American Institute of Chemical Engineers

16th Topical Conference on Ethylene Producers 2004

AA003008

Held at the 2004 AIChE Spring National Meeting

April 25-29, 2004
New Orleans, Louisiana, USA

Volume 1 of 2

Printed from e-media with permission by:

Curran Associates, Inc.
57 Morehouse Lane
Red Hook, NY 12571
www.proceedings.com

ISBN: 978-1-60423-459-6

Some format issues inherent in the e-media version may also appear in this print version.

American Institute of Chemical Engineers

16th Topical Conference on Ethylene Producers
2004

TABLE OF CONTENTS

Volume 1

Volume 2

Online Asset Management
Tools for Machinery Sytems

Glenn M. *Poché*
Regional Sales Director

Bentley Nevada LLC.

Prepared for presentation at the 2004 AIChE Spring National Meeting

Ethylene Plant Rotating Equipment – [T8007] – Rotating Machinery Reliability – Impact of Operational Practices

Overview

Asset management of machinery requires the use of tools/systems, which provide value beyond that of machinery protection systems. Protection systems, in essence, provide insurance – protection against significant machinery changes. By design they are reactive – the change must occur before the monitor system provides an alarm output. A management tool must be proactive – providing an output/alarm before the significant machinery change takes place. Additionally the asset management tool should provide information and not just data. The information needed to manage the machinery will vary by department/groups - Operations, Maintenance, Management, … . Therefore, flexibility in the dissemination of the information to meet their specific needs would be ideal. The ultimate objective of the asset management tool is to enhance equipment reliability while optimizing personnel effectiveness and increasing plant safety.

Information provided by an asset management tool should address two aspects: one, the enhancement of the protection system, and two, the proactive information or optimization that allows for proper action by the necessary personnel depending on the machinery/process issues.

- In the first instance, the software alarms allow pre-warning before the protection systems detect an alarm condition. This function provides an early indication of changes due to possible equipment degradation and/or process changes impacting the equipment, which could be a potential safety issue. Additionally, these software alarms examine parameters not typically monitored in the protection system. As a result, plant personnel could be made aware of changes that would normally go unnoticed.

- In the second instance, use of rule based advisories minimizes the need for personnel to continuously review data – most of which typically has not changed. The advisories can be specific for certain types of equipment/assets or generic for a particular operation or process. These advisories can be directed to the necessary personnel with specific text to the different groups. Custom rules also allow for preservation of corporate knowledge/intellectual assets.

The system ideally would be able to interface with a DCS and/or other system(s) to access the repository of data for managing assets and a communication link to departments/groups needing to be aware of changes and action to take.

Benefits/Features

Monitoring systems of many types are used extensively throughout the process industry providing alarming functions and in some cases shutdown functions. These systems are typically seen/used as “the last line of defense” for the monitored parameter particularly if the systems are used in auto shutdown service. An asset management tool with “soft alarm” features can be used as a “first line of defense” in providing earlier warning of parameter changes. Asset management tools are most often applied in applications associated with large critical centrifugal and reciprocating machinery therefore the examples provided in this paper will be directed toward those machines; however, the capability of these systems is much broader with the ability to manage any type of machinery/asset - including stationary equipment.

Most monitor systems measure a single parameter with one or two alarm levels provided with an increase in the value of the monitored parameter. There are variances to this measurement/alarm relationship, but these comprise in excess of 80% of the monitoring systems with some systems providing "banding alarms" such as for Rotor Position. The assumption, of course, is that an increasing value is moving in a least desirable direction (in fact, a case or situation that is not always correct). Additionally, the typically monitored parameter is more of a "macro" value while there are other more subtle measurements that are not monitored. "Software" alarms are set inside – typically below – the alarm levels established in the monitor systems. Additionally, the system provides the ability to detect reductions in the measured parameter as well as changes in some of the more subtle measurements. A few examples are provided to further explain the value.

Overall Vibration – A very common measurement on centrifugal rotating machinery is that of the overall vibration level. Most monitor systems – as specified by API standards – provide two "high" alarm levels as an indication of machine degradation. Machinery malfunctions such as degradation of the balance condition or increases in bearing clearances – for hydrodynamic type bearings - will generally be reflected in increases in the overall vibration and the monitor systems provide good annunciation/protection alarming capability. A preload malfunction due to a cocked bearing or an alignment abnormality can result in a drop in the overall vibration level with the monitoring system not providing any annunciation/protection alarming capability. Note on the following how a preload from the 10 o'clock direction pushes on the shaft reducing the amplitude in the one direction.

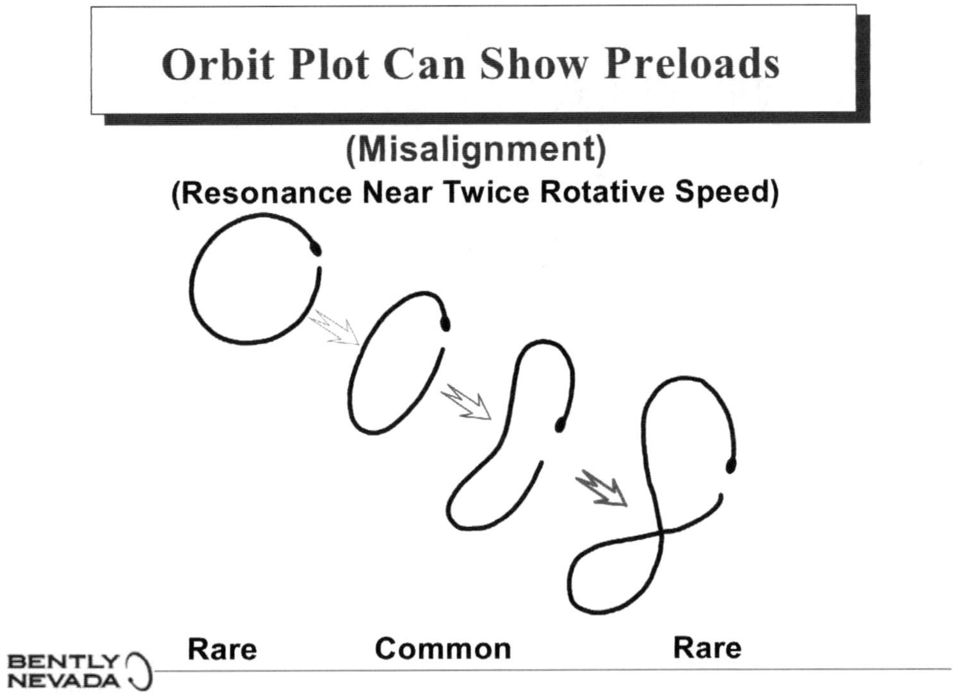

Subtle Measurements – The Overall Vibration value is comprised of all frequency components. A way of observing/monitoring the overall vibration level in greater detail is to look at the two most common frequencies generated by large centrifugal machines – the 1X and 2X components. The 1X component is amplitude of vibration occurring at the machines operating speed and the 2X that component occurring at twice the machine speed. Associated with these two frequencies are the phase angles that provide further subtle indications of changes in the overall vibration signal. Any one of these four variables can change independently having an impact on the overall vibration value. Detection and alarming of these four parameters provides enhanced management of the asset. Below is shown the definition of phase – distance from the keyphasor® to the next positive peak of the filtered signal.

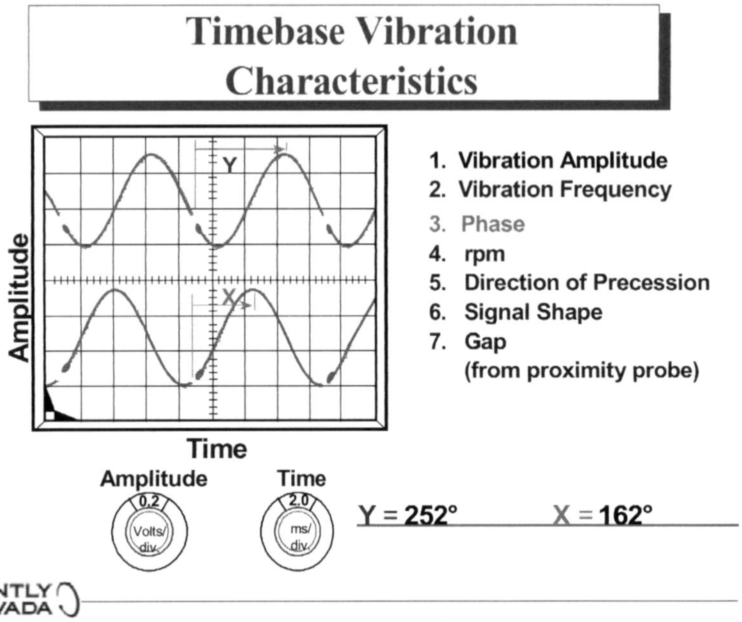

The following shows a method of putting an acceptance region around the 1X or 2X vector. The size and shape of the region is determined or based on the operational aspects of the machine where some may show variance in the phase component while

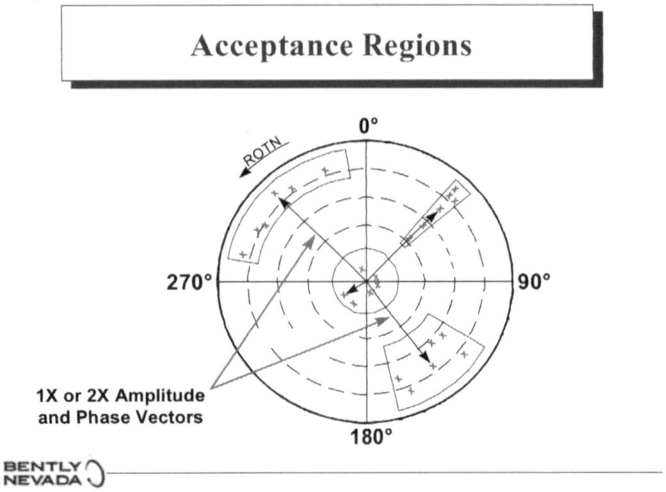

others may show variance in the amplitude component. Changes of the predetermined acceptance region would be cause for concern and merit further investigation of the data.

The following is showing a trend plot of the 2X amplitude and phase versus time on a rotor with a crack that gets progressively worse. The rotor is operating significantly below the first balance resonance – first critical. One of the characteristics of a shaft crack is that it can generate a significant 2X component due to the symmetry difference in the breaking shaft. As the crack propagates, the stiffness of the rotor decreases causing the resonance to decrease eventually coinciding with and moving through the 2X component. Hence the rise and then drop in the 2X amplitude and corresponding phase shift.

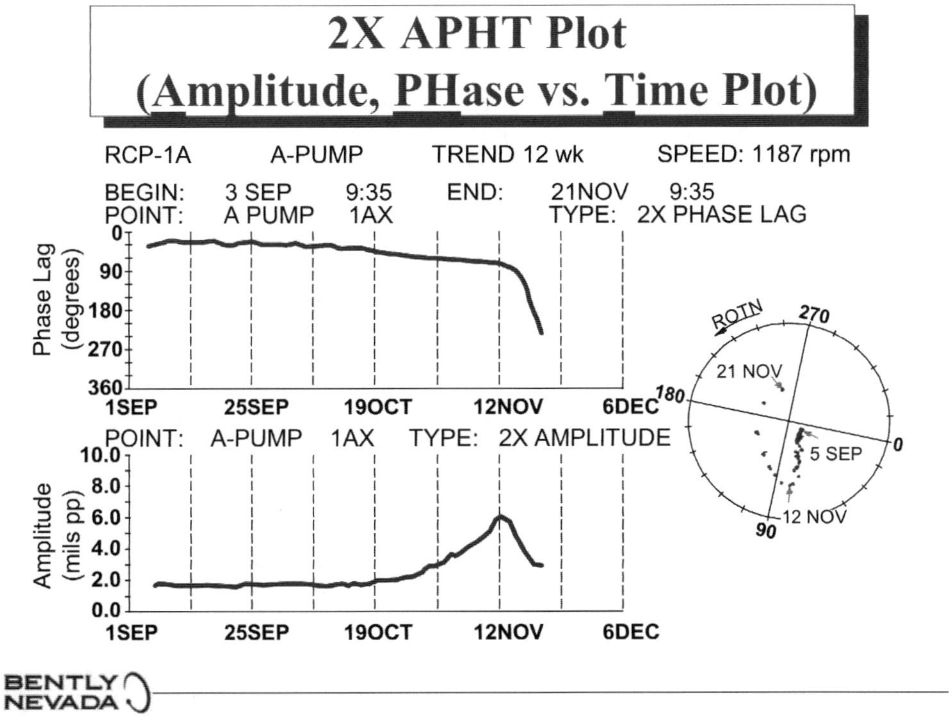

Another measurement that can be included in the monitoring/protection system, but typically isn't, is the radial position of the rotor in the bearing. This measurement is known as the DC Gap Position or the Shaft Centerline Position. By incorporating an XY set of proximity probes, properly installed, the position of the shaft in the bearing can be determined. This data is important in itself but provides greater value when used in conjunction with the rule-based software noted below.

As shown on the following, proximity probe systems have two components – an ac component and a dc component. The ac component represents the motion of the shaft/target relative to the probe and the dc component represents the average gap between the shaft/target and the probe.

Average Gap (dc Voltage)

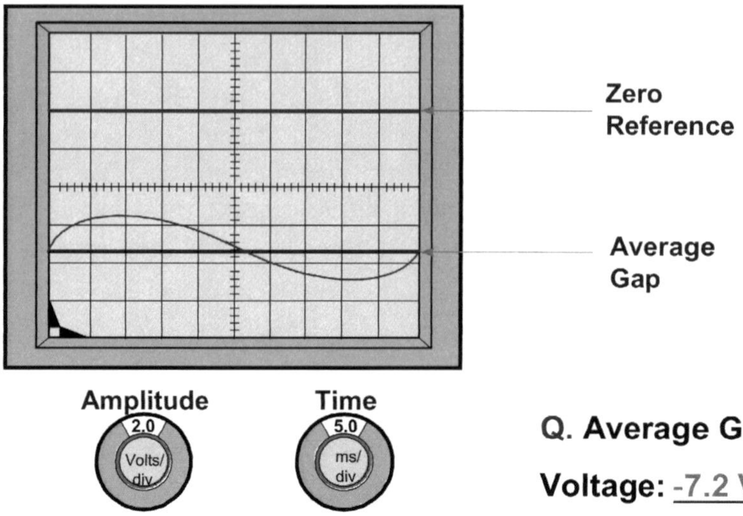

Q. Average Gap

Voltage: -7.2 Vdc

By using two radially mounted orthogonal proximity probes at a bearing, the position of the shaft in the bearing can be determined/monitored. The following shows the centerline movement of a shaft from startup to running speed.

Average Shaft Centerline

(Not Orbit or Polar Plot)

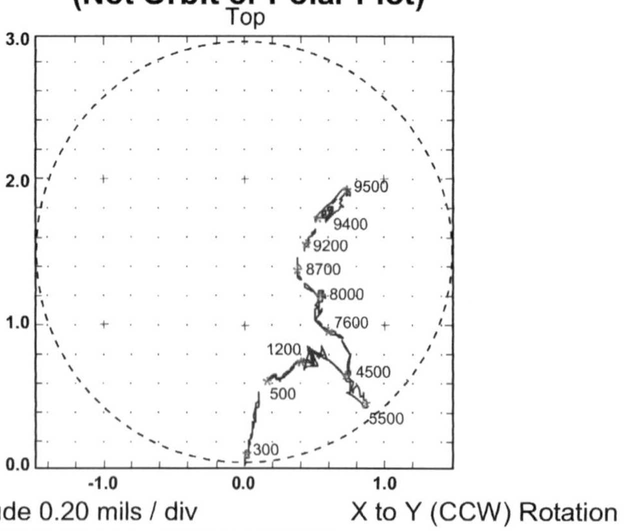

Changes in the overall amplitude as detected in the vibration monitoring system typically results in a call to someone knowledgeable in troubleshooting machinery by evaluating dynamic data from the various transducers on the machine. The necessary data can be viewed through various diagnostic systems if the transducer mix is correct and proper. An on line asset management system providing this dynamic data eliminates the need for any auxiliary test equipment. Further, if the database can be accessed remotely by off-site personnel – corporate experts, OEMs, selected consultants, ... efficient technical support can be resourced in a timely manner. Dynamic data from the transducers can be correlated to process data to determine whether there was an impact due to some process change or simply a change in the machine condition with proper integration to the DCS and/or historian.

Some reliability programs consist of taking periodic data – weekly, bi-weekly, monthly, bi-monthly, ... - from the installed transducers and then have the data "analyzed/reviewed" by a Reliability Engineer to detect the incidence of or the advent of problems. This effort can be quite time consuming, and in most cases, results in the data being "taken/analyzed/reviewed" being found as "good/ok/no change". This is to be expected as machines today run for significant periods of time with generally few problems, but because of the significant impact on the operation of the process, a user has to insure that their machinery is delivering and not deteriorating. And, hopefully, if the delivery drops and deterioration is detected, it is done early enough to correct the malfunction or as a minimum to begin proper planning to deal with the issue(s).

An on line asset management system offers a more efficient solution to the above. These systems' software continuously monitors all of the transducers at a sufficient rate to allow capture and storage of sufficient data to include all frequency information that might be needed for analytical purposes. By using software alarms along with selective filtering, the subtle changes can be detected and annunciated.

But, the real value comes in the area of software that provides a rule-based subset to run audits not only on static data, but dynamic data. These rules compare the data to defined machinery malfunction rules and determine not only if a malfunction exists, but also indicates the severity of the malfunction. This ability eliminates the need to take periodic data and to spend valuable engineering time reviewing what is typically "good" data. Notification of an identified problem can be sent via e-mail, phone call, pager, ... and to the appropriate people. The deliverable is in the form of text – text that directs one to take specific action – that is, not data, but precise instructions on what action to take and if multiple actions, in which order to take them. Additionally, the dynamic data storage ideally is such that should a machine trip, all the data from just prior to the shutdown is stored providing information critical to determining the cause, sequence of events, ...

The following is an example of a presentation of "Actionable Information". Such a presentation provides information as to the location of the problem, time the problem initiated, nature of the problem, severity of the problem and some direction to take to alleviate the problem.

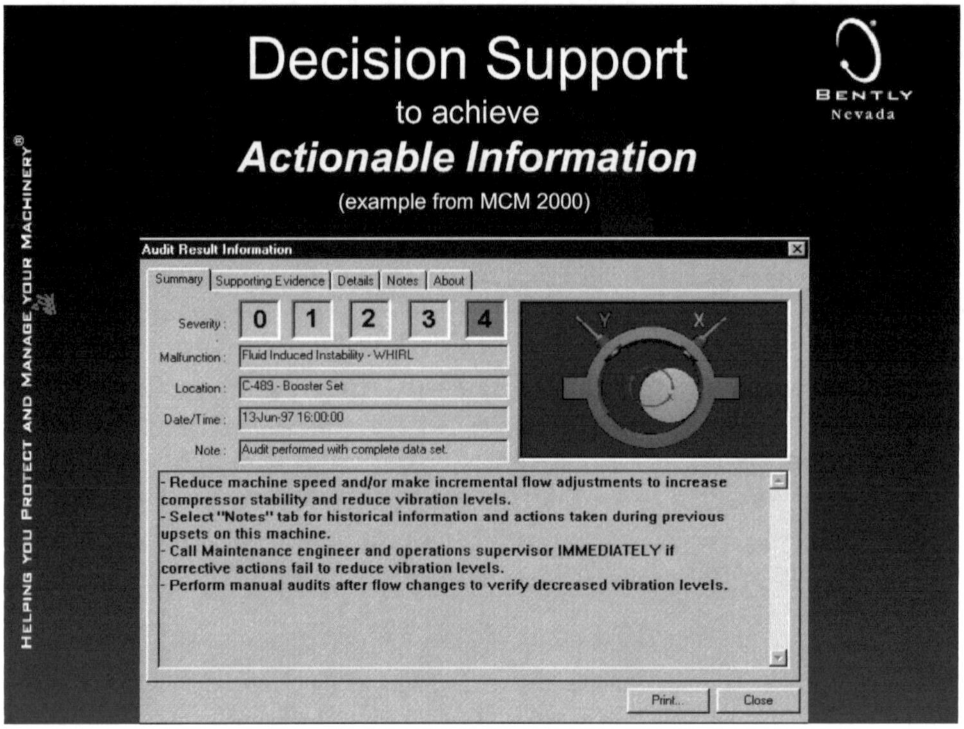

Preserving corporate knowledge/intellectual assets is critical to the continued success and competitive positioning of all companies. Being able to incorporate one's corporate knowledge/intelligence of machinery systems into an online system is of great value. Some systems provide this capability by providing a means for one to "write" rules based on their corporate knowledge. These rules could be an extension of standard rules by incorporating corporate knowledge into custom rules written based on process knowledge, machinery knowledge or combinations of both. The rules are easily written using a basic "drag and drop" methodology, therefore a writer does not have to know programming language in order to write and install these custom rules. Any data accessible to the online asset management system - the DCS, historian, or other can be used as inputs to the rules.

The following is a pictorial of how rules could be generated by using basic math functions to create the logic behind the information driving the condition and resulting in the appropriate action to be taken. The resulting presentation could be in the format as above – Decision Support – or by electronic notification – email, pagers, phones, ...

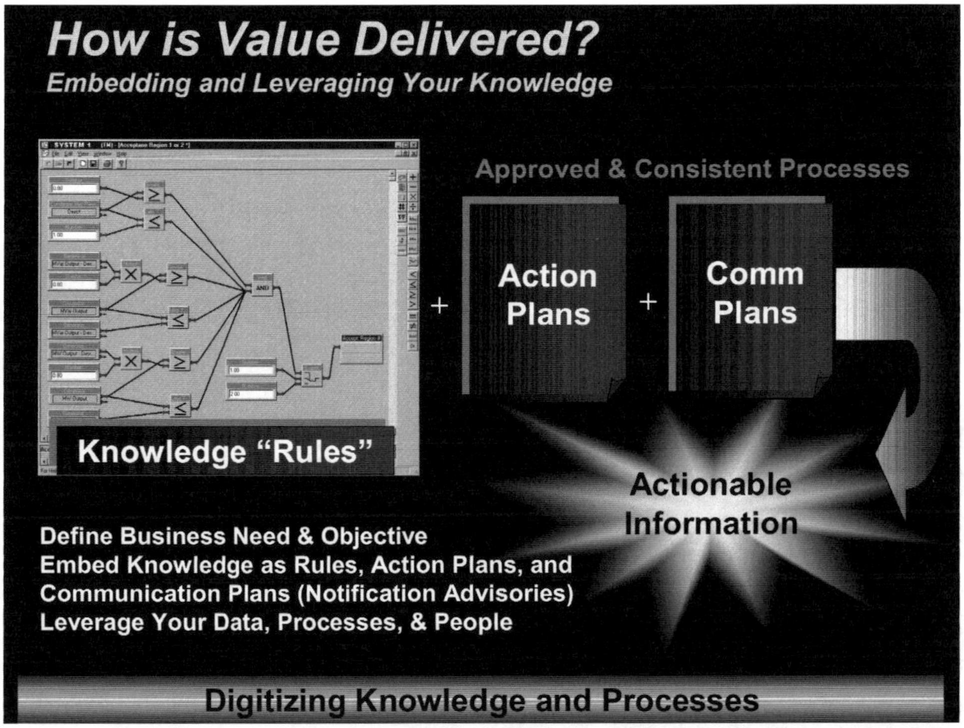

Other features of an Online Asset Management System may include the following:

Parametric or State Monitoring – There may be cases where machinery operates under different conditions – varied speed range or varied process parameters. In which case it might be desirable to have alarms varied based on the different conditions, if they have a significant impact on various monitored parameters. For example on the plot below, if a machine was running in a speed range that brought it closer to the natural resonance, one might want to consider adjusting soft alarms accordingly.

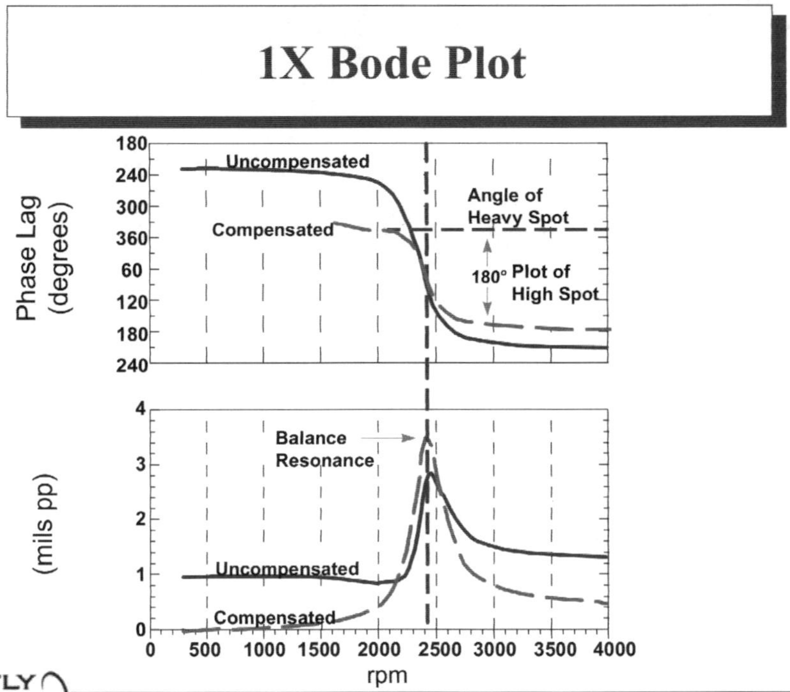

BENTLY NEVADA

<u>High Speed Scanning</u> – If process data along with machinery data is managed together in the system, scan times down to 1.0 second provides an avenue for data correlation to determine the sequence of events.

<u>Thermodynamic Performance Monitoring</u> – The Online Asset Management System can be used to continuously measure a unit's performance providing notification should degradation or changes to the norm occur.

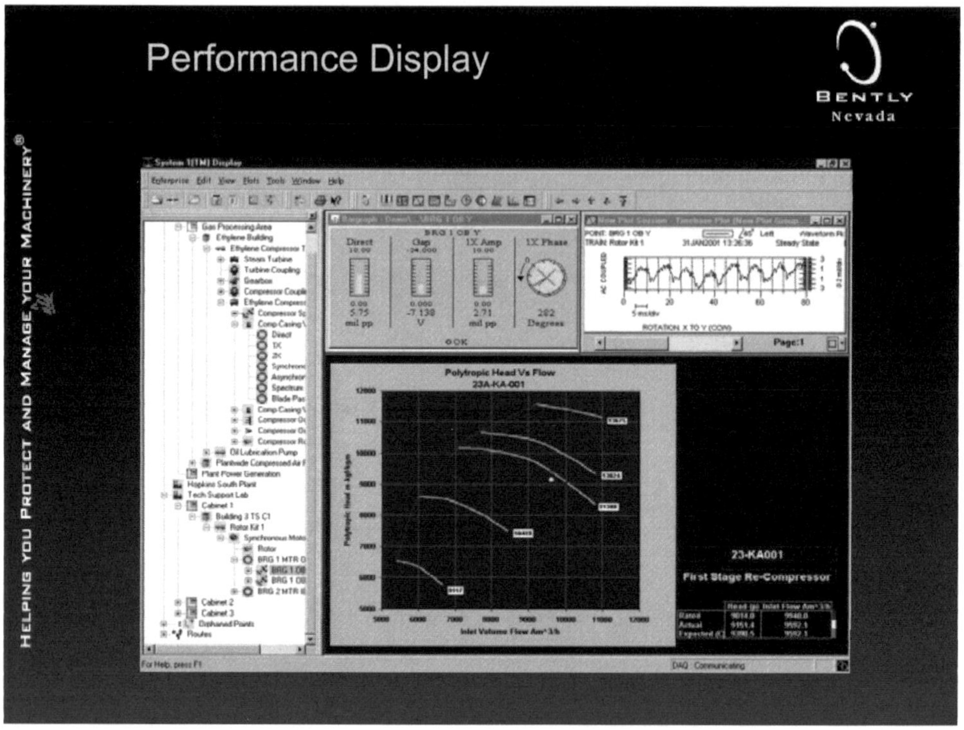

<u>Trending and Reference Data</u> – Historical trend data can be used to compare current abnormalities to previous data of the same or similar abnormalities. Or, current data can be compared to a reference data of a known condition/situation. Above, the "1X Bode Plot" under Parametric or State Monitoring shows a significant amount of "run out" – the 1X vector at low speeds. Noting the change/no change of the vector is an example of being able to use reference data.

Below is an example of gearbox data showing data from a cracked tooth condition with that of good gearing. The data with the cracked tooth shows the high 7X component as well as the high gear mesh frequency component.

Baseline Comparison

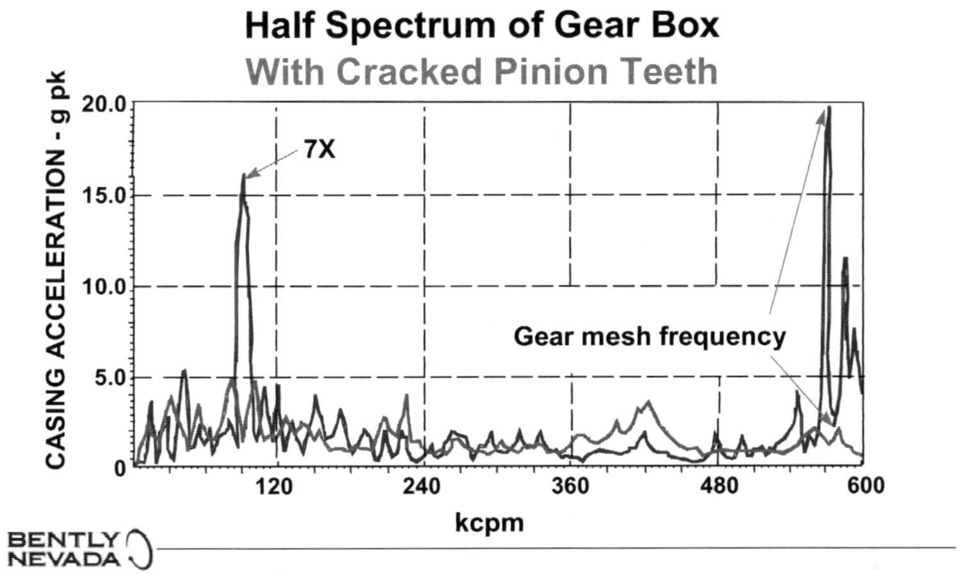

Conclusion

An Online Asset Management System, properly configured, integrated with a DCS and/or a historian and included with the rule-based software, offers a user:

- an enhanced monitoring system with the ability to "monitor" more detailed data on machines or fixed assets
- an "on-line machinery expert", so to speak, as the system can run continuous audits on the data determining if a specific problem exists
- the ability to include custom rules incorporating corporate knowledge/intelligence about the process and/or machinery
- have online access to such things as:
  - Panametric/State monitoring
  - High speed scanning for machinery/process data correlation
  - Thermodynamic performance monitoring
  - Trending and reference data
- Summary Perspective
  - Level 1 – Protection system only
  - Level 1+ – Protection system with off-line diagnostics
  - Level 2 – On-Line system – alarm triggered by soft alarms with on-line diagnostics
  - Level 2+ – Same as Level 2 but inclusive of process data
  - Level 3 – On-Line system utilizing rules to initiate early action
  - Level 4 – Same as Level 3 plus corporate knowledge/intelligence rules

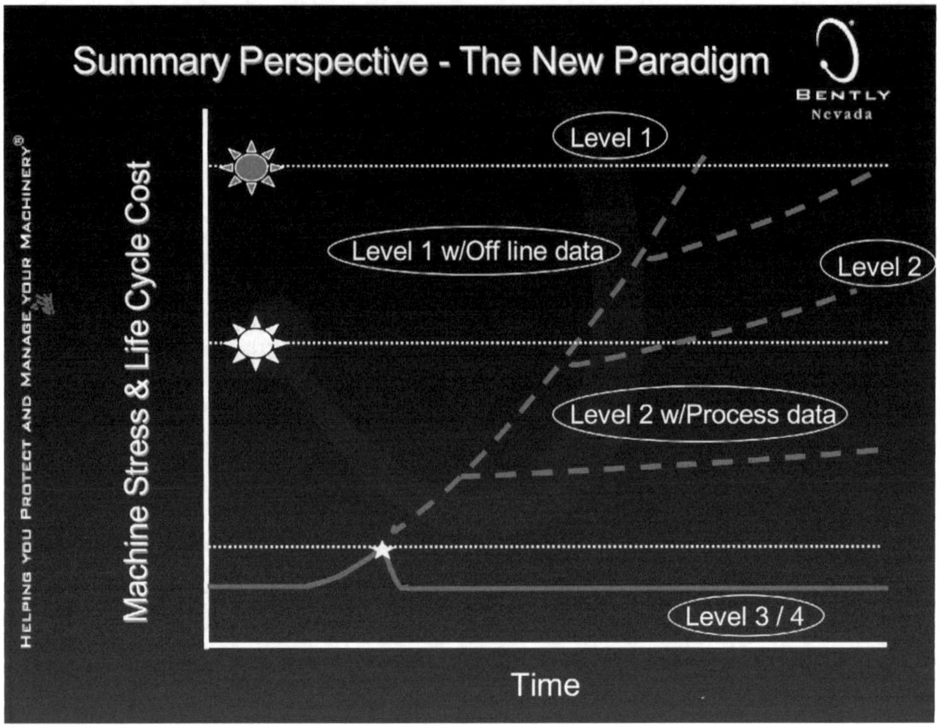
Summary Perspective - The New Paradigm
BENTLY
Nevada
HELPING YOU PROTECT AND MANAGE YOUR MACHINERY®
Machine Stress & Life Cycle Cost
Level 1
Level 1 w/Off line data
Level 2
Level 2 w/Process data
Level 3 / 4
Time

Surge:

What It Is, Why It Will Damage Your Compressor, and How To Avoid It, with Discussions on Choke.

Authors: M. Theodore Gresh
Bruce G. Heckel
Brian S. Spencer

Elliott Turbomachinery Company Inc.

Prepared for Presentation at
The 16th Ethylene Producers' Conference
April 25th thru 29th

[T8007] Rotating Machinery Reliability – Impact of Operational Practices

Copyright © Elliott Turbomachinery Company Inc.
January 2003

Unpublished

ABSTRACT

Surge

What It Is, Why It Will Damage Your Compressor and How To Avoid It, with Discussions on Choke.

Surge has been defined as peak head. However, actual "hard surge" is a condition where the discharge pressure of the compressor overcomes the dynamics of the compressor and flow is reversed from the discharge to the inlet of the compressor. Unless something is done to correct the situation, the system will continue to cycle until something breaks. Surge can be a mild stall or an intense heavy hard reversing flow. For centrifugal compressors, if the impellers are not damaged from the reverse pressure gradients, the diaphragms may crack. Surge is best avoided by a good anti surge system that accurately monitors compressor inlet and discharge conditions and flow rate. The surge point should be verified by actually surging the compressor in the field at several speeds (for variable speed machine) as the surge points supplied by the OEM is approximate and very hard to predict. In fact, the surge point is somewhat affected by the system in which the compressor is installed and location of the anti-surge line and check valve can be crucial.

Even very mild surging or stall can cause serious damage to a compressor, resulting in high rotor vibration or internal damage. Methods of detecting surge and incipient surge will be presented.

INTRODUCTION

As key components in Ethylene manufacturing, the compressors' reliability must be ensured. Among the more difficult problems in compressor control and operation is the ability to detect and prevent surge, incipient surge and how to identify operation in choke. This paper will cover these aerodynamic phenomena in an effort to inform the reader of the governing principals and associated characteristics. It will also describe, and present examples of, how they can damage components of the compressor and the implications the damage has on the operation of the machine. Surge (and incipient surge) prevention will be discussed in order to provide insight into the detection as well as the safe and effective avoidance of surge through the use of an anti-surge system. Finally, how to limit the compressor's operation beyond the end of the manufacturer's supplied curves in the choke region will be addressed.

AERODYNAMIC VECTOR ANALYSIS OF GAS MOTION

Surge has been defined as peak head. Below the surge point, head decreases with a decrease in flow (Figure 1).

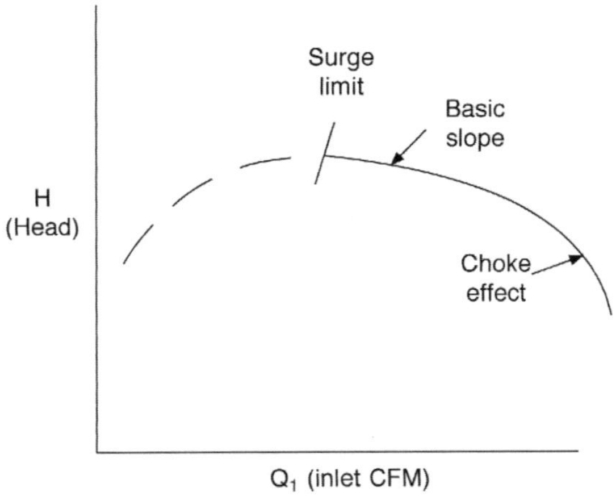

Figure 1: Head curve for a compressor stage.

This is better understood by first explaining the velocity vector diagrams for the impellers which lead to the characteristic curves for each stage in the machine, and ultimately the entire compressor.

V_{rel} (Figure 2) represents the gas velocity relative to the blade.

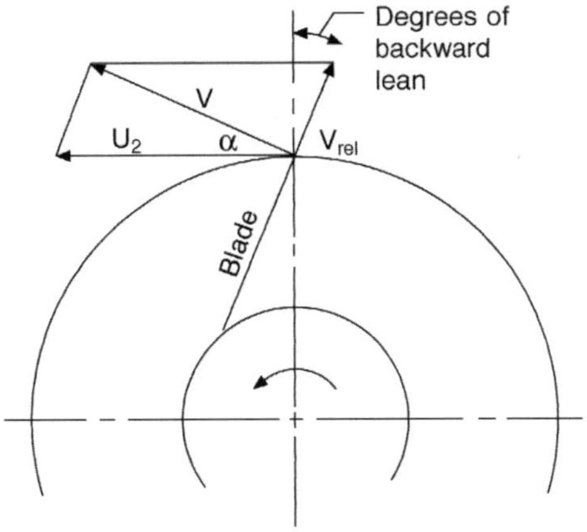

Figure 2: Vector diagram of the gas velocity relative to the impeller blade.

U_2 represents the absolute tip speed of the blade. The resultant of these two velocity vectors is represented by V, which is the absolute velocity of the gas ($U_2 + V_{rel} = V$). Knowing the magnitude and direction of this absolute velocity, we can break this vector into its radial and tangential components (Figure 3).

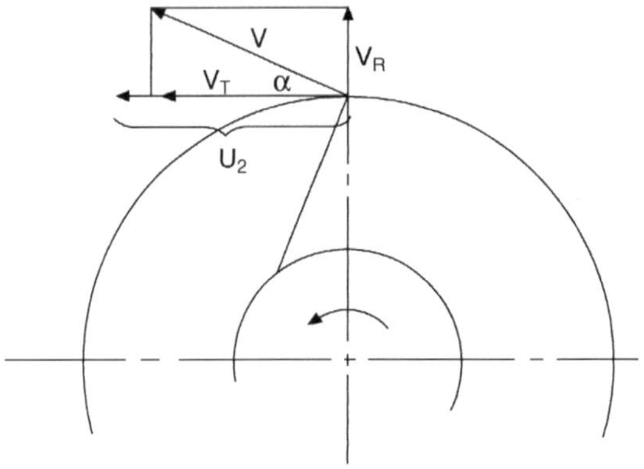

Figure 3: Vector diagram of the gas velocity shown in Figure 2 in radial and tangential components.

For a radial inlet impeller the head output is proportional to the product of U_2 and V_T. For a typical backward-leaning bladed impeller, as the flow decreases at constant speed, V_{rel} decreases. This makes V_T increase, which increases head output. This head increase with decreasing flow is what causes the basic slope to the compressors' characteristic curve.

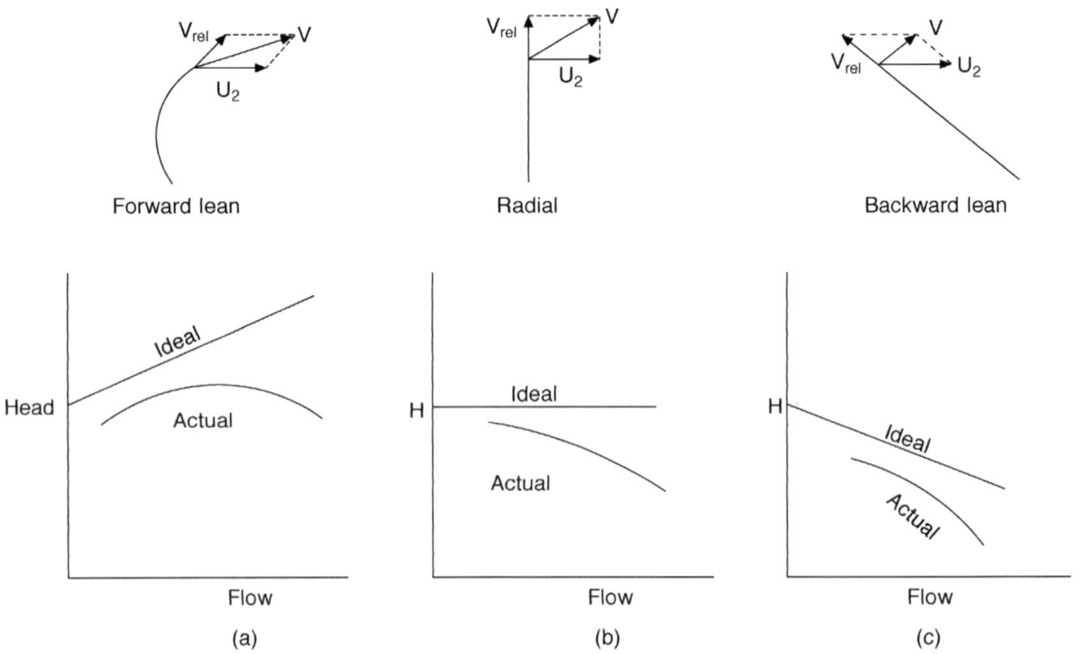

Figure 4: Three basic head curve shapes for centrifugal compressors.

Figure 4 shows characteristic curves for three basic configurations: forward-leaning, radial, and backward-leaning blade profiles. Note that the forward-leaning blades provide a positive sloping head curve and the maximum head output. This is because V_T is increasing with increasing flow.

A radial bladed impeller has a theoretical constant (flat) head curve since V_T does not change with flow.

Overall stage efficiency is highest for backward-leaning impellers, while efficiency is lowest for forward-leaning blades.

For best efficiency, most modern centrifugal compressors use backward-leaning bladed impellers. Directionally speaking, the greater the backward lean, the better the efficiency. However, as the angle increases, the head is reduced (Figure 4(c) and 5).

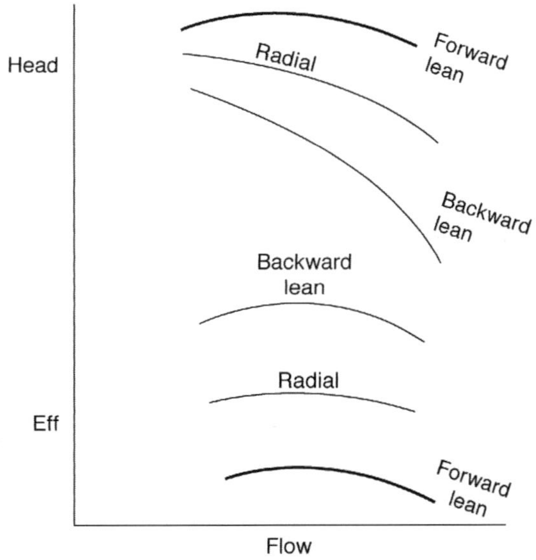

Figure 5: The effect of impeller blade angle on head.

Current designers use 3-D full-inducer blading to re-gain some of the head loss due to high backward leaning blades. A designer can select a blade angle and tip width to best fit the desired head and efficiency characteristics of the particular application.

WHAT IS CHOKE

Stonewall, or choke, is a condition at which increased capacity (flow) results in an excessive decrease in head (see Figure 6). This occurs because the Mach number is approaching 1.0.

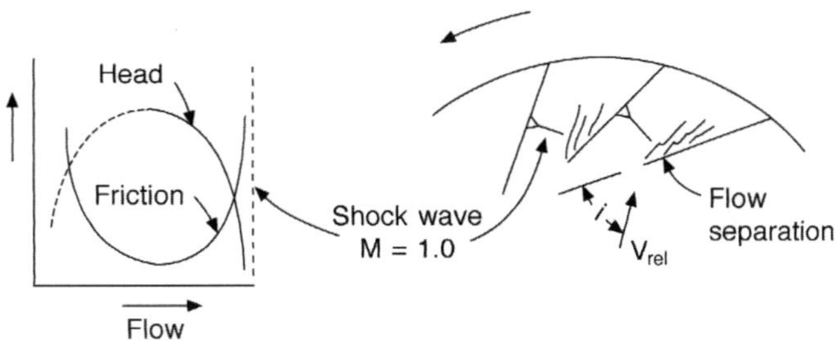

Figure 6: Stonewall.

Operating at a very high flow rate has very negative effect on the performance of a centrifugal compressor, and can sometimes be damaging. For axial compressors, high flow rates can also create blade flutter and result in serious damage to the blading. The stonewall effect of the centrifugal compressor stage with a vaneless diffuser is controlled by impeller inlet vector geometry.

U_1 in Figure 7 represents the tangential velocity of the leading edge of the blade.

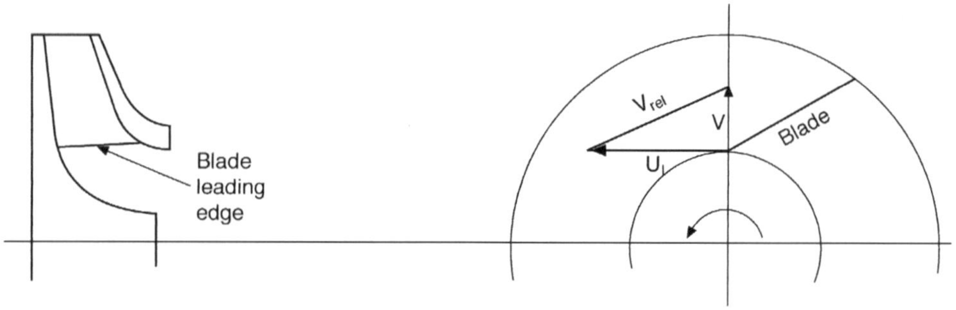

Figure 7: Stonewall.

V represents the absolute velocity of the inlet gas, which, having made a 90° turn, is now moving essentially radially (in the absence of prewhirl vanes)- hence the name radial inlet. By vector analysis V_{rel}, which is gas velocity relative to the blade, is of the magnitude and direction shown.

$$V = U_1 + V_{rel}$$

At design flow, V_{rel} lines up with the blade angles. As flow increases beyond design, V increases. As V increases so does V_{rel}. V_{rel} now impinges at a negative angle to the blade, a condition known as negative angle of attack. High negative angles of attack contribute to the stonewall phenomenon because of boundary-layer separation and a reduction of the effective area in the blade pack. This area reduction, in addition to the already high V_{rel}, brings on Mach 1 and a corresponding shock wave, as shown in Figure 6.

WHAT IS SURGE

As flow is reduced at constant speed, the magnitude of V_{rel} decreases proportionally, causing the flow angle to decrease. (See Figure 8 and 9.)

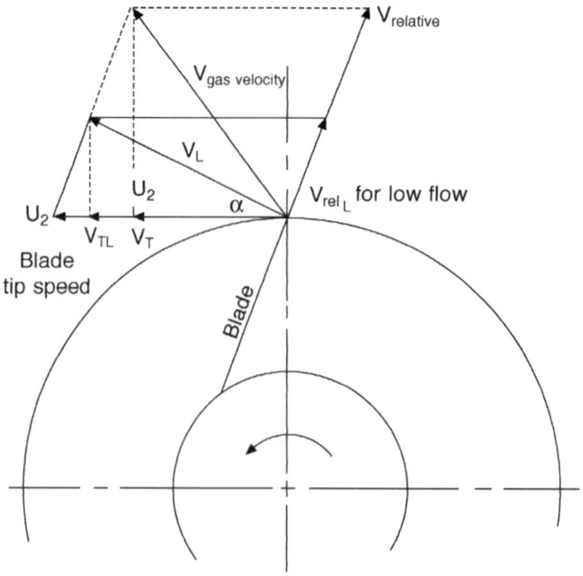

Figure 8: The effect of a change in flow rate on the vector diagram at the impeller O.D. is shown.

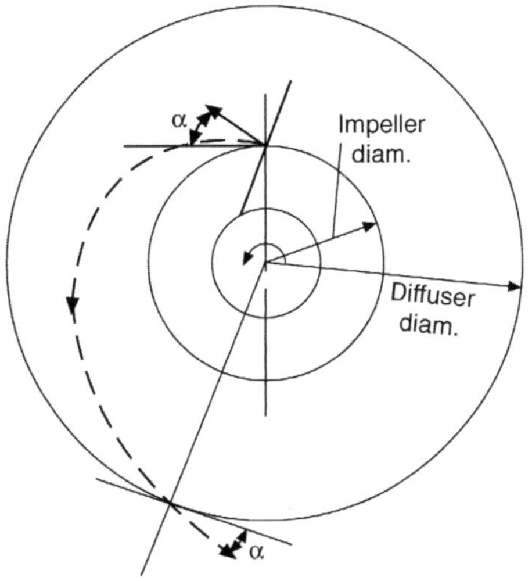

Figure 9: Flow through a diffuser.

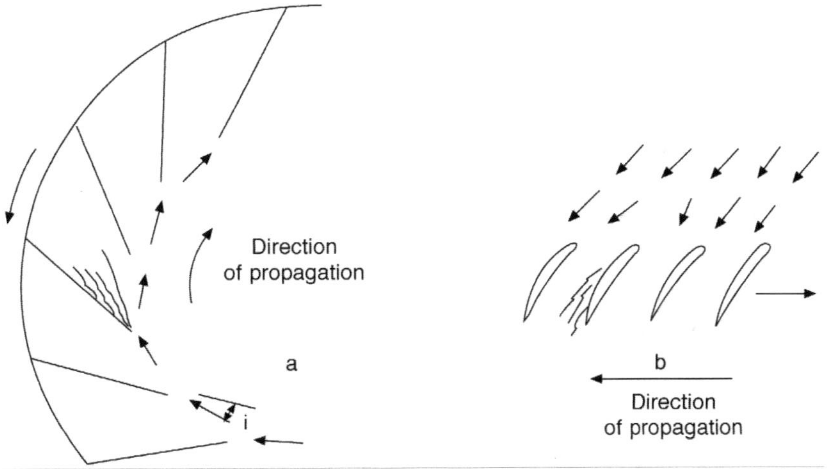

Figure 10: Rotating stall: a. Centrifugal compressor impeller. b. Axial compressor blade pack.

Additionally, the incidence angle is increased (Figure 10).

The smaller the flow angle α, the longer the flow path of a given gas particle from the impeller tip to the diffuser outside diameter. When angle α becomes small enough, and the diffuser flow path long enough, the flow momentum of the gas is dissipated by the diffuser walls by friction to the point where the frictional forces are increasing (versus flow reduction) faster than the head is increasing (versus decreasing flow).

The high losses associated with low flow (see Figure 10) are partly caused by a poor incidence angle *i*, which can result in flow separation at the low-pressure side of the blade leading edge. This flow separation, commonly referred to as rotating stall, frequently starts at one or more blades and continuously shifts around the impeller blades. This occurs at relatively low speeds just before full surge occurs. At higher speeds, the compressor generally goes directly from stable operation to flow separation on all blades and finally to full flow reversal.

The flow separation plus the higher frictional losses result in a positively sloped curve. Since system resistance curves are also positively sloped, the system is unstable.

Stalling and/or surging are most likely to occur during two specific times. During start-up and shutdown of the compressor and when operating at high pressure ratios, small deviations from design conditions cause large variations in performance.

Multi-stage compressors are designed to give the optimum performance at full-load conditions when the flow intersecting every blade is approaching as close to the correct angle of incidence as possible. However, at off-design low speeds, as in the case of start-up and shutdown, there is a much lower pressure rise across each stage causing higher velocities in the rearward stages of the machine. This causes the flow in the back end of the machine to accumulate in each impeller leading to choke. As the flow accumulates in the back end of the machine, the flow is restricted to the front end of the machine. The slower gas velocity in the front end of the compressor causes a severe positive angle of incidence leading to impeller stall.

In contrast to stalling at low speed, which begins in the first stages of the compressor, off-design high-speed stalling is more likely to begin in the last stages. Compressors operating at high speeds will tend to have higher pressure ratios across the machine. This implies that across each stage there is a large pressure rise and corresponding volume reduction and increase in gas density. Additionally, the volume reduction and increase in gas density become progressively greater from the inlet to the discharge in each section of a machine. Because of this, a small deviation in design conditions, such as reduced mass flow, will only have a minor effect on the first stages of a machine, but the effect will worsen as it propagates through the compressor to the final stages. Again, as in the low-speed scenario, the gas velocity in the effected stages is reduced leading to a high positive angle of incidence and ultimately to impeller stall.

This is why the characteristic curve for high-pressure-ratio compressors is so steep and the corresponding operating range is highly restricted.

By adjusting the diffuser area during the design of the compressor, the surge point can be controlled through the increase of V_R and flow angle α. The designer must balance between the desired surge point and stage efficiency due to the higher frictional losses associated with the higher gas velocities.

The surge point can be reduced by the addition of vanes in the diffuser (see Figure 11).

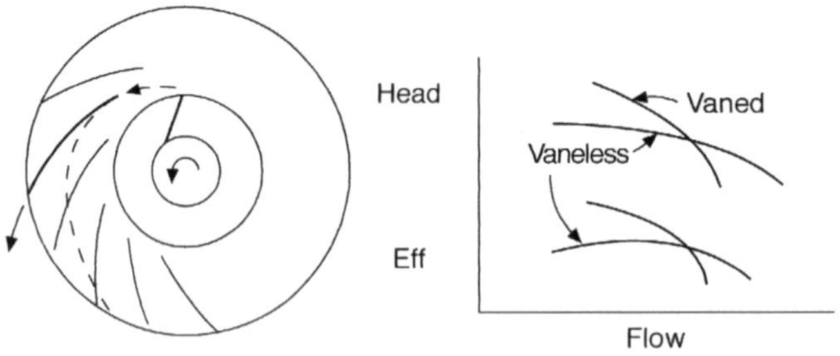

Figure 11: Diffuser vanes.

The vanes shorten the flow path through the diffuser, reducing frictional losses and controlling the radial velocity component of the gas. Due to the lower friction, head and efficiency are enhanced, but the operating range is reduced. When operating off-design in the reduced operating range, the incidence angle to the vanes is changed and flow separation occurs.

Adequate flow margin to surge must be ensured, especially at higher, off design speeds. When a given stage design is operated at a much higher impeller tip speed, the diffuser flow angle will decrease as speed increases. This can cause diffuser stall to initiate prior to impeller stall. To counteract this, the stage is designed with a narrowed diffuser width that increases the flow angle back to near the original design speed. This helps to improve the margin to surge in both flow and pressure rise. Consequently, the flow at which the diffuser stall occurs will be lowered, and the stall will initiate in the impeller first. Narrowing the diffuser will only have an impact on the location of stall initiation if test measurements don't already show the impeller initiates stall first, at higher Mach numbers.

For each new stage design the rotating stall characteristics must be analyzed to determine where incipient surge will originate. Experimental data is required to accurately assess whether the impeller or the diffuser is more likely to initiate stall. Tests involving multiple dynamic pressure probes are used to accurately define when a rotating stall pattern occurs. These probes are located at shown in Figure 12.

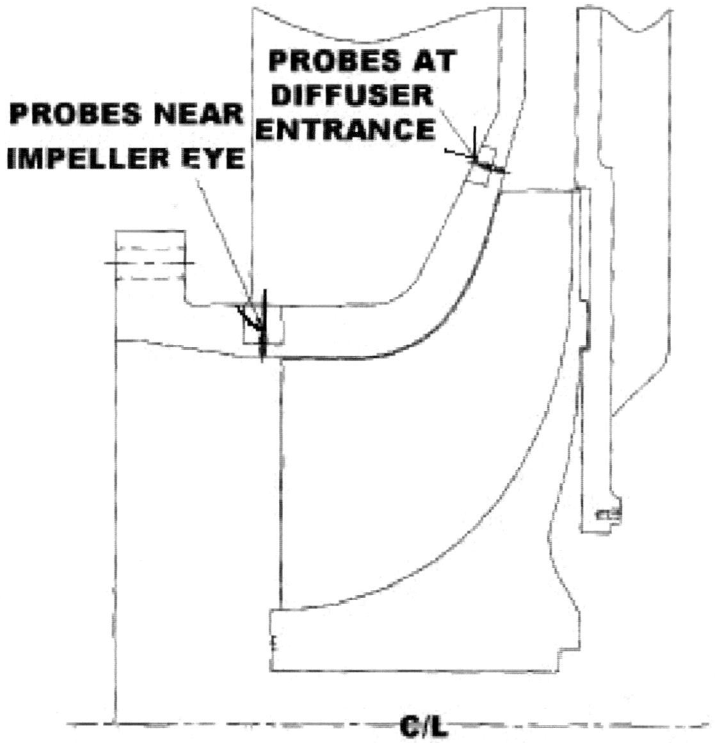

Figure 6: Schematic Showing Radial Location of Dynamic Pressure Probes; Typically Three Probes at Each Location @ 0, 45, and 60 Degrees for Phase Reference.

Frequency analysis is used to determine the number of stall cells and also the likelihood whether the rotating stall is caused by the impeller or diffuser, or due to an interaction of both. Rotating stall cells propagate around the impeller blades opposite the direction of shaft rotation at some measurable fraction of the rotor's spinning speed. The rate of propagation is referred to as 'stall speed'. Stall caused by the diffuser usually gives very low stall speeds, in the range of 3 to 20 percent of rotor speed. Diffuser-impeller interaction generally gives higher speeds in the range of 30 to 60 percent, and impeller stall at or near blade leading edges gives high speeds in the range of 60 to 95 percent of rotor speed. Evidence of the component that is causing the stall is usually given by changes in rotating speed of the stall as flow is reduced further past the initiation point. Typically, stall speed increases for diffuser stall as flow further reduces, and vice-versa for impeller blade incidence stall. It should be noted that these characteristics are typical of acceptable designs from this OEM; others could have somewhat different characteristics. Particular attention is paid to any occurrence of one-cell rotating stall patterns; number of stall cells also can change with either flow or speed variations. One-cell rotating stall is the only pattern capable of exciting a rotor by the unbalanced aerodynamic force. Thus, during these tests, the bearing vibration probe measurements are also monitored and recorded for any sign of increased rotor displacement.

An example is given in Appendix-A of two specific impellers for which frequency analysis was performed resulting in two different stall conditions.

Process gas compressors frequently require a wide operating flow range. This is needed to accommodate variations in process operation, field pressures and gas compositions, seasonal changes. To meet this need, designers today are employing the latest, state of the art computational fluid dynamics (CFD) computer programs. CFD is now used routinely to evaluate the compressor flow field with the goal of maximizing design point efficiency, and delaying the onset of flow separation at off design, lower flow rates. For a new impeller design, at a given tip Mach number, blade geometries are specified along the hub and cover, from inlet to tip, with the purpose of controlling the rate of change of the suction and pressure side velocities (or static pressure). Also, the hub and cover contours are determined based on flowrate, pressure ratio, and stage spacing requirements. Contour curvature and inlet to tip area schedule must be closely examined in order to reduce or eliminate regions of flow reversal. CFD calculations are run on new geometries at design and several off design flows with the intent of defining the performance qualities from maximum flow (choke) to surge flow. Since surge and rotating stall are unsteady, transient phenomena, CFD and other more empirical 1-D analysis methods cannot directly predict the flowrate where surge or stall will occur. Still, they can often give good estimates and provide the designer guidance on when a design is approaching a rotating stall situation. By utilizing proper indicators for stage design, hazardous one-cell stall due to diffusers has been able to be avoided for flows above the desired surge control line.

Also, surge is dependent on the specific system piping arrangement that is not normally modeled during the stage design process. However, as the compressor flowrate is reduced, CFD can indicate the growing presence of localized areas of low momentum fluid and regions of separated or reverse flow. These internal "pictures" of the compressor design have been successful in estimating where the minimum flowrate will occur, and the likelihood of rotating stall. Still, only detailed measurements during the prototype testing can determine the exact flowrate where rotating stall occurs, the number of rotating cells, their frequency, and the likely compressor component that initiated the stall. The location of full system surge and the peak of the head curve are also defined during the test. These data are then used to establish aerodynamic performance ratings, for families of similar stage designs that are applied in a wide range of industrial compressor applications. CFD analysis has helped to improve the probability of success for new designs, thus minimizing or eliminating any costly redesign activity.

In Figure 13, performance test results are compared for two stages designed for similar flow and head coefficients.

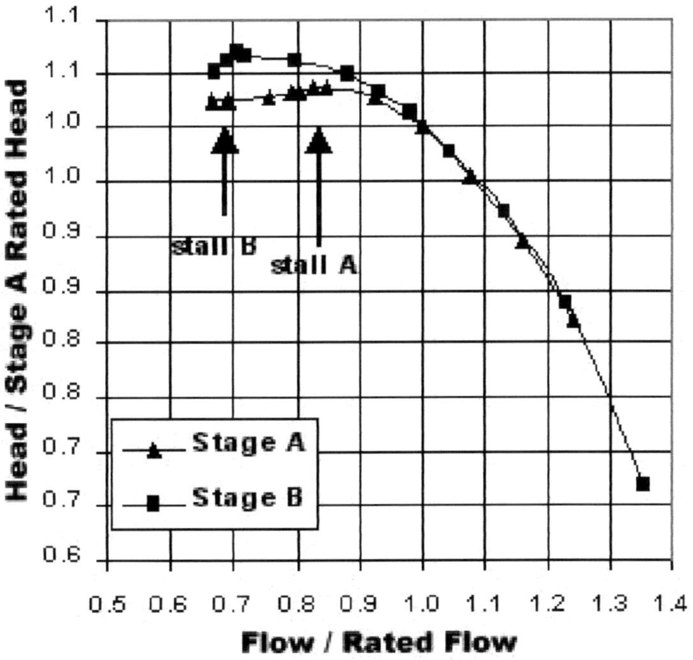

Figure 13: Tested Headrise for Rejected Stage A and Accepted Stage B at Tip Mach # = 0.7

Both had 3-D, ruled surface impeller blade geometry that attempted to control the hub to shroud incidence angle at the leading edge, and blade loading levels from inlet to tip. Stage A was designed without the benefit of CFD analysis but utilized proven, quasi-3D gaspath and empirical performance prediction programs. Stage B was designed with the same techniques as for stage A, but also benefited from concurrent analysis of the impeller with a commercial CFD code. The test results show a clear difference in the location of the peak head and also the flow coefficient where rotating stall was measured. During a later post-mortem evaluation of stage A, CFD analysis at the flow point where stall initiated on test, indicated a severe pattern of reverse flow along the shroud, suction side of the blade. Also, early signs of flow separation were evident near the design flow point. Had CFD been used during the design phase, it is obvious this stage would never have been built and tested in that configuration. Efforts would have been made to eliminate or minimize the reverse flow regions, especially at design flow. Figure 14 compares impeller A and B velocity maps at the stall flow point of each stage.

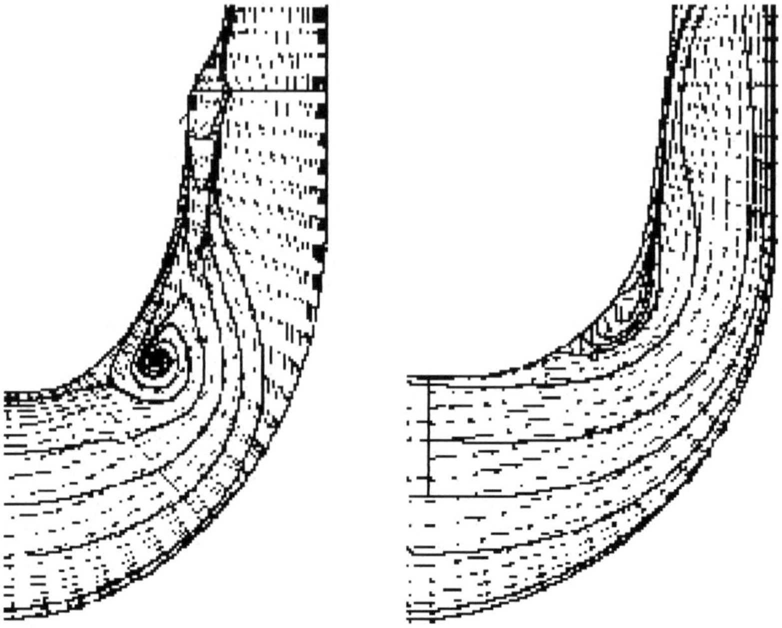

Figure 14: CFD Velocity Vector Comparison at Measured Stall Flow.

Note the much larger recirculation zone along the shroud wall for the rejected stage. It is believed the higher rate of impeller diffusion and the tighter shroud curvature caused impeller A to stall much sooner than impeller B. Blade loading was not considered a factor since the A impeller, with 21 blades, had lower loading rates than the B impeller, with 15 blades.

As previously discussed pertaining to most influencing factors that determine when a compressor surges, or experiences stall, off-design operation of a compressor can dramatically affect the characteristic curve shape. Any change in inlet conditions can change the discharge pressure and gas horsepower as shown in Figure 15.

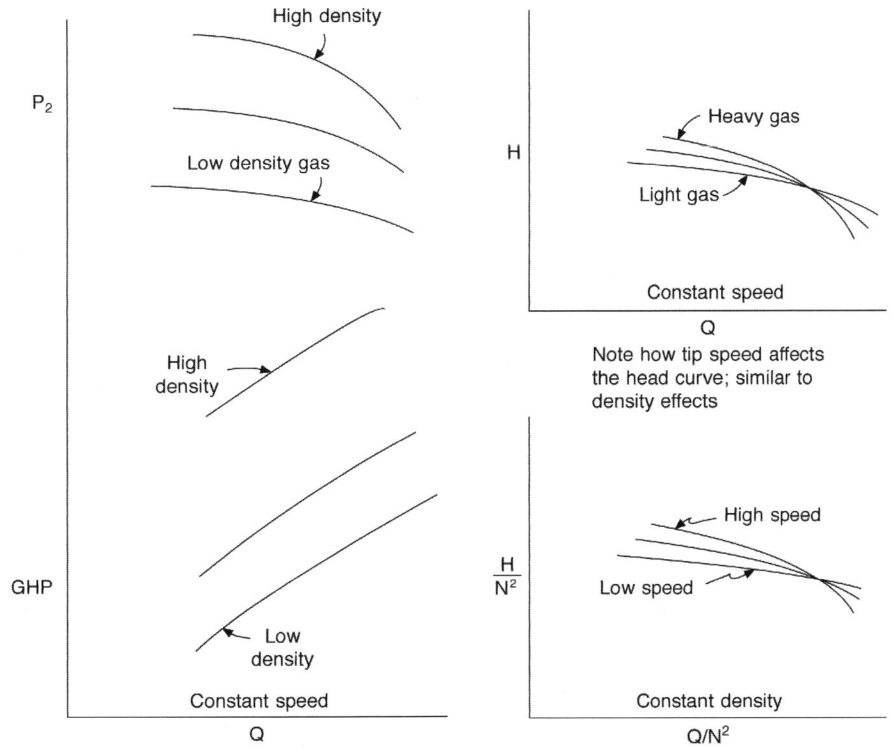

Figure 15: Effect of varying inlet conditions at constant speed for a single-stage compressor

In addition to changing the characteristic pressure and horsepower curves, the characteristic head curve also changes. This is due to volume ratio effects and equivalent speed effects.

If a constant discharge pressure is desired and gas conditions are changed (inlet pressure or temperature, mole weight change), a speed change is required. Since the curve shape changes with speed (higher losses at higher speeds), the head curve then changes (Figure 16). This effect is further compounded by volume ratio effects (Figure 17).

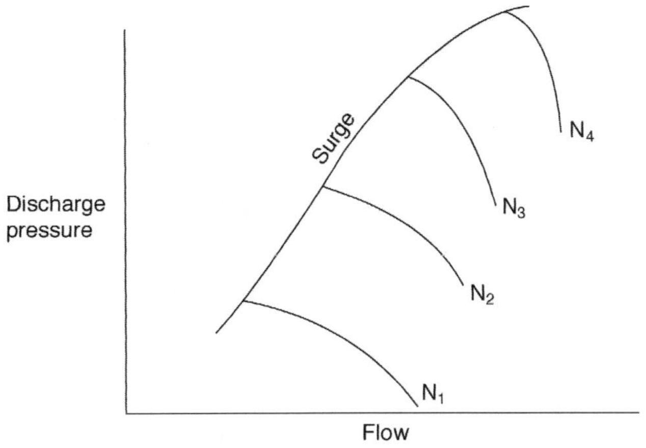

Figure 16: Effect of speed change on compressor curve shape.

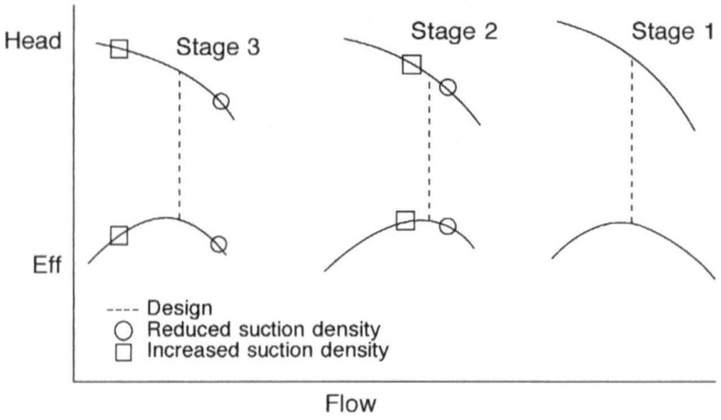

Figure 17: Volume ratio effects.

The head characteristics are a function of the acoustic velocity of the gas (see Figure 6). Knowing this, it is most convenient to refer to some "constant" gas, and obtain an "equivalent tip speed". This reference constant is typically air at 80°F, since this is what most "developmental" testing use as a test medium. A brief example of this scale-to-reference is presented in Appendix-B.

In a multi-stage compressor the "equivalent speed" effect is compounded by volume ratio effect (Figure 17). If the gas density varies, the pressure rise and volume ratio will also vary. This will feed a different flow rate to the second stage. The effect on following stages will be compounded and the end result is a premature choke and surge.

SURGE SCENARIO

To better relate these governing principals of Turbomachinery to a real-world scenario, visualize the simple system shown in (Figure 18).

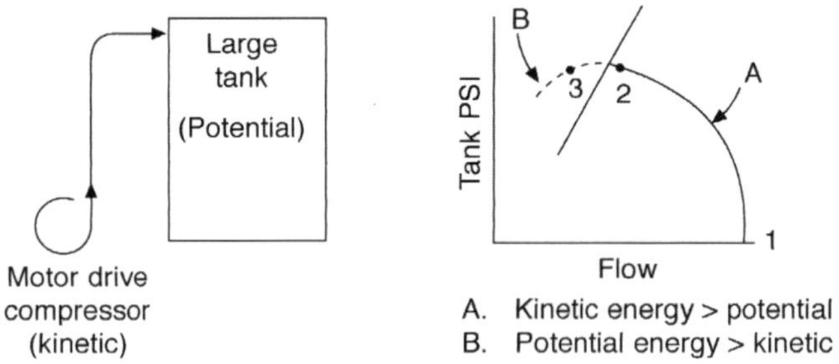

Figure 18: Surge Scenario.

The system consists of a small motor-driven compressor delivering air to a relatively large tank. While in an idle state, the entire system is at ambient conditions. The instant the unit reaches design speed, the pressure in the tank is still zero (Point 1). As time passes, the pressure in the tank builds and the flow is reduced due to increased resistance. Eventually Point 2 is reached where the pressure in the tank causes such a high backpressure on the compressor that flow through the impeller is significantly reduced. Much of the energy input is going to friction instead of building head. This is due to both the mismatch of inlet angle and the longer diffuser passage described earlier. Since this effect continues to build as flow is reduced, the slope of the head curve is reversed. As flow is reduced to Point 3, the head output of the compressor is also reduced. Since the pressure in the tank is still at Point 2, flow occurs from the tank to the compressor. Once the pressure in the tank is reduced (by reverse flow) to a level less than the head capability of the compressor, the process will then recover and the gas will flow from the compressor to the tank. This process will continue to repeat itself indefinitely.

Understanding what the governing principles are that exist in conditions that invoke stall and surge are as important to compressor designers as is the ability to identify the tell-tale signs of a surging unit in the field is to reliability, process, and service engineers. These detectable indicators are also what are used as the inputs to an anti-surge system, which will be discussed later.

Included in Appendix-C is a surge identification procedure. This procedure is written as a guideline for field surge testing of a unit and covers how to monitor audible and vibration characteristics, as well as fluctuating pressures and flow meter differentials at both the inlet and discharge of the machine.

HOW SURGE CAN DAMAGE YOUR COMPRESSOR

There are three main causes of damage to a compressor as a result of surge- temperature, alternating stresses, and vibration. Temperature, or more specifically, high temperature, is the result of the re-compression of gas as it cycles back and forth through the impeller during an excursion of surge.

The thermal growth consequences of the elevated temperatures are of greater importance to axial compressors than centrifugal ones. Both the rotating and stator blades suffer excessive thermal expansion and eventually the rotating components will contact the stationary ones leading to failed broken blades, a phenomenon commonly know as "corn-cobbing" the rotor.

In centrifugal compressors, the impact of increased temperatures in the machine presents adverse effects to temperature-limited materials, such as Lead interstage seals. Especially in refrigeration service, machines that were designed prior to the currently available abradable materials, such as fluorosint and memory-capable thermoplastics, such as Peek, utilized Lead interstage seals to obtain the tightest design clearances possible. The thermal restrictions of Lead, due to its low melting point, were not expected to present a problem in low-temperature applications. However in the case of surge, the re-compressed flow will increase in temperature proportionally to the pressure rise across the stages undergoing surge. If the temperature were to increase enough, the interstage material could become fluid and the pressure will cause severe damage to the abradable portion of the seal permanently opening the seal clearance. Another consequence of high-temperature gas recirculation is thermal stresses on the impeller materials. These may work in conjunction with some of the other alternating stresses, discussed below, to cause equipment failure.

Alternating stress caused by surge occur in mainly three forms- axial, centrifugal, and torsional. Lateral alternating stress is the result of the fluctuation of pressure differential across axial surfaces in the compressor, such as impellers and diaphragms. The results of these stresses are usually impeller cracks and failures. When cracked or failed impellers are examined after the fact, a type of fatigue cracking is observed, usually at the joint between the impeller hub or cover and the leading edge of the blade, called "bow wave" cracks (Figure 19).

Figure 19: Bow wave cracks.

Also examining the crack surface will reveal "beach marks" when the crack was caused by alternating stresses causing fatigue failure after a high number of cycles. Cracked diaphragms are also a possibility for the same reasons as the impellers, but usually their relatively robust design will outlast the impellers.

The alternating axial loading can also create damage to the thrust bearings in a compressor. The repeated pounding of the rotor's thrust disk on the thrust bearings cause the thrust button in the back of the bearing shoe to wear into the leveling blocks behind it. This is a result of the fluctuating thrust loads as stage differential pressures oscillate during surge. This will cause the total rotor's float to be increased.

Centrifugal alternating stresses are a result of the speed fluctuations associated with surge. In the case of hard surge of a large, turbine-driven compressor, the forces exerted on the rotor due to centrifugal alternating stresses may cause the same high-cycle fatigue cracking, and eventual failure, of the impellers as in the case of the axial alternating stresses [Kushner pg 54].

Torsional alternating stresses are usually not of high enough amplitude to cause any damage. Although, during an off-design speed condition if the machine is operating on the rotor's torsional critical frequency, excitation due to surge can prove disastrous.

As with alternating stresses, the high radial vibration associated with surge can cause damage and eventual failure of compressor components if allowed to persist for long periods of time. The effects of radial vibration, regardless of its origins, are usually the increase of the clearances in the journal bearings, interstage seals and balance piston seal. Another effect, similar to torsional alternating stresses, is the excitation of the rotor's lateral critical frequency. The machine should be designed to operate far enough away from its critical speed under normal conditions that this should not be a factor, but when transitioning through an off-design speed, as in the case of start-up and shutdown, any sustained excursions of surge can excite the lateral critical when close to the critical speed.

Some case studies describing failures associated with surge and stall phenomenon are presented below in Appendix-D.

HOW CHOKE CAN DAMAGE YOUR COMPRESSOR

Problems with centrifugal compressor operation in stonewall where a stage is actually choked are rare, but have been reported. The rarity may mainly be due to the fact that almost all compressor discharge systems choke before an internal component. In addition, the driver is often limited in power availability to reach stonewall. There have been experiences in which impeller failures have definitely occurred with a combination of high liquid ingestion along with choke. These are described in references given below. Failures in others are suspected, although not proven, including dry gas and air compressors. In addition the last reference given below, written by another vendor, describes a case for natural gas compressors where choke operation during trip-out of the sister parallel unit was reported to be the cause; in this case impellers at volute stages failed. The eventual solution was to install vaned diffusers so that the diffusers choked instead of the impeller throats.

"In the vast majority of cases, process system resistance aids in preventing operation with deep choke. Parallel operation where shutdown of one unit can drive the remaining unit on-line into choke is suspected for another plant, also with high potential of liquid ingestion, giving similar failures at the leading edge area. Opening blow-off valves at high flow, (instead of the typical function of opening to protect from surge), is suspected for the direct cause in another unit. The solution to both vibratory modes of failures for Case No. 3 was to eliminate excessive liquids and to operate in the confines of the designed operating map."

"For gas processes, system resistance normally prevents severe impeller overload at high flow. Compressors in parallel and other operations with reduced system resistance, including start-up, where an impeller can actually choke should be avoided for significant time periods accumulating high number of cycles. This is especially true if there is likelihood of liquid ingestion."

Kushner, Richard, and Strickland (2000)

"Operation of radial compressors in choke, especially for last stage impellers could present added interaction resonance concerns, especially for high gas density and loads. Running in deep stonewall should be avoided, only having transients with limited number of cycles, just as is recommended for surge. This is especially true with potential liquid ingestion, which also must be avoided since it intensifies engine order resonance, and has caused at least one verified case of loss of overall damping. Better on-line monitoring procedures should be developed..."

Kushner (2000)

HOW TO AVOID SURGE

In order to avoid damage from surge, as described above, some type of anti-surge control system must be used. As a minimum, an anti-surge system must include a check valve located on the compressor discharge and a valve to relieve discharge pressure between the compressor discharge and the discharge check valve. Figure 20 shows such a system.

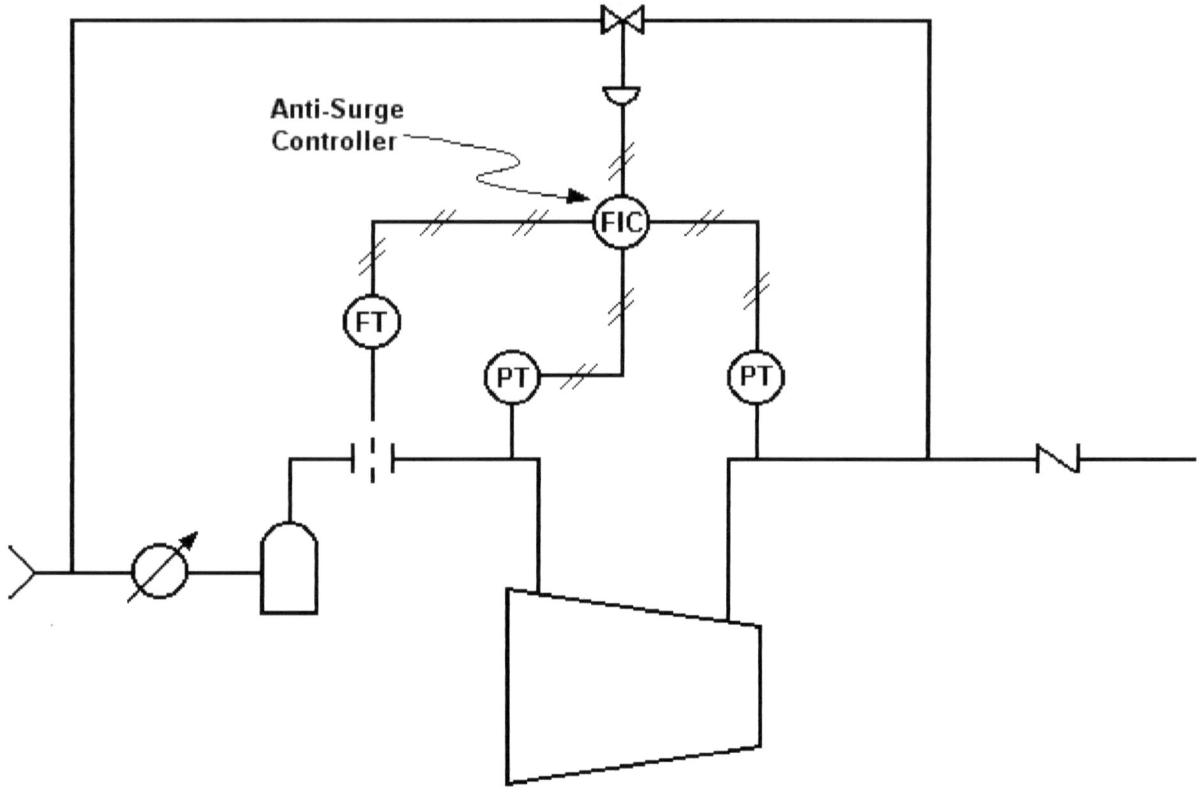

Figure 20: Typical anti-surge system.

For air compressors, the above valve would vent the discharge air to a silencer and then to the atmosphere. It would then be called a "Blow-Off" valve or "Vent" valve. If the compressor is a gas compressor the valve must return the gas to the compressor inlet (after cooling) and is called an "Anti-Surge" valve or "Recycle" valve.
The check valve must be located as near as possible to the compressor discharge to minimize the amount of gas that is trapped in this pipe segment which must be handled by the Blow-off or Anti-Surge valve during transient or off design operation. It is impossible to size the Blow-off or Anti-Surge valve properly if the check valve is not properly positioned.

The valve must be sized, as a minimum, large enough to handle compressor flow near surge. However, a valve sized to the minimum criteria would have no reserve capacity to account for some variation in actual surge flow versus predicted and no reserve capacity to handle the trapped gas in the discharge pipe described above. There is no "one size fits all" rule that covers these valves. Experience in field installations and in using dynamic simulations, has indicated that a valve sized to pass 150% of the flow at surge will be adequate to protect most installations of this type. However, if a cooler (or other process vessel) is located in the compressor discharge pipe upstream of the check valve, the additional volume may require a larger valve. The valve size can best be verified by using a dynamic simulation of the system. The valve stroking speed should be typically 1 to 2 seconds in both directions. This speed can usually be obtained with pneumatic actuators and positioners. Slower speeds may be adequate for larger compressors that typically have slightly longer surge cycles. This can best be verified with a simulation.

In addition to the check valve and Blow-Off or Recycle valves, the anti-surge system must have a means of measuring compressor flow, inlet pressure, discharge pressure, and an anti-surge controller. Figure 20 shows an orifice located at the compressor inlet and pressure transmitters at the inlet and discharge all connected to the anti-surge control. The anti-surge controller is a PID (Proportional, Integral, Derivative) controller with special features needed to implement anti-surge controls. It typically employs a means to open the valve quickly while restricting the closure speed (prevents instability), a means to open the valve more quickly in case the flow crosses the anti-surge line, and surge recognition to over ride the normal anti-surge control to prevent repeated surging. There are many methods used to configure anti-surge controls. The optimum method can best be determined by plotting the compressor surge line for expected changes in compressor inlet conditions.

Using the compressor shown in Figure 20 and with characteristics shown in Figure 21, various control algorithms are plotted (surge line) for changes in compressor inlet gas temperature conditions (Figures 22, 23, and 24).

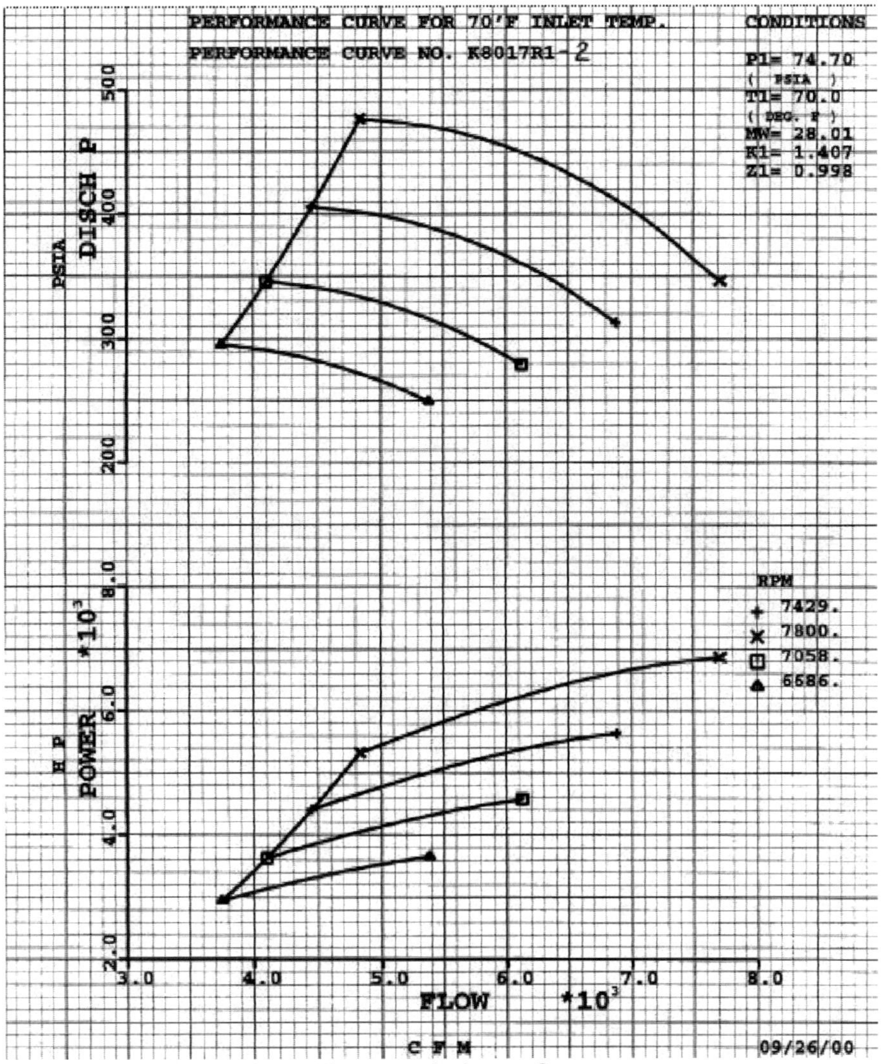

Figure 21: Characteristic performance curve.

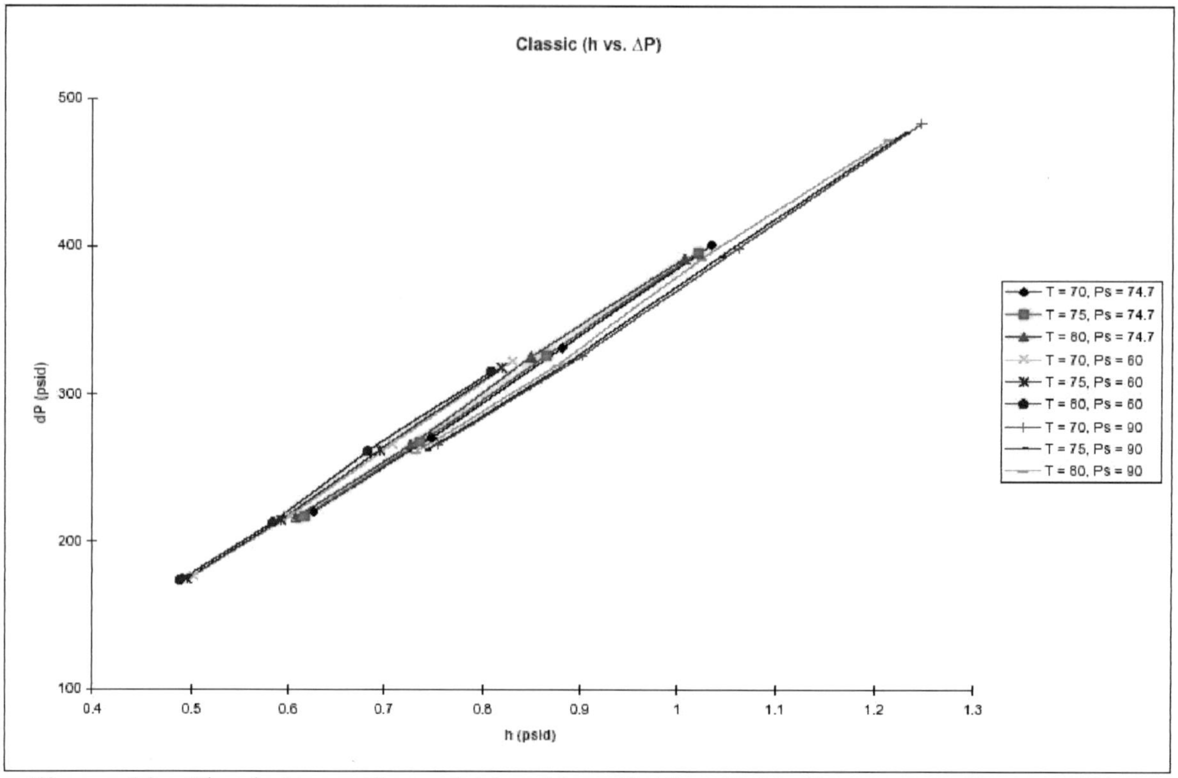

Figure 22: Classic Surge Control Algorithm- Orifice dP Vs Compressor Pressure Rise.

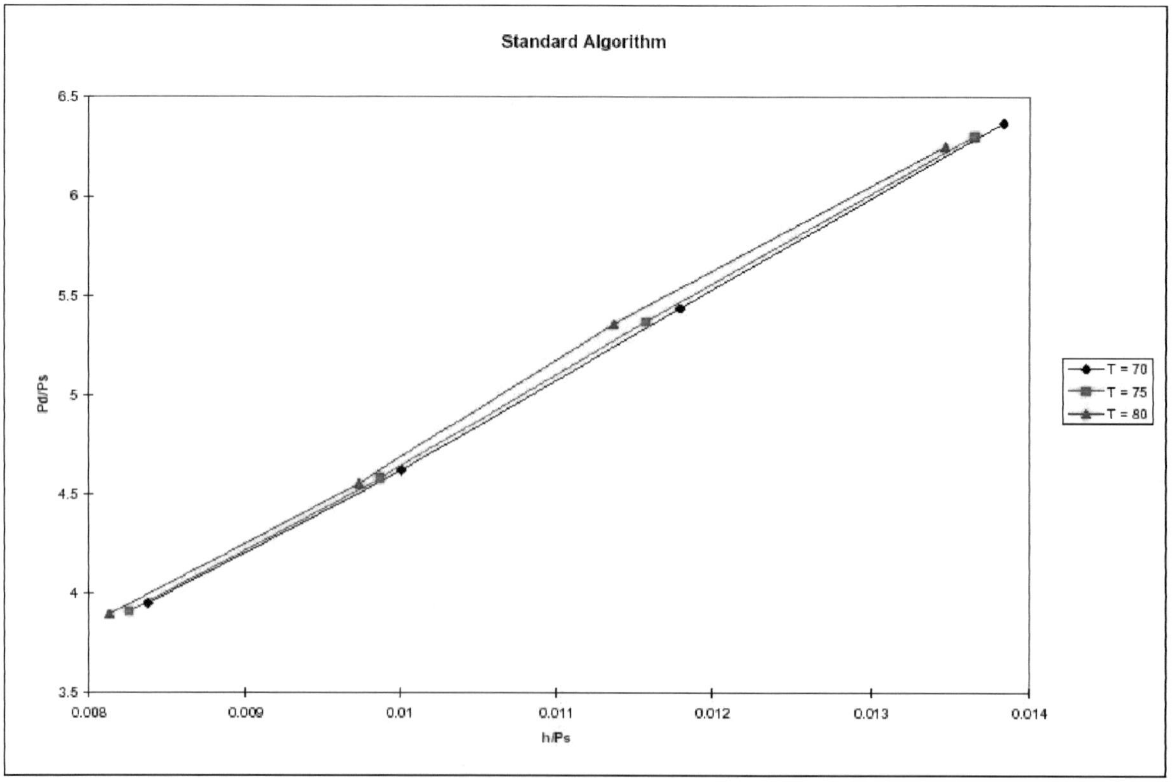

Figure 23: Standard Algorithm- Discharge Pressure/Suction Pressure Vs. Orifice dP/Suction Pressure

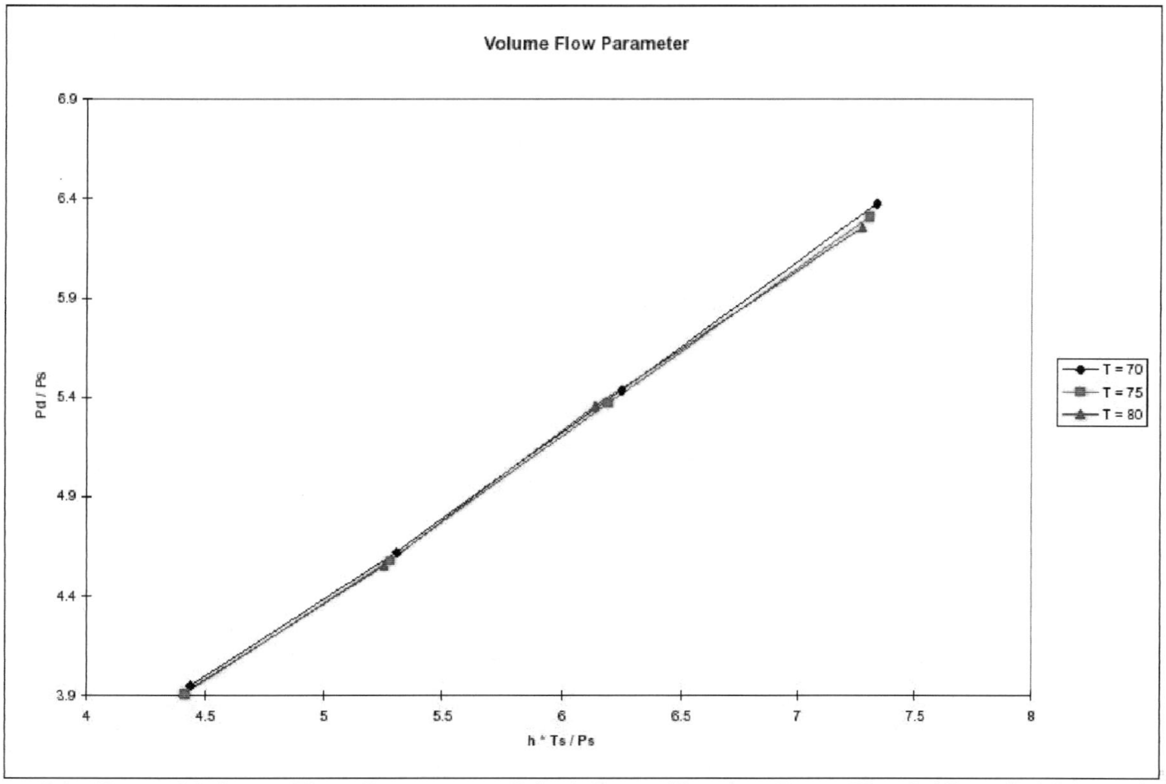

Figure 24: Surge Volume Flow Parameter- Disch Press/Suction Press Vs. Orifice dP*(Temp/Press)

Each surge line consists of four speed points. Similar plots can be made for changes in molecular weight and inlet pressure. The optimum algorithm is the one that shows the least deviation from a straight line when all surge lines are plotted together. This may require adding temperature measurement for compensation.

Once the control algorithm is established, a control line is calculated to provide a safety margin to the right of surge (see Figure 25).

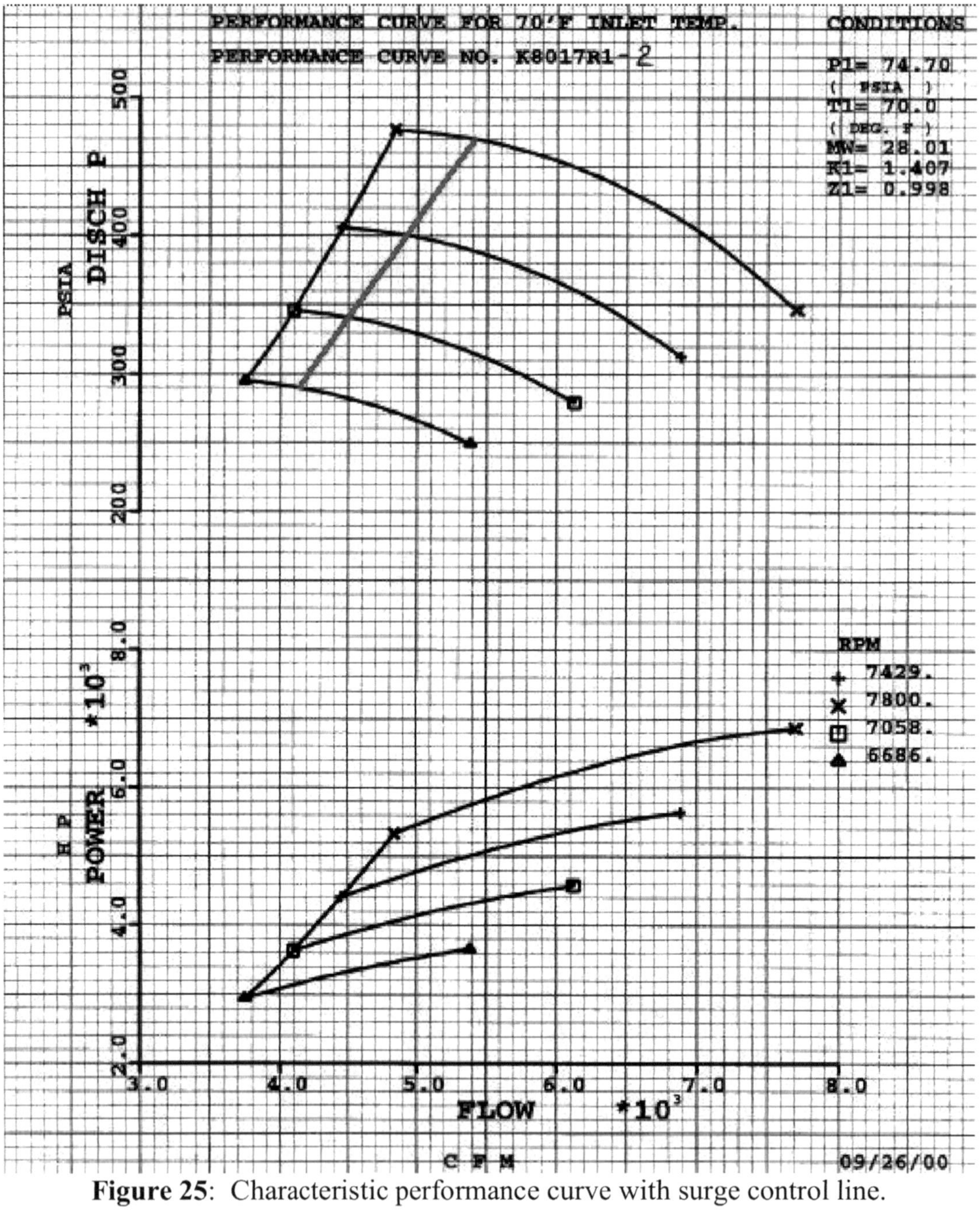

Figure 25: Characteristic performance curve with surge control line.

The control line is typically located 10% away from surge. This line is easily movable toward or away from surge depending on operating experience. Most systems will automatically increment the control line away from surge if surge is detected.

Surge may be detected in several ways. One method is to monitor the flow measurement rate of change. Since surge is defined as a flow reversal, the system can be set up to detect a rapid reduction in flow. If the flow reduction is above the preset amount, the incident is considered a surge. A similar method monitors discharge pressure. If discharge pressure drops rapidly it can be considered a surge. A slightly different method is to monitor flow and if compressor flow is reduced below surge flow, the incident is considered a surge. Incidents of surge are usually counted as a means to determine if the system is working properly.

In summary, the above discussion deals with a basic anti-surge system for a single section compressor. Multi-section compressors complicate this significantly in both control logic and in piping considerations.

Simulation is the best way to study anti-surge system dynamics. Using the system described in Figures 20 and 21 and assigning representative volumes and pressure drops, a model was constructed. Three valve sizes were studied (1.05, 1.5, and 1.8 X surge flow) as to their ability to prevent surge following a discharge flow blockage. These results are shown in the attached (Figures 26 and 27 using the 60 cu ft curves).

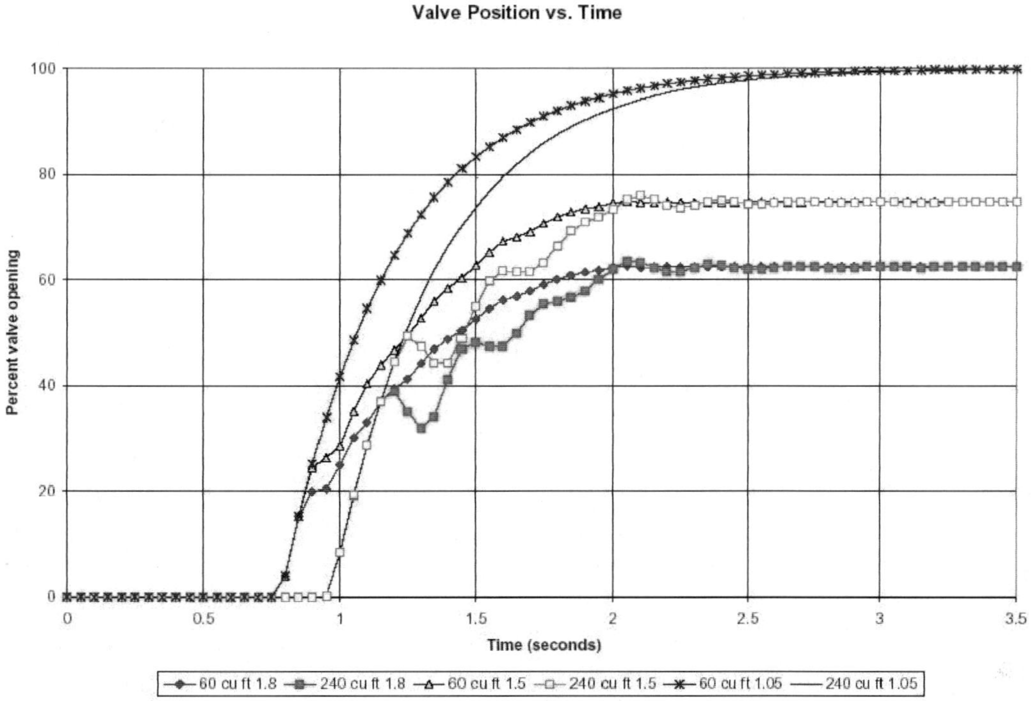

Figure 26: Valve Position Vs. Time

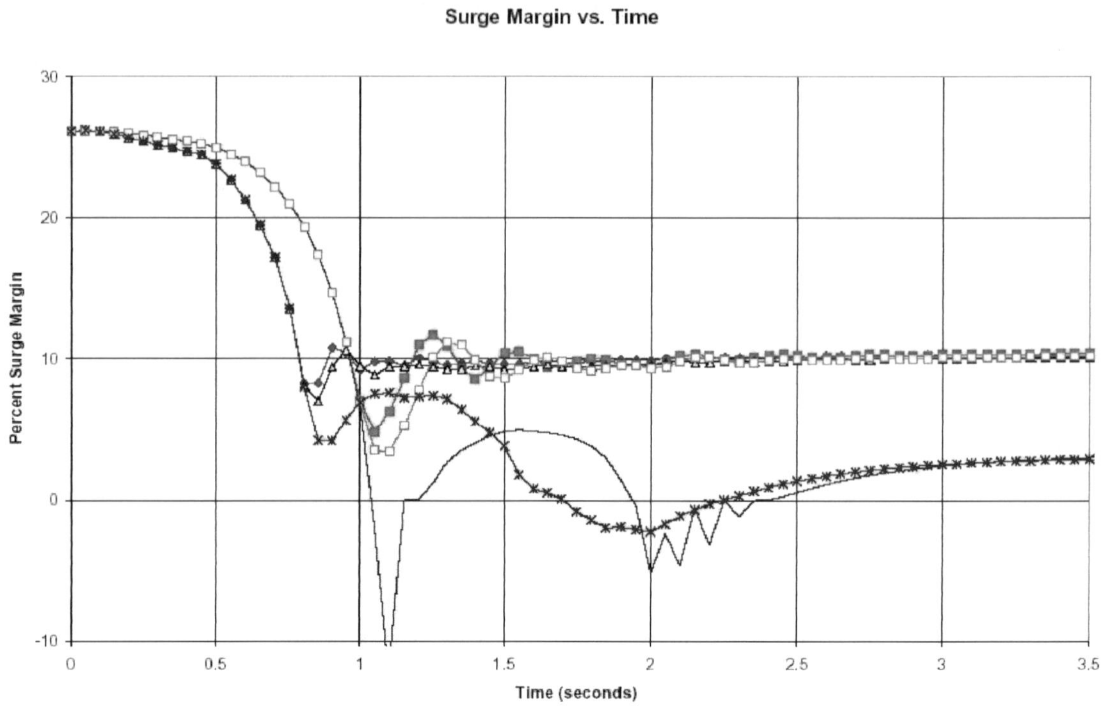

Figure 27: Surge Margin Vs. Time

The smallest valve (5% greater than surge flow) shows it is just adequate to prevent surge during this transient. The valve sized for 1.5 X surge flow is shown to be adequate and a valve sized for 1.8 X surge flow is shown to be larger than necessary. Valves sized excessively large could result in compressor operation in stonewall. Long-term operation in stonewall has been shown to cause compressor impeller damage.

If volume is added to the compressor discharge (60 cu ft increased to 240 cu ft), the job of the anti-surge valve to prevent surge is made more difficult. The simulation was then re-run using the same discharge flow blockage transient with the same valves that were used in the first study. These results are shown as the 240 cu ft curves. These show that the larger valves perform better than the smaller valves by relieving more of the trapped gas in the larger volume discharge segment. Surge is predicted for the 1.05 X surge flow valve but not for the 1.5 and 1.8 X surge flow valves.

While the above explanation covers only the basics of anti-surge systems, it is hoped that understanding has been enhanced.

HOW TO AVOID CHOKE

As stated above, choke is usually self-avoiding as most drivers lack the power required to operate the machine far out into the choke region of the performance curve. But avoiding operation beyond the end of the manufacturer's curve is important to minimize the possibility of damage especially if liquid ingestion potential exists at these high associated flows. The best way to avoid operating in a choke condition is to use an online performance monitoring system when available. Without an online performance system, the operator must make periodic manual field performance evaluations, especially when choke operation is suspected.

Operation of the machine in choke is not easily detectable, and there are no warning signs to the failure of internal components at this operating condition. As the machine is pushed beyond choke, the possibility of liquid ingestion in refrigeration machines increases and the corresponding risk of a failed impeller increases. By monitoring the machine's on-line performance, an actual operating point can be placed on the compressor's performance curve and the operator can limit the machine's operation. Generally, the ends of the performance curves are accepted to be the limit of safe operation for a given speed line.

IN CLOSING

While surge has long been recognized as being a source of equipment failures, it is just recently that choke has been understood to be a source of impeller failure as well. Respecting the machine's limits in regard to surge and choke are the key to ensuring long-term reliability. In most cases, the costs involved, and time required, recovering from a failure due to surge or liquid ingestion due to choke much outweigh the amount of profit gained due to the period of unsafe operation. It is the equipment manufacturers' responsibility to design and build a compressor that meets the purchaser's varying needs, but it is the end users' responsibility to operate the machine according to the intended design.

APPENDICES

Appendix-A

AN EXAMPLE OF FREQUENCY ANALYSIS TO DETERMINE ORIGIN OF STALL

Results of the dynamic pressure transducer measurements for the test stages discussed above indicated that both impellers exhibited one cell rotating stalls. In impeller A, at design tip speed, one-cell stall was verified by phase analysis, and indicated the cell was rotating at 65 percent of the compressor rotor speed. This confirmed that the stall initiated within the impeller. In Figure A-1, the pressure pulsation frequency at 50 Hz was indicated first; then as the flow was lowered, another pulsation at 4.5 Hz appeared.

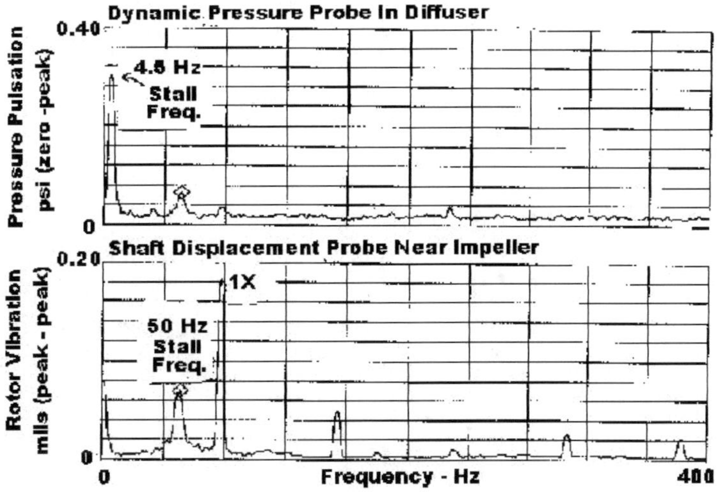

Figure A-1: Pressure Pulsation and Rotor Vibration Frequency Spectrum after Peak of Head Curve for Stage A.

The second pulsation was due to the diffuser, which had a two-cell rotating stall at 4.5 Hz, traveling near three percent of rotor speed. Note the significant rotor vibration at 50 Hz, but negligible response at 4.5 Hz frequency. CFD analysis after the test had shown that the stall initiated near the middle of the meridional passage and propagated back toward the eye as flow was reduced. This may be why the stall speed was lower than the 80 to 95% impeller stall that is typically found when stall is initiated at the blade leading edge.

During tests for accepted Stage B, at design tip speed, the dynamic pressure probes showed there was a two-cell rotating diffuser stall that developed, immediately followed by a one-cell impeller stall with frequencies as shown in Figure A-2.

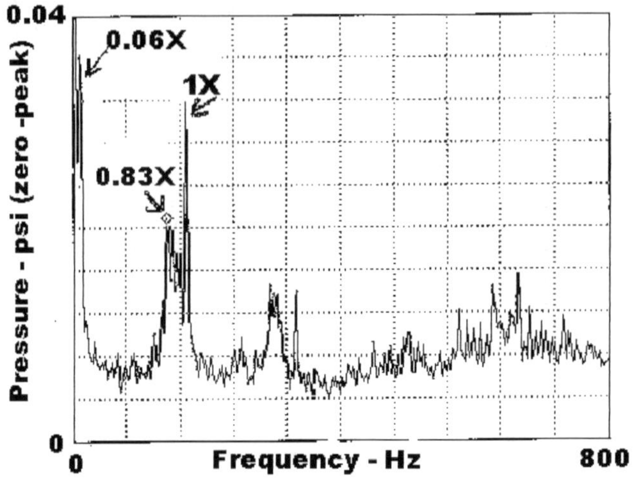

Figure A-2: Pressure Pulsation - Frequency Spectrum Analysis at Peak of Head Curve for Stage B.

The two-cell diffuser stall had a frequency of 9 Hz or 0.06 times speed, and thus was rotating near three percent of rotor speed (0.06 X / 2 = 0.03 X). The impeller one-cell stall, with frequency equal to 139 Hz or 0.83 times speed, was rotating at 83% of rotor speed. There was no significant rotor vibration at the corresponding frequency of pressure pulsation at 9 Hz, whereas the one-cell stall did show a mild response. At higher impeller tip Mach numbers; there was only multi-cell diffuser stall just before surge. Surge is the condition of full flow reversal that was proven by the fact that pressure pulsations at surge frequency were in phase for all three dynamic pressure probes.

Appendix-B

EXAMPLE OF CALCULATING EQUIVALENT TIP SPEED

We know the sonic velocity of air at 80°F is 1140 fps, and that of propylene at –40°F is 740 fps. If a compressor stage is operating at a mechanical tip speed of 780 fps on propylene at –40°F, the equivalent tip speed is:

$$U_{eq} = 780 \times \frac{1140}{740}$$
$$= 1200\ fps$$

The stage characteristic head curve shape at 780 fps on propylene is therefore the same as 80°F air at 1200 fps. (Note: This is not exact. There is also an impeller tip volume ratio effect based on gas density that causes head and work input to change somewhat. An air and propylene test will not result in exactly the same head curve at constant equivalent speed, but for all practical purposes the results are close enough to be considered the same.)

Appendix-C

SURGE IDENTIFICATION
The following is the preferred procedure for establishing the location of the surge point.

1) Slowly close the recycle or blow off valve, while monitoring the following parameters:
   a. Blow off or recycle valve position, % open.
   b. Audible sound level at the inlet of the compressor. Listen for a pulsing sound.
   c. Audible sound level at the compressor discharge. Listen for a pulsing sound, a low frequency 0 to 25 Hz.
   d. Compressor suction pressure immediately upstream of the compressor inlet flange. Monitor both the local pressure gage (for low pressure, a water manometer works well) and the pressure transmitter. Watch for a bouncing in the pressure level. Note that the pressure transmitter may not show this unless it is rated as a dynamic device, with a rise time below 0.1 sec. When the dynamic amplitude exceeds 20% of the gage static pressure or the compressor pressure rise, consider the unit to be in surge (Figure C-1)

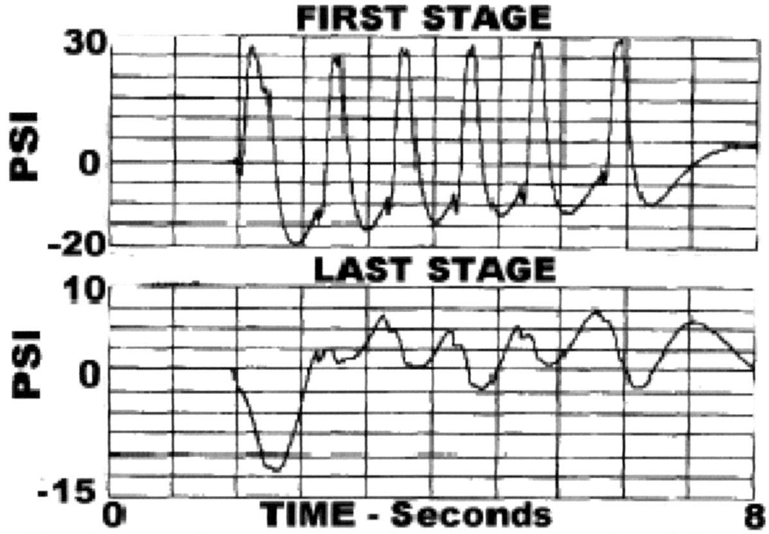

Figure C-1: Simultaneous Pressure Pulsations During Part-Load Testing at Front and Rear Stages for a 14-stage Air Axial Compressor.

   e. Compressor discharge pressure near compressor discharge flange (see Figures C-1, C-2, & C-3). As the blow off valve is slowly closed, the pressure will rise. Monitor both the pressure gage and the pressure transmitter. Watch for the pressure to bounce (see "d" above). Also watch for any drop in pressure. At the first indication of a drop in pressure (with decreasing flow), consider this to be surge, and record data.

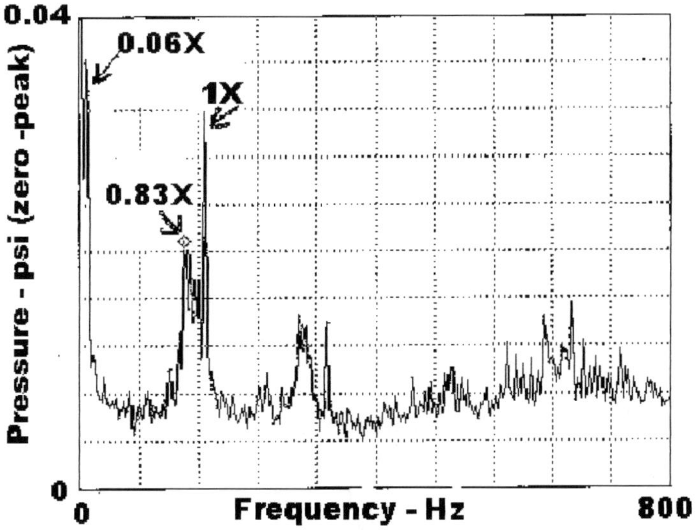

Figure C-2: Pressure Pulsation - Frequency Spectrum Analysis at Peak of Head Curve for Stage B.

f. Compressor flow rate. Watch for fluctuations in the flow meter differential pressure. Note that an electronic output on the flowmeter will not indicate surge unless the device is rated for dynamic conditions, with a rise time below 0.1 sec. It is best to locally attach a manometer (for low pressure) or differential pressure gage and monitor this. Any dynamic differential pressure in excess of 20% of the nominal (steady state) differential at the given flow rate is to be considered surge, if no other indications (c, d, e, or f) are observed.
g. Compressor vibration level. Pay particular attention to subsynchronous amplitudes. Very small increases or bouncing of amplitudes indicate possible onset of surge. An increase of 20% at the given speed of the overall vibration level, or 0.20 mils increase of the subsynchronous, while alone not a sign of surge, indicates the proximity of an instability. Use extra caution when exceeding these values.

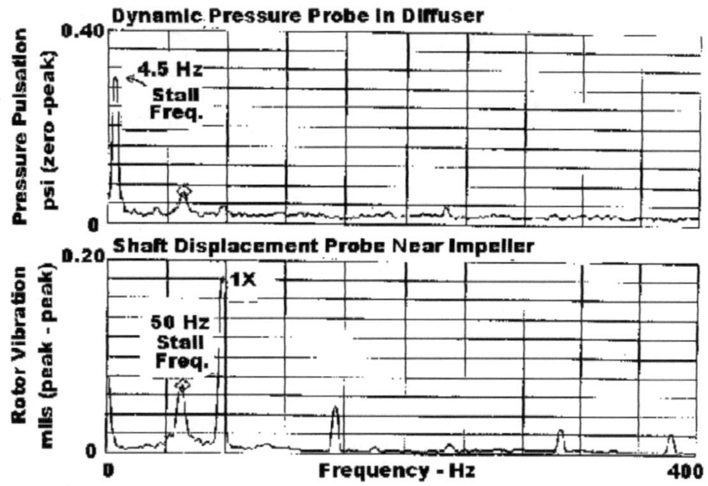

Figure C-3: Pressure Pulsation and Rotor Vibration Frequency Spectrum after Peak of Head Curve for Stage A.

2). When any of the above items (except the peak head condition, e above) indicates surge, the position of the blow off valve should be immediately noted, and then opened to the full open position.
3). Close the valve back to within a few percent of the point where the instability occurred. Example: The suction pressure began to bounce at blow off valve opening of 79%. The valve is immediately opened to 100% open. The blow off valve is then closed back to 81% open.
4). Additional time period until data is stable and record data to assure the system is stable.
5). Repeat steps 1 through 4 for the other speed lines, or inlet guide vane positions.
6). Record all data for future reference.

Note that the ideal method of detecting the point of aerodynamic instability is to monitor dynamic pressure probes near the inlet to the impeller and in the diffuser. Flow instability can develop in either location. In some units it appears in the inlet due to flow separation on the inlet of the impeller blades. The position of this point on the compressor head curve generally lines up with the point of peak head. On other units, stall will occur in the diffuser section. This is caused by the inability of the diffuser to overcome the compressor discharge pressure. This event may not fall in line with peak head.

Sophisticated instrumentation is not required to detect surge. The instability usually can clearly be heard and even felt when standing near the compressor. Sometimes the instability is subtle and you must listen very closely.

If you are standing in the compressor discharge area, you may not hear an inlet stall condition. Likewise, if you are in the control room observing the flow and pressure on the slow responding process monitor equipment, you may only see a deep hard surge condition, when it occurs.

Keep in mind that too much surging will eventually cause equipment failure. Surge the equipment hard enough and long enough and something will eventually break. When setting the surge line, the equipment should experience only one or two surge pulses. Allowing the unit to surge any more is only asking for trouble. In order to accomplish this the recycle or blow off valve must have a quick opening (1 to 2 sec.) response and a slow (10-20 sec.) closing time to keep the system stable.

Be safe and assume that the unit is very sensitive to surge and that the machine could easily wreck if surged very much.

Appendix-D

CASE STUDIES

Case Study #1

In March 1993, a single stage air compressor (Figure D-1) in a desulphurization plant in western Canada suffered a major failure after approximately three weeks of operation. Damage was extensive, not only to the impeller (Figure D-2) but to the casing, bearing housing, shaft and other associated parts as well.

Figure D-1: Failed impeller.

Root cause of the problem was eventually traced to operation in a stall condition or a condition that many would call "incipient" surge. During set up of the anti-surge system, the compressor was operated farther and farther beyond stable operation into an area where flow separation was occurring, while full hard surge never did appear. Eventually vibration levels increased due to the stall condition and this point was defined as the surge point for the sake of the anti-surge system.

The flow separation across the impeller blades caused the blades to flutter and eventually initiate a crack in one of the blades during the period the anti-surge system was being set up. Once this crack was initiated, the crack propagated with normal operation of the compressor that eventually led to failure of the blade. The impeller blade jammed in-between the impeller and casing causing the damage described above.

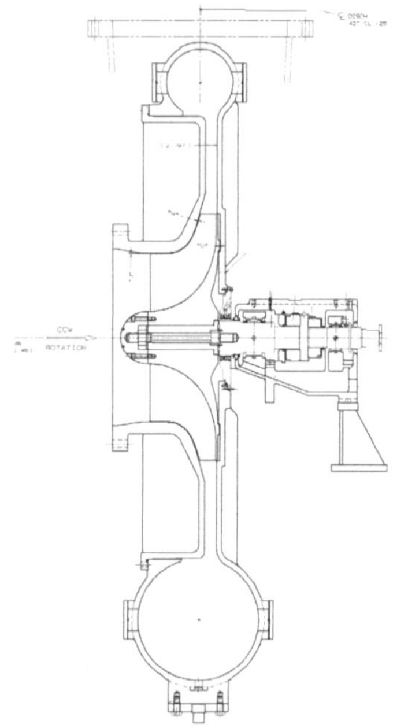

Figure D-2: Single stage air compressor.

Case Study # 2

In 1997, a newly installed propylene refrigeration compressor tripped on high vibration after being in service for just 20 hours (Ref. 1). During this period, the compressor was operating entirely on false load with low vibration levels. Nitrogen was being bled from the receiver drum for the first few hours of operation to remove noncondensibles from the closed loop system. The unit tripped on high vibration after 20 hours.

Inspection of the compressor found that the cause of the high vibration was a failure of the last stage impeller. As shown in Figure D-3, a triangular section of the impeller back plate was missing.

Figure D-3: Damage at tip of impeller after operating in choke for 20 hours.

A close look at the operating data prior to the failure indicated that the density of the propylene was very high at the first suction and side load suction. Both sections of the propylene compressor were operating at or beyond the "choke" point (end of the OEM performance curve, Figure D-4).

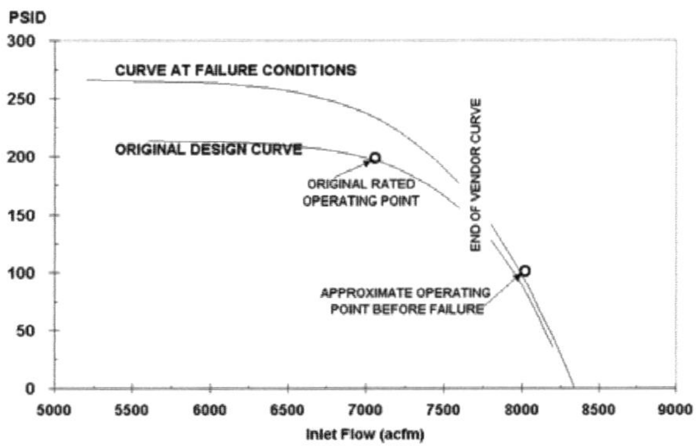

Figure D-4: Pressure rise vs. inlet flow. Note how the compressor operating point is well beyond end of vendor curve.

The users experience with operating multistage compressors with enclosed impellers in or near choke had not been considered detrimental to the compressor, unlike surge, thus the continued operation at these conditions for such a long period of time (20 hours).

Case Study # 3

In the early 1970s four duplicate 30,000 hp barrel compressors for mixed-refrigerant gas were experiencing numerous covered impeller failures at a natural gas liquefaction plant in Libya. The cause of the failures was eventually found after using a task force including many consultants. Many believed that the root cause was excessive surging. Strain gauge readings on the impellers eventually proved that crack initiation was due to liquid ingestion, while crack propagation involved excessive surging as well. Field data (Figures D-5, D-6, and D-7) show transients for flow reversal with corresponding pressure and thermal loads.

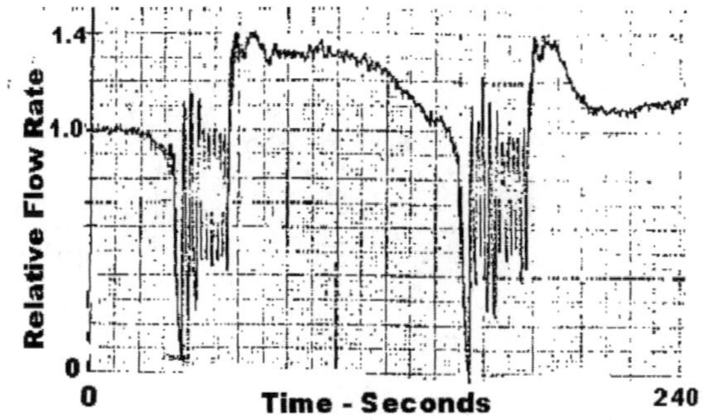

Figure D-5: Multiple Surges of Mixed Refrigerant Compressor Showing Gas Flow Variations

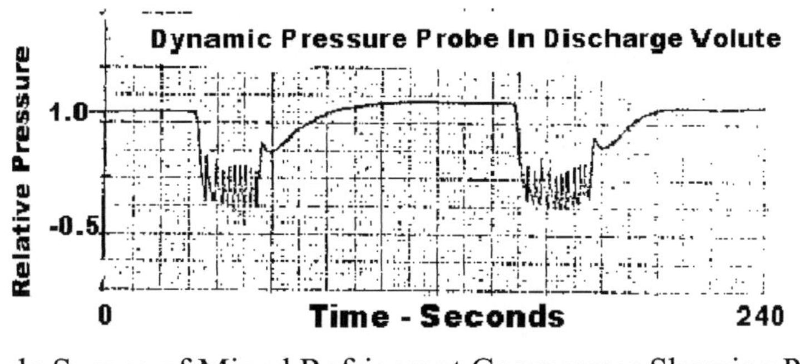

Figure D-6: Multiple Surges of Mixed Refrigerant Compressor Showing Pressure Pulsations in Discharge Volute

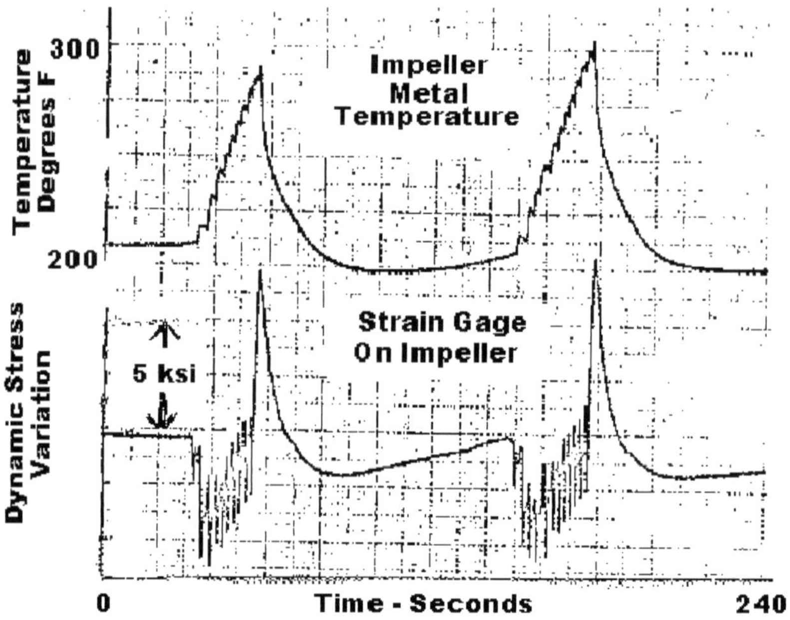

Figure D-7: Multiple Surges of Mixed Refrigerant Compressor Showing Last Stage Metal Temperature and Dynamic Stress near Crack Initiation Point.

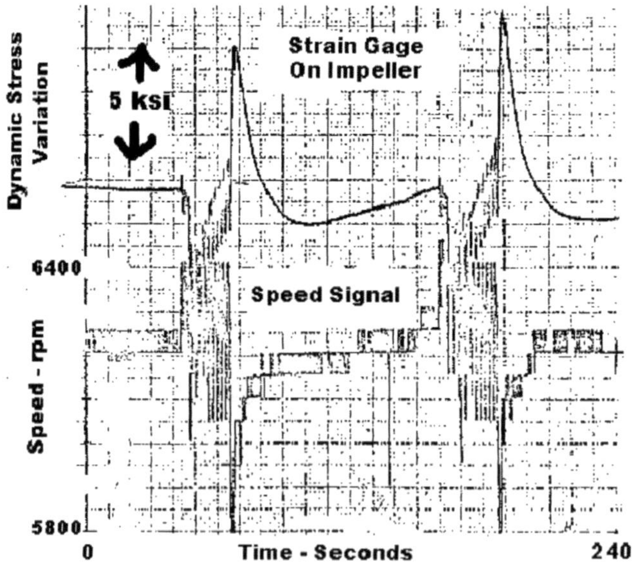

Figure D-8: Multiple Surges of Mixed Refrigerant Compressor Showing Last Stage Dynamic Stress Near Crack Initiation Point and Rotor Speed Variations caused by reverse flow and corresponding torque reversals.

References:

Gresh, M. T., 2001, *Compressor Performance: Aerodynamics for the User*, Second Edition, Boston, Massachusetts: Butterworth-Heinemann.

Kushner, F., 2000, "Compressor Blade And Impeller Rotating Disk Vibration Avoidance Parameters", ASME Conference: "Challenges and Goals in Industrial and Pipeline Compressors", Orlando, Fl, PID-Vol. 5; pp. 81-89.

Kushner, F., Richard, S., and Strickland R. A., 2000, "Critical Review of Compressor Impeller Vibration Parameters For Failure Prevention", Proceedings of the Twenty-Ninth Turbomachinery Symposium, Turbomachinery Laboratory, Texas A&M University, College Station, TX, pp. 103-112.

Kushner, F., Walker, D., and Hohlweg, W., 2002, "Compressor Discharge Pipe Failure Investigation with a Review of Surge, Rotating Stall, and Piping Resonance", Proceedings of the Thirty-First Turbomachinery Symposium, Texas A&M University, College Station, TX

Shepherd, D. G., 1956, *Principles of Turbomachinery*, New York, New York: Macmillan Publishing Co., Inc.

Minimum Flaring -

Impact on Compression System

Greg Dunnells and Kazim Akhtar

ABB Lummus Global
3010 Brairpark Dr
Houston, TX 77042

Prepared for Presentation at the 2004 AIChE Spring National Meeting

Greg Dunnells
Copyright © ABB Lummus Global

New Orleans, LA, April 25 – 29, 2004

Rotating Machinery Reliability – Impact of Operational Practices

Unpublished

April 26, 2004

AIChE shall not be responsible for statements or opinions contained in papers or printed in its publications.

Abstract

Recent environmental regulations have targeted air emissions caused by flaring, thus making flare minimization even more important. Reportable quantities of flared pollutants such as 1,3-butadiene and benzene are as low as 10 pounds per incident. In addition, NO_x generated by the combustion of the pollutants must also be reported. A recycle system can also be used to recycle off-specification or intermediate products to minimize flaring.

Thus, designing a system that minimizes plant cost and helps to minimize flaring during start-up or shutdowns is highly desirable. An example of this is to use the charge gas compressor to recycle inert gas to the cracking heaters in order to generate additional super-high pressure (SHP) steam. This minimizes the required start-up boiler size, which in turn reduces plant investment.

Introduction

One effect of designing a highly efficient ethylene plant is that most, if not all, of the steam required to operate the plant is normally generated by the cracking heaters. Import steam may not be required to operate the plant under normal operating conditions, but a large amount may be required for upset conditions when flaring occurs. In the past, ethylene plants had large capacity boilers designed for start-up requirements and these were operated at some lower rate during normal operation. During an upset, the extra boiler capacity could be used to maximize the smokeless capacity of the flare.

In the past, it was generally acceptable to start fresh feed to the heaters in order to generate steam to power the compressors; however, this resulted in long periods of charge gas flaring until the charge gas compressor was started. Recent environmental reporting requirements penalize charge gas flaring because the gas contains large quantities of both 1,3 butadiene and benzene.

Thus a system that minimizes both boiler steam and charge gas flaring during start-up and upsets is required.

Plant Design

ABB Lummus Global designed and built a gas cracker plant in Louisiana, USA, that started up in the late 1990's. This plant has a recycle system designed to both assist during plant start-up and minimize flaring during plant upsets.

The recycle system (Figure 1) utilizes piping that is present for normal operation (i.e. the ethane recycle and fresh feed piping). Some additional piping was added to allow for recycle of the charge gas compressor discharge as well as the deethanizer overheads (acetylene converter feed) and the acetylene converter effluent.

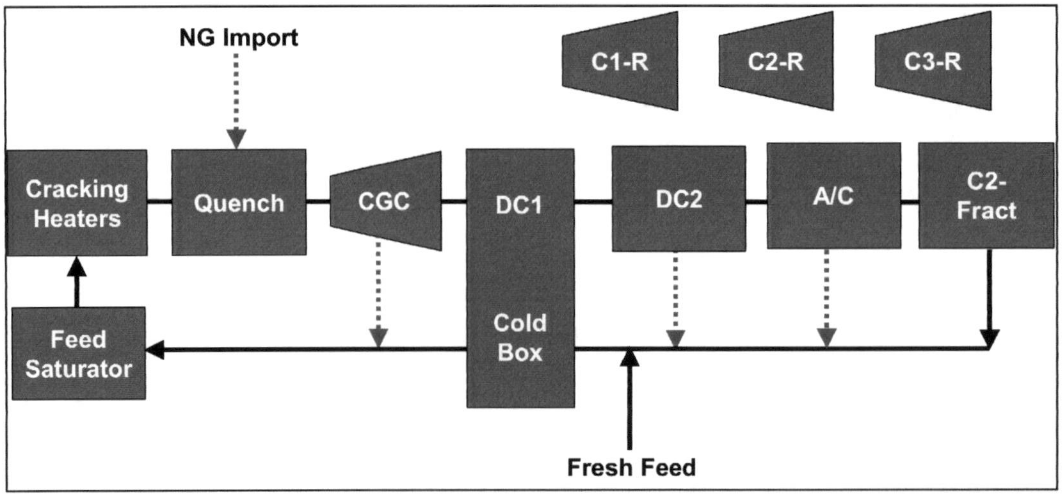

Figure 1. Recycle system.

The piping allows an inert gas, such as methane or nitrogen, to be circulated through the compressor to the feed piping/feed saturator and the cracking heaters. The advantage of using an inert gas is that it can be recycled through the heaters indefinitely without significant degradation. If methane is used, a water gas shift reaction occurs due to the catalytic effect of the radiant tube metal, but it is not significant and the caustic tower can remove any CO_2 produced. Natural gas can also contain significant amounts of CO_2 and this needs to be accounted for during the design. Allowing process gas containing CO_2 through the dryers will cause problems in the recovery section.

Most, if not all, ethylene plants are designed to augment start-up boiler steam by using the cracking heaters to generate some amount of super-high pressure steam. This is done by passing dilution steam through the heaters and waste heat exchangers to the quench tower, where it is condensed and then recycled back to the heaters after being converted back into dilution steam by heat exchangers using medium pressure steam. The disadvantage of doing this is that it takes almost one pound of super-high pressure/high pressure (SHP/HP) steam to generate one pound of dilution steam.

An inert gas has the advantage that it does not condense and therefore steam is not required to revaporize it: it only absorbs sensible heat while it is heated and cooled during circulation. An inert gas can be recycled at a high rate (limited by compressor capacity and system pressure drop) and thus increase firing and SHP steam production in the cracking heaters.

Methane has a higher heat capacity than nitrogen, so it generates the most steam. SHP steam is then used to power the compressors and is extracted at some lower level to generate dilution steam. The plant has a feed gas saturator and dilution steam is generated by heating a circulating water stream using low pressure (LP) steam. This maximizes the available work from the steam cycle. The only disadvantage of using methane is that its molecular weight is significantly lower than charge gas so the compressor cannot generate its normal discharge pressure. The mass flow is also reduced since the compressor is compressing volume, not pounds. Figure 2 is an example of a typical performance map for a charge gas compressor operating with methane.

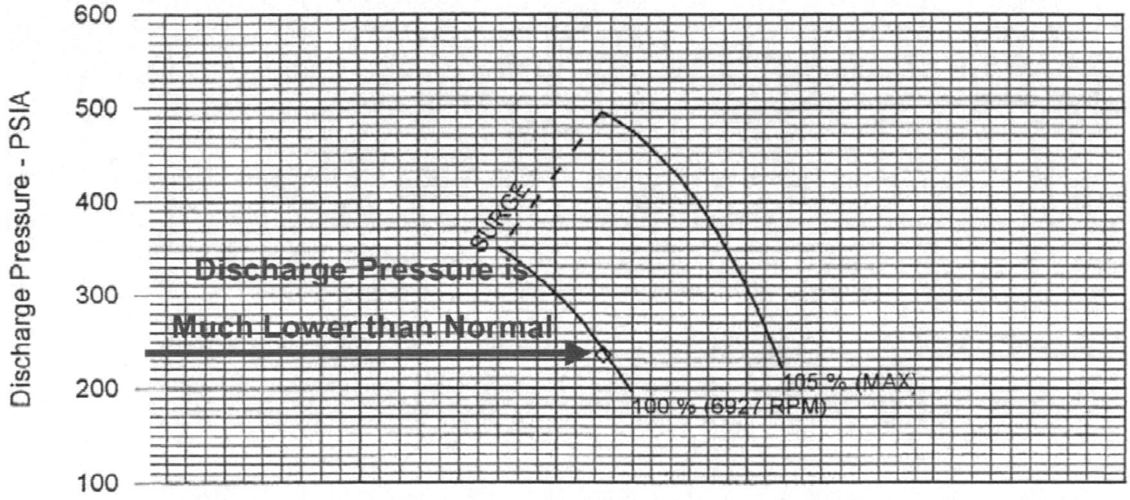

Figure 2 – CGC Performance Curve on Methane

In this case, the compressor will only develop a discharge pressure of around 240 psig, whereas the normal discharge pressure is approximately 500 psig (see Figure 3). This is not a problem if it is recognized during the design and the piping/control valves are designed accordingly. The power required is only about 60% of normal, which helps to reduce the steam demand.

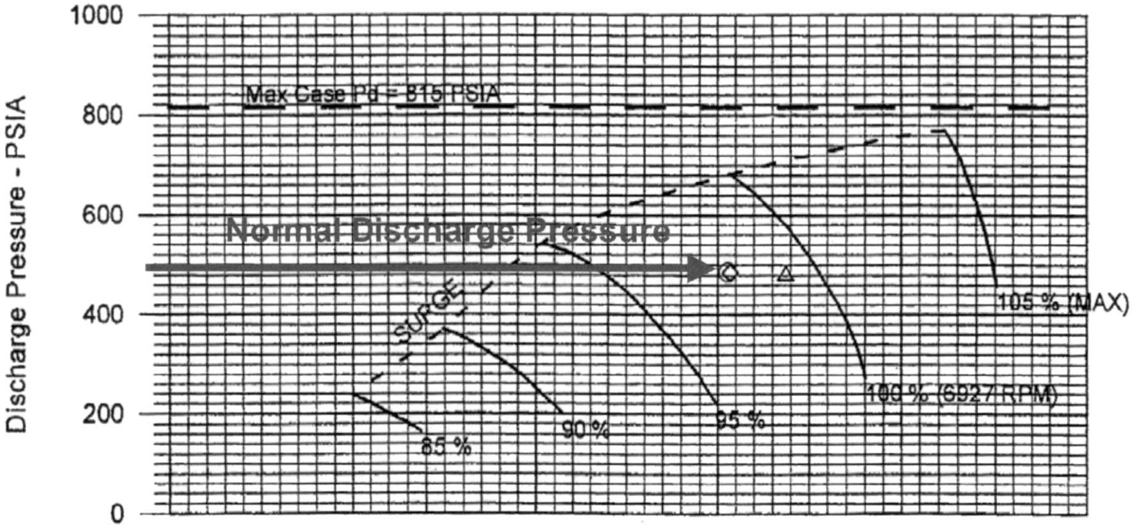

Figure 3 – Normal Performance Curve]

The art versus science of ethylene plant design requires that a realistic steam balance be calculated and that sufficient steam is available for all possibilities, including flaring. It is relatively easy to “save” too much money in a “low cost” plant by leaving out something that is really needed but not recognized (e.g. boiler capacity), or if the designers are too optimistic about such things as turbine efficiency, especially pump driver turbines, which can be notoriously inefficient.

During the design process, several start-up scenarios were investigated and, together with the plant personnel, it was decided that the heaters and start-up boilers would generate SHP steam at 1500 psig. This minimized start-up boiler capacity as the refrigeration compressor turbines operate using steam extracted from the charge gas compressor turbine at the HP level (Figure 4).

Steam for smokeless operation of the flare is generated at the SHP level, let down to the HP level, and further let down to the pressure required by the flare tip.

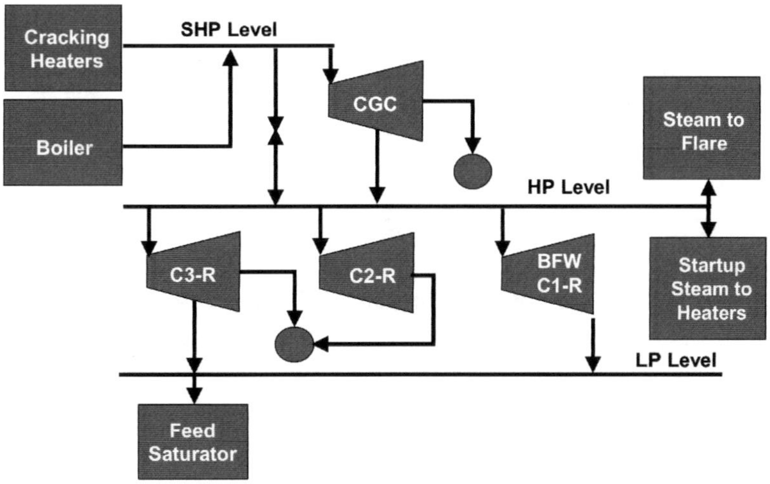

Figure 4 – Steam System Layout

Start-up Philosophy

As part of a minimum-flaring start-up, it is desirable to start the refrigeration compressors and the recovery section distillation towers on total reflux before starting the charge gas compressor and heaters. This minimizes recovery flaring because the recovery section is essentially "on-spec," waiting for feed forward. The plant has a start-up vaporizer to allow products imported from product and feed pipelines to be distributed as a vapor to pressurize the recovery towers before liquid is introduced. This avoids the problem of purging and inventorying the plant with cryogenic hydrocarbons in carbon steel vessels.

Following is an outline of a typical start-up sequence describing how the recycle system is used:

- Inventory and start the propylene refrigeration compressor (C3-R).
- Inventory and start the ethylene fractionator (C2-Fr).
- Inventory and start the ethylene refrigeration compressor (C2-R).
- Inventory and start the deethanizer.
- Inventory and start the depropanizer.
- Start natural gas flow into the quench tower/CGC system. Pressurize until the quench tower pressure is just below the first-stage-suction relieving pressure.
- Start at least two heaters on hot-steam-standby (HSSB). All other heaters should have warming fires but no steam flow. The heater effluent flow should be through the quench tower to start warming the transfer line and quench water system.
- Start the CGC and bring to minimum governor speed while maintaining positive suction pressure.
- Increase compressor speed slowly while maintaining positive suction pressure. When compressor is up to full speed, start flow to the feed header and bring up to normal pressure.

- Start circulation of methane to the heaters from the discharge by opening the heater feed valves. Increase the flow as quickly as possible, while maintaining positive suction pressure.
- Check the CO_2 content at the overhead of the caustic tower and if acceptable, start a small feed forward to the dryers and demethanizer feed system. Limit feed flow to the demethanizer feed system so that the system cool-down rate does not exceed 25°F per hour. This allows the equipment and especially the flanges and flange bolts to cool slowly and avoid leaks caused by a too rapid chill-down. If the system is under nitrogen pressure, the vent(s) should initially be routed to flare, not the fuel system.

When the system is stabilized/cooled and the plant is ready for fresh feed, introduce fresh feed to the heaters at a controlled rate in order to maintain pressure control in the system. The discharge pressure on the charge gas compressor starts to increase, as does the SHP steam production and consumption. Liquids start to condense in the demethanizer feed drums and in the demethanizer. When liquids have accumulated in the demethanizer bottom and the reboiler has started, start feed to the deethanizer.

To avoid flaring, the overhead vapor from the deethanizer reflux drum is routed to the recycle header under flow control. Heater Coil Outlet Temperatures (COTs) should be maintained at a relatively low value, say 1450°F, in order to minimize conversion of ethylene to acetylene. Otherwise, it may take additional time and flaring to achieve an acceptable inlet acetylene concentration to the acetylene converter. Once this recycle is established, add additional fresh feed and increase the plant load. Some flow can be started through the acetylene converter to warm the converter to its start-of-run (SOR) inlet temperature. This material is recycled to the heaters via the recycle line after the converter. H_2 should not be added to the acetylene converter feed until at least 70 percent of normal flow rate is established. This is done to minimize channeling through the bed and reduce the possibility of hot spots.

When the acetylene converter flow reaches 70 percent of the normal flow, add H_2 and adjust the operating conditions to make on-specification effluent. As mentioned earlier, it is necessary to control the heater COTs and manage the acetylene converter inlet concentration. It is a dynamic situation and needs operator skill to manage it smoothly, without flaring. Once the acetylene converter effluent is on specification, start feed flow to the ethylene fractionator and product flow will start almost immediately. At this time, shut in the recycle system and increase fresh feed and heater COTs quickly. Otherwise, since the feed is ethane, the system will be inundated with recycle ethane.

Plant Experience

The recycle system has been used successfully during start-ups and functions well. The deethanizer and acetylene converter recycles are used during upsets to reduce flaring. Experience indicates that judgment is required in order to decide when to use the recycle system, such as when the acetylene converter goes off-specification or some other upset occurs. It is extremely important to manage the heater cracking severity when recycling ethylene, otherwise the acetylene content in the converter feed can rise quickly and overwhelm the acetylene converter. This can result in heavy flaring. Typically, feed forward to the ethylene fractionator is not started until the on-line GC and lab sample both show acceptable acetylene concentrations.

The plant has modified the start-up sequence according to their experience, which is that boiler trips do occur at times, so it is best to minimize the plant import steam demand. For example, the C2-R compressor is not started until feed forward has started, which gives an additional margin on boiler steam production. In order to minimize steam consumption, they also do not start many of the steam turbine driven pumps.

The recycle of ethylene/ethane back to the heaters does not cause a dramatic increase in coking. To be safe, the heater decoke cycle is started earlier than normal after recycling for a significant amount of time.

Conclusions

The system is relatively easy to use; and has worked according to the design assumptions. The design intentions – i.e. minimum cost and minimizing emissions during startups and upsets – have been achieved.

AIChE Ethylene Producers' Conference

New Orleans
April 25-28th 2004

Knowledge-Based Predictive Maintenance For Olefins Turbocompressors [T8007]

By

Constantine Lau, Dresser-Rand Company
Martin Maier, Dresser-Rand Company

Knowledge-Based Predictive Maintenance For Olefins Turbocompressors

Introduction

The success of a predictive maintenance program is more than just using predictive technologies used to analyze data; it is a business process that relies on how the information is being used and what action can be derived from it. An analogy would be a physician using tools to diagnose a patient, and prescribing a course of corrective action to the patient based on the (health) data he collected from the tools. Some predictive maintenance programs are not effective because of the lack of available machinery expertise that can make full use of the information. Therefore, what is the use of tools without the doctor?

The creation of a knowledge-based system is not intended to replace the machinery expertise but to enhance the business process of a predictive maintenance program. Also, a fully qualified, reliability professional is very hard to find -- and usually very expensive. A knowledge-based system is an additional set of "expert brains" that can help diagnose a problem, and even recommend corrective action, depending on the robustness of the system. Best of all, it never sleeps—it is like having the best machinery expert working 24/7 around the globe for you.

This paper will examine the development of a machinery knowledge-based system that analyzes, diagnoses, and makes recommendations based on the condition data collected from compressors. The knowledge-based system is built on an expert system software application based on proprietary machinery information supplied by design engineers, field service engineers and condition monitoring specialists.

Expert System

Expert system technology is designed to capture human expertise on software. It is built on rules and procedures, applicable when experience can be stored in the knowledge base, and when no mathematical model is available. An expert system is a subset of artificial intelligence that was initiated with the invention of computer. It enjoyed some early success but has not been accepted into mainstream technology. Some initial deployment failures are attributed to the limitation of hardware processing power, but with the advent of computer technology, hardware is cheap and readily available off the shelf.

An expert system is commonly used as a decision-making tool in monitoring, control and optimization applications. Such a system can also be used to perform "what-if" analyses to help plant personnel make better decisions.

Intelligent alarming is one example in which an expert system would trace the cause of an alarm by back tracking rules in the system, then providing the operator with advice on what is the most probable cause of the problem. The system can even go further by suggesting a series of actions that can be performed to remedy the problem.

Unlike neural net technology approaches, expert system technology does not use mathematical models. While expert systems are very powerful and flexible, this power and flexibility requires a significant amount of customization effort.

The earliest application of an expert system was in the medical field, where the computer asks a patient or the physician a series of questions, then diagnoses what was wrong with the patient based on the captured expertise of renowned medical experts. This was never widely accepted, perhaps because building an expert system that will be equivalent to a human brain would be an enormous undertaking. The final decision is always made by a human ---the physician -- and the ultimate reason is that nobody wants to take chances with the human body.

Challenges

The biggest challenges with a knowledge-based application are technology and time investment. In the earlier days of expert system deployment, complex applications were restricted by speed and bandwidth issues, but this has gradually become a non-issue with the advent of technology and computer hardware becoming a commodity.

The next challenge is the effort required to develop an expert system application. The system is only as good as the amount of knowledge being put into the knowledge base. Most projects failed simply because people underestimate the effort to develop a robust, usable expert system. An expert system should be treated as an evergreen project that requires nurturing as it grows.

Neural Net Technology is not a Knowledge-Based System

Like neural cells in the human brain, neural net technology provides self -learning capability embedded in software. As one of the more commonly applied branches of "artificial intelligence," neural nets enable software applications to learn and make associations and decisions on their own. There are a few selective application areas where neural net technology has proved to be very successful, but its application is not yet widespread. The most powerful and intriguing aspect of neural net technique is based solely on cause and effect, and no intimate knowledge of the process is required. There are pros and cons of this approach, as an analogy -- if a neural net learns how to drive a car, it may not know how the engine works, but it can learn from past data that driving the car at 63 miles an hour will get the best mileage.

Machine or Human

Knowledge base building will be an ongoing process that requires discipline and patience. In the initial phase, most of the diagnostic work will be handled by humans; for example, 30percent machine, 70 percent human. As time goes by, as the knowledge base becomes more robust and complete, most of the diagnostic work will be handled by machine; for example, 70 percent machine, 30 percent human. Machines will never completely replace humans because there will always be cases that require some judgment and common sense that the machine will never have.

Expert System Protocol

A method is presented for evaluating vibration in turbomachinery although the same approach can be applied to other aspects of machine operation (performance, mechanical, etc.). Using information relating to the basic application, unit design parameters and real-time performance, mechanical and vibration information downloaded from a data acquisition system, the vibration signature of a rotor system is evaluated to identify abnormal operation that may be symptomatic of a potential problem. It then determines the probable cause(s) and recommends an appropriate course of action. Evaluations can be made based on whatever information is available. The greater quantity of pertinent information that is provided to the system, the more definitive will be the results. At the heart of the system is the rules base that contains the knowledge and experience of people from various sources within a company (test, field service, product design, rotor dynamics) and outside a company (universities, publications, consultants). The end result is a tool to enable a machine operator on site, or a remote condition system operator to diagnose and correct vibration problems in the timely manner.

The system evaluates data as both a "snapshot" in time relative to "base line" operation and within the context of previously recorded data to identify trends. If a problem (or potential problem) is discovered, the system will provide guidance as to what response should be taken, whether a root cause has been isolated, and how the problem can be resolved. The system also performs a risk assessment that can be used in the design phase to determine how challenging the selection is, based on an aggregate evaluation of each aspect of the design.

Model

Figure 2 is the model of an expert system. Application and design data, together with data downloaded from the data acquisition system, are normalized by converting them into "Intensity Factors." An example of this conversion process is shown in Figure 3. These factors are a measure of the severity of a parameter relative to a set of reference values. They range from 0 (low severity) to 1 (high severity) and are defined in the rules database. The rules database also contains a list of vibration phenomenon and a unique set of characteristics that define them and distinguish them from other phenomenon. These distinguishing characteristics are quantified using "Rating Factors" which range in value from –10 to +10. A rating factor of +10 indicates that a particular characteristic strongly points to the particular phenomenon. A rating factor of 0 indicates no correlation exists. A rating factor of –10 indicates a negative correlation and serves as dismissing evidence. The intensity factors are specific to machine operation, whereas rating factors are specific to one or more vibration phenomenon.

An example is shown in Figure 4. It references three rotor dynamic phenomenon: fouling, shaft bow and journal bearing degradation. A number of parameters are listed that are associated with each phenomena. This matrix is a small subset of the entire matrix that lists all rotor dynamic phenomenon in the data base and all parameters used to define each phenomenon. Raw data is non-dimensionalized and converted into intensity factors in the manner shown in Figure 3. Rating factors, which uniquely define each phenomenon in terms of their characteristics and attributes, are also shown. The intensity factors and the rating factors are then multiplied and added. The probability that the unit is exhibiting a particular vibration phenomenon is then calculated in the following way:

$$\text{Probability} = \frac{\sum(\text{Intensity Factor} * \text{Rating Factor})}{\sum(\text{Rating Factor})} \text{ x } 100\%$$

The probability is calculated for each phenomenon and they are then listed in descending order. Confidence factors are also determined which indicate the amount of pertinent data that was provided to the system. These are calculated in the following way:

$$\text{Confidence Factor} = \frac{\sum(\text{Rating Factor Of Parameters That Are Assigned Values})}{\sum(\text{Rating Factors Used To Define Phenomenon})} \text{ x } 100\%$$

If all the data necessary to define a particular phenomenon was provided, the confidence factor would be 100 percent. If only a small fraction of the data that the system needs are provided, the confidence factor approaches 0 percent. By making use of both the probability and confidence factors, the expert system can make reasonable evaluations with only limited amounts of data.

Rotor Dynamic Expert System Rules Database

This is essentially the knowledge base in the rawest form before being converted into rules. Figure 5 is an example which deals with polymerization and general fouling in cracked gas trains. It is documented in terms of phenomenon and described in qualitative and quantitative terms. In analogy, this is like a business plan before any work is conducted; without this documentation, it will just be creating rules in an *ad hoc* manner.

Detailed information can be accessed for each phenomenon as to its nature, factors that influence it most, additional things to investigate, corrective action, etc. Although much of this information is encoded into the rules database, it still requires human judgment to determine the best course of action.

Conclusion

Predictive maintenance ensures that machinery is operating at peak efficiency, maximizes availability and, most importantly, prevents catastrophic failures. A predictive maintenance business process requires the machinery expert to analyze process, performance, and vibration data that allows the user to make better decisions that can impact production and maintenance. A knowledge-based system can enhance this process by supporting the machinery expert with meaningful information. With the continuing trend toward diminishing machinery expertise, the knowledge-based system will be the primary means of maintaining these knowledge bases.

An expert system is a form of knowledge-based system that captures human knowledge. The effectiveness of a knowledge-based system depends on the quantity and quality of rules. The amount of effort required to create one should not be underestimated.

There are a few steps in transforming knowledge into software. The first step is to define the knowledge base, equivalent to creating a business plan. Without this important step, it would just be *ad hoc* implementation of rule base that would guarantee very little success. In this paper, the example used is the documentation of compressor polymerization and fouling phenomena in the rotor dynamic expert rule system database. With this qualitative information in hand, the second step is to transform and redefine it into a rating and intensity factor that contributes to the polymerization and fouling phenomena in a matrix format. Third step, a set of probability and confidence level numbers will be generated from this matrix to indicate the root cause phenomena. After all these steps are completed, data contained in the matrix will be programmed into rules forming the basis of the knowledge- based system.

References

1. API Standard 617, 7th Edition, July 2002, "Axial and Centrifugal Compressors and Expander-compressors for Petroleum, Chemical and Gas Industry Services", Section 2.6, "Dynamics"

2. API Publication 684, 2nd Edition, 2004, "Tutorial on API Standard 617, 7th Edition", Section 1.6, "Stability Analysis"

3 Haq, Inam and Bu-Hazza, Ahmed and Al-Baz, Khalid, "Multistage Centrifugal Compressor Fouling Evaluation At High Power Settings", 1998, ASME 98-GT-53.

4 Haq, Inam and Bu-Hazza, Ahmed, "Modeling and Computation of Fouling of a 36 MW Multistage Centrifugal Compressor Train Operating In a Cracked Gas Environment", 2001, ASME 2001-GT-229.

Acknowledgements

The authors would like to thank the following individuals for their valued support in providing much of the source material for this paper: Gary M. Colby, Inam U. Haq, James M. Sorokes

Figure 1: Knowledge-based Condition Monitoring

(Human Experts Supported by Expert Systems)

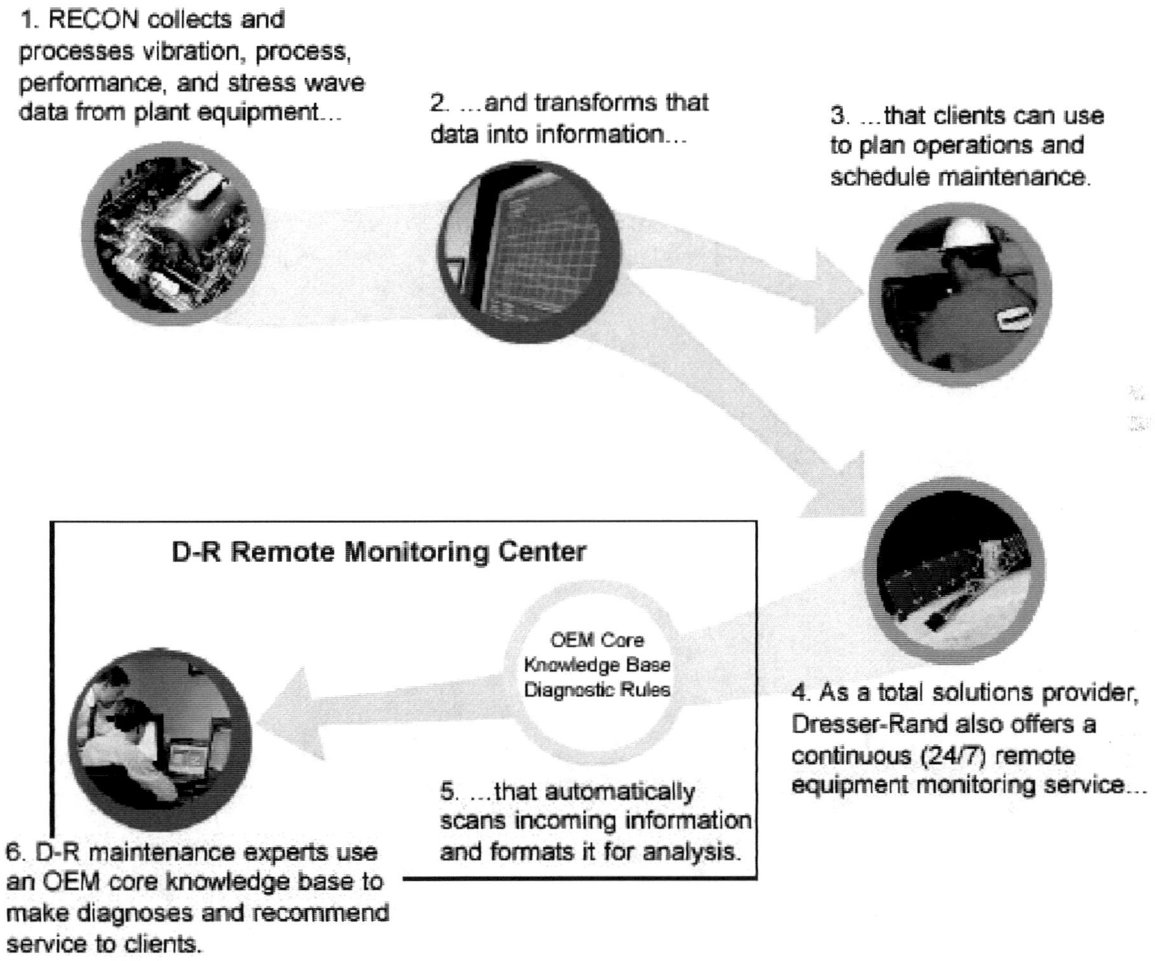

Figure 2: Rotor Dynamic Expert System Model

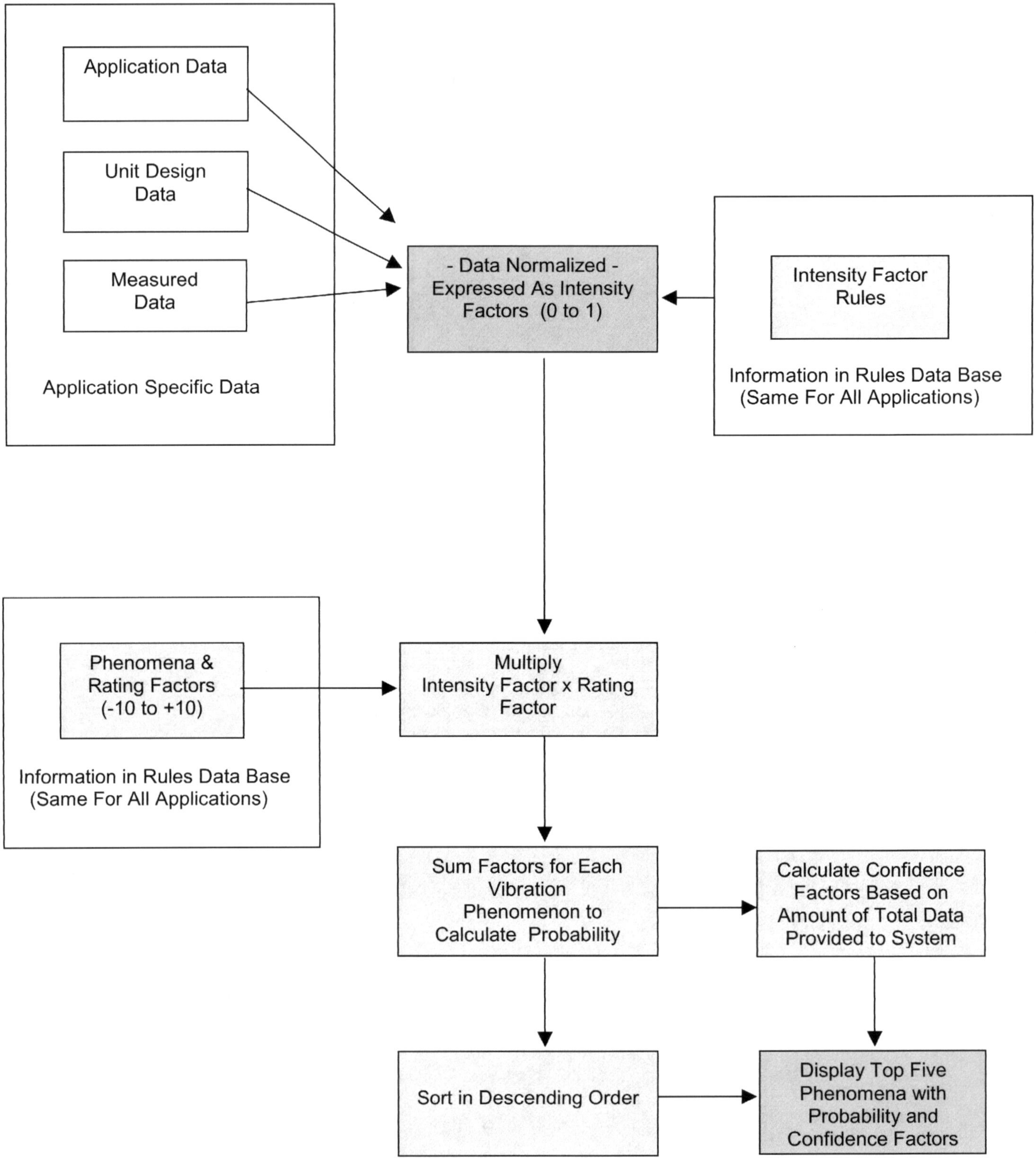

Figure 3: Intensity Factors

Data are normalized and expressed as Intensity Factors. Intensity factors are a measure of the severity of measured parameters relative to a set of reference values (normalizing factors). They range from 0 to 1.0 where 1.0 is the most severe. A linear relationship is assumed between two sets of points which are stored in the rules data base.

Driver Power (% Increase)

Driver power increases as fouling progresses

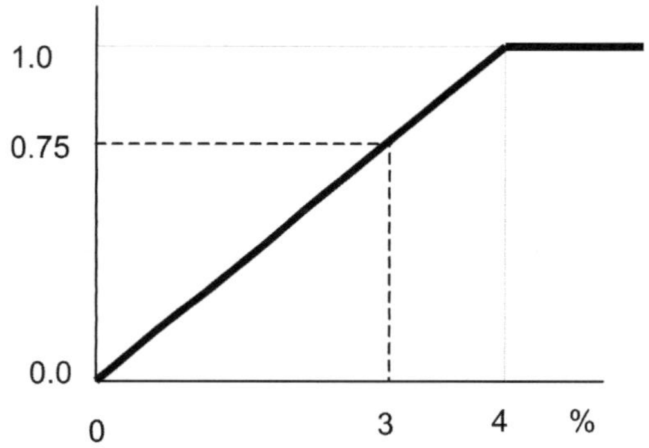

Increase In Suction Pressure (% Increase)

The first section suction control valve increases the suction pressure to offset the reduction in mass flow. The percent increase in suction pressure is based on absolute pressure.

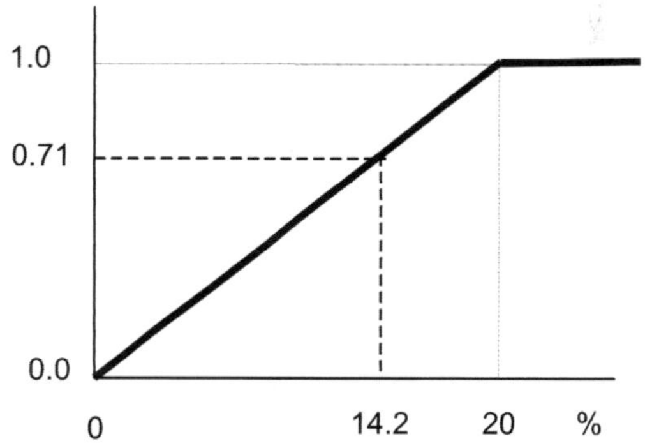

Discharge Temperature (Deg F)

Temperatures above 200 degrees are only considered relevant when assessing the risk of polymerization.

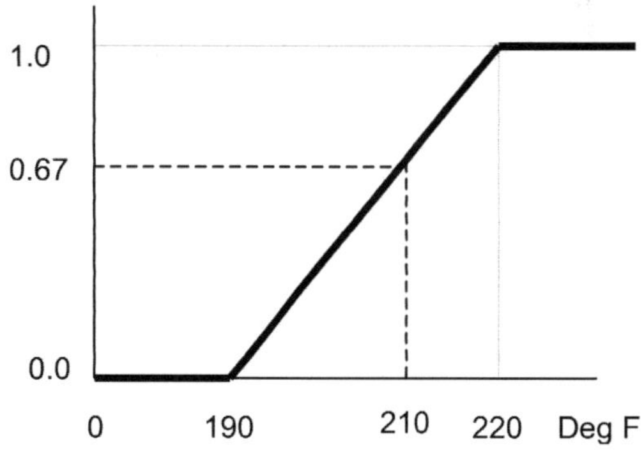

Synchronous Vibration (Mils)

Baseline vibration from the commissioning run is 0.5 Mil with the rotor in a new and clean condition. Reference vibration limit is 1.5* API 617 Specification Limit.

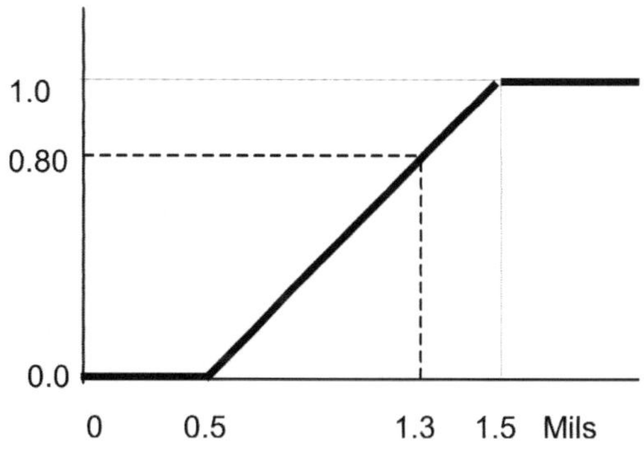

Figure 3: Intensity Factors
(continued)

Change In Nc1 Amplification Factor (%)

The amplification factor (AF) of the rotor's first critical speed (Nc1) is a measure of the damping of the rotor/bearing/seal system. Changes in rotor unbalance should not significantly alter system damping. Variations up to 20% are anticipated due to changes in acceleration rate and oil film temperature.

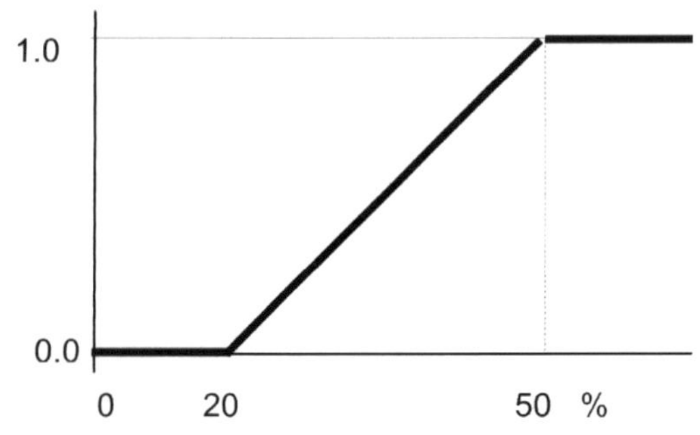

Change In Runout At Slowroll (Mils)

Electrical and mechanical runout are normally constant and independent of rotor unbalance.

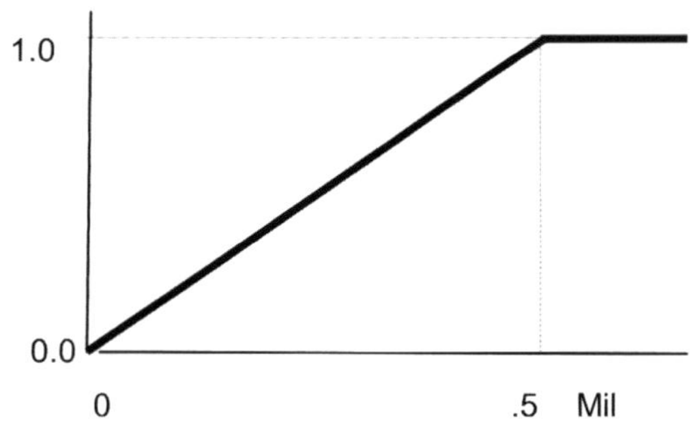

Figure 4: Example Illustrating Method (Note - Matrix shown below is a subset of actual matrix)

Raw Data – See Previous Figure For Normalization Of Data

	Observed Trend			
Driver Power (%)	Increase	3.0	3.0	3.0
Suction Pressure (%)	Increase	14.2	14.2	14.2
Discharge Temperature (Deg F)	Increase	210	210	210
Fouling Index (-)	-	0.80	0.80	0.80
Vibration Peak Amplitude (Mils)	Increase	1.3	1.3	1.3
Change In Nc1 Amplification Factor (%)	-	0	0	0
Change In Runout At Slowroll (Mils)	-	0	0	0

Intensity Factor Matrix [Application & Unit Specific]

	Observed Trend			
Driver Power	Increase	0.75	0.75	0.75
Suction Pressure	Down	0.71	0.71	0.71
Discharge Temperature	Increase	0.67	0.67	0.67
Fouling Index	-	0.80	0.80	0.80
Vibration Peak Amplitude	Increase	0.80	0.80	0.80
Change In Nc1 Amplification Factor	-	0	0	0
Change In Runout At Slowroll	-	0	0	0

Rating Factor Matrix [Same For All Applications and Services]

	Observed Trend	Phenomenon#1: Fouling	Phenomenon#2: Shaft Bow	Phenomenon#3: Journal Bearing Degradation
Driver Power	Increase	8	0	0
Suction Pressure	Down	9	0	0
Discharge Temperature	Increase	6	0	0
Fouling Index	-	10	0	0
Vibration Peak Amplitude	Increase	7	8	6
Change In Nc1 Amplification Factor	-	0	0	7
Change In Runout At Slowroll	-	0	9	0
Rating Factor Sum (Defined)		**40**	**17**	**13**
Rating Factor Sum (Total In Full Matrix)		**50**	**50**	**50**
Confidence		**80%**	**34%**	**26%**

Evaluation Matrix – Elements of Rating and Intensity Factor Matrices Multiplied

	Observed Trend	Phenomenon#1: Fouling	Phenomenon#2: Shaft Bow	Phenomenon#3: Journal Bearing Degradation
Driver Power	Increase	6.0	0	0
Suction Pressure	Down	6.4	0	0
Discharge Temperature	Increase	4.0	0	0
Fouling Index	-	8.0	0	0
Vibration Peak Amplitude	Increase	5.6	6.4	4.8
Change In Nc1 Amplification Factor	-	0	0	0
Change In Runout At Slowroll	-	0	0	0
Sum		**30**	**6.4**	**4.8**
Probability (Weighted Avg)		**75%**	**13%**	**10%**

Figure 5: Sample Rotor Dynamic Expert System Rules Database

Identification #	**P- 8010**
Phenomenon	**Polymerization and General Fouling in Cracked Gas Trains**
Probable Cause	**Elevated Gas Temperature and Ingestion Of Contaminants**

DESCRIPTION:

Polymerization is commonly seen in cracked gas service in ethylene plants. Ethylene plants manufacture ethylene by cracking ethane which is one of the primary feed stocks to the process. It is heated to approximately 840 degrees C in successive steps where it is broken down to ethylene (30 percent), ethane (23 percent), hydrogen (32 percent) and a variety of other gases (light and heavy hydrocarbons and non-hydrocarbon gases). The gas is then fed into a compression system. During this time, heavy hydrocarbons can solidify and coat the surfaces of the primary and secondary flow passages with a sticky polymer material. Depending on pressure, concentration and other factors, this process occurs when the gas temperature reaches 200 to 230 degree F. Hence, inter-cooling is required between sections to maintain the temperature below this limit. The suction temperature of each section can be adjusted by varying the cooling water flow rate. The same can be accomplished by injecting a small (<3 percent weight flow) amount of atomized water into the gas stream. This reduces gas temperature via the heat of vaporization. Buildup tends to occur in the latter stages of a section rather than in the early stages because of the elevated temperature. The buildup increases at an accelerated rate as the surface becomes more conducive to the formation of deposits and because of the increase in gas temperature as fouling progresses.

Fouling has also been known to occur at the inlet of the first body of cracked gas trains [Ref 3 and 4]. Catalyst dust from the furnace that feeds the process combines with heavy-end gases that liquefy at the suction end of the compressor and coat the first stage(s). This is in contrast to polymerization that generally can occur at the discharge of each compressor section.

Introduction of wash agents such as C9 oils or naptha into the gas stream via atomizing nozzles helps to reduce the accumulation of polymers, particularly on the rotating components. It can be done on a continual basis at speed or on an intermittent basis at reduced speed. Continuous injection is limited to 3 percent liquid solvent by weight flow of the total gas flow, although a rate of less than 1 percent has been found to be effective. Intermittent washing may not remove the material uniformly from a severely fouled rotor which can result in a significant increase in vibration. Washing agents can be injected into the process piping but is more effective if introduced between stages, particularly in the diffusers where deposits are more apt to accumulate. A thorough (steam) cleaning of the rotor and bundle should be a standard part of any maintenance program.

Polymer can accumulate in seal clearances. Rotating labyrinths are frequently used to minimize this.

Polymer can presumably fill the flow passages of a shunt system. If a traditional shunt design is used (without swirl brakes), the resulting preswirl can destabilize the rotor system.

Polymer can cause anti-surge systems to malfunction by fouling sensors and/or causing control valves not to operate freely.

There are a number of ways that fouling can be identified with regard to its severity, location and probable cause. There are also certain characteristics that distinguish it from other related phenomenon. These are discussed below.

CHARACTERISTICS:

1. VIBRATIONAL

- As deposits accumulate on the rotor, the synchronous rotor vibration may slowly increase. Although relatively large quantities of material can be deposited over time, it tends to be uniformly distributed which changes the net balance of the rotor at a slow rate. This effect is less prevalent if an effective injection wash is utilized.

- Sudden jumps in vibration can occur when material breaks off from the rotor. This presumably can excite the rotor's fundamental frequency at the instant this occurs. The space that is left then tends to fill in, reducing the vibration to some degree. Material can be dislodged from the rotor from the following causes:

    ➔ Process upsets such as surge and liquid slugging.
    ➔ Traversing the critical speed of the rotor during startup or shutdown. This is particularly true if the critical speed is severe.
    ➔ Contraction of the impeller as centrifugal growth is lost upon shutdown.

- Sudden jumps in vibration can also occur as material breaks away from the stationary flow path and is sucked into the successive stages. As debris impacts the rotor, this can excite the rotor at it's fundamental frequency and/or at a variety of frequencies.

- A pronounced increase in vibration can occur following (intermittent) injection wash when material is not uniformly removed from the rotor. This is particularly true if a wash system is installed after a rotor has become severely fouled.

- The location of fouling can be estimated from the shaft vibration data by comparing the change in vibration at each end of the machine. It is inconclusive, however, when dealing with double flow arrangements where equal amounts of fouling are assumed to occur at each end of the machine. Double flow compressors are frequently selected as the first body of cracked gas trains.

- The first critical speed (Nc1) may show a more significant change in amplitude than what is observed at the design speed depending upon how the deposits are distributed on the rotor.

- The level of damping associated with Nc1, as determined by the amplification factor (AF), should not change. This distinguishes it from various bearing or seal phenomenon where the system damping changes.

- The vibrational amplitude at slowroll (<10% MCOS) should not change. This distinguishes it from thermal or mechanical rotor bow.

- In straight-through units, fouling of the gas balance line can reduce thrust compensation (particularly if flow passages are undersized) which is evident by a gradual shift in axial rotor position and a gradual change in thrust bearing RTD temperature. The temperature can either increase or decrease depending on whether the load on the inboard/outboard thrust bearings is increasing or decreasing.

- As the seals fill with material, a (continuous) rub condition may develop which may result in minor levels of subsynchronous vibration.

2. IMPACT ON AERODYNAMIC PERFORMANCE

One of the first signs of fouling is the inability of the compressor to produce its rated pressure at normal mass flow. Or conversely, the inability of the compressor to pass its normal mass flow at design operating pressure. A simple way to confirm that fouling has occurred is to reduce the flow to the point of surge. At a given pressure ratio, if surge occurs at a higher flow relative to the prediction or a prior test of the unit in the clean condition, then fouling has occurred.

Fouling is characterized by a shift of the performance curve both down and to the left as illustrated in Figure 6. The downward shift is caused by the pressure drop associated with energy dissipation due to alterations of the rotating or stationary flow (path). The shift to the left is related to the reduction in passage area making the component's "smaller" in flow capacity. Additionally, the increase in gas temperature that takes place due to dissipative losses and the reduction in efficiency due to the progressive shift along the performance curves of successive downstream stages. Figure 6 also shows the shift in the surge line to the right.

The way the system responds to this shift in the performance curve is dependent on the type of control system used to regulate the process. The objective of the control systems is to maintain the mass flow and final discharge pressure at constant levels, within prescribed tolerances.

Systems with fixed suction pressure accomplish this by regulating rotational speed. As fouling progresses, the rotational speed increases to offset the reduction in mass flow and pressure.

Systems with variable suction pressure accomplish this by regulating the position of the suction pressure control valve. As fouling progresses, the suction pressure is increased to compensate for this change.

Fouling is also accompanied by the following changes in the system:

- An increase in power consumption.
- An increase in discharge temperature.
- An increase in inter-cooling water flow rate to offset the rise in gas temperature.

The interstage seal clearances can open up during operation because of normal wear, erosion or performance upsets (surge, liquid slugging, etc.). Additionally, in a "dirty" process, the sealing (tooth) clearances can fill up with debris and can reduce the effectiveness of the seals. Rotating labys reduce the likelihood of this occurring. In either case, interstage leakage can increase as the seals progressively deteriorate. In other instances, particularly when there are no excursions in rotor vibration, the seal clearances can fill up thereby reducing interstage leakage.

3. EFFECTS ON SURGE CONTROL SYSTEM

Surge control systems are intended to keep the compressor operating on the allowable region of its performance map. In process compressors, this is generally accomplished by suction throttling, recycling gas from the discharge to inlet, etc. One frequently used control system measures Delta P/Ps and compares it to a reference value to determine the amount of flow adjustment necessary to keep the unit out of surge. As fouling progresses and the surge line moves toward the right on the performance curve (Figure 6), the set point of the surge control system may need to be adjusted to compensate for this shift. Otherwise, the surge control system will progressively become ineffective.

4. APPLICATION RELATED:

- Fouling tends to be application specific. That is, it is commonly associated with certain types of service or process gas. In cracked gas service in ethylene plants, the catalyst from the furnace can enter the compressor and combine with the heavy ends of the hydrocarbon mixture which drop out in the first stage(s). This fouling is fundamentally different from the polymerization fouling that frequently occurs at the final stage(s) of each section when the gas temperature reaches 190-220 degrees F. General fouling and polymerization in cracked gas trains are primarily distinguished by the location in which they occur along the rotor.

5. PROCESS RELATED:

- An effective injection wash system should mitigate fouling, particularly on the rotor. If indications of fouling are observed either by increased synchronous vibration levels or performance degradation, then the wash system needs to be evaluated to determine if it is operating properly or if the process gas or conditions have changed in some way.

- Injection wash should be installed at the time of commissioning or following a turnaround when the rotor is in a clean condition. Installing a wash system once a compressor has become severely fouled may not be effective in removing deposits completely or uniformly resulting in increased levels of rotor vibration.

- Coatings applied along the rotating and stationary flow paths have been found to be effective in slowing the rate of fouling by making the surfaces less susceptible to buildup.

ADDITIONAL INVESTIGATION:

- Reduce the flow and observe the shift in the surge point. Doing this periodically will indicate at what rate fouling is occurring. A word of caution: Bring the unit into full surge can result in considerable damage to the rotor and seals. It can also cause deposits to break off and obstruct the flow path and/or increase rotor unbalance.

- Investigate long- term trends of rotor vibration increase and performance degradation and schedule maintenance accordingly.

CORRECTIVE ACTION:

- Reduce the suction temperature if fouling is suspected to be caused by polymerization.
- Use injection wash, change type and/or increase flow rate (within prescribed limits)
- Inject wash agent between stages, particularly in diffuser walls where buildup becomes most prevalent.
- After prolonged operation and/or after signs of severe fouling, remove and thoroughly clean rotor, bundle, piping, valves, sensors, etc.

Figure 6: Change in Aerodynamic Performance as Fouling Progresses

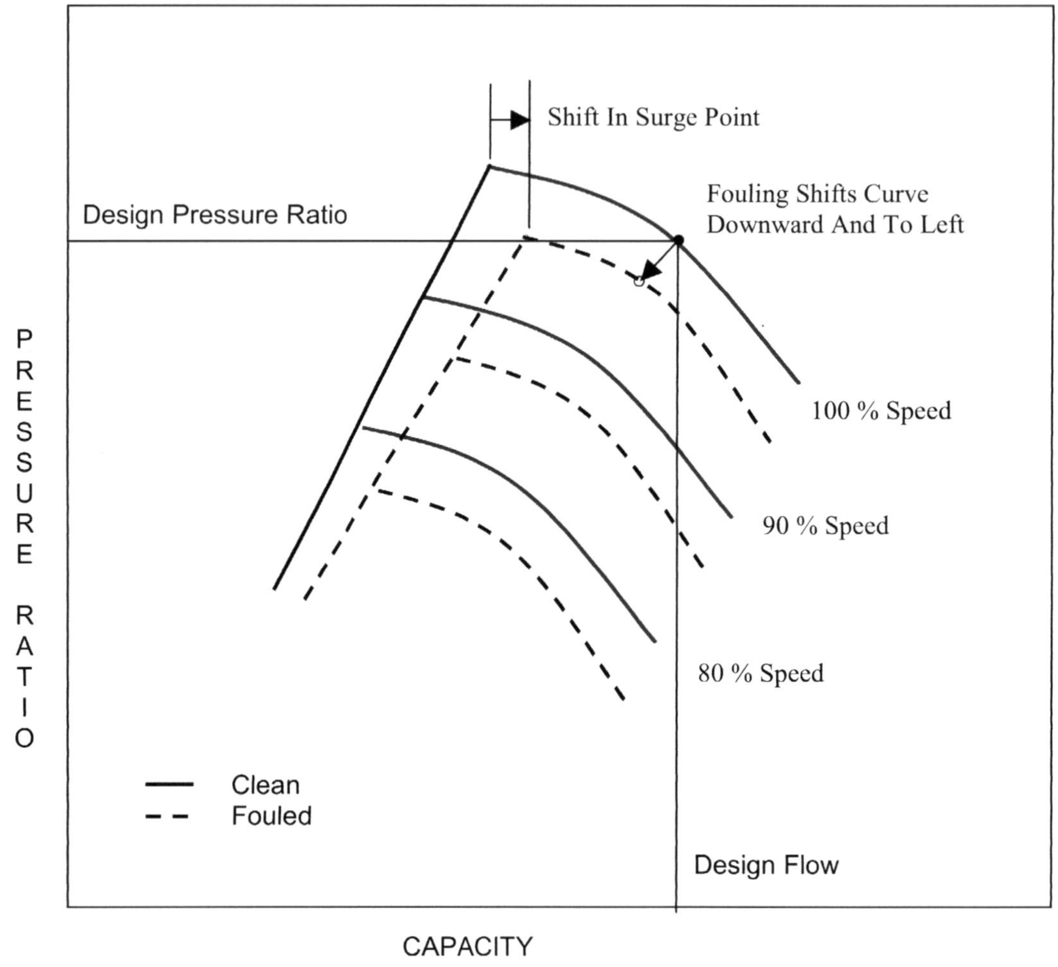

Using Improved Inspection Tooling to Perform Machinery Condition Assessments

Chris Brown
Foster-Miller - Waltham, Massachusetts, USA

and

John R. Dugas Jr.

DuPont Packaging & Industrial Polymers - Orange, Texas, USA

2004 Ethylene Producers Conference

April 25-29, 2004

[T8007] – Rotating Machinery Reliability – Impact of Operational Practices

Outline

- Turnarounds and Condition Assessment
- Remote Visual Inspection Tooling
- Case Studies
  - Worthington turbines at Sabine River Works (SRW)
  - Elliott turbine at SRW
  - Identifying the cause of high vibration at a commercial utility

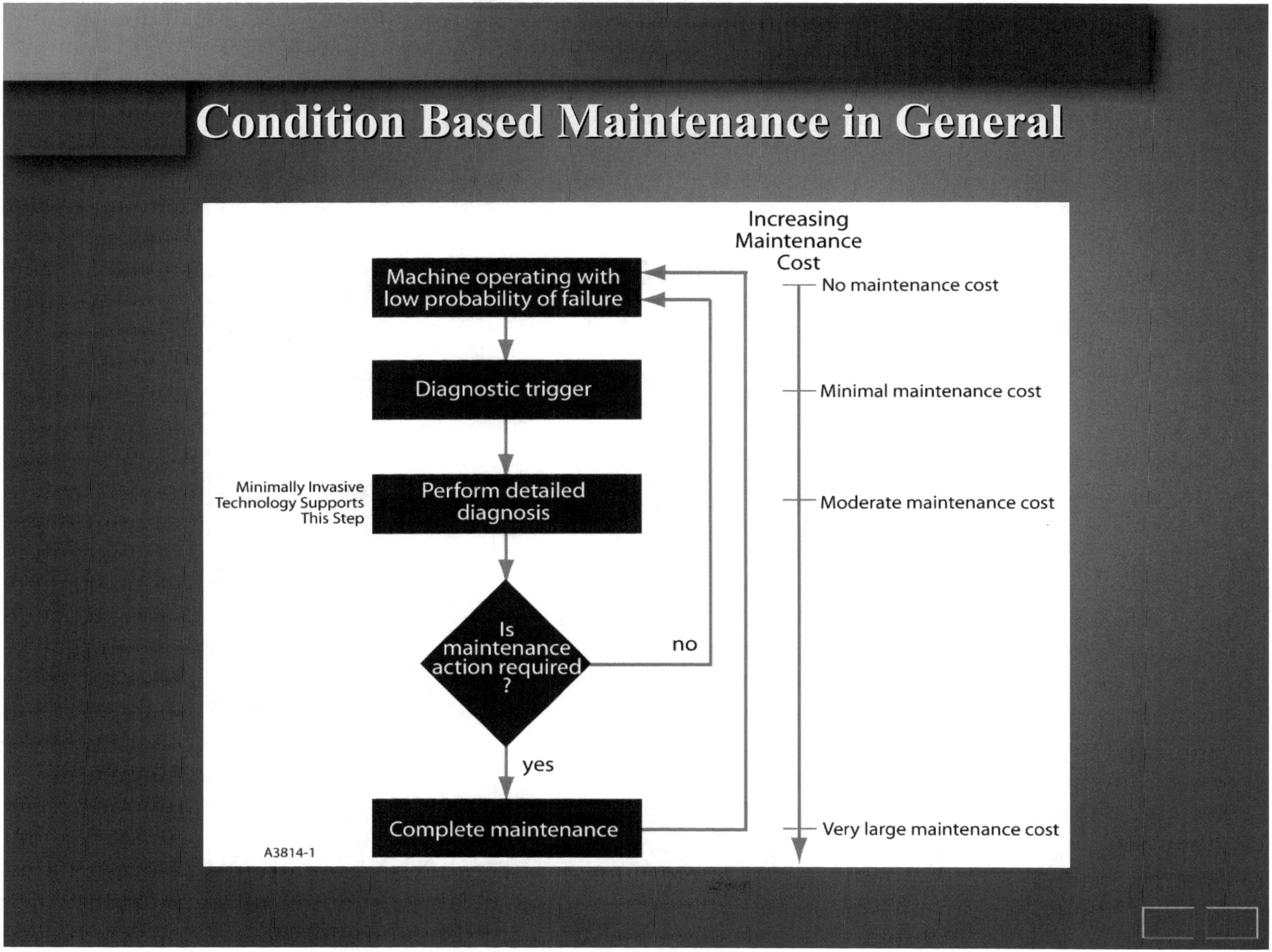
Condition Based Maintenance in General
Increasing Maintenance Cost
No maintenance cost
Minimal maintenance cost
Moderate maintenance cost
Very large maintenance cost
Machine operating with low probability of failure
Diagnostic trigger
Minimally Invasive Technology Supports This Step
Perform detailed diagnosis
Is maintenance action required ?
no
yes
Complete maintenance
A3814-1

Diagnostic Triggers and the Detailed Diagnosis

- Routine monitoring is the most common diagnostic trigger
  - Higher vibration
  - Changes in efficiency, steam rate, etc.
- Routine monitoring techniques generally indicate THAT you have a problem, but don't answer
  - What exactly IS the problem?
  - What needs to be done about it?
- Detailed diagnosis is the process of answering those two questions
  - In the case of rotating machinery, this generally involves analysis and internal inspection

Petrochemical Plant Turnarounds

- Opportunities for 'detailed diagnosis' of turbines and compressors in a single-train petrochemical plant are limited to turnarounds
  - More limited than even a base-load commercial utility plant
- The detailed diagnosis and the maintenance action (if any) need to occur in a short time period
- Detailed diagnosis is thus often skipped
  - Leads to unnecessary maintenance action
  - Leads to higher than necessary maintenance costs

Remote Visual Inspection Technology

- Used extensively and regularly for a variety of equipment
- Mostly off-the-shelf 'one size fits all' tools
- Not commonly made to go to particular places in particular equipment

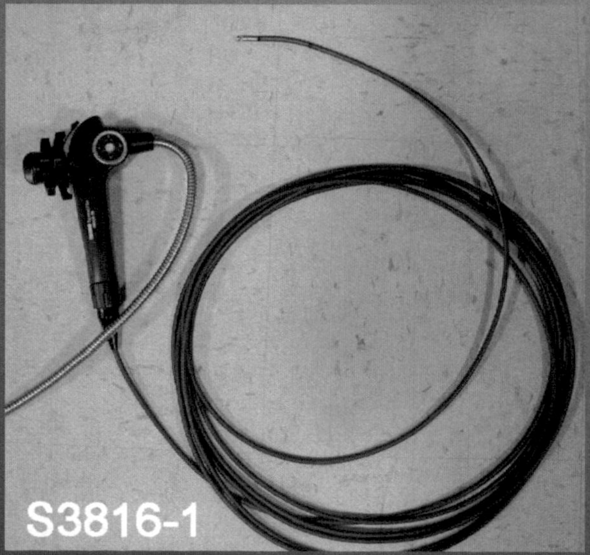

Typical Video Probe

Custom Tooling for Steam Turbines and Other Large Rotating Equipment

- A family of tools designed to go where off-the-shelf tools cannot
  - To reach 1st stage nozzles in large steam turbines
  - To go further down rows of turbomachinery
  - To cross long unsupported distances

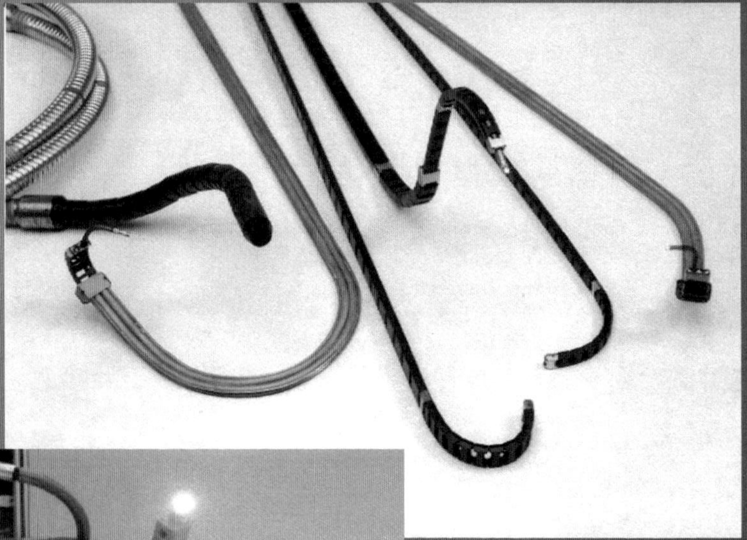

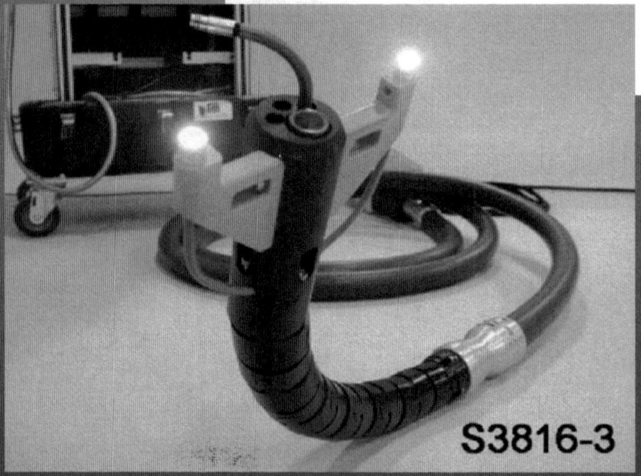

Case
Worthington Steam Turbines at SRW

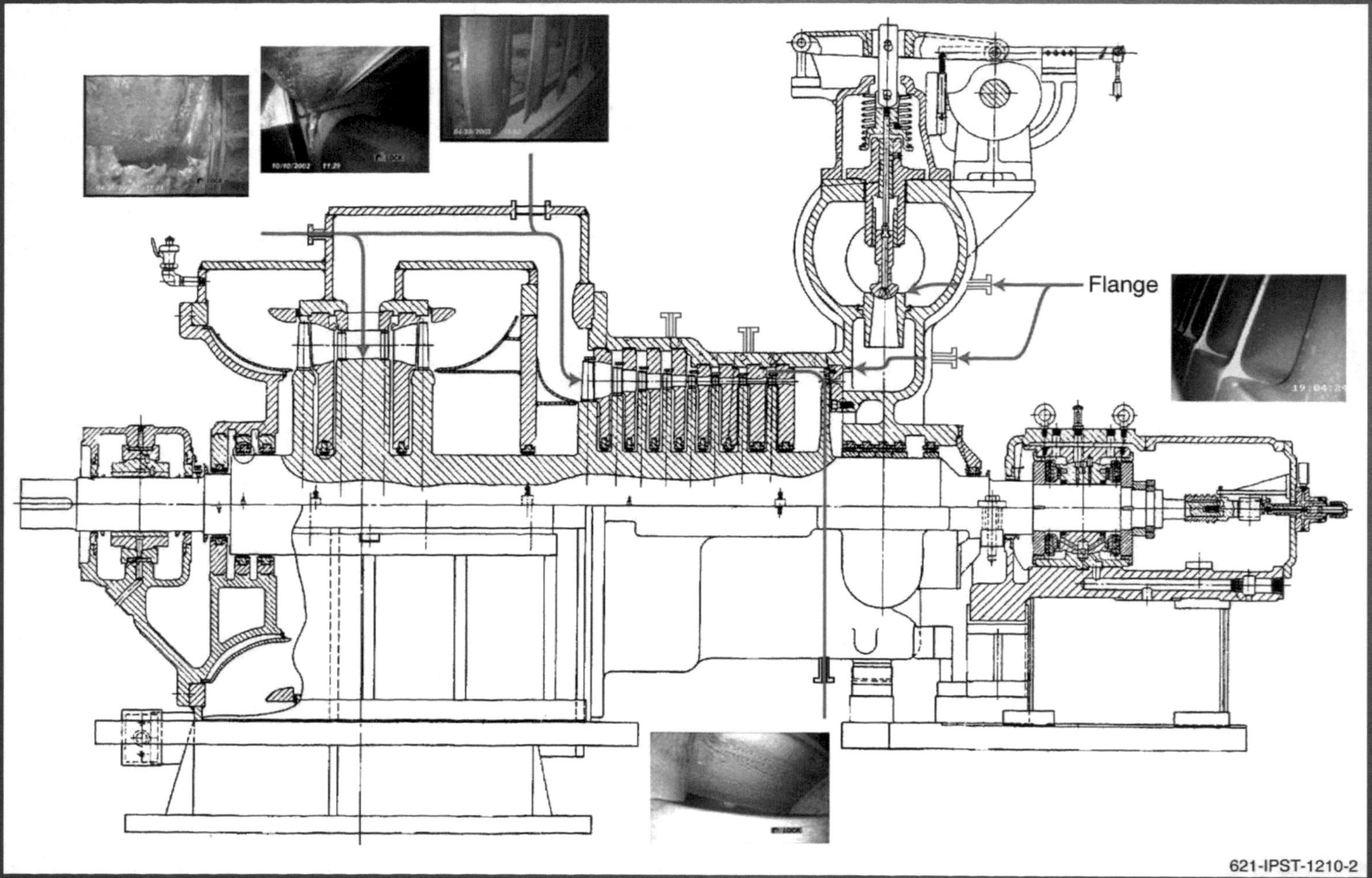

Turbine cross section: Inspection paths marked in red

Worthington
Inspection Description

- Two identical Worthington units inspected
  - GB201 and GB501
- Inspection took place in the first three days of a five week turnaround
  - Afforded maximum time possible for any maintenance action
- Both critical units fully instrumented:
  - Vibration monitoring
  - Efficiency, steam rate, and torque monitoring
  - Monthly oil analysis

Worthington
Diagnostic Triggers

- Loss of performance on one unit (GB201) due to fouling
  - GB201 typically ran at full capacity
  - Required monthly online water wash to maintain capability
  - Concern about resulting erosion/corrosion
- No mechanical issues (vibration, torque, etc) detected

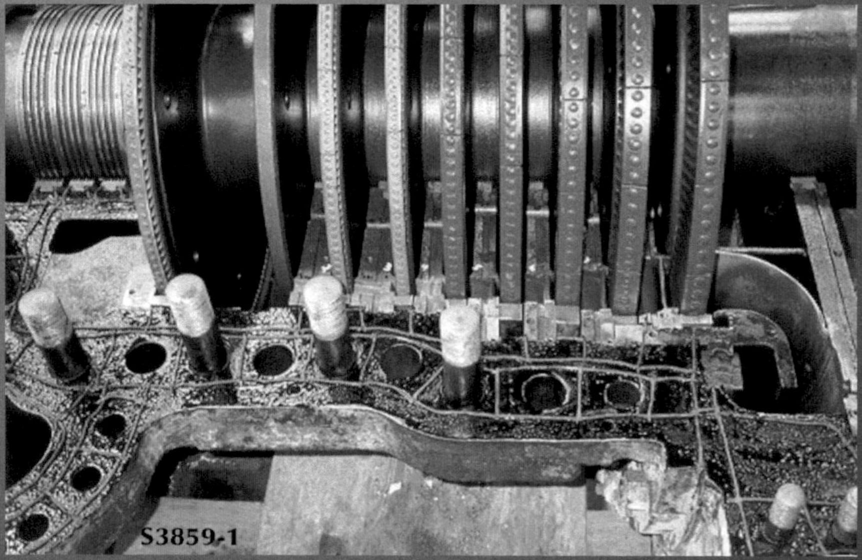

- Primary goal of inspection was to evaluate status of nozzles, buckets, and coatings

Worthington
Low Pressure Stage

- Access through pressure tap port
- Stiff tooling ‘parked’ at the leading edge of the first stage nozzles
- Flexible video scope extended through the nozzles and the radial guide nozzles to the second stage rotating buckets

Worthington: *Low Pressure Stage*
Key Find: *Eroded Radial Nozzles*

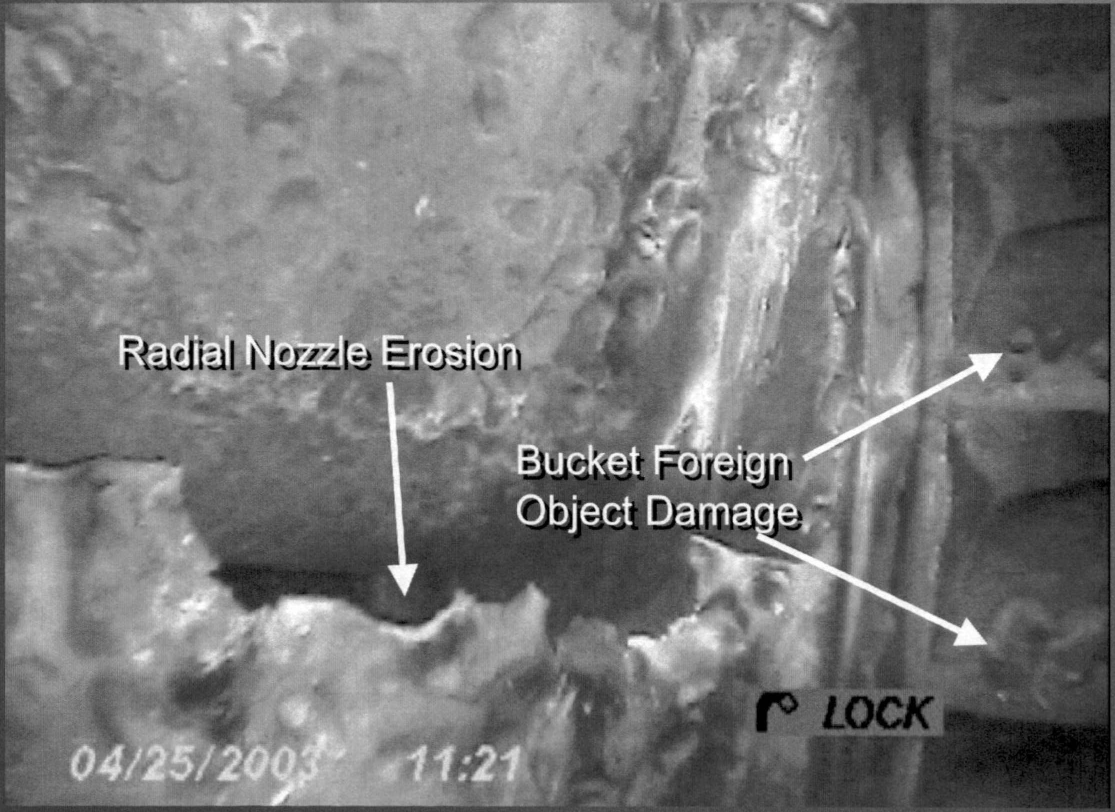

Worthington
Low Pressure Stage Eroded Radial Nozzle Stills

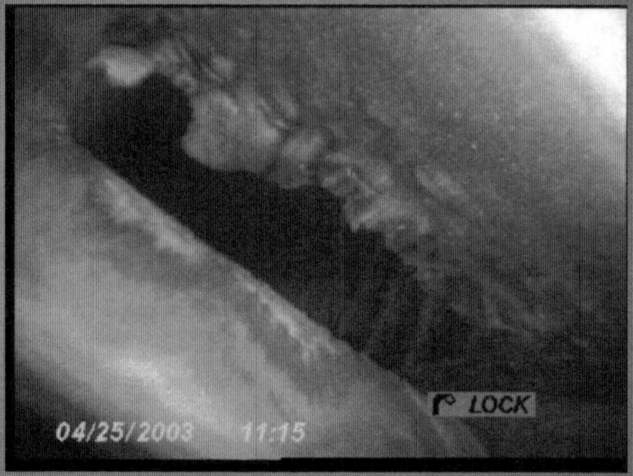

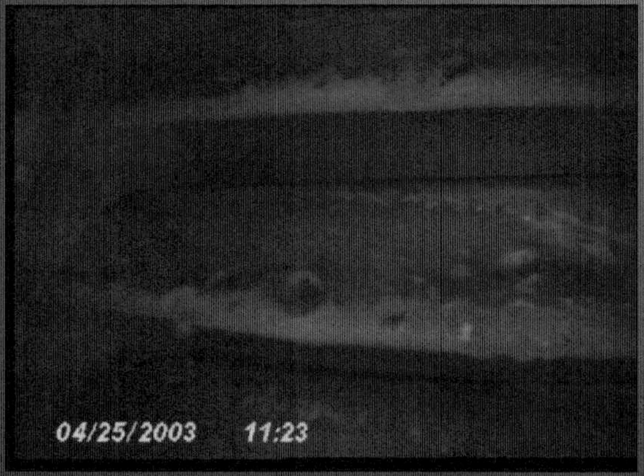

Worthington: ***Low Pressure Stage***
Maintenance Action: ***Immediate Maintenance Required***

- Erosion was severe enough to warrant opening casing and replacing exhaust end radial vanes on both turbines
- One spare was in stock
- The other set of vanes was rebuilt and reinstalled
- There was foreign object damage (FOD) on the buckets which required repairs to one of the removed rotors prior to reinstallation

Worthington: *Low Pressure Stage*
Maintenance Action: *Remove Casing to Replace Nozzles*

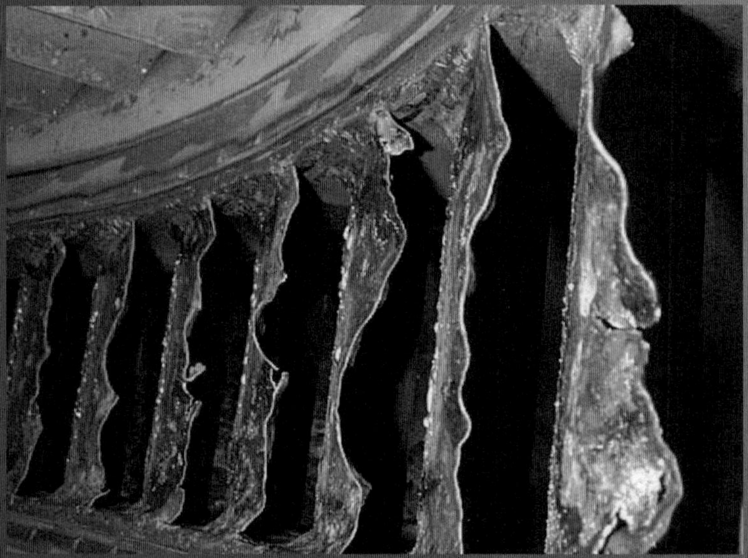

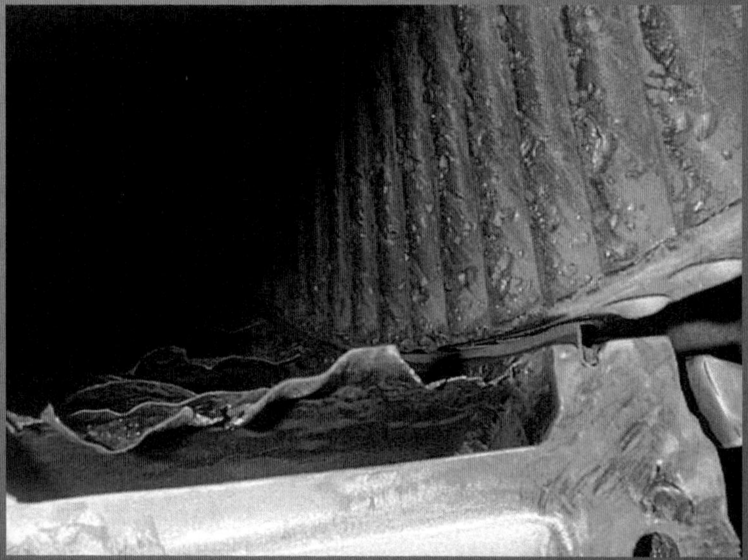

Pictures above were taken after casing was removed

Worthington: *Low Pressure Stage*
Maintenance Action: *Radial Nozzle Replacement*

Radial guide nozzles replaced with in-stock spares

Worthington
High Pressure Stage

- Aft stages visible from low pressure stage access
- First stage nozzles and buckets accessed through the governor poppet valves
- Intermediate stages accessed through drain port on the lower casing
  - Required the tooling to travel straight up to the top casing

Worthington: *High Pressure Stage*
Key Find: *Coating Evaluation*

- An antifoulant coating had been added to the entire steam path
- Inspection revealed it had eroded away on the first stage buckets
- No further damage was seen

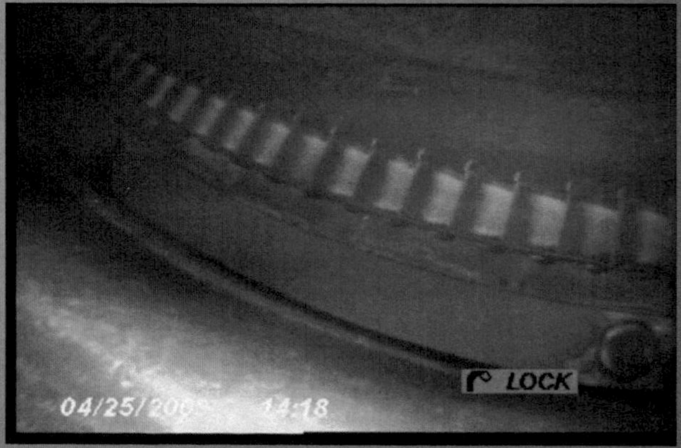

Worthington: *High Pressure Stage*
Maintenance Action: *No Immediate Maintenance Action Required*

- Blade coating status obtained from inspection
  - Top antifoulant coating was gone
  - Intermediate and base coats meant to prevent erosion/corrosion of base materials was still present
  - Antifoulant coating was gone due to poor steam quality
    - Poor steam quality due to surface condenser leaks and treatment chemicals
- No time to recoat or to install coated replacements
  - Planned for the next turnaround

Case
Elliott Steam Turbine at SRW

11 stage turbine

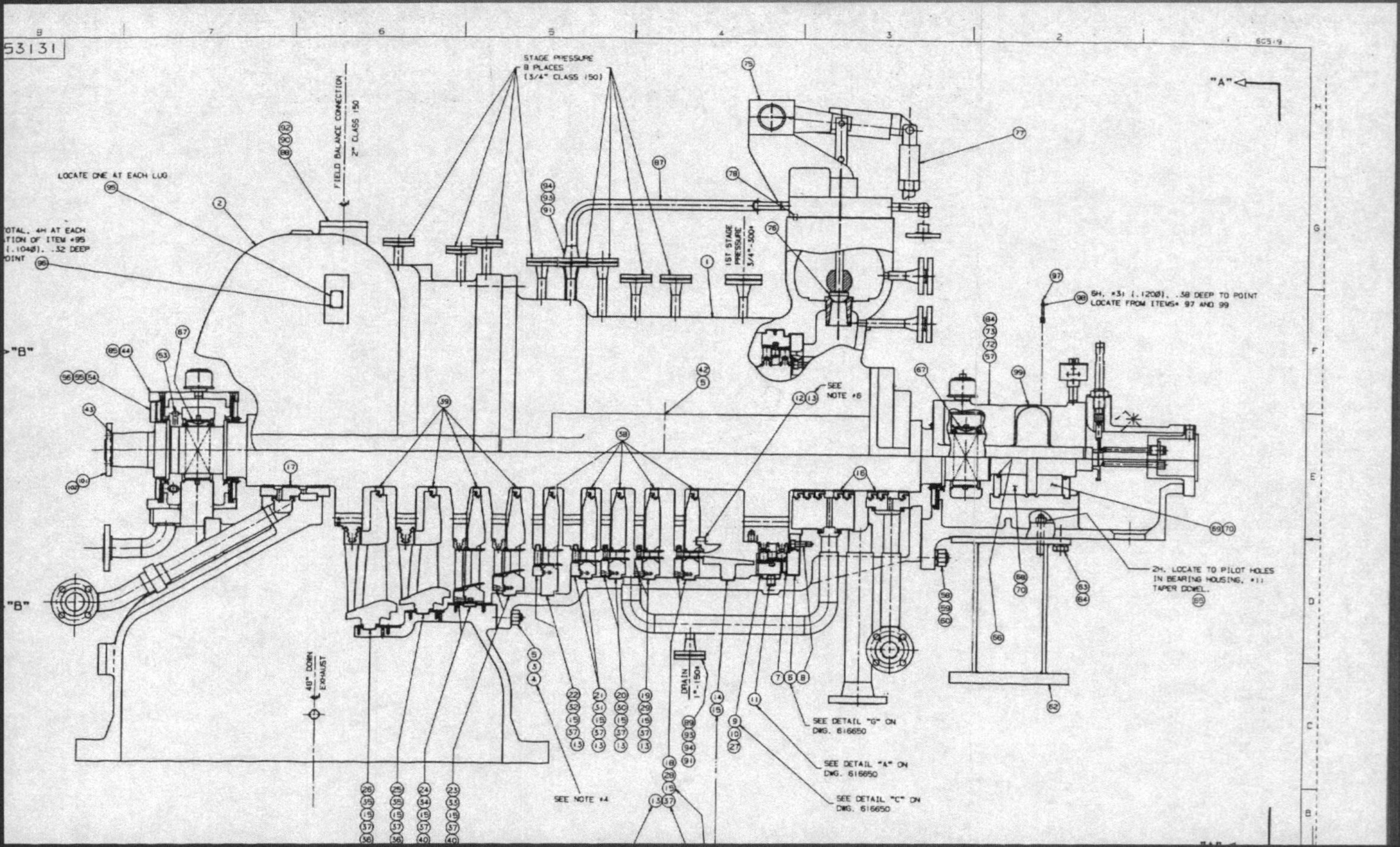

Elliott
Inspection Description

- Each stage accessed through stage-specific pressure ports
- Below: Unit with Casing Off in 1999

Elliott
Exhaust End Stages

- Fouling deposits visible

- Coating has eroded away
- Otherwise buckets are in adequate condition

Elliott
Inlet End Stages

- Nozzles and Buckets in better condition than on the exhaust stages
- Less evidence of corrosion or deposits

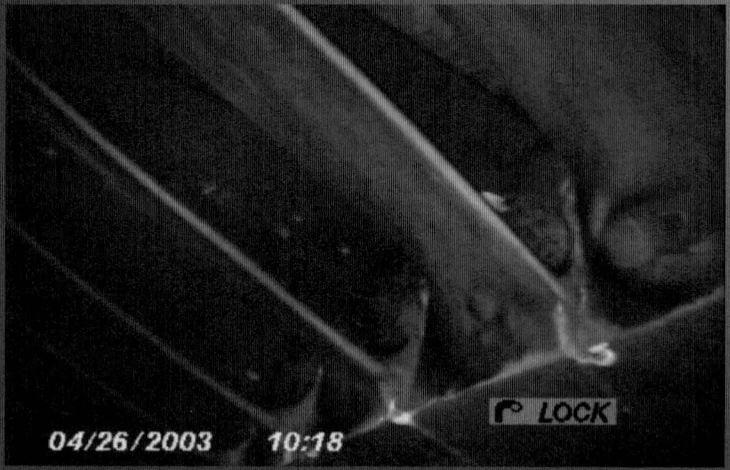

Elliott
Maintenance Action

- Condition not unexpected given
  - The unit's time in service
  - The online water washing that had been done
- No immediate maintenance action is required

Summary

- Turnarounds are a rare opportunity to assess the condition of large rotating machinery
- Tooling specially designed to maximize coverage inside the machinery makes the best use of the turnaround opportunity
- If the inspections are done early in the turnaround, maintenance can be performed if problems are found
- Spare parts and maintenance planning are enhanced even if no problems are found requiring immediate maintenance action

SESSION 57 -ROUNDTABLE DISCUSSION
OPERATIONAL IMPACTS ON EQUIPMENT RELIABILITY
PRESENTED BY THE ROTATING MACHINERY SUBCOMMITTEE

The Roundtable was an open forum for the audience to discuss non-proprietary issues around operational impacts on equipment reliability.

The following rotating equipment related topics were discussed , as well as other items of interest from the audience, in an informal question and answer type format:

* Equipment Start-up
* Equipment Shutdown- scheduled and non-scheduled
* Slowroll or Turning Gear Operation
* Equipment & Auxilliaries Safety Interlocks/Alarm & Trip philosophy
* Process Upset Conditions- operating discipline
* Operator experience and training
* Compressor anti-surge control systems
* Equipment Strainers- startup and permanent
* Suction knockout- demisters- liquid ingestion
* Cleaning of equipment piping
* Equipment performance monitoring/trending/fouling
* Equipment vibration/bearing temperature monitoring/trending
* Checking steam quality, lube oil quality

Paper: 55a

Report on Activities of the NOx in Ethylene Recovery Task Group

Dane C. Grenoble
Consultant

Prepared for Presentation at the 2004 Spring AIChE National Meeting
April 25-29 New Orleans, LA

Copyright © 2004 Dane Grenoble, All rights reserved.

UNPUBLISHED

AIChE shall not be responsible for statements or opinions contained in papers or printed in its publications

Abstract

In 1991, the AIChE Ethylene Producers Safety Subcommittee chartered a Task Group to investigate safety issues associated with accumulation of NOx compounds in ethylene plant cold boxes. The impetus for forming the Task Group was the 1990 explosion at the Shell Berre ethylene plant that was attributed to accumulation of NOx compounds in the Plant's ethylene recovery cold box. This report will summarize the principal findings of the Task Group and provide an overview of suggested guidelines for dealing with suspected accumulation of NOx compounds. The Task Group remains active and typically holds an annual meeting at the Spring AIChE conference.

Disclaimer:

The material given here is presented in the interests of sharing data related to safe operations. Neither the AIChE, the North American EPC of the AIChE, the author nor their employers make any guarantee as to accuracy or completeness of the written materials.

Introduction

Accumulation of NOx compounds in the Cold Box of an Ethylene Recovery Train is a serious safety issue. The most significant event yet related to accumulation of NOx in a Cold Box is the Shell Berre Plant explosion in 1990. The explosion completely destroyed the Cold Box but fortunately the explosion was confined to the Cold Box area with no subsequent explosion from the release of a large hydrocarbon vapor cloud. The following picture shows part of the damaged Cold Box.

Shell Berre Cold Box

This event initiated the formation of the Task Group on Nitrogen Oxides in Cryogenic Sections of Ethylene Plants as a Subgroup within the EPC Safety Section. The Charter of the Subgroup was to provide an open forum to all interested Companies to develop an improved understanding of the issues associated with the accumulation of NOx in Cold Boxes.

Since 1991 the NOx Subgroup has been an active forum for information exchange for Ethylene Producers with concerns about NOx accumulation in the cold box section of the ethylene recovery train. In the early years there were several information exchange meetings per year and there were also several subgroups set up to develop in depth information regarding NOx Literature and Analytical techniques for detection of minute quantities of NOx gases and NOx compounds that can accumulate in the Cold Box environment. Much of the early activities have been reported in detail by Roy Halle ("Potential Hazards of Nitrogen Oxide Compound Accumulation in Cryogenic Ethylene Recovery Facilities" presented at the 1993 Ethylene Producers Conference at the AIChE Spring National meeting). Later, David Simms ("Update on the Activities of the NOx Task Group" presented at the 1997 Ethylene Producers Conference at the AIChE Spring National Meeting) provided an additional update of the Task Group activities. In recent years, the Subgroup has been holding annual meetings at the Ethylene Producers Conference during the Spring AIChE meeting.

This current report will summarize the key issues with processing steam cracked hydrocarbon products contaminated with NOx compounds and suggest guidelines for managing potential accumulations of NOx compounds.

Background

Accumulation of NOx compounds in the cold box is due to trace amounts of NO and O_2 contamination in the cold box feed. Other contaminants like NH_3 can also contribute to the accumulation of NOx compounds. The source of NO has generally been linked to ethylene units that recover refinery gases and in particular FCC gas. Since the start of the subgroup it has become evident that there are other potential sources of NOx contamination associated with reprocessing of various byproduct streams from the ethylene plant.

The chemistry of NOx accumulation is complex and not completely understood. The reactions of main concern are:

NO (nitric oxide) + O_2 (oxygen) → N_2O_3 (dinitrogen trioxide, a.k.a. Blue Liquid)

N_2O_3 + Unsaturated Hydrocarbons (like butadiene) → NOx Gums

(NO + O_2) or N_2O_3 + NH_3 (ammonia)→ NH_3NO_2 and NH_3NO_3 (ammonium nitrite and nitrate)

It is important to note that in addition to NO; oxygen and ammonia are also required contaminants for the accumulation of NOx compounds and NOx salts to occur in the Cold Box. Removal of these co-contaminants would be one method to mitigate accumulation of NOx compounds but the Author is not aware of technology that has been successful for complete removal of these contaminants.

The key properties of the NOx compounds are summarized in the following Table.

Table 1

	NO	NO_2	N_2O_3	NH_4NO_2	NH_4NO_3
Melting Point:	-161 °C	-9.3 °C	-102 °C	60 °C (explodes)	170 °C
Boiling Point:	-151°C	+21 °C	+3.5 °C (decomposes)		210 °C
Appearance:	Colorless Gas	Red-Brown Gas	Blue Liquid	Solid White Crystals	Solid Colorless Crystals

Of the ammonium NOx salts, ammonium nitrite is of most concern since it is much less stable than ammonium nitrate. As noted, NH_4NO_2 can explode at temperatures not far from ambient conditions. In work cited by Halle and Simms, NH_4NO_2 has been shown to be unstable at temperatures as low as about 35°C. N_2O_3 is of concern since at ambient conditions it readily decomposes into NO and NO_2 gases that are both quite toxic. Also N_2O_3 as discussed above will readily react with unsaturated hydrocarbons like butadiene to form NOx gums even at the cryogenic temperatures in the Cold Box.

The properties of NOx gum are not included in the above Table due to incomplete information. NOx gums are complex hydrocarbon compounds that are highly unstable mixtures of nitrated oligomers of unsaturated hydrocarbons. Butadiene is perhaps the unsaturated hydrocarbon that is of most concern. Butadiene is known to react readily with N_2O_3 at cryogenic temperatures to form highly unstable and explosive gums that have been reported to decompose at temperatures well below ambient conditions.

The section of the cold box that recovers the coldest liquid feed to the demethanizer tower has been the main location where the NH_3NO_2 and NH_3NO_3 salts have been found. N_2O_3 and NOx gums have been found mostly in cold boxes that recover cryogenic hydrogen gas. The Figure shown below is typical of most cold boxes in ethylene recovery service.

Typical Areas for NOx Accumulation

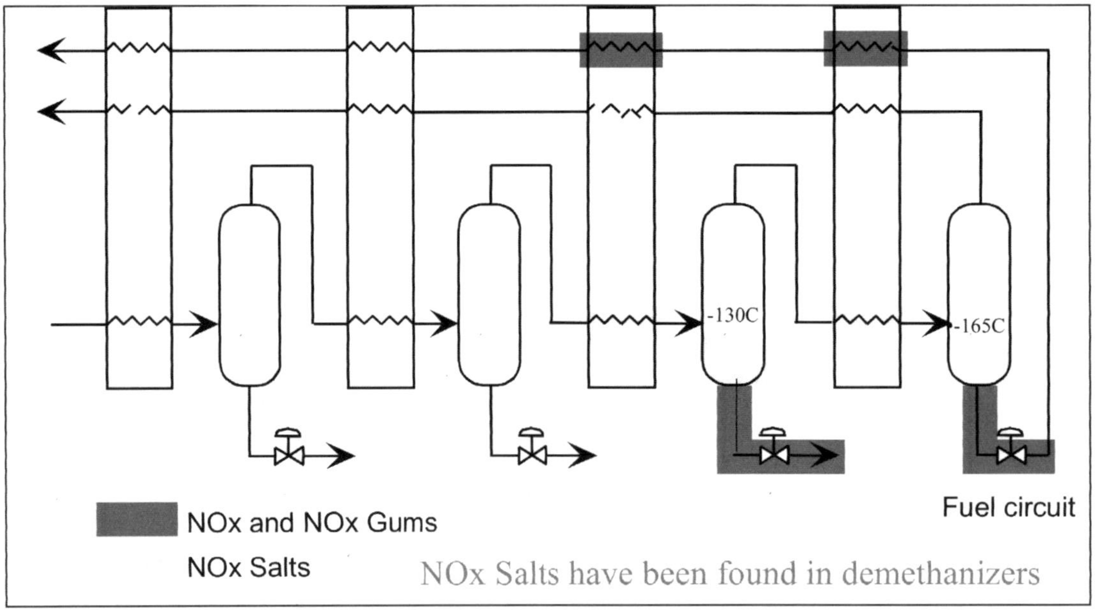

The ammonium nitrite and nitrate salts have been found mostly in the region of the drum that operates at about -130°C. N_2O_3 and NOx gums have been found mostly in the drum and surrounding areas that operate in the coldest section at about -165°C. Note that many units do not have this coldest section since they do not recover high purity Hydrogen in addition to fuel gas. As noted in the Figure, NOx salts have also been found in the demethanizer tower downstream of the Cold Box. It is not clear if these salts were formed directly in the demethanizer tower or were carried forward from the Cold Box with the liquid product from the -130°C drum.

Sources of NOx Contamination

Contamination of Cold Box feeds with NOx (mainly NO) was originally thought to be restricted to Crackers that processed Refinery Cracked gases especially FCC gas. Since the advent of the NOx Subgroup, additional sources of NO contamination have been identified. The following is a list of possible sources of NO contamination.

Refinery Gas Streams

- FCC Gas
- Coker Gas
- Saturated/non-cracked refinery gas streams

Other Sources

- Recycle cracking streams that contain Process Additives containing the '–NO' functional group
- Some Producers have reported NOx accumulation without processing any streams defined here.
- Other yet unidentified sources of NOx contamination?

Analysis for NO and NOx Compounds

Trace levels of NO are very difficult to detect. The Analytical Subgroup of the NOx Task Group has published procedures for measuring NO in Process Gas and for measuring NOx gums recovered in alcohol washes of the Cold Box. NOx salts recovered in alcohol washes can be measured using conventional anion and cation ion chromatography.

The preferred method for measuring NO in Cold Box feed is to use an online analyzer equipped with a Photoionization Detector (PID). Such an instrument is capable of detecting NO down to about 10 wppb. A downside for this detector is that the PID lamp has relatively short lifetime and must be replaced frequently. It is likely that most suppliers of online analyzers should be able to provide a custom built unit for analysis of NO in cracked gases.

Is there an acceptable level of NOx accumulation in Cold Boxes?

Perhaps the most common question posed to the Subgroup has been how much NOx accumulation is acceptable. Unfortunately there is no easy answer to this question. Both the ammonium NOx salts and the NOx gums have significant explosive energy equivalent to TNT and other military explosives. So even a small amount of accumulation can result in a significant incident should the accumulated NOx compounds explode. The incident at Shell Berre clearly illustrates the potential consequence of such an explosion.

Regarding the amount NOx compounds that is acceptable, a brief description of an incident reported by a Member Company may be helpful. During the warm-up of this Plant's cold box, an explosion occurred resulting in rupture of one of the tubes located in the coldest section of the cold box. Since this unit had a history of NOx compound accumulation, their standard procedure was to warm up the cold box using inert nitrogen gas. Thus there was no potential for a secondary explosion from the loss of containment of the process gas.

The following Figure shows the damaged pipe in the cold box.

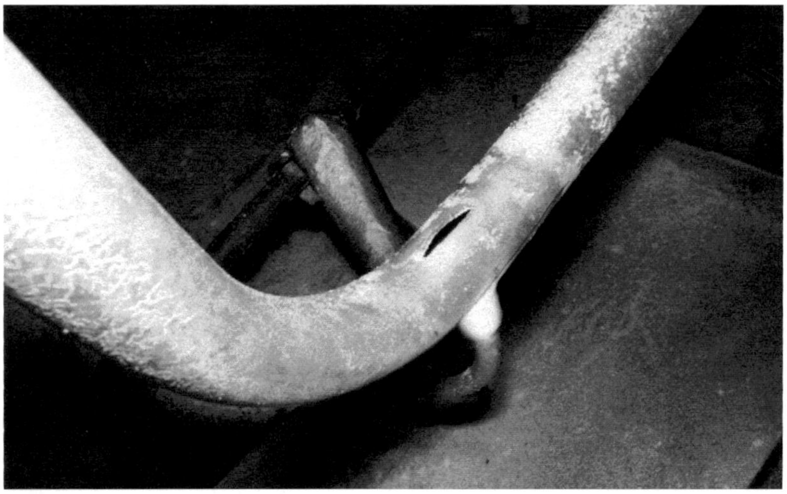

The investigation report concluded that the pipe rupture was due to detonation of a small amount of NOx gums that accumulated in the bleeder pipe just opposite to the rupture. The amount of NOx gum may have been only about 10 grams. As noted above, NOx gum is formed by reaction of N_2O_3 with unsaturated hydrocarbons like butadiene. Since this Plant was aware that N_2O_3 accumulates in this region of the cold box, they had procedures in place to prevent carrying forward process cracked gas during any occasion that the cold box would be subjected to thawing. Thus the presence of the NOx gum was a surprise. The source of the unsaturated hydrocarbon was eventually traced to the presence of a buffer gas line between the cold box and the plant compression train. Apparently during a period of upsets in the plant, cracked gas was able to back flow into the cold box resulting in accumulation of butadiene as well as N_2O_3 in the coldest section of the cold box leading to the formation of NOx gum and the subsequent explosion.

The following Figure is a schematic of the cold box showing the buffer gas line that has subsequently been removed. This incident also illustrates the importance to identify all possible sources of hydrocarbon flow to the Plant Cold Box.

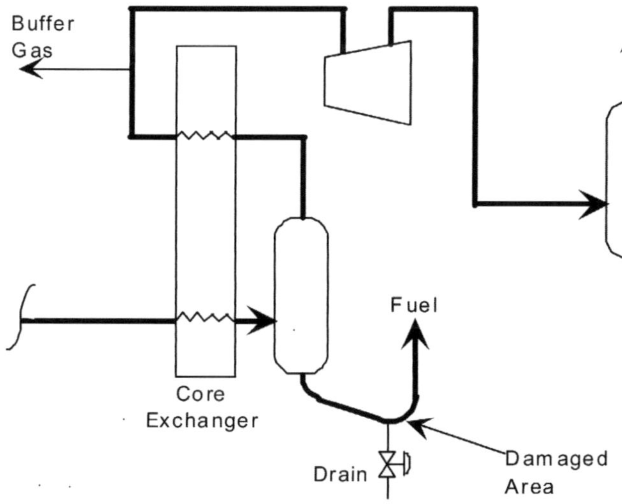

In this example only a small amount of NOx gum was sufficient to cause an explosion with sufficient force to rupture the piping in the cold box.

The prudent answer to how much NOx accumulation is acceptable is none to very little.

Most important is to recognize that NOx accumulation can or has occurred and then take steps to manage the risks associated with the possible or known accumulation. This Plant was aware of NOx accumulations and had taken a significant number of steps to manage and mitigate consequences of the presence of NOx compounds. Because of these steps, this incident was relatively minor compared to a possible incident where the NOx accumulation was not known and when such a pipe rupture could lead to release of a substantial amount of hydrocarbon vapor.

The following sections provide suggested Guidelines for managing potential hazards associated with the accumulation of NOx compounds in a Cold Box.

Operating/Safety Guidelines

Checking for NOx Accumulation

Since the measurement of low levels of NOx compounds in Cold Box feed can be difficult, it is potentially easier to look for NOx compounds in the Cold Box where concentrations have increased.

The following are some action steps to be considered:

- Determine likely locations for NOx accumulations such as:
  - ✓ Cold process gas vessels and piping (temperatures < -100° C)
  - ✓ Vaporizing methane streams
  - ✓ Low flow areas in low temperature methane rich streams
- Purge systems with 'inert' gas during shutdowns
  - ✓ Measure purge gases for NO/NO_2 from all cold process gas streams.
  - ✓ Use length of stain tubes
- Be alert for
  - ✓ 'Blue Liquid'
  - ✓ Yellow-brown Gases

Startup Procedure Issues

The following list suggests guidelines to consider when starting up or shutting down the Cold Box under planned or unplanned conditions.

- Unplanned Startups following an unplanned shutdown
  - ✓ Is the Cold Box a 'Dirty System' that may contain NOx and NOx compounds?
  - ✓ What is the condition of cold box?
    - ➢ Did NOx locations warm up?
    - ➢ Did the shutdown event carry reactive material forward to NOx locations?
- Unplanned Startup Procedures
  - ✓ Exclude butadiene if system may contain NOx compounds
  - ✓ Stabilize conditions before forward flow of process gas → bypass coldest sections until gas at normal operating conditions
- Planned Startups
  - ✓ Condition of Cold Box?
    - ➢ Clean system?
    - ➢ NOx measurements during shutdown?
- Startup Procedure, if answers are **NO** to above, is same as for **unplanned shutdowns**

The most important aspect is to prevent introduction of reactive compounds in the process gas from entering the regions of the cold box where NOx accumulation may have occurred.

Shutdown Procedure Issues

- Control Shutdown of NOx Areas
  - ✓ Purge gas composition → use inert gas (nitrogen) if possible. It is not recommended to use cracked gas to warm up the cold box.
  - ✓ Drain liquid from warm drums to avoid butadiene to NOx areas as system warms up
  - ✓ Warm up rate should follow the plate fin exchanger Vendor recommendations at minimum → typically ~20° C/hr
  - ✓ Monitor evolution of NO/NO_2 → use length of stain tubes.
    - ➢ Expect rapid rises in NOx concentration as purge gas reaches beginning of N_2O_3 decomposition temperature at about -40° C
- Trip or emergency shutdown or loss of refrigeration **requires** procedures/facilities to avoid butadiene getting to possible NOx accumulation areas

Deriming / Thawing Considerations

- Washing Considerations
  - ✓ Volume requirements
  - ✓ Wash temperature
  - ✓ Disposal of solvent
  - ✓ Weight design limitations when liquid filling
  - ✓ Soak or circulate
- Solvent(s) → use polar solvents to remove salts/gums
  - ✓ Methanol dry (99.5%) → Methanol corrosivity (dry + heat)
  - ✓ Methanol Freezing Point (-94° C)
    - ➢ Methanol + H_2O (eutectic mixture)
    - ➢ Methanol / Propanol mixtures

Routine Monitoring that may indicate accumulation of NOx

- System Performance
  - ✓ Unusual Pressure drops?
  - ✓ U value changes?
  - ✓ Recoveries/purities change?
- NOx Accumulation Monitoring
  - ✓ NO analyzer on cold box feed
  - ✓ Estimate NOx accumulation per amount of processed FCC gas or other suspect feed gas

Introduction of New Raw Materials

- NO produced from furnace feeds
  - ✓ Feeds containing NOx contaminants
  - ✓ Feeds / recycle streams with inhibitors that crack to NOx
  - ✓ Water washed feeds with water that may contain nitrates/nitrites (?)
  - ✓ Dilution steam water that may contain NO precursors
- NO in Cracked Gas Compressor feeds
  - ✓ Recycle from other units
  - ✓ Inhibitors / antifoulants

If NOx Accumulation is suspected:

- Have a plan in place to deal with NOx accumulation → don't wait to be surprised with the unexpected discovery of NOx compounds before developing a plan
- Considerations
  - ✓ Possible Personnel Exposure
  - ✓ Avoid butadiene migration to suspect NOx locations
  - ✓ Have a cleanup plan to purge and/or alcohol wash
  - ✓ Measure NOx at each shutdown in purge gases and in solvent if washed

Summary

- Be aware of the possibility of NOx, particularly if:
  - ✓ You process FCC gas (or other refinery gas).
  - ✓ You have certain recycle streams.
  - ✓ You have raw materials containing some inhibitors.
  - ✓ You have unexplained cold box operating problems.
- If you suspect NOx, test for presence
  - ✓ Best time is during thaw/deriming.
- If you suspect NOx gums or salts, treat them as potentially hazardous.
  - ✓ Do not subject to rapid heating or shock.
  - ✓ Consider solvent wash for cleaning.

The above "Guidelines" are intended to provide an outline of procedures and considerations for Ethylene Crackers that might have the potential for contamination of Process Gas with NOx compounds. It is not intended to be a complete description of the possible issues and safety implications if NOx accumulation occurs. Nor is it intended to be a complete listing of recommendations to manage accumulation of NOx in Ethylene Plant Cold Boxes.

For additional information or if there are questions regarding NOx issues, please contact the author at dgrenoble@houston.rr.com or Richard Rodriguez at richard.rodriguez@shell.com.

Paper 55b

Feedstock Contaminants in Ethylene Plants - an Update

Authors

Arthur J. Baumgartner, Shell Chemical Co.
Marilyn W. Blaschke, Baker Petrolite
Steven T. Coleman, Equistar Chemicals, LP
Robert Kohler, UOP
Timm E. Paxson, Shell Chemical Co.

Prepared for Presentation at the 2004 Spring AIChE National Meeting
April 25-29 New Orleans, LA

UNPUBLISHED

AIChE shall not be responsible for statements or opinions contained in papers or printed in its publications

Abstract

Ethylene plant operators in North America and Europe were polled for their responses on a wide range of feedstock contaminant issues. This paper presents the responses to this survey and an analysis of the gathered information. Comparisons with the 1992 Feedstock Contaminant Survey are also included.

Introduction

It has been 12 years since a survey regarding contaminants has been conducted by the AIChE Feedstocks Subcommittee, and the subcommittee thought it timely to execute another questionnaire. The 2003 survey also was constructed in part to give the group direction for future activities. This paper will present an analysis of the gathered information, and comparison with the 1992 survey will also be included where applicable.

A copy of the 1992 survey is included as **Appendix I**. That survey was divided into 2 sections; the first dealing with a list of almost 30 named contaminants, while the second dealt with the specific contaminant mercury. Detailed questions of the first section included:

- Indication of the level of concern (high, medium, low) of the list of contaminants.
- Consequences of allowing the contaminant into the unit.
- Quantification of the contaminant level.
- Problems caused by the contaminants, how the problem was solved, and willingness to share the solution with others in the ethylene industry.

2003 Survey -- General

The current survey was sent out in electronic form to facilitate mailing and completion. The subcommittee obtained 35 separate replies from ethylene producers. (Please note that the values presented in the following three charts will not completely agree since several respondents combined results.) As depicted graphically below, almost 70% of the respondents were from North America (NA). One quarter are located in Europe (E), and 2 respondents gave no response (No Resp).

Plant Locations

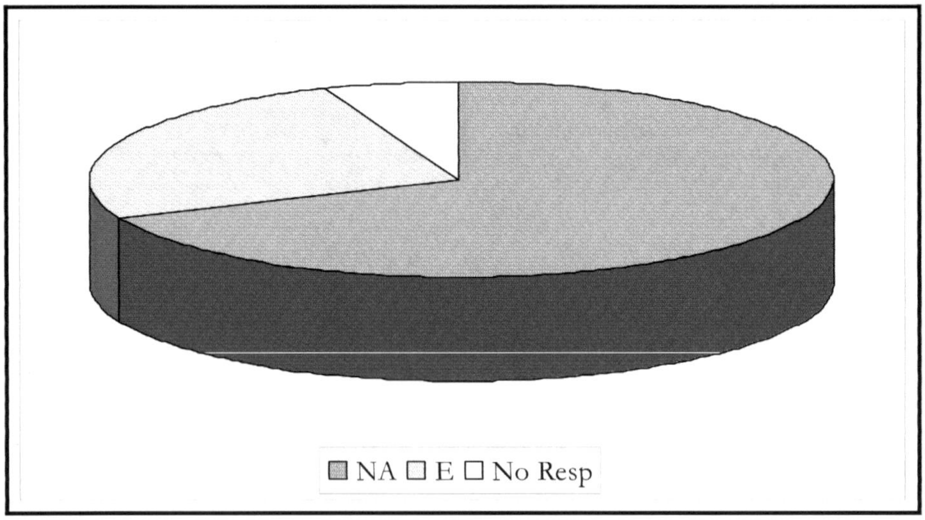

Seven (18%) of the plants represented in the survey were built before 1970. Almost half (49%) were built in the '70's, while 13% and 15%, respectively were built in the '80s or post-1990. (Two respondents did not indicate plant age.)

Plant Age

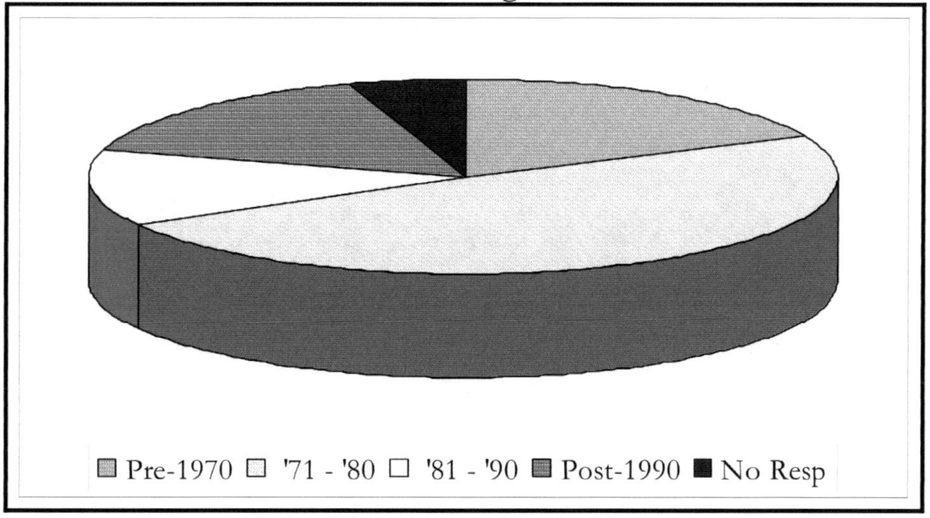

Cracker feedstocks were divided into 7 categories for analysis: Ethylene and/or C_2/C_3 and/or C_4 feedstocks (abbreviated EP), Gas Field Condensates and/or Natural Gasolines (GFC), RCO feedstocks, which were defined as Refined Crude Oil liquids (Naphtha, Gas Oils, *etc.*), and the various combinations: EP+GFC, EP+RCO, GFC+RCO, and All 3.

Fourteen plants or 40% of the population crack EP feeds, which represents the largest group by far. The next largest group (5 or 14%) are those cracking Gas Field Condensate (GFC), while four (11%) crack Refined Crude Oils (RCO). Ten producers, which represented 29% of the pool, crack multiple feeds; the **Table** below delineates that data:

Feed Category	Number	%
EP+GFC	1	3
EP+RCO	3	9
GFC+RCO	2	6
All	4	11

The combined data is shown pictorially in the pie graph below:

Cracking Feedstocks

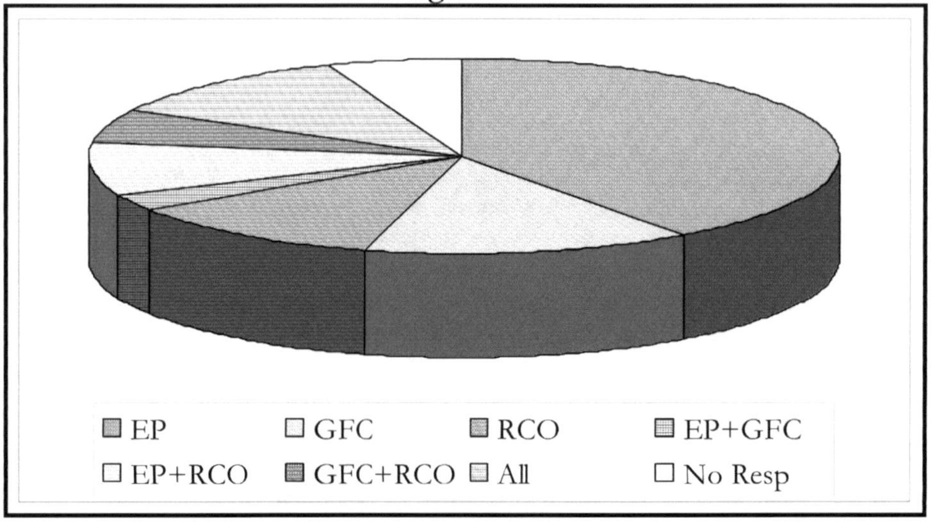

The final general category is plant size. The questionnaire included 3 categories – those plants smaller than 300 metric tons per annum (mta), those of intermediate range, 300 – 600 mta, and those greater than 600 mta. The data is displayed below both in tabular and graphic form.

Plant Size	Number	%
<300 mta	1	3
300 - 600	13	37
>600	19	54
No Resp	2	6

Plant Size Data

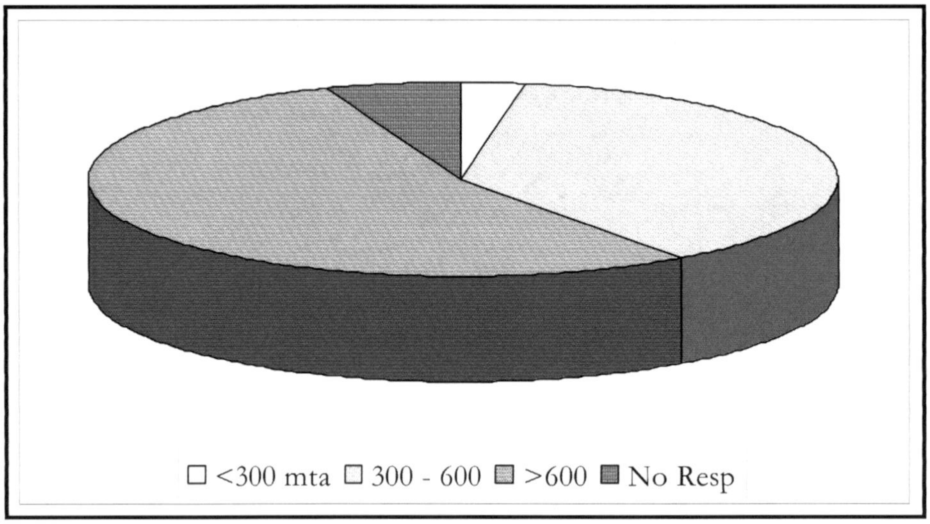

2003 Survey – Question One

The format of the current survey was also divided into a general section and a section asking questions about several specific contaminants. Question One was divided into 6 pages with each page focusing on contaminants found in various feedstocks: 1) Ethane or E/P (Ethane/Propane) mix, 2) Gas Oils, 3) Naphthas & Condensates, 4) Propane or Other NGLs (Natural Gas Liquids), 5) Raffinate Streams (C_4, C_5 recovery), and 6) Others, such as Refinery Gases. Each page of Question One lists 28 contaminants (with space for the addition of a few more). Detailed questions include for each of the listed contaminants:

- Level of concern
- Do you test for it?
- Do you detect it?
- Base Line/Normal Value?
- Maximum Permissible Level

Responses to Question 1 are organized into 3 tables:

1) Level of concern for specific contaminants
2) Contaminants tested for and detected
3) Normal baseline and maximum permissible concentrations of contaminants

The data are reported according to the six above mentioned feedstock categories with the 6th category being entirely devoted to Refinery Gases.

In **Table Q1-1**, the level of concern, or in other words, the amount of interest shown in each of the contaminants is listed. Responses are tabulated in terms of the percentage of plants responding where there was either a medium level of concern or a high level of concern (low responses ignored). A 3rd column was created for each contaminant, which was defined as the sum of the percentage of plants responding with medium and with high levels of concern, to identify the most important contaminants.

When the contaminants are ordered according to highest percentages of the plants where there was at least a medium level of concern for the four C_2-C_5 feedstock categories, the following ranking of important contaminants results, (note: ranking of 1 is the highest percentage value).

Ranking of C2-C5 Feedstock Contaminants*

Ethane or E/P Mix		Propane or Other NGL's		Refinery Gas		Raffinate Streams from C_4, C_5 Recovery	
Rank	**Contaminant**	**Rank**	**Contaminant**	**Rank**	**Contaminant**	**Rank**	**Contaminant**
1	CO_2	1	COS	1	COS	1	MTBE
2	Methanol	1	Fluorides	1	H_2S	2	Total Sulfur
3	COS	3	CO_2	1	CO_2	2	Organic Cl
4	H_2S	4	H_2S	2	Arsine	2	Inorganic Cl
5	Total Sulfur	4	Methanol	2	O_2	2	Lead
6	H_2O	4	Total Sulfur	2	NOx	6	Fluorides
7	Inorganic Hg	7	H_2O	2	CO	6	Methanol
8	O_2	7	Inorganic Cl			6	Inorganic Hg
8	Radon	7	Olefins			6	Organic Hg
8	Fluorides	10	Organic Cl			6	Silicon Deriv.
8	Arsine	11	Inorganic Hg			6	Olefins
8	NH_3	11	Radon			12	COS
8	RSH	13	RSH			12	Oxygenate
8	CO	13	Arsine			12	Transition Metals
15	Olefins	13	CO			12	Sulfolane
15	Inorganic Cl	16	Propylene			12	Vinyl Acetate
17	Organic Hg	16	Organic Hg			12	H_2O
17	Sodium	16	NH_3			12	NH_3 (Ttl Nitrogen)
17	Corrosion	16	Sodium			12	DME
17	Organic Cl	16	MTBE			12	Arsenic
*)	According to highest percentages of plants where there was at least a medium level of concern.						

The above **Table** shows similar concerns expressed in previous contaminant surveys. Light acid gases, such as CO_2 and COS that are difficult to remove, are of special importance in the C_2 and C_3 feedstocks. Also, methanol, mercury, fluorides, chlorides, sulfur compounds and arsine continue to be a troublesome contaminant to many plants cracking C_2-C_5 feedstocks.

Similar rankings for Naphthas/Condensates, (including BTX Raffinates) and Gas Oils are listed in the following table.

Ranking of Naphthas/Condensates, Gas Oils Feedstock Contaminants*

Naphthas/Condensates (including BTX Raffinates)		**Gas Oils**	
Rank	**Contaminant**	**Rank**	**Contaminant**
1	Lead	**1**	Total Sulfur
2	Total Sulfur	**2**	Inorganic Chlorides
2	H_2S	**2**	Inorganic Hg
2	Inorganic Hg	**4**	Silicon Derivatives
5	MTBE	**4**	Total Nitrogen
5	Sodium	**6**	Lead
5	Inorganic Chlorides	**6**	Organic Hg
8	Organic Hg	**8**	Organic Chlorides
8	Organic Chlorides	**8**	Sodium
10	Arsine (precursor)	**8**	Phosphorus
10	Methanol	**8**	Total Chlorides
10	Silicon Derivatives	**8**	Total Oxygenate
13	Olefins	**8**	H_2S
13	Florides	**8**	H_2O
15	Vanadium	**8**	Methanol
15	Total Chlorides	**8**	MTBE
17	Total Transition Metals	**8**	Vanadium
17	COS (precursor)	**8**	Arsenic
17	H_2O	**8**	Organic Chlorides
17	Total Oxygenate		

*) According to highest percentages of plants where there was at least a medium level of concern.

The above **Table** on contaminant rankings in naphthas/condensates and gas oils also shows similar concerns expressed in past feedstock contaminant surveys. Motor gasoline additives continue to be of special concern in naphtha/condensates but also of some concern in gas oils. Sulfur compounds, mercury, sodium, chlorides and arsenic continue to be problem contaminants for both naphthas/condensates and gas oils.

In **Table Q1-2**, contaminants tested for and detected are listed in terms of percentage of plants responding. As expected, there was a relatively large amount of testing reported for contaminants of highest concern such as CO_2 and methanol in C_2 and C_3 feedstocks.

In **Table Q1-3**, the normal or baseline concentration and the maximum permissible concentration of the contaminants in the survey are listed. Only contaminants listed as medium or high level concern are included in the table, along with the number of responses for the contaminant.

2003 Survey – Question Two

Question 2 was designed to determine how various contaminants impact the operation of the ethylene plant. The raw data, **Table Q2-1,** which is presented in survey question form, was a matrix with the contaminants of Question One listed in one column with a number of consequences forming the headers of the remaining columns of the matrix. The headers (consequences) include Quality, Corrosion, Catalyst Poison, Operability, Safety, Environmental, and Capacity Limit(ations). Boxes are to be checked regarding the consequence of introduction of each contaminant into the unit. Responses from all plants are shown as well as a breakdown of results by gas cracker (**Table Q2-2**) and liquid cracker (**Table Q2-3**). For the purposes of this survey, gas cracking plants are those who listed their feedstock as Ethane/Propane/Butane only. All plants that crack liquids exclusively or a mixture of gas and liquid feedstocks are designated as liquid crackers. A total of 35 plants participated in the survey.

Figure Q2-1 shows that of all responses, sulfur, methanol/alcohols, sodium, chlorides, CO_2 and mercury were the contaminants of highest concern. In fact 71% of all plants surveyed had experienced effects from sulfur and methanol/alcohol contamination in their feedstock.

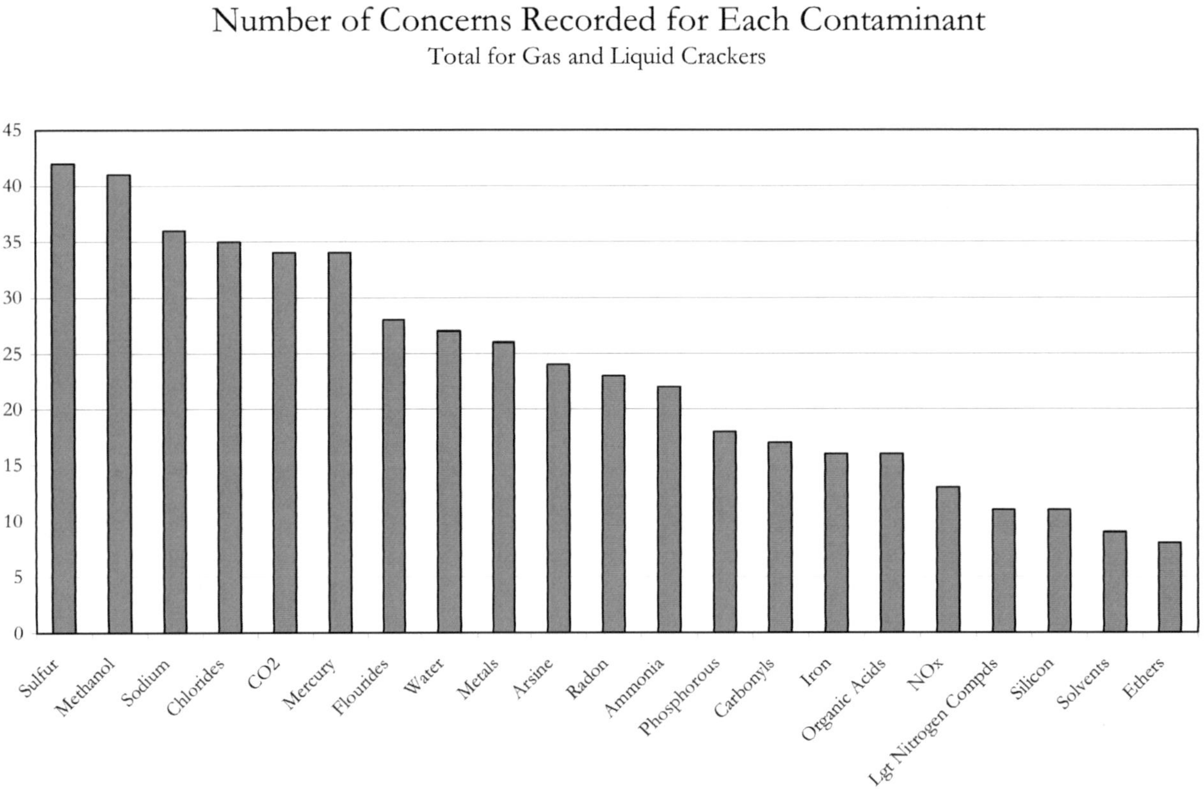

Figure Q2-1

An analysis of gas versus liquid cracker responses showed that the high interest contaminants were relatively consistent between the two types of plants. With methanol/alcohols, CO_2, chlorides, sulfur, fluorides and radon being the highest level of concern in gas crackers. The data shifted slightly to show mercury, sulfur, sodium, methanol/alcohols, arsine and chlorides to be of highest concern with liquid crackers. Ninety percent of all liquid crackers had a concern with sulfur contamination.

Figure Q2-2 shows the consequences by type. This data shows that the ethylene producers surveyed are most concerned about the effect of contaminants on product quality, corrosion and catalyst poisoning.

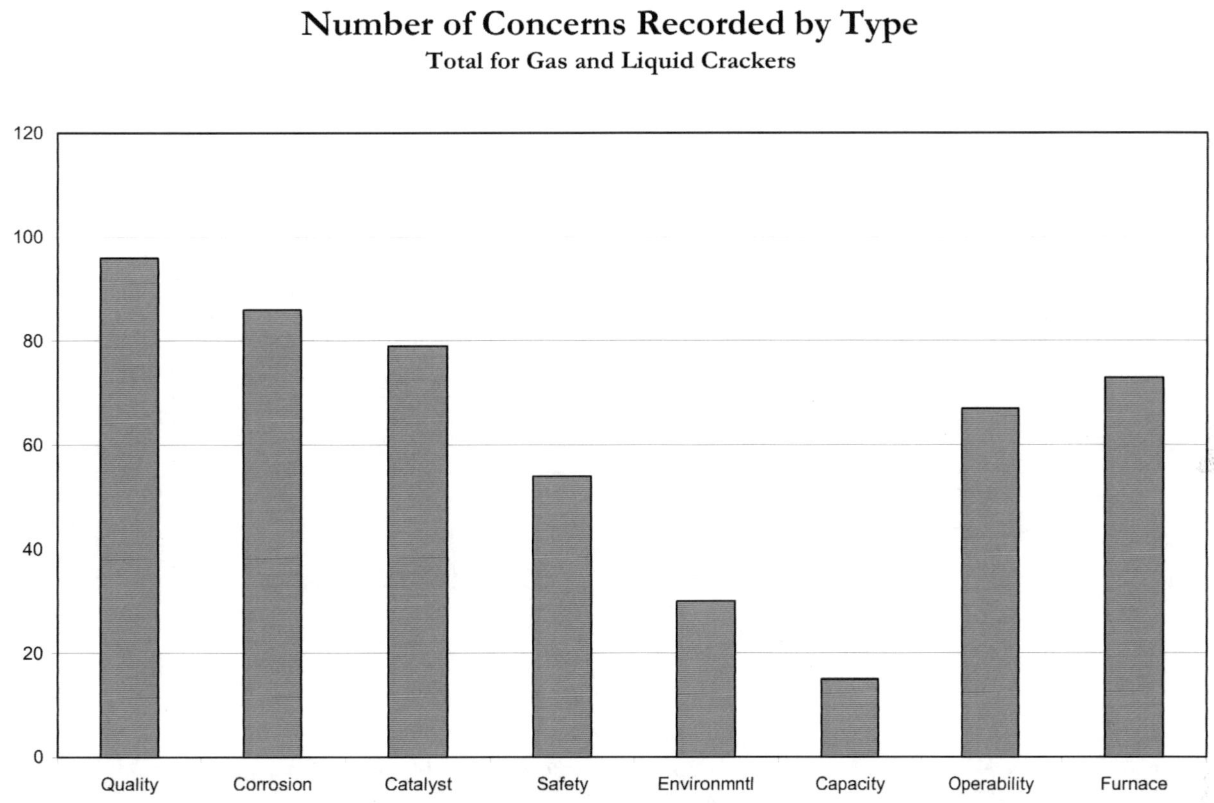

Figure Q2-2

This is consistent with the liquid crackers surveyed. The gas crackers, however, had less distinct areas of concern. The gas cracking data shows that the consequences most experienced were:

1. Product Quality and Operability – 22 responses
2. Corrosion and Furnace – 18 responses each
3. Catalyst Poison and Safety/Hygiene – 16 responses each

A final analysis comparing the older vintage plants with newer plants was conducted. For survey purposes, any plant listing their age as greater than 20 years old is classified as an older vintage plant. Twenty-four (24) of the survey respondents are classified as older vintage plants, nine (9) plants are classified as new plants and two plants did not list their vintage and therefore are not included in this data analysis.

Looking at the most problematic feedstock contaminants for each category shows that while the ranking may change, the problem contaminants are consistent within the gas cracking plants, old or new. The newer liquid cracking plants are more concerned about arsine, phosphorous and carbonyls than the older vintage liquid crackers.

Overall the fewest contaminant concerns per plant were reported by the new gas crackers. Older plants cracking liquids reported the highest number of concerns/consequences with feedstock contaminants.

Gas Crackers – Top Five Problem Contaminants		
All Survey Data	**Older Vintage**	**Newer Vintage**
Methanol/Alcohols	CO_2	Methanol/Alcohols
CO_2	Fluorides	Chlorides
Chlorides	Sulfur	CO_2
Sulfur	Chlorides	
Fluorides	Methanol/Alcohol	

Liquid Crackers – Top Five Problem Contaminants		
All Survey Data	**Older Vintage**	**Newer Vintage**
Mercury	Methanol/Alcohols	Mercury
Sulfur	Mercury	Arsine
Sodium	Sodium	Sulfur
Methanol/Alcohols	Sulfur	Phosphorous
Arsine	Water	Carbonyls

2003 Survey – Question Three

Question 3 is divided into four pages with each page dealing with a specific contaminant – Page 1), Mercury, 2), Salt (Sodium Chloride), 3), Arsenic, and 4), Ammonia/Amines/Nitriles. The survey data is tabulated in **Tables Q3-1 – Q3-5**. The 1992 survey focused only on the details of mercury contamination.

Mercury

Mercury typically enters a steam cracking plant as the element, but could also enter as an organometallic complex such as methyl mercury. It moves through the furnaces and distributes into the various hydrocarbon fractions as delineated in the following Table:

Olefins Unit Mercury Distribution[1]

Stream	Distribution (Relative %)
Tail Gas	2
Ethylene Product	8
Ethane Recycle	2
Propylene Product (MAPD Catalyst)	19
Propane Recycle	1
Butanes/Butylenes	53
Pyrolysis Gasoline	1
Unaccounted	Balance

The element can cause problems in several different ways. It can condense in the core exchangers or in low point drains or headers. Personnel opening the drain valves associated with these headers could risk mercury exposure. The core exchangers are typically constructed of aluminum alloys and if allowed to warm above -38°F (-39°C), severe pitting and intergranular cracking can occur with concurrent equipment failure. The cracking phenomenon may be catalytic in mercury (Hg continually migrates to crack propagation front) and occurs in the cm/sec velocity range. Failure of the coldbox is possible with the uncontrolled release of hydrocarbons. Mercury can concentrate on and deactivate the MAPD catalyst. Personnel are at risk to mercury exposure when the catalyst vessels opened and the catalyst deinventoried.

The survey data is shown in **Table Q3-1**; overall, mercury ranked high with respect to contaminant concerns in the current survey. Eighteen respondents (of 35) indicated that their cracking plants had no problems regarding mercury contamination; the majority of these cracked EP (Ethane and/or Ethane/Propane and/or Butane feedstocks). In contrast, 17 locations indicated concerns regarding mercury contamination, and all of these except one producer cracked heavier feeds than EP. This was also true of the 1992 respondents – those cracking heavier feedstocks had a higher level of concern.

Mitigation in 1992 focused primarily on feed specifications (<1 parts-per-billion to 10 ppb), and sulfur impregnated activated carbon. One 1992 respondent indicated the installation of a mercury trap. No one specified mercury traps in 2003, but specifications were in place for the majority of the 2003 survey participants[2]. Activated carbon beds and sulfur impregnated carbon beds are still popular, but also in use now are alumina and molecular sieve based adsorbents. One respondent used a proprietary sulfided metal adsorbent.

Crackers predominantly target mercury removal from the feedstock and before the cracked gas entered the chilling train section. Three respondents removed mercury from the hydrogenation

1 McPhaul, D. R., and Reid, J. A., (1995) "Ethylene Plant Feedstock Contaminants Treatment," Proceedings of the 7th Annual Ethylene Producers Conference of the AIChE, December. Feedstock mercury levels 10 –21 ppb. MAPD reactor removed 94% of the Hg in reactor C_3 feed. No Hg removal beds were in place during data collection period

2 The specification level was not an answer requested of the 2003 survey.

reactor feed. Mercury was not purposely removed from several plants, but investigation revealed deposition in process locations such as furnace piping and the process dryer mol sieves.

The majority of the locations that responded to the question (17 of 19) indicated that the method of removal was satisfactory. One unsatisfied respondent considered the adsorption capacity/efficiency of the bed used (alumina) to be too low; while another indicated that the plant did not treat all the feeds and would have to investigate further. Another unit had discovered hydrogenation catalyst deactivation caused by mercury; in response, a carbon adsorbent layer was loaded on top of the catalyst bed.

Sodium Chloride

Sodium chloride contamination is often found as a component of brine solutions in liquid feeds due to contamination arising from an incident during ocean transportation. Unfractionated feedstocks such as condensates may contain a residual amount of naturally occurring brine. Thirdly, salt cavern storage is often used for the lower mole weight hydrocarbons, and it is not impossible that a spike of brine might find its way into and down the transportation pipelines.

In the scenarios above, the salt will enter the cracking process as a solution, but the water will quickly vaporize leaving solid NaCl and other salts. At temperatures above 1000°F, these salts, and especially sodium chloride, will attack austenitic steels including the alloys used in the convection section shield coils. In addition to the inherent damage and weakening of the coils, the salts can form a "slag", which will migrate in the direction of process flow, and ultimately cause tube restriction bringing about flow maldistribution to the radiant coils. In an evenly fired box, this will in turn, result in over-cracking in the flow-restricted tubes and high tube metal temperatures. An added difficulty is that the slag deposition could cover already formed coke and render its removal during decoking problematic[3].

In the furnace radiant section, sodium chloride is converted over typical furnace tube alloys (Iron, Chromium, Nickel) into a variety of compounds. The interaction of NaCl with chromium leads directly to $(Na_2O)(Cr_2O_3)$, which will peel from the tube walls. The chloride generated generally appears as gaseous HCl, which will be swept out of the furnace area and end up in the process water. Sodium chloride is also converted to Na_2SO_4, NaOH, and Na_2S. A complex reaction scheme involving these three compounds results in the formation of additional $(Na_2O)(Cr_2O_3)$ thus increasing chromium loss. Loss of tube surface chromium increases coke make. Coking rates will be higher than normal for several furnace cycles since chromium will only slowly migrate from the bulk alloy to the tube surfaces. Chromium loss also leads to increased carbide formation, which causes loss of tube ductility. Carbided tubes are less able to flex and are more prone to failure during the temperature swings inherent to the decoking exercise.

The data for the 2003 survey regarding sodium chloride are tabulated in **Table Q3-2**. Comparison of the data from the two surveys reveal an increased concern level in the current survey. It indicated Sodium Chloride or Salt to be # 3 in order of concern for all and gas crackers, and #2 for liquid crackers (with MeOH as #1). The recent survey also indicates that Chlorides are a catalyst poison, which the 1992 survey does not mention as a "consequence" of this contaminant's presence. Of the 17 respondents in the 1992 survey, 9 indicated that they look for Chlorides. In the current survey,

---

3 Orriss, R., Address to an EPC conference held in Atlanta GA.

18 respondents indicated that salt was NOT a problem, with 16 indicating it was. Only 6 of the latter monitor salt in the cracking furnace feed. 10 do not. 12 of the 16 indicate the system used is meeting their expectations, while 4 of the respondents say there is salt in the feed sometimes.

Of the operators citing equipment or process problems with salt, 11 indicated the Cracking Heater, 2 mentioned the water quench system, one mentioned Mole Sieve Dryer degradation, 1 ethylene purity problems, and one cited problems with the Caustic system. 10 operators listed tank draining as a part of a fix to remove this contaminant, 12 listed feed specs, with some citing a combination of both. One operator uses ion exchange, 4 have desalters, 3 mention water wash and coalescers and one cites "operating discipline."

In the current survey, 16 out of the 34 respondents identified salt as a contaminant, so that the percentage of feeds where this is a problem has slightly decreased.

The 16 respondents who encounter salt as a problem, treat it *via* Feed Specs, desalter, plus tank draining (3) Feed Specs, coalescer plus tank draining (3), Tank Draining (1), Ion Exchange plus Tank Draining (1), Feed Specs only (3), Feed Specs plus Tank Draining (4) and Operating Discipline (1).

Arsenic

Compounds of arsenic have been found in certain natural gases, but the volatile arsenic species related to ethylene production is AsH_3 which comes in with FCCU (Fluidized Catalytic Cracking Unit) dry gas or is produced from arsenic compounds intrinsic to the cracker feed. Concentrations typically range from 50 - 1000 ppbw.

Arsine is extremely poisonous but the concentrations encountered in the ethylene cracking business are low and the likelihood of exposure is very low.

Under controlled conditions, arsine and the lower mole weight alkyl arsines thermally decompose to elemental arsenic, which is deposited on hot surfaces as a mirror. Water saturated arsine readily decomposes to the metal upon exposure to light. Many of the arsenic containing compounds that enter an ethylene-cracking furnace will also readily decompose and emerge from the furnace as arsine.

Arsine will deactivate the acetylene (AC) and methylacetylene/propadiene (MAPD) catalysts. Typically, the catalyst used is *ca* 300 Hydrogen Cyanide Pd on alumina. Note that only 20% of the metal is actually functioning as a hydrogenation catalyst; therefore, it takes very little arsenic to completely poison a catalyst.

The 2003 survey data regarding arsine is found in **Table Q3-3**. In the 1992 survey, Arsine is listed in # 4 spot of concern, with CO_2 as being highest. It fell to a significantly lower level in the current survey. It appears as # 10 in order of concern for all crackers, and #5 for liquid crackers (With MeOH # 1). As expected, it is a much more frequently encountered contaminant in liquid crackers, since it is of little concern in gas crackers. Catalyst Poison and Safety/Hygiene are the overriding impact of this contaminant.

23 respondents indicated that As was NOT a problem, with 12 indicating it was. 4 respondents indicated that As was introduced by AGO/VGO feed stocks, 7 mentioned the source as Refinery

Dry Gas or Propylene, 3 mentioned the source as naphthas or condensate feeds. Of the 12 who encounter As, 9 indicated that they do analyze spent catalyst for it. Typically, it is found in concentrations ranging from 300 Hydrogen Cyanide to 0.4 wt% on pygas catalyst (4), MAPD catalyst (3), Acetylene converter catalyst (1) and Acetylene guard bed (1)

Removal or control of the contaminant As is accomplished by Feed specifications (9), deposition on Hydrogenation catalyst (5), CuO bed (4), PbO bed (1), and Activated Alumina (1). The tabulated results are listed in **Table Q3-3**.

Ammonia/Amines/Nitriles

Elemental Nitrogen, ammonia and very small amounts of Hydrogen Cyanide can be the products of cracking Nitrogen-containing compounds such as aliphatic amines or aromatic rings carrying the $-NH_2$ group or more generally, the -NRR' moiety. Amine carrying feeds typically include naphthas, condensates and gas oils; refinery gases, ethane, and FCC and coker off-gases may contain ammonia itself. Amines may also be introduced *via* chemical additives. The most commonly encountered nitrile is acetonitrile, which is a solvent used in butadiene and isoprene recovery.

NH_3 is usually removed in quench water through blowdown or in an amine or water scrubber with final removal commonly done by mole sieve dryers. If converted into its chloride salt (NH_4Cl) *via* reaction with HCl, it can be very corrosive. Amines that are introduced after the cracking furnaces can also react with HCl to produce corrosive salts. Oftentimes, the fate of acetonitrile is hydrolysis into ammonia acetate, which is in equilibrium with ammonia and acetic acid. Thus, one may introduce corrosive acetic acid into the process if conditions are such that the ammonia can vent in an overhead leaving behind the less volatile acetic acid.

Excess NH_3 accumulates in the ethylene product. In addition to the corrosion problem, nitriles can affect the activity of hydrotreating catalysts and concentrate in raw pygas.

The summarized data regarding ammonia may be found in **Table Q3-4**. Of 35 respondents, 19 indicated that they had no ammonia issues while 16 answered that they did. Nine EP crackers checked the "no" box while 5 pointed to issues. Only one of 5 Gas Field Condensate crackers has ammonia issues. The responses were evenly divided (4, yes; 4 no) for those cracking refined crude oil feeds as were those 10 crackers that process mixed feeds – 5 responded that ammonia issues were encountered, 5 did not have any issues.

Ammonia introduction *via* treatment chemicals (14 crackers) was the largest single source of contamination. Contamination from feedstocks (8 plants) rated second. The other sources of contamination noted were vent gases and internal feeds.

Ammonia contamination caused off-spec ethylene product in 11 cases and adversely affected the process water in 5 cases. Ammonia contamination also was traced to furnace corrosion, the formation of ammonium nitrate in the chilling system, catalyst deactivation, and contamination of products other than ethylene.

Varied strategies were used to mitigate ammonia contamination. Ten facilities removed this contaminant from the process water, 6 dealt with the issues by adjusting feed specifications, while the same number were able to remove ammonia *via* the amine treater. Four producers were able to trap ammonia on the mol sieve dryer beds, while one had installed a low pH absorber.

The final question regarding ammonia contamination was regarding the issue of the employed mitigation meeting expectations. Fourteen (78%) of the 18 respondents who replied to this query indicated a positive response, while only 4 answered negatively.

Table Q3-5 tabulates the results regarding amine and nitrile contamination. Of eighteen respondents, the split was equally divided (9 and 9) between those indicating issues with amines and those reporting no problems. Amine issues did not seem to correlate with any particular feedstock, but only 2 European plants reported problems *vs.* 7 North American crackers. The most problematic area was ethylene quality, while dilution steam/quench water contamination was of concern in two cases. Ammonium nitrate formation was indicated on one case of amine contamination.

Only 4 respondents (3 of the 4 were North American ethylene producers) indicated issues regarding nitrile intrusion; 80% (16) indicated no nitrile issues. Similar to amine contamination, nitrile contamination did not correlate with any feed type. Only two responses were written in the "**Issue & Mitigation**" column. One plant tracked byproduct nitrile concentrations while the other plant indicated C_4 byproduct contamination. In response, a water wash tower was installed to scrub out the contaminating nitriles.

2003 Survey – Question Four

The final question of the survey addresses feedstocks that enter the ethylene unit downstream of the furnaces. Five common streams are identified: Flare drum knockout liquids, Vents from other plants, Refinery FCC dry gas, Wastewater stripper overhead, and Raw pyrolysis naphthas. An "Other" Category is provided to capture any other streams the respondents wish to comment on. For each of the streams, the survey has the following questions:

- Do you process (the particular feed)?
- Specify origin if applicable.
- Where does it (the feed) enter the process?
- Give a brief description of any problems caused by the stream.

Survey responses for Question 4 are summarized below.

Flare Drum Liquids

Of the 35 respondents to the survey, 16 reported processing flare knockout drum liquids in their plants. Thirteen of the sixteen (81%) were liquid crackers, well above the 60% liquid / flexible feed crackers in the survey population, indicating the liquid crackers are better able to handle, or perhaps produce more of, the liquids collected in the plants' flare drums. Larger crackers are more likely to process flare drum liquids, with about 70% of the plants processing flare drum liquids being >600K ton/year, vs. 60% in the general survey population. North American vs. European as well as plant vintage distributions were in line with survey population.

The flare drum liquids are typically reprocessed to the primary fractionator or water quench tower (9 of the 12 reporting the reprocessing locating with the oily water / NESHAPS stripper) and PGC being the other locations for bringing these liquids back into the plant. For the most part, there were no issues with processing this stream through the plant. Three sites mentioned concerns with Chlorides, one mentioned quench water fouling, and another product quality issues. A detailed breakdown of responses is shown in Table Q4-1.

Vents from Other Plants

Twenty three respondents reported processing vent streams in the recovery section of their ethylene plants, making it the most popular "non-cracking" feedstock. The largest source of vent streams is polymer units (12), followed by vents from butadiene recovery units (7), alcohol units (3) and aromatic processing units (2). In most cases, the vents are returned to the Cracked Gas Compression Train (16), with a few recovering the vent streams further downstream in product purification (4).. A few return the vents to the Primary Quench area, and one reported returning the vent gas to the cracking furnaces.

There were very few problems reported by those reprocessing vent streams. Three mentioned fouling as a concern, with two of these specifically mentioning vinyl acetate as the offending contaminant. Two reported Nitrogen as an issue: one indicated that the Nitrogen appears in the Hydrogen product, and the other that Nitrogen ties up compressor capacity. Unsteady plant throughput and emulsion in oil-water separation were the other issues mentioned. Detailed responses are presented in Table Q4-2.

Refinery FCC Dry Gas

FCC dry gas was processed by four of the respondents. Two of the four reported issues with this stream: acetylene converter catalyst poisoning from arsine and/or hydrogen sulfide, and Nitrogen oxide contamination resulting in formation of ammonium salts in the cryogenic section of the plant. Results are included in Table Q4-3.

Wastewater Stripper Overhead

Eight of our thirty five respondents process wastewater stripper overhead streams in their plants. Plant location, size, and vintage of plants processing this stream, were in roughly the same proportion as the general survey population. Only 2 were gas crackers, slightly below the 40% gas crackers in the overall survey response. The streams were reprocessed through the primary quench system. There were surprisingly few issues reported with this stream, with one mentioning this stream as a source of amines and organic acids which can upset quench water chemistry, and another commenting that polymer buildup in the process has been attributed to this stream. Additional information can be found in Table Q4-4.

Raw Pyrolysis Naphtha

Nine producers reported processing Raw Pyrolysis Naphtha in their olefins plants, 8 of which were North American. A significant source of this naphtha is the drip oils from gas crackers, which are more abundant in North America, so this is not surprising. The usual source of entry for this feedstock is the primary fractionation area (3) or debutanizer (3); one reported bringing the stream into the transfer line exchangers.

Respondents reported the following issues when processing this stream: equipment fouling (1), catalyst fouling (1) and upsets in quench water chemistry (1). Complete responses are captured in Table Q4-5.

Other External Streams Recovered in Olefins Plants

Additional streams fed to the recovery section of ethylene plants not covered in the above categories were contributed by six producers. Two appear to be FCC propylene, and mention arsine, sulfur, oxygen and guard bed (arsine) fouling as issues. Another two streams appear to be liquid recycle streams from butadiene recovery units, and the issues associated with these streams include serving as a source of unwanted Nitrogen and adversely affecting quench water chemistry. The other streams are a recycle from polymer plant (no issues) and an unidentified stream recycled to the cracking furnaces. The detailed responses to this section are presented in Table Q4-6.

2003 Survey – Closing Notes

The 2003 survey included 35 respondents with 70% of these located in the Americas. The balance were European respondents. This is in contrast to 18 respondents to the 1992 survey. Many of the contaminant concerns voiced in 1992 were also indicated in this survey; notably sulfur, methanol, sodium, chlorine, carbon dioxide, mercury, and fluorine although the order of concern changed somewhat.

The last question of the survey asked for suggestions that would improve it for future questionnaires. Only a few mechanical issues regarding manipulation of the electronic file were surfaced. The more significant request was that responders wanted more space for adding comments and more space for listing additional comments. These suggestions will be carefully studied during the creation of the next survey.

The authors provided some analysis of the data, but by no mean exhaustively examined and interpreted the responses. Thus, the reader is encouraged to continue this process; much more information can be extracted from the paper. The reader is also free to tailor any analysis to a particular focus or goal.

The authors appreciated all the input given by the responders. Without participation this paper would not have been possible.

Table Q1-1 - Part 1 of 2
Level of Concern for Specific Contaminants

Feedstock Class	Ethane or E/P Mix			Propane or Other NGL's			Raffinate *ex* C4, C5 Recovery		
Number of Plants	29			27			12		
Contaminant	% Medium	% High	% Medium +High	% Medium	% High	% Medium +High	% Medium	% High	% Medium +High
COS	31	7	38	26	15	41	17	8	25
DMDS	0	3	3	0	0	0	0	0	0
RSH	7	3	10	7	4	11	0	0	0
H_2S	24	10	34	26	15	41	0	8	8
DMS	0	0	0	4	0	4	0	0	0
CO_2	31	34	66	22	15	37	0	0	0
H_2O	7	14	21	11	11	22	17	0	17
Inorganic Hg	0	17	17	7	7	15	0	25	25
Organic Hg	0	10	10	0	7	7	0	25	25
Organic Chlorides	0	7	7	0	19	19	25	8	33
Inorganic Chlorides	0	14	14	7	15	22	25	8	33
Fluorides	0	14	14	4	33	37	0	33	33
Arsine	7	7	14	4	7	11	0	8	8
Phosphine	3	0	3	0	4	4	0	8	8
Methanol	45	10	55	19	15	33	17	8	25
NH_3	14	0	14	4	4	7	0	17	17
Sodium	0	7	7	0	7	7	0	8	8
NOx	0	3	3	0	4	4	0	0	0
Lead	0	0	0	0	0	0	25	8	33
Phosphorus	0	0	0	0	0	0	0	8	8
Radon	10	3	14	11	4	15	0	0	0
MTBE	0	0	0	7	0	7	8	33	42
DME	0	0	0	7	0	7	8	8	17
O_2	14	3	17	0	4	4	0	0	0
CO	14	0	14	0	11	11	0	0	0
Vanadium	0	0	0	0	0	0	8	0	8
Olefins	14	0	14	19	4	22	25	0	25
Silicon Derivatives	0	0	0	0	0	0	25	0	25

Table Q1-1 - Part 1 of 2
Level of Concern for Specific Contaminants
(Additional Contaminants - Written in Responses)

Feedstock Class	Ethane or E/P Mix			Propane or Other NGL's			Raffinate *ex* C4, C5 Recovery		
Number of Plants	29			27			12		
Contaminant	% Medium	% High	% Medium +High	% Medium	% High	% Medium +High	% Medium	% High	% Medium +High
Total Sulfur	17	0	17	22	7	30	33	0	33
Total Chlorides				4	0	4			
Total Nitrogen				4	0	4			
Arsenic							8	0	8
Total C_4	3	0	3						
Cor	0	7	7						
TVS	7	0	7	4	0	4			
1-3, BD	3	0	3						
Total Oxygenate				0	4	4	0	17	17
Total Transition Metals							17	0	17
CS_2				0	4	4			
Copper Strip - H_2S				4	0	4			
Propylene				4	4	7			
Oil				0	4	4			
Sulfolane							17	0	17
Vinyl Acetate							17	0	17
NMP							8	0	8
Diolefins							0	8	8
Amine based inhibitors							8	0	8

Table Q1-1 - Part 2 of 2
Level of Concern for Specific Contaminants

Feedstock Class	Naphthas/Condensates			Gas Oils			Dry Gas		
Number of Plants	22			10			3		
Contaminant	% Medium	% High	% Medium +High	% Medium	% High	% Medium +High	% Medium	% High	% Medium +High
COS	9	5	14	0	0	0	33	33	66
DMDS	0	0	0	0	0	0	0	0	0
RSH	9	0	9	0	0	0	0	0	0
H_2S	32	27	59	0	10	10	67	0	67
DMS	5	0	5	0	0	0	0	0	0
CO_2	0	0	0	0	0	0	67	0	67
H_2O	9	5	14	10	0	10	0	0	0
Inorganic Hg	14	45	59	0	60	60	0	0	0
Organic Hg	9	36	45	0	30	30	0	0	0
Organic Chlorides	14	32	45	30	0	30	0	0	0
Inorganic Chlorides	23	27	50	20	40	60	0	0	0
Fluorides	9	23	32	0	0	0	0	0	0
Arsine	9	27	36	0	20	20	0	67	67
Phosphine	0	0	0	0	0	0	0	0	0
Methanol	18	18	36	0	10	10	0	33	33
NH_3	5	0	5	0	0	0	0	0	0
Sodium	32	23	55	30	0	30	0	0	0
NOx	0	5	5	0	0	40	0	33	33
Lead	32	45	77	40	0	40	0	0	0
Phosphorus	9	0	9	20	0	20	0	0	0
Radon	0	0	0	0	0	0	0	0	0
MTBE	32	27	59	0	10	10	0	0	0
DME	0	5	5	0	0	0	0	0	0
O_2	0	5	5	0	0	0	0	33	33
CO	0	0	0	0	0	0	33	0	33
Vanadium	14	14	27	10	0	10	0	0	0
Olefins	27	5	32	0	0	0	0	0	0
Silicon Derivatives	23	9	32	60	10	70	0	0	0

Table Q1-1 – Part 2 of 2
Level of Concern for Specific Contaminants
(Additional Contaminants - Written in Responses)

Feedstock Class	**Ethane or E/P Mix**			**Propane or Other NGL's**			**Raffinate *ex* C_4, C_5 Recovery**		
Number of Plants	29			27			12		
Contaminant	**% Medium**	**% High**	**% Medium +High**	**% Medium**	**% High**	**% Medium +High**	**% Medium**	**% High**	**% Medium +High**
Argon+Oxygen							0	33	33
Total Sulfur	27	32	59	30	30	60			
Total Chlorides	9	14	23	10	0	10			
Total Nitrogen	5	0	5	40	0	40			
Arsenic	5	0	5	30	0	30			
Total C_4	5	0	5						
Total Oxygenate	5	5	9	0	10	10			
Total Transition Metals	0	14	14						
CS_2	5	0	5						

Table Q1-2 - Part 1 of 2
Contaminants Tested for and Detected

Feedstock Class	Ethane or E/P Mix		Propane or Other NGL's		Raffinate *ex* C_4, C_5 Recovery	
Number of Plants	29		27		12	
Contaminant	**% Testing**	**% Detecting**	**% Testing**	**% Detecting**	**% Testing**	**% Detecting**
COS	17	14	26	19	8	0
DMDS	7	10	0	4	0	0
RSH	10	7	7	7	0	0
H_2S	32	21	33	26	8	8
DMS	0	3	4	4	0	0
CO_2	79	55	33	26	0	0
H_2O	24	14	19	15	17	17
Inorganic Hg	14	14	11	7	25	8
Organic Hg	14	10	11	0	25	8
Organic Chlorides	7	7	15	7	33	17
Inorganic Chlorides	7	7	7	4	33	17
Fluorides	7	3	22	7	33	8
Arsine	10	3	7	7	8	0
Phosphine	3	0	4	4	8	0
Methanol	38	28	22	19	17	8
NH_3	0	0	4	0	17	0
Sodium	3	3	4	0	8	0
NOx	3	0	4	0	0	0
Lead	0	0	0	0	33	17
Phosphorus	0	0	0	0	8	0
Radon	7	0	4	0	0	0
MTBE	0	0	4	4	33	17
DME	0	0	4	4	8	0
O_2	0	0	4	4	0	0
CO	14	3	15	7	0	0
Vanadium	0	0	0	0	8	0
Olefins	24	21	44	37	25	25
Silicon Derivatives	0	0	0	0	17	0

Table Q1-2 - Part 2 of 2
Contaminants Tested for and Detected

Feedstock Class	Naphthas/Condensates		Gas Oils		Dry Gas	
Number of Plants	22		10		3	
Contaminant	% Testing	% Detecting	% Testing	% Detecting	% Testing	% Detecting
COS	0	0	0	0	33	33
DMDS	0	0	0	0	0	0
RSH	14	9	0	0	0	0
H_2S	55	18	10	10	100	67
DMS	5	0	0	0	0	0
CO_2	0	0	0	0	100	33
H_2O	23	9	30	10	0	0
Inorganic Hg	41	14	60	10	33	33
Organic Hg	32	18	30	10	0	0
Organic Chlorides	45	9	20	10	0	0
Inorganic Chlorides	45	18	60	10	33	33
Fluorides	9	5	0	0	0	0
Arsine	23	18	20	20	67	33
Phosphine	0	0	0	0	0	0
Methanol	27	14	10	10	33	33
NH_3	0	0	0	0	0	0
Sodium	41	14	40	20	0	0
NOx	0	0	0	0	0	0
Lead	82	32	10	10	0	0
Phosphorus	9	5	30	20	0	0
Radon	0	0	0	0	0	0
MTBE	50	18	20	10	0	0
DME	5	0	0	0	0	0
O_2	0	0	0	0	33	33
CO	0	0	0	0	33	33
Vanadium	27	14	20	20	0	0
Olefins	50	32	0	0	0	0
Silicon Derivatives	14	9	40	20	0	0

Table Q1-3: Contaminant Levels: Medium to High Level of Concern – Page 1 of 3

Feedstock Class	Ethane or E/P Mix					Propane or Other NGLs				
Number of Plants	29 (15 gas crackers, 14 flexible crackers)					27				
Contaminant Level		Normal Concentration		Maximum Permissible			Normal Concentration		Maximum Permissible	
Contaminant	Responses	Average	Range	Average	Range	Responses	Average	Range	Average	Range
CO_2, ppmm	23	100-500	0 – 40,000	1,000-2,000	100 – 60,000	10	<100	0 – 500	1,000	500 – 3,000
Total Sulfur, ppmw	8	<5	0 – 5	30	5 - 90	9	<20	0 – 200	50	10 – 200
COS, ppmm	11	<5	0 – 10	5	0.2 - 10	11	<1	0 - <10	1 – 2	1 - 20
H_2S, ppmm	10	5	0 – 10	10	5 - 25	9	<1	0 – 10	10	1 – 100
RSH, ppmm	4	5	0 – 10	1, 20	1 - 20	3	<1 – 10	0 – 50	50	10 – 120
DMDS, ppmw	1	10				1 (DMS)				
H_2O, ppmw	7			4, 90, No Free H_2O		6				No Free H_2O
Inorganic Hg, ppbw	5	< 0.1	0 – 5	0.02, 0.1, 5	0.1 - 5	4	1	0 – 5	5	5
Organic Hg, ppbw	2	< 0.1	0 - 0.1	0.02, 0.1	0.02 - 0.1	2	<0.1	<0.1	0.1, 10	0.1 – 10
Total Hg, ppbw						1	<0.1	<0.1	5	5
Fluorides, ppmm	4	< 1	0 - 1	1	1 - 4	11	<1	0 – 1	1 – 2	1 - 50
Organic Cl, ppmm	2	< 1				5	<0.1	<0.1	1, 5	1 - 5
Inorganic Cl, ppmm	3	< 1	0 – 1	1, 2	1 - 2	6	<0.1	<0.1	2	2 - 5
Total Cl, ppmm						1	<0.1		1	
Sodium, ppmw	2					2	<0.1	<0.1	1	1
Arsine, ppbm	4	<0.1	0 – 0.1	0.1, 10	0.1 – 10	3	<1	0 – 1	0.1, 10	0.1 - 10
Phosphine, ppbw						1				
CO, ppmm	4	<1	0 – 50	1	1 – 100	3	<1	0 - <10	1	0.1 - 1
Oxygenates, ppmw						1	<10		100	
Methanol, ppmm	16	1-10	0 – 200	5-10	1 - 50	9	<1	0-100	20	5-500
MTBE, ppmw						2			150	100 - 200
DME, ppmm						2			55	10 - 100
Oxygen, ppmm	4	< 0.1, <10	0 – <10	5	5	1				
Nitrogen, ppmm	1	< 100	0 – 100	100	100					
Olefins, wgt %	3	<.05, <2, 18	0 – 20	.05, 5, 25	0.05 - 25	6	<1	0 – 500	2	0.05 - 70
Diolefins, wgt %	1	<1	0 – 1	15	15					
NH_3, ppmm	4	<1	0 – 1	1	1	2				

Table Q1-3: Contaminant Levels: Medium to High Level of Concern – Page 1 of 3
(Continued)

Feedstock Class	Ethane or E/P Mix					Propane or Other NGLs				
Number of Plants	29 (15 gas crackers, 14 flexible crackers)					27				
Contaminant Level		Normal Concentration		Maximum Permissible			Normal Concentration		Maximum Permissible	
Contaminant	Responses	Average	Range	Average	Range	Responses	Average	Range	Average	Range
Propylene, wt%						2	1.5	1.5	5	5
Butadiene, ppmw	1	<1		50						
Methane, wgt%	1	2.5		6						
Hydrogen, ppmm	1	<1		10						
CS_2, ppmw						1	<1		5	
Radon	4					4				
Corrosion, Cu Strip	2	1A		1A		2	1A		1B, 2B	

Table Q1-3: Contaminant Levels: Medium to High Level of Concern – Page 2 of 3

Feedstock Class	Raffinate From C_4, C_5 Recovery					Naphthas/Condensates (Including BTX Raffinates)				
Number of Plants	12					22				
Contaminant Level		Normal Concentration		Maximum Permissible			Normal Concentration		Maximum Permissible	
Contaminant	Responses	Average	Range	Average	Range	Responses	Average	Range	Average	Range
Total Sulfur, ppmw	4	<100	0 – 500	500	20 - 500	12	100 - 200	100 – 2000	200 – 500	10 - 2000
COS, ppmw	2	<1	0-1	1	1					
H_2S, ppmw	1	<1	N/A	10	N/A	12	0-10	0 - 50	10-15	1 - 500
DMS, ppmw						1				300
H_2O	2				No Free H_2O	2				No Free H_2O
Inorganic Hg, ppbw	1					12	8	6 - 10	10	10
Organic Hg, ppbw	1					9	3	1 - 5	55	10 - 100
Total Hg, ppbw	2	<100	0 – 100	100	100	4	< 5	0 - 40		2 - 100
Fluorides, ppmw	3	< 1	0 - 1			6	<0.1	0-0.1	4	4
Organic Cl, ppmw	4	< 1	0 – 1	5, 5, 0.15	0.15 - 5	9	< 1	10	10	5 - 50
Inorganic Cl, ppmw	4	< 2	0 – 2	6	6	11	< 1	0 - 10	5	1 - 10
Total Cl, ppmw	2			5	5	7	< 1	0 - 5	20	5 - 30
Sodium, ppmw						11	< 1	0 – 1	1	0.1 - 4
Lead, ppbw	4	5, <500	0 – 500	50, 5000	50 - 5000	15	< 100	0 – 5000	50 – 100	50 – 10,000
Vanadium, ppbw	1					5	< 10	0 – 10	50	25 - 100
Arsine, ppbw	2	< 5	0 - 5			7	< 5	0 – 50	50	5 - 100
Phosphine, ppbw	1					2	<10, <500	0 - 500	10 , 500	10 - 500
Trans Metals, ppmw	2					1	< 2	0 – 2	2	2
Silicon, ppbw	3	< 1000	0 – 1000	1000	1000	7	< 1000	0 – 5000	500	100 - 1000
Methanol, ppmw	3	<10	0 - 10	35	20 - 50	7	< 5	0 – 10	100	2 - 500
MTBE, ppmw	5	< 0.1	0 – 0.1	65	50 - 100	11	< 10	0 – 50	50 – 100	20 - 1000
DME, ppmm	2	< 10	0 - 10	10	10	1			100	100
Oxygenates, ppmw	2	<1, <500	0 – 500	20, 1000	20 - 1000	2	< 10, <1	0 – 10	100	100
Nitrogen, ppmw	3	< 100	0 – 100	100	100					
Olefins, wgt %	3	<.05, <2, 18	0 – 20	.05, 5, 25	0.05 - 25	6	< 1	0 – 1	1 – 2	1 - 5
Diolefins, wgt %	1	<1	0 – 1	15	15					
Vinyl Acetate, ppmw	2	<1000	0 – 1000	1000	1000					
Sulfolane, ppmw	2	<150	0 – 150	150	150					
NMP, ppmw	1	< 1	0 – 1	100	100					

Table Q1-3: Contaminant Levels: Medium to High Level of Concern – Page 3 of 3

Feedstock Class	Gas Oils					Dry Gas				
Number of Plants	10					3				
Contaminant Level		Normal Concentration		Maximum Permissible			Normal Concentration		Maximum Permissible	
Contaminant	Responses	Average	Range	Average	Range	Responses	Average	Range	Average	Range
Total Sulfur, ppmw	7	<1600	<200-4000	2300	300-6000					
H_2O	2	No haze		No free water						
Inorganic Hg, ppbw	6	<50	<2-100	74	2-100					
Organic Hg, ppbw	3	<2		37	2-100	1	3	1-5	5	
Organic Cl, ppmw	1	<2			50					
Inorganic Cl, ppmw	6	<1.2	<1-2	3.7	1-5					
Total Cl, ppmw	1	<1		5						
Flourides, ppmw	1			1						
Lead, ppbw	3	<1		5						
Phosphorus, ppb	1			500						
Vanadium, ppbw	1	<100		100						
Arsine, ppbw	1					1	8	6-10	1000	
Silicon, ppbw	4	<2000	<100-5000	1700	100-5000					
Methanol, ppmw						1	5		20	
MTBE, ppmw	2	<10		300						
Oxygenates, ppmw	1	<1								
Olefins, m %						1	0.05		0.3	
Total Nitrogen, ppmw	4	<1500		2000						
NH3/amines, ppmm						1			10	
NOx, ppbw						1			50	
O2+Argon, ppm w						2	<1000		5000	5-10000
CO, ppmm						1			50	
COS, ppmw						2	<0.05		4	0.1-8
H_2S, ppmm						2	30	10-50	300	
CO2, ppmm						2	300	100-500	850	200-1500

Table Q2-1
Total Responses from Survey

Contaminant	Quality	Corrosion	Catalyst Poison	Safety/ Hygiene	Environmental (air/water)	Capacity Limit	Operability	Furnace problems	Other: please list
Methanol/ Alcohols	19	2	6	0	1	1	9	3	Front end hydrogenator upsets if MeOH to CO
Carbon Dioxide	17	6	0	1	0	4	6	0	
Radon	0	0	0	16	7	0	0	0	Polyethylene level devices
Mercury	0	5	13	9	6	0	1	0	
Iron	0	2	2	0	0	0	4	8	Polymers
Arsine	3	1	10	5	5	0	0	0	
Metals	1	2	7	0	0	1	6	9	
Sodium	3	7	3	1	0	0	6	16	
Fluorides	4	10	2	0	0	1	5	6	Polymers
Chlorides	2	14	3	1	0	0	7	8	
Sulfur	6	10	9	5	2	4	3	3	
NOx	0	0	0	8	2	1	2	0	
Water	1	7	1	0	1	2	7	8	Total plant trip/ice formation in piping
Phosphorous	3	1	7	1	2	0	1	3	
Ammonia	9	6	2	4	0	0	1	0	
Organic Acids	4	9	1	1	0	0	1	0	
Light Nitrogen compounds	5	1	1	1	0	1	1	1	
Carbonyls	5	1	4	1	2	0	4	0	
Silicon	1	0	6	0	0	0	0	4	
Ethers	6	1	1	0	0	0	0	0	Drier cycle shortened
Solvents	4	1	0	0	1	0	2	1	
Others:									
O_2	1	1	0	1	0	0	0	0	Polymers
MTBE	1	0	0	0	0	0	0	0	
Nitrogen	1	0	0	0	0	0	1	0	
Zinc	0	0	1	0	1	0	0	0	
Aluminum	0	0	0	0	0	0	0	1	
Highly unsaturated comps/olefins	0	0	0	0	0	0	0	1	
Particulate dirt	0	0	0	0	0	0	1	1	
Asphaltene/ polymer/resid									Erosion of tubes in convection section

Table Q2-2
Responses from Gas Cracking Plants
(15 gas crackers surveyed)

Contaminant	Quality	Corrosion	Catalyst Poison	Safety/ Hygiene	Environmental (air/water)	Capacity Limit	Operability	Furnace problems	Other: please list
Methanol/Alcohols	6	0	3	0	0	0	4	1	Front end hydrogenator upsets if MeOH to CO
Carbon Dioxide	5	1	0	1	0	3	4	0	
Radon	0	0	0	6	2	0	0	0	Polyethylene level devices
Mercury	0	0	1	1	0	0	0	0	
Iron	0	1	0	0	0	0	2	1	
Arsine	1	1	0	0	0	0	0	0	
Metals	1	0	2	0	0	0	1	1	
Sodium	0	1	0	0	0	0	2	3	
Fluorides	2	2	1	0	0	1	1	3	
Chlorides	2	3	2	1	0	0	1	4	
Sulfur	1	4	3	2	0	0	0	1	
NOx	0	0	0	2	1	0	0	0	
Water	0	1	0	0	1	1	3	0	
Phosphorous	1	0	1	0	0	0	1	0	
Ammonia	0	1	2	1	0	0	1	0	
Organic Acids	0	1	0	1	0	0	1	0	
Light Nitrogen compounds	0	1	0	1	0	0	0	0	
Carbonyls	2	1	1	0	0	0	0	0	
Silicon	0	0	0	0	0	0	0	3	
Ethers	0	0	0	0	0	0	0	0	
Solvents	0	0	0	0	0	0	0	1	
Others:									
Nitrogen	1	0	0	0	0	0	1	0	

Table Q2-3
Responses from Liquid Cracking Plants
(20 liquid crackers surveyed)

Contaminant	Quality	Corrosion	Catalyst Poison	Safety/ Hygiene	Environmenta l (air/water)	Capacity Limit	Operability	Furnace problems	Other: please list
Methanol/Alcohols	13	2	3	0	1	1	5	2	
Carbon Dioxide	12	5	0	0	0	1	2	0	
Radon	0	0	0	10	5	0	0	0	
Mercury	0	5	12	8	6	0	1	0	
Iron	0	1	2	0	0	0	2	7	Polymers
Arsine	2	0	10	5	5	0	0	0	
Metals	0	2	5	0	0	1	5	8	
Sodium	3	6	3	1	0	0	4	13	
Fluorides	2	8	1	0	0	0	4	3	Polymers/Freon
Chlorides	0	11	1	0	0	0	6	4	
Sulfur	5	6	6	3	2	4	3	2	
NOx	0	0	0	6	1	1	2	0	
Water	1	6	1	0	0	1	4	8	Total plant trip/ice formation in piping
Phosphorous	2	1	6	1	2	0	0	3	
Ammonia	9	5	0	3	0	0	0	0	
Organic Acids	4	8	1	0	0	0	0	0	
Light Nitrogen compounds	5	0	1	0	0	1	1	1	
Carbonyls	3	0	3	1	2	0	4	0	
Silicon	1	0	6	0	0	0	0	1	
Ethers	6	1	1	0	0	0	0	0	Drier cycle shortened
Solvents	4	1	0	0	1	0	2	0	
Other:	0	0	0	0	0	0	0	0	
O_2	1	1	0	1	0	0	0	0	Polymers
MTBE	1	0	0	0	0	0	0	0	
Zinc	0	0	1	0	1	0	0	0	
Aluminum	0	0	0	0	0	0	0	1	
Highly unsaturated comps/olefins	0	0	0	0	0	0	0	1	
Particulate dirt	0	0	0	0	0	0	1	1	
Asphaltene/ polymer/resid									Erosion of tubes in convection section

Table Q3-1
Specific Feedstock Contaminants - Mercury

Location	Plant Age	Size	Feed	Contamination	Feed Specification	Removal *via*	Removal Location	Meeting Expectations
NA	Pre-1970	<300	EP	No	-	-	-	Yes
NA	'81 - '90	300-600	EP	No	-	-	-	Yes
NA	Pre-1970	300-600	EP	No	-	-	-	-
NA	Pre-1970	300-600	EP	No	-	-	-	-
NA	Pre-1970	300-600	EP	No	-	-	-	-
NA	Pre-1970	300-600	EP	No	-	-	-	-
NA	'71 - '80	300-600	EP	No	-	-	-	-
NA	'71 - '80	300-600	EP	No	-	-	-	-
E	Post-1990	>600	EP	Yes	Yes	Syntex metal sulphide adsorbent	Feedstock	Yes
NA	'71 - 80, Post-1990	>600	EP	No	-	-	-	-
NA	Post-1990	>600	EP	No	-	-	-	-
NA	Post-1990	>600	EP	No	-	-	-	-
NA	'71 - '80	>600	EP	No	-	-	-	-
NA	Post-1990	>600	EP	No	-	-	-	-
E	Pre-1970	300-600	GFC	Yes	-	Alumina based adsorbent.	Hydrogenation Reactor Feed	No. Adsorption capacity/efficiency is low.
NA	'71 - '80	>600	GFC	Yes	Yes	-	None.	Yes
NA	'71 - '80	>600	GFC	Yes	-	Molecular Sieve	No purposeful removal of Hg. Deposits on Process Dryer Mole Sieve.	Yes
NA	'71 - '80	>600	GFC	Yes	Yes	-	Feedstock	Yes
NA	'71 - '80	>600	GFC	Yes	Yes	Adsorption onto molecular sieve	Dryer/Reactor Regeneration Train	Yes
E	' 71 - 80, '81 - '90, Post-1990	300-600	RCO	No	-	-	-	-

Table Q3-1
Specific Feedstock Contaminants - Mercury

Location	Plant Age	Size	Feed	Contamination	Feed Specification	Removal *via*	Removal Location	Meeting Expectations
NA	Pre-1970	<300	EP	No	-	-	-	Yes
NA	'81 - '90	300-600	EP	No	-	-	-	Yes
NA	Pre-1970	300-600	EP	No	-	-	-	-
NA	Pre-1970	300-600	EP	No	-	-	-	-
NA	Pre-1970	300-600	EP	No	-	-	-	-
NA	Pre-1970	300-600	EP	No	-	-	-	-
NA	'71 - '80	300-600	EP	No	-	-	-	-
NA	'71 - '80	300-600	EP	No	-	-	-	-
E	Post-1990	>600	EP	Yes	Yes	Syntex metal sulphide adsorbant	Feedstock	Yes
NA	'71 - 80, Post-1990	>600	EP	No	-	-	-	-
NA	Post-1990	>600	EP	No	-	-	-	-
NA	Post-1990	>600	EP	No	-	-	-	-
NA	'71 - '80	>600	EP	No	-	-	-	-
NA	Post-1990	>600	EP	No	-	-	-	-
E	Pre-1970	300-600	GFC	Yes	-	Alumina based adsorbant.	Hydrogenation Reactor Feed	No. Adsorption capacity/efficiency is low.
NA	'71 - '80	>600	GFC	Yes	Yes	-	None.	Yes
NA	'71 - '80	>600	GFC	Yes	-	Molecular Sieve	No purposeful removal of Hg. Deposits on Process Dryer Mole Sieve.	Yes
NA	'71 - '80	>600	GFC	Yes	Yes	-	Feedstock	Yes
NA	'71 - '80	>600	GFC	Yes	Yes	Adsorption onto molecular sieve	Dryer/Reactor Regeneration Train	Yes
E	' 71 - 80, '81 - '90, Post-1990	300-600	RCO	No	-	-	-	-

Table Q3-2
Specific Feedstock Contaminants - Sodium Chloride, Salt

Location	Plant Age	Size	Feed	Contaminant?	Issue, or Area Introduced	Mitigation	Monitor?	Meet Expectations
E	Post-1990	>600	EP	No	-	-	No	-
E	Pre-1970	>600	EP	No	-	-	-	-
NA	Pre-1970	<300	EP	No	-	-	No	Yes
NA	'81 - '90	300-600	EP	No	-	-	No	Yes
NA	Pre-1970	300-600	EP	No	-	-	-	-
NA	Pre-1970	300-600	EP	No	-	-	-	-
NA	Pre-1970	300-600	EP	No	-	-	-	-
NA	Pre-1970	300-600	EP	No	-	-	-	-
NA	'71 - '80	300-600	EP	Yes	Cracking Heater Opn, (CHO); Cracking Heater Reliability (CHR), Mol Sieve Degradation	Operating Discipline	No	Yes
NA	'71 - '80	300-600	EP	No	-	-	-	-
NA	'71-'80; Post-90	>600	EP	Yes	Caustic System	Tank Draining& Ion Exchange	No	Yes
NA	Post-1990	>600	EP	Yes	CHO, CHR, Water Quench System	Feed Specs, Water Wash & Coalescer	Yes	Yes
NA	Post-1990	>600	EP	No	-	-	-	-
NA	'71 - '80	>600	EP	-	-	-	-	-
NA	Post-1990	>600	EP	No	-	-	-	-
E	Pre-1970	300-600	GFC	Yes	CHO, Quality problems in ethylene	Tank Draining Practices	No	No, salt in feed sometimes
NA	'71 - '80	>600	GFC	Yes	Other Crude Unit	Feed Spec, Desalter Tank Draining	No	Yes- exceed limits occasionally
NA	'71 - '80	>600	GFC	Yes	Cracking Heater	Desalter	Yes	Yes
NA	'71 - '80	>600	GFC	Yes	CHO, CHR	Feed Specs	No	Yes
NA	'71 - '80	>600	GFC	No	-	Feed Specs, Tank Draining	Yes	Yes
E	'71 - '80	300-600	RCO	No	-	-	-	-

Table Q3-2
Specific Feedstock Contaminants - Sodium Chloride, Salt
(Continued)

Location	Plant Age	Size	Feed	Contaminant?	Issue, or Area Introduced	Mitigation	Monitor?	Meet Expectations
E	'71-'80; '81-'90 Post-90	300-600	RCO	Yes	Water Quench System	Feed Specs	Yes	Yes
E	'71 - '80	>600	RCO	No	-	-	-	-
E	'71 - '80	>600	RCO	Yes	CHO, CHR	Feed Specs, Tank Draining	Yes	No, operators fail to drain
NA	'71 - '80	>600	EP, GFC	No	-	-	-	-
E	'71 - '80	>600	EP, RCO	No	-	-	No	-
E	'71 - '80	>600	EP, RCO	Yes	CHO	Feed Specs	No	Yes
NA	'81 - '90	>600	GFC, RCO	Yes	CHO, CHR	Feed Specs, Desalter, Tank Draining	No	No, no robust control system
NA	'81 - '90	>600	GFC, RCO	Yes	CHO, CHR	Feed Specs, Tank Draining	No	No, no robust control system
NA	'71 - '80	300-600	All 3	Yes	CHO, CHR	Feed Specs, Desalter, Tank Draining	No	No, no robust control system
NA	'81 - '90	>600	All 3	No	-	-	No	-
NA	'71 - '80	>600	All 3	Yes	CHO, CHR	Feed Specs, Water Wash & Coalescer	No	Yes
NA	Pre-70; '71-'80;	>600	All 3	No	-	-	-	-
				Yes	-	Feed Specs, Water Wash & Coalescer, Tank Draining	Yes	Yes
				Yes	Other	Feed Specs, Tank Draining	Yes	Yes

Table Q3-3
Specific Feedstock Contaminants - Arsine

Location	Plant Age	Size	Feed	Contaminant?	Problems, or Area Introduced	Mitigation	Meet Expectations	Spent Analysis	Quantity As Found
E	Post-1990	>600	EP	Yes	Refinery Propylene	CuO bed	Yes	No	-
E	Pre-1970	>600	EP	No	-	-	-	-	-
NA	Pre-1970	<300	EP	No	-	-	Yes	No	-
NA	'81 - '90	300-600	EP	No	-	-	Yes	No	-
NA	Pre-1970	300-600	EP	No	-	-	-	-	-
NA	Pre-1970	300-600	EP	Yes	Refinery "Dry Gas", Refinery Propylene	Feed Specs, AA bed	Yes	Yes	Acetylene C guard bed
NA	Pre-1970	300-600	EP	No	-	-	-	-	-
NA	Pre-1970	300-600	EP	No	-	-	-	-	-
NA	'71 - '80	300-600	EP	Yes	Refinery Propylene	Feed Specs	Yes	No	-
NA	'71 - '80	300-600	EP	No	-	-	-	-	-
NA	'71-'80; Post-90	>600	EP	No	-	-	-	-	
NA	Post-1990	>600	EP	No	-	-	-	-	-
NA	Post-1990	>600	EP	No	-	-	-	-	-
NA	'71 - '80	>600	EP	-	-	-	-	-	-
NA	Post-1990	>600	EP	Yes	Refinery Dry Gas	CuO bed	Yes	Yes	-
E	Pre-1970	300-600	GFC	No	-	-	-	-	-
NA	'71 - '80	>600	GFC	Yes	Naptha, AGO/VGO	Feed Specs, Deposition on Hydr. Cat.	Yes	Yes	Pygas Cat. .09 - .15%
NA	'71 - '80	>600	GFC	Yes	Naphtha, AGO/VGO, expect in Pygas Hydr., and A-Conv. Cat.	Feed Specs, Deposition on Hydr. Cat.	Yes	Yes	MAPD Cat. 0.0103%
NA	'71 - '80	>600	GFC	No	-	Feed Specs	-	-	-
NA	'71 - '80	>600	GFC	Yes	Naphtha Condensate	Feed Specs, Deposition on Hydr. Cat. In gas/fueloil	Yes	Yes	Pygas Cat. 0.1-0.4 wt%
E	'71-'80; '81-'90 Post-90	300-600	RCO	No		-	-	-	-

Table Q3-3
Specific Feedstock Contaminants - Arsine
(Continued)

Location	Plant Age	Size	Feed	Contaminant?	Problems, or Area Introduced	Mitigation	Meet Expectations	Spent Analysis	Quantity As Found
E	'71 - '80	300-600	RCO	No	-	-	-	-	-
E	'71 - '80	>600	RCO	No	-	-	-	-	-
E	'71 - '80	>600	RCO	Yes	Refinery Propylene, Dry Gas	Feed Specs, CuO	Yes	Yes	MAPD Cat. 300-400 Hydrogen Cyanide
NA	'71 - '80	>600	EP, GFC	No	-	Form ppt with S clean out residue	Yes	Yes	2 ppm wt in Acetylene converter
E	'71 - '80	>600	EP, RCO	No	-	-	-	-	-
E	'71 - '80	>600	EP, RCO	No	-	-	-	-	-
NA	'71 - '80	300-600	All 3	Yes	Refinery "Dry Gas"	Feed Specs, PbO bed	Yes	Yes	-
NA	'81 - '90	>600	All 3	No	-	-	-	-	-
NA	'71 - '80	>600	All 3	Yes	Refinery Propylene	CuO bed	Yes	Yes	Pygas Cat. .05mg/L
NA	Pre-70; '71-'80;	>600	All 3	No	-	-	-	-	-
NA	'81 - '90	>600	GFC, RCO	No	-	Feed Specs	-	-	-
NA	'81 - '90	>600	GFC, RCO	No	-	Feed Specs	-	-	-
No Resp	No Resp	No Resp	No Resp	Yes	Refinery Propylene, Dry Gas, AGO/VG	Feed Specs, Deposition on Hydr. Cat.	Yes	Yes	MAPD Cat. <200wppm
No Resp	No Resp	No Resp	No Resp	Yes	AGO/VGO	Feed Specs, on Hydr. Cat.	Yes	Yes	Pygas Cat. 0.2 wt%

Table Q3-4
Specific Feedstock Contaminants - Ammonia

Location	Plant Age	Size	Feed	NH_3 Issues?	Introduction Mode	Issue Locations	Mitgation Strategy	Meeting Expectations
E	Post-1990	>600	EP	Yes	Treatment chemicals	Process water	Removal from quench/process H_2O; amine treater system.	Yes
NA	Pre-1970	<300	EP	No	-	-	-	Yes
NA	'81 - '90	300-600	EP	No	-	-	-	Yes
NA	Pre-1970	300-600	EP	No	-	-	-	-
NA	Pre-1970	300-600	EP	No	-	-	-	-
NA	Pre-1970	300-600	EP	No	-	-	-	-
NA	Pre-1970	300-600	EP	No	-	-	-	-
NA	'71 - '80	300-600	EP	Yes	Treatment chemicals	Ethylene product quality	No response	No Resp
NA	'71 - '80	300-600	EP	No	-	-	-	-
NA	'71 - 80, Post-1990	>600	EP	Yes	Treatment chemicals	Process water; furnace corrosion *ex* dilution steam	No response	Yes
NA	Post-1990	>600	EP	No	-	-	-	-
NA	Post-1990	>600	EP	No	-	-	-	-
NA	'71 - '80	>600	EP	Yes	Treatment chemicals	Catalyst	No response	No Resp
NA	Post-1990	>600	EP	Yes	Treatment chemicals	No response	No response	No Resp
E	Pre-1970	300-600	GFC	Yes	Treatment chemicals; vent gas	Ethylene product quality; ammonium nitrate in chilling train	Feedstock specs; NH_3 analyzer	Yes
NA	'71 - '80	>600	GFC	Yes	Other internal feeds	Ethylene product quality	Other	No

Table Q3-4
Specific Feedstock Contaminants - Ammonia
(Continued)

Location	Plant Age	Size	Feed	NH_3 Issues?	Introduction Mode	Issue Locations	Mitgation Strategy	Meeting Expectations
NA	'71 - '80	>600	GFC	Yes	Feedstock; treatment chemicals; NESAPS system	Ethylene product quality; other product quality; process water	Removal from quench water/ process water	No. Too much ammonium salts circulating in our quench H_2O system. Small quench H_2O pH swings can cause our unit to experience product quality problems
NA	'71 - '80	>600	GFC	No	-	-	-	-
NA	'71 - '80	>600	GFC	Yes	Feedstock; Treatment chemicals	Ethylene product quality	Removal via mole sieve or other adsorbents	Yes
E	' 71 - 80, '81 - '90, Post-1990	300-600	RCO	No	-	-	-	-
E	'71 - '80	300-600	RCOL	Yes	Treatment chemicals	Ethylene product quality; Process water	Removal from quench / process water; amine treater system.	Yes
E	'71 - '80	300-600	RCOL	No	-	-	-	-
E	'71 - '80	>600	RCO	Yes	Upset of BD plant could contaminate raffinate stream with solvent NMP. Contaminated stream has to be processed as cracker feedstock	Process water	Calculating maximum permissible feedstock rate on basis <5 Hydrogen Cyanide NH_3 formation, extended monitoring of pH-values in quench H_2O, increase of quench H_2O refresher	Yes

Table Q3-4
Specific Feedstock Contaminants - Ammonia
(Continued)

Location	Plant Age	Size	Feed	NH_3 Issues?	Introduction Mode	Issue Locations	Mitgation Strategy	Meeting Expectations
NA	'71 - '80	>600	EP, GFC	No	-	-	Feed specifications	Yes
E	'71 - '80	300-600	EP, RCO	No	-	-	-	-
E	'71 - '80	>600	EP, RCO	No	-	-	-	Yes
E	'71 - '80	>600	EP, RCO	Yes	Treatment chemicals	Ethylene product quality	Removal from quench/process H_2O; amine treater system.	No Resp
NA	'81 - '90	>600	GFC, RCO	Yes	Feedstock; treatment chemicals	Ethylene product quality	Feedstock specs; removal w/adsorbents; removal from quench / process water; amine treater system.	No; suffer periodic ammonia excursions
NA	'81 - '90	>600	GFC, RCO	Yes	Feedstock; treatment chemicals	Ethylene product quality	Feedstock specs; removal w/adsorbents; removal from quench / process water; amine treater system.	No; suffer periodic ammonia excursions
NA	'71 - '80	300-600	All 3	Yes	Feedstock; treatment chemicals	Ethylene product quality	Feedstock specs; removal w/adsorbents; removal from quench / process water; amine treater system	Yes
NA	'81 - '90	>600	All 3	No	-	-	-	-
NA	'71 - '80	>600	All 3	Yes	Feedstock; treatment chemicals	Ethylene product quality	Feedstock specs	Yes
NA	Pre-1970 '71 - '80	>600	All 3	No	-	-	-	-
-	-	-	-	No	Feedstock	No response	Removal from quench / process water; low pH water absorber	Yes
-	-	-	-	No	Feedstock	None	Removal from quench/process H_2O	Yes

Table Q3-5
Specific Feedstock Contaminants - Amines & Nitriles

Location	Plant Age	Size	Feed	Amine Issues?	Issue & Mitigation	Nitriles Issues?	Issue & Mitigation
E	Post-1990	>600	EP	Yes	Liquid carryover from absorber units into gas feeds. Have modified system to prevent.	No	-
NA	Pre-1970	<300	EP	No Resp	-	No	-
NA	'81 - '90	300-600	EP	No	-	No	-
NA	Pre-1970	300-600	EP	No Resp	-	No Resp	-
NA	Pre-1970	300-600	EP	No Resp	-	No Resp	-
NA	Pre-1970	300-600	EP	No Resp	-	No Resp	-
NA	Pre-1970	300-600	EP	No Resp	-	No Resp	-
NA	'71 - '80	300-600	EP	No	-	No	-
NA	'71 - '80	300-600	EP	No Resp	-	No Resp	-
NA	'71 – 80 Post-1990	>600	EP	Yes	Monitoring of the dilution steam system	No	-
NA	Post-1990	>600	EP	No Resp	-	No Resp	-
NA	Post-1990	>600	EP	No Resp	-	No Resp	-
NA	'71 - '80	>600	EP	No Resp	-	No Resp	-
NA	Post-1990	>600	EP	Yes	No Response	No	-
E	Pre-1970	300-600	GFC	Yes	Nitrate formed in furnaces gives ammonium nitrate in chilling train. Removal by flushing during plant outage.	No	-
NA	'71 - '80	>600	GFC	No	-	Yes	-
NA	'71 - '80	>600	GFC	Yes	Amines/NH_3 showing up in product streams. To prevent - improved quench water pH control -- closer monitoring of MEA addition.	No	-
NA	'71 - '80	>600	GFC	No Resp	-	No Resp	-
NA	'71 - '80	>600	GFC	No	-	No	-

Table Q3-5
Specific Feedstock Contaminants - Amines & Nitriles
(Continued)

Location	Plant Age	Size	Feed	Amine Issues?	Issue & Mitigation	Nitriles Issues?	Issue & Mitigation
E	' 71 - 80, '81 - '90, Post-1990	300-600	RCO	No Resp	-	No Resp	-
E	'71 - '80	300-600	RCOL	No Resp	-	No Resp	-
E	'71 - '80	300-600	RCOL	No Resp	-	No Resp	-
E	'71 - '80	>600	RCO	No	-	No	-
NA	'71 - '80	>600	EP, GFC	Yes	Converted to nitriles which contaminate byproducts. Analyze byproducts for nitriles.	Yes	Analyze byproducts for nitriles.
E	'71 - '80	300-600	EP, RCO	No Resp	-	No Resp	-
E	'71 - '80	>600	EP, RCO	No Resp	-	No Resp	-
E	'71 - '80	>600	EP, RCO	No	-	No	-
NA	'81 - '90	>600	GFC, RCO	Yes	Ethylene quality; no action because of rarity of occurrences.	No	-
NA	'81 - '90	>600	GFC, RCO	Yes	Ethylene quality; no action because of rarity of occurrences.	No	-
NA	'71 - '80	300-600	All 3	No	-	Yes	-
NA	'81 - '90	>600	All 3	No	-	No	-
NA	'71 - '80	>600	All 3	Yes	Amine carryover in refinery propylene fouled the arsine guard bed. Worked with suppliers to ensure free water and amines were removed prior to transfer.	No	-
NA	Pre-1970 '71 - '80	>600	All 3	No Resp	-	No Resp	-
-	-	-	-	No	-	No	-
-	-	-	-	No	-	Yes	Affects C_4 product quality; added wash water tower to remove ACN.

Table Q4-: Flare Drum Knockout Liquids

Location	Plant Age	Size	Feed	Process?	Origin	Process Entry	Issues
NA	Pre-1970	300-600	EP	No			
NA	Pre-1970	300-600	EP	No			
NA	Pre-1970	300-600	EP	No			
NA	'71 - '80	300-600	EP	No			
NA	'71 - '80	300-600	EP	Yes	Flare drum knockouts.	PGC suction.	No Response
NA	'71 - 80, Post-1990	>600	EP	Yes	No Response	Oily water system.	Contribution to quench water fouling.
NA	Post-1990	>600	EP	Yes	No Response	Quench.	No Response
NA	'71 - '80	>600	EP	No			
NA	Post-1990	>600	EP	No			
E	Pre-1970	300-600	GFC	Yes	No Response	Quench or feed.	No problems, specification for line-up to process/incineration.
NA	'71 - '80	>600	GFC	Yes	No Response	Quench tower *via* slops injection.	Product quality issues depending on source.
NA	'71 - '80	>600	GFC	Yes	Flare knock-out liquid (H_2O containing) are sent to NESHAPs	See wastewater stripper section.	No Response
NA	'71 - '80	>600	GFC	No			
NA	'71 - '80	>600	GFC	Yes	Olefins plant and alcohol plant	Water Quench Tower	Pyrolysis gasoline quality, alcohol in propylene.
E	' 71 - 80, '81 - '90, Post-1990	300-600	RCO	Yes	Only cracker products, not from external plant.	Quench column.	None.
E	'71 - '80	300-600	RCO	No			
E	'71 - '80	300-600	EP, RCO	No			
E	'71 - '80	>600	EP, RCO	Yes	No Response	No Response	No Response
E	'71 - '80	>600	EP, RCO	Yes	Origin.	Slops, fuel to boilers.	No probems.
NA	'81 - '90	>600	GFC, RCO	Yes	Ethylene plant.	Primary fractionator.	Some chloride concerns.
NA	'81 - '90	>600	GFC, RCO	Yes	Ethylene plant.	Primary fractionator.	Some chloride concerns.
NA	'71 - '80	300-600	All 3	Yes	Ethylene plant.	Primary fractionator.	Some chloride concerns.
NA	'81 - '90	>600	All 3	No			
NA	'71 - '80	>600	All 3	Yes	Aromatics unit. Light slops tank.	Primary fractionator. Oily water separator.	None.
NA	Pre-1970 '71 - '80	>600	All 3	Yes	Recycle stream in process.	Plant equipment cleaning; safety valves.	None.
-	-	-	-	Yes	No response.	Primary fractionator.	None.

Table Q4-2: Vents from Other Plants

Location	Plant Age	Size	Feed	Process?	Origin	Process Entry	Issues
E	Post-1990	>600	EP	Yes	Ethanol.	Cracked gas compressor.	
NA	'81 - '90	300-600	EP	Yes	Oxo alchohols; polyethylene.	3rd & 4th stage charge gas.	No response
NA	Pre-1970	300-600	EP	Yes	HD and LD polymer plants.	Deethanizer for HD and quench for LD.	HD isobutane can cause need for flaring to keep temperatures down on reboilers; LD can cause fouling in caustic tower if vinyl acetylene concentration high.
NA	Pre-1970	300-600	EP	Yes	Poly plant.	Purification.	No response.
NA	Pre-1970	300-600	EP	Yes	Poly plant.	Purification.	No response.
NA	'71 - '80	300-600	EP	Yes	Polyethylene vents.	Initial stages of compression.	Components used to produce polyethylene promote/cause fouling/corrosion in process equipment.
NA	'71 - '80	300-600	EP	Yes	LDPE/LLDPE/PP vent recovery.	PGC suction and discharge.	Vinyl acetates from LDPE cause polymerization problems in quench and caustic system.
NA	'71 - 80, Post-1990	>600	EP	Yes	Polyethylene & LAO purge streams.	Charge gas compressor suction.	Unsteady plant throughput.
NA	Post-1990	>600	EP	No			
NA	'71 - '80	>600	EP	No			
NA	Post-1990	>600	EP	No			
E	Pre-1970	300-600	GFC	Yes	Aromatics, butadiene, polyethylene plants.	Compression.	No problems.
NA	'71 - '80	>600	GFC	Yes	Polyethylene plant.	3rd stage compressor.	Nitrogen carry-over to hydrogen product.
NA	'71 - '80	>600	GFC	Yes	Butadiene acetylene vent stream.	Furnace feed.	No response.
NA	'71 - '80	>600	GFC	Yes	Derivative units. Ethylene recycle.	1st stage suction drum.	Nitrogen content from PE plants ties up compressor capacity.

Table Q4-2 : Vents from Other Plants (Continued)

Location	Plant Age	Size	Feed	Process?	Origin	Process Entry	Issues
NA	'71 – '80	>600	GFC	Yes	Alkylation, Butadiene Recovery	Cracked Gas Compression	None
E	'71 – '80, '81 - '90, Post-1990	300-600	RCO	Yes	Waste gas from butadiene extraction.	CG compression.	None.
E	'71 - '80	300-600	RCO	Yes	Ethanol.	Cracked gas compressor.	No response.
E	'71 - '80	300-600	RCO	No			
E	'71 - '80	>600	RCO	Yes	Polyethylene & polypropylene.	Pyrofractionator or quench water to gas compressor.	Emulsion in oil/water separation. Line plugging. Self ignition during cleaning.
E	'71 - '80	300-600	EP, RCO	Yes	Light gas from BD unit.	PGC.	No problems.
E	'71 - '80	>600	EP, RCO	No			
E	'71 - '80	>600	EP, RCO	Yes	Origin.	Partly in fuel gas; partly back to PGC.	No problems.
NA	'81 - '90	>600	GFC, RCO	Yes	BD recovery; hydrotreaters.	Pyrolysis gas compressor.	None; clean streams.
NA	'81 - '90	>600	GFC, RCO	Yes	BD recovery; hydrotreaters.	Pyrolysis gas compressor.	None; clean streams.
NA	'71 - '80	300-600	All 3	Yes	BD recovery; hydrotreaters.	Pyrolysis gas compressor.	None; clean streams.
NA	'81 - '90	>600	All 3	No			
NA	'71 - '80	>600	All 3	No			
NA	Pre-1970 '71 - '80	>600	All 3	No			
-	-	-	-	Yes	PE/PP plants.	PGC.	None.
-	-	-	-	Yes	PE. PP.	Cracked gas compressor, depropanizer.	None.

Table Q4-3: Refinery FCC Dry Gas

Location	Plant Age	Size	Feed	Process?	Origin	Process Entry	Issues
NA	Pre-1970	300-600	EP	No			
NA	Pre-1970	300-600	EP	No			
NA	'71 - '80	300-600	EP	No			
NA	'71 - 80, Post-1990	>600	EP	No			
NA	Post-1990	>600	EP	No			
NA	Post-1990	>600	EP	Yes	H_2 recovery plant.	Cracked gas compressor.	
E	Pre-1970	300-600	GFC	No			
NA	'71 - '80	>600	GFC	No			
NA	'71 - '80	>600	GFC	No			
NA	'71 - '80	>600	GFC	No			
E	' 71 - 80, '81 - '90, Post-1990	300-600	RCO	No			
E	'71 - '80	300-600	RCOL	No			
E	'71 - '80	300-600	EP, RCO	Yes	FCC unit. Gas treatment.	PGC.	Arsine has caused catalyst poisoning. Occasional H_2S peaks have resulted AC hydrogenation contamination.
E	'71 - '80	>600	EP, RCO	No			
E	'71 - '80	>600	EP, RCO	No			
NA	'81 - '90	>600	GFC, RCO	No			
NA	'81 - '90	>600	GFC, RCO	No			
NA	'71 - '80	300-600	All 3	Yes	Onsite refinery; purchased gas.	Before PGC.	None; clean streams.
NA	'81 - '90	>600	All 3	No			
NA	'71 - '80	>600	All 3	No			
NA	Pre-1970 '71 - '80	>600	All 3	No			
-	-	-	-	Yes	FCC.	PGC.	NOx contaminants forming ammonium NOx salts in cold box.
-	-	-	-	No			

Table Q4-4: Wastewater Stripper Overhead

Location	Plant Age	Size	Feed	Process?	Origin	Process Entry	Issues
NA	Pre-1970	300-600	EP	Yes	From process.	Quench water.	
NA	Pre-1970	300-600	EP	No			
NA	Pre-1970	300-600	EP	No			
NA	'71 - '80	300-600	EP	No			
NA	'71 - 80, Post-1990	>600	EP	Yes	No response.	Quench water.	No response.
NA	'71 - '80	>600	EP	No			
NA	Post-1990	>600	EP	No			
E	Pre-1970	300-600	GFC	No			
NA	'71 - '80	>600	GFC	Yes	NESHAPs stripper overhead water.	Primary fractionator.	Amine and organic acids source. Can upset quench water chemistry.
NA	'71 - '80	>600	GFC	No			
NA	'71 - '80	>600	GFC	Yes	Process water and rainwater runoff.	Quench Tower	Oxygen and oxygenates can affect quench water and caustic tower.
E	' 71 - 80, '81 - '90, Post-1990	300-600	RCO	No			
E	'71 - '80	300-600	RCOL	Yes	No response.	P/F distillate drum.	No response.
E	'71 - '80	>600	RCO				
E	'71 - '80	300-600	EP, RCO	No			
E	'71 - '80	>600	EP, RCO	Yes	No response.	No response.	No response.
E	'71 - '80	>600	EP, RCO	No			
NA	'81 - '90	>600	GFC, RCO	No			
NA	'81 - '90	>600	GFC, RCO	No			
NA	'71 - '80	300-600	All 3	No			
NA	'81 - '90	>600	All 3	No			
NA	'71 - '80	>600	All 3	Yes	Unit water stripper. Offsite water water stripping unit.	Primary fractionator.	None.
NA	Pre-1970 '71 - '80	>600	All 3	Yes	Recycle stream.	Wastewater stripper.	Occasionally, polymer build-up.
-	-	-	-	No			

Table Q4-5: Raw Pyrolysis Naphtha

Location	Plant Age	Size	Feed	Process?	Origin	Process Entry	Issues
NA	Pre-1970	300-600	EP	No			
NA	Pre-1970	300-600	EP	No			
NA	'71 - '80	300-600	EP	No			
NA	'71 - 80, Post-1990	>600	EP	No			
NA	Post-1990	>600	EP	No			
NA	'71 - '80	>600	EP	No			
NA	Post-1990	>600	EP	No			
E	Pre-1970	300-600	GFC	No			
NA	'71 - '80	>600	GFC	Yes	No response.	TLEs.	Increased fouling.
NA	'71 - '80	>600	GFC	Yes	Purchased on open market.	Primary fractionator.	Impurities (silicon, organic acids) can cause catalyst fouling or upset quench water chemistry.
NA	'71 - '80	>600	GFC	No			
NA	'71 - '80	>600	GFC	Yes	EP cracker drip oils,	Primary Fractionator	Potential source of oxygen; product quality issues.
E	' 71 - 80, '81 - '90, Post-1990	300-600	RCO	No			
E	'71 - '80	>600	RCO				
E	'71 - '80	300-600	EP, RCO	No			
E	'71 - '80	>600	EP, RCO	Yes	No response.	No response.	No response.
E	'71 - '80	>600	EP, RCO	No			
NA	'81 - '90	>600	GFC, RCO	Yes	Many sources.	Debutanizer.	Some quality issues.
NA	'81 - '90	>600	GFC, RCO	Yes	Many sources.	Debutanizer.	Some quality issues.
NA	'71 - '80	300-600	All 3	Yes	Many sources.	Debutanizer.	Some quality issues.
NA	'81 - '90	>600	All 3	No			
NA	'71 - '80	>600	All 3	Yes	Purchased shipments..	Primary fractionator.	None.
NA	Pre-1970 '71 - '80	>600	All 3	Yes	Recycle and by-product streams.	Quench system.	None.
-	-	-	-	No			

Table Q4-6: Other Streams Recycled or Reprocessed in Olefins Plants

Location	Plant Age	Size	Feed	Origin	Process Entry	Issues
NA	Post-1990	>600	EP	No response.	PGC 4th stage discharge, upstream of caustic tower.	Possible source of arsine. Process has arsine beds for removal. Possible source of O_2 and sulfur.
NA	'71 - '80	>600	GFC	Butadiene heavies stream (green oil, dimer stream).	Primary fractionator.	May contain Nitrogen compounds that can affect quench water chemistry.
E	'71 - '80 '81 – '90 Post-1990	300 - 600	RCO	Butadiene extraction plant.	CG header inlet quench column.	None.
E	'71 - '80	300 - 600	RCO	No response.	Furnaces.	No problems.
E	'71 - '80	300 - 600	EP, RCO	Polyethylene plant. Polypropylene plant. Polypropylene plant.	PGC. Debutanizer overhead drum. PGC.	Just flow fluctuations. No problems. No problems.
NA	'71 - '80	>600	All 3	Refinery.	C3 purification.	Arsine guard bed catalyst fouling.

Appendix I

1992 Feedstock Contaminants

Survey Results

AMERICAN INSTITUTE
OF CHEMICAL ENGINEERS

December 18, 1992

The Ethylene Producers Committee feedstock issues questionnaires are in and the results have been tabulated and are submitted for your reference and review. A lot of time has passed since May 26, when you first received this survey to fill out. The Ethylene Producers Committee of the AIChE appreciates your patience. There were some delays in receiving the responses, and a lot of time was spent in maintaining the integrity of the data. We feel confident that you will find this material to be useful, as all the respondents have expressed their concerns and shared solutions to the questions asked.

The Ethylene Producers Committee feels proud that over 80% of the members responded with their meaningful comments regarding impurities or concerns in their feedstocks. As promised, the results are being returned to all participating companies to be digested and disseminated as you see fit. This material will be published under the auspices of the AIChE at the National meeting held the spring of 1993 in Houston, Texas. The AIChE has approved making the results public during the Safety Session, the morning of March 30. The material will be presented by Mark Graham of Chevron Chemical Company.

Regarding the survey, every effort was made to ensure that the reported information appears in the same form and context as provided by the respondents. There was no attempt to interpret any of the respondent's comments or data. The subcommittee in charge of evaluating these results felt this was necessary to preserve the authenticity, since all the ethylene producers have unique supply and process situations. Confidentiality of the data has been maintained so as no individual member would be identified.

It does not end here. As a sequel, the subcommittee on Feedstocks, (where this survey was processed), does not plan on sitting still. A new group has been formed to address radon and fluorides in feedstocks. A questionnaire and published results should be expected in late 1993. If you have any other interests or comments regarding these or other topics, please contact your company representative or direct them to the undersigned.

It is our pleasure to serve the industry through the members of the Ethylene Producers Committee of the AIChE in this endeavor. Our mission is to be a clearing house of information for the ethylene producers, as well as other affiliated companies. We hope this survey accomplished that. We thank you for your patience and support, for without it, this type of information sharing could not happen.

If you have questions or comments concerning the attached survey results, contact Mark Graham at the following address:

Mark A. Graham
Chevron Chemical Company
1301 McKinney Street
Houston, TX 77010
Telephone (713) 754-2612

M. G. Brown
Chairman – Ethylene Producers Committee

R. C. Parnell
Chairman – Feedstocks and Products Issues Subcommittee

EPC FEEDSTOCK ISSUES SUBCOMMITEE OUESTIONNAIRE RESPONSES COMPILATION

Please indicate your level of concern (H-High, M-Medium, L-Low) that the following substances from your feedstocks may have an adverse impact on your process and/or your customers processes. Please also identify other contaminants not listed that concern you.

NOTE ON RESULTS: one respondent clarified sodium to be NaCl and NaOH.

CONTAMINANT	# RESP.	CONCERN		
		L	M	H
Methanol	16	3	8	5
Chlorides	16	4	0	12
NOx	15	2	5	8
Arsine	16	0	4	12
Carbon Dioxide	18	0	7	11
Sulfur	17	1	5	11
Sodium	16	8	2	6
Ammonia	16	3	8	5
Mercury	17	1	3	13
Water	17	9	5	3
Fluorides	14	3	2	9
Radon	14	7	6	1
Phosphorus	14	9	3	2
OTHERS:				
Metal	3	0	2	1
Oxygenates (MTBE)	3	0	2	1
HF	1	1	0	0
Iron Oxide	1	0	0	1
Oxygen	2	0	0	2
Lead	4	1	2	1
Carbon Monoxide	1	0	0	1
Vanadium	2	0	1	1
Antimony	1	0	0	1
Olefins	1	1	0	0
Styrene	1	0	1	0
C_{10+} in LPG	1	0	0	1
Vinyl Acetate	1	0	0	1
Bottom Sediment & Water	1	1	0	0

2) We have compiled the following possible consequences of these feedstock contaminants to ethylene crackers. Please let us know if you agree, if not please explain.

NOTE: In the tabulation of results, agree and disagree were combined.

CONTAMINANT	# RESP	AGREE/ DISAGREE	NO COMMENT	OTHER CONSEQUENCES
Methanol	15	8	7	Routinely injects into system; economics; Makes CO in furnaces; Emulsion; Short run on C_3 driers
Carbon Dioxide	17	10	7	Increased treatment costs, caustic usage
Mercury	14	10	4	Safety - cold box; Cat poison; Safety-cat poison; Safety, hygiene, environment; Safety; Cat poison; Safety and cat poison
Fluorides	14	7	7	Furnace damage; pH in quench water; Corrosion, waste treatment; Corrosion; Fluoride corrosion; Furnace tube problem; No experience
Chlorides	15	8	7	Corrosion; Furnace damage; pH of quench water; Quality Product quality; Furnace tube problem
Sulfur	11	8	3	Corrosion, cat poison; Cat poison; Cat poison; Environment, cat poison; Operability, cat poison; corrosion; cat poison; Lose front end reactor control; Cat poison; Corrosion, quality, environmental
Radon	18	7	11	
NOx	17	8	9	Cold box damage
Sodium	18	9	9	Quality
Water	14	7	7	Freezing; Corrosion; Operability; Inject steam and recover; Economics; Corrosion, safety
Phosphorus	16	6	10	Cat poison; Furnace damage

2) We have compiled the following possible consequences of these feedstock contaminants to ethylene crackers. Please let us know if you agree, if not please explain.

NOTE: In the tabulation of results, agree and disagree were combined.

CONTAMINANT	# RESP	AGREE/ DISAGREE	NO COMMENT	OTHER CONSEQUENCES
Arsine	18	9	9	
Ammonia	15	6	9	Corrosion; Operability; Cold box pluggage (ammonium salts); Quench water pH
OTHERS:				
Metal	2			Safety, Quality
Oxygenates (MTBE)	2			Product quality, Quality
HF	1			Furnace tube problem
Iron Oxide	1			Nuisance - plugs meters
Oxygen	2			Quality, Furnace damage, polymerization
Lead	4			Furnace damage; Corrosion; Cat poison; Corrosion
Carbon Monoxide	1			Operability
Vanadium	2			Corrosion; Corrosion
Antimony	4			
Olefins	1			Furnace coking
C_{10+} in LPG	1			Furnace coking

3) (A) Check if you look for these contaminants in your feeds? (B) Check if you have seen these contaminants in your feeds? If so, have you established baseline data (what is normal in PPM or PPB)? What minimum level of detect ability have you achieved (in PPM or PPB)? What maximum level will you tolerate? (C) Would you be willing to share your analytical methods with the ethylene industry? (YES/NO)

NOTE: The reader should not conclude any linkage of data by respondent between the rows and columns. Respondents made comments on some, but not necessarily all, issues.

CONTAMINANT	LOOK FOR	FOUND	BASELINE	DETECT.	MAXIMUM LEVEL (PPM/PPB)	SHOW METHOD
Methanol	8	12	5-20ppm; 10-200ppm; <100 ppm; 50 ppm; 20 ppm; 20-50 ppm; <1 ppm; 100 ppm	0.1 ppm; 2 ppm; 1 ppm; 1 ppm; 1 ppm; 1 ppm, 1 ppm; 0.2 ppm; 1 ppm	1000 ppm; 5 ppm; 100 ppm; 200 ppm; 100 ppm; 2000 ppm; 2 ppm; 20 ppm; 10 ppm	5 said yes
Carbon Dioxide	15	17	0.4%; no; <75 ppm; 5000 ppm; 1000 ppm as gas; 1000-3000 ppm; 500 ppm; 60 ppm; 100-200 ppm; 200 ppm; 150 ppm; 5-10 ppm; <100 ppm; 5-15 ppm	1 ppm; 1 ppm; 0.1 ppm; 2 ppm; <1 ppm; 100 ppm; 200 ppm; 1 ppm; 0.2 ppm; 1 ppm; 5 ppm; 0 ppm; <1 ppm; 0.1 ppm	1%; 200 ppm; 6000 ppm; 15000 ppm; 2000 ppm; 175 ppm; 1000 ppm, 1000 ppm; 3250 ppm; 750 ppm; 1000 ppm	6 said yes
Mercury	11	11	<2 ppb; 0.38 ppb; <0.01 ppm; .004 ppm; 0.001-0.006 ppm; <40 ppb; 20 ppb; 0-5ppb; 25 ppm; no	1 ppb; 0.001 ppb; 0.002 ppb; 0.001ppm; <0.001 ppm; 0.001 ppm; 10 ppb; 5 ppb; 1 ppb; 1 ppb	2 ppb; 1 ppb; 0.01 ppm; 0.01 ppm; 0.01 ppm; 10 ppb	5 said yes
Fluorides	9	8	1-3 ppm; 0.5-3 ppm; 1-2 ppm; <1 ppm; <1 ppm	0.5 ppm; <1 ppm; <1 ppm; <0.1 ppm; 0.1 ppm	5 ppm; 5 ppm; 5 ppm; 1 ppm	2 said yes

3) (A) Check if you look for these contaminants in your feeds? (B) Check if you have seen these contaminants in your feeds? If so, have you established baseline data (what is normal in PPM or PPB)? What minimum level of detect ability have you achieved (in PPM or PPB)? What maximum level will you tolerate? (C) Would you be willing to share your analytical methods with the ethylene industry? (YES/NO)

NOTE: The reader should not conclude any linkage of data by respondents between the rows and columns. Respondents made comments on some, but not necessarily all, issues.

CONTAMINANT (Continued)	**LOOK FOR**	**FOUND**	**BASELINE**	**DETECT.**	**MAXIMUM LEVEL (PPM/PPB)**	**SHOW METHOD**
Chlorides	9	8	<2 ppm; <2 ppm; <2ppm; <lppm; <1 ppm; l ppm; no	0.1 ppm; 1 ppm; 2 ppm; 2 ppm; 2 ppm; l ppm; (100 ppb in-organic) (500 ppb-organic); 0.2 ppm	5 ppm; 5 ppm; 5 ppm; 5 ppm; 5 ppm; 4ppm as NaCl	5 said yes
Sulfur	17	17	20 ppm; 50-100 ppm; 50-100 ppm; 10-50 ppm; 3 ppm; 1 ppm; 430 ppm; 10-50 ppm; 15 ppm; 0-900 ppm	0.1 ppm; 1 ppm; 1 ppm; ?; ?; 0.5 ppm; 1 ppm; <0.1 ppm; 0.25 ppm; 25 ppb; 5 ppm; 1-100 ppm	100 ppm; 200 ppm; 200 ppm; 200 ppm; 5 ppm; 4 ppm; 1000 ppm; 50 ppm; 30 ppm; 500 ppm; 5-2000 ppm	6 said yes
Radon	3	3	5-3000 CT/ 10 min.	5pCi/ml; 25 counts	800 counts; 2000 counts	2 said yes
NOx	5	2	20-50 ppb; no	not sure; 0-20 ppb	0.1 ppb	3 said yes
Sodium	7	7	0.1-03 ppm; 0.1-1 ppm; <.01 ppm	0.01 ppm; 0.01 ppm; 0.01 ppm	4 ppm as NaCl; 5 ppm; 5 ppm; 0.5 ppm	2 said yes
Water	11	14	<100 ppm; saturated; 50 ppm; 110-180 ppm; 1-30 ppm; 18 ppm; 0-20 ppm	2 ppm; 1 ppm; 1 ppm; 1-5 ppm; 1 ppm; 1 ppm; 1 ppm; <l ppm	200 ppm; 200 ppm, saturated; 300 ppm; 130 ppm; saturated; 5 ppm; 25ppm; 50 ppm; saturated; 0.1%	2 said yes

3) (A) Check if you look for these contaminants in your feeds? (B) Check if you have seen these contaminants in your feeds? If so, have you established baseline data (what is normal in PPM or PPB)? What minimum level of detect ability have you achieved (in PPM or PPB)? What maximum level will you tolerate? (C) Would you be willing to share your analytical methods with the ethylene industry? (YES/NO)

NOTE: The reader should not conclude any linkage of data by respondents between the rows and columns. Respondents made comments on some, but not necessarily all, issues.

CONTAMINANT	LOOK FOR	FOUND	BASELINE	DETECT.	MAXIMUM LEVEL (PPM/PPB)	SHOW METHOD
Phosphorus	5	0	<0.05 ppm; <0.05 ppm; <0.05 ppm	0.05 ppm; 0.05 ppm; 0.05 ppm		1 said yes
Arsine	9	11	No <0.002 ppm; <0.002 ppm; <0.002 ppm; 50 ppb; <1 ppm; 4 ppm	1 ppb; 10 ppb; 5 ppb; 0.002 ppm; 0.002 ppm; 0.002 ppm; 20 ppb	7 ppm; 0.01 ppm; 0.01 ppm; 0.01 ppm; 100 ppb; 1 ppm; 10 ppb	4 said yes
Ammonia	7	9	<160 ppb; <0.05 ppm; <0.05 ppm; 3 ppm; no	0.1 ppm; 160 ppb; 0.05 ppm; 0.05 ppm; 0.05 ppm, 3 ppb; 1 ppm	2000 ppb; 0.1 ppm; 0.1 ppm; 0.1 ppm; 5 ppm	2 said yes
OTHERS:						
Metal	1	1			2 ppm	
Oxygenates (MTBE)	2	2		50 ppm	100 ppm	2 said yes
HF						
Iron Oxide						
Oxygen	1	1	<0.1 ppm	0.5 ppm	2 ppm; 1000 ppm	1 said yes
Lead	5	5	0.02-0.05 ppm; 0.022 ppm; 0.013 ppm; <20 ppb; 5 ppm	?; 5 ppb; <5 ppb; 1 ppb	0.05 ppm; 0.05 ppm; 0.05 ppm; 25 ppb; 50 ppb; 50-100 ppb	2 said yes
Carbon Monoxide	1	2	0-3 ppm ; 1500 ppm	1 ppm; 2 ppm	4000 ppm	
Vanadium	1	1	<20 ppb	5 ppb	25 ppb	1 said yes
Antimony						
Olefins	1	1	0.600 ppm	10 ppm	20,000 ppm	
C_{10+} in LPG						

4) (A) Check if you have had problems with any of these contaminants in the past and solved your problem.
(B) If so, would you be willing to share this information with the ethylene industry? (YES/NO) What would you be willing to share? (Briefly indicate this information on a separate page.)

NOTE: Three respondents answered to "(B)" generically. One respondent answered saying: Contaminants are removed in the plant by caustic/amine units (CO_2 and sulfur), arsine bed or cleaning the process water system (fluorides). Another respondent said: In all cases specs were written, communicated, enforced. Impurities were removed at the Vendor. The last respondent said they would be willing to share safety related information on mercury, radon, NOx.

# RESP.	CONTAMINANT	(A)	(B)	Indicate which feedstock contaminant was found, (i.e. Refinery gas, NGL's, Naphtha Gas oil, Other)
9	Methanol	8	6	Naphtha & NGL; NGLs; Ethane, propane, barged feed; NGLs; Propane - installed heaters; EP and naphtha; E-P; E-P, P/P mix; Ethane Feed/Mol Sieve Dryer
13	Carbon Dioxide	10	9	E-P mix; Worked with suppliers to reduce Spec's; Refinery gas; Ethane; FCC off gas; NGL; NGL; NGLs; Ethane; NGL's E-P; E-P, FCC gas; Found in NGLs; Higher strength caustic
7	Mercury	5	2	<1 ppb in P/Pmix; Naphtha/Algerian condensate; Other; Condensate/Heavy oils; E-P
4	Fluorides	3	3	Refinery propane/butane; Propane, P-P mix; Refinery propane; Butane - didn't use it
3	Chlorides	2	1	Other; Barged feed; Naphtha

4) (A) Check if you have had problems with any of these contaminants in the past and solved your problem.
(B) If so, would you be willing to share this information with the ethylene industry? (YES/NO) What would you be willing to share? (Briefly indicate this information on a separate page.)

NOTE: Three respondents answered to "(B)" generically. One respondent answered saying: Contaminants are removed in the plant by caustic/amine units (CO_2 and sulfur), arsine bed or cleaning the process water system (fluorides). Another respondent said: In all cases specs were written, communicated, enforced. Impurities were removed at the Vendor. The last respondent said they would be willing to share safety related information on mercury, radon, NOx.

(Continued)

# RESP.	CONTAMINANT	(A)	(B)	Indicate which feedstock contaminant was found (i.e. Refinery gas, NGL's, Naphtha Gas oil, Other)
13	Sulphur	10	8	Ethane feed, H_2S okay, CS_2, COS problem; Refinery gas; Installed treater; P/P mix; E-P/Naphtha; Gas Oils; Gas Oil; Found in all feeds; Naphtha, LPG; Ethane/propane; FCC off gas; P/P mix; Naphtha and Gas oil
3	Radon	1	2	NGLs; Propane; We have detected radiation that could be consistent w/Radon presence
4	NOx	3	3	Refinery gas and?; Refinery gas; Refinery gas
3	Sodium	3	1	Barged feeds; NGLs/Naphtha; Ethane
5	Water	4	2	Ethane, propane; Salt dome storage; Ethane feed/Mol Sieve Dryer; Any of the feeds; E/P mix; found in NGL's
	Phosphorus	1	1	None
7	Arsine	6	4	Propane, P/P mix; P/P mix; FCC gas; Refinery gas; E/P; Refinery gas; Refinery gas; Refinery propylene

4) (A) Check if you have had problems with any of these contaminants in the past and solved your problem.

(B) If so, would you be willing to share this information with the ethylene industry? (YES/NO) What would you be willing to share? (Briefly indicate this information on a separate page.)

NOTE: Three respondents answered to "(B)" generically. One respondent answered saying: Contaminants are removed in the plant by caustic/amine units (CO_2 and sulfur), arsine bed or cleaning the process water system (fluorides). Another respondent said: In all cases specs were written, communicated, enforced. Impurities were removed at the Vendor. The last respondent said they would be willing to share safety related information on mercury, radon, NOx.

(Continued)

# RESP.	CONTAMINANT	(A)	(B)	Indicate which feedstock contaminant was found (i.e. Refinery gas, NGLs, Naphtha Gas oil, Other)
5	Ammonia	4	2	Refinery Gas; FCC gas; Refinery gas; Refinery gas; Ethane and E-P
OTHERS:				
1	Other		1	Propane/Butane
2	MTBE	2	1	Barged feeds; naphtha
2	Vanadium	2	1	Barged feeds; naphtha
1	Oxygen	1	1	Refinery gas
1	Carbon Monoxide	1		E/P mix
1	Lead	1	1	Naphtha

1. What are your concerns about mercury as a feedstock contaminant? (Continued)

NOTE: The third column indicates the number of respondents which acknowledged the area of concern with a check mark. Some respondents made comments irrespective of column three.

# Respondents			**Have you experienced any adverse effects as a result of mercury?**
8	Environmental	8	No, Oily sludge contamination disposal by incinerator as hazardous waste; None to date
6	Waste disposal	6	None to date

2. What specifications or guidelines for mercury do you have for your feedstocks?

# Respondents		
15	LPG's	We have no specifications or guidelines other than avoiding feedstock from sources known to have significant quantities of mercury; No spec; No spec; No Hg; N/A; None established; <1 ppb; Do not check; 12 respondents said "none"
12	Condensate	5 ppb; Prohibit use of known Hg containing condensate; No Hg; N/A; N/A; <2 ppb; Do not check; 5 respondents said "none"
12	Liquid feeds (Naphtha Gas Oil, Etc.)	5 ppb; No spec - "determine & report on all naphtha's; No Hg; N/A; N/A; <2 ppb; Do not check; 5 respondents said "none"

2. What specifications or guidelines for mercury do you have for your feedstocks? (Continued)

# Respondents		
11	Supplemental feeds (Pyrolysis gasoline, Etc.)	No specs; No Hg; No Hg; N/A; No formal specification limits; Do not check 5 respondents said "none"
1	Refinery Gas	Requested max 1.0 ppb

3. What treatments (existing or planned) are utilized to remove mercury from your materials? What are the typical mercury concentrations before and after treatment?

NOTE: Two respondents said "none" to question 3 in general. There were 13 respondents to part a).

a) Treatment of Feedstocks Outside your plant:

		Before	After
Outside your plant:	N/A; 9 respondents said "none"		
Inside your plant:	Planned absorption; None - will install equipment within next 2 years; 6 respondents said "none"		unknown

NOTE: There were 14 respondents to part b).

b) Treatment of process streams other than feedstock? (Cracked gas, product streams, etc.) Treatment

Treatment	Location in Process	Before	After
Calgon Carbon	Cracked Gas	13 ppb	0 ppb
Acetylene converters	Front end	25 ppm	0 ppm
Absorption	C_3 hydrogenation beds - (sacrificial)		<1 ppb

Treatment	Location in Process	Before	After
Sulfur impregnated, activated carbon bed	Downstream from mol sieve dehydration	0.38 ppb	0.006 ppb
Cracked gas	Downstream of driers	20-50 ugm/ m^3 (design)	1 ppb (design)

Seven respondents said "none";
one respondent said "N/A"

4. What precautions have been implemented to make your process resistant to mercury related problems? (Spare equipment, material of construction, equipment re-design, changed processing scheme, changed feedstock, etc.)

NOTE: There were 16 respondents to this question.

1. Installed connection to allow installation of mercury trap.
2. Existing process scheme removes mercury contamination before reaching aluminum equipment.
3. None
4. All liquid feedstocks analyzed for mercury before use. Cold boxes checked for mercury during shutdowns.
5. None
6. We are still working on ppb level detection methods of the source (feed?, caustic?, etc.?)
7. None to date - possible to work with Coretest Labs.
8. None.
9. None.
10. No precautions other than sulfur impregnated activated carbon bed for mercury removal.
11. Evaluation in progress for absorption from feedstock
12. Vigilance to keep it out of feed
13. None.
14. We have not made any changes to make the process more resistant to mercury other than installing mercury removal facilities. We avoid feeds that we believe may contain significant mercury.
15. None.
16. Install Calgon Carbon trap on cracked gas.

5. Do you utilize special operating procedures to address your mercury concerns? If so, what are they and what has been your success with them?

NOTE: There were 16 respondents to this question.

1. No.
2. No.
3. No.
4. No.
5. No.

6. N/A at this time.
7. None really - during shutdowns we monitor before opening lines and equipments.
8. None.
9. None.
10. Condition of activated carbon bed is tested quarterly.
11. No.
12. If spot samples taken at the cold box inlet showed over 1 ppm, any new feed suspected of containing mercury would be backed out.
13. No.
14. We have established guidelines on leak testing in the event of planned or unplanned shutdowns. We also have guidelines on flushing equipment with alcohol.
15. No.
16. Check low point drains. Special startup and shutdown of cold box.

6. Do you utilize special decontamination procedures to address your mercury concerns? If so, what are they and what has been your success with them?

NOTE: There were 16 respondents to this question.

1. No.
2. No.
3. No.
4. No - decontamination has not been necessary.
5. No.
6. N/A at this time.
7. None - possible solution with core test.
8. None.
9. None.
10. No procedures are in effect at this time.
11. No.
12. No, haven't found any accumulation of mercury.
13. No.
14. We look for mercury when analyzing gas used to purge and warm-up aluminum core exchangers. We have measured low concentrations. We have not found mercury deposits or mercury caused equipment damage.
15. No.
16. No.

7. What analytical methods do you utilize to detect the presence of mercury? What special sampling or sample handling procedures are utilized for samples to be analyzed for mercury? What are the limits of delectability using these methods?

NOTE: There were 17 respondents to this question.

1. None.
2. None.
3. None.

4. Liquids: Jerome 511 gold foil analyzer. Detection limit 0.3 mg/ 10-ml sample. Gases: Jerome 431, specially modified for process use. Detection limit approximately 0.1 ppb. Online analysis of cold box feed stream has been demonstrated using portable instrument.
5. Jerome gold foil for gas analysis.
6. Jerome instruments. Three filters are being used to remove sulfur to minimize sulfur interference with the mercury test. Min. detect ability = 5 ppb/wt. Liquid feed streams are not filtered, only gas streams.
7. See 4(#6) above.
8. Vapor - Jerome analyzer, liquid - extraction followed by ICP-AES.
9. We do not analyze for mercury at this time.
10. Outside lab analysis on feedstock only so far.
11. All analyses are performed by analytical services contractors (two groups are employed for these services. Minimum detect ability 0.002 ppb).
12. Jerome mercury analyzer, trap tubes and mass spec.
13. Jerome analyzer on cold box feed samples and new gas feeds. 0.5 ppb on Jerome. Dosimeters used to collect gas samples. See (12) below for liquid sample method.
14. Checked dryer regen gas once - hydrogen purification unit - gold filament - none detected.
15. We use a Jerome 411 for analyzing vapor samples and a Jerome 511 for liquid samples. We are concerned that those don't detect organic mercury in the liquid phase.
16. None.
17. Jerome analyzer with dosimeter. Liquid samples to outside laboratory.

8. Are you involved in developing new technology to detect mercury? Or to remove it from your process streams? Or to remove it from contaminated equipment?

NOTE: There were 16 respondents to this question.

1. No.
2. No.
3. No.
4. Evaluating available technology from other sources, including some under development.
5. No.
6. N/A at this time.
7. For removal - we favor removing mercury from incoming feed.
8. Not at this plant, although we have developed technology for mercury removal.
9. No.
10. No action items are in progress at this time.
11. Not in the Chemical company.
12. Yes, we worked w/ACS laboratory in Houston to develop in ICP-MS method for detecting mercury in liquids down to 2 ppb.
13. No.
14. No.
15. No.
16. No.

Mercury Removal from Cracked Gas and Liquid Streams

Steve Coleman and Jason Ries, Equistar Chemicals, LP
Channelview, Texas, USA
and
Don Campbell and Keith R. Clark, UOP LLC
Houston, Texas, U.S.A.

Prepared for Presentation at the AIChE 2004 Ethylene Producer's Conference

Abstract:

To maximize profits in today's business climate, ethylene producers are cracking heavier, more cost-effective "opportunity" feedstocks. Mercury is often present in these liquid feed streams, causing significant quantities to be carried forward into the cracked gas and liquids. Mercury is known to deactivate metal-based hydrogenation catalysts and damage aluminum heat exchangers and cold boxes to the point of catastrophic failure. Removal of this contaminant upstream will improve the safety of unit operation and reduce the potential for personnel exposure during routine maintenance or shut downs. HgSIV™ adsorbents are regenerable and remove both mercury and water. Since the sorption sites for mercury removal are separate from and additive to the dehydration sites, mercury removal is accomplished by replacing a portion of the dehydration grade molecular sieve with HgSIV adsorbents. This paper will discuss a case study where mercury is being successfully removed from cracked gas at an ethylene plant.

BACKGROUND

Mercury, present in nature, is found in trace concentrations in various ethylene feedstocks, in particular natural gas field condensates. Removal of this mercury is desirable to protect metal-based hydrogenation catalysts from deactivation. In addition, aluminum equipment and piping is subject to attack by mercury via a variety of mechanisms including liquid metal embrittlement, amalgamation and amalgam corrosion (Crippen, 1997). Such attacks have resulted in equipment failures in ethylene plants (English, 1989) and similar processing facilities (Wilhelm, 1990; Kinney, 1990). Moreover, the presence of mercury in process equipment being cleared and cleaned during plant shutdowns requires steps to prevent personnel exposure. For these reasons, ethylene plant operators prefer to exclude mercury from their facilities or reduce it to a level that can not be detected with available analytical capabilities. Currently, this means reducing the mercury to less than 10 ng/Nm^3, which corresponds to about 10 ppt by volume.

Copyright © 2004 UOP LLC and Equistar Chemicals, LP All rights reserved.
April 26, 2004
Unpublished
AIChE shall not be responsible for statements or opinions contained in papers or printed in its publications

UOP has been testing for the presence of mercury for decades, but historically, low-level mercury analysis had to be done in a laboratory setting. Initially, this work was done in UOP's labs in Des Plaines, Illinois. However, for over a decade, UOP has been measuring mercury in gas streams in the field, both to validate the effectiveness of its HgSIV adsorbent and to determine if mercury is present in new facilities. UOP started with natural gas streams and now has experience working with cracked gas streams from ethylene plants. Over time, UOP has upgraded its techniques and analytical equipment allowing even greater precision and accuracy in the field. UOP has learned much from practical experience and has incorporated these discoveries into perfecting field mercury analysis. Due to the complex hydrocarbon matrix, this latest endeavor in cracked gas measurement presented unique challenges.

FIELD MEASUREMENT

Obtaining precise and accurate measurements requires the utmost care, and a precision sampling system. In adsorbent bed outlets, levels of mercury are extremely low and any introduction of error can dramatically skew the results. Obtaining a representative sample from a process line requires a special sample tap (stainless steel quill), as shown in Figure 1, that has been thoroughly purged. This quill is imperative because pipe wall effects and condensed liquids in sample points will distort the mercury readings for a significant period of time.

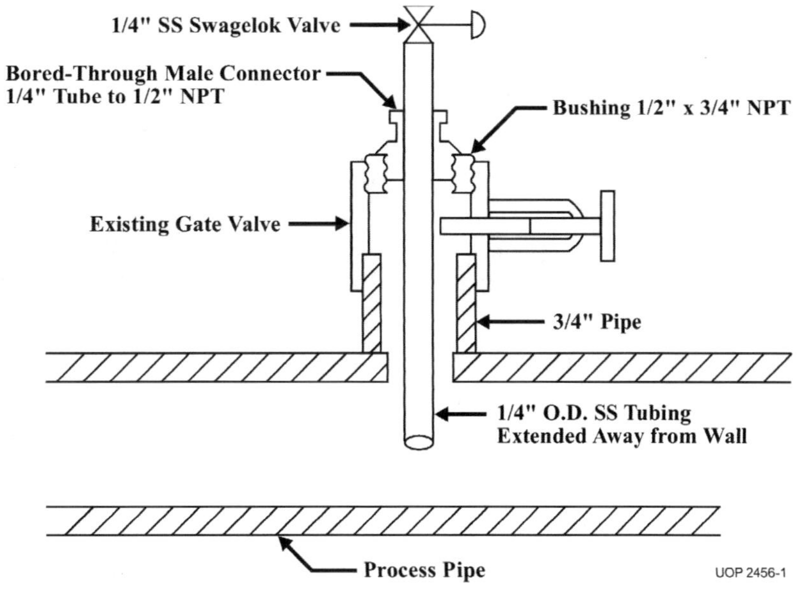

Figure 1: *Sample Point Configuration for Hg Analysis*

Field measurements are performed by first concentrating the mercury in the sample stream on a gold surface trap, a technique utilized by many available mercury analyzers. The traps are then thermally desorbed, and the mercury pulse is swept into the analyzer where a detector quantifies the amount of mercury.

The best technology to analyze low levels of mercury in hydrocarbons is cold vapor atomic fluorescence spectroscopy (CVAFS), but the CVAFS's size has historically limited its use to the laboratory. Complementing its long experience measuring mercury in the field, UOP added CVAFS technology to its testing capabilities about four years ago with the implementation of a new analyzer. As with other techniques, the sampling protocol is critical and it has been optimized through practical experience in hydrocarbon gas streams. At the plant site, UOP reliably measures mercury concentrations as low as 10 ng/Nm^3. Each sample can be collected and analyzed within an hour, compared to days or weeks with an off-site facility. Before the analytical team departs, the level of mercury in a treated or untreated stream has been quantified. This information is used to determine whether streams need to be treated for mercury or confirm that an existing mercury removal system is functioning as designed.

HgSIV ADSORBENTS

HgSIV adsorbents were created for effective mercury removal in existing molecular sieve adsorption units. The HgSIV products are molecular sieves that contain silver on the outside surface of the molecular sieve pellet or bead. Mercury from the process fluid (either gas or liquid) amalgamates with the silver, yielding a dry, mercury-free stream. Adding a layer of HgSIV 3 pellets to an existing cracked gas dryer results in the simultaneous removal of the design water and mercury load, without requiring an additional downstream vessel. Mercury and water are both regenerated from the HgSIV adsorbents using conventional gas dryer techniques.

Physically, HgSIV adsorbents have a similar appearance and integrity to standard molecular sieves and are available in a beaded or pelletized form. These HgSIV adsorbents are loaded into an adsorption vessel in the same manner. There is no need for special care, such as the use of nitrogen blanketing during the installation. For unloading, only the same precautions need to be taken as with conventional molecular sieves.

The disposal requirements are also the same as for dehydration grade molecular sieves, provided the material is properly regenerated prior to unloading. Analysis of fresh and used samples of HgSIV adsorbents have shown that they pass the EPA Toxicity Characteristic Leaching Procedure test (TCLP), and are thus classified as non-hazardous waste for disposal purposes.

COMMERCIAL EXPERIENCE

UOP first installed HgSIV adsorbents in the natural gas market two decades ago. Currently there are over thirty units operating successfully around the globe. Now UOP is applying this expertise to the ethylene market and has four units in operation. HgSIV 3 molecular sieve is successfully used to remove mercury from cracked gas streams, thereby preventing mercury from entering into the system downstream of the cracked gas dryers. Protection of the downstream cold fractionation section, aluminum heat exchangers and hydrogenation catalysts is thereby readily accomplished, without significantly altering the configuration or operating procedure of the dryers.

Mercury removal is accomplished by a "drop-in" substitution of adsorbent products, without affecting the dehydration performance of a cracked gas dryer, because the sorption sites for mercury removal are separate from and additive to the dehydration sites. The loading diagram for a typical cracked gas dryer is shown in Figure 2. This is a common configuration consisting of a compound main adsorbent bed with a segregated guard bed. A moisture probe is located in the plenum space between the main bed and guard section, and the bed is operated until breakthrough is detected by the moisture analyzer. The guard section sees minimal water and is only used in emergencies.

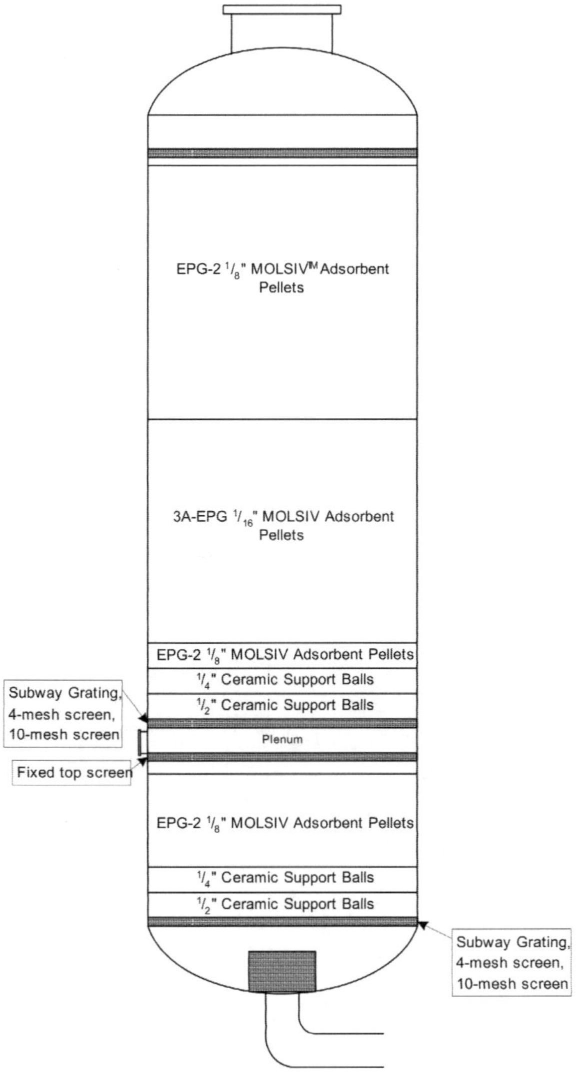

Figure 2: *Typical Cracked Gas Dryer Configuration*

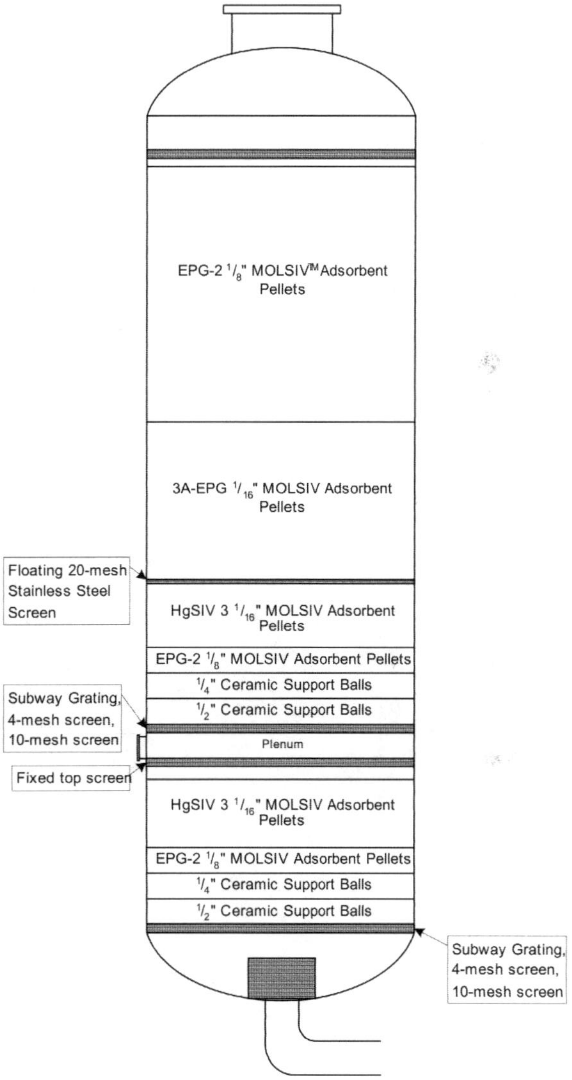

Figure 3: *Cracked Gas Dryer Configuration with HgSIV 3 Pellets*

Figure 3 illustrates a loading configuration that incorporates the HgSIV 3 product. The dryer is operated normally, and the mercury effluent levels achieved are 10 ng/m^3 or lower. Although HgSIV 3 pellets have the same dehydration performance as 3A-EPG MOLSIV™

adsorbent, it is positioned in the bottom portion of the bed to minimize the hydrothermal aging that occurs during regeneration, allowing for it to be reclaimed and reused. The use of a floating-screen in the main bed facilitates the reuse of the mercury adsorbent on subsequent recharges of the 3A-EPG molecular sieve.

CASE STUDY: EQUISTAR CHANNELVIEW, TX

Equistar processes a variety of feedstocks through its olefins plants in Channelview, Texas. Some of these feeds, including gas field condensates from Algeria, contain trace levels of mercury, some as high as 50 ppb. Recognizing the potential hazards mercury can present to aluminum exchangers, Equistar replaced its brazed aluminum core exchangers in 1989 with a mercury tolerant design and has developed procedures for clearing equipment where mercury might be present. This includes extensive monitoring of equipment for mercury during plant turnarounds and outages.

In March 2003, Equistar's Olefins 2 plant was shutdown for a scheduled turnaround. The plant had been on-line since November 1995. The plant was cleared in a typical manner to remove residual hydrocarbons in preparation for maintenance. During clearing activities, a significant quantity of mercury, both liquid and vapor phase, was found throughout the cold fractionation area of the plant (Figure 4).

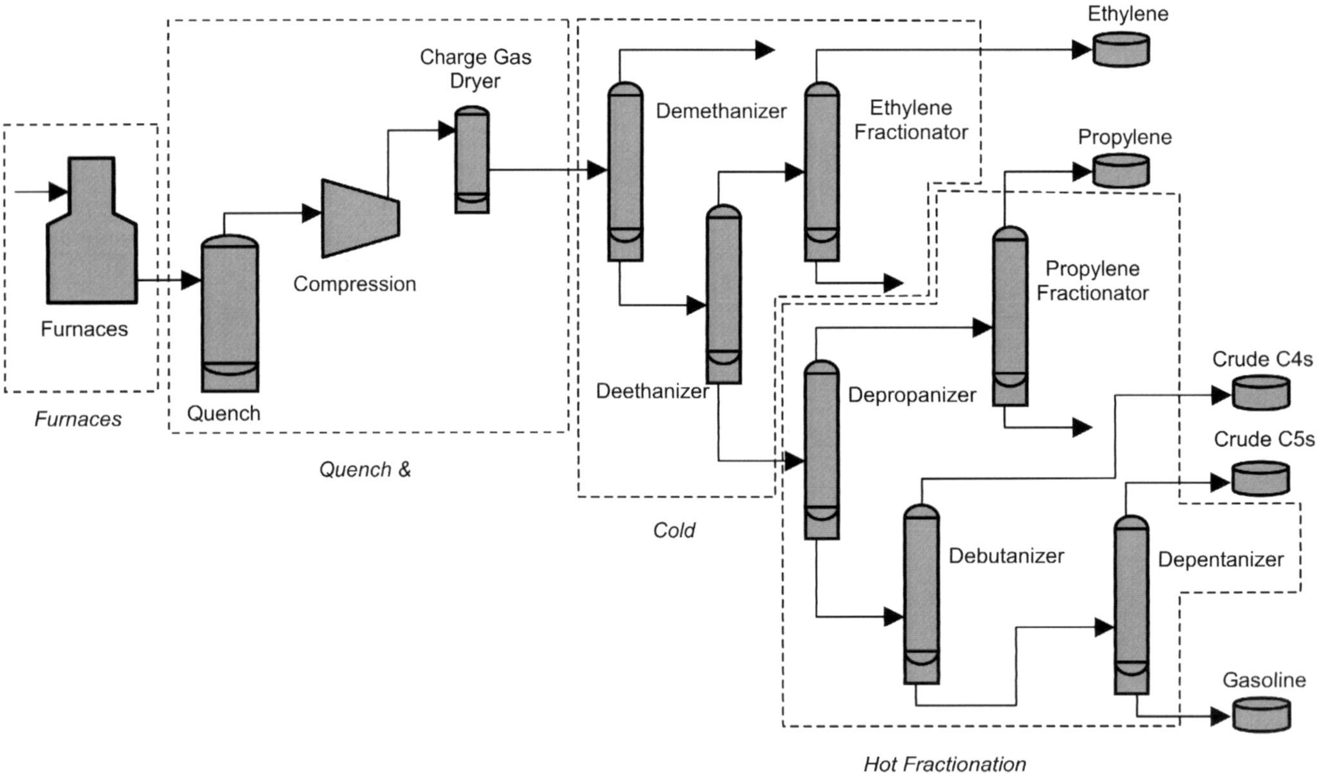

Figure 4: *Typical Olefins Plant Flow Diagram*

The presence of mercury in the cold fractionation area of the plant was expected based on processing feedstocks contaminated with mercury and previous turnaround experiences in 1989 and 1995. The main difference between the 2003 turnaround and historical experience was the location and quantity of mercury found. In previous turnarounds, the majority of mercury had been found in small concentrations upstream of the brazed aluminum core exchangers. During the 2003 turnaround, mercury was found downstream of the core exchangers and throughout the cold fractionation area of the plant.

Mercury concentrations were high enough in many areas to require additional remediation efforts and personal protection equipment to limit personnel exposure. The American Conference of Governmental Industrial Hygienists has assigned mercury vapor a threshold limit value (TLV) of 0.025 mg/m^3 as a time weighted average for a normal 8-hour workday and a 40-hour workweek (ACGIH). The mercury concentrations in the cold fractionation vessels ranged from less than 0.010 mg/m^3 to 0.700 mg/m^3. Coping with this unexpectedly high level of mercury delayed a number of turnaround activities.

Based on the high levels of mercury encountered in the cold fraction section of the ethylene plant and the desire to prevent mercury from contacting aluminum exchangers and deactivating selective hydrogenation catalyst systems, Equistar initiated an investigation of options for excluding mercury from the cold fractionation area of the plant. Because of the high mercury removal efficiency (>99.9%) and the speed and ease of implementation, Equistar decided to install UOP's HgSIV adsorbent in the Olefins 2 plant's process gas dryers during the March 2003 turnaround. The HgSIV pellets were a "drop in" solution requiring no additional vessels or equipment. A portion of the dehydration grade molecular sieve, which was coincidentally being replaced during the turnaround, was substituted for HgSIV material.

The HgSIV adsorbent was installed in the guard bed and the lower portion of the main bed, as shown in Figure 3. Loading the sieve in the outlet (bottom) portion of the dryer vessel will extend its life by minimizing degradation due to hydrothermal aging. While the conventional 3A-EPG pellets loaded above the HgSIV material will require replacement in 3 – 4 years, the HgSIV adsorbent could be left in place for an additional 3 – 4 years or until the next scheduled turnaround.

Several months after the dryers containing HgSIV adsorbent were put in service, UOP Field Technical service was on-site to conduct field mercury analyses of the cracked gas dryer inlet and outlet using the CVAFS instrument. Several trips were made to the Channelview facility, once while running Algerian condensate and once while not. The mercury concentration of the inlet to the cracked gas dryers ranged from 1,500 ng/nm^3 to 500 ng/nm^3, with significant quantities of mercury detected in the system even when mercury-free feed was cracked. Measurable amounts of mercury remain in the plant, an indication that mercury adheres to equipment and is not readily purged from the system.

Testing was also conducted on the regeneration gas inlet; Figure 5 illustrates evidence of mercury detected in the regeneration system approaching 100 ng/nm^3. This complicates matters by reintroducing mercury into the adsorber vessel on each regeneration cycle. Mercury contained in the regeneration gas, accumulates in the ceramic support material and grating, and then it is bled slowly back into the dryer effluent. This residual mercury presented analytical measurement challenges which were overcome with modified sampling and analytical techniques.

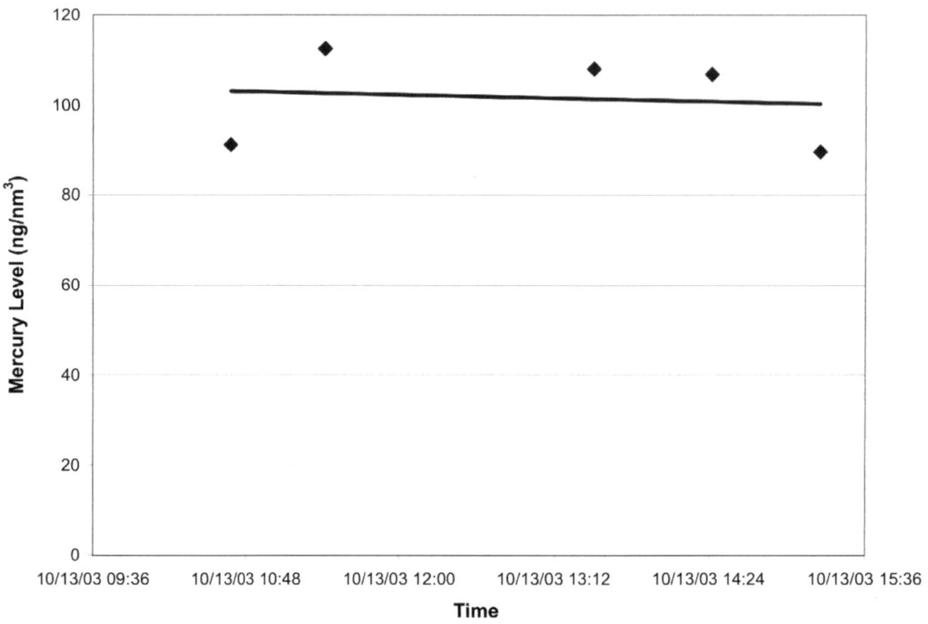

Figure 5: *Mercury Concentration of Bed A Regeneration Inlet*

To illustrate the complexity of the measurements and the utmost importance of achieving an equilibrium condition with respect to the mercury levels in the process, Figure 6 shows the effluent measurements of one cracked gas dryer bed through two adsorption cycles. Initial mercury readings were twenty times what we typically expect to see at the bed outlet due to sampling point contamination. Once the system equilibrated, the mercury levels were at or below the 10 ng/nm^3 detection limit. To facilitate future mercury determinations, a quill-type sample tap (Figure 1) was modified on the common outlet of the cracked gas dryers. Analysis of the cracked gas confirmed mercury levels at or below 10 ng/nm^3. In addition, there was no reported moisture breakthrough or elevated pressure drop. The results in Figure 7 verify that the HgSIV 3 adsorbent removes mercury from cracked gas streams down to very low levels, without loss of moisture capacity or pressure drop penalties.

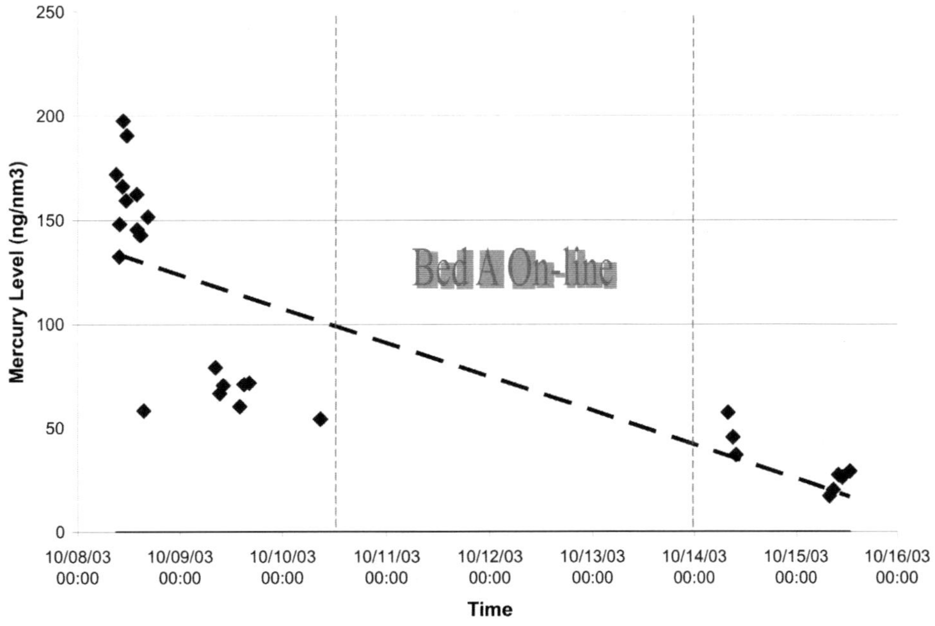

Figure 6: *Mercury Concentration of Bed B Outlet*

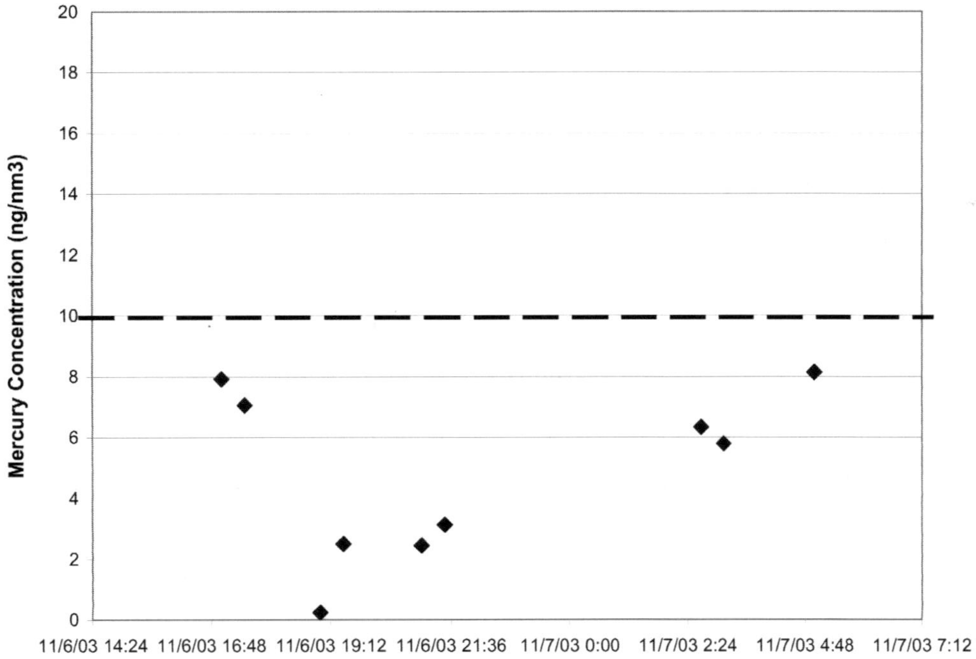

Figure 7: *Mercury Concentration of Cracked Gas Dryer Outlet*

In summary, the unit has currently been in service for one year. Equistar confirms that the dryers are performing well. Mercury continues to be removed, and cycle times and pressure drop are tracking with the design projections.

REFERENCES

American Conference of Governmental Industrial Hygienists (ACGIH), "1994-1995 Threshold Limit Values for Chemical Substances and Physical Agents and Biological Exposure Indices", Cincinnati, Ohio.

Crippen, K., and S. Chao, "Mercury in Natural Gas and Current Measurement Technology," 1997 Gas Quality and Energy Measurement Symposium, February 3-5, Orlando, Florida, pp. 1-16.

English, Jerome J., "A Mercury Induced Aluminum Alloy Piping Failure in an Ethylene Manufacturing Plant", Proceedings of the First Ethylene Producers Conference, presented on April 4, 1989.

Kinney, G. T., "Skikda LNG Plant Solving Troubles," *Oil & Gas Journal*, September 15, 1975.

Wilhelm, Dr. S. Mark, "The Effect of Elemental Mercury on Engineering Materials Used in Ammonia Plants", prepared for presentation at the 1990 Summer National AIChE Meeting in San Diego, CA, presented on August 1, 1990.

Adsorbent Reactivity: An Important Factor in Contaminant Removal

Vladislav Kanazirev and David L. Smith
UOP, LLC

Prepared for Presentation at the AIChE 2004 Ethylene Producer's Conference

Abstract:

Over several decades, adsorption processes have grown to be the most practical and cost effective means of controlling trace contaminants in various streams. Unlike catalysts, adsorbents by definition should not exhibit any appreciable catalytic activity towards the stream to be purified. Most adsorbents fulfill this requirement when used in non-reactive streams, but potential problems arise in the case of reactive feeds. Adsorbents acting as catalysts may sometimes promote highly exothermic reactions such as olefin oligomerization and polymerization, which are potential safety concerns. They can catalyze numerous side reactions involving both the main stream and original contaminants. This can have a negative impact on the adsorbent's performance and longevity.

Selected theoretical and practical aspects of an adsorbent's reactivity will be discussed, such as the stability of reactive intermediates and reaction pathways involving both the main process stream as well as the contaminants. This paper will also introduce a series of tests of increasing sensitivity to evaluate the reactivity of adsorbents. Knowing an adsorbent's relative reactivity versus other adsorbents helps the user understand any potential impact on his process and offers the opportunity to balance this adsorbent characteristic versus other performance criteria or operating conditions.

INTRODUCTION

For nearly half a century, molecular sieves, activated aluminas and other adsorbents have successfully served the chemical industry by removing trace contaminants from hydrocarbon streams. First applied to the sweetening of light hydrocarbons such as propane and butane (Clark, 1962), molecular sieves quickly proved valuable for the simultaneous dehydration and desulfurization of olefin-bearing LPG streams (Manchanda and Gilmour, 1973). Later, the use of molecular sieves and other porous solids expanded to the more demanding application field of trace contaminant removal from olefins. Today, advanced adsorbents and adsorption processes have grown to be the most practical, cost-effective means to control contaminants in olefin feedstocks and products (Reid and McPhaul, 1996).

The new generation of site-specific, polymerization catalysts requires nearly contaminant-free monomers. Even a small decline in the monomer purity adversely affects the polymer product quality and the process economics (Detrait and Grootjans, 1994). Thus, adsorbent producers are challenged once again to deliver solutions that are attuned to the newest catalyst developments to satisfy the complex needs of this market.

Copyright © 2004 UOP LLC All rights reserved.
April 26, 2004
Unpublished
AIChE shall not be responsible for statements or opinions contained in papers or printed in its publications.

The ethylene industry periodically reviews state-of-the-art contaminant removal technologies and issues reports on their findings (Shah and Coleman, 1996; Grose and Bowen, 1996; Coleman, 1998). While selected aspects of these technologies, such as arsine and arsenic compound removal methods, have been extensively discussed (Reid, 1997), the catalytic reactivity of adsorbents has only been briefly touched upon (Reid, 1999). There has not been a separate, detailed study on this topic to date.

The aim of this paper is to focus on adsorbent reactivity as a factor that must be considered along with other performance criteria when choosing adsorbents for contaminant removal and the test methods to measure that reactivity.

Can Adsorbents Act as Catalysts?

Generally, an adsorbent can act as a catalyst in certain situations. Similar to most catalysts, adsorbents are porous solids, which exhibit high surface area and distinct surface functionality. These features lead to high surface energy and active sites that play a crucial role in both adsorption and catalysis. Specialized, typically proprietary, technologies can turn a porous solid into either a good catalyst or a valuable adsorbent by fine tuning selected properties of the base material.

Unlike catalysts, adsorbents by definition should not have any appreciable catalytic activity. While being efficient in handling trace contaminants, adsorbents must not cause unwanted chemical changes to the main stream. This requirement is not difficult to satisfy when adsorbents are employed in non-reactive streams, such as paraffinic hydrocarbons or inert gases. However, several potential problems arise for reactive streams where adsorbent catalyzed reactions may lead to serious complications. Even worse, the combination of reactivity and elevated temperature may be harmful if reactive fluids are used for adsorbent regeneration or cooling.

Reactivity Background

The laws of thermodynamics and kinetics govern the transformations of all chemical substances. Catalysts help to speed up chemical reactions. The term reactivity applies to both chemical substances and catalysts to designate their tendency of readily participating in reactions that cause chemical change. By definition, a reactive substance is readily susceptible to chemical change. Thus, even “weak” catalysts or adsorbents can transform a reactive substance. Likewise, a reactive catalyst can crack “tough” molecules that would otherwise be stable in the presence of a less reactive catalyst.

Reactivity of Olefin Feedstock and Products

The thermal stability of olefins was the main concern in the early era of the ethylene industry (Reid, 1999; Briton et al., 1986 and the references herein). Table 1 summarizes data extracted from the NIST database, which show that acetylenes, ethylene and propylene are “endothermic” compounds that release significant energy upon decomposition.

Table 1
Standard Heat of Formation of Selected Hydrocarbons

Compound	Standard Heat of Formation, kJ/mol
Acetylene	226.73
Methylacetylene	185.4
Ethylene	52.47
Propylene	20.41
1- Butylene	- 0.63
2-Butylene, cis	- 7.7
2-Butylene, trans	- 10.8
Isobutylene	-17.9

Complex events of a chemical and physical nature can supply the heat / energy needed for the decomposition reactions. These events can lead to catastrophic consequences as described in the literature (Reid, 1999; Briton et al., 1986), especially in the case of the energy-rich "endothermic" compounds listed in the first part of Table 1.

In the absence of catalysts, thermal runaways in heated equipment are the principal cause of overheating that may trigger gas-phase oligomerization reactions, which generate more heat and eventually lead to uncontrolled decomposition phenomena. The initiation of these thermal runaways is non-catalytic in nature. The presence of oxygen and free radicals in the system greatly enhances the reactivity of the system. The reactivity of the olefin feed certainly plays a role, but the particularities of the initiating event are usually more important. Finally, it seems that the ethylene industry has grown to understand the chemical and engineering aspects of these free radical initiated thermal runaway reactions, which has helped to improve the safety of olefin plant operation.

In a manner different to the free radical initiated reaction pathways, other intermediates play a decisive role in the catalytic conversion of olefins. Carbenium ions are widely accepted as the principal reaction intermediates in acid-catalyzed reactions. These ions are easily formed by the addition of a proton to an olefin:

$$C_nH_{2n} + H^+ = C_nH_{2n+1}^+ \qquad [1]$$

Table 2 lists the heat of formation of different types of carbenium ions (Lossing and Semeluk, 1970). Inferred by the heat of formation is the relative stability and hence the reactivity of these ions (Pines, 1981).

Table 2
Heat of Formation of Selected Carbenium Ions

Compound	Heat of formation (298°K) of carbenium ions, kJ/mol
Methyl	1093
Ethyl	917
Propyl	871
Secondary propyl	804
Primary butyl	842
Primary isobutyl	833
Secondary butyl	766
Tertiary butyl	699

The differences in Table 2 imply the following sequence of reactivity of basic olefins in acid-catalyzed reactions:

Isobutylene >> Butylene > Propylene > Ethylene [2]

On the other hand, the ease of the most common reactions of hydrocarbons, which are catalyzed by solid acids, decreases (Pines, 1981; Manchanda and Gilmour, 1973) in the sequence:

Double Bond Shift > Skeletal Isomerization > > Oligomerization > Cracking [3]

While the double bond shift is a relatively easy reaction that even "weak" catalytic sites can facilitate, strong acid sites and elevated temperatures are usually needed to carry out more demanding reactions such as cracking and oligomerization.

Catalytic Effects of Purification Beds

The ethylene industry uses purification beds at numerous stages in the process. In addition to the traditional application of molecular sieves, activated aluminas for drying, other adsorbents and scavengers are used to remove contaminants ranging from arsine and phosphine to organic sulfur compounds and oxygenates. In many instances, guard beds based on metal or metal oxide scavengers are used to protect expensive selective hydrogenation catalysts (Voight et al., 1996; McClung and Novalny, 2002). Many of these adsorbents can cause unwanted catalytic activity that can adversely affect the process in a number of different ways.

A. Catalytic Conversion of the Main Stream

Olefin oligomerization and double bond shift are two examples of this group's unwanted catalytic reactions.

Oligomerization

The catalytic oligomerization of olefins can proceed if the adsorbent possesses a small concentration of strong acid sites. At low temperatures, this reaction produces a series of oligomers.

$$C_nH_{2n} \rightarrow C_{2n}H_{4n} \rightarrow C_{3n}H_{6n} \rightarrow C_{4n}H_{8n} \rightarrow C_{5n}H_{10n} \qquad [4]$$

Since the oligomerization reaction is highly exothermic, it can cause temperature excursions and potential safety concerns. In less severe cases, oligomers known as "green oil" are produced. "Green oil" has negative consequences downstream of the adsorbent bed. Heavy oligomers and polymers can deposit on the adsorbent as coke precursors, which ultimately decrease adsorbent efficiency and shorten service life, thus resulting in increased costs for the user.

The chemistry of the acid-catalyzed oligomerization reaction, practiced as an industrial process (Pines, 1981; Sanati et al., 1999), is well understood. Moreover, the first observation of butylene oligomerization over the natural molecular sieve *chabazite* occurred about 60 years ago. The fact that synthetic molecular sieves can promote, to a certain extent, olefin oligomerization has been recognized (Norton, 1964) since the first applications of molecular sieves as adsorbents. The order of catalytic activities for propylene polymerization,

$$10X > 13X > 5A > 4A > 3A$$

and for olefin reactivity in the same reaction,

$$(CH_3)_2C{=}CH_3 > CH_3\,HC{=}CH_2 > CH_2{=}CH_2$$

was published by Charles Norton in 1964.

Double Bond Shift

In contrast to oligomerization, the double bond shift olefin reaction should not affect adsorbent longevity or process safety.

$$CH_3\,CH_2\,HC\,{=}CH_3 \rightarrow CH_3\,HC{=}CH\,CH_3 \qquad [5]$$

However, this reaction can have a negative impact on product quality and economics in polymerization processes where an alpha olefin (butylene-1; hexene-1 or octene-1) is used as a co-monomer. The double bond shift is, in this case, an unwanted reaction that can potentially occur when adsorbents remove oxygenate contaminants from the recycled co-monomer stream. Remarkably, both acid and base catalysts can facilitate this unwanted reaction.

B. Catalytic Reaction between the Carrier Stream and the Contaminants

Catalytic reactions between the carrier stream and the contaminants rarely pose any safety concerns, but potentially decrease the efficiency and increase the cost of the purification process. Formation of mercaptans from olefins and H_2S is an example of this type of reaction.

$$CH_3\,HC{=}CH_2 + H_2S = CH_3\,HCSH\,CH_3 \qquad [6]$$

An example where there is a concern with this reaction is that the secondary contaminant produced, the mercaptan, is more difficult to remove by adsorption than the original contaminant, H_2S.

C. Catalytic Transformation of Contaminants

Sometimes, the catalytic action of an adsorbent or a metal oxide-based scavenger may cause reactions with the contaminants, which yield yet other contaminants. For example, methyl mercaptan can convert to dimethyl sulfide and hydrogen sulfide:

$$2CH_3SH \rightarrow CH_3\,S\,CH_3 + H_2S \qquad [7]$$

Since dimethyl sulfide is less polar than the mercaptan, its removal by adsorption is more difficult. Similarly, in mixed olefin streams alkyl mercaptans can convert into disulfides.

Occasionally, methanol is present along with H_2S and other contaminants. Known chemistry predicts that methanol would form mercaptans with adsorbents acting as catalysts at slightly elevated temperatures, high contaminant concentration and long contact times. This type of conversion can be easily observed during the adsorbent regeneration where desorbing fronts of different contaminants interact at elevated temperatures.

Methanol can also catalytically produce a "difficult" secondary contaminant such as dimethyl ether. In addition, another contaminant, water, is also released into the olefin stream upon methanol reaction.

$$2CH_3OH \rightarrow CH_3\,O\,CH_3 + H_2O \qquad [7a]$$

The well-known catalytic conversion of carbonyl sulfide into hydrogen sulfide is another example of contaminant reaction:

$$COS + H_2O = H_2S + CO_2 \qquad [8]$$

It is believed that basic hydroxide groups readily available on some adsorbents catalyze reaction [8]. Remarkably, the COS conversion is one of the rare examples of converting a "difficult" contaminant to an "easier" one. The reverse of reaction [8] happens often in the natural gas industry when gases containing both H_2S and CO_2 are treated with molecular sieve or activated alumina adsorbents.

D. Adsorbent as a Source of Contaminants

Most adsorbents do not produce contaminants while in service. However, changes in adsorbent properties and deactivation often occur due to formation of heavy residue and coke precursors on adsorbents exhibiting residual catalytic activity. The adsorbent deactivation often happens while in service with reactive feeds. This phenomenon has negative consequences for process efficiency.

Metal oxide or carbonate-based scavenger adsorbents acting as chemisorbents can produce some contaminants while in service. A metal oxide, for example, would remove trace hydrogen or CO, but release one mole of H_2O or CO_2 for each mole of contaminant scavenged. A scavenger based on a basic metal carbonate would release two molecules of contaminant for each original contaminant molecule that has been scavenged as illustrated in equation [9].

$$Me(OH)_2.MeCO_3 + 2H_2S = 2MeS + 3H_2O + CO_2 \qquad [9]$$

The interaction with some contaminants can turn a non-reactive adsorbent to a reactive one. Thus, while generic activated alumina is a good scavenger for HF and HCl, these impurities can enhance its catalytic properties and convert it to an acid catalyst, resulting in green oil formation and adsorbent deactivation.

Common Adsorbents – Origin of Their Catalytic Activity

Molecular sieves, activated aluminas and silica gels are three common classes of porous solids currently used for adsorption. All three groups of adsorbents potentially have, inherent in their structures, strong acid sites that cannot be completely neutralized by simple methods. One can further differentiate Broensted and Lewis acid sites in these porous solids as represented by protonized hydroxyl groups and electron acceptors or metal ions, respectively. The following examples should illustrate that small impurities may embed strong catalytic properties in porous solids, which originally do not have catalytic activity:

- According to R.L Burwell and co-workers (West, et.al, 1973), additional processing to remove Al from the raw material for silica gel production decreases by 30 times the gels' catalytic activity for hexene-1 isomerization. Moreover, doping the gel with Al to a ratio of Al/Si of 0.0002 causes a more than 100-fold increase in the catalytic activity in regard to the same reaction.

- HZSM-5 type zeolites exhibit strong acidic properties at Al levels as low as 20 ppm. The protonated tetrahedral aluminum atoms in the SiO_2 network have been identified as the catalytic active sites in aluminosilicate catalysts (Haag et al., 1984).

Numerous reports also discuss the source of the catalytic activity in activated alumina where the crystalline phase and activation conditions are important along with the presence of certain impurities and additives.

UOP Experience in Solving the Reactivity Problem

As the world's pioneer in molecular sieve adsorption technology (Sherman, 1999), UOP recognized many years ago that adsorbent reactivity is an important factor in contaminant removal technology. Modifications in the manufacturing of molecular sieve adsorbents, formulations and methods for attenuation of adsorbent reactivity at unit start up and during normal operation have led to industrial successes in very difficult applications. More recently, the concept of low reactivity adsorbents was successfully applied to the alumina and alumina-zeolite (AZ) adsorbents and other proprietary materials produced by UOP. UOP's experience in the development and evaluation of catalysts is directly applicable in solving reactivity problems in adsorption and scavenging type applications. Testing of reactivity in all stages of development and commercialization appears to be the most important factor in the implementation of the low reactivity concept.

Examples of UOP Testing of Adsorbents

UOP has developed a series of tests of increasing sensitivity to measure reactivity. The adsorbents selected to demonstrate the following tests depended on the sensitivity and type of test and do not in all cases compare two adsorbents used in the same application. They were chosen for the purposes of this paper to demonstrate the capability of each test. Now that the tests have been validated, UOP will more broadly apply these techniques across the portfolio of adsorbents typically used in olefin duty.

A. Testing on a Milligram Scale: Microbalance

These tests aim to initially evaluate the intrinsic catalytic activity of a given adsorbent or raw material for adsorbent manufacturing. Even though these tests only employ a 5-25 mg sample, they can detect significant differences in adsorbents. Figure 1 depicts the testing of two adsorbents via the TG (thermogravimetric) method utilizing propylene. After grinding the adsorbent, approximately 10 mg of powder is transferred into the pan of a microbalance. The material is then subjected to a thermal treatment in an inert gas flow to remove the oxygen from the system and restore the activated state of the adsorbent. Finally, the inert gas flow is replaced by the flow of the same inert gas containing about 1% by volume of propylene.

Figure 1 shows two different cases. Both adsorbents are capable of adsorbing similar amounts of propylene at a temperature of about 38°C. However, the UOP composite adsorbent AZ-300 quickly desorbs propylene upon an increase in temperature. In contrast, another commercial aluminosilicate adsorbent (commonly referred to as silica gel) we provisionally label as Silica Gel 1 demonstrates weight gain upon the same temperature protocol. Obviously, propylene converts to heavier oligomers, which remain trapped in the adsorbent's pores.

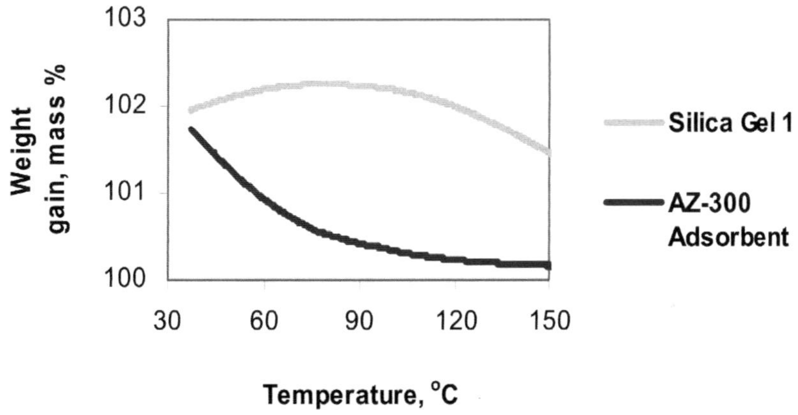

Figure 1: *TG microbalance test with 1 vol. % propylene stream*

B. Testing on a Milligram Scale: Flow Reactor with Hexene-1

This series of tests requires a micro reactor equipped with flow and temperature control and on-line gas chromatographic (GC) analysis of the feed and products. Small sized, crushed particles are used in these tests in an amount ranging from 20 to 250 mg.

Figure 2 shows the temperature dependence of hexene-1 conversion over two types of adsorbents. Selective Adsorbent 2 and UOP AZ-300 Adsorbent are both alumina-zeolite adsorbents. UOP Guard and Guard 3 are non-regenerative metal oxide-based materials typically used as scavengers. At the test conditions, the double bond shift is the most prominent route of hexene-1 conversion accounting for more than 90% of the converted hexene-1.

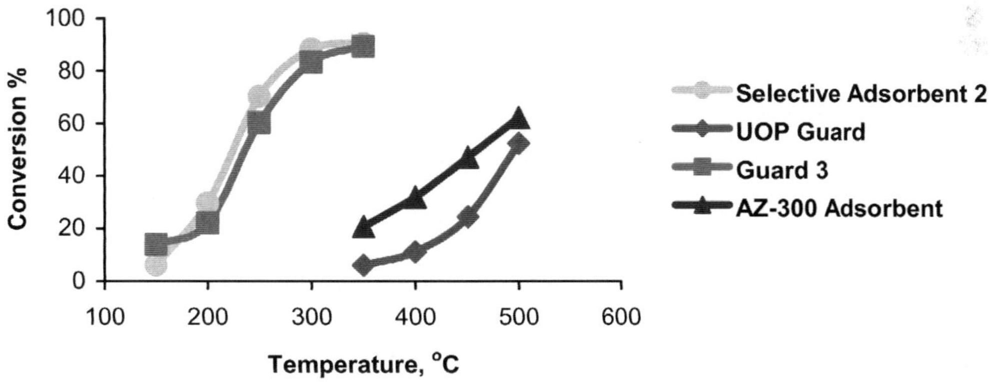

Figure 2: *Example of hexene-1 isomerization test with adsorbents*

It is clear that adsorbents and guard materials that convert hexene-1 at low temperatures are more reactive towards an olefinic stream than adsorbents that need a temperature in excess of 300°C to achieve appreciable hexene-1 conversion. The intrinsic catalytic activity of the latter is so low that they could easily tolerate less inert streams as regenerants in industrial service.

C. Testing on a Milligram Scale: Flow Reactor with Heptene

Another UOP micro-test employs heptene as a probe molecule and on-line post hydrogenation of the products prior to GC analysis. Thus, pronounced changes of the hydrocarbon skeleton such as isomerization, cyclization and cracking can be easily detected without the interference of the olefin components. Certainly, skeleton changes of a hydrocarbon require a higher level of catalytic activity of the tested solid material compared to the activity that would be needed to facilitate just the double bond isomerization reaction. The Heptene Conversion Test is carried out at a constant temperature of 425°C while variation of the feed flow rate permits gathering data at different contact times.

Figure 3 illustrates testing with the heptene probe by comparing different adsorbents at the same experimental conditions. The Silica Gel 1 adsorbent shows a very high reactivity towards heptene. The Selective Adsorbent 2 sample also causes high heptene conversion while AZ-300 Adsorbent essentially does not convert heptene over the complete range of contact times tested.

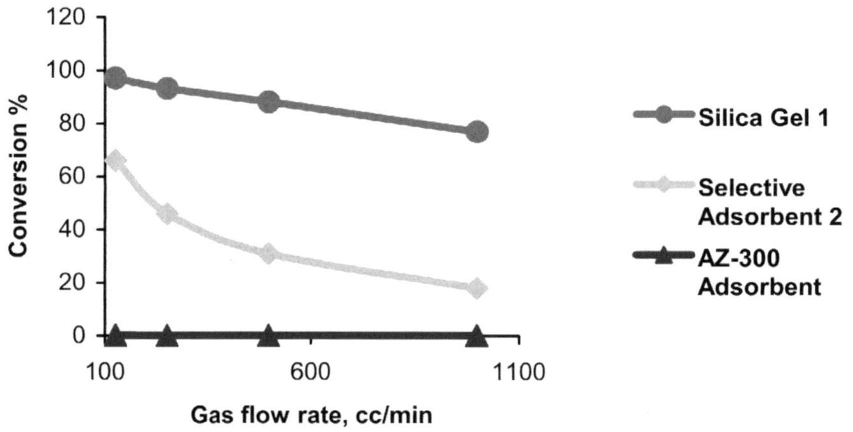

Figure 3: *Heptene Conversion Test with selected adsorbents*

D. Laboratory Testing at Process Relevant Conditions

While evaluating adsorbent reactivity with probe molecules at a micro scale is a useful laboratory method for characterization, one might argue that the conditions applied are not relevant to real process conditions particularly pressure, fluid phase, probe molecule species and adsorbent form (powder and small granules versus pellets or beads).

To assess the adsorbent reactivity at relevant process conditions, a temperature programmed reaction (TPR) method was developed. In the simplest setup, liquid isobutylene or propylene is slowly heated in a stirred autoclave in the presence of a similar volume of adsorbent. Both autoclave pressure and temperature are monitored over the test period and compared to a case where no adsorbent is present in the autoclave.

A decrease in pressure indicates the onset of an oligomerization reaction. The onset temperature is lower when isobutylene is used instead of propylene as a probe (Figure 4), indicating that an acid catalyzed oligomerization reaction is occurring. Indeed, the composition of the final liquid product is typical of acid catalyzed oligomerization.

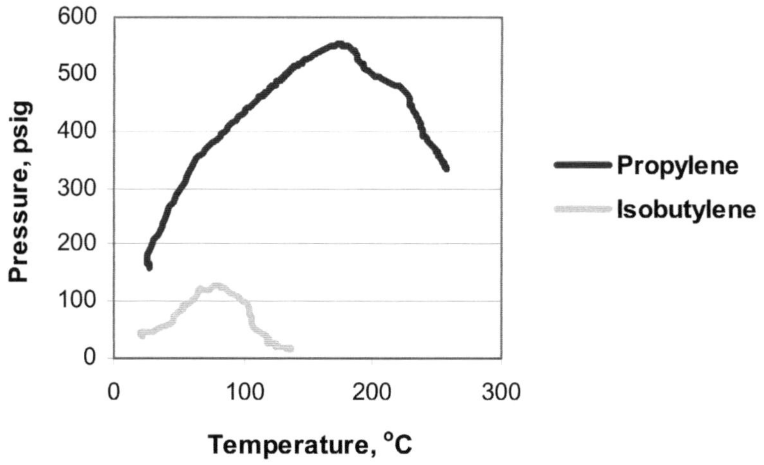

Figure 4: *Comparison of propylene and isobutylene in the TPR test on a reactive adsorbent*

Using this technique, corroborated by other methods, UOP has confirmed that commercial adsorbents used in the same application can differ vastly in terms of reactivity. Figure 5 shows the results for two commercial alumina-zeolite adsorbents, two commercial 13X molecular sieves, and a test using no adsorbent. No oligomerization takes place with AZ-300 adsorbent even at temperatures as high as 300°C while the other alumina-zeolite adsorbent, Selective Adsorbent 2, promotes the oligomerization reaction at near ambient temperature. Significant differences in reactivity can also be seen with the two 13X materials, one manufactured by UOP and the other designated as 13X-Source A. The onset of oligomerization for 13X-Source A begins at temperatures much lower than that required for the 13X MOLSIV™ sample. This demonstrates the fact that the same type of adsorbents made by different manufacturing methods can differ greatly in reactivity.

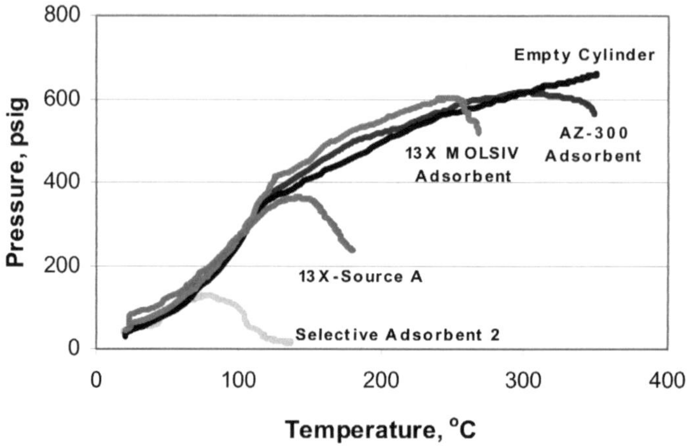

Figure 5: *Isobutylene TPR test on selected adsorbents*

The more reactive adsorbents will tend to produce green oil and deactivate more quickly. This translates into additional costs for the processor in the form of operating inefficiencies and more frequent recharges when using these reactive adsorbents.

In addition, we observe large differences in reactivity between some metal oxide-based guard bed adsorbents, which are generally regarded as less reactive than molecular sieve adsorbents. Since such guard beds are increasingly used to treat reactive C_4 feeds, their adsorbent reactivity patterns deserve special attention. Figure 6 shows that guard materials can differ widely in reactivity. Again, a safe guard material should not convert even reactive substances such as isobutylene at a near ambient temperature.

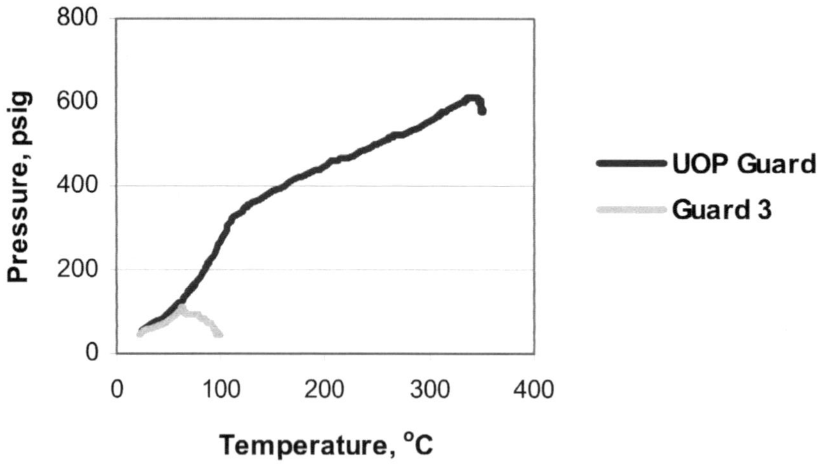

Figure 6: *Isobutylene TPR test on selected guard materials*

E. Dynamic Test at Process Relevant Conditions

The contact time of the olefin liquid to be purified in typical adsorbent beds is in the order of magnitude of a few minutes to tens of minutes. However, this time can increase during certain conditions such as startups and process upsets that cause low or stagnant flow. Sometimes, elevated temperature and long contact time combine, which may increase the danger of side reactions if reactive adsorbents are used.

To simulate these conditions, UOP employs a dynamic test where adsorbent particles confined in a basket, are soaked in excess isobutylene in a stirred autoclave equipped with a recycle pump, temperature control and on-line GC analysis. The reactor is kept at constant temperature while small liquid samples are withdrawn from the recycle loop and analyzed.

Figure 7 compares two commercial adsorbents that differ substantially in reactivity.

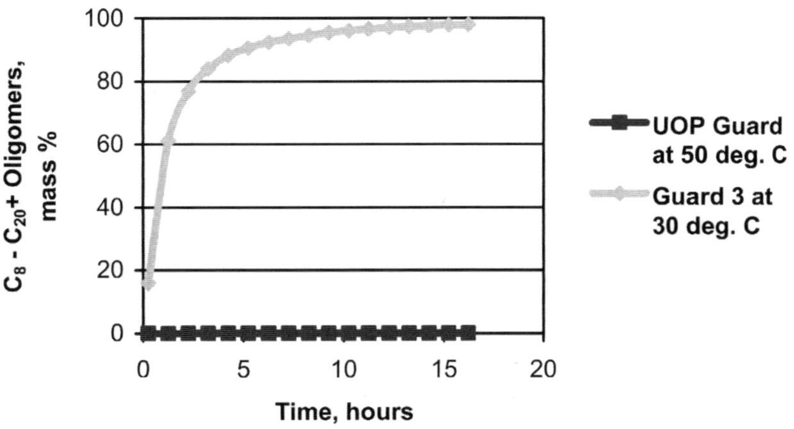

Figure 7: *Isobutylene dynamic test on selected guard materials*

The reactive guard bed adsorbent promotes an oligomerization reaction that produces a high boiling residue at a near ambient temperature. This residue can impact not only on the performance and longevity of the guard bed, but also the downstream equipment and processes.

There is a good correlation between the results of the micro test methods discussed earlier and the data from the dynamic test. For example, the high reactivity first predicted for the Selective Adsorbent 2 on the basis of micro data was later corroborated by the dynamic test shown in Figure 8. Figure 8 also evidences the isobutylene conversion over Selective Adsorbent 2 as a typical acid catalyzed reaction, which produces the expected product distribution of C_8 – C_{20+} oligomers.

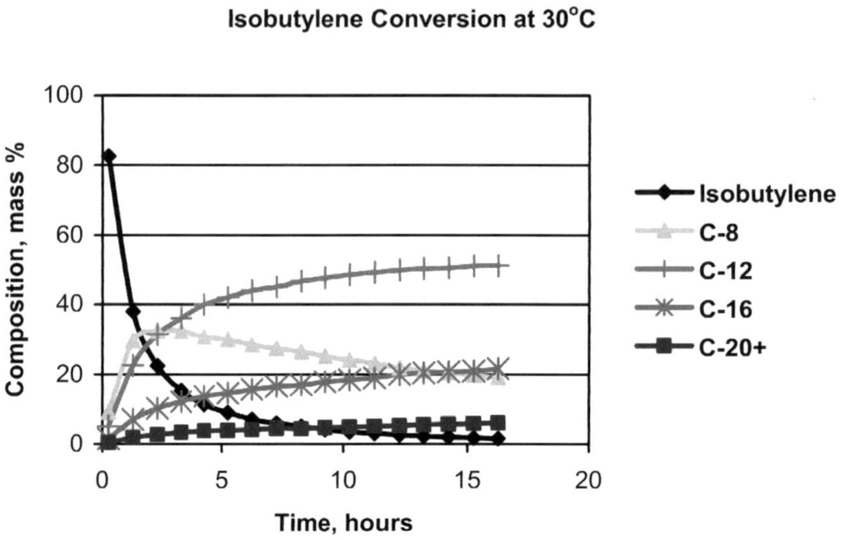

Figure 8: *Isobutylene dynamic test on a reactive adsorbent. Reaction product distribution versus time on stream.*

F. Basket and Field Testing

The evaluation of an adsorbent's reactivity does not end with the special tests described above. This is a process that continues, with analysis of spent samples from actual purification beds, to verify contaminant removal performance and validate the adsorbent's low reactivity at process conditions.

CONCLUSION

Adsorbent reactivity should be considered along with capacity and other performance criteria when choosing adsorbents for contaminant removal from olefin streams. Adsorbents acting as catalysts may sometimes promote highly exothermic reactions such as olefin oligomerization and polymerization, which are potential safety concerns. They can catalyze numerous side reactions involving both the main stream and original contaminants. This can have a negative impact on the adsorbent's performance and longevity, especially in the case of reactive streams and high-purity requirements for the final product.

UOP has developed a series of tests of increasing sensitivity to evaluate the reactivity of adsorbents. Knowing an adsorbent's relative reactivity versus other adsorbents helps the user understand any potential impact on his process and offers the opportunity to balance this adsorbent characteristic versus other performance criteria or operating conditions. These factors should all be considered collectively when choosing an adsorbent. Now that these techniques have been developed, UOP can apply this knowledge more broadly to commercial adsorbents and to the development of new adsorbents for contaminant removal.

REFERENCES

Briton, L. G., Taylor, D. A. and Wobser, D. C., "Thermal Stability of Ethylene at Elevated Pressures", *Plant/Operations Progress*, Vol. 5, No. 4, pp. 238 (1986).

Clark, E. L., "Molecular sieves perform well in sweetening light-hydrocarbon liquids," *Oil and Gas Journal*, pp. 178 (November 12, 1962).

Coleman, S. (Chair), "Sulfur Symposium", Proceedings of the 7th Ethylene Producer's Conference, pp.649 (1998).

Detrait, J. C. and Grootjans, J. F., "Adsorbent key to polypropylene catalyst activity", *Oil and Gas Journal*, pp. 50 (October 10, 1994).

Grose, R. (Chair) and Bowen, C. (Co-chair), "Effects of contaminants in ethylene plants", Session 22, Proceedings of the 8th Ethylene Producer's Conference, pp. 389 (1996).

Haag, W. O., Lago, R. M. and Weisz, P. B., "The Active Site of Acidic Aluminosilicate Catalysts", *Nature*, Vol. 309, pp. 589 (June 14, 1984).

Lossing, F. P. and Semeluk, G. P., "Free Radicals by Mass Spectrometry. XLII. Ionization Potentials and Ionic Heats of Formation for $C_1 - C_4$ Alkyl Radicals", *Canadian Journal of Chemistry*, Vol. 48, pp. 956 (1970).

Manchanda, K. D. and Gilmour, R. H., "Drying and desulfurizing light olefins with molecular sieves," AIChE Symposium Series, No. 134, Vol. 69, pp. 82 (1973).

McClung, R. G. and Novalny, S., "Choosing a Selective Hydrogenation System", *Petroleum Technology Quarterly*, pp. 147 (Autumn, 2002)

Norton, C. J., "Olefin Polymerization over Synthetic Molecular Sieves", *Industrial & Engineering Chemistry, Proc. Design Dev.*, Vol. 3, No. 3, pp.230 (1964).

Pines, H., "The Chemistry of Catalytic Hydrocarbon Conversions", Academic Press, New York, NY, pp. 5 (1981).

Reid, J. A., "Introduction to Arsine and Arsenic Compounds Removal Methods", Proceedings of the 9th Ethylene Producers' Conference, Vol. 6, pp. 781 (1997).

Reid, J. A., "An Overview of Olefin Unit Accidents and Safety Lessons Learned", Proceedings of the 11th Ethylene Producers' Conference, Vol. 8, pp. 353 (1999).

Reid, J. A. and McPhaul, D. R., "Control contaminants in olefin feedstock and products", *Hydrocarbon Processing*, pp. 128 (September, 1996).

Sanati, M., Hoernell, C. and Jaras, S.G., "The Oligomerization of Alkenes by Heterogeneous Catalysts", *Catalysis, Specialist Periodical Reports, The Royal Society of Chemistry*, Vol. 14, pp. 236 (1999).

Shah, R. (Chair) and Coleman, S. (Co-chair), "Effects of contaminants in ethylene plants", Session 21, Proceedings of the 8th Ethylene Producer's Conference, pp. 241 (1996).

Sherman, J. D., "Syntetic Zeolites and Other Microporous Oxide Molecular Sieves", *Proc. Natl. Acad. Sci. USA*, Vol. 96, pp. 3471 (1999).

Voigt, R.W., Merriam J.S. and Blankenship, S.A., "Evolution of Catalyst for Selective Hydrogenation of C_2 Acetylenes in Ethylene Streams", Proceedings of the 8th Ethylene Producers' Conference, Vol. 5, pp. 157, 1996.

Weisz, P. B., "Some Classical Problems of Catalytic Science: Resolution and Implications", *Microporous and Mesoporous Materials*, Vol. 35-36, pp.12000.

West, P. B., Haller G. L. and Burwell, Jr., R. L., "The Catalytic Activity of Silica Gel", *J. of Catalysis*, Vol. 29, pp. 486 (1973).

SESSION 56A – ETHYLENE PRODUCERS CONFERENCE KEYNOTE ADDRESS ETHYLENE PLANT ENVIRONMENTAL ISSUES

PRESENTED BY:

Sally Shaver (speaker), U.S. Environmental Protection Agency

This address was delivered by the Director of the EPA Emission Standards Division (ESD). The ESD is responsible for establishing emission standards under Section 112 of the Clean Air Act and managing federal programs for nationwide control of hazardous and criteria pollutant emissions from stationary sources. The ESD is part of the EPA Office of Air Quality Planning and Standards.

The Increased Focus on Episodic Air Emissions in Texas and Louisiana

Jennifer Keane
Baker Botts L.L.P.

©2004
Baker Botts L.L.P.

Prepared for Presentation at AIChE Ethylene Producers' Conference/April 26, 2004/Environmental Session

Unpublished

AIChE Shall Not Be Responsible For Statements or Opinions Contained in Papers or Printed in its Publications

Over the past several years, EPA and state environmental agencies have increasingly focused their attention on reducing episodic air emissions. Episodic air emissions are those events which can cause large amounts of air emissions in a relatively short period of time—such as upset events, maintenance events, leaks from cooling towers, and flaring. Efforts to reduce episodic emissions have been particularly marked in the Gulf coast region where most ethylene producers are located. In Texas, the Texas Commission on Environmental Quality (TCEQ)[1] has promulgated rules designed to reduce emissions from "highly-reactive" volatile organic compounds (HRVOCs) in addition to revising the State's upset and maintenance rules to increase stringency. While the changes in Louisiana have not been as sweeping to date, the Louisiana Department of Environmental Quality (LDEQ) is considering promulgating HRVOC rules of its own in the Baton Rouge area and has recently adopted emergency rules addressing upset emission reporting requirements. On the federal front, EPA Region VI has sent requests for information (pursuant to its authority under Section 114 of the federal Clean Air Act) seeking information on flaring emissions in the Beaumont/Port Arthur area. At least one Section 114 request has gone to an ethylene producer.

I. <u>Ozone Attainment Demonstrations and the Role of the State Implementation Plan in Increasing Scrutiny on Episodic Emissions.</u>

The Houston/Galveston (HGA) ozone nonattainment area is classified as a "severe" nonattainment area under the one hour ozone standard. With its high concentration of industry and its Gulf shore breezes, the HGA region has posed a difficult challenge for regulators attempting to bring the area into attainment with the 0.12 ppm limit by the 2007 deadline mandated by the federal Clean Air Act Amendments of 1990. Initial attainment demonstration efforts focused on reducing NOx emissions with rules requiring an up-to-90% reduction from existing NOx stationary sources by 2005. This culminated in submission of an attainment demonstration State Implementation Plan (SIP) to EPA on December 6, 2000. After litigation was brought by a coalition of industry sources on the December 2000 SIP, agreement was reached to make changes to the SIP and to conduct further evaluation of the science behind ozone formation in the HGA nonattainment area.

That effort, the Texas 2000 Air Quality Study (TexAQS), revealed that while NOx emissions from industrial sources in the HGA and east Texas regions of the State were generally correctly accounted for in the state emissions inventory, industrial VOC emissions appeared to be significantly understated.[2] Analysis of the data indicated that ozone formation can occur rapidly in HGA industrial areas because of the close proximity of large NOx and VOC emissions. In particular, the study showed that highly reactive VOCs (including ethylene, propylene, and 1, 3 butadiene) played a key role in ozone formation. Accordingly, in 2002, TCEQ promulgated rules to impose additional controls on routine industrial VOC emissions from sources that emit HRVOCs.[3]

The HRVOC rules address two needs: the need to reduce episodic emissions and the need for still more data to better understand how episodic events contribute to ozone formation. The

[1] In this paper, references to TCEQ also include TCEQ's predecessor agencies, including the Texas Natural Resource Conservation Commission and the Texas Air Control Board.

[2] The study also showed that surface monitors did not capture the formation of ozone plumes downwind of industrial facilities; aircraft monitors recorded ozone levels that were not recorded on the surface monitors.

[3] In Harris county (Houston), HRVOCs are defined as one or more of the following VOCs: 1, 3 butadiene; all isomers of butene, alpha-butylene and beta-butylene; ethylene and propylene. In Brazoria, Chambers, Fort Bend, Galveston, Libery, Montgomery and Waller counties, HRVOCs consist of ethylene and propylene.

rules require reductions from four industrial sources: fugitives, flares, process vents, and cooling towers. When the rules were proposed, TCEQ estimated that approximately 48% of the highly reactive VOCs came from fugitives, 30% from flares, 8% from vents and 7% from cooling towers.[4] The rules are performance-based, emphasizing monitoring, recordkeeping, reporting and enforcement, rather than establishing individual unit emission rates. Instead, the rules create a site-wide cap with compliance determined on a 24-hour rolling average.[5]

In general, flares are required to be tested and then equipped with continuous flow monitoring systems capable of measuring flow rates over the full potential range of operation by no later than April 30, 2005.[6] Cooling towers must similarly have a continuous flow meter on each inlet of each cooling tower as well as a system that will continuously determine the total strippable VOC concentration at each inlet. Speciated HRVOC concentrations must be determined at least once per month. Compliance is required by December 31, 2004.[7] Increased fugitive monitoring requirements are imposed, including separate tagging, the removal of certain exemptions from existing monitoring requirements (such as leak-skip provisions), and weekly inspection requirements for process drains equipped with water seals. Stringent leak repair and follow-up provisions are imposed along with extensive recordkeeping requirements. The rules also impose audit requirements, whereby an independent third-party organization must conduct an audit of each process unit subject to the rules at least once every two calendar years. The majority of the fugitive monitoring rules become effective on March 31, 2004.[8]

In addition to adopting stringent new standards designed to reduce HRVOC emissions, Texas changed the episodic emissions reporting requirements found in Texas' General Rules (discussed in more detail in Section II below) for HRVOCs. In the Houston/Galveston and Beaumont/Port Arthur ozone nonattainment areas, the reportable quantity (RQ) for butenes, ethylene, propylene, acetaldehyde and toluene were lowered to 100 lbs of unauthorized emissions in any 24 hour period. This ensures that TCEQ receives timely information about episodic events involving HRVOCs.[9]

II. Texas Episodic Emission Rules

Releases of air pollutants that result from episodic emissions traditionally have been of great concern and interest to both state and federal regulators because of the potential for adverse health effects. Increasingly, however, both EPA and TCEQ are focusing on episodic or unanticipated emissions associated with industrial processes to address chronic and worsening ozone problems in urban areas. A reduction in these types of emissions is one of the few remaining mechanisms to tackle this seemingly intractable problem that will not impinge on the rights of individuals living in the area (a politically unattractive solution). Therefore, this aspect of air pollution that once was a matter of

---

[4] 27 *Tex. Reg*. 5396 (June 21, 2002).

[5] The site-wide cap requires an average of an across-the-board 64% reduction from the 2000 inventory. This allowed TCEQ to propose an 80% reduction in NOx from stationary sources in lieu of the previously adopted 90% reduction requirement.

[6] *Highly-Reactive Volatile Organic Compounds, Division 1: Vent Gas Control;* 30 T.A.C. §§ 115.720 - .729.

[7] *Highly-Reactive Volatile Organic Compounds, Division 2: Cooling Tower Heat Exchange Systems*; 30 T.A.C. §§ 115.760 - .769.

[8] *Highly-Reactive Volatile Organic Compounds, Division 3: Fugitive Emissions;* 30 T.A.C. §§ 115.780 - .789.

[9] In addition to agency rules and materials, two overhead presentations were used as source materials for this section: "Event Emissions and their impact on ozone formation in the Houston Galveston Area" by David Allen, Yosuke Kimura, Cynthia Murphy, William Vizuete, Harvey Jeffries and Byeong-Uk Kim and "Houston Area Ozone Air Quality, Overview of Tools, Techniques, Communications and Improvements to Better Understand Ozone" by Doug Deason (November 6, 2003).

concern primarily for emergency response officials has become part of the debate over air quality and ozone formation.

A. 1997 Rules Incorporated the Concept of "Reportable Quantity" into the Rules.

In Texas, reporting of episodic releases is directly related to the air pollution emission limits that are imposed by regulations, permits, and orders. Prior to 1997, "major upsets" and certain maintenance events above authorized limits were required to be reported to TCEQ. In 1997, TCEQ adopted the concept of Reportable Quantities (RQ) for upset and maintenance rule reporting. A release was required to be reported only if unauthorized emissions of air contaminants exceeded an RQ in any 24-hour period. TCEQ adopted the federal RQs under EPCRA and CERCLA; however, because the Texas Clean Air Act regulates more than hazardous air pollutants, Texas promulgated state-based RQs for various non-hazardous air pollutants.[10] An RQ of 5,000 pounds was adopted for a limited number of hydrocarbons of particular interest in Texas, such as pentane.[11] For all other air contaminants, a default RQ of 100 pounds was established.[12]

Under the agency's upset and maintenance rules (U/M Rules), an owner or operator was required to report releases above the RQ as soon as practicable, but not later than 24 hours after the discovery of the upset. Any release below the RQ only required that the owner or operator maintained records. Proper recordkeeping (or reporting) was established as a threshold criteria for having upsets or maintenance events excused from the underlying permit or regulatory limits. The 1997 U/M rules were submitted to EPA for approval as part of the Texas SIP. Expressing concern that the State's new reporting rules were insufficient due to lack of public access to the records maintained at company sites and alleging the rules were unclear as to who had burden of proof that an event was "reasonably avoidable," EPA notified TCEQ in 1998 that the U/M Rules would not be approved as part of the Texas SIP. In particular, EPA objected to what it believed to be the granting of an automatic exemption if the objective criteria in the revised regulation were satisfied.

B. 2000 Rule Revisions Designed to Gain EPA SIP Approval

In light of EPA's disapproval of the 1997 U/M rules, TCEQ proposed additional revisions to the regulation that mirrored the criteria set out by EPA in its 1999 Excess Emissions Policy.[13] The proposal affirmatively required that sources seeking to justify excess emissions must make a much more detailed, and arguably greater showing than before, including proof that: (1) the excess emissions were caused by a sudden, unavoidable breakdown of technology, beyond the control of the owner or operator; (2) the excess emissions did not stem from any activity or event that could have been foreseen and avoided or planned for, and could not have been avoided by better operation and maintenance practices; (3) repairs were made in an expeditious fashion when the operator knew or should have known that the applicable emission limitations were being exceeded; (4) the owner or operator's actions in response to the excess emissions were documented by a properly signed, contemporaneous operating log or other relevant evidence; and (5) the excess emissions were not part of a recurring pattern indicative of

[10] Because "air contaminant" is broadly defined in the Texas Clean Air Act (Texas CAA) as virtually any substance discharged into the air, only releases of carbon dioxide, water, nitrogen, methane, ethane, noble gases, hydrogen, or oxygen are exempt from reporting requirements. [10]

[11] 30 T.A.C. § 101.1.

[12] *Id.* at § 101.1.

[13] 25 *Tex. Reg.* 530 (Jan. 28, 2000).

inadequate design, operation, or maintenance. The adoption of these new criteria signaled a dramatic change in TCEQ's treatment of upset and maintenance-related emissions. Emissions that had been almost routinely excused were now subject to intense scrutiny by the agency.

C. 2002 Rule Revisions to Incorporate Legislative Changes into the Rules

That scrutiny increased in 2001, when, in response to legislation passed by the Texas Legislature during TCEQ's Sunset Review process in 2001,[14] TCEQ again revised the U/M Rules.[15] These 2002 rule revisions (called the "Emission Events" rules) added an additional layer of complexity to the reporting and recordkeeping requirements, making it more difficult to meet the threshold reporting requirements. Inconsistent guidance to companies regarding the reporting requirements has further complicated the program. The 2002 revisions included rules to prevent or minimize non-routine emissions of highly reactive volatile organic compounds (HRVOCs) in the Houston/Galveston (HGA) and Beaumont/Port Arthur (BPA) ozone nonattainment areas through the development of corrective action plans, detailed case-by-case reviews of every emission event, enhanced reporting centered around substantially lower RQs for certain HRVOCs, in-depth root cause analyses, and enhanced enforcement. "Upsets" were renamed "emission events" as part of the 2002 rulemaking.

The new Emission Events rules also placed the burden on owners and operators to demonstrate why emission events are not "excessive" by requiring the following information to be reported to TCEQ: frequency of the facility's emission events; the cause of the emission event; the quantity and impact on human health or the environment of the emission event; and the duration of the emission event.[16] Emission events deemed "excessive" (which could occur on the basis of one event) cannot be excused from underlying emission limits—regardless of the circumstances. If an emission event is determined not to be excessive, TCEQ only will exempt an owner or operator from compliance with underlying emission limitations where certain stringent demonstration requirements are met, including: the episodic emissions were caused by a sudden breakdown of equipment or process beyond the control of the operator; the episodic emissions could not have been foreseen or avoided; and prompt action was taken to correct the problem.[17] TCEQ anticipated that a review of each emission event on a case-by-case basis would result in reductions in HRVOC emissions.

Reducing emissions during scheduled maintenance, startup or shutdown activities also were addressed in the 2002 Emission Events rules. Like emission events, TCEQ promulgated rules to exempt owners and operators from compliance with emission limitations during scheduled maintenance, startup or shutdown activities only where unauthorized emissions during maintenance, startup or shutdown could not have been prevented through planning and design; unauthorized emissions were not part of a recurring pattern indicative of inadequate design, operation, or maintenance; and operations were consistent with good practices for minimizing emissions.

---

[14] Act of May 28, 2001, 77th Leg., R.S., ch. 965, 2001 Tex. Gen. Laws 1933.
[15] 27 *Tex. Reg.* 8499 (Sept. 6, 2002).
[16] 30 T.A.C. § 101.222(a).
[17] *Id.* at 101.222(b).

D. More Revisions Required by EPA in 2003/2004

The 2002 Emission Events Rule exempted excess emissions from emission events (other than "excessive" emission events) and scheduled maintenance, startup and shutdown activities from compliance if the owner or operator properly recorded or reported them, as appropriate, in accordance with the ten initial notification requirements and twelve final notification requirements for emission events or the eleven initial and final notification requirements for maintenance events. If these reporting or recording requirements were not satisfied, the underlying event could not be exempted regardless of the circumstances. If the threshold reporting/recording requirements were met, then the event was evaluated in accordance with eleven demonstration criteria for emission events and nine for maintenance events to determine if the event was exempt from otherwise applicable emission limitations.

Despite the stringency of the rules, once again, EPA expressed its displeasure—this time threatening to disapprove Texas' Title V Operating Permits Program unless the rules were revised such that upset and maintenance events could not be "exempt" from underlying rules and regulations. In response, TCEQ proposed in 2003 to substitute an "affirmative defense" for the exemption. The proposal included several other changes that would have the combined effect of severely limiting any legal protection afforded by the defense. One of the regulated community's key objections to the proposal was that companies would be placed in the untenable position of notifying TCEQ that they were going to violate the law whenever they submitted an advance notice of scheduled maintenance, startup, and shutdown activities. (TCEQ traditionally has declined to affirmatively authorize emissions from such activities and, instead, has suggested that companies rely on reporting and the satisfaction of the demonstration criteria. Companies felt strongly that the proposal, if adopted, would eliminate this option and could expose them to significant liability for planned events.)

Although TCEQ commissioners did adopt revisions to the rule, they did so only after expressing concerns that stakeholders had not been afforded adequate opportunity to comment on the agency's approach to scheduled maintenance, startup, and shutdown emissions that were made in the final rule being presented for adoption.[18] (Rather than provide an affirmative defense, the rules simply state that emissions from a scheduled maintenance, startup or shutdown activity are required to be in a permit unless all nine of the demonstration criteria are satisfied.) In an unusual move, the rules were adopted, but with an expiration date of June 30, 2005, clearly in an effort to avoid what the commissioners perceived as a realistic threat that EPA otherwise would refuse to approve the emission events and scheduled maintenance, startup, and shutdown emissions rules as part of the SIP. As a result, this action will require almost immediate reconsideration of further revisions to the rule.[19]

E. Issues of Particular Concern to Ethylene Manufacturers

Several aspects of TCEQ's emission event and maintenance events rules are particularly troubling for ethylene manufacturers.

(1) *Threshold Reporting Issues*

---

[18] 29 *Tex. Reg.* 118 (Jan. 2, 2004).

[19] The current rules are found at 30 T.A.C. Chapter 101, Subchapter F. *See* http://www.tnrcc.state.tx.us/oprd/rules/index.html

Originally, the concept of RQs were incorporated into the Texas episodic emission rules as a way of ensuring that TCEQ was properly informed of large events that might pose a public health concern, but was not the recipient of numerous reports for small events that did not pose health concerns. Today, that concept has completely reversed. TCEQ is now taking the position that its rules require reporting on a facility-by-facility basis. That means that a company that experiences a major event at its manufacturing site that impacts many facilities is required to submit a separate electronic report for each facility. (Of course, application of the rules on a facility-by-facility basis also means that RQs are applied on a facility-by-facility basis, further complicating an already complicated program.) This can require large manufacturing sites to submit thirty or more reports for a single emission event.

Moreover, the rules require the initial notification to include the common name of the facility that incurred the emission event and the common name of the emission point where the unauthorized emissions were released to the atmosphere, while the final report requires the common name and the agency-established facility identification number of the facility that experienced the emission event and the common name and the agency-established emission point number where the unauthorized emissions were released to the atmosphere. Mistakes in properly identifying affected facilities and their authorized emissions limits have been used by TCEQ to disallow emission events. For example, if an emissions point does not have an emissions point number with an authorized emissions limit, the failure to write the word "zero" in the blank is grounds for a finding that the initial threshold reporting requirements were not satisfied. Similar disapproval can occur if TCEQ disagrees that the facility that "incurred" or "experienced" the emission event was not properly identified in the report. Failure to satisfy the threshold reporting requirements means that penalties can be assessed for the underlying event, regardless of the circumstances that led to the release.

(2) *Smoking Flares*

TCEQ's rules have separate requirements for releases of unauthorized emissions of air pollutants and for opacity. The issue of how to properly report smoking flares—if at all—has been the subject of conflicting guidance to the regulated community. When the concept of RQs was first incorporated into TCEQ's rules, the subject of smoking flares was discussed in the preamble and the regulated community was instructed to consider the degree of opacity caused by such flaring and report smoking flares if over the opacity RQ. TCEQ specifically wanted to know when a major smoking flare event was occurring so it could respond to public inquiries, should any occur. Later, the focus on flaring events switched to the unburned hydrocarbons and thermal NOx generally associated with flaring events. Many flaring events were not excused because companies reported only that the flares were smoking and not associated speciated emissions.

TCEQ then issued guidance stating that flares were subject to visible emission standards, not opacity. Most recently, we have been informed of a verbal agency determination taking the position that, while unauthorized emissions associated with smoking flares must be reported to TCEQ, the fact that the flare is smoking is not a required element of any report to the agency. Moreover, if a report were to include a statement that the flare is smoking, then that is an automatic violation of the Chapter 101 visible emissions standards since the emission event rules do not include any demonstration criteria for visible emissions (only opacity). It is unclear whether this is "official" TCEQ guidance but it does serve as evidence as to the confusion caused by the current rules.

(3) *Repeated Flaring Events*

One of the demonstration criteria requires the owner or operator to prove that the unauthorized emissions did not stem from any activity or event that could have been foreseen and avoided, and could not have been avoided by good design, operation, and maintenance practices.[20] Some within TCEQ are narrowly interpreting this provision—taking the position that an event can be excused if it occurs once, but repeated events cannot meet the demonstration requirement. When the 2000 rule changes were adopted, TCEQ responded to comments from the Sierra Club that the flaring of off-specification product from olefins plants should not be considered to be an upset under the rules.[21] TCEQ rejected the Sierra Club's blanket assertion, noting "an upset that is beyond the control of the owner or operator could result in the production of product that is not within specifications. The flaring of any off-specification product resulting from an upset that is eligible for exemption . . . would be examined in conjunction with the upset."[22]

Similar comments were raised again in 2003 when the HRVOC rules were promulgated. Environmental Defense (ED) and the Sierra Club requested that TCEQ include specific language in the final flare rules stating that flares which are not permitted process flares may be used only for emergencies, startups, shutdowns, and malfunctions. ED requested further "clarification" language explaining that flares may not be used to dispose of off-specification product or surplus on-specification product and that these products must be stored on site or recycled. TCEQ responded that it believes that some of these practices could be part of a comprehensive emissions management plan implemented by affected sources and that compliance with the rules will require companies to promote the use of other methods to dispose of materials commonly routed to flares.

To date, it does not appear that TCEQ has taken the position that repeated flaring events of off-specification product cannot be excused under the Chapter 101 demonstration criteria. However, given the position of the agency toward other "repeat" events and concern over flaring emissions, it is certainly likely that TCEQ Enforcement may take this position in the future. This could place ethylene manufacturers in the position of flaring off-specification product with the knowledge that the event will not be excused or having to provide very large storage capacity at the plants for such materials. Similarly, unless changed, the recent rule changes addressing scheduled maintenance, startup and shutdown (SMSS) emissions could be interpreted in a manner as to drive companies toward storing materials instead of flaring.[23]

The 2004 rule changes provide that SMSS emissions "are required to be included in a permit" unless the owner or operator proves nine criteria, including that "the unauthorized emissions from any scheduled maintenance, startup, or shutdown activity were not part of a recurring pattern indicative of inadequate design, operation, and maintenance."[24] In the preamble to the final rule, TCEQ included the statement that "Section 101.222(c)(3) [quoted above] requires the demonstration that the scheduled maintenance, startup, and shutdown activities was not part of a recurring pattern. Therefore, many routine scheduled maintenance, startup, and shutdown activities would not meet this criteria and

---

[20] 30 T.A.C. § 101.222.

[21] 25 *Tex. Reg.* 6731 (July 14, 2000).

[22] *Id.*

[23] These rule changes are among those scheduled for sunset on June 30, 2005.

[24] 30 T.A.C. §101.222(c)(3).

are subject to the requirement to obtain authorization for those emissions."[25] While the preamble appears to misstate the rule requirement (the rule does not require that a company prove that an activity is not part of a "recurring pattern," but rather that it is not part of a recurring pattern indicative of inadequate design, operation, or maintenance), an interpretation that such activities required permitting could pose difficulty for ethylene manufacturers. TCEQ requires full health effects review for SMSS emissions and other industries have found it difficult to meet the agency's stringent standards.

III. Louisiana's Increased Focus on Episodic Emissions

While Louisiana has not focused on episodic emissions to the same degree as Texas, the State is increasing its efforts to become better informed on the magnitude of certain episodic emission events within the State, particularly in the Baton Rouge nonattainment area. In November 1990, Baton Rouge[26] was designated as a serious ozone nonattainment area. Attainment of the ozone standard was to occur by 1999. Despite progress, the area failed to reach attainment by 1999, and the area was bumped up to a severe ozone nonattainment area effective June 23, 2003. Investigations of ozone episodes in the Baton Rouge area have indicated that most of the ozone events appear to be very localized and not regional in nature.[27] One monitor located near the Capitol building typically shows speciated VOC emissions typical of those emitted by mobile sources—particularly during the morning rush hour. However, an analysis of peak concentrations observed before and during ozone events has indicated that the high concentrations are coming from non-mobile VOC emissions. Specifically, the analyses have indicated peak HRVOC concentrations presumed to come from industrial sources.

As a result, LDEQ adopted an emergency rule lowering the reportable threshold (RQ) for HRVOCs, including ethylene and propylene[28]. The rule became effective on August 12, 2003 and was renewed in December. HRVOC RQs have been reduced to 100 pounds and the combined emissions of HRVOCs must be totaled to determine if an RQ has been exceeded. The rule is being credited with providing important information regarding upset and maintenance emissions in the Baton Rouge area. A rulemaking to incorporate these changes permanently into Louisiana's rules, *Requirements for Emergency Notification*, is currently underway.[29]

LDEQ also issued an administrative order last September, requiring companies to maintain and submit HRVOC data to LDEQ by December 23, 2003. While further study is still underway, LDEQ is exploring the possibility of requiring speciated data for ethylene, propylene and butenes in permits. In addition, a "HRVOC technical committee" has been convened to review additional data and oversee appropriate research. This will culminate in recommendations for control

---

[25] 29 *Tex.Reg.* 128 (January 2, 2004).

[26] The Baton Rouge Nonattainment area includes the parishes of Ascension, East Baton Rouge, Iberville, Livingston, and West Baton Rouge).

[27] Slide presentation "Baton Rouge Ozone: Cause and Effect" (October 2, 2003). Subpresentation "Ozone Exceedence Assessment" by Jim Hazlett.

[28] HRVOCs affected by the rule are: acetaldehyde, butenes (all isomers), ethylene, propylene, toluene, xylene (all isomers) and isoprene.

[29] *See* Notice of Intent OSO52.

measures, if any are deemed necessary.[30] It seems likely that LDEQ will look to the HRVOC rules in the Houston Galveston area for guidance, if needed.

IV. EPA's Efforts to Reduce Episodic Emissions

Not to be outdone by Texas and Louisiana, EPA has also entered the episodic emissions fray. Eric Schaffer, former head of EPA Enforcement, and his group, The Environmental Integrity Project (EIP) have focused attention on flaring events in the Port Arthur, Texas area. Specifically, EIP has focused on flaring incidents at six plants, including one ethylene producer. By issuing reports, sending letters to EPA, and garnering publicity, EIP has raised questions about whether the severity and frequency of the flaring events violate the federal Clean Air Act. This has culminated in the issuance of requests for information on flaring events by EPA pursuant to its authority to seek such information under Section 114 of the Act. EPA has requested information on all smoking flare events, all reports associated with such events, all estimated emissions associated with such events, and an explanation as to how the events have been legally excused or are legally excusable. While no enforcement actions have been brought to date, the Section 114 requests have served to further increase the scrutiny on episodic emissions.

V. Impact of Federal Release Reporting Rules on Ethylene Producers

In addition to the state episodic emission rules, companies remain subject to federal emergency release reporting requirements. Fortunately, there has been few changes to the CERCLA and EPCRA release reporting rules since EPA issued its guidance on "federally permitted releases" in April 2002.[31] A federally permitted release is one that is subject to a permit or control regulation under the following CAA programs: (1) New Source Performance Standards (NSPS); (2) National Emission Standards for Hazardous Air Pollutants (NESHAP) and Maximum Achievable Control Technology (MACT) standards; (3) Prevention of Significant Deterioration (PSD) permits for major sources in areas that attain the NAAQS; (4) Non-attainment New Source Review (NSR) permits for major sources in areas that violate the NAAQS; (5) SIPs submitted to and not disapproved by the EPA; and, (6) any "schedule or waiver" pursuant to any of the above.[32] These include most of the basic stationary source air pollution control programs under the CAA.

The 2002 guidance confirmed EPA's position that emissions in excess of permitted limits are not considered to be federally permitted and noted that "[p]ermit limits and control regulations usually do not control or limit unanticipated releases such as accidents or malfunctions and for that reason such releases generally do not qualify for the CERCLA section 101(10)H) federally permitted release exemption."[33] EPA also rejected the notion that accidental releases that occur when a facility is operating consistent with an accident and malfunction plan could be considered "federally permitted." EPA takes the position that releases due to accidents and malfunctions are, by definition, not anticipated,

---

[30] Slide presentation "Baton Rouge Ozone: Cause and Effect" (October 2, 2003). Subpresentation "Recent Actions & Next Steps" by Hall Bohlinger. *See also*, Baton Rouge HRVOC Workgroup webpage at: http://www.deq.state.la.us/evaluation/ozone/otcintro.htm.

[31] 67 *Fed. Reg.* 18899 (April 17, 2002).

[32] 42 U.S.C. § 9601(10)(H).

[33] *Id.* at 18902.

and therefore difficult to subject to controls to limit or eliminate emissions, therefore reasoning that "Congress did not intend to exempt unanticipated releases from CERCLA section 103 and EPCRA section 304."[34]

However, EPA did state that if a company is operating in compliance with an approved startup/shutdown plan that contains federally enforceable procedures which limit or control releases during startup or shutdown, then the release that occurred during such periods would generally qualify for the federally permitted release exemption. The requirement that sources subject to the Maximum Achievable Control Technology (MACT) standards must have a startup, shutdown and malfunction (SSM) plan[35] is cited as an example of a plan that might qualify. EPA was quick to add, however, that if a release of a hazardous or extremely hazardous substance is exempt from regulation or is otherwise not subject to emission limits or other controls during the startup or shutdown of an operation, then those uncontrolled releases would not qualify for the federally permitted release exemption. Such emissions must comply with CERCLA and EPCRA notification requirements. Thus, while not unlimited in scope, it does appear that ethylene producers may be able to categorize certain startup and shutdown emissions as "federally permitted" by virtue of the SSM plan that will be required as companies comply with the Ethylene MACT rules by 2005.[36] This might help reduce the number of emergency releases that would otherwise have to be reported under federal law—though, alas, it will have no impact on state reporting requirements.

VI. Conclusion

As EPA and state environmental agencies have increasingly focused their attention on reducing episodic air emissions, it has become more difficult for companies to avoid enforcement and new regulatory requirements designed to reduce the frequency and emissions associated with such events. Texas' upset and maintenance rules, along with proscriptive rules to address sources of HRVOC emissions in the Houston/Galveston ozone nonattainment area are the most stringent in the country. While the changes in Louisiana have not been as sweeping to date, the Louisiana Department of Environmental Quality (LDEQ) is considering promulgating HRVOC rules of its own in the Baton Rouge area and has recently adopted emergency rules to better obtain information on upset emissions. On the federal front, EPA Region VI has sent requests for information (pursuant to its authority under Section 114 of the federal Clean Air Act) seeking information on flaring emissions in the Beaumont/Port Arthur area. While the federally-permitted release exemption may provide some slight relief for controlled startup and shutdown emissions addressed in a company's SSM plan, in general it appears that companies should prepare themselves for continued focus in this arena.

---

[34] *Id.* at 18903.

[35] 40 C.F.R. § 63.6(e)(3).

[36] *See* 40 C.F.R. Part 63, Subpart XX, *National Emission Standards for Ethylene Manufacturing Process Units: Heat Exchange Systems and Waste Operations* and Subpart YY, *National Emission Standards for Hazardous Air Pollutants for Source Categories: Generic Maximum Achievable Control Technology Standards.*

Paper No. 56c

ENVIRONMENTAL SELF-ANALYSIS AND SELF-REPORTING UNDER TITLE V AND SARBANES-OXLEY

Matthew Kuryla and Whitney Swift
Baker Botts L.L.P.
Houston, Texas

Copyright © 2004 Matthew Kuryla and Whitney Swift

Prepared for presentation at AIChE Ethylene Producers Conference/April 26, 2004/Environmental Session

Unpublished

AIChE Shall Not Be Responsible For Statements or Opinions Contained in Papers or Printed in its Publications

ABSTRACT

The U.S. Environmental Protection Agency's ("EPA's") audit policy and the many state environmental audit statutes provide incentives for the voluntary self-analysis and disclosure of environmental compliance issues. The Title V operating permits that many of the country's most complex industrial sites are now receiving, however, raise questions about what self-analysis is "voluntary." Title V permits require sites to identify and to report promptly all deviations from air quality requirements, and to certify compliance (other than the deviations that have been reported) with all applicable air quality requirements on an annual basis. A responsible official of the permit holder must certify that the deviation report or compliance certification is true, accurate and complete, based on a "reasonable inquiry." The U.S. Security and Exchange Commission's ("SEC's") disclosure requirements add to public companies' *in*voluntary self-analysis and disclosure obligations. Under the Sarbanes-Oxley Act of 2002, public companies must establish "disclosure controls and procedures" designed to raise to senior management any material environmental violations or liabilities. Title V and Sarbanes-Oxley require companies to systematize their internal procedures for environmental compliance and self-reporting. This paper will discuss how companies can structure environmental self-analysis systems to ensure that those systems are sound and defensible under Title V and Sarbanes-Oxley. This paper will also evaluate what, if any, legal protections remain for environmental self-analysis and self-reporting in light of the mandatory compliance evaluation and disclosure requirements, and discuss how companies can structure self-analysis post-Title V permit issuance in an attempt to take advantage of those legal protections.

I. Environmental Self-Analysis: Benefits and Incentives

A. Benefits of Environmental Auditing and Self-Analysis

The ever-increasing volume and complexity of federal and state environmental regulations continue to elevate the challenges associated with achieving and maintaining environmental compliance at complex industrial sites. The owners and operators of those sites have increasingly relied on environmental self-analysis and voluntary environmental audits as vital elements of their compliance programs.

The U.S. Environmental Protection Agency ("EPA") defines environmental audit as a "systematic, documented, periodic and objective review by regulated entities of facility operations and practices related to meeting environmental requirements."[1] A voluntary self-analysis or environmental audit program provides a number of benefits. Most significantly, environmental audits will help a company achieve greater compliance with applicable environmental requirements. Audits evaluate how a site, or part of a site, is complying with applicable requirements and provide an excellent opportunity to review the mechanisms in-place to confirm and track compliance. Self-analysis will identify potential compliance issues at a site, allowing companies to act in a manner that will prevent a potential or threatened noncompliance from becoming an actual violation. An audit will also provide an opportunity to correct violations before a site is inspected by state or federal authorities. All of these factors combine to improve compliance and to reduce the risk of fines from state or federal authorities and to help reduce a company's liability to potential toxic tort plaintiffs.

B. Incentives for Environmental Auditing, Self-Analysis and Self-Reporting

EPA has long recognized the benefits of environmental self-analysis. In proposing its Audit Policy in 1995, EPA stated that environmental compliance "can be achieved only with the voluntary cooperation of thousands of businesses and other regulated entities subject to these requirements."[2] EPA repeated that sentiment when revising its Audit Policy in 2000, stating that "universal compliance cannot be achieved without active efforts by the regulated community to police themselves."[3] Consistent with its recognition of the value of environmental self-analysis, EPA has provided incentives for the voluntary identification and disclosure of noncompliance issues. EPA's civil penalty policies for media-specific programs provide varying degrees of penalty mitigation for disclosure.[4] In addition, companies that satisfy the conditions of EPA's Audit Policy qualify for the following benefits:

---

[1] U.S. EPA, "Incentives for Self-Policing: Discovery, Disclosure, Correction and Prevention of Violations," 65 Fed. Reg. 19618, 19625 (April 11, 2000) ("EPA Audit Policy"), at Section II.B.

[2] 60 Fed. Reg. 16875, 16875 (Apr. 3, 1995).

[3] 65 Fed. Reg. at 19619.

[4] *See, e.g.,* U.S. EPA, "Clean Air Act Stationary Source Civil Penalty Policy" at p. 16 (Oct. 25, 1991).

- Elimination or substantial reduction of "gravity-based penalties."[5] Gravity-based penalties are the portion of a civil penalty imposed by EPA over and above the economic benefit component that "reflect the egregiousness of the violator's behavior and constitute the punitive portion of the penalty." EPA will waive or substantially reduce the gravity-based component of a penalty associated with a noncompliance, giving companies that have not obtained an economic benefit from noncompliance an opportunity to avoid penalties entirely.

- No recommendation for criminal prosecution.[6] EPA will forego recommending to the U.S. Department of Justice ("DOJ") that criminal cases be brought against regulated entities that voluntarily disclose and promptly correct the violations identified through self-policing efforts, unless it finds that the violation is part of a pattern or practice demonstrating "potentially culpable behavior that merits criminal investigation."

- No routine requests for audit reports.[7] EPA will not request environmental audit reports (for example, during an inspection) to trigger an enforcement investigation.

Many state governments have joined EPA in establishing incentives for self-analysis of environmental compliance. According to the National Conference of State Legislatures, 25 states have enacted audit privilege or immunity laws and 11 states have issued audit policies that encourage the use of environmental audits.[8] Like EPA's audit policy, the state laws generally provide limited immunity from fines and penalties if a company voluntarily discloses the results of the audit and corrects the violation in a timely manner. [9] Many of the state audit acts also protect information generated through a qualifying audit with a privilege against disclosure in administrative or judicial proceedings. [10]

C. Key Limitation: *Voluntary* Self-Analysis and Reporting

The significant benefits provided by EPA's Audit Policy and the various state environmental audit laws do not extend to *all* self-analysis activities. For self-analysis to qualify for a

---

[5] EPA Audit Policy Section II.C.1.&2., 65 Fed. Reg. at 19625.

[6] EPA Audit Policy Section II.C.3., 65 Fed. Reg. at 19625. The U.S. Department of Justice ("DOJ") has also recognized the value of audits, taking the position that the performance of an audit will be viewed as a "mitigating factor" in DOJ's exercise of enforcement discretion and the selection of penalties. See U.S. Dep't of Justice, "Factors in Decisions on Criminal Prosecutions for Environmental Violations in the Context of Significant Voluntary Compliance or Disclosure Efforts by the Violator" (July 1, 1991).

[7] EPA Audit Policy Section II.C.4., 65 Fed. Reg. at 19625. EPA may, however, seek information from an audit that is relevant to identifying violations or determining liability or the extent of harm if it has independent reason to believe that a violation has occurred.

[8] National Conference of State Legislatures, Environment, Energy and Transportation Program, "State Environmental Audit Laws and Policies: An Evaluation," available at: http://www.ncsl.org/programs/esnr/auditsum.htm.

[9] *See, e.g.,* Tex. Rev. Civ. Stat. Ann. art. 4447cc, Texas Environmental, Health, and Safety Audit Privilege Act ("Texas Audit Act") § 10(a).

[10] *See, e.g.,* Texas Audit Act § 5.

penalty reduction or privilege, it must have been *voluntary*.[11] EPA's Audit Policy defines "voluntary discovery" as follows:

> The violation was discovered voluntarily and not through a legally mandated monitoring or sampling requirement prescribed by statute, regulation, permit, judicial or administrative order, or consent agreement.[12]

Similarly, the Texas Audit Act states that the privilege provided by the act does not apply to "a document, communication, datum, or report or other information *required by a regulatory agency* to be collected, developed, maintained, or reported under a federal or state environmental or health and safety law."[13]

The remainder of this paper will review the *mandatory* self-analysis and self-disclosure requirements imposed by Title V and the U.S. Securities and Exchange Commission ("SEC") disclosure requirements, including the Sarbanes-Oxley Act of 2002, and discuss the mechanisms and systems that companies use to satisfy these requirements. Following a review of what companies *must do* under Title V and Sarbanes-Oxley, the paper will return to the environmental audit laws and policies, and discuss how companies can structure self-analysis to take advantage of those benefits in light of those mandatory requirements.

II. Title V: Self-Analysis, Self-Disclosure and "Reasonable Inquiry"

A. Title V Permits

The Title V operating permit program was established by the 1990 amendments to the Federal Clean Air Act.[14] States act as the permitting authorities under Title V, developing state permit programs and issuing Title V permits that must meet the minimum federal requirements of Title V and EPA's Title V regulations. As a result of the time delays associated with (1) EPA's promulgation of the federal Title V regulations, (2) states promulgating Title V regulations that satisfy federal requirements, and (3) state permitting authorities developing and issuing Title V permits, many complex industrial sites have recently been issued, or will soon be issued, initial Title V operating permits.

Title V permits are designed to be comprehensive documents that identify all of the applicable air quality-related requirements (the "applicable requirements") for a particular site. The stated purpose of the Title V program is to provide regulators, members of the public, and the permit

---

[11] EPA Audit Policy Section II.D. 2., 65 Fed. Reg. at 19625. "Voluntary disclosure" is one of nine criteria established in EPA's Audit Policy. *See also* Texas Audit Act § 3(a) (defining a qualifying audit as a "systematic *voluntary* evaluation, review or assessment") and 10(a) (limiting the penalty immunities provided by the act to persons who make "voluntary disclosure").

[12] EPA Audit Policy Section II.D. 2., 65 Fed. Reg. at 19625.

[13] Texas Audit Act § 8(a)(1).

[14] 42 U.S.C. § 7401 *et seq.*

holders themselves a clear picture of all of the permit holder's applicable requirements. In this respect, the permits are "non-substantive"; that is, Title V permits consolidate *existing* applicable requirements.

The fact that Title V permits do not create new emissions limits does not mean that life *after* Title V permit issuance is the same as life before Title V. In what many Title V permit holders would call a tremendous understatement, EPA in its Title V regulations states that, "while Title V does not impose substantive new requirements, it does require . . . that certain procedural measures be adopted especially with respect to compliance."[15] The effect of these "certain procedural measures" is tremendous, as Title V permit issuance shifts the burden of compliance demonstration squarely onto the permit holder. This shift is accomplished through two mechanisms, deviation reporting and the annual compliance certification, that establish mandatory self-analysis and self-reporting obligations for every Title V site.

B. Deviation Reporting

EPA's Title V regulations at 40 C.F.R. Part 70 require prompt reporting of all "deviations" from permit requirements.[16] Part 70 does not define deviation. Deviation *is* defined, however, in the federal Title V program regulations[17] at 40 C.F.R. Part 71:

> Deviation means any situation in which an emission unit fails to meet a permit term or condition. A deviation is not always a violation. A deviation can be determined by observation or through review of data obtained from any testing, monitoring, or recordkeeping. . . . Included in the meaning of deviation are any of the following:
>
> *(1)* A situation where emissions exceed an emission limitation or standard;
> *(2)* A situation where process or emissions control device parameter values indicate that an emission limitation or standard has not been met;
> *(3)* A situation in which observations or data collected demonstrates noncompliance with an emission limitation or standard or any work practice or operating condition required by the permit;
> *(4)* A situation in which an exceedance or an excursion, as defined in part 64 of this chapter, occurs.[18]

[15] 40 C.F.R. § 70.1(b).

[16] 40 C.F.R. § 70.6(a)(3)(iii)(A)&(B).

[17] While the Part 70 regulations establish minimum requirements for state Title V programs, the Part 71 regulations establish the Title V program requirements when EPA is the permitting authority. *See* 40 C.F.R. § 71.1(a). Part 70 and Part 71 are identical in most respects.

[18] 40 C.F.R. § 71.6(a)(3)(iii)(C). Part 64 establishes EPA's compliance assurance monitoring ("CAM") requirements.

In accordance with this definition of deviation, Title V permit holders are required to self-report *any* situation in which an emission unit fails to meet a permit term or condition.

Part 70 does not adopt a "one size fits all" approach with respect to what constitutes prompt reporting of a deviation, allowing state permitting authorities to define prompt "in relation to the degree and type of deviation likely to occur and the applicable requirements."[19] State permitting authorities have generally implemented this provision by requiring expedited reporting of emissions-related deviations (e.g., upsets) and semiannual or quarterly reporting of deviations not associated with unauthorized emissions (e.g., a failure to report or to keep records).[20]

C. Compliance Certification

The annual compliance certification is the companion to deviation reporting under Title V. More than self-reporting, the compliance certification is self-*certification* of the permit holder's compliance status over the annual reporting period. Part 70 states that the compliance certification must identify:

> The status of compliance with the terms and conditions of the permit for the period covered by the certification, based on [the monitoring and related recordkeeping and reporting required by the permit]. The certification shall identify each deviation and take it into account in the compliance certification. The certification shall also identify as possible exceptions to compliance any periods during which compliance is required and in which an excursion or exceedance as defined under part 64 of this chapter occurred.[21]

Part 70 further requires that the compliance certification identify both the permit terms and conditions being certified and the method used by the permit holder for determining compliance status, though the certification may cross-reference the permit for that information.[22] Through the annual compliance certification, Title V self-reporting obligations extend beyond the requirement to report noncompliances and require the permit holder to certify in the affirmative its compliance status over the prior year.

D. Responsible Official Certification

The Title V program uses potential corporate officer liability to ensure that Title V permit holders take their self-analysis and self-reporting obligations seriously:

> Any application form, report, or compliance certification submitted pursuant to these regulations shall contain certification by a responsible

---

[19] *See* 40 C.F.R. § 70.6(a)(3)(iii)(B).

[20] *See, e.g.,* 30 T.A.C. § 122.145(2) (Texas Title V deviation reporting requirement).

[21] 40 C.F.R. § 70.6(c)(5)(iii)(C).

[22] 40 C.F.R. § 70.6(c)(5)(iii)(A)&(B).

> official of truth, accuracy, and completeness. This certification and any other certification required under this part shall state that, based on information and belief formed after reasonable inquiry, the statements and information in the document are true, accurate, and complete.[23]

The requirement that a responsible official certify the truth, accuracy and completeness of deviation reports and compliance certifications is a clear attempt to encourage permit holders to more closely scrutinize the reports submitted under Title V.

Part 70 defines a narrow class of persons who can qualify as a "responsible official" for a corporation:

> A president, secretary, treasurer, or vice-president of the corporation in charge of a principal business function, or any other person who performs similar policy or decision-making functions for the corporation, or a duly authorized representative of such person if the representative is responsible for the overall operation of one or more manufacturing, production, or operating facilities applying for or subject to a permit and either:
>
> (i) The facilities employ more than 250 persons or have gross annual sales or expenditures exceeding $25 million (in second quarter 1980 dollars); or
>
> (ii) The delegation of authority to such representatives is approved in advance by the permitting authority.[24]

Under Title V, the duty to certify reports cannot be delegated *below* a person "responsible for the overall operation of one or more manufacturing, production, or operating facilities." Someone meeting this criteria *must* certify the truth, accuracy and completeness of every Title V deviation report and compliance certification.

E. Reasonable Inquiry

Section 70.5(d)(10) quoted above contains the "reasonable inquiry" language that must be included in any responsible official's certification. Part 70 does not describe what EPA considers a "reasonable inquiry" with respect to a Title V permit holder's obligation to identify all deviations that have occurred over the reporting period (for a deviation report) or to determine the compliance status with all of the applicable requirements identified in the permit over the certification period (for the compliance certification). In the preamble to the 1991 proposal of the reasonable inquiry requirement, EPA explained that the language was modeled on Rule 11 of the Federal Rules of Civil Procedure.[25] The limited additional explanation states that the certifying official must make a reasonable (under the circumstances) inquiry before attesting to the document's truth, accuracy and completeness.

---

[23] 40 C.F.R. § 70.5(d)(10) (emphasis added).

[24] 40 C.F.R. § 70.2.

[25] 56 Fed. Reg. 21712, 21734 (May 10, 1991).

EPA provided a great deal more guidance as to the kind of inquiry it expects for a responsible official's certification when it prepared draft revisions to the federal Title V regulations in 1998. The draft revisions—which were never formally published as proposed revisions in the *Federal Register*—delete the reference to "reasonable inquiry" in the certification and replace it with the following, more explicit statement:

> The responsible official shall certify, by his or her signature, the following statement: "I certify under penalty of law that this document and all attachments were prepared under my direction or supervision and in accordance with a system designed to assure that qualified personnel properly gather and evaluate the information submitted. Based on my inquiry of those individuals who manage the system, or those individuals directly responsible for gathering the information, I certify that the information submitted is, to the best of my knowledge and belief, true, accurate, and complete. I am aware that there are significant penalties for submitting false statements and information or omitting required statements and information, including the possibility of fine or imprisonment for knowing violations."[26]

While EPA intended that the above requirements replace, rather than define, the term "reasonable inquiry," the steps set out by EPA are instructive regarding what EPA likely expects from a responsible official prior to certifying a Title V submittal. The draft rule language would require that the deviation report or compliance certification: (1) be prepared under the responsible official's direction, (2) by qualified personnel, who (3) properly gather and evaluate the information. The responsible official must then inquire into the accuracy of the report or certification with the persons who prepared or supervised preparation of the final document.

EPA provided additional guidance in its explanation of the draft rule language. Recognizing that submittals such as compliance certifications can contain a large amount of detailed information, EPA stated that a requirement to "personally examine" or "be familiar with" the information submitted would be beyond the normal scope of an responsible official's knowledge and responsibilities.[27] Rather, the draft language allows responsible officials "to base their certifications on the opinions of technical staff who may be subject matter experts in the areas for which information is being collected and reported."[28] EPA's explanation confirms that a responsible official can conduct a reasonable inquiry by discussing the contents and accuracy of the deviation report or compliance certification and the system used to prepare the document with the persons who prepared the document.

The compliance demonstration obligations imposed by the Title V program, as well as the obligation for a responsible official of the permit holder to certify that it conducted a reasonable

---

[26] Draft 40 CFR § 70.5(d) (Feb. 17, 1998 Draft) (available at http://www.epa.gov/ttn/oarpg/t5pgm.html).

[27] Preamble to draft 40 CFR Part 70 revisions, at p. 92 (Feb. 18, 1998 Draft) (available at http://www.epa.gov/ttn/oarpg/t5pgm.html).

[28] *Id.* at p. 92-93.

inquiry in preparing those documents, impose significant, mandatory self-analysis and self-reporting obligations on Title V permit holders.

III. Federal Securities Disclosures and Sarbanes-Oxley Certification Requirements

A. Federal Securities Regulations: Self-Reporting of Material Environmental Matters

Federal securities regulations require public companies to self-report "material" legal proceedings, as well as material financial impacts of environmental compliance and environmental contingencies that may reasonably have a material impact. These environmental matters are to be disclosed in the quarterly Form 10-Q and annual Form 10-K filed with the SEC.

The SEC's Regulation S-K, Item 103 requires that companies registered with the SEC disclose "material environmental proceedings and litigation" in Forms 10-Q and 10-K.[29] While there is no bright-line test with regard to what is material,[30] SEC guidance classifies a proceeding as material if (1) a governmental authority is a party to the proceeding and the company has a reasonable belief that the proceeding will result in monetary sanctions of $100,000 or more, or (2) any other proceeding involving a claim for damages, potential monetary sanctions, or costs that exceeds 10 percent of the current assets of the company. Companies registered with the SEC are also required to disclose, under Regulation S-K, Item 101, any capital expenditures related to compliance with (or failing to comply with) environmental requirements that are "material" to the capital expenditures, earnings and competitive position of the company and its subsidiaries.[31] Similarly, Regulation S-K, Item 303 requires company management to disclose environmental contingencies that may reasonably have a "material" impact on revenue or income from continuing operations.[32]

B. Sarbanes-Oxley: Certification of Reports Filed with the SEC

The Sarbanes-Oxley Act ("Sarbanes-Oxley") was passed in July 2002 to bolster federal securities laws in response to a series of high-profile corporate governance scandals. Among the many aspects of Sarbanes-Oxley aimed at increasing corporate accountability are provisions that enhance the financial disclosure certification requirements and require corporations to implement a series of disclosure controls and procedures. While Sarbanes-Oxley did not change any requirements with respect to the disclosure of environmental matters, its focus on disclosure controls will impact the self-disclosure of environmental matters in federal securities reports and, in all likelihood, other certifications.[33]

---

[29] 17 C.F.R. § 229.103.

[30] Material information is that information "to which there is a substantial likelihood that a reasonable investor would attach importance in determining whether to purchase the security registered"—that is, a reasonable investor would find the information important in deciding whether to invest in the company. 17 C.F.R. §§ 230.405.

[31] 17 C.F.R. § 229.101.

[32] 17 C.F.R. § 229.303.

[33] *See* Judson W. Starr *et al.*, The New Rules in Environmental Criminal Enforcement, ALI-ABA Course of Study: Criminal Enforcement of Environmental Laws (2003) (theorizing that Sarbanes-Oxley may come to be viewed as establishing a "best

Any environmental self-disclosure contained in a Form 10-K annual report or Form 10-Q quarterly report filed with the SEC will be subject to the Sarbanes-Oxley certification requirements. Section 302 of Sarbanes-Oxley requires the signing officer to certify that the officer has reviewed the report and that, based on the officer's knowledge, the report "does not contain any untrue statement of a material fact or omit to state a fact necessary in order to make the statements . . . not misleading."[34] Section 302 also requires that the certification explicitly reference the company's disclosure controls. The required certification must provide that the signing officers (1) have designed internal controls to ensure that material information relating to the company is made known to the officers by others within the company and (2) have evaluated the effectiveness of the internal controls within the 90 days prior to filing the report. The certification must include the officers' conclusions about the effectiveness of the internal controls.[35]

III. Compliance with Title V and Sarbanes-Oxley Certification Requirements

The Title V reasonable inquiry requirement and Sarbanes-Oxley certification requirements mandate that responsible officials and corporate officers be assured that they are signing accurate and complete environmental disclosures. While the self-analysis and self-reporting required for Title V and SEC reporting purposes are designed to identify very different matters, there are striking similarities between a reasonable inquiry conducted in accordance with EPA's 1998 draft rule and the system of disclosure controls required to support an SEC certification.

A. Title V: Identification of Potential Deviations through an Audited System

Deviation reports identify all noncompliances that have occurred over the reporting period. Compliance certifications describe the permit holder's compliance status over the reporting period, identifying those periods in which the permit holder was not in compliance. For both deviation reporting and compliance certification purposes, companies need a system to assure that qualified personnel gather and evaluate the compliance-related information that is required to be collected by the permit. A comprehensive compliance evaluation system should incorporate all of the testing, monitoring and recordkeeping required by applicable requirements and the permit itself, as well as any other "credible evidence" of the permit holder's compliance status.[36] A properly functioning system will be designed to identify potential deviations, bring them to the attention of persons charged with determining whether a deviation has occurred, and record and/or report the proper information if a deviation is confirmed.

---

practice" for corporate governance and that its standard may extend to cover other reporting mechanisms and that "Sarbanes-Oxley changes what was once considered a 'reasonable inquiry.'").

[34] Sarbanes-Oxley Act § 302(a)(1)-(2).

[35] Sarbanes-Oxley Act § 302(a)(4).

[36] *See Annual Compliance Certification Guidance*, Samuel Coleman, P.E., Director, Compliance Assurance and Enforcement Division, EPA Region VI, at p.4 (Dec. 22, 1998) ("If a source becomes aware of *other material information* that indicates that an emission unit has experienced deviations (as that term is defined in State programs) or may otherwise be out of compliance with an applicable requirement even though the unit's permit-identified data indicates compliance, the source must consider this information, identify and address it in the compliance certification, and certify accordingly. See, e.g., 62 FR 8314, 8320 (Feb. 24, 1997) [EPA's Credible Evidence Rule].").

In accordance with the guidance provided by EPA's 1998 draft rule, the responsible official who certifies that the deviation report or compliance certification is true, accurate and complete is charged with confirming that the system used to prepare the document is working. The certifying official must ask the persons responsible for managing the system, or those who directly gather the information processed in the system, whether the compliance information generated for the report or certification is true, accurate and complete. It is based on these questions, and information and belief that the permit holder has a functioning system designed to identify and evaluate potential noncompliance, that the responsible official can certify that the deviation report or compliance certification was prepared after a reasonable inquiry.

A responsible official's inquiry of the persons who conduct the compliance evaluation can take multiple forms. Some Title V permit holders have used meetings in which the responsible official, after reviewing the report or certification, requests verbal confirmation from those persons involved in the collection and evaluation of compliance-related information that the results of the evaluation are true, accurate and complete. Other permit holders have opted for a more formal "sub-certification" approach, under which the persons involved in preparing the deviation report or compliance certification prepare an *internal* certification of the truth, accuracy and completeness of that person's role in the compliance evaluation, on which the responsible official can rely in certifying the document. The sub-certification approach provides the benefit of creating a "paper trail," should problems with a certification later arise.

B. Sarbanes-Oxley: Identification of Material Environmental Matters through an Audited System

The disclosure controls mandated by Sarbanes-Oxley require a similar system, albeit one designed to identify different matters. While a system designed for a Title V reasonable inquiry should identify and evaluate potential deviations from applicable air quality requirements, a Sarbanes-Oxley disclosure control system should identify, track and evaluate the materiality of potential environmental liabilities that may trigger SEC disclosure requirements. The disclosure control system should incorporate a variety of factors, including: anticipated expenditures associated with environmental matters; compliance and enforcement risks; regulatory trends; operational constraints; and self-certifications and reports filed under substantive environmental requirements. The disclosure controls must identify and help quantify the potential liabilities and bring them to the attention of the certifying officers. In certifying the SEC filings, the officers represent that the company has implemented a system of disclosure controls that *ensures* that material information matters that may require reporting are brought to the officers' attention.[37]

Sarbanes-Oxley requires the signing officer to certify that the officer has recently evaluated the effectiveness of the company's internal controls. Like the inquiry into a Title V compliance evaluation system, an audit of the disclosure controls can take multiple forms, from a face-to-face discussion to a series of sub-certifications prepared by the persons charged with implementing the disclosure controls.

---

[37] Sarbanes-Oxley Act § 302(a)(4)(B).

IV. Impact of Mandatory Self-Analysis and Self-Disclosure on Audit Protections

This paper opened with a discussion of the benefits of environmental self-analysis and the incentives that EPA and many states have provided to encourage self-analysis and disclosure. Following Title V permit issuance, however, the protections for self-analysis and disclosure of potential violations of air quality requirements[38] is greatly impacted. The Title V self-analysis and certification requirements overlap with traditional objectives of an environmental audit and will reduce, if not eliminate, the available audit protections. What self-analysis or disclosure could be considered "voluntary" when a Title V permit holder is required (1) to report all deviations and (2) to conduct an ongoing "reasonable inquiry" into its compliance with air quality requirements?

Title V permit holders are required to report all deviations. After Title V permit issuance, the fact that a deviation is identified through self-analysis—whether required by the permit or not—will *not* protect that deviation from disclosure.[39] While the audit report itself may still be protected, any deviations discovered during the audit should be identified and addressed in the permit holder's deviation report and compliance certification. The failure to disclose such deviations risks the knowing submission of a false compliance certification.

The penalty reduction benefits provided by EPA's Audit Policy and the state audit laws will also be impacted by Title V permit issuance, as the penalty reductions are limited to violations that are voluntarily discovered and disclosed. In its Audit Policy Interpretive Guidance, EPA takes the position that a violation *can be* voluntarily discovered post-Title V, and that the Audit Act penalty benefits can apply, if the permit holder can "demonstrate that its inquiry *exceeded* its [reasonable inquiry] obligations."[40] EPA states that this determination would be made on a case-by-case basis, but has not provided guidance with respect to how self-analysis can exceed a reasonable inquiry.

A Title V permit holder hoping to take advantage of the immunity provided by the Audit Policy or a state audit statute should structure its audit activities in a manner that will differentiate between (1) the inquiry required by the permit and (2) the compliance evaluation performed as part of the audit. Factors that may help distinguish between the required reasonable inquiry and a voluntary audit include:

- the audit must involve testing, monitoring or the review of records that is not already required by a term or condition of the Title V permit

---

[38] Because Title V permits include only air quality requirements, Title V permit issuance will not impact a Title V permit holder's ability to take advantage of audit protections for other media.

[39] *See, e.g.,* Texas Audit Act § 8(a)(1) (excluding from the audit privilege any "report or other information required by a regulatory agency to be . . . reported"). This disclosure obligation raises another consideration: after Title V issuance, an audit should not be performed unless there is a firm commitment by management to correct any environmental problems discovered, and thereby avoid creating a knowing, unresolved violation.

[40] U.S. EPA, Audit Policy Interpretive Guidance at q.2 (January 1997) (emphasis added); *see also* Texas Audit Act guidance at p.12 (September 1997).

- the audit is performed by different personnel than those charged with day-to-day compliance evaluation for the emissions unit(s), such as corporate environmental staff or outside contractors

- the audit is a multimedia audit and is not restricted to compliance with air quality requirements found in the Title V permit

- the audit is conducted in a narrow timeframe

- timing of the audit is independent of any certification or reporting obligation under the Title V permit

Self-analysis *must* meet the first criterion listed above to qualify for audit protections. Whether the audit is judged to have exceeded a reasonable inquiry will be a subjective, case-by-case evaluation, and will depend on how well the Title V permit holder is able to distinguish between permit obligations and the voluntary audit activity.

GAS IMAGING
A NEW ALTERNATIVE TO METHOD 21

Michael Smylie
ENVIRON International Corporation
101 Rowland Way
Novato, California 94945

Michael Koonce
Gas Imaging Systems
1 Post Street, Suite 2515
San Francisco, CA 94104

Prepared for Presentation at AIChE Ethylene Producers Conference
April 26, 2004
Environmental Session

No Copyright
Unpublished

AIChE shall not be responsible for statements or opinions contained in papers or printed in its publications.

INTRODUCTION

At present, federal and state fugitive emission monitoring programs are based on EPA Method 21[1], which involves the use of a portable hydrocarbon analyzer to monitor for an emission leak at the leak interface of equipment such as valves, pumps, compressors, and connectors. Monitoring is performed so that leaks can be identified – this is done by comparing the hydrocarbon analyzer reading, or screening value, with the leak definition in the applicable regulation. Data gathered from monitoring is then used to determine what actions are needed to decrease and/or eliminate the leak concentration. Generally if a component is found to be leaking, an attempt to repair the component must be completed within a specified time frame. If the component cannot be repaired within this time frame, it must be repaired at the next available plant shutdown.

Under Method 21, an operator is required to visit and screen each regulated component on a defined frequency to determine if the component is leaking. Given the variations in state fugitive emission regulations, and between the federal New Source Performance Standards (NSPS) and National Emission Standards for Hazardous Air Pollutants (NESHAP) requirements, leak definitions for different components and monitoring frequencies can vary substantially, even within the same refinery/chemical plant complex. The actual number of components to be tested in a refinery or chemical plant can be quite large, making Method 21 monitoring both time intensive and expensive – in a typical U.S. refinery with over 200,000 traditional components, the annual cost for an LDAR program can exceed $1,000,000. Furthermore, a study in 1994 sponsored by API found that over 90% of controllable fugitive emissions come from only about 0.13% of the piping components[2]. As a result of concerns over the cost and effectiveness of Method 21, efforts at finding more productive ways to identify and repair high leakers are underway in both the public and private sectors.

OPTICAL GAS-IMAGING DEVICES

An emerging class of technology, generally referred to as optical gas-imagers, offers an operator the ability to monitor traditional and nontraditional components from a distance and identify – in some cases instantaneously – leaking components (of a sufficient mass) within the line of sight of the optical imager. The remote sensing and instantaneous detection capabilities of optical gas-imaging technologies allow an operator to scan areas containing tens to hundreds of potential leaks, thus eliminating the need to visit and manually measure all potential leak sites.

Between May and August 2002, two field studies were conducted to measure and detect fugitive emissions using a portable optical gas-imaging device. The portable optical gas-imaging device (CO_2 laser) evaluated during the field studies is based on the Backscatter Absorption Gas Imaging (BAGI) technique that makes visible invisible gases. Rather than monitoring each individual component by measuring the concentration of any leaking gas at the component's surface as current Method 21 procedures require, the CO_2 laser can detect leaks instantaneously from a distance. The technology is similar to "night vision goggles" in that it views surroundings in infrared light; for the demonstration conducted in the field studies, the area viewed was illuminated with infrared light that allowed images of chemical leaks to be viewed live. Leaking gas appears in real-time as a "black cloud" on a viewfinder or an attached TV screen.

The technology is not yet developed to provide a quantitative measure of the gas leak; however, the CO_2 laser shows the two-dimensional size of the leak and the darkness and dimension of the gas cloud image is relative to the thickness of the leak plume. During the two field studies, a TVA-1000B portable hydrocarbon analyzer was used to obtain a screening value of all leaks detected by the CO_2 laser. Although commercially available, the device is not yet certified intrinsically safe.

Backscatter Absorption Gas Imaging (BAGI)

The principle of operation of the BAGI technique is the production of a video image by reflected (backscattered) laser light, where the laser wavelength is such that it is strongly absorbed by the gas of interest. Figure 1 shows a schematic of how the BAGI technique works.

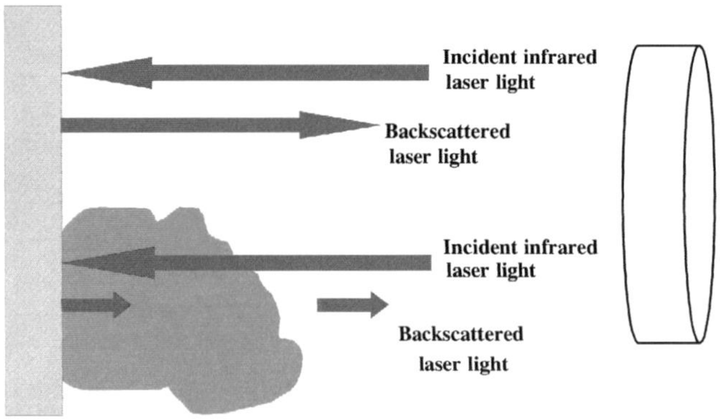

Figure 1: Schematic description of the BAGI process.

BAGI illuminates the scene with infrared light. A video camera-type scanner picks up the backscattered infrared light. The camera converts this backscattered infrared light to an electronic signal, which is displayed in real-time as a black and white image on both the camera's viewfinder and a TV monitor. Since the scanner is only sensitive to illumination from the infrared light source and not the sun, the camera is capable of displaying an image in either day or night conditions. The camera can be switched between visible and infrared views, which provide an important attribute when the operator needs to differentiate between steam plumes and gas plumes (see discussion of atmospheric window below).

STUDY RESULTS

The first field study was conducted at Site A between May 13 and 16, 2002; the second was conducted at Site B between August 5 and 8, 2002. Both sites are located in the greater Houston-Galveston area. Study results from these two field tests are presented below.

Site A Study Results

The first field study was conducted at Site A between May 13 and 16, 2002. At site A, approximately 7,187 components were monitored in three process blocks. This represents about 18 percent of Site A's traditional LDAR components in these process blocks.

Screening was conducted on both traditional and nontraditional fugitive emission components including piping manifolds, valves, flanges, threaded fittings and other connectors, heat exchangers and other process vessels, open-ended lines, pumps and compressors. Much of the equipment in the Cold Section was coated in heavy insulation and sheet metal cladding that covered virtually the entire vessel and piping flanges.

Mass Rate Calculations

At Site A, emissions of propylene and ethylene were known to be present in the areas screened with the CO_2 gas-imaging device. 70 of the 95 components found leaking during the field study were traditional LDAR components. Applying EPA correlation equations, the mass rate was calculated for each detected leak from a traditional component. Table 1 shows the mass rates of ethylene and propylene from each component category.

Table 1. Calculated mass rates from traditional sources at Site A.

Component Type	Number of Components		Total Mass Rate (Kg/hr)		Percentage of Total Mass		Average Mass Rate (Kg/hr)	
	E	P	E	P	E	P	E	P
Open-ended line (oel)	3	0	0.03	0.00	0.94%	0.00%	0.01	Not Applicable
Compressor seal	1	0	0.09	0.00	3.32%	0.00%	0.09	Not Applicable
Connector	27	3	1.96	0.11	72.73%	20.47%	0.07	0.04
Valve	24	12	0.62	0.41	23.01%	79.53%	0.03	0.03
TOTALS	**55**	**15**	**2.70**	**0.52**			**0.20**	**0.07**

Notes:
E= ethylene
P = propylene

Comparison of Mass Rates from Bagging and Correlation Equations

12 of the 70 traditional components recorded as leaking were bagged during the field study. Leaks at two of the 12 components were not detected by the CO_2 laser (tags 644 and 769) and were known to be leaking propylene. Table 2 compares the mass rates determined by bagging and those calculated using EPA's correlation equations.

Because the calculation of mass rate assumed all fluids except cracked gas are pure compounds of a single constituent, the calculated results are for either ethylene or propylene. This method certainly overestimates the quantity of the dominant constituent in a mixed stream; however it is the most reasonable approach based on data readily available.

In all cases, the calculated mass rate were several orders of magnitude larger than the bagging results. Bagging results show the smallest detected leak had a mass emissions rate of 5.63E-07 kg/hr, compared to a correlated mass rate five orders of magnitude higher at 0.012 kg/hr. The largest detected leak was emanating from a connector (tag 3815) with mass emissions of 2.32 E-02 kg/hr.

Table 2. Comparison of bagging results with calculated mass rates at Site A.

Component Data		**Plant Area**	**Seen by Laser?**	**Bagging Result (Kg/hr)**		**Correlation Result (Kg/hr)**	
Tag #	**Type**			**E**	**P**	**E**	**P**
510	Connector	Cold Area	Yes	7.24E-05	ND[a]	0.006	NA
1535	Valve	Cold Area	Yes	4.70E-04	ND	0.006	NA
0090	Connector	Cold Area	Yes	3.66E-05	5.76E-07	0.01[d]	NA
644	Valve	Cold Area	No	4.33E-06	2.60E-04	NA	0.05
769	Valve	Cold Area	No	1.20E-05	2.10E-04	NA	0.01[e]
3496	Connector	Cold Area	Yes	7.41E-06	6.65E-07	0.03	NA
3815	Connector	Compressor	Yes	2.32E-02	ND	0.127[f]	NA
3451	Connector	Compressor	Yes	9.05E-04	ND	0.05[g]	NA
4278	Valve	Compressor	Yes	2.80E-03	2.64E-06	0.009[h]	NA
4279	Connector	Compressor	Yes	1.95E-02	7.30E-06	0.017	NA
Bull Plug[b]	Connector	Compressor	Yes	1.62E-03	6.39E-07	0.01	NA
Seal Oil Breather Vent[c]	Other	Compressor	Yes	5.63E-07	ND	0.09	NA

Notes:
E= Ethylene
P= Propylene
[a] ND = not detectable
[b] This bull plug was not tagged, nor was the valve near it.
[c] This seal oil breather vent was not tagged.
[d] Average of the three screening values taken on 5/14 and 5/16.
[e] Average of three screening values taken on 5/13 and 5/14.
[f] Average of two screening values taken on 5/15 and 5/16.
[g] Average of two screening values taken on 5/15 and 5/16.
[h] Average of three screening values taken on 5/15.

Site B Study Results

The field study at Site B occurred between August 5 and 8, 2002. Site B is a large producer of ethylene with operations conducted in two on-site locations, a base plant and an expansion plant. The field study focused on those areas of the plant where high concentrations of ethylene were present in the process pipes and vessels. The first two days of testing were spent in the base plant; as the week progressed, field study activities moved to the expansion plant. Monitoring was conducted in three primary process areas.

The types of equipment and components screened in these areas included piping manifolds, valves, flanges, threaded fittings and other connectors, heat exchangers and other process vessels, compressors, open ended lines, and pumps. In all, approximately 1,178 components in the three areas were screened and field data was recorded from 52 leaking sources.

Mass Rate Calculations

At Site B, 49 (or 94%) of the 52 components detected as leaking during the field study were traditional components. All but one of the 49 sources was leaking ethylene. Applying EPA's correlation equations, the mass rate was calculated for each detected leak from a traditional component. Table 3 shows the calculated mass rates of ethylene and propylene from each component category. As seen in the table, although 69% of the leaking components were valves, connectors accounted for 66% of the calculated mass rate.

Table 3. Calculated mass rates from traditional sources at Site B.

Component Type	Number of Components		Total Mass Rate (Kg/hr)		Percentage of Total Mass		Average Mass Rate (Kg/hr)	
	E	**P**	**E**	**P**	**E**	**P**	**E**	**P**
connector	15	0	1.413	0.000	69.0%	0.0%	0.094	0.000
valve	33	1	0.635	0.004	31.0%	100.0%	0.019	0.004
TOTALS	**48**	**1**	**2.048**	**0.004**				

Notes:
E= ethylene
P = propylene

Comparison of Mass Rates from Bagging and Correlation Equations

Twenty of the 52 leaking components were bagged during the test. Table 4 compares the mass rates determined by bagging and those calculated using EPA correlation equations. Because the calculation of mass rate assumed all fluids to be pure compounds of a single constituent, the calculated results are for one species – ethylene. Again, this method may overestimate the quantity of the dominant constituent in a mixed stream; however, it is the most reasonable approach based on data readily available from Site B.

A comparison of the bagged and calculated mass rates of ethylene emissions shows mixed results as summarized below:

- In nine cases (e.g. tags 15792 and 15859), the bagged and calculated mass rates are of the same order of magnitude.
- In two of the nine cases where the bagged and calculated mass rates are of the same magnitude (nontraditional valve and tag 15910), bagging mass rates were larger than the calculated mass rates.
- In twelve cases, the results differ by one or more orders of magnitude. The calculated mass emissions for these twelve cases are larger than the mass rate determined by bagging.
- It is unclear why the CO_2 laser was unable to detect two leaks of about 2x10-4 kg/hr at components 15792 and 15846 while it detected leaks of similar and smaller size at other locations.

STUDY FINDINGS

From analysis of the field study data conducted as part of this project, the primary findings from this effort are as follows:

1. The field studies demonstrated that the portable optical gas-imaging device (CO_2 laser) was able to identify leakers while monitoring both traditional and nontraditional components under normal petrochemical plant operating conditions and good weather conditions (light wind, clear sky, summer temperatures).

Table 4. Comparison of bagging results with calculated mass rates at Site B.

Component Data		Plant Area	Seen by Laser	Bagging Result (Kg/hr)			Correlation Result (Kg/hr)		
Tag #	Type			E	P	M	E	P	M
valve near 1454[a]	Valve	Base Plant C_2= Prod. Pump	Yes	7.37E-03	ND[b]	4.87E-05	2.99E-03	NA	NA
1358	Connector	Base Plant C_2= Prod. Pump	Yes	3.02E-03	ND	1.31E-05	2.59E-02	NA	NA
15792	Valve	Base Plant	No	2.44E-04	1.36E-07	7.57E-07	8E-04	NA	NA
15884	Valve	Base Plant – Off-spec Line	Yes	6.20E-04	4.50E-06	1.41E-07	4.63E-03	NA	NA
15859	Valve	Base Plant – Off-spec Line	Yes	2.03E-03	1.97E-07	ND	4.1E-03	NA	NA
15846	Valve	Base Plant – Off-spec Line	No	2.00E-04	ND	4.94E-08	1.28E-03	NA	NA
15910	Valve	Base Plant – Off-spec Line	Yes	4.44E-03	ND	ND	1.25E-03	NA	NA
15918	Valve	Base Plant – Off-spec Line	Yes	1.813-02	ND	2.48E-07	3.66E-02	NA	NA
16639	Connector	Base Plant – Cold Ends	Yes	7.46E-03	3.12E-07	1.91E-07	2.85E-01	NA	NA
16854	Flange	Base Plant – Compressor	Yes	1.34E-02	7.20E-07	1.25E-08	2.45E-02	NA	NA
16358	Flange	Base Plant – Cold Ends	Yes	7.08E-04	1.20E-07	ND	3.78E-01	NA	NA
11737	Valve	Base Plant – Cold Ends	Yes	1.31E-02	7.53E-07	7.03E-07	1.423E-02	NA	NA
9930	Valve	Expansion Plant – Prod. Manifold	Yes	2.41E-04	2.47E-07	ND	4.26E-03	NA	NA
9933	Valve	Expansion Plant – Prod. Manifold	Yes	1.71E-02	1.29E-06	1.67E-06	6.19E-02	NA	NA
9951	Valve	Expansion Plant – Prod. Manifold	Yes	6.82E-04	1.23E-07	ND	1.03E-02	NA	NA
9934	Valve	Expansion Plant – Prod. Manifold	Yes	2.08E-02	ND	4.47E-07	1.02E-02	NA	NA
9988	Valve	Expansion Plant – Prod. Manifold	Yes[c]	1.16E-02	1.27E-02	1.50E-06	2.77E-02	NA	NA
10575	Valve	Expansion Plant	Yes	1.33E-05	ND	1.34E-07	6.19E-02	NA	NA

Component Data		Plant Area	Seen by Laser	Bagging Result (Kg/hr)			Correlation Result (Kg/hr)		
Tag #	Type			E	P	M	E	P	M
10577.2[d]	Connector		No	4.95E-04	ND	1.37E-07	NA[e]	NA	NA
10597.2	Connector	Expansion Plant	Yes	1.75E-03	ND	ND	4.47E-02	NA	NA
12404	Valve	Expansion Plant C_2= Prod. Pump	Yes	2.64E-03	ND	ND	2.18E-02	NA	NA
15496	Valve	Expansion Plant 2nd Level Platform	Yes	2.34E-02	1.74E-05	5.34E-03	2.68E-02	NA	NA

Notes:
E= Ethylene
P= Propylene
M= Methane
[a] **Nontraditional component**
[b] **ND = not detectable**
[c] Leak was through a pinhole on the bonnet and was only clearly visible when a corkboard was used as a background.
[d] This component was bagged although no leak was detected from component when area was screened with the CO_2 laser and no data was recorded in the field notes for this source. The data is retained because it is instructive of the mass rates that the CO_2 laser can detect.
[e] Not applicable as neither methane nor propylene was the species of interest or a dominant specifies. As such, no emission rate was calculated.

- At Site A, the CO_2 laser identified 95 leaking components (70 traditional and 25 nontraditional) in three process blocks; at Site B, the CO_2 laser identified 52 leaking components (49 traditional and 3 nontraditional) in 28 equipment areas within three operating units.

- At Site A, the mass rate of leaks, as determined by bagging, ranged from 5.63×10^{-7} kg/hr to 2.32×10^{-2} kg/hr. At Site B, the mass rate of leaks, as determined by bagging, ranged from 1.3×10^{-5} kg/hr to 2.34×10^{-2} kg/hr.

- At Site A, six of the ten bagged leaks (60%), detected by the CO_2 laser, had a mass emission rate in excess of 0.1 g/hr. Four of these six components (67%) were above 1 g/hr. At Site B, 18 of the 19 bagged leaks (95%) detected by the CO_2 laser were above 0.1 g/hr. 15 of the 19 components (79%) were above 1 g/hr. This is illustrated in Figure 2. As seen in this figure, it can be concluded that the CO_2 laser detects all leaks above 1 g/hr.

- Ethylene was the only species detected during testing at both Sites.

2. The majority of components detected as leaking had screening values above 1,000 ppmv. This result is in keeping with trends seen in other industrial studies.

   - At Site A, 96.7% of the 95 detected leaking components (traditional and nontraditional) had screening values over 1,000 ppmv; 63.3% had screening values over 10,000 ppmv.

   - At Site B, 83% of the 52 detected leaking components (traditional and nontraditional) had screening values over 1,000 ppmv; 56% of the 52 leakers were above 10,000 ppmv.

3. Additional testing of the CO_2 laser is warranted to more fully establish the sensitivity of the instrument to factors such as distance, reflective background, wind speed, sight lines, angle-of-view and hydrocarbon species, and conditions that might be encountered during routine use at a petrochemical facility.

   - In the field studies, leaks from as close as 4 feet to as far as 20 feet were detected. The component with the lowest mass rate was detected from a distance of greater than 10 feet. The highest leaking component was detected from a distance of 8 feet 1 inch. No attempt was made to determine the viewing threshold of any of the detected leaks.

   - The CO_2 laser successfully detected leaks against a variety of typical reflective backgrounds found at petrochemical facilities – smooth and rough surfaces; shiny, flat, painted and corrugated metal surfaces; concrete; and convex and concave surfaces.

   - Leaking components were detected at eye-level, beneath equipment, and at elevations above normal Method 21 reach.

   - The CO_2 laser successfully detected leaks with wind speeds ranging from 0 meters/second to 3.9 meters/second.

Figure 2. Emission rate of bagged components from Sites A and B.

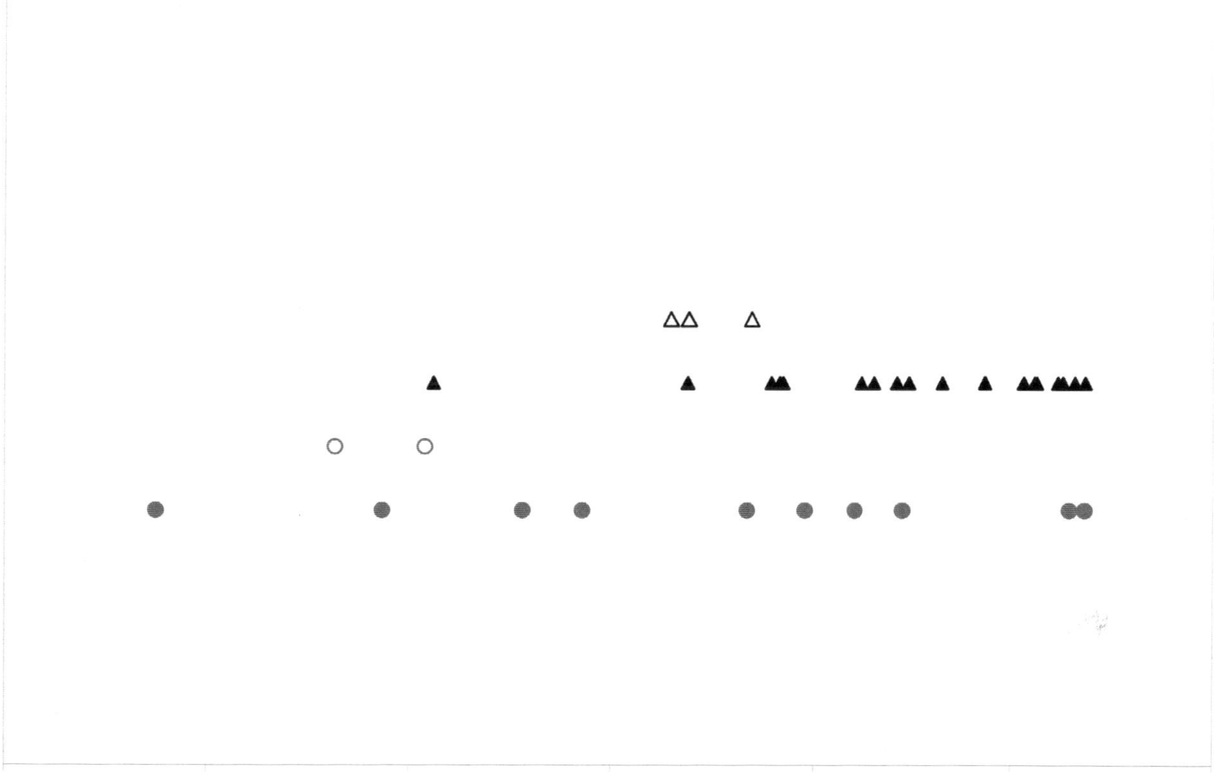

Emissions Rate (g/h (gram/hour)

○ Site A - not seen ● Site A - seen △ Site B - not seen ▲ Site B - seen

CONCLUSIONS

Analytical evaluations of optical gas imaging devices in the laboratory and in field studies have successfully demonstrated that these devices can locate leaking fugitive emission components in refineries and chemical plants under normal operating conditions. While these evaluations provide some promise that a more effective, cost-efficient method for identifying fugitive emissions may be available in the near future, additional work is required to further assess the capabilities of these devices. In particular, work is needed to further evaluate optical gas-imaging devices to determine the detection sensitivity of these devices to the broad range of chemicals and to various factors that would be encountered during routine use at petroleum and chemical plants, and to develop and demonstrate the ability of these devices to estimate fugitive mass emissions.

Ultimately, the end goal of these analytical evaluations – namely to obtain regulatory approval for the use of optical gas-imaging devices as an alternative work plan for fugitive emissions monitoring – provides an opportunity for a "win-win-win" situation, a situation which is often an elusive goal of environmental regulations. For industry, the approval of an alternative monitoring work plan provides a potentially more cost-efficient and effective method to monitor fugitive emissions. For regulatory agencies, the use of optical gas-imaging devices as an alternative or supplement to Method 21 provides a more credible means for reducing a primary source of emissions from industrial facilities. For the public, the reduction of ozone precursors and other air toxic emissions from refineries and chemical plants through a more e effective and credible monitoring procedure improves the health and well-being of those in communities near these facilities and others that may be exposed to these pollutants.

REFERENCES

U.S. Environmental Protection Agency, Method 21, Code of Federal Regulations, Title 40, Part 60, Appendix A – Test Methods.

American Petroleum Institute. *Analysis of Refinery Screening Data,* API Publication 310. 1997.

Smylie, M., et. al., *Measurement and Assessment of Equipment Leak Fugitives in Industrial Ethylene and Other Chemical Source*s. 2003

McRae, Tom. GasVue: A Rapid Leak Location Technology for Large VOC Fugitive Emissions. (Presentation at the CSI Petroleum Refining Sector Equipment Leaks Group, Washington DC, September 9, 1997.)

What Can We Do About HRVOC's?

By

Stephen Lang
Principal, Environmental Engineering
KBR

Richard Churns
Olefins Process Design
KBR

A Paper Prepared for the

2004 Spring National Meeting
American Institute of Chemical Engineers
Ethylene Producers' Meeting, New Orleans, LA
Environmental Session
26 April 2004

ABSTRACT

Presented in this paper is the background on environmental regulations in the Houston Galveston Area for highly reactive volatile organic compounds (HRVOCs) and the affect of these regulations on ethylene producers. Quantified are the small continuous ethylene plant vents typically flared, including pump and compressor seals, flare purge gas, control valve and relief valve leaks, periodic equipment blowdowns and other miscellaneous sources. The path forward to minimize these emissions or engineer recovery of these streams has to be based upon site-specific analytical work, as emission factor based estimations may not properly characterize these sources or provide a proper basis for cost effectiveness determinations.

INTRODUCTION

In December 2002 the Texas Commission on Environmental Quality (TCEQ) adopted regulations [TCEQ 2002] for control of highly reactive volatile organic compounds (HRVOCs) for the 8 county Houston-Galveston Area (HGA). HRVOCs are currently defined in Harris County as ethylene, propylene, 1,3-butadiene, and all isomers of butenes, and as only ethylene and propylene for the seven surrounding counties. The HGA is composed of Harris County where Houston is located and surrounding Brazoria, Chambers, Fort Bend, Galveston, Liberty, Montgomery, and Waller Counties. This highly industrialized area with world scale refining and chemical manufacturing sites has had a long-standing designation as a severe non-attainment area for the 1-hr ambient air quality standard for ozone. The HGA's non-attainment designation will also continue under the upcoming 8-hr ozone standard according to the EPA's assessments made in 2003 and as shown in Figure 1.

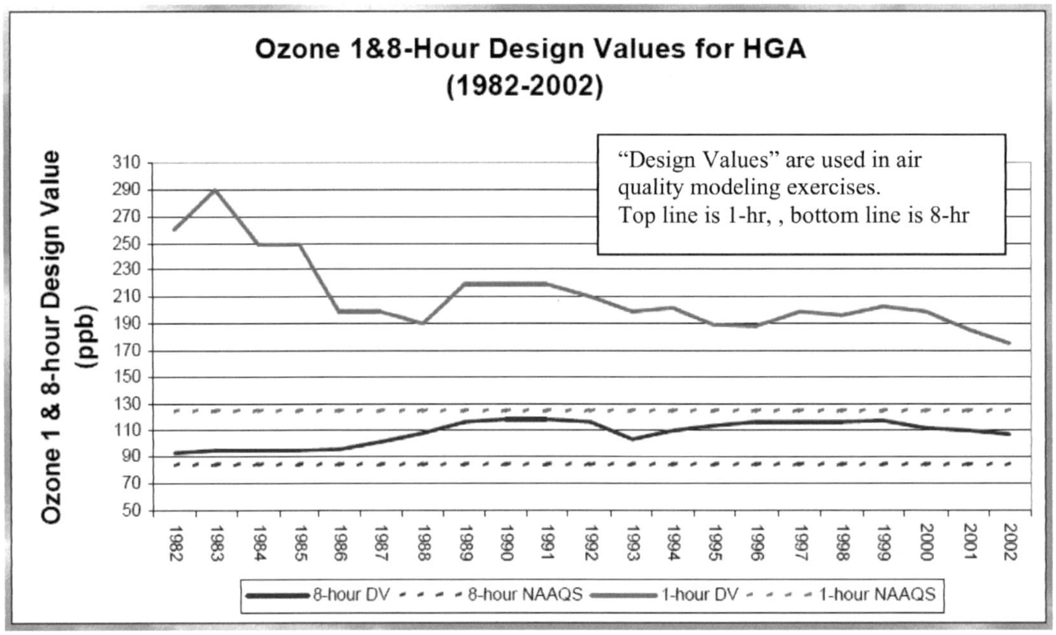

Figure 1. Long Term Ozone Levels versus Ambient Air Quality Standards for HGA

The HGA's new HRVOC regulations address emissions from four key industrial sources: fugitives, flares, process vents, and cooling towers and are a direct outgrowth of the Texas Air Quality Studies [TexAQS 2000] conducted in the summer of 2000. The TCEQ currently believes the distribution of HRVOC sources is as shown in Figure 2 [Forbes 2003] until new monitoring and inventory data is collected under these rules.

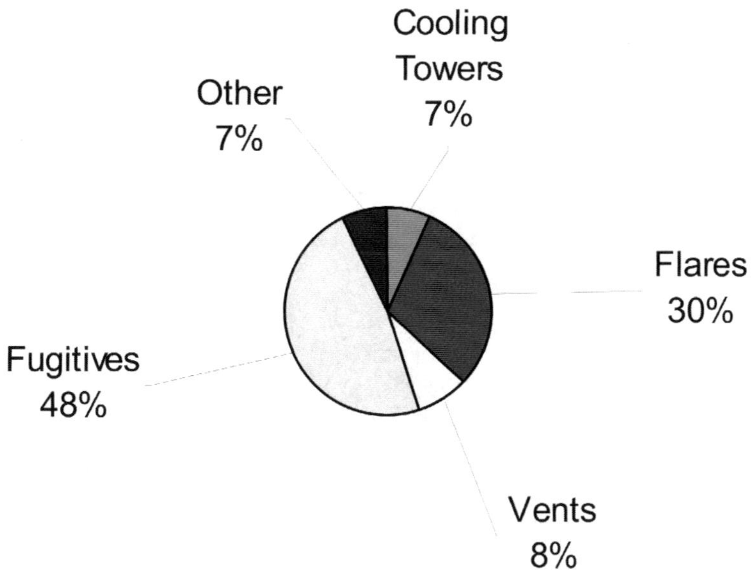

Figure 2. Current TCEQ HRVOC Source Distribution per Emissions Inventory

The TexAQS identified HRVOCs as responsible for the unique character and distribution of ozone formation in the HGA by an extensive scope of chemical and meteorological sampling at ground level monitoring stations, as well as at elevation via aircraft laboratories flying over the regional air-shed to capture air quality data. In summary, HRVOCs were found to induce prompt ozone formation when NOx and sunlight were present and then regional concentrations would build to higher levels from the slower photochemical smog reactions from other VOC and NOx sources such as mobile (traffic) and stationary industrial sources [Kleinman 2002]. Shown in Figure 3 is the relative reactivity of VOC and HRVOCs and in Figure 4 the high concentrations of numerous HRVOCs evident at a central Houston sampling location coincident with a high ozone peak [Allen 2003]. The HRVOC reactivity and emissions inventory is believed to account for what was otherwise considered to be "missing VOC" needed to generate the ozone levels found in the HGA.

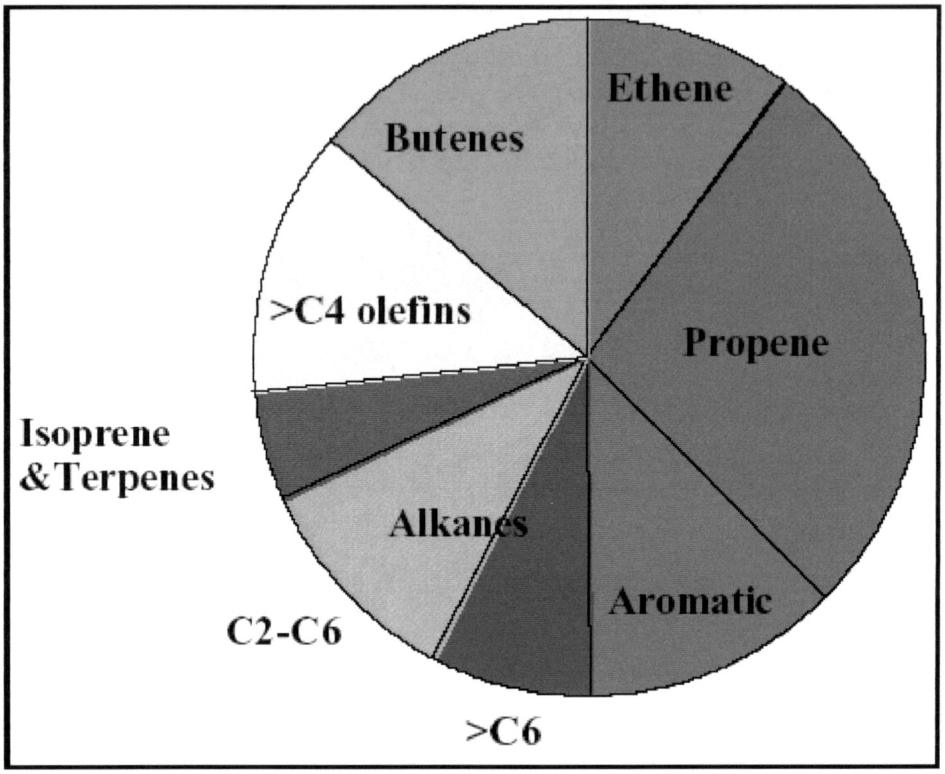

Figure 3 Relative Reactivity of VOCs versus HRVOCs in HGA

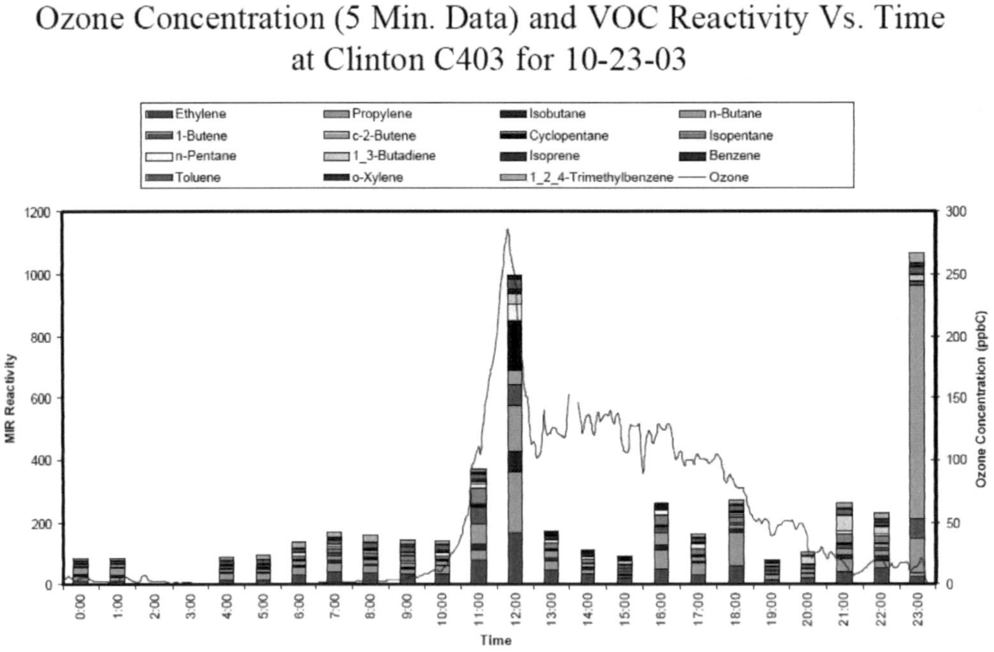

Figure 4. HRVOC Concentrations at An Air Quality Sampling Location

Shown in Figure 5 [Allen 2003] is a time-concentration snapshot view of an ozone episode developing across the HGA in October, 2003 originating along the Houston Ship Channel, and its extent driven by early day onshore winds and later day offshore winds.

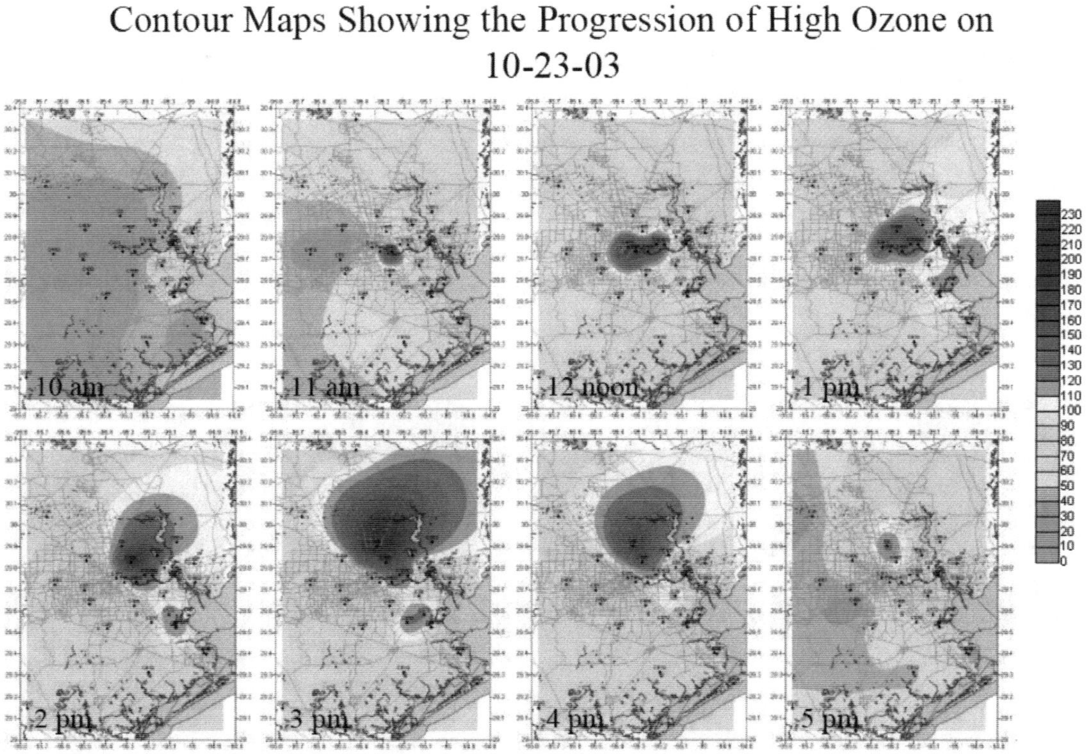

Figure 5. Ozone Episode in HGA

Because of the interrelationship of NOx and HRVOC emissions in the formation of ozone, the first round of HRVOC regulatory controls adopted in 2002 also offered TCEQ an approach for moderating the ongoing nitrogen oxide emission controls from a nominal area-wide 90% reduction to an 80% reduction. This allowed the TCEQ to relieve certain legal challenges to its recently implemented NOx cap and trade rules for the State Implementation Plan (SIP). However, this is believed to be a short-term relief for industrial sources, as TCEQ must develop mid-term corrections to its rules and fashion a program to come into attainment status for the 8-hr ozone standard. Since it is expected that other similar industrialized ozone non-attainment areas will use the HGA regulations as a potential model for solving their air quality problems, the initial impacts and experiences with the HGA rules deserve special attention.

HRVOC REGULATION

To understand the challenges faced by ethylene producers in complying with the HGA HRVOC rules, it is first important to understand how emission inventories are being used in air quality modeling studies related to the HRVOC control program. **S**ummary charts of baseline inventory data versus episodic release events [Allen 2003] are shown in Figure 6. It can be seen that large amounts of HRVOCs may be emitted on an intermittent basis from HGA sources. These emissions typically last for less than 24 hour durations and have varying frequency as shown in Figure 7.

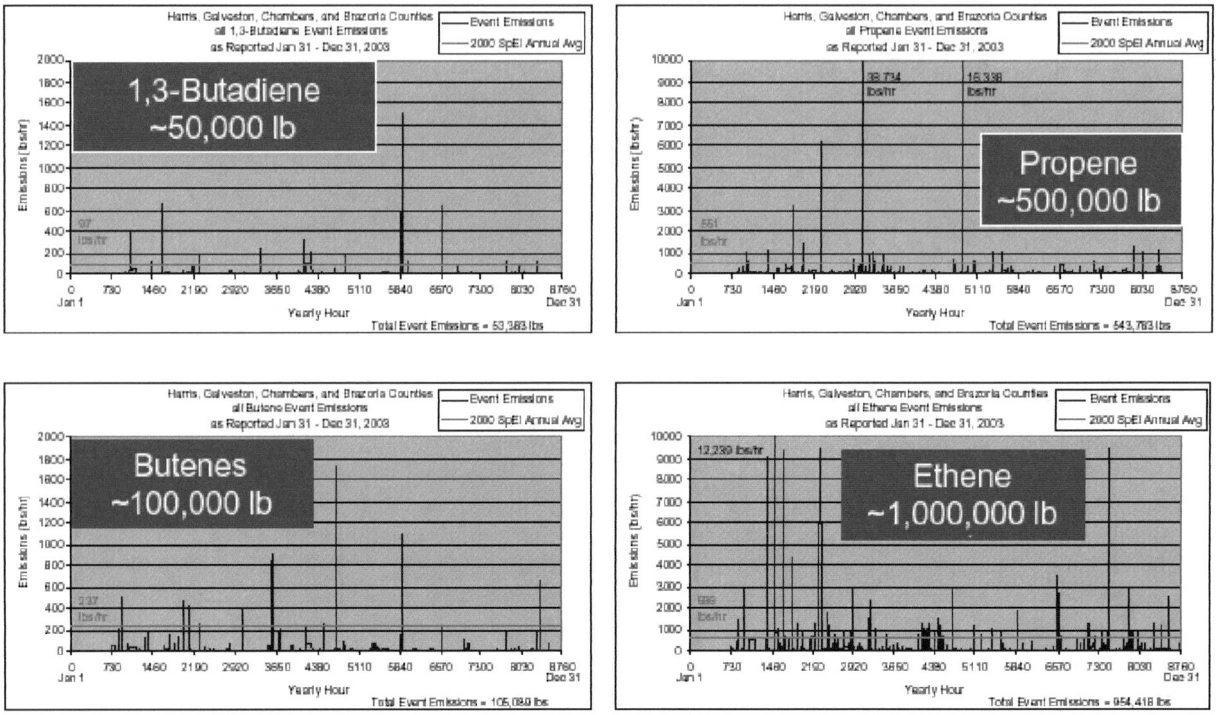

Figure 6. HRVOC Release Events versus Identified Emission Inventories

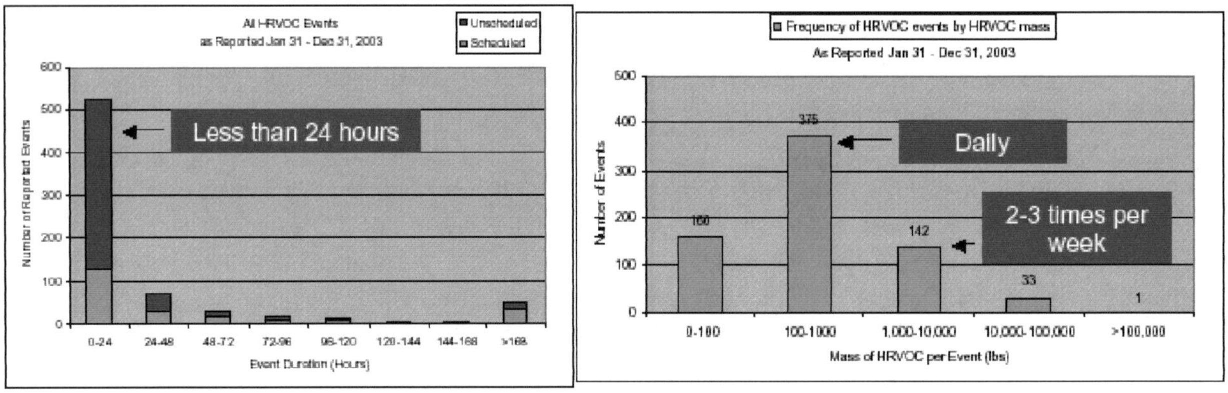

Figure 7. HRVOC Event Characterizations

While intermittent releases certainly have a short-term impact on regional air quality, it is difficult to account for their contribution to long-term annual ozone levels without using many simplifying assumptions in air dispersion modeling studies used for regulatory planning. As a result, high ambient ozone concentration episodes are identified from monitoring records and then utilized to calibrate dispersion modeling inputs and approaches. These short time scale studies become central to the air quality planning exercises and analyses. One such episode noted during the TexAQS [Smith 2004] is shown in Figure 8.

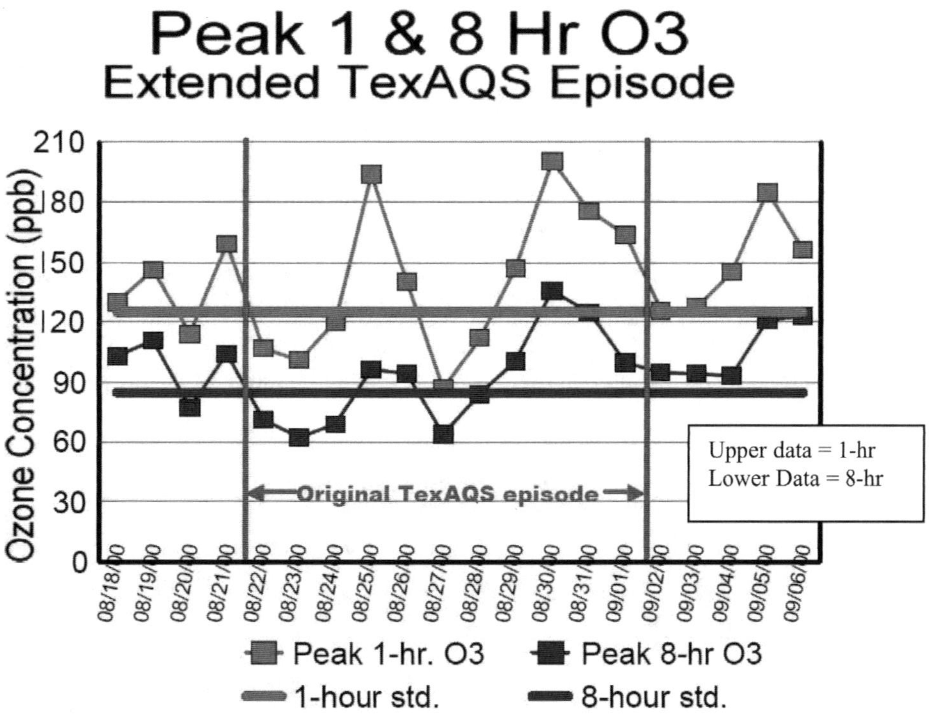

Figure 8. TexAQS Ozone Episode

To address the need for input to the steady-state emission type air dispersion models and ultimate development of HRVOC controls, the TCEQ established its Site-Wide Cap and Control regulatory program for process vents and flares including flow monitoring, testing, record keeping and reporting provisions. Under Section 115.722 rules , emissions of highly-reactive volatile organic compounds (HRVOC) at each permittee account relating to Vent Gas Control or Cooling Tower Heat Exchange Systems are limited to a 24-hour rolling average as specified in Table 6-2.1, Initial HRVOC Site-Cap Allocations: Harris County, and Table 6-2.2, Initial HRVOC Site-Cap Allocations: Seven Surrounding Counties, of the *Post-1999 Rate-of-Progress and Attainment Demonstration Follow-up SIP for the Houston/Galveston Ozone Nonattainment Area* adopted on December 13, 2002. Additionally, all flares shall continuously meet the requirements of 40 Code of Federal Regulations §60.18(c) - (f) as amended through October 17, 2000 (65 FR 61744) when vent gas containing volatile organic compounds is being routed to the flare.

Under Section 115.725 of the rules, Monitoring and Testing Requirements, each vent gas stream at a permittee account site must be tested by applying the appropriate reference method tests and procedures specified in Section 115.125 to establish maximum potential HRVOC hourly emission data expected during any operation not defined as an emissions event or a scheduled maintenance, startup, or shutdown activity. The data shall be used in accordance with the test plan required under Section 115.726 of the rules (relating to

Recordkeeping and Reporting Requirements) to demonstrate compliance with the control requirement of Section 115.722(a). This data will be critical in documenting applicability determinations if exemptions are sought under Section 115.727 related to rule thresholds of 5.0 wt% HRVOC and those containing less than 100 parts per million by volume (ppmv) HRVOC at all times.

In the TCEQ's SIP Tables 6.2.1 and 6.2.2 there are 17 ethylene plants identified in the HGA with an average adjusted HRVOC emission from cooling towers, flares and vents of 221 lb/hr. Data extracted from these tables for this set of ethylene plants is presented below in Table 1. Note that the TCEQ's Site-Wide Cap Program is implemented by plant site location and in Harris County, the HRVOC's are defined as ethylene, propylene, 1,3-butadiene and all butenes while in the seven surrounding counties, only ethylene and propylene count as HRVOC's. Thus this regional data can only be generally compared. The TCEQ is very interested in HRVOC speciation and looking for the monitoring and analytical portion of its rules to confirm inventories and provide greater detail for air quality modeling input.

Table 1. Control of Highly Reactive VOCs in Houston Galveston Area (1)

		Emissions for Site-Wide Caps, lb/hr				
Plants	**No. Plts.**	**Total**	**Adjust-ment**	**Adjusted Baseline**	**Control Level**	**Target**
HARRIS COUNTY						
Ethylene plants	9	2736	0.81	2208	0.68	706
Others (2)	83	4116	0.83	3402	0.64	1231
Total Harris County	92	6852	0.82	5610	0.65	1938
Ethylene Plants Average	9	304	0.81	245	0.68	78
Brazoria, Chambers, Fort Bend, Galveston, Liberty, Montgomery, and Waller Counties						
Ethylene plants	8	1754	0.89	1556	0.68	496
Others (2)	53	2154	0.89	1911	0.64	684
Total Seven Counties	61	3908	0.89	3467	0.66	1180
Ethylene Plants Average	8	219	0.89	194	0.68	62
All Ethylene Plants	17	4490	0.84	3764	0.68	1203
Average All Ethylene Plants	17	264	0.84	221	0.68	71

Notes to Table 1:

(1) Based upon TCEQ's HGA SIP Tables 6.1-1 & 6.2.2

(2) Primarily refinery, polymer, and specialty chemicals plants.

The target HRVOC emissions caps for the Harris County ethylene sites represent a 40.9% reduction from their adjusted emission inventory while in the surrounding 7 counties, ethylene sites represent HRVOC reductions of 46.3% . If the HRVOC emission distributions shown in Figure 1 are representative for an average ethylene plant, a challenge will be to find the most cost effective distribution of controls across the sets of fugitive, flare, vent, and cooling tower HRVOC emission sources to approach roughly a 50% HRVOC reduction. First however, the operators must clearly establish where their baseline emissions are compared to the TCEQ caps.

ETHYLENE PLANT FLARE SYSTEMS

To the engineer designing it, an ethylene plant flare system has several basic features:

- There is a warm flare header collecting vapors that are wet and are warm enough that they won't cause water that is present to freeze. Sometimes there is a warm liquid header. These two headers go to the warm flare drum where the liquids can be vaporized or pumped back to the process and the vapor can go on to the flare.

- There is a cold flare header collecting dry vapor and a cold liquid blowdown header collecting light, dry liquid. These two headers go to the cold flare drum where the liquids are vaporized and the combined vapor goes to the flare. In newer plants, this vapor is heated to above 32°F so it can be combined with the vapor from the warm flare and go to the flare in a single carbon steel line. This saves money compared to having a warm carbon steel line and a cold stainless steel line to the flare.

- Combined streams go to the main plant flare where the hydrocarbons are burned.

During the normal design contingency, total cooling water failure, the flare system loadings for a world scale generic ethylene plant may look like this to the design engineer:

Warm Flare	Flow, lb/hr
Quench Tower	620,000
Acetylene Reactor	90,000
C_3 Splitter	710,000
Ethylene Compressor	90,000
Debutanizer	50,000
Total Warm Flare	1,560,000
Cold Flare	
C_2 Splitter	240,000
MAPD Reactor	20,000
Total Cold Flare	260,000
Total Flow to the Flare	1,820,000

To the normal process engineer, essentially all of the hydrocarbons sent to a flare will burn to produce carbon dioxide and water. However, to the environmental engineer there are emission factors applied to flares that account for incomplete combustion and generation of carbon monoxide, nitrogen oxides, carbon based particulate matter, and other trace unburned hydrocarbons and products of incomplete combustion, such shown on Table 2 below.

Table 2. FLARE EMISSION CALCULATIONS BASED UPON: AP-42 Table 13.5-1. EMISSION FACTORS FOR FLARE OPERATIONS (1)

Component	Emission Factor (lb/MMBtu [HHV])
Total Hydrocarbons	0.14
Carbon Monoxide	0.37
Nitrogen Oxides	0.068
Soot	(2)

Notes to Table 2:

(1) 80&% propylene, 20 % propane stream feed to flare

(2) (2) Soot in concentration values, 0 micrograms per liter (ug/L); lightly smoking flares, 40 ug/L; average smoking flares, 177 ug/L; and heavily smoking flares 274 ug/L.

It should be appreciated that the notes regarding the AP-42's soot emission factor are of continuing interest due to the potential products of incomplete combustion and the fine size distribution of produced soot. Large steam assisted type flares, such as those found at ethylene plants, may only be truly smokeless up to 10-15% of their total capacity, due to the very large design loads and limits to induced air flow from steam injection.

Hopefully, the design contingency (~36,000 MMBtu/hr) never occurs. So to the plant operator, the flare system could look something like this, if every relief valve was tight:

Warm Flare	Flow, lb/hr
Quench Tower	0
Acetylene Reactor	0
C_3 Splitter 0	
Ethylene Compressor	0
Debutanizer0	
Total Warm Flare	0
Cold Flare	
C_2 Splitter 0	
MAPD Reactor	0
Total Cold Flare	0
Total Flow to the Flare	0

In this case, since there are no lost hydrocarbons, the flare produces nothing. However, since flares must however comply with the velocity, heating value, pilot and visible emission requirements of 40 CFR 60.18 (c)-(f), if nothing else is tributary from a plant, the flare header sweep gas flow, typically of natural gas, will contribute minor emissions per the AP-42 emission factors given in Table 2.

Where Do Flared HRVOC's Come From and How Much is Released?

The following items typically discharge into an ethylene plant flare system:

- Relief valves
- Control valves
- Relief valve bypasses and other vents
- Compressor seals ("sour" oil pots)
- Flare header purge gas

Relief Valves - A potential source of HRVOCs to the flare is leaking relief valves. However, relief valves are subject to API Standard 517 – Seat Tightness of Pressure Relief Valves. This sets the leakage at 0.60 scf/day for 'D' orifice and smaller and 0.30 scf/day for 'E' orifice and larger. 0.30 scf/day is less than 0.001 lb/hr so it takes more than 1,000 relief valves to leak 1 lb/hr. This is not to say that individual relief valves don't leak more than this. Hopefully the HRVOC rules' required monitoring programs will find leaking relief valves over time.

Control Valves - Another potential source of HRVOCs to the flare is control valves which leak to the flare. While there aren't nearly as many control valves discharging to the flare as relief valves, there are some large ones and the leakage standards aren't as strict. A typical ethylene plant has the following control valve dumps to the flare:

- Quench Tower Overhead Vapor – full plant flow
- Acetylene Reactor Effluent – net forward flow
- Acetylene Reactors – depressuring / purge flows
- C_2 Splitter – ethylene product flow

Control valve leakage is governed by ANSI/Fluid Controls Institute standard 70-2. This sets the following maximum leakage rates:

- Class II – 0.50-percent of full flow - air
- Class III – 0.10-percent of full flow - air
- Class IV – 0.01-percent of full flow - air
- Class V – 0.005-ml/min/psi/in – water - (about 0.001-percent of full flow)
- Class VI – 0.1-ml/min/in^2 - air - (about 0.0001-percent of full flow)

Although newer plants may be built to tighter standards, earlier plants considered "Tight Shutoff" to be Class IV leakage. Then the control valves mentioned above could be estimated by a reasonable person to have the following total and HRVOC leakage, for a 1.5 billion lb/yr, ethane based, ethylene plant:

	Leakage lb/hr	HRVOC's lb/hr
Quench Tower Overhead	39	20
Acetylene Reactor Dump	47	27
Acetylene Reactor Effluent	14	8
Off-spec Ethylene Product	19	19
Total	118	74

A 1.5 billion lb/yr, naphtha based, ethylene plant could have the following estimated leakage:

	Leakage lb/hr	HRVOC's lb/hr
Quench Tower Overhead	66	38
Acetylene Reactor Dump	51	37
Acetylene Reactor Effluent	17	14
Off-spec Ethylene Product	19	19
Total	152	108

Even if the plant were built with Class V or Class VI valves, after 20 or 30 years in service the seals must get brittle and they have to be considered to act more like Class III or Class IV valves.

Relief Valve Bypasses and Other Vents - Another potential source of HRVOC's is leaking relief valve bypasses or other vents to the flare. The leakage standard for block valves is API Standard 598 – Valve Inspection and Testing. This sets the leakage at 0.06 scf/day for 2-inch and smaller, 0.36 scf/day for 2½-inch to 6-inch, 0.60 scf/day for 8-inch to 12-inch and 0.84 scf/day for 14-inch and larger valves. 0.84 scf/day is less than 0.003 lb/hr so it takes more than 300 valves to leak 1 lb/hr. Again, this is not to say that individual valves don't leak more than this. Again, it is expected that under the TCEQ's HRVOC rules, the required monitoring programs will find any leaking valves.

Compressor Seals - Compressor seals at the cracked gas, ethylene, and propylene refrigeration compressors (CGC, ERC, PRC) are expected to pass 3-12 acfm per shaft seal depending on shaft size, and seal condition with age. Upwards of 3200 lb/hr of VOCs and 2600+ lb/hr of HRVOC emissions can reasonably be estimated at a world scale ethylene plant with six seals at the CGC, and two each at the ERC & PRC. While these type estimations are very equipment specific, they point at seals being prime candidates for early review and continual monitoring. While older compressor seals would be expected to approach the 12 acfm per seal rate, these seals are usually vented back to the process rather than sent to flare.

There is a small, usually unmeasured flow of de-entrained gas from the seal oil ('sour oil' pot vent) on the lube oil system which is often simply vented to the atmosphere. These emissions are often considered problematic due to the solubility of hydrocarbons at their operating conditions versus seal oil pot and oil console storage tank pressures, where HRVOC flashing is expected to occur through restriction orifices and vents. RO's on seal pots are sometimes found to have been sized under different economic conditions, when feedstock and thus product prices were much, much different than today. Seal pot vents have historically been simply routed to flare or recycled for hydrocarbon recovery. Oil console tanks have often been ignored or otherwise conservatively controlled. The key to managing HRVOCs in these vents cost effectively are to understand the true amount of flashing that can occur and any regulatory exemptions that may be available. As there are errors inherent in making estimations of these sources from assumptions, new pressure and vent composition data should be acquired. An estimation of flow through a set of 6 CGC, ERC & PRC sour oil pot vents (2 for each compressor) was made, resulting in emissions on the order of 143 lb/hr VOC with 128 lb/hr of HRVOCs.

HRVOCs From COOLING TOWERS

In addition to the site-wide cap requirements in the HGA for HRVOCs, there are monitoring, testing, record keeping and reporting requirements for cooling towers. A continuous flow monitor and continuous analysis to determine the strippable VOC concentration are required for each cooling tower inlet for systems of greater than 8000 gpm. For cooling water systems of less than 8000 gpm, continuous flow monitoring is still required but the total strippable VOC concentration may be determined by sampling at least twice a week. Monthly samples must be taken to determine the composition of the strippable VOCs. If the strippable VOCs are greater than 50 parts per billion by weight (ppb), daily samples must be taken until the concentration drops below 50 ppb. All of this is required by December 31, 2004.

The "leak rate equivalency" of this 50 ppb regulatory threshold may be looked at from several perspectives given that pickup of VOC from air scrubbing and leakage from process heat exchange systems may be expected to be contributing factors to the presence of VOCs in cooling water systems. A world scale ethane based ethylene plant may have a cooling water inventory of about of 1.5 million gallons and a recirculation rate of 125,000 gpm, while a naphtha based plant would have a cooling system approaching twice as large. Thus, 50 ppb of VOC within the cooling water inventory is equivalent to 625-1250 lbs of VOC, while a 50 ppb leak into the cooling water return water line that is totally stripped into the cooling tower exhaust is equivalent to a loss of 75-150 lbs/day of VOC.

Beyond historical knowledge of leaking exchangers, operators are encouraged to document their inventory of exchangers and note via equipment data sheets where cooling water supply and process pressure differentials are the greatest, to focus on potential leak locations. It is noted that exemption language under TCEQ rules may be available if minimum cooling water pressures are above process pressures by 5 psig. Monitoring downstream of individual problem locations should provide operators with higher concentrations of VOCs and thus better accuracy of leak rate determinations.

FUGITIVE EMISSIONS

Process unit components must be monitored and inspected quarterly, with a leak defined as 500 ppmv above background. Pumps, compressors and agitators must be visually inspected for leaks weekly. Equipment must be upgraded to new standards for bypass lines, pressure relief valves, pumps, compressors, agitators, closed vent systems and control devices. These changes were required by December 31, 2003. Independent third-party audits are required once every two calendar years with the first audit completed by December 31, 2004.

It is instructive to understand the emission factors used for emission inventories within the various chemical industry sectors, including factors addressing ethylene impacts. Table 3 provides emission factor values that are typically used for permitting and are generally accepted by regulatory authorities. Given the new attention to HRVOCs, and the proportion of the HRVOC inventory associated with fugitive emissions, it is expected that these factors will be revisited with special attention paid to component sizes and operating conditions. As it is common for ethylene plants to have tens of thousands of fugitive piping sources, this is no easy undertaking.

Table 3. Fugitive Emission Factors for Piping Equipment (lb/hr/source)

Source	Stream Composition	SOCMI Average Emission Factor (1)	SOCMI Without Ethylene Emission Factor	SOCMI With Ethylene Emission Factor (2)	SOCMI Non-Leaker Emission Factor	Typical Control Method	TCEQ 28VHP Control Efficiency, %
Valves	Gas/Vapor	0.0132	0.0089	0.0258	0.0003	A	97
	Light Liquid	0.0089	0.0035	0.0459	0.0004	A	97
	Heavy Liquid	0.0005	0.0007	0.0005	0.0005	--	0
Flanges	Gas/Vapor	0.0039	0.0029	0.0053	0.0002	B	30
	Light Liquid	0.0005	0.0005	0.0052	0.0002	B	30
	Heavy Liquid	0.00007	0.00007	0.00007	0.0002	B	30
Pumps	Light Liquid	0.0439	0.0386	0.1440	0.0041	C	100
	Heavy Liquid	0.0190	0.0161	0.0046	0.0046	C	100
Compressors	Gas/Vapor	0.5027	0.5027	0.5027	0.1971	D	85
Relief Valves	Gas/Vapor	0.2293	0.2293	0.2293	0.0986	E	100
Open-Ended Lines/Valves	All	0.0038	0.0040	0.0075	0.0033	F	100
Sampling Connections	All	0.0330	0.0330	0.0330	0.0330	G	100

Notes to Table 3:

1. Ethylene content less than 11%
2. Ethylene content between 11 – 85%

Control Methods for Table 3:

A. Quarterly leak detection and repair program. Leak definition is 500 ppmv
B. Weekly physical inspection of flanges.

C. All pumps are leakless design and are either dual seal, magnetic drive, or canned pumps
D. Quarterly leak detection and repair program. Leak definiiton is 2000 ppmv
E. Relief valves routed to control device or equipped with rupture disc.
F. Open ended lines/valves plugged,capped or blinded.
G. Closed-loop sampling station.

Given the API standard for RV leakage is 30 scf/day/rv, this is equivalent to ~ 0.0007 lb/hr/rv which appears negligible. Likewise, the API standard for valve leakage applied to vents and RV bypasses is 4 to 40 bubbles per minute per valve, equivalent to ~0.0002 lb/hr/valve which also appears negligible. These values approach the fugitive SOCMI Non-Leaker definitions in Table 3, and could lead one to questioning whether more definition of emission factors for different services aren't truly warranted. Table 4 provides order of magnitude estimates for VOCs and HRVOCs based on Table 3 type emission factors for a set of world scale ethylene plants. The ranges in emission values are significant but are indicative of the larger HGA group emissions and lead to the TCEQ's need for a consistent analytical basis for reporting emission inventories.

Table 4. Order of Magnitude VOC & HRVOC Emissions for Flares, Cooling Towers, Fugitives

	VOC, tpy	C2=, tpy	C3=,tpy	Valves, #	Flanges, #
Flares	1 - 225	10+	1 - 25	---	---
Cooling Towers	20 - 45	5 – 10	2 - 5	---	---
Fugitives	60 - 300	15 – 80	5 - 100	6,500 - 15,000	13,000 - 60,000

ANALYTICAL STANDARDS FOR FLARE EMISSION CALCULATIONS

Quantifying the small continuous vents to ethylene plant flares for environmental permitting needs must use equipment counts and emission factors or present other rational bases as described above for obvious reasons. However, for development of emission inventories at existing plants, there is no better definition than actual sampling of tributary streams and application of combustion efficiency factors. The TCEQ's new rules require monitoring stream properties of flows to flares by the end of 2004 with 2% accuracy for stream temperature, within 5 mm Hg accuracy for stream pressure, and within 5% accuracy for stream flow rate. These standards for flow determinations are coupled requirements for compositional sampling at least once every 15 minutes for basic HRVOC speciation and of any other components necessary to assure stream molecular weight determinations are within 5% accuracy.

TCEQ also has specified that the monitoring instrumentation be available 95% of the time on a yearly basis. Most sites will use their speciation data to compute heating value of flare tributary streams to assure compliance with EPA flare standards for that parameter under 40 CFR 60.18. When exit velocity and heating value standards are met, ethylene and propylene destruction efficiency may be assumed to be 99%, while other VOCs and HRVOCs may assume 98% destruction. If exit velocity and heating value standards are not met, then 93% destruction efficiency is to be used for emissions inventory calculations.

Also, to assure that flare tributary stream heating value meets minimum standards of 300 Btu/scf for steam or air assisted flares or 200 Btu/scf for non-assisted flares specified at 40 CFR 60.18(b)(3), natural gas or process purge gas may be used. With flare purge gas set to provide 0.1 ft/sec sweep velocity, in a 48-inch or 60-inch flare header this may become an appreciable amount of gas usage and potential HRVOC emission source depending on purge gas supply.

CONCLUSIONS

Because of uncertainty in historical HRVOC emission inventories, consistency in monitoring and emissions calculations is among the first regulatory steps being taken by the TCEQ. Information derived from open literature, as described above, is summarized in Table 5 to estimate potential overall VOC and HRVOC losses from world scale ethylene plant operations.

Table 5. Estimated Ethylene Plant VOC and HRVOC Losses

Emission Source	TCEQ Regional HRVOC Percentage Distribution (1)	Estimated VOC Losses lb/day	Estimated HRVOC Losses lb/day
Cooling Towers	7%	100 - 250	40 – 100
Flares	30%	5 - 1200	60 – 200
Fugitives	48%		
Relief Valves	---	300 - 1500	100 – 1000
Control Valves	---	115 - 155	75 – 110
RV Bypasses & Other Vents	---	---	---
Compressor Seals	---	3200	2600
Compressor Seal Oil Pots	---	145	130
Vents	8%	---	---
Others	7%	---	---

Note to Table 5:

(1) TCEQ Regional Average HRVOC Emissions Percentages are not source specific. They are presented for consideration of potential order of magnitude losses.

Normally if there are enough losses of VOCs or concentrated HRVOCs, a plant would have identified them and sought to recover them for product value first or for fuel value as a second priority. Using Table 5's midrange values, approximately 5000 lb/day of VOCs and 1200 lb/d of HRVOCs may be subject to determinations for control or recovery systems. Assuming the HRVOC losses and value are too small to justify product recovery systems, assigning a minimum heating value of 20,000 btu/lb VOC and $4.00/MMBtu fuel cost, this would be equivalent to 4.2 MMBtu/hr or roughly $150,000/year of fuel value.

It is believed that most if not all operators are controlling their compressor seal losses by recovery designs with only secondary purges going to flare. The next largest emission sources, relief and control valves are typically routed to flares, but operators may not have an appreciation for the magnitude of these or seal oil pot HRVOC losses. Operators will thus likely be faced with options for recovery that are economically unappealing. Definition of control cost effectiveness, in terms of net annual $ per ton HRVOC controlled will be required for decision support between control schemes such as flare header recovery compressors or oil scrubbers to enable recycle to process or fuel systems or alternatively installing small, enclosed, staged ground flares for ~1,000 - 10,000 lb/hr duty with all their regulatory instrumentation versus routing to their existing large flare systems.

As the site inventories of HRVOCs are established and certified, it is expected that there will be new site-wide caps in the future and a more certain basis for defining engineered controls. Thus there are high stakes for assuring emission inventories are correct and fully understood as soon as possible. Additionally, given there are always operational changes occurring between turnarounds, studying the impact of operations and maintenance on emission inventories should be a continuing major focus for operators.

RECOMMENDED ACTION PLAN FOR COMPLIANCE WITH SITE-CAP

The following steps are recommended for HRVOC sources to address their site-wide caps:

1. Review Historical Emission Inventory and Process Unit Data Bases
2. Conduct On-Site Systems Review including Screening Sampling and Analyses
3. Front End Engineering – Systems, Furnaces, Process, Control Systems, and Estimating
4. Develop Monitoring & Control Options and Cost Estimates
5. Develop Compliance Strategy, Sorting Emission Source Controls by Cost Per Ton HRVOC Emission Reduction
6. Prepare Report with Recommended Control Strategy to Attain Required Reductions
7. Prepare Required Air Permit Applications
8. Detailed Engineering
9. Field Work for Construction / Installation
10. Field Validation Studies

References / Bibliography

Allen et. al. 2003. Event Emissions and their impact on ozone formation in the Houston Galveston Area, Work supported by the Texas Environmental Research Consortium, Project H13

Breitenbach 2002. Accelerated Science Evaluation of Houston Ozone, Regional Air Quality Planning Committee, March 21, 2002

Forbes 2003. TCEQ HRVOC Rules Overview, SIP Development Team, 3/18/03

Kleinman et. al. 2002. GEOPHYSICAL RESEARCH LETTERS, VOL. 29, NO. 0, 10.1029/2001GL014569, 2002, Ozone Production Rate And Hydrocarbon Reactivity In 5 Urban Areas: A Cause Of High Ozone Concentration In Houston

Smith 2003. Houston/Galveston/Brazoria Modeling Update, TCEQ Technical Analysis Division, 1/21/04

TCEQ 2000. Revisions To The State Implementation Plan (Sip) For The Control Of Ozone Air Pollution Post-1999 Rate-Of-Progress And Attainment Demonstration Follow-Up Sip For The Houston/Galveston Ozone Nonattainment Area Texas Commission On Environmental Quality, December 13, 2002, Rule Log No. 2002-046a-Sip-Ai

TCEQ 2002 Enhancements to the State-of-the-Science Photochemical Modeling**,** at URL http://www.tnrcc.state.tx.us/air/aqp/airquality_impscience.html#section2

TCEQ 2002. Conceptual Model For Ozone Formation In The Houston-Galveston Area, Appendix A To Phase I Of The Mid Course Review Modeling Protocol And Technical Support Document, Technical Support Section, Technical Analysis Division TCEQ, December 13, 2002

TCEQ 2002. HRVOC related SIP rules under 30 TAC 115.10 (17)(A) & (B)

TCEQ 2003. Revisions To The State Implementation Plan (Sip) For The Control Of Ozone Air Pollution Complete Attainment Demonstration Sip For The Houston/Galveston Ozone Nonattainment Area, Updated: September 8, 2003

TexAQS 2000. at URL http://www.utexas.edu/research/ceer/texaqs/

ETHYLENE PLANT COOLING TOWER MONITORING AND

HEAT EXCHANGER REPAIR METHODS

Lisa LeBlanc

Huntsman Petrochemical Corporation
4241 Savannah Avenue
Port Arthur, Texas 77640

Prepared for Presentation at: AICHE 2004 Spring National Conference
Session T8002 – Ethylene Plan Environmental Issues

© Lisa S. LeBlanc, Huntsman Petrochemical Corporation

April 2004

Unpublished

AIChE shall not be responsible for statements or opinions contained in papers or printed in its publications

Abstract

This paper will begin with a discussion of cooling water monitoring for an ethylene plant cooling tower that has been stringently regulated since 1998 by a Texas Commission On Environmental Quality (TCEQ) air permit. The air permit contains cooling tower monitoring requirements similar to the heat exchange systems section of the Ethylene MACT. Details will be presented for heat exchanger leak episodes, including extraordinary measures that have been taken over the past five years to keep the ethylene plant online and reduce the shutdown flaring emissions associated with the exchanger repairs.

Historical Background

The Huntsman Light Olefins Unit (LOU) is part of the Huntsman Aromatics and Olefins (A & O) Facility in Port Arthur, Texas. The LOU began producing ethylene and propylene upon startup in the late seventies as part of Texaco Chemical Company. The facility was purchased in 1994 by Huntsman Petrochemical Corporation and now operates as part of the Huntsman Jefferson County Operations (JCO) Complex in Jefferson County, Texas.

The olefins unit cooling tower operates at a water circulation rate of approximately 120,000 gallons per minute. The first structured periodic monitoring program for the LOU cooling tower was benzene monitoring which began in early 1995. A structured monitoring program for other Volatile Organic Chemicals (VOC) began in late 1997. Prior to that time VOC monitoring was conducted on an as-needed basis. The cooling tower monitoring program has continued since that time with a variety of monitoring methods. Figure 1 below shows a simplified flow diagram for the LOU Cooling Tower.

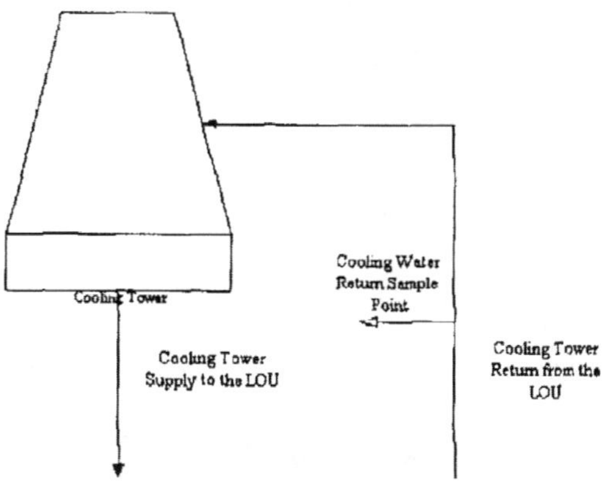

Figure 1
Light Olefins Unit (LOU) Cooling Tower

The air permit for the LOU cooling tower was amended in 1998 to include permit requirements for VOC monitoring and heat exchanger repair similar to the Ethylene MACT Heat Exchange System requirements in 40 CFR Subpart F. At that time the definition for a leak into the cooling tower from a heat exchanger was set at 50 ppbv benzene and 100 ppbv VOC. The following leak detection methods were required by the 1998 permit amendment:

- Daily benzene monitoring with a method similar to EPA Method 8260 performed by the A & O lab
- Weekly monitoring of VOC by EPA Method 8260

- Weekly monitoring of VOC by EPA Method TO14 and TO15
- Weekly VOC monitoring by the El Paso Method.

The monitoring program was altered again in late 2002 when cooling water monitoring requirements were reduced to daily benzene and weekly El Paso tests. The leak definition remains the same as originally set in 1998. A provision in the current permit also requires a sample to be taken for EPA Method TO14 and TO15 analysis in the event that the weekly El Paso reading exceeds a concentration of 30 ppbv in the cooling water return from the LOU. The cooling tower monitoring methods used at A & O are described in the following section. A copy of the current monitoring permit provisions is included in an attached appendix.

Monitoring Methods

In early 1995 the Huntsman A & O lab developed a Gas Chromatography (GC) test method for benzene based on EPA Method 8260 (i.e., "The Purge and Trap Method"). The method detection limit for the A & O lab benzene test is 3 ppbv. The plant has been using this method daily since January 11, 1995. This has provided an extremely reliable way to detect trends and leaks and to track the efficiency of repair methods for over nine years. The greatest advantage of having an in-plant benzene monitoring program is that we know almost immediately when we have a benzene leak. It is also a very inexpensive test method and allows rapid troubleshooting for the location of a leaking exchanger.

The El Paso air stripping method was developed by the El Paso Chemical Company to detect the presence of strippable VOC in a water stream. The El Paso sampling apparatus simulates the VOC stripping from the cooling water that would occur in a cooling tower by flowing air counter to the current of a stream of cooling water through a large cylinder This test method is described in the TCEQ Air Quality Permit Technical guidance for Chemical Sources: Cooling Towers, February 2001, Draft RG-108.

The El Paso Method has both positive and negative attributes. First the positives; the El Paso method gives an immediate result. This allows a plant to begin leak source location tracking immediately, allowing for more timely repairs and reducing emissions to the atmosphere. Additionally, the conversion calculation used to calculate concentration value in the cooling water from the El Paso test reading gives you a good estimate of the size of the leak into your cooling water system. Also, the test is reasonably repeatable. And, in Texas, the El Paso test is a recognized method of cooling tower leak detection in air permits. It's also an inexpensive test. A facility can, if desired, construct an El Paso test board and can, if they have a monitor, take numerous El Paso readings for little cost other than manpower.

However, there are some drawbacks to the El Paso method.. First, it gives a result in unspeciated VOC that includes hydrogen, methane and other non-VOC. This makes tracing the source of the leak into the cooling tower more difficult, and it can be misleading. For example, if an olefins facility has a methane leak from an exchanger around the demethanizer, the El Paso reading would indicate evidence of a VOC leak into the cooling water system. Also it's not an EPA recognized test method. So, while it is good for TCEQ permit compliance, another method must be used to demonstrate compliance with the Ethylene MACT heat exchange system requirements.

As part of a permit requirement, Huntsman began working with the TCEQ in 1998 to develop a method for cooling tower monitoring that would give repeatable speciated VOC results. There are standard methods that are very reliable for detecting heavy components, such as benzene, in a cooling tower return. EPA Method 8260 has been used since 1995, as previously mentioned, for detection of benzene and other heavy components in the LOU cooling water return. The problem, particularly for an olefins cooling tower, is trying to find a method of sampling and analysis that will ensure that the light hydrocarbons are not lost before they can be analyzed.

In 1995 Huntsman started using the "Air Force Method" for detection of light hydrocarbons in water. To provide a sample for this test, a cooling water sample was collected in a zero head space vial and then capped quickly. The analysis for this method is a proprietary method of an outside lab, SPL of Houston.

The primary advantage of the Air Force Method is that very light hydrocarbons can be identified. The most important chemicals for an olefins unit that can be identified by the Air Force Method test are ethylene and propylene. Methane can also be identified by this method. Upon comparison to other methods, probably due to sampling methodology, this method was not considered repeatable enough to provide ongoing assistance in heat exchanger leak detection. It is still used if needed for chemical speciation of the cooling water return.

A Gas Chromatograph Mass Spectrometer (GCMS) was purchased for the A & O Lab in early 1999 for cooling water analysis. Along with the GCMS, a method to sample and introduce the water to the GCMS for analysis that would prevent the light hydrocarbons from escaping was developed. For this method a sample is taken of the cooling water in a pressurized sample cylinder, then injected directly into the GCMS. The GCMS method is good for speciation of the components that are in the cooling water and the sampling method met the "closed loop" requirement such that no light hydrocarbons were lost during sampling. However, there were some problems that caused the lab to abandon this as an acceptable cooling water sampling and analysis method.

The initial investment for GCMS equipment was high and the standards for the GCMS for the cooling water method were difficult and expensive to obtain. Also, since the method detection limit for most of the compounds were so low there was interference from airborne hydrocarbons in the lab. A separate room was constructed to correct the problem. However, enough problems persisted that the GCMS was abandoned as a cooling water monitoring method. It was determined that other methods could provide speciation more easily and less costly than the GCMS.

In order to demonstrate the repeatability of the Huntsman benzene lab test, EPA Method 8260 performed at an outside lab was used in a side-by-side comparison to the Huntsman lab method for three years. Samples for 8260 analysis were taken once weekly at the same time the samples for the lab test were taken. The A & O lab was able to demonstrate that the Huntsman method was a reliable, repeatable test method for benzene that produced results statistically matching the results from the 8260 test. Therefore, during the August 2002 permit amendment the TCEQ removed the requirement for the 8260 test.

The methods that have proved to be most efficient at yielding repeatable speciated VOC results have been EPA Methods TO14 and TO15. These methods were first required by a 1998 TCEQ permit amendment. The sampling procedure uses an El Paso apparatus as a sampling device with the exhaust of the El Paso apparatus captured into a vacuum canister for a specified period of time. The canister is then sent to an offsite lab for analysis.

The list of analytes for TO-14 and TO-15 include most species of hydrocarbons present in a olefins unit cooling tower; therefore, it is a very good way to speciate VOC present in cooling water return. One problem with the TO14 and TO15 tests include the length of time it takes for sample results to be turned around which can be several days to a week or more. Also, since the sample is taken with an El Paso apparatus, any inaccuracy that is inherent in the El Paso method is reflected in the results of the TO-14 and TO-15 test.

After completing more than nine years of cooling water monitoring, for our facility the least expensive monitoring program that gives results satisfying the need to quickly locate and repair leaks occurring in the olefins cooling tower system is a daily test for benzene, a weekly El Paso test for VOC and the TO14/15 for VOC speciation as needed.

Locating Exchanger Leaks into the Cooling Water System

When the VOC level exceeds 100 ppbv or the benzene level exceeds 50 ppbv in the LOU cooling return as indicated by the monitoring program described previously, the plant must locate and repair the exchanger leak according to A & O air permit provisions. The time table for repair of a heat exchanger leaking into the LOU cooling water system is forty-five days from the date any analysis results are obtained indicating the presence of a leak. However, typically when an increase above baseline is noted, operations personnel will begin to look for suspected leaking exchangers.

The olefins unit has a Cooling Water System Troubleshooting guide that systematically allows for location of a leaking exchanger. Figure 2 below shows a simplified drawing of the LOU cooling water system. The system contains sample points on the trunks of the cooling water lines and at each of the exchangers throughout the LOU to allow for a stepwise approach to locate the leaking exchanger as shown in the following plot plan.

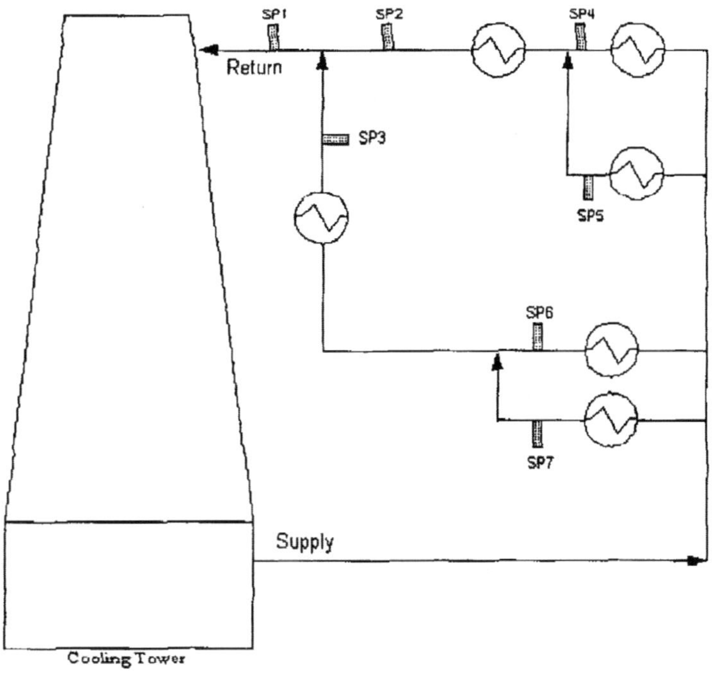

Figure 3
Simplified Troubleshooting Diagram

Upon verification of elevated levels of benzene or VOC compounds in the cooling tower return sample point, SP-1, troubleshooting is initiated. This would involve tracing back through the unit cooling water system by testing at each of the sample points, SP-2 and SP-3, with an El Paso test to indicate which side of the cooling water trunk contains the leak. This approach is continued until the leaking exchanger is located. Note: The LOU contains more than fifty exchangers in VOC service and the exact leaking exchanger can normally be located within one to two days. When the leaking exchanger is located the plant will take steps to eliminate the source of the leak.

Episodic Heat Exchanger Leak Events

There have been approximately twenty separate heat exchanger leak episodes into the LOU cooling water system since 1995. Some repairs of leaking exchangers have been as simple as isolating exchangers from the process, but most have been more complicated requiring the shutdown of all or part of the LOU. The plant staff has also found ways to repair exchangers online and keep the LOU operating. The following is a history of three episodes of LOU leaking heat exchangers including equipment repair. Some of the leaks occurred prior to the implementation of a monitoring program at the LOU, but the experience gained is now used to help meet the more stringent permit repair requirements that have existed since late 1998.

In early 1995 a heat exchanger leak into the LOU cooling water return was identified. Elevated VOC in the LOU cooling tower was noted by a chemical treatment vendor when quantities of treatment chemicals increased. The leak was then verified by further testing. The leak was traced to eight exchangers, four sets of two, in the propylene refrigeration section of the LOU.

Valves were not available to isolate the exchangers from the process. It was determined that the pair of exchangers that were the largest contributors of VOC to the cooling tower return could be repaired at reduced process rates without completely shutting down the LOU if appropriate isolation equipment was installed. Locations were selected and stopples were installed via hot tap. Once the stopples were installed the selected exchangers were repaired.

The propylene refrigeration exchangers continued to have leaks, with episodes occurring through 1997. During a major plant shutdown in 1997 the bundles in all eight exchangers were replaced with welded tube sheets.

Another leak episode occurred in July of 1998 when a VOC leak was noted in the LOU cooling tower by the El Paso method. The leak was traced to an exchanger in the propylene refining section of the LOU, the propylene tower overhead condenser bank. No isolation valves existed around the leaking exchanger to allow the exchanger to be isolated from the rest of the unit.

In order to address the exchanger leak the plant installed temporary piping to bypass the propylene tower, allowing propylene tower feed to be downgraded and routed to a cavern normally used to store refinery grade propylene (RGP). The propylene refining section of the LOU was then taken offline for the exchanger to be repaired.

During the most recent major plant shutdown in late April 2003, bypass valves were added to allow the exchanger to be isolated for maintenance. Additionally the temporary piping that routed propylene to the storage cavern was upgraded to permanent piping.

In early August 2000 benzene was identified in the LOU cooling water return by the daily GC analysis. The total VOC in the cooling tower return was also elevated as indicated by the weekly El Paso reading. An interstage cooler in the cracked-gas compression train was identified as a leaking exchanger. By mid-August 2000, the exchanger leak increased to the point that a permit action leak definition was exceeded. There were no existing bypass valves to allow the exchanger to be isolated from the process. The LOU had forty-five days to find a way to repair the leak on line or, as required by the air permit, the plant would have to shut down in order to remove the exchanger from service.

A booster pump was added to the cooling water side of the leaking exchanger system, increasing the cooling water pressure above the process pressure, thus preventing VOC from leaking into the cooling water. This arrangement was very effective in slowing the rate of VOC leakage into the cooling tower. The VOC concentration in the cooling tower dropped below the permitted action level shortly after the new equipment was placed on line. In early 2001 the exchanger leak increased to the point that cooling water was leaking so rapidly into the process that it became necessary for the plant to shutdown for exchanger repair.

During the most recent plant shutdown for maintenance in April 2003, the design of the cracked-gas compressor interstage coolers was improved to reduce erosion between the floating head and the impingement plate. Also, the exchangers were upgraded to welded tube sheets.

Conclusion

In conclusion, the emissions from the LOU cooling tower have dropped dramatically from several thousand tons per year in the mid nineties to less than twenty tons per year in each of the last five years. The monitoring program and the exchanger design improvements practiced at the LOU have allowed the facility to greatly reduce the emissions from the LOU cooling tower from heat exchanger leak episodes. The facility will continue to track emissions from the cooling tower in order to ensure that any past improvements have been effective and any new improvements needed to reduce emissions are implemented.

Appendix 1
A & O Facility Cooling Tower Permit Conditions

Flexible Permit Nos. 16989 and PSD-TX-794

27. Cooling Tower and Heat Exchanger System in VOC Service **(7/02)**

The holder of this permit shall perform sampling and make repairs as necessary to demonstrate ongoing compliance with the emission limits for the Cooling Tower(s) (EPNs LOUCOOLTWR and AUCHXUCLTR). The results of all monitoring, maintenance, and repair efforts affecting the cooling tower and heat exchanger system(s) shall be recorded and such records shall be maintained for a period of five years. The records shall be made available to TNRCC personnel or other air pollution control agency upon request.

This permit does not authorize any emission rates from the cooling tower and heat exchanger system(s) other than those listed on the attached maximum allowable emission rate table (MAERT). Emissions from repairs and delay of repair are not authorized for emission rates in excess of those listed on the attached MAERT.

A. Monitoring and Detection Limits: The VOC emissions associated with cooling tower water shall be monitored weekly with an approved method. The sampling method shall be the air stripping method known as the "El Paso" method, which is described in the TNRCC Publication: "Air Quality Permit Technical Guidance for Chemical Sources: Cooling Towers, TNRCC Publication No. RG-108, May 1997 (Revised)." The sample port for the water returning to the cooling tower shall be located on the top of the horizontal section of the water line returning to the cooling tower.

The minimum detection level of the overall testing system shall be 0.030 ppmw or 30 ppbw VOC (concentration VOC in water entering the cooling tower). The minimum detection limit for the air stripped VOC shall be 0.500 ppmv (concentration VOC in the stripping air). Calibration standards shall include at least zero ppmv and 10 ppmv VOC in air (as methane). The concentration of VOC in the stripped air may also be determined by EPA Methods TO14 and TO15. The testing, analysis, and recordkeeping shall include the air contaminant speciation, physical characteristics, and instrumentation response factors used to convert from the of air contaminant concentration in the stripping air (ppmv) to air contaminant concentration in the cooling water (ppmw).

If the El Paso method is used to determine the concentration of strippable VOC in the water returning to the cooling tower and the water returning to the process, the net concentration of strippable VOC in the water shall be used to determine emissions from the cooling tower (this concentration is the difference between the concentration of VOC in the water returning from the process and the water supplied to the process). If the El Paso method is used to determine the concentration of strippable VOC in the water returning to the cooling tower without sampling the water returning to the process, the concentration of VOC in the water supplied to the process shall be taken as zero.

Alternate methods may be approved by the TNRCC Engineering Services Team.

B. Action Level: The action level shall be a net concentration of strippable VOC in the water returning to the cooling tower of 0.100 ppmw or 100 ppbw. If the action level is exceeded and cooling tower emissions are less than the maximum allowed by the MAERT, the permittee shall comply with the repair requirements of paragraph C below, except as provided in paragraph D.

C. Repair: If the concentration of VOC in the water exceeds the action level as defined by paragraph B, and cooling tower emissions are less than the maximum allowed by the MAERT, then the cooling tower and heat exchanger system shall be repaired so that the concentration of VOC is reduced to below the action level. The repairs shall be made as soon as practical but not later than 45 calendar days after the permittee receives results of the monitoring tests indicating an exceedance of the action level, unless the permittee demonstrates that the results are due to a condition other than a leak.

D. Delay of Repair: For the purposes of this permit condition, delay of repair means not performing the repairs within 45 calendar days after the owner or operator receives results of monitoring tests indicating a concentration of VOC in the water exceeding the action level described in paragraph B. The permit holder may delay repair as provided for in paragraphs (1) through (5) below. Records of a decision to delay repair shall state the reasons repair was delayed and shall provide details to demonstrate compliance with paragraphs (1) through (5) below. Records shall be maintained on site for at least 5 years and shall be provided to the TNRCC or other air pollution agencies upon request.

(1) Delay of repair of heat exchange systems for which leaks have been detected is allowed if the leaking equipment is isolated from the cooling tower system.

(2) Delay of repair of heat exchange systems for which leaks have been detected is allowed if the pressure of the cooling water throughout the leaking exchanger is increased to a pressure greater than the process side of the exchanger, thus reversing the direction of the leak.

(3) Delay of repair may be allowed if repair is technically infeasible without a total unit shutdown or a partial unit shutdown and a shutdown is expected within the next two months.

(4) Delay of repair may be allowed for up to 120 calendar days if necessary parts are not available.

(5) Delay of repair may be allowed if repair is technically infeasible without a shutdown and the shutdown would cause greater emissions than the potential emissions from delaying repair. The permit holder may delay repair until the next shutdown of the process equipment associated with the leaking heat exchanger.

28. In addition to the sampling required in Special Condition No. 27, sampling using EPA Methods TO14 and/or TO15 is required to speciate the VOC emissions from the air stripping apparatus only in the event of a VOC concentration in the water greater than 0.030 ppmw or 30 ppbw as indicated by an El Paso reading. **(7/02)**

29. In addition to the requirements of the Special Condition No. 27 on Cooling Tower and Heat Exchanger System in VOC Service, the permit holder is also required to monitor for benzene in the cooling water. The concentration of benzene in the cooling water shall be monitored daily with EPA Method 8260 or the previously approved Huntsman lab method, at a minimum detection limit of 0.010 ppmw or 10 ppbw (concentration of benzene in the water). Alternate methods may be approved by the TNRCC Engineering Services Team. The action level for benzene shall be a concentration of benzene in the cooling water of 0.050 ppmw or 50 ppbw. If the concentration of benzene in the cooling water is greater than or equal to the action level, the permittee shall comply with the repair requirements of paragraph C, except as provided in paragraph D, of the Special Condition No. 27 regarding Cooling Tower and Heat Exchanger System in VOC Service. (7/02)

Energy Efficient Burner Design and Maintenance in the Olefins Industry

Darin R. Foote

John Zink Company, LLC
Tulsa, OK

Abstract

Energy efficiency has long been at the forefront of the design of fired equipment in the olefins industry. This paper will first discuss the complexities of Ethylene Cracking Furnaces, which should be considered in energy efficient burner design and preventative maintenance. General burner design considerations will be discussed. Advantages to next generation technology will be noted. General operating considerations for fuel efficiency will be covered. In addition, the importance of fuel gas filtration will be discussed. Lastly, recommended inspection and maintenance practices designed to keep fired equipment operating fuel efficient will be presented.

Introduction

Energy efficiency has long been at the forefront of the design of fired equipment in the olefins industry. Organic chemicals (ethylene, propylene, and many others) represent the highest energy consumption of all chemical sectors, accounting for about 49% of total industry energy use. Generally, 50% of energy consumed in organic chemicals manufacture is fuel for endothermic reactions. The remainder of energy consumed is in the form of feedstock (natural gas and LPG).[i]

Ethylene producers have long understood the need for burner maintenance to operate efficiently. In recent years, stack emissions regulatory requirements have become more stringent. This, coupled with a greater understanding of the relationship of burners and their effect on the rest of the system has lead many producers to place greater emphasis on burner maintenance as a preventative measure for problems such as radiant tube failure, tube coking, and fuel efficiency.

This paper will first discuss the complexities of Ethylene Cracking Furnaces, which should be considered in energy efficient burner design and preventative maintenance. General burner design considerations will be discussed. Advantages to next generation technology will be noted. General operating considerations for fuel efficiency will be covered. In addition, the importance of fuel gas filtration will be discussed. Lastly, recommended inspection and maintenance practices designed to keep fired equipment operating fuel efficient will be presented.

Complexities of the Ethylene Cracking Furnace in Burner Design

The environment in the Ethylene Cracking Furnace presents a set of challenges that add to the difficulty of design and maintenance of burners. Some of these complexities are unique to the olefins industry and some common to most of the refinery and petrochemical industry.

First, the kinetics of ethylene cracking is very much temperature and residence time dependent. The feedstock needs to reach a design process temperature and remain at temperature for a specific residence time. Burners heat a refractory wall, which in turn radiates to the process tube coil. This process is dependent on the heat flux through the process tubes. Any disturbance in the process flow or heat flux through the tubes can cause a disturbance in yields. If allowed to continue unchecked, disturbances in the process flow or heat flux can cause the cracking reactions to go to completion or coke. Coking is classified as a severe thermal cracking reaction were the end products are gases such as H_2, H_2O, CO_2, light hydrocarbons, and a solid carbonaceous material known as coke. The coke is deposited on the inside surface of the tube coil. The rate of coke formation is one of the main parameters for determining run length, tube life, firing rate, and other parameters relating to energy usage. Proper burner design, operation, and maintenance are critical to maintain design flux rates to the process tubes.

Ethylene Cracking Furnaces operate at a much higher temperature than most other refinery and petrochemical process heaters. Most Ethylene Units have a design temperature of 2000-2350°F. This severe operating environment must be considered in the design and material selection of burners.

Many ethylene complexes put undesired components from the product stream back into the fuel train for the process heater. Depending on the feedstock, the product stream can be high in olefin content. Olefin cracking rates are much higher than those of corresponding paraffin compounds.[ii] Liquid carryover in fuel gas systems is also a great concern with heavier constituents in the fuel gas system. Cracking rates of liquid hydrocarbons are also much greater than those in the gas phase. In addition, cracking begins at a much lower temperature in liquid hydrocarbons. Since the burner tips are exposed to the high temperature environment, having a fuel that has a high propensity for cracking can lead to coking of burner tips.

In addition to high olefin content, the ethylene complex fuel gas can also be high in Hydrogen content. Hydrogen is a product of the cracking process. Some ethylene complexes can put as much as 80% H_2 by volume into their fuel gas system. Hydrogen and Olefins both are more prone to flashback in pre-mixed burners. Flashback will be discussed later in this report. However, flashback in pre-mixed burners generally results in moderate to severe damage to burner components.

Ethylene cracking furnaces in the 1980's and early 1990's were typically designed to 4,000 – 6,000 Btu/(hr·ft^2). New ethylene cracking furnaces are typically designed to 7,000 – 10,000 Btu/(hr·ft^2). This general increase in heat density presents a new challenge to meet stringent emissions regulations and design burners for long term operation in this severe furnace environment. At these higher heat densities, furnace designers and burner vendors must now look more closely at dimensional constraints on flame envelopes to avoid burner to burner interaction, roll over, and tube impingement. Older generation cracking units are now being retrofitted with new, low NO_x burner technology. Low NO_x projects are usually coupled with furnace expansion projects to increase the fired duty 25-40% over the initial design, increasing the heat density to current design standards.

General Burner Design Considerations

A burner is a fuel and air metering device. Its purpose is to provide an environment for the proper fuel and air mixture ratio to mix and react. Much time and energy is spent to find the proper tip drillings, tile design, venturi design (pre-mixed burners), and air inlet design. This section will discuss the proper material selection and design considerations to keep the burner operating efficiently over a long term operating period.

Gas tip and riser design must be resistant against coking, scaling, and corrosion. Inside the furnace is a high temperature, reducing and oxidizing environment. Corrosion and scaling resistant materials must be used for tip casting and gas risers. In general, this means high quality 300 series stainless steel risers or better and CK20 (310 stainless steel cast equivalent) gas tips or better. As stated earlier, the cracking reaction is temperature and residence time dependent. Since gas tips can reach gas cracking temperatures it is important to control the residence time in the burner tips. In general, riser velocities of

100-250 ft/sec should be used as a standard for burners in ethylene service. As a further measure to prevent overheating of gas tips and the problems that result, burner designs should embed external gas tips inside the burner refractory to shield them from radiation.

The burner tile plays a large role in shaping the flame and providing a means for air and fuel mixing. Tile failure can lead to uneven flame patterns, which can result in uneven heating of the process tubes and coke formation. Tile material should be selected such that the maximum recommended service temperature is 3000°F as a minimum and the alumina content should be 60% or greater. Resistance to thermal shock and reducing atmospheres are other considerations. Proper material selection will ensure that the tile holds its shape and serves its purpose during long term, continuous operation.

Burner flashback is one of the main considerations in the design of premixed burners for service in Ethylene Cracking Furnaces. Flashback occurs when the flame speed overcomes the gas / air mixture flow velocity exiting the premixed tip. When this condition is satisfied, the flame propagates back into the premixed tip. It is generally accompanied by a loud popping sound followed by a prolonged acoustic resonance. Premixed burners can sustain irreparable damage if flashback occurs. This can lead to a reduction in firing capacity, flame quality, maldistribution of heat, emissions excursions, and other operational problems which effect the fuel efficient operation.

Figures 1 through 3 list the burning velocity of various hydrocarbon mixtures as a function of mixture ratio of air and fuel. Notice first that the flame speed of paraffin hydrocarbons is the lowest, olefins are slightly higher, and constituents such as Hydrogen and Acetylene are among the highest burning velocities. Next consider the air to fuel ratio. As shown in Figures 1 through 3, the peak flame velocity generally occurs when fraction of stoichiometric fuel is above 1.0, or in the fuel rich conditions. Therefore, the design of premixed burners against flashback must consider the following.

1. *Exit Bulk Velocity and Velocity Profile*: Flashback occurs when the burning velocity overcomes the exit velocity of the fuel / air mixture. The bulk velocity is a function of the exit area and flow. Since the bulk velocity is generally fixed, the exit area and venturi design should be sized for the optimum eductor performance and bulk exit velocity. Also important is the velocity profile. The peak to average velocity should be as close to unity as possible. This is done through minimizing the entrance effects to the premixed tip slots with baffles and proper slot design.

2. *Fuel Composition*: Consider the maximum swings of Hydrogen, Olefins, and other constituents with high burning velocities.

3. *Firing Rate*: Generally, the lowest firing rate where the excess air is controlled will be worst case for burner flashback. Therefore, control of excess air may not be achievable at turndown conditions unless specifically considered in the design of the burner.

4. *Stoichoimetry Exiting the Premixed Tip*: As shown in Figures 1 through 3, the peak burning velocity occurs in fuel rich conditions. Most pre-mixed burners that employ older designs operate with 10-30% secondary air. Secondary air is pulled into the furnace through air slots external to the venturi. Therefore, the stoichiometry of the fuel / air mixture exiting the tip is in the fuel rich region, making it more prone to flashback. The latest ultra Low NOx burners employ Lean Premix Technology. In lean premix technology, 100% of the combustion air is entrained with a fraction of the fuel. The remaining fraction of the fuel is introduced later as secondary fuel. Therefore, the stoichiometry of the fuel air mixture is in the extremely fuel lean region. Note the burner velocity as a function of mixture ratio for Hydrogen shown in Figure 3. The slope of the curve in the lean region is much greater than the slope in the rich region. This makes the newest Lean Premix technology less prone to flashback, particularly with hydrogen rich fuels.

Utilization of these design considerations will help ensure that the burner will operate flashback free.

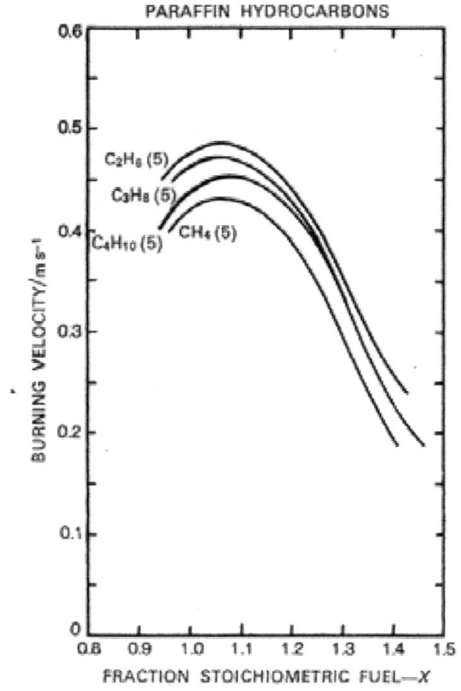

Figure 1: Burning Velocity vs. Stoichiometry for Paraffins.[iii]

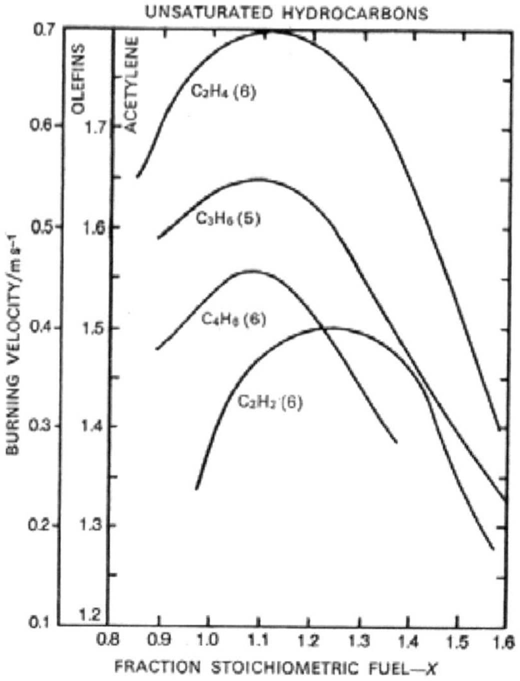

Figure 2: Burning Velocity vs. Stoichiometry for Unsaturates.[iii]

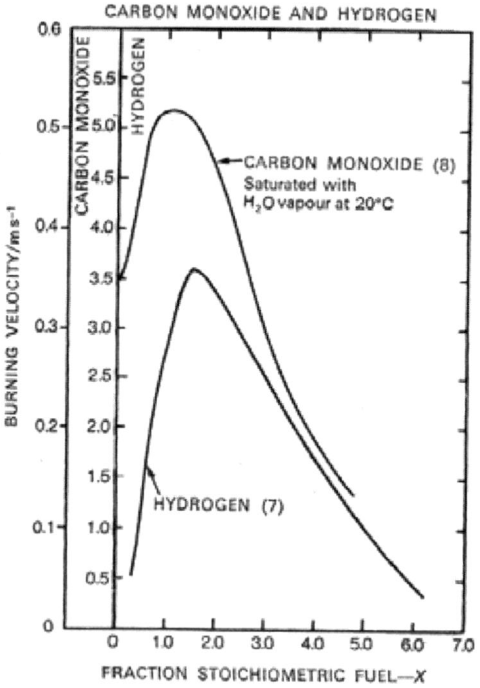

Figure 3: Burning Velocity vs. Stoichiometry for CO and H_2.[iii]

General Burner Operating Considerations

Current burner technology for Ethylene Cracking Furnaces is generally designed for 10% excess air operation. Through the years furnace manufacturers and burner vendors have worked closely to ensure that the furnace specifications match the burner specifications for capacity. This ensures that the burners will operate in a range where the excess air is controllable. Controlling tramp air leakage by sealing the firebox and closing dampers of burners that are out of service is important. Additionally, controlling heat losses to conduction by having adequate furnace insulation are also paramount in energy efficient operation. These items represent basic elements of fired equipment efficiency. This section will discuss some additional operating practices, which may help to ensure efficient burner operation in the long term.

One of the most stressful conditions on burner components is the no flow condition. Burners are designed for continuous operation. Taking burners out of service in this severe environment removes the cooling flow of combustion air and fuel and can effect the life of some burner components. It is recommended to rotate burners in and out of service every few days to help spread the load.

Coking and plugging of fuel gas tips are the most common fuel side problems that can effect burner efficiency. When tips are partially or fully plugged this can effect flux profiles, burner stability, capacity, and many other parameters. Plugged and coked tips are generally plugged with a complex blend of deposits that have accumulated over time. In many cases the deposits have spent time being heat treated and the molecular structure has physically changed. For this reason, it is difficult to analyze and determine the exact mechanism coking and/or plugging. However, there are three main mechanisms for a tip to coke.

One mechanism involves the tip being initially plugged with a foreign material such as pipe scale, rust, or other particulate matter. When gas orifices are partially plugged the flow to the gas tip is reduced. As the flow is reduced, the gas tip heats up and the gas residence time increases. Once the gas reaches cracking temperature it begins to lay down coke until the tip becomes completely plugged.

Another mechanism is liquid carryover. Fuel gas tips generally run at temperatures in the liquid cracking range. As liquid droplets come in contact with the inside surface of the fuel gas tip they will crack and lay down coke. Flow is reduced and the gas constituents begin to crack and plug the tip identical to the previous mechanism.

The third mechanism is a gas cracking mechanism. This can occur whenever conditions in the fuel gas tip favor gas cracking. Ethylene Furnaces have standby and startup conditions where a portion of the burners are taken out of service and exposed to high temperature. In this condition, fuel gas tips can glow white hot. Since conditions in these tips are in the range of gas cracking temperatures, it brings into question what may occur when gas is reintroduced to these tips.

John Zink Company investigated this condition in a coking test performed in 2002, where a single gas tip was installed among three other burners operating at cracking furnace temperatures. The tip was exposed to a no flow condition and allowed to heat up to full furnace temperature. An olifinic hydrocarbon mixture of 85% Natural Gas, 5% H_2, and 10% C_3H_6 was used for the coking test. Two flow conditions were tested.

In the first test, the tip was allowed to heat up to full furnace temperature. Approximately 300 scfh gas mixture was introduced until the tip was visibly cool. The gas was then shut off and the tip was allowed to heat up to furnace temperature. The cycle was repeated fifteen (15) times over a 2 hour period. At the conclusion the riser was removed from the furnace and the tip was examined. Surprisingly, the tip was free of any significant coke buildup.

In the second test, a low flow condition was investigated. The tip was again allowed to glow white hot. The fuel mixture was introduced incrementally until the tip cooled from white hot to red hot, which corresponded to a flow rate of 30 scfh. The fuel flow was then fixed at this rate left to run for one hour. At the completion of testing the tip was examined and found to be completely plugged with coke. The internal surface of the riser was coated with a coke layer approximately 1/8" thick.

The results of the testing give some insight into a few of the operating conditions and the propensity to form coke during these conditions. The first test is a good simulation taking a burner offline for maintenance and then bringing it back online. The second test simulates the effect of a leaking fuel gas block valve to a burner that has been removed from service. Many ethylene cracking furnaces employ separate fuel gas headers for primary and staged fuel gas in hearth burners. This design allows burners to achieve higher turndown during "primary only" operation. In this condition, the staged tips will approach the furnace temperature. As the process is brought back online flow is sent back to the staged tips. Given the results of the low flow test, staged tips can plug if gas is reintroduced at a low rate or if the isolation valve leaks.

Burners which mix the combustion air and fuel prior to introducing it into the furnace are defined as premixed burners. In addition to tip plugging from coke buildup, premixed burners are also prone to tip plugging through another mechanism. Premixed burners use the fuel energy to entrain combustion air through a jet pump assembly. Particulate matter such as dust, ceramic fiber insulation, and other particulates can be entrained into the fuel / air mixture and plug the exit slots of the premixed tip and/or the inside surface of the venturi. This can lead to operational problems such as a reduction in capacity, flame quality, emissions, and others if allowed to continue. Premixed burner tips can also sustain irreparable damage after the particulate matter is heat treated inside the exit slots of the premixed burner tip.

<u>Fuel Gas Filtration</u>

Fuel gas filtration is one of the most effective ways to control burner tip plugging. As stated earlier, the two more common causes for tip plugging is particulate matter and liquid carryover. Conventional filters are generally used to capture particulates. They can be installed in the fuel gas manifold or at individual burners. Coalescing filters are the most common way to eliminate liquid carryover. Coalescing filters employ a medium of hydrophilic fibers that coalesce smaller liquid droplets into larger ones. Eventually, the liquid droplets fall due to gravity or are separated by inertial forces and are collected at the outlet of the filter system. Most coalescing filters have two stages to prolong the life of the coalescing media. The first stage is a conventional filter, which filters out any particulates. The second stage is the coalescing filter. Generally, the best fuel gas filtration designs employ a two stage filter at the inlet of the furnace manifold and individual strainers at each burner inlet to control particulates which are generated downstream of the filter system. Filtration systems do take time to maintain, however, the time spent in filter maintenance generally is less than the alternative of continued burner tip maintenance and replacement.

Online Burner Inspection for Efficient Operation

There are a few simple ways to inspect burners for efficient operation while the heater is online.

1. *Heat Patterns:* Visual inspection of the heat patterns on the wall should be one of the first indications that the burner is not operating efficiently. Much can be learned when inspecting the heat patterns on the wall. Non-uniform heat patterns on the wall are an indication of maldistribution of heat. Leading causes are tip plugging / coking and tile failure.

2. *Tramp Air Sources:* Tramp air sources are one of the leading causes to inefficient operation. The most common places for tramp air are around tube seals, site doors and mounting locations for burners and auxiliary equipment. In extreme cases, sources of tramp air are visible by a cooling streak on the furnace refractory local to the tramp air inlet.

3. *Visual Inspection of the Flames:* Flames should be free of soot and uniform in height, width, and projection out from the fired wall. Any non-uniformity in flame dimensions and appearance should be investigated further.

4. *Fuel Composition Monitoring:* Fuel composition should be regularly monitored. The supply temperature at the burners should be kept at a safe margin above the dew point to avoid liquid carryover and tip plugging.

5. *Manifold Pressure:* The burner supply pressure, temperature, flow, and composition should be monitored on a regular basis. Increases in pressure are an indication of tip plugging. It is important to record flow, composition, and temperature in addition to pressure as this will give you a true indication if the fuel pressure is in fact on the rise. If tips are allowed to continue to plug until a no flow condition, the tip will eventually fall off leaving an open ended gas riser injecting fuel in to the furnace. This can lead to overheating of process tubes.

6. *Air Inlet Inspection:* Burner inlets should be monitored on a regular basis for restrictions. Premixed burners are particularly prone to plugging of the air inlet. Heavy fuel gas constituents can lay down a thin hydrocarbon film on the cool surface of the venturi. This film attracts dust and particulates which accumulate over time. It is not uncommon to plug off the entire inlet with the exception of a small hole where the fuel is allowed to pass through. Therefore, regular inspection of burner inlets should be practiced.

7. *Tile Inspection:* Burner tiles should be visually inspected on a regular basis. Pitting, cracking, glassing, and an increase in tile porosity are indications that the tile not in good operating condition. Glassing and porosity can be difficult to see in while the burner is online.

<u>Offline Inspection for Efficient Operation</u>

When the heater is offline and entrance is allowed to the furnace, there are some additional inspection that can be done.

1. *Tile Inspection:* The burner tile is integral to the proper operation of the burner. As stated earlier, pitting, cracking, glassing, and an increase in tile porosity are indications of tile failure. Pitting occurs in cases of prolonged jet impingement and overheating. Burner refractory should not pit in normal operation. If pitting occurs, the tile should be replaced. Surface cracks are expected in the normal life of the tile. However, if the surface cracks propagate through tile thickness this can lead to tile failure. As a general rule of thumb, if a surface crack propagates more than ¼" into the tile then the tile should be replaced. If the tile becomes overheated then the surface of the tile will vitrify or glass. This is an indication of physical change in the tile composition and the tile should be replaced. An increase in tile porosity is another indication of physical change in the tile composition. Tiles that show an increase in porosity may fail if allowed to stay in service and should be replaced.

2. *Furnace Refractory Inspection:* The same criteria for burner tile is true with furnace refractory. Many furnace designs employ ceramic fiber modules to insulate parts of the heater. These modules shrink over time leaving exposed steel. If left unchecked, improper insulation coverage may effect fuel efficiency and also lead to furnace structural damage from overheating.

3. *Fuel Gas Header Maintenance:* Fuel gas headers should be inspected and cleaned during scheduled turnarounds.

4. *Fuel Gas Tip Inspection:* Fuel gas tips should be inspected while the heater is online. Normal inspections of tip plugging / coking should be done while the heater is offline also. In addition, tip elevations and tolerances should be verified against the burner general arrangement drawing.

5. *Burner Damper Operation:* All burner dampers should be inspected for proper operation. Replace dampers that do not operate through the complete range of motion.

6. *Tramp Air Leakage:* If possible, the furnace should be smoke tested under positive pressure during scheduled maintenance to identify any sources of tramp air leakage. Tramp air sources should be sealed.

Conclusions

With an increasingly competitive marketplace and a reduction in emissions regulations, fuel efficient operation of fired equipment in the olefins industry is becoming increasingly important. The environment of the Ethylene Cracking Furnace has many unique aspects that present a formidable challenge to fuel efficient operation. However, proper design, routine inspection, and maintenance will ensure that fired equipment operates fuel efficient in the long term.

References

[i] Energy Information Administration, *Manufacturing Energy Consumption Survey* (1997).

[ii] Gary, James H. & Handwerk, Glenn E., *Petroleum Refining* (Marcel Dekker, Inc., 1984), p. 111.

[iii] Rose, J.W. & Cooper, J. R., *Technical Data on Fuel* (John Willey and Sons, New York, 1977, 7th Edition)

"Utilities Leak & Steam Trap Inspection and Repair Program"

David B. Johnston
Assistant Department Superintendent, Utilities & Feedstock Maintenance Department
Texas Eastman Division, Eastman Chemical Company
Longview, Texas

ABSTRACT

Managing repairs to utility line leaks (steam, condensate, air, nitrogen) and steam traps has always presented a challenge in chemical processing plants. Although the potential exists for reducing utility costs by several millions of dollars annually, these efforts are often limited by staffing availability and the higher priorities placed on process equipment maintenance. However, higher costs associated with utilities production now provide increased emphasis on the effective identification and repair of all utility leaks and problem steam traps. This paper includes a discussion of recent efforts by Eastman Chemical Company in Longview, Texas to establish centralized management of utility leak and steam trap inspections/repairs plant-wide. One Eastman mechanic, chosen for suitable experience and job skills, is assigned full time to coordinate program activities. Selected vendors are scheduled periodically to perform plant-wide surveys. Once jobs are prioritized and categorized, higher skilled level repair work is assigned to Eastman maintenance personnel, where existing work processes are initiated for job execution. The remaining work (approximately 80% of all program activities) is performed by on-site continuous use contractor personnel, which basically includes database management and lower skill level leak/trap repair execution. Associated contractor staffing level is managed to fit seasonal workloads. Discussion will focus on these and other key program elements, along with details and results of pilot efforts. The fundamental program goal is to provide continuous, centralized management of all site utility leaks and steam trap inspections/repairs.

INTRODUCTION

Eastman's Longview, Texas Operations includes approximately 25 separating operating units, which manufactures over 60 major chemical and plastic products. Virtually all products are manufactured from two primary raw materials-propane and ethane. As will all petrochemical manufacturing facilities, this site has a complex network of utility systems to support all operating areas. Managing utility system leaks and steam traps at Texas Eastman is an integral part of minimizing utility costs.

The purpose of the Utilities Leak and Steam Trap Inspection/Repair Program is to provide centralized management of steam/condensate/air/nitrogen leaks and steam trap inspections/repairs site-wide. System/component improvements, weld repairs, and more complex steam, air, nitrogen leak or steam trap repairs are still performed by Eastman personnel. The program maintenance contractor manages program documentation and a largest percentage of repair work, specifically less complex repairs (i.e. small steam/air/nitrogen tracing or steam trap repairs). A Program Coordinator, assign to Eastman's Contractor Services Group, coordinates surveys, program logistics, and works with Contractor Services Group supervision to manage the Program Maintenance Contractor staffing levels and budgeting requirements. The Program Coordinator also provides technical assistance as required to troubleshoot and recommend repair scopes or improvements to more complex utility system problems.

PROGRAM DEVELOPMENT

Background

The success of previous utility leak and trap inspection/repair efforts was limited by staffing availability and higher priorities placed on process equipment maintenance. Repair resources such as equipment mechanics, welders, and pipe fitters utilized for leak and trap repair work were not always available to support this effort. As a result, productivity of utility leak and trap inspection and repairs suffered. However in early 2003, the higher costs associated with utilities production presented a renewed emphasis on the effective identification and repair of all utility leaks and problem steam traps. It was also evident that a dedicated group would need to be organized to manage and staff this effort.

Eastman has implemented several successful site-wide maintenance programs, including an asbestos abatement, paint maintenance program, road maintenance program, and most recently, a refractory maintenance program. Managed by Eastman's Contractor Services Group, these programs utilize contract labor as dedicated resources and well defined procedures to achieve program objectives. Based on existing program successes, and limited success with previous utility leak and steam trap inspection/repair efforts, it became evident that a new centralized program was needed to manage utility leaks and problem steam traps.

In January, 2003, a Steering Team was formed and resultant subcommittee organized to achieve the following charge: ***"Develop a Plant-Wide Utilities Leak/Trap Inspection & Repair Program. Initially piloted in the Utilities and Feedstock Division, this program***

would provide centralized ownership and coordination of all utilities leak and steam trap inspections, work prioritization, planning/scheduling, and repair execution. Plans would include site-wide program expansion, beginning January 1, 2004."

The Steering Team further defined the scope of the new utilities leak/trap maintenance program would include the following utility systems:

- Steam leaks
- Instrument air leaks
- Plant air leaks
- Nitrogen leaks
- Steam Trap Failures

In order for the implementation subcommittee to accomplish the program objectives they determined their efforts should focus on the following key elements during program development:

- Program Justification
- Staffing & Responsibilities
- Accounting & Budgeting Logistics
- Utility System Survey Logistics
- Work Categories & Associated Assignments
- Repair Pre-work/Planning/Scheduling
- Program Documentation...*Database, Supporting Measures, and Management Control System*

During early 2003, the team convened frequently to develop program logistics, addressing the key elements listed above. One Eastman mechanic, chosen for suitable experience and job skills, was assigned full time to coordinate program activities. A continuous use contractor was selected as the program maintenance contractor to provide what would become the majority of program staffing duties. A program rough draft was completed, and the program piloted in one operating area to test program logistics on a manageable level, refine the program justification, and resolve potential issues before site-wide implementation.

Program Justification

The significant potential in achieving program objectives was evident, but team members realized definitive savings had to be documented for site-wide program approval. So to support program justification, an initial steam leak survey was performed in selected operating areas by the Program Coordinator, and factored up to provide a site leak rate estimate. An outside service was contracted to perform a plan wide air and nitrogen leak survey. Eastman's steam trap supplier also conducted site-wide trap surveys.

Steam loss associated with traps is fairly well defined since there is a known pressure and orifice size. The contractor that surveyed our nitrogen and air lines used an acoustics emissions detector to quantify leaks. Initially, steam leak loss estimates were quantified based on estimated plume size (ft) and reference tables listing estimated pounds per hour for a given plume length. However, to verify accuracy of this method, the program coordinator conducted experiments to condense a small leak and measure the actual condensate rate. Figure 1 shows this experiment, which is a leaking valve seat. Estimated loss (2 ft plume), based on plume length and reference tables is 17.5 lb/hr. Actual condensation rate determined by experiment was 25 lb/hr. And since this experiment was repeatable with several similar small leaks, we felt our current steam loss estimates were a bit conservative, but consistent. The team is continuing to evaluate steam loss estimate methods to determine the most accurate and consistent technique for our program. However, based on current loss estimate methods used, program costs and economics were determined. Throughout most of 2003, the ratio of savings to program expenses exceeded 15:1 consistently. Projected 2004 net benefits are expected to exceed $3M.

[Figure 1 – Steam Leak Quantification Experiment]

Program Accounting and Budgeting Logistics

Separate cost centers are assigned to accrue program charges, specifically including program coordination, leak surveys, and repairs (inside and outside battery limits) managed by the managing contractor. The program database is used to track cost history. Repairs executed by Eastman personnel are accrued on separate maintenance work orders issued specifically for these leak/trap repairs and charged to the appropriate operating area cost center.

Similar to other Contractor Services managed program budget plans, an on-going 3-year budget plan has been developed and will be maintained by the program managing contractor. Historical data will be logged into the program database and used to establish and update this budget. The budget will have the following components: Condition Assessment, Survey/Leak Inspection, Preventive Maintenance, and Small Repairs. In addition, a separate list of larger Improvements and Specials will be maintained and communicated for inclusion in the area specific budgets. These larger jobs, typically involving Eastman resources, will generally be charged to separate work orders generated through existing maintenance practices. All of the program work will be prioritized based on the consequences of not performing the work. Accompanying the budget will be documented reduction strategies, including expected consequences for each in the event that budget reductions are requested.

Utilities Leak & Steam Trap Maintenance Program

Staffing and Responsibilities

Throughout the remainder of 2003, program efforts were expanded to include all utilities and olefins plant operating areas, and the managing contractor staffing level increased to accomodate the increased work load. Figure 2 below represents the basic roles of program participants.

In order to take advantage of lower cost contractor labor, the team defined program job duty opportunities that would match the skill set of our on-site continuous use contractor personnel. As a result, it was determined that the contractor would manage the program database and other program related documentation through Eastman's Contractor Services Group. In addition, the majority of program repair work, specifically less complex steam, condensate, air, and nitrogen leak repair work, as well as trap repair work would be staffed by the continuous use contractor. All told, the team estimated that approximately 80 to 90% of all program work would be staffed by the continuous use contractor. Other program responsibilities were identified, and are outlined below:

1. Program Coordinator : Eastman General Maintenance Mechanic is assigned to the Contractor Services Group to coordinate program processes, with emphasis on steam leak surveys, more complex repairs, and associated program improvements. The Program Coordinator is the focal point for compiling survey data, including leak surveys he performs himself (pre-turnaround surveys), surveys performed by contractors, or leak/traps problems discovered and reported to him by various operating area personnel. The Program Coordinator also provides technical assistance as required and executes the Management of Change process on any changes resulting from repairs and/or improvements.

2. Program Maintenance Contractor The approved Continuous Use Contractor will
   a) Provide competent staffing for less complex repair work to match the fluctuating workload.
   b) Manage the program database containing an inventory of the involved utility systems and historical documentation (including value) of work performed and related program measures.
   c) Supervise their employees on repair execution for their assigned work.
   d) Serve as the focal communication point for phone/office mail reports of utility system leaks and steam trap problems.

3. Steam Trap Supplier: Performs plant-wide steam trap surveys annually and provide survey reports to the Program Coordinator, who will distribute to the Continuous Use Contractor database manager (for database entry) and the appropriate operating department supervision.

4. Utility System Survey Contractor: Performs annual surveys on steam, air, and nitrogen systems site-wide.

5. On-line Utility Leak Repair Contractor: Will perform all clamp, wrap, and leak pump jobs by request, per on-line process and utility leak repair procedure.

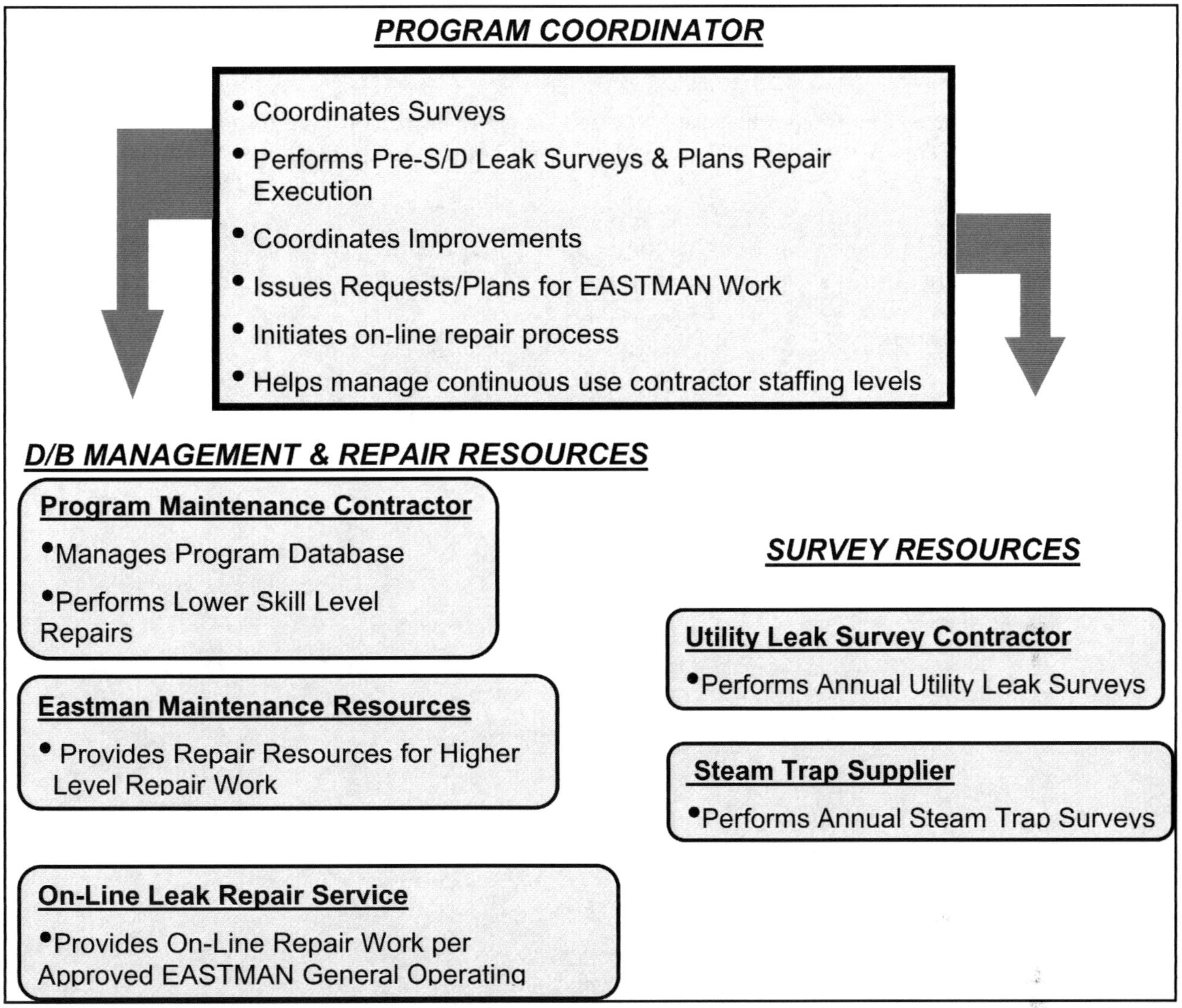

[Figure 2 – Basics Roles of Program Personnel]

Utility System Surveys

1. Steam/Condensate Leak Surveys: The Program Coordinator will assure that steam and condensate leak surveys are performed prior to plant shutdowns and as needed at other times. The Program Coordinator will perform the surveys himself or otherwise arrange leak surveys be performed by the Utility Leak Survey Contractor. Figure 3 shows the Yellow leak tags that will be used to identify each leak, and a corresponding Job Plan form filled out to capture pertinent leak data. The bottom portion of each leak tag and corresponding Job Plan Form will be submitted to the database manager who will log the survey data into the Program Database.

2. Steam/Condensate/Air/Nitrogen/Hydrogen Surveys: The Program Coordinator will schedule Utility Leak Survey Contractor to perform annual utilities surveys site-wide. Once completed, the Program Coordinator will issue the survey report to the database manager, who will log survey data into the Program database.

3. Trap Surveys; The Program Coordinator will schedule the Steam Trap Supplier to perform steam trap surveys annually. The Program Coordinator will issue trap survey reports to the database manager.

4. Steam Line Insulation Surveys; Program Maintenance Contractor will use existing survey processes to identify insulation needs and log energy loss and associated job details into the program database.

5. Area requests; operating area personnel may call the program managing contractor or Program Coordinator to report leaks or trap problems at any time.

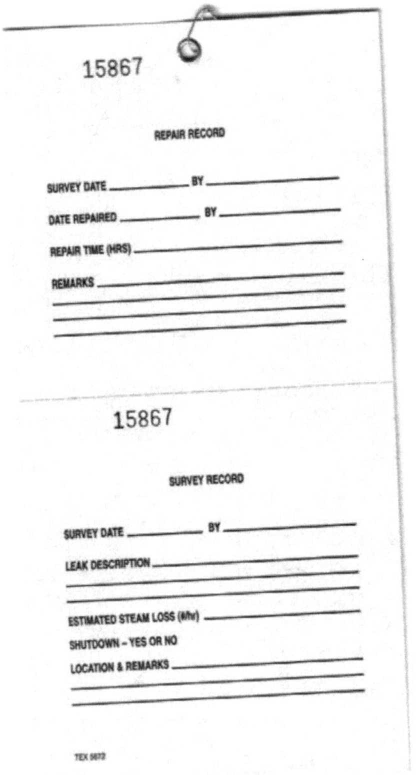

[Figure 3 – Yellow Leak Tag]

Database Development

In order to manage the information generated from surveys, to aid in categorization and assignment of repairs, and to capture data for supporting measures, a method was needed to store, retrieve, and analyze information. The team chose to use Microsoft Excel for the database application since the database manager was familiar with this software. Eastman's preferred trap supplier uses database is Microsoft Access to report trap survey results. This information is cut and pasted into the program database by the database manager. Database results are office mailed to the Program Coordinator, who prepares a weekly work schedule for the managing contractor lead repair person. Shown below is a sample work list taken from the program database. Several additional data categories are available and used as required for pre-work planning.

Tag #	Survey Date	Repair Category	Est Rep. Hrs	Bldg. No.	Prty. Rating	Trap Type	Leak Type	Pressure	Location	Scope
15597	9/29/03	Clam/Wrap/Pump repair	10	68	High		Steam	300	PR south D-10 @ B-68	Replace line. Re-pump until line S/D
676	1/27/03	Small steam/air/nitrogen tracing or steam trap repair	1	68	Low		Air		POSITIONER GASKET, SE OF 3T-57 (GRD)	C/O positioner gasket
15611	10/6/03	Small steam/air/nitrogen tracing or steam trap repair	3	68	Low		Condensate	30	South metal bldg 68-1 along side road	Replace trap. Use UTD-52L.
15590	9/29/03	Small steam/air/nitrogen tracing or steam trap repair	3	68	Low	Thermostatic	Steam	100	NE corner 4th Ave & 8th St at stanchion	Replace trap. Re-pipe using all 3/4 fittings & steel gate valve.
15613	10/6/03	Small steam/air/nitrogen tracing or steam trap repair	6	68	Medium		Steam	80	CM 2 at B-68	Repair leaking threads on manifold tie-in at top flange.
15249	8/22/03	Large or high pressure steam/air/nitrogen or leak or steam trap repair	4	74	High		Steam	1500	South EA-452 at RS	3/4 NR valve to be c/o
429	1/27/03	Small steam/air/nitrogen tracing or steam trap repair	1	74	Low		Nitrogen		PACKING, 1" VALVE, AT UTILITY STATION, NE OF FB-917 (GRD)	Repack valve

[Figure 4 – Sample Database Work List Report]

Planning&Scheduling Repairs

Repair Prework
The program database manager enters survey information into the database. Jobs are categorized by safety and repair value, divided into separate work lists per repair assignment, and issued to the program coordinator. This process is performed weekly (Thursday) to include new leaks reported, to update program measures, and to allow the program coordinator to organize work assignments for the week.

Repair Categories and Associated Assignments: Repair work is categorized as follows:

Repair Category	Job Execution Assignments
Small steam/air/nitrogen tracing or steam trap repairs	Repair execution generally staffed by managing contractor. Operating areas also have the option of using Ops mechanics.
Steam line insulation	Repair execution always performed by managing contractor.
Large or high pressure steam, air, nitrogen, or leak or steam trap repairs	Repair execution generally to be performed by Eastman personnel.
Weld related repairs	Repair execution always to be performed by Eastman personnel
Clamp/wrap/pump repairs	Repair execution always to be performed by Eastman's on-line utility leak repair contractor
Improvement or changes Requiring an MOC	MOC prepared by the Program Coordinator or other Eastman Personnel, Repair execution will depend on job category above.

Repair Priority: Once survey data is logged, the Program Database will prioritize leak/trap repair jobs according to perceived safety risk and repair value. Work lists will be generated by job execution assignments (table above) and grouped as follows:

- *Within 1 Week*: Repairs to be executed within 2 to 3 days, which include leaks/traps that present a perceived safety hazard; those that have negative affects on operating area production or equipment; or steam leaks that have a relatively higher repair value (5 ft or more in plume height).
- *Within 3 Weeks*: Repairs to be executed as soon as reasonably possible (within 2 weeks), which include minor leaks or small traps that do not impact safety or production, have a moderate repair value, but do not present a perceived safety hazard that would warrant immediate repair execution.
- *Backlog Jobs*: Nuisance leaks or small trap applications, including small leaks/traps with a relatively low repair value (typically seasonal or in remote locations), and do not present a perceived safety hazard.
- *Shutdown Jobs*: The estimated value of the leak (utility loss) and safety risk will determine whether on line repair work is warranted.

Repair Planning/Scheduling:
Job planning for leaks/trap problems identified on leak surveys will be detailed on the Job Plan form. Managing contractor repair personnel will use material from area stocking bins or from stock available on the Program lift truck. For all other repair categories existing planning/scheduling processes will apply.

Job Execution

1. Permitting and Job Prep: Repair personnel (specifically the job leader) will notify the Operating Department, at least the day before job execution, of upcoming repair plans to allow operators to prepare equipment ahead of time, thus avoiding repair delays. On the day of the job execution, repair personnel will request a Safe Work Permit from the Operating Department for each job. If the repair crew is assigned for a week or more, the program coordinator will request the operating department assign an operator, at least part time, to the repair group for job preparation and permitting. Operating Department personnel will issue the Safe Work Permit, and work with repair personnel to prepare the job per "Lockout/Tag out of Energy Sources" procedures.
2. Repair Work Assigned to Managing Contractor: A prioritized work list (using prioritization criteria above) will be generated from the database for all jobs to be assigned to Program Maintenance Contractor each Thursday and submitted to the Program Coordinator. The Program Coordinator will compile the work list, associated Job Plan forms, and leak tags (bottom portion of tag), and issue the work list packet to Program Maintenance Contractor supervision for job assignment Monday morning. A copy of the work list will also be issued electronically to the operating department Maintenance Coordinator and Production Assistant. Program Maintenance Contractor will execute repairs, by priority, remove the top portion of the leak tag from the repair site, record completion date per tag number, and submit top portion of the leak tag back to the database manager, who will log completion dates into the program database.
3. Repair Work Assigned to Eastman Resources or Preferred On-line Utility Leak Repair Contractor: Following job categorization and prioritization, the database manager will issue a leak/trap job list to the Program Coordinator per operating department each Thursday. Via office mail system, the Program Coordinator issue work notifications to begin existing Eastman maintenance processes. The Program Coordinator will also forward associated Job Plan forms and leak tags to the appropriate maintenance planner for planning/scheduling information.

   a. The preferred on-line utility leak repair contractor will perform all on line repair work; including clamp, wrap, and leak pump jobs by request, per existing general operating procedures.
   b. Jobs assigned to Eastman resources will follow existing maintenance processes.

Upon job completion, contractor repair personnel or Eastman personnel assigned to the repair will remove the top portion of the leak tag, record completion date per tag number, and submit top portion of the leak tag back to the database manager, who will log completion dates into the program database.

Program Effectiveness Measures

1. Program Savings - tabulated by managing contractor via program database and issued monthly. Also includes cumulative value of outstanding work remaining.
2. Program Cost Vs. Savings - tabulated by managing contractor via program database and issued monthly.
3. Productivity Measure – Repair hours per leak and job delays.

Management Control System

1. Overdue Job Repairs: Prior to the regularly scheduled monthly review meeting, the program database manager will issue a program job backlog report, along with results of all program measures to the program coordinator. The Program Coordinator will issue email notes to the appropriate operating department and maintenance department Superintendents and Assistant Superintendents, to inform them of the overdue repair jobs and associated value.
2. Monthly Program Review: The Program Coordinator and Contracting Specialist will meet with Program Managing Contractor supervision monthly to review program status and supporting measures. Managing contractor staffing levels will be assessed for the upcoming months and adjusted as necessary to balance work requirements with budget goals.
3. The managing contractor database manager will issue a monthly report to document program status and results of supporting measures, basically summarizing review details in item 2 above.
4. Quarterly Program Review: Contractor Services Group Leader, Contracting Specialist, and Program Coordinator will meet quarterly with Program Managing Contractor supervision to review program status and supporting measures. Based on measure results, Eastman Contractor Services will reinforce or provide guidance to program personnel.
5. Annual Program Review: The original program Team will meet each January with the Steering Team to discuss program status, results of supporting measures, and proposed program enhancements.

ISSUES TO OVERCOME

Setting up a centralized utility leak and steam trap inspections/repairs plant-wide program can pose several problems. Program effectiveness can be hindered by issues such as job preparation delays (Lock-out/Tag-out, permitting), repair material availability, weld repair or fabrication delays, and defining utility loss for program justification.

To address job preparation delays, the team is requesting an operator be assigned, at least part time, to devote a necessary resource to prepare each job while the repair group is scheduled in that operating area. This will prevent delays associated with an entire repair crew waiting around for their jobs to be prepared.

Concerning repair material availability, on-site vendor stocking of all typical repair materials was already in place at this site. In addition, a truck equipped with tool cabinets and stock with the more commonly used materials has helped facilitate material requirements.

Weld repairs and/or jobs requiring pipe fabrication is planned and managed by the program coordinator. This eliminates dependency on area maintenance planners, who would normally place program work at a lower priority.

Defining utility loss for program justification will most likely be questioned by management. However, proper research of available methods now utilized by reputable survey contractors will provide sound justification for your program.

CONCLUSION

Establishing centralized utility leak and steam trap inspections/repairs plant-wide program can be very rewarding to a manufacturing facility. There are several key elements and issues that must be addressed for your site when developing and implementing such a program. The fundamental program goal should be to provide continuous, centralized management of all site utility leaks and steam trap inspections/repairs. Management must also be committed to this effort, and allocate the necessary resources upfront to capitalize on what promises to result in significant savings long term.

ACKNOWLEDGEMENTS

I would like to recognize the significant contributions of the all Eastman Chemical Company team members involved in program development: Shayne Cowan, Will Dierksen, David Dodgen, Leland Hanson, Steve Lewis, and Kenneth Pitt.

I would also like to recognize the contributions of the following key contractor personnel for their efforts in program implementation: Gary Cassity and Bobby Hill with Gregg Industrial Insulators, Mike Mabry with Spirax Sarco, and Bobby Brooks and Jay Shoesmith with EAD's.

THE AUTHOR

David B. Johnston is an Associate mechanical engineer with the Eastman Chemical Company – Texas Eastman Division in Longview, Texas. He has worked at Texas Eastman Division for 23 years assigned to a variety of maintenance responsibilities for repair and reliability improvements of both fixed and rotating equipment, with emphasis on rotating equipment reliability. His current assignment is Assistant Department Superintendent of the Plant Maintenance Department, responsible for equipment maintenance in the Utilities and Olefins Feedstocks operating division. David graduated from Lamar University with a BS Degree in Mechanical Engineering in 1980.

Utilities Leak & Steam Trap Maintenance Program

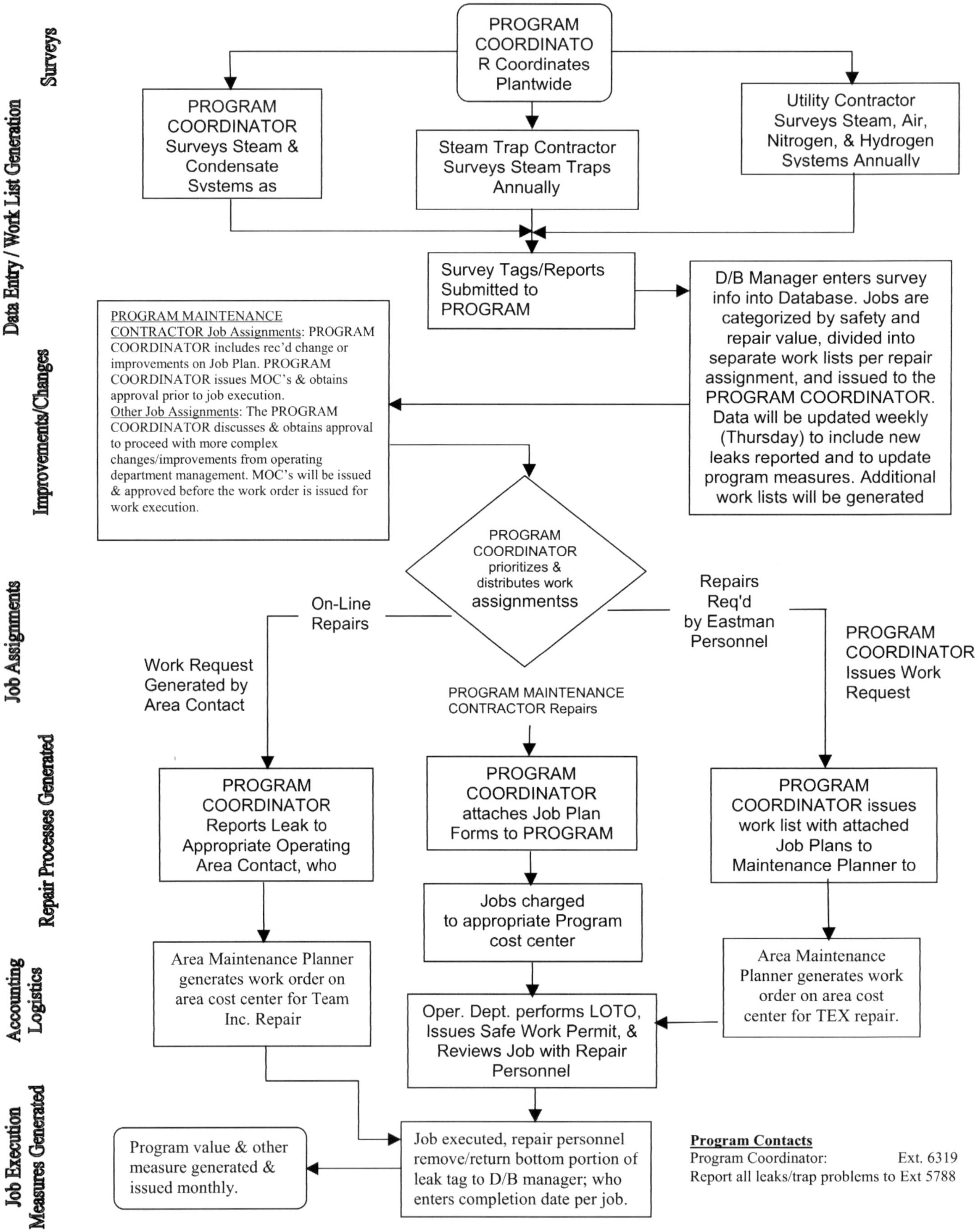

The Use of Torque Meters for Monitoring Turbine/Compressor Performance

Louis J. Trahan
Sr. Mechanical Engineer
E. I. DuPont

Lee W. Allison
Maintenance Area Superintendent
E. I. DuPont

Prepared for Presentation at the 2004
AICHE - Ethylene Producers Conference
New Orleans, April , 2004

Dupont Sabine River Works Arrangement

- Elliott Turbine and Compressor for booster system
- Dresser Rand Turbine and Elliott Compressors for Charge Train
- Only the Charge experience will be shared

Charge Gas Compressor

- Charge Compressor Steam Turbine
  - Dresser-Rand
  - 11 rateau stages (11, 12 stages double flow)
  - inlet steam - 620 psig @ 734 F
  - exhaust steam - condensing @ 5.8" hgA
  - 36,000 horsepower
  - 285,000 lb/hr steam required

Charge Gas Compressor

Charge Compressor Train (4 stages):

Rated for 367 M PPH process gas from 23 psig and 104F (40C) to 555 psig @ 217F (103C)

- That requires:
- 36M horsepower total
- 258,000 lb/hr steam used
- $1995/hr @ $7.00/1000 lbs in 2003
- $17,476,200/yr.

Charge Compressor Train

Our Goal

- •Production demands require reliable performance
  - •at max capacity, no room for losses
  - •Reliability?
- •Need a system to monitor performance
  - •We use continuous data monitoring
- •What do we do?
  - •Look at data periodically/special occasions for adverse trends
  - •Looking for losses - trigger what data to look at -can either be in driver or driven equipment

Troubleshooting Guides

- Deviations from design must be detected:
  - Look @ several pieces of data
    - Production rates
      - Mass flow rate out the 4th stage
    - Loss of horsepower
      - 800-1000 flag
    - Pressure at the exit of the turbine 1st stage
      - 620 –430 normal drop
      - 620 -435 abnormal drop
    - Other items to consider
      - turbine speed, efficiency, turbine cam position, exhaust vacuum

Discharge Mass Flow

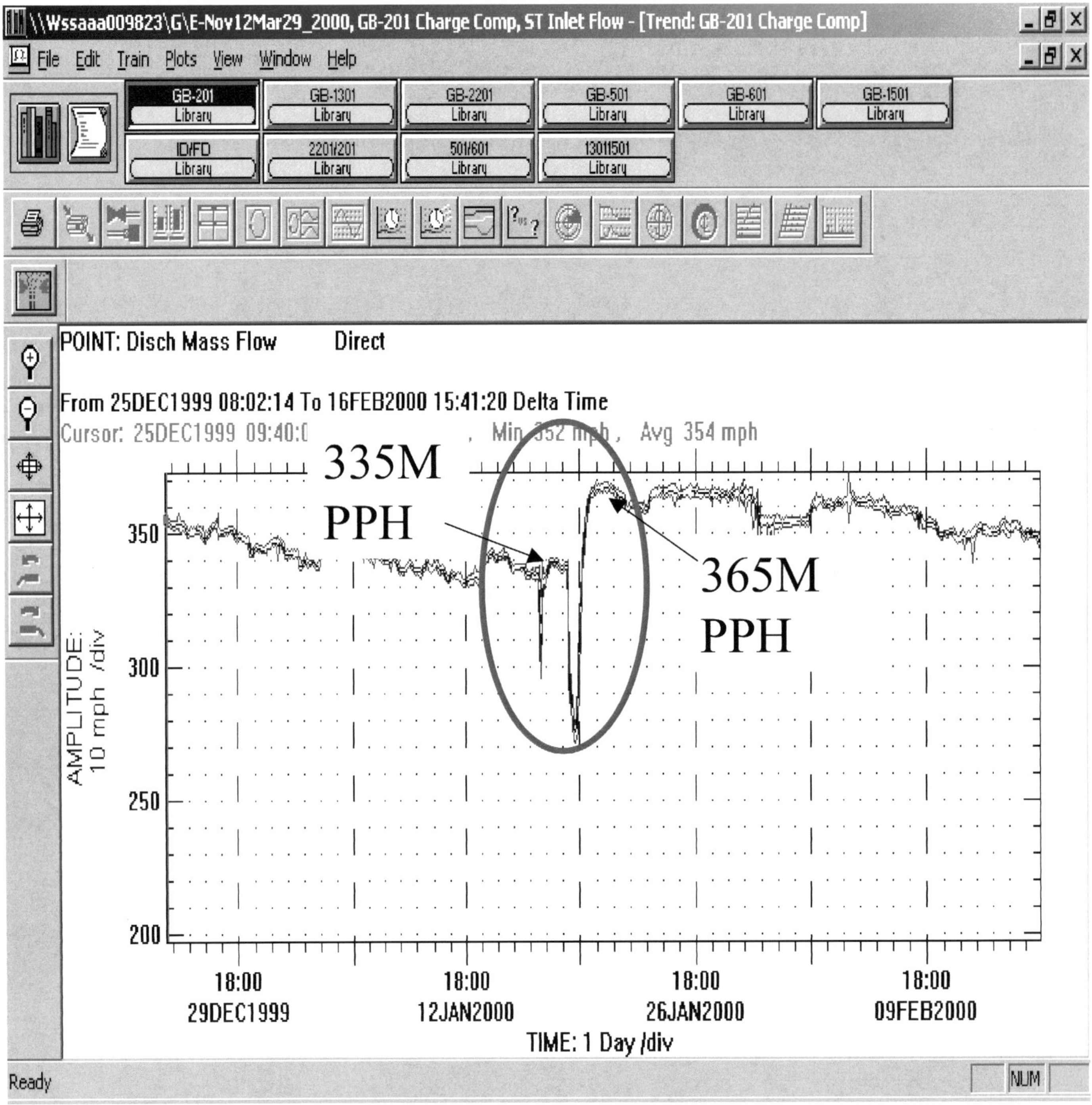

Steam Turbine Horsepower

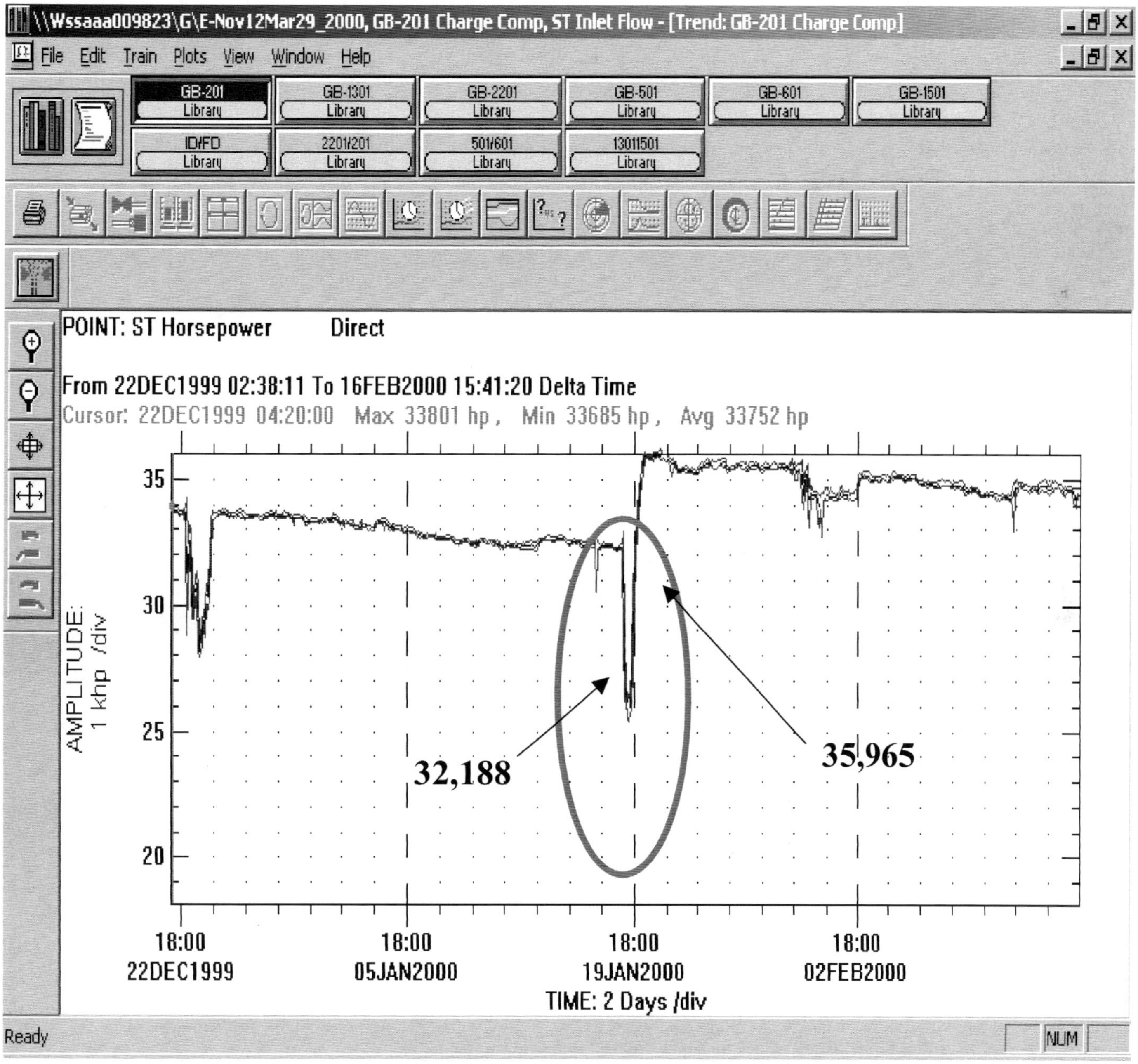

Steam Turbine 1st Stage Pressure

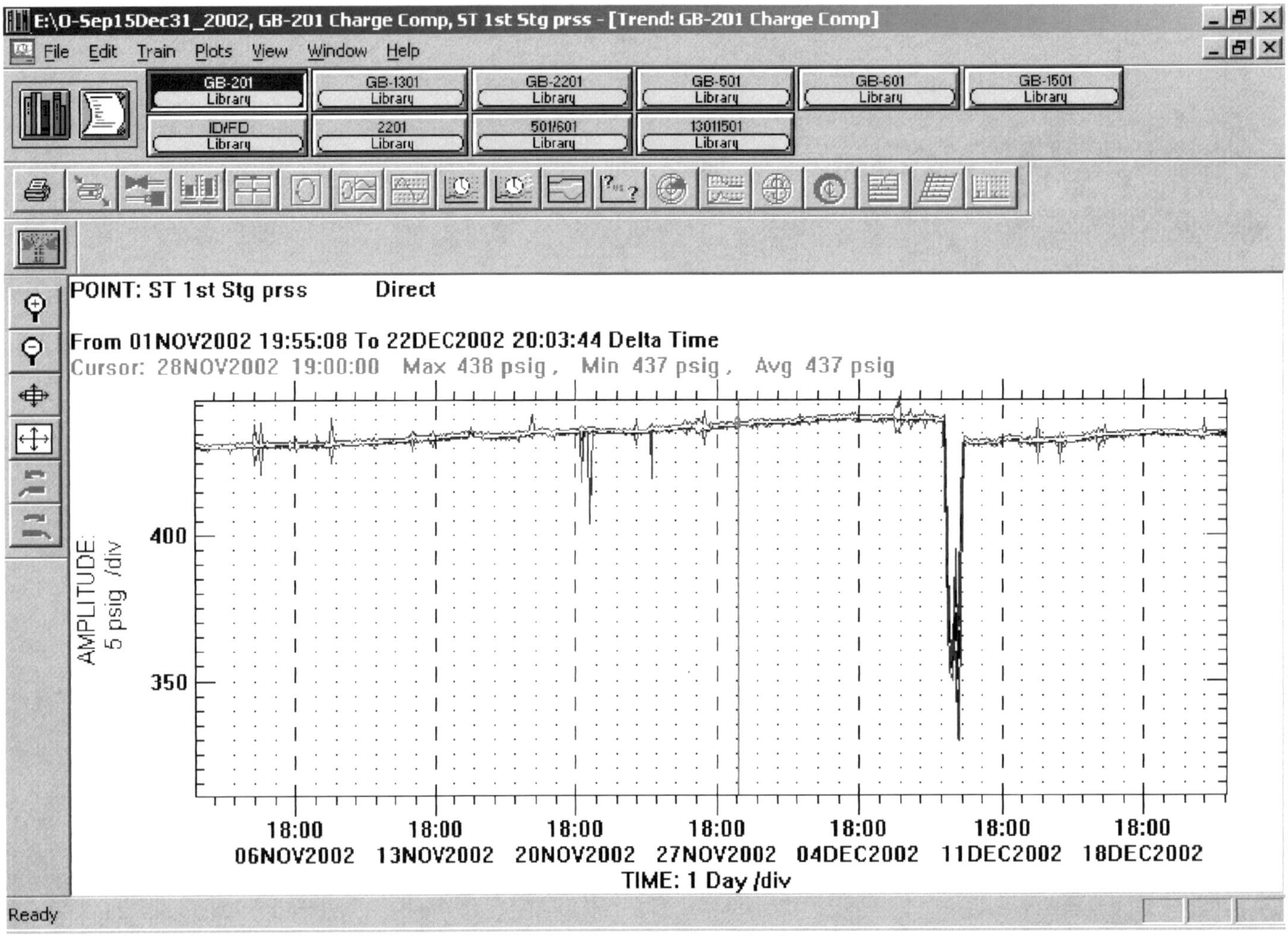

Second-Fourth Stages Speed

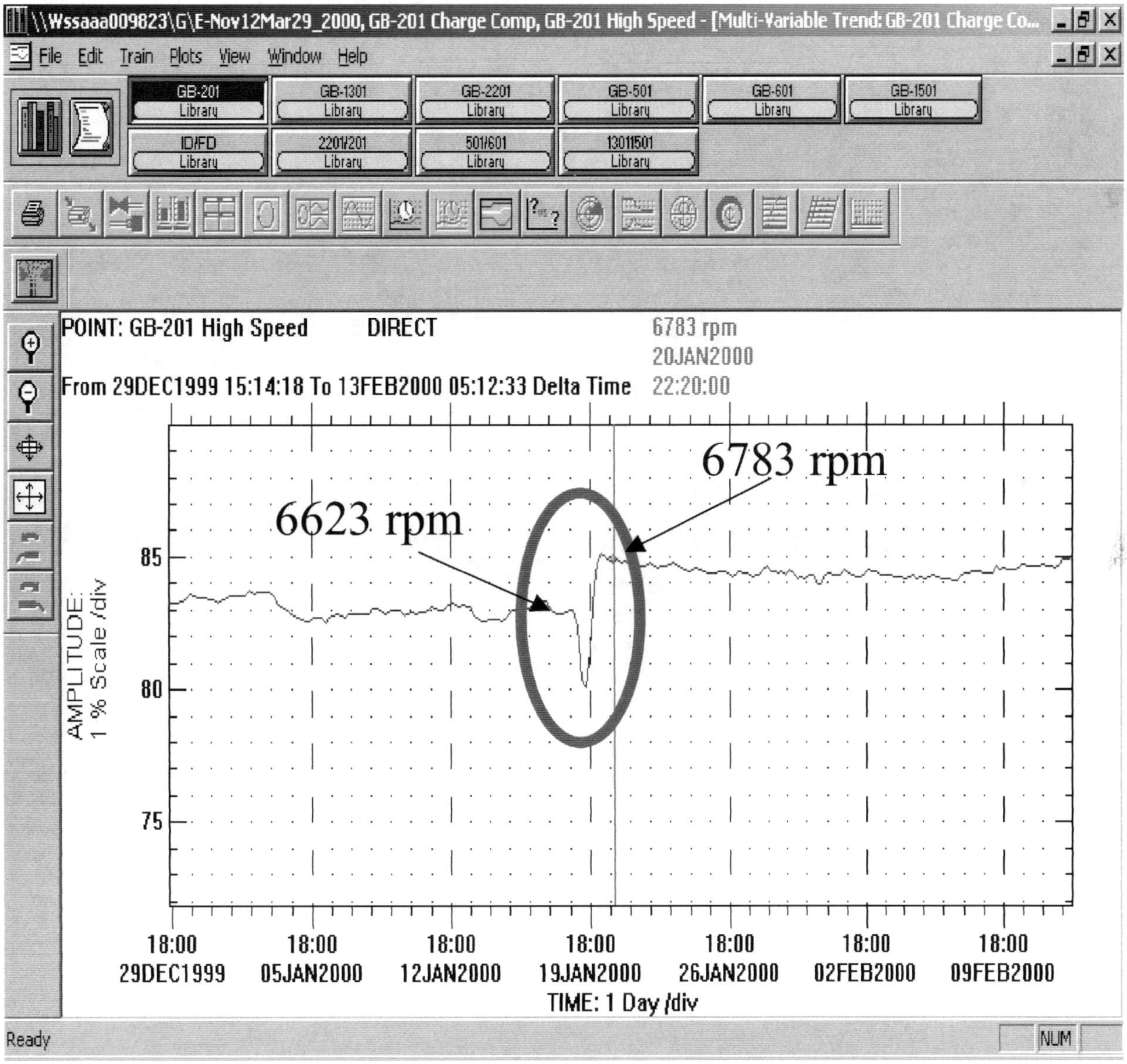

Before TA

•Relied upon continuous monitoring data

•Used a strain gage type of meter

After TA

Post TA Torque Meter Problem

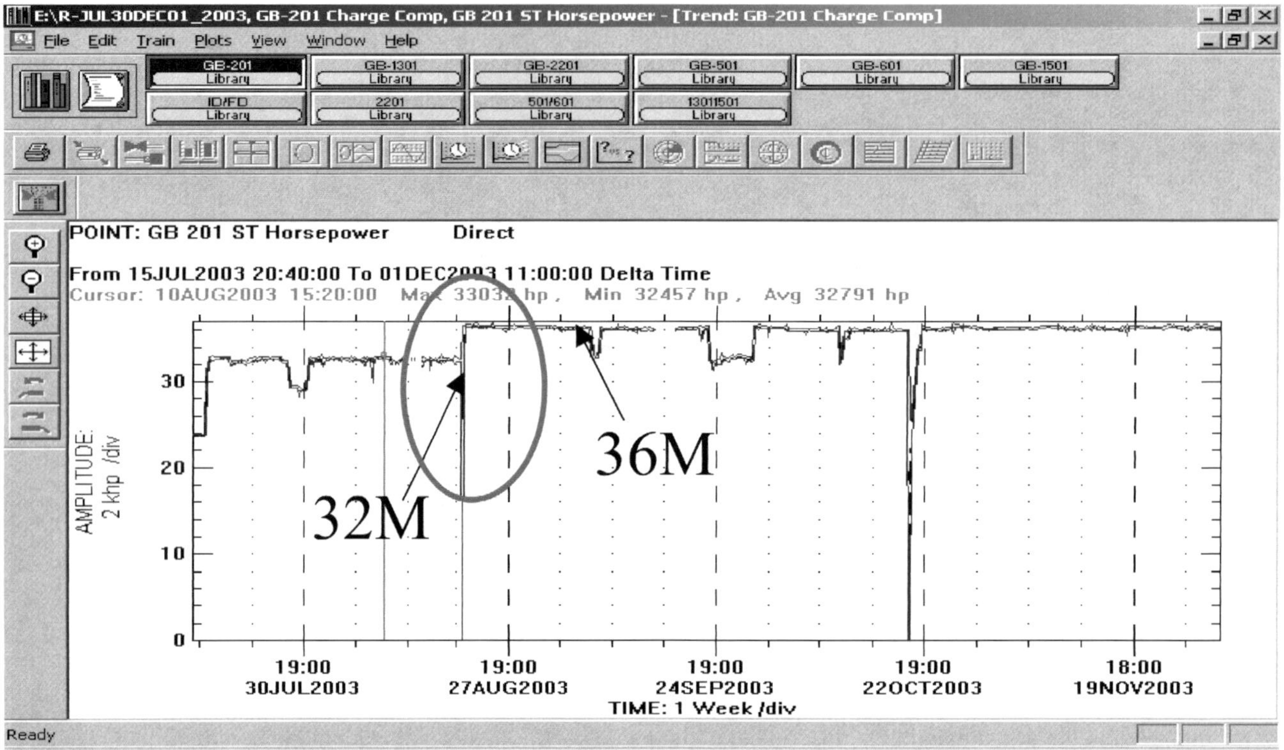

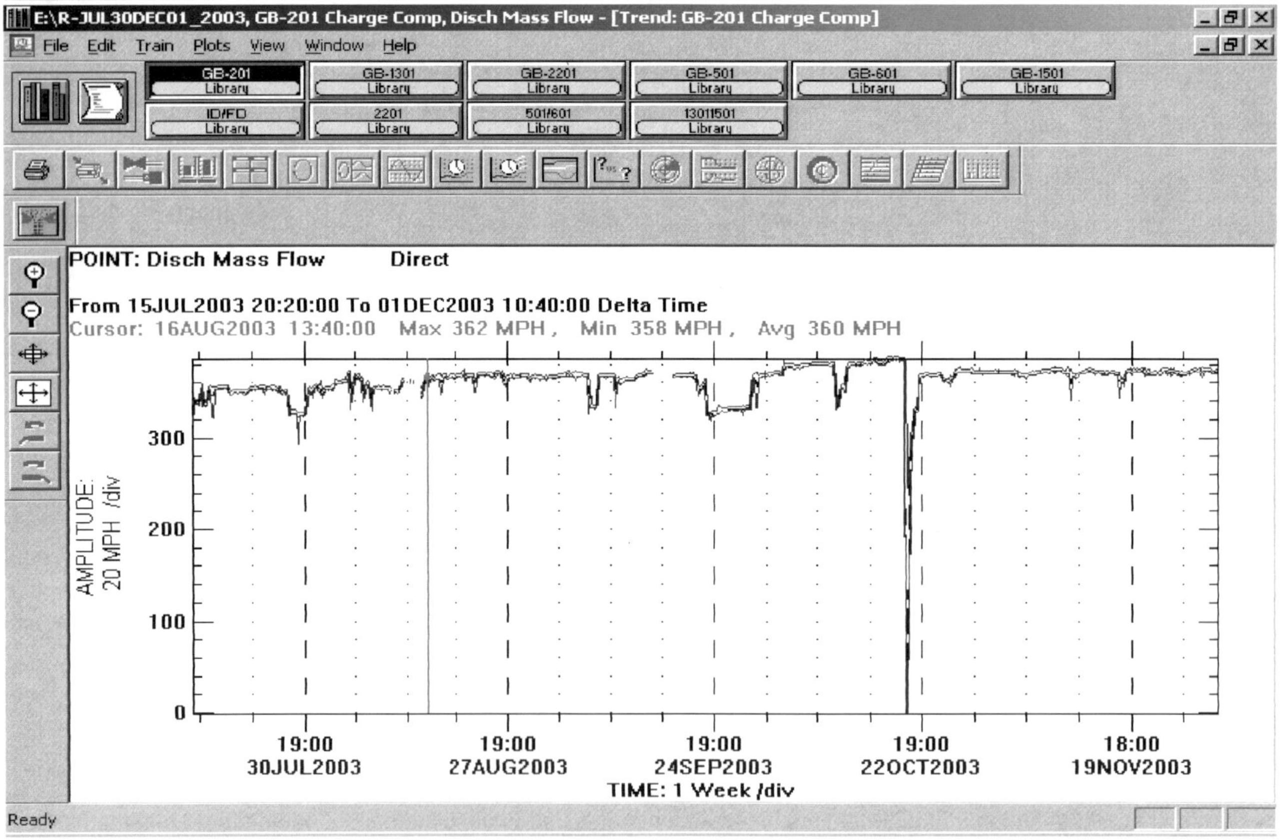

Torque Meters

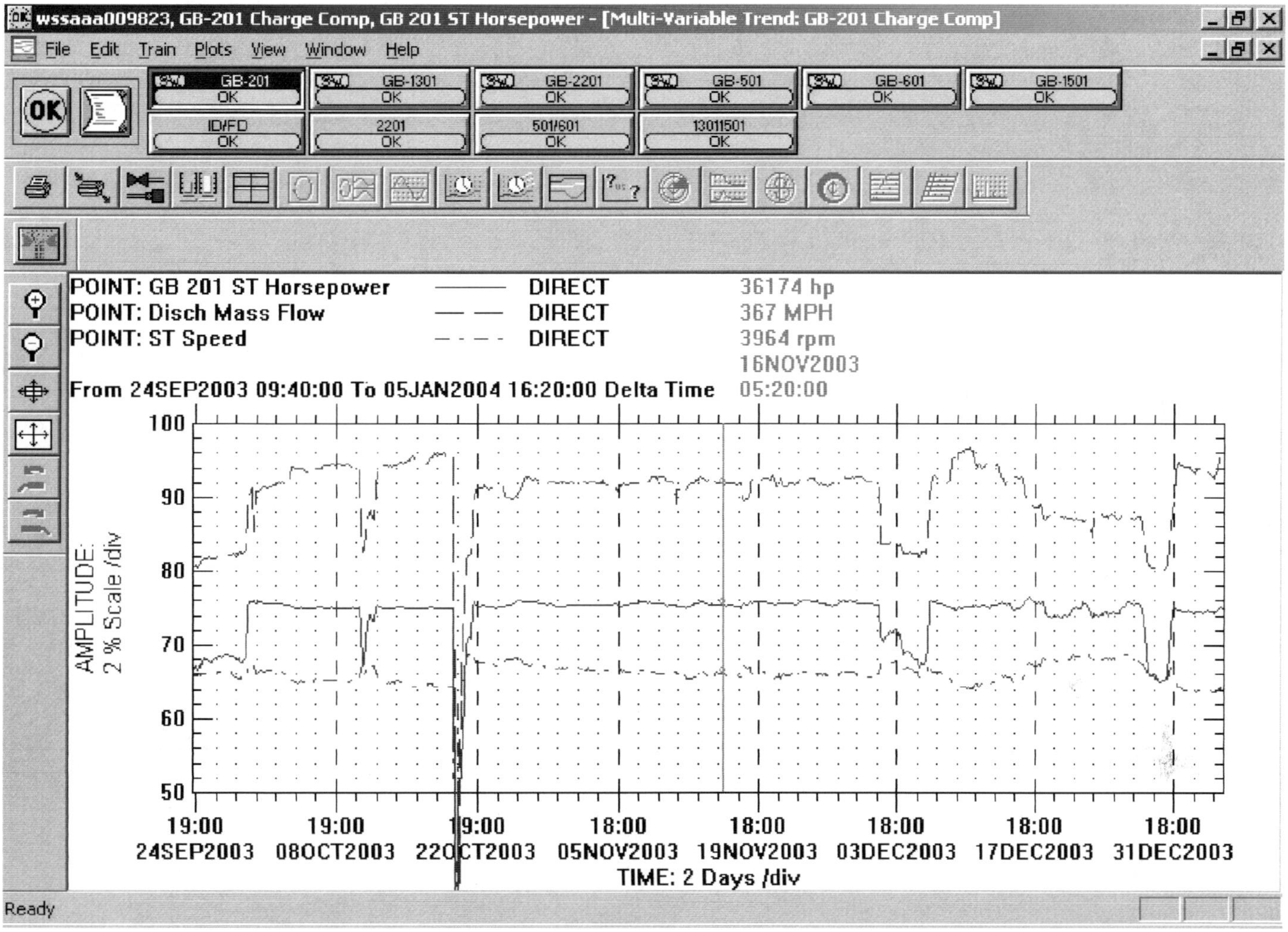

Turbomachinery Efficiency Gains Through Reliability Improvements

Patrick J. Cooper
Senior Engineer
Shell Chemical Company

Prepared for Presentation at the 2004 Spring National Meeting
New Orleans, LA, April 25-29, 2004
Maintaining Olefins Plant Energy Efficiency

UNPUBLISHED

April 2004

AIChE shall not be responsible for statements or opinions contained in papers or printed in its publications.

ABSTRACT

The purpose of this paper is to provide an end users view of ethylene turbomachinery upgrades available to improve reliability and at the same time improve efficiency. Turbomachinery used within ethylene process plants consume a significant amount of energy to operate. A number of hardware revisions can be implemented within steam turbines and compressors to improve reliability and also improve efficiency. The hardware upgrades discussed can be planned and implemented during routine plant turnarounds with some requiring no hardware modifications to install while others will require advance fabrication of spares ready for installation.

There are numerous upgrade options available that are worthy of consideration. As with any upgrade, each situation has to be analyzed to determine the expected benefit. This paper will limit the discussion to the following proven upgrade options available and how it affects reliability and efficiency.

Steam Turbines

- Nozzle box (snout) rings
- Interstage seals
- Tip Seals
- Blade and steam flow path designs

Compressors

- Interstage seals
- Coatings
- Fluid Injections
- Impeller and gas flow path designs

INTRODUCTION

The petrochemical industry is constantly challenged to improve health, safety and environmental performance along with reducing operational and maintenance costs. Improving turbomachinery reliability and efficiency can result in reducing unplanned flaring, reducing boiler emissions and ultimately increased profitability. End users of turbomachinery are constantly looking for ways and working with suppliers to improve machinery reliability. There are a number of commercially available hardware upgrade options worthy of consideration to improve steam turbine and centrifugal compressor performance in terms of reliability and efficiency.

Ideally machines capable of operating from turnaround to turnaround with no unplanned shutdowns or loss of efficiency should be the primary goal. Machinery reliability improvements usually require the installation of redesigned or totally new hardware to address a reliability issue. Reliability issues can be associated with repetitive failures, fouling, rubs, etc. In the process of addressing these, often times a gain in the machine's efficiency can also be realized. Also, these same reliability improvements could provide the necessary efficiency improvements to prevent replacing the equipment if it is of marginal performance. The efficiency improvements could be used for additional horsepower requirement or increased throughput. Ethylene production ranks high in regards to energy consumed amongst all chemicals manufactured and therefore efficiency improvements are worthy of consideration.

It goes without saying, that implementation of any of these upgrades requires a close working relationship with the supplier and its' engineering group. It also requires discussions way in advance of the actual installation to address all of the "what if" scenarios to prevent any unexpected delays. Previous repair history and documentation along with photographs are invaluable in this exercise. *A picture is worth a thousand engineering hours to figure out.* This paper will discuss several of the options available to consider for improving reliability and efficiency.

Steam Turbines

Steam turbines within ethylene manufacturing plants are used as the primary driver for the cracked gas and refrigeration compression systems. The turbines range in size up to 100,000 horsepower. Inlet steam pressures range up to 1500 psig. Since turbines are subjected to relatively high temperatures, enough clearance needs to be provided to prevent contact between the stationary and rotating components. If the clearances are too much, then a loss in efficiency will occur. It is always a challenge to try to harness this amount of energy in such a way as to optimize both reliability and efficiency.

The following will be a discussion pertaining to the components within the turbine whereby technology exists to improve reliability and efficiency.

__Nozzle box rings__

This upgrade applies to turbines that use a nozzle box. Inlet valves control the amount of steam that enters a turbine. The inlet valves fit into valve ports located in the nozzle box. The nozzle box directs the inlet steam to the 1st stage wheel. It is necessary to provide a sealed passage for steam between the outer shell and the inner shell (figure 1). An inlet pipe, referred to as a "snout", surrounded by a series of seal rings accomplishes this. The seal rings have to be capable of allowing thermal expansion between the parts in multiple directions. They must also maintain the minimal clearance designed to prevent steam leakage.

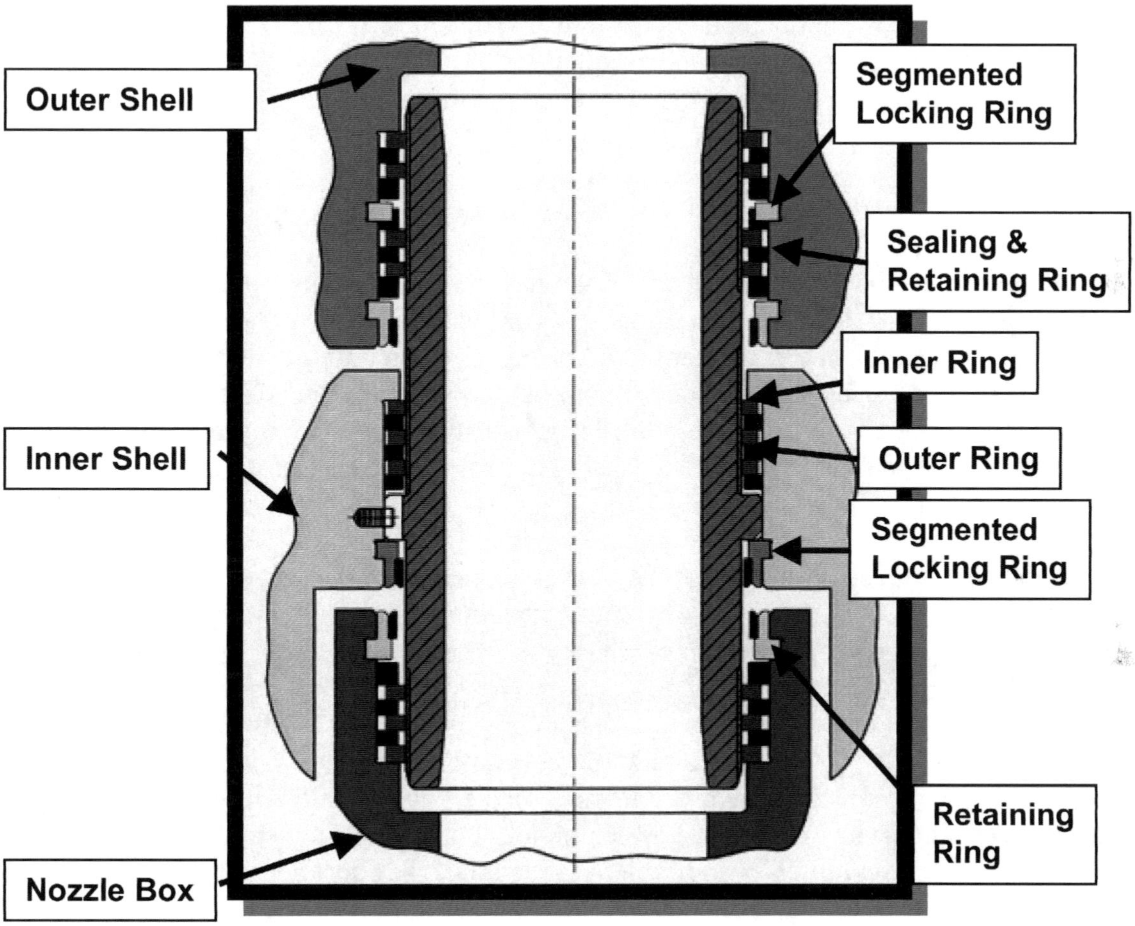

__Figure 1__

Conventional seal rings are susceptible to oxidizing causing the seal rings to experience high friction. Galling of the seal rings is common and presents difficulties during maintenance to remove them without damaging other parts. Once the seal rings gall, maintaining the designed clearance will no longer occur thus increasing steam leakage.

In order to improve on the reliability of these sealing rings, the rings are made of different materials that allow the rings to expand at different rates. In addition the rings are resistant to oxide formation and have a low coefficient of friction that allows the rings to slide freely to accommodate thermal gradients. This improved reliability results in about a 0.5% in efficiency improvement by reducing the steam leakage.

Interstage Seals

Labyrinth seals exist between the wheels of the turbine and the shaft ends to prevent steam losses along the shaft. Conventional seals are manufactured out of bronze or stainless steel material. The seal clearance is designed to accommodate thermal gradients and to prevent rubs during transient conditions such as start-ups and shutdowns. If the clearance is not enough, a rub may occur during start up and essentially open up the clearance since the seals are relying on a spring to provide the flexibility to absorb the energy from the rub.

Another concern with seal rubs is that the heat generated from the rub will cause the rotor to bow and create an unbalance. Allowing the machine to slow roll and bringing it up slowly to allow it to expand will help prevent rubs.

Newer seals are available that allows the seals to be positioned away from the rotor during start up (figure 2). The seals gradually close off as steam pressure increases and the rotor comes up to speed. This prevents rubs on start up thus allowing the seal to function as designed when running. The seals should only require inspection during turnarounds and can be reused if no problems are found. The springs are the only parts requiring replacement during maintenance. These seals can contribute to about 1.7% in efficiency improvement while improving reliability by preventing rubs.

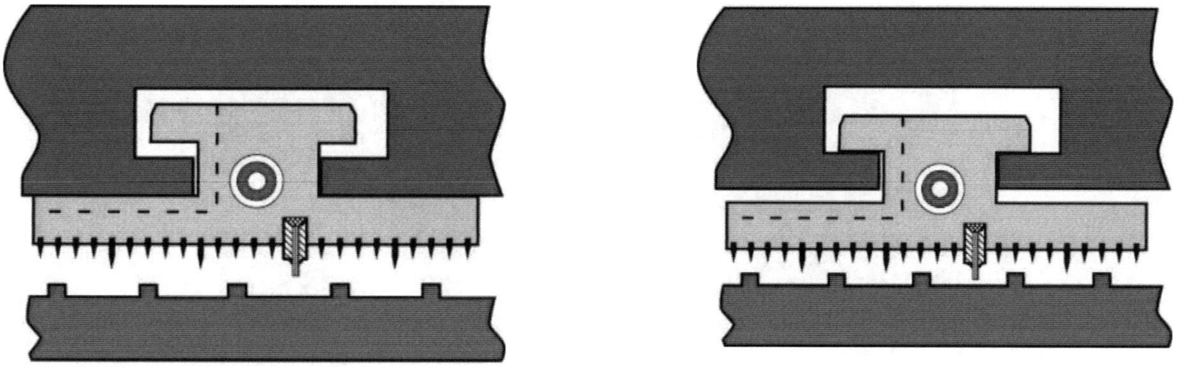

Open Position Closed Position

Steam Flow ⟶

Figure 2

Depending on the turbine design, it may be possible to include brush seals into the packing. The brush seals are another close clearance seal that is designed to take rubs but is flexible enough to regain it's shape. This feature alone is worth 0.3% in efficiency improvement.

Tip Seals

Tip seals are a single or double close clearance labyrinth attached to the diaphragm (figure 3). This seal minimizes the steam leakage around the rotor shroud thus forcing more steam to travel through the blades. Although tip seals provide no reliability improvement, and in fact if not installed correctly can actually introduce a potential reliability issue, it is worthy to mention since they could contribute to about 0.75% in efficiency gain. Not all turbine designs can accommodate this feature since some are limited by enough real estate to install them.

Figure 3

Blade and Steam Path Flow Designs

The redesign of blades and diaphragms can benefit turbines manufactured more than several years ago since the computer software available today can optimize the design for long-term reliability and improved efficiency. Given the variable speed operating conditions of the plant's turbines, operation on resonance during some periods can occur. The failure mechanism is high cycle fatigue initiating in a stress concentrator in the root attachment. Through the use of finite element analysis (FEA), a thorough analysis of the blade stresses and vibration characteristics can be determined. The blade design is further optimized through an aerodynamic evaluation. This evaluation provides a blade design that will also maximize efficiency. The diaphragms can also be evaluated and harmonically tuned to reduce the stimulus on the blade.

This upgrade option is generally pursued if reliability issues such as blade failures have occurred. Redesigning blades and diaphragms is quite time consuming. The upgrades will require the spare rotor to be modified and ready for installation during a turnaround. Any diaphragm changes may require new diaphragms to be fabricated prior to a turnaround if no spare diaphragms are available. It goes without saying that preventing blade failures during operation is a significant benefit and improving the efficiency is another added benefit.

Steam Turbine
Component Upgrade Summary

	Reliability Effect	Efficiency Effect	Comments
Nozzle Box Rings	Prevents rings from hanging up during operation. Minimizes damage during removal of the inlet pipes.	Provides improved sealing. Contributes to 0.5% efficiency gain.	Allow extra time to hone nozzle box bores and to finish machine rings to size.
Interstage Seals	Prevents rubs during startups and transient events. Seals can be reused and should reduce spare parts costs.	Provides improved sealing from startup to shutdown. Contributes to 1% - 3% efficiency gain.	No major modifications required fitting these.
Tip Seals	No significant benefit to reliability improvements.	Provides improved sealing on turbines without any originally. Contributes to 0.75% efficiency gain.	May not have much space to install them. Need to pay close attention on securing seals.
Blade and Steam Path Flow Designs	Addresses the causes for blade failures.	Depends on the workscope as to the actual benefit realized.	Will typically involve upgrading the spare rotor and manufacturing a new diaphragm to be installed during shutdowns

Centrifugal Compressors

Centrifugal compressors are used within ethylene plants for cracked gas and the propylene and ethylene refrigeration process. The cracked gas compressor consumes approximately 50-60% of the total compression power within the plant. It usually consists of three casings with four to six stages of compression. Each stage is cooled external to the compressor and returns to the next stage. The propylene compressor consumes 25-30% of the total compression power. It has one casing comparable to the low pressure or mid pressure cracked gas compressor in terms of size. The ethylene compressor consumes approximately 10-20% of the total compression power. It has one case also and is the size of the high-pressure cracked gas compressor. The ethylene compressor is the highest speed of the three services.

The following will be a discussion pertaining to the components within the compressors whereby technology exists to improve reliability and efficiency.

Interstage Seals

As with turbines, the compressors utilize labyrinth seals to seal gas leakage along the shaft and across the impellers. The seals consist of an eye and shaft seal for each stage (figure 4). Aluminum labyrinths have been used extensively in the past for these seals. During startups or other transient events the rotor can rub the seals resulting in increased clearance. The aluminum seals are not flexible when rubbed and can cause rotor damage as well. So as much time as was spent in achieving the designed clearance during installation, it can be quickly lost during startup resulting in reduced efficiency.

The use of engineered thermoplastics has proven to be effective in improving reliability and efficiency. The seals are fitted with tighter clearances to start with. The seals are flexible and regain their shape after rubs occur thus maintaining their installed clearances. In terms of reliability, it prevents seal and rotor damage. In terms of efficiency, the seals reduce gas leakage by having a tighter clearance that is maintained throughout the run. The balance piston seal is another option to consider for upgrading to thermoplastics.

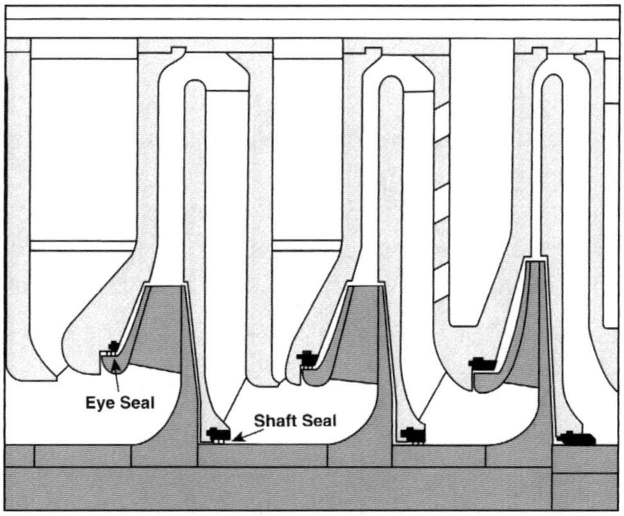

Figure 4

Fluid Injections

Fluid injections as discussed here involve the use of a wash oil, water or specialty chemicals to inhibit fouling. Fluid injection is used on the cracked gas compressors and not the refrigeration compressors. Water has found it's way into the line up since it is used to prevent the discharge temperatures from getting too hot and polymerizing the process. The fluid type, flow rates, continuous or non-continuous option vary depending on the process. Compressor fouling can cause significant reliability problems by building up around the shaft and actually machining the shaft and/or impellers. In terms of efficiency, fouling probably contributes the most to efficiency losses that affects throughput and energy consumed.

Some compressors use injection nozzles in the suction line while others inject between the impellers that offer advantages of ensuring good coverage. Retrofitting a compressor without nozzles between the impellers requires advanced planning to assure the quality of the installation along with accomplishing it during a turnaround window. The efficiency gain with performing fluid injection depends on the application but it is not uncommon to experience 1-2% efficiency loss per year. So regaining or preventing any loss due to fouling has a significant payback.

Coatings

Another way to combat fouling in the cracked gas compressors is to coat the rotors and diaphragms with an anti-stick coating. The coatings are a multi-step application process. While the coatings do provide benefits there are a number of issues to think about.

One is good coordination between the user and supplier is paramount to incorporate coatings to compressors that don't currently have it. The difficulties arise since diaphragms are not typically stocked. Coating diaphragms can be done during turnarounds but requires good upfront planning to accomplish during the turnaround window. Another issue pertains to coated compressor internals may require revising the decontamination procedure. Some users have experienced blistering of the coating that appears to be related to decontamination methods. The last issue relates to how to clean the coated parts after a run without damaging the coating. A number of cleaning methods are available. Blast cleaning with frozen CO_2 pellets works well and minimizes coating damage. The durability of the coating is dependent on the aggressiveness of the fouling and fluid injections. It is best to prepare for a recoat during turnarounds.

Impeller and Gas Flow Path Designs

As with turbines, computer programs have significantly enhanced designing compressor internals to improve reliability and efficiency. The use of FEA and computational fluid dynamic (CFD) programs to design compressor internals has provided improvements in mechanical integrity and minimizing flow losses. Manufacturing the improved designs is no longer a limiting factor. The use of 5-axis milling has overcome this hurdle.

You can imagine that this upgrade is not high on the list unless a rerate or a reliability issue pertaining to the compressor internals exists. The costs involved will play a large part in justifying. Again, each application has it's own issues and this upgrade can provide significant benefits with regards to efficiency. It is one of those upgrades that is considered when throughput increases are required and this option may prevent having to purchase a new compressor.

Compressor
Component Upgrade Summary

	Reliability Effect	Efficiency Effect	Comments
Interstage Seals	Thermoplastic seals can tolerate rubs and maintain clearance from startup to shutdown. Reduces impact to rotor shafts. The thermoplastic materials can tolerate the process better than aluminum.	Provides improved sealing from startup to shutdown. Contributes approximately 1% - 2% per compressor case in efficiency gain.	Larger compressors may not benefit as much from the efficiency gain compared to smaller ones.
Fluid Injection	Depending on the application, can significantly improve reliability by preventing and removing fouling.	Provides a handle to maintain and regain design efficiency. Without it a loss of 1%-2% annually can occur.	Injection between the compressor wheels is preferred. Erosion and coating removal can be an issue.
Coatings	Addresses the accumulation of polymer deposits which could lead to increased vibration.	Depends on the machines operational history as to how much it impacts efficiency.	The use of interstage fluid injection may impact the life of the coating. May have to revise decontamination procedures for maintenance shutdowns to maintain coating integrity.
Impeller and Gas Flow Path Designs	Improved mechanical integrity.	Depends on the application but will provide improved efficiency.	May be difficult to justify unless a reliability issue exists or a rerate is planned.

Conclusion

What this paper has provided is a brief discussion on some upgrade options to consider. More detailed information is available through previous papers written on each option along with supplier discussions. It is important for those of us in the ethylene production business to keep abreast of the latest technology so that we can continue to improve machinery reliability and efficiency. A good working relationship with suppliers is important to understand the benefits and risks associated with any change. Retrofitting machinery with new technology will always be a viable alternative to purchasing new equipment. *It all depends on the cost to benefit.*

Acknowledgements

The author would like to thank the machinery engineers and inspectors at Shell Norco Chemical and Shell Deer Park Chemical for their contributions in the evaluation and implementation of many of the options discussed.

Thanks to Shell Chemical Company for the opportunity and permission to publish and present this paper.

Thanks to Turbocare, Inc. and Elliott Inc. for the turbine and compressor graphics.

Steam System Optimization

Robert Aegerter, Equistar Chemicals, L.P., (815) 942-7390

Abstract

Refinery and chemical plant steam systems are complex and the fuel required to produce the steam represents a major expense. The incremental cost for generating a 1,000 lb./hr. of steam is typically $45,000 - $60,000/year. Most plants have numerous low/no cost opportunities to save several thousand pounds/hour of steam. Implementing these low/no cost projects can result in significant savings.

Reduce Steam Demand and Cost

The ultimate goal in optimizing the steam system is to minimize the steam generation costs. Most projects are dependent on the steam balance and can only be justified if low-pressure steam is being vented or if steam is being let down. Some projects are independent of the steam balance, such as eliminating high-pressure (HP) steam leaks, insulating HP steam piping, optimizing the boiler operation, and improving the performance of condensing turbines. These projects can be implemented without a complete knowledge of the steam balance.

Insulating bare pipe or replacing damaged insulation on the HP steam system can usually be justified with a quick return. An infrared thermography survey can help identify areas of damaged insulation and will help justify insulation projects. The 3E Plus insulation program provided free by the Department of Energy can also be used in justifying insulation projects.

Steam leaks on the HP steam system should be repaired quickly and can be justified with quick returns. Steam traps in HP steam service should be rated for the service.

Repairing air leaks around the boilers is also important. On negative draft boilers, air leaks waste fuel, damage refractory, and cause erroneous excess oxygen readings. On positive draft boilers, air leaks waste fuel and can cause personnel injury. Fan and boiler capacity can also be lost with significant air leaks. Air leaks should be fixed at the first opportunity.

To operate fuel-fired boilers at the minimum oxygen level, the fuel and air controls need to be calibrated and tuned to optimum performance. Part of that work should include verifying that the fan louvers and the stack damper function properly.

To maintain efficiency, the boiler's heat transfer surfaces need to remain clean. On the water side, sites should work closely with their water treatment company to prevent hardness deposits from forming. On the process side surfaces must also remain clean. Economizers collect dirt and products of combustions and must be periodically cleaned. Water washing the fins is the most effective method of economizer cleaning and should be done about every five years when boilers burn clean fuels. If dirty fuels are used, then soot blowing should be frequently performed and the economizers cleaned on a more frequent schedule.

For sites with condensing turbines, the turbine blades and the surface condensers must remain clean to maintain efficient operation. Solid deposits on the turbine blades can be removed by online water washing. Surface condensers can be continuously online cleaned with systems that circulate brushes or sponge balls through the tubes that gently clean the tubes. Although continuous online cleaning is preferable, surface condensers are often designed such that half of the condensers can be taken out of service and hydro-blasted while the turbine is still running with the other half of the surface condenser in service. This may require the turbine to run at reduced load while half of the condenser is out of service.

Develop a Steam Balance

Most steam system optimization projects require an understanding of the steam balance. A steam balance is an excellent tool for understanding the steam system. The more accurate and detailed the steam balance, the more successful a plant will be in reducing its steam costs. Once a steam balance has been developed, opportunities for steam savings become apparent, and project savings can be quantified.

Steam Venting or Letting Down?

Typically a plant will be either venting excess low-pressure steam or letting down steam to meet the low-pressure steam demand. If a plant has multiple operating units, it is possible that one unit may be venting low-pressure steam, while the other unit is letting down steam to the low-pressure level.

The strategy for reducing the steam demand is different when steam is being vented or let down. Reducing the steam demand should be a continuous process of eliminating sources of excess low-pressure steam until there is a low pressure steam demand and then reducing the steam demand to create an excess of low-pressure steam. As changes are made to the steam system, the balance should be updated. The steam balance should be used as a guide that identifies the most economic solution to reducing the HP steam demand. As the steam balance changes due to changes in feedstock, production levels, weather conditions, etc., the steam balance should be updated and opportunities implemented as they become available.

Eliminating Excess Steam

When the amount of steam that is entering the header is in excess of the amount of steam required to maintain the pressure controller's set point, steam is either vented to the atmosphere or condensed in fin-fan coolers. Often plants that are experiencing excess steam first look at ways of utilizing the steam. This often takes the form of spending capital to install equipment or hiding the excess steam in distillation columns or steam to the flare. A better solution is to find no/low cost solutions to eliminate or reduce steam entering the header.

The quick-fix solution to eliminating excess steam is to shut down steam turbines that exhaust into the header and to start up the motor driven spare equipment. Often times this step will be enough to eliminate the venting. Although shutting down steam turbines is the easiest solution, it may not be the most cost-effective solution because an electric motor is now being operated. In addition if the plant's electrical rate schedule includes heavy penalties when new peak demands are set, then consideration should be given to when turbines are shut down and motors started up. If the excess steam can be eliminated without shutting down steam turbines, then other solutions should be pursued.

Steam Let Down Valves

To eliminate the excess steam condition, all sources of steam that are contributing to the excess steam condition must be identified. One of the best places to start looking is at steam let down valves. If a let down valve is open from a higher pressure header and steam is being vented at a lower pressure level, then steam is at an excess at the higher steam pressure level and sources of steam supplying the higher steam pressure header must be investigated. If the steam let down valves are closed and steam is being vented, then the let down valves may be leaking through and contributing to the excess steam. Many times steam letdown valves operate in a pinched position, which can cut the valve seat. It is not unusual for a control valve to leak through 10,000 LB/hr or more of steam. The likelihood of a control valve leaking through is directly dependent up the length of time since the valve has been serviced, the differential pressure across the valve, the amount the valve is throttled, and the materials of construction.

If there is a desuperheating station downstream of the steam letdown valve, one way of checking if the steam valve leaks through is to check the BFW valve that is used to desuperheat the steam. If the BFW valve is open, then the steam valve is leaking through.

Another way to check if a steam let down valve is leaking through, is to close the upstream and downstream isolation valves. The effect of this change should be observed on the upstream and down stream systems (boiler steam production and upstream steam letdown valves and vents). Depending on the size of the system, it may take 15 – 30 minutes until the system fully responds to the change. After waiting an appropriate time, place the valve back into service and check again for a response to this change. If the system responds both times to the step changes, the control valve is leaking through. The leak flow rate can be estimated by the change in steam production or the changes in the valve positions in the upstream and downstream control valves.

If the let down valve leaks through it should be repaired or replaced. Consideration should be given to replacing leaking steam valves with ANSI class V control valves, that have tighter shutter off and longer life than a standard valve. The additional cost of installing a better quality valve can justified with the savings realized with a longer valve life.

Steam traps that discharge into a steam header should be checked for proper operation. Badly leaking steam traps can over pressure a steam header.

Steam Turbines

Another potential source of excessive steam into the header is the single-stage steam turbines that are used to drive pumps, fans, and air compressors. The efficiency on these turbines can be improved by operating the turbines with the hand valves closed (when they are not required) and restoring the efficiency to the original design.

Often hand valves are provided by the turbine manufacturer to provide additional horsepower. When the hand valves are open they leave all the nozzles in the turbine nozzle block open providing for the maximum available power. Operating with the hand valves open when not required reduces the pressure in the nozzle block by forcing the throttle valve to pinch back. Closing the hand valves blocks some of the nozzles, causing the throttle valve to open more. Since there is less pressure drop across the throttle valve, more work is done through each remaining nozzle making the turbine operate more efficiently.

Figure 1. The hand valves extend into the steam chest and can be accessed by valve stems that are located beneath the outboard bearing housing.

Hand valves should be operated either fully open or closed. The hand valves can be checked while the turbine is operating by monitoring the turbine speed while closing each hand valve. The hand valve farthest from the inlet throttle valve should be closed first. If the turbine can maintain speed with any of the hand valves closed, the hand valves should remain closed.

If the hand valves are closed and the turbine is maintaining the design speed, then the turbine should be checked for further efficiency improvements. Opportunities may exist to restore the turbine to its original design efficiency or rerate the turbine to a more efficient design.

The nozzles and turbine blades can wear over time causing the turbine to use more steam than designed. Often these components are ignored when maintenance is done on the bearings and seals. The blades and nozzles are typically not a concern unless the turbine cannot maintain its speed. The turbine efficiency can be checked by measuring the inlet and outlet pressures and temperatures. From these readings the actual turbine efficiency can be calculated by ratioing the actual and ideal enthalpies as follows:

$$\text{Turbine Efficiency} = \frac{h_{inlet} - h_{outlet\ actual}}{h_{inlet} - h_{outlet\ ideal}}$$

The original design data sheets has the turbine water rate (or steam rate) that is expressed in terms of LB/HP-Hr. The design water rate can also be converted to the design efficiency by calculating the ideal amount of steam the turbine would use to produce 1 HP of work (assuming constant entropy) and dividing the theoretical by the design water rate.

The theoretical amount of steam required to do work is calculated as follows:

$$\text{Water rate}_{ideal} = \frac{(2545\ \text{BTU/HP-Hr})}{(h_{inlet} - h_{outlet\ ideal})\ \text{BTU/LB}}$$

If there is a significant difference between the actual and design efficiency, consideration should be given to replacing the nozzle block and turbine blades.

Often turbines are over designed and can be rerated for additional efficiency. If the pressure drop across the governor valve is more than 10% of the steam inlet pressure, then the turbine is over designed. The turbine can be rerated to operate more efficiently. Usually this will require that a new nozzle block be installed. Rerating a steam turbine is relatively inexpensive and can be justified if the turbine is causing steam to vent to the atmosphere. The plant should work with the steam turbine's manufacturer to rerate the turbine.

Although more expensive than rerating an existing turbine, it may be cost effective to replace a steam turbine with a more efficient turbine or an electric motor driver to obtain the amount of steam flow reduction that is desired. When installing a new steam turbine, efficiency should be the prime concern. While most turbines operate at either 1,800 or 3,600 rpm, single-stage steam turbines operate more efficiently in the 5,000-6,000 rpm range. Operating a turbine at a higher speed with a gear reducer to the driven equipment may be justified with the increased turbine efficiency.

Another option to replacing a steam turbine that drives a fan or horizontally split case pump or fan, is to modify the rotating assembly so that that shaft extends out on both ends of the driven equipment. A motor driver and a steam turbine can be installed on opposite ends of the driven equipment. Either the motor or the turbine can easily be selected as the main driver by increasing or decreasing the speed of the steam turbine above or below the motor's synchronous speed. If the turbine is operating at above the motor's synchronous speed, the turbine will be carrying the load. If the turbine's governor is set to operate at below the synchronous speed, then the load will be transferred to the motor driver.

To enhance this strategy, an electronic governor with a speed pickup can be installed on the turbine to split the work being done between the turbine and motor drivers. The turbine's speed can be tied into the steam header pressure controls so that steam would be preferentially letdown through the turbine instead of the letdown valves. This type of installation adds flexibility to controlling the steam balance.

Steam Headers

Varying steam header pressures can affect the steam rate through turbines. If lower turbine steam rates are desired, then either the inlet steam pressure can be increased or the exhaust pressure decreased. Lowering the exhaust pressure will have more impact on turbine steam rates than raising the inlet pressure. The same technique can also be used to obtain more horsepower from a steam turbine that has a fully open throttle valve.

Varying steam header pressures can also help transport steam between battery limits, which can help eliminate excess steam conditions. When steam headers are too small, varying the steam header pressures is an effective alternative to installing larger or parallel steam headers. The best way to determine the amount of variation available in your steam headers is to conduct a test. The steam header pressures should be adjusted by only 1-2 psi for each stage of the test and then left for a couple of days to determine the effect. If no problems are experienced, the test should be continued until the bottlenecks are identified.

If a plant has several operating units, some units may have an excess low-pressure steam and others letting down. To optimize a multi-unit steam system, the units should have adequately sized headers connecting the units.
The header pressures should be controlled to optimize the site's steam balance instead of the individual area's balances. All the steam system pressure controllers should receive input from an optimizing program that adjusts the steam header pressures to minimize letting down and venting throughout the site. To avoid upsets the upper and lower pressure limits need to be determined for each steam header and changes in header pressures must be done gradually.

If it is not possible to eliminate excess low-pressure steam, then effectively utilizing the steam is the next best alternative. The deaerator offers the opportunity to recover excess low-pressure steam by increasing the deaerator's pressure. Increasing the pressure will admit more steam to the deaerator and heat up the BFW to near the saturation temperature of the deaerator's operating pressure. The hotter BFW will generate more waste heat boiler steam and will reduce the amount of fuel used in fired boilers.

The deaerator's pressure can most likely be raised if the operating pressure is more than 5 PSIG below the MAWP of the storage drum and the scrubbing section.

To optimize the deaerator's pressure, the deaerator's pressure control can be tied into the low-pressure steam header's pressure control. The existing steam header pressure control valves should respond quickly to handle upsets, and the deaerator's pressure control should respond slowly to seek an optimum pressure setting.

If the excess steam is at too low of a pressure to be used by most users, the excess steam can be used in a higher pressure service by:

- Boosting the pressure using a thermo-compressor and
- Injecting the steam down stream of a user's temperature control valve. The injected steam is added at a fixed flow and the temperature control valve is required for precise control.

A survey of heat exchanger chest pressures will identify potential candidates for this technique.

Minimize Steam Let Downs

If steam is constantly being let down, the short-term solution is to start up steam turbines that exhaust to the proper steam header and to shut down the motor driven equipment. Care should be given not to vent steam by operating too many turbines. The economics of shutting down motors is quickly eliminated when steam is being vented.

If a site has an extraction turbine, pressure controlling the steam headers with extraction steam is more efficient than operating smaller single-stage turbines. For sites with extraction turbines and multiple operating areas, the extraction steam controls can be connected into other area's steam header pressure controls by using the unit's existing steam pressure control valves as the primary pressure control and using the extraction controls to respond more slowly to close the unit's pressure control valves.

If an operating area is not venting steam, the goal should be to reduce the steam demand. Eliminating steam leaks and recovering condensate usually has a very high return. Condensate and steam leaks should be repaired soon after they are detected because they can quickly grow larger. If a leak cannot be isolated, a leak repair service should be used.

Steam Traps

Faulty steam traps can contribute a significant steam demand. Too much emphasis has been placed on the manufacturer and type of steam traps used. There are numerous suppliers of quality steam traps. Standardizing on a specific trap that functions well, maintaining a consistent steam trap testing program, and repairing faulty steam traps soon after they are identified will minimize steam trap energy costs.

The most effective method of testing steam traps is done using a computerized steam trap management system. The steam traps' pertinent information is entered into a data base that includes the trap's manufacturer, model number, and operating conditions. A trained technician uses an ultrasonic and temperature sensing instrument that stores the field measured data. The data is later downloaded into a software program that analyses the results. From the program's results, failed traps can then be identified, prioritized, and then repaired.

A computerized steam trap management system will help identify traps that frequently fail. Often the failures are due to using the wrong trap for the service, too long of tracing or return lines, traps that have return lines that are tied into the supply lines, etc. Resolving these issues will reduce the number of repeat trap failures.

Using more steam than the required amount for process needs can significantly increase the steam demand. Too often set points are determined by "what we've always run before" and not what is most efficient. Over use of steam will cause more heat to be rejected to cooling towers or fin-fan coolers. Sometimes this additional load can cause plants to be rate restricted during the summer months by because of limited cooling system capacity. Using the plant steam balance and plant design information, the actual versus plant design steam use set points for all major steam users should be compared. If there are large discrepancies that cannot be accounted for by changes in plant operation, there may be savings opportunities.

The operating targets on distillation columns should be reviewed to verify that additional energy is not being used obtaining a higher purity than is required. Sometimes the product specifications can be relaxed to reduce the energy requirements.

Distillation

When the operating rates are reduced, often the amount of steam required for numerous users can be reduced. The steam use for most users should be varied with the production rates to obtain the most efficient operation.

Reducing the operating pressure of distillation columns increases the volatility of the components being separated and can reduce the reboiler steam demand. Tests can be conducted to determine if the pressure can be reduced. This is especially an opportunity when operating at reduced rates.

Steam to Flares

Most plants can decrease their steam demand by having automatic steam control flow to the flare. Typically board operators manually control steam to the flare using camera surveillance. This method results in excessive steam use that can reduce the flare's destruction efficiency from 98% to 60%. The decrease in destruction efficiency is caused by the steam snuffing out the flame which is re-ignited by hot zones.

The ideal steam flow needed for smokeless operation is not linear with the hydrocarbon flow. Higher gas flows cause higher exit velocities, which entrains more air into the flared gas and reduces the required steam flow. The gases' molecular weight, the hydrogen/carbon ratio, the heating value, and other factors also affect the ideal steam rate.

To optimize the amount of steam used on a flare, an automatic steam controller should be used. There are several types of steam controllers, but the most common uses infrared technology to measure the flame's opacity and optimizes the steam flow to eliminate smoking. Automatic steam controllers often have a one-year payback.

Frequently flare steam flow control valve's seats are worn from operating in a throttled condition. Steam valves with worn trim can leak over 10,000 LB/Hr of steam to the atmosphere. The steam valve, when it is in the closed position, can be checked for leakage by isolating the valve and watching the flare tip for a change in steam flow. If the steam plume is reduced when the valve is isolated, the valve leaks through and should be repaired or replaced. Upgrading the valve to a ANSI class V control valves will prolong the valve's life.

Insulation

Insulating bare equipment and piping in hot service or repairing damaged insulation can reduce the steam demand. Sites that have a refrigeration compressor driven by a steam turbine, can also benefit from insulating bare cold surfaces and repairing damaged refrigeration insulation by reducing the system's refrigeration demand and the turbine's steam flow rate. Almost all plants have numerous insulation opportunities. Often flanges, control valves, steam turbines, man ways, sections of piping, heads on vessels, etc. are bare and can significantly increase the steam demand.

An insulation survey should be conducted of the steam, condensate, and refrigeration systems. Typically there is a quick payback for insulating bare surfaces. Bare piping in hot or cold service should be insulated and damaged insulation and metal jacketing should be repaired to stop further damage and corrosion under insulation. Although removable insulation covers are excellent for flanges, man ways, and heat exchangers heads that need periodic opening, permanent insulation is required for cold insulation and is preferable for hot insulation because it has less energy loss.

Waste Heat Recovery

If all of the steam users are efficiently using steam, then waste heat recovery opportunities need to be explored. The duties and temperature profiles on services that are cooled by air or water should be compared to services that are heated by steam. If the duties and temperature profiles compare favorably, consideration should be given to projects that recover waste heat energy. Deaerator make-up water is an excellent heat sink for waste heat recovery. When the deaerator make-up water is preheated, then the deaerator's steam demand is reduced. When installing deaerator make-up water preheat projects, the following precautions should be addressed:

1. When water that has not been deaerated is heated above a specific temperature, then oxygen will be released and corrosion in the carbon steel piping will occur. Your water treatment representative should be consulted for the exact temperature. If your project will recover enough preheat to the deaerator make-up water that oxygen will be released, it will be necessary to install a stainless steel return system to the deaerator.
2. If the make-up water is heated too much and there is no stripping steam needed in the deaerator to maintain the set operating pressure, then good oxygen stripping will not occur in the deaerator. Your deaerator manufacturer should be consulted to determine the deaerator's minimum stripping steam requirements.

If after reducing the demand on all steam users is exhausted and all economically attractive waste heat recovery projects have been implemented and steam is still being let down to meet the demands of the LP steam header, then consideration should be given to installing steam turbine drivers to replace electric motor drivers. Efficiency needs to be a prime concern when specifying steam turbines.

Adding Flexibility

Steam systems are dynamic. Changes in the process can change the steam balance (the amount of steam that is venting and/or being let down). Flexibility needs to be built into the steam system to allow for process changes. The following items should be considered to add flexibility to your steam system:

1. Extraction steam should be used instead of letdown steam to pressure control the steam headers. The extraction controls can be tied into other operating unit's steam let-down controls to further optimize the extraction.
2. Steam turbines and motor drivers should be utilized to minimize steam vents and let down flows.
3. Steam header pressures should be adjusted to allow steam to be transported to other locations or to reduce the steam flow through turbines.
4. The deaerator pressure should be varied slowly to eliminate steam venting but not to cause excessive steam being let down.

Significant savings can be accomplished with good communications between operating areas, but control schemes that continuously optimize the system are preferable.

Summary

This article is a guide to reducing the steam demand in refineries and chemical plants. Most of the opportunities presented are low or no cost approaches. By understanding how your steam system operates and by using the techniques presented in this article, most sites should be able to find numerous opportunities to reduce their site's steam demand.

MAINTAINING AND INCREASING COOLING TOWER EFFICIENCY

by

James L. Willa
Willa, Inc.
18 Fortune Lane
St. Louis, MO 63122

ABSTRACT

The lack of adequate cooling tower thermal efficiency often adversely effects the total output of a plant. This paper outlines the various methods available to increase efficiency. However, it primarily covers specific maintenance and operating procedures, which can significantly increase performance without major capital expenditures. Often this approach will equal or even exceed the efficiency improvement to be gained by such major expenditures. This list of specific actions is accumulated from the authors' over fifty (50) years experience in cooling towers, and in particular from the last seventeen years, during which time the author has averaged testing and inspecting over seventy-five (75) towers per year.

INTRODUCTION

The Cooling Technology Institute (CTI) spends half or better of its efforts helping industry to improve the design, and to test cooling towers to an accuracy of one-tenth of one degree Fahrenheit. Specifying engineers, likewise, go into great detail to obtain the proper cooling tower. However, the fact remains that once these towers are placed in service, they become the "step-children of industry", and as such, begin to lose much more than mere tenths of a degree in performance, because of operation and maintenance procedures ranging from good to unbelievable. Loss in thermal performance of cooling towers is unnecessary.

The current emphasis is on "debottlenecking" or getting more production out of existing units. This process quite often reveals the cooling tower as the or one of the limiting factors. Therefore, "uprating" cooling towers has become the "buzz word" in the cooling tower industry for the past few years. It appears that this trend will continue for some time into the future.

"Uprating" means getting colder water back from the tower under current operating conditions, or maintaining current cold water return temperatures under increased heat load, or various combinations of both. There are several valid approaches to this problem, many of which require substantial changes and expenditures.

After the cooling tower is constructed, there are three factors that affect the performance. These factors are the amount of water being pumped over the tower (L), the amount of air being pumped through the tower (G), and the "tower characteristic" (KaV/L), which is a measure of the tower's heat transfer ability. Therefore, a change in any of these three factors will change the tower's thermal performance.

Total water flow can be altered by changes in pumping rate, giving a water flow rate either above or below the design rate. In general, if this difference is excessive the distribution system is adversely affected, either by flooding with high water flow or by improper spray patterns with low water flow.

Airflow can be affected basically by changing static pressure against which the fan is pumping, fan speed, pitch, or efficiency. Once a unit is in operation, there is usually no way in which the fan speed can be changed. However, pitch and efficiency can be influenced. Of course, a change in static pressure is the most likely cause of a change in airflow.

A change in the "tower characteristic" or ability to cool water is brought about primarily by changes in physical configuration in the wetted surface area.

The performance of the tower is not only dependent upon maintaining the proper L/G or water/air ratio, but it also is based upon the assumption that they are thoroughly mixed or both properly distributed.

The following specifics cover thermal performance only. Structural integrity, safety items, etc. are not included.

LOUVERS

Keep louvers free of algae, moss, and scale, to prevent increased static pressure and resulting reduction of airflow. One tower inspected had silica scale obstructing one third of the total air inlet area, even though the vertical distance between louvers was 36". New counterflow towers are built without louvers. On existing counterflow towers, two of every three louvers of the old (½" x 6" x 3') style can be removed. On old straight wall crossflow towers, two out of every three louvers can be removed.

Keep steam exhaust and/or other heat sources from entering the air inlets. Where steam driven pumps are used, be certain the exhaust steam stack is tall enough to prevent its discharge into the air inlets.

On crossflow towers, maintain the one foot tall plywood air seal below the bottom louvers to prevent air from bypassing under the fill.

CASING

Any hole, crack or leak through the casing causes loss of air through the fill. The closer the leak is to the top of the tower the more damaging the effect.

Missing corner rolls near the top of a counterflow tower can be the source of an appreciable bypass air leakage into the plenum area.

Quite often, a new tower will have the casing installed "leak free" with caulking, etc. Then the fire protection contractor will cut a one foot square hole through the casing for a 6" sprinkler supply line. Patch all holes and cracks.

End wall entrance doors on crossflow towers must be kept intact and closed.

BASIN

Total cooling ability depends not only on proper cold water temperature, but also upon getting this water to the right places in the right amounts. Therefore, the sump screens become an important and almost universally neglected item in the tower's cooling ability. All towers must have a double screen at every outlet to sump or pumps. The cooling tower, due to its size and negative pressure at the air inlets, acts as a huge vacuum cleaner/air washer sucking up dust, sandwich baggies, styrofoam coffee cups, paper, cardboard, plastic ribbons/sheets, etc. The screens are designed to keep this material out of the heat exchanger tubes. However, over 90% of all screens do not have a "cup" or "catch" on the bottom leading edge of the first screen (the screens should be installed in sets of two). This means that when the first screen is removed for cleaning, all the debris falls to the bottom and goes under the first screen to the second screen. When the second screen is removed for cleaning, the debris escapes under the bottom and is headed through the pump and into the exchanger, where one "baggie" can plug four to five tubes. The lead screens must have a 6" wide x 6" deep screen lined "cup" or "catch" on the bottom edge, to catch debris when screens are lifted for cleaning. Use S.S. framing and ½" mesh S.S. screen liner for this "cup".

These screens must be cleaned regularly. A hoist, either hand operated or electric, should be installed over the screens, and adequate hooks or handles on the tops of the screens for attachment to the hoist. Be careful where these screens are cleaned. An inspection of a hyperbolic natural draft tower revealed excellent stainless steel screens with “cups”, hoists, etc. However, they were moved by the hoist on a supporting rail to one side, leaned against the louver face and washed with a pressure hose. This location resulted in all the wood fiber and other debris being added right back into the tower through the louvers. The hoist support rail must extend past the sump to a “cleaning area”.

Extraneous streams are often dumped into the cooling tower basin. Detrimental streams include steam condensate from perhaps the steam powered circulating water pumps, or heated water from the steam condenser on the cold water pumps. Hot boiler blowdown is sometimes added directly to the basin for makeup.

The most common extraneous stream, and usually the most detrimental, is side stream filter water. This water is generally taken from the hot water line to the tower, and run through the side stream filter and from there back into the basin, having insufficient remaining head to be put over the top of the tower. In one instance, in a ten cell six cube doubleflow crossflow, a stream of water was being taken off the hot water header and run through an elaborate side stream filter setup, which of course is good. However, the effluent from the side stream filter was then piped back to the cold water basin. The thermal performance test on this unit revealed the fact that a 0.7°F rise in the cold water temperature in this 107,000 GPM cooling tower, was due to this bypassing of hot water directly into the basin. This same company had just paid approximately $250,000 to lower the cold water temperature by 0.9°F in a four cell crossflow tower, to increase their production efficiency. Whereas in the ten cell tower, 0.7°F was being given away strictly for the lack of a line being run to the top of the tower and the side stream filter water added into the distribution basin, rather than in the cold water basin. Even the addition of a small pump if necessary to accomplish this would be a minor expenditure, compared to the difference in cooling tower performance. All extraneous streams at above cold water temperature should go over the top of the tower, where the effect will be minimal.

Do not allow hot water to bypass into the cold water basin. Often a bypass line opened in the winter will be left on during the summer. Many counterflow risers have drain lines from above the valve, into the basin. These lines are for winter use to drain "out of service" risers, to prevent freezing in the riser. However, many of these lines are left open during the summer.

Add cold water to the acid dilution trough, not hot water. Remember, the "product" of the cooling tower unit is cold water - don't pollute it with hot streams, and don't waste it. Many plants blowdown cold water, whereas hot water should be used for blowdown.

Chlorine (or other algicide/slimicide) injection location should be selected where it is most needed. If slime in the exchanger is the most critical problem, then the chlorine should be added into or somewhere near the sump. However, if algae growth on the distribution deck of a crossflow tower is causing poor distribution and loss of performance, or film pack fill is being plugged, then the chlorine should be added to the hot water supply line.

Water distribution pattern in the unit is most important. Many times the spare cooling tower pump is run continuously, as one condenser needs the pressure. This results in more water over the tower, which raises the cold water temperature. This rise in cold water temperature often outweighs the gain in heat exchange from the increased velocity. Avoid the "send more water syndrome".

FILL

This is the area of major heat transfer. The fill must be kept clean from algae, slime, scale, product contamination, mud and debris, particularly film pack type fill.

Splash Fill

Splash fill must be kept level and in position. The first six foot of crossflow fill near the air inlet should be kept in place with plastic clips to the support grid, and should never exceed 6" overhang past the last support grid near the louvers. Ice damaged fill slats near the perimeter must be repaired or repositioned. Crossflow and counterflow fills should not be utilized interchangeably. When icing threatens to be a problem, heavy duty redwood fill should be used rather than plastic. Missing fill should be replaced. Crossflow splash fill should be positioned parallel to airflow, in most cases.

Film Pack Fill

Film pack fill has a very large surface area in comparison to splash fill, and is much more efficient per foot of height, which allows increased efficiency and reduced pumping head. However, film pack fill does have limitations. It is much more susceptible to decrease in performance from poor distribution patterns, and is more susceptible to clogging or plugging.

During the 1950's and early 1960's, it was very difficult to convince industrial users to accept film pack fill. The pendulum has now swung completely the other way, as often happens with a new product. Now film pack is being used in many, if not a majority, of cooling towers. Many of these applications are really not suited to film pack fill due to water conditions and/or product contamination, which will result in its fouling or plugging. Therefore, a thorough investigation of the water and possible contaminants should be made, prior to using film pack fill. Evidence of this "over use" is the abundance of technical literature on "cleaning film pack fill", and "non clogging" film pack fill. All film pack fill will clog or plug under adverse conditions.

The tower owners' best defense against loss of performance due to fill problems is a good water treating program that will prevent accumulation of mud, algae, slime and scale, and a good screen design and operation to prevent debris.

FAN STACKS

Holes in the fan stack beneath the fan cause air to bypass the fill. The most common type of holes in this area is a "slot" in the fan stack at the drive shaft, oil line and mechanical equipment support steel. The hole for the drive shaft should not be a slot, but only a hole approximately two inches larger in diameter than the coupling on the drive shaft. The oil line, vent line and any other line or beam penetrating the stack should have holes only large enough for a tight fit around the line or beam. Vertical joints must be kept tight, and caulking or putty where used must be kept in place.

The "back pressure" against the fan can be reduced and the air delivery increased at the same horsepower, by using a tall "velocity recovery" fan stack that flares out after the "throat" area, thereby increasing the discharge area and reducing the "back pressure".

The fan blade tip clearance should be kept to a minimum such as 1" – 1½" for a 28' diameter fan. If the tip clearance is greater than this, a tip seal should be installed.

FAN DECK

Any hole in the fan deck will leak air which bypasses the water. These holes are common under the fan motor, and somewhat less common around the perimeter of a counterflow tower between the fan deck and the outside casing. The access hatches must be kept closed.

Often, when the fan deck is repaired, boards will be left extending into the fan stack area. These must always be cut off flush with the inside bottom flange of the fan stack to avoid obstruction to airflow. On counterflow towers, a ladder should extend from the fan deck through the drift eliminators to the top of the fill, to facilitate periodic checking of the distribution. Pultruded glass reinforced polyester plastic panels should be used for any new or replaced fan deck.

FANS

The fans must be pitched evenly between blades. The degree of pitch must be high enough so that the motor draws full load amps or very near full load during summer operation. They may be left at this pitch all year. Even though the colder (heavier) air of winter will increase the load on the fan, there are two compensating factors. First, the air going through the fan is the hot air coming off the hot water, and thus never varies nearly as much as ambient air. Second, the motor is being cooled by outside ambient (cold) air, which will prevent the motor from overheating during winter operation, even if it is drawing somewhat more than rated amps.

A very common mistake is to leave the center air disk or air seal off the fan. This allows easier access to the fan blade attachment area for mechanics, but it also allows the air near the center to flow backwards or down through the fan. This can be a significant loss (up to 15%), which is reflected directly as a 15% loss in total performance.

Fan blades must be kept clean and free of algae and scale with leading edges smooth. The fans must be kept at full speed for maximum performance. Therefore, mechanical equipment must be kept in good order, and vibration kept below triggering the vibration switch. The vibration switch should be set to allow starting torque, and sensitive enough to prevent excessive vibration, but not so sensitive that it causes unnecessary shut down. The vibration switch should be mounted on the fan motor, or the motor end of the mechanical equipment support steel.

MECHANICAL EQUIPMENT SUPPORT

The fan stack must be sealed tight against the support steel. Torque tubes must be sealed on one end, preferably the outboard end, to prevent major air loss. Plastic caps on the outboard end are normal and often missing. A steel plate welded on the inboard end keeps saturated air out of the inside of the tube.

CROSSFLOW - SHORT WALL (Between Fan Deck and Distribution Deck)

Any air leak into the tower reduces performance. The closer the leak is to the fan the more air will pass through a given size leak, due to increasing negative pressure nearer the fan. Missing boards or panels and holes in this wall, will significantly reduce airflow through the fill. Seals are often left out between fan deck joists, when installing new mechanical equipment supports, causing major leaks. Access panels through this wall to the plenum area must be kept closed.

DISTRIBUTION SYSTEMS

All cooling tower calculations and rating systems must presume the air and water to be evenly and thoroughly mixed. However, in the "real world" they very seldom are evenly and thoroughly mixed.

The problem of proper distribution is the area of greatest concern effecting thermal performance of a cooling tower.

Crossflow distribution systems are simple and easy to clean. Therefore, you would expect them to be maintained. However, the large majority of these open basins are not maintained, and often have as many as 50% or more of the nozzles plugged with mud, algae, wood, scale, sand, rust particles, plastic, debris, etc.

These basins must periodically be swept clean, with the mud and debris being removed from the system, and each individual plugged nozzle cleaned. Normally, a minimum of 5% to 10% will be plugged, and often over half. These plugged nozzles cause a rise in water level in the distribution basin, resulting in excessive drift loss. Plugged nozzles also create extreme imbalance in the water flow pattern through the tower, and hence, appreciable loss in performance. These nozzles can be cleaned quickly and easily, and should be checked and cleaned regularly. Replace all missing nozzles.

In a system without proper algae control, the distribution deck will be the primary area of algae growth and must be cleaned periodically. Plugged nozzles should be cleaned by hand or with a small tool. They should not be opened with the "popular" steel rod or broom handle, because invariably the bottom distribution plates are knocked off, and the distribution nozzles merely become down take orifices, which lowers thermal performance. In some towers, these distribution nozzles are washed out of position by the velocity of the water. In these instances, they must be fastened down using two stainless steel nails on each nozzle, which catch the edge of the nozzle under the head of the nails. Some nozzles have a tendency to lose their inside orifice plates, which increases the diameter of the nozzle, lowers the water level in the distribution basin, and results in poor distribution. Sometimes, outside or inside rows of nozzles may be plugged or removed, to help prevent ice formation during winter operation. Quite often, these plugs or missing nozzles are forgotten, and this portion of the distribution system behaves poorly during the summer.

Often, covering these basins with an opaque cover will appreciably retard algae growth and consequent pluggage. However, if the cover is too low for a man to walk under the distribution basin will be checked either less often, or not at all, and the unit will perform worse than without a cover. Likewise, if the cover changes the area to a "confined space" rating, again the frequency of inspection and cleaning may drop precipitously.

Many times the number of nozzles plugged becomes so great that water pours over the inside and/or outside basin walls. The inside water turns to drift and is the base cause of the majority of excessive drift problems on crossflow towers. The outside overflow damages the fill near the louvers. Both types of overflow greatly reduce thermal performance.

The distribution basins must all be kept at the same depth or equal water flow (within $\pm$ ½" of average depth). The depth should be at least 3" to allow water to spread over the deck to the far corners, and not over 8" to prevent overflow. If the depth does not fall in this range when the water flow is balanced between cells and nozzles are clean, then the orifice diameter of the nozzles should be changed to a larger or smaller size, to accomplish a depth in this range (3" - 8").

Crossflow distribution valves must be kept operable, in order to properly balance the flow between basins. The locknuts must be set when the balancing is completed. It is a great time saver to count and record the number of exposed threads on each distribution valve. This facilitates the ability to quickly reset the valves after any upset or change.

The flow control box, usually built out of plywood or redwood, equally distributes the flow around the distribution basin, and prevents areas of excessive velocity. Check for delamination of the plywood and for proper sealants, so that water does not come out the top of the box. A properly constructed and operating flow control box will have a minimal area where the velocity over the nozzles is so fast that they do not flow properly. Look for large turbulent areas extending from the flow control box to the back wall of the distribution basin. High water level and/or turbulence can cause splashing over the top of the basin wall onto the top of the drift eliminators. This water then exits the tower as drift. If large pieces of rust, scale or other debris is plugging nozzles, then install "fence" around flow control boxes. Nail 2" x 2" redwood strips to the distribution deck (being careful not to cover any nozzles) around all four sides of the flow control boxes, 6" out from the perimeter of the boxes. Attach ½" mesh S.S. wire, 10" tall (vertically), to the inside edge of these 2" x 2"'s, to form a "fence" around the boxes. This will greatly reduce nozzle pluggage by trapping all the larger pieces of rust, scale and debris within the fence. This debris can then be cleaned out by hand by kneeling down on the flow control boxes, rather than having to walk in the water on the distribution deck.

Many crossflow towers have a section of the distribution basin that extends behind the short vertical wall, between the distribution deck and the fan deck. This results in the last 12" to 18" of the distribution basin being out of sight. Nozzles are often stopped up or nonexistent in this back area. Determine whether or not there needs to be nozzles in this area, to adequately cover the fill. It is necessary to remove the access door in the short wall in order to see this part of the distribution basin, and to determine whether or not any water is flowing over onto the drift eliminators. Remember, this overflow is the most prevalent cause of excessive drift in crossflow cooling towers. It often goes unrecognized to the extent that all new drift eliminators will be installed, in an attempt to control the problem, without realizing the actual source. If the cause of the splash over is excessive turbulence coming from the flow control box, then the installation of one or two baffles to slow the velocity will solve the problem. If the cause is excessive water depth in the distribution basin, then the most likely cause is plugged distribution nozzles. If the water is still too deep after the nozzles are cleaned, then the distribution flow control valve must be adjusted to obtain a more satisfactory operating level, or additional or larger inside diameter nozzles installed.

The installation of handrails and kneerails around the outside of the distribution basin will encourage the regular checking and cleaning of crossflow distribution nozzles.

All holes or cracks in the distribution deck floor or walls must be patched. Cant strips on the perimeter of the distribution decks must be kept leak proof with proper caulking and nailing. Often, sprinkler supply pipes will be brought across the distribution basins. Sometimes these pipes are set so low that cuts must be made in the distribution deck walls to accommodate them. These invariably start leaking, adversely effecting performance and adding to the drift. These pipes should be raised to <u>above</u> the level of the distribution basin walls. The distribution deck must be level. Sagging between joists often results in up to 1½" to 2" difference in depth, reducing performance. Any new or rebuilt distribution deck should be constructed of pultruded glass reinforced polyester plastic panels.

Heat is not removed from the water in an even fashion as it drops through the tower. When the hot water leaves the nozzles in the top of the tower, within a very few feet approximately one third of the heat is transferred to the air because of the large driving force (Δ enthalpy) between the water and air streams. The next one third of the heat load is transferred by approximately halfway down through the tower. The last half of the tower is transferring less than one third of the total heat load, due to the continuing decrease in driving force. The cold water temperature is asymptotically approaching the wet bulb temperature.

Low approach towers (cold water temperature minus wet bulb temperature) are more expensive because the water must be held in contact longer, in order to compensate for this decreasing difference between wet bulb and cold water temperatures. The net result of this effect is that the water is being cooled through the tower in a logarithmic manner. When it reaches the basin and mixes with other water of different temperature, this mixing is an arithmetic mixing. Therefore, cooling is lost because the water was cooled logarithmically and is mixing back together arithmetically. An area or a cell making hotter water can negate the last two or three degrees that were so hard won by the area or cell making colder water. This is most important because it is the largest single and most common factor in the loss of thermal performance in industrial cooling towers today. Having different flow rates in various areas or cells of the tower results in an overall decrease in thermal capacity. Therefore, the best overall tower performance is when all portions of the tower produce the same cold water temperature, or as close as possible. Hence, the distribution basins or cells must be balanced to achieve maximum performance. Strive to balance water depth in the crossflow distribution basins to within the recommended ± ½" of the average depth.

Counterflow distribution systems are somewhat more difficult to maintain properly as they are in an area definitely considered a "confined space". The importance of maintaining even flow between cells and/or portions of the tower is, of course, as important as in a crossflow tower.

A coarse "rule of thumb" is that a plugged nozzle costs you the equivalent of twice the area covered. First, you obviously lose any cooling in the area where no water is flowing. Second, this area with no water has less static and the air will take the path of least resistance, thereby robbing air from adjacent areas. This "rule of thumb", of course, breaks down at high pluggage conditions because at 50% plugged you still obviously get some cooling.

The majority of plugged nozzles almost invariably start at the far side of the tower from the riser, in the last two to three laterals, as this is where the majority of debris will settle. The end nozzles on the laterals will plug and the effect will migrate back towards the main distribution pipe. This lack of water flow can be seen in the lower air inlet area, and on the surface of the basin. Checking the cold water basin while the tower is in operation will not only reveal areas with plugged nozzles, but also any broken or missing nozzles, and broken or leaking pipes.

These nozzles must be cleaned with the water on. This is necessary both to determine which nozzles are plugged, and to have the water pressure to blow the debris out while cleaning. Most nozzles can be cleaned by inserting a finger, screwdriver, rod, etc. into the nozzle opening and letting the pressure wash the nozzle clean. If the pluggage is severe or with some nozzle designs, the nozzle must be removed from the lateral, cleaned, and reinstalled. Under severe pluggage conditions, a "sewer tape" or similar tool must be inserted through a nozzle opening into the lateral to clear the lateral of debris. Plugging of nozzles and laterals on the far side from the riser can be significantly reduced by installing a 2" PVC line out of the bottom of the ends of the mains on the far end from the risers. Extend this line vertically down inside the casing to 5' above grade, with a PVC valve at the bottom end for flushing trash out of the ends of the mains. Replace all missing nozzles. It is essential that the nozzles be checked and cleaned, if necessary, at least at the beginning of the hot weather season each year.

Counterflow distribution valves on the risers must be kept operable so the cells can be balanced. A counterflow tower can be balanced between cells by the following method.

First, clean all nozzles. Install a stopper with copper tube in the first nozzle nearest the access panel in each cell. Attach a piece of clear tubing to the copper tube and bring the tubing through the access panels, and attach to the top handrail. Adjust valves until water is at same level in all tubes.

Be certain the riser is leak free and firmly attached to the main header. If the header is wood, keep all bands (wood, metal, fiberglass) secure. A block at the end of the lateral is helpful to prevent longitudinal motion. Keep bands on wooden laterals secure. On PVC laterals, cover ends with removable plugs or caps to facilitate cleaning when necessary. Check connections between laterals to mains, and nozzles to laterals. Every few years, where rubber "O" rings are used, replace them with new ones as they become brittle with age.

Water flow rates appreciably ($\pm$ 15%) different from design will adversely effect distribution patterns, particularly in a counterflow tower. Over pumping will cause excessive overlapping of spray pattern. Under pumping will prevent interlocking spray pattern. A proper interlocking spray pattern is essential, especially in a film pack tower. A "full cone" nozzle is essential in film pack cooling towers.

Another common operational procedure concerning total water flow is the situation in which a unit is operating at only a partial load. In this case, only a portion of the cooling tower is needed. However, almost invariably, the entire pumping capacity on the total tower will be utilized with a greatly reduced range (hot water temperature minus cold water temperature). This results in the inefficient utilization of the cooling tower. Much could be saved by reducing the water flow in proportion to reduced load on the unit, and taking the appropriate number of cells out of operation on the cooling tower. This will give efficient cooling tower operation and save the power consumption on the fans and pumps out of operation.

DRIFT ELIMINATORS

Any pluggage in the drift eliminators can cause uneven or reduced airflow. This pluggage can be caused by algae, slime, moss and mud accumulation, which should be controlled by chlorine or other algicide and dispersants. It can be caused by scale, which can be controlled by proper pH, blowdown and dispersants. Objects on the drift eliminators such as scaffold boards, work platforms, collapsed partition walls, walkways, etc. can restrict flow.

Missing blades or sections of drift eliminators not only cause increased drift, but also induce uneven airflow patterns and resultant loss of performance.

In counterflow towers, an access panel must be made in the drift eliminators to allow inspection and work on the distribution system.

Be certain the seal at the top of the drift eliminators in crossflow towers is intact, to prevent air bypass.

Proper chemical treatment is the best way to keep drift eliminators clean.

PARTITIONS

Wind checks in the bottom of counterflow towers are important, particularly in towers with open end walls. These partitions reduce windage (water blown out of bottom of tower by wind) and help direct the airflow properly. Keep them intact by renailing when necessary.

Partitions between cells do not play an active role in performance as long as all cells are in operation. However, maintenance and load conditions often call for one or more fans to be out of service. Complete intact partitions with doors closed (crossflow towers), are then important to prevent active fans from obtaining their air from the plenum of adjacent inoperative cells, thereby bypassing the fill.

WATER TREATING

A proper water treatment program which will prevent scale and control algae and slime, and reduce suspended solids, is essential to maintain maximum cooling tower performance.

CONCLUSIONS

The point must be emphasized that all of the foregoing ideas on improving the efficiency of a cooling tower are not valid unless they are actually implemented by the operator and/or maintenance. Therefore, insist that the operator make a complete visual inspection around and on top of the tower at least once every eight hour shift. Be certain that a logbook is maintained covering the various important aspects of the tower. Have a check list for the operator to utilize and confirm that he does not overlook any important items. Further, have a check list for your inspection department for their periodic inspection of the cooling tower. Every one or two years, an outside professional cooling tower inspector should be utilized to ascertain the condition of your cooling tower.

Be certain that you are getting the maximum thermal efficiency out of your cooling tower by checking the above items, <u>before</u> you consider major expenditures for performance improvement.

MAINTENANCE AND ENERGY ECONOMICS

Claire L. Cagnolatti

Senior Consultant
Solomon Associates
Dallas, Texas 75240
claire.cagnolatti@SolomonOnline.com

Prepared for 2004 AIChE Ethylene Producer's Conference

Unpublished

AIChE shall not be responsible for statements or opinions contained in papers or printed in its publications.

Maintenance and Energy Economics

Maintenance Performance Analysis

Solomon Associates (Solomon) has been performing competitive performance analyses of fuels refining since 1980 and olefin plants since 1987. Currently, Olefin Studies are performed on a 2-year cycle, conducting the data analysis for the odd years' performance during the even years.

Our methodology calculates a Maintenance Cost Index, which is total maintenance cost as a percent of replacement asset value. The total maintenance cost includes an annualized portion of the turnaround (referred to as T/A in the following figure) cost. Annual non-turnaround costs are averaged for the 2 years of the Study cycle.

$$\text{Maintenance Cost Index} = \frac{\text{Routine + Annualized T/A Maintenance}}{\text{Replacement Value}} \times 100$$

$$\text{Reliability Indicator} = \frac{\text{Downtimes and Slowdowns + Annualized T/A Downtime}}{\text{Ethylene Capacity}} \times 100$$

$$\text{Maintenance Effectiveness Indicator} = \frac{\text{Routine Maintenance + Annualized T/A + Cost of Unreliability + Flaring}}{\text{Replacement Value}} \times 100$$

Figure 1. Maintenance Indices

We measure reliability in a manner similar to the Maintenance Cost Index to develop our Reliability Index. Lost production during the Study year is documented by cause and duration. Annualized turnaround downtime is calculated from turnaround duration and interval data. These two items are summed and expressed as a percent of ethylene production capacity.

Our Maintenance Effectiveness Index combines the measure of the cost of maintenance and the cost associated with lack of reliability. We convert the Reliability Index into cost of lost production, using each plant's actual net cash margin per ton of ethylene calculated from that year's Study data. We add this amount to the total maintenance cost (2-year average of non-turnaround costs plus annualized turnaround cost) and add the cost of flaring, valued at average feedstock cost. The sum is then divided by replacement value to represent the combination of maintenance cost plus the cost of unreliability.

Study year downtime losses
\+ Study year slowdown losses
\+ Annualized turnaround downtime
Total lost ethylene production

Total lost ethylene production x
Actual net cash margin per ton of ethylene
= Cost of Unreliability

Figure 2. Cost of Unreliability

Years of data analysis in the refining and petrochemical industries have shown that maintaining reliability is the key to good performance in many areas, especially with respect to maintenance cost. It is difficult, if not foolish, to try to control maintenance costs without the proper focus on reliability.

In an unreliable plant or a plant whose reliability is not managed but is left to chance, control of maintenance costs is impossible without compromising ongoing operations. If maintenance costs are arbitrarily controlled, things may be okay for a year or two. But eventually, arbitrarily controlled maintenance costs will compromise the plant's operability. Our recommendation to clients is: Manage reliability and maintenance costs will decrease automatically.

In addition to reduced maintenance costs, other advantages of managed reliability include improved quality of life for personnel (with less plant upsets to handle) and a greatly reduced impact on the environment since most environmental infractions occur during plant upsets. Further, as much as 70 percent of all safety incidents occur during non-steady-state operations.

In the 2001 Olefin Study, the first quartile in maintenance effectiveness consumed only 88.5 percent of the energy consumed by the average plant in the Study. The other three quartile averages are very close to the total Study average in energy consumption. Coincidence? We think not. Better reliability, less flaring, more steady-state operations, and more predictive and preventive maintenance (as opposed to reactive or emergency maintenance) allow the plant to operate at optimum levels and allow maximized energy efficiency.

North American Competitiveness

Historically, North American olefin plants have been competitive with their European and Asian counterparts with respect to energy costs, as shown in the energy expense trend chart in Figure 3.

While pre-2000 energy costs were competitive, energy consumption per ton of high-value chemicals produced is higher in North America than in other regions. As a percent of the 2001 Olefin Study average energy consumption, North American consumption was 108 percent of average, Europe was 97 percent, and Asia was 94 percent. Prior to 2000, this higher energy consumption was offset by a much lower energy cost, making North American plants' average energy cost exactly the same as Europe or Asia.

Part of the higher energy consumption is due to a predominance of light feed plants in North America. Per the 2001 Study, ethane crackers consume 121 percent of the Study average energy consumption, where naphtha crackers only consume 97 percent of that average. However, even after allowing for feed type differences, average energy consumption rates are significantly higher for the North American region, as shown in Figure 4.

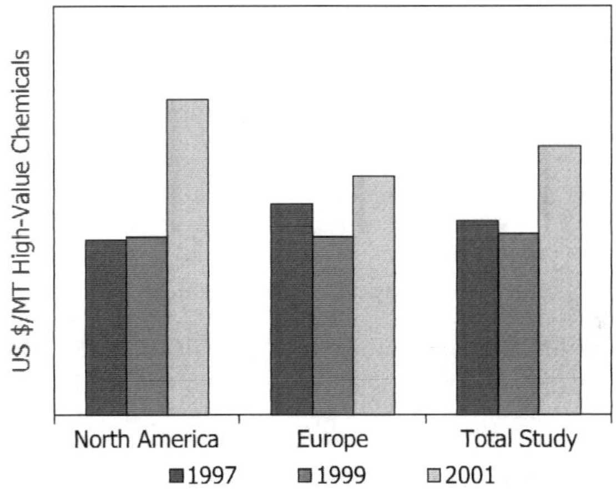

Figure 3. Energy and Utility Expense

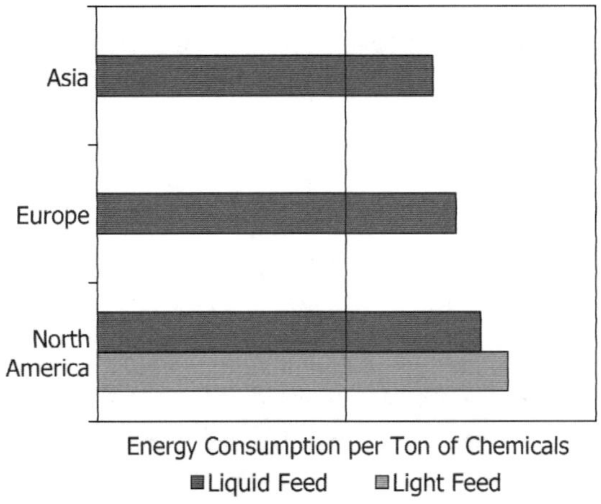

Figure 4. 2001 Regional Energy Consumption

How Can North American Olefin Plants Stay Competitive?

Could North American olefin producers have reduced costs by pursuing increased energy efficiency? It would seem so. The problem is that the low energy costs enjoyed by North American manufacturers prior to 2000 made it difficult to realize the required return on investment to justify capital spending for reducing energy consumption. A 5 percent reduction in consumption, as might be realized after capital investment of $20 million to improve waste heat recovery, would have a payout of approximately 8.6 years for an average-sized olefin plant using 1999 average energy prices. Using 2001 average energy prices, that payout is reduced to 4.4 years. This drastic change demonstrates that any previously shelved capital projects related to energy conservation should be dusted off and revisited.

Since 2000, energy prices in North America have more than doubled, certainly justifying more capital for energy conservation. However, the capital investment required to achieve this energy consumption reduction is not installed overnight. Any short-term energy consumption improvements must therefore be no- or low-cost measures. This is where maintenance becomes of the utmost importance.

Steam leaks, condensate leaks, inefficient furnaces or boilers, or less-than-optimum performance of any piece of energy-consuming equipment (almost everything in an olefin plant) must now be re-examined with an entirely new attitude. What are the economics of repairing a steam leak or tuning the burners on a furnace? If your company has not **recently** evaluated the potential cost savings in these maintenance-related activities, your maintenance department may be missing some opportunities to have an impact on reducing energy consumption.

Importance of Process Control

Also important are instrumentation maintenance and properly functioning process control systems. Good process control is key to maintaining steady-state processes operating at their optimum efficiency and optimum energy consumption. There are several levels of questions that need to be answered with respect to maintaining optimum energy consumption through process control:

- If your plant does not have advanced process control due to lack of economic payout, have you re-evaluated the economics in light of the recent energy and feed price changes?
- If your plant has advanced process control, when was the last time your plant evaluated the system for relevant energy economics and optimized energy consumption? Are model economics, including energy prices, being updated in real time?
- If your plant does not have advanced process control, are your regulatory controls tuned for optimum energy consumption? There are products available that constantly scan the tuning of regulatory controllers and produce reports on the worst ones.
- Are process-control setpoints evaluated or audited by engineers? Audits can ensure that no extra "cushion" is built into a control setpoint, which might be very costly from an energy standpoint.

Energy is No Longer Just an Afterthought

While it is best to have a plant design that facilitates energy efficiency, most ethylene plants in North America were built when natural gas cost $2.50 per million Btu or less. It will take some time before capital investment to lower energy consumption can be installed. In the meantime, a major shift in the mindset of operations and maintenance personnel must occur at North American plants. Energy consumption, energy cost, and energy conservation must be brought into the forefront for consideration in all decisions. This mindset change may take years, but the economic picture is not likely to change soon. We should all get used to it.

SESSION 61
ROUNDTABLE DISCUSSION
MAINTAINING OLEFIN PLANT ENERGY EFFICIENCY
PRESENTED BY THE MAINENANCE SUBCOMMITTEE

The Roundtable was an open forum for the audience to discuss non-proprietary issues around maintaining and/or improving the energy efficiency of olefins plants. No papers were presented, however, the authors from Session 58, Maintenance Paper Session relating to Energy Efficiency, served as panelists for the Roundtable along with other topical industry experts.

The panel consisted of the following individuals:

- Claire Cagnolatti – Solomon Associates
- Patrick Cooper- Shell Chemical Company
- Darrin Foote – John Zink Company LLC
- Mike French – Chevron Phillips Chemical, LP
- David Johnston – Eastman Chemical Company
- Gabe Nath- Pittsburg Corning Corp
- Robert Aegerter – Equistar Chemical, LP
- Louis Trahan-E.I. Dupont Sabine River Works
- Jim Willa – Willa Inc.

The Roundtable covered the following efficiency related topics as well as other energy efficiency related items of interest from the audience in an informal question and answer type format:

Electrical Power Usage

Analyzers/Meters/Control Systems

Utilities/Water Cooling Towers/Steam Boilers

Thermal Insulation

Furnace Refractory Systems

Turbines/Compressors

Pumps

Heat Exchangers

Cracking Furnaces (General)

Propylene: Olefin of the Future?

Dr. Jeffrey S. Plotkin
Director, PERP Program
Nexant ChemSystems

Over the last 15 years or so the growth of polypropylene has been very strong and is forecast to remain so. This has caused the growth rate for propylene to overtake the growth rate of ethylene. And because propylene has historically been produced as a by-product of ethylene, this has caused concern that there will be a propylene supply gap in the future. This concern has led to the development of a number of "on-purpose" propylene process technologies:

- Propane dehydrogenation
- Olefin metathesis
- C_4/C_5 cracking
- Deep catalytic cracking
- Methanol to propylene

Propane Dehydrogenation

Propane dehydrogenation is simple in concept, but from an engineering perspective there are some fairly complex process design issues. Propane dehydrogenation is very energy and capital intensive. The high-energy consumption is due to the endothermic nature of the dehydrogenation reaction and the high capital cost is due to the need to operate the reaction at high temperature and low or even sub-atmospheric pressures. UOP's Oleflex and Lummus' Catofin processes are the leading propane dehydrogenation technologies for license. There are several plants already in operation and several more under construction. The problem with this approach from a business viewpoint is that it requires a source of very low cost propane to make the economics workable.

Olefin Metathesis

Olefin metathesis was first used in 1964 by Phillips to convert excess and low cost propylene to ethylene and butene-2. In the mid-1980s, Lyondell used the reverse approach and converted ethylene and butene-2 to propylene. In the late 1990s, Lummus bought the licensing rights to this technology from Phillips and is trying to broadly license this technology as one solution to the propylene supply gap. BASF/Fina is reportedly going to implement this technology in their new Port Arthur, Texas steam cracker. Once implemented, this plant will actually produce more propylene than ethylene. The downside to this approach is that 1/3 of the carbon atoms in the propylene produced is derived from higher priced ethylene. Thus, the economics of this method is highly dependent on the valuation of the butene-2 and the forecasted long-term price differential between ethylene and propylene. Nevertheless, Lummus has successfully licensed a few more metathesis units.

C4/C5 Cracking

Another approach for increasing propylene supply is to use specially formulated zeolitic type of catalysts to crack C4 and C5 olefins at elevated temperatures. ExxonMobil, UOP, Lurgi, and Kellogg are all offering versions of this approach. One of the potential advantages of this method, as compared to olefin metathesis, is that ethylene is not consumed but instead is produced in small amounts. On the downside, the amount of propylene produced is smaller than what can be made by using olefin metathesis. While there are not yet any commercial units running this technology, at least one license has reportedly been sold.

Deep Catalytic Cracking

Deep catalytic cracking is a technology that was developed in China. It is based on a unique fluidized cracking catalyst and operating conditions that allows heavy oil feedstocks to be cracked and afford relatively high amounts of propylene. Typically, propylene yields are increased from 4.8% to 17%. Stone and Webster is the exclusive licensor of this process outside of China.

Methanol to Propylene

Monetization of stranded gas is currently receiving a lot of attention. Since methane is difficult and expensive to transport, approaches are being explored to convert methane to items of commerce that can be easily shipped to developed markets. Gas to liquids (GTL) is one route that is potentially economic. In this approach the gas is reformed to synthesis gas and then, using Fischer-Tropsch catalysis, converted to liquid fuels. Another approach under investigation is to convert the synthesis gas to methanol and then convert the methanol to ethylene and/or propylene. Methanol to olefins (MTO) first semi-commercial experience was in Mobil's methanol to gasoline plant in New Zealand in the mid-1980s. They found that their ZSM-5 zeolite catalyst was capable of converting methanol to light olefins. This technology laid dormant until the mid 1990s when UOP and Norsk Hydro built a pilot plant to convert methanol to ethylene. The UOP/Hydro process used SAPO-34 catalyst and gave better selectivity to light olefins than the Mobil process. Lurgi has developed their own version of this process that is geared to make mostly propylene. The downside to this approach is that a large amount of water (56%) is unavoidably made in this process. This contributes to the large capital investment. No MTO processes are commercial yet, however, projects in Nigeria and China are under study.

Summary

Several different approaches for solving the impending propylene shortage have been developed. There is no one approach that will be best; rather it is likely that all of the technologies discussed will ultimately find a home depending on the specific situation. Once again innovation in the chemical industry has come to the fore to solve a supply problem.

Paper No.59B

Options for Meeting the Propylene Supply-Demand Gap

James M. Andersen, Robert B. James, and Daniel H. Wei

UOP LLC
Des Plaines IL, USA

&

Walter Vermeiren

ATOFINA
Feluy, Belgium

Key words: Olefin Cracking, Ethylene, Propylene, Steam Cracking

Prepared for Presentation at the AIChE Spring National Meeting
Olefins Production: Fundamental and Technological Aspects: Creativity and Technological Innovation) of the Ethylene Producers Conference
April 27, 2004, New Orleans, LA,

© 2004 UOP LLC and ATOFINA All rights reserved
March/2004

Unpublished

AIChE shall not be responsible for statements or opinions
contained in papers or printed in its publications

Abstract:

Demand for propylene as a monomer building block is expected to increase in the future faster than the need for ethylene. Expected increased use of ethane feed to produce ethylene will cause additional propylene supply - demand gap, generating market pressures and creating opportunities for the entrepreneurial petrochemical supplier. Propylene can be produced and recovered from a variety of feedstocks using various chemical and refining processing operations and catalytic materials. This paper surveys the traditional routes used for the production of polymer grade propylene and introduces new technology integrations that an alert producer can use to increase propylene production in the most efficient and cost effective manner.

Traditional technologies used for propylene production include recovery from fluidized catalytic cracking (FCC) units operating on atmospheric and vacuum gasoils and by-product production from ethylene thermal cracking units operating on propane, butane, and naphtha through gas oil range petroleum fractions. Propylene is also produced directly from propane via catalytic dehydrogenation.

More recent developments employ special catalytic systems operating on unconventional feedstocks. One large project in the design phase will convert stranded natural gas to methanol followed by conversion to ethylene and propylene. Additional projects are presently in development that will produce propylene either directly as a single product or as a co-product with ethylene. Another announced project utilizes the metathesis reaction to catalytically react by-product butenes with ethylene to produce propylene.

To extend the production of propylene from other low value hydrocarbon feed streams UOP and ATOFINA have recently announced a next generation catalyst system that converts C_4-C_7 olefins to propylene. The ATOFINA/UOP Olefin Cracking Process is a robust, catalytic process that converts low value, higher molecular weight olefins from a variety of sources to propylene at good conversion and equilibrium yields. The authors examine the traditional routes and feedstocks to propylene and characterize the technical and economic conditions favouring each application and integration into refining, petrochemical, and stranded gas facilities.

Options for Meeting the Propylene Supply-Demand Gap

By James M. Andersen, Robert B. James, and Daniel H. Wei; UOP, Des Plaines IL, USA
Walter Vermeiren, ATOFINA, Belgium

Introduction

Propylene is one the fastest growing petrochemicals as a result of high growth rates of polypropylene. Although steam crackers are the largest current source of propylene, this source is not expected to keep pace with propylene demand due to the expected move toward lighter steam cracker feedstock.

Figure 1: Steam Cracker Feedstocks Worldwide

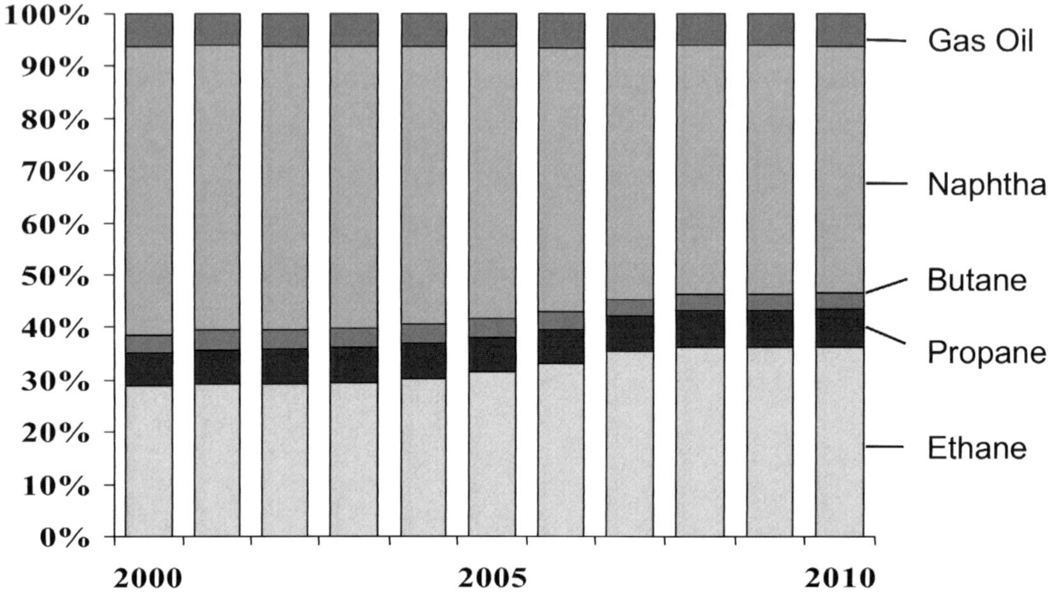

Data Source: CMAI 2002

Fig.1 shows expected trend of Steam Cracker feedstock

The trend to lighter feedstock reduces the propylene to ethylene production ratio. At the same time the relative demand of propylene to ethylene is increasing; causing a propylene gap (see Figure 2). This predicted propylene gap presents an opportunity for an alert producer using alternate new technologies to increase propylene production in the most efficient and cost effective manner.

Other available sources for propylene production are:

- Fluidized catalytic cracking
- Propane dehydrogenation
- Natural gas or methanol to olefins
- Olefin conversion
  - Metathesis

- Olefin cracking

Figure 2: World Light Olefin Ratios

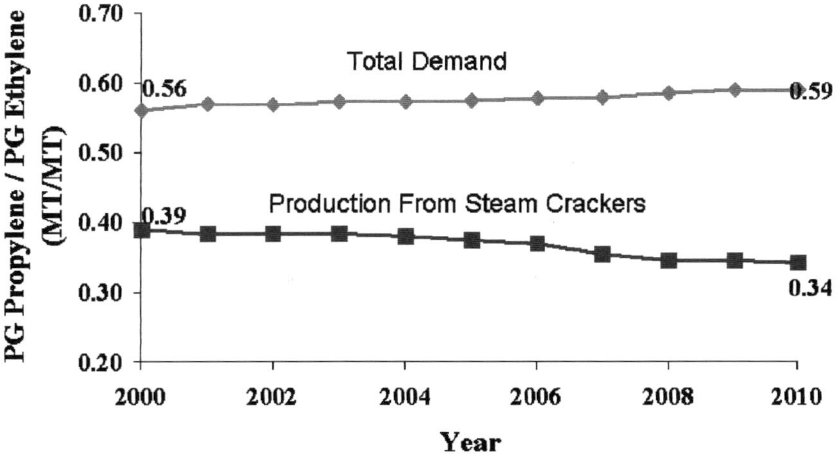

Data Source: CMAI 2002

Fig. 2 plots the trend of relative Propylene demand and relative supply from Stem crackers

The second largest source of propylene supplied for petrochemical applications comes from a propylene recovery unit (PRU) added to fluidized catalytic cracking (FCC) units in refineries. The primary function of the FCC unit is to convert gas oil and heavier streams to lighter products such as gasoline, LPG, and light cycle oil. Substantial amounts of propylene are contained in the light by-product streams. Many refineries utilize the FCC derived propylene for motor fuel production via alkylation or catalytic condensation processes. Some FCC refineries employ PRU units to recover propylene as chemical or polymer-grade propylene products. These refiners often utilize specially formulated FCC catalysts and/or catalyst additives, which increase the yield of propylene but at the expense of gasoline C_5^+ product. Propylene recovery from FCC units is especially prevalent in locations in close proximity to petrochemical facilities, such as in North America, Western Europe, and Northeast Asia.

Since 1990, propane dehydrogenation (PDH) has been providing a growing source of propylene for petrochemical applications. There are currently seven plants in operation producing 2% of the world's petrochemical propylene. Six of these plants use the Oleflex™ process developed and licensed by UOP. PDH offers three specific advantages to propylene derivative producers. First, a PDH plant makes a single product, propylene. A company specifically interested in producing a propylene derivative may not want to produce ethylene or C_4^+ by-products that are made within naphtha crackers. PDH focuses an investment specifically on propylene capacity. Second, production costs for a PDH plant are tied to the cost of propane. Propane prices are not tied directly to naphtha prices or the propylene market; therefore, PDH allows large propylene derivative producers to diversify the overall cost structure of their feedstock. Finally, the best locations for propylene derivative plants may not have good access to by-product propylene. Given the high cost of shipping and storing propylene, PDH is generally more cost-effective than buying propylene for these locations.

New technologies such as the UOP/HYDRO MTO process have been developed and are near to commercialization to convert natural gas to light olefins by first converting the gas to methanol and then in a second stage, converting the methanol to olefins (MTO).[i] MTO projects are driven by the desire to monetize stranded gas and by the market demands for ethylene, propylene and their derivatives. Stranded natural gas prices are generally independent of crude oil and naphtha market prices so MTO provides another means for olefin derivative producers to diversify the overall cost structure for their feedstock. MTO offers high selectivity to light olefins with flexibility to adjust the ratio of propylene to ethylene production over a fairly broad range. MTO units can be located in integrated facilities that convert natural gas to polyolefins or near olefin derivative facilities with methanol shipped from a remote plant.

Olefin conversion technologies produce light olefins from other olefins. There are two main types of olefin conversion technologies available; metathesis and olefin cracking. Metathesis processes produce propylene by reacting ethylene with normal butenes. The process mechanism only reacts ethylene with 2-butene and must depend on isobutene removal and a 1-butene to 2-butene isomerization step. It also consumes 1/3 ton of ethylene per ton of propylene produced. In situations where ethylene is valued similar or higher than the propylene product the costs of production for propylene via the metathesis route can be very high.

Olefin cracking processes produce ethylene and propylene by cracking C_4 and heavier olefinic feedstocks. The reactions conserve the alkene bond but transfer the olefinic bond to a three and two member carbon chain. Because C_4^+ olefinic feedstocks are produced as by-products olefin conversion process units can be economically combined or integrated into the other main olefin producing technologies mentioned above to provide higher yields of light olefin products. To extend the production of propylene from other low value hydrocarbon feed streams UOP and ATOFINA recently announced a next generation catalyst system that converts C_4-C_8 olefins to propylene. The ATOFINA/UOP Olefin Cracking Process is a robust, catalytic process that converts low value, higher molecular weight olefins from a variety of sources to propylene at good conversion and yield.

ATOFINA/UOP Olefin Cracking Process

Following initial work by ATOFINA in the mid 1990s, UOP and ATOFINA formed a joint-development alliance in late 2000. The development activities included successful operation of a demonstration unit, catalyst performance testing in pilot plants, feed-yield determination, catalyst manufacturing scale-up and process design development. The demo unit, started in 1998 at a Fina (now Total) industrial facility in Antwerp, processes commercial feedstocks from operating plants. The demo unit includes feed pretreatment, reactor section, catalyst regeneration facilities, and internal recycle capabilities.

Figure 3: ATOFINA Demo Plant

Fig. 3 is a photograph of the Olefin Cracking Demo plant facilities in Antwerp

The heart of the ATOFINA/UOP Olefin Cracking process (OCP) features fixed-bed reactors utilizing a proprietary zeolitic catalyst to produce high yields of propylene. The OCP reactor section consists of three main process blocks. A feed diolefin saturation block uses conventional technology to convert diolefins to mono-olefins. The conversion block operates at temperatures between 500 and 600 °C and pressure between 1 to 5 bars gauge. Reactor size and operating costs are minimized by operating at high space velocities and high conversions and selectivities without an inert diluent stream. A swing reactor system block is used for catalyst regeneration. The specific product recovery and separation section configuration depends on how the OCP unit is integrated with other processing plants. In general the separation section will produce a light olefins product stream, and a heavy C_5^+ product stream. The separation section will also prepare a C_4-C_5 OCP recycle stream of unconverted olefin and produce a drag product stream of olefin depleted, paraffin rich C_4-C_5 range material.

Figure 4 Atofina/UOP Olefin Cracking Process

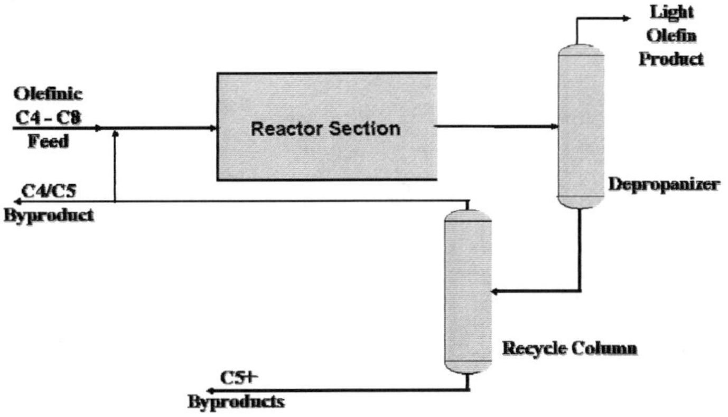

Fig. 4 is a schematic block flow diagram of the OCP

The ATOFINA/UOP Olefin Cracking process was developed to utilize low value by-product streams containing C_4 to C_8 olefins from steam crackers, cokers and fluidized catalytic cracking (FCC) units in refineries, and/or methanol-to-olefins (MTO) plants. When combined with steam crackers, the process allows the steam crackers to achieve higher propylene/ethylene ratios to help meet the growing gap between propylene demand and supply from conventional steam cracking plants. For refineries, the technology utilizes FCC by-product streams to increase propylene and ethylene production while reducing the olefin content of gasoline blending streams with little or no loss of octane. For MTO complexes, the technology increases the yield of light olefins for a given amount of methanol feedstock and reduces the C_4^+ by-product streams.

Steam Cracker Integration

A typical block flow diagram for a steam cracker without butene recovery is shown in Figure 5. The C_4 and C_5 olefin streams are recycled to the cracker furnace after diolefin saturation to maximize light olefin production from a fixed naphtha feed quantity. For this example the quantity of naphtha assumed is 2000 kmta with a P/E ratio of 0.6. Figure 6 shows the integration of an OCP unit with the steam cracker. Low value C_4 through C_6 by-product streams produced in the cracker furnaces are charged to an OCP unit where additional light olefins are produced with high P/E ratios. Internal recycle is used to optimize conversion of olefins. The OCP light olefin product stream is sent to the naphtha cracker recovery section, while the paraffin-rich residue is charged as feed to the naphtha cracker furnace. A small stream of higher boiling material is produced by the OCP and joins the pygas product stream from the cracker. The differential yield for the complex is shown in Figure 7.

Figure 5: Steam Cracker Basis

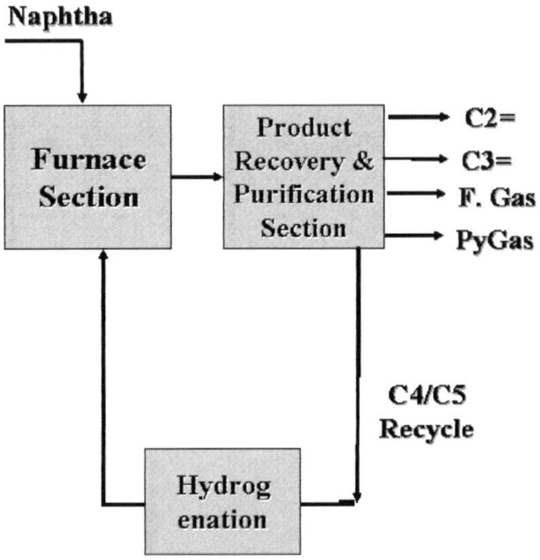

Fig. 5 shows the steam cracker basis for the integration study.

Figure 6: OCP Integration

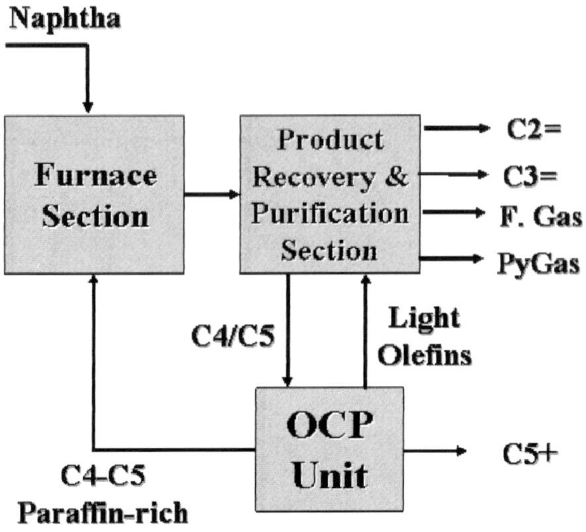

Fig. 6 shows the block flow of the integration of OCP with the cracker

Figure 7: Differential Yields Steam Cracking Integration with OCP

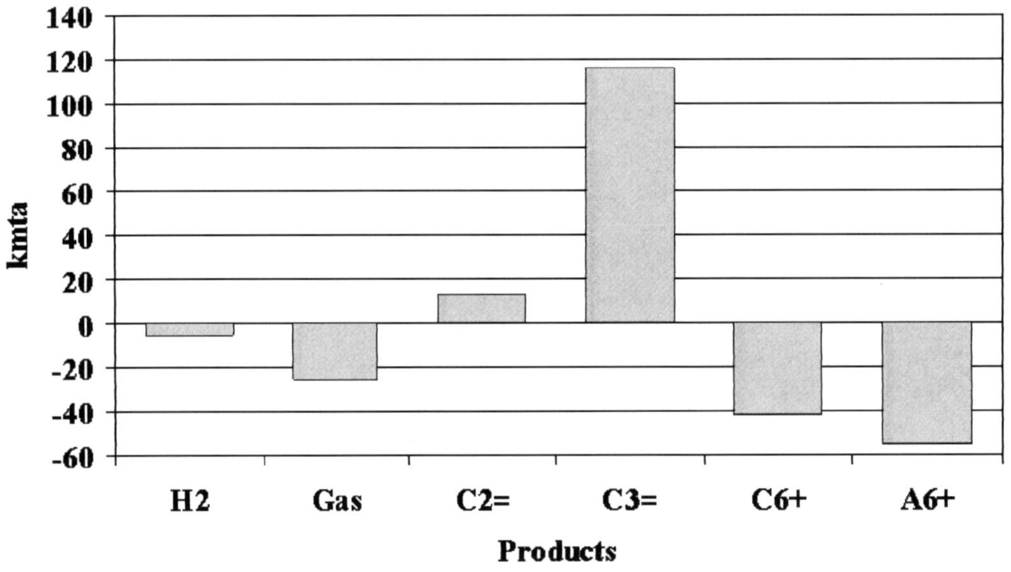

Fig. 7 shows the differential yields for the Cracker plus OCP minus the cracker alone

Figures 8 and 9 show economics for steam cracking integration with OCP. The grass roots naphtha based ethylene plant processing 2,000 kmta naphtha feed with medium operating severity is compared with the same cracker integrated with an OPC unit. The differential yield gain is calculated from Figure 7 with the values given in Figure 8. The integrated plant produces 30% more propylene compared to the base case with no loss of ethylene (P/E=.8). The capital charges for the integration are shown in figure 9. Simple payout period for this case is dependent on the relative value of propylene/ethylene but in all cases is expected to be less than 4 years.

Figure 8: Integration Study Values

Study Basis:
Grass Roots Naphtha Based Ethylene Plant Complex with normal recycle - 2000 kmta naphtha feed (670 kmta C2=), medium severity compared with same cracker integrated with Olefin cracking

Naphtha	$190/mt	Fuel	$10/MMkcal
Ethylene	$400/mt	Electricity	$.05/kWh
Propylene	$400/mt	Steam	$8.50/mt
Hydrogen	$500/mt	Cooling W.	$.02/m3
Fuel Gas	$120/mt		
Gasoline	$190/mt		

Fig. 8 lists study values

Figure 9: Economics for Steam Cracking Integration

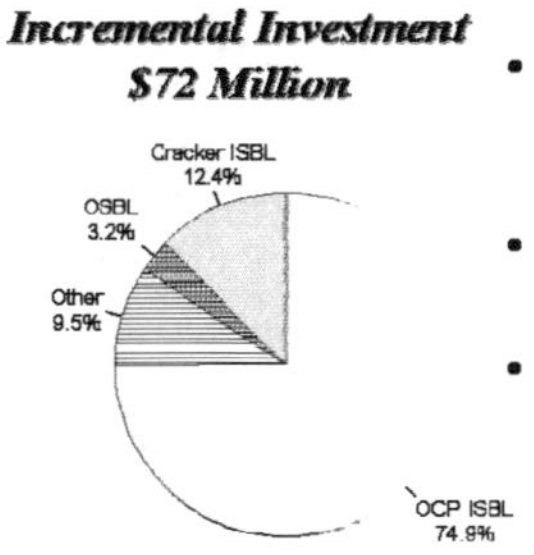

- 30% higher propylene production compared to base case
- 0.8 overall C3=/C2= ratio
- Less than 4 year payout on incremental investment

Fig. 9 shows estimated investment costs

FCC Refinery Integration

When integrated with an FCC refinery (Figure 10), the OCP unit converts C_4-C_8 olefin rich streams from the FCC and coker units to high-value light olefins mostly propylene. The differential yield of the FCC integration with OCP is shown in Figure 11.

Figure 10: OCP Integration with FCC

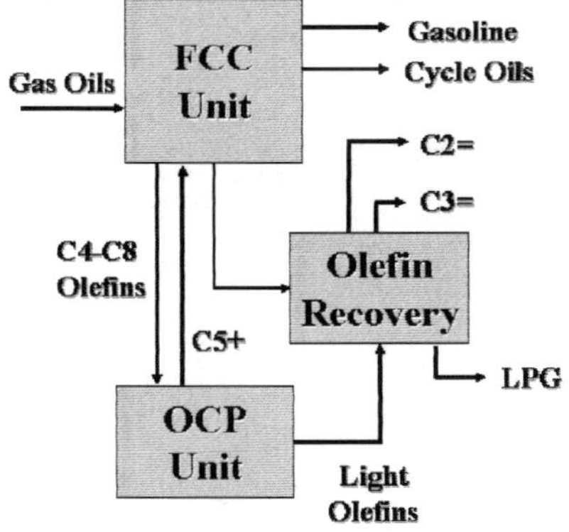

Fig. 10 is a block flow of the OCP-FCC integration

Figure 11: Integration Differential Yields

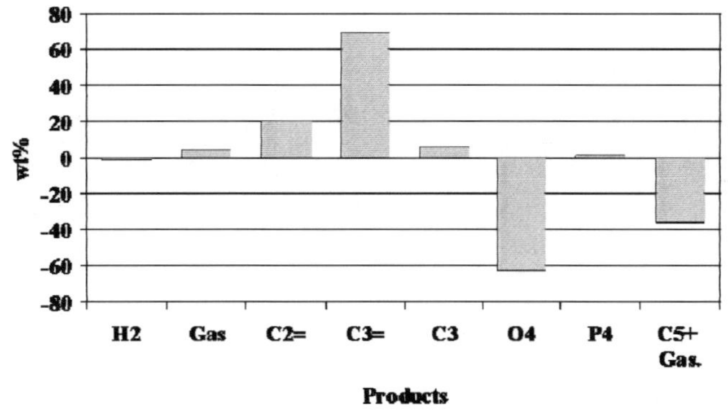

Fig. 11 gives the differential yields for the OCP-FCC minus FCC

Although the gasoline blend from the FCC after OCP integration is lower in olefin content the octane number remains the same due to a small amount of aromatics produced and skeletal isomerization that occurs in the OCP (Figure 12). Most integrations with FCC will be as retrofits and expansions of existing facilities. The economics are more difficult to define than for a grass roots case but the simple payout of integrating an OCP unit into a refinery should certainly be less than 4 years.

Figure 12: FCC Gasoline Composition

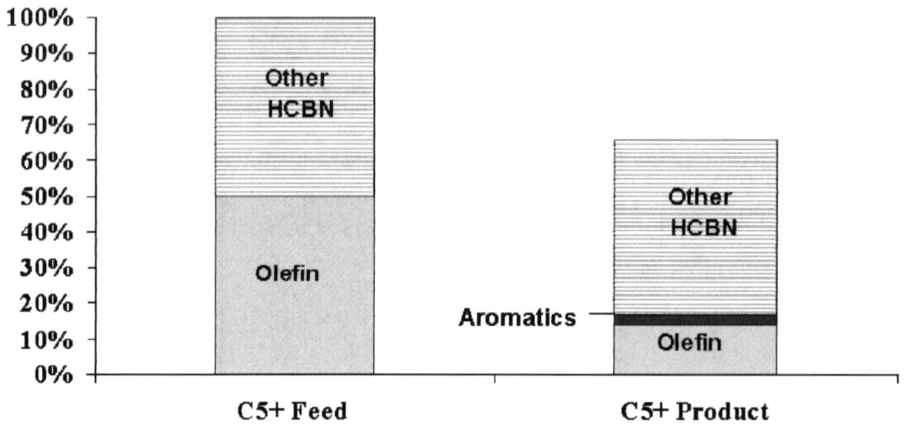

Fig. 12 gives the composition of FCC gasoline with FCC alone and with integration with OCP

MTO Integration

Although the MTO reactions are quite specific, by-product C_4 and C_5 streams are produced. Since some MTO plants are expected to be based in remote locations, efficient utilization of by-products can be difficult. By integrating an OCP unit into an MTO complex (Figure 13), the overall yield of ethylene and propylene on feed to the complex can be greatly increased. Figure 14 shows the yield of methanol feed that goes to light olefins (carbon basis) for an MTO complex and for an MTO complex augmented with an OCP unit. Because of the low value of remote MTO by-product C4+ olefin streams the simple payback for the incremental cost of adding an OCP unit to the complex is close to two years.

Figure 13: OCP Integrated with MTO

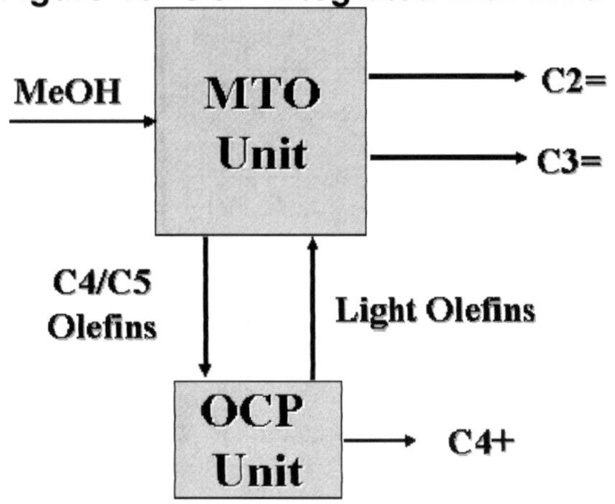

Fig. 13 is a block flow of MTO/OCP integration

Figure 14: MTO Integrated Yield

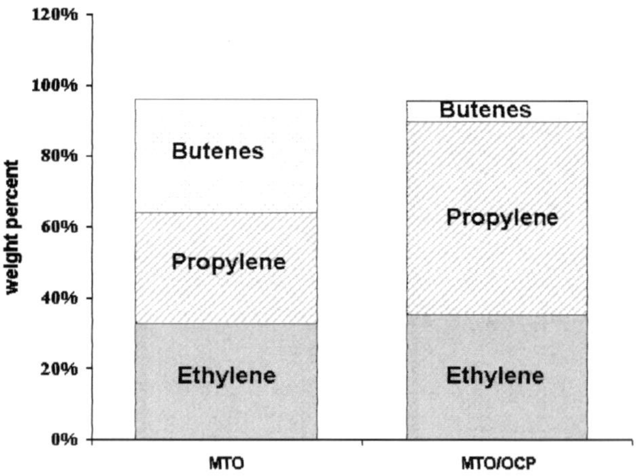

Fig. 14 shows the MTO yield before and after integration with OCP

Summary

Increased demand for polypropylene together with a decreased production ratio of $C_3^=/C_2^=$ from conventional steam cracker sources will result in an increasing propylene supply-demand gap in the near future. The ATOFINA/UOP Olefin Cracking process was developed to meet the increasing demand for propylene, by converting low-value by-product streams to propylene and ethylene at high propylene to ethylene ratios. OCP units are capable of processing a wide range of C_4 to C_8 olefins from steam crackers, FCC refineries, and MTO plants. The technology is available for license from UOP.

---

[i] J. Andersen, S. Bakas, S. Kvisle, H. Reier Nilsen, et al, "MTO: Meeting the Needs for Ethylene and Propylene Production", ERTC Petrochemical Conference, Paris, March, 2003

Progress report on MTP® with focus on DME

Harald Koempel, Waldemar Liebner, Martin Rothaemel
mg engineering - Lurgi AG
Lurgiallee 5
D-60295 Frankfurt am Main, Germany

Prepared for Presentation at
"AIChE Spring National Annual Meeting (April 25-29, 2004)
Session: Olefins Production: Fundamental and Technological Aspects"

© Waldemar Liebner, Lurgi AG, February 2004

AIChE Shall Not Be Responsible For Statements or Opinions Contained in Papers or Printed in its Publications

Introduction

At the 2003 AIChE spring meeting Lurgi had the opportunity to present a paper "Demonstrating The Methanol to Propylene (MTP®) Process". Since the paper aroused considerable interest we considered it appropriate to provide the continuation and conclusion report on the Demo Unit operation. For coherence the process and Demo Unit basics are repeated here – for details the reader is referred to the earlier paper.

In addition the current paper will elaborate on the first step of the MTP® process, the dehydration of methanol, a.k.a. dimethylether (DME) synthesis, explaining why this "old" two-step process is still attractive. The route from natural gas through methanol (via Lurgi MegaMethanol®) to DME and propylene offers a way for monetizing stranded gas and oil associated gas. The propylene from MTP® can be processed into even higher value polypropylene (PP) or into other chemicals. This constitutes a gas-based route to "petro-chemicals" which are usually based on steam cracking technologies.

Figure 1 MTP ®: Simplified Process Flow Diagram with production figures

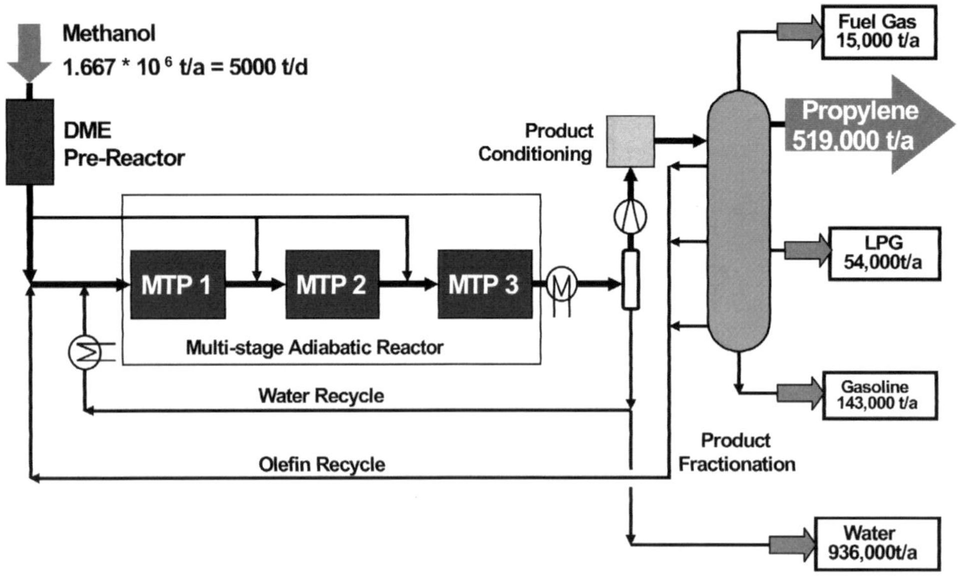

The MTP® Process

Lurgi has developed a process for converting methanol to olefins that exhibits a high selectivity to propylene. Ethylene, butylenes and C_5-C_6 olefins, which are by-products, are recycled and converted to propylene. The remaining by-product is a high-octane gasoline that can be blended into a gasoline pool. Figure 1 shows the process flow and the basic material balance of the MTP® process. For a feed rate of 5,000 metric tons of methanol per day (1.667 million metric tons annually), approx. 519,000 metric tons of propylene are produced per year. Saleable by-products are LPG and sulfur-free gasoline. Fuel gas is used internally. Process water can be used for irrigation after simple biological treatment. The amount of coke produced during normal operation is extremely low.

The Demo Unit

The Demo Unit was installed at the Statoil methanol plant at Tjeldbergodden (Norway) in November 2001 following a cooperation agreement with Statoil ASA. Later in 2002 Borealis joined the cooperation. The main purpose of the unit was to demonstrate that the catalyst lifetime meets or exceeds the commercial target of 8000 hours on stream – under "real-world" conditions of a large-scale methanol plant.

Description of the Demo Unit and Scale-Up. All reactors are fixed-bed types and the MTP® reactors are loaded with a tailor made proprietary zeolite catalyst. The operating conditions of the MTP® reactors in terms of bed height, linear gas velocities, pressure drop and adiabatic temperature profile are identical to the proposed commercial reactor design, which will ensure a risk-minimized scale-up to commercial-size reactors. The plant is equipped with a simplified work-up section for separating the product into gas, liquid hydrocarbons and water. The water produced and the $C5^+$ fractions from the gasoline distillation column are continuously recycled. If ethylene and butylene have to be recycled, these are added from gas cylinders in the appropriate mass flow rates.

MTP Demo Unit Highlights

Demo Unit Status and Results

As of September 8th 2003, the Demo Unit completed the scheduled 8000 hours life-cycle test. In the following the main results are presented and discussed. Directly after this successfully completed test a new series was initiated with a second batch of the original catalyst. These additional tests shall serve for the verification of the first series' results and shall explore some parameters which were found of interest in the first run but did not fit into its schedule. Some highlights of the recent tests will be reported here.

Methanol Conversion. Figure 2 depicts the dependence of the methanol conversion on the total run-time of the Demo Unit as well as its changes during every cycle. While the catalyst activity for methanol conversion to all products including DME is rather stable, the breakthrough of DME increases in the course of the run-time of the Demo Unit resulting in shorter cycles (see below).

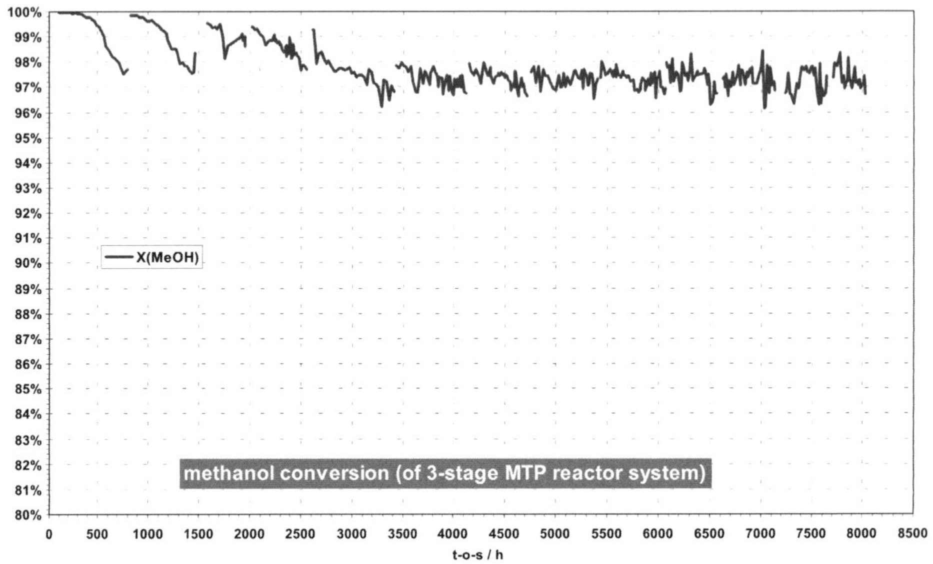

Figure 2. Dependence of methanol conversion on the total time-on-stream.

Cycle Length. The target value for cycle length was 500 – 600 hours. The criterion for the cycle end is mainly determined by the requirement of the commercial plant: if the methanol conversion is too low, a larger methanol recycle has to be applied which can be limiting. Furthermore, the DME content in the product gas can also be critical since DME has to be removed from the propylene product to less than 1 ppm. Both factors were taken into account when the cycles in the Demo Unit were ended. The resulting cycle lengths are depicted in Figure 3.

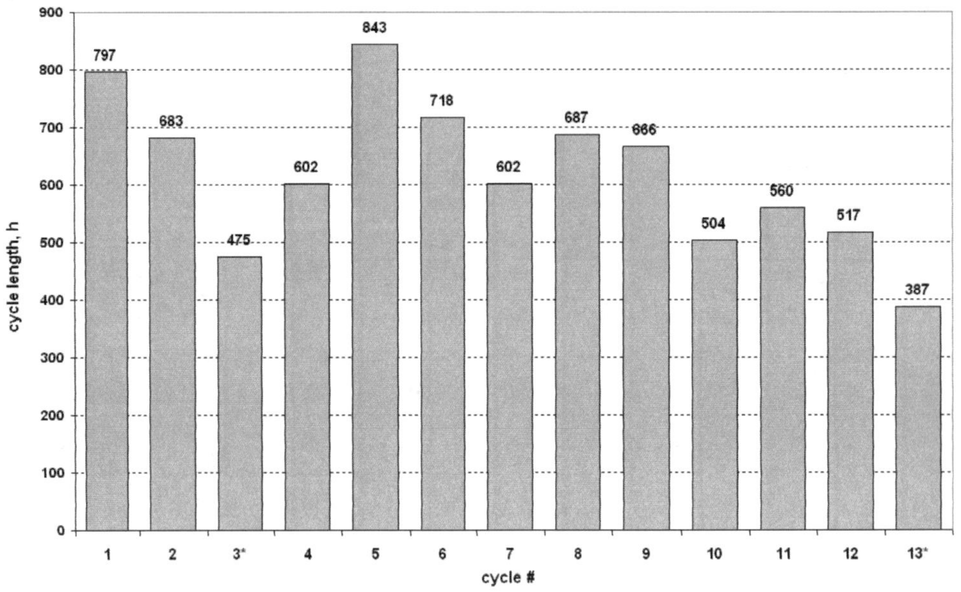

Figure 3. Cycle length in the Demo Unit (NB: the 3rd cycle was ended prematurely due to a turnaround of the methanol plant, 13th cycle was ended after 8000 hours of total on-stream time)

The results of the Demo Unit operation confirm previous expectations that the cycle length exceeds the values measured in the smaller lab-scale unit. Cycle lengths of more than 600 hours are obtained most of the time, decreasing only towards the end of the life test.

Propylene Selectivity and Yields. As indicated by the propylene-to-C3 ratio (see Figure 4), the propylene selectivity has steadily increased in the first half of the run-time of the Demo Unit and reached a stable level in the second half, of around 97-98%.

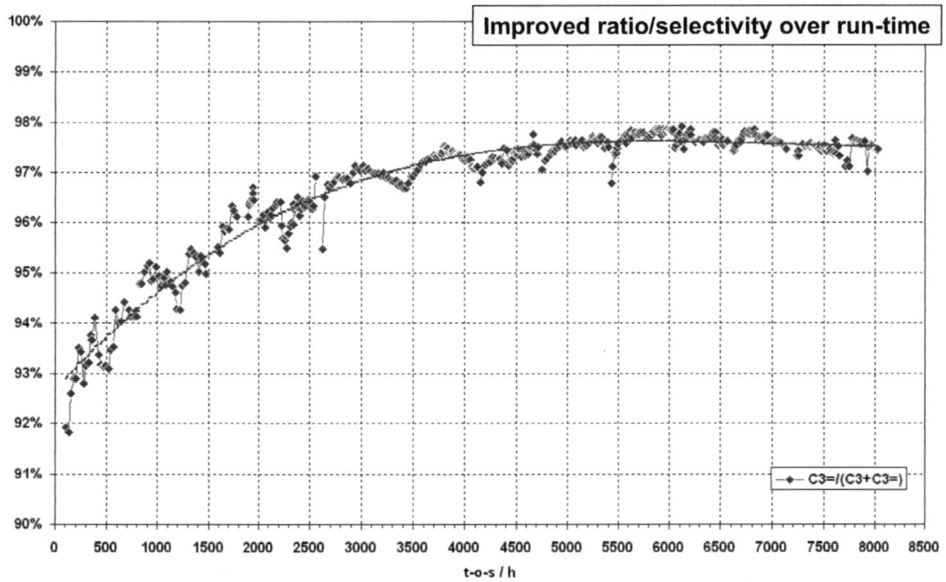

Figure 4. Propylene-to-total C_3 ratio as a function of the time-on-stream of the Demo Unit.

It has to be stressed that the demonstration of the propylene yield was not the objective when designing the Demo Unit. The main objective was to test catalyst lifetime at the selected reaction conditions. However, based on the tests with the addition of ethylene, butylene and pentenes (simulated recycle) as well as the recycle of a C5/C6 fraction from the gasoline column ("live recycle"), a yield of 65% propylene was achieved in the Demo Unit. The commercial MTP® unit will have a complete recycle which contains also inerts that will (indirectly) improve propylene selectivity by lowering the adiabatic temperature increase which results in a more favorable temperature profile in the reactors. Based on the analysis of these effects, the propylene yield of the commercial plant is expected to amount to about 71%.

Product Quality. Specific acetylene trace analyses by Borealis have proven that the product gas of the Demo Unit does not contain acetylene nor methyl-acetylene/propadiene. An initial quality test of the gasoline product by the Statoil refinery in Mongstad indicated its high quality in terms of extremely low benzene content, low overall aromatics content, high percentage of branched paraffins and olefins and a rather high RON of >95. Since it is also sulfur-free it is a high-value blending stock.

Further, Borealis used a batch of purified MTP -propylene for a polymerization test. A polypropylene product was obtained with all properties to specification, i.e. identical to standard PP, proving that the MTP -propylene was totally free of possible catalyst poisons. Finally, Borealis produced cups from this polypropylene. These cups are now the first PP -products made from natural gas.

Demo Unit - Lessons from Life-cycle Test

- catalyst lifetime is above 8000 h (depending on end-of-life restrictions)
- measured total propylene yield under the restricted conditions of the Demo Unit was 65 % which translates to an expected 71% for a commercial plant with recycle of all relevant streams
- gasoline product quality is very good (Statoil)
- purified propylene from MTP can be polymerized with similar yields and product properties as from any other C3= source (Borealis)
- An approach to an isothermal reaction pass favorably increases selectivities and yields
- commercial design of MTP plant is revised and now based on Demo Unit results

The commercial design will now implement a more staged reactor which enables a reaction pass nearer to the isotherm. Together with the fact that the restrictions inherent in this Demo Unit no longer prevail in the commercial plant, this ensures that yields and selectivities will be at their best.

Demo Unit - Second Operation Phase

A new operating phase already began in December 2003. These additional tests with a second batch of the original catalyst shall serve for verifying the results of the first series and shall explore some parameters which were found of interest in the first run but did not fit into its schedule. Figure 5 shows the improved propylene selectivity caused by the more careful run-in of the second batch.

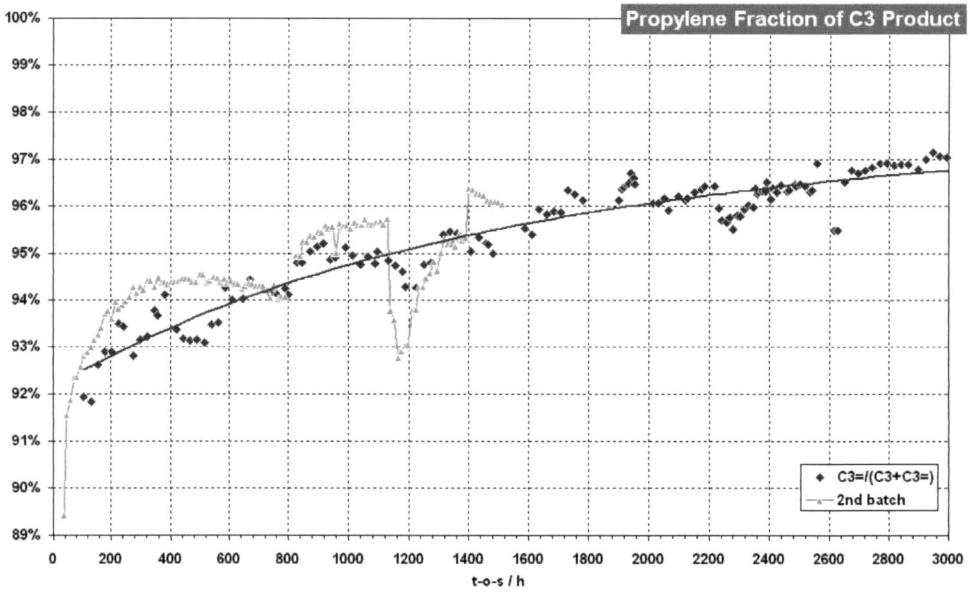

Figure 5. Comparison 1st versus 2nd operating phase

The preliminary results after the first two cycles of the second operation phase with another batch of the same catalyst production are as follows:

2nd batch has...

- ...comparable activity in terms of methanol conversion
- ...higher once-through selectivities for propylene and olefins

- ...comparable/higher propylene/C_3 ratio

It is assumed that these improvements stem from the milder start-up procedure as compared to the very first one, i.e. lessons were learned and thus

<u>2nd batch did not feature...</u>

- ...exotherms
- ...marked fluctuations of temperature/pressure
- ...extended periods of steam treatment

This confirms the above assumption that the results of the first phase can be improved.

Demo Unit Outlook

- Second-phase operation is to continue for a total of 3000 hours.
- Furthermore, at the Lurgi R&D labs in Frankfurt, a new MTP pilot plant was installed that includes equipment for olefin recycles and propylene purification to produce propylene samples that can be tested in bench-scale polymerization tests by potential customers. Estimated start-up is the first quarter of 2004.

Conclusion

The operation of the Demo Unit has provided sufficient data for full commercialization of the MTP process. A key lesson learned is that a near-isothermal reaction path considerably improves the propylene yield and selectivity.

DME - Ye Olde Two-Step

Taking the opportunity we would like to focus for a moment on the very first step of MTP, the production of dimethyl ether (DME) from methanol. On an industrial scale, DME is produced by the dehydration of methanol which in turn is manufactured from natural gas via the intermediary production of synthesis gas. Thus, DME is obtained from synthesis gas by an "old-fashioned two-step process". Or, at least this is what proponents of single-step processes from Japan and Denmark say.

We would like to briefly explain here why the old way is more economical for the production of DME as such, as e.g. as energy carrier or diesel fuel, or in the context of MTP. Firstly, the two steps give the opportunity to split a complex in two building the methanol plant at the natural gas site and the DME and/or propylene plant near the consumers. While this is a logistical argument which may or may not hold true for a concrete project there are fundamental thermodynamic reasons why a single step "gas-to-DME" process is inferior to "the old two-step".

DME Production / Thermodynamic Principles

Methanol synthesis. Formally, methanol synthesis is described by the equation

$CO + 2 H_2 \Leftrightarrow CH_3OH$ $\Delta H_{R\,(298\,K)}$ = - 90.84 kJ/mole (I)

Inexorably linked with methanol synthesis is the CO conversion

$CO + H_2O \Leftrightarrow CO_2 + H_2$ $\Delta H_{R\,(298\,K)}$ = - 41.27 kJ/mole (II)

The combination of both reactions leads to an additional reaction equation which is dependent on (I) and (II) above

$CO_2 + 3 H_2 \Leftrightarrow CH_3OH + H_2O$ $\Delta H_{R\,(298\,K)}$ = - 49.57 kJ/mole (III)

The heat of this reaction therefore also corresponds to the difference of the heat of reactions (I) and (II) above. These equation systems provide the explanation why methanol can be produced through the hydrogenation of both CO and CO2.
The equilibrium constants dependent on the partial pressure for the above reaction equations are defined as follows

$$K_{p(I)} = \frac{p_{CH_3OH}}{p_{CO} * p_{H_2}^2}$$

$$K_{p(II)} = \frac{p_{CO_2} * p_{H_2}}{p_{CO} * p_{H_2O}}$$

$$K_{p(III)} = \frac{p_{CH_3OH} * p_{H_2O}}{p_{CO_2} * p_{H_2}^3} = K_{p(III)} = \frac{K_{p(I)}}{K_{p(II)}}$$

As with all exothermic reactions, the throughput rises (reaction from left to right) with decreasing temperature. The temperature rise resulting from reaction under adiabatic conditions therefore leads to a continuously deteriorating equilibrium.

Methanol dehydration. The conversion of methanol to obtain DME follows the reaction equation

$2 CH_3OH \Leftrightarrow CH_3OCH_3 + H2O$ $\Delta H_{R\,(298\,K)}$ = - 23.5 kJ/mole (IV)

This reaction, too, is exothermic, but the heat of reaction is clearly lower causing the equilibrium to be less strongly influenced by the temperature. Unlike methanol synthesis, the formation of DME as per equation (IV) is basically not affected by pressure.

The equilibrium constants are defined as follows

$$K_{p(IV)} = \frac{p_{CH_3OCH_3} * p_{CO_2}}{p_{CH_3OH}^2}$$

Combined methanol/DME synthesis ("single-step" DME). At least in formal terms, methanol synthesis and methanol conversion to DME can be combined. This gives

$$3\,CO + 3\,H_2 \Leftrightarrow CH_3OCH_3 + CO_2 \qquad \Delta H_{R\,(298\,K)} = -\,246.45\ \text{kJ/mole} \qquad (V)$$

The corresponding equation constant is

$$K_{p(V)} = \frac{p_{CH_3OCH_3} * p_{CO_2}}{p_{CO}^3 * p_{H_2}^3} = (K_{p(I)})^2 * K_{p(II)} * K_{p(IV)}$$

While in methanol synthesis the formation of methanol and the conversion of CO occur simultaneously on the same catalytically active centers, the generation of DME requires separate catalytic centers. Therefore, catalysts for the conversion of synthesis gas to DME (direct synthesis) are also referred to as bifunctional catalysts. Methanol and DME are basically generated one after the other. The advantage of methanol conversion to DME not being affected by pressure is lost when combining these reactions.

At first glance, the direct synthesis of DME apparently has clear advantages when compared with the 2-step production. In particular, the favorable equilibrium of reaction (IV) even at temperatures above 300 °C provides the opportunity of circumventing the unfavorable equilibrium of the MeOH synthesis reaction at elevated temperatures.

At a second glance, however, serious disadvantages show up. Methanol and DME catalyst operate in clearly different optimum temperature ranges which requires a trade-off that results in high catalyst consumption. The equally combined CO conversion inevitably leads to a substantial formation of CO2 as a byproduct.

As the direct synthesis of DME, too, requires gas recirculation to achieve a sufficient conversion rate, the formation of CO2 results in the corresponding rise of the CO2 content in the reactor feed gas which substantially affects the kinetics of methanol production. This in turn also leads to increased catalyst consumption.

One of the core problems of DME direct synthesis is the physical properties of DME. DME is gaseous at ambient temperatures and can only be partially liquefied at the standard pressures and condensation temperatures prevailing in MeOH synthesis. Therefore, an increase in the synthesis pressure or a reduction of the condensation temperature will be required to achieve sufficient conversion rates.

Economics of Single-step versus Two-step DME Production

Comparing the costs of the two alternatives for large-scale plants like 3000 to 5000 t/d of DME, as suggested for energy or MTP applications, thorough analyses have proven that single-step processes may have a slight advantage in the synthesis section but have a marked disadvantage in the work-up section and through the necessary CO_2 -recycle. This translates into a 10-20% lower total installed cost for the two-step route which again means 5-10% lower production costs.

Hence, we prefer "the old two-step", not only as a dance but also in the production of DME, going through methanol production and subsequent separate dehydration.

Planning and Designing the Modernization of the Recovery Area of a Flexible Cracker

Andre Bernard, Process Design Engineer, NOVA Chemicals Corunna site
P.O. Box 3600, Sarnia, ON, Canada N7T 8C7
bernara@novachem.com

Richard Hayden, Process Manager, KBR
601 Jefferson Ave., P.O. Box 3, Houston, TX, 77001-0003
richard.hayden@halliburton.com

Abstract

NOVA Chemicals' Corunna site operates a flexible cracker with a current nameplate capacity of 160% of original design. Aging compressors often limit operation flexibility and plant throughput. In order to enable the plant to consistently and efficiently meet its current nameplate capacity for the different feed slates processed, replacement of the Charge Gas, Ethylene Refrigeration and Propylene Refrigeration compressors are required.

NOVA funded a feasibility study by KBR to determine the best options to replace the compressors, given NOVA's overall objectives of maximizing throughput with minimal capital investment while achieving improved efficiency and energy savings. As a result of this study reconfiguration of the cold section and the introduction of Liquid Driers were recommended.

KBR also proposed the implementation of its patented Liquid Reinjection Process within the cold section of the plant. This process scheme eliminates a significant recycle stream to the Charge Gas Compressor and will also provide a significant refrigeration load reduction by mixing dry condensed hydrocarbons with dry charge gas upstream of the chilling train. Further modifications are required to the existing Demethanizer to convert the tower into a Prestripper and a new Demethanizer will be constructed. The new Demethanizer pre-fractionation scheme will also include changing the Deethanizer to a dual feed configuration. This will allow a significant reduction in the Deethanizer reflux requirements.

This paper discusses the key process changes that will achieve the modernization objectives with regards to throughput, efficiency, capital and a tight turnaround window.

Introduction

A project for modernizing NOVA Chemicals Corunna ethylene facility is currently underway. The project is driven by the need to replace aging compressors that currently limit the plant's ability to consistently achieve its maximum capacity. Low efficiencies of the existing equipment also impose constraints on plant utility systems (steam and treated water).

In order to have a successful project of this magnitude, it is not enough to simply replace the older equipment with bigger or parallel units. It is also desirable to change the process, making it more energy efficient, while shifting flows and duties throughout the recovery section in order to maximize the reuse of existing equipment items and minimizing replacement services.

When KBR evaluated NOVA's goals for the modernization project, they determined that the existing NOVA process was a candidate for including new Liquid Dryers, working with the addition of a Demethanizer Prestripper as well as KBR's patented Liquid Reinjection technology to reduce specific energy requirements of the process while re-using both the existing Demethanizer and Deethanizer. NOVA selected KBR's proposed new process scheme as it was found to best fit the project objectives. At this point, basic engineering has been completed by KBR and detail design is currently in progress. Shut down and startup activities are scheduled for the fall of 2005.

Project Objectives

After project completion, the recovery section of the ethylene unit will not limit furnace utilization for the different combination of feeds processed at the site. Opportunities with regards to unit reliability and overall energy efficiency were to be considered by the design. The design also takes into consideration a 45 days (spec to spec) turnaround schedule.

Given the feedstock flexibility, 2 design feed cases believed to properly define the unit turndown requirements were developed. As in any modernization project plot space is critical. As such, the new compressors must fit the existing footprint to avoid structural work on the compressor deck.

Plant History

NOVA Chemicals Corunna Site was North America's first fully integrated refinery and petrochemical complex when it came on stream in 1977. The original facility was based on a naphtha technology olefins plant coupled with a front-end Crude Unit and back-end Aromatics Unit. Originally only naphtha was being processed in the cracking furnaces, but over the years the feedstock has been diversified and the plant now has the capability of processing a wide range of feeds. The plant has undergone a few debottleneck projects expanding its ethylene nameplate by 160% since its original start-up. Along with ethylene, the Corunna Site produces a full range of primary petrochemicals, including propylene, butadiene, iso-butylene, n-butylene, benzene, toluene and xylenes. Co-products produced at the Corunna Site include gasoline components, hydrogen, isoprene, dicyclopentadiene, resin oil, diesel fuel and industrial fuel.

Current Configuration Process Description

In the current process scheme, as shown on Figure 1, recovery of olefins from furnace effluents starts with the quench area where charge gas temperature is reduced and heavies such as pyrolysis fuel oil (PFO) and heavy gasoline (RPG) are extracted. The Charge Gas (CG) effluent from the quench area is compressed through 5 stages of compression. Condensed water streams from the compressor train knock-out drums are recycled to the quench area along with condensed hydrocarbons from the first 3 stages of compression. Condensed hydrocarbons from the last 2 stages are processed in a high-pressure Condensate Stripper where overhead vapors are recycled to the Charge Gas Compressor (CGC) 4th stage and the tower bottoms stream forwarded as feed to the Depropanizer. An acid gas unit is integrated within the CGC train for removal of hydrogen sulfide and carbon dioxide. Compressed Charge Gas is dried and is then processed in the Chill Train/Demethanizer area for separation of methane and hydrogen. Chilling is achieved with ethylene refrigerant along with hydrogen and methane streams recycled to the chill train. The Demethanizer bottoms stream is processed in the Deethanizer where C3+ hydrocarbons are separated and forwarded as feed to the Depropanizer in the C3/C4 area. The Deethanizer overhead vapor stream is first treated in Acetylene Converters where its acetylene content is being hydrogenated. It is then processed in the C2 Fractionator for recovery of dilute and polymer grade ethylene. The ethane stream leaving the bottom of the C2 Fractionator is recycled to the furnace for cracking.

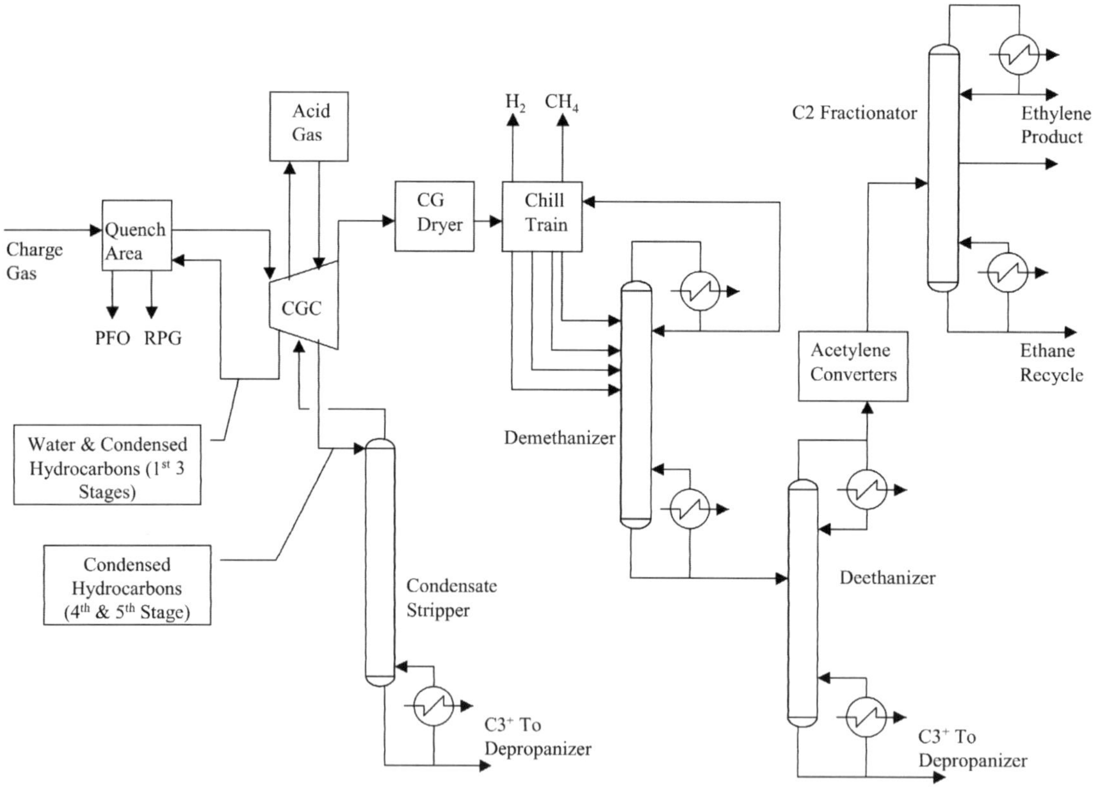

Figure 1: Current Process Scheme

Current Plant Constraint

The recovery section of the plant is currently limited by the capability of the CGC. It is not unusual to also encounter hydraulic constraints in the Chill Train as well as limitations with the existing Ethylene Refrigeration Compressor (ERC).

From a mechanical standpoint, excessive pressure drops between the CGC 5th stage and the Demethanizer could lead to pressure exceeding the relief valve setting when considering higher production rates. The existing Demethanizer tower bottom section is made of carbon steel and requires safety management systems to prevent occurrence of low temperatures during a plant upset.

Another current plant constraint is in the plant utility systems, particularly in steam supply and water treatment areas. Increasing the demand for treated water or steam is achievable but at the expense of plant reliability. Any increase in capacity of these utilities would severely affect capital cost and are constrained by plot space in the case of water treatment and steam generation facilities.

KBR Patented Liquid Reinjection Concept

The concept of Liquid Reinjection involves the injection of liquid hydrocarbons into dry charge gas to enhance condensation against propylene refrigeration. The liquid hydrocarbon stream is preferably obtained within the recovery section of the plant but could also come from outside sources such as importing propane and/or propylene into the process. Typically dry condensed hydrocarbons from the CGC area, the Deethanizer bottom or the Depropanizer or combinations of the above are used. Injection of C2 lean liquid hydrocarbons in the charge gas raises the stream dew point temperature, lowering refrigeration energy usage and shifting cooling requirements from ethylene to propylene refrigeration. The overall amount of liquid injected should be sufficient to enhance condensation against warmer refrigeration, but limited by impacts on downstream separation equipment. In fact, for this project the hydrocarbon condensate from the 4th and 5th stages of compression are dried and sent forward for reinjection.

Technology Selection & Description

Different process schemes, which would limit the size and power requirements of the new compressors were evaluated at the early stage of the project and were leading to similar conclusions. Along with a new CGC and refrigeration compressors (ethylene and propylene), integration of the existing Demethanizer with a Prestripper was recommended. Removal of the existing parallel cold box and replacement with a new unit was also justified.

KBR's Liquid Reinjection technology incorporated many of the above changes but reduced the overall scope of replacement. Furthermore, the power savings realized by integrating the technology reduces the demand on the water and steam system, which is known to limit the flexibility of certain operating activities at various times. The technology could also be easily integrated into the existing process scheme.

KBR has achieved the project objectives sought by NOVA through the application of its unique Liquid Dryer, Liquid Reinjection, Demethanizer Prestripper system to the Corunna Olefins Unit. This scheme opens up capacity in existing equipment by eliminating recycles such as that produced by the high-pressure Condensate Stripper. It also reduces refrigeration loads by condensing charge gas components at warmer temperatures against less expensive refrigeration levels.

As illustrated on Figure 2 the Liquid Reinjection scheme will be implemented by eliminating the condensate stripper. Hydrocarbon condensate from the 4th and 5th stage compressor knock out drums will now be collected and fed to a Liquid Coalescer to remove any free water. The hydrocarbon effluent from the Coalescer will be pumped to a Liquid Dryer. This Dryer contains the same type of molecular sieve that is used in the Charge Gas dryers. The dried condensate will then be combined with the effluent from the Charge Gas dryers upstream of the chillers in the front end of the demethanizer feed chilling section. The injection of dry liquid hydrocarbon in the charge gas raises the dew point of the mixture by 40 to 50 ^{o}C (72 to 90 ^{o}F) and allows more condensation across the 1st propylene chiller. As a result, the volumetric vapor flow to the chill train drops by 8 to 11%, which reduces the chilling requirements from the ethylene refrigeration system. The ethylene compressor power needed is then reduced by about 10%. The lower flow to the chill train also benefits the hydraulics of the unit by reducing pressure drops.

Elimination of the recycle vapor stream from the condensate stripper to the CGC 4th stage suction reduce compression power on the last 2 stages by 3.5 to 5% for an overall compressor power reduction of about 2%. The volumetric vapor flow to the Charge Gas Dryers is also reduced by 4 to 6%, which contribute to lower pressure drops between the CGC 5th stage discharge and the Demethanizer.

The most significant change in the demethanizer feed chilling area is the installation of a new Demethanizer Prestripper. Actually, the existing Demethanizer will be retrofitted as the Prestripper, and a new stainless steel Demethanizer will be added. The liquid from the bottom of the first demethanizer feed separator feeds the top of the Prestripper, rather than flowing directly to the Demethanizer. As the bottom of the existing Demethanizer is made of carbon steel, its conversion to a Prestripper reduce the safety management systems require to protect the vessel from brittle fracture if cold temperature were to occur during a propylene refrigeration system trip or failure. In the new configuration the vessel will only see the warmest stream exiting the chill train.

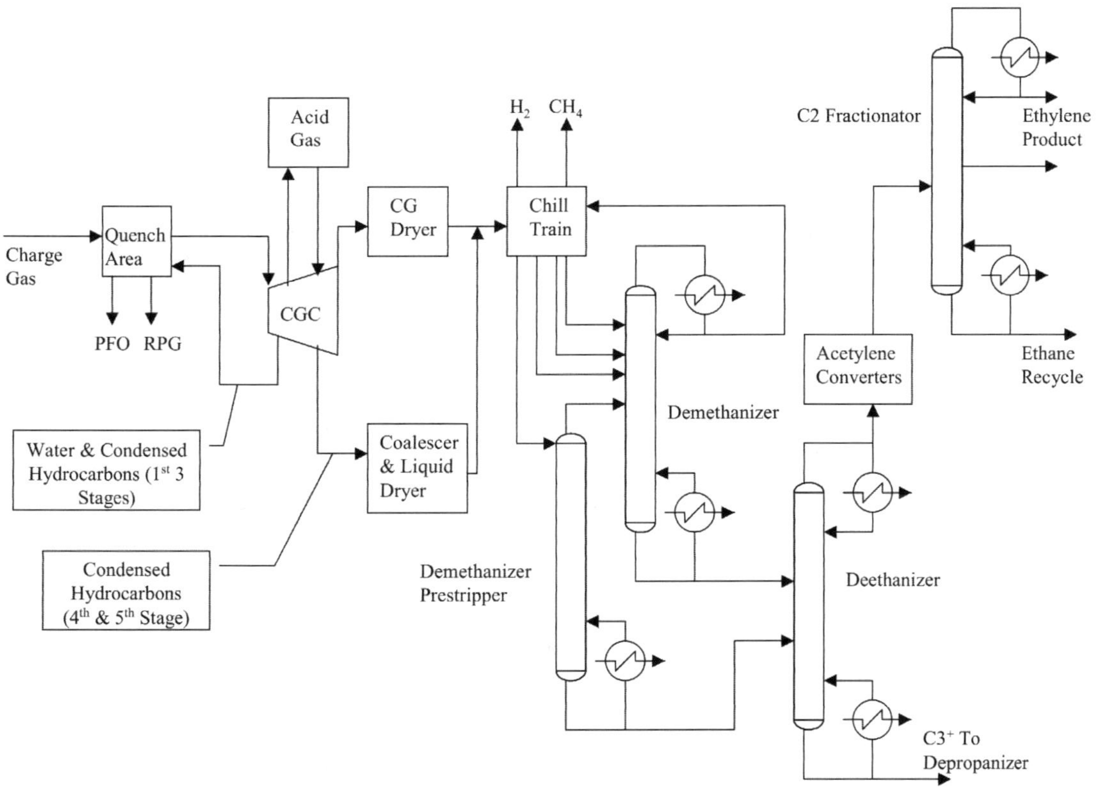

Figure 2: Recommended Process Scheme

The Liquid Reinjection process scheme causes the Deethanizer to process the additional heavies, which are currently fed from the Condensate Stripper to the Depropanizer. As shown on the McCabe & Thiele diagram illustrated on Figure 3, the new feed line moves to the left and cause the operating line in the rectifying section to be slightly steeper. The 5 % increase in slope cause a 7 % increase in condenser duty. In the current plant configuration a portion of the feed is preheated prior to entering the Deethanizer, which creates 2 feeds of identical composition but different energy levels. A Prestripper provides 2 feeds of different compositions with the top feed significantly lighter. As shown on Figure 4, the lighter feed allows a pinch point that enables an operating line that would otherwise cross the equilibrium curve. Without reconfiguring the tower with regards to number of stages and feed locations, the operating line is found to run closer to minimum reflux in the rectifying section. The condenser duty is then reduced by about 22% when compared to the current process scheme and by 27% when implementing Liquid Reinjection without the Demethanizer Prestripper. The bottom feed line provides an efficient pinch point that allows reasonable reboiler duty and tower hydraulic loadings. This feed is being preheated through integration with the CGC area. However, care must be taken in the amount of preheat as excessive preheat could cause tower instability and/or premature flooding.

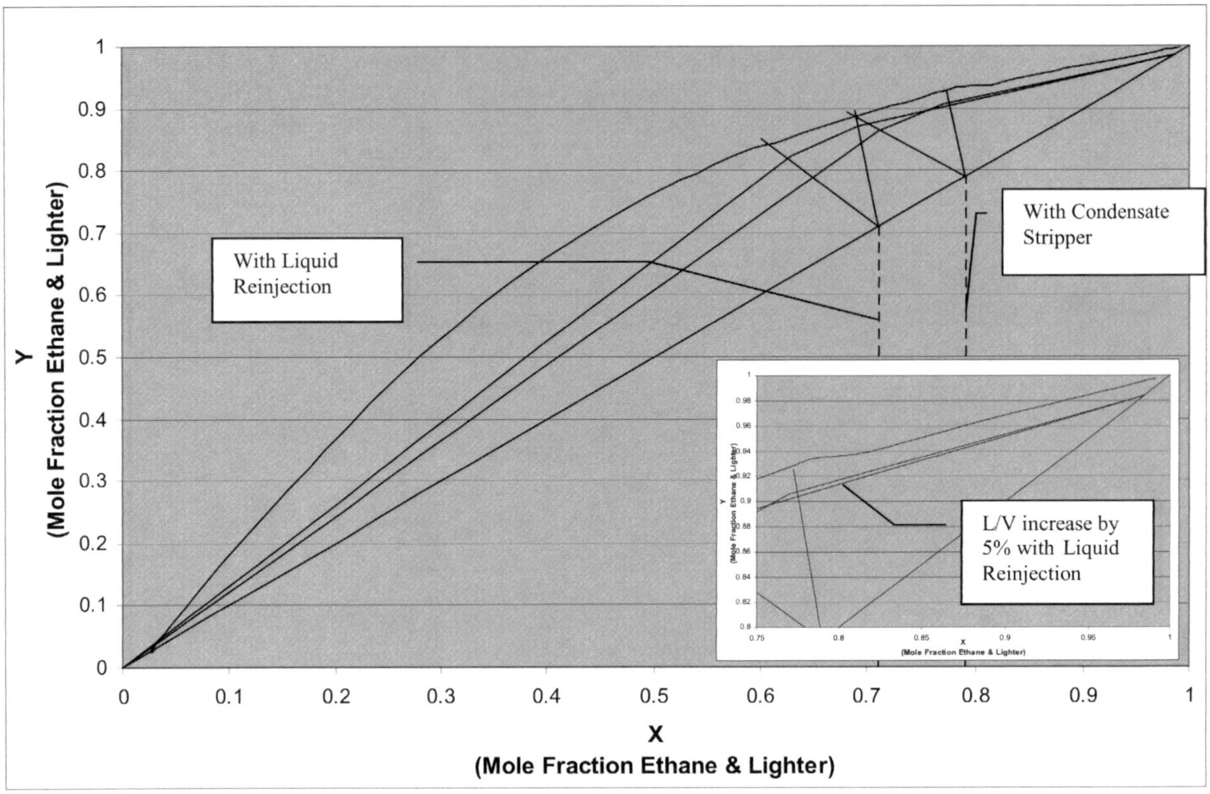

Figure 3: McCabe Thiele Diagram – No Prestripper

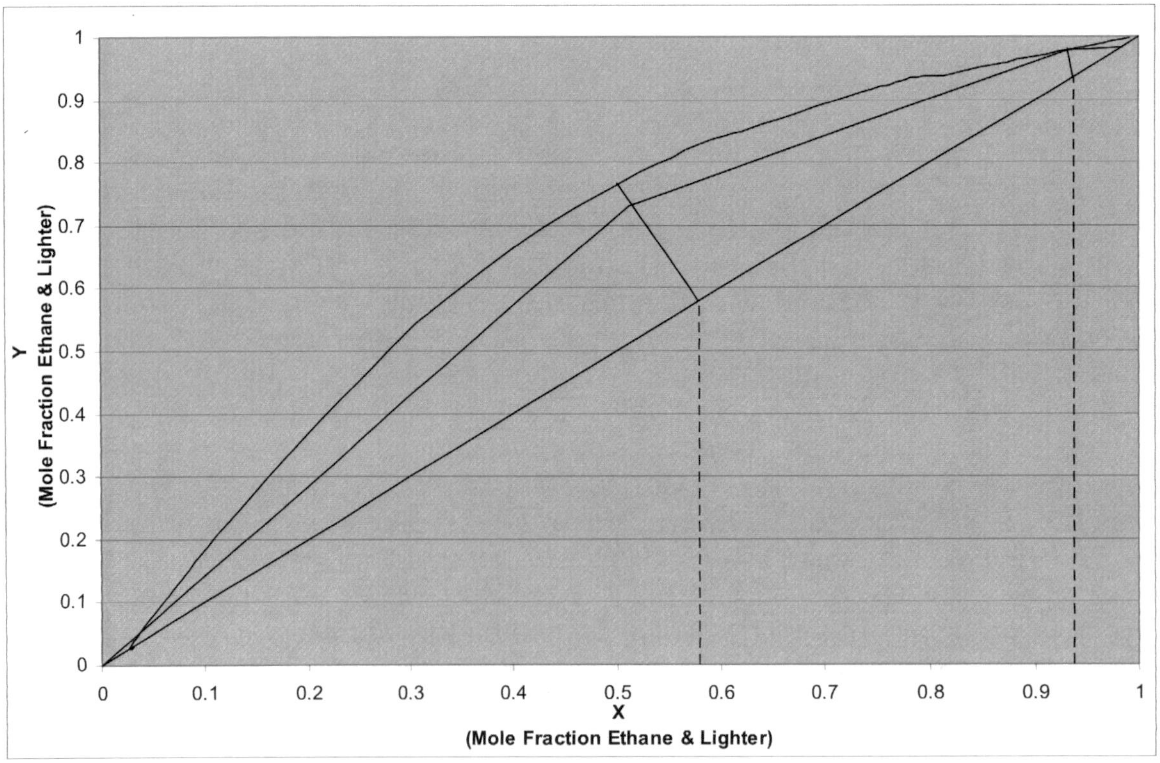

Figure 4: McCabe Thiele Diagram – With Prestripper

The net impact on the propylene refrigeration system is about even. There is an increase in demand at the coldest level of propylene refrigeration, due to the shifting of loads caused by the Liquid Reinjection and Demethanizer Prestripper. There is also a decrease in refrigerant recuperation available from the Demethanizer Reboiler. However, this is offset by the reduction in Deethanizer Condenser requirements, caused by the use of the Demethanizer Prestripper, which provides two feeds that enable easier fractionation.

Conclusions

Replacement of the compressors (CGC, Ethylene & Propylene Refrigeration) along with the introduction of KBR's unique Liquid Dryer, Liquid Reinjection, Demethanizer Prestripper system means that this part of the process will no longer limit the cracking heaters capacity for the different feed slates being processed.

The modifications will also improve the site energy balance significantly, by reducing Charge Gas Compressor power by 2% and Ethylene Refrigeration Compressor power by 10%, which will benefit the utilities systems operating margins.

Paper No. 59e

Improve Butadiene Recovery with the KLP™ Process

Timothy D. Foley, Steven L. Krupa, and Stephen A. McColl
UOP LLC

and

Byung-Goo, Cho
LG Petrochemical Co. Ltd.

Key words: Butadiene, selective hydrogenation

Prepared for Presentation at the
AIChE Spring Meeting
New Orleans, Louisiana
April 2004.

© 2004 UOP LLC and LG Petrochemical Co. Ltd. All rights reserved.
Unpublished.

AIChE shall not be responsible for statements or opinions contained in papers or printed in its publications.

Introduction

Recent improvements to the KLP process, which is a cost-effective alternative to two-stage extractive distillation processes, have reduced the costs for petrochemical producers looking to either expand existing butadiene recovery capacity or install new capacity to meet increasing demand for high purity butadiene. Since acquiring the KLP process licensing and catalyst manufacturing business from Dow in the early 1990s, UOP has continued to improve the process. These improvements have been successfully commercialized.

Increasingly, butadiene producers are lowering the acceptable amount of acetylenes in the product butadiene. Typical product specifications can now be as low as 20 wppm total acetylenes with a maximum of 5 wppm vinylacetylene, which is a lower limit than two-stage extraction processes can achieve. The two-stage extraction processes extract the acetylenes, producing a by-product stream containing approximately 25% acetylene and approximately 75% butadiene. The acetylene content of the waste stream needs to be limited to reduce safety concerns associated with high purity acetylene streams.

As an alternative to the two-stage extraction processes, the KLP process can be coupled with a single-stage extraction process, collectively referred to as a KLP/One-stage extraction process, to yield very high purity butadiene with minimum loss to waste streams. In the KLP process, acetylenes are selectively hydrogenated to the corresponding olefin or di-olefin, and the reactor product contains less than 5 wppm total acetylene. The reactor product can then be fed to a single-stage extraction unit for separation of butadiene from other C_4 hydrocarbons. Of particular benefit is the significant quantity of the acetylenes converted into valuable butadiene product. Therefore, while in the two-stage extraction processes, up to 3-4% of butadiene is lost to waste streams, there is a net production of butadiene in the KLP/One-stage extraction process. Additionally, there is no waste acetylene stream and therefore the safety concerns with handling this stream are eliminated.

UOP has continued to optimize the process, commercializing several improvements recently, all aimed at reducing capital and operating costs. The principal improvements discussed in this paper are the replacement of the carbon burn regeneration system with a rejuvenation system, and the 25% reduction in catalyst requirement by modifying the reactor system.

KLP Chemistry

The KLP-60™ catalyst used in the KLP process is a copper-nickel formulation on a spherical alumina support. The catalyst system operates at low temperature. Hydrogen is added to the process in an amount slightly greater than stoichiometric for conversion of acetylenes. The process is operated at a pressure great enough to ensure the reaction takes place in the liquid phase. The acetylenes are hydrogenated with minimal conversion of butadiene, according to the following primary reactions:

$$CH_3C{\equiv}CH + H_2 \rightarrow CH_3CH{=}CH_2$$
Methyl Acetylene → Propylene

$$C_2H_5C{\equiv}CH + H_2 \rightarrow CH_3CH_2CH{=}CH_2$$
Ethyl Acetylene → 1-Butene

$$CH_2{=}CHC{\equiv}CH + H_2 \rightarrow CH_2{=}CHCH{=}CH_2$$
Vinyl Acetylene → 1,3-Butadiene

The total amount of acetylenes in the reactor product is less than 5 wppm, with a vinylacetylene content of less than 1 ppm. Because vinylacetylene is typically 65-70% of the total acetylene, conversion to butadiene significantly increases the yield of product butadiene.

The by-product of primary concern is green oil, which is formed in the reactors. The formation of this by-product is limited by optimized operating conditions. The green oil is removed from the reactor product by distillation. This green oil is a small fraction of the feed to the unit and is typically recycled to a steam cracker for further conversion.

The following figures illustrate the performance of the KLP-60 catalyst system. These figures contain data from a commercially operating unit and illustrate not just the acetylene conversion and product purity, but also the stability of the system:

Figure 1	Vinylacetylene in the feed and product
Figure 2	Ethylacetylene and methylacetylene in the product
Figure 3	Net production of butadiene across the KLP unit

Process Flow and Improvements

A KLP unit operates with two reactors in parallel, while a third reactor is either being rejuvenated or is in stand-by mode. Each reactor contains 50% of the total catalyst needed for the design conditions of the unit. Previously, two reactors were specified, one operating while the other would be in regeneration or stand-by mode. In the old design, each reactor would contain 100% of the catalyst need for the design conditions of the unit. With the new design, the total amount of catalyst required is reduced, thus reducing the operating cost of the unit.

Feed enters the lead reactor at a controlled temperature where it is mixed with a controlled amount of hydrogen, which is dependent on the concentration of acetylenes in the feed. The product of the lead reactor is temperature adjusted, mixed with hydrogen again, and charged to the lag reactor. The product of the lag reactor is sent to a small fractionator which removes the green oil by-product. The overhead of the fractionator is then sent to the single-stage extraction unit either as a vapor or a liquid.

The reactors are operated in such a way as to carefully limit the temperature rise due to the exothermic hydrogenation reactions. Operating at optimized conditions reduces the production of by-product, but cannot eliminate it. Some of that by-product is deposited on the catalyst in the form of heavier hydrocarbons. Over time, this deposition will degrade catalyst performance. Therefore, periodically, the reactors require rejuvenation. One reactor will be rejuvenated while the other two operate either as the lead or the lag reactor.

Previously, the KLP process employed a carbon-burn regeneration, which required a fired heater and other associated equipment. The current rejuvenation procedure is quicker and more cost effective, producing the same results. The rejuvenation procedure uses a solvent that is commonly available. During the rejuvenation step, the solvent is stripped of the heavy material and recovered and recycled.

KLP Process Example

The KLP process can be used in a revamp scenario to increase the capacity of an existing two-stage extraction unit, or it can be used in a grass roots scenario. In either case, the KLP process offers a cost effective solution for the production of butadiene.

Grass Root Example

Compare with a conventional two-stage extraction unit, a KLP/One-stage extraction unit offers significant advantages for the production of butadiene. A simplified material balance comparing the operation of the two alternatives is provided in Table 1.

Table 1
Performance Comparison
Two-stage Extraction Unit v. KLP/One-stage Extraction Unit

	Two-stage Extraction	KLP/One-stage Extraction
	MTA	MTA
Total C_4 Feed	200,000	200,000
Acetylenes	2,000	2,000
Butadiene	**92,500**	**92,500**
Other C_4s	105,500	105,500
KLP Reactor Product Butadiene		94,000

Butadiene loss in reject streams and fractionation	**(2,500)**	**(1,000)**
Net Butadiene Product	**90,000**	**93,000**
Total reject stream loss	5,800	600
Raffinate Product	104,200	106,500

It can be seen that for the same feed rate, the KLP/One-stage extraction unit produces 3% more butadiene product rate than the two-stage extraction unit. Additionally, the reject streams are dramatically reduced.

The economics of the two alternatives are estimated in Table 2:

Table 2
Economic Comparison
Two-stage Extraction Unit vs. KLP/One-stage Extraction Unit

		Two-stage Extraction	KLP/One-stage Extraction
Fixed Investment, ISBL $US MM		28	26
Total Investment Cost (TIC), $US MM		40	37
		$ MM / year	
Butadiene Product Value,	$420/MT	37.8	39.1
Raffinate Value,	$200/MT	20.8	21.3
Raw Materials Cost,	$180/MT	36.0	36.0
Utilities		2.7	2.4
Other Operating Costs		3.6	3.4
Depreciation		3.3	3.0
Margin		13.0	15.6
Pretax ROI, %		32.5	42.2
Simple Payback, years		3.1	2.4

Table 2 shows that the KLP/One-stage extraction unit provides a significant advantage over the two-stage extraction unit. This advantage increases significantly as the value of butadiene varies from $420/MT, which is an average historical value.

Revamp Case

The KLP process can be used to increase the capacity of existing two-stage extractive distillation units. In this scenario, the two-stage extraction unit is converted into two single-stage extraction units which are then placed downstream of a new KLP unit. Using this technique, the capacity of the existing unit can be increased up to 15-25%, depending on the original design, with minimal changes to the existing extraction unit equipment. The KLP process can be placed upstream of any extraction technology, therefore the application is not limited by previous technology selections.

In the most recent KLP application, LG Petrochemical installed a KLP unit upstream of the existing two-stage extraction unit in their Yeosu, Korea, facility. The KLP unit was of the latest design, incorporating the three reactor system and the latest catalyst rejuvenation system

The goal for the LG Petrochemical project was to increase capacity at minimum cost. As such, equipment was re-used wherever possible. For example, the existing vaporizer was converted into the fractionator for the removal of green oil from the KLP unit reactor product.

While the basic engineering for the KLP unit was completed by UOP, LG Petrochemicals and their contractor provided the basic and detailed engineering for the revamp of the two-stage extraction unit to parallel one-stage units. At this time, for various reasons, LG Petrochemicals are operating only one of the parallel one-stage extraction units. The total project time, from project execution decision to start-up of the unit was 10 months. The KLP unit and the revamped extraction unit came on-stream in July of 2003, meeting the targeted goals:

Capacity Increase	6%
Utility Reduction	10%

Greater capacity increase could be achieved by operating both of the parallel one-stage extraction units. With this option, greater than 15% capacity increase is possible.

The total investment for the LG Petrochemical project, including UOP's license, engineering, and catalyst fees, was $8,000,000.

Conclusions

The KLP process, now in operation in 8 units worldwide, has provided an economic method for the recovery of butadiene from crude C_4 streams. Used in conjunction with any single-stage extraction technology, the KLP process produces the highest purity butadiene product. That technology has now been improved to provide even greater economic benefits. In summary, the advantages of the KLP process are:

- Nearly 100% conversion of acetylenes
- Product acetylene concentration of 5 ppm total acetylenes
- Lower capital and utility requirements than two-stage extractive distillation
- Elimination of concentrated acetylene waste streams
- New catalyst rejuvenation system for reduced cost catalyst activity maintenance
- Reduced catalyst inventory.

Reference: Bipin V. Vora, Bernd Heida "A New Cost Effective Process for Producing 1,3 Butadiene," 5th EMEA Petrochemicals Technology Conference, Paris, France, 25-26 June 2003

Figure 1
Vinylacetylene in the KLP Unit Feed and Product
Commercial Operation

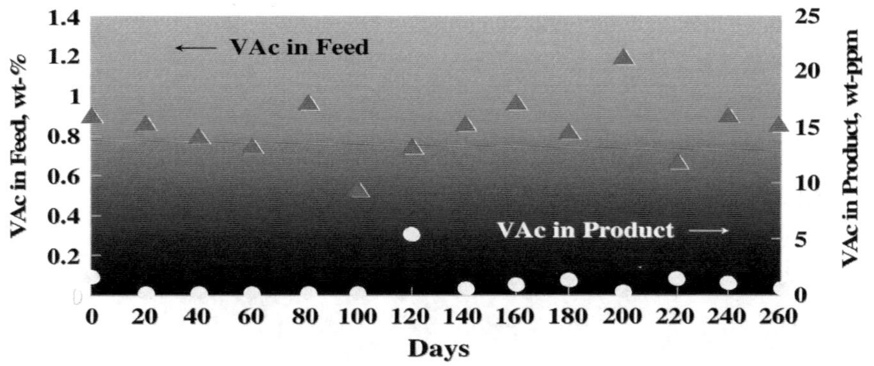

UOP 3706-9

Figure 2
Ethylacetylene and Methylacetylene in the KLP Product
Commercial Operation

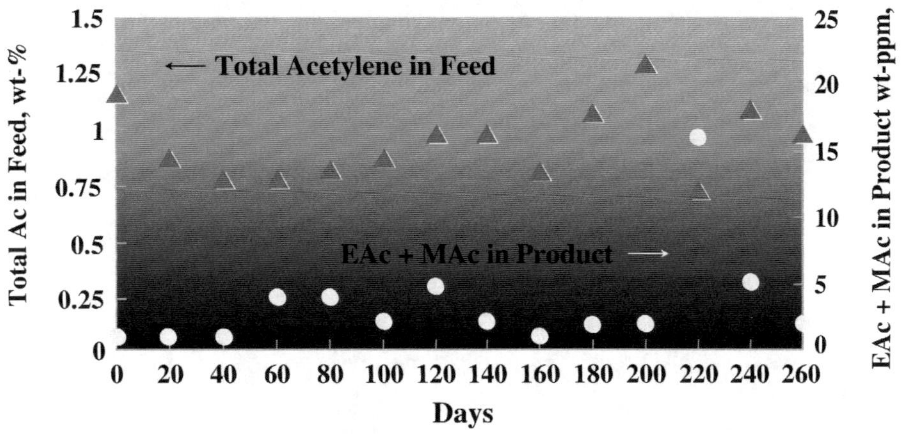

UOP 3706-1(

Figure 3
Net Production of Butadiene in the KLP Process
Commercial Operation

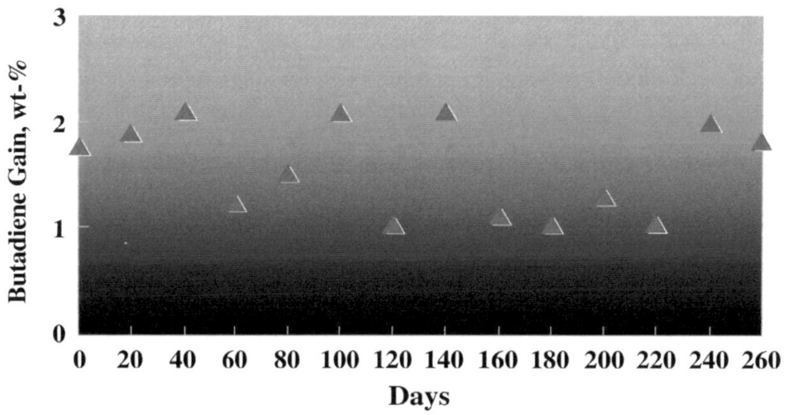

UOP 3706-11

Figure 4
KLP/One-stage Extraction Process

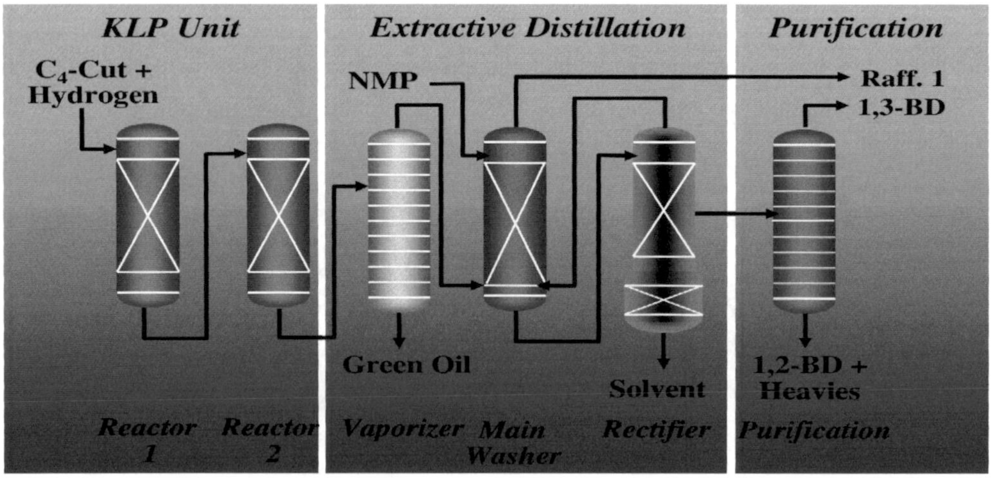

UOP 3897-08

Coking Prediction Capability Enhances Steam-Cracker Economics

By

Philippe Mouillé
BP Lavéra (France)

Brian H. Cooke
BP Grangemouth (United Kingdom)

Erin M. Surdo
BP Naperville (IL, USA)

Odile Josse
Naphtachimie (France)

Nicolas Sumian
BP Lavéra (France)

Bernard Descales
BP Lavéra (France)

1. Introduction

Coking prediction is of primary importance to the olefins business, because coke deposition inside the steam-cracking furnaces tubes requires regular furnace shutdown for de-coking. The lost margin due to lost production for de-coking purposes for the Lavéra (France) 730 ktpa ethylene steam-cracker alone is about 4 millions USD per annum. As a consequence, coking often dictates the operating severity and the choice of feedstocks.

Due to the highly complex physics and chemistry governing coke deposition, the coking rate is generally poorly predicted by state-of-the-art commercial models and, as a consequence, so is the maximum tube metal temperature. To tackle this problem, a pragmatic approach has been developed, based on rigorous modelling using extensive actual plant data.

This paper briefly presents the basis of the model development and focuses on its various applications at different BP olefins sites in Europe and in the USA.

2. Predictive coking model

Although it is not the aim of this paper to enter deeply into the details of the model equations, this paragraph provides an overview of this model. The predictive coking model is a mix of theoretical and semi-empirical modelling, guaranteeing robustness, fitted to actual plant data, guaranteeing accuracy for on-line process control. It is divided into three “elementary bricks”.

The first brick is the coking rate model – which is integrated over time to predict the coke thickness inside the radiant coil. Apart from the furnace operating conditions (hydrocarbon flow rate, dilution steam ratio, coil outlet temperature) and the coke already deposited, the feedstock type and quality are of primary importance to the coking rate prediction.

This information is summarized into a single quality marker called the “coking index”. The coking index is a relative scale reflecting the coking potential of each of the pure components of the feed, as per microcracker laboratory trials. From a gas chromatography analysis of the feed, it is then possible to determine its coking index. Based on gas chromatography and near infra-red (NIR) analysis of various feeds, the coking index has also been modelled in the NIR on-line analyser and can be used as an input for the on-line coking model.

The second brick of the model is the tube metal temperature (TMT) prediction. This is a heat transfer calculation from the incident heat flux through the tube material, the coke layer (as per the coking rate model, integrated over time) and the convective heat transfer on the process side of the coil. An important characteristic of the coke layer for the heat transfer calculation is its thermal conductivity, which has been measured from coke samples obtained from different feed types (ethane, C4s, naphtha) on several plant furnaces.

The third brick of the model is the pressure drop calculation, which is especially useful for Venturi Ratio (VR)-constrained furnaces. Based on the coke layer thickness (again as per the coking rate model), classical chemical engineering equations provide an estimation of the pressure drop in the radiant coil, hence a prediction of the VR.

Altogether, the first model including the three components was tuned on a large amount of actual data from the Lavéra plant: 80 runs from 2000-2001 and 120 runs from 1996-2000 were used for calibration and validation respectively, covering four largely different furnace technologies and feedstocks such as recycle ethane, fresh and recycle C4s and a wide range of naphthas.

Slight modifications on a limited set of tuning parameters were then required to adapt the model to furnaces located at other BP sites, based on only little particular plant data: Grangemouth (United Kingdom), where four different types of furnaces are cracking gas, C4s and naphtha; Chocolate Bayou (TX, USA), where three different types of furnaces are cracking ethane-rich and propane-rich mixtures as well as light liquid streams.

3. Coking model applications

This paragraph presents the main applications of the predictive coking model over the last two years at BP olefins sites around the world. Day-to-day usage includes on-line process control of TMT and monitoring of furnace tube life. Occasionally the model is also used over a few months to evaluate the performance of anti-coking technologies.

3.1. On-line process control

The application of the predictive coking model to on-line process control has been set up at the Lavéra plant. It addresses two crucial issues for operating staff.

The first one is the continuous prediction of the coke thickness inside the radiant coil. This is especially important since, for safety reasons, the operating envelope of the furnaces defines a maximum allowed coke thickness. Indeed, if the coke layer is too thick, there is a significant risk of tube rupture in the event of a thermal shock linked to abnormal operation because of different dilatation behavior between the tube material and the coke. Picture 1 hereafter shows an example of such a catastrophic rupture.

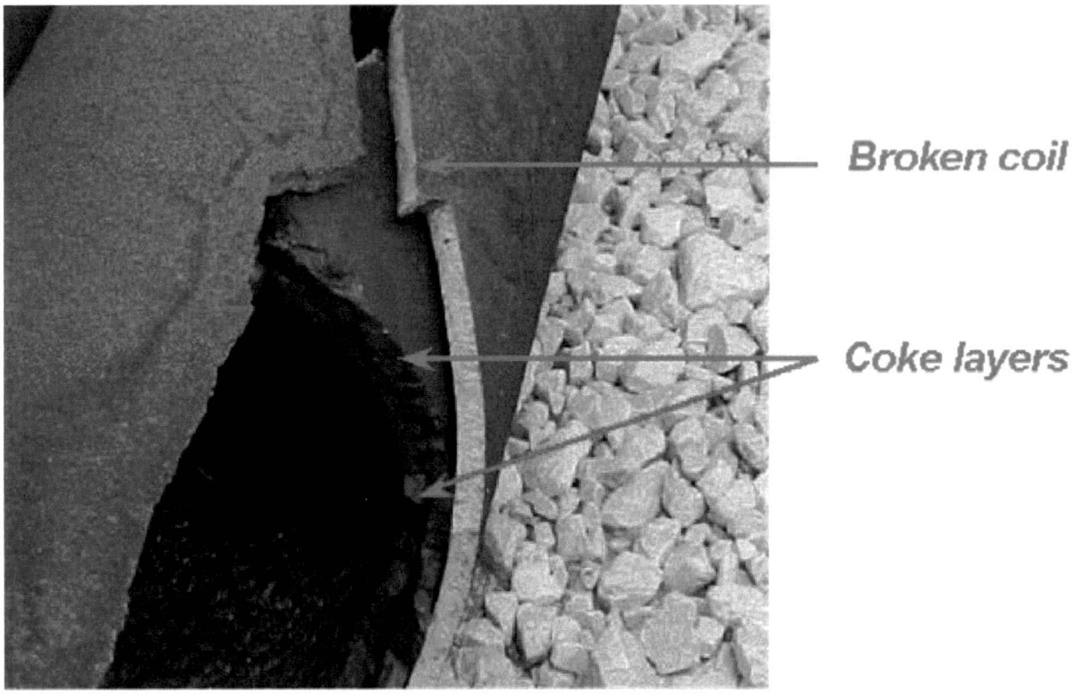

Picture 1 – Coke layers inside a broken radiant coil after abnormal furnace operation

Using the coking model to back calculate the coke thickness from TMT measurements and current furnace operating parameters, and updating the prediction over time between two measurements thus improves plant reliability and contributes to asset conservation.

The second area where accurate process control is required is the TMT prediction. Again, the operating envelope defines a maximum allowed TMT for radiant coils. However, the TMT measurements are only discrete in time, typically once every 2-3 days at Lavéra. Between measurements, the TMT model acts as a virtual TMT sensor whose values are used as feedbacks to the control loops of the main furnace settings.

Picture 2 hereafter shows an example of TMT control for a furnace in operation at the Lavéra plant. The maximum allowed TMT is reached for the first time after about 15 days on-stream[1]. From that moment till the end of the run, the hydrocarbon flow rate and the coil outlet temperature are continuously adjusted according to the model predictions to prevent the TMT from going off-limits, while maintaining severity.

---

[1] For chart clarity and confidentiality reasons, a constant offset has been applied to the TMT values, but scale is preserved.

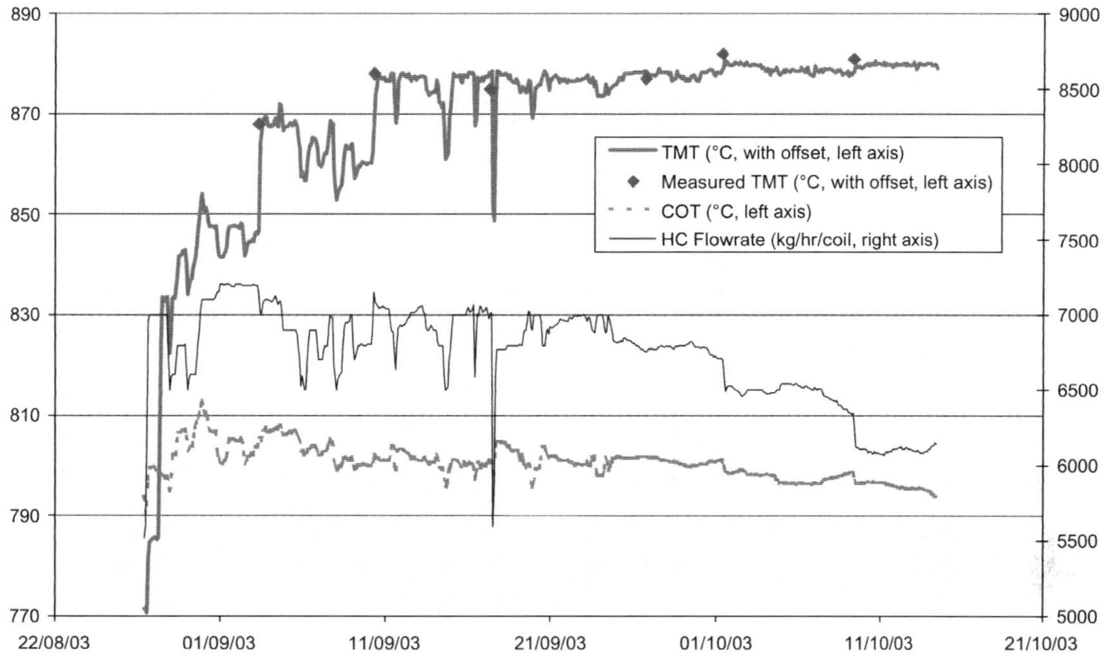

Picture 2 – On-line control of tube metal temperature for a Lavéra furnace

Taking into account the time spent by the plant under furnace constraint and the time spent by the furnaces at or close to their TMT constraint, the estimated profit associated to the process control application of the coking model is about 0.5 million USD per annum.

Picture 3 hereafter shows the evolution of measured and predicted VR over time for a furnace that can sometimes hit that constraint. The increase in coke thickness during the run length increases the pressure drop and hence the VR. However picture 3 shows that decreasing the flow rate for constant coil exit pressure also significantly increases the VR. This can result in hitting a VR constraint as the hydrocarbon flow rate is reduced in an attempt to avoid a TMT constraint.

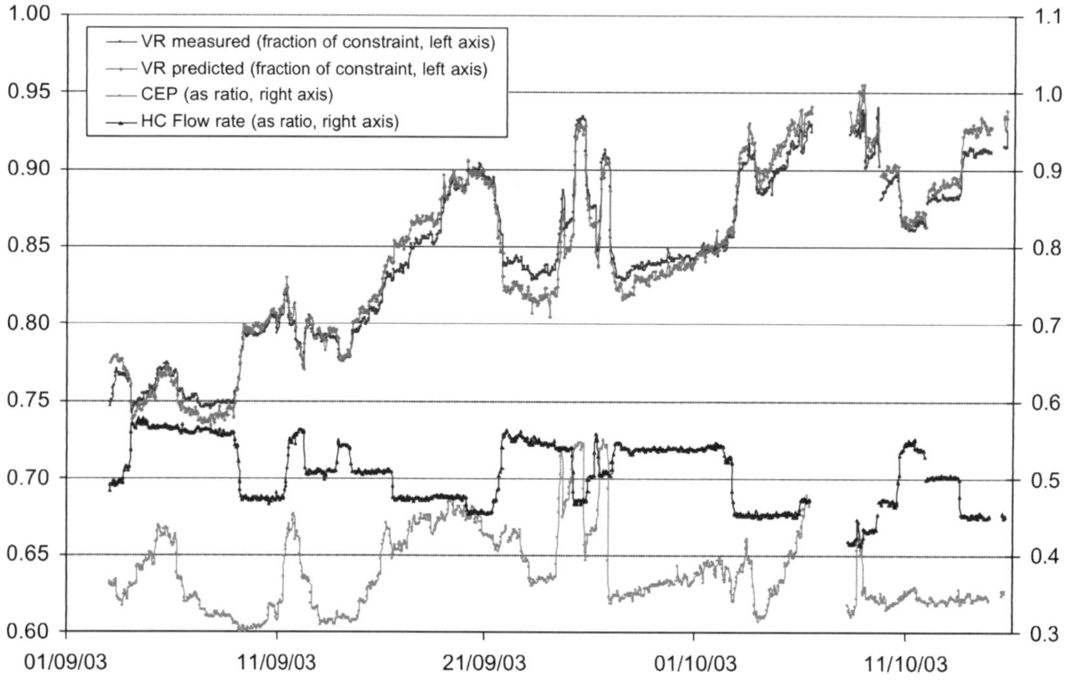

Picture 3 – Measured and predicted VR for a Grangemouth furnace

3.2. Estimation of furnace coil life

Furnace radiant tube life is a function of tube material and operating conditions and can be either shortened or preserved according to how well the carburisation and creep phenomena are managed.

For each type of material used in the manufacture of radiant coils, the Larson-Miller equations provide an estimation of the number of hours that the material can bear in terms of operating temperature at different stress constraints.

From those equations, specifying furnace operating conditions and using the predictive coking model to evaluate the TMT – considered as the operating temperature – it is possible to model furnace tube life and take it into account in off-line studies aiming at assessing various operating policies.

Using actual plant data and the TMT calculated by the model, it is also possible to monitor on a weekly basis the consumption of furnaces tube life over time and identify with the maintenance teams which furnaces are the most fragile and should possibly undergo early maintenance. This monitoring is especially relevant when a furnace trips for a short period of time with no damage. The coil life model is able to estimate how much of the coil lifetime the incident has consumed.

3.3. Assessment of new anti-coking technologies

Coking is the number one steam-cracking constraint. As a consequence, more and more companies are offering different solutions to try and reduce the coke formation in radiant coils of pyrolysis furnaces: original coil geometries, additives and coatings designed either to reduce the catalytic activity of the inner coil surface or to ease the coke gasification.

In the past years, BP has been trying at least two such new technologies at two different sites. In both cases, the predictive coking model has been used as reflect of the baseline situation, i.e. when no anti-coking solution is in place. The model then comes as a substitute for a "twin" anti-coking technology free control furnace that would have to be run at exactly the same settings as that under new technology trial.

The first anti-coking technology, named "technology A" in the sequel of this document, has been tried on an ethane recycle furnace. Vendors claimed that run length could be increased at same ethane conversion, or that conversion could be increased at same run length. This basically meant that, at given operating conditions, TMTs were expected to be lower than before installation of technology A on the furnace.

Both charts on picture 4 hereafter show two typical TMT profiles predicted for the furnace equipped with technology A[2]. The discrete plots correspond to TMT measurements.

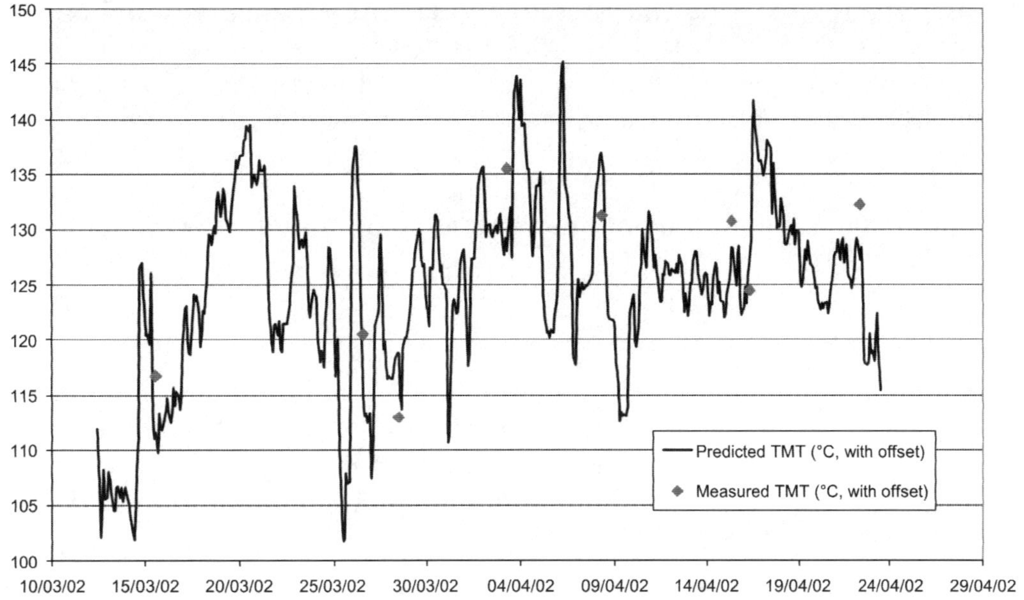

[2] For confidentiality reasons, the same constant offset has been applied to the measured and predicted TMT values, but the scale is preserved.

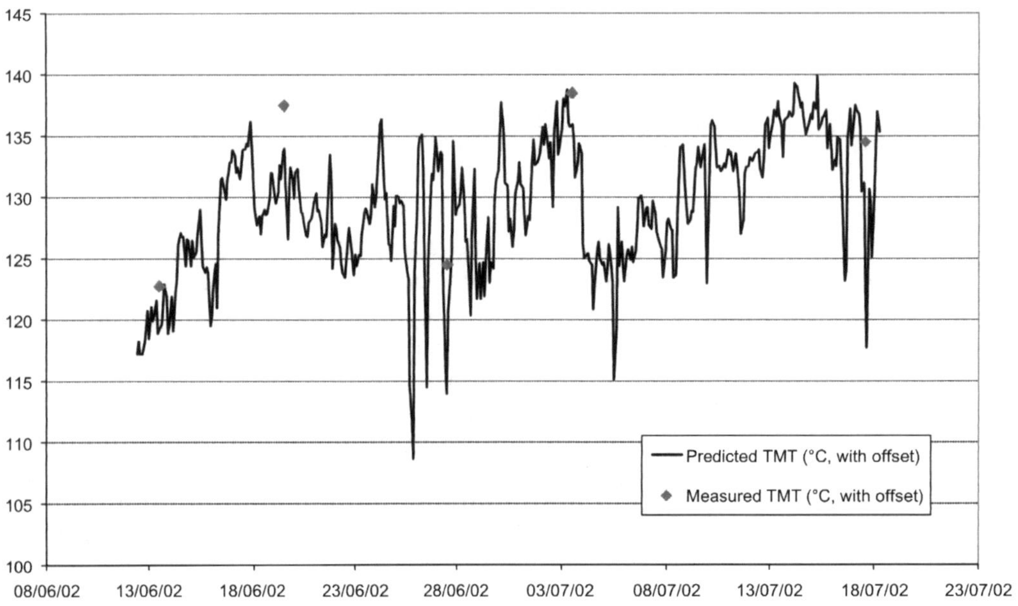

Picture 4 – Monitoring of measured and predicted TMT against time for two ethane runs on a furnace equipped with technology A

Those two runs are representative of what has been observed on several runs over more than one year on this ethane recycle furnace equipped with anti-coking technology A. The measured TMT almost perfectly matches the model prediction which itself reflects the operation of a furnace not equipped with that technology. The measured TMT has never been significantly and consistently lower than the predicted one at same run lengths and conversions than prior to technology A installation. It was therefore concluded that technology A had not matched expectations and installation of that technology on other furnaces was discontinued.

The second anti-coking technology under trial at BP, named "technology B", has been tried on a gas-cracking furnace.

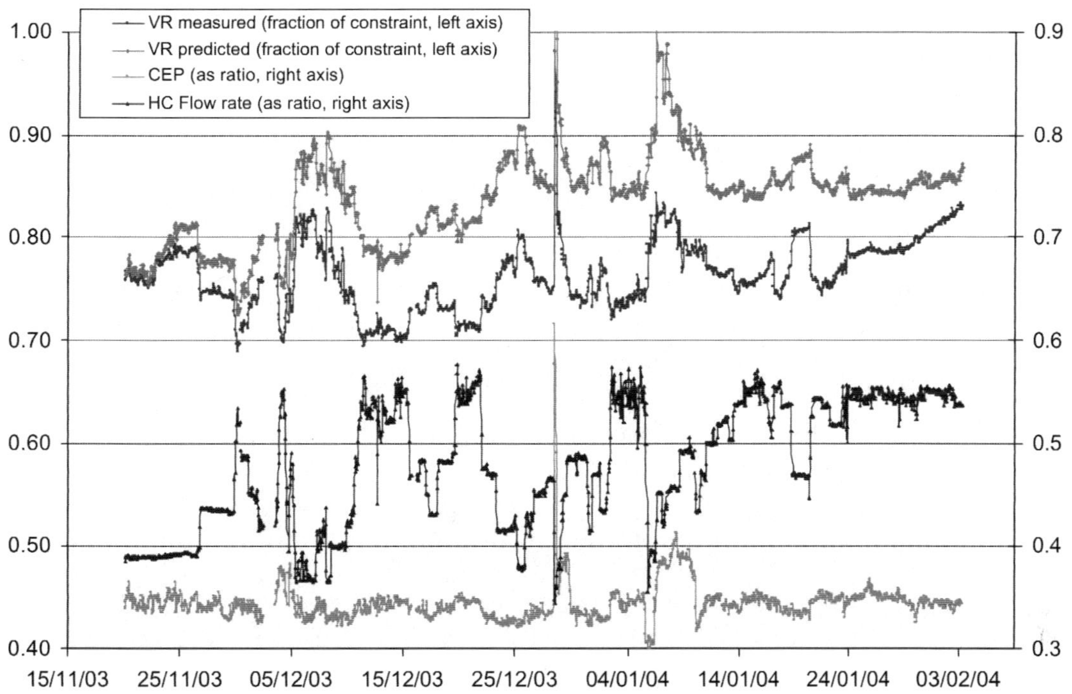

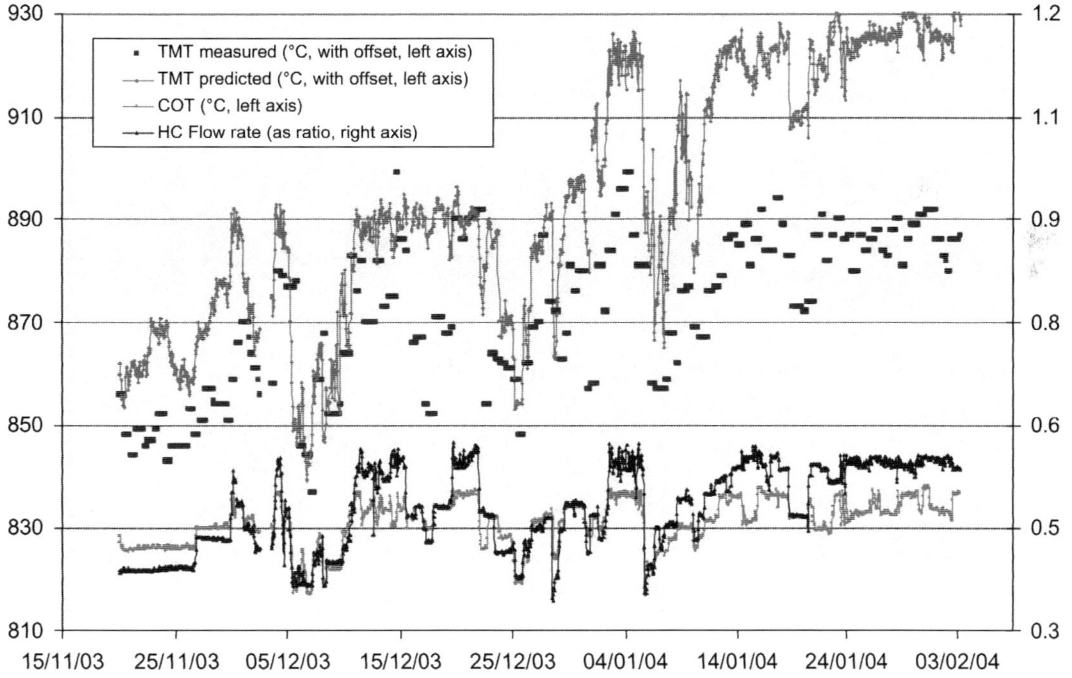

Picture 5 – Monitoring of measured and predicted TMT and VR against time for a furnace equipped with technology B

The runs shown in picture 5 are typical for furnaces equipped with technology B where measured VRs and TMTs are significantly and consistently lower than the model predicted values for furnaces not equipped with technology B. The lower TMTs for similar operating conditions mean that the furnace run length is typically doubled, and in this case the furnace was de-coked without reaching a VR or TMT constraint.

Acknowledgements

The authors are grateful to the Analytical teams at various BP locations around the world for their experimental support, to the plant support people who extensively shot the furnaces for collection of TMT data at the time of model tuning, and finally to the plant staff for their confidence in the on-line application of the predictive coking model.

AUTHOR INDEX

AUTHOR INDEX

AUTHOR INDEX

American Institute of Chemical Engineers

16th Topical Conference on Ethylene Producers 2004

Held at the 2004 AIChE Spring National Meeting

April 25-29, 2004
New Orleans, Louisiana, USA

Volume 2 of 2

Printed from e-media with permission by:

Curran Associates, Inc.
57 Morehouse Lane
Red Hook, NY 12571
www.proceedings.com

ISBN: 978-1-60423-459-6

Some format issues inherent in the e-media version may also appear in this print version.

American Institute of Chemical Engineers

16th Topical Conference on Ethylene Producers
2004

TABLE OF CONTENTS

Volume 1

Volume 2

Sessions 63 and 64 - Ethylene Plant Technology - Furnace Burner Technology Review and Roundtable

Chair: Chris Wallsgrove
Vice Chair: Chell Chellappan

Session 65 - Ethylene Plant Process Control

Chair: R. Donald Bartusiak
Vice Chair: James E Hackney

Session 66 - Ethylene Plant Technology - Advances in Distillation Technology for Ethylene Plants

Chair: Steven Borsos
Vice Chair: Brian Bahr

Paper T8012 A

Ethylene Plant Start-up Strategy for Flaring Minimization

Sasha Vragolic
Huntsman LLC
2701 Spur 136
Port Neches, TX 77651

Prepared for Presentation at the 2004 Spring National Meeting, New Orleans LA, Apr 25-29
16th Annual Ethylene Producers' Conference
Session T8012 – Ethylene Plant Operations

Copyright © Huntsman, LLC
April 2004

UNPUBLISHED

AIChE shall not be responsible for statements or opinions contained in papers or printed in its publications.

Ethylene Plant Start-up Strategy for Flaring Minimization

For many chemical engineers, the aspect of our job that is most difficult and the one for which I was least prepared, is community relations. Its easy for us to understand chemical reactions, they are based on scientific principles and highly predictable. But public reactions and attitudes aren't predictable. The facilities we design, build, operate and manage often become the center of public attention and engineers are expected to help the public understand plant operations and answer their questions.

One of the Huntsman Petrochemical Corporation chemical plants is a feed flexible ethylene plant[1]. This ethylene plant was originally part of a single large complex that included a large oil refinery. Over the years, the residential neighborhoods near the plant have grown. Our neighbors have expressed their desire to know more about our facility and its possible impact on the community.

Working with community and neighborhood leaders has been time consuming but rewarding[2]. We have worked hard to build trust relationships with our neighbors because we understand that having a strong working relationship with the community is vital to our future. One of the greatest environmental concerns expressed by our neighbors has been the ethylene plant's flare. This safety device can produce a huge flame, black smoke, high noise levels and noticeable vibrations. By regularly meeting with our neighbors, we have taken steps to help demystify flares. Because of our sensitivity to our neighbors' concerns we have achieved a wonderful success story that has produced cost savings and substantially reduced the negative community relations impacts associated with flaring during plant startup.

For many years, our management, technical and operating staffs have sought ways to minimize flaring and flaring losses during start-up. In the early-1980s, we installed a waste gas system at our ethylene plant. This system permitted some off-specification material to be routed to furnace feed instead of to the flare. In the nineties, we made control system modifications and added access to purchased hydrogen. Previous attempts at flaring minimization during startups also included the use of natural gas as the startup gas for the cracked gas compressor. Although each of these investments improved plant start-up performance and helped reduce flaring, we felt that there was room for significant improvement. During our 2003 turnaround, we developed a new breakthrough startup strategy that resulted in an 87% reduction in gases flared during the startup. This successful strategy also significantly reduced our emissions and costs.

Ethylene Plant Basics

A simplified version of Huntsman's ethylene process flow diagram is presented in Figure 1. Ethylene is made by the pyrolysis of hydrocarbons at very low pressures and at temperatures that often exceed 1,500 degrees F. This reaction yields a mixture of mostly unsaturated hydrocarbons. This mixture is rapidly cooled, compressed to pressures above 500 psig, chilled, and the liquid fractions distilled. The compression is achieved using a large multi-stage compressor. The nominal size of this compressor, at Huntsman's Port Arthur plant, is 55,000Hp. The chilling and distillation is done to separate: a) Hydrogen, b) Methane, c) A C_2 fraction consisting of acetylene, ethylene and ethane, and d) A higher molecular weight fraction. As part of this processing, the cracked gas is chilled to temperatures below -200 degrees F. This temperature extreme is necessary to achieve the desired hydrogen-methane separation. The chilling is accomplished through the use of a propylene refrigeration system, an ethylene refrigeration system, and through Joule-Thompson expansion of methane. The heat-integrated set of heat exchangers that are used to chill the cracked gas is often referred to as the "cold train." The C_2 fraction, exiting as the overhead of the deethanizer, flows through a set of C_2 reactors where acetylene is selectively hydrogenated to ethylene. This stream is then fed to the C_2 fractionator for the separation of ethylene and ethane.

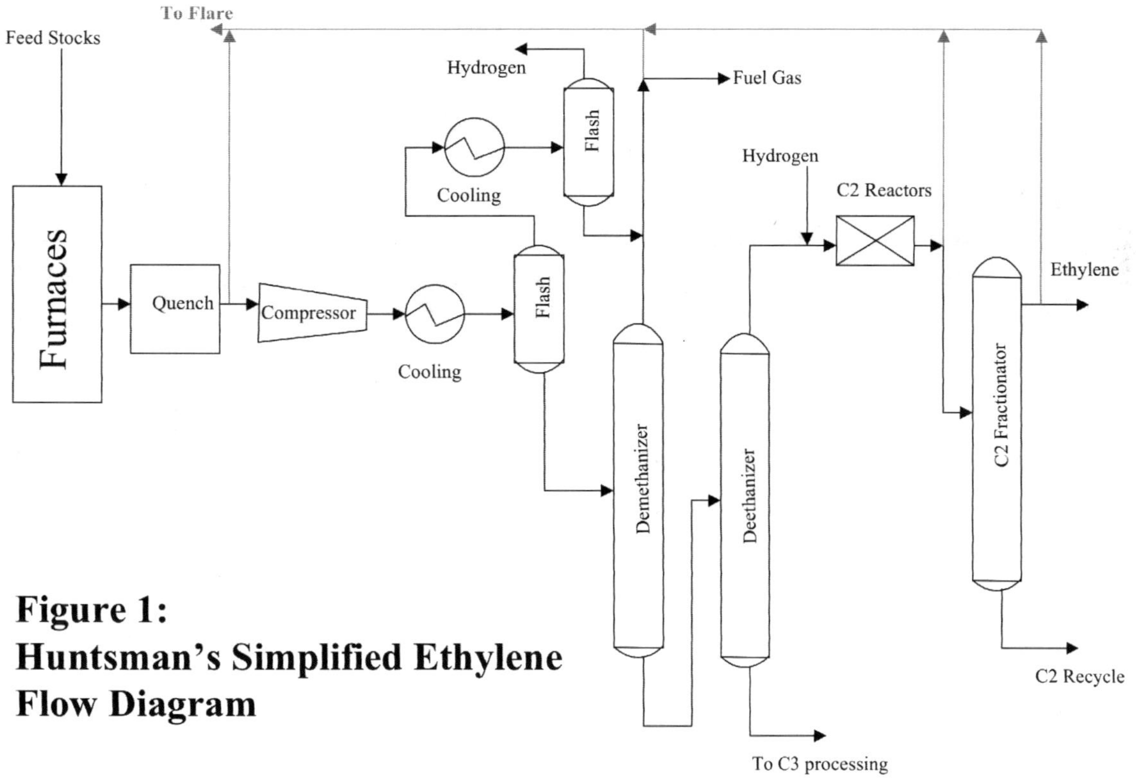

Figure 1:
Huntsman's Simplified Ethylene Flow Diagram

Start-Up Challenges

The primary start-up challenges faced by the technical and operating staffs of ethylene facilities are: a) Providing adequate cracked gas compressor feed; b) Cooling the cracked gas at a controlled rate; c) Removing the acetylene from the C2 fraction; and, d) Ensuring that the final product meets all customer specifications. Each of these start-up challenges may result in flaring of valuable hydrocarbons and emitting large quantities of the residual combustion by-products into the air[3].

The initial startup of the multi-stage centrifugal cracked gas compressor (CGC) involves slow rolling the compressor followed by rapid acceleration to the minimum governor speed. This rapid acceleration is necessary to rapidly pass through the critical speeds for the compressor. However, a significant volume of cracked gas must be available at the suction of the compressor when this rapid acceleration occurs. Without the gas, the compressor would pull a vacuum on the upstream equipment and result in considerable equipment damage. However, prior to fully commissioning the compressor, the cracked gas has nowhere to go except the flare. In an effort to reduce flaring and save money, ethylene manufacturers typically time the introduction of hydrocarbon feed to the furnaces as close to the startup of the compressor as possible. When feed is introduced, it pressures-up the equipment upstream of the CGC and is usually routed to the flare via CGC suction pressure control. The amount of feed required is tempered by the fact that there are minimum flow controllers on the compressor to provide loading, but even so, a significant amount of feed must be introduced and flared prior to startup.

Once the compressor is running and flow is established to the downstream equipment, the cracked gas must be chilled in a controlled manner. Rapid chilling of the gas will cause the metal equipment and piping to shrink faster than the flange bolts. This rapid temperature change will create leaks. Some of these leaks may be severe enough to require a unit shutdown followed by repairs and restart. Typically, ethylene manufacturers attempt to chill down the cold equipment at a controlled rate to minimize potential emissions. This means that less than normal amounts of the cracked gas are liquefied in the process and, although some of the gas can be used as fuel in the pyrolysis furnaces, initially a significant amount of this gas must be disposed of through the flare. The amount of hydrocarbons flared shrinks as the temperatures drop and more of the higher molecular weight hydrocarbons are condensed.

As the C_2 fraction is separated in the deethanizer, it begins flowing through the C_2 reactors. Hydrogen and an adequate reactor temperature are required to hydrogenate the acetylene. Until the acetylene is removed, the C_2 fraction must not be allowed to enter the C_2 fractionator. As a consequence, until the reactors are fully commissioned and all of the acetylene is removed from the stream, the hydrogenated C_2 fraction must be flared. Although hydrogen is a by-product of the ethylene manufacturing process, it takes a significant amount of time to obtain hydrogen of the proper purity for use in this part of the process[4]. In the nineties, Huntsman constructed piping that allows us to purchase high-purity hydrogen and inject it into the C_2 reactor feed. This approach eliminates the time required to obtain hydrogen at the necessary level of purity. As a result, the flaring of the C_2 fraction was reduced to the amount of time it takes to warm up the reactors, to the correct start-of-run temperatures, and to subsequently sweep out the acetylene in the downstream equipment, after the reactors are fully commissioned.

Since very low concentrations of acetylene are typically specified by polyethylene manufacturers and the removal of acetylene from downstream equipment exhibits exponential decay behavior, the acetylene removal step of the startup process can account for a significant portion of the total gas flared. Once acetylene has been removed from the C_2 fraction, the stream is routed to the C_2 fractionator. Until a proper composition profile is established in the tower, the distillate that doesn't meet pipeline quality is typically flared.

Huntsman Strategy to Reduce Flaring

Huntsman applied the following startup strategy improvements to save time, money and minimize flaring: 1) Pre-commissioning the C_2 reactors and tower with pipeline ethylene; 2) Using the absolute minimum hydrocarbon feed for initial CGC commissioning; 3) Timing the introduction of feed to the furnaces; 4) Starting up on an ethane-propane (EP) feed; and, 5) Starting up with minimum conversion of the EP feed.

The pre-commissioning of the C_2 system was accomplished by first introducing pipeline propylene into the deethanizer. We filled the bottom sump, as well as the re-boiler circuit, with propylene. Pipeline ethylene was introduced into the deethanizer through startup ethylene piping as shown in Figure 2. The deethanizer condenser and re-boiler were commissioned and reflux was established. Flow from the deethanizer, through the C_2 reactors and into the C_2 fractionator was established. The reactors were rapidly heated to normal operating temperatures. The ethylene vapor entering the C_2 fractionator was condensed and refluxed back into the tower. Once liquid ethylene flow was established in the C_2 fractionator re-boiler, the re-boiler was commissioned. Reflux was increased up to the normal operating reflux rate. The distillate was routed through the product pumps and piping, and back to the deethanizer through the ethylene startup piping. This pre-commissioning allowed the operating staff to prove out the mechanical integrity of all of the equipment in this section of the unit. Leaks that previously would have required a unit shutdown were identified and repaired prior to startup. This approach saved time and money because repairing leaks merely temporarily delayed startup. We didn't have to shutdown the entire unit to make repairs and then once again begin restarting the unit with the accompanying flaring. By having the reactors warmed up and pipeline hydrogen available when feed was introduced, we were assured of no initial acetylene breakthrough and, thereby, eliminated the associated flaring of the contaminated C_2 fraction.

Figure 2: C2 Section Pre-Commissioning

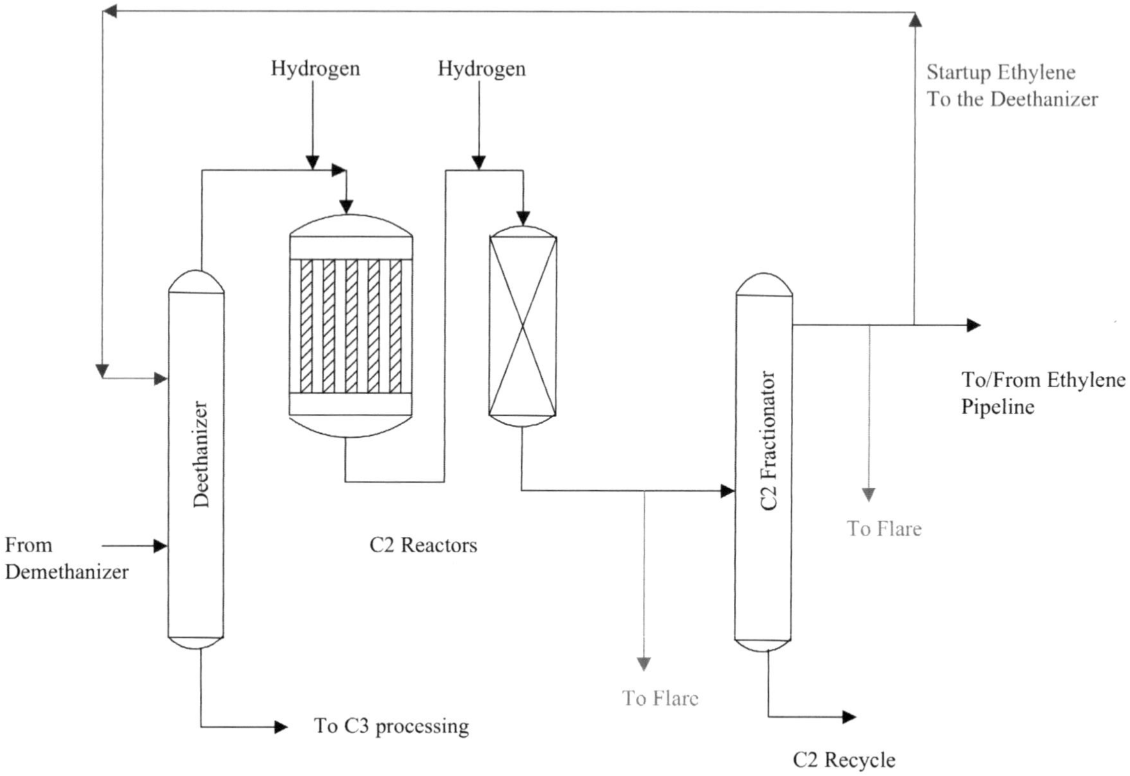

Using the absolute minimum feed prior to the commissioning of the CGC, also resulted in reduced flaring. This is because all of the gas that is used in this step is eventually incinerated by the flare. We were able to startup the CGC with only 25% of the furnace capacity. As previously stated, timing the feed introduction as close to the compressor acceleration is crucial to minimizing flaring. In order to determine the timing of feed introduction, the volume of major piping and vessels was estimated and dynamic simulations performed. Since this is a low- pressure system, where pressure drop must be minimized, the equipment is large diameter and therefore, high volume. The main equipment is shown in Figure 3. In the figure, N refers to the number of cracking furnaces, which are the main chemical reactors for the facility. This is where the pyrolysis reaction occurs. It is a minor point but when choosing furnaces for startup, it is preferred to choose the furnaces on the extreme ends of the feed and effluent piping to displace the maximum amount of inert gas from the system.

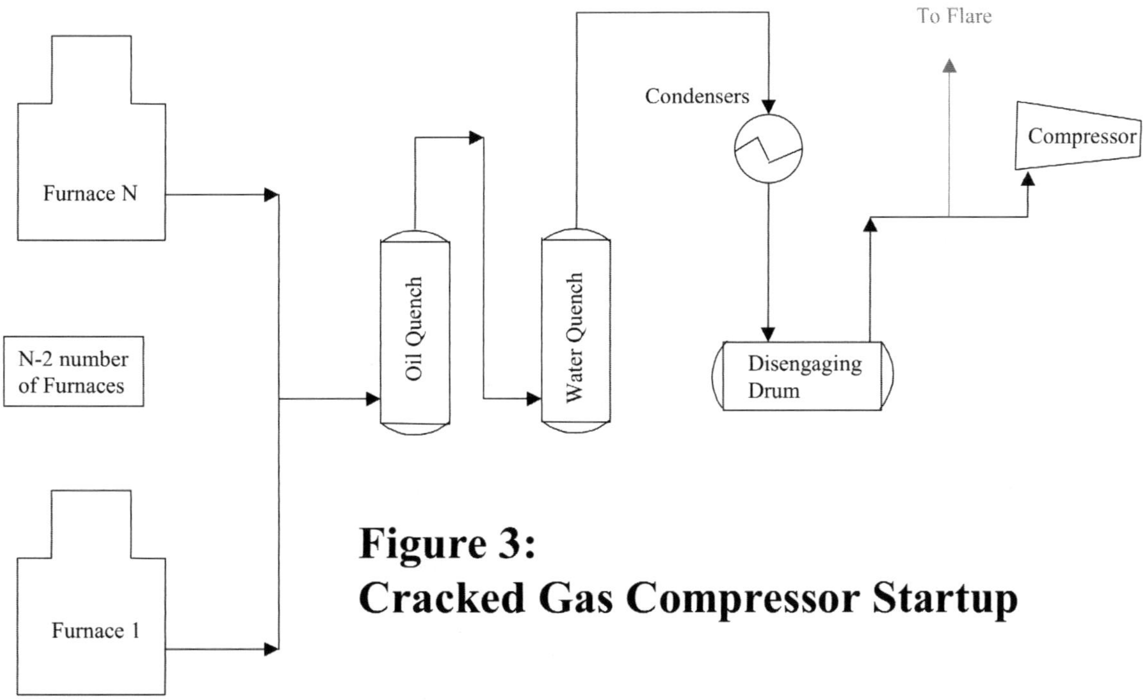

Figure 3: Cracked Gas Compressor Startup

Aspen Dynamics[5] was used to calculate the suction pressure rise for a feed introduction profile. This profile was used to determine the timing of feed introduction so that no flaring occurred. However, practical considerations to ensure equipment integrity and personnel safety were taken into consideration, so the timing was slightly extended. This resulted in some flaring but dramatically less than experienced during this step of past startup efforts. With EP feed and minimum conversion (20%) there were minimal coking consequences to allowing the furnace coil outlet pressure and, therefore, CGC suction pressure to rise close to the pressure relief valve (PRV) settings for the upstream equipment. Huntsman allowed the pressure to rise to within 20% of the PRV settings. This minimized flaring and also ensured a greater quantity of gas available for CGC startup.

The CGC was accelerated to minimum governor speed and fully commissioned. As the CGC began operating it surged a few times; but with adequate minimum flow, it stopped surging without additional furnaces being commissioned.

Initially, none of the ethylene refrigerant users (chillers) were commissioned. Only propylene refrigerant users were required to lower the cracked gas temperature. By starting with low conversion EP, we were able to have less hydrogen and methane in the cracked gas. The resulting mixture had a higher liquid fraction at these relatively high temperatures. As a result, a greater percentage of the gas was liquefied. The majority of the remaining gas was used as fuel gas in the furnaces. As the temperature in the cold train was lowered, EP conversion was increased and additional furnaces were commissioned. This balancing of EP conversion and cold train temperature, to minimize excess gas, resulted in significant flaring reductions.

Using a furnace coil design common to Huntsman's facility we simulated furnace yields at various conversions[6]. Flash calculations were performed on the effluent stream at various temperatures, using AspenTech's AspenPlus[7] simulator, to determine the liquid fraction as a function of conversion and the demethanizer feed flash drum temperature. This data is shown in Figure 4 and was used to determine the appropriate EP conversion during each phase of the startup.

Figure 4: Liquid Fraction as a Function of Temperature and Conversion

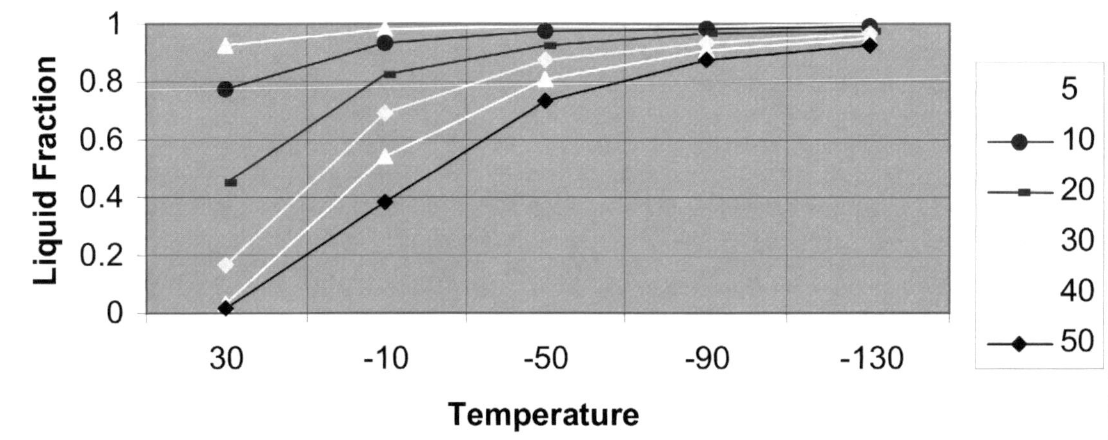

Once the CGC was commissioned and liquid feed to the demethanizer established, a low flow of hydrogen was fed to the C_2 reactors. Once the C_2 fraction flow doubled from the initial startup ethylene rate, ethylene flow to the deethanizer was blocked. From this point, all produced ethylene flowed directly into the pipeline. Since the product section had already been commissioned, all of the existing on-line analyzers, that assure product quality, were already functioning properly. The ethylene product had met all customer specifications and there was no need to flare any off-specification product during this final phase of the startup.

Conclusion

Although several built-in precautions and unforeseen events did result in some flaring, those instances will be analyzed and solutions will be incorporated in future startups. However, even with these precautions and miscues, flaring was reduced by 87% and startup time was reduced. In addition, several potential problems that would have normally forced a shutdown in the C_2 section were identified prior to startup and fixed.

We recently presented our new strategy and the remarkable results to the Citizen's Advisory Panel. Although they may not have fully understood all of the technology, they certainly appreciated our effort and the results. Most of our near neighbors expressed that they hadn't even realized that we had re-started the ethylene unit because of the limited flaring. This new strategy was more than an economic success; it was a success that was appreciated by our fence-line neighbors.

Footnotes

[1] The plant was designed by Stone & Webster and constructed in the late 1970's. It was modified by ABB Lummus-Crest in the late 1980's. Its current nominal capacity is 1.5 billion pounds of ethylene per year.

[2] For years, Huntsman has worked with a Community Advisory Panel composed of close neighbors to our plants in Port Arthur and Port Neches. In 2001, Huntsman was one of initial members of the Port Arthur Industrial Group composed of leaders in the city, neighborhoods and industries in Port Arthur.

[3] During the 1997 turn around and inspection of our ethylene unit, we flared typical quantities of start-up hydrocarbons; however, our management challenged the technical and operating staff to find ways to reduce the volume of hydrocarbons sent to the flare. We developed a comprehensive plan designed to reduce hydrocarbon typically sent to the flare system for disposal at each of the four start-up challenge steps.

[4] Some plant designs utilize front-end hydrogenation and avoid this difficulty but with other operational and financial trade-offs.

[5] Aspen Dynamics is a trademark name of a dynamic simulation program licensed from Aspen Technology, Incorporated, of Cambridge MA. Version 11.1 of this software was used for the referenced calculations.

[6] Visual EFPS was used to generate the yields. Visual EFPS is a trademark name of a pyrolysis kinetic simulation program licensed from Technip USA Corporation's Pyrotec division. Version 1.06 of this software was used for the referenced calculations.

[7] Aspen Plus is a trademark name of a process simulation program licensed from Aspen Technology, Incorporated, of Cambridge MA. Version 11.1 of this software was used for the referenced calculations.

About the Author

Sasha Vragolic is the Acting Technology Resources Team leader with Huntsman Petrochemical Corporation's Jefferson County Operations. He has 15 years of experience, mostly in ethylene manufacturing. He has had various assignments in process and process control engineering as well as in operations and simulation and optimization. He has provided technical assistance to four different ethylene plants. He earned a BS ChE from the New Jersey Institute of Technology and a Masters degree in ChE from Lamar University. He is a member of the AIChE and serves as Huntsman's representative on the Process control subcommittee of the Ethylene Producers Committee of AIChE. He can be contacted at sasha_vragolic@huntsman.com or (409) 723-3502.

EROSION/CORROSION FAILURE OF COMPRESSOR INTERSTAGE COOLERS IN AN ETHYLENE PLANT

Luiza M. Jorge [1]*, José Antônio C. Cunha [2], George Nowowiejski [3]

COPESUL – Companhia Petroquímica do Sul
(Unidade de Engenharia – Time de Processo)
BR386 - Km. 419 - Triunfo – RS - CEP 95853-000 - Brasil

---

1 – INTRODUCTION

COPESUL have two ethylene plants located in the South Petrochemical Complex, Triunfo, Brazil. The first plant was commissioned in the late 70's, and the second plant which has a capacity of 450,000 MTA was started up in June of 1999. This plant experienced tube failures on the fourth stage cracked gas interstage coolers (equipment name - 112P04A/B) in November 2000 and in June 2001. The plant design includes a 5 stage cracked gas compressor with a front end depropanizer located on the discharge of the fourth stage. Water injection is used on stages 1 through 4 to control discharge temperature and fouling.

This is not a common problem faced in ethylene plants and Stone Webster was contacted in July 2001 to help COPESUL determine the cause of failures and assist in providing a permanent solution.

2 – BACKGROUND

2.1 - SHUTDOWNS

November / December 2000 Shutdown:

- COPESUL experienced a tube leakage in exchanger 112P04A in November 2000, which necessitated a plant shutdown. The leakage was noticed by the presence of hydrocarbons in the cooling water. The tube bundles were replaced in December 2000 (after 18 months in operation).
- General Inspections: There was a granulated solid residue among the tubes in the bundle with a similar consistency as coke and had formations of 20mm diameter "stones" on average;
- Shell 112P04A/B: There was a small amount of process residue, a black powder, similar to the coke on the tubes. The exchanger shell showed some evidence of erosion and corrosion (active metal). In the areas close the baffles it had brightness consistent with a polished material similar in color to the tube bundle described below.
- Tube Bundle 112P04 A/B:

---

[1] COPESUL (luiza@copesul.com.br)

[2] JP Engenharia (jose.antonio@jpessa.com.br)

[3] Formerly with Stone and Webster currently with Equistar (george.nowowiejski@equistar.com.)

*To whom all correspondence should be addressed

- ✓ A process residue was found on the bundle which was similar to coke deposited on the tubes and shell, more concentrated in the lower part of the bundle. However, most of residue was in powder form. In some places this was in the shape of stones up to 20mm in diameter;
- ✓ Tubes of the first rows, close to the entrance of each baffle in the shell, showed severe erosion demonstrated by a bright aspect of the active metal and by the flat and polished surface, contrasting with the remaining of the bundle. This was the area where the tube failure occurred. In these areas no deposit was found as it was probably removed by the high-speed fluid hitting the tubes;
- ✓ At the entrance of the tube side there was little residue deposited from the cooling water, and this did not adversely affect the heat exchanger performance.

- The bundle of these heat exchangers had severe erosion at the shell entrance.
- The third stage CGC intercooler was also inspected and the tube metal loss was slight.

June /July 2001 Shutdown:

- The same hydrocarbon leakage to the cooling water occurred in the fourth CGC intercoolers 112P04A/B.
- General Inspection: The equipment was found to be almost clean, with a small amount of residue at the front head, rear head and tube sheet.
- Shell 112P04 A/B: The shell had severe erosion near the inlet process nozzle for an approximate length of 2 meters. Erosion is in the form of parallel lines in the same flow direction. In several places, there was need for solder filling, to maintain minimum wall thickness.
- Tube Bundle 112P04 A/B: The tubes had a bright and flat aspect, immediately after the opening. The tubes located in the area of the inlet nozzle had severe erosion with several tubes completely failed.
- There was more metal loss in the 112P04B than in the 112P04A, despite symmetrical piping at the entrance of heat exchangers.
- There was some metal loss on the baffles.
- Flow meters (annubar type) in the CGC third, forth and fifth stage were full of coke.

2.2 – MAINTENANCE and REPAIR

December 2000:

- ✓ Tube bundles were replaced in December 2000 (after 18 months in operation).
- ✓ Impingement plates were relocated closer to the bundle.
- ✓ Many tubes, located near the entrance (in the area of the inlet nozzle) where the erosion/corrosion was expected to be the worst, were plugged as a preventive measure against the possibility of further erosion.
- ✓ The modifications did not solve the problem as these tube bundles failed within 7 months.

July 2001:

- ✓ The tube bundles were completely replaced. The impingement plate was changed. Two perforated plates were installed, with holes offset from plate to plate as a way of reducing the inlet gas flow. The new impingement plates had larger dimensions than the old ones, enlarging the protected area. A total of 42 tubes located near the inlet nozzle were plugged, as an additional measure of prevention to the erosion.
- ✓ A new third exchanger of the same design as the existing exchangers to be operated in parallel, for each third and fourth stage was specified as a tentative solution to avoid the chance of another shutdown in a short period of time.
  - ✓ The new coolers would take approximately 1/3 the flow to reduce the velocity at the exchanger inlet
  - ✓ The inlet configuration of the new exchangers was changed to include three rows of triangular rods at the bundle inlet.

✓ Block valves were installed on process inlet and outlet side to allow maintenance in any of the exchangers while the plant remained in operation.

2.3 –INVESTIGATION OF FAILURE

2.3.1. Metallurgical Investigation

a) Copesul Laboratory

Samples of tubes were sent to COPESUL Laboratory and to an outside Lab for tests to identify failure mechanism and to determine if the failure was caused by: erosion, corrosion or combination of both erosion/corrosion. The conclusion from Copesul Lab by SEM (Scanning Electron Microscope) was that the metal loss was due to erosion. The outside lab (UFRGS) concluded that the failure was due to enhanced erosion by particles contained in gas flow.

b) IRIS (Internal Rotary Inspection System) Report

Report by IRIS determines metal loss along of tube and the cause of loss. The report showed:

112P04A/B - 4th stage discharge: Extensive metal loss on the largest number of tubes.
112P03B – 3rd stage discharge: Many tubes with significant metal loss.
112P02B – 2nd stage discharge: Showed some metal loss
112P01B – 1st stage discharge: No noticeable metal loss.

The metal loss was in the area for tubes closest to the process fluid inlet.

2.3.2. Chemical Analysis / Data Collection

Following the exchanger leak in June 2001, Copesul began a comprehensive testing program to determine the cause of the problem.

a) Iron content, Suspended Solid and pH were checked in fourth stage suction and discharge drums, respectively 112V04 and 112V05.

Table 1 – Average results for the period in September 2001

	Suspended Solids, mg/l	Fe, mg/l	pH
112V04	1,2	10	3,6
112V05	9,6	44	4

pH values in the second and third stage suction drums were in the range of 3,5 to 4. pH values in Plant 1 are in the range of 6 and are maintained by amine injection in the cracked gas compression circuit.

b) Vent stream from Polyethylene plants

The effect of the vent stream from Polyethylene plant was tested by stopping this flow. The pH was slightly lowered when this occurred, but this was not the cause of erosion / corrosion problem.

c) BFW Injections to the Cracked Gas Compressor:

Plant 1 uses a closed system for water injection where the water is recycled from the second stage suction drum to low pressure and medium pressure casings and water is recycled from the fifth stage suction and discharge drums to the high pressure casing. An amine injection system maintains the pH at a value in the range of 5 to 6.

Plant 2 uses an open system where boiler feed water is continuously added and excess water leaves the plant via dilution steam blow down. The pH value in Plant 2 is considerably lower than in Plant 1. The boiler feed water was analyzed for conductivity, iron content and pH for 15 days in September 2001. The results were somewhat scattered due to a difficulty in sampling and handling, but in general showed that boiler feed water contaminants were not the source of the corrosion / erosion problem.

d) QWT / Chemical Injection

COPESUL uses caustic injection to QWT. Much of the industry uses amine injection. Both will adequately control pH in circulating quench water. The caustic however has a very low vapor pressure and remains entirely in the quench water, thus the condensate from the cracked gas suction drums will not be buffered. The pH of this condensate will then be acidic due to the presence of H_2S, CO_2 and organic acids. When amines or ammonia is used to control pH in circulating quench water, a portion of these compounds leave the quench system with the cracked gas and condensate in the intercoolers. This raises the pH of the condensate and reduces the corrosion effects of acidic environment.

3 – *RESULTS*

3.1 – EXCHANGER DESIGN REVIEW

S&W checked the design of CGC second, third and fourth stage intercoolers. The second stage intercooler is designed with higher velocities than the fourth stage intercooler, and yet had minimum erosion problems.

The design of these exchangers is consistent with many other designs in this service, but other plants have not experienced similar failures.

Nozzle to shell inlet/impingement velocity and Rho-V- squared were well below TEMA allowable limits for all exchangers

A new third cooler, in parallel, for each third and fourth stage was specified as a solution to minimize the velocity in the shell (as the new cooler would take approximately 1/3 the flow). The start-up of the new third cooler for the fourth stage took place in May 2002.

3.2 - RESULTS OF TEST PROGRAM IMPLEMENTTED

In late September 2001, a test program was implemented to investigate the cause of failures, primarily to determine if the failures were due to erosion or corrosion. The program consisted of the following:

3.2.1. Particulate Filter

A probe was installed on the discharge of the third stage of the cracked gas compressor upstream of the 112-P-03A/B exchangers. The purpose was to determine the solids content of the gas leaving this stage as the downstream coolers had seen quite a bit of metal loss. This was installed in early 2002 and the results after several weeks of sample collection showed an extremely small (3 g of solids) amount of solids. The material was 80% organic as determined by weight loss on combustion at 700° C. These results strongly indicated that erosion by solids in the gas was not the cause for the exchanger tube leaking.

3.2.2. BWF Injection in casing of Cracked Gas Compressor (112-C-01)

A review of the plant historical operating data showed that the discharge temperature of the fourth stage was at or below 75°C for much of the time. Computer simulations indicated that this is approximately the dew point of the gas at the operating pressure. This provided strong evidence that water was present at the compressor discharge which would contribute to the tube failure.

The problem was solved by reducing BFW injection to the fourth stage CGC and controlling the temperature in the range of 78 to 82°C which ensured a condition well above dew point.

As a result iron content in forth stage discharge drum (112-V-05) was reduced from 44 mg/l (September 2001) to 21 mg/l (February 2002). Figure 1 shows this trend.

3.2.3. Polyethylene Plant Vent

The vent from the polyethylene plant was analyzed for a variety of components and bubbled through water to determine the pH to confirm if it had any highly acidic species. The componential analysis indicated about 335 to 1200 ppm weight of oxygenates predominantly acetone, ethylene oxide and Tert-butyl alcohol.

The pH of the stream after contact with water was about 6. The conclusion was that the PE vent aside from producing some coke due to the presence of oxygenates was not a contributor to the erosion/corrosion problem.

The PE vent stream were processed only in Plant 1 after September 2001 as a precautionary measure

3.2.4. Amines injection into fourth stage

The cause for the metal loss from the exchanger tubes was believed to be most likely due to corrosion caused by low pH. In order to confirm this theory, an amine injection package was designed to inject amines into the fourth stage cracked gas compressor discharge stream. This was finally implemented in March 2002. The amount of Amine injection was sufficient to keep pH in fourth stage discharge drum in the range of 5 to 7.

Previously, in September 2001, laboratory analysis showed that the condensate from third and fourth discharge drums had values of acetic acid in the range of 159 a 427 ppm and pH values in the range of 3.3 to 4.6.

Additionally, a program was implemented to analyze condensate from fourth stage suction and discharge drums for pH and iron content. This data was collected from September 2001 to July 2002. This data shows that the amine injection raised the pH and reduced the iron content in the condensate substantially. This data is in Figure 1. The iron content was reduced from 21 mg/l (March 2002) to less than 10 mg/l (July 2002).

Start-up of the new third cooler in May 2002 increased iron content results, but in June 2002 iron content returned to the lower values due to the amine injection and the purging from the system of the loose metal associated with a new piece of equipment and piping.

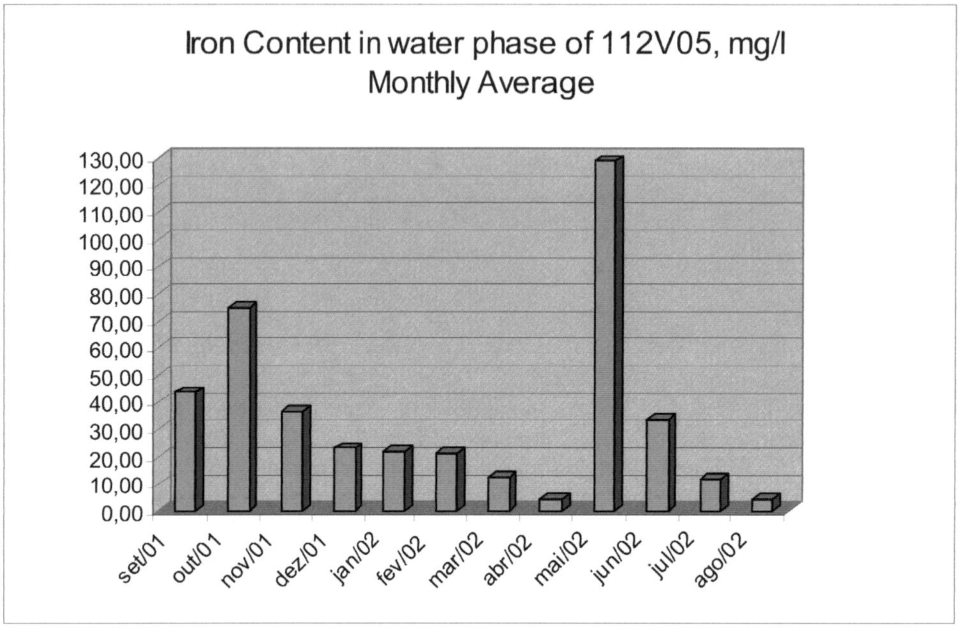

Figure 1 – Presence of Iron in forth stage compressor discharge drum (112-V-05)

4 - CONCLUSION

The general conclusion is that the tube failures in the CGC fourth stage coolers (112-P-04 A/B) were caused by a combination of erosion and corrosion. The low pH provided a mechanism for corrosion and the liquid entrained in the cracked gas rapidly eroded the corrosion layer on the inlet tubes. This appears to be an erosion problem but is in fact a velocity accelerated corrosion problem.

5 - RECOMMENDATIONS

The following recommendations were implemented for long-term successful operation of the plant.

- Maintain amine injection to control pH between 5 - 7 in CGC fourth stage suction and discharge drums (112-V-04 and 112-V-05).

- Periodic sampling of condensate in CGC fourth stage suction and discharge drums for pH and iron content and adjust the amine injection as required.

- Keep BFW injection(s) to the cracked gas compressor so that the temperature at the CGC fourth stage discharge will be maintained in the range of 78° to 82°C.

References

French, Eddie C. (1993). New inhibitor reduces crude-unit corrosion problems at lower pH. Oil & Gas Journal, May, Vol. 24, 45-53.

Troubleshooting a Cooling Water System with a Rigorous Flow and Pressure Network Model

By

André Bernard, Process Design Engineer, NOVA Chemicals (Canada) Ltd
P.O. Box 3060, Sarnia, ON Canada N7T 8C7
Bernara@novachem.com

Abstract

Cooling water is supplied to the entire NOVA Chemicals' Corunna Ethylene facility by a single 16 cells system, circulating 160,000 gallons/minute. Cooling water temperature is a production constraint during hot summer days.

In June 2003, the cooling water supply pressure was observed to increase. Ultrasonic flow measurements and temperature surveys were used as a means of troubleshooting the problem. The results from these surveys suggested that the cause had originated in the olefins quench area. In order to validate the field survey, an existing rigorous flow and pressure network model of the cooling water system was modified to reflect the current quench area cooling water flows and other suspected conditions. The simulated cases indicated that the site was experiencing a widespread fouling problem. All simulated exchangers outside of the quench area had to have an 80% increase in pressure drop in order to match plant conditions. This evaluation of the entire cooling water network allowed the root cause of the pressure increase to be identified as system wide fouling. As a result, the final solution proved to be different than the action plan that had been developed based only on the results from the field surveys.

This paper discusses the methodology that has been used to build the cooling water system model and how it has been used to diagnose the extent of fouling in the entire system. The model includes approximately 145 exchangers in a network that involves common supply and return headers, as well as 2 intermediate headers.

Introduction

NOVA Chemicals Corunna Ethylene facility is integrated with a front end Crude Unit and back end DPG Unit served by a single cooling water system. Since the original start up, new cooling water headers and exchangers were added as a result of plant de-bottlenecks and integration of new process units. The hydraulic distribution of the system has been significantly affected by those changes. As such the operating cooling water supply pressure decreased from the original 65 psig to the current 58 psig with water circulation rate increasing from the original design 145,000 gallons/minute to 160,000 gallons/minute.

System fouling had been experienced in the early days of plant operation due to wood debris coming from the cooling tower fill. Significant reduction of exchanger fouling has been observed over the years through replacement of the fill with fiberglass material. Currently only 10% of the tower still has fill made out of wood. The Cooling water exchangers were found to be relatively clean during the last 2 major plant turnaround (1997 & 2001) with fouling concentrated mostly on the exchangers that are located at the ends of cooling water headers.

Speculations on fouling mechanism on exchangers known to foul, operation desire to increase the cooling water supply pressure back to the original design along with process design need to better understand the system have lead to the development of a cooling water system hydraulic model. When first built, the model provided valuable answers with regards to plant operation as well as becoming an important process design tool. The model was recently used to investigate the reason why a sudden rise in cooling water supply pressure had occurred.

Cooling Water System Description

A single cross-flow cooling tower with 16 cells removes the heat that has been extracted from the process by cooling water. The heat removal process is limited by ambient conditions, tower efficiency and hydraulic capability. Water is made-up to the tower to compensate for evaporation and blow down losses. The water makeup supply comes from the river after clarification. Water is blown down at a rate that ensures good water chemistry in the system. Injection of chemicals into the system ensures protection against bacterial growth, exchanger fouling and corrosion.

A simplified schematic of the cooling water distribution is illustrated on Figure 1. Five operating pumps with one spare ensure the circulation of cooling water in the system. A common supply and return header ensure distribution to the exchangers and collection of the water back to the tower. The Olefins Cold-side area is distinct from the other areas because it has an intermediate temperature header that supplies two surface condensers before joining the site common return header. Chemicals are added at a single location in the cooling tower sump. Water is blown down from the system at the supply header at a rate manually set by operators. Cooling water supply temperature is controlled by a programmable logic controller (PLC), which selects the number of operating fans and their speed (low or high). Normally all fans are operated at their maximum speed during summer conditions where the tower is constrained by ambient conditions.

Figure 1: Cooling Water System

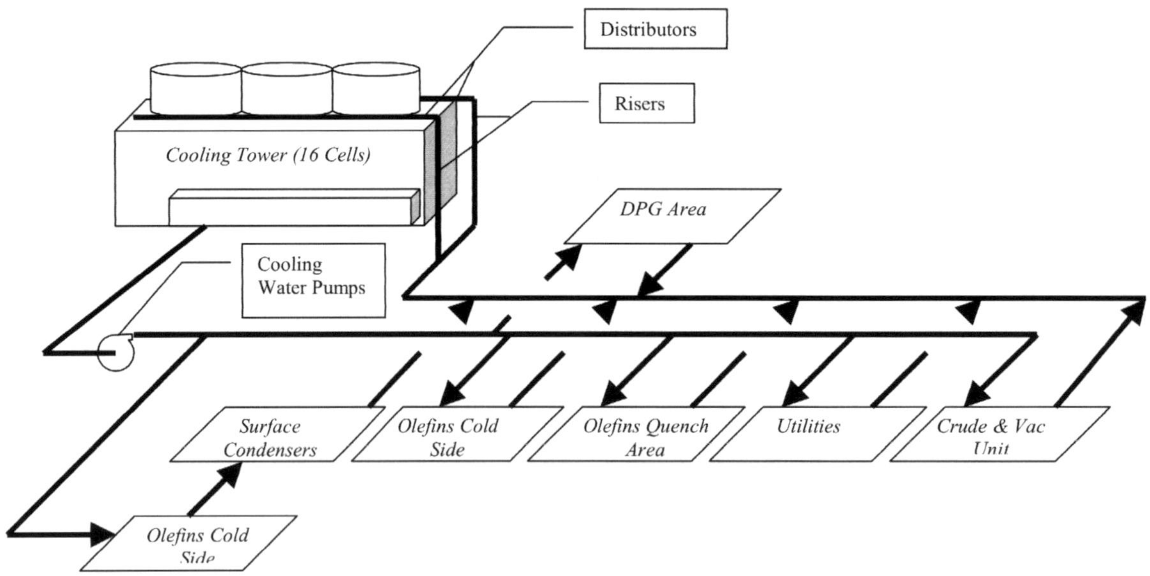

Modeling Methodology

Depending on the size of a cooling water system, modeling the hydraulic network could become a tedious exercise. A review of similar modeling initiatives had indicated that there might be a problem with achieving convergence. Building such a model would also require a significant investment of time and resources. The methodology described below was developed in order to build the model within a 4-month period using limited resources.

The model was built on a commercial software platform that is designed to model looped cooling water systems. Piping isometric configurations, exchanger hydraulic resistance expressed with the resistance coefficient K and pump curves are the typical inputs required to the software. The following simplifications were made in order to accelerate the modeling effort and ease convergence.

- Exchangers with inlet nozzle smaller than 2" were excluded.
- Because of software limitations, 45^0 elbow were modeled with 90^0 elbows.
- Dead-end pipe and minimum flow bypass were excluded.
- The cooling water system was assumed to be clean on the first generation model.
- Exchangers were assumed to operate without being pinched back.

As mention above, heat exchanger pressure drops are based on the resistance coefficient K evaluated at a given equivalent diameter. The following expression provides the relationship between pressure drop and the K value.

$$\Delta P = \frac{8\rho K Q^2}{\pi d^4} \tag{1}$$

Calculating K-value at a given equivalent diameter from exchanger data sheet (as calculated by exchanger software) would have provided consistency between the cooling water model and the exchanger data. However, given the number of exchangers involve and the number of changes expected for process design activities; this methodology would have been impractical and difficult to manage. This led to the development of the following simplified method:

Heat exchanger K-values were calculated from the following equations which only required the number of tubes (N_t), the tube ID (d_{tube}), number of tube pass (N_p), nozzle size (d_{nozzle}) and the equivalent diameter (d_{eq} nozzle diameter used as basis) as inputs on exchanger for which cooling water flows on the tube side.

The term K1 represents friction losses across the tubes, K2 losses to enter and leave the tubes and K3 losses across exchanger inlet and outlet nozzle. The exchanger K value is the sum of the above 3 terms. f in the K1 term represents the friction coefficient, which depends on tube roughness and diameter. The values of f were established from the CRANE Handbook for pipe roughness of 0.00015 (commercial steel pipe).

The correlation was found to predict reasonable exchanger pressure drops, which were however slightly underestimated when compare to those predicted by heat exchanger design and rating software. Differences of 0.5 psi were noticed in some exchangers when comparing the 2 methods. Those were judged to have little impact in evaluating the overall system hydraulic distribution.

Heat Exchanger K value Evaluation

$$\Delta P = \frac{K\rho v^2}{2g_c} \qquad K = fL/d$$

$$K1 = \frac{12 f d_{eq}^4 L N_p^3}{N_t^2 d_{tube}^5}$$

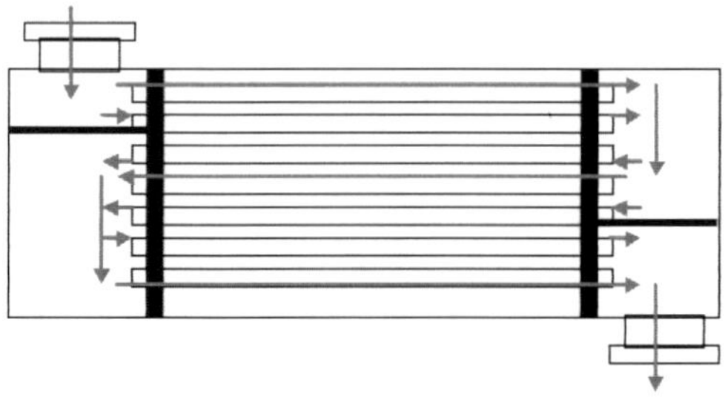

$$K2 = \frac{1.8 N_p^3 d_{eq}^4}{N_t^2 d_{tube}^4}$$

$$K_{Exchanger} = K_1 + K_2 + K_3$$

$$K3 = 1.25\left(d_{eq} / d_{nozzle} \right)^4$$

Given that the number of exchangers that had cooling water on the shell side is limited, it was decided that their K-values would be evaluated from pressure drops that had been calculated by heat exchanger software at the design flow.

There are two possible approaches for modeling the cooling water system hydraulic network. It could either be modeled as one single model or it could be broken down into sub-models, which would then be integrated together. The approach selected should be based on the size of the model. In this case 143 exchangers from 5 different operating areas were modeled. Given the size of the area being modeled, it was broken down into 9 sub-models with 20 different boundary conditions. Breaking down the model enables easier convergence, however it introduces another complication with regards to integrating the sub-models together. A trial and error approach would be tedious and time consuming. A method, which requires 2 iterations, was developed. Each sub-model is first run with an initial estimate of inlet and outlet pressure. The results of the first run are used to evaluate the overall resistance coefficient K of the entire area so it could be simplified to a single exchanger. Those K-values are then input into a site model that includes the cooling water pumps, the main supply and return headers and all sub-models represented by a single exchanger. Sub-models single exchanger inlet and outlet pressure calculated on the site simplified model are re-input in each individual sub-models. Those sub-models are re-run and provide estimated flows for each individual exchanger. The pressure profile across the cooling water system can also be extracted from the model results. The material balance of each sub-model is compared to the simplified model and normally closed within 1 to 2 %.

A Tool for Troubleshooting.

The primary function of the model is for use as a process design tool. However, it has been used on several occasions to assess operating strategies, contingency plans and to perform cooling water system troubleshooting. The model has been used to evaluate the cooling water supply pressure against the number of pumps online in order to determine the critical number of operating pumps that would prevent a cooling water flow-based flare safety interlock being energized. Cooling water flow shedding contingencies were also tested as a means of increasing the margin from the interlock trigger point when operating at or below the critical number of pumps. Overall cooling water circulation rate evaluations with regards to the number of operating pumps were used to evaluate the reaction forces on the cooling tower riser after an incident where one of the tower distribution headers joint had moved and caused a leak. These are some of the examples of how the model has been used to support the plant operation. The following section describes in more detail how the model has recently been used to evaluate the cause of an unexpected increase in cooling water supply pressure.

Problem Description

Cooling water supply is normally ensured by 5 operating centrifugal pumps with one spare (total of 6 pumps). Except for the drivers, which are equally split between electric motor and steam turbines, the pumps are identical and operated in parallel. Throughout the month of June 2003 the cooling water supply pressure ramped up from its normal 58 psig up to 65 psig. It reached a peak of 69 psig then dropped down and stabilized at 64 psig. This increase in pressure occurred approximately 19 months after the last plant turnaround. Flow surveys were done with an ultrasonic portable flow meter on critical exchangers. All flows were equal to, or above, those for the flow survey that was performed after the last plant turnaround, with the exception of the Quench Water exchangers. One of the Quench Water exchangers was suspected to be badly fouled as almost no cooling water flow was detected. The Quench area was using approximately 15% of the overall cooling water and was suspected to be the cause of the increase in pressure. Operations requested that the model be used to validate if this was indeed the cause.

Figure 2: Cooling Water Supply Pressure Trend

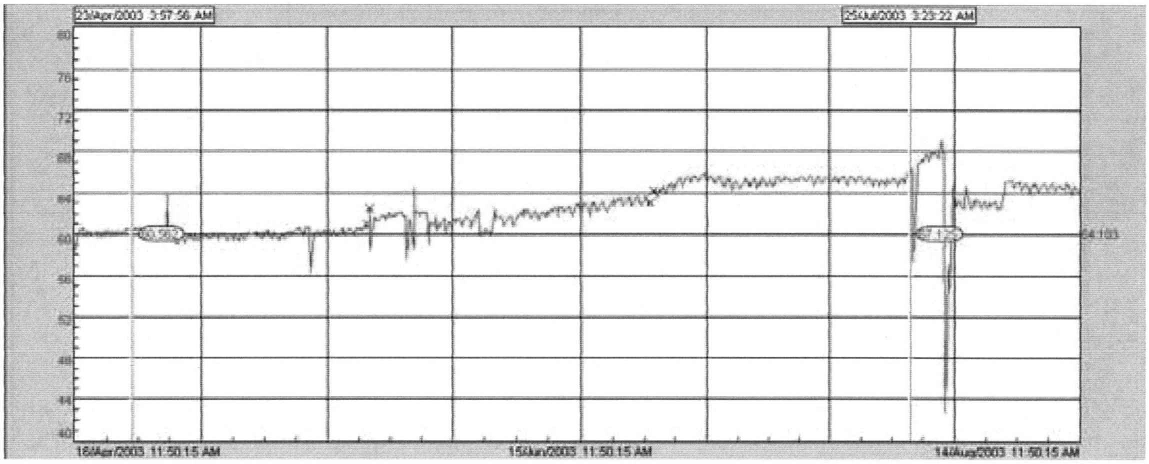

Troubleshooting Activities

Based on the information and field data supplied by Operation, the hydraulic resistance of the Quench area was increased to reflect the observed flow. As such the Kvalue of the single exchanger representing the Quench area on the site simplified model was increased until the flow of the area matched the field survey. The cooling water supply pressure calculated by the model was recorded and compared to the actual operating pressure. Based on the model the extent of fouling in the Quench area would only explain a 1 to 2 psig pressure increase. The impact of the Quench area on the overall circulation was also checked by completely blocking the flow to the area. That case still did not fully explain the system pressure increase (2 psig short). It became evident at that point that the Quench area was not the only source of the problem.

Potential causes of high cooling water supply pressure were laid out to define different case studies as described in Table 1. A validation methodology that was based on the use of the simulation model was developed for each potential cause.

Table 1: Potential Cause & Validation Methodology

Potential Cause	Validation Methodology
Faulty pressure transmitter	Validate the circulation rate with the cooling tower performance as extrapolated from the cooling tower performance curves.
Pump speed set higher then normal	Validate that the circulation rate calculated by the model is consistent with the tower performance.
Other areas fouled	Increase the Kvalues of other areas on the site simplified model by trial and error to match the cooling water supply pressure. Keep the quench area fouling extent as per initial evaluation.
Site Wide Fouling	Assume that the extent of fouling is equally distributed across the entire network by increasing the Kvalue of each area on the site simplified model until the supply pressure matches the actual plant operating value. Keep the Quench area fouling extent as per initial evaluation.

The cooling tower performance was found to be consistent with the calculated circulation flow. However the sensitivity of the calculation is questionable as 8 % reduction in cooling water circulation rate would only cause 0.6 °C difference in cooling water supply temperature. In this case, a field check of the pressure transmitter was the only valid method to confirm that those two causes did not apply.

Limiting the fouling to a limited number of areas (1 or 2) did not appear to explain the problem. As observed on the model, when one or more area fouled the flow redistribute to the clean areas. A small increase in supply pressure was observed but not to the extent of what had occurred in the system.

By considering site-wide fouling, the supply pressure was matched by increasing the Kvalues of each area by 80% of the estimated clean Kvalues while maintaining the Quench area to the initial fouling evaluation. The full model (including all the sub-models) was run with each individual exchanger Kvalue increased by 80%. Exchanger individual flows were then compare to the field flow survey. As shown on Table 2, 60% of the exchangers verified had calculated flows within 20% of the measured flows. The large deviations on some exchangers were explained by higher heat duty being likely to cause high cooling water outlet temperatures that were outside the level of protection offered by the chemical program. Those exchangers would therefore foul at a faster rate and would have more than the simulated 80% increase in hydraulic resistance. Other exchangers were known to operate with pinched cooling water valve. In reality it is unlikely that the exchangers will have exactly the same degree of fouling. In this case it is fair to assume that the bulk of the exchangers exhibit the behavior evaluated by the simulation. Some exchangers will have a higher degree of fouling while others will have less. However, the results support the suspicions of a site-wide fouling problem and, on average, the exchangers exhibit an 80 % increase in hydraulic resistance.

Table 2: Cooling Water Flow Comparison

Exchanger	Field Measurements (usgpm)	Model Predictions (usgpm)	Relative Difference (%)
EA-2118A	3.2	0	N/A
EA-2118F	4200	3455	-17.7
EA-2119A	4068	4014	-1.3
EA-2119B	4544	4050	-10.9
EA-2119C	3514	4075	16.0
EA-2119D	3804	4474	17.6
EA-8102	8903	7232	-18.8
EA-2501A/B	6842	6972	1.9
EA-2501C/D	7027	6923	-1.5
EA-2501E/F	5838	6917	18.5
EA-1008	977	2824	188.9
EA-2203B	2219	2368	6.7
EA-2202A	1320	2141	62.1
EA-2202B	1936	2135	10.2
EA-2202C	2090	2115	1.2
EA-2466	4042	2994	-25.9
EA-2453B	1683	3185	89.2
EA-2425	1746	2563	46.8
EA-2421	1273	1393	9.4
EA-5003	1625	1767	8.8
EA-4004	1585	3109	96.1
EA-3308	581	807	38.8
EA-3317	36	64	78.5
EA-3314	17	13	-21.8
EA-1506	660	371	-43.8
EA-1012	1730	1527	-11.8
EA-1015	1770	2322	31.2

The extent of the problem was also validated by sampling deposits that were in exchangers that could be taken off line without causing a total plant outage. These exchangers were also located in different areas.

Fouling Cause

Once the extent of fouling was established the cause of its occurrence needed to be understood so that proper action could be taken to stop it and if possible cure it before the next plant turnaround. Table 3 summarizes some of the significant changes done on the system since 1998.

Table 3: Cooling Water System Operational Changes

Change	When	Why	Comments
Implement Summer Operating Strategy of 6 vs 5 pumps	Summer of 1998	Supply more pressure to the elevated exchanger in the Crude & Vac Unit	No other significant benefits expected
Automated pH Control	1999	To Optimize Chemical Consumption	
Plant Turnaround	Fall 2001	Normal Plant Turnaround	Majority of the cooling water exchangers inspected & hydro pressure cleaned.
Rebuilt Cooling Tower Splash Boxes & Replace Distribution Valves	Fall 2001	Part of the Plant Turnaround Scope	
5 Pumps operation in the Summer	May 2002	Distribution Header Stand pipe overflow on 6 pumps	
Increase speed of one of the turbine pumps	Summer of 2002	Provide maximum achievable pressure for the Vac unit elevated exchangers	
Leak on distribution header	September 2002	Pressure spike caused during a pump swing had caused excessive reaction forces on the Cooling Tower risers	
Power Failures	July 2000, April 2003 & July 2003		

One of the significant changes implemented in 1998 was the operating strategy of running 6 pumps during the peak of the summer period. The intent at the time was to provide more pressure to elevated exchangers in the Crude and Vacuum unit. Cooling deficiency on the jet ejector system was causing the Vacuum tower to loose vacuum. Operating the additional cooling water pump was found beneficial to that system during the hot summer days. During the 2001 plant turnaround the cooling tower splash boxes and distribution valves were all replaced. This repair had caused a small increase in backpressure on the cooling water system. Even though small, the backpressure would cause the standpipe at the far end of the tower distribution header to overflow when operating 6 pumps. The 6 pumps operation was then no longer possible. To alleviate the problem, Operations had increased the speed on one of the turbine-driven pumps so that the circulation flow would operate at the limit of what the distribution header could handle. This practice had caused an incident on a pump swing where a sudden spike in pressure had cause a joint on the distribution header pipe to move back and cause leak. Reaction forces on the tower risers were evaluated to be excessive for the available support. A recommendation was made not to exceed 5-pump operation and the pump-swing procedure was reinforced. Since this was the first summer since 1997 where only 5 pumps supplied the cooling water system, it was quickly identified as a potential cause of the fouling. Less cooling water circulation would have led to temperature and tube velocities outside of the level of protection offered by the chemical program. However, after thoroughly analyzing the differences between 5 and 6 pumps configuration, this cause no longer seemed to be credible

From process simulation, the additional circulation of water when operating 6 vs 5 pumps was found to have negligible and sometimes slightly negative impacts. The increase circulation rate results in an increase of ½ °C in cooling water supply temperature. In most cases the warmer temperature offsets the benefits of incremental flow to the exchangers. The outlet cooling water temperature measured on the Pyrolysis Gas Compressor (PGC) and the Propylene Refrigeration Compressor (PRC) surface condensers also supports those conclusions. The cooling water effluent of approximately 70% of the exchangers in the olefins area supplies those two surface condensers. It is therefore unlikely that the primary exchangers would have cooling water outlet temperature above those measured at the outlet of the surface condensers, which was within the limit of the chemical program.

The timing of the pH control automation implementation makes it an unlikely cause. Similarly the power failures experienced since 2000 would have cause this problem to occur before.

Indication of a reduction in cooling water cycle which coincided with the time where the pressure rise stabilized suggest that the system may have been operated outside a safe operating range with regards to water chemistry. Even though no strong evidence could be found, the symptoms are consistent with the nature of the problem.

Corrective Actions

Critical exchangers that can be isolated and cleaned without a total plant outage were cleaned with high-pressure water. An acid wash was done on some exchangers that cannot be taken offline. The cooling water cycles were reduced. After implementation of those changes the operating pressure dropped down to 61 psig. A fair number of exchangers may still be significantly fouled and could become bottleneck to plant rates during the next summer season. Contingencies and other potential cleaning methods are being investigated to ensure that the plant runs effectively until its next schedule turnaround.

Conclusion

Monitoring only the flow rates across exchangers is not enough to assess the degree of fouling. To evaluate the incremental hydraulic resistance caused by fouling, both flow rates and pressure drops need to be measured. Unfortunately pressure taps on exchanger inlet and outlet are often unavailable. Pressure gauges do not offer much accuracy either. Furthermore, relying only on field measurements would also be tedious as the analysis has to be done on each individual exchanger. Using a model that represent the flow pressure network of the cooling water system requires a limited number of field measurements. In this case the system supply pressure along with a flow survey done on a limited number of exchangers was enough to perform a detail analyses of the system. The model also provides a detail pressure profile across the system, which could be used to further evaluate individual exchangers. In this case the model was successful in evaluating the extent of the problem and also provided key information in validating the potential causes.

It would be unrealistic to build such a model at a time where a problem occurs. The time require to build a model far exceeds the available response time before the problem becomes untreatable and leads to a plant outage. The model should be built by the process design group as they are likely to be responsible for changes to the system. Operation should then request support from the process design group when face to a similar problem. This investigation demonstrates the synergy that may exist between process design and operation in troubleshooting significant plant problems.

Nomenclature

ρ : Density (lb/ft^3)
ΔP : Pressure Drop (psi)
v: Velocity (ft/sec)
d_{eq}: Equivalent diameter (in)
d_{nozzle}: Exchanger nozzle diameter (in)
d_{tube}: Exchanger inside tube diameter (in)
f: Friction factor (as define in reference 1)
g_c: Conversion factor between lb_f and lb_m = 32.174 ft lb_m /(sec^2 lb_f)
K: Resistance coefficient
K1: Resistance coefficient for flowing across the exchanger tubes.
K2: Resistance coefficient for entering and exiting exchanger tubes.
K3: Resistance coefficient for flowing across the exchanger inlet and outlet nozzle.
$K_{Exchanger}$: Overall exchanger resistance coefficient
L: Tube length (ft)
N_p: Number of tube pass
N_t: Number of tubes
Q: Flow rate (USGPM)

References:

1. Crane Technical Paper No. 410M, 4th Printing, Crane CO. New York, 1982.

2. Nicholas P. Chopey, Tyler G. Hicks, Handbook of Chemical Engineering Calculation, McGraw-Hill, 1984.

3. Warren L. McCabe, Julian C. Smith, Peter Harriott, Unit Operation of Chemical Engineering, 4th Edition, McGraw-Hill, 1985.

4. T.W. Cochran, Simplifying Piping Network Analysis, Chem. Eng. April 1995, pp. 104-106.

5. Jim MacDonald, George Hanks, Cooling Systems Use of the C-Factor for Exchanger Monitoring, Industrial Water Treatment May/June 1993.

Paper No. 60d

Repeat Issues With High Pressure Steam Control Valve

Brian Baca
Operations Specialist
Chevron Phillips, LP
Sweeny Complex
PO Box 866
Sweeny, TX
bacabd@cpchem.com
979-491-2514

Prepared for Presentation at

2004 Spring National Meeting
April 25 - 29, Hyatt Regency, New Orleans, LA

Introduction

Chevron Phillips (CPC) operates a petrochemical complex located in Sweeny Texas. This complex, consists of NGL fractionation and ethylene production units. One of the three ethylene units is Unit 24. This unit was built in 1977 and produces 1450-psig steam as the driver on the cracked gas compressor train. This steam system consists of pressure and flow controllers which maintain a stable operation of the cracking furnaces, high pressure steam super-heater, and the cracked gas compressor turbine. These three main systems define the integrated high-pressure steam system for the unit. One of the components of the pressure control system is a pressure operated pneumatic valve. This valve when open, will vent the 1450 psig steam system to atmosphere. During normal operation, this valve remains closed and is set in the automatic mode in the distributed control system. Only during large pressure increases is this valve utilized, this normally occurs during startup and shutdown activities. These activities seldom happen, therefore, the valve is rarely exercised.

The pressure control valve housing is equipped with a plug type sealing mechanism. Plug travel through the valve body is achieved with pneumatic pressure aided by internal springs. Behind the plug body is a void or chamber containing steam and or condensate which has leaked by the rings or seals on the plug body. This valve when operated, utilizes an internal pilot piston, which will open initially to drain the void behind the plug and remove stored steam and or condensate (Figure 1). When the piston has traveled a set distance, the main stem of the valve will begin to push the plug in the open position. The chamber behind the plug should be free draining if the valve is oriented in such a way where the chamber is above the plug. Reversing the orientation of the valve would result in steam and water remaining in the void and chamber and would not allow for free draining.

Background

Since original installation in the unit, this valve has had numerous failures. Many times, the internal parts of the valve were found to be damaged. At times extensive damage made the valve inoperable. This valve is not spared, and repair needed to made immediately. Prolonged downtime of the unit was often required resulting in ethylene production lost. Internal inspection of the valve revealed damage usually occurred to the sealing surface of the valve body, and extensive plug damage. This type of damage usually resulted in the valve not sealing off and becoming stuck in the closed position. During these times, the unit relief valve system functioned as required to maintain safe operating pressures. Because these failures were spread out over many years and valve repairs were made quickly, previous investigations performed did not determine the root cause of the failures. This control valve discharges into an atmospheric stack. The exhaust stack also houses the 450-psig steam vent valve. The 1450 psig steam controller is arranged where the actuator is located below the valve (Figure 2), the inlet to the valve is located on the side, and the discharge is located above the valve and discharge is in an upward flow.

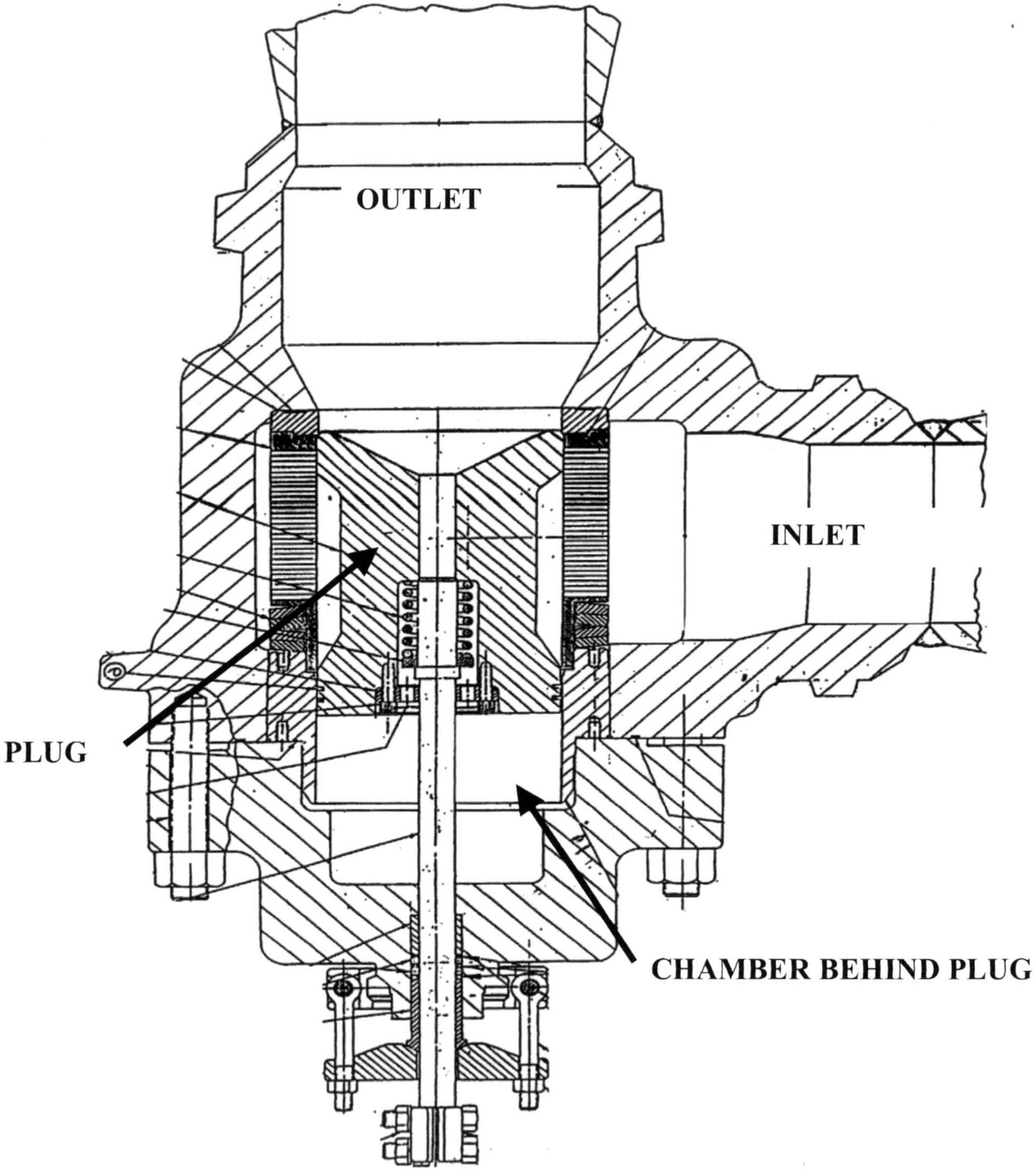

FIGURE 1

Recent Event

The last issue with the high-pressure steam control valve occurred on 06/01/03. During a non-routine shutdown, the valve was opened to vent the 1450 psig steam to the atmosphere after a trip occurred on the cracked gas compressor. An extended downtime was required to repair manual valves in the boiler feed water system elsewhere in the unit. During this downtime, the steam pressure control valve was left open to depressure the 1450 psig steam system to atmosphere. After boiler feed water valve repairs were made, the unit was prepped for startup. During startup activities, the valve was operated to warm up the 1450 psig steam system. For warm up, the valve is closed partially to allow the system to pressure up, but yet allows a slip stream of high pressure steam to be vented to atmosphere thus warming up the entire system. The operations group discovered that field verification on the valve output did not match the control room output indication. It was determined that the valve would not open up to the position requested by the control system, but was in fact closed and stuck in that position. During repair attempts by the instrument department, the personnel in the field requested to open and close the valve in an effort to troubleshoot the failure of the valve. The valve could not be opened and removal of the valve was necessary. Once again, extensive repairs were required on the internals of the valve resulting in additional downtime of the ethylene unit (Figure 3).

Investigation

A cross functional team was created to investigate the failure of the valve and determine a root cause of this failure. After dismantling the valve, it was discovered the valve had actually slammed shut with great force, and the valve plug had become lodged in the pilot seat, damaging both the plug and the seat. During the investigation, it was determined that due to the orientation of the valve in the field, steam had condensed on the top side or exhaust side of the plug, and drained into the body of the valve during the opening sequence. This draining occurred into the void or chamber behind the plug, which due to orientation, was below the plug. This did not allow for free draining of the valve body. With this configuration, when the pilot opened, the condensate draining downward into the steam chamber flashed and extreme steam pressure slammed the valve back into the closed position. After reviewing manufacturers drawings, the investigation team determined the valve was incorrectly installed in the field during original installation in 1977. If the valve were turned 180 degrees, the discharge would be in a downward flow, and when the internal piston opened the steam chamber could vent or drain properly. Also if the valve were in the downward flow position, the discharge side of the plug would not hold condensed steam on the exhaust side of the valve and create a flashing problem (i.e., would be self draining). A determination could not be found as to why the original installation of the valve was in the wrong position. However, assumptions were made that the designer of the valve installation felt that this orientation would aid in access to the valve actuator.

Solution

After the investigation team presented its recommendation, a project was opened to design piping such that the valve could be rotated and a downward discharge of the valve obtained.

Conclusion

This investigation on this controller concluded that original design and installation errors occurred. Care should be taken during review of design and installation phases to ensure manufacturers requirements and recommendations are taken into consideration. Recurring failures of equipment can often point toward these types of errors and should be closely reviewed to ensure that original installation is correct.

FIGURE 2

FIGURE 3

Ethylene Producer's Conference

April 27, 2004

F-P3-1180 FURNACE FAILURE

Bryce Kagay – Shell Chemical Company, Deer Park, Texas

Abstract

During a routine steam-air decoke (SAD) of ethane furnace F-P3-1180, the 1250 psig steam superheat coil ruptured. Upon isolation of the furnace it was observed that there was also extensive damage to the furnace radiant coils. A team comprised of operations personnel and process engineers investigated the failed SAD.

The investigation revealed that during the third attempt to decoke the furnace, steam and air flow were compromised. The furnace was firing based on coil outlet temperature and since there was no flow to carry heat from the firebox to the thermocouple, firing automatically increased. Before operations could intervene, all tubes in the radiant section failed as did the 1250 psig steam coil in the convection section. The investigation team concluded that a faulty pressure indication at the steam let-down station caused the steam controller to close and interrupt steam flow to the furnace. Loss of steam pressure tripped the air valve closed. This paper will cover both the equipment and systems level issues causing and contributing to the failure.

Process Description

Furnace F-P3-1180 is a two-cell ethane furnace that is designed to crack either ethane or butane. Shell's typical design is different than F-P3-1180. This furnace has two independently fired radiant cells, cell A and cell B. The two radiant cells share a common convection section. Figure 1 shows a simplified flow diagram of F-P3-1180.

When on feed, cold feed is preheated in the feed preheat section. Clean, medium pressure steam is let down to dilution steam pressure via a split-range controller. Dilution steam is added to the feed downstream of the feed preheat coil. The mixture of preheated feed and steam exits the convection section and enters the radiant section. Furnace effluent is quenched in the Transfer Line Exchanger (TLX). The effluent from the TLX is routed to a secondary quench exchanger.

Firing of the firebox is normally controlled by coil outlet temperature control cascaded to fuel gas pressure control. The thermocouple of the temperature controller is located outside of the firebox.

Flue gas exits the firebox and enters the convection section. Heat from flue gas is recovered by preheating feed and generating superheated 1250 psig steam as it travels from the bottom to the top of the convection section. The superheated steam temperature and degree of superheat is controlled by an attemperator where BFW is injected to de-superheat the steam before it re-enters the 1250 psig steam superheat section.

During steam-air decoke, steam enters the furnace via feed process passes as well as dilution steam passes. Decoking air is added to the furnace downstream of the dilution steam tie-ins to the process feed passes. This combined steam and air flow travels through the furnace passes and vaporizes the coke, carrying the vaporized coke out the furnace through the transfer line exchangers (TLX).

Decoking air cannot be added to the furnace unless the decoking air trip system is satisfied. One of the requirements is that the effluent isolation valve is closed to prevent decoking material from entering the scrubber column. The isolation valve cavity must meet a minimum pressure requirement to ensure it is closed. As shown in Figure 1, the cavity pressure is supplied by dilution steam.

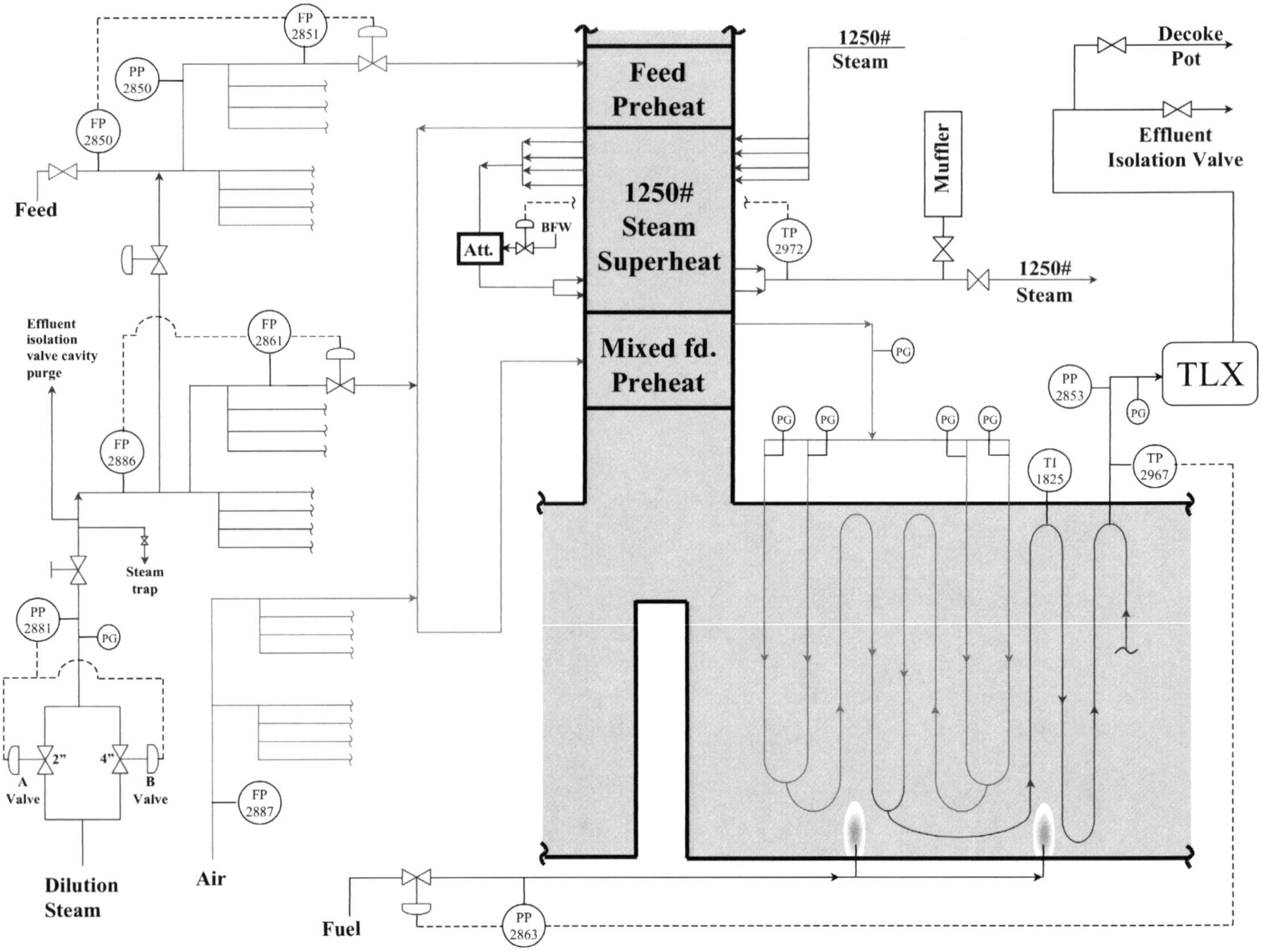

Figure 1. Simplified process flow diagram of F-P3-1180.

Incident Description

During steam-air decoke (SAD) of F-P3-1180, the 1250 psig steam superheat coil suddenly ruptured. Upon isolation of steam to the furnace it was observed that there was also extensive damage to the furnace radiant coils.

No injuries occurred. The local alarm was activated and non-essential field personnel in the Olefins area were evacuated. There was no impact to Olefins operations.

Contributing Factors

F-P3-1180 experienced excessive heat input causing the superheat coil to suddenly rupture. The excessive heat input was caused by rapid firing due to the loss of process flow through the furnace. In the morning of the incident, air and steam flows were erratic due to a faulty dilution steam letdown pressure transmitter. During the final excursion, both steam and air flow stopped. The thermocouples that control furnace firing are outside the firebox at the TLX inlets; when process flow stopped, heat was no longer carried from the firebox to the controlling thermocouples. The controlling thermocouples registered a temperature below setpoint and subsequently fired up the furnace and resulted in failure.

The current control system configuration for F-P3-1180 allows over-firing in the event process (i.e. dilution steam) flow stops. A single pressure transmitter failure (PP2881) lead to loss of both dilution steam and air flow to the furnace. Other than operator intervention, there is no automated system in place to protect the furnace from over-firing during SAD.

Through the root cause analysis (RCA) for the F-P3-1180 SAD failure, the investigation team found that malfunctions in the alarm and information system contributed to confusion prior to the furnace failure. Alarm values were not readily available to operations. It was not universally understood why limits were defined nor was clear and concise corrective action readily available for use. It was also not well understood when the existing limits were applicable (i.e. during SAD, standby, steam decoke, shutdown).

Summary of Learnings

1. F-P3-1180's design is such that loss of dilution steam pressure will result in loss of steam and air flow through the furnace. This loss of flow will cause the TLX inlet temperatures to be cooler than the tubes in the radiant section. Since the thermocouples that control furnace firing are at the TLX inlets, the controller will increase fuel gas flow to the firebox. The thermocouples at the TLX inlet will not respond to the increased fuel gas flow, and the temperature controller will continue to increase the fuel gas to the firebox.

2. Because the inlet taps to the dilution steam letdown pressure transmitter were plugged, steam pressure was compromised for two 20-minute periods in the final four hours of the SAD of F-P3-1180.

3. Other than operator intervention, there is no automated system in place to protect the furnace from over-firing during SAD; operator is the last line of defense to protect the furnace.

4. A malfunction in the alarm and information system confused operations prior to the furnace failure. Alarm values were not universally known to operations and thus impaired their ability to diagnose the malfunction.

Action Plan

1. Evaluate design of other furnaces to determine whether single-point failure could result in complete loss of flow to a furnace. Consider mitigation and increased operator awareness.

2. Improve robustness of alarm and information system. Define clear & concise corrective action to be considered when alarms activate. Make alarm limits and justification documentation readily available for operator's use.

Photo 1a. Failed 1250 psig steam coil.

Photo 1b. Pressure Transmitter PP2881

Photo 1c. Failed radiant tubes

Auto-Refrigeration / Brittle Fracture Analysis of Existing Olefins Plants – Identification of Potential Excursions

Ralph E. King, P.E.

Lyondell Chemical Company and Equistar Chemicals, L.P.
Channelview Complex
P.O. Box 30
Channelview, TX. 77530-0030

Prepared for Presentation at Ethylene Producers Conference / Session T8012 Ethylene Plant Operations
2004 AIChE Spring National Meeting, New Orleans, LA

Copyright Equistar Chemicals, LP
January 2004

Unpublished

AIChE shall not be responsible for statements or opinions contained in papers or printed in its publications.

ABSTRACT

This paper describes the use of Process Hazards Analysis (PHA) techniques and API 579 Recommended Practice for Fitness-For-Service, Assessment of Existing Equipment for Brittle Fracture methods to evaluate existing olefins plants. The paper describes some of the identified transient process excursions that can result in operation below vessel Minimum Allowable Temperature (MAT), creating the potential for brittle fracture. The methods of the evaluation are described. The importance of identifying transient process conditions and making materials-of-construction selections based on these conditions is emphasized.

Key words: Auto-Refrigeration, Olefins, Brittle Fracture, API 579, Process Hazards Analysis (PHA)

NOMENCLATURE

Auto-Refrigeration – The UNITENTIONAL and UNCONTROLED phase change from a liquid to a vapor of a hydrocarbon that results in refrigeration.

Brittle Fracture – The sudden and often catastrophic fracture of a material with little or no plastic deformation. Occurs with minimum energy absorbed in the material prior to fracture. Rapid crack propagation through the material with no prior indication (leak-before-break).

Critical Exposure Temperature (CET) – The lowest temperature experienced by the piece of equipment as a result of auto-refrigeration.

Minimum Allowable Temperature (MAT) – The lowest temperature allowed at a coincidental pressure to ensure the vessel remains ductile.

Maximum Allowable Working Pressure (MAWP)– The maximum pressure the vessel may experience so as to keep within Code Allowable Stress.

Minimum Design Metal Temperature (MDMT) – The MAT at MAWP – The lowest temperature allowed at MAWP to ensure the vessel remains ductile.

Post Weld Heat Treat (PWHT) – Post welding thermal heat treatment of the vessel to refine the weld metallurgy and to reduce residual stresses.

INTRODUCTION

Auto-refrigeration and the low temperatures that can result in brittle fracture of carbon steel equipment have been a concern since industry experienced a number of failures. The American Society of Mechanical Engineers (ASME) revised the Boiler and Pressure Vessel Code (BPVC), Section VIII, Division 1 in 1987 to reflect limitations of low temperature for new pressure vessels. An industry accepted method to analyze and evaluate existing pressure vessels for brittle fracture was not available until the American Petroleum Institute (API) issued API 579 Recommended Practice for Fitness-For-Service in January 2000. As a part of ongoing review of process risk, the potential for auto-refrigeration and associated brittle fracture was highlighted as one of the major potential risks associated with the Lyondell/Equistar olefin plants. A program to analyze, evaluate and mitigate the risk of auto-refrigeration / brittle fracture in existing plants was initiated in 2001. After having evaluated several olefins plants, this paper reflects the techniques and methods used and also describes potential transient conditions that were identified during the analysis. Figure 1 reflects the major steps in the Auto-Refrigeration / Brittle Fracture Prevention Program.

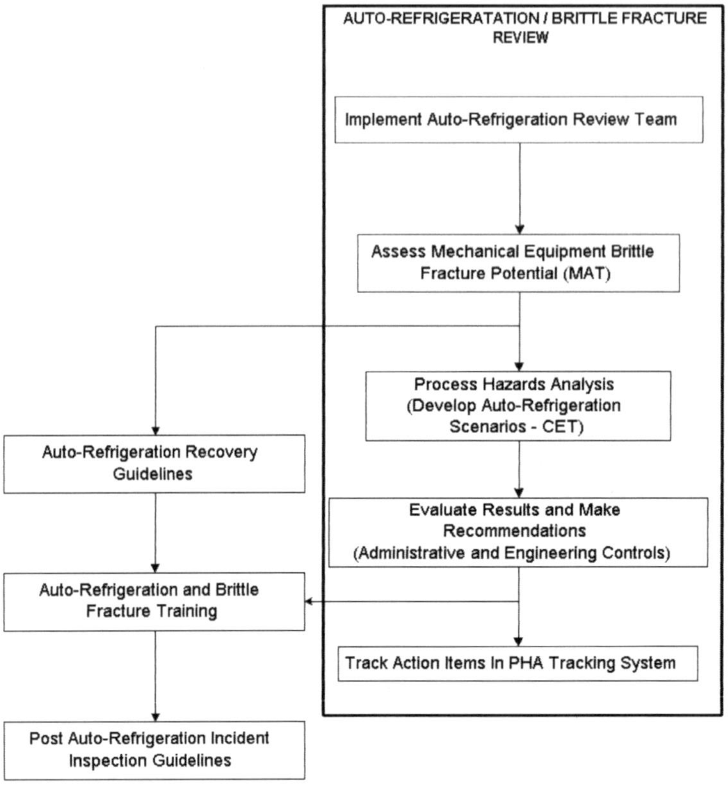

Figure 1 Auto-Refrigeration / Brittle Fracture Prevention Program

BRITTLE FRACTURE BASICS

Carbon steel and other ferritic steels become susceptible to brittle fracture with decreasing temperatures. Brittle fracture is a particularly undesirable failure mode because it can occur without warning, and the damage may manifest itself in complete rupture of the equipment. Brittle fracture occurs in a break-before-leak mode rather than the preferred leak-before-break mode of failure. For brittle fracture to occur, all three of the following elements must be present simultaneously:

a) A susceptible steel (susceptibility increases as temperature decreases),

b) A stress riser, such as a crack or a notch (as often is found in weld defects), and

c) A sufficient applied stress above a minimum stress level (~7,000psi for carbon steels)

These three elements are often represented in the form of a Brittle Fracture Triangle (Figure 2). All three sides of the triangle must be present for brittle fracture to occur. Remove any side and brittle fracture is not possible. For most pressure vessels under normal operating conditions, the stress is almost always above 7,000psi. And since no vessel is fabricated perfectly, there are always some weld flaws (cracks or notches). The net effect is that two sides of the triangle are always present during normal operation. The best protection against having a brittle fracture is not to get carbon steel vessels cold enough to "close" the Brittle Fracture Triangle.

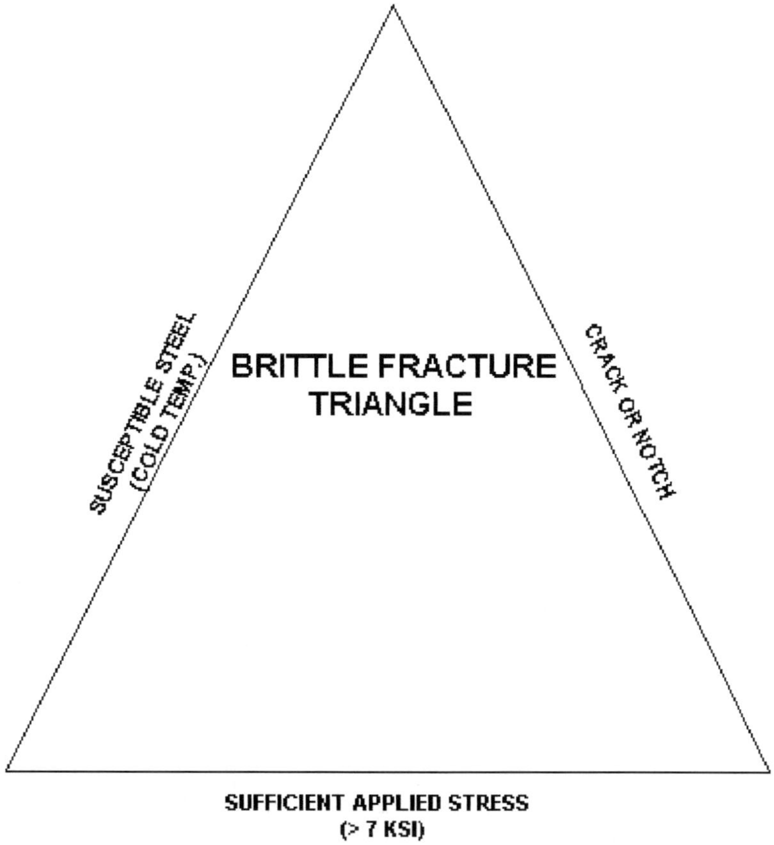

Figure 2 – Brittle Fracture Triangle

The susceptibility of carbon steel to brittle fracture is related to temperature. As temperature decreases, the susceptibility to brittle fracture increases. Auto-refrigeration can provide the mechanism for low temperature exposure. When a particular grade of steel becomes susceptible depends on its grain size and the melting practices used. The best grades of carbon steel used for pressure vessel construction can be applied at the allowable stress down to approximately –50°F. These steels are specially processed to obtain good resistance to brittle fracture. Carbon steels that are not specially processed for brittle fracture resistance may become susceptible at room temperature and above. The specific temperature where a pressure vessel becomes susceptible to brittle fracture can be captured in the concept of Minimum Allowable Temperature (MAT). The MAT is the lower temperature boundary at all possible vessel pressures (stresses) to ensure brittle fracture does not occur. The MAT at Maximum Allowable Working Pressure (MAWP) is defined as the Minimum Design Metal Temperature (MDMT).

Other types of ferritic steels such as chrome-moly and carbon-moly steels do behave similarly to carbon steels that are not specially processed for resistance to brittle fracture. Even nickel steels can become susceptible to brittle fracture at low enough temperature. Austenitic stainless steels, nickel-base alloys, aluminum alloys and copper alloys are essentially resistant to brittle fracture.

One key concept for complete understanding of brittle fracture is that a minimum level of applied stress is needed to propagate a brittle fracture. For the crack tip to propagate through the carbon steel, it must have sufficient energy. For carbon steel, this minimum level of applied stress (energy) is about 7,000psi. This equates to about 40% of the MAWP for pressure vessels built to the ASME Sec. VIII, Div. 1 1998 or earlier Code, (35% for vessels built to the 1999 or later Code). The potential for catastrophic failure is reduced when the pressure is brought below the 7,000psi stress level (<40% MAWP). If an auto-refrigeration excursion has occurred and the Brittle Fracture Triangle is "closed", the vessel is at risk of failure. The only fail-safe response is to "open" the triangle by dropping the pressure on the vessel, thereby lowering the stress to less than 7,000psi. This may induce additional auto-refrigeration, resulting in lowering the vessel temperature, but as long as the stress is reduced, the potential for a catastrophic failure is diminished. The vessel may crack and leak from secondary stresses, but the primary stresses, the stresses that cause vessels to rupture, have been reduced.

ANALYSIS APPROACH

The Auto-Refrigeration / Brittle Fracture Review process involves two major analysis methods:

1. Mechanical evaluation – In the mechanical evaluation the equipment is analyzed for brittle fracture potential using API 579, Section 3. For most equipment, a Level 2 Method A, B or C analysis is used. The result of this evaluation is to generate a curve that reflects a vessel's Minimum Allowable Temperature (MAT) for all pressure conditions.

2. Process Hazards Analysis (PHA) – The PHA is undertaken to evaluate the process for possible auto-refrigeration excursions. The purpose of this analysis is to generate a Critical Exposure Temperature (CET) for all vessels that are subject to exposure to light hydrocarbon materials. The CET is represented by a curve of process temperature conditions for pressures below the MAWP of the vessel.

Mechanical Evaluation:

The mechanical evaluation is undertaken using the methods described in API 579. Since most auto-refrigeration excursions occur under upset / non-steady state conditions, the MAT is calculated for all pressure conditions of the vessel. API 579 allows the MAT to decrease as the primary (pressure) stress on the vessel falls. The MAT is allowed to decrease to as low as –155°F for carbon steels. To facilitate the analysis, a spreadsheet was developed that takes vessel information (material of construction, MAWP, corrosion allowance, weld joint efficiency, impact test data, etc.) and calculates the MAT. The spreadsheet also has pure component pressure / temperature equilibrium curves for the more common light hydrocarbon materials encountered in an olefins plant. The pure component curves can be used to approximate the expected process temperatures (CET) for equipment. Input of process simulation results (flash temperature and pressure) can also provide a more exact picture of the expected process conditions. If the CET curve is at or below the MAT curve, then the vessel is at risk of brittle fracture. Figure 3 represents the MAT curve calculated for a non-PWHT SA516-70 carbon steel vessel with an MAWP of 305psig. The MAT at MAWP is 8°F. For the example vessel, the vapor pressure (CET) curve for the ethylene goes below the MAT for the vessel. In this case the vessel is acceptable for containing liquid propylene but is unacceptable for containing liquid ethylene.

Brittle Fracture Analysis Report for Example Vessel of SA-516-70

Fail Level 2 - Maximum difference between the MAT and vapor pressure curves or User entered points = 23°F.
Difference for selected curves: Ethylene = 23°F
Minimum acceptable (coldest) temperature (MAT) when the pressure is at MAWP is 8°F.
Maximum permitted pressure when the temperature is at -155°F is 143 psig
Recommended minimum field hydrotest temperature = 38°F

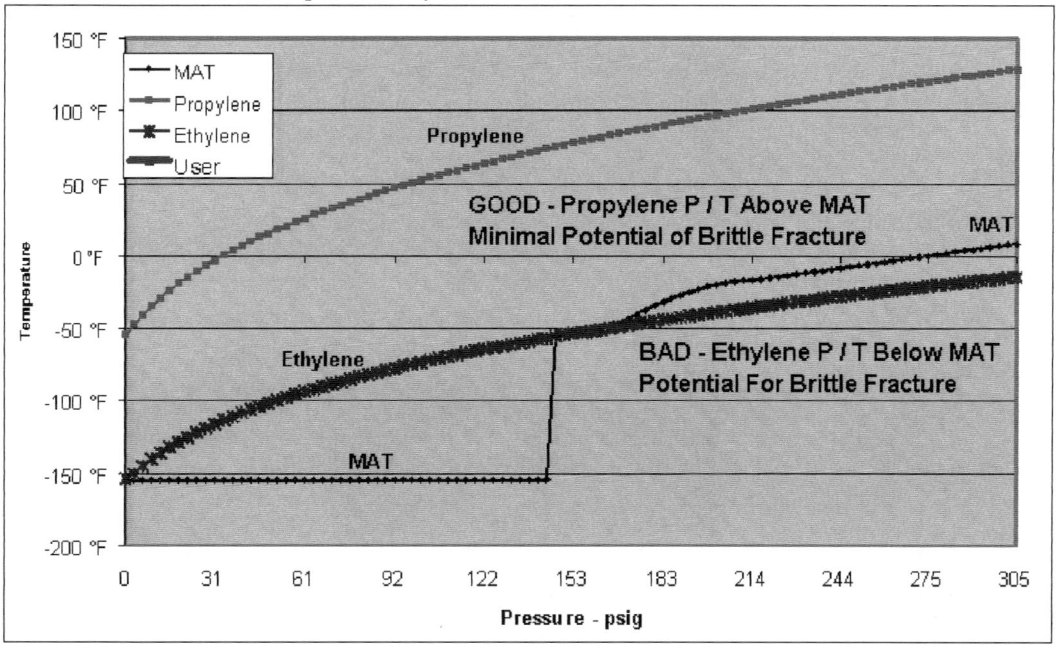

Figure 3 - Example Vessel MAT

Process Hazards Analysis (PHA):

The purpose of the PHA is to identify all possible scenarios that could result in auto-refrigeration of the light hydrocarbon process material. The analysis may use any of the OSHA / EPA accepted PHA methods. Because auto-refrigeration most often occurs during transient conditions, the analysis must evaluate the unit during each of the following operating modes and particularly the transition between operating modes.

a) Normal Operation

b) Upset Conditions

c) Normal Startup

d) Normal Shutdown

e) Emergency Shutdown

f) Air Freeing / Nitrogen Freeing

g) Inventory

h) De-Inventory

i) Not in Operation / Maintenance-in-Progress

j) Commissioning / Leak Testing

Because of the transient, unsteady state conditions that may exist during all operating modes, the PHA team must have an extremely good understanding of the operation of the plant and knowledge of operational steps that occur when making transitions from one mode to another. For each scenario identified, the Critical Exposure Temperature (CET) and coincident pressure are determined. The CET is the lowest expected temperature possible given the scenario developed and the process material contained in the vessel. The scenarios must take into account the possibility of having light hydrocarbon materials in vessels that do not normally contain such materials. Overflowing of separation drums, recycle streams, cross connections via common headers and cross heat exchange in exchangers must all be considered. Subsequent actions executed by the process controls or by operating personnel must be anticipated and evaluated.

Of particular concern is the ability to increase operating pressure after an auto-refrigeration excursion prior to the equipment warming. This has the potential to place the vessel well below the MAT. Scenarios involving increasing pressure were evaluated by considering a limiting case scenario where the equipment temperature was considered to remain constant at the CET as the pressure increases. Separator vessels (compressor suction and discharge drums, reflux drums, chilling train separators) that normally contain liquid inventory are vulnerable to excursions below the MAT as a result of rapid re-pressurization. When such a vessel is de-pressured, the liquid in contact with the vessel wall vaporizes, creating a very effective mechanism to rapidly cool the vessel wall to the process equilibrium temperature. As the vessel is re-pressurized, the temperature of the portion of the wall in contact with the liquid lags well behind the process equilibrium temperature because heat transfer is limited by less effective convective and conductive mechanisms.

The PHA techniques most often used during the analysis are the Event (Fault) Tree and Guide-List methods. In all cases, the two methods were used in combination to ensure complete identification of potential process auto-refrigeration excursions. The Event Tree method was implemented only to the extent that the PHA Team had an adequate grasp of the initiating events and the process sequence that potentially could result in auto-refrigeration / brittle fracture. The Event Tree method allows the PHA Team to identify and assess the measures in place to minimize the likelihood of occurrence of initiating events and to make judgements about the adequacy of protective measures. The Guide-List provides a structured analysis of all operating modes of the plant and ensures that all transient conditions are evaluated. The developed Guide-List is frequently updated to reflect the "learning's" from the previously completed evaluations.

FINDINGS

Based on the experience gained from several reviews, a number of generalized findings can be made in the evaluation of an existing olefins plant. While these generalized findings are based in part on the Lyondell/Equistar studies, they do not represent the exact detailed findings of the studies. The findings described below and the associated vessels are purely examples of some of the possible excursion scenarios that may be encountered when undertaking an auto-refrigeration / brittle fracture PHA.

Acceptable Vessels:

Many of the vessels / scenarios evaluated did not result in any substantial findings or recommendations. These vessels were acceptable for continued operation for the process services being contained. The SA516-70 carbon steel example vessel illustrated by Figure 4 reflects the typical MAT results and upset scenarios encountered.

Brittle Fracture Analysis Report for Acceptable Vessel

MAT is acceptable for the vapor pressure curves and the User operating pressure/temperature combinations

Min. acceptable (coldest) temperature (MAT) when the pressure is at MAWP is 17°F.
Maximum permitted pressure when the temperature is at -155°F is 130 psig
Recommended minimum field hydrotest temperature = 47°F

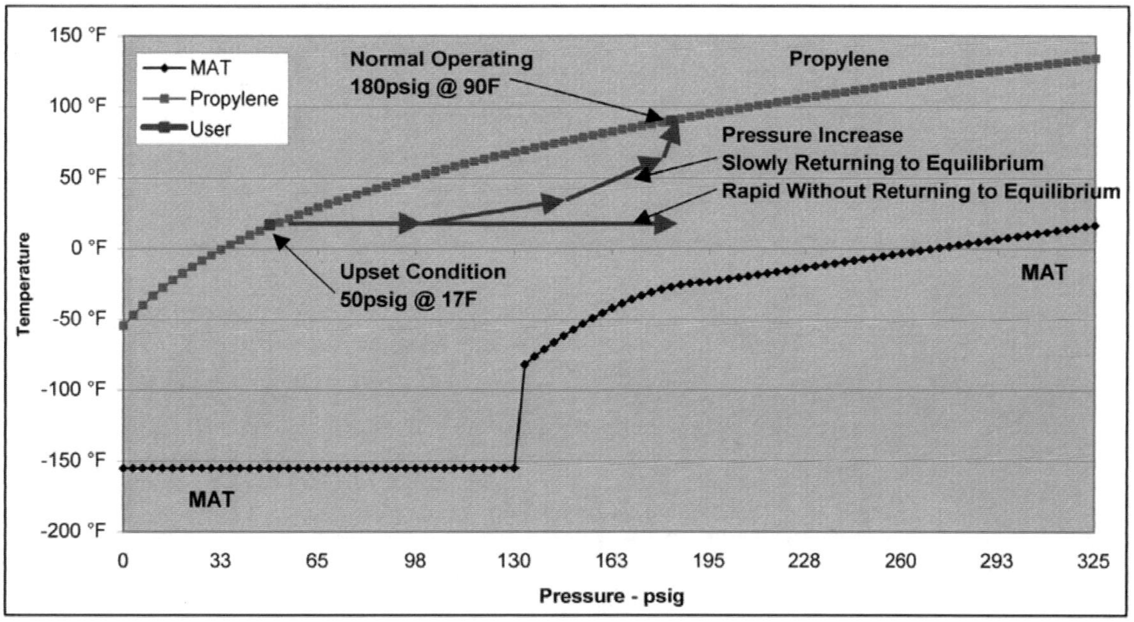

Figure 4 – Example Acceptable Vessel

Loss in system pressure with a subsequent return to normal operating pressure (or even up to the MAWP) does not result in conditions at or under the MAT. The rate at which the pressure increases determines the transient upset conditions. Large systems or where pressure increase is based on heat transfer results in a slow pressure increase with vessel wall conditions tracking near equilibrium. Systems like compression trains results in rapid pressure increase and the vessel walls can deviate substantially from equilibrium temperature. In most cases, the final upset conditions do not approach the MAT and no mitigation is required. This represents the majority of vessels evaluated.

3½% Nickel Vessels in Liquid Methane Service:

Vessels fabricated from 3½% nickel low temperature steels (SA203 Gr. B & E, SA350 – LF3) have substantial impact resistance even at very low temperatures. These alloys, however, are still limited to –155˚F MAT. When containing liquid methane, as in the de-methanizer feed separator and the de-methanizer reflux drum, the auto-refrigeration CET can be well below the MAT. See Figure 5 for an example of MAT analysis results for a nickel vessel.

Brittle Fracture Analysis Report for Example 3 1/2% Ni Vessel in Liquid Methane Service

Fail Level 2 - Max. difference between the MAT and vapor pressure curves or User entered points = 105°F.
Difference for selected curves: Methane = 105°F
Minimum acceptable (coldest) temperature (MAT) when the pressure is at the MAWP is -150°F.
Maximum permitted pressure when the temperature is at -155°F is 475 psig
Recommended minimum field hydrotest temperature = -120°F

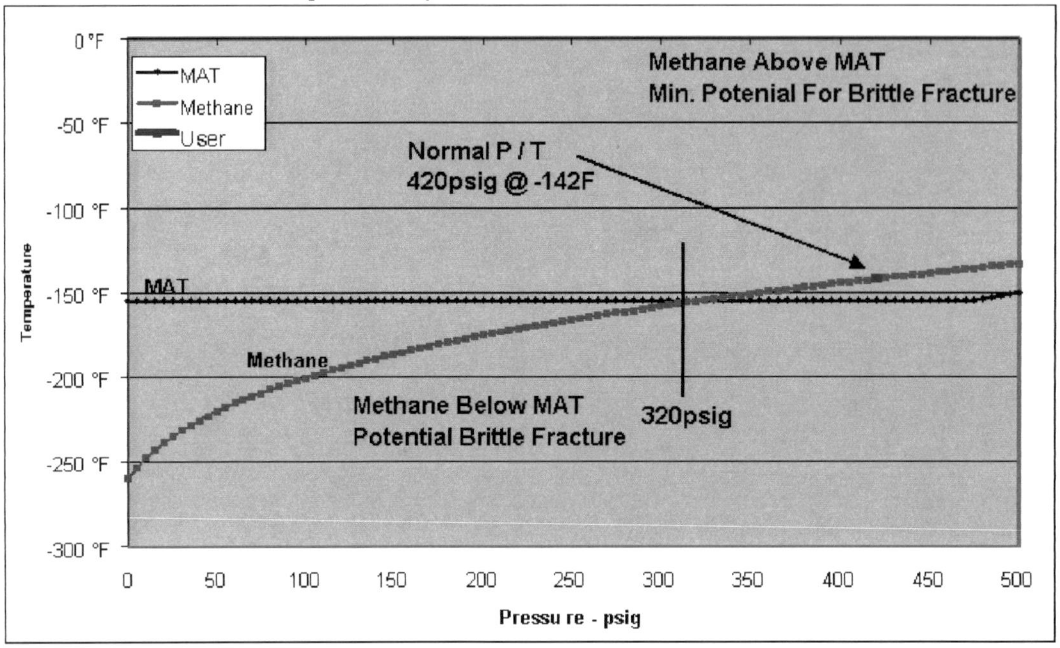

Figure 5 – Example 3½% Nickel Vessel MAT – Liquid Methane Service

With liquid methane in this example, if the pressure is dropped much below 320psig, the CET will be lower than the MAT of the 3½% nickel vessel. The vessel is acceptable for "normal" process conditions but is not acceptable under "upset" conditions where the pressure is lost and auto-refrigeration occurs. Replacement of the vessel using stainless steel is one option for correcting this situation.

Chilling Train Vessels:

Figure 6 represents a simplified diagram of an example Process Gas Chilling Train. Loss of forward process gas flow due to a trip of the Process Gas Compressor or from flaring process gas going forward (lower streams flowing from left to right) results in low-flow / stagnant flow of the process gas in the core exchangers. If the liquid / vapor methane and hydrogen chilling streams (upper streams flowing from right to left) continue flowing through the core exchangers, process gas is condensed at a temperature lower than normal. This condensed process liquid either gravity-flows to the next drum or in some instances may back-flow into the previous drum, depending on the physical arrangement of the cores and separator drums. Due to process gas composition and the associated reduced pressures in the chilling train during the excursion, liquid temperatures can be suppressed significantly below normal operating conditions. This auto-refrigeration excursion has the potential to have 3½% Nickel drums below the –155°F limit and / or vessel conditions could be below MAT if normal process pressures are re-established (restart Process Gas Compressor) prior to warming of the vessels. Figure 7 reflects an example vessel where the vessel is chilled followed by subsequent reestablishing of normal process pressure.

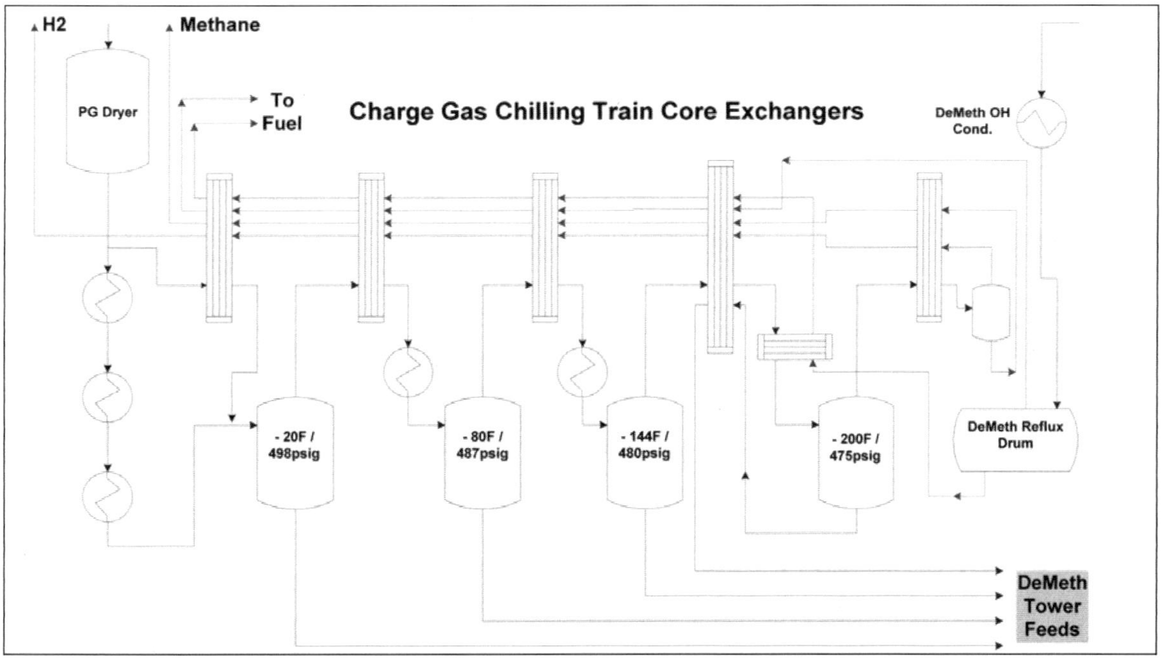

Figure 6 – Simplified Process Gas Chilling Train

Brittle Fracture Analysis Report for Demethanizer 1st Precooler Separator

Fail Level 2 - Max. difference between the MAT and vap. pres. curves or User entered points = 117°F.
Difference for selected curves: Methane = 105°F; User entered points = 117°F
Min. acceptable (coldest) temperature (MAT) when the pressure is at MAWP is -50°F.
Max. permitted pressure when the temperature is at -155°F is 264 psig
Recommended minimum field hydrotest temperature = -20°F

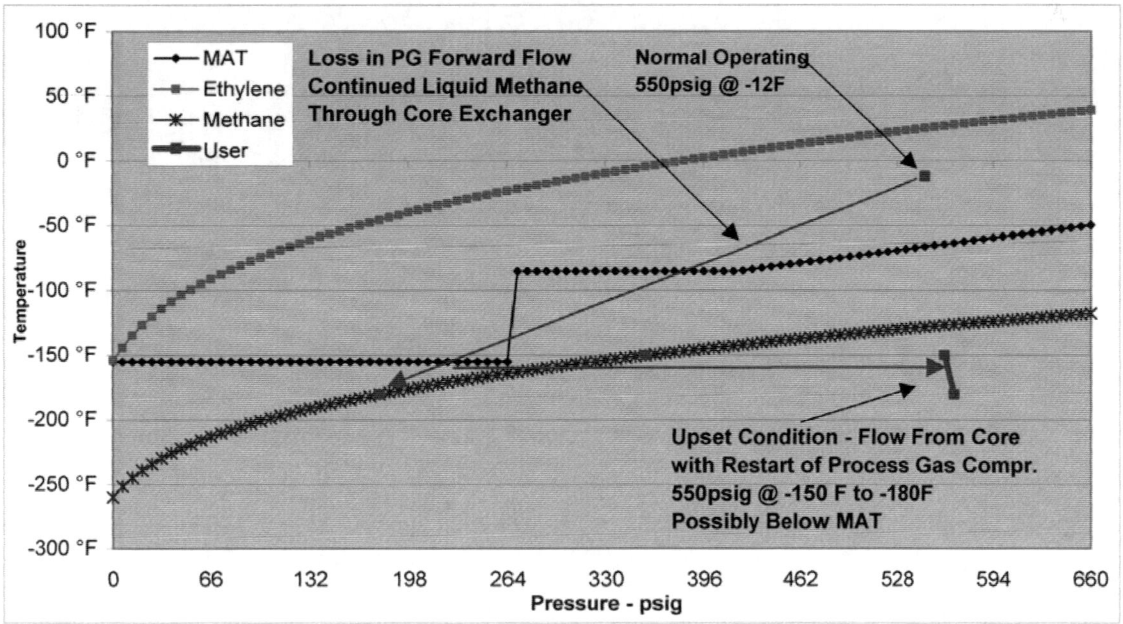

Figure 7 – Example Chilling Train Vessel MAT – Continued Chilling Streams Through Core Exchangers

The vessel is acceptable for "normal" process conditions but is not acceptable under "upset" conditions where forward process gas flow is lost and the cold liquid methane / hydrogen streams continue flowing through the core exchangers. Possible options for correcting this situation are replacement of the vessel using stainless steel or installation of protective interlocks. This same scenario can occur on standard shell and tube heat exchangers in other process positions where an auto-refrigeration excursion on one side of the exchanger has the potential to chill the other side to conditions below the MAT.

Distillation Towers and Peripherals:

Process upsets in distillation towers may result in the potential for exposure of the tower or any of its peripheral equipment (re-boiler, overhead condenser, reflux drum) to conditions below the equipment MAT. Figure 8 represents an example of a De-Ethanizer Reflux Drum.

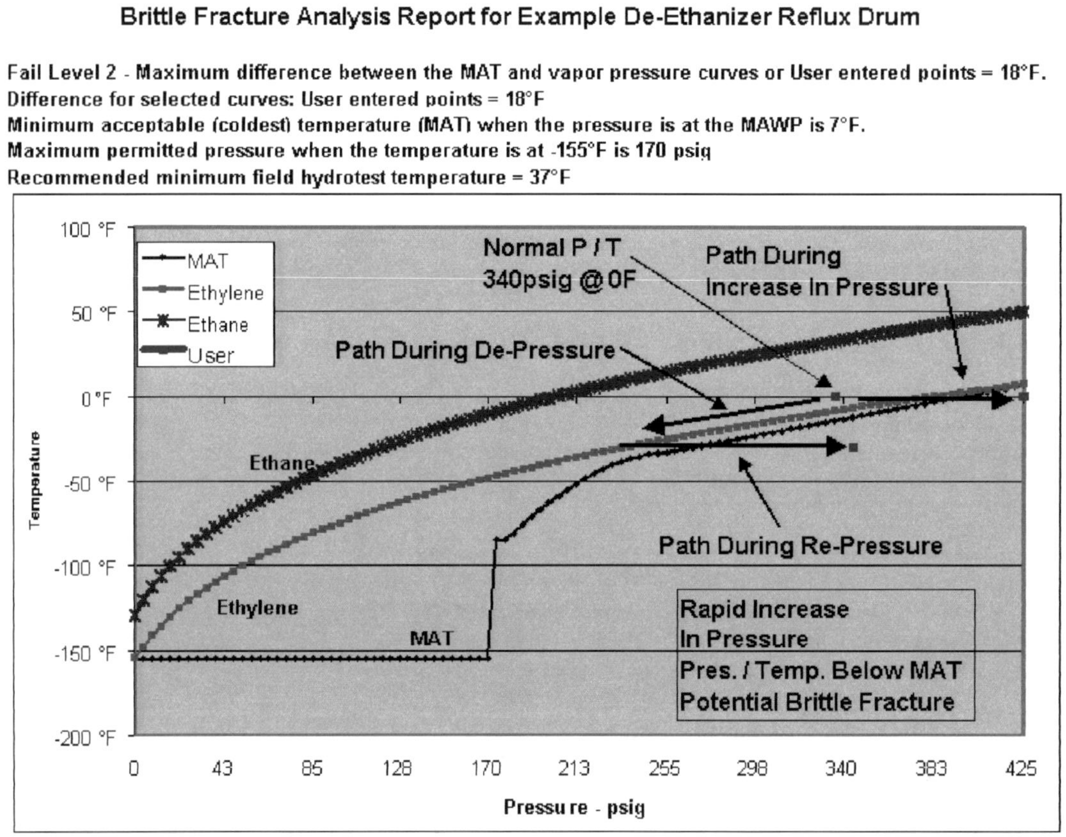

Figure 8 – Example De-Ethanizer Reflux Drum MAT

Under normal operating conditions, the vessel operates above its MAT. Upsets in operating pressure to values either higher or lower than normal may result in excursions below the vessel MAT. Rapid increase in pressure can result in excursions below the MAT. Possible initiating events include loss of condensing or pressure controller failure. An excursion to low pressure does not initially result in exposure below the MAT. Initiating events include loss of re-boil or pressure controller failure. However, if the Operations response to the upset is to attempt to quickly return to normal pressure, the CET may be lower than the MAT. The vessel wall temperature will always lag the process temperature. One alternative for avoiding this scenario is to install controls / interlocks to limit the source of the pressure (re-boil), thus preventing operation below the MAT.

Bimetallic De-Methanizer:

De-Methanizers made from two different materials present a significant challenge in terms of auto-refrigeration and brittle fracture potential. Vessel designers have historically handled the wide differences in tower process temperatures with an overhead temperature of –140°F and a bottoms temperature of +20°F by using a bimetallic construction. The top section of the tower is typically made of a 3½% nickel steel while the bottom may be fabricated from a Charpy impact-tested carbon steel. During process upsets that result in loss of vapor flow up the tower (loss of re-boil, etc.) the tower slumps and the cold process material from the upper section of the tower cascades down into the lower section. Depending on the process composition and the materials-of-construction, this tower upset can represent a significant potential for brittle fracture. Figure 9 represents an example analysis of the lower carbon steel section of a bimetallic de-methanizer. The "User" points shown just below the ethylene equilibrium curve represent the process material vapor / liquid equilibrium (CET) if all of the tower inventory, both upper and lower sections, were combined and well mixed. The "User" points near the methane equilibrium curve represent the vapor / liquid equilibrium (CET) for the average composition of the material in the upper section of the tower. If the process material from the upper section is allowed to drain into the lower section, there is a potential for conditions to be below the MAT of the lower section.

Brittle Fracture Analysis Report for Example - Lower C.S. Sect. Of a De-Methanizer

Fail Level 2 - Maximum difference between the MAT and vap. pres. curves or User points = 105°F.
Difference for selected curves: Methane = 105°F; User entered points = 55°F
Minimum acceptable (coldest) temperature (MAT) when the pressure is at the MAWP is -75°F.
Maximum permitted pressure when the temperature is at -155°F is 220 psig
Recommended minimum field hydrotest temperature = -45°F

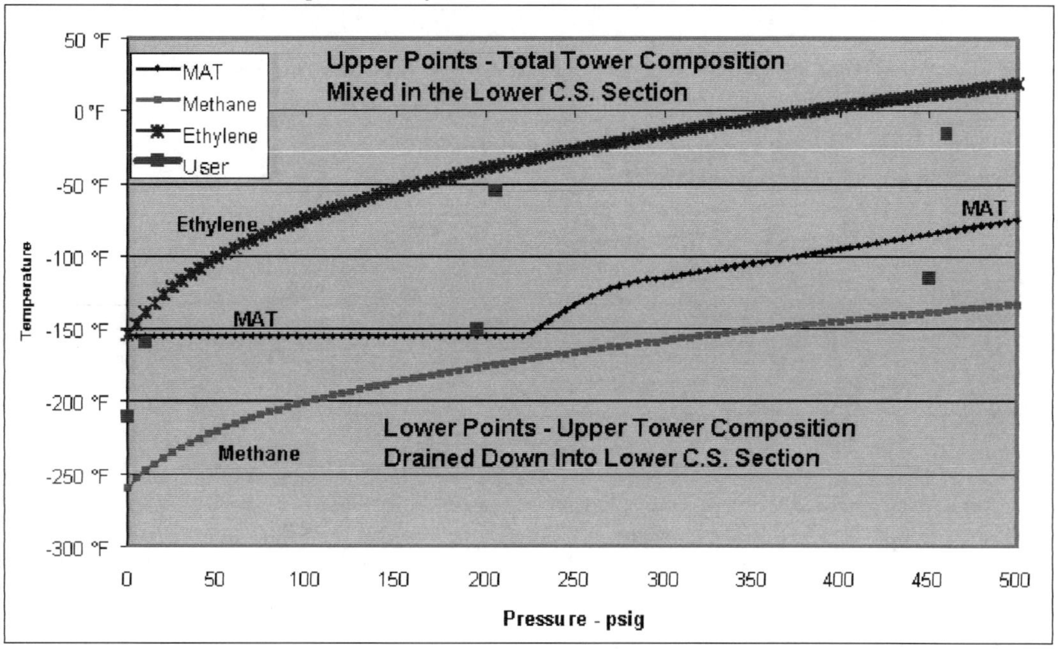

Figure 9 – Example De-Methanizer - CS Bottom Section MAT

API 579 uses a bimetallic de-methanizer as the Level 3 Analysis example. This method uses fracture mechanics, statistical application of NDT methods and finite element analysis to evaluate and inspect the vessel for continued service. While certainly an API 579 Level 3 analysis may provide assurance for fitness-for-service, there is no guarantee that a particular vessel will pass such a rigorous analysis. Another approach to mitigating the risk of operation below the MAT is to prevent the cold material in the upper section of the tower from reaching the lower carbon steel section of the tower. This can be accomplished by modifications to the tower internals and the installation of controls / interlocks that divert the upper section process material out of the tower if the lower section conditions approach the MAT.

Nitrogen Freeing of Equipment:

During inventory of the unit, light hydrocarbon liquids can be introduced into a process environment inerted with nitrogen. The nitrogen contained in the process is free of any hydrocarbon. The effect is to lower the partial pressure of the hydrocarbon below the expected flash temperature (CET) for the hydrocarbon at the measured process pressure. Process simulation indicates that ethylene can flash to as low as –190˚F when it is at equilibrium with nitrogen at atmospheric pressure. This is 40˚F lower than the normal atmospheric flash temperature of –150˚F. Propylene can flash to as low as –110˚F in nitrogen, compared to its normal atmospheric flash temperature of –53˚F. These substantially suppressed temperatures present lower CETs than would be expected based solely on the light hydrocarbon material being contained. Because the temperature reduction is due to the presence of the nitrogen, the CET can occur while substantial pressure is applied to the vessel. Such suppressed CETs at substantial pressure can present a brittle fracture potential for the exposed vessel as the pressure builds during the introduction of the light hydrocarbon. Figure 10 shows an example using propylene as the light hydrocarbon. The MAT is for a vessel made from SA516-70 non-PWHT. It can be seen that introducing propylene into the nitrogen-filled vessel padded at 100psig results in a flash temperature of –56˚F. As the vessel builds pressure and the nitrogen is vented off, the propylene moves to its vapor / liquid equilibrium temperature but not without being substantially below the MAT. Freeing the vessel of nitrogen by purging with propylene vapor prevents exposure below the MAT.

Brittle Fracture Analysis Report for Example: Nitrogen in a Vessel

Fail Level 2 - Maximum difference between the MAT and vapor pressure curves or User entered points = 37°F.
Difference for selected curves: User entered points = 37°F
Minimum acceptable (coldest) temperature (MAT) when the pressure is at the MAWP is 37°F.
Maximum permitted pressure when the temperature is at -155°F is 148 psig
Recommended minimum field hydrotest temperature = 67°F

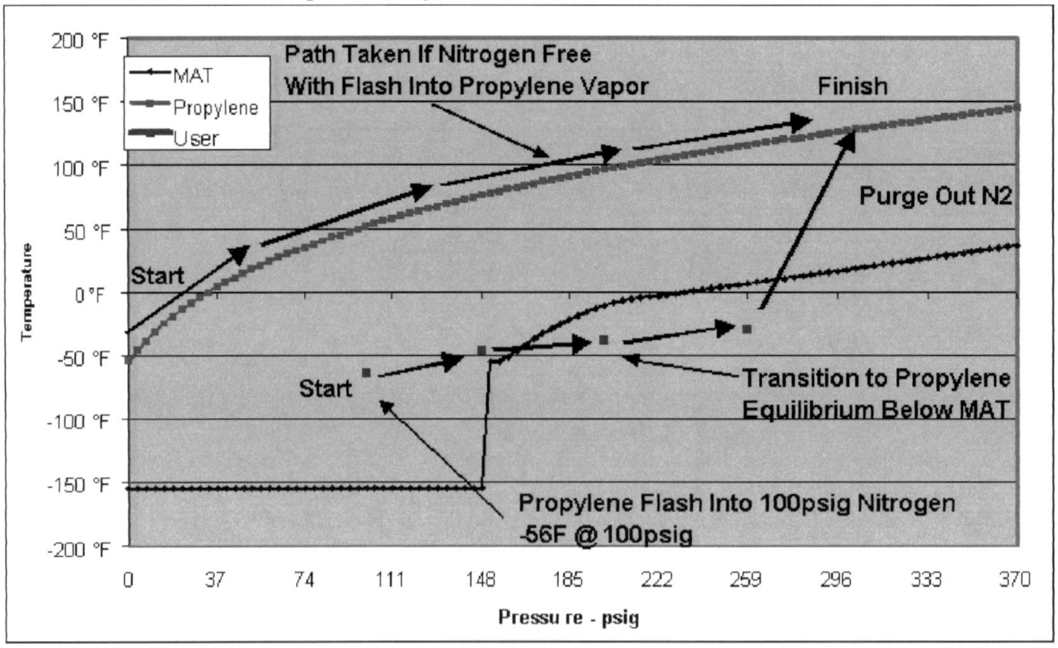

Figure 10 – Example of Nitrogen Filled Vessel during Inventory

This same effect can occur during other process steps where nitrogen is introduced into the process and where there is sufficient mixing to allow the light hydrocarbon and the nitrogen to reach equilibrium. Process equipment such as dryer or catalyst beds is at risk if nitrogen or other non-condensable vapors such as methane or hydrogen are used as a purge or drying media. Such auto-refrigeration excursions can occur at substantial pressures and place the vessel at risk of brittle fracture.

Compressor Suction Drums:

Multi-stage compressor trains represent an unusual case for evaluation. The normal operating pressure can be substantially different than the "settling out pressure" that occurs when the compressor is stopped. For compressor suction drums, the normal operating temperature can be rather low at a correspondingly low pressure. When the compressor stops, the settling out pressure can be substantially higher as the pressure equalizes across all of the compressor stages. With the pressure rise being rather rapid and the suction drum being filled only with vapor, there is insufficient time to allow the vessel wall temperature to maintain equilibrium with the rise in pressure. If the vessel material-of-construction is not specified correctly, this increase in pressure can result in vessel conditions being below the MAT for the vessel. Two examples are presented that reflect these potential results.

Figure 11 represents the transient shutdown for a compressor suction drum made of material specified with a –150°F Charpy impact temperature. It can be seen that the sudden rise in the pressure conditions results in a CET above the MAT of the vessel, making this example acceptable for service. Figure 12 represents the same design but specified with a –100°F Charpy impact temperature. For the vessel in Figure 12, the CET is below the MAT at the settling out upset conditions, making the vessel susceptible to brittle fracture.

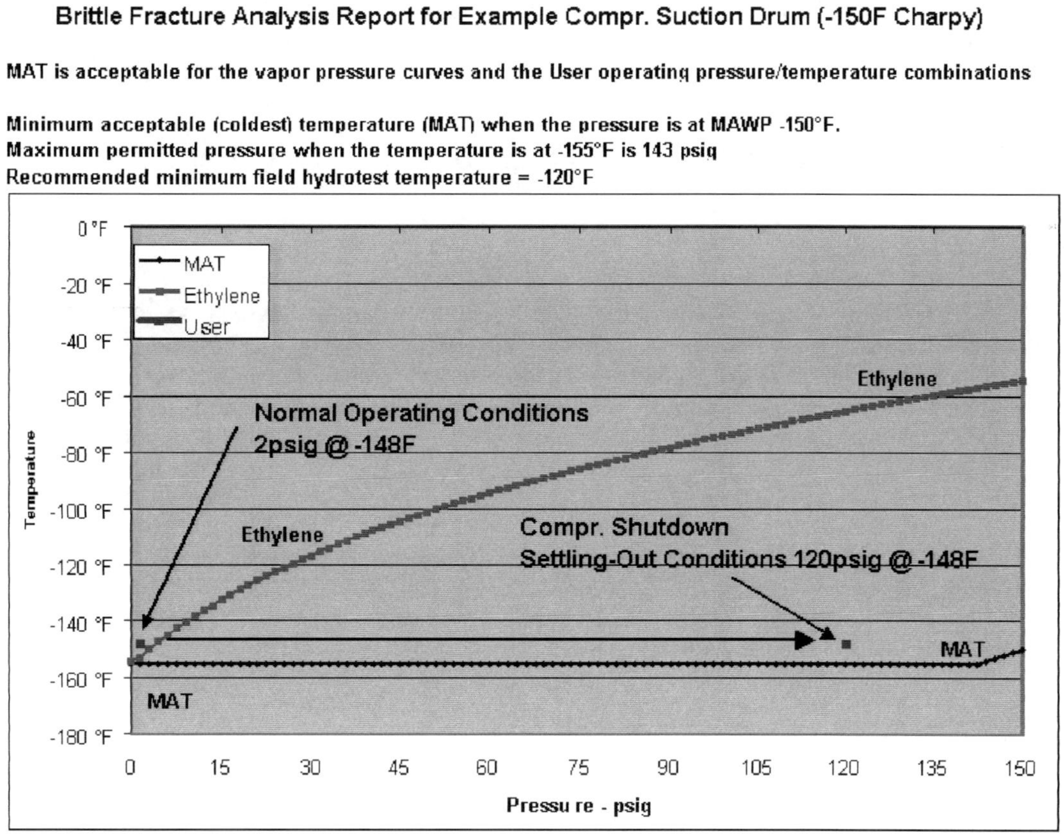

Figure 11 – Example Compressor Suction Drum MAT (–150°F Charpy)

Brittle Fracture Analysis Report for Example Compr. Suction Drum (-100F Charpy)

Fail Level 2 - Maximum difference between the MAT and vapor pressure curves or User entered points = 28°F.
Difference for selected curves: User entered points = 28°F
Minimum acceptable (coldest) temperature (MAT) when the pressure is at MAWP is -100°F.
Maximum permitted pressure when the temperature is at -155°F is 77 psig
Recommended minimum field hydrotest temperature = -70°F

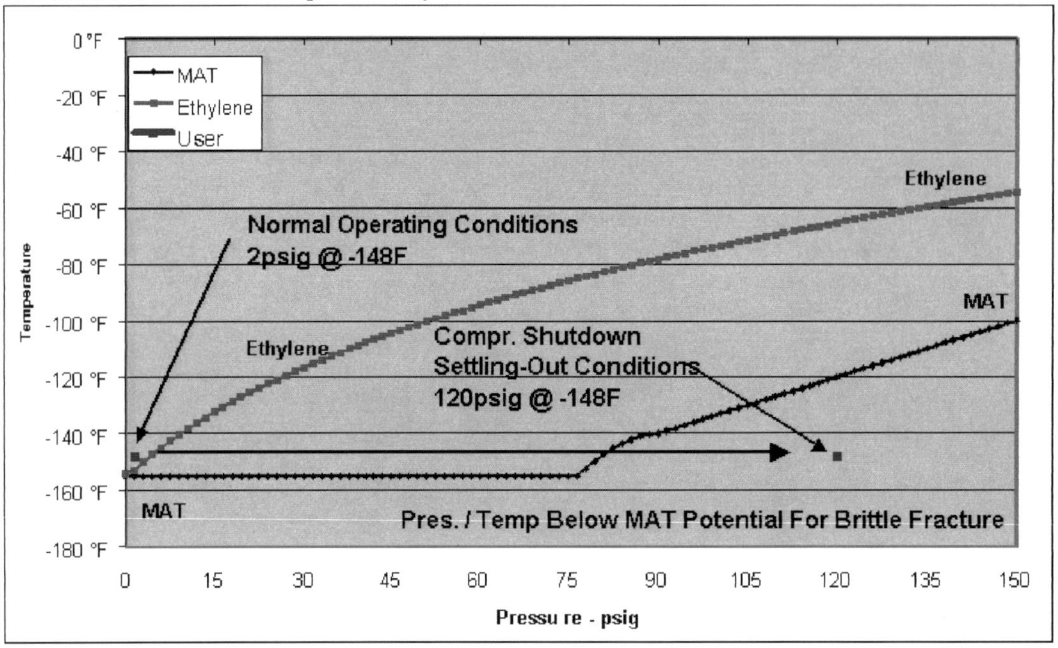

Figure 12 – Example Compressor Suction Drum MAT (–100°F Charpy)

Compressor Discharge Drums:

Compressor discharge drums present a similar challenge except that at normal operating conditions these drums contain condensed light hydrocarbon liquid at a substantial pressure. Upon shutdown of the compressor, the discharge side of the system will slowly lose pressure, resulting in auto-refrigeration of the liquid thus chilling the vessel. Upon restart of the compressor, the discharge pressure rises very rapidly, much faster than conductive heat transfer can warm the vessel wall. Under these conditions, steps must be taken to rid the vessel of the auto-refrigerating liquid or a different material of construction should be specified to prevent having the CET below the MAT. Figure 13 represents an example compressor discharge drum undergoing a slow de-pressure event with subsequent compressor restart. It illustrates how the discharge drum ends up at CET conditions below the MAT once the normal compressor discharge pressure is reestablished.

Brittle Fracture Analysis for Example Compr. Disch. Drum - S.D. / De-Pres. with Restart

Fail Level 2 - Maximum difference between the MAT and vapor pressure curves or User entered points = 26°F.
Difference for selected curves: User entered points = 26°F
Minimum acceptable (coldest) temperature (MAT) when the pressure is at the MAWP is 11°F.
Maximum permitted pressure when the temperature is at -155°F is 240 psig
Recommended minimum field hydrotest temperature = 41°F

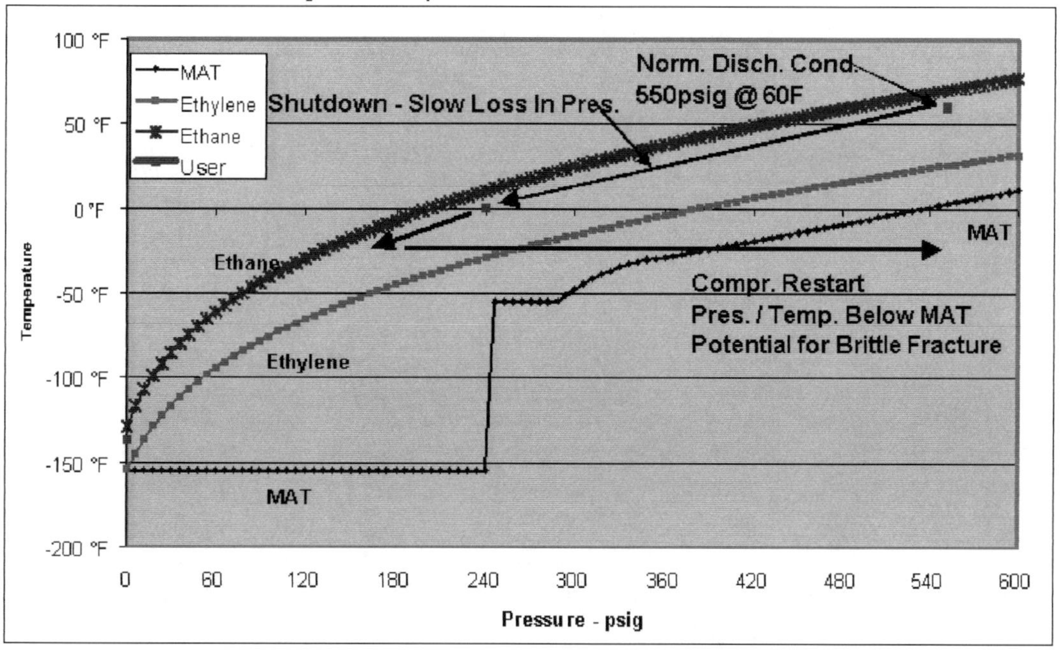

Figure 13 – Example Compressor Discharge Drum MAT - Loss in Pressure with Restart

Loss in Heat Input

Vessels that normally receive a heated process stream may not be designed to accept the non-heated stream without conditions being below the MAT. Vessels such as acetylene guard beds / reactors and product surge drums are susceptible to being chilled if the heat source is lost to the primary heat exchanger. Figure 14 represents an example product surge drum that depends on a preheat exchanger to keep process conditions above the MAT. Failure of the heating source has the potential for process CET to be under the MAT.

Brittle Fracture Analysis Report for Product Surge Drum

Fail Level 2 - Max. difference between the MAT and vapor pressure curves or User entered points = 81°F.
Difference for selected curves: User entered points = 81°F
Min. acceptable (coldest) temperature (MAT) when the pressure is at MAWP is -50°F.
Maximum permitted pressure when the temperature is at -155°F is 136 psig
Recommended minimum field hydrotest temperature = -20°F

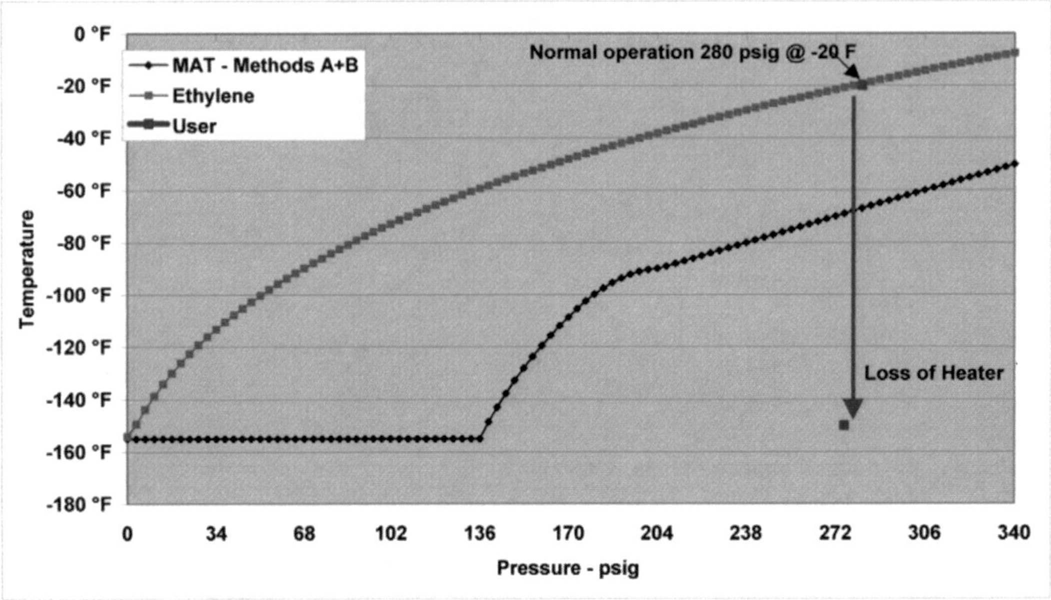

Figure 14 – Example Product Drum MAT - Loss in Heat Input

The vessel is acceptable for "normal" process conditions but is not acceptable under "upset" conditions where heat input is lost. Options for mitigating this potential scenario are replacement of the vessel with one having the appropriate MAT or installation of protective interlocks.

Overflow / Transport of Liquid:

Another potential scenario involves the overflow or transport of a light liquid hydrocarbon into sections of the process that are not supposed to contain liquids. Typically, these type scenarios occur due to multiple failures during a chain of events, culminating in an auto-refrigeration excursion. The carry over of liquid into downstream systems can be an initiator or a consequence of the chain of events. A particular area of concern in an olefins plant equipped with a back-end acetylene converter is the carry over of liquid material out of the de-ethanizer reflux drum into the normally dry, acetylene converter pre-heat exchanger train. For a brittle fracture potential to exist in this scenario, all three of the following conditions must occur:

1. The level in the de-ethanizer reflux drum must increase, overflowing the drum to the downstream exchanger train.

2. The pressure of the acetylene converter exchanger train must be reduced due to operations response or due to automated control actions, resulting in auto-refrigeration of the exchanger.

3. Through operator actions, normal process pressure is restored to the exchanger train after the exchangers are chilled due to auto-refrigeration.

If any one of the above items does not occur, the potential for brittle fracture, even under low temperature, auto-refrigeration conditions, is reduced.

Figure 15 represents an example of the MAT for a typical back-end acetylene converter preheat exchanger. For this example, the pressure / temperature (CET) conditions can be well below the MAT after operations reestablishes normal operating pressure.

Brittle Fracture Analysis Report for Example Acetylene Preheat Exchanger

Fail Level 2 - Maximum difference between the MAT and vapor pressure curves or User entered points = 92°F.
Difference for selected curves: User entered points = 92°F
Minimum acceptable (coldest) temperature (MAT) when the pressure is at the MAWP is 8°F.
Maximum permitted pressure when the temperature is at -155°F is 212 psig
Recommended minimum field hydrotest temperature = 38°F

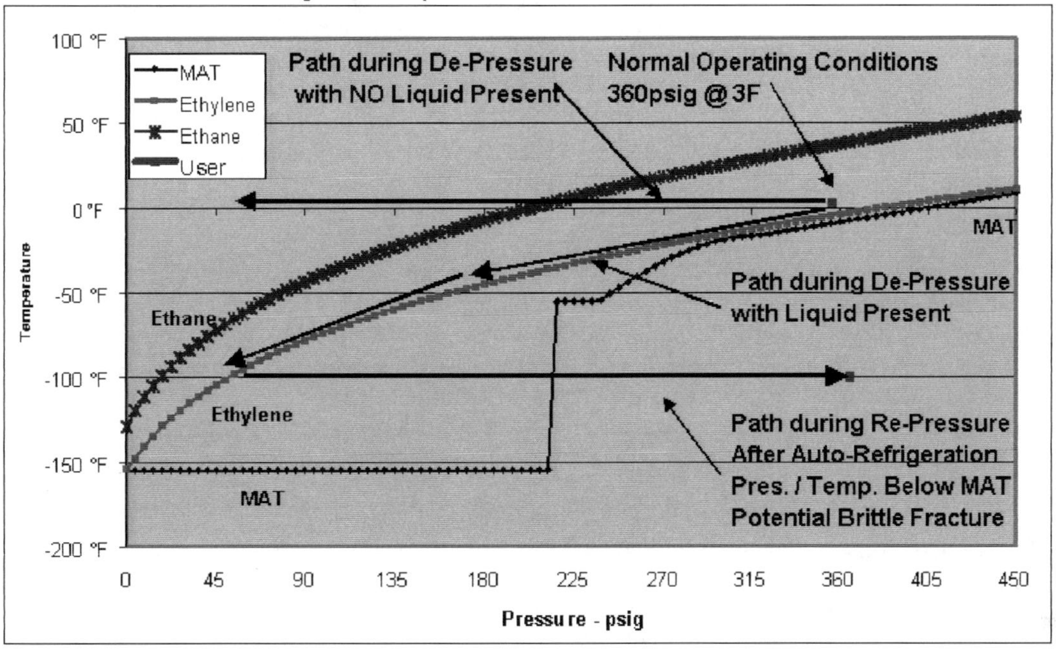

Figure 15 – Example Acetylene Converter Preheat Exchanger MAT

Prevention of this type of excursion can take the form of high level alarms and / or overhead isolation valves to prevent allowing the liquid light hydrocarbon from entering the downstream equipment. This example is particularly important from an Operator response / training perspective. This is a rather typical vessel made of SA516-70 non-PWHT carbon steel. Thousands of these types of vessels are in light liquid hydrocarbon services. Operations personnel should be trained that anytime a de-pressure excursion occurs, the vessel temperature must be confirmed ***prior*** to reestablishing normal operating pressure. If it is determined that the vessel is cold due to auto-refrigeration, the pressure is ***not*** to be increased on the vessel until it has been warmed. Operations personnel should be trained to avoid the auto-refrigeration excursion. Failing that, they are ***never*** to increase the pressure on a cold vessel.

General Observations

Based on the experience of executing several Auto-Refrigeration / Brittle Fracture Reviews of existing olefins plants, a number of general observations can be made about the design and operation of a typical Olefins plant and about the evaluation process to identify potential scenarios.

1) Plants were designed for steady state operating conditions: Since many olefins plants were built prior to full understanding of auto-refrigeration and potential brittle fracture, the current materials-of-construction do not reflect these now known temperature limitations. In many instances the startup / shutdown or upset conditions were not fully evaluated when making vessel materials-of-construction decisions. In many instances the mere act of removing heat input from the process exposes vessels to a temperature potentially below the MAT.

2) A single event block: Using the described PHA Methods indicates that in many instances, operating mode sequences can result in auto-refrigeration. Often only one event block separates conditions from being at potential risk of brittle fracture. Additional protective measures to avoid the excursion or to prevent certain sequence of events are required to ensure adequate levels of protection.

3) Materials-of-construction selection for new vessels: In many instances the materials-of-construction do not provide adequate separation between process CET and vessel MAT. For a vessel to be intrinsically safe, the MAT (MDMT) should be set at the lowest possible CET. For small vessels, there are minimal cost implications for this selection. For larger vessels, a more detailed analysis may allow selection of a lower cost alloy as long as adequate separation is provided between the CET and MAT under all operating modes. Adequate separation must be provided to allow vessel wall temperature to return to equilibrium conditions with the process without reaching the MAT.

4) Key points for a successful Auto-Refrigeration / Brittle Fracture Review Program:
   a) Management Support: Support of both the review process and the follow up recommendations.
   b) Analyze Equipment: **All** vessels that contain or have the potential to contain a light liquid hydrocarbon must undergo an analysis.
   c) Simulate Process Conditions: In cases where the normal operation is close to the MAT curve, process simulation results may be necessary to accurately predict the behavior of the hydrocarbon mixture.
   d) Get the Right Team: Key to the success of the team is to have an Operational Expert. The Operational Expert is that key, highly experienced operations representative (operator, supervisor, specialist) who understands how, when and why certain operational actions are taken.
   e) Have Dedicated Resources: A high intensity, short-term effort is required to maintain focus and to bring about completion of the review.
   f) Acceptable Risk: Predefine the level of acceptable risk (Risk Matrix – consequence and predicted frequency of occurrence) prior to the review.
   g) Risk Ranking: During the PHA, the team should rank each scenario that results in a recommendation so as to determine when sufficient levels of protection / mitigation have been attained. Ranking post PHA results in substantial confusion by uninvolved project team members.

5) Evaluate Abnormal Operation: Auto-refrigeration excursions by their nature do not occur during steady state operation. In order to identify possible auto-refrigeration excursions, the team must consider process upset scenarios that are far removed from normal operation and likely outside the past operating history or experience. In addition to normal operation, the team must consider:

   - Normal Operation
   - Upset Conditions
   - Normal Startup
   - Normal Shutdown
   - Emergency Shutdown

- Air Freeing / Nitrogen Freeing
- Inventory
- De-inventory
- Not in Operation / Maintenance-in-Progress
- Commissioning / Leak Testing

The challenge to the review team members is to identify all of the ways the pressure and / or temperature excursions can occur and to develop appropriate action items to prevent or mitigate the risk. Use of the Process Data Historian to search for temperatures at or below the affected vessel MAT can assist in identification of past excursions that potentially could have resulted in brittle fracture.

6) Provide Training: Key to a safe recovery from an auto-refrigeration excursion is for operations personnel to be well trained in the phenomenon of auto-refrigeration. How to prevent it from occurring, how to identify that it is occurring, what to do if confronted with auto-refrigeration and, just as important, what ***not*** to do if confronted with auto-refrigeration. Training provides the first line of defense in preventing auto-refrigeration excursions.

7) Inspection Guidelines: Provide guidance to inspection personnel on the method and extent of post auto-refrigeration excursion inspection. The extent of the inspection is based on the excursion CET compared to the vessel MAT.

Concluding Remarks

This description of an Auto-Refrigeration / Brittle Fracture Review process is provided to encourage owner / operators of light liquid hydrocarbon processing units to evaluate their existing units for auto-refrigeration / brittle fracture potential. The "Findings" and "General Observations" are provided as thought provoking examples of some of the potential scenarios that can be identified while undertaking such a review. Making appropriate process changes, installing alarms and interlocks and training operating personnel in how to properly respond has been emphasized so as to avoid excursions and ensure that the CET is above the MAT under all possible operating conditions. Replacement of equipment with upgraded materials-of-construction / metallurgy is suggested when an appropriate level of protection cannot be obtained by other means.

Acknowledgments:

Having executed several Auto-Refrigeration / Brittle Fracture Reviews, the author has had the privilege to work with several very talented and knowledgeable engineers. Their support in this endeavor and their assistance in writing this paper are greatly appreciated.

Thomas E. Kelly, Consulting Engineer, Process Engineering – Olefins, Lyondell / Equistar
Craig Thompson, Consulting Engineer, Process Engineering – Olefins, Lyondell / Equistar
James H. Walker, Process Supervisor, Process Engineering – Olefins, Lyondell / Equistar
Walter Pinto, Manager Stationary Equipment – Specialty Engineering, Lyondell / Equistar
J. Robert (Bob) Sims, Becht Engineering Co., Liberty Corner, NJ

References:

1. API 579 Recommended Practice For Fitness-For-Service, Section 3 Assessment of Existing Equipment for Brittle Fracture

2. McLaughlin, J.E., and Sims, J. R., "Assessment of Older Equipment For Risk of Brittle Fracture", ASME - PVP-Vol. 261, Service Experience and Life Management: Nuclear, Fossil, and Petrochemical Plants

3. Messinger, John R., "Cold Service Risk Assessment of Existing Bimetallic Demathanizers", AICHE 2001 Spring National Meeting, 13th Ethylene Producers Conference, Houston, TX April 22-26

4. Jacobs, Wesley S., "ASME Code Material Requirements for Low Temperature Operation, Section VIII, Division 1 & Division 2, 1998 Edition, 1999 Addenda"

EEPC

EEPC Working Group

Issue Group: Operator Training

Presentation by: H. Kruman, Huntsman
K. Funk, Linde

HSE Conference, Portugal 2003

EEPC

Several ethylene plants face a situation where experienced operators retire or leave, taking a lot knowledge with them.

The purpose of the issue group is to collect
and benchmark the experiences and practices
of the different companies, how to capture this knowledge.

EEPC

The following companies were involved:

- N. Vilhelmsen / A. Svendsen **Noretyl / Norway**
- D. Bretelle **AtoFina / France**
- R. Ophorst **Sabic / Netherlands**
- W. Van engelandt **BASF / Belgium**
- K. Funk **Linde / Germany**
- K. P. Hambrock **BP-Cologne / Germany**
- R. Krumpen **Basell-Wesseling / Germany**
- E. Viviano **Technip / Italy**
- H. Kruman **Huntsman / UK**

EEPC

11 main points have been identified for benchmarking:

- Staff Organization - Training Resources
- Documentation - Access
- Procedures: Routine - Start-up - Troubleshooting
- Media - Video - Training Simulators
- Lesson Learning
- Validation - Documentation
- How to capture knowledge: on-the-job training + others
- Recruitment - Qualification - Retirement of sr. operators
- Plant modifications / Revamp: training
- How to motivate for being trained and/or give training
- Main priority / Major challenges / positive experience

EEPC

1.) Staff Organisation –

Training spare on shift

Multiskilled operators

Apprentices on shift

Dedicated daytime trainer

Additional days per year for training

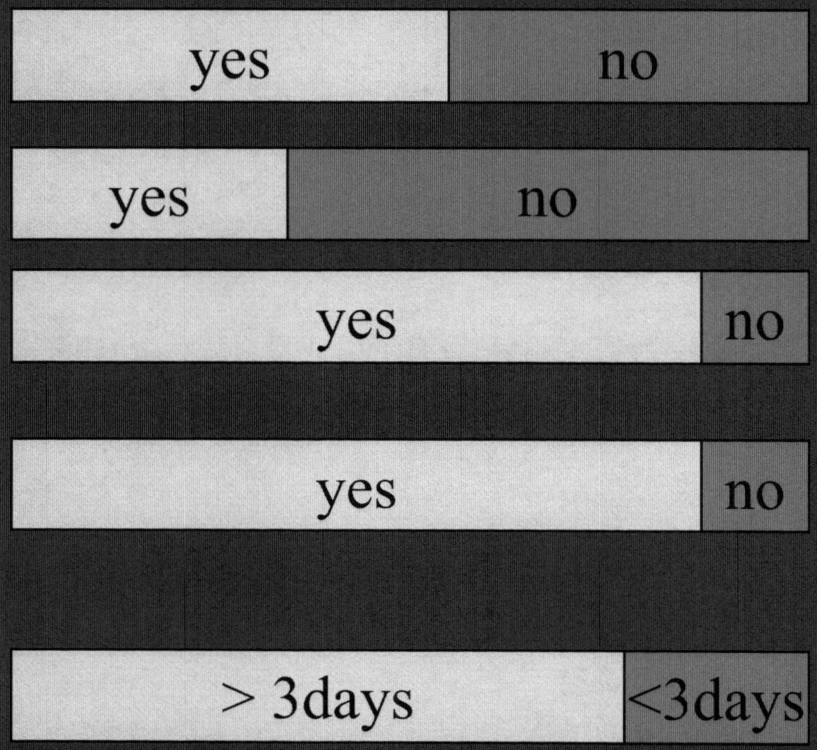

EEPC

Main Conclusion:

The major challenge to everyone is to sustain enough manpower to allow for spare operators on shift, this being recognized as the only way to practice on the job training.

EEPC

2.) Access to Documentation –

All of the companies have paper documents used for training purposes.

However in most cases electronic access to documents is available.

It is recognized as “best practice” to grant full electronic access to documents / databases.

Only in a few cases do companies allow individual internet and / or e-mail accounts.

EEPC

3.) Procedures –

Routine / incident / environmental and safety — yes

Start-up / shut-down / upset response / health / management skills training — yes | partly

Trouble shooting / maintenance preparation / general knowledge — yes | partly | no

EEPC

4.) Tools Used -

Documents / drawings / electonic access / classroom desktop	yes
Digital Photos	yes / partly
Interactive documents / video / CD / outside simulation / process data storage	yes / partly / no
Generic or Tailor made simulator	partly / no

EEPC

5.) Lesson Learning (incidents, upsets, etc.)

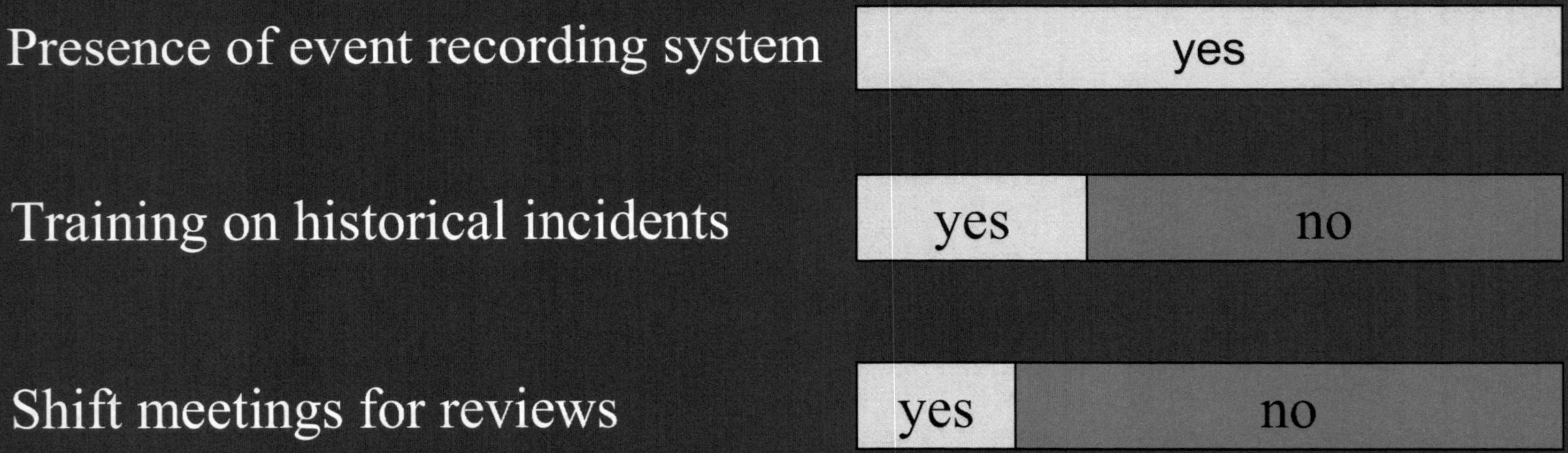

EEPC

Main conclusion:

Root cause analysis is a standard practice
Communication is given across the shifts on all topics

Major events (SHE related) are communicated across businesses and sites

Not one company is fully satisfied with their present system and see it as a challenge to improve the system

EEPC

6.) Validation / documentation of training –

Written test — yes | no

Test by interview including registration — yes

Validation of experienced operators — yes | partly | no

Not one company is fully satisfied with their present system

EEPC

7) How to Capture Knowledge –

On job training	yes
Defined content for on job training	yes / no
Development of procedure by experienced operators	yes / no
Special programs to deal with retirement	no

EEPC

8.) Recruitment – Qualification

- Formal qualifications: Technical / Process school
- In most cases, there is no overlap period
- As the origin of the companies is so diverse, details can be found in the hand-outs

EEPC

9.) Plant Modifications –

- Operators are involved only in large scale projects
- For large scale projects, formal training with class room and formal procedures are used
- For small projects, the training is scaled down to procedures

EEPC

10.) Motivation of the Trainer

- Is it a problem?
- Job rotation
- Reward, promotion
- Interest

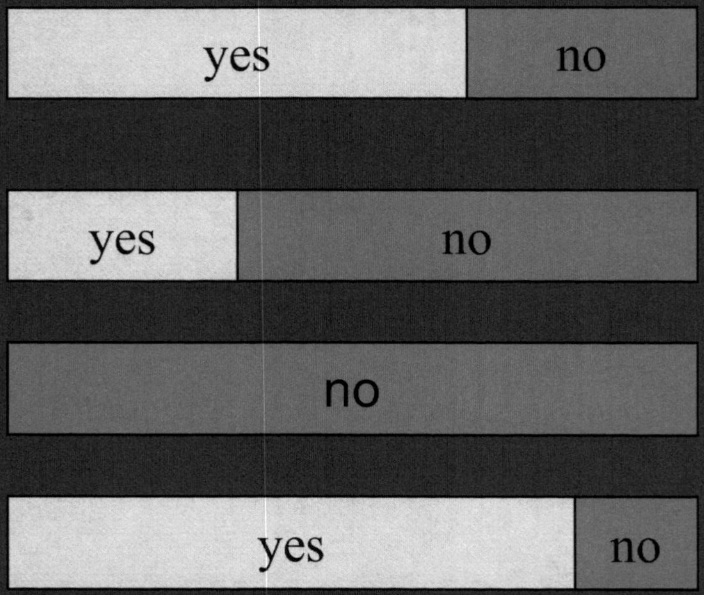

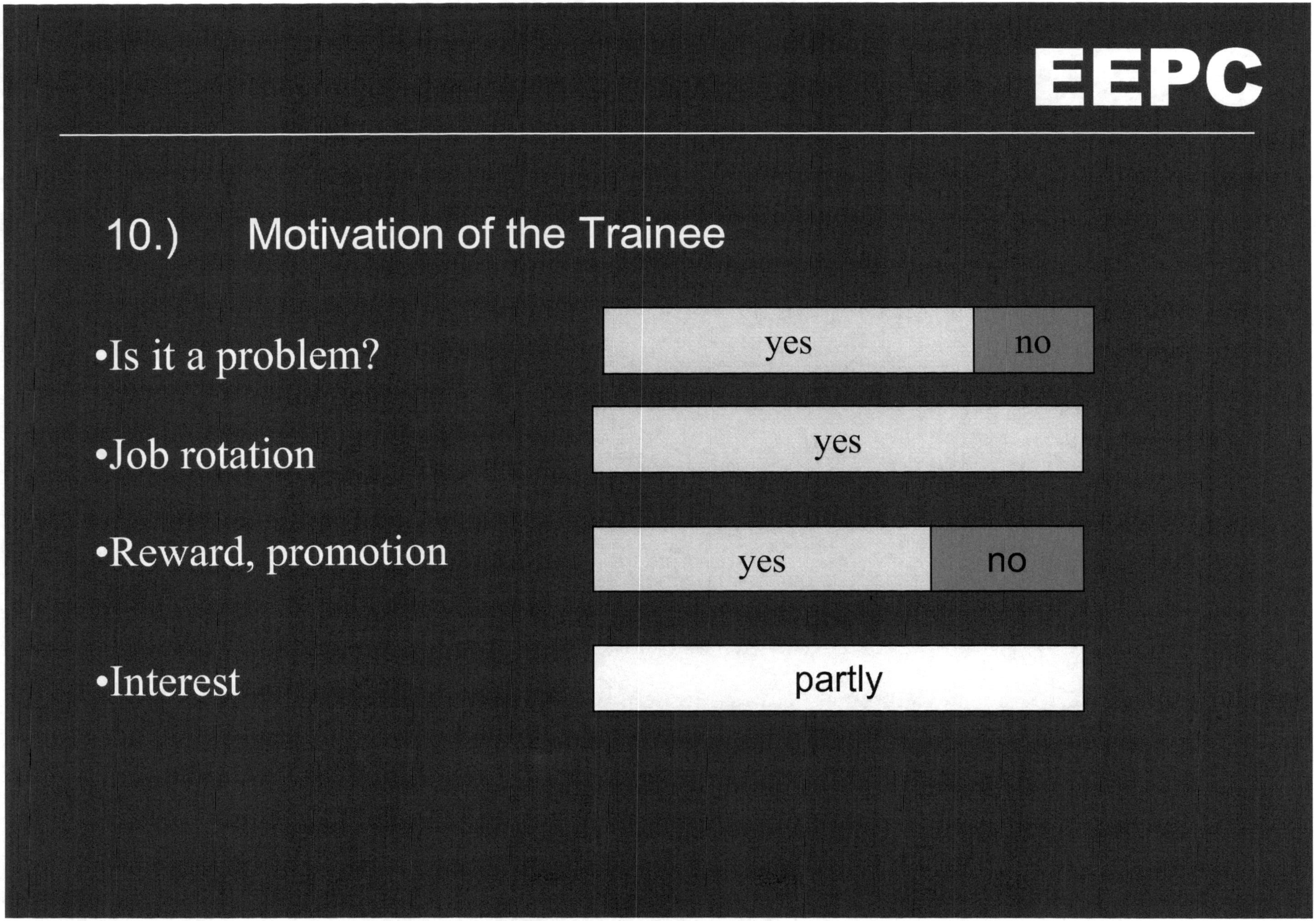
EEPC
10.) Motivation of the Trainee
•Is it a problem?
yes
no
•Job rotation
yes
•Reward, promotion
yes
no
•Interest
partly

EEPC

11.) Main Priorities for the Plants of the Team Members

- Written operating instructions, checklists, immediate response lists
- Capturing knowledge by working experienced and unexperienced together
- Training operators for troubleshooting and abnormal situation handling
- Start-up / Shutdown training
- Getting operators motivated for (self) training
- Validation of operator knowledge and skills

EEPC

The findings of the 11 bullet points discussed are all included within the handouts distributed as part of this presentation

EEPC

The Work Group looked in depth at all of the topics during their meetings & felt that a lot could be gained by inviting other Ethylene producing companies to contribute to the bench marking process

Many “best practice” methods could also be put in place if the group were to continue

EEPC

Shared Experiences

EEPC

1. Capturing Knowledge
2. Apprentice Schemes
3. Job Methods

EEPC

Capturing Knowledge

A database system whereby the knowledge of a Technician, gained over many years , can be retained for use by other personnel on the Plant.

A video in which plant start-up is trained (presented at the Rome 2001 EEPC meeting)

Plant instructions written by technicians themselves e.g. with linked photos

EEPC

Data base

Prior to retirement the Technicians are given quality time to put his knowledge into the data base

Putting the knowledge into the data base can take many formats

The data base is user friendly and has access for everyone involved with running the plant

EEPC

The inputs are written in such a way that everyone can understand the presentation

The inputs are updated as & when modifications / changes are made to the Plant

Everyone is encouraged to contribute to the system

EEPC

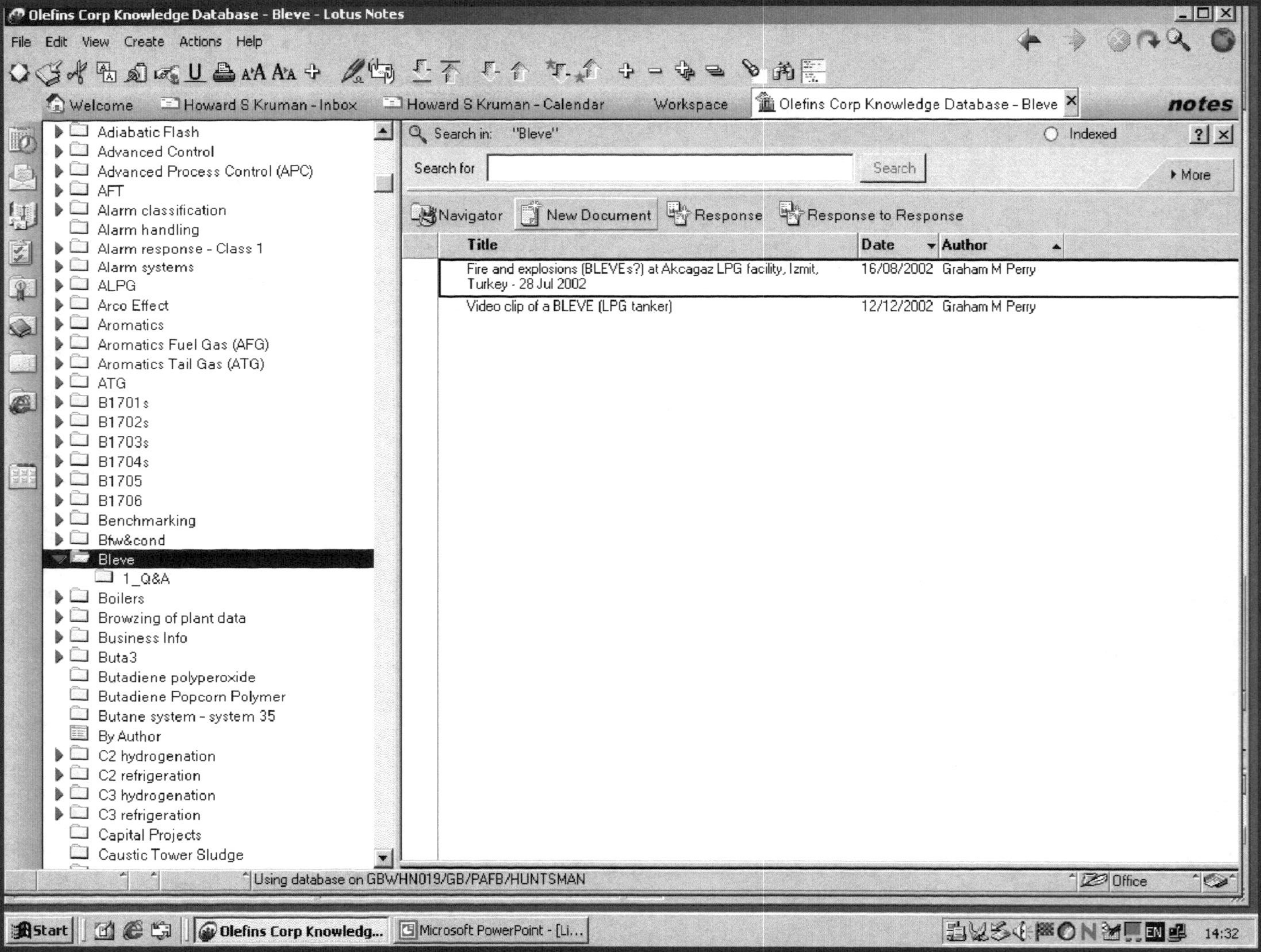

Olefins Corp Knowledge Database - Bleve - Lotus Notes
File Edit View Create Actions Help
Welcome
Howard S Kruman - Inbox
Howard S Kruman - Calendar
Workspace
Olefins Corp Knowledge Database - Bleve
notes
Adiabatic Flash
Advanced Control
Advanced Process Control (APC)
AFT
Alarm classification
Alarm handling
Alarm response - Class 1
Alarm systems
ALPG
Arco Effect
Aromatics
Aromatics Fuel Gas (AFG)
Aromatics Tail Gas (ATG)
ATG
B1701s
B1702s
B1703s
B1704s
B1705
B1706
Benchmarking
Bfw&cond
Bleve
1_Q&A
Boilers
Browzing of plant data
Business Info
Buta3
Butadiene polyperoxide
Butadiene Popcorn Polymer
Butane system - system 35
By Author
C2 hydrogenation
C2 refrigeration
C3 hydrogenation
C3 refrigeration
Capital Projects
Caustic Tower Sludge
Search in: "Bleve"
Indexed
Search for
Search
More
Navigator
New Document
Response
Response to Response
Title
Date
Author
Fire and explosions (BLEVEs?) at Akcagaz LPG facility, Izmit, Turkey - 28 Jul 2002
16/08/2002
Graham M Perry
Video clip of a BLEVE (LPG tanker)
12/12/2002
Graham M Perry
Using database on GBWHN019/GB/PAFB/HUNTSMAN
Office
Start
Olefins Corp Knowledg...
Microsoft PowerPoint - [Li...
14:32

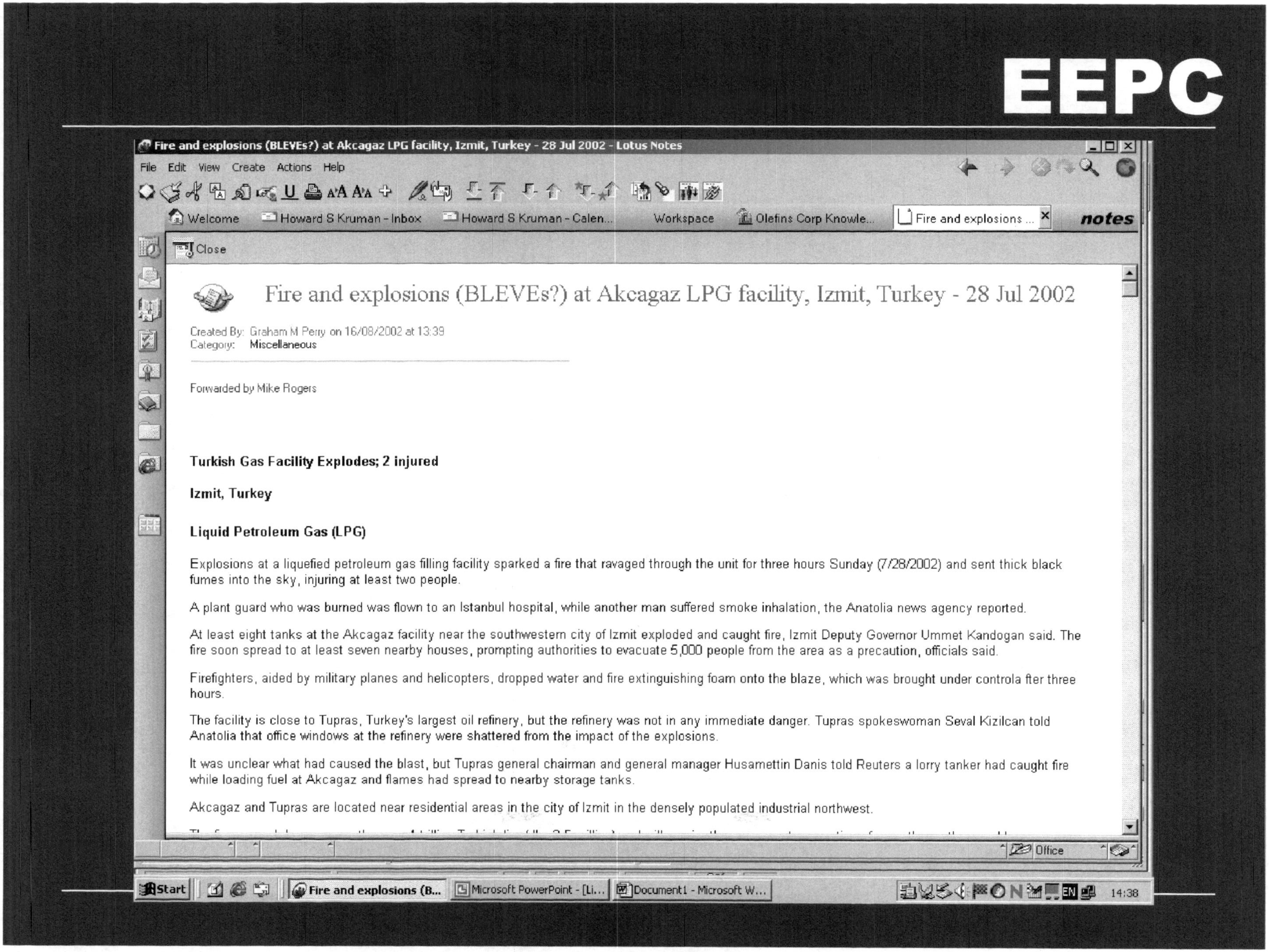

Fire and explosions (BLEVEs?) at Akcagaz LPG facility, Izmit, Turkey - 28 Jul 2002

Created By: Graham M Perry on 16/08/2002 at 13:39
Category: Miscellaneous

Forwarded by Mike Rogers

Turkish Gas Facility Explodes; 2 injured

Izmit, Turkey

Liquid Petroleum Gas (LPG)

Explosions at a liquefied petroleum gas filling facility sparked a fire that ravaged through the unit for three hours Sunday (7/28/2002) and sent thick black fumes into the sky, injuring at least two people.

A plant guard who was burned was flown to an Istanbul hospital, while another man suffered smoke inhalation, the Anatolia news agency reported.

At least eight tanks at the Akcagaz facility near the southwestern city of Izmit exploded and caught fire, Izmit Deputy Governor Ummet Kandogan said. The fire soon spread to at least seven nearby houses, prompting authorities to evacuate 5,000 people from the area as a precaution, officials said.

Firefighters, aided by military planes and helicopters, dropped water and fire extinguishing foam onto the blaze, which was brought under controla fter three hours.

The facility is close to Tupras, Turkey's largest oil refinery, but the refinery was not in any immediate danger. Tupras spokeswoman Seval Kizilcan told Anatolia that office windows at the refinery were shattered from the impact of the explosions.

It was unclear what had caused the blast, but Tupras general chairman and general manager Husamettin Danis told Reuters a lorry tanker had caught fire while loading fuel at Akcagaz and flames had spread to nearby storage tanks.

Akcagaz and Tupras are located near residential areas in the city of Izmit in the densely populated industrial northwest.

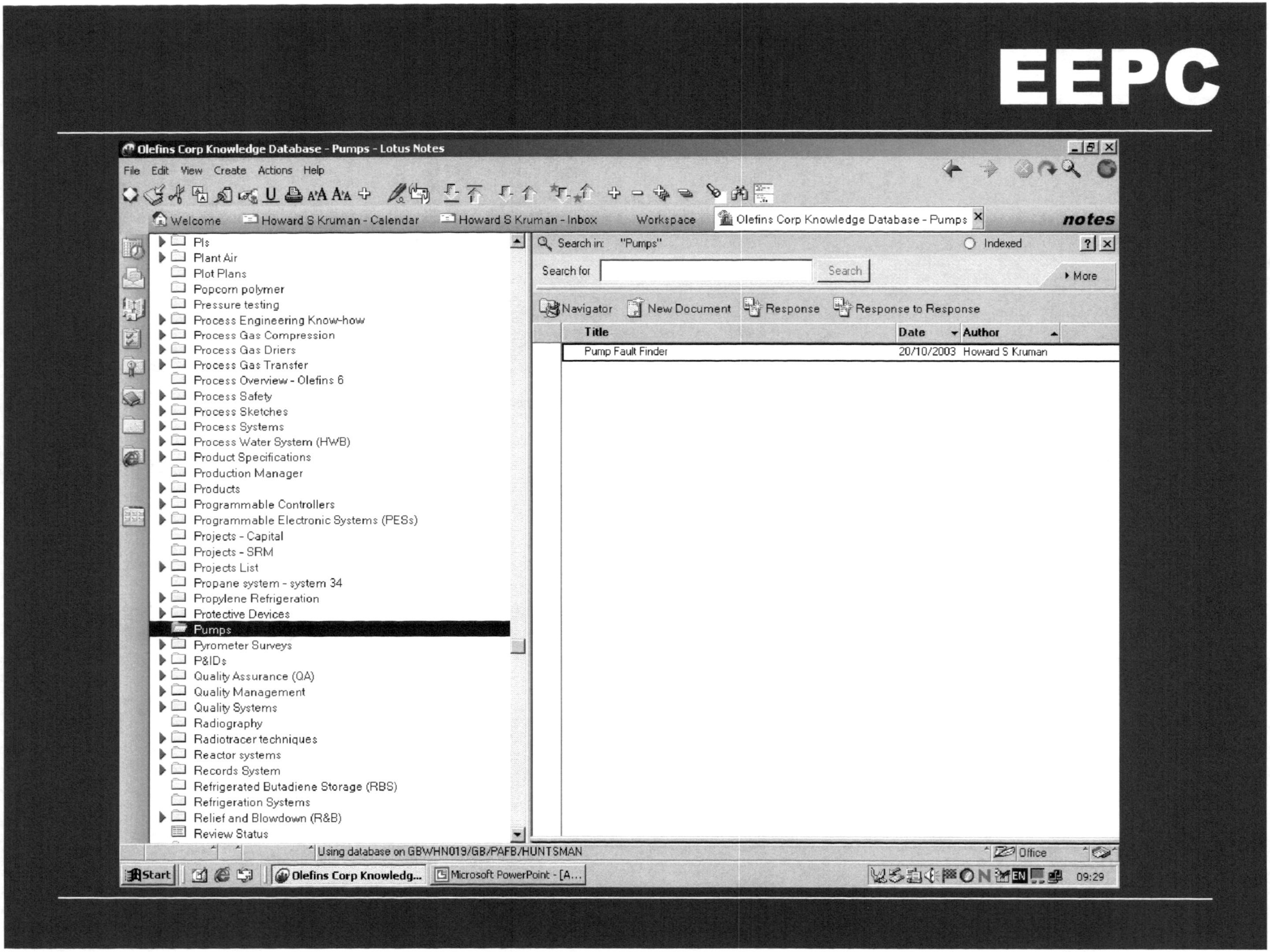
EEPC
Olefins Corp Knowledge Database - Pumps - Lotus Notes
File Edit View Create Actions Help
Welcome
Howard S Kruman - Calendar
Howard S Kruman - Inbox
Workspace
Olefins Corp Knowledge Database - Pumps
notes
PIs
Plant Air
Plot Plans
Popcorn polymer
Pressure testing
Process Engineering Know-how
Process Gas Compression
Process Gas Driers
Process Gas Transfer
Process Overview - Olefins 6
Process Safety
Process Sketches
Process Systems
Process Water System (HWB)
Product Specifications
Production Manager
Products
Programmable Controllers
Programmable Electronic Systems (PESs)
Projects - Capital
Projects - SRM
Projects List
Propane system - system 34
Propylene Refrigeration
Protective Devices
Pumps
Pyrometer Surveys
P&IDs
Quality Assurance (QA)
Quality Management
Quality Systems
Radiography
Radiotracer techniques
Reactor systems
Records System
Refrigerated Butadiene Storage (RBS)
Refrigeration Systems
Relief and Blowdown (R&B)
Review Status
Search in: "Pumps"
Indexed
Search for
Search
More
Navigator
New Document
Response
Response to Response
Title
Date
Author
Pump Fault Finder
20/10/2003
Howard S Kruman
Using database on GBWHN019/GB/PAFB/HUNTSMAN
Office
Start
Olefins Corp Knowledg...
Microsoft PowerPoint - [A...
09:29

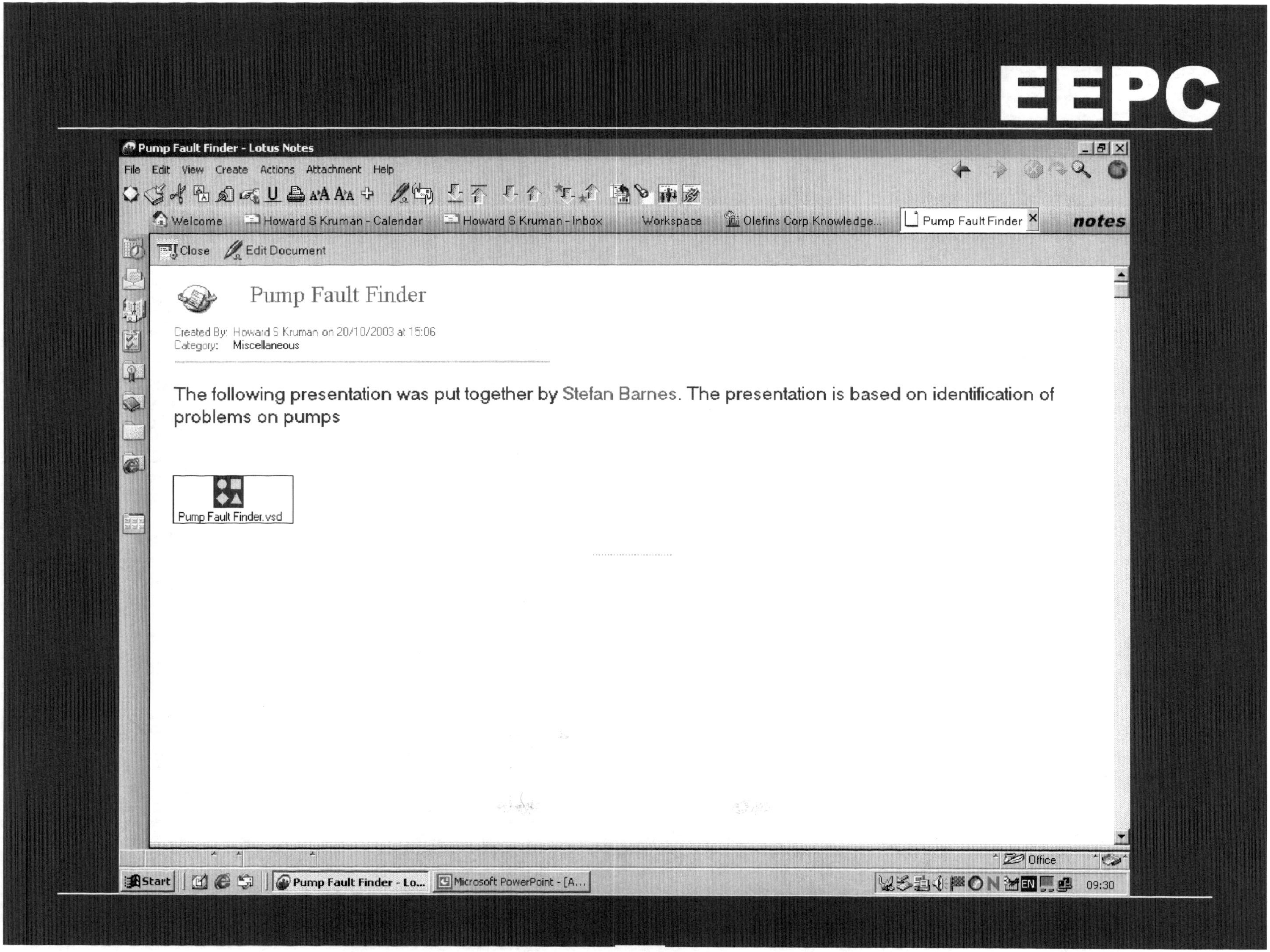
EEPC
Pump Fault Finder - Lotus Notes
File Edit View Create Actions Attachment Help
Welcome
Howard S Kruman - Calendar
Howard S Kruman - Inbox
Workspace
Olefins Corp Knowledge...
Pump Fault Finder
notes
Close
Edit Document
Pump Fault Finder
Created By: Howard S Kruman on 20/10/2003 at 15:06
Category: Miscellaneous
The following presentation was put together by Stefan Barnes. The presentation is based on identification of problems on pumps
Pump Fault Finder.vsd
Office
Start
Pump Fault Finder - Lo...
Microsoft PowerPoint - [A...
09:30

EEPC

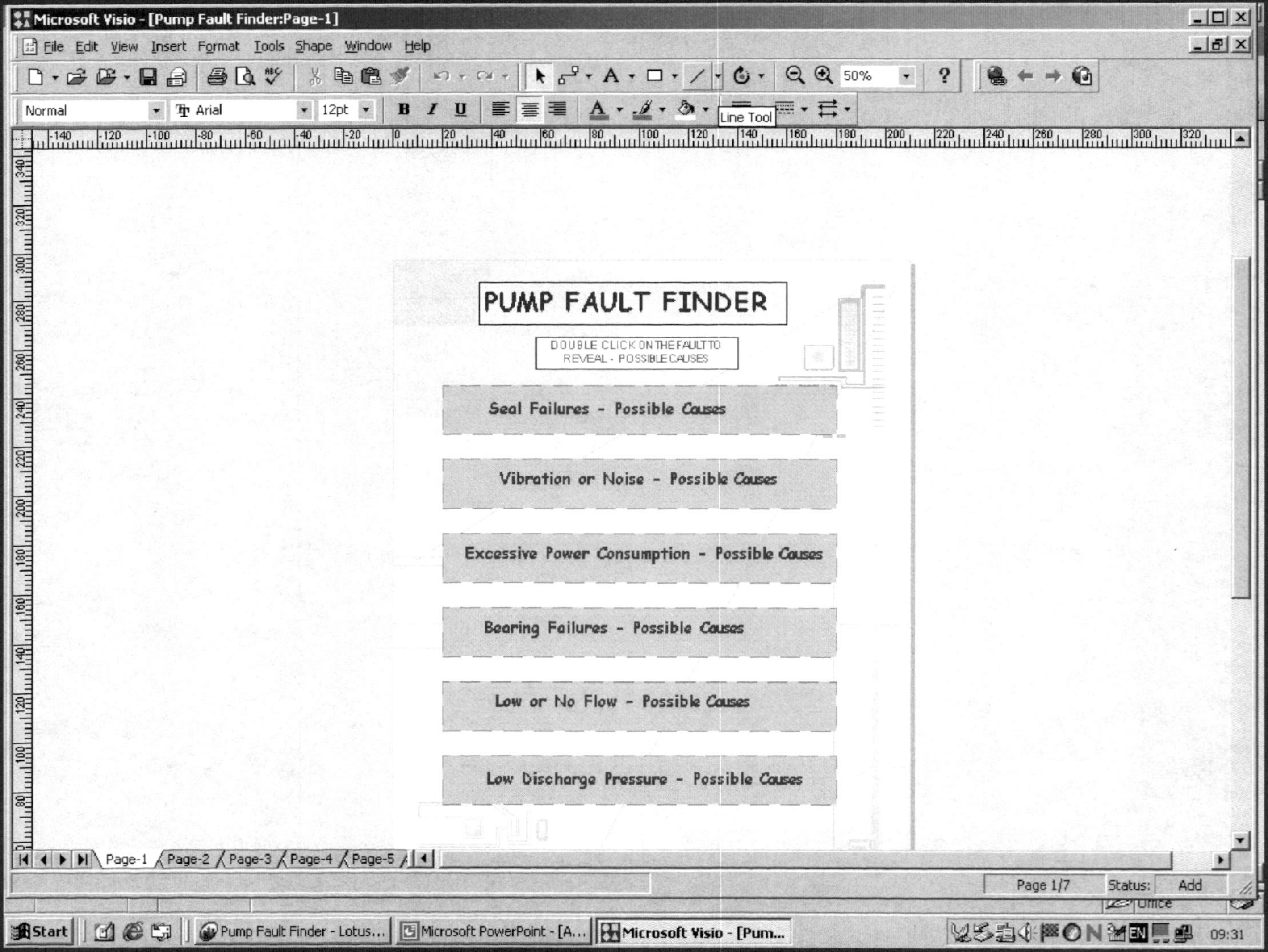

Microsoft Visio - [Pump Fault Finder:Page-1]
File Edit View Insert Format Tools Shape Window Help
PUMP FAULT FINDER
DOUBLE CLICK ON THE FAULT TO REVEAL - POSSIBLE CAUSES
Seal Failures - Possible Causes
Vibration or Noise - Possible Causes
Excessive Power Consumption - Possible Causes
Bearing Failures - Possible Causes
Low or No Flow - Possible Causes
Low Discharge Pressure - Possible Causes
Page-1 Page-2 Page-3 Page-4 Page-5
Page 1/7
Status: Add
Start
Pump Fault Finder - Lotus...
Microsoft PowerPoint - [A...
Microsoft Visio - [Pum...
09:31

EEPC

The one thing common to all of the systems is the use of experienced Operators in generating the material

EEPC

Apprentice schemes

Some of the participating Companies are running Apprentice schemes.

These schemes are required to supply labour due to an aging workforce some of which will be retiring within the next few years

Some Companies are training their own apprentices whilst others are are drawing from a pool of labour that is being trained by a Central Training Resource

EEPC

If the apprentices are trained on the plant on which they will work they will gain experience not only from the Training Officer but also from the experienced Technicians in the area

The greater the exposure to the plant the young trainees have the easier the transition once they have served their apprenticeship

EEPC

Job methods

A good Job Method not only tells you how to do the task but also gives you the thinking processes whilst carrying out each stage of the task …. These are known as Key Points

The best Job Methods are those written by the Technicians

EEPC

	Action	Key Point	Signed
1	Confirm that all WCPs are signed off	Your supervisor will advise	
2	Check around the pump for completeness	Check that all vents & drains are shut	
3	Open the suction isolation valve on the pump	Keep your eyes open for leaking joints Make sure that all of the inerts are vented to flare	
4	Crack open the delivery valve on the pump	Inform the Panel Technician that you are about to start the pump	
5	Start the pump & slowly open the delivery valve	The amps should read 15A & the Delivery pressure should be in the region of 300psi	
6	Shut down the spare pump as per Job Method 1234	Remember to tell the panel man	

EEPC

Questions ?

IMPLEMENTATION OF THE HAZARDS AND EFFECTS MANAGEMENT PROCESS (HEMP)

Natalie Salter
Shell Global Solutions (US) Inc.

Mark Siddoway
Shell Chemical LP

Westhollow Technology Center,
PO Box 4327,
Houston, TX, 77210

Prepared for Presentation at the 2004 Spring National Meeting, New Orleans LA, Apr 25-29
16th Annual Ethylene Producers' Conference
Session T8001 – Ethylene Plant Safety

Copyright © Shell Global Solutions (US) Inc.
April 2004

UNPUBLISHED

AIChE shall not be responsible for statements or opinions contained in papers or printed in its publications.

IMPLEMENTATION OF THE HAZARDS AND EFFECTS MANAGEMENT PROCESS (HEMP)

Natalie Salter
Shell Global Solutions (US) Inc. – Westhollow Technology Center

and

Mark Siddoway
Shell Chemical LP – Westhollow Technology Center
Houston, Texas, USA

Abstract

The Shell Group of Companies has developed the Hazards and Effects Management Process (HEMP) to identify the Health, Safety, and Environmental (HSE) hazards at a manufacturing facility. The HEMP process is an analysis technique that reviews the identified hazards and uses an internal risk assessment matrix to rank the risks based on consequence and likelihood. The hazards and associated risk rankings of high, medium or low are documented in a hazard register. The hazards with a high risk ranking are reviewed in detail using the "bowtie" technique. The bowtie technique combines a fault tree analysis with an event tree analysis into a pictorial representation. The top event is identified at the center of the bowtie. Threats that could lead to the top event are identified on the left, and the consequences from the top event are identified on the right. The control measures that are identified as preventing the top event from occurring are reviewed for each threat and placed on the line from threat to top event, while recovery measures are identified for each consequence and placed on the line from top event to consequence. The resulting pictorial representation resembles a bowtie. Tasks and activities are identified for each control measure, and responsibility for maintaining those tasks and executing required activities are assigned. As a result, the site has a complete list of tasks and activities to manage the "high risks" that are identified as part of the HEMP process.

Background

The Royal Dutch/Shell Group of companies introduced the Hazard and Effect Management Process (HEMP) in 1994, as a new element in their revised Health Safety and Environment Management System (HSE-MS) to address risk assessment. The HEMP methodology identifies the hazards and demonstrates how those risks are managed. The process requires key or critical tasks and/or activities to be identified with a responsible individual who is accountable for maintaining the risk controls.

HEMP Process

The HEMP process has three parts:

1. Hazard Register
2. Bowtie Analysis
3. Gap Analysis

Once these parts are completed, the work is combined into an HSE Case that is used as reference documentation for future work at the sites. This documentation is maintained via the management of change, incident investigation, and other ongoing processes. A complete revalidation of the HSE Case is done on a five year interval.

Hazard Register

The Hazard Register is used to document the hazards identified, the consequences related to the hazard, the risks associated with the hazard, and what actions are used to control the risk. Hazard identification consists of reviewing all of the activities on site and determining the hazards associated with the activity. Typically, hazards are chemicals or physical phenomena such as heat, impact or height. Examples are high-pressure gas, LPG, hot surfaces, specific bacteria, and parasites. Hazards have the potential to cause harm, including ill health and injury; environmental impact; company reputation damage; damage to property and products; production losses; or increased liabilities. Once a hazard has been identified, the "top event" that leads to the consequence is discussed. The most common top events are "loss of containment" for a process fluid or "loss of control" for physical events. Consequences of the hazard are documented in the hazard register, and then each consequence is risk ranked using an internal risk assessment matrix. Risks are ranked in four categories: People (personal impact); Assets (loss of assets and consequential business loss); Environment (affect on the environment and the community); and Reputation (of the company). A sample list of the hazards categories that may be reviewed can be found in Table 1. See Table 2 for a sample from an olefins unit.

The team of individuals used to complete the hazard identification process may include process specialists, HSE Specialists, industrial hygienists, hardware discipline specialists, and operational specialists. The team then reviews the hazards, and all the hazards identified as "high risks" are collated for further evaluation using the bowtie methodology.

Bowtie Analysis

The bowtie analysis is a risk assessment tool that generates the following three outcomes:

1. A framework for each hazard that identifies incident scenarios and the associated threats that lead to one or more consequences. The framework associates control measures to the threat and consequence lines and relates these to the critical tasks and activities. This is done at a higher level than a Process Hazard Analysis (PHA) review in that one bowtie analysis might combine multiple threats with multiple consequences into numerous theoretical incident scenarios and hundreds of control measures, critical tasks and activities. A bowtie analysis is different from other risk assessment or review tools that do not generate such a framework.
2. An action list that addresses the status of controls and/or critical tasks and activities in the actual project, site or production unit.
3. A link to the management system of the site or production unit for assurance that the necessary controls and/or critical tasks and activities are in place.

The bowtie starts with the hazard that has been identified in the hazard identification and investigates the threats that could lead to the top event. Here a threat is a possible cause that will potentially release the hazard and produce an incident. Examples of threats related to a hydrocarbon hazard are high temperature, corrosion, runaway reactions, fatigue, overpressure, freezing, and opening of equipment. The top event is the way in which the hazard can be released to do harm. Examples of top events are loss of containment, loss of control, oxygen deficiency, and uncontrolled release of energy.

Controls measures are used to reduce the probability of either the hazard's potential for harm and/or its consequence. These control measures are designed to prevent the threat from becoming an event or prevent a particular event from reaching a particular consequence. Examples of controls are materials of construction, protective devices, segregation, procedures, inspection, and training. The controls are represented on the main line of the bowtie between the threat and the top event in Figure 1.

The consequence is the final result of the hazard being released but not being controlled. Examples of consequences are fire, explosion, injury or illness, and vegetation or fish kills. Controls on the main line from the top event to the consequence are called recovery measures. These reduce the risk that the release leads to the consequence or mitigates the consequence. Examples of these include dikes, gas detectors, emergency responders, and fire suppression. These can reduce the likelihood that the top event will develop into further consequences and/or provide life saving capabilities should the event develop further. Figure 1 shows the complete bowtie. The resulting bowtie is a combination of a fault tree analysis with an event tree analysis into a pictorial representation.

Each of the controls, whether on the threat side or consequence side of the bowtie, requires a critical task or activity to be defined to maintain the effectiveness of the barrier. At least one critical task or activity is required for each control, but there may be more, and they may be cross-functional. Engineering, operations, and maintenance may all have activities to support a single control on a given threat line. It is important for the task or activity to be documented so that accountability for the activity can be realistically placed with a single individual (e.g. department head, inspector, operator, etc.) The critical tasks and activities identified during the bowtie analysis are then linked to the HSE-MS of the unit and/or site so that the complete set of risk reduction controls is maintained and effective.

Gap Analysis

Once the bowtie analysis is complete and the critical tasks and activities are identified for the production unit or site, the risk reduction controls have only been identified and must then be implemented. The next step is for the production unit or site to complete a gap analysis. The gap analysis helps to determine if each of the identified required controls are in place and if the individuals accountable to maintain the control are aware of their responsibility. These gaps are then documented, and remedial action plans are developed if necessary. Implementation is tracked via the HSE-MS management review process.

HEMP Results

The time it takes to complete the HEMP work at a facility can vary depending on the size of the site. The first week consists of the hazard identification phase for the site. The group completing the hazard identification may consist of process personnel, HSE specialists, public relations staff, pressure equipment specialists, and industrial hygienists, as well as the facilitator. The process units are broken into smaller process blocks with similar chemical and physical properties. An overview of what happens in the process block is given so the whole group understands the basis for the area. Then the top event and consequences are discussed for the process block. Each consequence is then risk ranked per the risk assessment matrix, and the risk is documented in the hazard register. Those items with high risks are identified and clearly marked. For a sample site, there can be on the order of 300 hazards identified and risk ranked, of which one quarter may require bowties from a process unit standpoint and less than 5% may require bowties to address personal safety issues. Table 2 shows an excerpt of the hazard registry from a sample olefins plant with the risk assessment completed for People, Assets, Environment and Reputation.

The bowtie analysis for the sample olefins plant unit results in bowties being developed, several of which are very similar due the nature of the equipment. For example, the propylene and ethylene refrigeration systems are very similar in their design, with the exception of materials of construction and temperature of operation. Therefore, one bowtie could be developed for the ethylene refrigeration system and then modified to address the propylene system. The bowtie analysis brainstorms the threats that could lead to the loss of containment. The controls that are in place for the prevention or reduction of the threat that leads to the top event are then documented.

As an example, internal corrosion is a threat to loss of containment of a process gas compressor. Controls that could mitigate this are:

- Design - Corrosion allowance, metallurgy, including the need to stress relief those systems in caustic service.
- Mechanical Integrity – Inspection of the pressure equipment and piping.
- Procedures – Decontamination of the lines in caustic service when down for maintenance.

Figure 2 has an excerpt from a Process Gas Compressor Bowtie – Corrosion Threat Line. This process is completed for all the threats identified for the loss of containment scenario.

The assumption is then made that loss of containment or the top event occurs and then the controls for the recovery or mitigation of the consequence are reviewed. In a Process Gas Compressor case, one of the consequences could be a Vapor Cloud Explosion (VCE). The controls that mitigate or minimize the risk of a VCE from occurring are:

- Design – Control of ignition sources (electrical area classification, unit spacing, grounding requirements, etc.).
- Design – Fixed fire water system, deluge, gas detection, etc.
- Permit to Work – Control of ignition sources (vehicle entry, low energy/hot work, etc).
- Initiate Emergency Response – Call the plant emergency number.
- Emergency Preparedness and Response – Drills, pre-plans, staging, people, equipment, etc.

See Figure 3 for an excerpt from a Process Gas Compressor Bowtie – Vapor Cloud Explosion Consequence Line

For each of the controls, the critical tasks and/or activities associated with the control are determined, and the responsible individual is assigned to the task. In this manner, a list of critical tasks is developed for each individual at the site from the operations/crafts specialists to the plant manager. These tasks are then pulled together and used to identify individual competencies required for each job role and to make the individual aware of the tasks that they are responsible to complete. For example, on the corrosion threat line, the procedure for decontamination of caustic lines was identified as a control. The critical tasks for this control are:

- Operations Specialist – Follow the procedure for decontaminating all lines in caustic service to determine that the temperature of the piping does not exceed design.
- Maintenance Foreman - Make craftsman aware of the lines to be worked on that may have caustic contamination and require stress relief after repair work.

This list of critical tasks and activities is now a powerful tool for the site to use to manage risks. It identifies key competencies that are needed by individuals and helps in training and development plans for the site. It also becomes a tool in the management of organizational change. When re-organization takes place, the key tasks are identified that need to be completed by the new organization. It also provides an auditing tool for reviewing how well the site is managing risk or the health of the HSE-MS. Most tasks and activities that are identified by this process are not new, and the site is already completing most of the tasks. The power of this tool is to put all this into a framework that can be used to show that the risks are being managed to an acceptable level.

Conclusions

- The Hazards and Effects Management Process or HEMP is a way to build a framework for the management of all identified risks at the site.
- HEMP provides a systematic consistent method of risk assessment that can be compared from site to site.
- The bowtie analysis provides a framework for demonstrating that identified high risks are managed to acceptable levels.
- HEMP allows the facility to compile a list of critical tasks and activities that allow the employees to understand what tasks are important and what competencies are required to complete those tasks.

Table 1 – Sample of Hazards for Hazard Identification

No.	HAZARD DESCRIPTION
H-01	**Process Hydrocarbon Feed and Streams**
	Specific to process units streams
H-02	**Refined Hydrocarbons**
H-02.01	Lube and seal oil
H-02.02	Hydraulic oil
H-02.03	Diesel fuel
H-02.04	Gasoline
H-02.05	Hydrocarbons above auto ignition temp
H-02.06	Asphalt/Bitumen
H-03	**Other Flammable Materials**
H-03.01	Cellulosic materials
H-03.02	Pyrophoric materials
H-03.04	Carbon fibre reinforced material
H-03.04	Dry vegetation
H-03.05	Hydrogen
H-04	**Explosives**
H-04.01	Detonators
H-04.02	Conventional explosives
H-04.03	Perforating gun charges
H-04.04	Explosive gases
H-05	**Pressure Hazards**
H-05.01	Hydrocarbons under pressure
H-05.02	Bottled gases under pressure
H-05.03	Water under pressure
H-05.04	NonHC gas under pressure in pipeworks
H-05.05	Air under high pressure
H-05.06	Hyperbaric operations
H-05.07	Decompression
H-05.08	Trapped pressure in equipment
H-05.09	High pressure equipment
H-06	**Hazards associated with differences in height**
H-06.01,2	Personnel at height
H-06.03	Overhead equipment
H-06.04	Personnel under water
H-06.05	Personnel below grade
H-06.06	Falling ice/snow
H-07	**Objects under induced stress**
H-07.01	Objects under tension
H-07.02	Objects under compression
H-08	**Dynamic Situation Hazards**
H-08.01	On land transport (driving)
H-08.02	On water transport (boating)
H-08.03	In air transport (flying)
H-08.04	Boat collision/offshore structures
H-08.05	Equipment with moving/ rotating parts
H-08.06	Use of hazardous hand tools,,knives etc.
H-08.07	Transfer from boat to offshore platform
H-08.08	Slips, Trips and Falls
H-09	**Physical Environment Hazards**
H-09.01	Weather
H-09.02	Sea state/river currents
H-09.03	Tectonic activity (earthquakes)
H-10	**Hot Surfaces**
H-10.01	Process piping equipment 60-150°C
H-10.02	Piping equipment > 150°C
H-10.03	Engine & turbine exhaust systems
H-10.04	Steam piping
H-11	**Hot Fluids**
H-11.01	Temperatures 100-150 °C
H-11.02	Temperatures >150 °C
H-12	**Cold Surfaces**
H-12.01	Process piping -25 to -80°C
H-12.02	Piping equipment < -80°C
H-13	**Cold Fluids**
H-13.01	Oceans, seas,rivers & lakes < 10°C
H-14	**Open Flame**
H-14.01	Heaters with fire tube
H-14.02	Direct fired furnaces
H-14.03	Flares
H-15	**Electricity**
H-15.01	Voltage > 50 - 600V in cables
H-15.02	Voltage > 50- 600V in equipment
H-15.03	Voltage > 600V
H-15.04	Lightning discharge
H-15.05	Electrostatic energy
H-15.06	Batteries
H-15.07	Stared charge (e.g. capacitors)
H-16	**Electromagnetic Radiation**
H-16.01	Ultraviolet radiation
H-16.02	Infra red radiation (e.g. fired eqpt)
H-16.03	Microwaves
H-16.04	Lasers
H-16.05	E/M radiation: high voltage ac cables
H-17	**Ionizing Radiation - Open Source**
H-17.01	Alpha, Beta - open source
H-17.02	Gamma rays - open source
H-17.03	Neutron - open source
H-17.04	Naturally occurring ionizing radiation /NORM
H-18	**Ionizing Radiation - Closed Source**
H-18.01	Alpha, Beta - closed source
H-18.02	Gamma rays - closed source
H-18.03	Neutron - closed source
H-19	**Asphyxiates**
H-19.01	Insufficient oxygen atmospheres
H-19.02	Excessive CO_2
H-19.03	Drowning
H-19.04	Excessive N_2
H-19.05	Halon
H-19.06	Smoke
H-20	**Toxic Gas**
H-20.01	H_2S, sour gas
H-20.02	Exhaust fumes
H-20.03	Ethyl Chloride
H-20.04	SO_2
H-20.05	Ammonia
H-20.06	Chlorine
H-20.07	Welding fumes
H-20.08	Tobacco smoke
H-20.09	CFCs (Old Freons0
H-20.10	HCFCs (New Freons0
H-21	**Toxic Liquid**
H-21.01	Mercury
H-21.02	PCBs
H-21.03	Biocides
H-21.04	Methanol
H-21.05	Brines
H-21.06	Glycols
H-21.07	Degreasers/Solvents
H-21.08	Isocyanates
H-21.09	Sulphanol
H-21.10	Amines
H-21.11	Corrosion inhibitors
H-21.12	Scale inhibitors
H-21.13	Styrene
H-21.14	Odorant additives
H-21.15	Alcoholic beverages
H-21.16	Recreational drugs
H-21.17	Used engine oils
H-21.18	Carbon tetrachloride
H-21.19	Grey and/or black water
H-21.20	Polycyclic Aromatic Hydrocarbons
H-21.21	MTBE or MMT
H-21.22	BTEX
H-21.23	Hexane
H-21.24	Ethyl Benzene
H-21.25	MEK
H-22	**Toxic Solids**
H-22.01	Asbestos
H-22.02	Man made mineral fibre
H-22.03	Cement dust
H-22.04	Sodium hypochlorite
H-22.05	Powdered mud additives
H-22.06	Sulphur/Sulphur dust
H-22.07	Pig trash
H-22.08	Oil based muds
H-22.09	Pseudo oil based muds
H-22.10	Water based muds
H-22.11	Cement Slurries
H-22.12	Dusts
H-22.13	Cadmium compounds & other heavy Metals (eg. Lead etc.)
H-22.14	Oil based sludges
H-22.15	Catalysts (fresh)
H-22.16	Catalysts (spent)
H-23	**Corrosive Substances**
H-23.01	Hydrofluoric acid
H-23.02	Hydrochloric acid
H-23.03	Sulphuric acid
H-23.04	Caustic soda
H-23.05	Spent caustic
H-23.06	Others
H-24	**Biological Hazards**
H-24.01	Poisonous plants
H-24.02	Large animals
H-24.03	Small animals
H-24.04	Food borne bacteria
H-24.05	Water borne bacteria (e.g. Legionella)
H-24.06	Parasitic insects
H-24.07	Disease transmitting insects
H-24.08	Cold & flu virus
H-24.09	HIV
H-24.10	Other communicable diseases
H-25	**Ergonomic Hazards**
H-25.01	Manual materials handling
H-25.02	Damaging noise
H-25.02	Loud, steady noises >85dBA
H-25.03	Heat stress
H-25.04	Cold stress
H-25.05	High humidity
H-25.06	Vibration
H-25.07	Work stations
H-25.08	Lighting
H-25.09	Incompatible hand controls
H-25.10	Awkward location of workplaces and machinery
H-25.11	Mismatch of work to physical abilities
H-25.12	Mismatch of work to cognitive abilities
H-25.13	Long & irregular working hours/ shifts
H-25.14	Poor organization and job design
H-25.15	Work planning issues
H-25.16	Indoor climate
H-26	**Psychological Hazards**
H-26.01	Living on the job/away from family
H-26.02	Working and living on a live plant
H-26.03	Post traumatic stress
H-27	**Security Related Hazards**
H-27.01	Piracy
H-27.02	Assault
H-27.03	Sabotage
H-27.04	Crisis
H-27.05	Theft, pilferage
H-28	**Use of Natural Resources**
H-28.01	Land take/ Land Use
H-28.02	Surface/Ground Water Contamination
H-28.03	Air Emissions
H-28.04	Soil Issues (Gravel)
H-28.05	Habitat (trees, Veg.) and Wildlife
H-29	**Medical**
H-29.01	Medical unfitness
H-30	**Hazardous Goods**
H-30.01	Dangerous goods in transport activities

Sample

Table 1 Continued:

No.	HAZARD DESCRIPTION
H-98	**Emergency Response/HAZmat**
H-98.01	BLEVE/ potential BLEVE
H-98.02	VCE/ unconfined VCE
H-98.03	Storage tank boil over
H-98.04	Storage tank explosion
H-98.05	Gas pressure release/ escalation (ignited/unignited)
H-98.06	Liquid inventory release/ escalation (ignited/unignited)
H-98.07	Pool fire
H-98.08	Jet fire
H-98.09	Toxic smoke, heat and combustion products
H-98.10	Toxic release
H-98.11	Spill/ clean up
H-98.12	ER equipment operations
H-98.13	General ER tasks

No.	HAZARD DESCRIPTION
H-99	**Environmental Aspects**
H-99.01	**Hazardous Air Emissions**
H-99.01.1	Sox
H-99.01.2	Nox
H-99.01.3	H2S
H-99.01.4	VOC
H-99.01.5	Toxics
H-99.01.6	Fine Particulates
H-99.01.7	Heavy Metals
H-99.02	**Other Air Emissions**
H-99.02.1	Greenhouse Gases
H-99.02.2	Noise
H-99.02.3	Odours
H-99.02.4	Visual
H-99.02.5	Ozone Depleting Substances
H-99.02.6	Light pollution
H-99.03	**Surface Water**
H-99.03.1	Impact on Water Source
H-99.03.2	Impact of effluent (vol. temp. quality)
H-99.03.3	Toxics in effluent
H-99.04	**Soil and Groundwater**
H-99.04.1	Potential for Contamination
H-99.04.2	Sensitive Area Neighbors
H-99.04.3	Existing Contamination
H-99.04.4	Contamination risk from Neighbours
H-99.04.5	Unique Aspect
H-99.05	**Waste**
H-99.05.1	Volume Generated
H-99.05.2	Routine/Non-Routine
H-99.05.3	Hazardous Waste

Sample

Table 2: Sample Olefins Plant Hazard Identification – Partial List

Hazard #	Stream	Top Event	Consequence	Risk Assessment			
				P	A	E	R
H-01.01	Fuel gas to furnace:	• Loss of containment	• Unignited release in open area (up to max 160 ppm H_2S in gas during start up)		L	L	L
			• Jet fire	L	M	M	M
		• Explosive mixture in firebox	• Vapour cloud explosion (VCE, furnace re-ignition)	H	H	H	H
			• Explosion on relight	H	H	H	H
H-01.02	Process Gas:	Loss of containment	• Fire or explosion • pump seal fire • flange fire • Vapour Cloud Explosion	H	H	H	H
			• Spills/Release (unignited release)		M	L	L
H-01.03	Green oil overhead to guard dryers	Loss of containment	• Fire or explosion • flange fire • Vapour Cloud Explosion	H	H	H	H
		Uncontrolled energy release	• Spills/Release (unignited release)	*	M	L	L
			• rupture, violent failure of equipment, mechanical damage, missile, pipe-work	H	H		
H-01.04	Dilution steam system	Loss of containment	• Spills/Releases	*	M	L	L
H-01.05	Ethylene refrigeration compressor cooling system	Loss of containment	• Fire or explosion • Vapour Cloud Explosion	H	H	H	H
			• Spills/Releases (unignited release)	*	H	H	M
H-01.06	Ethylene to pipeline – product	Loss of containment.	• Fire and Explosion	H	H	H	H
			• Pipeline explosion (detonation)	H	H	H	H
		Exposure to	• Spill release (unignited release)	*	H	H	H
		Hazardous Decomposition in Pipeline	• Release to flare	-	-	H	H
H-01.07	Vents and drop outs to flare	Loss of Containment	• Liquid fire at the flare tip and ground.	H	M	M	H
			• Explosion in Flare line.	H	H	H	H
		Smokey flare	• Unignited vapour cloud, liquid release	*	M	M	M
			Compliance incident (full unit relief to flare)			M	H

P – People, A – Assets, E – Environment, R - Reputation

H – High, M- Medium, L – Low

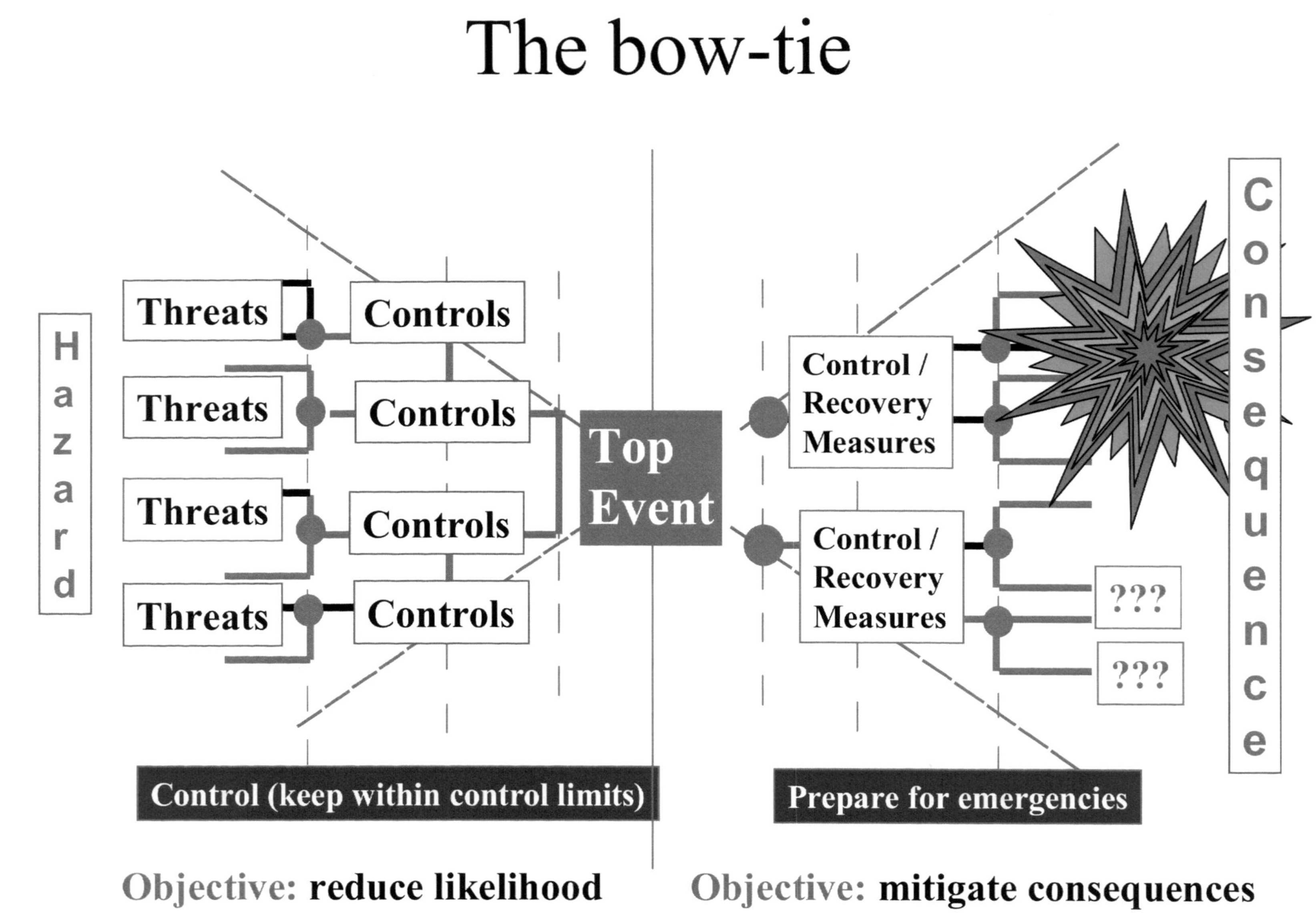

Figure 1: The Bowtie

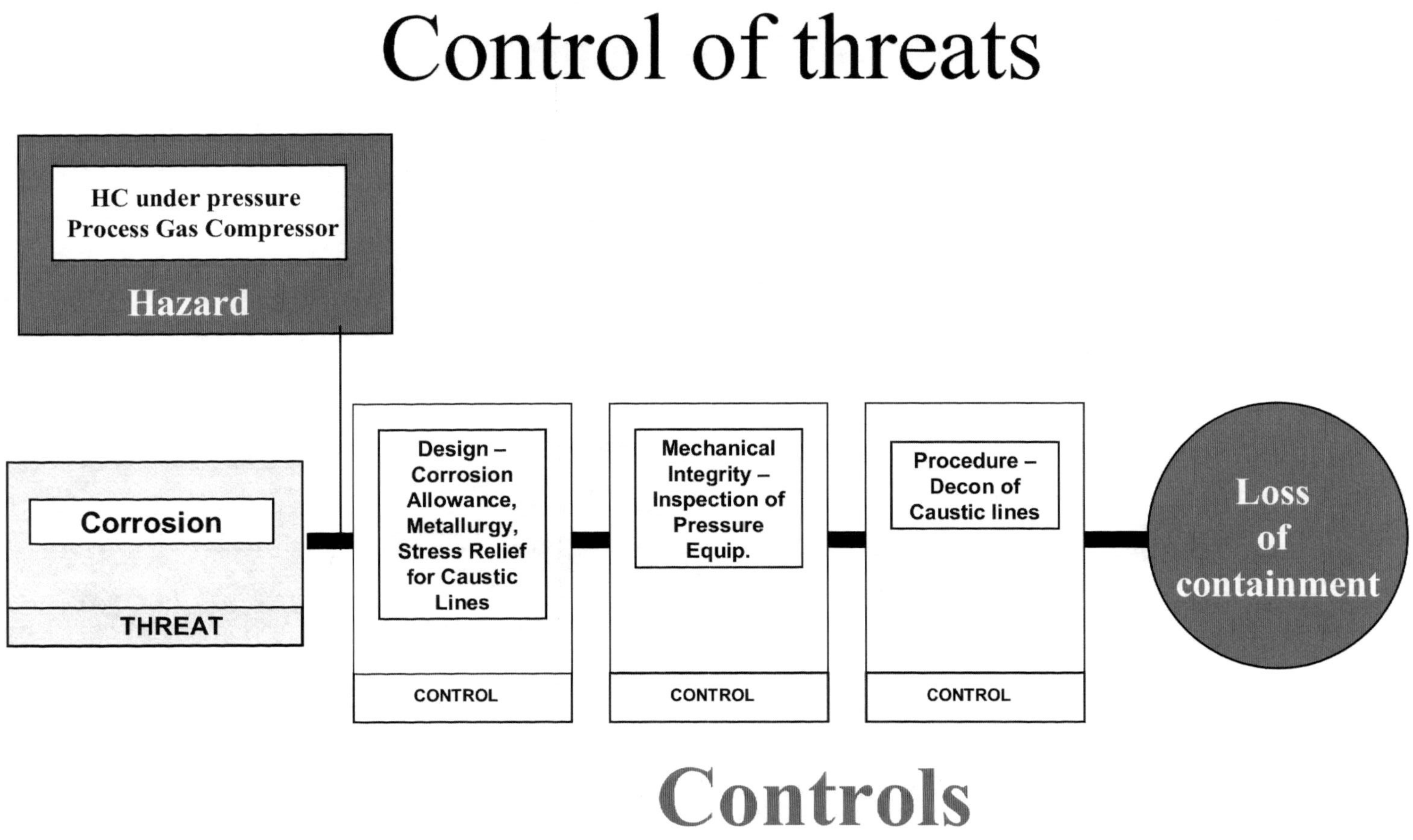

Figure 2 – A Sample Process Gas Compressor Bowtie excerpt – Corrosion threat line

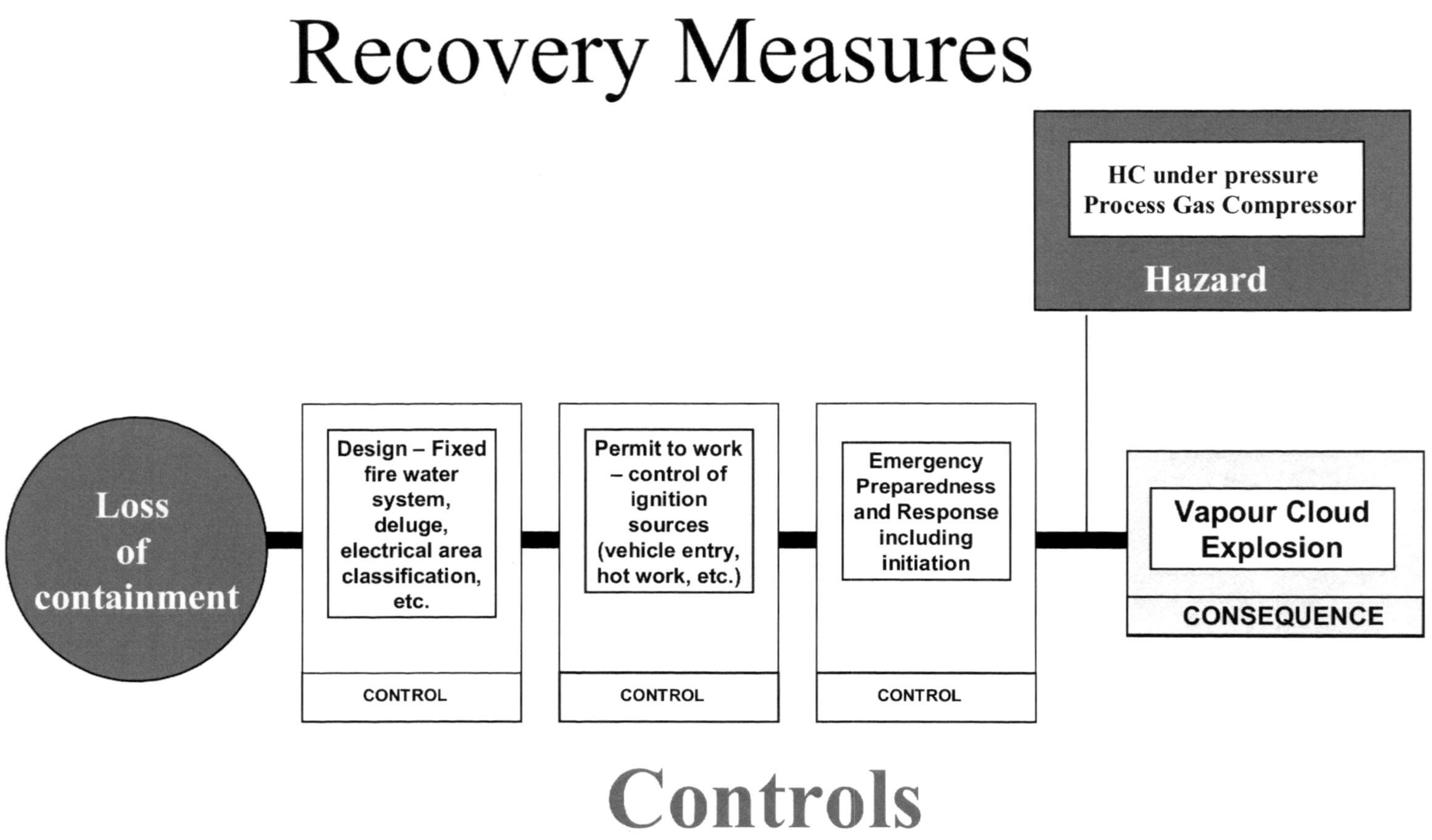

Figure 3 – A Sample Process Gas Compressor Bowtie – Vapour Cloud Explosion Consequence Line

End

PIPELINE INTEGRITY MANAGEMENT PROGRAM IMPLEMENTATION OBSTACLES

Reggie Harris
Eastman Chemical Company
Texas Operations
Hwy. 149, Kodak Blvd.
Longview, TX 75603

Prepared for presentation at the 2004 Spring National Meeting, New Orleans LA, Apr 25-29
16th Annual Ethylene Producers' Conference
Session T8001 – Ethylene Plant Safety

Copyright © Eastman Chemical Company
April 2004

UNPUBLISHED

AIChE shall not be responsible for statements or opinions contained in papers or printed in its publications.

Abstract

Eastman Chemical Company and Mustang Pipeline Company own and operate 780 mile of olefins and feedstock pipelines in Texas. The Integrity Assessment and Management Plan of 2001 was initially viewed as an unnecessary costly government regulatory program. Our exemplary past record with the Texas Railroad Commission inspections was ample evidence that our pipelines were well maintained and operated safely. Initial work on the new program was driven solely by the need to comply with the new rules. Now that we have overcome some of the obstacles, run the in-line inspection tools, received the results of these runs and completed some excavations of anomalies, we realize the added safety value of the Integrity Assessment and Management Plan. With two-thirds of the initial phase of this program remaining, we are encouraged by the results and the opportunity to continue our exemplary performance in this field. Eastman plans to be a leader in pipeline operations and utilize the new regulations to reinforce our commitment to maintaining a safe and reliable pipeline system.

Eastman Chemical Company and Mustang Pipeline Company own and operate 780 mile of olefins and feedstock pipelines in Texas. Department of Transportation Rule 49 CFR 195, "Pipeline Safety: Pipeline Integrity Management in High Consequence Areas" became effective May 29, 2001. This rule required pipeline operators to:

1. Identify pipeline segments that could affect high consequence areas.
2. Develop a written Integrity Management Program, and
3. Perform an assessment of each pipeline system either by using an in-line inspection tool (smart pig) or hydrostatic test. The rule required each operator to have at least 50% of line pipe in high consequence areas assessed by September 30, 2004.

The new rule was initially viewed as an unnecessary costly government regulatory program. Our exemplary past record with the Texas Railroad Commission inspections was ample evidence that our pipelines were well maintained and operated safely. Initial work on the new program was driven solely by the need to comply with the new rules. As we began, we believed the DOT schedules could be met without much trouble. However, we would soon learn that this was not going to be the case. Through our risk assessment analysis we selected the 6", 150 mile propylene pipeline as the first objective.

The search for a smart pigging supplier began and it became evident from our discussions, that our pipeline posed many unique opportunities. There were three areas of concern: (1) Length of the run – our sections were 89 and 61 miles, the pigging suppliers were use to less than 50 miles, (2) Speed – their optimum speed was 5 – 7 mph, our speed would only be 1.5 – 2.5 mph, and (3) Dryness – propylene is extremely dry and they were use to running in lines that have some water or oil in them to act as a lubricant for the plastic cups. Each one of the obstacles taken separately had a solution, but all three together caused several of the suppliers to recommend someone else. We delayed making a decision in hopes that the pigging technology would catch up with our needs. The delay gave us time to modify the pipeline to accept a smart pig. Since the line was built in 1960, the launching and receiving traps were not long enough. New traps were fabricated and installed. To compound our concerns, we had not performed any type of pigging on this line since 1988 when the service was changed from propane to propylene. We were entering unknown territory.

The pigging program began with us launching a foam pig in one of our shorter pipeline sections (44 miles) to evaluate pig performance and get an estimate of how much debris we could expect on the longer runs. All went well with the initial run. We then launched a foam pig in the 89 mile section with good results and followed with another foam pig. We were starting to feel confident. When we ran a foam pig in the 61 mile section we received a bit of a surprise. There were two pigs in the receiver trap instead of one and the pig we launched a day earlier was the first one removed from the trap. Go figure. Two foam brush pigs were then run in each section of line. The pigging was progressing well and we continued to gain experience with each run. However, the more frequent operation of the forty year old valves at our intermediate trap site resulted in several packing leaks. These eventually got to the point where we had to shutdown the pipeline, depressure about two miles, and replace four valves. We were now ready to select a supplier.

Before we could run a smart pig, the supplier required us to run a series of different type pigs in each section of line. These included (1) a more aggressive brush pig, (2) a cleaning pig, (3) two gauging plate pigs, and finally a geometry pig (measures out of roundness problems and dents). On several occasions, the pigs did stop (one time for about 4 hours) at short radius elbows or areas with heavy wall pipe, but we were always able to free the pigs before significant supply problems occurred. The supplier said that it was not unusual for these pigs to temporarily stick. The line was now qualified to run a smart pig. We decided to evaluate the shorter 61 mile section first.

After five (5) runs we achieved success. The dryness of the product and the drive section (pull pig) of the smart pig caused us most of the problems. The first pig was launched and within two miles we knew we were having problems. The speed of the pig substantially decreased due to a worn pull pig (propylene velocity of 3.2 mph and ILI tool velocity of 2.1 mph). When the unit was received and removed from the trap, the pulling section was badly damaged and the wires from the sensors to the computer were frayed or broken. No data was received. The pull pig was modified by the supplier and a different loading technique was used when the pig was pushed into the launcher. The pull pig was again damaged and one of the U-joints was leaking oil, but the rest of the pig looked good. Again no data. This time, propylene entered the computer compartment causing a computer failure. More modifications were made to the pull pig and oil was added to the product at the time of launch. We received 59 miles of good data during this run. This time propylene had entered the battery compartment and prevented data collection for the last 2 miles. Once again the pull pig was damaged and a U-joint was leaking oil. Since we had received data for 59 miles, the next pig would be launched in the "sleep" mode and not begin recording data until it had traveled 50 miles. More changes to the pull pig, more oil at launch and another long 61 mile run resulted in another failure. A magnet from the smart pig became lodged in the RTJ flanges on the last valve before entering the receiving trap causing it to break into two pieces. The receiver trap had to be removed from the pipeline to extract the back half of the pig. No data was received because of a computer failure. We were not making a lot of progress. On Run #5 we finally achieved success. The revised pull pig seemed to work better and it gave us a little more confidence as we planned to inspect the 89 mile section of pipeline.

We had now fallen behind schedule due to all the problems encountered on the first section of line. We hoped that a new year would bring us better fortune and help us get back on track. In mid-January, 2003, we launched an ILI Tool in the 89 mile section of our propylene pipeline. Everything went well and it appeared as though we had a successful run. The supplier determined that we received data for 69.7 miles before the computer stopped recording. A u-joint had again failed allowing propylene to enter the computer compartment. The next pig was launched asleep and was set to come on after 50 miles of travel. Damaged wires resulted in no data being recorded. The 3rd smart pig was also launched in the sleep mode and this time everything worked. The data for the rest of the line was obtained. Even though we had finally completed the 150 mile pipeline, we could not afford any further delays. The next two sections of propylene pipeline were scheduled for completion by mid-year 2003.

In late May, 2003, we received the final inspection report on the first pipeline section. We now had to locate and repair the anomalies that had been reported as well as continue with the in-line inspection program. Fortunately, there were no "immediate repair" conditions or any that had to be completed within six months. We began inspecting the most severe anomalies immediately and making repairs as needed. Through the end of December, 2003, we had inspected over 430 anomalies and 3341 feet of pipe. We had also installed 15 composite wraps over areas where severe pitting was found. These repairs sometime caused additional problems. One of the digs was at one of our valve stations. Shortly after the anomalies were inspected and repaired, a small leak developed in the valve body of the mainline valve. Several miles of pipeline had to be depressured and the mainline valve replaced.

Bids were received for the second round of inspections and another ILI supplier was selected to run the next two sections. One section was 44.5 miles long and the other was 33.9 miles. The 44 mile section would be our next task. This line was built in 1988, and we believed it to be in excellent condition. We had already run several foam pigs and foam brush pigs in the line and no problems were encountered. This would also allow us to replace the launcher and receiver traps on the older 33 mile pipeline (built in 1960). The new supplier required us to run only a gauging pig before their smart pig was launched. The gauging pig was run without incident and everyone was satisfied that we were ready to run the inspection tool. The smart pigged was launched in late August, 2003, and after 40 miles became lodged in the pipe. Over the next several days, attempts to free the pig were unsuccessful. We finally had to depressure 6 miles of line and cut the tool out. Upon inspection, it was determined that metal slivers (apparently from the manufacturing process at the factory) had become lodged under the magnets on the ILI tool and when it reached a heavy wall elbow, the pig stuck. Over the next 2 ½ months, we ran 43 pigs of various types (majority were a combination brush/magnet pig) before the line was cleared of all the metal debris.

During this time, the new traps had been installed on the 33.9 mile section of pipeline and we were anxious to complete the in-line inspection. The supplier's gauging pig was run in early November, without incident. The line was declared ready for the smart pig. During the first week of December, the pig was launched and two feet from the receiver trap it stuck. A foam pig was launched several hours later in an attempt to free the pig. When it arrived the next night, not only did it not free the smart pig, all flow into the plant was shutoff causing several operational problems. The pig was cut out and it was determined that a misaligned elbow and bad weld had reduced the ID of the pipe enough to prevent the magnets on the pig from passing through. Since valid data was recorded for all but the last 20 feet, the run was considered a success.

In January of this year, we attempted to rerun the smart pig through the 44 mile section. The pig was launched and received without incident. However, for some undetermined reason, only 22 miles of data was recorded. The in-line inspection tool was rebuilt and in early February, 2004, the line was rerun without incident.

This will be another trying year. In order to meet DOT regulations, we must run an in-line inspection tool through our 12", 150 mile propane pipeline by September 30. This line does not have any intermediate traps and several obstacles lie ahead. We do not know how long it will take to clean the line or even if the smart pig can record data over that distance (the supplier assures us they can). Many questions remain and at the moment, we don't have the answers. This year we will also conduct a hydrostatic test on our 4", 33.9 mile propane/butane pipeline since it cannot be pigged.

Over the next couple of years, we will also have to find ways to inspect our 10", 195 mile ethylene pipeline, and 10", 44.5 mile propane pipeline. Each line poses unique problems. We realize that the in-line inspections are a valuable tool in assessing the integrity of our pipeline system, but we now realize it will be very a costly and difficult task to accomplish.

Paper 62c

Safety Instrumented System design for a Grassroots World Scale Ethylene Complex

Alastair Anderson
Shanghai Secco Petrochemicals Company Ltd.
A6 Westwood Villas
299 Cheng Jia Qiao Road
Shanghai 201103
PRC

Mark M. Moderski
ABB Lummus Global Inc.
3010 Briarpark Drive
Houston, Texas 77042

Prepared for Presentation at the 2004 Spring National Meeting, New Orleans LA, Apr 25-29
16th Annual Ethylene Producers' Conference
Session T8001 – Ethylene Plant Safety

Copyright © ABB Lummus Global Inc.
April 2004

UNPUBLISHED

AIChE shall not be responsible for statements or opinions contained in papers or printed in its publications.

Authors:
Alastair Anderson - BP Chemicals / SECCO JV Process Control Manager

Mark M. Moderski - ABB Lummus Global Inc. / Chief - Process Safety

ABSTRACT

Safety Instrumented System design for a Grassroots World Scale Ethylene Complex

In the early stages of planning for a world scale chemical complex to be built in China, SECCO decided to utilize equation based risk methods for determining the need and the robustness required for safety instrumented systems. This paper will address the safety system lifecycle approach adopted for this project by ABB Lummus Global and SECCO. Real project examples will be used to illustrate the equation based methods used, specifically; calibration of the methods to corporate risk criteria, accounting for individual and societal risks, the concept of ALARP (As Low As Reasonably Practicable) and the concept of occupancy mapping.

SECCO, a Sino-foreign equity joint venture recently formed by BP Chemicals East China Investments Limited ('BP') (50%), China Petrochemical Company Limited ('Sinopec') (30%) and Sinopec Shanghai Petrochemical Company Limited ('SPC') (20%) is currently developing a petrochemical complex to be located in the Shanghai Chemical Industrial Park. The project shall include world scale cracker and derivative processing units. When the complex comes onstream in 2005 it is expected to be the largest ethylene production facility in China.

INTRODUCTION

This paper will address at a high level the process that was undertaken in the Safety Instrumented System design of a grassroots chemical complex. The approach is one that had to meet key word requirements such as:

- Systematic
- Efficient
- Reliable
- Consistent
- Non-conservative
- Documentable

This approach has been applied at a chemical complex level and we have completed approximately 800 calculations. The approach to date has proven robust and has met all project requirements.

BACKGROUND

The Project:

The Shanghai Secco Petrochemical Co. Ltd. facility in Caojing (Approx. 50km South of Shanghai), China comprises the following units:

- A 900,000 MTA Ethylene Unit to produce polymer grade ethylene, polymer grade propylene, chemical grade propylene, and associated byproducts.
- Olefins Conversion Unit and Selective C4 Hydrogenation Unit to produce polymer grade propylene by the catalytic metathesis reaction of ethylene and n-butenes. The hydrogenation unit is used to remove butadiene from the mixed C4's and the C4 raffinate stream from the BD extraction unit before being used as feed to the OCU.
- Gasoline Treating unit to produce a C6 – C8 Heart Cut that is suitable for use in the Aromatics Extraction Plant and a fully hydrogenated C5 cut.

The ABB Lummus Global Inc. technology is being used for the Ethylene, OCU, Gasoline Treating and Butadiene units.

The Project Objectives:
Secco:

- Project Safety (HSE) – Best in China
- Schedule – 38 months to Beneficial Production
- Advanced Technology
- Single point Responsibility from ABB Lummus Global Inc.
- Maximum Chinese Content
- High Level of China Experience

- Quality
- Integration with all Units

ABB Lummus Global and Sinopec Engineering:

- Best in Class Facility
- Ensure schedule is met
- Deliver a facility to the client that meets or exceeds the project requirements.

The project execution strategy is as follows, the PDP package was done by Lummus Technology Division in Bloomfield, New Jersey. The extended basic engineering has been done by ABB Lummus Global Inc. in the Houston, Texas office. Sinopec Engineering Inc. in Beijing, China is doing detailed engineering.

This execution strategy presented unique opportunities to the design teams in each of the execution centers. The majority of the Process Safety / SIS design activities were carried out in the Houston, Texas office of ABB Lummus Global.

At the onset of the project SECCO developed a specification that explains how the project will manage the functional safety requirements of the instrumented protection systems across the complex. This specification titled "SIS Functional Safety Management Strategy" is closely modeled after the IEC 61508 / 61511 standards. As with the international standards the Secco strategy utilized a lifecycle approach to the project.

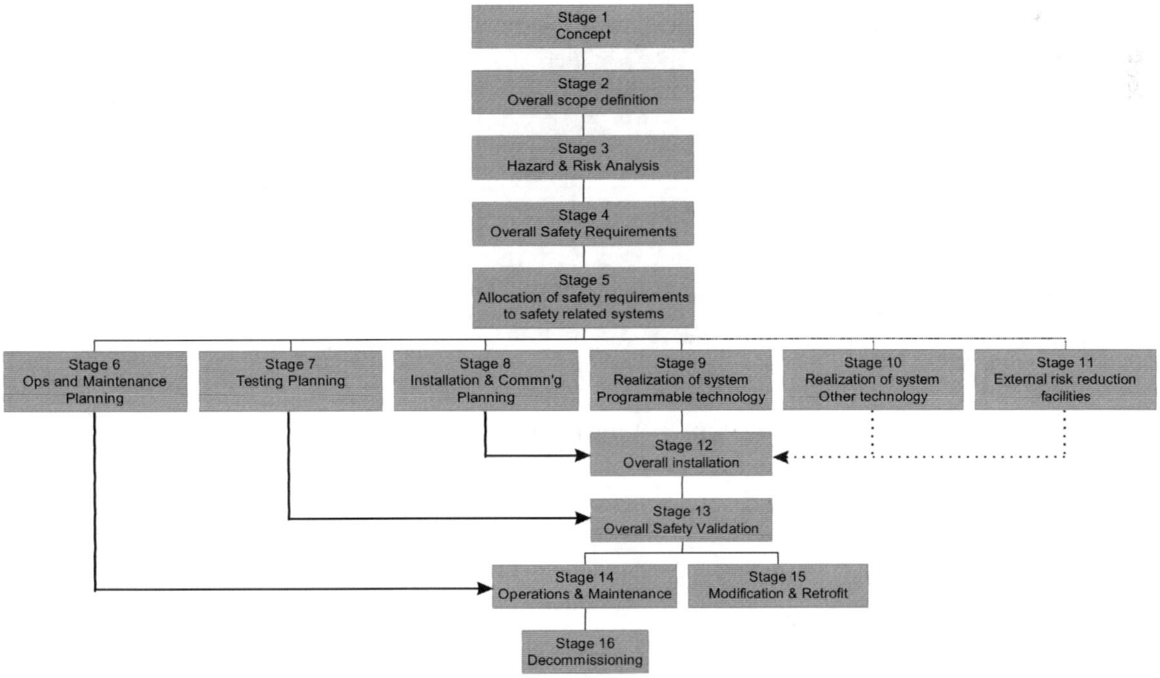

The stages are defined below with the applicable core responsibilities and timing.

Stage 1 Concept
Core Responsibility: ABB Lummus Global / Secco
Timing: During Basic Engineering Design
The requirement of this stage is to develop a level of understanding of the EUC and its environment sufficient to enable the other safety lifecycle activities to be satisfactorily performed. A statement of project assumptions are also required. Items that are included are things such as acceptable individual and societal risks and target proof test intervals.

Stage 2 Overall Scope Definition
Core Responsibility: ABB Lummus Global / Secco
Timing: During Basic Engineering Design
The requirement of this stage is to determine the boundary of the EUC and to define the scope of the hazards and risk analysis.

Stage 3 Hazard and Risk Analysis
Core Responsibility: ABB Lummus Global
Timing: During Basic Engineering Design
The requirement of this stage is to identify the hazards of the EUC (equipment under control) and its control system. A Preliminary Hazards Listing was developed. At this stage the risk / consequence assessment was qualitative.

Stage 4 Overall Safety Requirements
Core Responsibility: ABB Lummus Global / Secco
Timing: During Basic Engineering Design
The requirement of this stage is to develop the overall safety requirements specification in terms of safety function requirements and safety integrity requirements. The major deliverables at this stage are the trip description documents, occupancy maps, and competency training and records.

Stage 5 Allocation of Safety Requirements to Safety Related Systems
Core Responsibility: ABB Lummus Global / Secco
Timing: During Basic Engineering Design
The major deliverables at this stage are, allocation of each safety function to a system, determination of integrity level ratings, external project compliance audit, a control systems review and the implementation of change control procedures.

Stage 6 Operations and Maintenance Planning
Core Responsibility: ABB Lummus Global
Timing: During Basic / Detailed Engineering Design
The requirement of this stage is to develop a clear plan as to how the functional safety of the systems are to be maintained.

Stage 7 Testing Planning
Core Responsibility: Secco and SIS System Vendor
Timing: During Detailed Engineering Design

The requirement of this stage is to develop a clear plan up front as to how the validation of the safety system shall be achieved.

Stage 8 Installation and Commissioning Planning
Core Responsibility: ABB Lummus Global / Secco
Timing: During Detailed Engineering Design / Construction
The intent of this stage is to ensure that the safety systems are installed and commissioned in a controlled manner.

Stage 9/10 & 11 Realization of System
Core Responsibility: ABB Lummus Global / Secco
Timing: During Detailed Engineering Design
The intent of this stage is to verify that the end – to - end design of the system demonstrates compliance with the required Integrity Level Rating. Deliverables at this stage include design validation calculations, QRA reports, IL determination sheets.

Stage 12 Overall Installation
Core Responsibility: Constructor
Timing: During Construction / Commissioning

Stage 13 Overall Safety Validation
Core Responsibility: Constructor
Timing: During Construction / Commissioning

Stage 14 Operations and Maintenance
Core Responsibility: Secco
Timing: During Construction / Commissioning

Stage 15 Decommissioning
Not applicable

Integrity Level Determination Flowchart

The flowchart attached shows the method that was utilized by the project to determine the IL, hardware and testing arrangements for the protective systems.

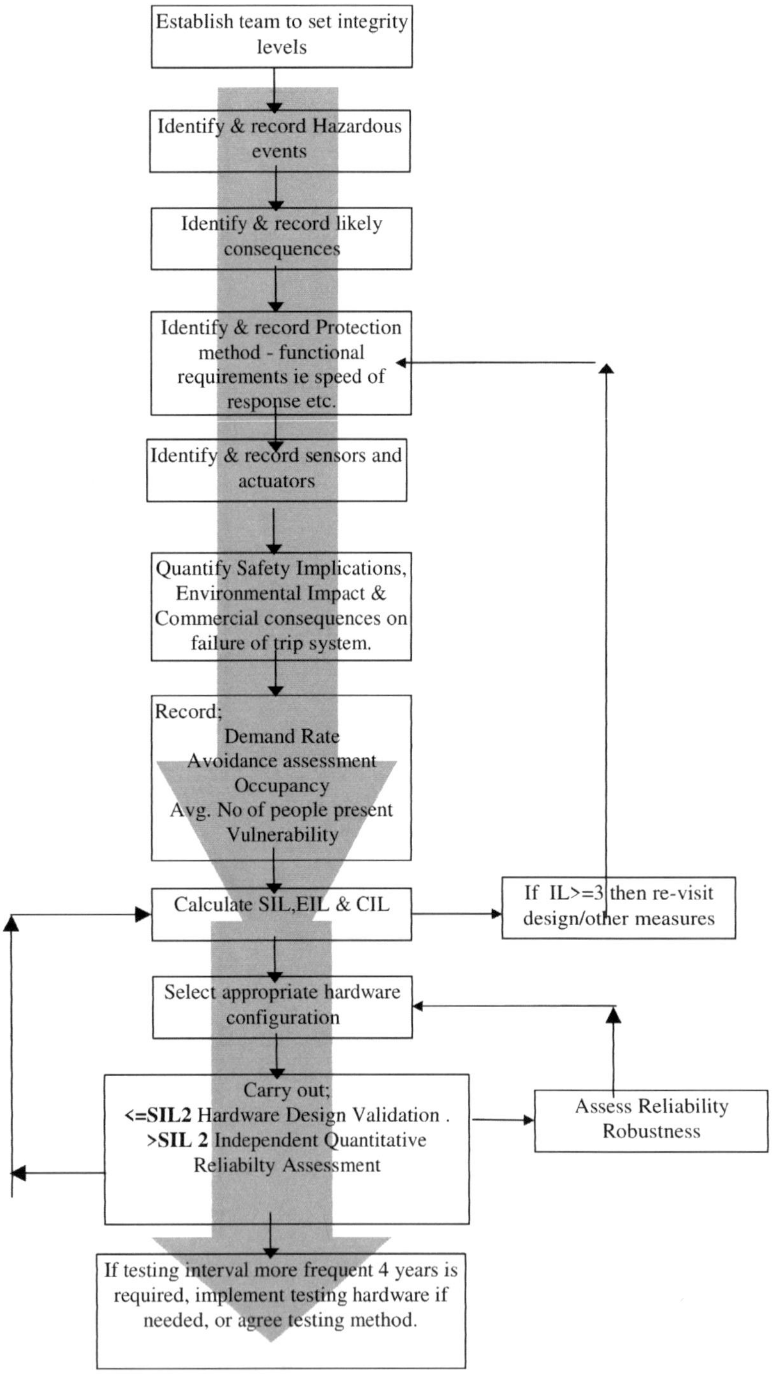

Determination of Integrity Levels utilizing ALARP

The approach taken by the Secco Project was to utilize a team for determination of Integrity Levels. The minimum team composition was as shown below:

- Lead Process Engineer Secco
- Lead C&I Engineer Secco
- Lead Process Engineer EP
- Lead C&I Engineer EP
- Team Chairperson Safety EP
- Operations representative

Secco has developed risk graphs that are based on the corporate risk criteria for the company. These risk graphs are calibrated and embedded into a Microsoft Excel spreadsheet. This spreadsheet is used to calculate the integrity level for the particular hazard being analyzed.

The four parameters that are calculated by the spreadsheet are The Safety Integrity Level (SIL), Environmental Integrity Level (EIL), Commercial Integrity Level (CIL) and the cost of unreliability in the input system through the asset life.

The calibration of the risk graphs results in a SIL rating that is higher than if one had used just the graphs defined under the IEC standards alone. The SIL rating generated by the calculation takes into account the ALARP practice. Therefore no further ALARP study is required.

The stage 3 Hazards Analysis is utilized to identify all equipment on the P&ID's that contain hazards. A sheet is generated for each hazard identified that lists the consequences, prevention method and sensors / actuators. The consequences of the protective system failing is then examined, documented and input into the spreadsheet. Based on the calculated risk of the hazard, an appropriate SIL, EIL and/or CIL rating was generated.

The SIL Calculation equation can be expressed as follows:

IR = DR * PFD * Iocc * V where;

- IR = Individual Risk to a given hazard
- DR = Demand Rate
- PFD = Probability of Failure on Demand
- Iocc = Occupancy of individual in the hazard area
- V = Vulnerability of the individual to the hazard

SR = DR * PFD * Aocc * V where;

- SR = Societal Risk to a given hazard
- DR = Demand Rate
- PFD = Probability of Failure on Demand
- Iocc = Occupancy of individual in the hazard area
- V = Vulnerability of the individual to the hazard

The following approach is then used to determine if the SIS design reduces the risk to "As low as reasonably practical (ALARP)".

The equation used to determine the maximum justifiable spend for reducing the risk by a factor of 10 (equivalent to improving the SIS design by one SIL) is;

Maximum Justifiable Spend = Reduction in Number of Fatalities * xMSUD * 10 * N^1.5 where;

- Reduction in Number of fatalities is given by (SR-(SR/10))
- 10 = number of years over which cost is spread
- xMUSD is the cost of a life defined by SECCO
- N is the number of fatalities.
- 1.5 power is set in the corporate risk criteria.

The maximum justifiable spend is then compared to the incremental cost of a SIL and if it is greater then the initial SIL is incremented by one until the spend is no longer justified. This SIL is the ALARP SIL that is displayed on the spreadsheet.

The Societal Risk is also evaluated to ensure that it does not exceed the maximum tolerable limit for employees.

The maximum tolerable limit has been defined by SECCO as;

Log10 (SRmax) = 1.5 * Log10 (N) + 1.5 where;

- N = Number of fatalities (Aocc * V)

x % of the maximum tolerable risk limit is allocated to the SIS design. The actual SR calculated is then compared to this value and if it is higher, a message is displayed on the spreadsheet that the SR Target has been exceeded.

The CIL Calculation equation can be expressed as follows:

It is assumed that the investment will payback in 10 years so the incremental cost is converted into the incremental cost per year over the payback period.

Incremental cost per year = incremental cost / payback period.

The cost of the incident per year is;

Cost of incident per year = cost of the incident * Demand rate * RRF

If the cost of the incident per year is greater than the incremental cost per year then the increase in RRF is valid and the integrity level is increased by one.

The EIL Calculation equation can be expressed as follows:

SECCO has defined an upper limit on the likelihood of incidents which result in major environmental damage. Four categories have been defined along with their respective intolerable cumulative frequency / yr.

The frequency of environmental events is determined by the equation;

Frequency = PFD * DR

The required PFD is then determined by the spreadsheet to meet the intolerable cumulative frequency. x % of this PFD is then taken as the portion of the total intolerable frequency allocated to SIS failure. The Integrity level required to meet this PFD is then displayed.

See APPENDIX A - GENERIC TRIP CAT SHEET

During the exercise of the team filling out the Trip Categorization Sheets, the demand rate is estimated by the team and documented on the sheet. This is done via direct team experience and via use of generic failure rate data. All generic data was agreed upon and documented prior to starting the project. In some cases the estimation of the demand rate can be quite complex. Simplified fault tree's were drawn to assist in calculating these rates.

The occupancy number as show on the trip categorization sheet is determined via the utilization of an occupancy map that was developed by Secco during the project. This map is a 2D map that shows the expected individual occupancy and the average occupancy of process areas. This is used as a guideline for the review team.

See APPENDIX B - OCCUPANCY MAP

This approach is aimed at providing a consistent approach for a project, by reducing the amount of 'judgment' that needs to be made at each SIL rating, e.g. occupancy and demand rate. By using the map and the generic failure data this leaves the team only having to consider consequences and vulnerability. Many of the other methods available and that have been used in the past are highly subjective and details of decisions were rarely recorded in detail.

This sheet encourages rigor in recording basis for values used, particularly demand rated and costs. Ultimately the aim is to prove, if required say in a court of law, that the approach used was systematic. In would be difficult to build a case which says "you could have done more".

It is also worth mentioning that this method is not necessarily better than any others but it offers an approach that meets the requirements of this project.

Design Validation

Where an integrity level of 3 or higher is generated, and independent Quantified Risk Assessment (QRA) was conducted which calculated the overall hazard rate. This hazard rate was then evaluated against the Secco corporate risk criteria to ensure that the risk was acceptable. If the risk did not meet the Secco criteria, then appropriate changes to the design were made. As an additional check, the results of the QRA were then put into the SIL spreadsheet to verify the results.

Secco Ethylene Statistics

The following are some statistics for the Safety Instrumented Systems Design for the Secco ethylene plant.

Total number of IL Categorization Sheets = 359
Total number of IL Rating Team Meetings = 30

The table shows the number of IL ratings by category for the facility. Note that a process hazard can have multiple ratings and the highest rating is chosen as the design case. For example, a hazard can be a CIL1 and a SIL 2. The SIL 2 design would be the case chosen.

IL Rating	1	2	3
CIL	103	38	8
EIL	0	0	0
SIL	14	20	17

At a SECCO complex level we have probably completed 800 calculations and the approach has proven robust. The results of the calculations have also been compared to the designs of similar facilities worldwide with similar results.

APPENDIX C - EXAMPLE PROBLEM

2001-1 LL Pressure CGC 1st Stage Suction

References:

1. Shanghai Secco Petrochemicals Company Limited, SIS Functional Safety Management Strategy, Document Number ST-00-TIJ-CI-0001, dated 5-31-02.
2. International Electrotechnical Commission (IEC) 61508, 1998.
3. International Electrotechnical Commission (IEC) 61511 – 2003.

APPENDIX A

GENERIC TRIP CATEGORISATION SHEET

赛科 SECCO

Trip Categorisation Sheet

Revision:	1
P&ID & C/E Chart No:	
Event ID:	1
Equipment No:	
Description:	

Hazardous Event and Demands

Likely Consequences:

Prevention Method:

Sensors & Actuators:

Initial Assessment:

Serious safety implications	Yes	Yes or No
Substantial Environmental Impact above 500K$	no	Yes or No
Commercial consequences above 500K$	Yes	Yes or No

Demand Rate Assessment (W):

How often is the protection likely to be needed?	0.0025	Demands per year

Avoidance Assessment (P):

On failure of the protection how likely is it that the event will be avoided?	0	%

Consequence Assessment (C):

Average number of people exposed?	2.4	Number
Vulnerability?	0.5	Probability

Exposure Assessment (F):

Occupancy of the most exposed person?	14	%
Calculated SIL:	**2**	

Environmental Assessment:

If the protection fails, what is the likely environmental cost?	0	MUSD

Calculated EIL Level: a

Commercial Assessment:

If the protection fails, what is the likely commercial cost?	4.3	MUSD

Calculated CIL Level: 0

Input Reliability Assessment:

Spurious failure rate (MTTFS) of the input system, once per	0	years
Cost of potential production loss	0	K US$
Cost of unreliability through asset life:	a	KUSD

Record basis for all values used above and any comments

Completed By:

APPENDIX B

OCCUPANCY MAP

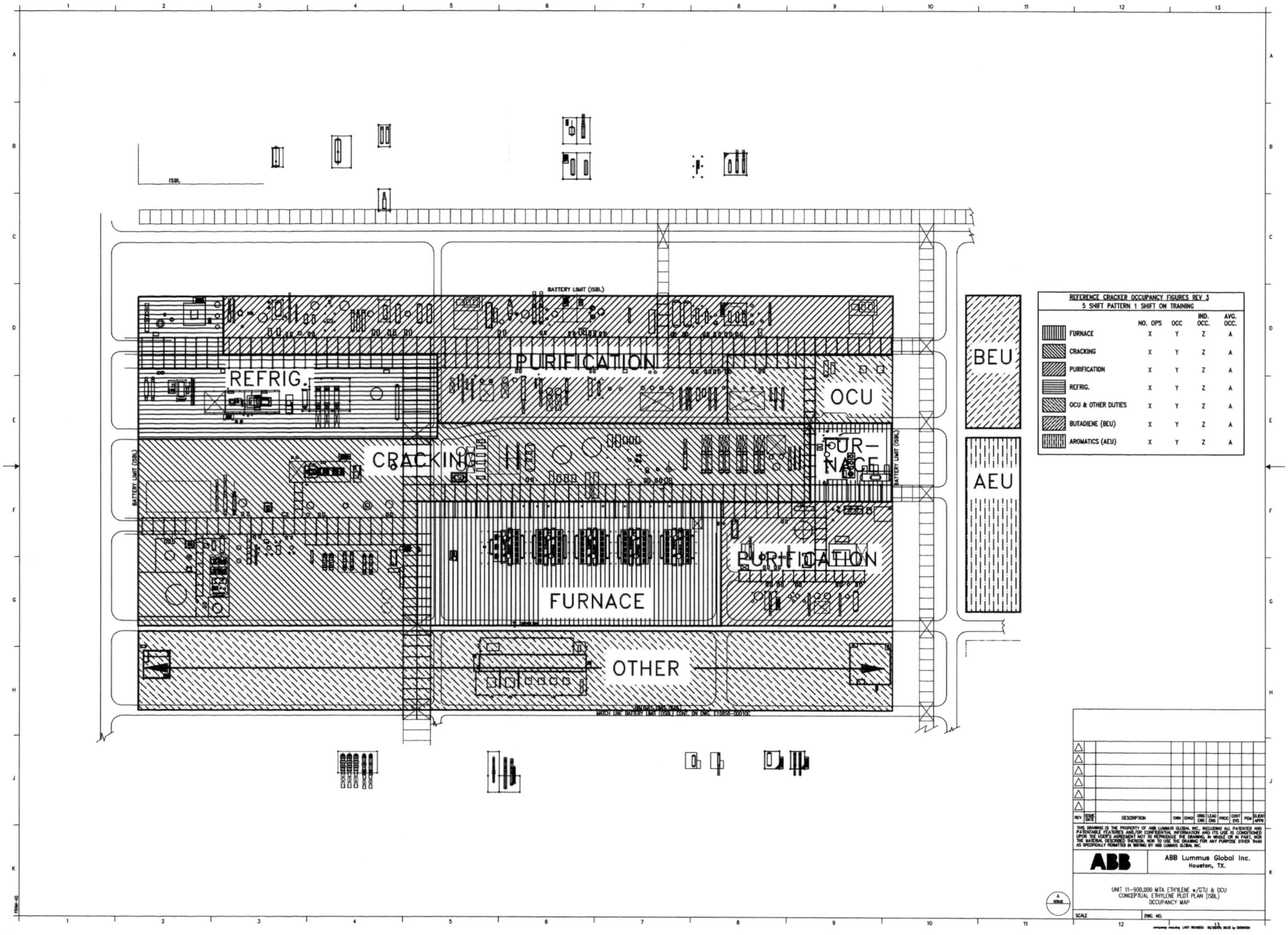
REFERENCE CRACKER OCCUPANCY FIGURES REV 3
5 SHIFT PATTERN 1 SHIFT ON TRAINING
NO. OPS OCC IND. OCC. AVG. OCC.
FURNACE X Y Z A
CRACKING X Y Z A
PURIFICATION X Y Z A
REFRIG. X Y Z A
OCU & OTHER DUTIES X Y Z A
BUTADIENE (BEU) X Y Z A
AROMATICS (AEU) X Y Z A
BEU
AEU
ISBL
BATTERY LIMIT (ISBL)
REFRIG.
PURIFICATION
OCU
CRACKING
FUR-NACE
FURNACE
PURIFICATION
OTHER
REV ISSUE DATE DESCRIPTION DWN CHKD ORIG ENG LEAD ENG PROC CONT SYS PEM CLIENT APPR
THIS DRAWING IS THE PROPERTY OF ABB LUMMUS GLOBAL INC., INCLUDING ALL PATENTED AND PATENTABLE FEATURES AND/OR CONFIDENTIAL INFORMATION AND ITS USE IS CONDITIONED UPON THE USER'S AGREEMENT NOT TO REPRODUCE THE DRAWING, IN WHOLE OR IN PART, NOR THE MATERIAL DESCRIBED THEREON, NOR TO USE THE DRAWING FOR ANY PURPOSE OTHER THAN AS SPECIFICALLY PERMITTED IN WRITING BY ABB LUMMUS GLOBAL INC.
ABB
ABB Lummus Global Inc.
Houston, TX.
UNIT 11-900,000 MTA ETHYLENE w/GTU & OCU
CONCEPTUAL ETHYLENE PLOT PLAN (ISBL)
OCCUPANCY MAP
A ISSUE
SCALE
DWG. NO.

APPENDIX C

EXAMPLE PROBLEM

LOW LOW PRESSURE CHARGE GAS COMPRESSOR 1ST STAGE SUCTION

赛科 SECCO

Trip Categorisation Sheet

Revision: 0

P&ID & C/E Chart No: 2001

Event ID: 1

Equipment No: C-2000 Charge Gas Compressor

Description: 1st stage inlet pressure

Hazardous Event and Demands

{Low low 1st stage inlet pressure to C-2000 caused by higher compressor speed if on manual control (operator error) (1in10) and vendor compressor controls in manual (1in10)} or {control loop PIC-15030 fails causing PV-15030A or B on T-1501 to vent to flare(1in50) and failure of vendor compressor controls(1in50)} or {FIC-20003 (min flow) causes FV-20003 to close (loss of spill back)(1in50) and failure of vendor compressor protection (vibration trip)(1in50)} or {furnace trips (4 heaters trip)(1in50) and failure of vendor compressor controls(1in50)}

Likely Consequences: Damage to compressor

Prevention Method: Low low pressure

Sensors & Actuators: PZT-20003ABC, stop turbine via closure of XZV-21020A/B and SZV-21001.

Initial Assessment:

Serious safety implications	no	Yes or No
Substantial Environmental Impact above 500K$	no	Yes or No
Commercial consequences above 500K$	Yes	Yes or No

Demand Rate Assessment (W):

How often is the protection likely to be needed?	0.022	Demands per year

Avoidance Assessment (P):

On failure of the protection how likely is it that the event will be avoided?	0	%

Consequence Assessment (C):

Average number of people exposed?	1.3	Number
Vulnerability?	0	Probability

Exposure Assessment (F):

Occupancy of the most exposed person?	14	%

Calculated SIL: a

Environmental Assessment:

If the protection fails, what is the likely environmental cost?	0	MUSD

Calculated EIL Level: a

Commercial Assessment:

If the protection fails, what is the likely commercial cost?	25	MUSD

Calculated CIL Level: 2

Input Reliability Assessment:

Spurious failure rate (MTTFS) of the input system, once per	0	years
Cost of potential production loss	0	K US$

Cost of unreliability through asset life: a KUSD

Record basis for all values used above and any comments

Damage to rotor, spares are held onsite, assume X week downtime (Xdays@$X) = $XMM +$XMM contingency.

Completed By:

Session T8001
Paper 62d

Software Upgrade Triggers Events That Lead To Plant Shutdown

David Gent
Nova Chemicals Corporation
P.O.Box 5006
Red Deer, Alberta, Canada T4N 6A1

Prepared for Presentation at the 2004 Spring National Meeting, New Orleans LA, Apr 25-29
16th Annual Ethylene Producers' Conference
Session T8001 – Ethylene Plant Safety

Copyright © Nova Chemicals Corporation
April 2004

UNPUBLISHED

AIChE shall not be responsible for statements or opinions contained in papers or printed in its publications.

Software Upgrade Triggers Events That Lead To Plant Shutdown

By:

David Gent (Nova Chemicals)

Abstract

When a plant undergoes a system failure, or major upset condition, there can be many consequences. In the case of a new plant, some of these consequences may be unforeseen.

This paper describes a series of problems that occurred as the result of the failure of an apparently simple software upgrade to a DCS control system. These problems occurred during the early phases of operation at the new NOVA Chemicals® Ethylene 3 plant at Joffre, Alberta, Canada.

The background that led to the initial DCS control failure is discussed and the consequences are described. These included a severe temperature excursion in the acetylene hydrogenation reactor, a loss of process containment at a brazed aluminum heat exchanger and the catastrophic failure of a compressor check-valve disc. The following vulnerabilities that were associated with these problems are also reviewed:

- Vulnerability to software problems in modern control systems
- Inability of the reactor emergency shutdown system to mitigate the temperature excursion
- Materials of construction for the check-valve service
- Integrity of the brazed aluminum equipment

Finally, some of the actions that were taken to avoid these issues in the future are discussed.

INTRODUCTION

The incidents that are described in this paper took place in the Ethylene 3 plant, which is located at the Nova Chemicals operating site in Joffre, Alberta, Canada. There are three olefins plants at the site and they have a combined ethylene capacity of more than 6 billion pounds per year.

The E3 facility is the newest of the three plants. It was completed in 2000 and has a design capacity of 2.8 billion pounds per year, based on an ethane feedstock.

The initial startup of the plant took place in August 2000 and it had been producing specification ethylene at a variety of operating rates up until the time of this incident in December 2000.

BACKGROUND

Process Configuration

E3 can best be described as a deethanizer-first plant. The acetylene hydrogenation step is in the C2-minus stream from the deethanizer overhead. This is then followed by the demethanization step and a C2 splitter heat-pump system. The low energy design of the plant has required a large degree of heat integration using process-intensive heat transfer equipment.

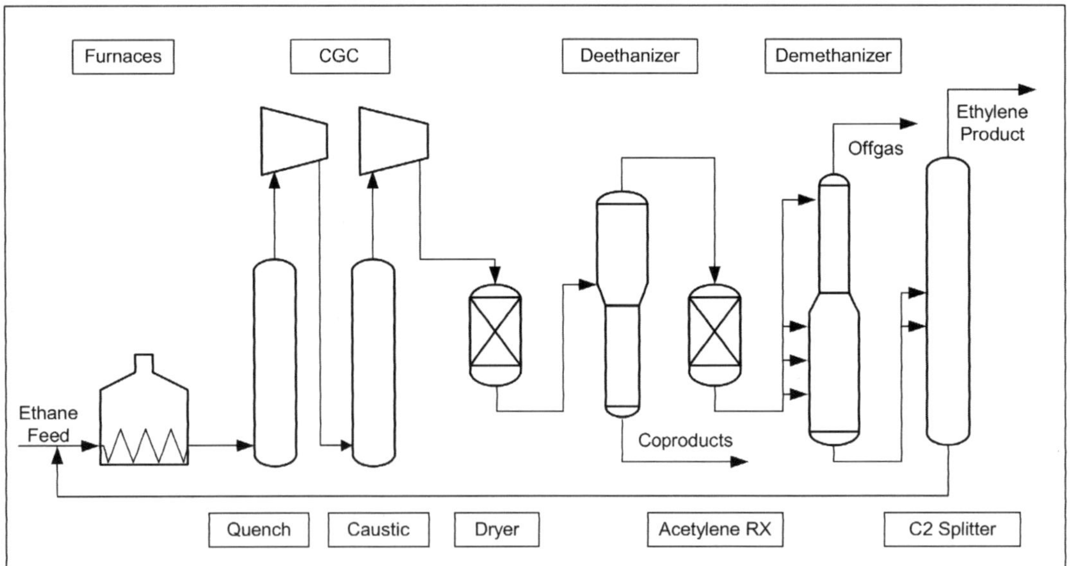

E3 Simplified Flowscheme

DCS Control System

The E3 control system is composed of a number of purpose-specific computers that are inter-connected by a dedicated network. Each control computer contains a redundant control module that processes the control algorithms that manipulate the process valves as well as providing communication to the operator display stations. The control computer also includes the input/output processing and signal conversion to communicate with the field devices. During normal operation, one control module provides the control and communication functions, while the second works in a 'shadow' (or hot backup) mode. In the event of a hardware or software failure in the primary module, the backup module automatically picks up the control and communication functions. The backup module is constantly undergoing diagnostic and communication checks to ensure that it is able to assume the primary module's functions as required. If neither module is providing the control and communication functions then all of the valves controlled by this module are configured to move to their fail-safe, or loss-of-air, position. All communication from the operator display stations to the field devices is also lost.

The control input/output terminations for this plant are marshaled in remote buildings that are located strategically throughout the plant. This geographic distribution did have some impact on the location of input/output within the process control computers. For example, one computer controls many of the most critical control valves that are associated with the operation of the cold section of the plant.

Changes to Control System

Several of the process control computers were highly loaded and this had led to performance issues such as response time. After further analysis of these issues, some recommendations for reconfiguration of the system were developed. These included both the addition of a fourth computer (three existing) to control the cold section of the plant and a modification to the system software. The system is designed to allow this type of reconfiguration while remaining in control of the plant. The reconfiguration work was completed in October 2000 without incident.

The proposed software change was implemented into the next release of the system software. This became available for installation in the E3 system in December 2000. This new release had already been installed in five other plants and had undergone extensive reliability testing in offline systems.

The installation process for the new software was agreed with the plant and it was decided to proceed on December 5th.

SEQUENCE OF EVENTS

<u>December 5th, 2000</u>

7:30 hrs:	The software migration commenced. A single operating station was migrated.
10-12:30 hrs:	7 out of 10 control computers were successfully migrated.
12:40 hrs:	During the migration of the 8th control computer, the swap failed. The result was a loss of view at the Operator control consoles and the movement of 82 affected control valves to their fail-safe positions. A swap back to the original primary control module was initiated. Valve control was restored within 5 minutes. A process upset occurred during the period when the control module was unavailable. The upset ultimately resulted in a plant trip.
12:55 hrs:	Flowrate to the acetylene reactors dropped from 285 Mg/h to 80 Mg/h. The recycle valve was opened to 100% in response to the low flow. The inlet temperature to the lead reactor increased from 82 °C to 120 °C over a 4-minute period. Deethanizer pressure safety valve was lifting.
13:02 hrs:	Reactor SD-1 trip activated. This shut-off the steam supply to the preheater and opened the bypass. Reactor SD-2 trip activated. This isolated and depressured the reactors. Nitrogen flowed into the reactors once the pressure was low enough.
13:03 hrs:	The charge-gas compressor was tripped manually due to high thrust.
13:06 hrs:	Some of the bed temperatures in the lead reactor had reached 550 °C This was the limit of the transmitter scale.
19:30 hrs:	A gas detection alarm rang-in. This was in the vicinity of the deethanizer overhead/ reactor effluent cross heat exchanger. On further investigation, hydrocarbon vapour was visibly leaking from underneath the insulation around the outlet flange of the heat exchanger.

December 6th

The leaking exchanger outlet flange (24-inch) was tightened and the leak stopped.

December 6-9th

The catalyst was removed from the reactors. There were significant quantities of carbon covering the catalyst. It was screened, graded and placed back into the reactors.

December 9th

A plant re-start was attempted.

Sustained operation of the charge gas compressor could not be achieved because the machine kept tripping due to excessive thrust (three trips occurred).

Gas was detected in the neighborhood of the same heat exchanger. The insulation was stripped from the plate-fin heat exchanger and gas was found to be jetting from a failure in the brazed face.

The re-start was aborted in order to allow repair of the leaks and to investigate the charge-gas compressor problem.

December 10th

Further analysis of the plant data from the December 5th incidents showed that the high-pressure compressor case had lost pressure quite rapidly (20 minutes) after the machine was shutdown. This suggested that the check-valve located in the suction line to this case was ineffective. A decision was made to inspect the valve.

December 12th

The check-valve was found to have catastrophically failed.

December 14th

Leak repairs had been completed and a full pneumatic test verified that the heat exchanger was ready for operation.

December 17th

The check-valve parts were replaced.

December 18th

Plant successfully re-started

REVIEW OF PROBLEMS

Troubleshooting the DCS Failure

A critical step during the software migration of a control module occurs when the backup control module assumes control of the valves from the primary module. Prior to performing this swap, a waiting period is built into the migration procedures in order for the control module to perform self-diagnostics. The backup control module will not allow a swap to be initiated if it detects any fault in its diagnostic cycle. If the control swap fails, all of the valves controlled by this module will move to their individually configured fail-safe position.

In this instance, all indications were that the backup control module was ready to assume control of the valves. However, this did not occur and all of the valves associated with this module went to their fail-safe conditions.

The error codes indicated that an input/output register overflow had occurred on a data network within the control module. This error caused the control module to branch to its error handling routines. There was no routine within the control program that was designed to handle this specific error, thus the module failed.

As the control had been successfully swapped back to the primary module, there was speculation that a transient condition might have caused the error. Based on this, it was decided that another migration would be attempted. This second attempt failed and produced the same error codes as the first.

Although all of the hardware diagnostics indicated that there was no hardware problem, it was decided to replace the backup control module. A third migration was attempted and this was successful. This pointed to some type of undetected hardware failure in the backup control module. It was sent to the vendor for detailed analysis.

A review of the software code was also initiated. On December 11th, 2000 the vendor communicated that they had found a problem with the software. The problem was in a section of recently changed code that scanned for hardware status information. This code change allowed a status register to be read before it had been properly initialized. Previous versions of the software had ensured that this would not occur.

The problem was not confined to the new software code that was being implemented in E3. It was also present in two earlier software releases in 2000. This meant that there were a number of plants running software containing the same code. It raised the question as to why did E3 encounter the problem when an estimated 2 million control hours had been accumulated without incident?

Firstly, although the control computer is designed to be re-configurable while remaining in control, few control computers actually undergo this specific type of re-configuration once installed. Secondly, not all control computer re-configurations would necessarily encounter problems with the software error. Indeed, two similarly re-configured control computers had been successfully migrated on the same day as the failure.

The reason that the third migration was successful after the module had been replaced was not because there was anything wrong with the original backup module. It was because the replacement involved powering down the module. The status register was initialized with default values when power was restored. This meant that the register no longer had an invalid value. It was confirmed that there was no problem with the hardware that had been removed.

Overall Consequences of Control Valve Failure Actions

The majority of the control valves that were affected by the loss of the control module failed to a closed position. These included all of the valves that control forward flow into the cold section of the plant. The result was that the compressor discharge was essentially blocked-in. This happened very quickly because many of the valves can go to their air failure mode in a few seconds. The loss of forward flow into the cold section of the plant caused a reduction in flow to the acetylene reactors and at the same time caused the safety valve on the deethanizer overhead to lift. There was some evidence of pressure surges and high thrust conditions were experienced at the charge-gas compressor.

The loss of the process control computer function meant that there was a loss of view to the affected parts of the process from the operator control consoles, including a loss of ability to interact with the associated control valves.

It should be emphasized that none of the safety systems were impaired by this incident. All hard-wired alarms, trips and pressure relief devices were fully functional and did respond when intended.

The following chart shows some of the key operating parameters over the timeline of the incident. It should be emphasized that the scan time for the instrument history was typically 1 minute (sometimes 2 minutes) and hence some transient conditions such as the PSV lifting are not well represented.

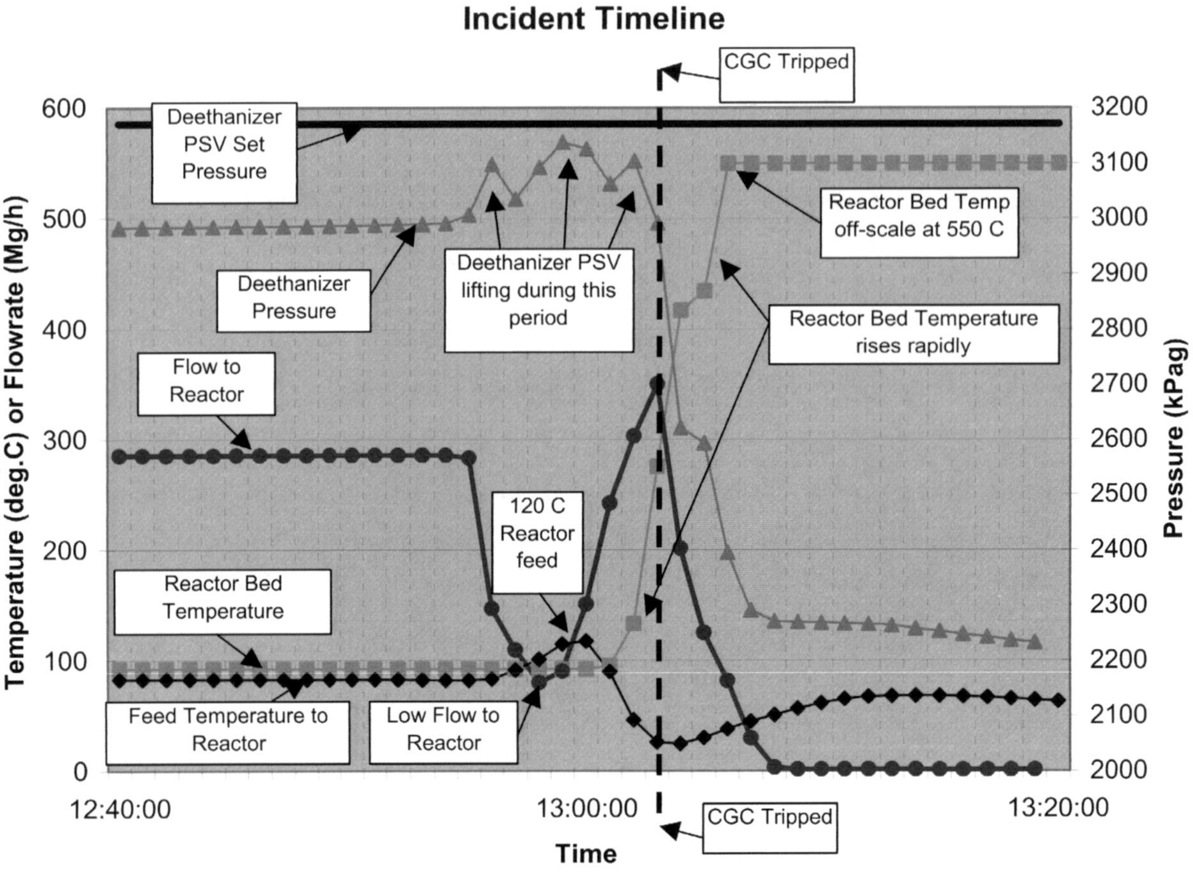

Acetylene Reactor Excursion

The closure of the control valves that allow feed to go forward into the cold section of the plant resulted in a significant drop in flowrate through the acetylene reactors. The space velocity guideline at the time called for this flow to be above 200 Mg/h. During this incident the flow dropped to as low 75 Mg/h for a period of 5 minutes or more.

In addition to the reduced flow, the inlet temperature, which had been running at 82 °C, rose to 120 °C. After about 4 minutes the reactor bed temperatures started to climb dramatically and even though the shutdown system was activated at 180 °C they eventually went off-scale at 500 °C. Data obtained later from the instrumentation asset management system indicated that the maximum readings on two specific thermocouples were 858 °C and 939 °C. Physical evidence based on inspection of both the reactor and the catalyst supports this observation.

Acetylene Reactor Inspection

The catalyst was removed from both reactors. A significant percentage of the pellets were covered with good quantities of carbon. The catalyst was screened to remove the carbon fines. It was then characterized by visual colour as 34% white, 30% grey and 36% black. The catalyst was re-installed in the reactors based on the expected activity that was assumed to be associated with this colour characteristic.

A replacement catalyst charge for one reactor was purchased and installed 9 months after this incident. The catalyst from the second reactor was moved into the lead reactor and the new catalyst was installed in the second reactor. Operating performance and catalyst selectivity has been very good since this change was made.

The reactor vessels were inspected for damage. Metal temperature indications in the range of 180-210 °C confirmed that the vessel wall metallurgy limits had not been exceeded. However, there was some minor internal mechanical damage. A hole was burned through the mesh screen above the bed support and some bolts were sheared off the outlet pipe support.

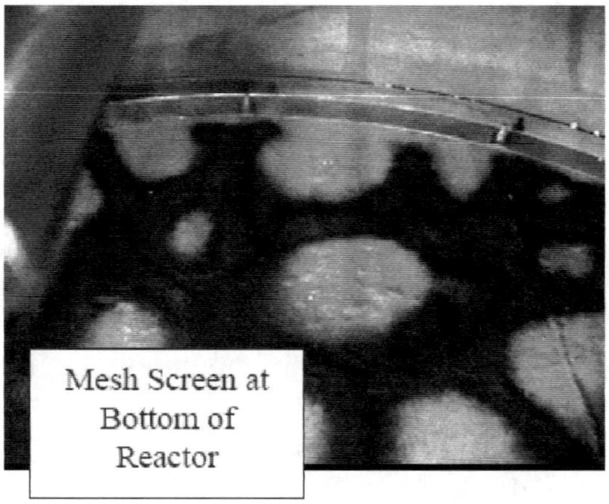

Acetylene Reactor Issues

Clearly the low flow to the acetylene reactors has the potential to cause a significant temperature excursion. However, the intent of the reactor trip systems is to prevent such an excursion from damaging the catalyst and the reactor. It was evident from the results of this incident that the trip systems were unable to fully accomplish this intent.

The first response to a temperature excursion is the SD-1 trip to emergency cooling mode. This trip has the following actions:

- Closes steam to the feed heater to the lead reactor
- Opens a bypass around the feed heater. This introduces cold deethanizer overhead into the reactor
- Closes a bypass around the intermediate cooler in order to send all of the flow through the cooler

The second response to a temperature excursion is the SD-2 trip to isolate and depressure the reactors. This trip has the following actions:

- Isolates the reactors by closing the inlet and outlet valves
- Opens the bypass valves around the reactors
- Opens the valves to the flare header
- Opens high pressure nitrogen valve to the reactor once the pressure is low enough

Unfortunately, and as was the case with this incident, these trip systems had not been able to prevent significant temperature excursions once they were initiated (the bed thermocouple measurements went off-scale on each occasion).

Catalyst Damage

The damaged catalyst that had been removed from the reactor was analyzed in terms of the ability to regenerate it. Sample batches were regenerated and were then tested for their activity and operability. Two types of operating tests were conducted. The first was to establish a temperature 'operating window' by running tests in a laboratory reactor. The second was to operate the catalyst on an actual plant C2-minus stream using a slipstream test reactor.

The first test showed that the regeneration had restored some activity to the catalyst. However, the operating window was significantly reduced when compared with a fresh catalyst sample.

The second test demonstrated a faster than normal rate of decline of catalyst activity. This was such that the catalyst could not be considered for continuous operation over the 5-year period that is required between plant turnarounds.

Response to Reactor Trip Limitations

Some immediate actions were taken to reduce the potential for a temperature excursion. These included;

- Lowering the temperature settings for the SD-1 and SD-2 trips
- Adopting an operating strategy that would risk having off-specification product rather than having a temperature excursion

- Implementing a low-flow trip

The basic philosophy was that it would be preferable to trip the reactors prematurely rather than risk further damage to the catalyst or to impact the integrity of the equipment. The low-flow trip was in itself a risk to plant operation because it was based on a single flow measurement element.

It was felt that the remedy to having an effective reactor shutdown system lay in further investigation of the design and operation of the depressuring and cool-down process. One key concern was the time taken for the reactor pressure to drop once the trip was initiated. The design intent was for the trip system to depressure the reactors to 50% of operating pressure within a period of 2 minutes. Actual operating data seemed to indicate that it was in fact taking 3 minutes or more to accomplish this degree of depressurization. Discussion with other operators suggested that 90 seconds to depressure to 25% of operating pressure might be a more appropriate target. Further analysis of the piping and valve hydraulics for the blowdown lines produced estimates in the range of 3 to 5 minutes to depressure to 25%.

Based on this information, it was decided that the capacity of the blowdown system should be increased in order to reduce the time required to depressure reactors. The most significant requirement was to replace the existing 3-inch blowdown valves with 6-inch valves. Data taken after this modification was made indicates that a time of 60 to 90 seconds to depressure to 25% is now being achieved.

The following chart contrasts the depressuring times that were experienced during the incident against those of the re-sized system. It should be noted that the instrument scan time for the incident was 1 minute whereas the scan time for the re-sized case is 1 second.

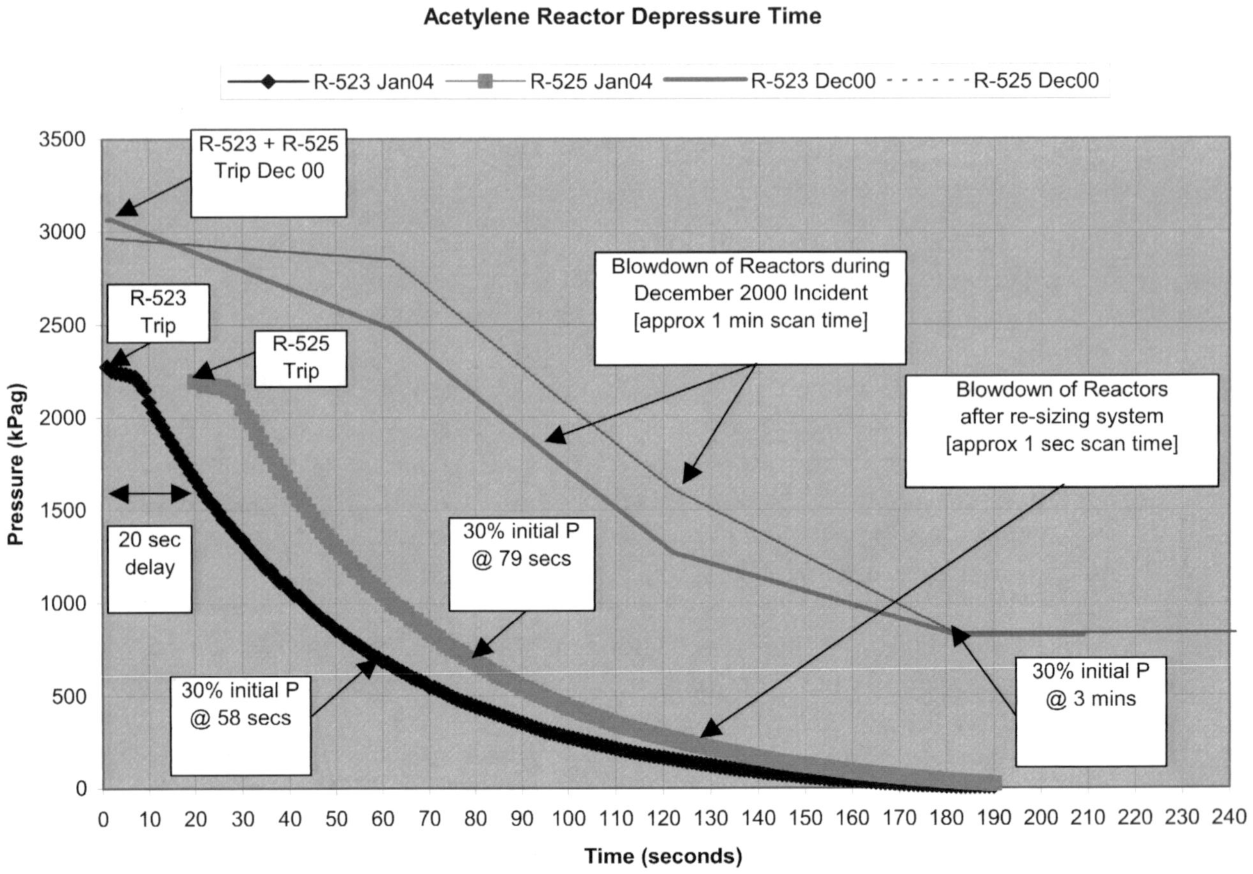

Platefin Heat Exchanger Failure

The platefin heat exchanger (E-520) is located in the deethanizer overhead flow and cross exchanges heat with acetylene reactor effluent flow.

A gas leak was detected in the vicinity of E-520 after the initial incident on December 5th. At that time the leak was visible from underneath the insulation at the outlet flange. This flange was tightened in order to stop the leak. The failure of the E-520 exchanger itself was not known until another gas leak was detected during the attempted re-start on December 9th. Once the insulation had been removed, this leak was clearly visible as gas jetting from the side of the exchanger core.

Exchanger Inspection and Repair

Exact leak locations were identified by pressurizing the flow passes with 700 kPag nitrogen. Initially two split bar-to-sheet leaks were found; one was 15 inches and one was 8 inches in length. These were ground-out to a depth of 0.5 inches as a weld preparation. They were then repaired by welding using a three-pass procedure.

Further pressure testing revealed that a new split bar-to-sheet had opened up while repairing the original splits. More split leaks developed as this was repaired and some small fissure-type leaks were also discovered. In the end, six separate weld repair sessions were conducted on leaks on the west side of the core and two on the east side. These took place over a one-day period. Once there was no detectible leakage at 700 kPag (100 psig) of nitrogen pressure, a pneumatic pressure test was conducted at 1.25 times design pressure. This verified the integrity of all of the exchanger passes.

Reasons for Platefin Heat Exchanger Failure

As is the case with most platefin heat exchangers, this unit is in a service that requires relatively close temperatures to be achieved. It is generally accepted that for a typical brazed aluminum core geometry under steady-state conditions, the maximum permissible temperature difference between streams is approximately 50 °C. For more severe cases such as two-phase flow, transient or cyclic conditions the difference should be lower, typically 20-30 °C. In addition to this, the Aluminum Platefin Exchangers Manufacturers Association (ALPEMA) guidelines recommend that the differential relative to the local metal temperature should not exceed 30 °C prior to flow being established. This condition had possibly been exceeded during previous startups of the plant. Furthermore, ALPEMA recommends that the rate of temperature change should be limited to 2 °C/ minute or less after flow is established. Again, this guideline had not necessarily been achieved during the early operation of this unit. These guidelines consider the impact to the brazed joint integrity.

The incident itself resulted in some significant pressure surges (spikes) that may well have exceeded the design pressure rating of the equipment. This was illustrated by the impacts on some of the other parts of the piping system (e.g. the flange leak and the check-valve failure).

The overall conclusion was that some of the thermal cycles that this platefin exchanger had seen prior to this incident might have already weakened joints where the brazing was more marginal. The pressures surges that occurred during the incident probably resulted in stresses that were above design and caused any weaker areas to fail.

Future Operation and Design

Operating and startup procedures for the plant have been adapted so that the temperature differential and rate of change on the E-520 unit is kept within the ALPEMA guideline. No further problems have been observed over 3 years of operation.

Future designs will need to consider the appropriateness of this type of equipment for the proposed service. The reason for specifying platefin equipment at this location was to avoid using two very large shell and tube exchangers and the plot space that they would have consumed.

Platefin heat exchangers are typically used in the cold sections of the plant and operate with very close temperature approaches. The alternative would be extremely large (and expensive) shell and tube heat exchangers made from stainless steel. The likelihood of operating outside of the ALPEMA guidelines tends to be somewhat less in many of these services. Nevertheless, startups and abnormal operations might still result in a temperature differential issue for any service. There are other similar units in the E3 plant where the temperature differential issue was recognized and the piping design and operating procedures were adapted in order to cope with it.

Troubleshooting the Check-Valve Failure

It was evident during the attempt to re-start the plant on December 9th that something was significantly different with the operation of the charge-gas compressor. The compressor tripped three times due to excessive thrust as it was being brought up to full operating speed.

The inability to successfully complete the start-up of the machine led to questioning what had changed. It was decided that the operating data from the original trip incident should be reviewed more closely.

Normally when the machine is shut down the stage system pressures equalize ('settle-out'). The E3 compressor is a four-stage, two casing machine. There is a check-valve in the suction line to the high-pressure case and this closes as forward flow stops and the compressor is shut down. The intent of the check-valve is to avoid rapid depressuring of the high-pressure case into the low-pressure case and the potential for reverse rotation of the shaft. The high-pressure case normally equalizes to a pressure of around 1380 kPag. This pressure gradually decays and after about 12 hours is about 420 kPag (30% of the original). A review of the settle-out situation following the original incident showed an initial pressure of 1600 kPag, which then decayed to 480 kPag (30%) within 17 minutes. This relatively rapid loss of pressure suggested that there was a problem with the check-valve.

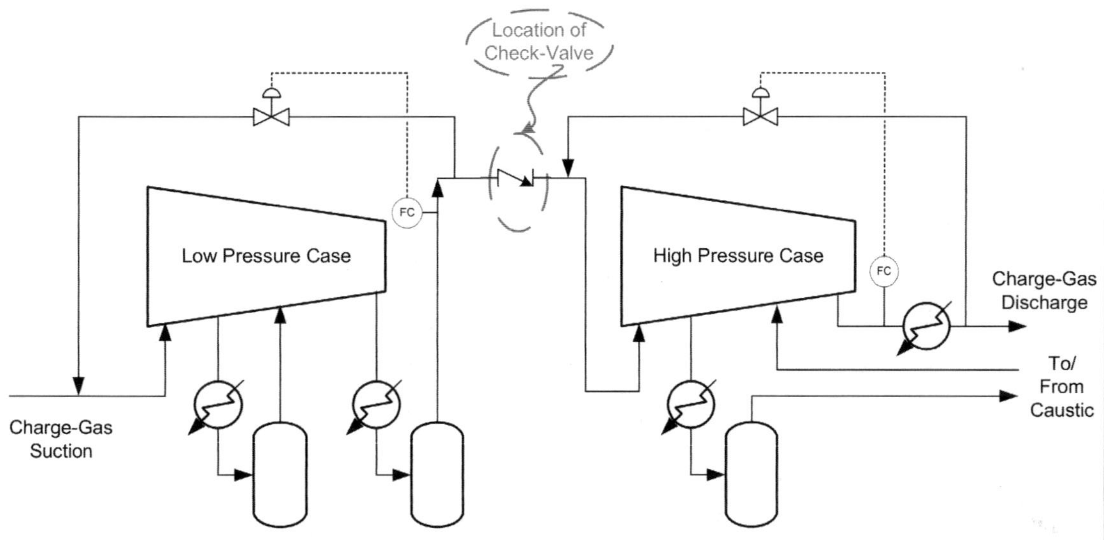

Charge-Gas Compressor - Anti-Surge Circuits

The following two charts illustrate the compressor settle-out pressures and the pressure decay times. The first chart shows the data for the incident where the check-valve has failed and the second chart shows the typical data when the check-valve is working normally.

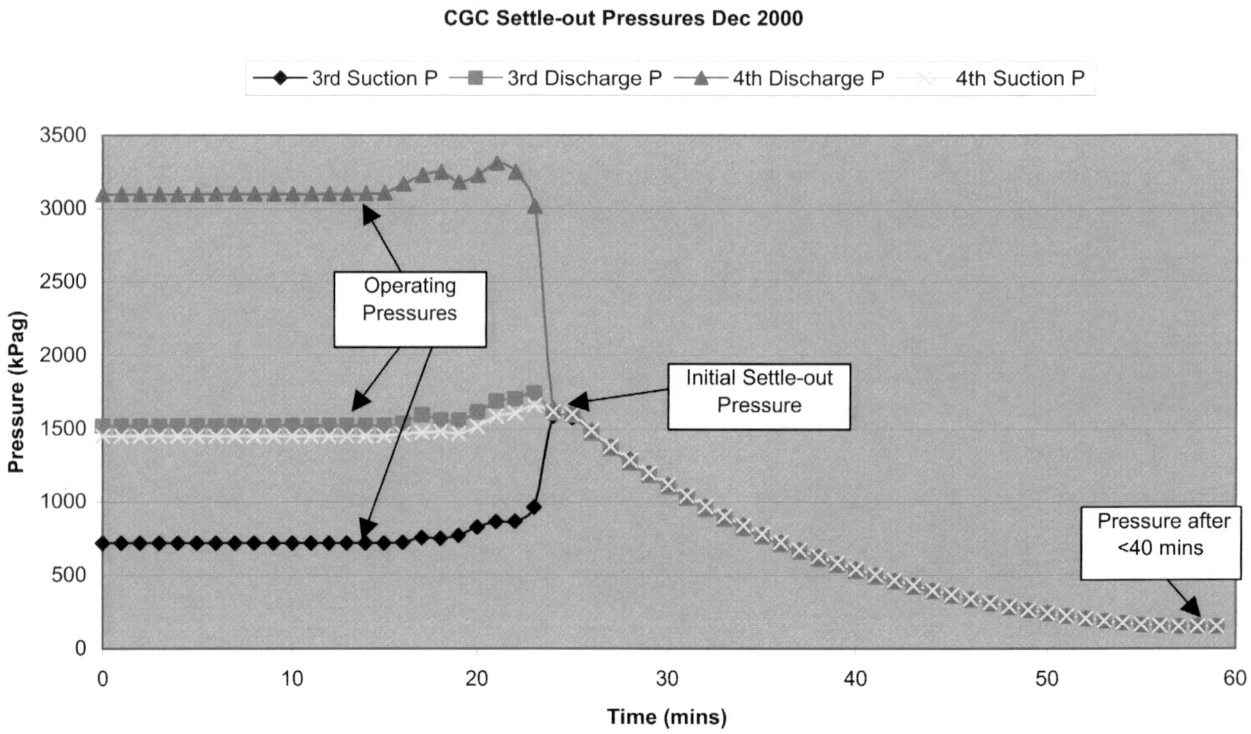
CGC Settle-out Pressures Dec 2000
3rd Suction P
3rd Discharge P
4th Discharge P
4th Suction P
Pressure (kPag)
3500
3000
2500
2000
1500
1000
500
0
Operating Pressures
Initial Settle-out Pressure
Pressure after <40 mins
0
10
20
30
40
50
60
Time (mins)

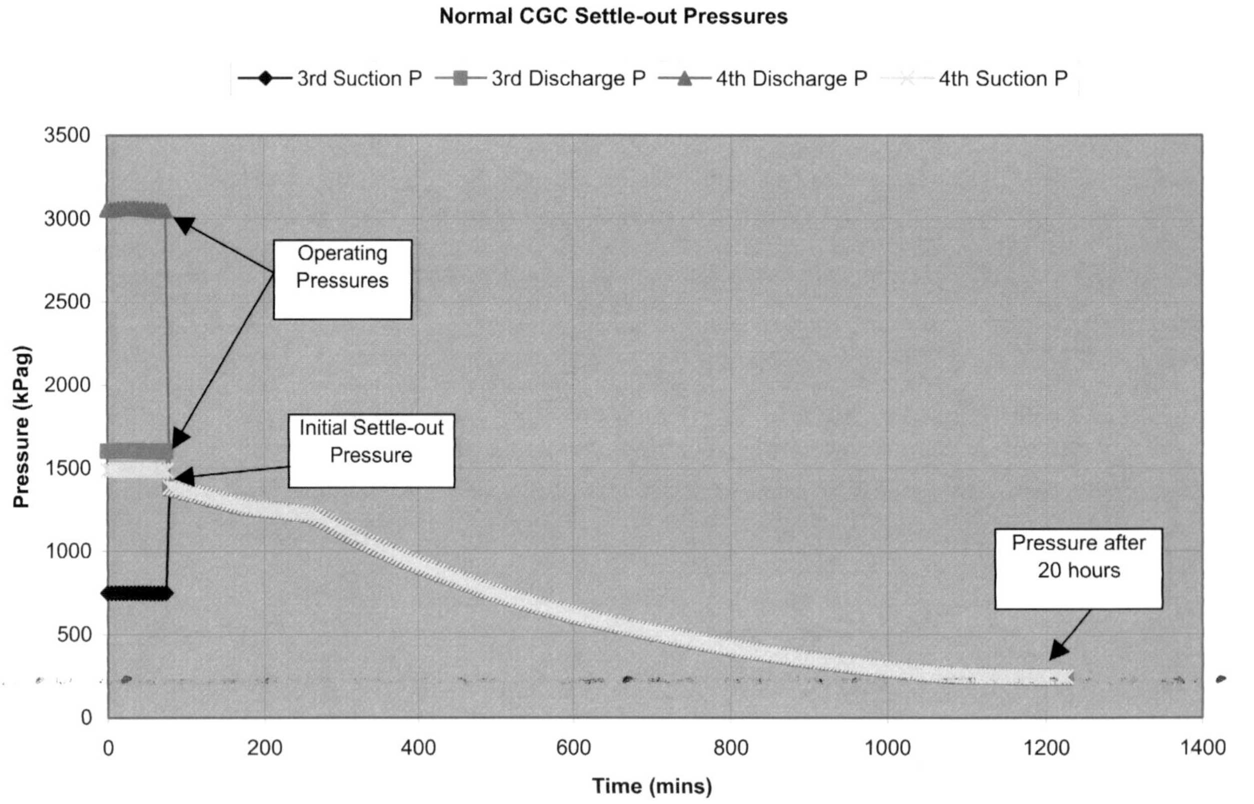
Normal CGC Settle-out Pressures
3rd Suction P
3rd Discharge P
4th Discharge P
4th Suction P
Pressure (kPag)
3500
3000
2500
2000
1500
1000
500
0
Operating Pressures
Initial Settle-out Pressure
Pressure after 20 hours
0
200
400
600
800
1000
1200
1400
Time (mins)

Check-Valve Inspection

The 42-inch check-valve is a dual-plate design. The two semi-circular plates are spring-operated and are hinged on a central pin. On opening the suction line a substantial part of one of the plates was found lying in the bottom of the pipe. The failed components were returned to the manufacturer for analysis.

Chemical analysis, photomicrographs and mechanical testing of the ASTM A217 CA15 material all indicated that the material met specification. The yield strength and ultimate tensile strength were determined to be 38% and 8% respectively above the minimums for the ASTM A216, WCB specification. Based on this, the manufacturer did not feel that the failure was a feature of the tensile strength of the material but rather it's inability to absorb the impact energy that it was subjected to. The failure mode was determined to be one of brittle fracture due to high impact.

The check-valve plates were replaced with more impact-resistant plates made from ASTM A350, LCC material.

Consideration of Failure Conditions and Future Designs

In the case of this failure it is likely that the valve saw greater reverse flow forces than might have been expected in this type of service. Piping systems that are exposed to very high reverse flow velocities can develop high overpressures. This overpressure is referred to as the Joukowsky pressure rise and is a result of the column of fluid, which is flowing backwards, being brought to an abrupt halt by the check-valve closure. The pressure wave emanates from the surface of the check-valve and travels back up the pipe at the speed of sound. In cases of unexpected high surge conditions this pressure rise can often result in the piping being exposed to pressures above the design pressure.

The valve manufacturer had not experienced this type of failure with this type of material in their check-valves previously. However, they do manufacture anti-slam (or anti-surge) versions of the dual-plate check-valves. These can be used in high deceleration systems where pressure surge or water hammer is expected. Another alternative is to specify a larger size high weight nozzle check-valve. This has the disadvantages of being a higher pressure drop and considerably higher cost device. In the case of the design of this system, neither of these types of check-valve had been considered because the potential severity of the service had not been recognized.

The overall conclusion was that high reverse flow forces resulted in a high impact force at the check-valve and that this, coupled with the use of a less impact-resistant material, led to the catastrophic failure.

In addition to changing the material specification for this check-valve, it was also decided to conduct a review of all similar check-valve installations within the plant. Three valves were identified but only one of these was considered to be in a service with a similar potential for high deceleration forces. This was a 30-inch check-valve located on the final discharge of the same compressor. It was decided to replace the plates in this valve with ones made from the more impact resistant material at the next plant turnaround. This has now been done. Future plant designs will look more closely at considering the potential for rapid deceleration (or surge) type conditions and specifying the type of check-valve and materials of construction accordingly.

Hazard and Operability Reviews (HAZOP)

Several safety and operability reviews took place during the course of the design of the plant. These included a formal HAZOP review that involved teams that were composed of experts who were drawn from the owner companies and from the engineering contractor. The HAZOP methodology is a well-established approach to structure creative thought about potential process deviations and their consequences. It is relatively thorough and detailed in its level of analysis. Nevertheless, the outcome is highly dependent upon the skill, knowledge and experience of the participants. In the case of E3 project, the quality of the input from both the owners and the contractor was felt to be very good.

In terms of the specific issues that are described in this incident, some aspects were considered during the HAZOP:

- The potential for low flow to cause an acetylene reactor excursion was known about and was discussed specifically in terms of the need for a low flow trip. It was decided that this was not required and that the protection should be focused on detection of the ensuing temperature excursion. However, a trip of the charge gas compressor was identified as the most likely cause of low flow and, for that reason, the reactor trip was interlocked with the compressor trip.

- The potential for reverse flow and depressurization was identified and was covered by the installation of check-valves at appropriate locations. The severity of the service that these check-valves might see was not necessarily considered.

- The potential for large temperature differences to impact the integrity of platefin heat exchangers was considered during the design. In some cases the piping design and the operating procedures were changed to account for this. In the case of the E-520 evaluation the focus had been on what might happen during a reactor temperature excursion. It was concluded that this would not appreciably impact the temperature differential at this exchanger because of the cooling provided by the reactor aftercooler.

It has to be remembered that these reviews took place before the plant had been in operation and as operating experience is gained there is an opportunity to incorporate new understanding. One of the approaches that has been taken to process safety management in the E3 plant is the maintenance of an 'evergreen' HAZOP database. This has been founded on the output that was available from the original design HAZOP and has been supplemented by adding relevant information on deviations that have been discovered during actual operation. In addition to this, each process system periodically undergoes a formal HAZOP review in order to ensure that any new information or understanding that has been derived from plant operation is included in the database.

Control Systems HAZOP (CHAZOP)

A separate safety and operability review was conducted for the design of the instrumentation and control systems. The intent was similar to the process HAZOP in that the review was structured around creative thought about what might go wrong with these systems.

One consideration for this type of review was whether it should incorporate the potential process impacts that might be associated with various possible control system failures. In the end, it was decided that the control systems safety review should be kept separate from the process reviews. The logic was that the control systems review would look to mitigate any risks that were identified to a degree that the process would be relatively unaffected. Similarly, the intent of the process design was to have fail-safe situations such as pressure relief for any abnormal operations that might result from instrumentation and control issues such as valve failures. The view that the two reviews should be separate was somewhat formed on the basis of the available schedule (control systems design lagged process design by many months) and somewhat on the complexity of coupling the two elements into one comprehensive review.

The results from the CHAZOP did identify many of the ways in which failures might occur and what the impacts might be. One example that is relevant to this incident is that the potential for a loss of view at the process operator consoles was identified and procedures were provided to handle this. However, the premise for these procedures was that basic process control would continue at the control computers during such an event.

Another possible consideration in the design of instrumentation and control systems is the grouping of elements within each system. It is recognized that components have a finite life and will fail. Hence the systems are designed with redundant backups and the ability to perform online maintenance. Nevertheless, it is still possible to have failures whether they are due to equipment, design, or human error. What are the process consequences when failures do occur? The process results and the risk to the plant might be very different depending upon what components are included within the failed system. It is typical that the basic safety systems of the plant such as trip interlocks and pressure relief valves independently offer the necessary additional layers of safety protection. Hence, in the end, while the type of control system failure that occurred might cause a significant process upset and even a shutdown, it would not be expected to jeopardize the integrity of the plant.

There may be an opportunity to limit the degree of process upset and economic losses that could occur from a control system failure by a judicious approach to what parts of the process are linked to what systems. As was described earlier, the E3 unit geography was a significant factor in determining the controller grouping for the cold section of the plant, as was the number of control computers that had been provided. If another computer had controlled some of the affected valves, then the charge gas compressor flow would not have been totally blocked. Of course the dilemma here is that, while this might have mitigated the impacts of this particular incident, a different distribution may be equally as bad in another type of incident. Nevertheless, these possibilities may be worthy of some type of structured review.

In the case of this incident there was really one initiating failure, that being the loss of the redundant process control computer. Thus once the computer failure occurred it was certain that all of the valves under its control would go to their failure positions. However, the equipment failures that then occurred can be considered to be separate and independent vulnerabilities that surfaced during this specific incident.

Improvements have been made to the E3 control systems and to the procedures for making changes to them. These are directed toward avoiding similar system failures and hence any process upset. They are not really intended to address the specific equipment consequences of the incident. The focus of the responses to this incident has been to address the individual process vulnerabilities by making changes to the design and to the procedures so that they are eliminated or avoided in the future.

Final Thoughts

Modern instrumentation and control systems are based on computer technology and they rely on software code to function. Troubleshooting these complex systems can be difficult, particularly if the problem lies within the software code. Given the complexity, it is perhaps surprising that these systems are as reliable as they are in most instances. Problems or failures may cause process upsets in the affected plant. However, these situations should not typically impact the integrity of the plant because there are independent safety systems such as high-integrity trips and pressure relief devices that are designed to provide protection.

In the case of this E3 incident, the software failure did create a plant upset, but the severity of the consequences was rooted in the individual equipment vulnerabilities. Other circumstances could have resulted in the same consequences. For example, closure of the wrong manual valve could have blocked-in the charge-gas compressor and led to the same type of scenario. The plant ultimately relies on its fundamental safety protection systems, which are not dependent on the process control system.

Some further analysis as to how failures and problems with the control systems could affect the process operation might be worthwhile. This might lead to some design changes and could allow the development of some appropriate response procedures.

The fact that the plant was new and included some new applications of equipment technology meant that some events were being experienced and responded to for the first time. The design has been modified and the operating procedures have been improved as the plant experience and knowledge base has grown. This should make similar experiences less likely in the future.

The E3 plant operated well for an extended period following this incident. During this time it met all of its guaranteed performance criteria.

ACKNOWLEDGEMENTS

E3 Manufacturing Department

Mike Carley (Senior Control Systems Engineer):
-Report of the diagnosis of the E3 controller failure

Marc Nobert (Senior Mechanical Engineer):
-Reports on the failure of the E3 compressor check-valve

Blake Switzer (Senior Inspector):
-Photographs and inspection reports of equipment damage

Dan Janzen (Operations Specialist):
-Report on the catalyst removal and inspection
-Troubleshooting and coordination of operating issues

Rick Beleutz (Senior Process Engineer):
-Analysis of operating events and recommendations for acetylene reactor catalyst
-Troubleshooting and coordination of operating issues

REFERENCES

1. The Standards of the Brazed Aluminum Plate-Fin Heat Exchanger Manufacturers Association Second Edition 2000

2. Goodwin International Technical Literature on Dual-Plate Check-Valves

Paper 62e

Morris Plant Furnace Start-up Incident
&
Lessons Learned

Bala S. Devakottai, Denise R. Karstensen, Bruce R. Raeburn

Equistar Chemicals, LP
Morris, IL / Channelview,TX

Prepared for Presentation at the 2004 Spring National Meeting, New Orleans LA, Apr 25-29
16th Annual Ethylene Producers' Conference
Session 62 – Ethylene Plant Safety

Copyright © Equistar Chemicals, LP
April 2004

UNPUBLISHED

AIChE shall not be responsible for statements or opinions contained in papers or printed in its publications.

Abstract

In March 2002, the Morris Equistar plant had an internal furnace firebox explosion. The incident occurred during the furnace start up following a routine maintenance shut down of the furnace. There were no injuries associated with the incident, but there was significant damage to the furnace and the adjacent furnace. A formal investigation of the incident was undertaken. This paper will review the details of this incident and describe actions taken to prevent a recurrence. An outgrowth of this incident has been a comprehensive review/critique of burner lighting practices utilized at all Equistar sites for ethylene furnaces, as well as for all fired equipment within the Lyondell / Equistar enterprise.

Introduction

Cracking furnaces can be considered to be the heart of an ethylene plant. The feed stock entering an ethylene plant is thermally cracked in the furnaces before the products are separated and recovered down stream. Proper operation of these upstream furnaces is essential for operation of downstream units. Typically, ethylene plants built in the 1970's were built with a large number of furnaces, with relatively small capacity. The trend has been gradually changing, with the plants built in the late 1990's having fewer and larger furnaces. Operator work force has been gradually on the decline, in part because of the reduction in the plant equipment and more automation. This makes it all the more important to make sure the right procedures are in place, procedures are being followed, and no compromise is made on safety. With the large number of cracking furnaces in an ethylene plant, there is a continuous set of activities in the furnace area such as decoking, exchanger cleaning, shut down for maintenance, burner re-light during start-up and transition to normal operating conditions. The furnace cycle is dynamic, with continuous operator involvement. While many of these are routine, repetitive activities, a small lapse can lead to a major incident.

Background of furnace

BA-107 is a Lummus SRT-1 furnace with natural draft stack. It has four radiant coils, A,B,C, D, with feed to each coil regulated via its own feed control valve. Dilution steam, also fed through 4 control valves, combines with the hydrocarbon feed downstream of the feed control valves, before entering the convection section. Each furnace has one stack damper with two furnaces sharing a common stack.

Fuel gas is supplied to the furnace through four fuel control valves, each provided with a bypass with manual block valve. Each furnace has a main fuel trip valve upstream of the four fuel gas control valves. The furnace is equipped with 120 wall burners in 6 rows. The furnace has a low fuel gas pressure interlock and a low steam drum level interlock.

Sequence leading to the event

On March 28, 2002, BA-107 was being prepared for startup following a maintenance shut down. BA-108, the adjacent furnace, which shares the stack with BA-107, was in operation. The other adjacent furnace, BA-106, was down for maintenance. Between 21:00 and 21:50 hours, pre-start-up activities such as stroking feed, dilution steam and fuel control valves were performed. The pre-startup checklist includes a step to make sure all individual burner cocks are closed. The furnace firebox was tested for flammability by measuring the Lower Explosive Limit (LEL), while the main fuel gas trip valve was still closed.

At 21:50 hours, the main fuel gas trip valve was latched open by the field operator and the bypass of all four fuel control valves were opened slightly to pressure up the fuel gas header to the individual burner cocks, as per the usual procedure. At this point, the lighting torch could not be found by the field operator. When the operator returned with a torch from another furnace, about 4-5 minutes had lapsed. The typical procedure is to light one burner associated with each radiant coil's fuel control valve, at the lowest row of wall burners.

At 21:55 hours, one burner in D coil and one burner in C coil were lit at the bottom row, with a lighting torch. At 21:56 hours, the torch was inserted into the lighting port to light B coil, and there was an explosion inside the firebox.

BA-107 furnace was isolated immediately and the area secured.

Impact of furnace incident

There were no injuries and there was no fire outside the firebox. There was significant damage on furnace BA107 and the adjacent furnace BA-106 to the walls, floors, radiant roof and structural columns, as seen in Figures 1 and 2 below. The stack damper on furnace BA 108, which shares a common stack with BA-107, was also damaged.

Figure 1 Walkway above firebox of furnace BA 107

Figure 2 Underside of the firebox on furnace BA 107

Root cause

An investigation of this incident revealed that the 4th row of 20 wall burners was left open. When the main fuel gas interlock valve was latched open by the field operator and the bypass valve around the four fuel control valves opened, the firebox was filled with fuel gas through the open burners.

DCS indication of fuel gas flow showed a sudden increase in fuel flow as soon as the bypass was opened, resulting in a fuel flow equivalent to a firing rate of 35 MMBTU/Hr. Stack oxygen indication dropped from 21 percent to zero and the arch draft pressure increased from –0.20 to 0.0 inches water gauge.

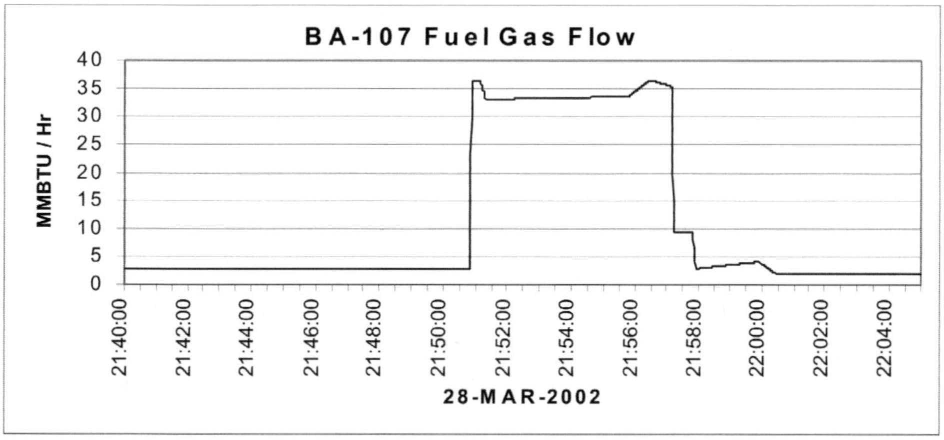

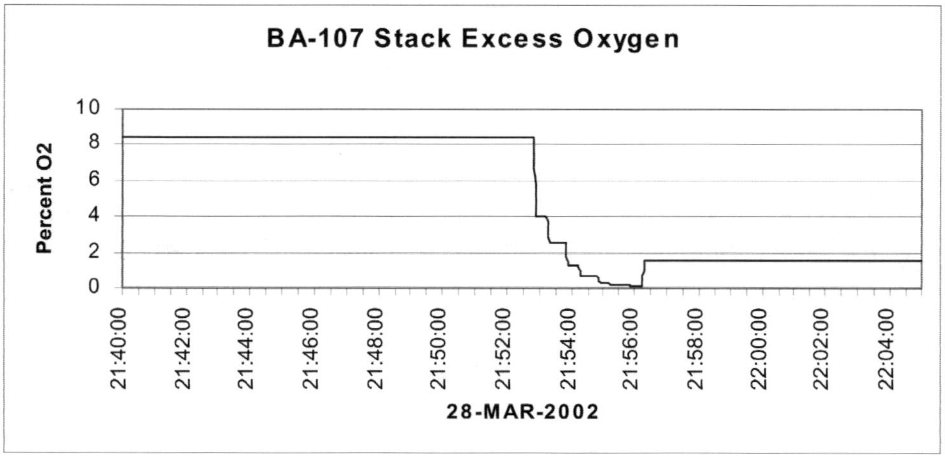

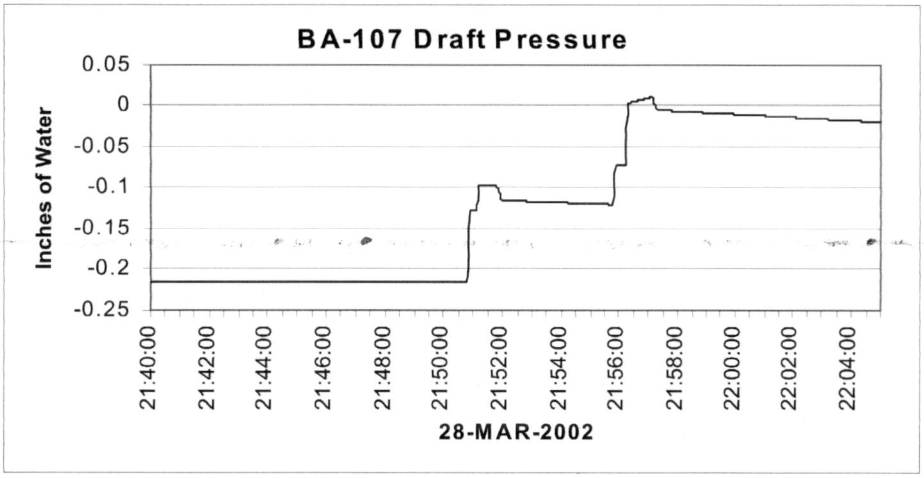

Lessons Learned

- In the procedure in place at the time of the incident (the "old procedure"), the operator opens the fuel gas bypass in the field and walks up to the first deck to light the first burners by opening the individual burner cock and inserting the lighting torch. During the incident, as soon as the operator opened up the bypass around fuel gas control valves, fuel gas started flowing into the firebox through the open burners. The fuel gas pressure, measured locally at grade downstream of the bypass valve, would have dropped indicating open burners. But this indication was not available at the first burner deck where the operator was lighting the burners.

  In the revised procedure, the bypass around fuel gas control valves will be opened to get 15 psig downstream of the control valve, then closed and watched for pressure decay. The first burner will be lit only after this header pressure check, ensuring all burner cocks are closed, has been completed.

- In the old procedure, typically, only one operator is involved in the first burner light up. With the fuel gas control valves and the local pressure gauge being at the grade level, and the first burner to be lit being on the first deck, it is not possible for one field operator to monitor all indications.

  The revised procedure requires two field operators for initial light off, one for watching the pressure at grade and the other for lighting the burners on the first deck.

- In the old procedure, the control room operator does not take control until the field operator lights the first four burners. The DCS indication on fuel flow, stack oxygen and arch pressure all would point to a fuel rich firebox, but these indications were not being monitored in the control room during the incident. There were two separate procedures – one for control room operator and one for field operator.

  In the revised procedure, the control room and field operator will coordinate all activities, including initial lighting of burners. The two separate procedures were merged into one to make sure both operators follow the same procedure.

- The old procedure requires LEL testing of the firebox only before the main fuel gas trip valve was latched open. Also, the wand used for LEL testing was short, which may not allow a representative sample of fire box gas to be obtained.

  The new procedure requires LEL testing to be done a second time, just before the first burner is lit. Since there are six rows of wall burners, LEL testing will be done using a long wand at all three decks of the firebox. Being a natural draft stack furnace with no ID fan, this step is critical to assure sufficient purging of the firebox before lighting the first burner.

- In the revised procedure, the stack damper will be open at least 25% to ensure adequate draft. This is all the more critical at Morris because of the natural draft stack.

- A DCS alarm based on a combination of low oxygen, high firing rate and low draft was set up to warn the control room operator of dangerous conditions in the fire box.

Olefin Furnaces at Other Equistar Sites

After the investigation of the Morris incident was completed, a brief survey of olefins furnace lighting procedures at other Equistar sites was conducted. The following are the preliminary findings from this analysis. A detailed review is currently in progress.

Furnaces in Equistar's six other olefins plants are equipped with ID fan to maintain furnace draft, and Morris ethylene plant has natural draft stack.

Fuel gas header pressuring to satisfy low fuel gas pressure interlock was done with nitrogen in 3 plants. In another 3 plants, the fuel gas header was pressured with fuel gas using a spring loaded dead man valve to bypass the fuel gas control valves and satisfy the low fuel gas pressure interlock. One plant is pressured with fuel gas using a manual bypass around the control valve.

In all plants, pressure testing the burner header using nitrogen or fuel gas ensures that all individual burner cocks are closed.

In 5 plants, low fuel gas pressure interlock permissive is required to latch open the main fuel gas trip valve. Of the other 2 plants, one bypasses the interlock and the other does not require a permissive, for initial burner light off.

LEL testing of firebox for flammable hydrocarbon is done in two plants. A permissive requiring the ID fan to be running to latch open the main fuel gas trip valve is in place in all plants. With ID fan in operation, the necessity of LEL testing is debatable.

Depending on operator availability, currently, initial burner light off at different sites involves either one or two operators.

Enterprisewide Fired equipment Review

In parallel with the ethylene furnace review, a review of initial burner light off procedure for all fired equipment in Lyondell/Equistar was initiated. The fired equipment review included boilers, steam super heaters, regeneration heaters, and other process heaters in various units.

Most of the large fired equipment such as boilers and steam super heaters were equipped with flame scanners, and had sufficient safety interlocks to avoid a fuel rich atmosphere without flame. Some flame scanners had reliability issues. However, a few of the small process heaters, with only a few burners and natural draft stack, were found to have insufficient time to purge the box before lighting a burner. These issues are being addressed along with procedural changes to minimize the occurrence of a fuel rich firebox.

Paper 62f

FLARE LINE FAILURE CASE, WHAT HAVE WE LEARNED?

Abram Kuo
Ryan Pitt
Westlake Group
900 Hwy 108
Sulphur, LA 70664

Prepared for Presentation at the 2004 Spring
National Meeting, New Orleans, LA
April 25-29
16th Annual Ethylene Producers' Conference
Session T8001 – Ethylene Plant Safety

Copyright © Westlake Group
April 2004

UNPUBLISHED

AIChE shall not be responsible for statements or opinions contained in papers or printed in its publications.

Overview

- Background
- Flare System Overview
- Event Sequence
- Root Causes / Corrective Actions
- Brittle Fracture Discussion
- Other Follow Up Actions
- Learning / Conclusions

Background

- The Westlake Petro 1 plant is a "demethanizer first" ethane cracker with a back end acetylene converter

- An off-spec event on 1/4/02 at the acetylene converter led to flaring of liquid ethylene product via the unit cold flare drum

- Through a sequence of events, the cold flare drum overhead line fell to below its minimum design metallurgy temperature

- On 1/5/02, the cold temperatures led to brittle fracture of the cold flare drum overhead line, loss of hydrocarbon containment, and ultimately an explosion and fire

Westlake Petro 1 Flare Drum System Overview

- The cold flare drum contents are vaporized and superheated with a closed loop propanol system
- Heat is supplied to the propanol system with 70# steam (~270 F)
- The vaporizer and superheater heats the cold flare drum material from cryogenic temperatures to above the minimum design metal temperature of the cold flare drum carbon steel overhead piping @ -10 F

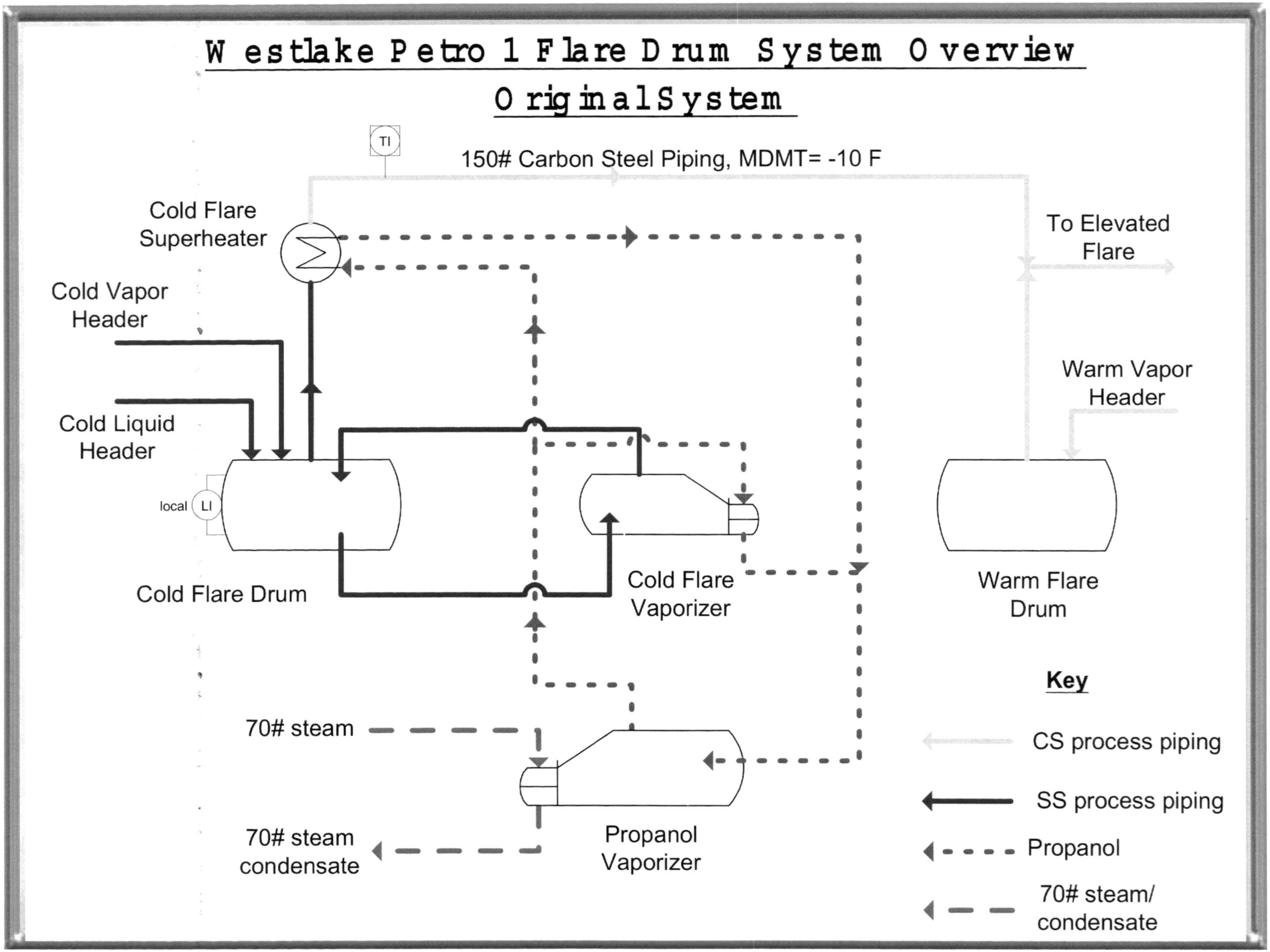
Westlake Petro 1 Flare Drum System Overview
Original System
TI
150# Carbon Steel Piping, MDMT= -10 F
Cold Flare Superheater
Cold Vapor Header
Cold Liquid Header
local
LI
Cold Flare Drum
Cold Flare Vaporizer
To Elevated Flare
Warm Vapor Header
Warm Flare Drum
70# steam
70# steam condensate
Propanol Vaporizer
Key
CS process piping
SS process piping
Propanol
70# steam/ condensate

Event Sequence

- The Westlake Petro 1 ethylene product went off spec on acetylene, initiating flaring of liquid ethylene product
- Acetylene converter outlet analyzer was in error which allowed the ethylene splitter inventory to be contaminated with acetylene prior to corrective action being taken
- A portion of the off spec liquid ethylene product was consumed by internal Westlake customers, with the balance being flared via the cold flare drum
- Malfunction of the cold flare drum vaporizer and superheater allowed the cold flare drum overhead line temperature to fall sharply

Event Sequence, Cont...

- A low temperature alarm sounded as the overhead line temperature fell to 0 F, and the thermocouple “went bad” at a value of -13 F

- With the cold flare drum overhead line running below its minimum design temperature of -10 F, the pipe ruptured, resulting in loss of hydrocarbon containment

- The hydrocarbon released found an ignition source, resulting in an explosion and fire

Cold Flare Drum Overhead Line Fire

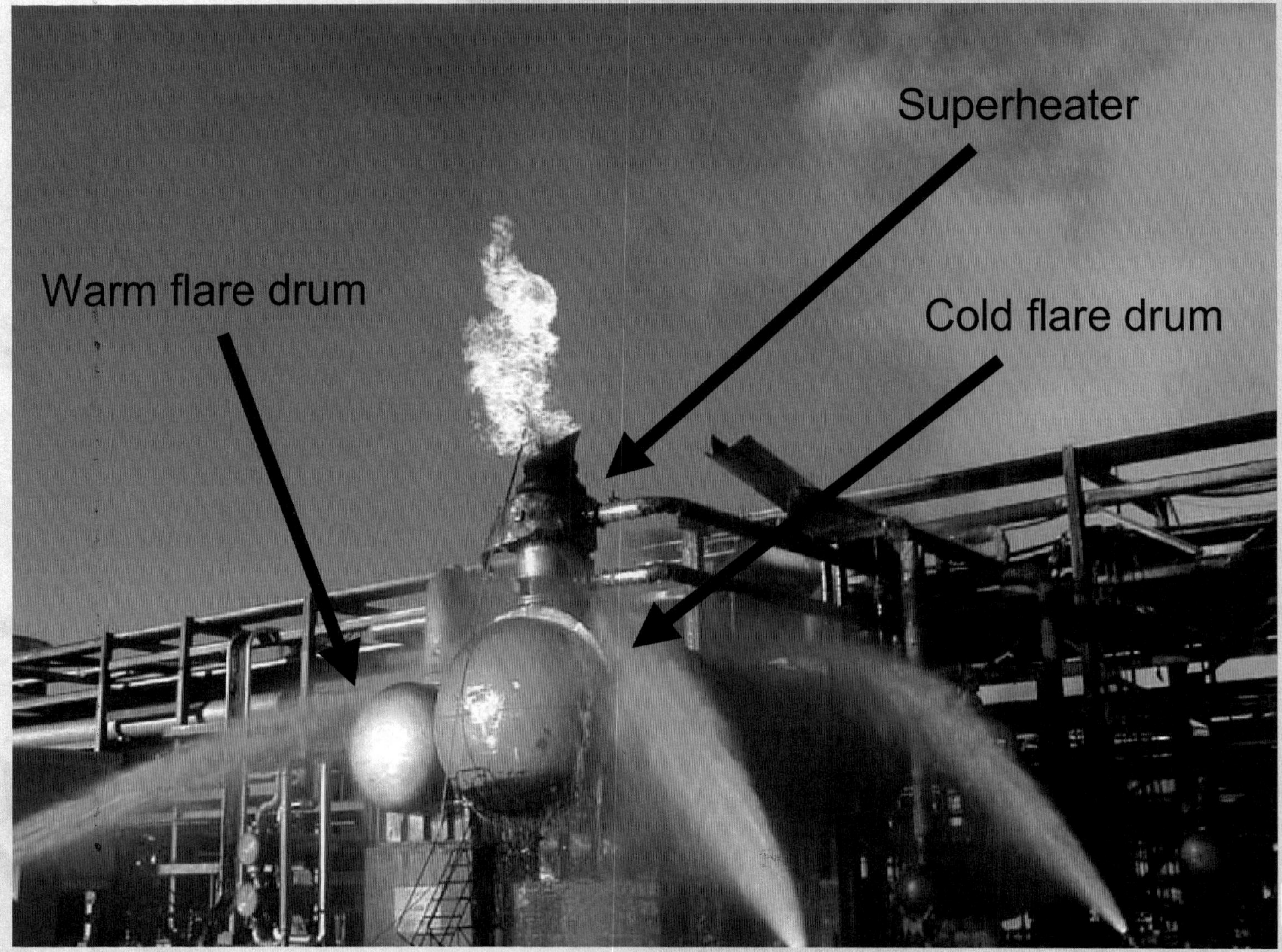

Root Causes of the Fire

- Vaporizer/superheater exchanger fouling had reduced heat transfer capacity of the cold flare system
- Once flaring began, the cold flare drum overhead line experienced low temperature
- Brittle fracture of cold flare drum overhead piping due to operation below the minimum design temperature of the carbon steel line

Root Cause #1: Vaporizer/Superheater Malfunction

- After the event, it was determined that fouling had compromised the operation of the vaporizer and superheater exchangers

Corrective Actions:

- 2 LI's installed on cold flare drum w/ DCS hi alarms @ 10%
- Heavy oil pull down system installed to allow removal of heavy oil from the cold flare drum
- Cold flare drum superheater replaced with 50% larger exchanger and propanol piping was replaced to handle higher throughput

Root Cause #2: Cold Safety Awareness/Management

- Once flaring began, the cold flare drum overhead line low temperature was not recognized

Corrective Action:

- Plant wide training conducted to re-emphasize cold safety awareness, including root causes and corrective actions for cold flare line event

Root Cause #3: Brittle Fracture

- The “final stress” that ultimately caused the brittle fracture of the piping has not been identified, but could have been any number of internal or external stresses
- External stress - Hard rain that came at the time of event?
- Internal stress - Contraction of the cold flare line due to temperature gradient?

Corrective Actions:

- Carbon steel flare system piping replaced with stainless steel, ~100' downstream of warm/cold flare drum tie-point
- Thermocouple installed on new stainless steel piping upstream of SS to CS transition
- Stress analysis on the new flare line, FEA on flare drum
- See “Flare Drum System Modifications” diagram

Overhead Line Brittle Fracture

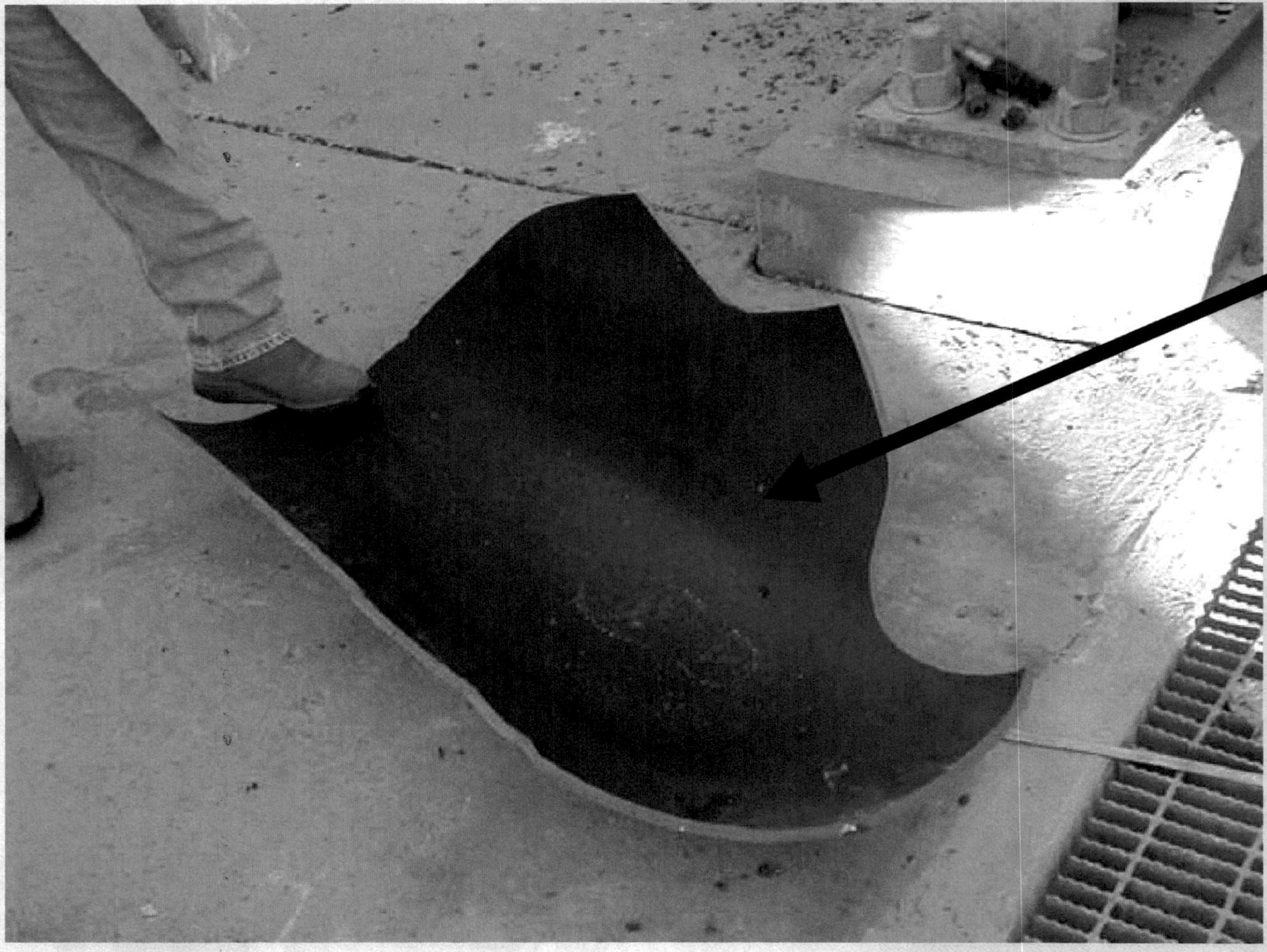

30" cold flare drum overhead line failure due to brittle fracture

Carbon Steel Brittle Fracture Discussion

- Carbon steel piping is typically used in services w/ temperatures above -10 to -20 F
  - Impact testing can certify the use of carbon steel piping in services as cold as -55 F (i.e. "killed carbon steel")
- At temperatures below -10 to -20 F normal carbon steel loses ductility and strength
  - The metal becomes brittle and can be susceptible to brittle fracture
- Three main factors influence the onset of brittle fracture:
  - Temperature
  - Flaw size (notches, scratches, etc…)
  - Stress

Impact Energy vs. Temperature for Carbon and Stainless Steel

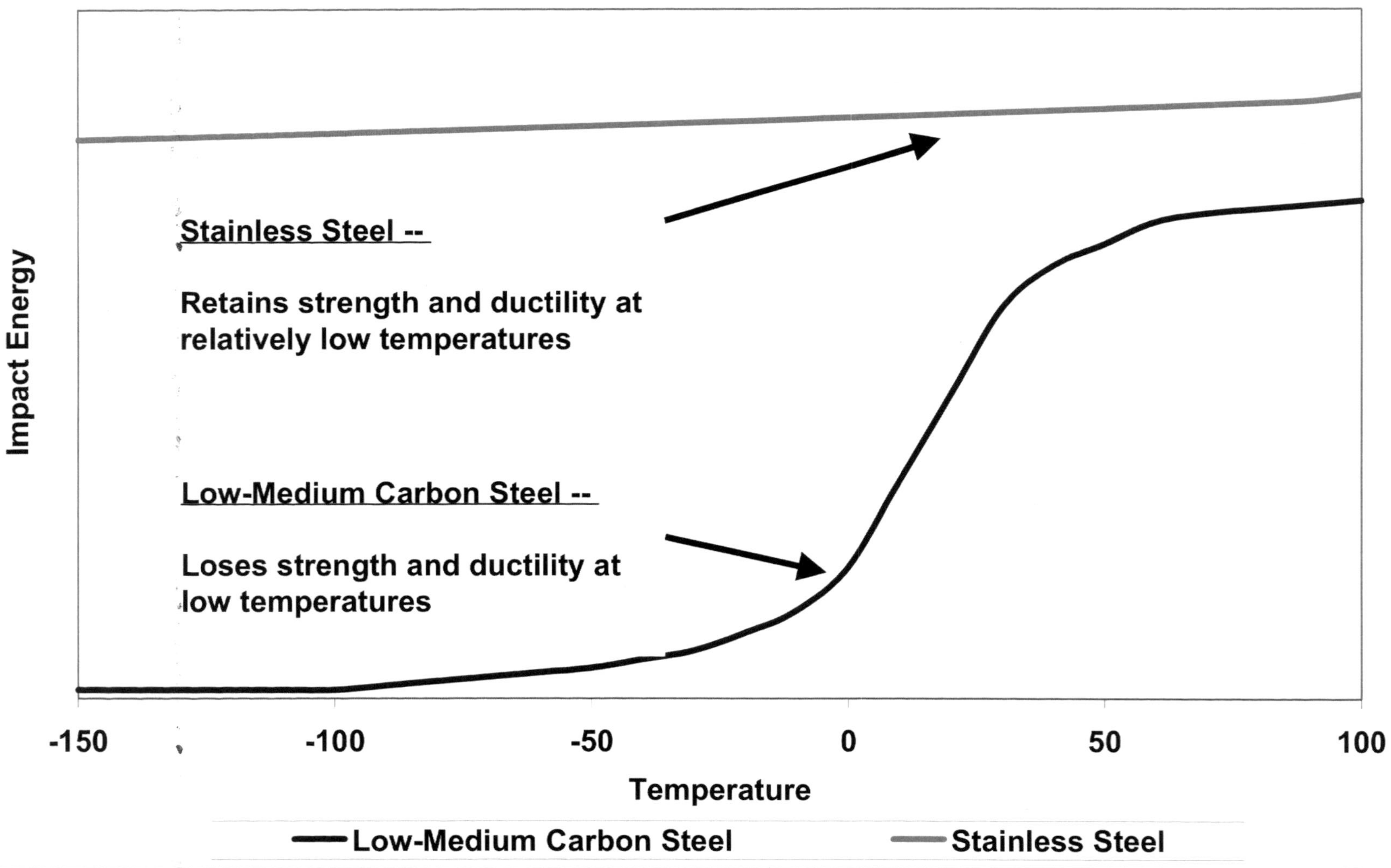

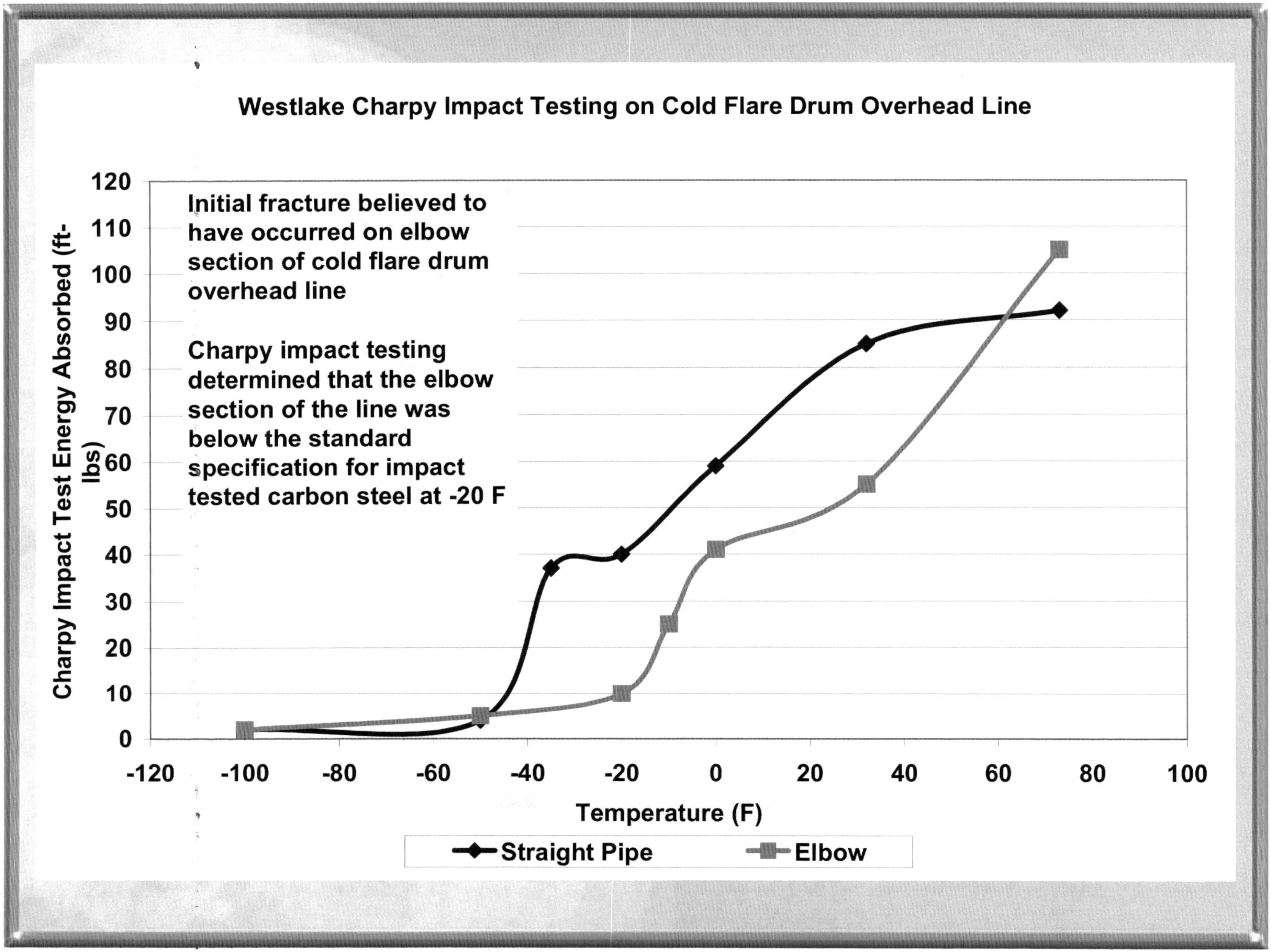
Westlake Charpy Impact Testing on Cold Flare Drum Overhead Line
Charpy Impact Test Energy Absorbed (ft-lbs)
120
110
100
90
80
70
60
50
40
30
20
10
0
-120
-100
-80
-60
-40
-20
0
20
40
60
80
100
Temperature (F)
Initial fracture believed to have occurred on elbow section of cold flare drum overhead line
Charpy impact testing determined that the elbow section of the line was below the standard specification for impact tested carbon steel at -20 F
Straight Pipe
Elbow

Other Follow Up Actions

- Study conducted to identify and replace carbon steel components in cryogenic service
  - Bimetallic demethanizer tower upgraded to stainless steel during 2003 T/A
  - Demethanizer tower reboiler piping upgraded from killed carbon steel to stainless steel during 2003 T/A
  - Ethylene splitter bottoms piping upgraded from killed carbon steel to stainless steel during 2003 T/A

- Install new cold flare line overhead thermocouple with a minimum range to -200 F

Westlake Petro 1 Flare Drum System Modifications

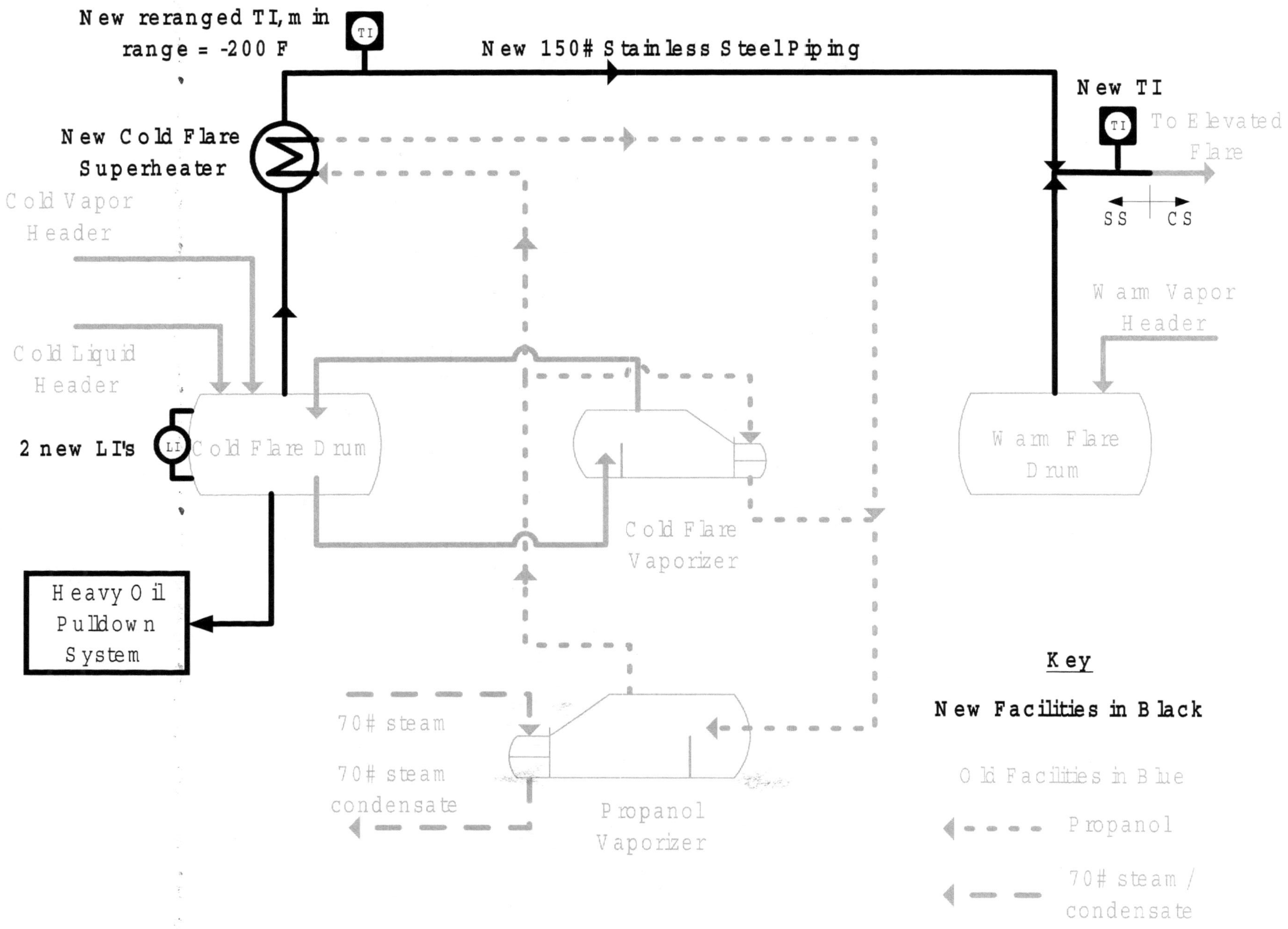

Safety/Environmental Impact

- The state police, the Local Emergency Planning Commission (LEPC), and the National Response Center (NRC) were notified of the event.

- No first aid or recordable injuries

- The fire was contained within a few hours, but it was not completely extinguished for 1 week

- The plant was shut down for 1 month while repairs were made to the unit

Key Learnings

- Event specific
  - Heavy liquid introduced to the cold flare drum should be removed in a timely manner
  - Cyclopentadiene (CPD) and other components in deethanizer bottoms streams can polymerize to foul cold flare heat exchangers
  - Cold flare system fouling may not be readily identified due to infrequency of use
- General
  - Unit upsets will tend to find the “weak link” in management systems, facilities, and personnel
  - Cold safety should remain a focus for all ethylene plants

SESSION 63A – BURNERS: WHERE WE ARE AND HOW WE GOT HERE
PRESENTED AT ETHYLENE PLANT TECHNOLOGY
FURNACE BURNER TECHNOLOGY REVIEW

PRESENTED BY:

R. Waibel (speaker), John Zinc Co LLC

This presentation was intended to provide a brief background regarding the evolution of burner design over the past 25 years. Changes in furnace and burner design over this period and the impact of these changes were discussed. Topics included:

Progression from premix burners to raw gas and now the re-introduction of premix burners,

Wall burners, hearth burners and combinations,

Going from welded fabricated parts to high alloy castings,

Changes in refractory burner blocks from dense solid refractory to light composite designs and all the stops in between,

Advances in design techniques from "cut and try" test firing to advanced computer aided simulations,

Advances in computer simulations and the integration of detailed burner models with detailed furnace models,

Fuel composition sensitivity, flame stability, and measures to enhance burner performance in these and other respects.

Flame Rollover and Other Flame Shape Problems

Ron Just

NOVA Chemicals Corporation

Prepared for presentation at 2004 Spring National Meeting,
New Orleans, LA, April 25-29

Copyright © NOVA Chemicals Corporation

March, 2004

Unpublished

AIChE shall not be responsible for statements or opinions contained in papers or printed in its publications.

Project Background

The troubleshooting activities described in this paper were conducted in support of the operation of the Ethylene 3 (E3) plant, which is located at the Nova Chemicals operating site in Joffre, Alberta, Canada. There are three olefins plants at the site and they have a combined ethylene capacity of more than 6 billion pounds per year. At the time of its start-up, E3 was the largest ethylene plant in the world, with a nameplate capacity of 2.81 bpy, based on ethane feedstock.

On-spec production of ethylene from E3 commenced at the end of August 2000. A trial run was performed on one of the cracking furnaces a week prior to the planned start-up of the entire plant, in order to verify furnace operation and tune controllers. It was during this trial run that challenges with burner performance were first identified.

General Design Information

E3 was designed with seven furnaces, but can maintain its nameplate capacity with one furnace in decoke. Each furnace, then, has a design capacity of 470MMlbs / yr of ethylene production. However, since the plant spends much of its time operating with all seven furnaces on-line, there was a second distinct design point for the furnaces at an ethylene production capacity of 400 MMlbs / yr.

The furnaces are a twin radiant cell design, with a shared convection section. A combustion air preheat (CAP) system was supplied, necessitating a forced draft fan for supply of combustion air to the individual burner assemblies. Both the forced draft and induced draft fans on the furnaces were equipped with variable frequency drives, in order to provide the controllability required across all the operating scenarios for the furnaces. The furnaces were also designed for on-line decoke of the radiant coils, with the effluent incinerated in the furnace firebox. Decoke nozzles are also used for incineration of acid gases and hydrocarbon tank vents within the plant.

The plant fuel gas supply system is designed to supply a variety of fuels to the furnace burners: a hydrogen-rich offgas produced within E3; a methane-rich waste gas from an on-site hydrogen recovery plant; natural gas make-up; and if required, E3's mixed C3 coproduct can be burned as fuel (this stream hasn't been used as a fuel for the E3 furnaces).

Furnace Commissioning Activities and Start-up Challenges

Commissioning activities for the furnaces included a limited number of activities directly related to the combustion system. Internal visual inspections of the combustion air supply ductwork were performed to ensure that all large debris capable of plugging or restricting air flow had been removed. Velocity and rupture air blows were performed on the plant's fuel gas supply headers, and velocity air blows were conducted on each furnaces' fuel gas piping for removal of debris, corrosion products, liquids, etc. General commissioning activities on the furnaces also included dry-out of the radiant box refractory at low firing levels. During this activity, the Operations staff had great difficulty in keeping all of the pilots on the furnace lit, while they started lighting the primary burners. Once primary burners were lit, and the refractory tile associated with that burner was hot, the pilot remained lit. However, it was noticed at that time that the pilot flame wasn't straight, and in fact was leaning heavily away from the wall and into the center of the furnace.

The second issue with the burners was identified during the individual furnace test run conducted one week prior to the planned start-up of the entire plant, using natural gas as fuel. The furnace was gradually ramped up to its design operating conditions, however a visual inspection of the firing patterns through the peep doors of the firebox revealed a tremendous amount of flame rollover and impingement on the radiant coils. The opposite end of the firebox and many of the radiant coils could not be seen due to the flame rollover! The combustion performance within the furnaces was clearly unacceptable, and the furnace designer and burner manufacturer were immediately engaged to resolve the operational problems.

Based on experiences from past furnace start-ups, the burner manufacturer provided recommendations for adjustments to the secondary burner elevations, which were promptly implemented on one furnace and retested. This change resulted in moderate improvements in flame quality, but performance was still not acceptable. The burner manufacturer recommended a second change to the secondary burners, which required changes to the orifice drillings on the tip. New tips were expedited and installed in combination with the initial elevation change. While this second change still didn't meet expected performance, it was now possible to operate the furnace at 64% of design throughput, with modified operating parameters as outlined later in this paper. The decision was made to replace the secondary burners on all seven furnaces, in order to allow for the start-up of the ethylene plant at reduced operating rates, while a focused troubleshooting effort commenced to resolve the long term operation of the burners.

At this point, the flame was pulling away from the refractory wall at an elevation of only 8 to 10 feet above the burner, and was then rolling over violently into the process coils. The resulting direct impingement of the flame and hot flue gas onto the coils was creating localized overheating of the radiant coils, which would lead to high rates of local coke formation inside the coils, shortened run lengths, and excessive carburization of the coils. Clearly, a concentrated effort was required to determine the root cause of the problems, and rectify combustion within the furnace in order to allow the E3 furnaces to support operation of the ethylene plant at its design production rate.

Troubleshooting Methodologies

Since the manufacturer's immediate suggestions didn't result in acceptable flame patterns within the furnace, a more significant and thorough troubleshooting effort was required. It was unknown whether the problems were the result of burner design, furnace design, operating parameters, or a combination of these factors, so it was felt that the troubleshooting effort needed to draw on expertise from all three of these areas. A team was formed which included members from NOVA's plant support and research personnel, as well as senior furnace designers from the engineering contractor, and senior combustion engineers, field personnel and computational fluid dynamic specialists from the burner manufacturer. A comprehensive plan was developed, and then readjusted weekly based on results received from the various team members. A variety of troubleshooting tools were used, and are described in the following subsections.

Visual Observation in E3 Furnaces

With initial operation on natural gas, the burners produced a very visible flame, making it easy to describe the degree and location of flame rollover within the firebox. For initial furnace trials and the early burner modifications, these visual observations formed the basis for all judgements of the flame quality.

However, as the plant transitioned to the design fuel, containing a majority of hydrogen, the flame was no longer visible. At this point, a tracer was required to make the flame / flue gas visible and allow comparative evaluations of the impacts of burner changes. A standard dry powder chemical fire extinguisher was used to illuminate the flow patterns within the furnace. Short injections of the dry powder into the combustion air ducting to a single burner created a brief (ie. 1 to 2 seconds) visualization of the flue gas patterns from that burner. The powder quickly mixed throughout the firebox, making further observations difficult. This method did allow for an evaluation of the elevation and amount of rollover taking place, and was particularly effective at highlighting when extreme rollover resulted in direct impingement onto the radiant coils.

An operating furnace experiences a natural down draft in the region of the process coils, because these coils are "cool" relative to the other surfaces and flue gas within the radiant box. This effect will ultimately cause some of the hot combustion gas to pull away from the refractory wall and "roll" toward the coils. Judging what degree of rollover is acceptable becomes highly subjective. As performance of the E3 burners gradually improved, visual observations of the flame / flue gas patterns became less effective, and it became extremely difficult to make comparative judgements between the final burner modifications.

These early visual judgements were critical in establishing operating parameters that allowed E3 to maximize the initial operating range of the furnaces. Observations made by Operations personnel during furnace start-ups led to the discovery that shutting in some of the primary or secondary burners in an evenly distributed pattern dramatically improved flame quality. This resulted in development of interim operating strategies in which some of the primary and sometimes the secondary burners on a furnace were blocked in to improve overall flame quality and extend the furnaces' operating capacity. These observations also directed the focus of modelling work to consider the interaction of adjacent burners as a possible cause of E3's problems.

Observation of tube colouration and pyrometer measurements of tube metal temperature were used to supplement indications of flame / flue gas patterns within the radiant box. Large, localized differences in radiant coil temperature were used as indicators that excessive flame and / or hot flue gas was still contacting the coils. As flame quality improved, this became the dominant method for rating combustion performance.

Visual observations also focused on relative hot and cool patterns on the refractory walls immediately adjacent to each burner, and between the burners. This allowed inferences to be drawn about the location of "cold" gas streams within the furnace, which were either signs of cold combustion air not properly mixing with fuel, or "cool" recirculation zones.

Manufacturer's Test Facility Furnace Trials

As part of the original project to construct the furnaces, burner testing was performed in a test furnace at the burner manufacturer's facility in 1998. The series of tests was performed on a single burner, and personnel from NOVA and the project's engineering contractor participated in the tests. Testing provided no indications of the flame rollover that was to be encountered during start-up of the E3 furnaces.

As part of the troubleshooting effort in late 2000 and early 2001, new burner tests were performed at the manufacturer's test facility. Several new approaches were implemented in this second round of testing in an attempt to replicate E3's combustion problems in the test furnace:

- Testing was conducted in a larger furnace that could accommodate installation of two adjacent burners. It was felt that this could help determine if burner to burner interactions were contributing to the problems.
- Sodium bicarbonate was injected into the combustion air stream to make flame / flue gas patterns more visible and determine to what extent rollover was taking place.
- Attempts were made to adjust operating temperatures within specific areas of the test furnace to more closely approximate the temperature distribution within E3's furnaces.

Despite all these efforts, the rollover problem from E3's furnaces couldn't be replicated in the test furnace. It was possible to draw comparisons between the test furnace and the E3 furnaces with respect to the temperature patterns on the refractory wall in the immediate vicinity of the burner. This allowed some conclusions to be drawn in the test furnace with regards to the degree of rollover that various burner redesigns were experiencing. For the most part, however, the test furnace couldn't be used to determine whether roll-over problems had been resolved, and testing of modified burner tips focused on two additional criteria prior to any installation in E3's furnaces:

- Radiant heat flux was measured, to ensure that revised burner designs met minimum requirements as specified by the engineering contractor.
- Flue gas NOx measurements were taken, to ensure that the project's performance guarantees could be met.

In this fashion, the test facility was used to screen potential burner modifications, prior to the expense of installing and testing a set of burners in one of the E3 furnaces.

Computational Fluid Dynamics Modelling

Much recent work has been undertaken in the field of modelling combustion processes and industrial heaters through the use of computational fluid dynamics (CFD). Since both the engineering contractor and the burner manufacturer had been involved in furnace CFD modelling, CFD efforts were undertaken to try to identify the root cause of the problems with the E3 furnaces.

The burner manufacturer's CFD modelling group created a near field CFD model that consisted of a slice across the width of one of the radiant cells, and included half of two adjacent burners. The slice consisted of a portion of two adjacent radiant coils, one half of the burners associated with these two radiant coils, as well as the inlet combustion air ducting associated with the burners. Detailed resolution on the model allowed specific modelling of the impacts of each burner tip, and the multiple orifice drillings on each tip. Modelling adjacent burners provided the ability to explore and evaluate possible interactions between adjacent burners. Inclusion of the air supply ductwork also allowed impacts of air distribution restrictions to be evaluated.

A full radiant cell CFD model was developed by a third party CFD modelling consultant. This model was built with a coarser grid than the burner manufacturer's model, and included a complete radiant cell of the furnace. The model also included the entrance to the convection section. This model simplified the specific design details associated with each burner, but was more capable of evaluating burner interactions and the overall effect of furnace geometry and flue gas currents.

An additional third party CFD modelling consultant was engaged to build a model of the combustion air ductwork, and evaluate the distribution of air flow to the individual burners on a furnace.

Results of the CFD studies contributed to the overall rationalization of the combustion problems within the E3 furnaces. However the tools as applied in E3's troubleshooting seemed limited in their ability to independently and proactively predict combustion phenomenon in the furnace. Once observations had been made in the field, the CFD teams were able to theorize the possible causes based on their models. The models didn't, however, prove capable of identifying the root cause, or estimating improvements in flame quality for the final sets of burner modifications.

Cold Flow Velocity Survey

In support of the various CFD models being prepared, a cold flow velocity survey was conducted on the burners in one of the E3 furnaces. This test focused on the impact that an air distribution plate located in the throat of the burner was having on the velocity distribution of combustion air within the individual burner.

The furnace was tagged-out and isolated following normal procedures for preparation of a furnace prior to entry into the firebox, with the exception that the forced and induced draft fans were still operational. Plant personnel were then able to enter the firebox, and use an anemometer to measure air velocities across the burner throat.

Velocities were measured using a 5 X 3 matrix, for a total of 15 velocity measurements per burner. The test was performed on burners with the original air throttle plate in place, and also on burners for which the plate was removed. Testing confirmed the team's belief that the throttle plate was resulting in severe maldistribution of air. Air was jetting out along the front and back edges of the burner throat, while the area above the plate was experiencing very low, and even reverse flows of air.

In-Plant Trials of Modified Burners

As described in previous sections, as the operation of E3's burners improved, it became increasingly more difficult to judge the relative performance of modifications, and the team struggled to identify measureable parameters that could be used to define success. With the inability to replicate E3's combustion problems either in the manufacturer's test furnace or through CFD models, it became clear that testing of modified burners in E3's furnaces was the only method for determining whether subsequent changes successfully resolved the problems. Ultimately, determination of successful burner performance was dependent on achieving design expectations for the furnaces as a whole. Targets for success, then, were based on two criteria:

1. Achieving repeatable minimum design run lengths when operating the furnaces continuously at design conditions.
2. Meeting the original project specification for NOx, of emission levels less than 0.0936 lbs NOx / MMBtu.

While proposed changes to the burners drew on the results of all the troubleshooting efforts described earlier, the team depended heavily on the knowledge and experience of the burner manufacturer's combustion engineering expertise to suggest variations to the burner design that would improve performance. The biggest challenge arose from the fact that changes that reduced the flame rollover typically resulted in less preferable radiant heat flux profiles and higher NOx generation. As noted earlier, the manufacturer's test furnace provided a reliable method for pre-screening heat flux and NOx impacts, but the final flame quality and NOx performance was measured in E3's furnaces.

Due to the large cost and maintenance effort associated with changes to the burners, it was desireable to minimize the number of burner modifications in a single furnace in order to confirm the performance of new modifications.

The minimum option considered was the change-out of the burners in one half of a single radiant cell. However, since some of the CFD results and in-plant tests indicated that interaction between burners was potentially contributing to the combustion issues in the E3 furnaces, the troubleshooting team decided that for the initial E3 furnace tests, all of the burners in one radiant cell would be modified. This was also beneficial since it allowed side-by-side comparisons of the old and new burners through observations and measurements in the two radiant cells on the same furnace. Some of the in-plant burner tests required that all burners in the entire furnace had to be modified. This was necessary during testing of air side modifications, since there was no other way to ensure equal air flow to all burners on the furnace. In addition, it was necessary to replace all the burners in a furnace prior to any verification of NOx performance.

Stack Testing for NOx Performance

As the multiple changes eventually resulted in burner performance that was deemed acceptable from the point of view of furnace operation, the final criteria that the burners had to achieve was the project NOx guarantee. Initial NOx estimates were obtained using portable flue gas monitoring equipment, but once these tests indicated that acceptable performance was being achieved, final tests were performed by a stack testing consultant, following the Alberta Environment stack testing protocol.

Burner Design Changes and Final Performance

Modifications to the E3 burner assemblies essentially fell into three main categories:

- Secondary burner tip changes.
- Air side distribution changes.
- Reallocation of the fuel split between primary and secondary burners, accompanied by primary and secondary burner tip drilling changes.

Changes to the secondary burner tips were the first series of changes attempted in order to resolve the E3 firing problems. Since the secondary burners supply a majority of the fuel to the combustion zone, these changes were directed at trying to force greater mixing between the combustion air and the bulk of the fuel gas. Elevation of the secondary burner tip relative to refractory floor was the first modification tested, and resulted in a modest improvement in flame quality. This was followed by modified fuel orifice drillings. A series of different orifice angles were tested, before arriving at a burner that produced significantly less flame rollover, at the expense of somewhat higher NOx levels and less ideal radiant heat flux profiles. These initial changes were performed on the basis of the observations in the operating E3 furnaces, and on the recommendation of the burner manufacturer's previous experiences with this burner type.

The second series of burner modifications focused on the throttle plate that had been installed to control the flow of air to each burner. The furnace was originally designed for induced draft. A forced draft fan was incorprated when a combustion air preheat system was added to the furnace later in the design. Final design of the burner attempted to achieve equal distribution of combustion air to each burner by installing three separate throttling devices in the combustion air supply to each burner. Two of these were fixed orifices: a perforated throttle plate installed in the throat of the burner; and a rectangular orifice in the ducting to the burner. The third device was a set of louvres, which are normally operated in the wide open position, but are throttled when the furnace is in start-up, shut down or decoke operation.

Air side changes focused on the belief that the burner throat plate was creating severe maldistribution of the combustion air exiting the burner tile, resulting in poor mixing of the air and fuel. This was supported by the observations made in the operating E3 furnaces, the cold flow velocity testing, and some of the early CFD modelling. The throat plate was removed, to provide an unrestricted entrance for the air flow into the burner tile, and the rectangular orifice was replaced with a perforated plate that had been sized to provide the combined pressure drop of the two original fixed orifices. This was one of the key changes to the E3 burners, and resulted in a significant improvement in burner flame pattern, a major reduction in flame impingement, and very little impact on NOx.

The burner throat restriction plate had also clearly had a significant impact on the operation of the raw gas pilot tip, due to the extremely high local velocities that were occurring in the vicinity of the pilot head cut out in the plate. Removal of the plate resulted in an immediate improvement in pilot stability, eliminating the early problems with the leaning flame and instability in a cold furnace.

Performance was still not meeting expectations relative to the degree of flame / flue gas impingement on the radiant coils, and so the final, and longest, portion of the troubleshooting effort focused on additional changes to the fuel distribution. During this phase, a series of tip drillings were tested in coordination with changes in the split of fuel between the primary and secondary tips. Much of the work was to reduce the potential for any interactions between burners, based on in-plant observations and CFD model results. Efforts focused on redistributing fuel; a series of redistributions was tested at the CFD level and in the manufacturer's test furnace. It appeared that significant improvements in flame quality could be achieved, however major increases in NOx accompanied the most successful changes.

The team decided to test one of the most successful tip redesigns in one of the E3 furnaces to confirm the CFD and test furnace conclusions. The plant test confirmed that the revised burner delivered exceptional performance from the point of view of flame quality, and as expected, resulted in NOx levels that were greater than the project guarantee values. Results of the plant testing resulted in one final modification, which attempted to compromise between the flame quality gains and the NOx requirements of the project. The final modification was tested in an E3 furnace, and was ultimately judged to have attained the required performance criteria. Furnace run lengths on two consecutive cycles easily exceeded the design basis for the project. Stack testing also confirmed that the furnace was exceeding the projects guarantees for NOx emissions, with estimated performance below 0.075 lbs NOx / MMBtu when operating at design conditions, versus the project performance guarantees of 0.0936 lbs NOx / MMBtu.

Following completion of the tests on the E3 furnace in April 2001, a plan was implemented which resulted in the final installation of new primary and secondary burner tips over a period of several months. By September 2001, the final E3 furnace was retrofitted, and a full year after the initial start-up concerns had been identified, E3 had achieved acceptable combustion performance across all of its furnaces.

Operational Impacts

Upon initial furnace operation, the high degree of flame rollover and impingement on the radiant coils described earlier resulted in localized overheating of the coils. Furnace operation had to be adjusted to minimize the local temperature differentials on the coils. This was accomplished through the following operational strategies:

- Furnaces were operated on natural gas fuel, rather than E3's produced offgas product. Flame temperatures were much lower on natural gas than on the offgas, which can contain over 80 mol% H2. This resulted in much less localized overheating of the radiant coils.
- Excess O_2 levels were increased to 5%, from the design level of 2%. This seemed to have the effect of reducing the amount of rollover, and again cooling the flame temperatures.
- Fired duty was limited by reducing the feedrates to the furnaces, and the target ethane conversion.
- As discussed earlier, individual, evenly-spaced primary or secondary burners were blocked in to reduce rollover and impingement on the radiant coils.

Ultimately, these strategies resulted in furnace operation being limited to only 64% of design production rates. In addition, the offgas that wasn't being consumed as fuel was flared continuously, incurring significant costs to the plant in additional fuel make-up. These restrictions were in place for much of the first month of plant operation.

The first two months of operation saw numerous changes in tip design, but the first major series of burner tip modifications was completed by the end of the second month of operation, and allowed furnace capacity to be increased to just over 90% of design production. However furnace operation was still restricted by only partial use of the hydrogen-containing offgas as fuel, and continued operation at elevated excess O_2 levels.

Modification of the air side ducting and distribution plates were validated early in the third month of operation, and resulted in the ability to finally operate on full offgas as fuel and at design excess O_2 levels. This was a major maintenance effort to implement, and required each furnace to be down cold for at least 10 days to allow disassembly of the combustion air ducting. Coordinating this maintenance with decoke schedules in order to maximize production levels resulted in the full implementation of all air side modifications extending over a three month period. Individual furnace production rates were still limited to just over 90% of design production during this period.

Final resolution of the combustion problems was attained in one of the E3 furnaces in April 2001, with the final burner change-out completed in September 2001. During the initial year of operation of E3, the furnaces experienced gradually decreasing levels of flame / flue gas impingement on the coils. The long term effect of this operation is unknown, although the radiant coils have been experiencing premature failures due to localized carburization of the coils, which may be the result of damage incurred during the initial operating period. Also of concern is the subjective nature of the final assessment of acceptable combustion performance. There was still a certain amount of rollover occurring in the firebox after the final tip configuration was installed. It is possible that the coil failures are being caused by ongoing impingement. NOVA is continuing to investigate existing performance and potential improvements to the current burners.

Plant Maintenance Impacts

The program to achieve acceptable performance of the burners on E3's furnaces resulted in significant impacts on plant maintenance. During the period from August 2000, until the final set of burners was changed out in Sept. 2001, over 25 partial or full furnace burner component change-outs took place (including air side modifications). Aside from the obvious cost and manpower implications for E3's maintenance team, the burner change-outs carried a number of personnel safety issues.

E3's furnaces sit approximately 10 feet above grade, which was driven by the need to provide appropriate personnel egress under the furnace, and the size of the combustion air supply ducting. As a result, however, the removeable burner assemblies can't be accessed directly from grade by maintenance personnel. Access to the burner assemblies is also complicated by the presence of three sets of fuel gas supply piping (pilot, primary and secondary) to each burner assembly, combustion air supply ducting to each burner, and decoke effluent piping that runs underneath the furnace, and along most of its length. In addition, burner replacements took place while furnaces were on-line or in decoke, meaning that the tips of the burner assemblies were red-hot when they were removed from the firebox. These issues resulted in a sizeable potential for injury to personnel during the numerous tip change-outs that occurred.

In order to reduce the risks associated with the burner replacements, continuous scaffolding was erected along the length of the furnace firebox to provide proper access to the burner assemblies (ie. four scaffolding assemblies under each furnace). Burner change-outs were also aligned with furnace turnarounds as much as possible in order to avoid the risks associated with removal of high temperature tips from the box. While this added additional cost and time to each set of burner change-outs, the work was accomplished without any major recordable injuries.

Management of burner change-outs has continued to be an issue within E3 over the past three years. Pluggage of individual burner tips, especially the secondary tips, is a chronic problem, with replacement of numerous tips required during one operating cycle on a furnace (ie. decoke to decoke). The pluggage appears to be a result of the collection of corrosion products in the tip. Two separate efforts have been iniated to reduce burner tip pluggage. The manufacturer has proposed a new tip style, which has proven successful at reducing pluggage at other facilities, in which the burner head is smaller, with less residence space inside the tip. This should result in a lower potential for cracking of the fuel leading to coking inside the tip. As well, new tip metallurgy is being tested to reduce the high temperature corrosion potential of the tip and riser. Resolution of the burner tip pluggage is viewed as a critical step to achieving stable and reliable performance of the burners in E3.

Conclusions

Initial combustion problems in E3's furnaces significantly impacted furnace operation for the first three months of operation. One of the keys to resolving the performance issues was the formation of a multi-disciplinary troubleshooting team, drawing on resources from NOVA, the furnace designer and the burner manufacturer. A wide variety of troubleshooting techniques was required in order to achieve successful performance, and roughly eight months after the initial on-site furnace test identified the problems, the final successful burner modifications had been proven. Final furnace performance exceeded design run lengths at design conditions, and produced NOx levels ~20% below guaranteed performance levels. Opportunities for further performance improvements are centering on reductions in burner tip pluggage.

63b

IMPACT OF TEXAS NO_X REGULATION ON THE U.S. ETHYLENE INDUSTRY

Vikki Medley
Chemical Market Associates, Inc. (CMAI)
Houston, Texas, U.S.A.

Prepared for Presentation at:
16th Annual Ethylene Producer's Conference
2004 AIChE Spring National Meeting
Ethylene Plant Technology – Furnace Burner Technology Review

Copyright © Chemical Market Associates, Inc. (CMAI) 2004

Original version published in proceedings from
CMAI's 2003 World Petrochemical Conference
Houston, Texas, U.S.A.
March 26-27, 2003

AIChE shall not be responsible for statements or opinions contained in papers or printed in its publications

Abstract:

NOx is a group of air pollutants that are formed during combustion. The reaction of NOx and volatile organic compounds, such as ethylene, creates ground-level ozone, one of the air pollutants specified for reduction under the United States Clean Air Act amendments of 1990. This paper takes a 50,000 foot view of one nonattainment area in the United States, the Houston/Galveston area in Texas, and investigates what regulations are impacting this area, how the area plans on meeting the regulations, probable changes to the state's implementation plan, and finally the impact of the regulations on the U.S. ethylene market. It does not attempt to investigate all potential methods of meeting the regulations or all potential issues around the regulations, including buying and selling emissions credits, alternate technologies, or other areas impacted by the Clean Air Act.

What is the Clean Air Act of 1990?

The Federal Clean Air Act of 1990, which amended the original Clean Air Act of 1970, was signed into law by then-President George H. W. Bush to combat acid rain, urban air pollution, and toxic air emissions and to mitigate their negative impacts on the environment and on human health. Title 1 of the Clean Air Act of 1990 called for the United States Environmental Protection Agency to establish National Ambient Air Quality Standards for six common air pollutants; ground-level ozone, particulate matter, lead, nitrogen dioxide, sulfur dioxide, and carbon monoxide. Upon establishment of the standards, the United States Environmental Protection Agency classified metropolitan areas of the United States as either being in "attainment" of the air pollution standard, meaning it is below the maximum level of the pollutant, or it is in "nonattainment", meaning that the area's air exceeds the maximum pollutant level set by the United States Environmental Protection Agency. The accompanying table entitled "Selected Clean Air Act Nonattainment Areas" details several nonattainment areas in the United States. For sulfur dioxide or lead, an area is either in attainment or not, but all other pollutants have a graduated rating system that defines the degree of nonattainment each area has been assigned. In this chart, the lowest level of nonattainment is moderate, followed by serious, severe, and ending with the highest level of nonattainment, extreme. This rating system also dictates the amount of time an area is granted to reach compliance with the law. Los Angeles is a nonattainment area for three pollutants; ground-level ozone, carbon monoxide, and particulate matter. It is also the only United States

Selected Clean Air Act Nonattainment Areas

Metropolitan Area	Pollutant					Total Categories
	Ozone	Carbon Monoxide	Sulfur Dioxide	Particulates	Lead	
Phoenix	Serious	Serious		Serious		3
Los Angeles	Extreme	Serious		Serious		3
New York	Severe-17	Moderate		Moderate		3
Pittsburgh		Nonattainment	X	Moderate		3
El Paso	Serious	Moderate		Moderate		3
Anchorage		Serious		Moderate		2
Chicago	Severe-17			Moderate		2
Las Vegas		Serious		Serious		2
Provo		Moderate		Moderate		2
Spokane	Moderate	Serious				2
St. Louis	Moderate				X	2
Denver				Moderate		1
Washington, D.C.	Serious					1
Providence	Serious					1
Houston/Galveston (HGA)	**Severe-17**					**1**
Milwaukee	Severe-17					1
Dallas	Serious					1

Source: US EPA, 7/19/02

metropolitan area that is classified as extreme, the highest degree of nonattainment for ground-level ozone. The New York area is in nonattainment of the same three pollutants as Los Angeles, but it is classified as severe-17, one step below extreme, for ground-level ozone. Severe-17 means that the area will have 17 years to come into compliance with the Clean Air Act of 1990, so New York must demonstrate compliance for ground-level ozone pollution by the end of 2007. The Houston/Galveston area, the nonattainment area that is the focus of this paper, is in nonattainment for one pollutant, ground-level ozone, and is classified as severe-17, establishing a compliance deadline of November 15, 2007.

Why is the Houston/Galveston nonattainment area so important?

The Houston/Galveston Area is the region in southeast Texas containing Harris and Galveston counties, and the six surrounding counties of Waller, Montgomery, Liberty, Chambers, Brazoria, and Fort Bend. This region has significant concentrations of refining and petrochemical assets, not only for the United States but on a world-wide basis as well. The Houston/Galveston area is home to 3 percent of the world's refining capacity, but 13 percent of the United States' refining capacity is within the nonattainment area. 7 percent of the world's benzene capacity is built in the Houston/Galveston area, but 30 percent of the United States' benzene capacity is inside this region. 13 percent of the world's refined butadiene capacity is in this region, but over 59 percent of the United States' refined butadiene capacity is in this region. 14 percent of the world's polymer and chemical grade propylene capacity is in this region, but over 51 percent of the United States' polymer and chemical grade propylene capacity is in this area. Finally, 12 percent of the world's ethylene production capacity is in the Houston/Galveston area, but over 44 percent of the United States' ethylene capacity is within this region. This region also contains the Houston Ship Channel area with the surrounding large refining and chemical plants, including ExxonMobil Baytown, Lyondell-Citgo Refining, Equistar Channelview and LaPorte, Shell Deer Park, as well as the Port of Houston. The extensive Mont Belvieu storage, trading, and pricing hub for natural gas liquids and chemicals as well as its natural gas fractionation assets are within the Houston/Galveston nonattainment area. Texas City's refineries and chemical plants, including refineries owned by BP, Valero, and Marathon, and chemical assets owned by Dow and Sterling are also within this region. Chocolate Bayou's Equistar and BP petrochemical complexes and the extensive refining and petrochemical facilities owned by ConocoPhillips and ChevronPhillips at Sweeney are also within this region. Freeport, home to petrochemical plants owned by Dow and BASF, is also in the Houston/Galveston nonattainment area. Clearly, any major environmental

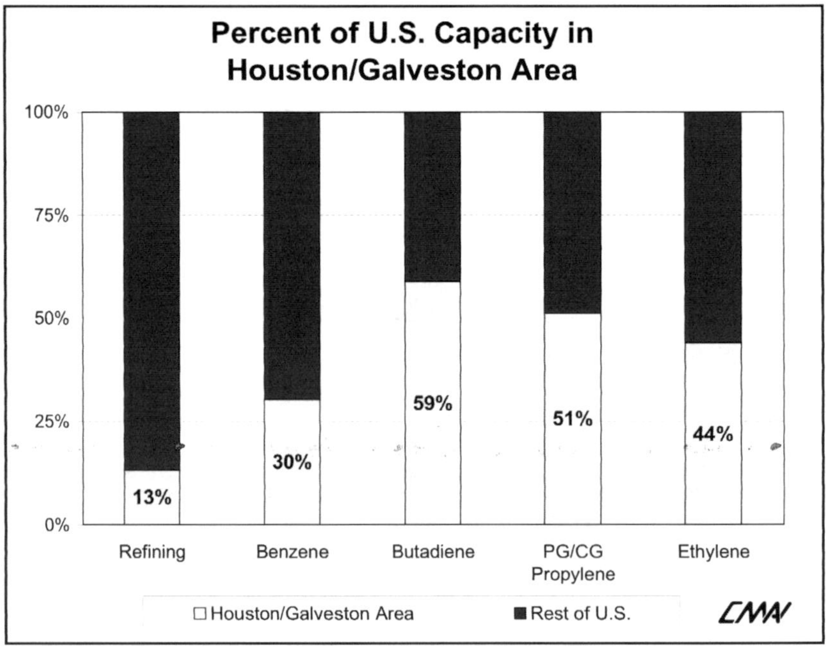

regulation that has such a broad impact on the refining and petrochemical industry can have far-reaching consequences.

What is the Environmental Protection Agency's approved plan for meeting the Clean Air Act requirements for the Houston/Galveston area?

The state of Texas is responsible for developing a State Implementation Plan, or a list of requirements and regulations that will allow the Houston/Galveston area to achieve attainment for ground-level ozone by 2007. The Texas Commission on Environmental Quality, which used to be known as the Texas Natural Resource Conservation Commission, is the Texas agency responsible for developing the State Implementation Plan. The United States Environmental Protection Agency has approval authority over individual State Implementation Plans, and can impose sanctions on areas that do not submit appropriate and detailed plans. Sanctions can also be assessed if the United States Environmental Protection Agency disapproves a State Implementation Plan, or if they find an area is not implementing the plan. The first level of sanctions they can enact is a "two-to-one offset," under which large, stationary or "point source" NOx emitters would be required to eliminate two tons of NOx for every one ton added in order to receive a new or modified construction air permit. The United States Environmental Protection Agency can also withhold federal highway funds for area projects as a secondary level of sanctions for areas that do not comply with the Clean Air Act regulations. Finally, the United States Environmental Protection Agency can develop a Federal Implementation Plan, and take over enforcement of the Clean Air Act for the state. It is clearly in the state's best interests to development a plan that will accomplish the environmental goals of the Clean Air Act, without severely compromising business growth.

The currently approved State Implementation Plan for the Houston/Galveston area requires a reduction of total NOx emissions from all sources by 75 percent. There is a mid-course review scheduled with the United States Environmental Protection Agency on May 1, 2004 to review progress towards attainment, and make adjustments as necessary to ensure compliance by 2007. The current United States Environmental Protection Agency approved plan includes a 90 percent reduction of NOx emissions from industrial and utility sources by 2007, with a staged compliance schedule. The mandated reduction targets are 44 percent of the required reduction by March 31, 2004, 89 percent of the required reduction by March 31, 2005, and the remainder of the reductions accomplished by March 31, 2007. These "point source" controls comprise the largest category for NOx reductions in the

Approved State Implementation Plan For The Houston\Galveston Nonattainment Area

HGA State Implementation Plan (SIP) Strategies:

- **Reduction of 90% of NOx emissions from industrial and utility sources**
  - **Point Source Reductions Targets**
    - **44% by March 31, 2004**
    - **89% by March 31, 2005**
    - **Rest by March 31, 2007**
- **Motor vehicle emission inspection and repair requirements**
- **Low sulfur diesel for East Texas**
- **California standards for large ignition engines**
- **Reduction in NOx emissions from airport ground equipment**
- **Motor vehicle speed limit reductions from 70 MPH to 55 MPH (May 2005)**

plan, accounting for over half of the reductions currently identified. However, this regulation impacts each and every citizen living in the eight county area. There are speed limit reductions to 55 miles per hour on all roads with a current maximum above 55 miles per hour (recently postponed to 2005 from 2002), enhanced motor vehicle inspection and repair requirements, and low aromatic content and higher cetane number diesel requirements for 110 counties in east Texas in 2005. The plan also requires California emission standards for all off-road gasoline engines greater than 25 horsepower across the entire state beginning in 2004. Agreements have been reached with two major carriers and the City of Houston to reduce NOx emissions from airport ground support equipment at all Houston airports through 2005. The state has also started providing grant funds to replace or retrofit heavy-duty diesel engines to lower NOx emissions under the Texas Emission Reduction Plan.

The state is also counting on voluntary emission reductions from alternative fuels and ultra-low emission public and private fleet vehicle purchases and commercial marine projects such as limiting transit speed on ocean-going vehicles in Galveston Bay and reducing emissions on tugboats, tow boats, and ferries. The state is anticipating that commuting options will reduce NOx emissions by 10 percent of the workforce avoiding one commute per week, and will be investigating the benefits of planting additional trees in urban areas. The Houston\Galveston area plan also takes credit for ongoing federal NOx reduction measures such as low sulfur diesel standards, national low emission vehicle standards, reformulated gasoline requirements, and marine recreation vehicle engine standards. The State Implementation Plan for the Houston/Galveston area as approved by the United States Environmental Protection Agency is still short 56 tons of NOx emission reductions, which the state has agreed to identify for the mid-course review, a review of achievement progress towards attainment with the United States Environmental Protection Agency in May of 2004. This current approved plan was submitted by the state to the United States Environmental Protection Agency on December 22, 2000, and the United States Environmental Protection Agency approved the State Implementation Plan on October 15, 2001.

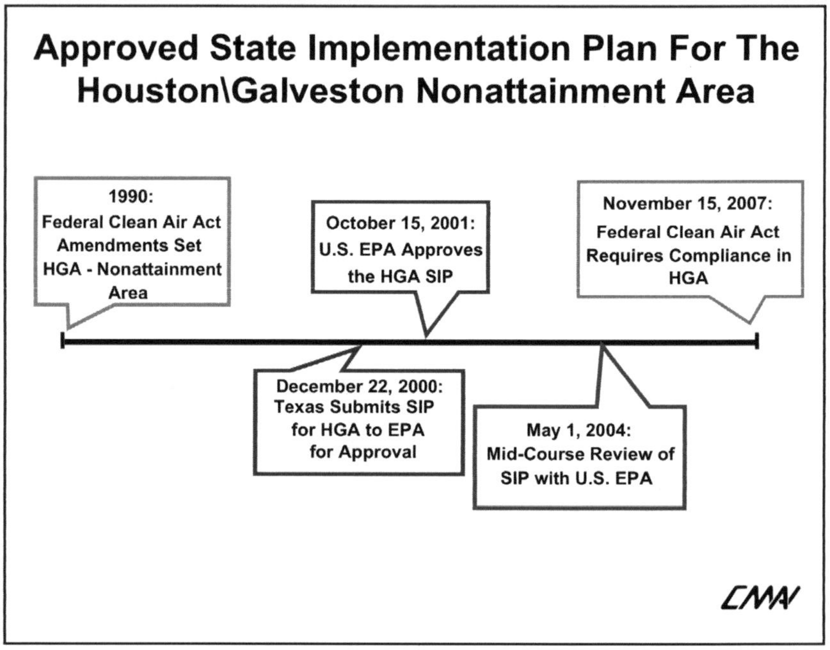

Why is the 90% reduction in NOx emissions so challenging for olefins plants?

Olefin plants can have several sources of NOx emissions, but the major source in most plants is the furnaces. Each cracking furnace carries raw material inside cracking sets, or special alloy tubes through multiple sections of the cracking furnace, but the firebox is where the reactions within the cracking sets accelerate. Furnace designs vary, but most furnaces have burners on the walls and the floor of the firebox. The furnace burners typically burn a methane rich fuel stream, often sourced primarily from the coproduct methane stream produced in the olefins unit. This stream typically has a higher content of hydrogen than purchased natural gas, and as hydrogen burns at a higher temperature than methane, this stream produces more NOx than a typical natural gas flame produces. Because of the higher temperature flame, steam cracking furnace fireboxes are some of the most severe in industry, with combustion occurring above 2000 degrees Fahrenheit or 1090 degrees Celsius. The industry invests in sophisticated control systems around furnaces in order to optimize furnace operating conditions and eliminate conditions that might cause premature failure of furnace components. As a result, there is a strong desire to control the heat flux within the firebox to ensure optimal cracking conditions. In the United States, industry has invested significant capital in feedstock flexibility, and producers utilize this flexibility to maximize profits. Because the feedstock mix in flexible crackers changes frequently, the product mix out of the plant changes, and the composition of the fuel stream firing the furnaces also changes. This changing fuel composition also makes reducing NOx emissions very challenging for U.S. ethylene producers.

The required 90 percent reduction in NOx emissions is also very expensive to implement. Based on a paper written by J.D. McAdams and S.D. Reed of John Zink, the current generation low NOx burners can be installed in a furnace for a little over $100,000 per year, and achieve a NOx reduction of 75 percent. For $200,000 to $300,000, a producer can use the conventional burners already installed on the furnace, and dilute the fuel gas stream with flue gas or other inerts prior to combustion. Fuel dilution will only achieve an 80 percent reduction in NOx emissions from the furnace. Installation of the next generation of low NOx burners will achieve an 88 percent reduction in NOx emissions. Recirculating the flue gas back through the firebox reduces the oxygen available in the furnace firebox and lowers the temperature of combustion, thus lowering the NOx production. Flue gas recirculation combined with conventional furnace burners allows a NOx reduction of only 70 percent. For around $400,000 per furnace, a combination of flue gas recirculation and current generation low NOx burners will only achieve an 80 percent reduction in NOx emissions. In order to achieve the 90 percent reduction required by the regulations, installation of a selective catalytic reduction system is required. A selective catalytic reduction system injects ammonia or urea into the flue gas stream before the flue gas flows across a catalyst bed, reacting the NOx to

Control Technologies for NOx Emission Reductions

	Current Generation Low NOx Burners	Fuel Dilution on Conventional Burners	Next Generation Low NOx Burners	Flue Gas Recirculation on Conventional Burners	Flue Gas Recirculation (FGR) & Low NOx Burners	Selective Catalytic Reduction System
Capital Cost Estimate	$130,000	$220,000	$280,000	$300,000	$430,000	**$860,000**
Additional Operating Cost (per year)	$0	$0	$0	$12,000	$12,000	**$68,000**
NOx Reduction*	75%	80%	88%	70%	80%	**90%**
For 90% NOx Reduction	Insufficient	Insufficient	Insufficient	Insufficient	Insufficient	**Sufficient**

*** Based on 100 MMBtu/hr process heater.**

Source: "Minimize NOX Emissions Cost-Effectively", by J.D. McAdams and S.D. Reed, in Hydrocarbon Processing, June 2001, Vol. 80, No. 6.

nitrogen and water. This option approaches $900,000 in capital investment per furnace, and nearly $70,000 in annual operating expenses associated with the selective catalytic reduction system, and is the only option that will meet the 90 percent reduction in NOx emissions.

A company-based approach to emission reduction and the system of emissions banking and trading, could render even less efficient NOx control devices viable options for companies. For example, credits for NOx reductions gained through the shutdown of older, inefficient facilities may be applied to other units, thereby lessening their NOx reduction requirements. Several plants reportedly have invested in new cogeneration facilities equipped with state of-the-art emission controls that would replace old boilers. Section 126 of the State Implementation Plan allows the trade of NOx emission allowances, providing companies with a more flexible approach to meet United States Environmental Protection Agency's emission reduction requirements. Despite the availability of these flexible approaches, companies have already begun to install NOx emission control devices. New facilities and equipment, such as ExxonMobil's new olefins furnace in Baytown, and ChevronPhillips/Solvay Polymer polyethylene plant in Cedar Bayou, Texas, as well as already existing units, like BP's Texas City refinery and ChevronPhillips' Sweeney complex have information in the press regarding their decisions to install NOx control devices. At the same time, engineering companies compete to provide more efficient and economical NOx reduction technologies. The next generation of ultra-low NOx burner also promises the attainment of the most stringent NOx requirements at more reasonable costs.

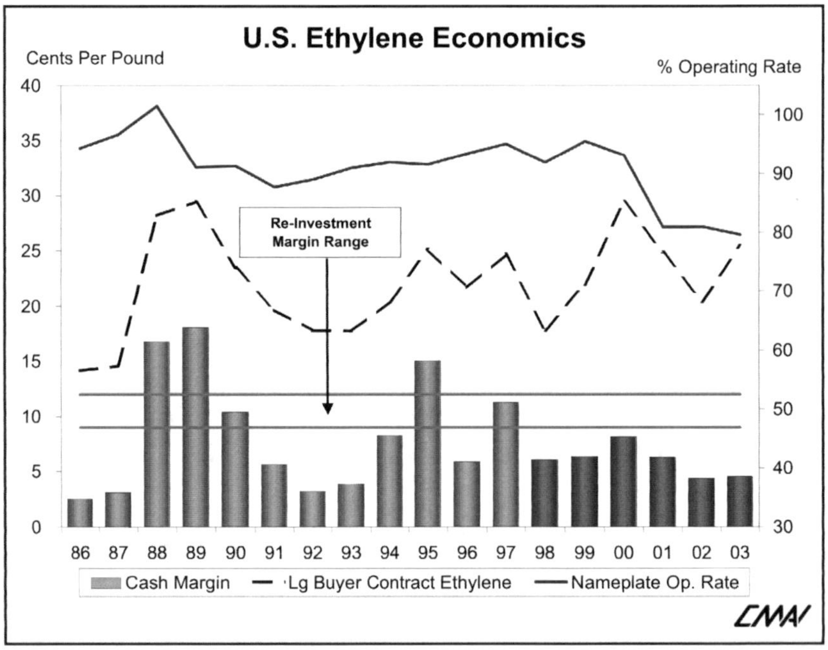

However, capital expenditures on environmental requirements under current market conditions are exceptionally difficult. The graphic entitled "U.S. Ethylene Economics" portrays CMAI's view of the average acquisition ethylene price, the weighted average feedslate cash margin, and the nameplate operating rate from 1986 through 2003. The two straight lines across the graphic represent approximate margin levels that justify re-investment in the business, the lower line for re-investment in an ethane cracker, the higher line representing the margin level for re-investment in a naphtha cracker. The United States industry has not seen re-investment level economics since 1997, leaving United States producers with 6 straight years of below re-investment level economics. Large capital expenditures on zero return environmental projects are clearly not good news to the industry at this time.

The Legal Challenge to the Houston/Galveston Area State Implementation Plan

Faced with implementing very costly and technically challenging options for meeting the regulations, a group of companies impacted by the State Implementation Plan called the Business Coalition for Clean Air – Appeal Group filed a legal challenge in January of 2001 against the state challenging the 90 percent NOx reduction requirements. In the lawsuit, the group claimed that there were significant technological impediments to achieving 90 percent reductions in NOx emissions. The group also claimed that the capital costs to the area businesses were greatly underestimated by the state, and would be closer to 9 to 12 billion dollars, with annual operating costs of 200 to 400 million dollars. The group also believed that the currently approved plan was based upon an incomplete scientific analysis of the air quality problems in the Houston/Galveston area. The group felt the 90 percent NOx reduction regulations overcontrolled NOx emissions, and largely ignored highly-reactive volatile organic compound emissions. On June 8, 2001, the court signed a consent order requiring the Texas Commission on Environmental Quality to consider results from an on going air study analyzing ozone events to determine whether volatile organic compound emission controls could be substituted for the last 10 percent of industrial NOx point source controls.

The air quality study referenced in the lawsuit was the Texas 2000 Air Quality Study, a 20 million dollar study conducted in August and September 2000 that set out to answer why the Houston/Galveston nonattainment area was different from other ground-level ozone nonattainment areas in the United States. The Houston/Galveston area has high concentrations of ground-level ozone that form very rapidly, and this phenomenon doesn't exist anywhere else in the United States. The study conducted experiments utilizing over 250 scientists from all over the United States, and has concluded that emissions of highly-reactive volatile organic compounds have been significantly underestimated in the past. The study concluded that these emissions were not intentionally underreported by operating companies, but that the current requirements for monitoring and reporting volatile organic compounds did not capture all sources of volatile organic compound emissions. In December of 2002, the Texas Commission on Environmental Quality revised the State Implementation Plan and has reduced requirements for controlling NOx emissions, but has substituted volatile organic compound emission monitoring and control requirements.

Revisions to the State Implementation Plan for the Houston/Galveston Area

The three-member commission that oversees the Texas Commission on Environmental Quality approved the revised State Implementation Plan on December 13, 2002. This plan trades a relaxation in NOx reduction requirements for industrial point sources from 90 percent to 80 percent with a 36 percent overall reduction of highly-reactive volatile organic compounds. The new regulations define highly-reactive volatile organic compounds as ethylene and propylene in the entire 8 county area, and adds butadiene and all butenes to the definition within Harris county; all of these chemicals are products in an olefins plant. The state has taken the approach of a "site" or plant emissions cap to allow some flexibility in emissions between units inside the site, and required reductions are identified based upon estimated total highly-reactive volatile organic compound emissions from flares, cooling towers, and vents. If the estimated inventory of emissions is greater than 500 pounds per hour of highly-reactive volatile organic compounds, the site cap requires a 70 percent reduction of highly-reactive volatile organic compound emissions. Sites that are estimated to emit between 125 pounds per hour and 500 pounds per hour are assigned a 68 percent reduction in emissions, and sources emitting between 10 and 125 pounds per hour are required to adhere to a site cap that is 60 percent below their estimated emissions. All sites between 2.28 and 10 pounds per hour of highly-reactive volatile compound emissions are required to reduce emissions by 50 percent.

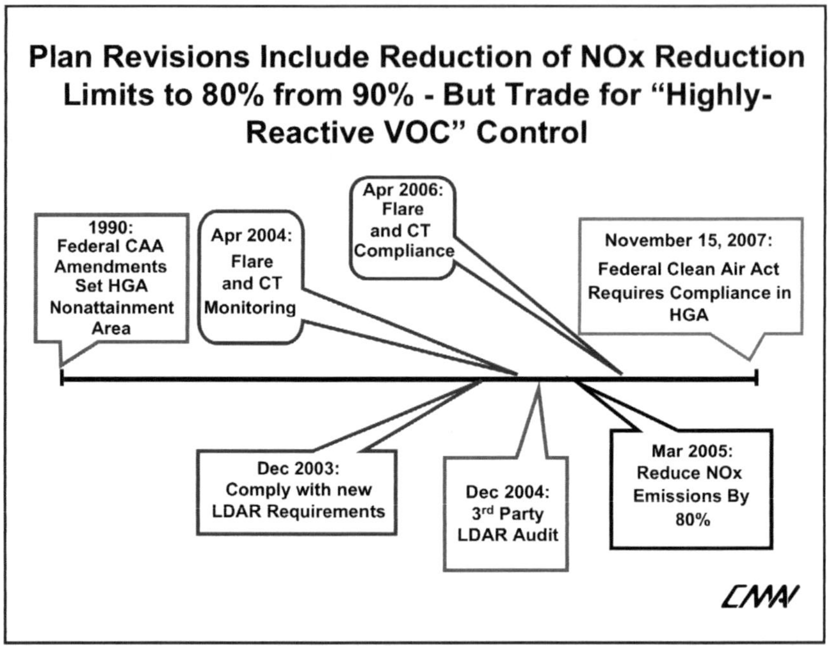

The regulations also require all large cooling towers that have the potential to emit highly-reactive volatile organic compounds to continuously monitor the inlet flow rate and the total volatile organic compound concentration in the inlet water to the cooling tower. Once a week, a sample must be taken to identify a minimum of 90 percent of the volatile organic compounds contained in the inlet water stream. If the water sample contains greater than 50 parts per billion of volatile organic compounds, daily samples are required until the concentration falls back below 50 parts per billion. For flares in the Houston/Galveston nonattainment area that contain highly-reactive volatile organic compounds, continuous flow monitoring and analysis of the flow once every 15 minutes is required. Extensive record keeping is required around both cooling towers and flares, including an hourly update of each site's 24 hour rolling average highly-reactive volatile organic compound emissions for the site wide cap. An owner cannot substitute emission reduction credits for implementing the requirements of this regulation. Operators in the Houston/Galveston nonattainment area are given until December 31, 2004 to implement the monitoring requirements, and April 1, 2006 to meet the site emissions cap.

The revised regulations also enact a new and more stringent leak detection and repair program than is currently required for units in which highly-reactive volatile organic compounds are present. More unit components than before are required to be monitored quarterly, including manway covers and heat exchanger heads, and the regulations require "extraordinary measures" to be undertaken before a leaking component can have maintenance deferred to a shutdown. An example of "extraordinary measures" is drilling and injection of sealant into a leaking component. All new pumps, compressors, and agitators must be equipped with shaft sealing systems. An independent third party audit of each company's leak detection and repair program is required every two years. Compliance with all aspects of this regulation was to be required by December 31, 2003 with the exception of the third party audit that must be completed by December 31, 2004.

Revisions to the NOx emission control requirements revise requirements for NOx emissions from all point sources down to 80 percent from nearly 90 percent. Emission reduction requirements for olefins plant furnaces, called "pyrolysis reactors" by the Texas Commission on Environmental Quality, have been relaxed to a more achievable 64 percent, or a NOx emission limit of 0.036 pounds of NOx per million Btu. Overall, the requirements for the category of process heaters and furnaces has been relaxed to a 70 percent reduction in NOx. The regulation requires measurement of total fuel flow and installation of either continuous emission monitoring systems or predictive emissions monitoring systems to measure NOx emissions. Carbon monoxide emissions are also limited and measurement is mandated, and ammonia emission monitoring is required if ammonia or urea is injected into the flue gas to control NOx emissions. Companies are given the option of complying with the NOx emission limitations by complying as a plant-wide site, and not necessarily control every NOx point source within the plant. The limitations on NOx emissions remain the same, whether a company chooses to address reductions on a component-by-component basis, or as an entire plant site. Compliance with all aspects of this regulation is required by March 31, 2005.

U.S. Ethylene Industry – Impacts of the Houston/Galveston Area State Implementation Plan

The accompanying chart entitled "Ethylene Capacity in the Houston/Galveston Nonattainment Area" displays each ethylene producing company that has olefins plants within the Houston/

Ethylene Capacity in the Houston/Galveston Nonattainment Area

Company	Location	Last Major Planned Maintenance Outage	Annual Capacity MM Lbs
BP	Chocolate Bayou, TX (#1)	September 1999 (27 Days)	1,400
	Chocolate Bayou, TX (#2)	October 2000 (35 Days)	1,800
Chevron Phillips	Cedar Bayou, TX	April 2004 (60 Days Planned)	1,770
	Sweeny, TX (#12)	*Unit Shutdown in 2001*	-
	Sweeny, TX (#22)	*Current (Idled)*	-
	Sweeny, TX (#24)	June 2002 (32 Days)	1,500
	Sweeny, TX (#33)	October 2003 (25 Days)	2,075
Dow	Freeport, TX (LHC-7)	October 1999 (30 Days)	1,350
	Freeport, TX (LHC-8)	January 2000 (57 Days)	2,198
	Texas City, TX	*Unit Shutdown in 2003*	-
Equistar	Channelview, TX (OP-1)	October 2003 (52 Days)	1,925
	Channelview, TX (OP-2)	March 2003 (56 Days)	1,925
	Chocolate Bayou, TX	October 2002 (59 Days)	1,100
	La Porte, TX	Idled for 74 Days Starting in June 2003	1,650
Exxon Mobil	Baytown, TX	September 2003 (73 days)	2,550
	Baytown, TX	April 2002 (35 Days)	2,293
	Houston, TX	Furnaces Remain Idled	225
Shell	Deer Park, TX (OP-2)	Furnace Restart Scheduled to Start up March 2004	1,120
	Deer Park, TX (OP-3)	April 2001 (43 days)	1,900
	Total 2004 Capacity in Houston/Galveston Nonattainment Area		**26,781**
	Total 2004 U.S. Capacity		**61,459**
	% in Houston/Galveston Nonattainment Area		**44%**

CMAI

Galveston area, the locations of the facilities, CMAI's record of the last major planned maintenance outage for each site, and the annual ethylene producing capacity for each facility. The Houston/ Galveston area has a total of 27 billion pounds, or 12 million metric tons of installed ethylene capacity within the region. Earlier in the paper a graphic entitled "U.S. Ethylene Economics" revealed that the United States ethylene industry has sustained below re-investment level economics for six consecutive years. Operating rates have fallen from 93 percent in 2000 to 80 to 81 percent in 2001 through 2003, indicative of very weak market conditions. This highlights the significant overcapacity present in the U.S. ethylene market. As a result of the very poor market conditions, several olefins units in the U.S. have been either shutdown completely, or temporarily idled. ChevronPhillips shutdown Sweeney #12 in 2001, and has since idled Sweeney #22. ExxonMobil had idled their Houston olefins plant, but has since restarted the facility at reduced operating rates; the furnaces at this site remain idled. Dow shutdown their Texas City olefins unit in June of 2003. All of these facilities are within the Houston/Galveston nonattainment area, and permanent closure of these units would certainly assist these companies in meeting their NOx reduction requirements. As a result of the weak operating conditions, Equistar as also idled their Lake Charles, Louisiana steam cracker, and Huntsman has idled their Port Neches, Texas cracker. Dow also permanently closed its Seadrift, Texas olefins unit in September of 2003. These actions have taken a total of 5.3 billion pounds, or 2.4 million metric tons of capacity out of the market.

CMAI's base case United States ethylene demand scenario is published in the ***2004 World Light Olefins Analysis***, and has been updated in preparation for CMAI's **2004 World Petrochemical Conference** in March of 2004. Our demand forecast anticipates United States ethylene demand growing at an annual average growth rate of 2.6 percent from 2003 through 2008. One of the most frequently asked questions from our client base is what if we are wrong? What if ethylene demand does not recover to our forecast percent growth rate and is less than what we forecast? What does the market look like at 1.5 percent demand growth? The accompanying graphic shows the base case forecast of 2.6 percent ethylene demand growth and an alternate forecast of 1.5 percent ethylene demand growth. The bars are surplus ethylene capacity in billions of pounds, read off the left axis of the graph. Surplus capacity is defined by the amount of nameplate ethylene capacity that exceeds total demand, or nameplate capacity – total ethylene demand = surplus capacity. The lines are read off of the right hand axis, and represent the surplus capacity as a percentage of total ethylene demand.

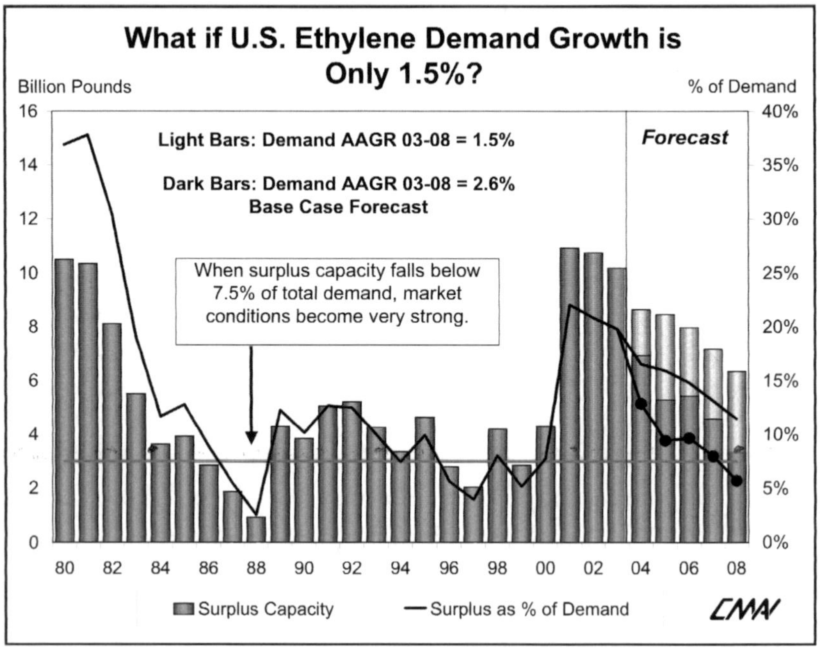

When looking back historically, surplus ethylene capacity was extremely high last in the early 1980s, when surplus capacity exceeded 10 billion pounds, or 4.5 million metric tons, and the surplus capacity reached 38 percent of total United States ethylene demand in 1981. Producers recovered from these abysmal market conditions by shutting down and idling excess capacity, and deferring re-investment in the industry. As a result of these actions, a very strong market peak occurred in the late 1980's when surplus capacity fell to less than 1 billion pounds. This is essentially an industry running effectively at 100+ percent, completely sold out of ethylene product. Looking at 1999, the last strong ethylene market year in the United States, Y2K lead to a build up of inventory throughout the chain, from the ethylene monomer producer, all the way through to the retail shelves as concerns regarding old computer system failures ran rampant throughout the world. In 2000, the United States economy slowed, the manufacturing industry entered a long and deep recession, and the inventory built up over 1999 was slowly consumed. By the end of 2000 and the beginning of 2001, extremely high natural gas prices in North America had caused olefins feedstock prices to soar, making North American ethylene derivative products largely uncompetitive in global markets. Material from new capacity in Western Canada began to enter the United States markets, further curtailing domestic ethylene demand, and imports of nondurable end use products from around the world entered the high priced United States markets. All of these conditions caused the surplus capacity in 2001 to exceed 10 billion pounds again, but because of the remarkable growth of the ethylene market during the 1990s, surplus capacity was only 21 percent versus 38 percent twenty years earlier. At the end of 2001, two new world scale facilities were added in the United States Gulf Coast, BASF/Atofina in Port Arthur, Texas, and Formosa in Point Comfort, Texas, causing market surplus capacity to grow even further. Our base forecast is represented in the graphic by the dark bars, calling for market conditions to tighten noticeably by 2004, and continue to tighten throughout the forecast period. After six consecutive years of below re-investment economic conditions in the United States ethylene industry, we anticipate that re-investment in this market will be considerably constrained as it was in the 1980s, especially considering the tremendous capacity build up in both China and the Middle East just on the horizon. However, the light bars on the graph represent a demand growth of only 1.5 percent annual average growth in U.S. ethylene demand. Under these conditions, surplus capacity only falls to just over 6 billion pounds, or 2.8 million metric tons by 2008, and the U.S. market never reaches another peak before the significant Middle East and China capacity additions are onstream. This scenario would likely entail further rationalization of ethylene capacity at a minimum, and might cause some producers to execute business exit strategies.

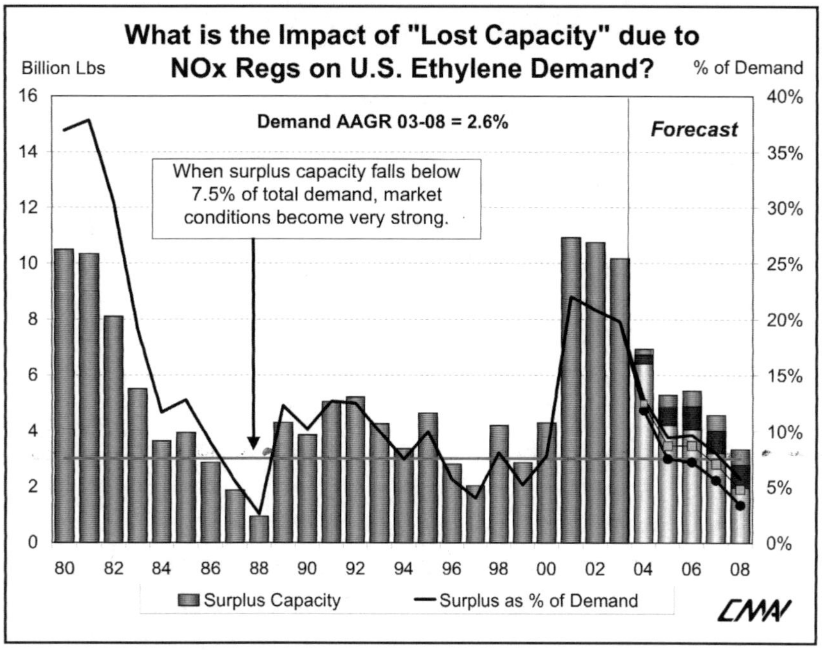

Returning to our base forecast of 2.6 percent annual average growth rate, the graphic entitled "What is the Impact of "Lost Capacity" due to NOx Regs on U.S. Ethylene Demand?" has our base forecast surplus capacity again in the tallest dark bars. Our preliminary feedback from talking to several producers that have installed and are operating on low NOx burners is that many are witnessing lower operable capacity out of their furnaces than they observed before switching to low NOx burners. We have overlayed a forecast of 2 percent capacity curtailment from each plant in the Houston/ Galveston area due to decreased capacity from low NOx burners, represented by the darkest bars. The light colored bars represent a secondary forecast of 5 percent capacity curtailment from each plant in the Houston/Galveston area due to decreased capacity from low NOx burners. It is likely that the curtailment on Houston/Galveston area capacity will likely be bracketed between the 2 to 5 percent loss, but the Houston/Galveston area is not the only area in the U.S. that is installing low NOx burners to meet Clean Air Act regulations. This additional curtailment of capacity comes to fruition at the very time our balances suggest the industry will be at the peak of both demand and earnings, and any curtailment of capacity could tighten market conditions considerably.

CMAI's price forecast has a cycle peak occurring in 2006 and 2007 with an integrated high density polyethylene chain margin reaching 20 cents per pound of HDPE. In 1995, the integrated chain margin for HDPE exceeded 22 cents per pound, but the ethylene market surplus capacity was 10 percent of total demand. Based on the surplus capacity analysis shown above, where we forecast surplus capacity to be somewhere below 10 percent starting in 2005, and range between 8 and nearly 10 percent of total demand in 2006 and 2007, margins certainly have the potential to be much higher than our base forecast.

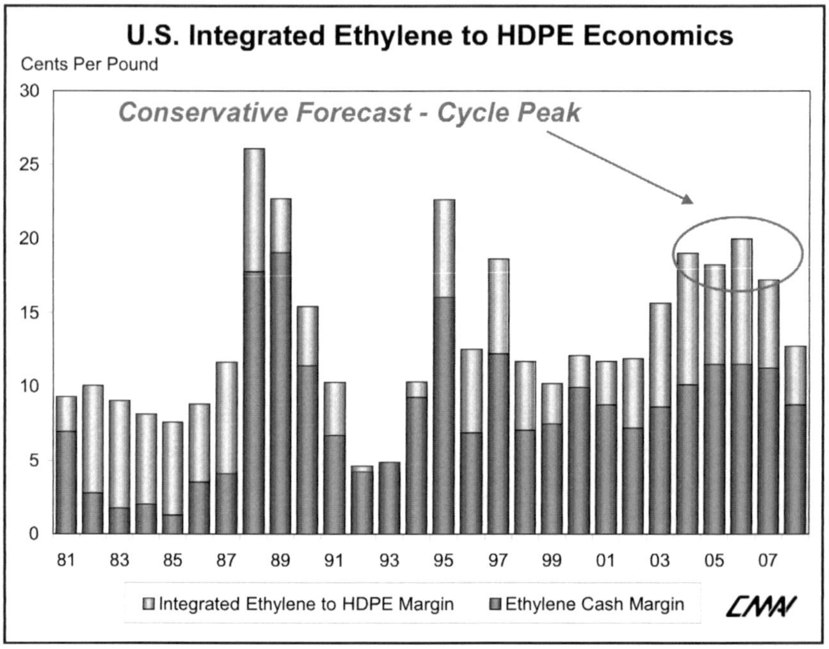

Future issues regarding the Houston/Galveston nonattainment area's State Implementation Plan

The Houston/Galveston area still has 56 tons of NOx emissions to identify for reduction by the mid-course review submittal to the United States Environmental Protection Agency planned in May 2004. While the revisions discussed above to the State Implementation Plan will undoubtedly be approved by the Environmental Protection Agency, the plan still lacks identifiable reduction measures for a portion of the emissions. Potential control technologies under consideration applicable to the petrochemical industry include requiring dockside vessel emissions to be subject to state New Source Review permits, including best available control technology review, emissions limits, and monitoring and testing requirements. Current regulations only require dockside emissions review when onshore facilities are new or modified. The state is also studying episodic emissions, or emissions occurring during upsets and maintenance activities, to determine the extent of these emissions, and how the state can begin to account for these emissions. Outreach programs are already underway to help facilities reduce episodic releases. The state also intends to continue studying results from the Texas 2000 Air Quality Study, to determine if other highly-reactive volatile organic compounds should be limited. The draft proposal of the new proposed regulations included many compounds other than ethylene, propylene, butenes, and butadiene that have since been removed from the regulation.

The current Federal Clean Air Act standard that the Houston/Galveston nonattainment area is required to achieve by 2007 is the one hour ground-level ozone standard, which the United States Environmental Protection Agency has set at 120 parts per billion. An area violates this standard if any monitor reads 125 parts per billion more than three times during any three year period. The Environmental Protection Agency has also set an eight-hour ground-level ozone standard that the Houston\Galveston area has also violated. The eight-hour standard is set at 80 parts per billion, and an area is in violation of this standard when the three-year average of each year's fourth highest ozone reading equals or exceeds 85 parts per billion. The Houston\Galveston area has also been designated in nonattainment of the eight hour ground-level ozone standard. Compliance measures towards the eight-hour standard will be addressed as part of the mid-course review submittal to the United States Environmental Protection Agency planned in May 2004.

SESSIONS 63D & 64A – THE LATEST GENERATIONS OF ETHYLENE FURNACE BURNERS A MARKET PERSPECTIVE FROM VENDORS PRESENTED AT: ETHYLENE PLANT TECHNOLOGY – FURNACE BURNER ROUNDTABLE

PRESENTED BY:

Scott Reed (speaker), Zeeco
Doule Bishop, Callidus Technologies LLC
Michael Wignal, Hamworthy Combustion Engineering
John Grever, Seals Corporation of America
Ashok, Patel, John Zink Co LLC

The principal vendors of burners for ethylene furnace applications each presented their latest technologies, references, experiences and expertise. The presentations emphasized the challenges being faced by the industry and the developments that have been offered to meet these challenges. NOx limitations are of particular concern for both new installations and for retrofit into existing furnaces. Low NOx and "ultra-low NOx" burner types were explained and illustrated. Modeling and design techniques were explained, and each vendor discussed test furnace and actual plant testing results.

Commercialization of Ultra-low NOx Burners for Ethylene Heaters

Robert J. Gartside – ABB Lummus Global
Peter R Ponzi – ABB Lummus Global
Frank D. McCarthy – ABB Lummus Global
S.G. (Chell) Chellappan – Equistar Chemicals, LP
Paul J. Chapman – Alstom Power Inc
R. T. Waibel – John Zink Co., LLC

Abstract

After more than fifty years of development, the thermal cracking or pyrolysis of hydrocarbons has matured into the single most important process to produce ethylene. It is carried out inside tubular coils that are installed vertically in large fired heaters where the heat for the endothermic cracking reaction is supplied by a large number of hearth burners and/or wall burners. High heat fluxes are required to supply the heat for the endothermic reaction. These heaters are typically fired with gaseous fuels derived from the ethylene plant itself containing primarily hydrogen and methane. Within this fired environment, thermal NOx is the primary source for NOx. Recently, the environmental restrictions for NOx release have become progressively tighter leading many producers to investigate means for reducing NOx. The primary options are selective catalytic reduction (SCR) and combustion technology.

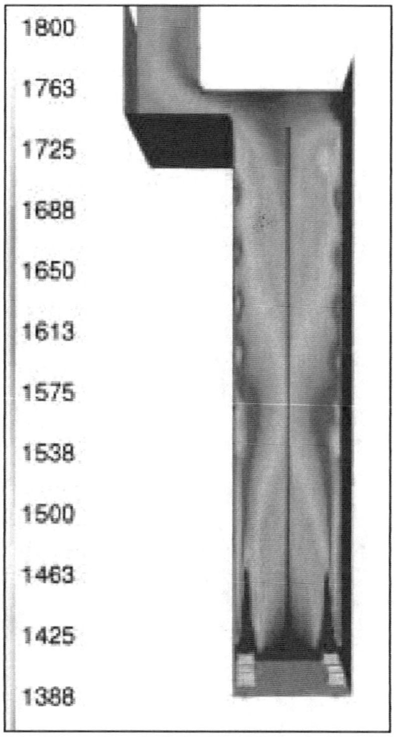

While any program to reduce NOx incorporates all fired services within a complex, the ethylene plant cracking heaters represent a major fired service. Incorporating catalytic units on all heaters represents a very costly option given the number of heaters within a plant and the complexity of the heat recovery in the convection section of those heaters. If a major reduction in thermal NOx from the ethylene plant cracking heater could be achieved utilizing burner technology, this would present plant operators with an interesting option.

In order to reduce thermal NOx, combustion technology has opted to utilize a combination of flue gas recycle, lean combustion, and fuel staging to reduce flame intensities and hence NOx. Staging, however, can lead to loose flames within the firebox that could roll over onto the process coils leading to increased fouling and mechanical damage. It also impacts the heat release patterns within the firebox and the overall process and heater performance.

Over two years ago, ABB Lummus Global, John Zink, and Equistar Chemicals LP began a program to commercialize the application of the Zink® patented* COOLMIX™ burner technology for this service. This program combined the furnace design and modeling expertise of Lummus, the combustion technology expertise of Zink, and the vision and operating experience of Equistar. Earlier attempts to implement ultra-low NOx burner technology achieving 0.03 lb NOx/MM BTU resulted in decreased heater performance caused by flame impingement on the cracking coils. Utilizing computational fluid dynamics (CFD), a unique combination of burner technology and firing patterns has been identified.

This program has resulted in the full modification of a commercial cracking heater at the Equistar Chocolate Bayou plant with ultra-low NOx burner technology. The cracking heater is a 60:40 hearth:wall-fired unit with nominal fuel hydrogen of nine percent and a duty of approximately 250 MM BTU/hr. This unit has been operating using these new COOLMIX hearth and wall burners with exhaust NOx values 28-32 ppmvd. As a result of the successful implementation of this technology, Lummus and Zink have decided to offer this technology jointly.

One of the keys to the successful implementation of this technology was the use of Computational Fluid Dynamic analysis to optimize both the burner and firing patterns. Lummus has spent the last four years developing a fully integrated and validated model of an ethylene plant cracking heater. This was described in detail last year in this session. This paper will present the results of this commercialization program and the CFD analysis critical to its success.

Introduction

Since 1999, ABB Lummus Global has been working on the development and validation of an integrated computational fluid dynamic (CFD) model of the ethylene plant cracking heater.[1, 2] This effort utilized the ethylene heater and CFD expertise of ABB Lummus Global and the burner testing and combustion expertise of John Zink. Lummus completed a number of CFD models of the test firebox at the John Zink test facilities in Tulsa, OK and together, Lummus and Zink gathered extensive data under a wide range of combustion conditions. This data, including in-flame measurements of temperatures and species, were used to develop a validated CFD model of an ethylene heater. The model has since been proven in numerous commercial installations. (The model was described in detail at last year's AIChE Spring National meeting.)

Now that the model has been fully validated, Lummus and Zink have formed a marketing and development alliance for the application of ultra-low NOx burner technology in cracking heaters and the application of Lummus's CFD capability in cracking heater firebox designs in general.

John Zink's patented COOLMIX hearth and wall burners are a new type of ultra-low NOx burner [3]. Since its introduction in 2000, this type of burner has been used in a variety of refinery process heaters and steam methane reformer applications. The burner technology utilizes a lean premix primary combustion zone combined with a staged secondary combustion zone, a design that significantly reduces both thermal and prompt NOx. Note that thermal NOx is a result of the direct reaction of oxygen with nitrogen under high temperature conditions while prompt NOx is a result of reactions at the flame front under fuel rich conditions.

In 2001, ABB Lummus Global and John Zink, working with Equistar Chemicals, began a series of CFD and process studies to define the operability of both John Zink's COOLMIX™ hearth and wall burners in a Lummus-designed short residence time (SRT®III) heater at the Equistar Chocolate Bayou ethylene plant. This paper will discuss the development program and CFD studies leading to the successful application of the ultra-low NOx burner technology.

Discussion

As part of a larger project to evaluate all of Equistar's NOx compliance in the Houston area, in 2001 Lummus, John Zink, and Equistar began a series of CFD, burner, and process studies to define the operability of the COOLMIX hearth and wall burners in an SRT III heater at Chocolate Bayou. Earlier in 2001, one of the base case SRT III cracking heaters at Chocolate Bayou (see below) had been modified with different ultra-low NOx burners (not COOLMIX technology) without utilizing CFD. The result was significant flame rollover that required the furnace to operate at reduced capacity. Attempts were made to mitigate this flame rollover problem, but any success was at the expense of higher NOx. The NOx levels did not meet Equistar's requirements when acceptable flame shape was achieved. Based upon Equistar's evaluation of the capability of Lummus's CFD model, Equistar felt that this tool would provide valuable input to achieve a successful implementation of ultra-low NOx burner technology.

Base Case – It is critical in any modeling evaluation to develop a validated base case. As the first task in the program, a full model of half the SRT III heater was developed. Initially installed in the early 1980s, all of the SRT III heaters were subsequently modified in 1995 to utilize the John Zink PLXMR-45 hearth burners to achieve a first level of NOx reduction. This heater contains six four-pass coils and operates at a design capacity of approximately 250 MM BTU/hr LHV. There are 12 hearth burners installed with an orientation roughly across from the inlet coil passes. This results in a higher heat flux to the inlet coils and provides "passive" zone firing. The heater also has 108 wall burners (LUM-4 type).

Initially, a base-case CFD analysis was completed using data from the "basis of design" information provided by Equistar. The figures below summarize the important criteria.

In Figure 1, the base-case flames are straight, as evidenced by the temperature profile. Note that some interference (and resulting high temperatures) is evident as a hearth burner plume reaches the first row of wall burners.

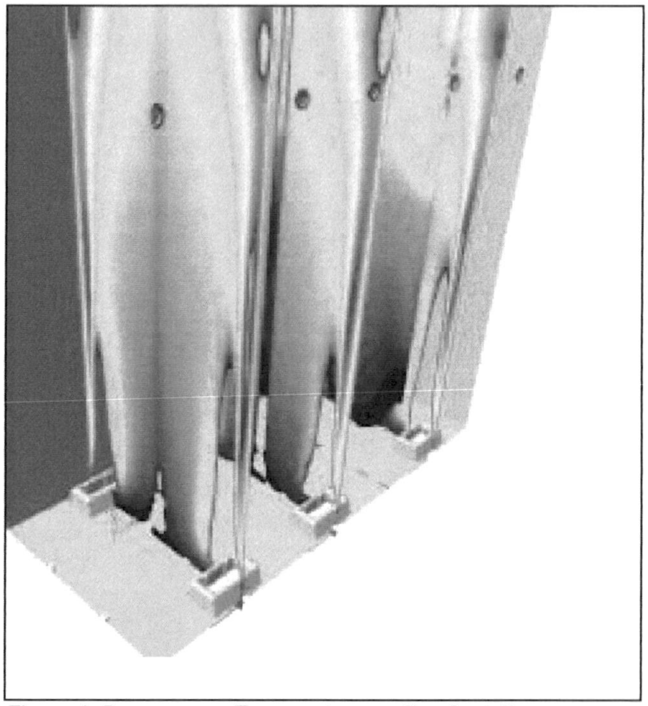
Figure 1. Base case – Temperature profile of existing burners

The wall temperatures show a cool lower portion of the firebox resulting from the high vertical nature of the flame (Figure 2). Note the similarity between the actual profile and the CFD profile.

Figure 3 shows the base-case flux and metal temperature profiles. The flux profile illustrates zone firing. Note that the outlet passes (3 and 4) have lower flux than the inlet passes. This zone firing is considered "passive" in that it is determined by burner layout, not fuel flow variations between burners. The coil metal temperatures show a maximum at the point of maximum heat release. The highest metal temperature is in the inlet pass, consistent with zone firing. At the lower conversion levels at this point in the coil, higher temperatures can be tolerated without fouling. Passive zone firing is a feature that should be retained in this design.

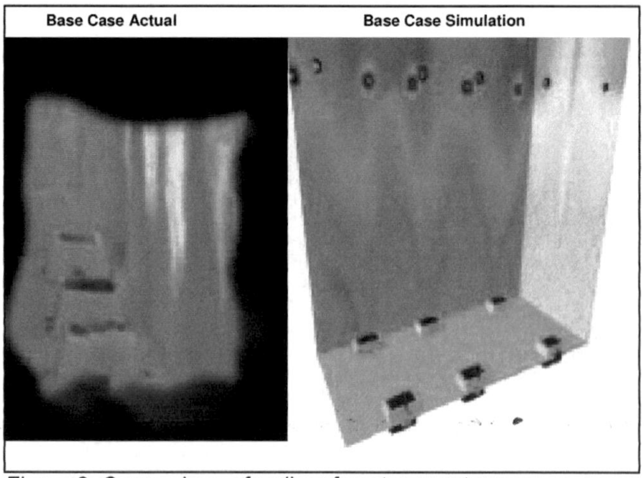

Figure 2. Comparison of wall surface temperatures

It is critical in any revamp that process performance be maintained. Thus, while it is important that the flames remain stiff and against the wall, the run length of the heater should not suffer. The operating heater utilizes a fuel consisting of nine percent hydrogen with the balance methane. The base-case furnace run length with the PLXMR burners was determined using statistical data for later comparison with operation using the new COOLMIX ultra-low NOx burner layout. The historical NOx measurements from this base heater were between 80 and 90 ppmvd.

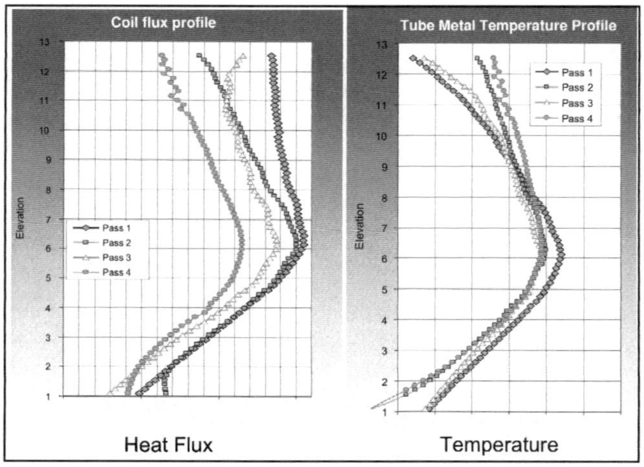

Figure 3. Base-case flux and TMT profiles

It is also important to ensure that the detailed model of the burner is correct. Lummus and John Zink conducted numerous firing tests using the COOLMIX burner, including detailed in-flame species measurements similar to those done in the initial validation studies used for the development of the CFD model. CFD simulations were also completed based upon those firing tests. The incident flux profile and oxygen concentration profile show excellent agreement between the data and the simulation, as shown in Figure 4. This validation confirms the accuracy of the CFD calculated heat release profiles.

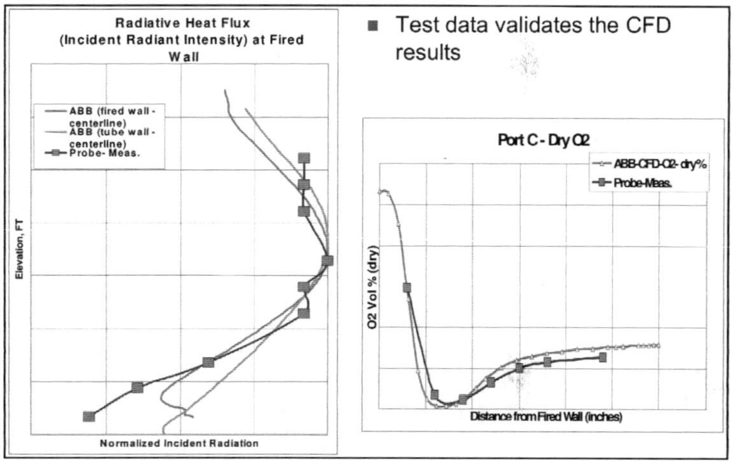

Figure 4. Burner testing – model validation

Ultra-low NOx Burner Retrofit Case - The design team then began to investigate the COOLMIX hearth and wall burner options for the Chocolate Bayou heater. John Zink had previously completed some initial developmental studies on the COOLMIX concepts and had installed a number of these for less severe service.

The COOLMIX hearth burner consists of four fuel-inspirated venturis for the airflow. The air (premixed with fuel) enters the firebox vertically. There are secondary and tertiary fuel tips to stage the fuel around the primary combustion zone. The COOLMIX wall burner also utilizes a fuel-inspirated venturi for the air flow. The air (premixed with fuel) enters the firebox parallel to the walls while a secondary fuel tip injects additional fuel into the firebox through the center of the burner tile.

The initial approach to the application of the COOLMIX technology was a 'drop-in" solution. However, unlike the previous attempt, the 12 PLXMR burners were first replaced "virtually" with 12 COOLMIX hearth burners using CFD. The wall burners remained as the conventional type in the initial simulations. The firing capacity of the selected COOLMIX hearth burners was higher than the existing PLXMR-45 burners, enabling four of the six operating rows of wall burners to be turned off. The design team thought that by allowing for a greater vertical distance, the inherently longer NOx flames would remain against the wall. This proved not to be the case.

The CFD analysis showed that the hot flames impinged upon the coil as shown in Figure 5. It was concluded that the capacity of the burners was too high. Excessive tube metal temperatures (>1100°C) are shown in Figure 6 in the lower portions of the coil and clearly this solution was unacceptable. John Zink had other versions of the COOLMIX technology operating in less severe service; however, these burners had approximately half the capacity

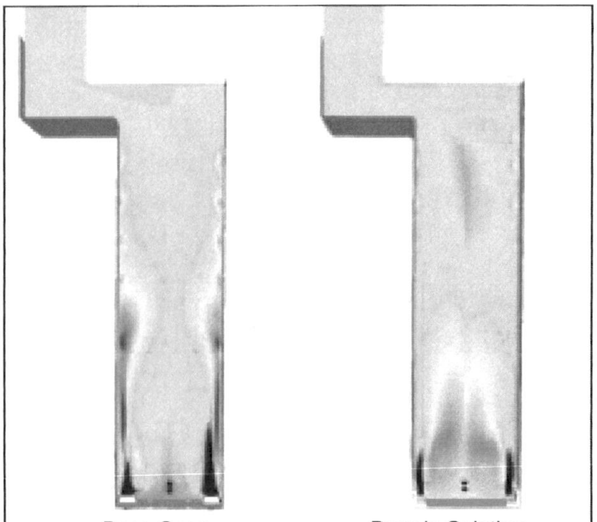

Figure 5. Gas temperatures at burner centerline for original design versus initial COOLMIX burner drop-in solution

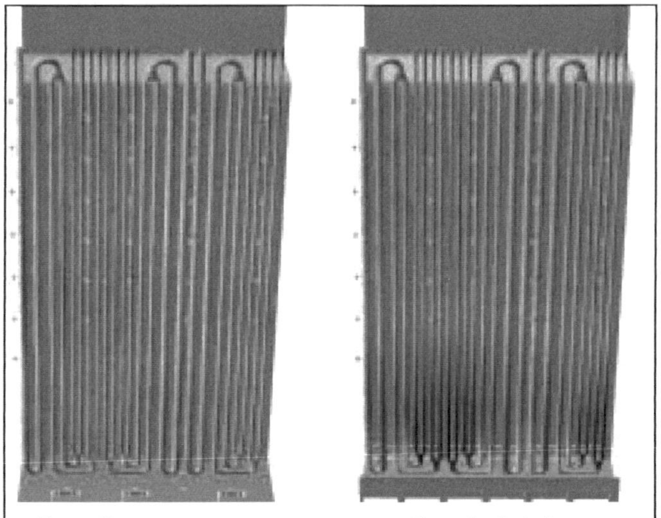

Figure 6. Tube wall temperature (OD) of original design versus initial COOLMIX burner drop-in solution

of the burners utilized in the first case. Thus, instead of the 12 COOLMIX hearth burners used, a case was run with 24 lower capacity COOLMIX hearth burners. This too showed severe flame rollover with flame impingement on the cracking coil.

At this point, Lummus and John Zink completed a number of smaller CFD studies to attempt to identify the root cause of the problem. Based upon those studies, it was determined that the critical factor in controlling flame shape is controlling the volumetric expansion profile of the flame. As a simple illustration, consider a firebox with a nominal temperature of 2000°F. The inlet air stream expands in volume by a factor of four as it combusts the fuel. Since one side is the furnace wall, there are only three directions for gas expansion: the vertical dimension up the wall (desired), to the side, or toward the coil plane (not desired). By allowing larger volume for expansion to the sides of the burners (by increasing burner spacing), some of the expansion can be accommodated in that direction, thus easing the tendency for the flames to roll over toward the coil plane. However, when high heat release in the furnace is required, the ability to increase the burner-to-burner spacing is limited. In order to achieve vertical flames, it is critical to maximize the vertical expansion. A careful balance must be achieved between initiating the combustion while the air inlet still has reasonable vertical momentum (thus having it accelerate vertically) and achieving too much mixing of fuel and air, thus creating high temperatures and increasing NOx.

CFD simulation provides an indispensable tool for studying this complicated hydrodynamic expansion/combustion problem. The Lummus models include all of the minute details of the burner itself, including the multiplicity of fuel injection ports on each tip. Each port has a distinct fuel flow and direction, and small changes can dramatically influence flame shape. The sensitivity of the flame to tip geometry is a fact known to anyone who has studied cracking heaters in detail. Further, in order to capture the burner-burner interaction, a model of a large portion of the heater is necessary.

Based upon this knowledge, a number of alternative solutions were examined. The final solution consisted of 18 "strategically placed" COOLMIX hearth burners with the firing tip profiles optimized for flame stiffness. The "strategic placement" allows for volumetric expansion control while optimizing the flux profile on the coil. All six rows of wall firing were retained but were now modeled as COOLMIX wall burners.

Figure 7 uses the temperatures at the burner centerline to show the general flame shape. It is important to note that the flame shape for ultra-low NOx burners is not as "tight" as for higher NOx burners. All low NOx flames show a more diffuse combustion on the outer edges of the flame envelope, a characteristic of ultra-low NOx combustion.

Figure 8 compares the CFD predicted wall surface temperatures with the actual temperature distribution within the firebox. The agreement is excellent. Note that the application of the COOLMIX hearth burners changes the temperature distribution within the firebox. In comparison to the base case (Figure 2), the bottom of the box is hotter, a result of the inherently lower flame height for the COOLMIX burners. One can also see the effect of the COOLMIX wall burners. Since the fuel is staged in these burners, the wall temperatures in the region of the wall burners are slightly lower than in the base case.

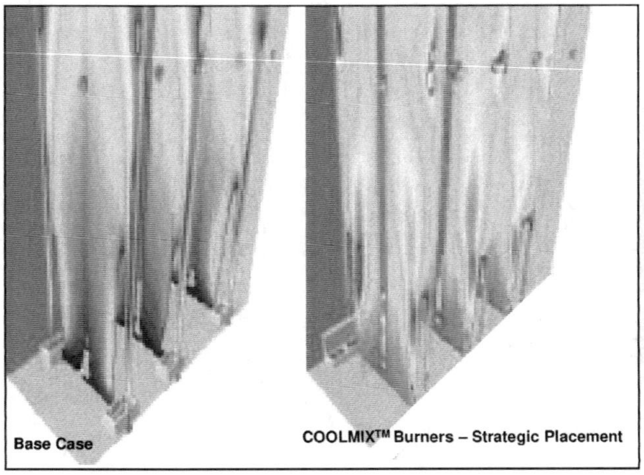

Figure 7. Gas burner temperature, burner centerline

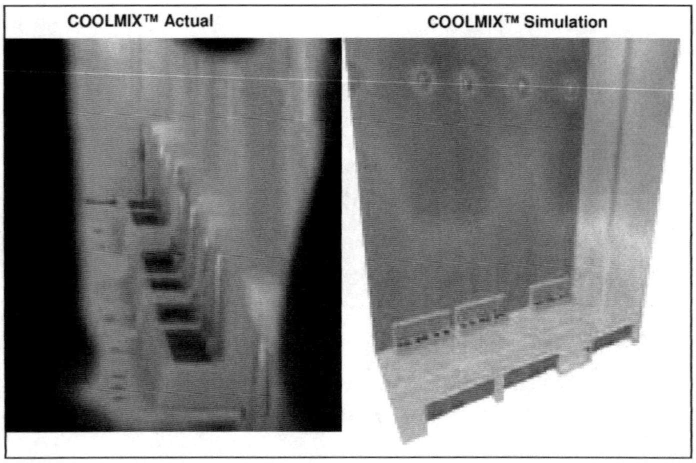

Figure 8. Comparison of wall surface temperatures

The performance characteristics of the coil are improved over the base case. The predicted maximum outlet pass coil metal temperatures for the base case and revamp case are essentially identical and the overall coil utilization is predicted to be slightly higher, reflecting the more uniform flux profile and hence metal temperature profile achieved by the burner layout. This is shown in Figure 9.

In order to confirm the predicted performance, extensive testing was carried out at the John Zink test facilities in early 2003. Firing was carried out with fuels ranging from 9 to 30 percent hydrogen. In the test heater, two COOLMIX hearth burners and 10 COOLMIX wall burners were installed with a total firing capacity of 25 MM BTU/hr. Those tests showed that under maximum firing and normal firing conditions respectively, the NOx values were 26 and 22 ppmvd corrected for three percent oxygen and 2200°F firebox temperature.

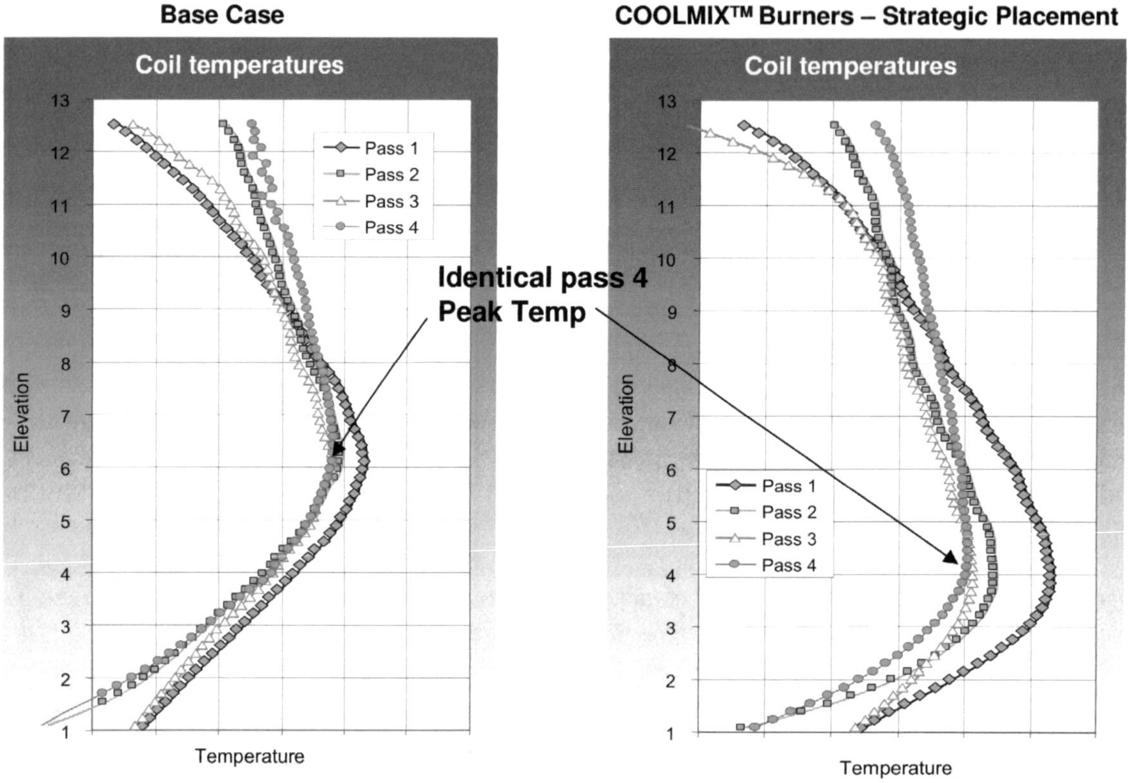

Figure 9. Comparison of TMT profiles

Based upon the results of these studies, one of the Equistar SRT III heaters was revamped to incorporate the COOLMIX hearth and wall burners. The unit was started up in July 2003 and has been operating ever since. The initial flame shapes were as predicted, showing generally stiff flames with the occasional diffuse combustion ("puffs") characteristic of low NOx combustion technology. Initial field measurements showed NOx emission levels of 25 to 40 ppmvd, depending on both ambient conditions and heater operating conditions. Although other furnace performance results were very encouraging, the NOx levels needed to be improved.

The extensive multi-burner array testing at the John Zink test facility confirmed the CFD prediction that flame interaction between the hearth burner and the radiant wall burners was significant. This interference created a "hot spot" at the first row of wall burners that was believed to be a major contributor to the NOx. This testing led to refinements in the relationship of the wall burners to hearth burners within the heater and to the wall burner staged fuel injection. When these modifications were recently tested on the Equistar Chocolate Bayou furnace, the NOx emissions were reduced to the 28 to 32 ppmvd level under low humidity and high firing rate conditions (full heater capacity); conditions that are normally adverse to low NOx emissions. Further, based upon the performance over the last seven months of testing, there has also been an improvement in furnace run length. The combination of lower NOx and improved performance represents an outstanding success.

This development work represents the first step change in combustion technology for ultra-low NOx performance. Lummus, John Zink and Equistar continue to develop improvements to increase process performance as well as reduce NOx. These studies are utilizing a combination of CFD studies and field testing, and the results of this ongoing development work will be reported in the near future.

Conclusions

The success of this program illustrates the power of the combination of Lummus's firebox design expertise and CFD model, John Zink's combustion system technology expertise and know how, and a motivated and technically enlightened client, all working together. As a result of this very successful commercialization program, Equistar is currently evaluating the possibility of installing the COOLMIX ultra-low NOx burner technology in all of the remaining SRT III heaters within their system and studies are underway for application to other heater types as well.

---

[1] Ethylene Furnace Modeling: An Integrated Approach: Presentation at the AIChE Spring National meeting 2003 New Orleans

[2] Ethylene Furnace Modeling: An Integrated Approach to Low NOx, Presentation at the EEPC Conference October 2003,Lisbon,Pt.

[3] U.S. Patent No. 6616442, 6607376; Patents Pending.

SESSION 64
ROUNDTABLE DISCUSSION
ETHYLENE CRACKING FURNACE BURNERS
PRESENTED BY THE TECHNOLOGY SUBCOMMITTEE

The Roundtable was an open forum for the audience to discuss non-proprietary issues around the latest technologies relating to ultra low NOx burners being employed to meet the latest stringent environmental regulations in parts of the United States as well as further discussion on the challenges the Olefins Industry faces. No papers were presented, however, the authors from Sessions 63 and 64, relating to Cracking Furnace Burners, served as panelists for the Roundtable along with other topical industry experts.

The panel consisted of the following individuals:

- Doyle Bishop – Callidus Technologies LLC
- Robert Gartside – ABB Lummus Global USA
- John Grever – Selas Corporation of America
- Ron Just – Nova Chemical
- Vikki Medley – Chemical Market Associate, Inc
- Scott Reed – Zeeco
- R. Waibel – John Zink
- Michael Wignal – Hamworthy Combustion Engineering Ltd

Paper 65a

Optimizing Ethylene Plant Production Using Model Predictive Control and Dynamic Real Time Optimization

Michael Boell
Basell Polyolefine GmbH
Wesseling, Germany

Rob Vasbinder, Nic Castelijns
Honeywell Process Solutions
Amsterdam, The Netherlands

Prepared for the presentation at the 2004 Spring National Meeting, New Orleans LA, Apr 24-29
16th Annual Ethylene Producers' Conference
Session T8009 – Ethylene Plant Process Control

Copyright © Basell Polyolefine GmbH
April 2004

UNPUBLISHED

AIChE shall not be responsible for statements or opinions contained in papers or printed in its publications.

ABSTRACT

Basell Polyolefine GmbH owns and operates two ethylene plants at Wesseling, Germany. To maximize the profit from one of its ethylene plants, Basell implemented a large-scale advanced process control (APC) and optimization project named "Schmetterling". The project was awarded in January 2002 to Honeywell Process Solutions. Use has been made of a Dynamic Plant Optimizer running on top of an APC level and a third party DCS system. Special attention was paid to obtain acceptance by operations as this was identified as the most important key success factor. This paper highlights the overall project and details the optimization application. Moreover it focuses on key success factors and lessons learned.

I. Process Description

Figure 1 shows an overview of the MG-4 Olefins plant.

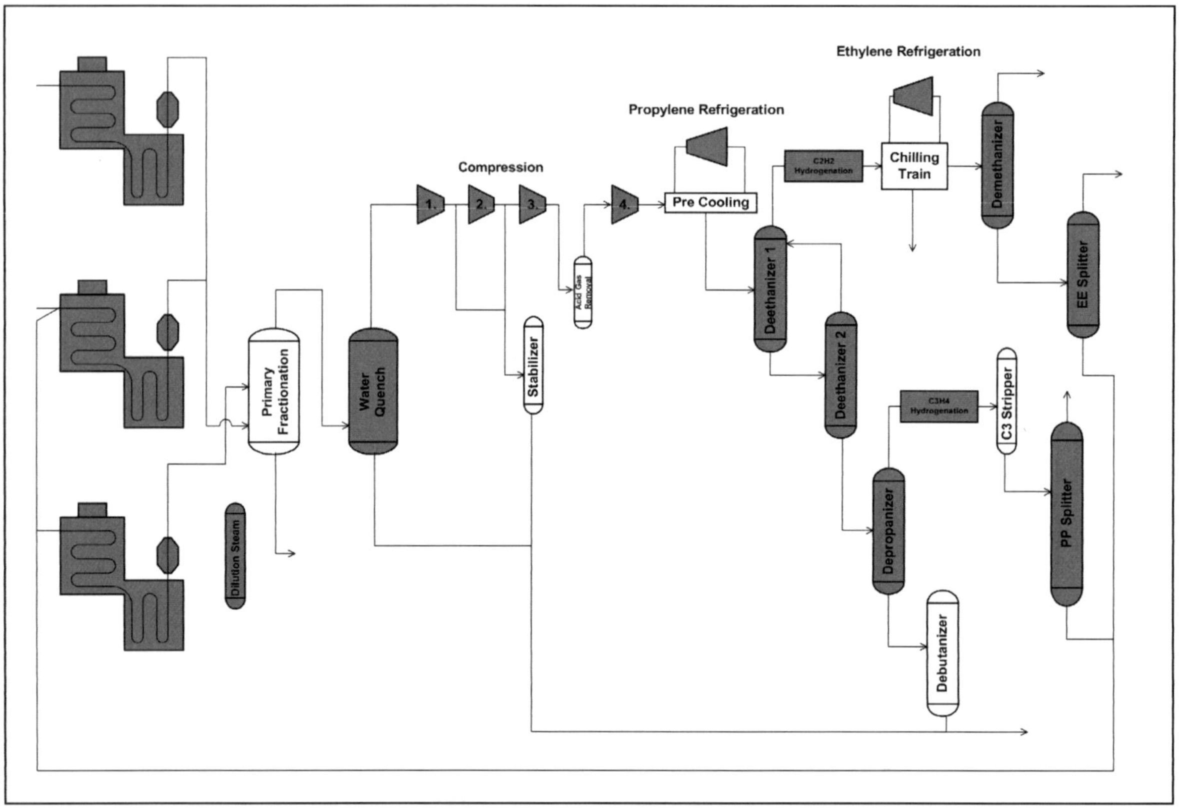

Figure 1: Flow sheet of MG-4 Olefin plant

It comprises the following units:

- 6 Furnaces and feed system
- Quench area
- Cracked gas compression and stabilizer
- Pre-cooling, Deethanizer
- Acetylene Converter
- Deep-cooling and Demethanizer
- Ethylene Fractionator and Ethylene Refrigeration
- Depropanizer
- MAPD Converter
- Propylene Fractionator, and
- Debutanizer

FURNACE AND FEED SYSTEM

There are six Technip furnaces, five liquid furnaces and one gas furnace. Hydrowax (a hydrogenated gas oil) from the upstream refinery is the primary feedstock with a relatively constant quality. All furnaces may also use naphtha as a feedstock, and the optimizer solution must support this significantly different feed type. Ethane and propane recycle are cracked to extinction. There is no on-line feedstock analyzer, nor furnace effluent analyzers in place. Furnace decoking is a campaigned operation.

QUENCH AREA

The quench area consists of a quench oil tower and a quench water tower. It separates the raw gasoline and fuel oil from the furnace effluent and recovers heat to produce dilution steam.

CRACKED GAS COMPRESSOR

The cracked gas compressor consists of four compression stages with inter-stage cooling water exchangers and knockout drums. The compressor is driven by a steam turbine. A caustic wash is placed between the third and the fourth stage. A stabilizer collects the C5+ from this section, which is mixed with the gasoline stream from the quench area.

PRE-COOLING, DEETHANIZER

The cracked gas from the fourth stage of the compressor is fed to the pre-cooling section, where propylene refrigerant is used. This results in four feed streams to the deethanizer. The deethanizer is a two-column system. The overhead of the second column acts as reflux to the first column. The overhead stream is sent to the acetylene converter, the bottom stream to the depropanizer.

ACETYLENE CONVERTER

The acetylene converter uses a single reactor, methanol cooled, to remove the acetylene from the process stream.

DEEP COOLING AND DEMETHANIZER

The effluent from the acetylene converter passes four stages of deep-cooling resulting in four feed streams to the demethanizer. Both the deep cooling and the demethanizer use ethylene refrigerant. Hydrogen and methane rich streams leave this section as product.

ETHYLENE FRACTIONATION AND ETHYLENE REFRIGERATION

The ethylene fractionator receives its feed flow from the bottom of the demethanizer. The overheads of the column include both a liquid and a gas product. The gas product is sent to a header feeding the polyethylene plants. The liquid flow is sent to storage. The bottom flow of the column is sent as ethane recycle back to the gas furnace.

The ethylene refrigerant compressor is 3 stage and electrical driven. The compressor is typically operated at full capacity and is a major plant constraint.

DEPROPANIZER

The depropanizer is fed by the bottoms of the second de-ethanizer column. The overhead goes to the MAPD converter and the bottom goes to the debutanizer. The tower is condensed with propylene refrigerant and reboiled with steam.

MAPD CONVERTER

The MAPD converter uses a single reactor, propylene refrigerant cooled, to remove the MAPD from the process stream.

PROPYLENE FRACTIONATOR

The propylene fractionator is preceded by a stripper to remove the light ends from the feed. The overhead propylene product is sent to tank farm and export. The bottom flow is sent as propane recycle to the gas furnace. The tower is condensed with cooling water and reboiled with quench water.

DEBUTANIZER

Depropanizer bottoms enter the debutanizer. The overhead C4 product stream is sent to storage. The bottoms gasoline product is mixed with the flows from the quench area and the stabilizer and is sent to storage.

II. Highlights of Technology Used

Honeywell's Profit Controller (Robust Multivariable Predictive Controller - RMPCT), Profit Optimizer and Profit Bridge Technologies are used in this project. Details of these technologies are briefly described below. For a detailed comparison between traditional and Honeywell's APC and Optimization Technologies please refer to [1].

PROFIT CONTROLLER TECHNOLOGY

The Profit Controller incorporates a model of the process dynamics that it uses to predict future behaviour of the process and to determine how to adjust the controller's outputs in order to bring all process variables to set points or within limits. If there are any degrees of freedom remaining, the controller adjusts the process to optimize operations, for example by maximizing total product value. Profit Controller is considered a fourth generation Model Predictive Controller [2] Its technology contains a number of differentiating features as reported in [3]. Some of the key features are described below.

Robustness

Firstly, the controller is mathematically robust to model uncertainty. All multivariable controllers are based on an empirical model of the plant gained through plant identification tests performed around the normal operating region. The rule is that the controller is only as good as the model that was derived from the plant test. The model can turn out to be bad because either not enough effort was expended in identification or the plant moved to a different operating point. Profit Controller recognizes this as a shortcoming and uses its Min/Max algorithm to find the best controller, which minimizes the sum of the maximum cumulative deviations from target for all practical process models. In essence the controller stays on-line and continues to perform well when catalyst activity decays, during feedstock switches or as cooling capacity changes.

Secondly, this controller minimizes movement in manipulated variables to achieve its control objective. It is therefore both the plant manager's and operator's best friend because it introduces the least additional disturbance to the process to achieve new targets or to get back on target. In addition to the operating and maintenance benefits of minimizing manipulated variable movement, this also means this controller remains on control in certain situations where other multivariable controllers would go unstable.

Product Value Optimization (PVO)

Utilizing the linear and/or quadratic objective functions, which come standard with Profit Controller, PVO based optimization strategies are easily implemented. The economic objective of the optimization strategy can be changed on-line. As a dynamic rather than steady state solution, optimization objectives are implemented in conjunction with the control solution to potentially offer a more precise final solution.

The general form of the objective function is:

$$\text{minimize} = \sum_i b_i \times \text{CV}_i + \sum_i a_i(\text{CV}_i - \text{CV}_{0i})^2 + \sum_j b_j \times \text{MV}_j + \sum_j a_j(\text{MV}_j - \text{MV}_{0j})^2$$

where the constants that you set to define the objective function are as follows:

b_i are the linear coefficients on the CV's

b_j are the linear coefficients on the MV's

a_i are the quadratic coefficients on the CV's

a_j are the quadratic coefficients on the MV's

CV_{0i} are the desired resting values of the CV's

MV_{0j} are the desired resting values of the MV's.

PROFIT OPTIMIZER TECHNOLOGY

Profit Optimizer has given a new definition to Real Time Optimization (RTO). Where traditional RTO is based on steady-state models of the process, executing at half an hour or slower frequency, Profit Optimizer has a dynamic model of the process and executes at minute frequency. Where traditional RTO had to rely on steady-state models developed from scratch, Profit Optimizer uses the collection of Profit Controller models of all the process areas within the scope of the optimization. Next to the controller models one has to spend only limited time to connect these models using the additional model capabilities in the optimizer. The user can introduce dynamic bridge models, source-clone models and combined constraint models. Details on this type of modelling can be found in [4]. The results of this modelling leads to a large matrix as shown in Figure 2, which is solved by the optimizer.

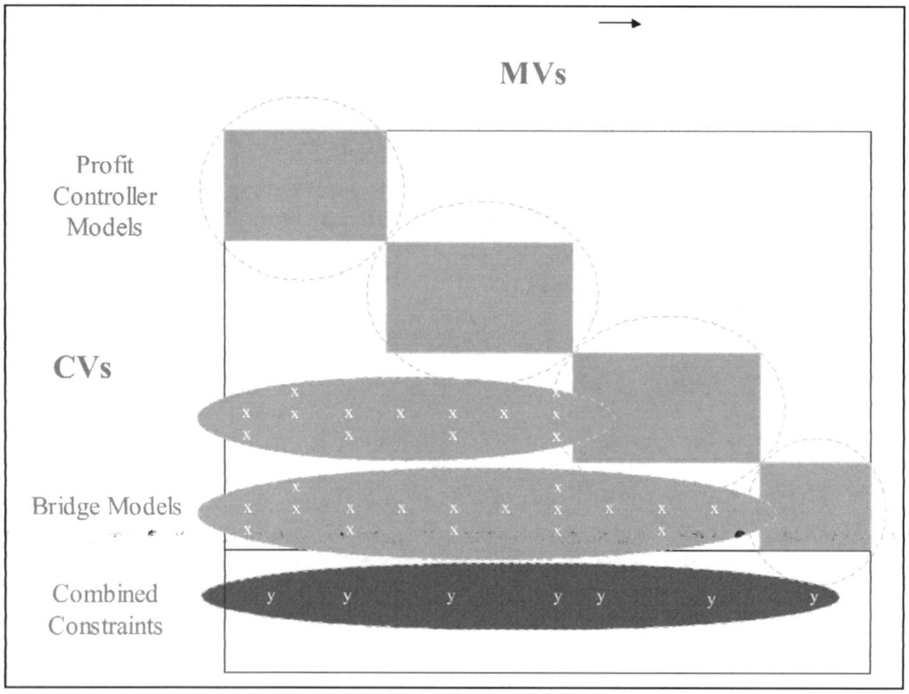

Figure 2: Profit Optimizer model structure

PROFIT BRIDGE TECHNOLOGY

Both Profit Controller and Profit Optimizer employ linear dynamic process models. In general, all processes are non-linear to some extent within the normal operating region. Traditional RTO tries to capture all these non-linearities by developing a full-scale rigorous flow sheet model of the process. For process areas that show a highly non-linear behaviour, gain updating has been proven to be an effective way to improve the controller performance over a wider operating range. With the Profit Bridge technology this is now extended to the optimizer layer as well.

Typically new gains are determined by numerical perturbation of non-linear process models. Minimal effort is needed to adapt process models for gain updating since no basic change in functionality of the process model is needed. Because only process derivatives are needed, rigorous parameter fitting is not required at each execution. This provides the framework for a very robust and reliable application. More information on interfacing steady-state process models with a dynamic optimizer can be found in [5]. Profit Bridge comes with dedicated process sockets as interface to steady-state models.

The controller and the optimizer allow for on-the-fly gain updating. In the controller these are the MV/DV-CV relationships. In the optimizer these are the gains in the Bridge Models and in the Combined Constraints. The mechanism of the Ethylene Socket is shown in Figure 3.

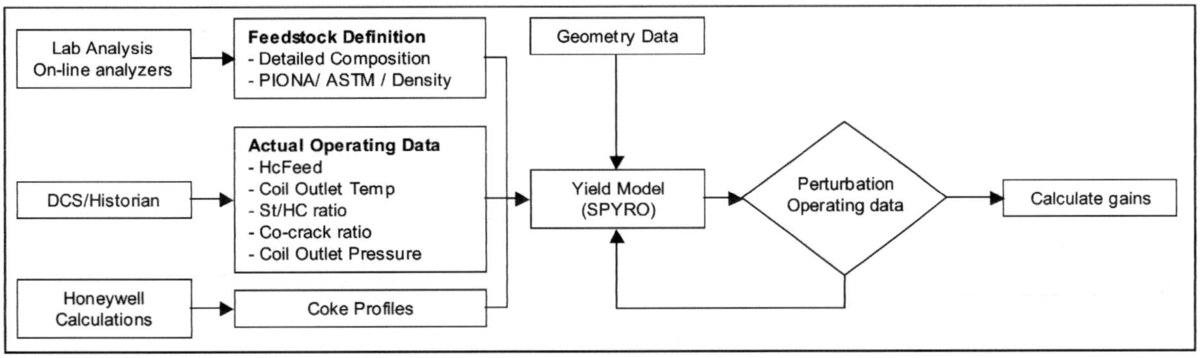

Figure 3: Profit Bridge Ethylene Socket

III. The Project

SCOPE

The APC project comprises 16 Profit Controllers: 6 for the furnaces, 1 for the quench tower, 1 for the dilution steam generator, 6 for the cold side separation and one each for the acetylene converters. In designing the controllers it was decided to keep related units under a single controller to improve controller performance. With the presence of the Profit Optimizer technology, much more flexibility is available to decide on the size of the controllers. Individual controllers can always be 'connected' again within the structure of the optimizer.
The units included in the APC scope are indicated in Figure 1.

Table 1 summarizes the scope of the APC project.

Table 1: Profit Controllers overview

Process Area	MV	DV	CV
Hotside:			
Furnace 1-3	26	1	68
Furnace 4-5	27	3	70
Furnace 6	10	3	34
Quench tower	4	2	5
Dilution steam	1	0	84
Coldside:			
Cracked Gas compressor	2	3	5
Deethanizer	4	3	7
Demethanizer	3	5	10
Ethylene Fractionator/C2= refrig	6	3	11
Depropanizer	2	1	4
Propylene Fractionator	2	1	5
Acetylene converter	2	2	5
MAPD converter	2	2	2
TOTAL	248	34	729

The scope of the real-time dynamic optimizer is comprehensive and includes the entire olefin plant, except for the acetylene units and the dilution steam system. For global constraints not covered by individual Profit Controllers, one single controller was added to the application. This controller contains CV's that deal with the total plant's throughput and product streams, allowing operations to set plant wide targets and constraints for the optimizer. This controller performs no direct control.
Bridge Models were defined for the cracked gas compressor and the column feeds.

The Profit Bridge application is used to interface to SPYRO models through the Ethylene Socket. Its resulting furnace yields is used to update the gains for severity and Tube Metal Temperature (TMT) in the furnace Profit Controllers and Bridge Models in the optimizer.

SYSTEMS OVERVIEW

The application architecture is shown in Figure 4.

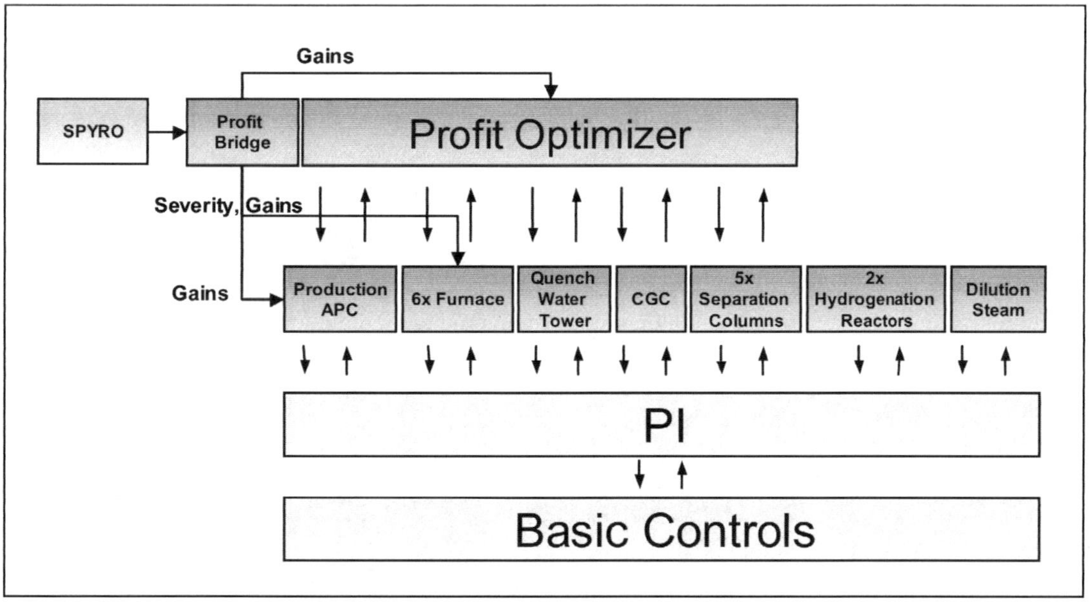

Figure 4: Application Architecture

The Profit Controllers are not directly connected to the DCS but all communicate through OSI's PI historian with the DCS. The Profit Optimizer has no direct communication with the historian, instead taking all its required data from the Profit Controllers. This minimizes load on the interface to the DCS, as well as avoiding duplication of process variable validation that already exists at the Profit Controller level. The two controllers on the hydrogenation reactors and the one for the dilution steam system are stand-alone controllers, and have no interaction with the Profit Optimizer.

Profit Bridge is connected both to the controller layer and the optimizer to update the gains. Profit Bridge is connected to the SPYRO program. The SPYRO program receives its information on-line from PI as well.

The systems architecture is shown in Figure 5.

As mentioned, all communication with the DCS layer is through the existing historian, for which an interface server was already in place. For the project, one additional server and 3 workstations were added to the system network. The APC/RTO server contained all the application software shown in Figure 1, including a PI-API for communication with the PI server and the SPYRO yield model. Three workstations were used for display of Profit Suite's standard operator graphics and some custom displays prepared in PI. One workstation was used for engineering purposes.

Remote support and maintenance could be provided with a direct access to the engineering station.

The controllers and the optimizer run at a 1-minute frequency, the gain update to these layers takes place at 5 minute and 1 minute frequency respectively. In this way the optimizer can respond directly to a gain change. The solution has no need for the plant to be at steady state to execute the optimisation runs, resulting in a high frequency, flexible approach that consistently optimizes to global objectives.

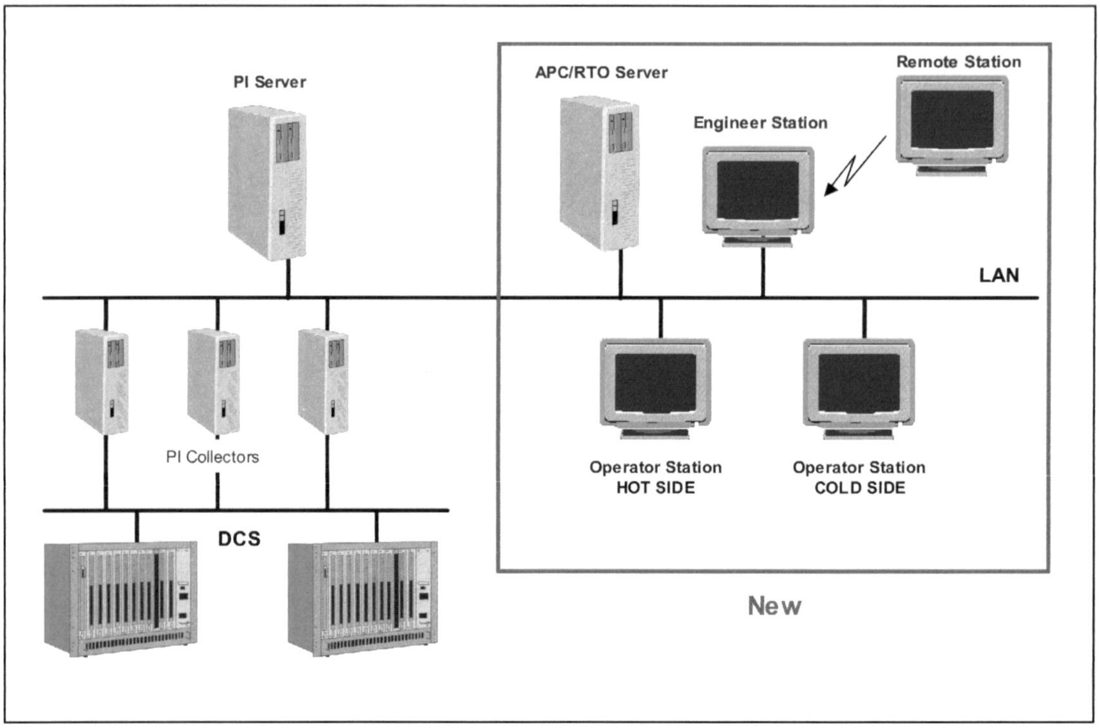

Figure 5: System Architecture

PROJECT SCHEDULE

The project was executed on a standard schedule, with completion within 18 months, including acceptance testing. The following is the actual schedule:

January '02	Project Kick-Off
April '02	Design Review
May '02-May '03	Controller Step testing and commissioning
Mar-June '03	Optimizer commissioning
July-Aug '03	Acceptance Testing

Commissioning of the optimizer was done in two steps: first the complete furnace section was brought into the optimizer scope followed by downstream section.

OBJECTIVES

The objective of the Profit Controllers is to maintain control of the process and in case of degrees of freedom remaining, control the process to a preferred local objective. Table 2 shows an overview of the local objectives for the controllers in place. These objectives are over-ridden by the Profit Optimizer objectives when the optimizer is in use.

Table 2: Local control objectives

Controller	Local Objective(s)
Furnaces	Maximize throughput Minimize O2% stack Pass Temperature balancing
Cracked gas compressor	Minimize suction pressure
Deethanizer	Stabilize Column Operation during load changes
Demethanizer	Minimize C2= loss top Maximize CH4 leakage bottom
Ethylene fractionator	Maximise C2- in C2= product Minimize C2= loss in bottom
Depropanizer	Stabilize Column Operation during load changes
Propylene fractionator	Maximize C3- in C3= product Minimize C3= loss in bottom
Acetylene converters	Maximize selectivity towards olefins

The objective of the Profit Optimizer is to optimize the complete process. Although many objectives are possible, the following two are the most common used:

- Ethylene Product maximization,
- Olefins Product maximization

IV. Key Success Factors

During the implementation of the project, the following three key success factors were identified.

- Operator acceptance
- Management support
- Selection of project team

The most important factor for project success is operator acceptance. Only if the project team manages to make the operators fully comfortable with the new technology will the APC applications stay online on a continuous basis. As it is the operators that have to work with the APCs in their day to day routine, they are the main customers of the project team. As this factor is so vitally important, a following chapter will deal with how to generate operator acceptance.

As the implementation of APC affects plant operation, especially during step testing, full support of the plant management is also essential.

The implementation of APC and RTO is a challenging task. Therefore the selection of the individual project team members is crucial. Resulting from the high complexity and the unit interdependencies in an ethylene plant, the design of APC and RTO solutions for such a plant require people who not only have a great understanding of APC in general, but also have performed similar projects in other ethylene plants. Only people with enough experience, who know all the potential challenges, will succeed in developing an optimal solution. Moreover, as any plant has its special challenges, people who have a sound understanding of the operation of the specific plant have to be included in the team. Also the view of the operator as the future customer has to be involved in the team. Figure 6 highlights the Basell side of the "Schmetterling" team.

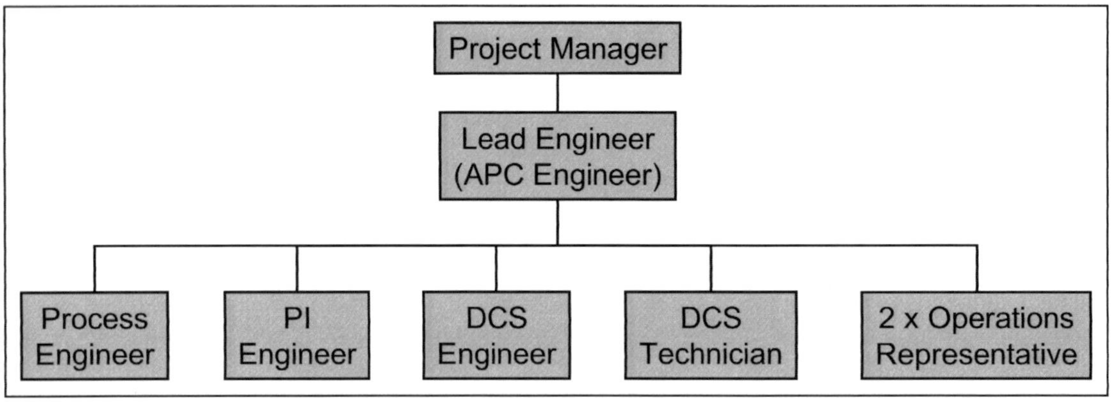

Figure 6 : Project team customer side

How to generate operator acceptance

The most important issue to generate operator acceptance is to **involve operations from the very beginning** on in the design of the applications. Requirements of operation can directly be accounted for during the design phase. As mentioned earlier, the Basell team included two operations representatives to perform this task. Moreover two focal points per shift were assigned to form the main connection to the APC team. These focal points received a special training before the start of the work in the control room, to generate sufficient knowledge about the design and implementation of APC. The focal points were the main contact persons for the team within the control room and they also gathered all feedback from the shifts to ensure that it is fed back into the APC design team. Moreover it turned out that the step testing activities could be performed more efficiently, because at least some operators already knew the requirements of the tests. A significant number of tests, especially later in the project, could be performed based on a step test plan by the shifts alone, without the full-time supervision of the project team. Step testing could be performed without any negative effect on the operation of the plant.

Sufficient **training** of the operators in the operation of APC and RTO is also essential. Operator training was mainly performed on the job, after some initial presentations of the project team. The training was supported by the shift focal points who participated in a 3-day operator course offered by Honeywell. The focal points were able to support their colleagues in more general questions, more specific topics were dealt with by the project team.

It is also important to **carefully choose the first application**. The implementation of APC and RTO is a big step for operations as it significantly changes the way the plant is operated. It is therefore not only important to organise sufficient training for the operators, it is also important to slowly familiarise operations with the new technology and make them confident that it really works. So the right choice of the first APC installed is important. If there is a freedom of choice, the Basell project team would suggest following three simple rules.

Choose an application with significant benefits for operators. To create operator acceptance it is crucial that it is proven to the operators that the technology will give them benefits in their day-to-day work. To achieve that, we would suggest realizing benefits for the operators even before implementing the most beneficial applications.

Choose APC with guaranteed first time success. It is important to prove that APC works. If the first application is commissioned smoothly and works perfectly from the beginning on, this buys confidence from the operators for the more challenging applications, which may not work perfectly right away. Initial problems will be accepted more easily.

Choose an easy application. The first application should ideally be small and simple with a straightforward concept. As mentioned earlier, in general APC is a new concept for operations with significant differences in plant operation. A simple application is much more easily understood.

Moreover it is crucial that the APC applications have a high online time from the beginning on. Only if they are running can operator acceptance be created. It is therefore very important to **solve problems without delay**. During the Schmetterling project, one team member was present in the control room almost all the time during working hours. During this time, all potential problems could be identified and solved without delay. Moreover, all questions concerning APC operation could be answered right away. Also the intensive on the job training for the operators took place during this time. For off working hours, the project team established a 7 day, 24 hour support, especially during the initial commissioning phase. One team member was on standby to react on APC problems immediately. The experience with such a support was very positive. The number of off-hour calls was limited and the extra workload on the team members was acceptable. As a result, all APC problems could be solved immediately and the applications showed a high online time from the beginning on.
With all the activities described above, the Schmetterling project team was able to generate a high operator acceptance for the new applications. The support that the team received from the operators was very helpful and the extra work the team spent to support operations paid off later in the project.

V. Acceptance Test

By implementation of APC and RTO on the MG4 plant, Basell aimed at reaching the following targets:

- Increase in ethylene production by 3%.
- Improvement of the quality of control on the key units in the plant by reducing the standard deviation of key controlled variables.
- Achieve an online time of the APC application of at least 90%.

After completion of the commissioning of the APC and RTO applications, Basell performed an acceptance test with duration of 4 weeks in the third quarter of 2003. During the test period, all targets of the project were achieved.

Improvement in quality of control

By implementation of APC on the key units of the plant, the quality of control could be significantly improved. As an example, the results achieved at the PP Splitter are presented in this paper. With APC in operation, it was possible to significantly reduce the standard deviation of the top and bottoms quality and therefore the column could be operated closer to the constraints, minimising the propylene loss in the bottoms and shifting the propane content in the propylene product closer to product specification. The following results were achieved:

- Reduction of standard deviation in top quality of 52%
- Operation of top quality closer to specification
- Reduction of standard deviation in bottoms quality of 58%
- Minimisation of Propylene loss in the bottoms

Figure 7 shows the Histogram of the top quality of the PP-Splitter before and after APC implementation. It clearly shows the reduction in standard deviation and a shift in the average closer to the product specification.

Figure 8 shows the Histogram of the bottoms quality of the PP-Splitter before and after APC implementation. It clearly shows the reduction in standard deviation and the minimisation of the propylene loss.

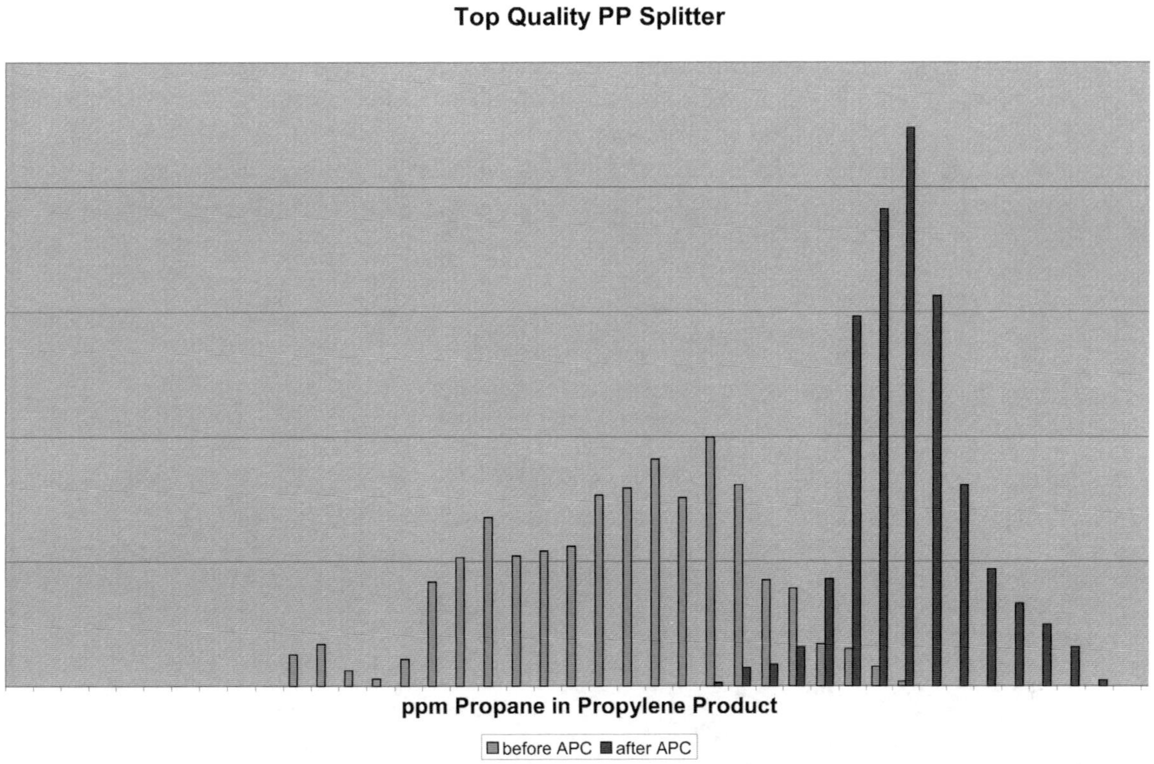

Figure 7, Histogram of top quality PP-Splitter before and after APC implementation

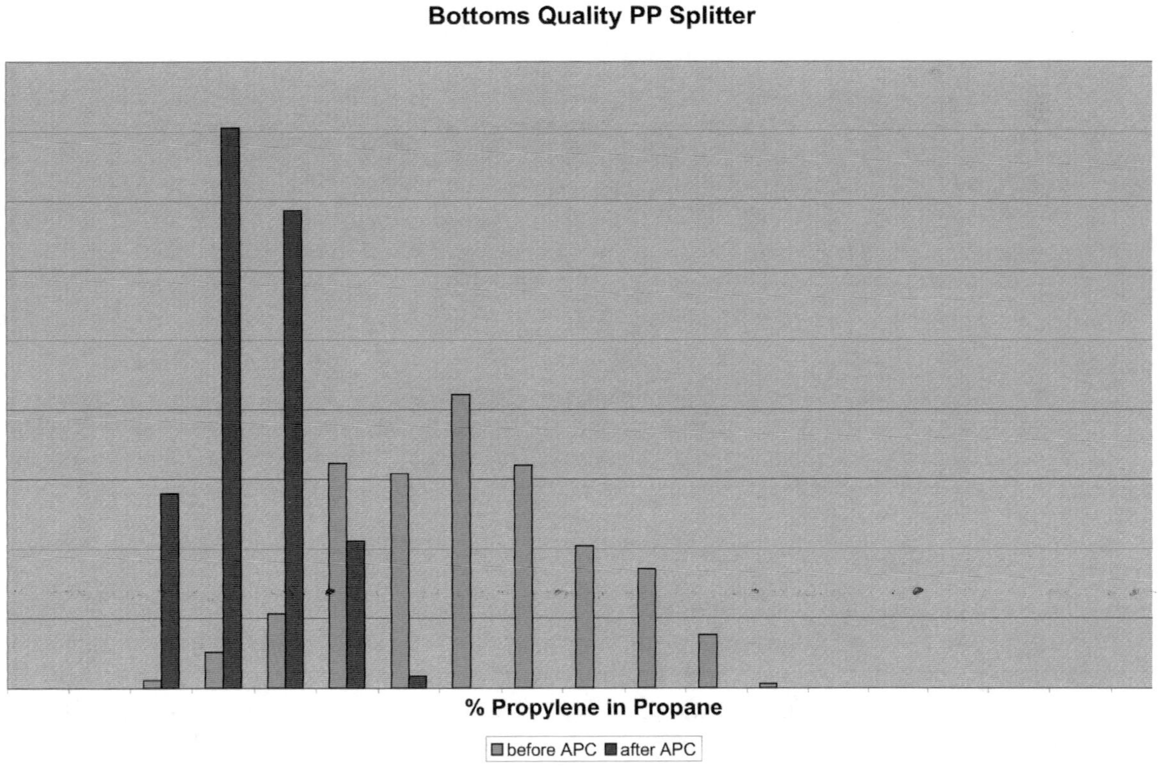

Figure 8, Histogram of bottom quality PP-Splitter before and after APC implementation

Online Time

To guarantee that the benefits from APC and RTO are captured on a continuous basis, Basell aimed to achieve a high online time of the applications. Figure 9 shows the online time for a typical month. Normally the online time is well above 90%, with only a few applications not reaching the 90%. Offline time of these applications mainly results from instrumentation maintenance or operations reasons that required the APC to be deliberately out of operation. The average corrected online time in the reference months was 94,6%, which is well above the target of 90%.

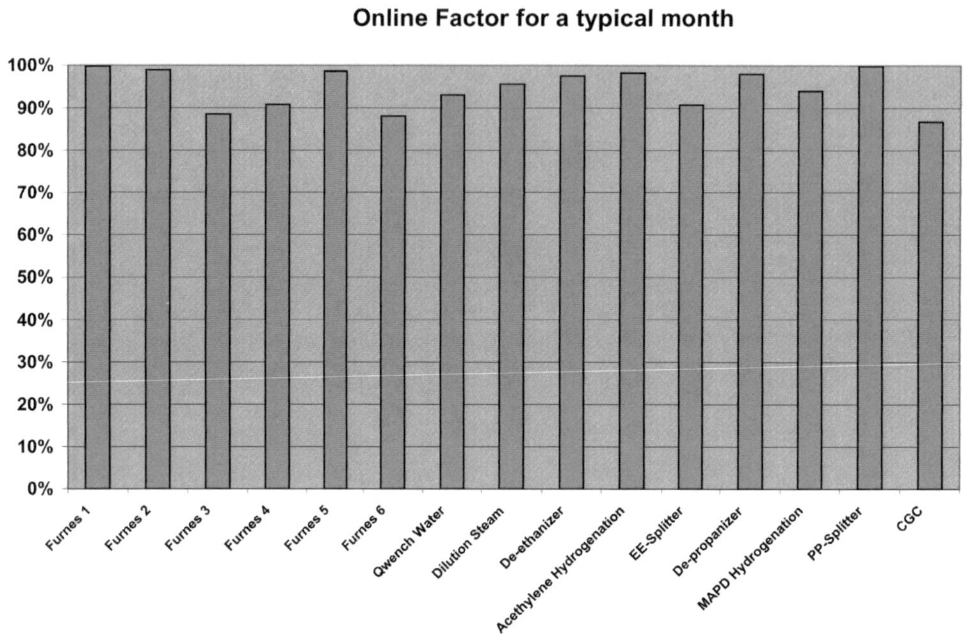

Figure 9: APC Online Time

Increase in ethylene production

During the acceptance test period, the optimizer was configured to maximise ethylene production. With the Optimizer in operation it was possible to operate the plant at various constraints. Overall operations was able to increase the ethylene production by 3,2% compared to the base case defined at the beginning of the project. Operations were also able to beat the production record that was valid before the start of the project by 1,5%.
These numbers are in-line with the experience of other projects with the same type of application [6].

VI. Conclusions

An Ethylene plant is one of the most complex process plants with a high number of interdependencies between different plant units. Implementing APC and optimisation tools from Honeywell's Profit Suite has enabled Basell to significantly improve the quality of control of the plant and help to continuously operate the plant at the true constraints to maximise profit.
All targets set for project " Schmetterling" were achieved and Basell is confident that these benefits can not only be sustained in the future, but can be increased as the operation's experience with APC and RTO grows.

VII. Acknowledgements

We would like to thank the all members of the "Schmetterling" project team (H. Bollig, M. Bleijenbergh, G.-R. Brychzy, K. Huthmacher, W. Lis, R. Werner, F. Wrede, F. Butoyi, R. Mahabeer, M. van Wijck) for their contribution to the project. We would also especially like to thank the whole operations crew of MG-4 for their support, which made the success of the project possible. We would also like to thank the management of Basell and Honeywell for their support and the permission to publish this work.

VIII. REFERENCES

[1] Nath, R., Alzein, Z., Pouwer, R. and Lesieur, M., "On-line Dynamic Optimization of an Ethylene Plant using Profit® Optimizer", paper presented at the 12th Annual Ethylene Producers Conference, Atlanta, GA, (March 2000).

[2] Qin, S.J., Badgwell, T.A., "A survey of industrial model predictive control technology", Control Engineering Practice 11 (2003), pp. 733-764.

[3] Lu, J., Escarcega, J., RMPCT : "Robust MPC technology supports APC", AICHE '97 Spring Meeting.

[4] Lu, J.,Escarcega, J., "Introduction to Dynamic Economic Optimization", Honeywell Internal Technology Document (1998).

[5] Vasbinder, R., "Combining Steady-State Models with a Dynamic Plant Optimizer", ERTC Petrochemical Conference, Mar 3-5, 2003, Paris, France.

[6] Nath, R., Halberg, M., Yap, P., Tjokro, S., Khair, M., Huat Ang, K., Wai Ng, C., "Maximizing Ethylene Plant Profits with Robust Multivariable Predictive Controller and Dynamic Plant Optimizer", ARTC Petrochemical Conference, Feb. 13, 2001, Kuala Lumpur, Malaysia.

Paper 65b

Alternate Methods of Process Model Identification: Benzene Distillation

Sasha Vragolic
Huntsman LLC
2701 Spur 136
Port Neches, TX 77651

Prepared for Presentation at the 2004 Spring National Meeting, New Orleans LA, Apr 25-29
16th Annual Ethylene Producers' Conference
Session T8009 – Ethylene Plant Process Control

Copyright © Huntsman, LLC
April 2004

UNPUBLISHED

AIChE shall not be responsible for statements or opinions contained in papers or printed in its publications.

Abstract

Huntsman LLC operates an aromatics unit in Port Arthur, Texas. The extract from the extraction unit is distilled to separate Benzene from Toluene and heavier aromatics. In December of 2001, Huntsman's technical staff decided to build a multivariable controller to minimize energy consumption in that distillation tower. It was decided to test alternate approaches to establishing the process dynamics for this column to prove out the concept, which would be more useful for processes with longer dynamics.

The initial approach was to take an existing Aspen Plus[1] steady state simulation of the tower and convert it to an Aspen Dynamics[1] dynamic model. This dynamic model was step-tested to obtain data for process identification using Aspen Technology's DMCplus[2] Model software. Analyzer deadtimes were estimated by inspection of online data that is gathered as part of the existing management information system. These deadtimes were added to the process models and a DMCplus[2] controller was built and tuned. The application was commissioned in January of 2002 with minor on-line tuning.

In July of 2002, the DMCplus[2] controller was perturbed in a closed-loop mode. That data was used to re-identify the process dynamics. These results are compared with the original model. Some of these models were used to update the controller model dynamics and improve performance.

In October of 2003, the current model was used as the seed model for a Aspen SmartStep[3] application. The resulting model is compared to the previous two models.

Introduction

During the past decade, multivariable controllers have been extensively applied in ethylene plants. These control applications have increased process efficiency and plant capacity. The benefits of these controllers have been well documented and the subject of numerous papers presented at the Ethylene Producers Conference. By now, the technology is well understood and widely accepted. The only obstacle to further implementation and better maintenance is the cost and the majority of the cost lies in the model identification step.

Traditionally, the process model is derived through the open-loop perturbance of the process. This procedure is known as a step-test and is part of any model based control project implementation. The step-test is a manpower intensive process that often requires around-the-clock engineering coverage and can last for several weeks. The duration is dictated by the quality of the test procedure, which dictates the number of perturbances required and by the duration of the process dynamics, which dictate how frequently the tester can impart a pertubance. The cost of the test is dictated by the duration and consists of the obvious manpower costs and the opportunity cost of sub-optimal operation of the process, during the test. Furthermore, with cost cutting efforts of the process industry, in recent years, the engineering staffs have been significantly reduced. Some engineering staffs have been reduced to the point where in-house step-testing is no longer possible. Therefore, new step-tests have to be contracted to outside firms, which further increases the cost of step-testing and thereby, raises the hurdle for new implementations and re-testing of existing applications.

Although it is relatively easy to calculate potential benefits for a new application, for an existing application, it is difficult to assign the percentage of the opportunity loss due to performance degradation, resulting from model inaccuracy. Because of the cost considerations, existing applications are often not re-tested, for model improvement, until the controller performance deteriorates to the point where the application is turned off. Besides the existing multivariable control (MVC) applications, there are a number of other potential applications where the economics are less compelling but can still yield significant benefits. To eliminate the efficiency degradation and to improve the cost effectiveness of the economically less compelling applications we need to reduce the cost of model identification. With smaller engineering staffs, we need to reduce the manpower requirements for MVC implementation, which will also serve to reduce the cost.

Application

One possible procedure for control model identification involves:

1) Building a first principles dynamic model of the process,
2) Perturbing the model rather than the process,
3) Building the control model for an MVC application from that data,
4) Commissioning the MVC controller,
5) Closed-loop testing the process for model refinement.

This procedure was implemented for a new MVC application on a Benzene/Toluene fractionator at Huntsman, LLC 's Aromatics and Olefins plant in Port Arthur, Texas.

The Benzene/Toluene fractionator is part of an aromatics extraction unit. The aromatics are extracted from an aromatics rich stream using liquid/liquid extraction. The benzene is separated from the aromatic extract in a distillation column, shown in figure 1. The light and heavy key components are benzene and toluene, respectively. The relative volatility for these compounds is approximately 2.5, making this a separation with a moderate reflux to feed ratio required. This means that incremental benefits of better control are moderate and justifying a major control project for such an application can be a challenge. But, the benefits of better control are not insignificant, especially with the high cost of energy, in recent years.

Figure 1:
Huntsman's
Benzene Fractionator

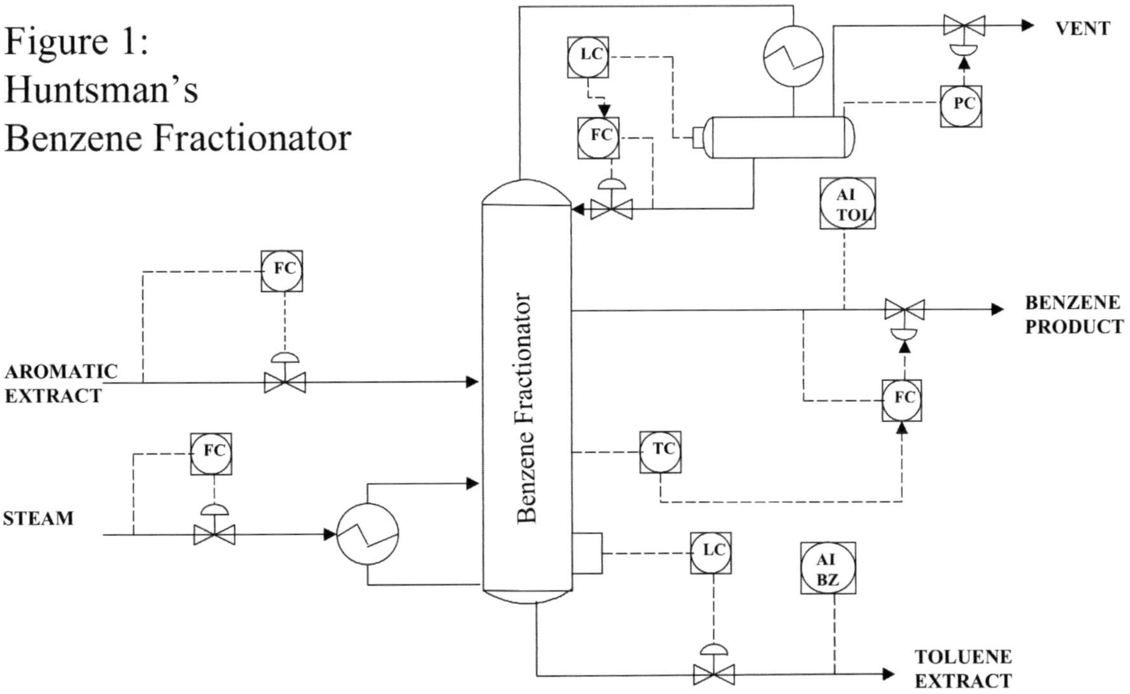

As shown in figure 1, the feed entering the column is controlled with a flow controller. Pressure control is used to remove any light hydrocarbons that are present in the extract. The benzene product is withdrawn from the distillation column as a sidedraw product. The benzene flow is set by a control tray temperature controller, which resets the sidedraw flow controller setpoint. The toluene and heavier aromatics exit the column as the bottoms stream that is controlled with a bottoms sump level controller. Reflux is controlled with a reflux drum level controller resetting a reflux flow controller set point. The heat input, to the column, is controlled with a reboiler steam flow controller. With this base level control scheme, the column operator has three Manipulated Variables (MV's) that he can set. He sets the setpoint for the feed flow controller to match the extract production from the liquid/liquid aromatics extractor. The temperature controller setpoint is set to achieve the desired benzene recovery and the reboiler steam flow controller setpoint is set to achieve the desired benzene product purity.

This is a fairly simple system but it is a control challenge. Due to the fact that it is a fairly easy separation (low cost) but with disparate values of the two products, it is a high recovery as well as a high product purity separation. This high purity of both major products means that this is a highly non-linear system. This implies high process sensitivity, at the constraints. The further one goes from the constraints, the lower the sensitivity and therefore, the easier the control. With low energy penalty for departure from constraints and easier control, it is easy for a human operator to drift away from constraint. This is where the automatic control opportunity lies.

Huntsman began a project to recover that energy cost in December of 2001. The first step was to develop a representative steady-state simulation of this distillation column. An existing Aspen Plus[1] model was used. The model was tuned to match a set of process data. The equipment geometry data was entered and the simulation imported into Aspen Dynamics[1]. The base level controllers were added to the simulation and tuned. There was no effort made to duplicate analyzer lags and deadtimes, at this point in the process.

To demonstrate the non-linearity of the system, figure 2 shows the steam rate required as a function of the amount of toluene in the benzene product, at constant feed rate and control tray temperature. The steam rate is expressed as a percentage of the steam required at the average operating point and the toluene concentration in the benzene product is shown as a fraction of the Huntsman constraint on that impurity. This figure clearly demonstrates the value of running closer to constraint and the shape of the curve shows the diminishing returns, at the higher impurity levels.

Figure 2: Steam Rate vs Benzene Composition

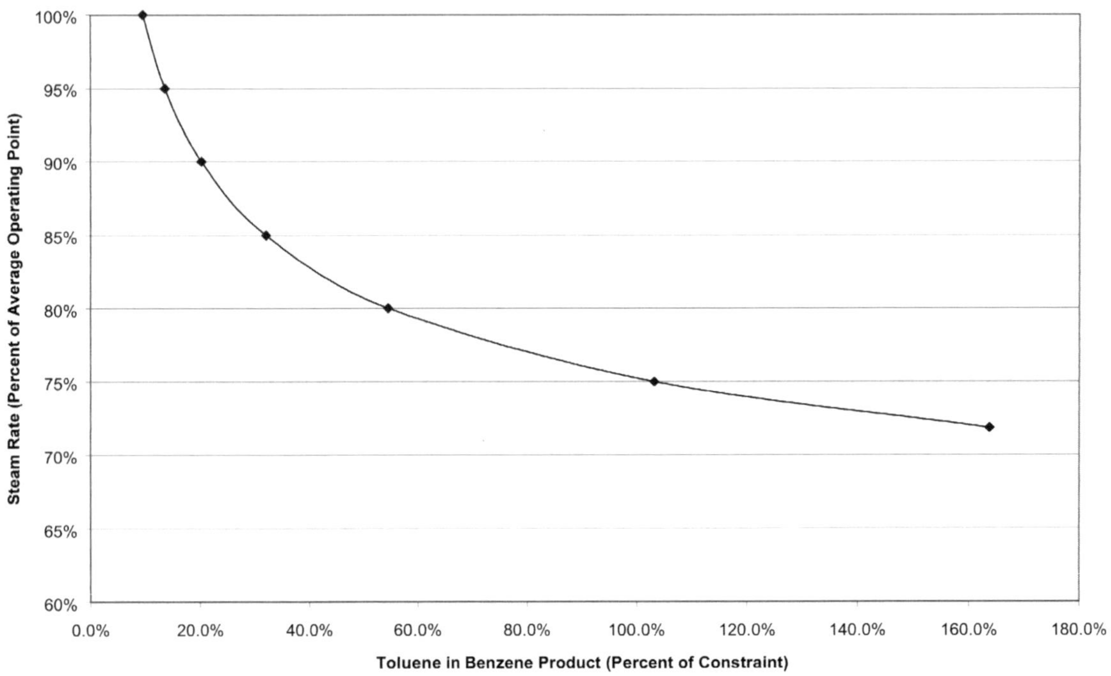

This data was used to calculate process gains of toluene in the benzene product to steam rate as a function of the benzene product composition in figure 3. This figure shows the high process sensitivity at the higher impurity levels and demonstrates that the process gain is very much dependent on where the process is tested. Of course, the point at which the model must have the highest degree of accuracy is at the constraints. From figure 3, it should be clear that it would be difficult to conduct a practical test at the constraints because of the high process sensitivity. The high non-linearity implies that the test must be carried out in a very narrow band of controlled variable (CV) values but with a real process, the test must impart sufficient CV variation to extract the true process dynamics from the noisy process data. Testing a dynamic simulation, with no noise, means that a very small perturbance can be imparted and an accurate process gain measured at the constraint.

Figure 3: Toluene/Steam Gain vs Benzene Composition

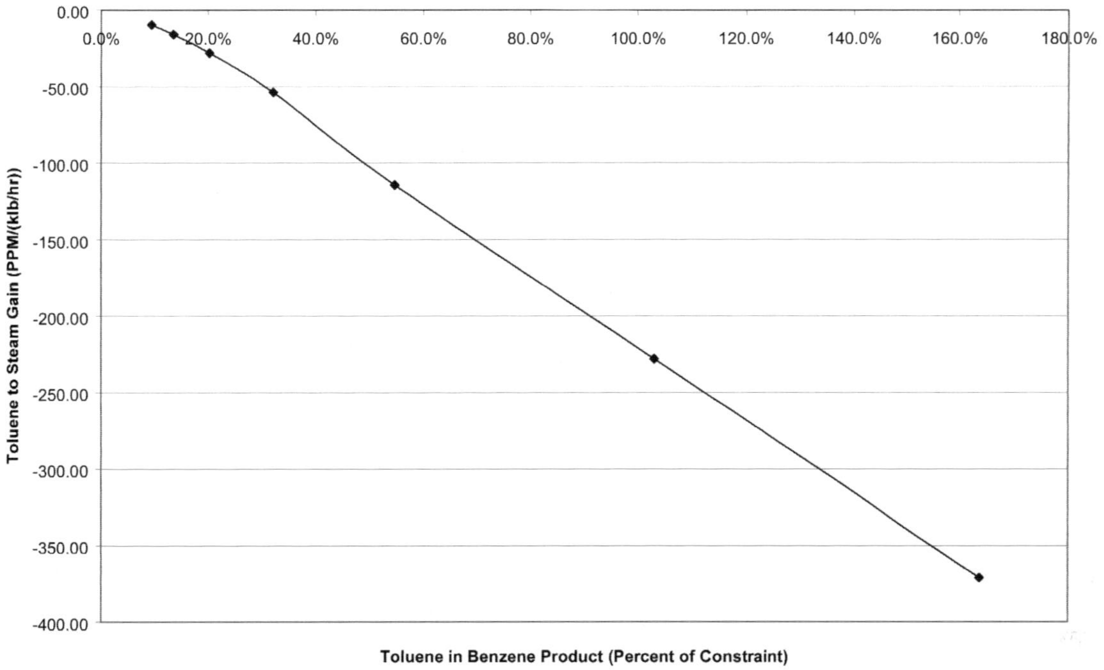

To determine the right set of constraints, the Aspen Plus[1] steady-state simulation was used to calculate the optimum operating point. The dynamic simulation was moved to the optimum point, and was perturbed. The data from Aspen Dynamics[1] was exported to an ASCII text file that was manipulated to resemble an AspenTech DMCplus[2] data collection file. That file was imported into the DMCplus[2] process model identification software. The DMCplus[2] identification was trivial since the data contained no noise and had perfect pure step responses. At this point, analyzer deadtimes were estimated, by inspection of process data. This time was added as pure deadtime to the process models. The resulting model is presented as figure 4. The controller configuration file was built and the controller tuned. With this procedure, there was a high degree of confidence in the process gains, but a very low confidence in the process dynamics. Therefore, the controller tuning was far less aggressive than normal.

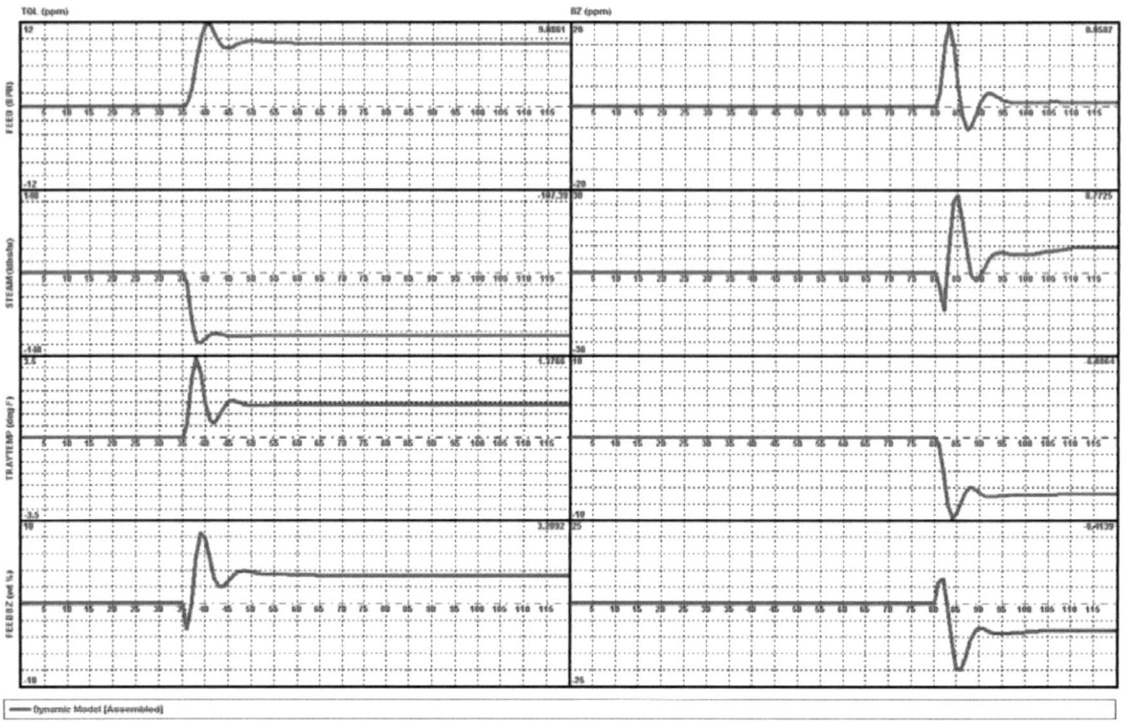

Figure 4: Control Model from Dynamic Simulation

At this point, a presentation was made to the operations team. The optimization results were used to show the maximum possible benefits of controller implementation. The methodology used to generate the control model was reviewed along with the risks of implementing a controller with this model. The risk minimization plan was reviewed along with the plan for closed-loop testing for process model improvement. Once the operation team understood the risks and benefits, the controller interface points were built and the controller commissioned in early January of 2002.

Results

The controller commissioning was extremely smooth with minimal on-line tuning required. In fact, the very first evening, the controller was left on-line without engineering coverage. The majority of the benefits were achieved on the first day of commissioning and amounted to a 15% reduction in steam usage for the same feed rate. The control performance was such that the plan to conduct the closed-loop testing was abandoned. However, over time, there were periods where the controller under-performed and was taken out of service. Therefore, the closed-loop testing was performed. The process dynamics were identified from that data and the new model, overlaid on the original model, is presented as figure 5. The new model was substituted into the controller in July of 2002 and the controller performance problems eliminated. After this iteration, any time that the controller was found to be out of service, the explanation, from the operators, was that no one knew why it was taken out of service and it was immediately re-commissioned. If more than a day had passed since the controller had been in service, the controller reduced steam consumption by 10% to 15% as it moved the process to the appropriate constraints.

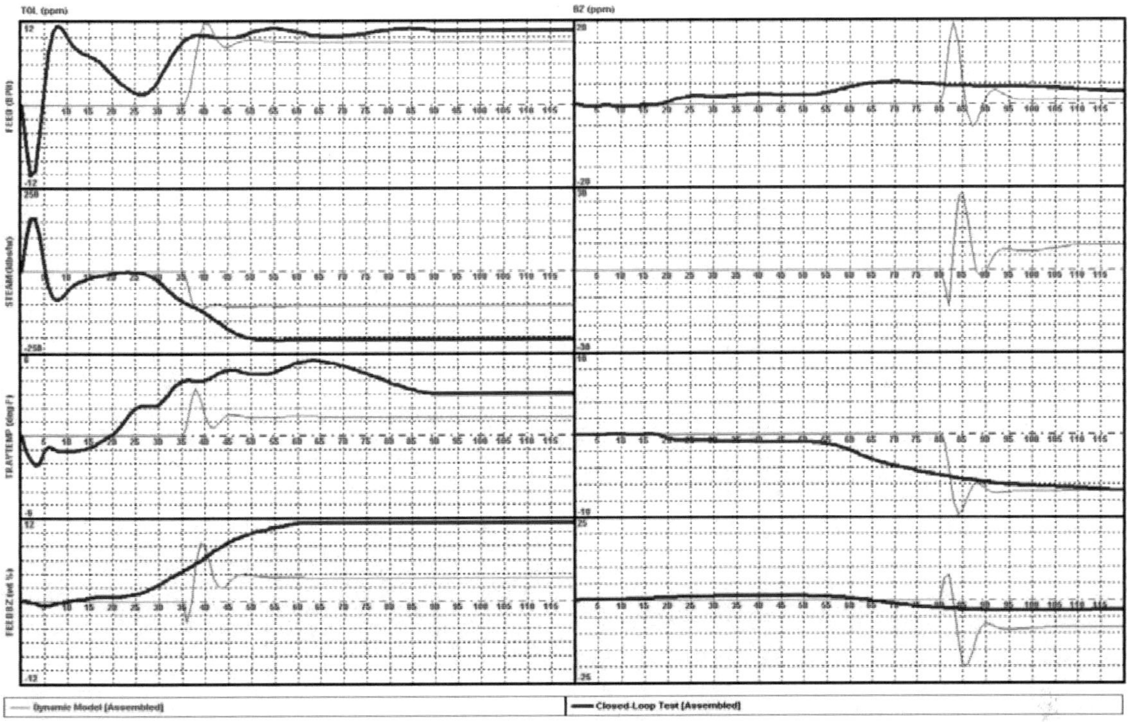

Figure 5: Control Model Comparison – Aspen Dynamics and Closed Loop Test Data

To demonstrate the benefits, the service switch position is plotted with the reflux to feed ratio for the past 3 years, as figure 6. The only manipulation of the raw data was to remove the process unit downtime periods. For the purpose of this figure, the reflux to feed ratio was normalized by dividing the actual ratio by the maximum reflux to feed ratio for this period.

Figure 6: Performance

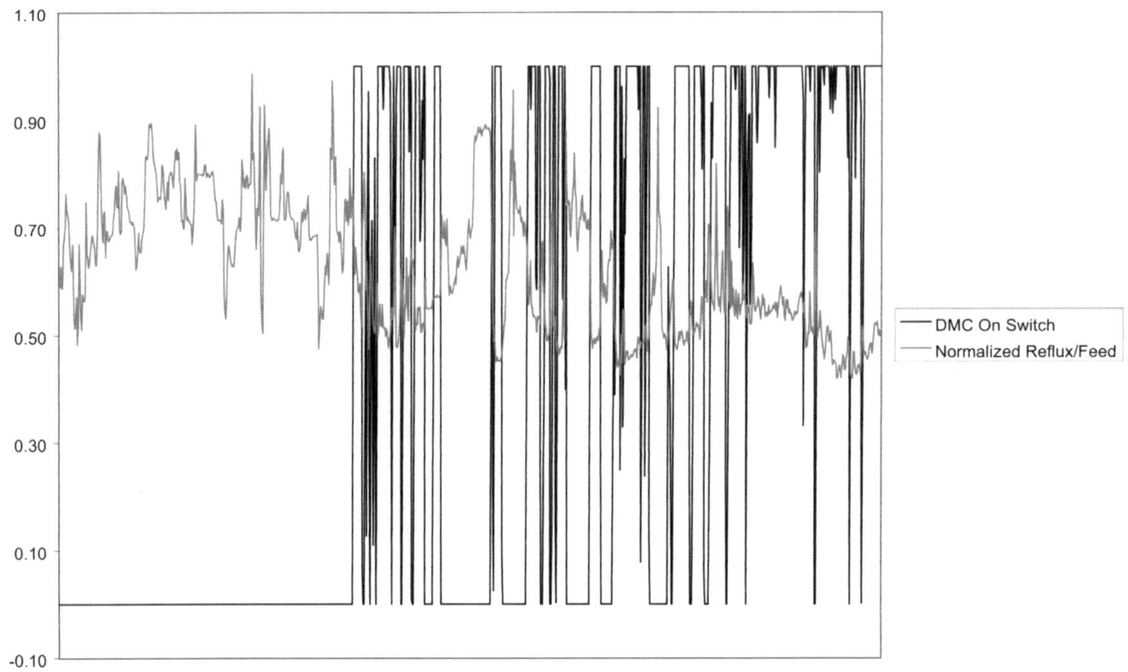

Alternative

In 2003, Huntsman licensed the use of AspenTech's automated step-testing software, Aspen SmartStep[3]. To test the step-testing software, this benzene distillation column application was chosen. The Aspen SmartStep[3] software requires a "seed model" which is an imperfect model that the tester can use to calculate the possible size of the step such as to not violate user specified CV limits. If the CV limits are violated, this "seed model" is used, by the Aspen SmartStep[3] tester, to move the process within the user specified limits. This happens when CV limits are violated as a result of either the testing routine (steps too large) or the action of unmeasured disturbances. This software presents an alternative approach to the closed-loop testing step in the procedure used by Huntsman. Instead of commissioning a controller with an imperfect model and closed-loop testing, one can now use the simulation model as the "seed model" for the automated tester. The Aspen SmartStep[3] application was configured and commissioned, using the current control model. Since the "seed model" was a model that had been used for control for over a year, there was minimal risk. The automated step-test was executed without incident. The new model was derived from the data and is presented, overlaying the previous two versions of the model, in figure 7. As you can see, an even better model was obtained and substituted into the controller.

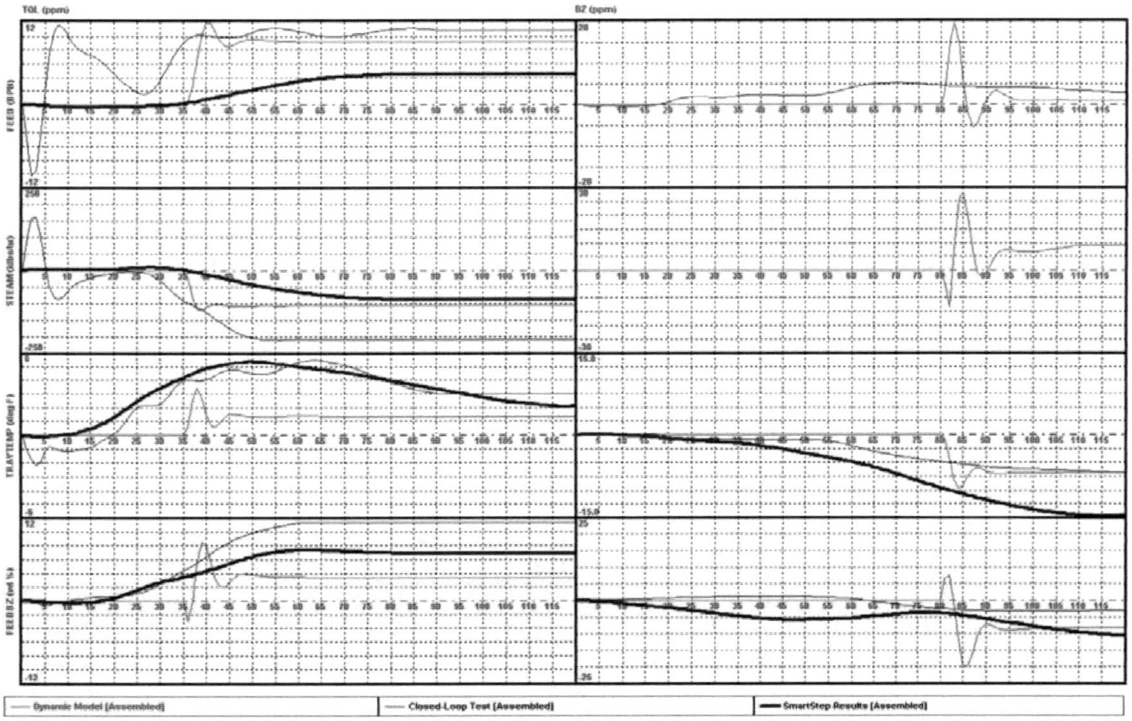

Figure 7: Model Comparison – Dynamics, Closed-Loop, and Aspen SmartStep[3]

Conclusions

A reasonable process model can be obtained from a process simulator. That model can be used as a starting point for a multivariable controller. The model can be improved with either closed-loop testing or, better, yet, with automated step-testing software such as AspenTech's SmartStep[3]. One engineer can carry out this entire procedure since it eliminates the need for continuous engineering coverage during the process testing phase of the process model identification. As such, it reduces the cost of re-testing existing applications for performance improvement and makes smaller applications such as this benzene distillation column application economically viable. The procedure is most useful for simple applications where it is easy to develop a representative first principles dynamic model but with long times to steady-state and therefore require very long step-tests.

Footnotes

[1] Aspen Plus and Aspen Dynamics are trademark names of process simulation programs licensed from Aspen Technology, Incorporated, of Cambridge MA. Version 10 of this software was used for the referenced calculations.

[2] DMCplus is trademark name of a set of multivariable control programs licensed from Aspen Technology, Incorporated, of Cambridge MA. Version 2 of this software was used for the referenced calculations.

[3] Aspen SmartStep is a trademark name of an automated step-testing program licensed from Aspen Technology, Incorporated, of Cambridge MA. Version 2.0 of this software was used for the referenced calculations.

About the Author

Sasha Vragolic is the Acting Technology Resources Team leader with Huntsman LLC's Jefferson County Operations. He has 15 years of experience, mostly in ethylene manufacturing. He has had various assignments in process and process control engineering as well as in operations and simulation and optimization. He has provided technical assistance to four different ethylene plants. He earned a BS ChE from the New Jersey Institute of Technology and a Masters degree in ChE from Lamar University. He is a member of the AIChE and serves as Huntsman's representative on the Process control subcommittee of the Ethylene Producers Committee of AIChE. He can be contacted at sasha_vragolic@huntsman.com or (409) 723-3502.

Paper 65c

Furnace Optimizer in a Naphtha Cracker

Hitoshi Shindo
Maruzen Petrochemical
3, Goi Minami-kaigan
Ichihara-shi, Chiba 290-8503 Japan
Hitoshi-Shindo@chemiway.co.jp

Kheng Lau and Ryuji Kaneko
Aspen Technology, Inc.
1293 Eldridge Parkway
Houston, TX 77077
Kheng.Lau@aspentech.com
Ryuji.Kaneko@aspentech.com

John S Ayala
Applied Manufacturing Technologies
3200 Wilcrest Drive
Westchase III Suite 400
Houston, TX 77042
John.Ayala@applied-mt.com

Prepared for Presentation at the 2004 Spring National Meeting, New Orleans LA, Apr 25-29
16th Annual Ethylene Producers' Conference
Session T8009 – Ethylene Plant Process Control

Copyright Maruzen Petrochemical/Aspen Technology/AMT
April 2004

UNPUBLISHED

AIChE shall not be responsible for statements or opinions contained in papers or printed in its publications.

Abstract

With changing availability and prices of feedstock as well as changing product demands, many operating olefins plants have to keep searching for the most optimum way of running their unit with several key constraints active. Rigorous real-time closed loop optimizers have been accepted by many operating olefins plants around the world as the preferred automation enabling application for optimizing the operations of the plant working in unison with advanced controllers. Depending on the plant constraints, the skill set of the engineers, the availability of technical resources and various other key factors, one has to make a decision on the economic viability of the rigorous plant wide optimizer. There have been only a few running optimizers reported where the furnaces are modeled rigorously and the rest of the process is modeled with component splitters. A furnace optimizer has been implemented recently at the world-scale naphtha cracker of Keiyo Ethylene in Japan. The optimizer has been able to move furnace severity to maximize propylene production while honoring many key plant constraints and a target ethylene production rate. The optimizer runs at a much higher frequency than a typical rigorous plant-wide optimizer. Hence, the optimizer can respond to naphtha feed composition changes much faster than a rigorous optimizer. Benefits and implementation issues will be discussed in this paper. With over capacity of ethylene, many olefins plants are trying to operate in a mode where propylene production rate is maximized. The simplified backend furnace optimizer offers a very attractive alternative in maximizing plant profitability.

Introduction

Using rigorous first principle process models to optimize the operation of olefins plants has been applied successfully at various operating plants worldwide. These optimizers play an important role in maximizing the profitability of the plants. The majority of these optimizers use SPYRO from Technip to model the cracking coil and the rest of the process is modeled using readily available rigorous models from the simulation software. These optimizers also use equation-based modeling technology. The typical size of such a process model representing a liquid cracker is well over 150 thousand equations and the number of non-zeros in the Jacobian matrix is about 2 million depending on the number of furnaces and components used in various sections of the process model. During the optimization cycle, two cases are typically solved; the first case is to tune the model to match the current operating conditions and the second case is the actual optimization case starting from the tuned process model. A model of this size took about an hour to complete the whole cycle with a very high end OpenVMS server five years ago. In the past 2 to 3 years, many optimizers were implemented with PC as the main computing platform. Also, the speed of the processor continues to improve dramatically. With a high end PC today, the optimization cycle can be completed in about 30-40 minutes depending on problem size. One main benefit of using the PC as the computing platform is the cost of the hardware. The cost of the hardware is not a significant fraction of the total cost of the optimizer anymore. However, the total cost of such an optimizer is still a significant investment.

One important task that has to be continued to achieve the economic benefit of the system is the maintenance of the optimizer post commissioning in order to maintain a high service factor. The maintenance work can take between 20 to 40 percent of a process engineer's time depending on the robustness and offline application of the model. The process engineer who was involved in the original optimization project typically performs this task. To be able to work with the optimizer effectively, several key skills are needed. These are as follows:

- Familiarity of the actual process and the control configuration (basic and advanced)
- Understanding of the process model representing the plant
- Familiarity of equation-based technology and the solver used to solve the equations
- Understanding of the modeling software
- Understanding of the online closed-loop configurations and how the optimizer interacts with the advanced controller

During commissioning of the optimizer, the project team typically strives to push the online model to the highest state of robustness possible by building many validity checks on key process tags that are brought in from the DCS or plant historian to fit the model parameters. In addition, for plants that can process various feed types, there may be a need to build initialization files so that the model can move from one feed scenario to another without any human intervention. The online model needs to be as robust as possible in order to maintain a high service factor thus delivering benefits to operating plants.

Furnace Optimizer

The concept of a furnace optimizer is not new. There have been a few such optimizers implemented around the world. Most of them were employed in gas crackers. The reported results were good and the furnace optimizer definitely delivered economic benefits to the plant. The furnace optimizer essentially only optimizes the furnace severity/conversion due to the rigor of the modeling approach taken. For the furnaces, SPYRO is used to represent the cracking coil. For the rest of the plant, only simple flow splitters, component splitters and mixers are used in the process model. Hence, the model only performs material balance for the plant sections after the furnaces. In addition to the material balance, key constraints of the process that are not captured by the process model have to be represented in the optimizer. The linear gain model from the advanced controller (DMCPlus) is typically used to capture such constraints. Since simplification is needed in implementing furnace optimizer, there can be several variations of the furnace optimizer. That is, the rigor of the model after the furnaces can vary from one optimizer to the other depending on the key active constraints.

Keiyo's olefins plant is furnace limited during decoking and back end limited with all furnaces in operation. This plant is the largest and one of the most modern olefins plants in Japan. The primary feedstock is naphtha. The naphtha composition changes regularly. Hence, there is a strong need for the automation of setting furnace severity while running the backend of the process as close to the process constraints as possible. With the demand of ethylene and propylene changing so rapidly, the plant quite often runs in maximizing propylene mode while maintaining a certain desired ethylene production rate. However, the plant can be backend limited if all of the furnaces are online. Keiyo decided that significant benefits could be captured by implementing a furnace optimizer.

For the furnace optimizer implemented at Keiyo Ethylene's olefins plant, SPYRO was used to model the cracking coil. The rest of the process was represented using simple component and flow splitters, mixers and linear gain models from DMCPlus. As one of the key initiatives in improving the plant profitability, DMCPlus and Feed Maximizer technology from Aspen Technology were implemented providing the backbone for implementing the furnace optimizer. DMCPlus is a multi-variable controller software package and Feed Maximizer is the composite linear program that combines all of the key DMCPlus controllers to maximize the feed to the furnace while honoring all of the key constraints in the backend of the process. More information about these technologies can be found in several papers published in the literature. For the furnace control, SPYRO was also used to predict furnace severity. For the naphtha feed, there is an online analyzer providing a carbon number based PINA composition of the feed. This is obviously very important; otherwise, there is no point in using SPYRO to predict furnace severity since the feed composition affects the furnace cracking prediction directly. The principle of garbage-in-garbage-out still applies. Although furnace effluent analyzers are present, they are not used to update the model in a closed loop fashion. The overall structure of the system is shown in Figure 1. A+ Optimizer is the equation-based environment within the standard AspenPlus modeling tool. This is the modeling tool that was used to implement the furnace optimizer. The optimizer is a layer on top of the advanced controller.

Implementation Details

While the controller was being implemented, the process model needed to implement the furnace optimizer was developed. It is very important that both of the SPYRO models (control and optimization) use the same furnace tuning parameters. Since the problem size of the furnace optimizer was very small compared to a typical full-scale rigorous optimizer, the optimization cycle could be completed in less than 5 minutes using a Pentium 3 processor running at 1GHz with 1GB of memory. The furnace optimizer has about 15 thousand equations. Averaged values of process tags for all of the furnace related process measurements were used. For the backend of the plant, the steady state predictions from DMCPlus linear program were used as measurements as shown in Figure 1. Since the furnace optimizer was configured to wait for only 30 minutes after implementation of new targets, it was not possible to see the changes in the backend of the process if the actual process measurements coming from the backend of the plant were used. Also, using the DMCplus steady state predictions allowed the typical steady state check which starts the optimization cycle to be tuned very loosely. The optimizer was able to send more than 40 solutions a day to the controller. During the commissioning of the furnace optimizer, findings indicated that it was imperative to include feed forward effects to the backend process variables for changes in furnace severity in the controller; otherwise, the feed maximizer would cut too much feed and put the feed back again later. Also, since the furnace optimizer ran at a much higher frequency compared to a rigorous optimizer, if there was any inconsistency between the controller and the optimizer, the problem would manifest itself quickly when the optimizer was switched on. This was particularly true as the plant was approaching the critical constraints.

Besides being consistent in process active constraint prediction, the behavior of the optimizer must be consistent with the feed maximizer as well when increasing or decreasing coil outlet temperatures. The coil outlet temperature effects of different furnaces were balanced in the optimizer just like the feed maximizer. That is, coil outlet temperatures were increased or decreased in groups. Also, the optimizer was configured to cut feed as well when key constraints were violated because the feed maximizer would do the same thing. Feed rates to furnaces were also balanced the same way the feed maximizer managed the furnace feed rates.

Results and Discussions

The furnace optimizer delivered the nonlinear trade-off between severity and yield while honoring the key constraints in the process. The feed maximizer alone could not deliver the nonlinear trade-off benefit since it was solving the linear problem. In addition, the combination of the furnace optimizer and the feed maximizer allowed more constraints to be pushed simultaneously. The feed maximizer alone can only adjust feed rates. The severity adjustments provided by the furnace optimizer allowed more benefits to be achieved. Also, since the furnace optimizer ran at a higher frequency than a typical rigorous plant wide optimizer, it was able to respond to feed composition changes quickly. If the optimizer was switched off, one would see that the operators were much more conservative in setting the furnace severity while pushing the plant constraints during naphtha tank swap.

The furnace optimizer required much less engineering effort in configuring the process model than a typical rigorous plant wide optimizer. Also, the process model can be upgraded to a full-scale rigorous optimizer as the need arises on a step-by-step approach. This makes it possible to balance the implementation cost and the potential benefit. The offline benefit of the furnace optimizer is clearly limited compared to the one delivered by a rigorous optimizer. Findings also indicated that the furnace optimizer requires much less maintenance since most of the process models are splitters and mixers. The overall process model obviously has less complexity due to simplification of the separation sections of the process. Very few convergence problems were encountered during the commissioning of the optimizer.

Since the furnace optimizer used the controller gain model in predicting backend constraints and the steady state prediction of backend process variables were used in the furnace optimizer model as measurements, it was very critical that the controller gain model was accurate. For instance, if the steady state prediction were cycling, the move calculated by the optimizer would cycle as well. As mentioned earlier, the furnace optimizer ran at a higher frequency than a rigorous plant-wide optimizer. These cycling problems would have shown up quickly if they did exist during commissioning. Due to the tight integration between the furnace optimizer and the controller, experience shows that it is critical that both the advanced control team and the optimization team to be at the site at the same time during commissioning.

Figure 2, 3, 4 and 5 have about 3 days of operating data at Keiyo ethylene plant. All figures cover the same period of plant operations. During that period, the furnace optimizer was switched on. The furnace severity used in the figures is the effluent propylene to ethylene (P/E) ratio. In light of confidentiality, plant naphtha feed rate, ethylene production rate, CGC power, naphtha specific gravity and P/E ratio were scaled using typical nominal value. The plant had 10 furnaces running initially and one furnace was brought down for decoke at the end of operating data shown. There were two feed naphtha composition swaps in the data as shown in Figure 4 where naphtha specific gravity increased twice. Figure 2 shows the severity adjustment made by the furnace optimizer when the plant was approaching the feed constraint in the hot distillation area.

When there was no room in the hot distillation area, the furnace optimizer reduced the P/E ratio to maximize ethylene production. As indicated in Figure 2, the high limit of the hot distillation feed was raised. The furnace optimizer immediately took advantage of the extra capacity by increasing P/E ratio to maximize propylene production and hence pushing more gasoline feed to one of the columns in the hot distillation area. Figure 3 shows the charge gas compressor (CGC) power as the furnace severity was adjusted by the optimizer. Looking at both Figure 2 and Figure 3, it was clear that the optimizer moved the plant to approach both the hot distillation section feed constraint and the charge gas compressor constraint. Ethylene production is shown in Figure 4. When the plant was approaching the hot distillation section feed constraint, the furnace optimizer increased ethylene production by reducing P/E ratio while maintaining the plant at or close to the hot distillation area feed high limit. Plant total naphtha feed is shown in Figure 5. Working with the CLP, the total feed naphtha was maximized while the backend constraints (hot distillation section feed and CGC power) were active. Also shown in Figure 5 is the total charge prediction from the furnace optimizer and the actual total charge. The optimizer prediction agreed with the actual charge quite well. This illustrates that the optimizer's prediction on backend constraints was very good and the prediction was consistent with the control (DMCPlus). The prediction of two major backend constraints was shown in Figure 2 and Figure 3. DMCPlus' steady state prediction and the optimizer prediction were very close to each other. This is a key factor in the success of the furnace optimizer guiding the plant successfully to its constraints. During the period of 10-furnace operation, the plant was either very close to or at both the hot distillation section feed constraint and the charge gas compressor power constraint. When a furnace was brought down for decoke, the plant was furnace limited. Then, the furnace optimizer increased the P/E ratio to increase the yield of propylene.

Conclusions

The furnace optimizer was implemented successfully at a naphtha cracker. There is an upgrade path for the optimizer as the need arises to include more rigor in the backend of the process. The payback of the implementation cost for the optimizer is very good even in today's investment requirements demanded by many manufacturing companies. For olefins plants that have many feed changes and are operating against back end constraints strongly affected by cracking severity, the furnace optimizer is an attractive option to consider. The furnace optimizer was able to capture more than 80 percent of the potential benefits of applying rigorous online optimization associated with severity adjustment. These benefits were captured with a significantly lower implementation cost and maintenance requirement. The furnace optimizer was not able to capture energy conservation benefits due to the simplified models employed for the recovery section of the ethylene plant.

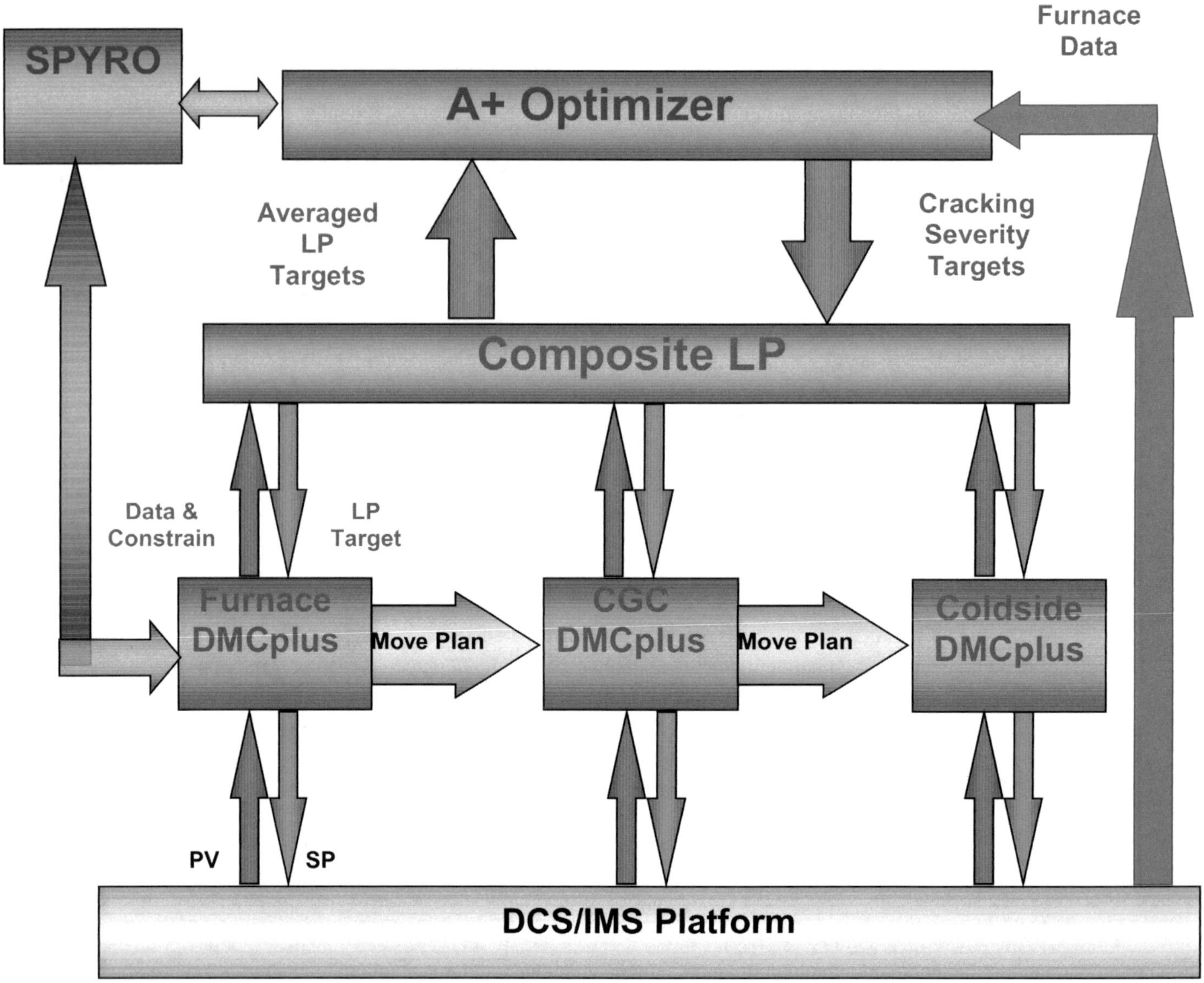

Figure 1 Overall Structure of Furnace Optimizer

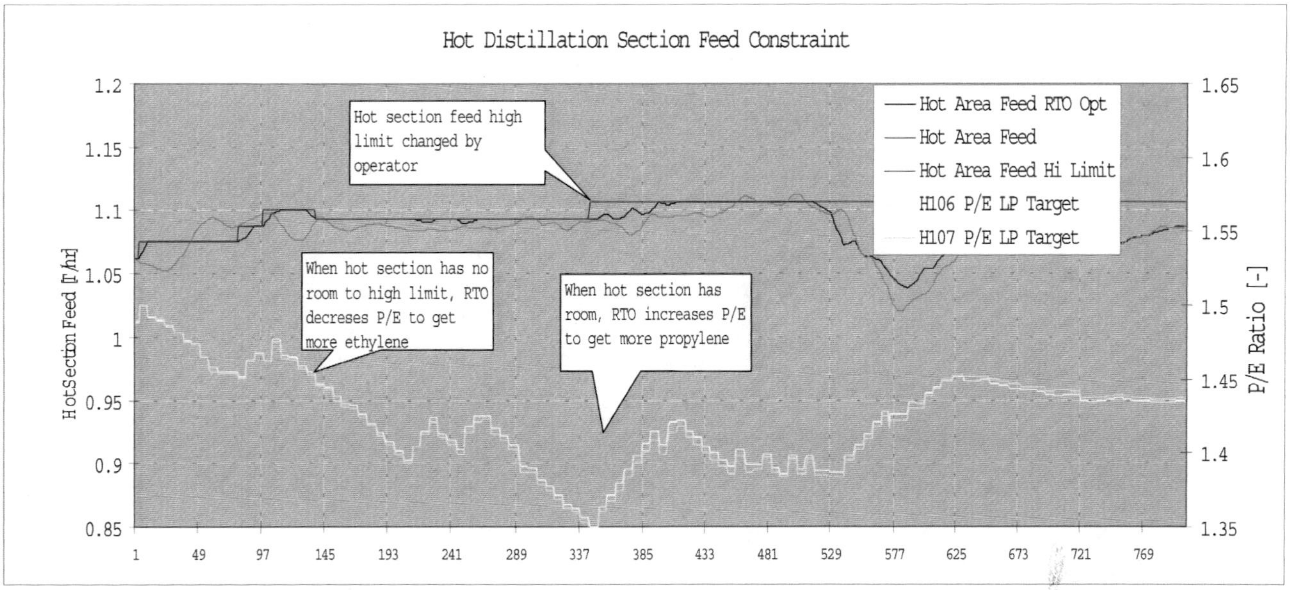

Figure 2 Hot distillation section feed constraint and furnace severity

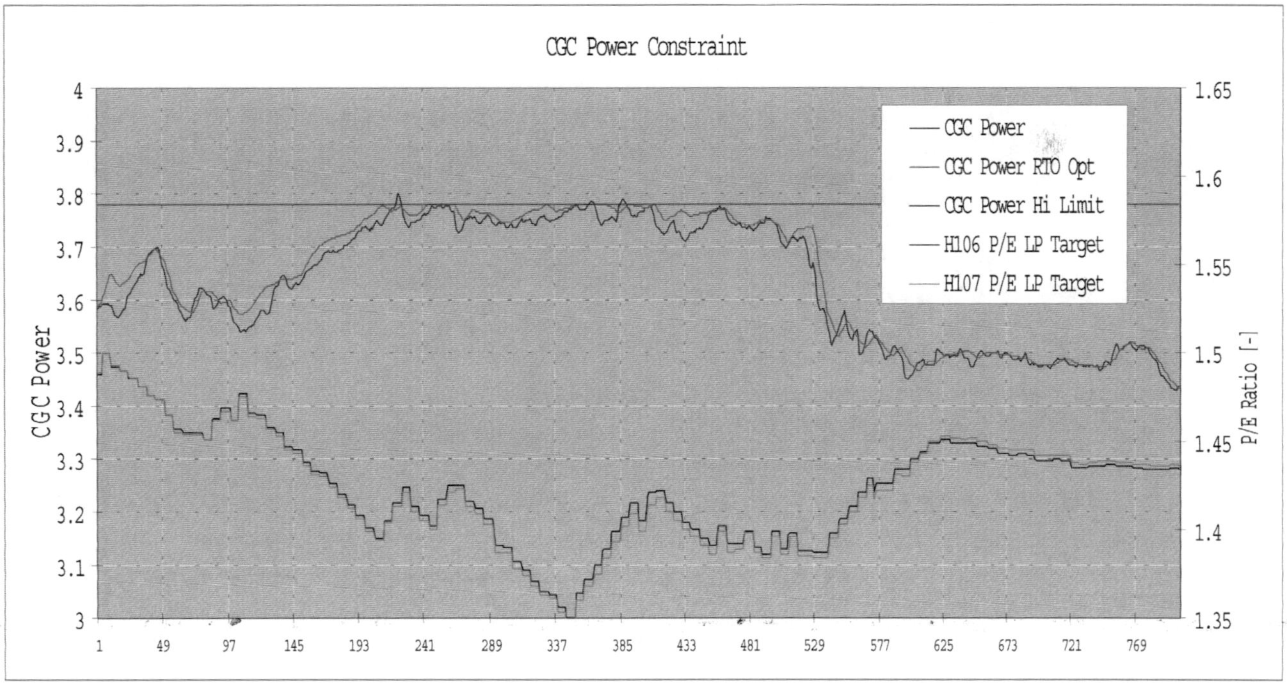

Figure 3 CGC power constraint and furnace severity

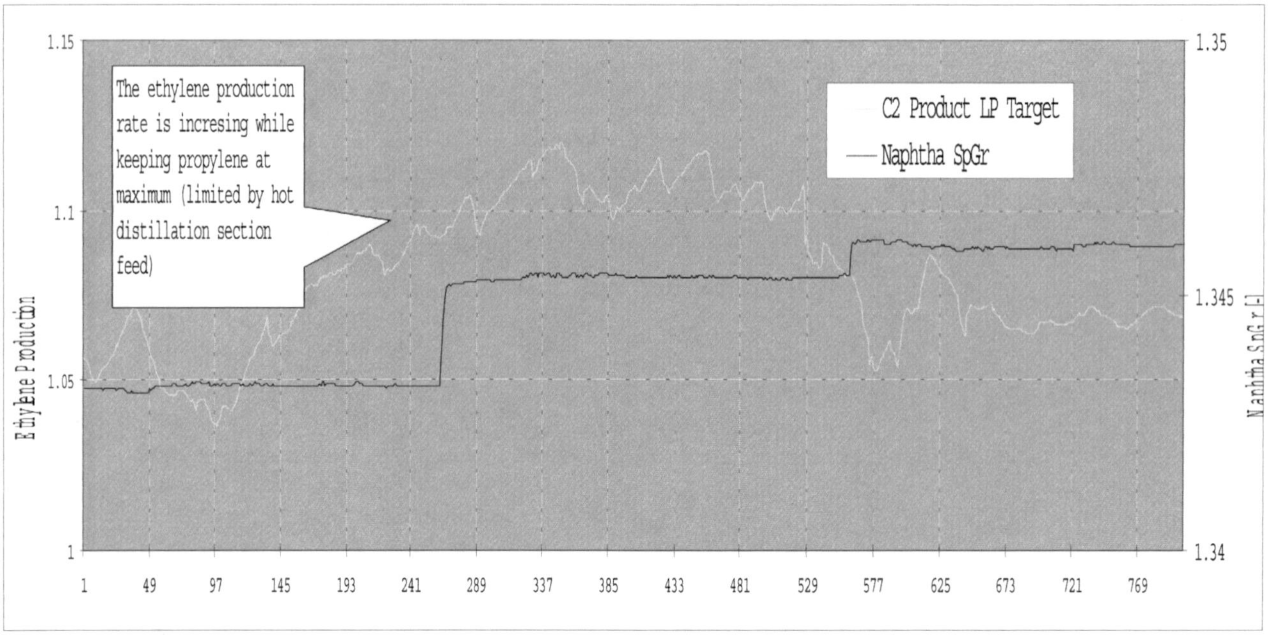

Figure 4 Plant ethylene production and feed naphtha specific gravity

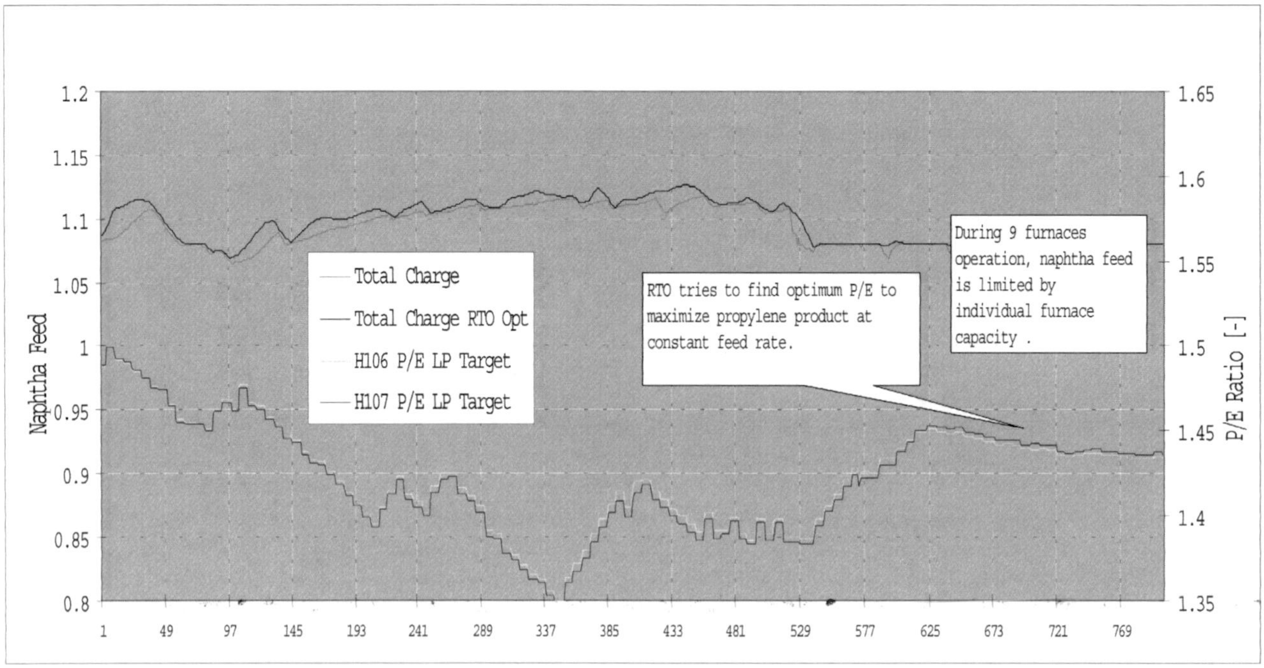

Figure 5 Total plant naphtha feed rate and furnace severity

Acknowledgement

The authors would like to thank the management of Maruzen Petrochemical especially Mr. Kensuke Iuchi for providing us the opportunity to work in both the control and optimization projects and allowing us to share the implementation details of the projects. Also, the operations stuff at Keiyo Ethylene played a key role in making the projects successful. Mr. Akio Miura, Mr. Ryushi Toki and Mr. Katsuyuki Funahashi were all very instrumental in many technical aspects of the projects.

After the Alarm Rationalization – Managing the DCS Alarm System

Jamie Errington, P. Eng.
Human Centered Solutions

Tim DeMaere, P. Eng.
NOVA Chemicals®

Dal Vernon Reising, PhD
Honeywell

Prepared for Presentation at the 2004 Spring National Meeting, New Orleans LA, Apr 25-29
16th Annual Ethylene Producers' Conference
Session T8009 – Ethylene Plant Process Control

Copyright © Human Centered Solutions, NOVA Chemicals Corporation, Honeywell
April 2004

UNPUBLISHED
AIChE shall not be responsible for statements or opinions contained in papers or printed in its publications.

Abstract

EEMUA[1] has issued recommendations for alarm system performance targets that might appear overly aggressive to some in the petrochemical industry. Results from an Abnormal Situation Management®[2] Consortium study that benchmarked alarm system performance across over 30 member sites suggests that alarm rationalization and achieving EEMUA recommendations for the number of configured alarms is not the sole driver - or "silver bullet" - of solid alarm system performance. NOVA Chemicals, one of the participants in the study, has been able to achieve those performance targets in several plants. This paper discusses what efforts have gone into achieving those alarm system performance targets in NOVA Chemicals' Ethylene 3 plant. While NOVA Chemicals has performed alarm rationalization on these exemplary performing consoles in Ethylene 3, perhaps as important have been the continuous improvement / lifecycle maintenance activities that NOVA Chemicals has performed since the rationalization and which will be described in the paper.

---

[1] *EEMUA Publication No. 191 – Alarm Systems A Guide to Design, Management and Procurement* (1999). London: EEMUA.

[2] ®ASM and Abnormal Situation Management are U.S. registered trademarks of Honeywell International, Inc.

I. Introduction

Alarm Management has been a problem for control room operations for many years, gaining in notoriety since the introduction of the distributed control system in the early 1980's. It is only now becoming a system that is receiving the attention it deserves due to an increased understanding of the impact of alarm management on plant performance, reliability and safety.

The Abnormal Situation Management® (ASM®)[3] Joint Research and Development Consortium is a group of 11 companies and universities that are finding approaches and solutions to mitigate – or where possible avoid – abnormal situations from occurring in production facilities. Abnormal Situation Management is often described as an integrated approach to improve reliability and operational integrity. Major aspects of the ASM Consortium's work has included (a) prevention, (b) early detection, and (c) mitigation of abnormal situations in industrial facilities (including petrochemical, refining, oil and gas processing, power generation, pulp and paper plants).

Alarm management is one area of particular interest and significant research for the ASM Consortium because it was the initial driving force for the Consortium's formation in 1993 as well as NOVA Chemicals Corporations' ("NOVA Chemicals") participation. Recently the Consortium has added to its detailed knowledge from prior research about alarm management with the completion of three major activities:

1) A two-phase alarm performance benchmarking study across 30 of its members' plants,
2) A human performance modelling project using high fidelity plant simulators,
3) An Alarm Management Effective Practices guideline from the collective experiences of the Consortium members.

This paper will show some of the alarm performance benchmarking results and some of the initial conclusions that the research team has been able to derive from them. The ASM Consortium will be continuing its research in this important area.

One of the participants in the benchmarking study was NOVA Chemicals' Ethylene 3 plant in Alberta, Canada. This world scale plant has been able to demonstrate that the EEMUA performance goals can be met for months at a time, though alarm performance issues still present challenges for unplanned major plant disturbances.

Ethylene 3 started production in Aug. 2000 and is capable of producing 2.8 billion pounds of ethylene each year. It is a facility jointly owned by Dow Chemicals Canada Inc. and NOVA Chemicals and operated by NOVA Chemicals along with other ethylene, polyethylene and utilities plants at its major Joffre, Alberta facility.

NOVA Chemicals has presented 2 papers[4,5] at previous AIChE conferences which described how it used a "Designing for Abnormal Situation Management" approach to several elements of its control systems including control room design, operator console, operator user interfaces. Alarm Management was a part of that project and has been essential to the success of the operator user interface.

---

[3] ®ASM and Abnormal Situation Management are U.S. registered trademarks of Honeywell International, Inc.

[4] Errington, J. & Bullemer, P., *Designing for Abnormal Situation Management*, AIChE 1998 Ethylene Producers Conference, New Orleans, LA.

[5] Errington, J. & Nimmo, I., *Designing an Ethylene Plant Control Room and Operator User Interface Using Best Practices*, AIChE 2001 Ethylene Producers Conference, Houston, TX.

II. Role of Alarms in Situation Awareness and Response

Of the many roles and responsibilities of console operators, perhaps the most challenging and time consuming is that of situational awareness and response. In this role the console operator is proactively monitoring the plant "to establish and maintain their awareness of the current state of the process relative to plant production goals, process operational limits and operating constraints."[6]

The states or modes of the plant can vary from normal, planned off-normal, abnormal or emergency shutdown. In normal mode the objective of the operator is to monitor the process and be aware of potential problems that can rapidly move them to a different state. In normal modes the operator is also looking for suboptimal conditions that require attention and adjustment to optimize the process. In a planned off-normal period, the operator will be executing the planned procedural operation and their duties are to monitor the process to ensure that the correct trajectory or plan is followed. "Once an abnormal or emergency situation is detected, the console operator is typically the focal point in the efforts to mitigate, compensate or recover from the undesirable situation."[7]

Alarms should play a key role in the situation awareness and response activities of console operators though the role of alarms needs to vary to match the different plant states to meet the changing needs of the operator. Designing the alarm system for the different plant modes and operator needs allows it "to assist the operator in detecting process problems and prioritizing their response."[8] Table 1 outlines the varying roles of the operator and the alarm system to the different plant states.

Plant State	Operator's primary role	Alarm system's primary role
Normal	Plant monitoring, optimization of production, and coordination of maintenance.	Inform operator of minor operating adjustments needed
Planned off-normal	Manage and coordinate the execution of the changes as the plant may be transitioning to a new production grade, starting up or shutting down pieces of equipment	Identify to the operator that execution is NOT occurring as planned or required
Abnormal situation	Situation management by intervening and augmenting the regulatory control system actions when the control system can no longer cope with the situation	Identify what is wrong and when operator intervention is required
Emergency shut-down	Ensure a safe shutdown	Identify when safety actions are required

Table 1: Roles of Operators and Alarms for Differing Plant States[9]

[6] Bullemer, P., Reising, D.V., Hadjukiewicz, J, and Errington, J. *Interaction Requirements Methodologies – ASM Consortium Technical Report*, Internal ASM document. Jan. 2004. p. 21

[7] Bullemer, P. (2004) p. 21

[8] Campbell Brown, D., & O'Donnell, M. (1997). *Too much of a good thing? Alarm management experience in BP Oil. Part 1. Generic problems with DCS alarm systems*. IEE Colloquium on Stemming the Alarm Flood (Digest No: 1997/136), pp. 5/1-5/6. 17 June.

[9] Table was developed in part from information from Campbell Brown, D., & O'Donnell, M. (1997). *Too much of a good thing? Alarm management experience in BP Oil. Part 1. Generic problems with DCS alarm systems*. IEE Colloquium on Stemming the Alarm Flood (Digest No: 1997/136). 17 June 1997

III. ASM Alarm System Benchmarking

The ASM Consortium supported the work of the Engineering Equipment and Materials User Association (EEMUA) in the development of its EEMUA Publication 191[10], which had defined some aggressive alarm system metrics that needed to be confirmed as being attainable. These levels included recommendations such as:

- Less than 1 alarm/10 minutes average alarm rate during normal operations; and
- Less than 10 alarms in first 10 minutes of a major plant upset

In 2001, the ASM Consortium initiated a series of research projects to benchmark the performance of alarm systems in member's facilities throughout North America. The motivation for this work was to assess whether members were achieving the EEMUA recommendations for alarm system performance as well as to contextualize that performance relative to plant complexity, degree of alarm rationalization, nature of plant operation (batch, semi-batch, continuous), and so on.

The study was conducted in two phases. In the initial stage, the participating companies agreed upon the set of metrics, duration of the data collection periods and committed to providing at least 3 months of data to Honeywell for each console in the study. Honeywell collated and analyzed the information and presented the results anonymously back to the project team. Participants in the 2nd phase were asked to provide additional console data and also anecdotal information about their alarm management practices which was used to examine the possible critical elements to explain the varying alarm performance results.

The project collected a total of 90 sets of monthly data from 37 unique consoles (three consoles appeared in both studies, which accounts for the 40 entries in Figures 1 and 2). Each console represented a portion or area of a plant that was typically operated by a single console operator under normal plant conditions. Some of these consoles will add additional console operators for planned major events such as a plant start-up. Static alarm configuration information was collected for all consoles and included number of analog points, number of control valves, digital points, and so on plus all their related alarm configuration information.

Dynamic alarm information was also provided from each console for the monthly periods. The dynamic information represented elements such as average and peak alarm rates over 10 minute windows. Figure 1 shows one example of that information in a plot of the overall average alarm rate from the 40 consoles.

Each bar represents the lowest, average and highest monthly average alarm rate for a single console that submitted 3 months of data. Single points represent consoles in the second phase that were only required to submit one month of data. The consoles have been grouped according to the proportion of configured alarms in each console's control system versus the EEMUA configured alarm recommendations (see Figure 41 on page 104 of EEMUA Publication No. 191). For example, the first 2 consoles at the left side of the chart had configured alarm totals of over five times that recommended by the EEMUA guideline. On the right hand side of the chart is a scale representing the 5 alarm performance categories described in Table 2 below.

[10] *EEMUA Publication No. 191 – Alarm Systems A Guide to Design, Management and Procurement* (1999). London: EEMUA.

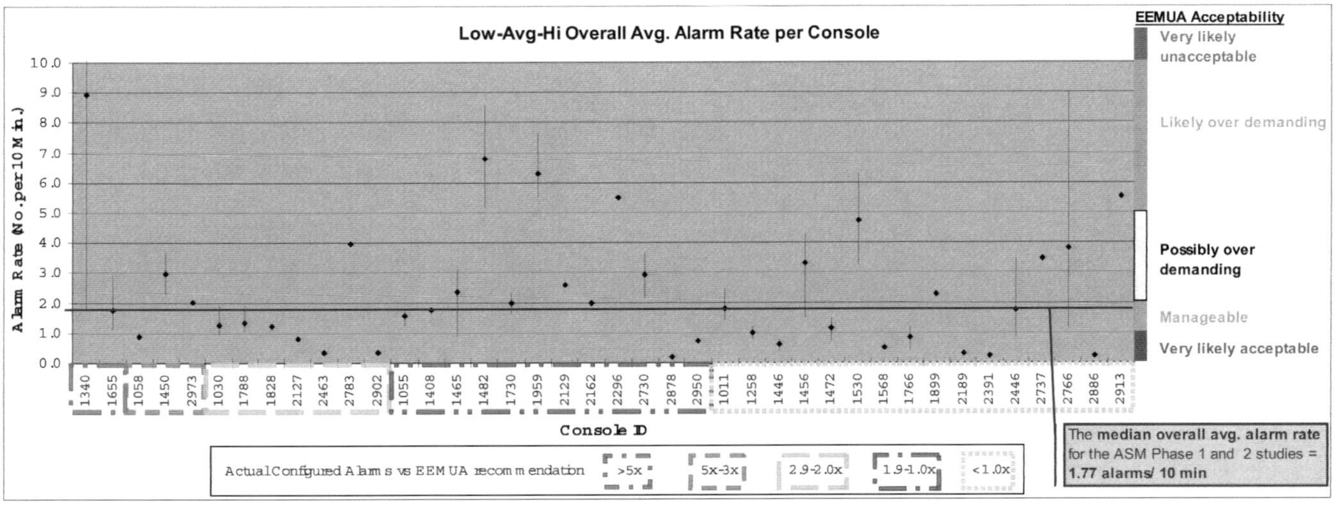

Figure 1: Overall Average Alarm Rate[11]

Long Term Average Alarm Rate in Steady Operation (per 10 minute window)	Acceptability
More than 10 alarms	Very likely to be unacceptable
More than 5 but less than 10	Likely to be over demanding
More than 2 but less than 5	Possibly over demanding (**Note:** *category not defined in EEMUA guide*)
1 or more but less than 2	Manageable
Less than 1 alarm	Very likely acceptable

Table 2: – EEMUA Benchmarks for Assessing Average Alarm Rates[12]

Overall average alarm rates were calculated for the entire month of each data set and included all modes of plant operation (normal, planned off normal, abnormal and emergency). The EEMUA acceptability categories are designed for only normal plant operation, which biases against those consoles in the ASM studies that had non-normal operation during the month. The average alarm rate across the ASM studies was 2.3 alarms per 10-minute period compared to the British Health and Safety Executive (HSE) study of 5 alarms per 10-minute window[13] that led to the EEMUA guidelines document.

What the chart in Figure 1 demonstrates is that the "very likely acceptable" level of performance is being achieved by one quarter of the consoles. The plot also shows that there appears to be very little correlation between the alarm rate and the number of configured alarms in the various consoles ($r = 0.103$, $p > 0.50$).

---

[11] Reising, D., & Montgomery, T. (2004). *Alarm System Performance Metric Benchmarking – Phase 1 and 2 Preliminary Analysis Results*, Internal ASM document. Jan. 2004. p. 4

[12] EEMUA (1999), Figure 42, p. 105.

[13] Bransby, M., & Jenkinson, J. (1998). *The management of alarm systems*. HSE Contract Research Report 166. Suffolk, UK: HSE Books.

The chart in Figure 2 displays the average peak alarm rate distribution to examine how much of the time each console was experiencing a specific alarm rate.

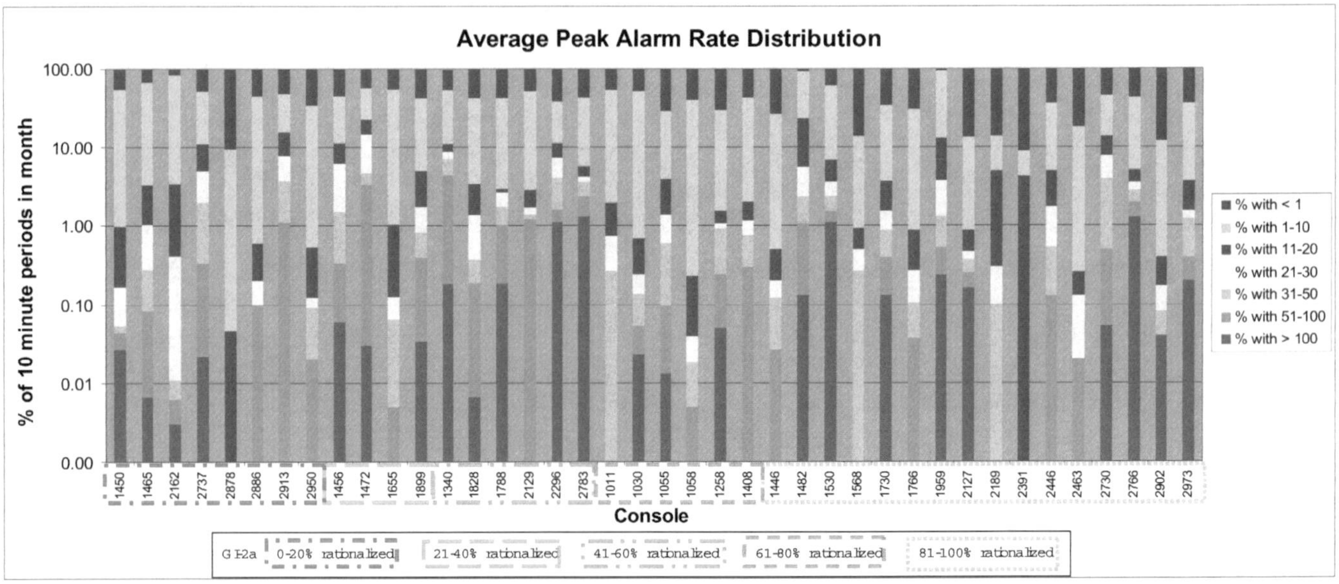

Figure 2: Average Peak Alarm Rate Distribution[14]

The contribution chart was generated by averaging values across each console's set of thee month data, if applicable, for the Peak Alarm Rate Distribution intervals (e.g., "% with < 1", "% with 1-10", and so on). The % of 10 minute periods for the various alarm count intervals are plotted on a Log scale to highlight those periods in which there were peak alarm rates and increased operator workload (e.g., Consoles 1530, 2296, 2766, and 2783 clearly had at least one *bad* day, on average, per month). However, care is needed in interpreting the Log-plotted results. As a general rule, bar lengths can not be compared between consoles (e.g., Console 2878 has 9.5% for 1-10 alarms and this colored bar looks as long as Console 1730's 1-10 alarms bar, which is actually representing 34.0%.

Figure 2 shows that 24 of the 40 consoles (60%) had a least one significant alarm flooding incident (over 100 alarms in a 10 minute window). In the case of 4 consoles, they averaged over 1 % of their time per month (over 7 hours per month on average) when their alarm rates were exceeding 100 alarm messages in each 10 minute window.

From a more positive aspect, 2 consoles had 3 months in which they never exceeded a peak alarm rate of 20 alarms per 10 minute window, which indicates that the EEMUA standards for peak alarm rates, though very aggressive, are within reach.

One other aspect that is highlighted in this plot is that the degree of rationalization, which has been used to group the consoles on the X axis, appears to have only minor relevance to the actual peak alarm rates. (Each of the plants were requested to indicate the percentage of alarms rationalized on that console (e.g., 0 – 20%, 21-40 %, and so on).)

[14] Reising, D.V. *Alarm System Performance Metric Benchmarking – Phase 1 and 2 Preliminary Analysis Results*, Internal ASM document. Jan. 2004. p. 4

"Some of the key findings from the Benchmarking Study include:

- Sites can meet and sustain the "Manageable" and "Very Likely Acceptable" EEMUA levels of overall average alarm rate performance in normal operation
- 70% of the consoles, on average, had overall average alarms rates of more than 1 alarm every 10 minute window (i.e. had more alarms than the "Very Likely Acceptable" level)
- Top ten Worst Actor tags often account for over 50 % of the alarm rate
- Alarm bursts (greater than 30 alarms per 10 minute window) occurred at least one time, on average, each month for 38 of the 40 consoles during the evaluation period
- Significant periods of peak alarm rates exceeding 100 alarms per 10 minute window were experienced at least once by 60% of the consoles during the evaluation period
- The degree of alarm rationalization or configured alarm levels does not appear to be the single significant factor in determining alarm rate performance."[15]

What the benchmarking information reveals is that the initial alarm specification and rationalization activities in the design of the alarm system which limit the number of alarms configured in the system do not guarantee good alarm system performance and conversely, good alarm system performance can be achieved without having a fully rationalized console. To better understand how a plant can achieve the recommended alarm performance levels, the following sections of this paper will examine the work that is underway at NOVA Chemicals' E3 facility where they have been able to consistently meet the recommended alarm levels.

IV. Alarm Management Lifecycle

Alarm management is not a one time project that can be executed and closed up in a tidy manner with a close-out report. Instead alarm management requires a continuous lifecycle approach regardless of whether it is a new or an existing plant. Figure 3 shows a simplified diagram illustrating the major elements of the E3 Alarm Management Life Cycle Model.

In the case of NOVA Chemicals' E3 plant, there have been several key elements which have led it to be successful. Some of these aspects were executed as part of the design project while others have been through the continuous improvement efforts, and a third group (Management Principles) has been necessary for all aspects.

Management Principles

From the outset of the E3 project in 1997, there has been tremendous support for an evergreen alarm management approach and the entire "design for ASM" philosophy. This philosophy meant that there was a clear understanding of the roles of the operator, the operator user interface and the alarm management system and how they interacted, as well as a willingness to commit the necessary resources to realize the philosophy. It was certainly not an afterthought by a project team getting ready for start-up to check to see if any alarms or the user interface had been created. It was a core element of the project from the outset.

[15] Errington, J. & Reising, D.V. (2004). *ASM Consortium Guidelines: Effective Alarm Management Practices*, Minneapolis, MN: Honeywell Laboratories

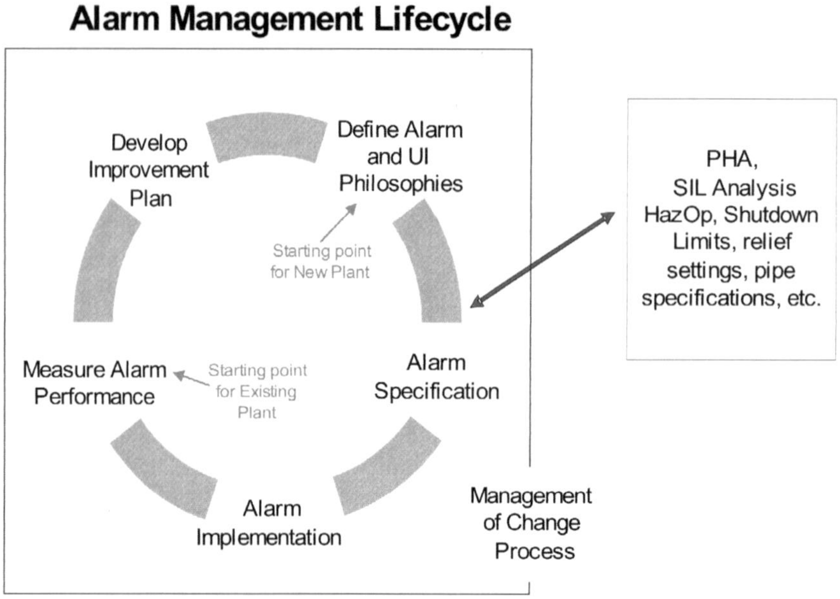

Figure 3: Alarm Management Life Cycle Model

Design and Implementation

The initial project design and implementation aspects of the E3 project included the early development of an alarm and operator interface philosophy. In 1997, NOVA Chemicals was in a unique position to apply the direct concepts about operator user interface designs that had been developed in ASM Consortium research projects and which had tightly integrated alarms into the graphics as navigation drivers. NOVA Chemicals' approaches to alarm management were drawn from the broad experiences of the other user member companies such as ChevronTexaco and BP Amoco. With elements such as these, coupled with management support, NOVA E3 could begin to develop a life cycle approach to alarm management, knowing what the alarm system had to deliver.

The E3 alarm philosophy defined for all plant personnel the expectations of the alarm system, the operators and the supporting tools and applications. Alongside the Alarm Philosophy, a matching Operator User Interface Philosophy was developed that described the principles of the operator interface and the way that alarm information, trends, online information and advanced applications would be integrated. Together these two documents describe the core of the console operator's situation awareness, and response support tool set.

Following the Alarm and Operator Interface Philosophy developments the project focused on capturing the alarm specification information from the series of plant reviews (PHA, SIL Analysis, etc.) and plant design information, as shown in Figure 3. This information was used in the alarm rationalization reviews (also known as Alarm Objectives Analysis) where multi-discipline teams performed the tag-by-tag review to determine the appropriate alarm settings. At the conclusion of this rigorous review process, E3 had significantly fewer configured alarms than the older NOVA Chemicals' E1 and E2 ethylene plants. E1 and E2 then used some of the design approaches implemented in E3 to improve their configurations though due to resource limitations they did not perform the rigorous tag by tag rationalization review. Today E3 still has from 16 to 43 % fewer absolute alarms and over 50 % fewer alarms configured per total points than the E1 and E2 ethylene plants which illustrates some of the difference that the rigorous approach can achieve. All three plants each employ two console operators to manage their alarms.

	Total Points	Total Alarms	Reduction in # Alarms (%)	Total Alarms / Total Points	Relative Improvement (%)	# Controllers	Total Alarms / Controllers
E1	8330	5847	16.6	0.70	51.3	532	11.0
E2	9779	8641	43.6	0.88	61.3	613	14.1
E3	14262	4874		0.34		740	6.6

Table 3: Joffre E1, E2 and E3 Configured Alarms

With the rationalization information captured in an electronic database where it could be retrieved by the operator interface, the alarm implementation process occurred. The implementation included the initial alarm settings along with alarm support applications to address known modal changes in plant operation.

Alarm Performance Monitoring

After the alarm system had been implemented and the plant started up, the performance monitoring of the alarm system began in E3, where any existing plant would enter the Alarm Lifecycle process (see Figure 3). Effective performance monitoring required the selection of key performance indicators for the alarm system. Monthly alarm metrics for E3 were set as follows:

- Average # of Alarms/10 Minute window
- Time in burst conditions (>10 alarms/10minute interval)
- Time weighted Average of Number of Standing Alarms

The average number of alarms for normal operation had been well defined in the EEMUA guide shown earlier in Table 2.

From the ASM Alarm performance research and from the data collected at NOVA, we have seen that alarm rates of less than 1 alarm per 10 minute window are achievable for normal operation. Upsets including major equipment trips or entire plant trips still often create an alarm burst of more than 10 alarms per 10 minute window for one or two periods. NOVA Chemicals is addressing this area and some of these activities will be described later in the paper.

Performance monitoring should also include a worst actors list of tags on a daily, weekly and monthly basis. This is a list of alarms that have occurred and are ranked by their frequency of occurrence. On existing plants, this is often the very first performance area to examine as often the worst 10 or 20 tags can account for a large (>50%) portion of the total alarm rate. A corrective action program that initially focuses on these tags can often make a significant improvement to the overall alarm rates at a console.

With the performance monitoring systems in place, the information can be collected and used to develop a continuous improvement plan that is described in more detail in following section. The plan uses the performance information to assess what types of alarm problems exist in the system and potential attack strategies to address them.

Figure 3 also illustrates how a comprehensive and effective Management of Change (MOC) program is required to manage the entire Lifecycle process. Without change control of the alarm settings, the system will be adjusted to the point where its performance and usefulness to the operators is very low. The MOC process is used whenever plant modifications impact the control system or the operator user interface.

V. Continuous Improvement Plans

The Alarm Life Cycle model represents a continuous improvement program. In the previous section we defined the process that NOVA used for its E3 plant, a new grassroots plant to design and implement its alarm system and reach the performance measurement step in the Lifecycle. Today E3 is an existing or brownfield plant and it continues to use the Life Cycle model. Presently E3 has very good alarm performance under normal and planned off-normal modes of operation. However by monitoring its alarms during major plant upsets, E3 still has opportunities to significantly improve its alarm performance even after its extensive alarm rationalization efforts during the design stages.

Figure 4 shows the average monthly alarm rates for the past 2.5 years in E3 on the 2 control consoles. The months where there have been significant plant upsets or outages stand out clearly in the plot as large spikes in the data against the typical lower base levels of alarms. There is also a significant difference between the more batch-oriented operations of the cracking console as compared to that of the finishing console. The cracking console manages the hot end of the plant (furnaces) and also includes the utilities (i.e., steam, fuel gas, quench water, etc.) units of the plants which present more difficulty in terms of alarm management particularly during plant upsets as utility loads vary.

Caution must be used in comparing one major plant upset against another, as they are never matching events in nature or duration. Both of these factors can significantly effect the alarm distribution and monthly average information.

The major improvement factors that have been implemented after the E3 plant start-up include:

- Improved setting of alarm dead-bands and delay-timers
- Use of difference alarms on redundant measurements
- Use of common maximum or minimum alarms
- Design for purpose alarms (i.e., mass balance, heat balance)
- Increased use of dynamic modal alarming applications and logic programming

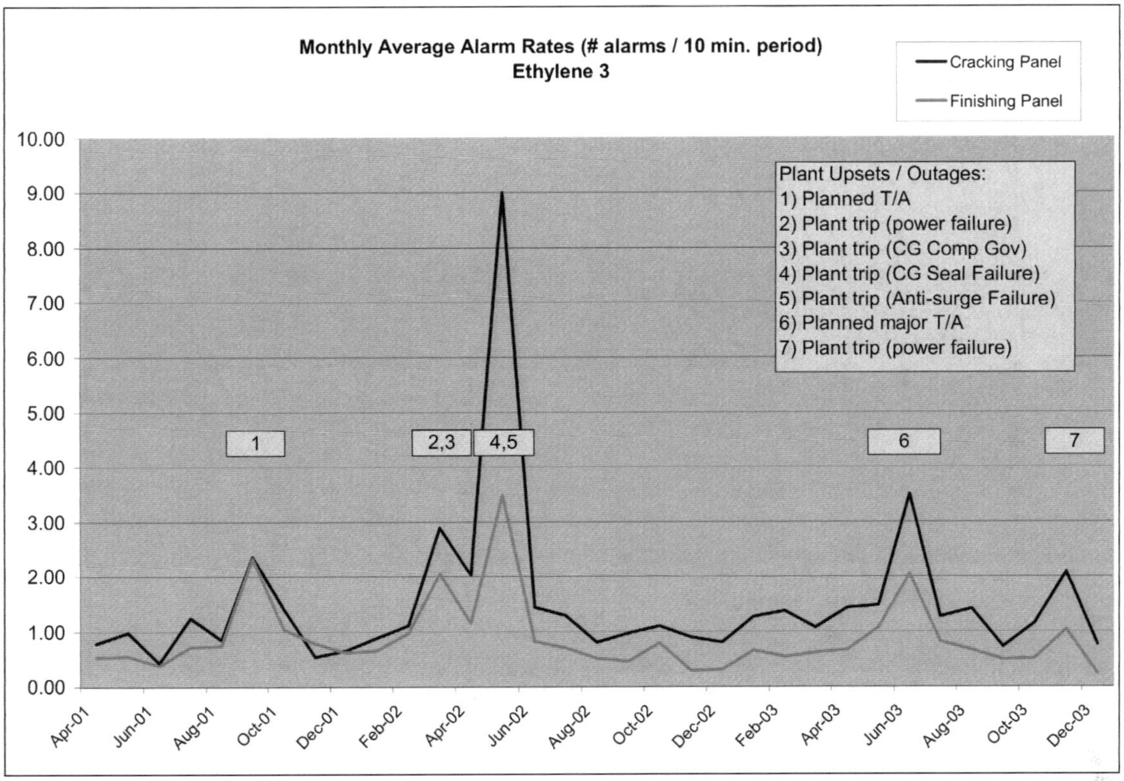

Figure 4: E3 Monthly Average Alarm Rates

Figures 5 and 6 show the alarm rate distributions for the 2 consoles in E3. For both consoles, a significant portion of their time (approximately 58 % for cracking and 69 % for finishing) there are no new alarm messages for the console operator. The peaks shown in Figure 4 from the plant upsets correspond to the extended red bars in Figures 5 and 6 where more than 10 alarms were received for a portion of the 10 minute periods.

However these are still exceptional results especially compared to the ASM benchmarking survey results presented earlier where many consoles were experiencing peak alarm rates 10 times higher than those in E3.

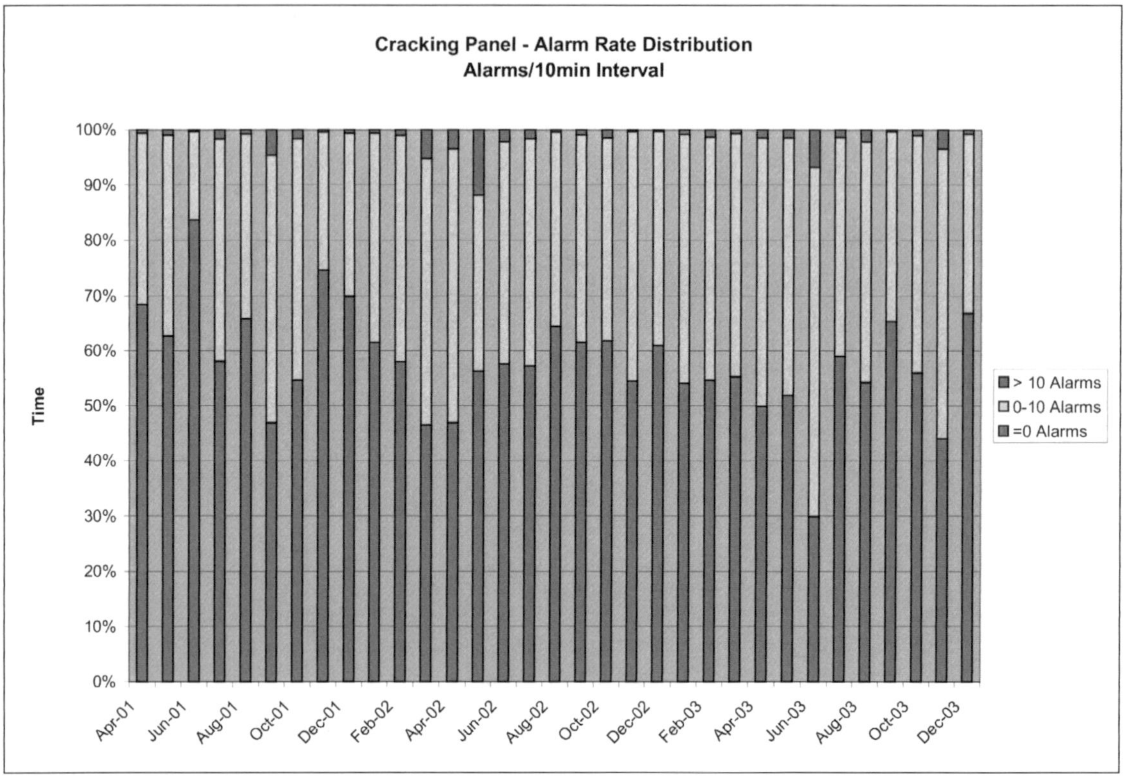

Figure 5: E3 Cracking Console Alarm Rate Distribution

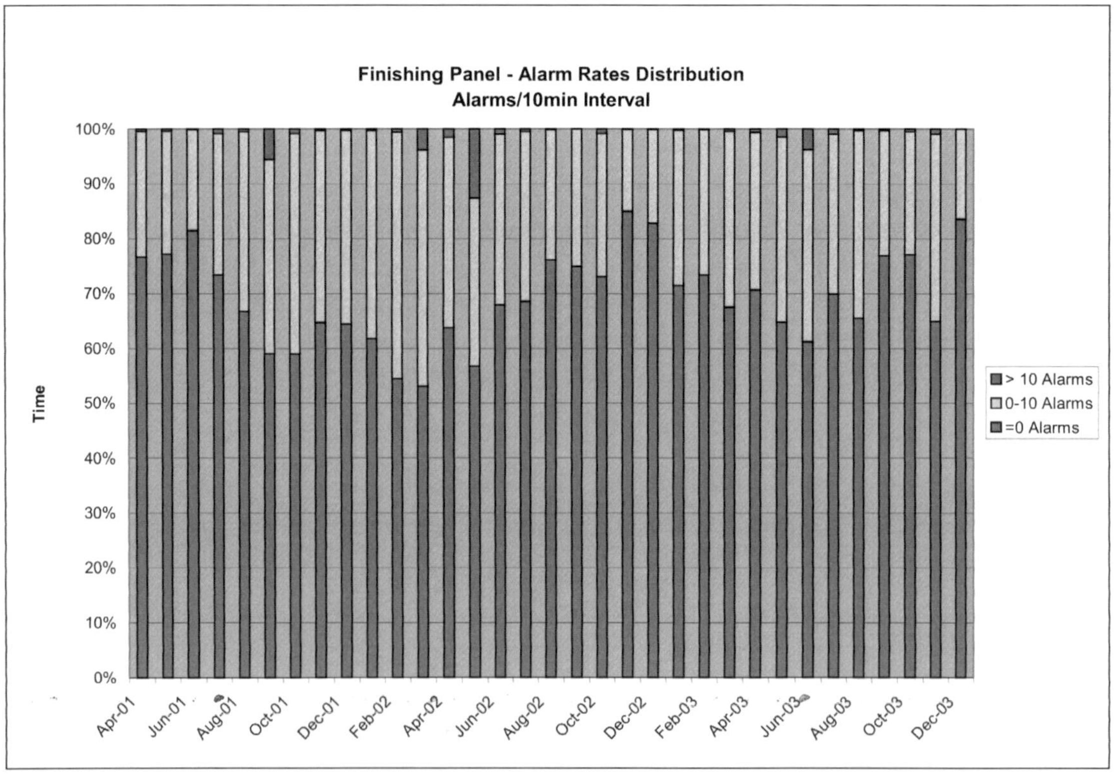

Figure 6: E3 Finishing Console Alarm Rate Distribution

The alarm performance assessment step shown in the Life Cycle model would be the starting point for any brownfield plant that was looking to begin an alarm improvement program. Measure where you are presently at, examine the gaps between your current performance and where you need to be, and then assess what actions will move you to the new state in the easiest and shortest manner.

To achieve the changes to the alarm system may require updates to the alarm and operator interface philosophies. This would be followed by a re-specification of alarms, then by new alarm implementations, which would lead the plant back to the assessment stage of the Life Cycle.

With alarm management – you will never, ever be finished.

Improvement plans can take several forms. For plants that have continuously poor alarm performance results during all operating modes, the worst actors list is often a good trigger for which type of improvement plan is required. If the 10 worst actors continuously represent a significant portion of the alarms (>50 %), then the focus should be on understanding and correcting these specific problems. If the problem is not centered on a small group of tags but is still characterized by high average alarm rates at all times, then the improvement plan may need to consider a broader and much more time-consuming alarm rationalization review.

For other plants that generally have good alarm performance in normal operation like E3, a more detailed improvement plans are required to address the specific events that cause the alarm problems.

Another valuable activity that E3 has employed in its continuous improvement program is the re-validation of its alarm rationalization information. E3 performs an alarm rationalization update on a three-year rotation (33 %/year) using their alarm review team as part of an evergreen safety program. This substantial commitment of resources and effort demonstrates management's support for the Alarm Management at this plant. This revalidation work is in addition to the alarm reviews are undertaken as required by the MOC process for all process changes. The revalidation exercise updates any information that might have been overlooked and acts as a training exercise for individuals who were not involved in the initial plant reviews.

VI Alarm Management Success Factors

Some of the factors that have contributed to the success of Alarm Management in E3 include:

Management Understanding and Support
The efforts to continuously refine and improve the alarm management system require a great deal of time and effort from the Operations and Technical personnel. This would not be possible without very strong management understanding and support.

Operator User Interface and Alarm Management Integration
Performing the best alarm design and rationalization that is possible in a plant will not guarantee good alarm performance. The operator user interface and its components (i.e. graphics, trends, online information, display hierarchy, etc.) must be designed to support and augment alarms by providing the console operators with good situation awareness and response capabilities.

E3 designed the operator interface and alarm management program with just that in mind. Figure 7 shows how the alarm specification and alarm performance information is brought to the operator from a context sensitive menu via a right mouse click on any tag in the operator interface. The operator's confidence and support of the alarm management system are significantly increased when they can have ready access to this information,

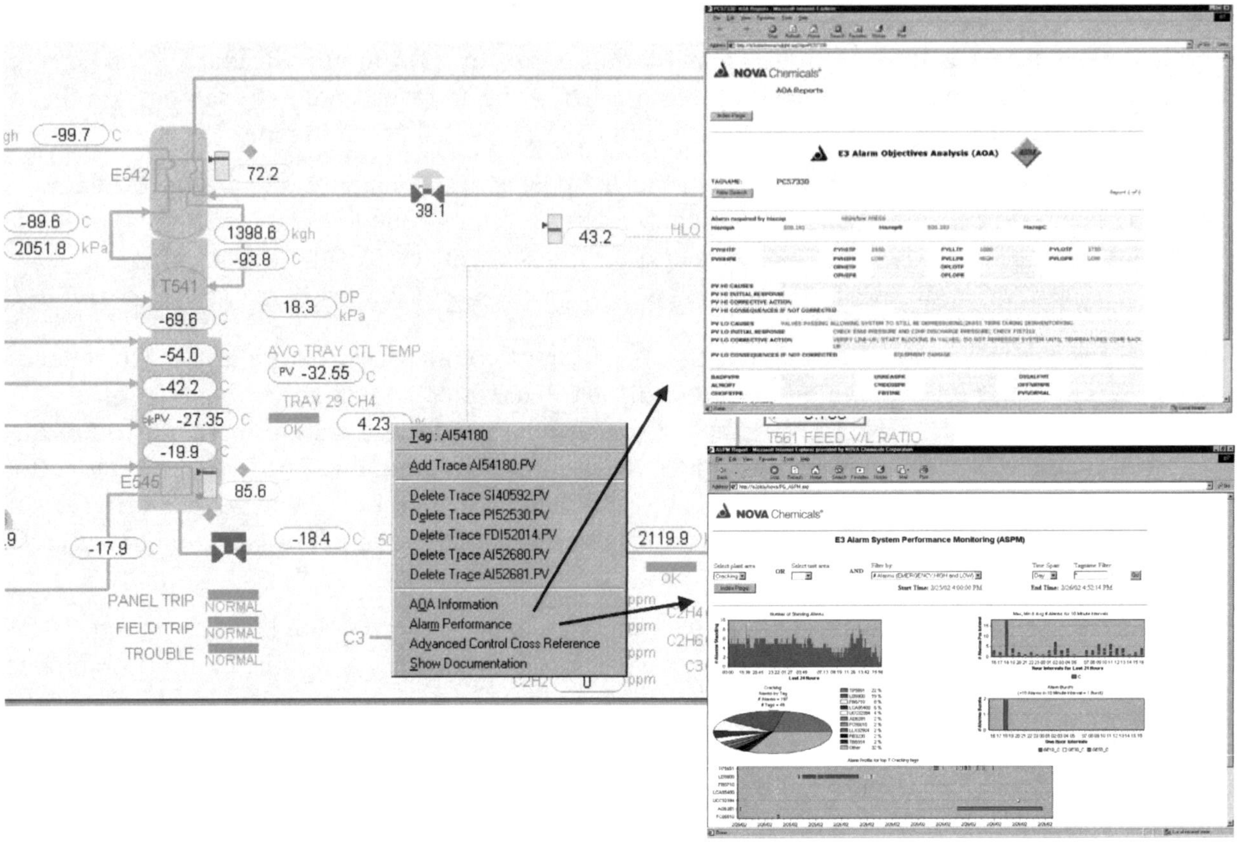

Figure 7: Operator Access to Alarm Specifications and Metrics

Alarm Performance Benchmarking

The alarm performance monitoring developments that have been made with the alarm and event historian have been essential to the continuous improvement efforts at NOVA. Key elements of this have been:

- Worst actors monitoring,
- Average and peak alarm rate monitoring,
- Standing and stale alarm analysis,
- Disabled/Inhibited or Shelved alarm monitoring,
- Excellent Systems Engineering support

Operator Training Program

Operator training is an important element to the continuous improvement plan for alarm management in E3. Operators must be given the knowledge and skills to effectively use the alarm management systems and its support applications. In E3, the dynamic training simulator and the overall training program play a large role in ensuring that all staff understand and respect the alarm management system as a tool to support situation awareness and response. The E3 training program includes:

- Scenario reviews and "what-if" training exercises,
- Use of the situation support tools and applications,
- Development, testing and use of all dynamic modal alarm strategies,
- Use of the alarm summary features during periods of high alarm rates.
- Routine alarm management duties for the console operator (i.e. shift turnover, etc.).

VII. Conclusions and Summary

The ASM Alarm Performance research has shown that there are significant challenges for many plants to meet the EEMUA alarm performance recommendations for both normal and off-normal plant modes of operation. The research has also shown, that though this performance is not being met at most consoles, several consoles in the industry are coming close to or are achieving these levels for a significant portion of their operating periods.

The research and the E3 results also indicate that good alarm design and rationalization are important and will assist a plant in terms of its alarm management, however they are not sufficient or the sole drivers to lead to good sustained alarm performance. Probably more important for long-term excellence in alarm performance is a life-cycle approach that includes performance monitoring and continuous improvement.

VIII. References

Bransby, M., & Jenkinson, J. (1998). *The management of alarm systems.* HSE Contract Research Report 166. Suffolk, UK: HSE Books.

Bullemer, P., Reising, D.V., Hadjukiewicz, J, and Errington, J. *Interaction Requirements Methodologies – ASM Consortium Technical Report*, Internal ASM document. Jan. 2004.

Campbell Brown, D., & O'Donnell, M. (1997). *Too much of a good thing? Alarm management experience in BP Oil. Part 1. Generic problems with DCS alarm systems.* IEE Colloquium on Stemming the Alarm Flood (Digest No: 1997/136). 17 June 1997

EEMUA Publication No. 191 (1999). *Alarm Systems: A Guide to Design, Management and Procurement.* The Engineering Equipment and Materials Users Association ISBN 0 85931 076 0.

Errington, J. & Reising, D.V. (2004). *ASM Consortium Guidelines: Effective Alarm Management Practices*, Minneapolis, MN: Honeywell Laboratories. Jan. 2004.

Errington, J. & Bullemer, P., (1998) *Designing for Abnormal Situation Management*, AIChE 1998 Ethylene Producers Conference, New Orleans, LA.

Errington, J. & Nimmo, I., (2001) *Designing an Ethylene Plant Control Room and Operator User Interface Using Best Practices*, AIChE 2001 Ethylene Producers Conference, Houston, TX

Reising, D.V. (2004). *Alarm System Performance Metric Benchmarking – Phase 1 and 2 Preliminary Analysis Results*, Internal ASM document. Jan. 2004.

Reising, D.V., & Montgomery, T. (2003). ASM Technical Report: Alarm System Performance Benchmarking Study - Phase 1 and Phase 2 Report. Minneapolis, MN: Honeywell Laboratories.

Paper 65e

MAXIMIZING THE YIELD AND THROUGHPUT OF ETHYLENE PLANTS THROUGH ADVANCED TURBOMACHINERY CONSTRAINT CONTROL SYSTEMS

Krishnan Narayanan
Anurag Gupta
Paul Fisher

Compressor Controls Corporation
4725 121st Street
Des Moines, Iowa 50323 U.S.A.
Knarayanan@cccglobal.com

and

Jagadeesan Kumar
Sankara Ramakrishnan
Khalid N. Al-Rushoud

Petrokemya
Al-Jubail, 31961, Kingdom of Saudi Arabia.

Prepared for Presentation at the 2004 Spring National Meeting,
New Orleans LA, April 25-29
16th Annual Ethylene Producers' Conference
Session T8009 – Ethylene Plant Process Control

Copyright © Compressor Controls Corporation
April 2004

UNPUBLISHED

AIChE shall not be responsible for statements or opinions contained in papers or printed in its publications.

Abstract

Control of an ethylene process presents numerous challenges and opportunities because the plant's production capacity depends largely on the availability, reliability, and efficiency of the turbomachines. Among these challenges are: minimizing flare, automation of plant startup, effective surge control for all operating modes; integrated load control, refrigeration compressor quench control, and effective process constraint control. In addition, operating safely at the lowest possible pressure on the suction of the Cracked Gas (CG) compressor to increase ethylene yield; and operating closer to the CGC final stage discharge pressure limit to increase production of the plant are some of the major issues faced by different producers. Turbomachinery control systems with proven capabilities to solve these challenges can have significant economic impact on Ethylene plant operating revenues and bottom line.

The safe operating regime of a turbo compressor is always constrained by the machinery and process limitations. Figure 1 depicts the safe operating envelope for a turbine driven compressor showing the various constraints superimposed on the compressor performance map. Advanced Surge control systems have been applied very successfully in ethylene plants to prevent compressor damage from surge related incidents. However, the same cannot be mentioned about other constraints imposed on a compressor especially due to fast load changes, or process upsets that cause the suction pressure to go rapidly below sub-atmospheric levels or an inter-stage pressure to exceed the pressure rating of machinery casing or process vessels.

The paper discusses solutions to address the aforementioned challenges using Integrated Turbomachinery Control Systems (ITCS). ITCS is a multivariable control system that provides antisurge control; capacity/throughput control for cracked gas and refrigeration compressors; governor controls for turbines, and quench control for refrigeration compressors.

Conventional control systems

Conventional turbomachinery control systems generally rely on standard closed loop control methods to constrain the operating point within the defined set points. CCC's extensive experience in the Ethylene process industry has shown that traditional constraint control actions are by and large inadequate to prevent excursions of critical process variables caused from major process upsets. As a result, the control system is configured with conservative safety margins and this invariably reduces the available operating zone of the compressor (Figure 1). The consequence of such action is a decrease in ethylene yield or compressor throughput with significant impact on overall plant production.

Accordingly, the paper discusses specific solutions to overcome the process constraint problems. Newly developed constraint control methods for ITCS will be shown to expand the operating envelope (beyond what is conventionally available) of a turbo compressor in a safe manner. Analysis of potential ethylene yield and capacity throughput increase will be presented for Petrokemya Olefins-1 plant. Dynamic simulation of a 6-stage CG compressor operating at Petrokemya will be used to demonstrate the effectiveness of the system. Responses to typical process disturbances such as furnace trips will be analyzed. Additionally, the article also addresses the handshaking requirements and role of ITCS in relationship with Plant Supervisory APC systems. A further objective of the paper is to highlight how a purpose-built turbomachinery control system (such as ITCS) can supplement the value of a Plant profit maximization system and provides enhanced benefits to an end user.

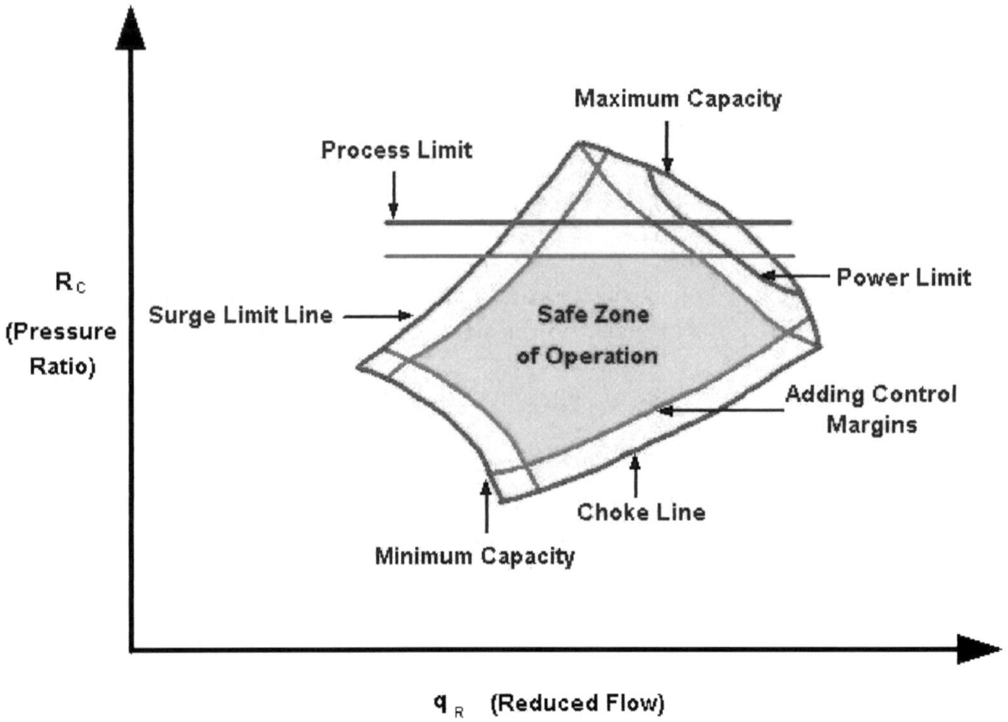

Figure 1: Compressor map with defined boundaries of operation

Olefins-1 process layout at Petrokemya

The paper will mainly focus on the hot section of the process with emphasis on CG compressor unit. At Petrokemya, Ethylene is produced by the pyrolysis of Ethane feedstock based on the process technology from CF Braun shown in Figure 2. The nameplate rating of the plant is 800,000 TPA. Eight furnaces are available out of which seven are in service to meet the production requirements. The cracked gas leaving the quench tower flows to CG compressor, where its pressure is increased from 0.9 bar g to 37.0 bar g. The gas is compressed in six stages and cooled between each stage. Any condensed hydrocarbons and water is removed in the knock out drums down stream of each cooler. Surge protection of CG compressor is through recycle of gas from third stage discharge to the first stage suction and from the sixth stage discharge to the fourth stage suction. CGC is driven by a condensing steam turbine.

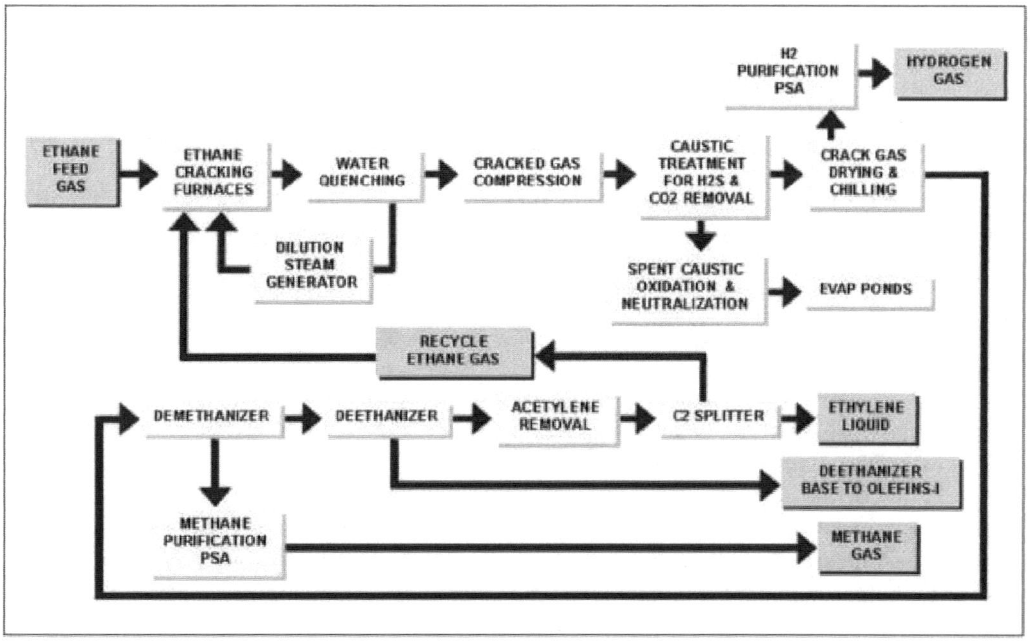

Figure 2: Schematic of Olefins-1 process at Petrokemya

Control philosophy and CG compressor process constraints

Main control philosophy is to regulate the first stage suction pressure of CG compressor to meet the plant loads. ITCS from CCC is employed to meet the suction pressure set point by varying the turbine speed. The suction pressure is set between 0.5 bar g to 0.9 bar g as per the plant capacity to handle additional gas volume.

Major constraints of the hot section are the limiting vapor velocities through the upstream quench tower, and suction drum overhead to avoid liquid carry over. In addition, the CG compressor inter cooler capacity is limited and the inter stage operating pressures are close to the equipment design limits thereby constraining the possibility of increasing the final stage discharge pressure. After the compressor revamp in May 2003 turnaround, the steam turbine has margins to handle the present plant load. However, it is not designed to handle the full compressor design load. The relief valve for blocked discharge is located at the 6th stage outlet, upstream of the Caustic wash tower which is located downstream of the 6th stage. The set point for the relief valve is 43 bar g. Low 1st stage suction pressure trip is set at 0.9 bar A. High stage inter-stage discharge temperature for stages 2 and 6 is 130 deg c. Constraint control loops of ITCS were configured to limit the compressor 1st suction pressure from going below 0.25 bar g and 6th stage discharge pressure from going above 40 bar g.

The steam turbine of C3 and C2 refrigeration compressors is constrained by the design mechanical limits of the casing. However, the governor valves for both the refrigeration compressors tend to be 100% open most of the time. In general, the plant is limited by hot section limits during summer and cold section during winter season.

Compressor stage	Design limits (bar g)	PSV setting (bar g)
1st stage suction	3.0	3.0
2nd stage suction	4.5	N/a
3rd stage suction	6.5	N/a
4th stage suction	10.0	N/a
5th stage suction	18.5	18.5
6th stage suction	27	N/a
6th stage discharge	43	43

Table 1: CG Compressor stage design limits

Integrated Turbomachinery Control System (ITCS)

Irrespective of the process technology used in Olefins plant design, integrated turbomachinery control is essential for meeting a plant's production objectives in a safe and efficient manner The fundamental objective is to efficiently operate the turbomachinery within a safe operating regime (as shown in Figure 1) by properly integrating the compressor and driver controls.

An ITCS (Figure 3) should be implemented by applying state-of-the-art microprocessor technology. Advanced control algorithms using adaptive open- and closed-loop techniques must be employed to prevent compressor surge and turbomachinery constraints. An ITCS must be versatile enough to adapt to diverse turbomachinery train configurations, such as driver, side-stream loads, and stage designs for CG and refrigeration compressors.

To ensure optimum performance, control algorithms should be executed in a maximum of 40 ms for an antisurge system, and 20 ms for a governor system. The control system must be designed for fail-safe operation with built-in strategies that allow safe and reliable on-line operation upon failure of transmitter signals. A minimum of duplex, fault-tolerant, control architecture is recommended to increase system reliability and availability.

Antisurge Control System Requirements

Algorithm Invariance

Suction conditions for refrigeration compressors change continuously. A surge-limit line, defined by an antisurge algorithm, must not vary with operating conditions [1,2]. The surge-limit line for CG compressor is a function of several parameters. The varying parameters are speed, gas composition, suction temperature, and suction pressure. The use of a surge-limit line without compensating for these parameters, in the antisurge controller, results in loss of turndown (as a result of unnecessary recycling), or operating in the unstable zone that can lead to surge.

Surge Parameter for Side stream refrigeration Machines

Side stream flows of a C3 and C2 refrigeration compressors present challenges to an antisurge control design. A compressor's internal design prevents the measurement of flow in the side stream-stage inlet. As a result, special algorithms (using principles of mass balance) are required to calculate this flow. CCC implements a unique algorithm to define surge parameters for a side stream compressor by using dynamic similitude theory [2].

Dynamic Decoupling between Series Stages

During plant start-up and part-load operation, compressor recycle valves are modulated under antisurge control action. For multi-stage CG compressor operating in series, opening the low-pressure (LP) stage recycle valve reduces flow to the downstream stage and pushes the high-pressure (HP) stage towards surge. Conversely, closing the LP stage recycle valve will drive the downstream stage away from surge. In order to prevent the HP stage from rapidly approaching surge, it is imperative that its recycle valve be opened by a necessary amount (as a feed-forward response) that is a function of the LP stage's recycle-valve opening. This feed-forward (decoupling) action stabilizes the interacting antisurge control loops.

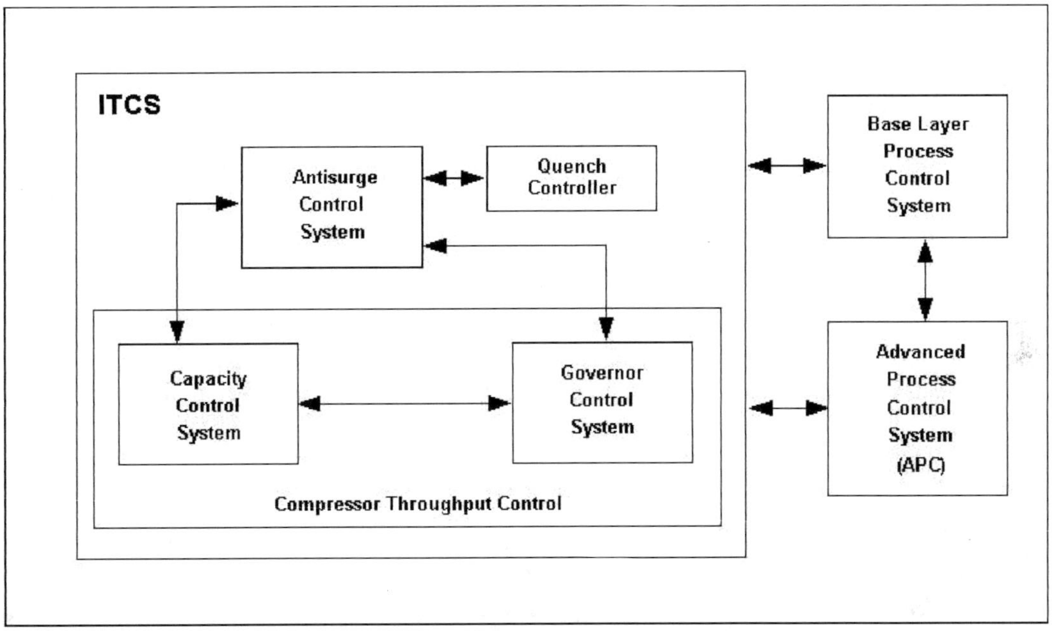

Figure 3: Integrated turbomachinery control system (ITCS) for Ethylene plants

For multistage C3 and C2 refrigeration compressors, the downstream stage (for example, the third stage) does not have a major influence on the upstream stage (first stage); these decoupling (feed-forward) gains must then be configured according to varying degrees of interaction between compression stages.

Automated Start-Up and Shutdown Algorithm

An antisurge control algorithm must be energy efficient for all operating modes of the plant. Automated recycle-valve response ensures that the compressor is automatically loaded and unloaded in an efficient, safe, and controlled manner during start-up and normal shutdown. In case of major, process upsets or emergency conditions, quick opening of the recycle valves unloads the compressor rapidly.

Constraint Control

An antisurge control system may be required to open the recycle valve even when operating conditions are away from the surge limit. Process and equipment design constraints may require the recycle valve to be opened in order to lower the discharge pressure at the compressor casing to prevent structural damage or process upsets resulting from sudden openings of the flare valves.

Suction pressure at the inlet drum to the compressor must not be allowed to drop below atmospheric conditions, owing to process and machinery constraints. The first-stage recycle valve should be opened by constraint-control action to maintain suction pressure at a safe level.

The design should guarantee that the compressor's operating point stays within the safe envelope shown in Figure 1. Frequent opening of the recycle valve results in diminished production caused by excessive power usage and reduction in CG compressor throughput or refrigeration compressor capacity. An inefficient antisurge control system will prevent a plant from operating at the desired production rates.

Requirements for Capacity Control System

A capacity control system's primary function is to adjust flow rates through the compressor by varying the turbine speed in accordance with the load imposed by production requirements.

Main Control Variable

The main process variable is suction pressure at the first-stage drum. The compressor's inlet drum pressure is maintained at a desired value by either operator-adjusted set point or from supervisory APC system for meeting profit based objectives.

Constraint Control

Constraint control algorithms will be required to limit turbomachinery performance within the boundary of a safe-operating envelope, as illustrated in Figure 1. This implies that a capacity control system must guarantee that safe operation is never sacrificed (during increased production demands) if the operating point of the turbomachinery is close to any performance limits. An ITCS ensures that performance is optimized within the safe-operating envelope; thus allowing maximum possible compressor throughput for a given constraint.

Process and equipment design constraints may require a capacity control system to restrict compressor throughput, thereby preventing structural damage to compressor casings; or preventing process upsets caused by sudden opening of the flare valves; or overloading the drivers if throughput exceeds the driver power limit.

Decoupling with an Antisurge Control System

Antisurge and capacity control systems for C3, C2 and CG compressors must be integrated through dynamic decoupling. At reduced rates, recycle valves are in operation and conflict with the capacity control system. This, in turn, results in unstable operation if the antisurge and capacity control systems are not integrated. For example, if production were to be reduced, an independent capacity controller would decrease flow through the CG compressor by reducing throughput. This could result in the recycle valves being opened by the antisurge controllers. Inlet-drum pressure would then increase and the capacity controller would react by increasing throughput. Antisurge controllers would close the recycle valves as a consequence of the increase in throughput, thereby causing suction pressure to drop, and so on.

This cycle would repeat itself and continue (causing unstable operation or making the response more oscillatory) unless the operator switches the throughput control to manual operation, or the capacity and antisurge control system gains are reduced. This reduction in gains is termed *loop detuning* and prevents unstable cycling (hunting) of control loops. However, loop detuning is undesirable because it results in loss of tight control during normal operation. But loop decoupling in an ITCS alleviates this problem and results in stable operation without having to detune the offending loops.

Automated Loading and Unloading

Refrigeration and CG compressors are generally brought up to minimum governor speed in LOCAL mode by employing a turbine's governor system. An operator can load the compressor by changing the governor speed set point, or by putting the governor control system in REMOTE mode and letting the capacity control system load the compressor automatically (in a controlled manner) using the remote-speed set point to the governor.

Successful loading of compressors in MANUAL mode relies heavily on operator experience and skill. For multistage compressors in series, interaction between the stages and recycle valves makes manual loading an extremely challenging task. Even seemingly simple mistakes can prove costly and severe; resulting in delays of process start-up and possible compressor damage from surge.

Accordingly, it is important that the capacity control system employ automated, loading and unloading algorithms. This will ensure that compressor throughput is regulated in a controlled manner and integrated with the antisurge control system to bring the compressor on-line smoothly, or take the compressor off-line to operate in a standby mode.

Governor Control System Requirements

The capacity control system interacts with the governor (in REMOTE mode) by adjusting the speed set point in accordance with the compressor suction pressure. In order to maximize production, the capacity control system will drive the governor to push the turbine to its power limit. The capacity control system should have windup prevention capability to prevent integral windup when the governor is constrained by the turbine power limit or operated in LOCAL mode. The governor and capacity control system should be able to detect a reduction in process load and drive the turbine away from a power-limit condition.

At Petrokemya, the automatic startup sequence of antisurge & capacity controllers allowed smooth startup of process not seen before. Tight constraint control combined with aggressive tuning due to decoupling strategy, allowed the load on CG compressor to be reduced to 100

Ton/hour with the recycle valves still closed. In addition to improved process control, ITCS provides significant energy savings to Petrokemya.

Quench Control System Requirements

A quench control system's main objective is to maintain the suction temperature of the refrigeration compressor stages by throttling the refrigerant coolant flow through the quench valve. The main process variable is typically the suction temperature of the C2 and C3 compressors.

Manual operation of quench control on refrigeration machines can cause a lot of problems. At Petrokemya, before the commissioning of CCC's control system, the compressors had a tendency to trip out on high suction drum level or surge. It was particularly severe on the ethylene refrigeration machine. The C2 compressor first stage suction drum size is very small and therefore floods very quickly. Invariably, the machine used to surge during start up. Automated quench control during process start up alleviated this problem.

Automatic sequencing provision in the CCC controllers enabled the quench valves to be opened only after the compressor speed is above a pre-defined threshold. Provision of variable ramp output during automatic startup sequence helped to introduce quench flows quickly, the rate being as fast as 4% per second for some compressor stages. The temperature set point for quench controller is typically adjusted based on stage suction pressure.

Quench controller's feed forward decoupling with antisurge control system minimized the loop interactions observed during fast openings of recycle valves.

Potential for maximizing the yield and throughput at Petrokemya

The problem

Maximizing the throughput during favorable market conditions and improving the yield for low market demand periods are some of the commonly accepted strategies employed by ethylene producers. Throughput increase results in operating the CG compressor at elevated stage pressures and closer to the discharge pressure limits. For example, at Petrokemya the CG compressor throughput increase is constrained by 6th stage discharge pressure PSV setting of 43 bar g. On the other hand, lowering of CG suction pressure leads to yield maximization. Producers are constantly faced with the potential risk of running the compressor in vacuum conditions while controlling the CG compressor at lower suction pressure levels. For Petrokemya, the CG compressor is configured to trip when the suction pressure goes below 0.9 bar A. Experience has shown that conventional constraint control systems are by and large inadequate to prevent excursions of critical process variables from major process upsets seen in an ethylene plant. In addition, upstream or downstream process constraints also prohibit running the compressor at an elevated discharge or lowered suction pressure levels. In such situations, the process units become the bottleneck in the plant.

Typical process disturbances

The following are main sources of process disturbances for CGC system at Petrokemya:

1. Furnace feed failure/furnace trip: Furnace feed failure results in the loss of approximately 27 t/h (7.5 kg/sec) of effluent with average Mol. Wt of 19 equivalent to

31850 Mm3/hr. The feed reduction takes place over a period of 30 to 60 seconds. Furnace trip will result in similar disturbance to CG compressor. In the worst-case scenario, a maximum number of four feed streams can fail or four furnaces can trip out.

2. Drier MOV malfunctioning: The closure of MOV will take place in approximately 30 to 60 seconds with sudden stoppage of forward flow of 195 t/h (54.2 kg/sec) from CG compressor to cold box.

3. Failure of differential pressure controls on excess non-condensed crude H2/tail gas: Causes reduction in forward flow of 90 t/h (25 Kg/sec) or 50% of normal feed over a period of 30 seconds.

4. Caustic wash tower off-spec operation resulting in stoppage of forward flow to cold box: This is same as MOV malfunctioning with the exception that there will be no sudden stoppage of forward flow, since the operator will be venting the gases from caustic wash tower overhead downstream of the 6th stage discharge.

Faced with the challenges discussed above along with the hot and cold section process constraints, the control system is configured with conservative safety margins to prevent the CG compressor from tripping out. This invariably reduces the available operating zone of the compressor (Figure 1). The consequence of such action is a decrease in compressor throughput with significant impact on Ethylene yield or overall plant production.

Economic justification and benefits for improving the constraint control solutions

Based on our findings, it was quite evident that there was a compelling need for developing a new and improved constraint control solution to expand the turbo compressor-operating envelope. This would have significant economic impact on the bottom line of an ethylene plant. To quantify the increase in production resulting from rise in discharge pressure, 6-stage compressor performance curves were used. A series of operating curves (mass flow versus stage discharge pressure) were developed for each stage running at the design speed with a defined backpressure downstream of CG compressor. Using these curves it was found that an average rise of 0.1 bar g in 6th stage discharge resulted in overall increase of 0.8 kg/sec CG compressor mass flow.

As an approximation, the resulting increase in CG compressor mass flow will linearly impact the ethylene production. Using the design CG mass flow of 55 kg/sec at Petrokemya, this translates to a minimum annual production increase of about 1.5% for every 0.1 bar increase in 6th stage pressure (assuming that yield changes due to increased suction pressure are negligible).

Yield prediction calculations due to changes in coil outlet pressure (or equivalently CG suction pressure) were estimated using software from the process licensor. Details of results are beyond the scope of this article. In summary, for Petrokemya ethane cracker, typical annual production increase of 1% was estimated for every 0.1 bar g reduction in coil outlet pressure.

The solution

CCC's expertise in advanced surge controls technology was leveraged to develop new solutions for addressing the challenges in ethylene plants. Problems encountered with traditional constraint control systems were overcome using fast acting adaptive control response ***(patent pending)*** employing recycle valve as the control element. The effectiveness of the solution is contingent upon stringent control system design such as recycle valve sizing, valve actuation, recycle piping configuration, system volumes, and field tuning. The details of ITCS implementation for CG compressor are shown in Figure 4.

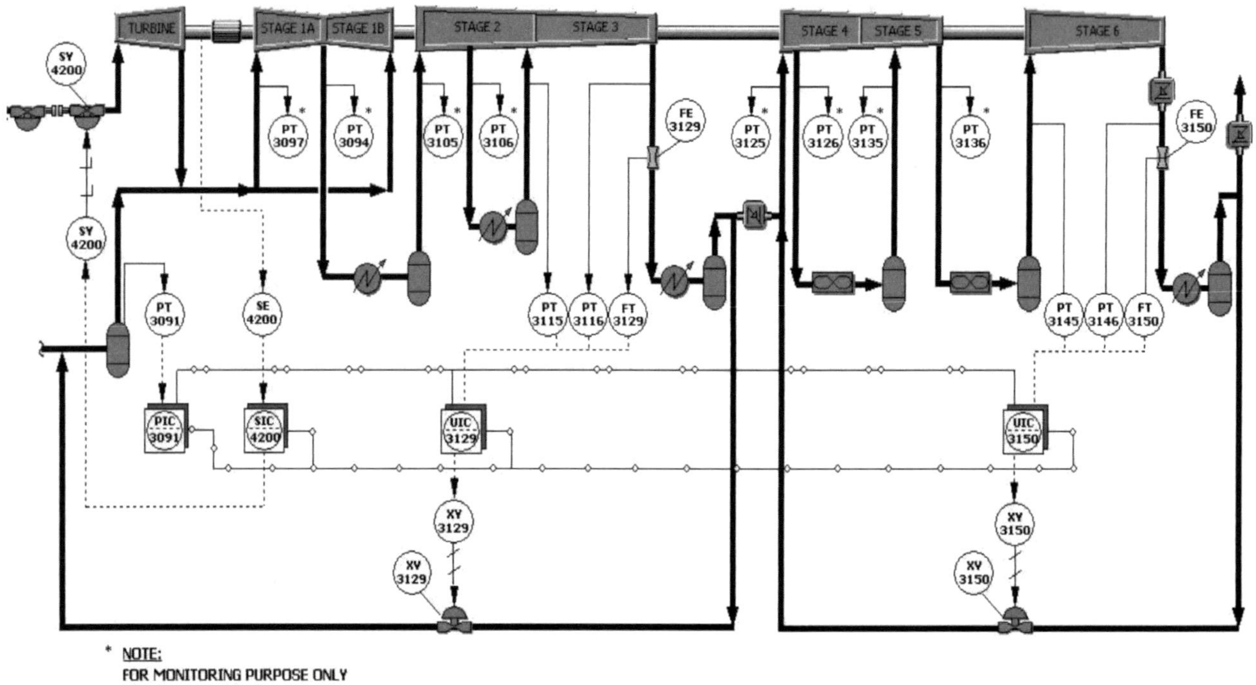

Figure 4: ITCS for CG compressor at Petrokemya

Validation of advanced ITCS

Newly developed advanced ITCS was verified to be successful using a real-time dynamic simulation of the 6-stage CG compressor at Petrokemya. Process disturbances simulating the furnace trip on suction and drier MOV malfunction in discharge were applied on CG compressor model to validate the advanced ITCS.

Dynamic Simulation of CG compressor system

A schematic of the dynamic simulation model used to describe the various elements of the CG compression system including the steam turbine driver is presented in Figure 5. The entire system was broken down into compressor, turbine, control valves, and a number of volumes for associated pipes, knockout drums, and the gas-cooler for constructing the dynamic model. Each volume element is numbered as shown in Figure 5. The parameters of every element were determined from the data provided by the equipment manufacturers, piping isometric drawings, and process operating data.

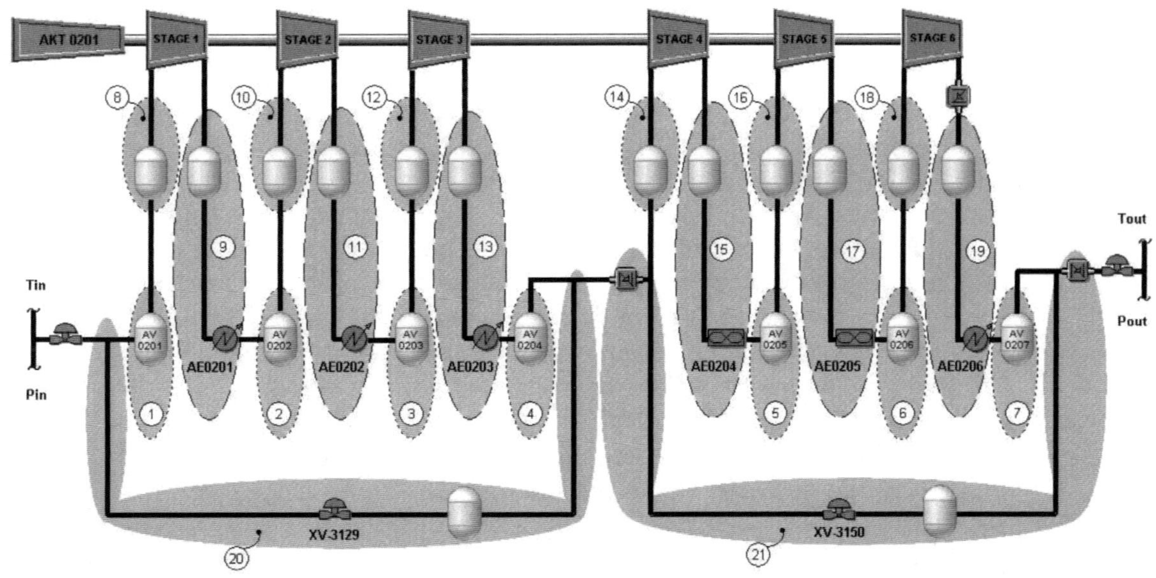

Figure 5: Dynamic model schematic of 6-stage CG compressor

In addition, the simulation model was connected to real controllers in an "in-the-loop" setting. This enabled direct comparison of traditional control system employing the standard PID-type constraint control algorithms with a purpose-built advanced ITCS employing newly developed control methods and predictive algorithms.

Model Assumptions

The following is a list of the major features of the model and associated simplifying assumptions.

- Cracked gas parameters — molecular weight, pressure, and temperature — in the suction and discharge headers [Pin, Pout, Tin, Tout] are assumed constant and are set up as boundary conditions for the model.
- Three invariant coordinates — compression ratio, reduced-speed, and reduced-flow — are used to model the CG compressor performance map [2]. Compressor reduced-speed and compression ratio are used to calculate inlet flow. Compression ratio and inlet flow are used to calculate shaft power.
- All pipes are modeled using a lumped-parameter approach to calculate pressures, temperatures and flows along the process. Pipe resistances are included into resistance of associated valves, and pipe volumes are split equally between the main upstream and downstream volumes connected by the pipe.
- Pressures and temperatures in process volumes are calculated using ordinary differential equations from the mass and energy balance in the volumes.
- The volume associated with the gas cooler in the compressor discharge section is split equally between adjacent upstream and downstream volumes.
- Process valves are modeled using standard ISA sizing equations, and take into account the actual valve size and flow conditions.

- Gas flow from furnace to CG compressor suction was regulated by means of a suction block valve. Process related upsets from furnace trips or feed failure were simulated by required closure of suction valve. Suction valve actuator stroke time and valve trim characteristics were modeled to match the flow reduction profiles (magnitude and time) typically seen during furnace upsets.
- Gas flow from CG compressor discharge to Caustic tower was regulated by means of a discharge block valve. Process related upsets from drier MOV mal-functioning or failure of differential pressure controls on excess non-condensed crude H2/tail gas were simulated by required closure of discharge valve. Discharge valve actuator stroke time and valve characteristics were mode-led to match the flow reduction profiles (magnitude and time) typically seen during drier MOV upsets.
- Actuator stroke-time for full opening and closing of both recycle valves was set as 2.0 seconds with nominal actuator dead time based on actual field operating set up.
- Actuator stroke time of governor valves and flow characteristics were matched with field data.
- Shaft rotational speed was calculated using ordinary differential equations from power balance between steam turbine and CG compressor stages.

Model validation

Once the model was developed, a comparison was made between the design data provided by the OEM and the compressor system model results under steady-state conditions with the compressor at the rated operating point. The values of speed, suction pressure, discharge pressure, discharge temperature, compressor power, and inlet flow were compared, with an error between design values and model-calculated values to be less than 1%. The small error between the values validates the model under steady-state conditions, and provides good initial reconciliation for the transient analysis of process disturbances. In addition, gas flows through the recycle valves were compared. The model-generated flows exactly matched the design-calculated flows in steady-state conditions, which validated the recycle valve model with respect to valve size and trim.

Simulation Results

Results for each test scenario are presented as time plots of compressor 1st stage suction pressure, 6th stage discharge pressure, and recycle valve outputs. For each process disturbance test, excursions in process variable above the constraint control set point will be analyzed to determine the margin of improvement. The control system parameters were tuned to the best possible extent using conventional field practices and best tuning principles. We believe that any potential variability in tuning parameters will have minor impact on the findings.

Scenario A – Drier MOV malfunction

Drier MOV malfunction results in full closure of discharge valve downstream of CGC 6th stage. Reduction in mass flow of about 54 kg/sec occurs over a period of 30 seconds. Compressor operating parameters at the start of test were:

- 6th stage discharge pressure: 40.75 bar g
- Suction pressure: 0.90 bar g
- Operating speed: 3900 RPM
- Mass flow: 65 kg/sec

Control system was tested to validate the new constraint control solution. Readers may recall that relief valve for blocked discharge is located at the 6th stage outlet, upstream of Caustic wash tower. The set point for relief valve is 43 bar g. The limit set point for high 6th stage discharge pressure was set at 42 bar g (instead of the traditional 40 bar g) to evaluate potential increase in CG compressor operating envelope. Figures 6 and 7 show the response of 6th stage discharge pressure using conventional constraint control systems. In Figure 6, the recycle valve ASV 3150 is used to limit the pressure excursions. On the other hand Figure 7 shows the pressure response using turbine speed to suppress the pressure rise. Figure 8 shows the significant improvement brought about by the advanced ITCS in reducing the pressure overshoot by a factor of 5 times (see Table 2 below) with respect to conventional system.

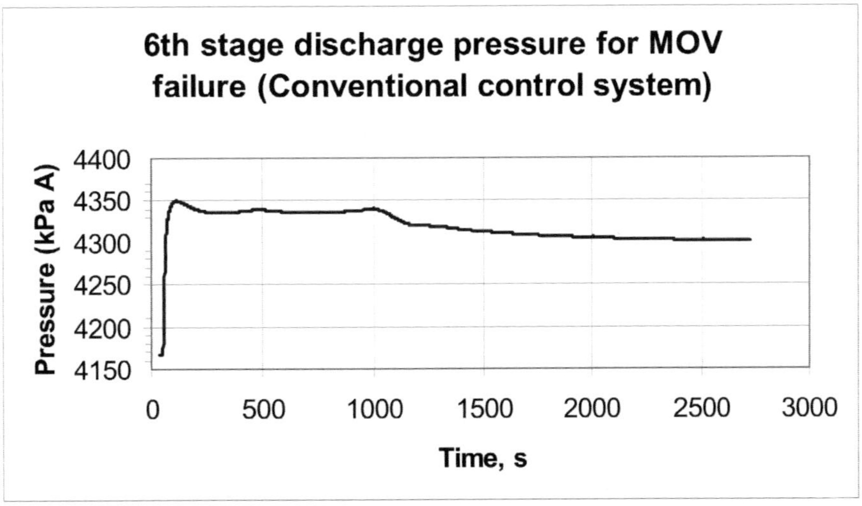

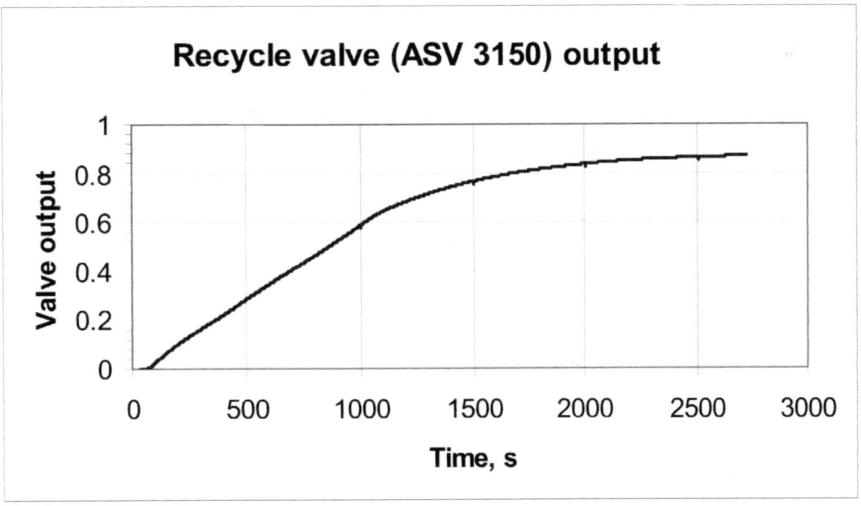

Figure 6: Typical 6th stage discharge pressure response for conventional control system

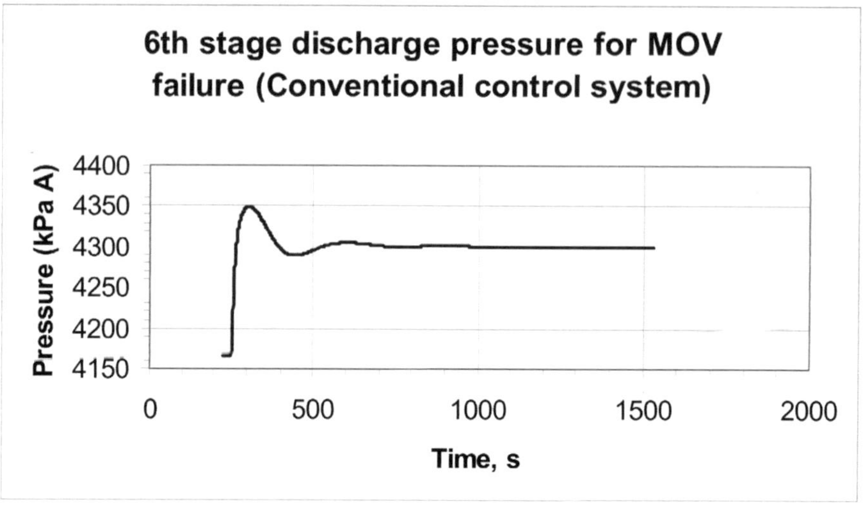

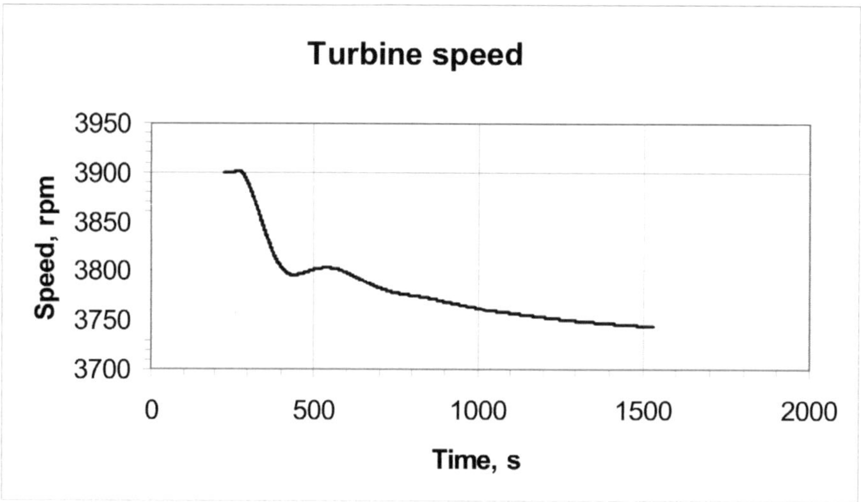

Figure 7: Typical 6th stage discharge pressure response for conventional control system

	Pressure set point	Maximum pressure response	Pressure over shoot	Overshoot ratio*
Conventional control system	42 bar g	42.5 bar g	0.5 bar g	1.0
Advanced ITCS	42 bar g	42.1 bar g	0.1 bar g	5.0

* with respect to conventional control system

Table 2: Comparison of discharge pressure control response

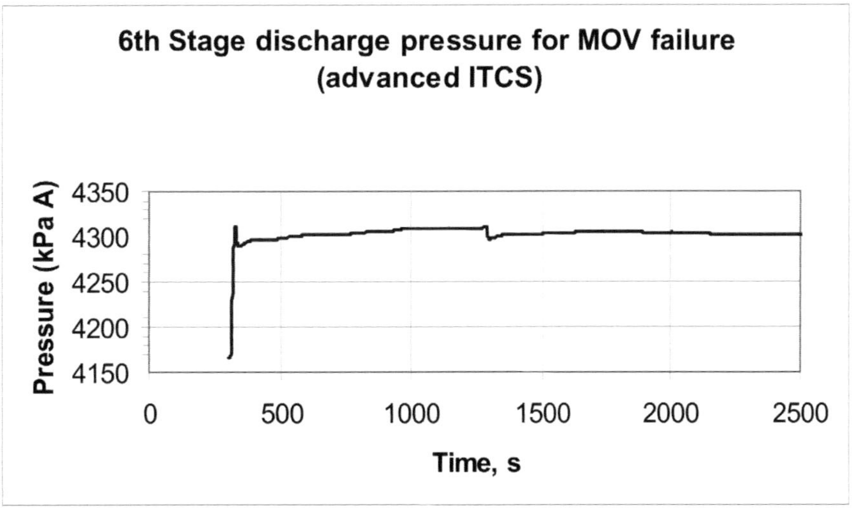

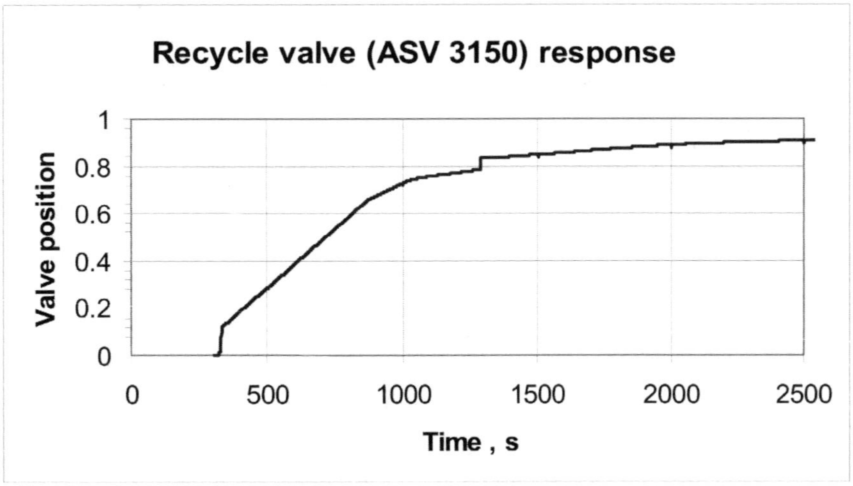

Figure 8: Typical 6th stage discharge pressure response for advanced ITCS

Scenario B – Furnace feed failure / furnace trip

Furnace feed failure or furnace trip results in rapid reduction of mass flow to CGC suction. Reduction in mass flow of about 30 kg/sec occurs over a period of 120 seconds. The flow reduction simulates taking four furnaces off line. Compressor operating parameters at the start of test were:

- 6th stage discharge pressure: 40 bar g
- Suction pressure: 0.25 bar g
- Operating speed: 3960 RPM
- Mass flow: 40 kg/sec

Control system was tested to validate the new constraint control solution. Readers may recall that CG compressor is configured to trip on 1st stage low suction pressure level of 0.90 bar A. The limit set point for low suction pressure was set at 0.15 g (instead of the traditional 0.25 bar g) to evaluate potential increase in CG compressor operating envelope. Figures 9 shows the response of 1st stage suction pressure using conventional constraint control system. The recycle valve ASV 3129 is used to prevent the pressure from going low. Figure 10 shows the significant improvement brought about by the advanced ITCS in reducing the pressure undershoot by a factor of 4 times (see Table 3) with respect to conventional system.

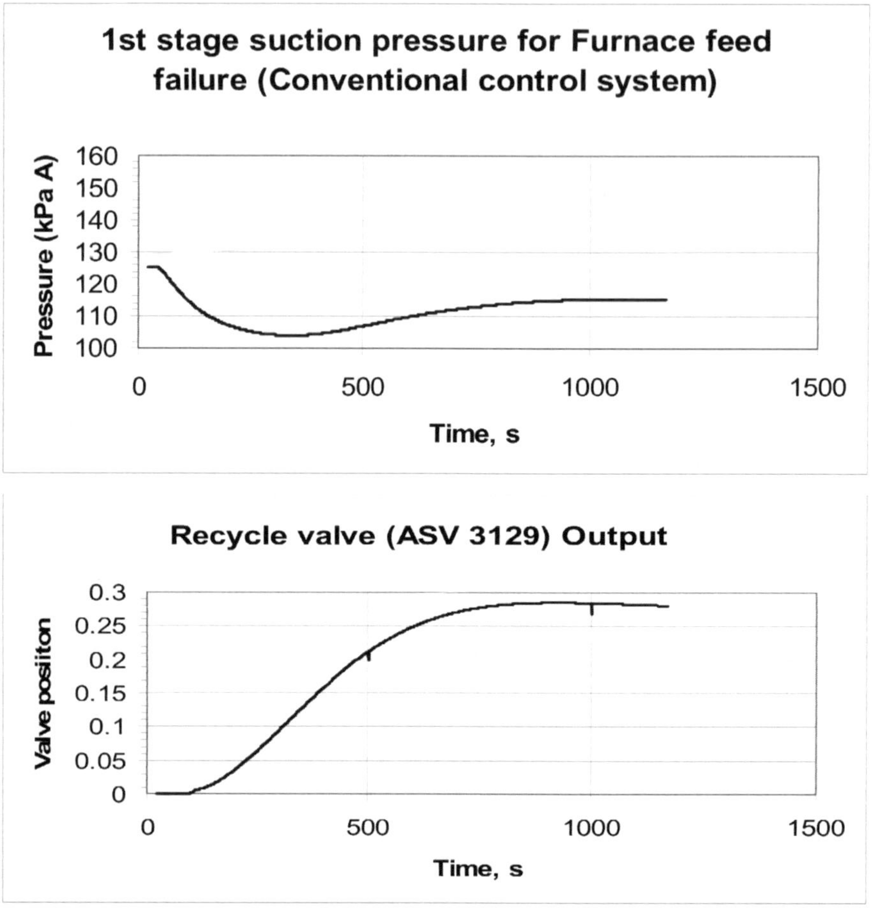

Figure 9: Typical 1st[h] stage suction pressure response for conventional control system

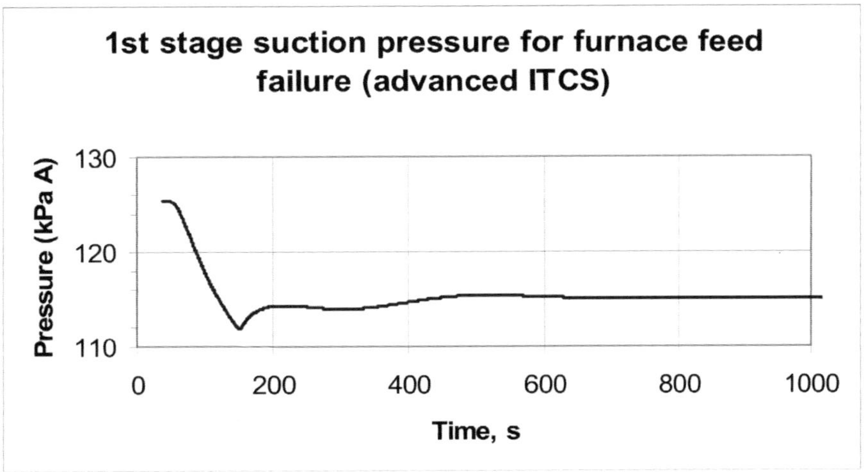

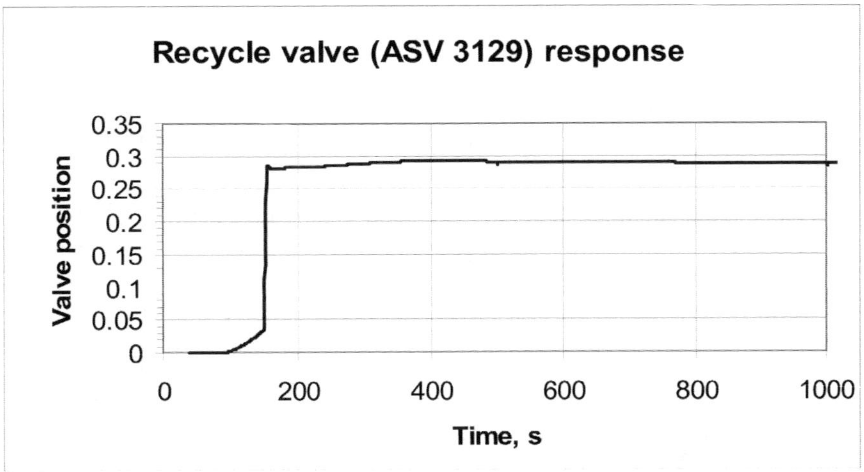

Figure 10: Typical 1st^h stage suction pressure response for advanced ITCS

	Pressure set point	Minimum pressure response	Pressure over shoot	Overshoot ratio*
Conventional control system	**0.15 bar g**	**0.03 bar g**	**0.12 bar g**	**1.0**
Advanced ITCS	**0.15 bar g**	**0.12 bar g**	**0.03 bar g**	**4.0**

* with respect to conventional control system

Table 3: Comparison of Suction pressure control response

Expanded operating envelope

We can see from the Figures 6 and 7 that conventional control system is not capable of suppressing discharge pressure increase due to downstream process upsets. 6th stage discharge was found to come very close to the relief valve pressure setting. Pressure response was found to be highly dependent on the magnitude of disturbance. In fact our studies proved that any marginal increase in the strength of downstream disturbance would cause the pressure to over-shoot above the PSV set point. On the other hand, response of advanced ITCS was found to be deterministic and repeatable to prevent the discharge pressure from going above the configured limit set point of 42 bar g. Irrespective of the magnitude of process disturbances, the adaptive nature of advanced ITCS effectively countered the process upsets effectively to minimize discharge excursions as seen in Figure 8. Now this gives Petrokemya an opportunity to potentially increase the 6th stage limit set point from 40 bar g towards 42 bar g thereby increasing the operating envelope of CG compressor. Direct benefit of this finding is an increase in CG throughput leading to overall production increase. However other considerations such as cold section constraints (refrigeration compressors), quench tower problems arising from increased suction pressure need to be carefully analyzed to determine the overall benefits.

Similar conclusions can be made for process upsets coming from suction of the CG compressor. Figure 9 shows that conventional control system is not capable of arresting the suction pressure collapse due to furnace related upsets. 1st stage suction pressure was observed to drop very close towards the trip set point of CG compressor. Consistent with the discharge pressure tests, any marginal increase in upsets from furnace would cause the suction pressure to go below atmospheric levels. On the other hand, response of advanced ITCS was found to be deterministic and repeatable to prevent the discharge pressure from going below the configured limit set point of 0.15 bar g. Irrespective of the magnitude of process disturbances, the adaptive nature of advanced ITCS effectively countered the process upsets effectively to minimize suction pressure decrease as seen in Figure 10. Again, this gives Petrokemya an opportunity to potentially decrease the 1st stage limit set point from 0.25 bar g towards 0.15 bar g thereby increasing the operating envelope of CG compressor. Direct benefit of this finding is an increase in ethylene yield leading to overall production increase. However other considerations such as problems in hot section arising from lower suction pressure, and turbine power limits need to be carefully analyzed to determine the overall benefits.

The potential increase in CG compressor operating envelope is illustrated using Figures 11 and 12.

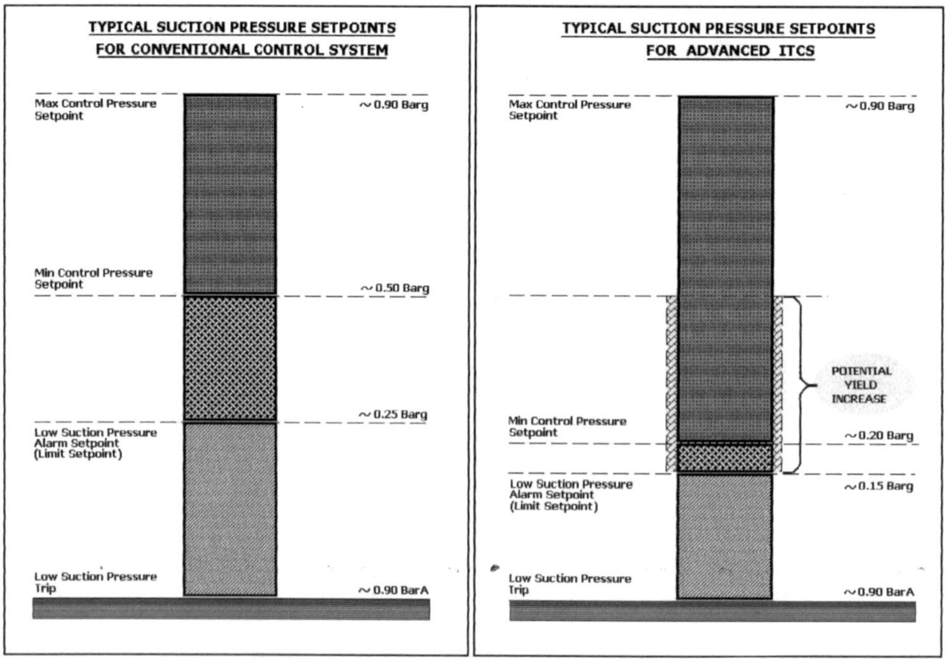

Figure 11: Comparison of suction pressure set points between conventional control system and advanced ITCS

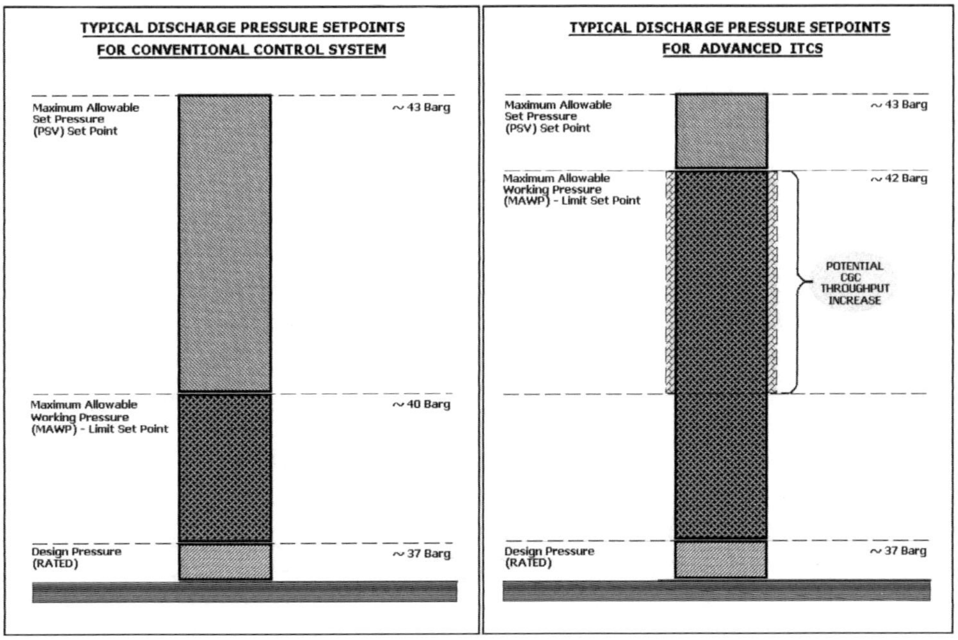

Figure 12: Comparison of discharge pressure set points between conventional control system and advanced ITCS

Interface between APC and ITCS

An advanced process control system (APC) runs in a traditional, distributed control system (DCS) to optimize production. The function of an advanced control system is to generate set points to the ITCS and to the base layer process control system in response to the process unit model constraints for maximizing the plant profitability. Since the plant design simulation & responses are built into APC system, it is expected to control the process sections close to the constraint limit.

The prime objective of ITCS is to provide surge protection during process upsets and turbine speed control during normal production run. Typically the recycle and quench valves should be closed during normal run, thus leaving only the capacity controller and limiting controller set points to be optimized by APC. Plants are generally designed without consideration for interface between an ITCS and base layer process control system that regulate other process sections impacting the performance of compressors.

The integration of APC could be done for example in setting the capacity controller set point for various furnace loads on CG compressor, which the ITCS cannot determine. If the SPYRO® program is integrated with the APC for furnace yield prediction, then the optimization of the capacity controller set point for CG compressor would become part of plant optimizer scope. This would definitely help in improving the furnace run length and yields.

Similarly adjusting the capacity controller set point for 1st stage suction pressure based on plant loads could optimize the C2 and C3 compressor. Another example is adjusting the 3rd stage suction pressure limit control set point of C3 compressor from APC as a function of the reboiler vapor supply control valve opening. This feature has helped Petrokemya for the first time to avoid field operator from adjusting manual jump over from 3rd discharge to 2nd discharge header, during winter operation. Maximizing the 6th stage discharge pressure of CG compressor would allow APC to perform what-if analysis for higher discharge pressure limit set points established in advanced constraint control scheme of ITCS.

The success of APC is hinged upon the stability and robustness of the base layer regulatory control system and ITCS. Increasing the operating envelope of a compressor widens the controllable range of ITCS. The wider the controllable range of an ITCS, the greater is the potential for an APC system to meet profitability objectives by operating at set point scenarios never included before. Thus the role ITCS and APC systems are complementary in nature to further supplement the value of a Plant profit maximization system and providing enhanced benefits to an end user.

Summary and conclusions

CCC's expertise in advanced surge controls technology was leveraged to develop new solutions for addressing the challenges in ethylene plants. Due to the excessive dead times or large time constants of CG compression process, traditional constraint control actions are usually ineffective to prevent excursions of critical process variables caused from process upsets. Moreover, constraint control response is initiated only if the process variable approaches closer to the limit set point. Increasing the gains of the controller does not necessarily mitigate the problem as the control action is found to be inadequate due to overall system sluggishness. As a result, the control systems are usually configured with conservative safety margins (process or turbomachinery) to overcome the problem. This invariably reduces the available operating zone of the compressor. The consequence of such action is a decrease in compressor throughput with significant impact on the plant production. The solution described in the article was shown to overcome the aforementioned problems by testing a dynamic model of 6-stage CG compressor system for various types of process upsets. It was shown that the newly developed fast acting adaptive control response of advanced ITCS is capable of minimizing the excursions of critical variables on a compressor, turbine or its associated process.

The turbomachinery control system at Petrokemya was retrofitted with CCC's ITCS during the May 2003 turnaround. Rigorous systems integration and testing of ITCS during the acceptance phase of the project along with conformance to stringent design requirements allowed the startup to be completed five days ahead of the normal schedule. Petrokemya is currently evaluating the possibility of upgrading the CCC system to seamlessly integrate the newly developed advanced constraint control module into the ITCS platform. This will allow us to take full advantage of the potential benefits offered by the advanced constraint control solution. Hot and cold section process constraints are being carefully reviewed to further refine the estimates of yield or throughput increase reported in this paper. We envision that advanced ITCS will play an important role in further enhancing the value of an APC system by expanding the envelope of compressor in a safe, reliable and efficient manner. APC systems can now expand the range of target set points to ITCS or base layer process controls by operating the compressors at suction or discharge pressures never considered before. The recommended strategy has potentials for significant economic impact on plant revenues and operating bottom line. Retrofit of control system to install the advanced ITCS from CCC is easily justified by production increases in millions of Dollars resulting from yield or throughput maximization.

Acknowledgements

The authors wish to thank their respective employers Petrokemya and Compressor Controls Corporation, for giving the permission to publish the article. We also would like to express our gratitude to the EPC Process Control Committee members for their support and opportunity given to present this paper. All the assistance received from various employees of CCC and Petrokemya during preparation of the article is sincerely appreciated.

References

[1] Compressor Controls Corporation Patent; No. 5,508,943. *Method and Apparatus for Measuring the Distance of a Turbocompressor's Operating Point to the Surge Limit Interface*. April 1996.

[2] Compressor Control Corporation. *Series 3 Plus Antisurge Controller for Centrifugal and Axial Compressors*. Instruction Manual, Publication IM301, February 1999.

Paper 65f

Implementation of Alarm Management Software in a DCS-controlled Ethylene Plant

Doug Rode
Eastman Chemical Company
PO Box 7444
Longview, TX 75607-7444

Karim Moti
Matrikon Inc.
1330 Post Oak, Suite 1600
Houston, TX 77056

Allen Monholland
Eastman Chemical Company
PO Box 7444
Longview, TX 75607-7444

Richard Bykowski
Eastman Chemical Company
PO Box 7444
Longview, TX 75607-7444

Prepared for Presentation at the 2004 Spring National Meeting, New Orleans LA, Apr 25-29
16th Annual Ethylene Producers' Conference
Session T8009 – Ethylene Plant Process Control

Copyright © Eastman Chemical Company April 2004

UNPUBLISHED

AIChE shall not be responsible for statements or opinions contained in papers or printed in its publications.

Abstract

In 2000, Eastman retrofitted an older pneumatically controlled ethylene plant with a DCS. Since the retrofit, operators have been swarmed by alarms. Efforts were made to reduce the excessive load of alarms through operator input. This did not significantly affect the alarm count.

In early 2003, Eastman entered into a pilot agreement with Matrikon to use their "ProcessGuard" alarm management software. Initially, the software was used to rank alarm data by alarm count. This quickly identified the most frequent alarms. Plant Management and DCS Technicians met daily and reviewed the top 20 alarms from the previous day. The team used this list to identify the obvious "nuisance" alarms and addressed them by changing alarm limits, deadbands, etc as needed. Once the majority of these "chattering" alarms were corrected, the software was then used to identify "parent/child" relationships between alarms. Each "parent/child" group was then analyzed, and long-term process improvement projects resulted.

The pilot effort reduced the alarm count by 80% in the first 3 months. Eastman has decided to implement this alarm management process on all of its older ethylene plants and is evaluating implementation in other operating areas.

Introduction

The main purpose of an alarm system is to provide an operator with critical process information at the right time. For older plants using panel-mounted instruments, alarms are typically reserved for extreme situations. This results in a "low" number of alarms. However, when an older plant is converted to DCS, the alarms available to the engineer are increased substantially. What was once just a TI (temperature indication), now has the ability to create an alarm. In fact, it can create a multitude of alarms – High, Low, HighHigh, LowLow, Deviation, etc.

A newly installed DCS, plus a basket of new alarms available for the engineer, will result in a host of new alarms configured on the DCS. When creating an alarm for a point that never had an alarm before, an engineer will be making their "best guess" at what the alarm values should be for that point. If the DCS point deals with capacity, the alarm might be configured based on maximum production rates. If the point deals with safety, the alarm range might be configured based on safety concerns. Sometimes, an engineer might not know what to base the range on. All of these scenarios typically result in configurations that are too "tight".

An alarm *point* configured incorrectly can result in nuisance alarms. Operators begin to ignore the alarm, or worse, suppress it. An alarm *system* configured incorrectly can result in a "flood" of alarms during a process upset. Operators can receive so many alarms that the true problem with the process hides among the flood. Both of these situations negate the purpose of the alarm system.

ProcessGuard

ProcessGuard is a software package provided by Matrikon, Inc. ProcessGuard receives alarm and event data from the DCS, parses the data based on a set of rules, initiates action, and histories the data in a database. ProcessGuard allows the study and manipulation of alarm data. ProcessGuard consists of the following components:

- *A&E Collector* is the point of entry for alarm and event data to the ProcessGuard system.
- *Rules Builder* helps create a set of rules. These rules are used by ProcessGuard to look at the alarm data and segregate the alarm type, time, value, etc.
- *A&E Archiver* receives the raw alarm data from the *A&E Collector* then applies the rules from the *Rules Builder*. It can then forward the data to the *A&E Viewer* and store the results in the ProcessGuard database.
- *A&E Viewer* is a tool for real-time monitoring of the alarm data.
- *A&E Analysis* is a Microsoft Excel Add-in that pulls data from the database. This lets you view predefined reports or perform ad hoc queries and statistical manipulation on the data in Excel. These reports are based on EEMUA (Engineering Equipment and Materials User Association) guidelines – publication 191.

The A&E Analysis tool is the main window to the alarm data. With the use of various pre-defined reports, ad hoc queries, filters, etc, the alarm data is extracted from the database and deposited into an Excel workbook. The reports can be run on a specific DCS console, process area, even a specific tag. Once the data is in Excel, it can be manipulated with all of the functions Excel has to offer.

Network Architecture

The following diagram outlines the architecture of collecting, storing, and accessing the alarm and event data.

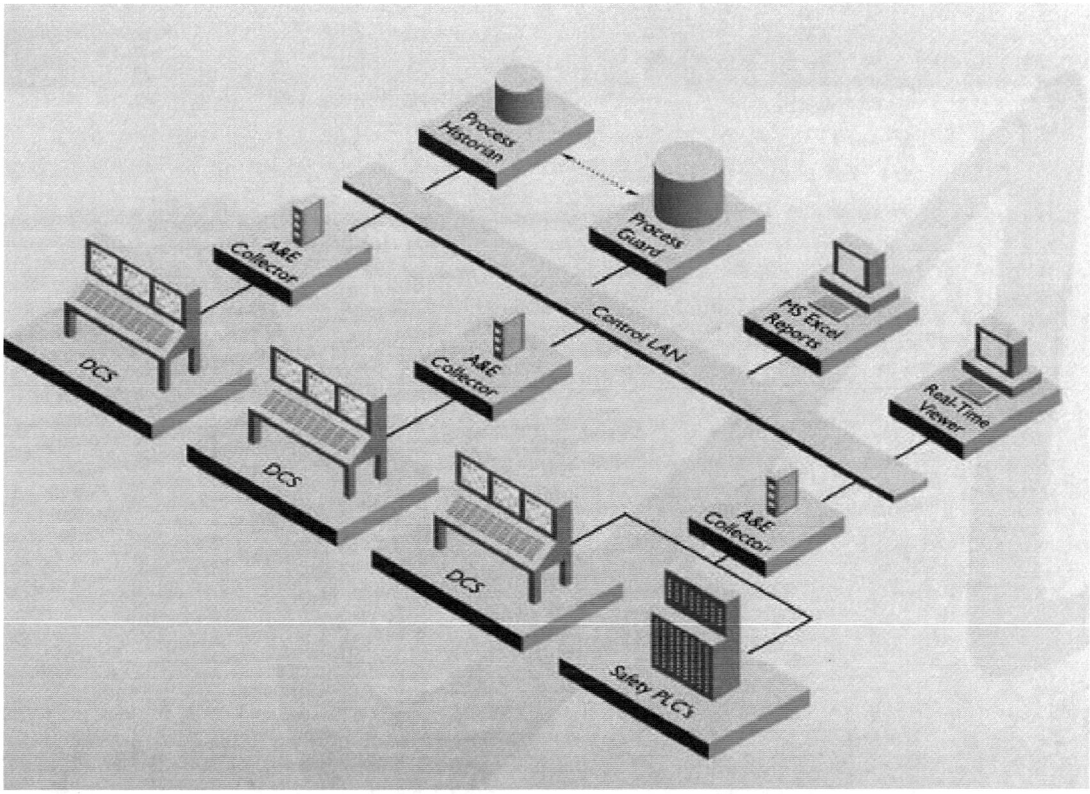

Matrikon initially installed the A&E Collector on a collection PC. The collection PC was connected to one Fischer Provox Pro Vue console. The collected data was transferred to the ProcessGuard server through TCP/IP, and stored within the database on the server. The client applications were installed on various PC's for data access and analysis.

Eastman's Experience

In 2000, Eastman retrofitted one of its older pneumatically controlled ethylene plants with a DCS. Since the retrofit, there has been a constant flow of alarms to the console. During process upsets, the operators are "flooded" with alarms. Various programs were implemented in order to curb the alarm load. These included a DCS logbook, a DCS alarm work request database, etc. None of the programs had the desired impact on the alarm load. With a constant barrage of alarms, it was hard to pinpoint which alarms needed attention.

In early 2003, Eastman entered a pilot program with Matrikon. The ProcessGuard server and software were installed by Matrikon to monitor the alarms on one DCS console. After initial installation, data was collected for two weeks in order to establish a baseline.

Matrikon advises that a well-configured alarm system should generate no more than 1 to 2 alarms every 10 minutes (150-300 alarms per day). This benchmark is intended to give the operator enough time to respond rationally to each alarm that occurs over the course of a shift. With this goal in mind, Eastman and Matrikon began to analyze the alarm data.

The first part of the analysis was to look at the *Alarm Distribution Over Time* report. This report quantifies the amount of alarm traffic that the operator sees during a specific timeframe. Figure 1 is a graph of the number of alarms received in a 10-minute period over 7 days. For the week, the total alarm count was 6073, for a daily average of 868. Compared to a benchmark of 1 alarm every 10 minutes, the load on the operator is generally high but is sometimes extremely high.

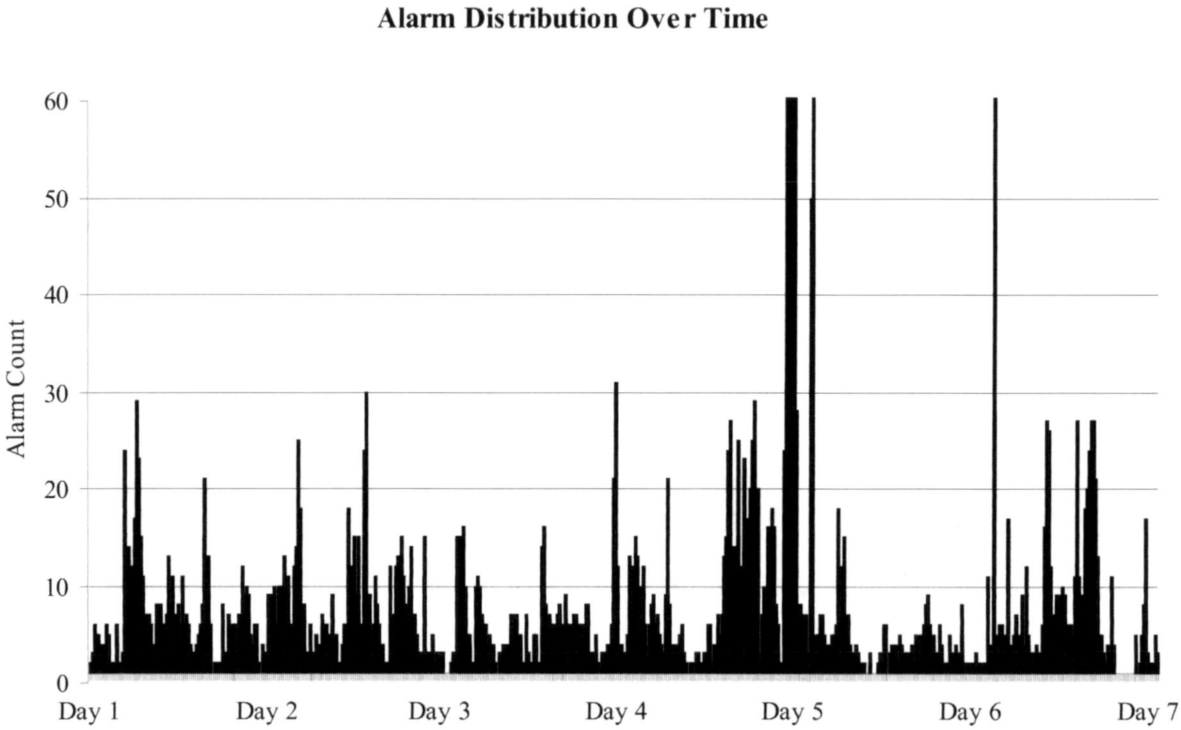

Figure 1 – 10 minute alarm count over 7 days (zero counts excluded)

After reviewing the *Alarm Distribution Over Time*, smaller periods of interest were picked out for further review. These were typically the peaks in the trend. Once a period was selected, the *Alarm Count by Tag* report was reviewed. This report shows the number of alarms each tag created during a specific timeframe. From Figure 1, it is obvious that there was a heavy alarm load on Day 5. Figure 2 shows the top 20 alarms during that period of interest.

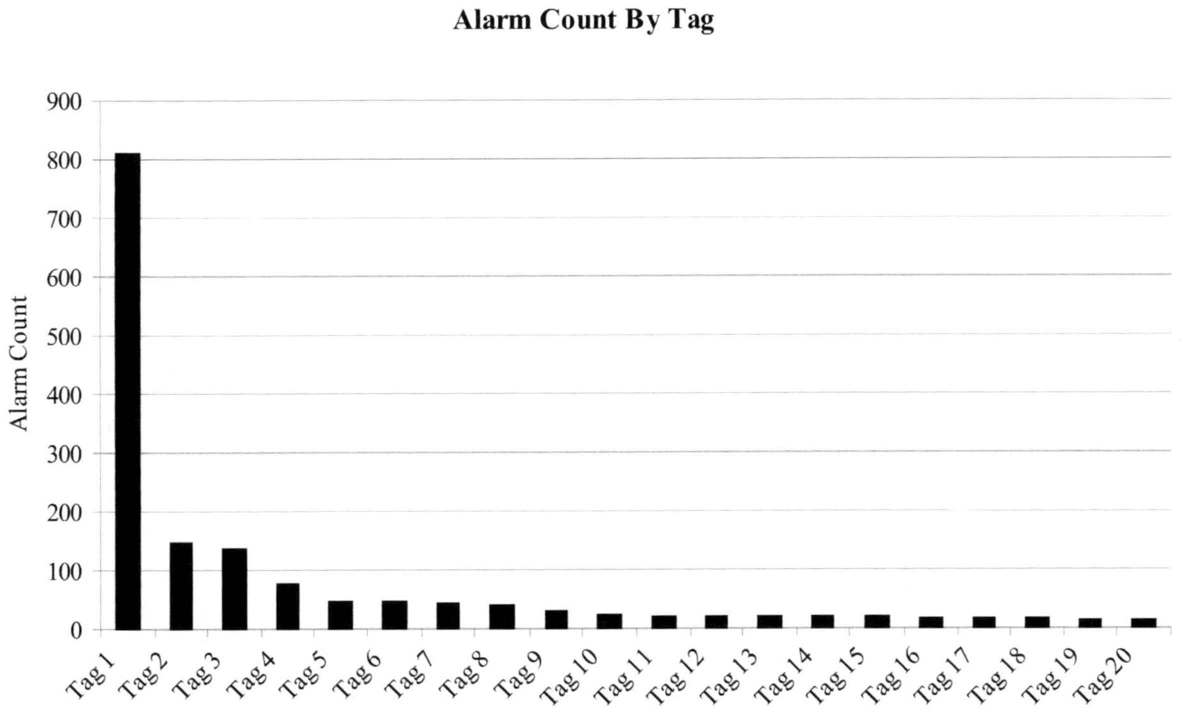

Figure 2 – Alarm Count by Tag for Day 5

Figure 2 shows that a large majority of the alarms were generated from just a few tags. After a few suspect tags had been selected, further review of the data was needed. *Chronological Reports*, *Ad Hoc Queries*, and process trends were used to determine the sequence of events that led to the frequent alarms. Table 1 shows a portion of the Tag 1 activity.

Tag	Timestamp	Alarm Identifier	Value	Message Type
Tag 1	8:18:20 AM	LO PRES	418.28	ReturnToNormal
Tag 1	8:18:22 AM	LO PRES	357.13	Alarm
Tag 1	8:18:25 AM	LO PRES	419	ReturnToNormal
Tag 1	8:18:26 AM	LO PRES	399.79	Alarm
Tag 1	8:18:30 AM	LO PRES	422.48	ReturnToNormal
Tag 1	8:18:30 AM	LO PRES	389.69	Alarm
Tag 1	8:18:34 AM	LO PRES	422.93	ReturnToNormal
Tag 1	8:18:37 AM	LO PRES	356.17	Alarm
Tag 1	8:18:40 AM	LO PRES	423.69	ReturnToNormal

Table 1 – Ad Hoc Query for Tag 1

Table 1 shows that Tag 1 is jumping around the low pressure alarm point. The *Operator Intervention* report was then used to determine that this tag had zero interventions during the period of alarming. Using the information from the reports, along with process knowledge about Tag 1, the controller tuning was modified and this "nuisance" alarm was eliminated.

After the initial 2 weeks of baseline data collection, modifications to the alarm configuration were started. Plant Management and DCS technicians met daily to review the alarm counts from the previous day. The top 3-5 alarms were selected, and the above methodology was incorporated to identify possible modifications. Modifications included controller tuning, changing dead bands, alarm limits, etc. Figure 3 shows the results after 2 weeks worth of modifications. For the week, the total alarm count was 1811, for a daily average of 258.

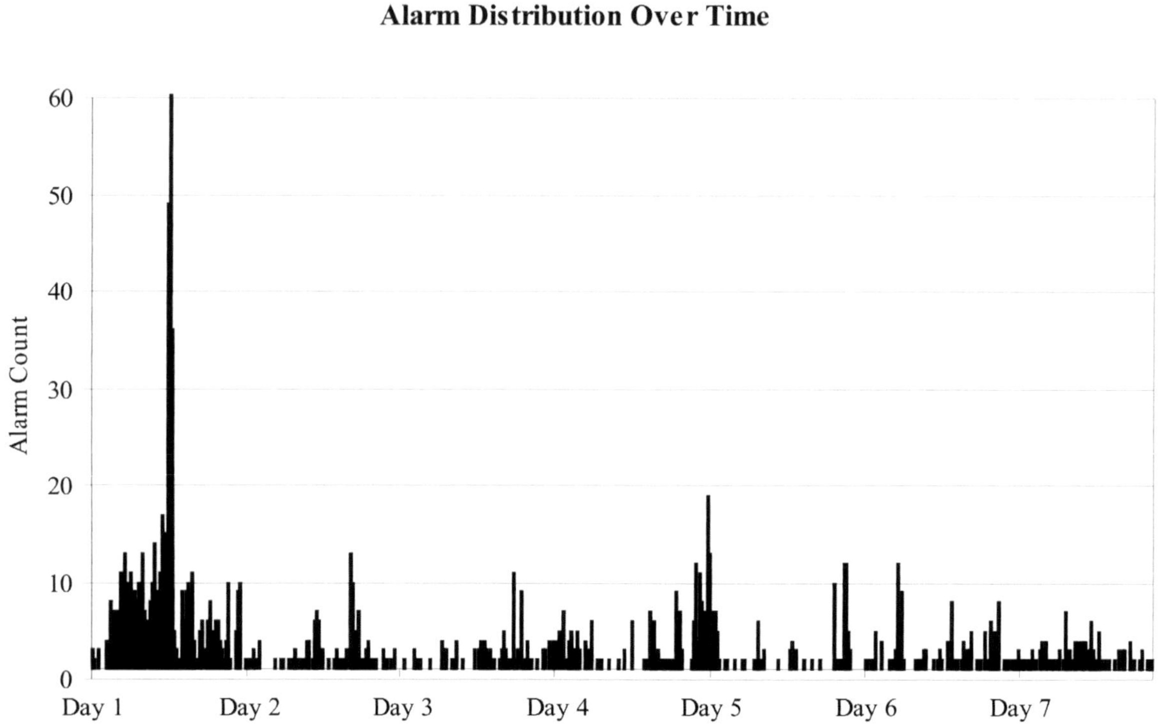

Figure 3 – Improved 10 minute alarm count over 7 days (zero counts excluded)

Table 2 details the performance before and after the initial modifications.

	Before	After	Change
Daily Average Alarms	868	258	-70%
Peak 10-minute Alarm Count	166	63	-62%
Peak Hourly Alarm Count	527	165	-69%
10-minute Intervals With no Alarms	13% of total	42% of total	+29% of total

Table 2 – Performance before and after

Once the initial modifications were complete, the majority of the "nuisance" and chattering alarms were eliminated. With the "noise" eliminated from the data, various other reports were then used to further study the system.

The *Standing Alarm – Duration* report looks at alarms that have occurred but have not returned to normal. This report was useful for checking how long conditions remained outside of the desired operating range. Because this report includes the start, end, and duration of an alarm, it quickly identified redundant and related alarms. For many of these related alarms (typically a pair), one of the alarms was eliminated.

As mentioned before, the *Operator Intervention* report shows the number of operator moves made on each tag over a specified time interval. This report was helpful in identifying alarm tags and process areas that required multiple operator adjustments. These tags were typically associated with control loops that required tuning adjustments.

The Symptomatic Analysis report shows correlations between an alarm (Parent) and the subsequent alarms that follow (Child). This report was useful in identifying how equipment and process areas affect one another. The various Parent/Child groups were studied, and typically resulted in long-term process improvement projects.

Figure 4 shows the results after 2 months worth of modifications. For the week, the total alarm count was 1124, for a daily average of 160.

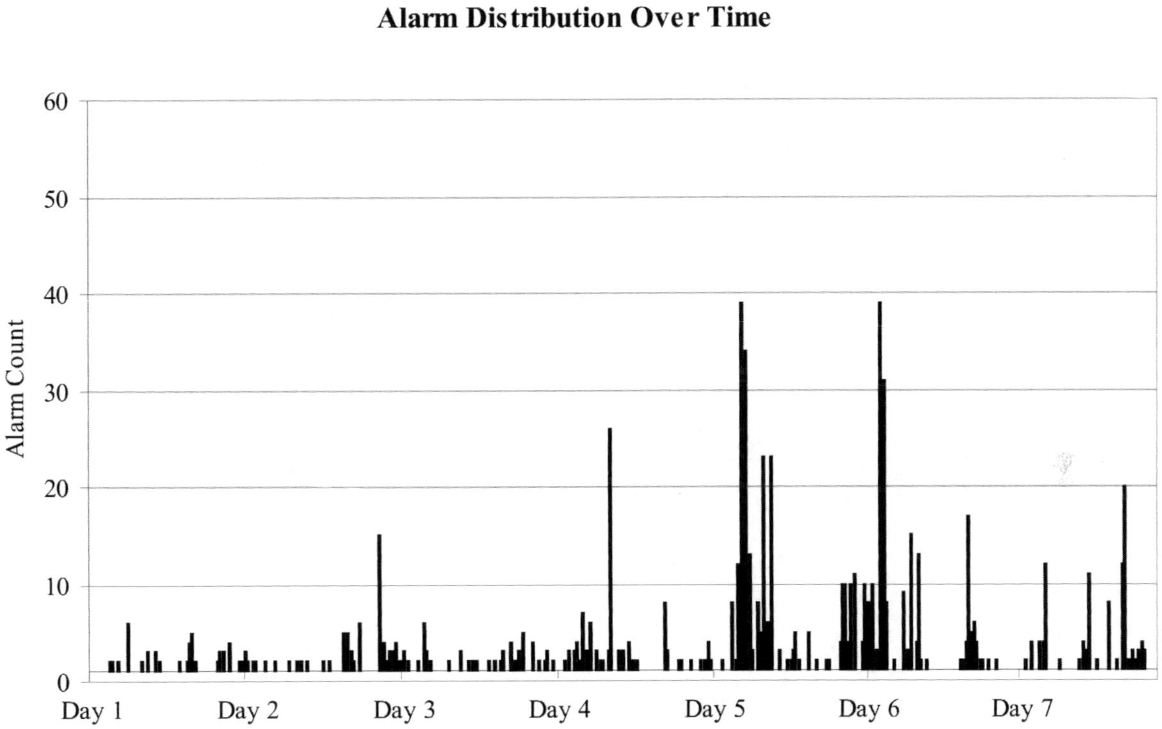

Figure 4 – 10 minute alarm count after 2 months (zero counts excluded)

Table 3 details the performance after 2 months worth of modification.

	Before	2 Weeks	2 Months	Total Change
Daily Average Alarms	868	258	160	-82%
Peak 10-minute Alarm Count	166	63	39	-77%
Peak Hourly Alarm Count	527	165	81	-85%
10-minute Intervals With no Alarms	13% of total	42% of total	62% of total	+49% of total

Table 3 – Performance after 2 months worth of modification

Management Process

After the initial optimization of the alarm system, a management process was developed and implemented in order to maintain the improved performance.

During the daily production meetings, the top 20 alarms from the previous day are reviewed. Any problematic alarm tags are assigned to engineers for further review. The engineers use the various ProcessGuard reports and process data to analyze the tags. The engineers will then present any potential changes at the next day's meeting.

Approved changes are given to the DCS technician who makes the change and enters it into a database. The database includes temporary and permanent changes as well as date, time, reason for change, etc. The DCS technician also maintains measures aimed at tracking the alarm system performance.

Plant upsets are reviewed in detail as they occur. Since upsets usually produce the "flood" of alarms, the goal is to look for ways to improve the alarming of that particular incident as opposed to individual tags.

Other Observations

In the later part of 2003, this ethylene plant had a shutdown and subsequent capacity increase. Shortly after startup, the alarm count jumped back up to previous alarm levels (>1000 per day). The same progression of analysis had to be repeated in order to re-optimize the alarm system at the higher production rates. The process is not a one-time fix. It does require continuous effort, albeit less than what is required initially.

The software has identified equipment/instrument problems. For example, a faulty limit switch.

Most of our older equipment does not have "first-out" type indication. The *Chronological Report* is helpful in determining what might have caused a piece of equipment to shutdown.

Summary

An alarm system that is not optimally configured can lead to excessive alarms that are a nuisance to the operator. This can mask the true problems with the process, ultimately hindering effective operator response.

In early 2003, Eastman installed Matrikon's ProcessGuard alarm monitoring and analysis software on one of its older ethylene units in order to optimize the alarm system. The software was initially used to quantify the alarm traffic seen by the operator. Then, smaller periods of interest were chosen, and individual alarm tags were studied. Modifications to the alarm system were made after the software identified obvious nuisance alarms. Further study of standing alarms, Parent/Child alarms, etc identified various other alarm tags that needed modification.

After 3 months of modification to the alarm system, the average daily alarm count was reduced from 868 to 160 (82% reduction). This alarm management process has since been installed on all of the older ethylene units and is being evaluated in other operating areas.

Design Guidelines for Distillation Columns in Fouling Service

Karl Kolmetz
Sulzer Chemtech, Singapore
kkolmetz@yahoo.com

Andrew W. Sloley
VECO USA, Inc.
andrew.sloley@veco.com

Timothy M. Zygula
Houston, Texas
tmz21@hotmail.com

Wai Kiong Ng
Sulzer Chemtech Singapore
waikiong.ng@sulzer.com

Peter W. Faessler
Sulzer Chemtech Singapore
peter.w.faessler@sulzer.com

Copyright 2004 K. Kolmetz, A. Sloley, T. Zygula

All rights reserved.

UNPUBLISHED

Prepared for

American Institute of Chemical Engineers

The 16th Ethylene Producers Conference

Section T8005 - Ethylene Plant Technology
Advances in Distillation Technology for Ethylene Plants

26 April 2004
New Orleans, Louisiana, USA

Abstract

Hydrocarbon Producers are exploring avenues to extend the on-stream time between outages for maintenance. Key equipment that can determine the end of run includes: catalyst life, cyclone erosion, and compressor and tower fouling. Critical equipment that has been shown to be a limiting factor can be duplicated to extend run length: for example parallel pumps, reactors and reboilers. This is a successful method to extend on-stream time, expensive and in fact, at times cost prohibitive. Incorporating design guidelines that increase the on-stream time of the key pieces of equipment is a better economic decision for most plants.

Currently Refiners are planning four-year run lengths and Ethylene Producers greater than five-year run lengths. These targets present challenges for distillation column design. Potential problem areas include refining vacuum wash oil beds, ethylene plant quench and saturator towers, and butadiene and other polymer producing distillation columns. Each of these applications has some common characteristics. A review of successful and not so successful designs can help develop key design criteria. Design guidelines developed from successful applications can improve the on-stream time of each of the applications.

Introduction

Designing mass transfer equipment for fouling service requires first an understanding of the fouling mechanism, the process in which the fouling occurs, and behavior of the process when the fouling is present. An understanding of these items needs to be developed in advance of designing mass transfer equipment for fouling service.

The challenges of fouling columns can result in;

1. Increase energy consumption due to heat transfer and efficiency issues.
2. Reduced column capacity, which may lead to production loses.
3. Increased down time for cleaning and disposing of fouling wastes
4. Potential need for the use of chemical additives

General Guidelines

Fouling service in a distillation column is a broad term that encompasses many fouling phenomena. The fouling phenomena can be localized in either the vapor phase or liquid phase. It the fouling is localized in the liquid phase an inhibitor can be utilized. Fouling phenomena can include;

1. Vaporization of volatile components
2. Polymerization
3. Condensation
4. Sedimentation / Precipitation / Crystallization
5. Foaming
6. Chemical Reaction
7. Corrosion

Complex fouling problems are a result of a combination of two or more of these phenomena occurring at the same time, such as in an ethylene caustic tower. They are nearly always mutually reinforcing.

Factors that assist fouling phenomena can include;

1. Residence time
2. Stagnate zones
3. Sharp transitions
4. Emulsion issues

Consideration must also include the fact that fouling gets worse over time. Fouling is not static.

Fouling Mechanisms

Vaporization of Volatile Components

In many applications fouling can result if the volatile components are allowed to vaporize. This fouling is found in refinery vacuum and de-asphaltene oil towers and ethylene quench oil towers. If the process is placed on recirculation without the addition of fresh feed with volatile material, the column can foul as the volatile components are removed. In one example an ethylene quench oil tower was placed on circulation and viscosity of the bottoms product increased beyond its pour point. The unit was shut down and required several days of off-line cleaning.

Polymerization

Polymerization is the linking of double bonds to form long chain molecules. Examples of polymerization include, polyethylene, polypropylene, polystyrene, poly butadienes and poly MMA. These products are undesirable in the monomer distillation column. These products can lead to reduced capacity and unit outages.

Condensation

Condensation is a reaction where two or more small molecules combine to from large, stable structure molecules. Extreme condensation is the formation of coke at high temperature and long residence times. Coke forms when thermal cracking removes hydrogen and light material from the fluid. A layer of coke is left behind. Thermal cracking can occur in both the gas phase and the liquid phase. However, the most common situation is liquid-phase cracking when low liquid rates extend residence time in high-temperature operations.

Examples of condensation can be found in ethylene furnace transfer lines to the quench oil tower and in refinery vacuum towers.

Sedimentation / Precipitation / Crystallization

Sedimentation is the accumulation of solids that are deposited in low velocity areas in process equipment. The equipment can include heat exchangers, tower distributors, distillation trays, random packing, and structured packing. If the trays are fouled below the feed point the cause might be sedimentation.

Many processes contain suspended solids, which can settle out on the mass-transfer surface. Products of sedimentation can include salts, metal oxides, catalyst fines, and coke fines. Precipitation and crystallization of dissolved salts can occur when process conditions become super saturated, especially at mass transfer surfaces. Ammonia salt deposition resulting from both water vaporization and direct solid deposition from the gas phase is a common refining problem.

Sometimes the deposits formed do not adhere strongly to the surface and are self limiting, the thicker a deposit becomes, the more likely is to be removed by the fluid flow and thus attain some asymptotic average value over time. Sedimentation fouling is strongly affected by velocity and less so by temperature, however a deposit can bake on a surface and become very difficult to remove.

Precipitation and supersaturation are fouling issues. Certain salts such as calcium sulfate are less soluble in warm water than cold. If such a stream encounters a wall at a temperature above that corresponding to saturation for the dissolved salt, the salt will crystallize on the surface.

Crystallization will begin at specially active points (nucleation sites) such as scratches and pits, often after a considerable induction period and then spread to cover the entire surface. The build up will continue as long as the surface in contact with the fluid has a temperature above saturation. The scale is strong, adherent and can require vigorous mechanical or chemical treatment for cleaning.

The following processes may cause super saturation:

1. Evaporation of solvent
2. Cooling below the solubility limit for solution with normal solubility, solubility increases with temperature.
3. Heat above the solubility of solutions with inverse solubility such as; $CaCO_3$, $CaSO_4$, $CA_3(PO_4)_2$, CaSiO3, $Mg(OH)_2$, $MgSiO_3$, Na_2SO_4, Li_2S0_4 and Li_2Co_3 in water.
4. Mixing of streams with different composition.
5. Variation of pH, which affects the solubility of CO_2 in water.
6. Ammonia salt deposition in refining

Additionally, solidification fouling can occur due to cooling below the solidification temperature of a dissolved component, such as solidification of wax from crude oil.

Knowing the properties of the fluid and suspended solid, flow conditions and mixing effects one can develop guidelines to predict sedimentation of solids from liquids. Petroleum fluid is a mixture of various compounds. Solubility balances among all such components make a petroleum fluid stable. Changes in that balance can cause precipitation from a petroleum fluid and result in fouling (asphaltene precipitation).

Foaming Systems

Some systems have a propensity for fouling in the form of foaming. In many systems an operating froth can be observed on the liquid phase. In distillation systems, the decrease in surface tension as equilibrium temperature rises promotes foaming. High gas velocities allow liquid to be entrained into the vapor phase. In non-foaming systems liquid disengagement from the vapor stream is relatively easy. Foaming makes liquid disengagement from vapor flow difficult. When foaming becomes severe it can lead to a reduction in capacity and loss of efficiency. In order to counter this effect distillation equipment that operate in foaming systems are run a lower vapor and liquid rates to reduce the amount of froth generated. Sometimes anti-foam additives are added to distillation columns to decrease the amount of foam generated in the column. There are drawbacks to using anti-foam chemicals. Sometimes they may contaminate the end product and produce product that is does not meet production specifications. (7)

Chemical Reactions

In distillation desired or undesired chemical reactions can occur. In an ethylene caustic tower there are competing chemical reactions. The caustic absorbs the CO_2 (desired) by electrostatic interactions and Van der Wall attraction. The acetaldehydes can be polymerized by the caustic (undesired) to form first a yellow oil polymer, and then a red oil polymer by aldol condensation reactions, which can then lead to emulsions.

Corrosion

When corrosion occurs the increase surface roughness may promote fouling from the other fouling mechanisms. A comprehensive theoretical discussion of the different fouling mechanism is given in [4].

Operating Conditions affect on Fouling Phenomena

Stagnant Zones

Stagnant zones are areas where fouling mechanisms can propagate: vaporization of volatile components, polymerization, condensation, sedimentation and chemical reactions. To eliminate stagnant zones a small slope can be added to collection devices and flow promoters to distillation trays. Stagnant zones increase the residence time of the process and lead to precipitation and build-up of polymer and coke. Stagnant zones can also promote unwanted chemical reactions because of high residence time and little or no movement of material.

Sharp Transitions

In the higher fouling cases sharp transitions and corners need to be avoided. Transitions and corners are areas where polymer and solids can seed and grow. Packed tower internals such as feed pipes and trough distributors can be areas for polymer and solids build-up. The fouling potential of packed tower internals can be reduced or eliminated through proper design practices.

If the fouling potential is in the vapor phase, the overhead vapor may be drawn from the side of the tower versus the tower over head to eliminate transition line length and an additional corner.

Emulsion Issues

In hydrocarbon service where water is present, as in ethylene quench and saturator towers, emulsion issues can create opportunities for fouling. If the pH of the water is allowed to depart from neutral ranges, emulsions can occur that create mixtures that can carry hydrocarbons along.

Trace contaminates like oxygen, nitrogen compounds, sulphur compounds, mercury, and chlorides have a significant effect on the performance of chemical process fouling and this need to be recognized if present in the process. Contaminates with surfactant properties should be reviewed carefully.

Measures to Mitigate Fouling Phenomena

Process Review

The first step in mitigation of fouling phenomena is to thoroughly review the process for fouling potentials. Investigations of previous applications need to be reviewed to develop the best solution to the specific process.

Equipment Selection

The most suitable mass transfer equipment for fouling service may also be the least efficient for mass transfer. Grid packing and shed decks can handle nearly every fouling service, but have low efficiencies when compared to sieve trays, random, and structured packings.

Packing

For packed towers the key fouling factors revolve around liquid distribution and packing residence time. The longer the residence times the less suitable. Low-pressure drop, smooth surface, low residence time packings perform best in fouling service. The order of preference is:

1. Grid
2. Structured packing
3. Random packing

Packing Distributor Concerns

In fouling service distributors are areas where residence time is increased and fouling phenomena can occur. In high-fouling services trough v-notch or other type of trough distributors are recommended over pan type distributors. A review of feed piping is also warranted.

Trays

The industry prefers trays in fouling service because of the long history of success trays have had in fouling service applications. The first continuous distillation column with bubble cap trays was developed in 1813 and structured packing was developed in 1964. The database and application know how is much larger with trays. The best trays to use in fouling services are sieve trays and dual flow trays. Moveable and fixed valve trays are less resistant to fouling because the valves are areas where fouling can seed and propagate. Distillation is part art, part science. As in any science, as you explore the limits you will find them

Example of Trays in Ethanol Mash Service.

Tray Downcomer Concerns

In fouling services downcomer design requires attention to detail. A major problem with downcomers is the creation of dead spots: in areas near the tower wall, in areas opposite the downcomer outlet, areas near outlet weirs, and at the ends of the downcomer. Using modified downcomers that decrease the downcomer volume helps to reduce the dead spot by keeping more of the downcomer volume full of agitated liquid. For quick revamps of existing equipment an alternative is to install metal or ceramic shapes to modify the downcomer volume. Consider the influence of weir height and avoid the use on inlet weirs when possible.

Baffle Trays

1. Shed Decks

Shed decks are essentially angle iron beams of various sizes from two to ten inches that are placed in rows across the column. They typically are at on 24 inch tray spacing. They may be set in overlapping rows or rotated 90 degrees from tray to tray. The open area on the tray is typically 50%. The positives of shed deck are that the fouling potential of the decks is almost zero because there are no stagnate zones and low residence time. Unfourtunaly, the efficiency of the tray almost matches the fouling potential, particularly if wide shed decks are utilized. Shed decks work well in fouling applications where their application is essentially for heat-transfer purposes.

Shed Trays

2. Disk and donut trays

Disk and donut trays are slightly sloped trays that allow the liquid to splash from inner circle ring to outer circle ring. Fouling potential of this tray is low along with the efficiency.

Disk and Donut Trays

3. Side to side trays

Side-to-side trays are trays that allow the liquid to splash from side to side. Fouling potential of this tray is low, as with efficiency

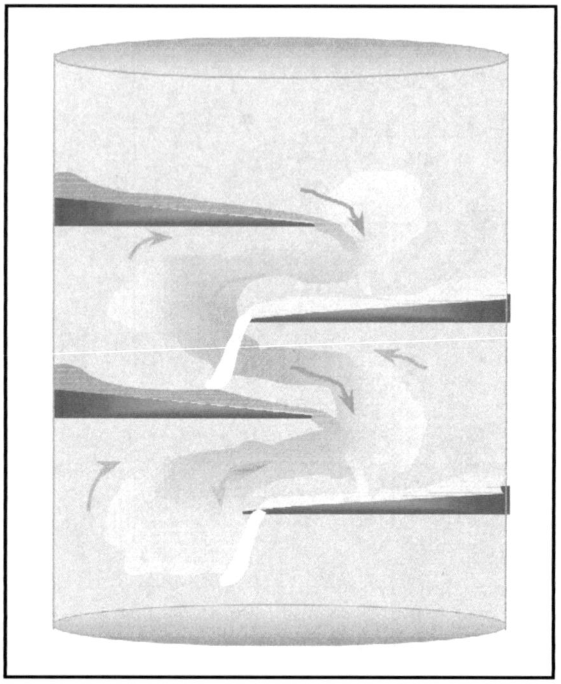

Side-to-Side Trays

Dual Flow Trays

Dual flow trays are the trays of preference for heavy fouling services. Dual flow trays have no downcomers, where products of fouling phenomena can accumulate or where polymer and solids can seed and propagate. Dual flow trays are designed with enough open area on the tray decks to eliminate stagnation and promote back mixing. A disadvantage to dual flow trays is that they are limited in turndown potential.

The vapor and liquid transfer up and down the column through the holes on the tray deck. This is an advantage in combating vapor-phase fouling as the underneath of the tray is washed. The continuous agitation of the liquid on the topside of the trays combined with continuous underside wetting/washing action makes this tray suitable for fouling services

The challenge of dual flow tray is mal-distribution in larger diameter towers. The top of a column will move in a typical storm as much as six inches. This movement will cause the hydrologic load to migrate in the column. If hydrologic flow instability is developed it propagates down the column. Improper feed, reflux or vapor distribution can also create mal-distribution problems,

Two types of dual flow trays are available; standard deck and rippled deck. The standard deck has a flat plate, and the rippled deck has sinusoidal waves

Spray nozzles vs. distributors

There is much discussion in the industry between the advantages and disadvantages of spray nozzles in fouling services. There are groups that advocate only spray nozzles with no packing or shed decks. There are groups that advocate only distributors with packing.

The best trough distributor will have superior distribution performance to the best spray distributor. In fractionation services, the trough distributor may be mandatory to meet separation requirements. In solids services, such as refinery FCC slurry zones, specially designed trough distributors have proven performance.

However, as long as the nozzles on a spray distributor have a large enough minimum flow passage to protect against plugging, spray distributors resist plugging. They have performed well in extremely severe services including refinery vacuum wash zones and delayed coker main fractionators.

Material of Construction / Surface Finishing

It may be possible to reduce fouling by the choice of material selection, or more importantly by surface finish. Smooth surfaces shed deposits easier than rough surfaces.

A corroded surface is more prone to other fouling mechanisms. The following guidelines on material selection can be considered on services that are very prone to fouling. (3)

1. Standard austenitic stainless steel such as 316 (18Cr 10Ni 3Mo) can be utilized for resistance to acids and reasonable resistance to pitting corrosion. Type 304L (18Cr 10Ni) stainless steel has a good resistance to nitric acid. Austenitic stainless steels have relatively low strength, poor anti erosion and abrasion properties and do not possess the ability to resist stress corrosion cracking.

2. Super austenitic stainless steels with relatively high nickel content (approximately 20 Cr 29 – 34 Ni), sometimes referred to as alloy 20 are more costly than standard austenitic steels but provide excellent resistance to acids and some acid chlorides.

3. Duplex stainless steel offer high strength, coupled with resistance to abrasion, erosion and stress corrosion cracking.

4. Nickel based alloys, such as Hastelloy, have out standing corrosion resistance in resisting acids, mixed acids, and acids at high temperatures. The principle restriction in their use is the high cost.

Use of Additives

An additive may act to resist fouling in one or more of the following ways.

1. Chemically react with the fouling species to modify it's fouling potential.

2. Change the physical interaction between the foulant and the equipment surface.

3. Modify the deposit residing on the surface, so that is can be removed by fluid flow.

It is also possible to reduce the fouling potential by pre-treating the metal with passivating solutions.

Case Studies

Case Studies can be utilized to refine design practices and standards for fouling services.

Refinery Vacuum Towers

Refinery Vacuum Towers are an excellent case study because many of the fouling phenomena are present in this application. The most challenging part of the vacuum column is the wash bed, which removes the entrained bottoms from the distillate products. Both trayed wash sections and packed wash sections have problems. The lower pressure drop of packing has overwhelming shifted the industry to packing for this service.

Liquid rates in the wash zone are very low. The low liquid rate, couple with high temperature, results in long residence times in stagnant sections of the packing. This occurs in the packing surface treatment, the interface between packing layers, on packing supports, and in dead portions of the packing (for random packing).

Asian Refinery Vacuum Tower Wash Bed

The industry average wash bed run length might approach 5 years. The average life of a wash bed is based on a variety of factors, which include;

1. Design of inlet feed system to reduce entrainment and vapour mal-distribution.
2. Design of inlet transfer line to reduce velocity and resulting entrainment
3. Design of furnace to prevent high flux areas which initiate cracking and coking
4. Design of collector tray, this can be sloped to reduce residence time
5. Rate of wash bed oil flow
6. Amount of vacuum – deep cut vacuum will result in more velocity gradients leading to more entrainment and local low-flow areas
7. Selection of packing and internals

Vacuum Tower Design Options

Baffle Trays

Baffle trays, particularly shed decks have seen excellent service factors in Refinery Vacuum Towers. The positives of shed deck are that the fouling potential of the decks is almost zero because there are no stagnate zones and low residence time. The negatives are the low stage efficiencies. Baffle trays are still in use in some older units.

Grid Equivalents

Grid Equivalents have proven service in wash zones. Capacity is high and entrainment removal is average. With sufficient liquid wetting, coking and fouling is unusual. Grids fit the design guidelines for packing in fouling service, which include low-pressure drop, smooth surface, low residence time.

Random Packing

Random packing has had successes in vacuum towers. However, due to the increased possibility of stagnant liquid pool formation, many installations have changed to structured packing or grid.

Structured Packing

Structured Packing may be used in wash beds if designed correctly. The challenge of structured packing in vacuum tower service is increased efficiency at slightly increased delta p compared to grids along with rougher surfaces and increased residence time.

If the packing is allowed to dry, by not properly designing the wash bed liquid rates with optimized liquid distributors, the packing can foul quickly causing very high-pressure drops and poor product quality and reduced capacity.

Current practice is a mixed bed of grid and structured packing combining fouling resistance at the lower sections and high separation efficiency at the higher sections. Computerized Fluid Dynamics (CFD) studies should be evaluated to optimize inlet vapor flow.

Examples

1. An Asian Refinery had a challenge maintaining wash oil rates due to plugging spray nozzles and the result is shown in the previous picture. The run length of this tower was less than one year. The spray nozzles were replaced with a liquid distributor and since the change was made there has been no sign of fouling. The same refinery had a second challenge of napthanic acid attack to the internals. The tower was revamped and metallurgy of the internals was upgraded.

2. An Asian Refinery wash bed was changed from random packing to grid packing in 1995, this tower is still in service today.

An example of the grid installed in the Asian Refinery

Lessons Learned

1. Grid equivalents may be an improved packing choice over structured packing at a reduced efficiency price in wash beds lowers sections.

2. Wash bed rates are difficult to simulate correctly and need to be carefully specified to insure long bed life with reduced fouling.

Ethylene Quench Oil Towers

Ethylene quench oil towers have the potential to encounter several of the fouling phenomena described in this paper. The distillation equipment selection needs to consider each of the fouling scenarios. The industry average Ethylene Unit run length might approach 6 years.

A challenge for quench oil towers is the formation of coke on the tower internals and fines. The tower internals must be designed, especially velocities and the residence time, to reduce the formation coke and fine accumulation on the internals.

An additional design criterion for ethylene quench oil towers is the concern for naphthalene concentrations. On heavy naphtha's the naphthalene concentration can cycle up in the tower. The naphthalene is too heavy to exit in the tower overhead, and can be stripped back into the tower from the fuel oil stripper, causing the liquid traffic in the tower to flood.

Quench Oil Tower Design Options

Baffle Trays

Baffle trays, particularly shed decks, have been used successfully in Ethylene Quench Oil Towers. The positives of shed deck are that the fouling potential of the decks is almost zero because there are no stagnate zones and low residence time. The negatives of baffle trays and shed decks are the low efficiency of the tray.

Ripple Trays

An Asian Producer has seen excellent fouling resistant service from Ripple Trays. They have been in service ten years will no issue of fouling.

Random Packing

A small negative of random packing is the horizontal area where liquid can pool. Fouling can begin to propagate from this area.

Lessons Learned

1. One on the lessons to learn is to utilize caution when commissioning quench oil towers. Normal guidelines are to purchase #2 Diesel Oil to start the circulation fluid. If the furnaces are on steam stand by to the quench oil towers, the steam can strip the light ends out of the circulation loops, increasing the viscosity of the circulation loops. If the viscosity is allow to increase above the pour point the flow from the tower can cease, requiring several days to clear the fouled lines. The vaporization of volatile components is a concern in Ethylene Quench Oil Towers.

2. The naphthalene concentration can cause high liquid loads in the bottom of the quench oil tower. Naphthalene is too light to exit with the fuel oil, and too heavy to exit the tower overhead.

Ethylene Quench Water Towers

Ethylene quench water tower in naphtha crackers that have quench oil towers can be relatively trouble free because the oil tower removes the coke fines and heavy hydrocarbons. Quench water towers for ethane crackers are a different story. There can be numerous challenges in light gas crackers. These can include;

1. Coke fines
2. Pyrolysis Tar
3. Emulsions

Quench Water Tower Design Options

Baffle Trays

Baffle trays, particularly shed decks, have seen excellent service factors in ethylene quench water towers. The positives of shed deck are that the fouling potential of the decks is almost zero because there are no stagnate zones and low residence time.

Ripple Trays

An Asian Producer has seen excellent fouling resistance service from Ripple Trays. They have been in service ten years will no issue of fouling.

Random Packing

A small negative of random packing and grid equivalents is the horizontal component, which is an area for the fouling phenomena to begin to propagate.

Examples

1. US Gas Cracker. A quench water tower was commissioned with a combination of spray nozzles, and random packing with pan distributors. The random packing was replaced in 1999 and 2000 because of high-pressure drop and loss of capacity. By February 2001 the tower was again showing signs of high-pressure drop and heat transfer losses. In July 2001 the tower was converted to v-notch liquid distributors and grid packing. The tower, which has been operating for 2 1/2 years, is now performing with higher capacity, lower pressure drop and improved heat transfer capacity than was experienced when the tower was first commissioned.

2. US Gas Cracker Quench Water Tower. The original design of the tower had random packing in the top bed and structure packing in the bottom bed. The tower was upset during start-up and the packed bed collapsed. The tower was then hot tapped and spray nozzles were inserted through ring of hot taps around the top of the tower. With the ring of hot taps and spray nozzles unit the tower was able to run at full rates, however, the tower's top temperature was higher than design, about 115 to 120 °F. The temperature in the bottom of the column was 140 to 160 °F. The plant should be commended for designing and installing a workable system in 20 days without bringing the plant down.

   In 1995 the tower was revised with random packing in the top and a draw pan for spray circulation in the tower bottom. The draw pan was upset during start up. Later inspection found hold-down clamps were not properly installed and bolts were missing. The ring of spray nozzles was reinstalled.

   In 2000 the circulation rate was suddenly reduced, as if a line plugged overnight. The tower was shut down for inspection and a loose vapor riser hat from the draw pan was found in circulation draw. The vapor riser hats were welded in place improving the previous bolting design.

   In September 2003 the tower was converted to v-notch liquid distributors and grid packing. The tower is now performing with lower pressure drop and improved heat transfer capacity.

Lessons Learned

1. The inspection of tower internals installation is important part of a successful column revamp. This is especially true of a column that has a system with hydrocarbons and water. The volumetric expansion of water is 1100:1. Any potential for water to vaporize will greatly increase the volume of vapor flow. Combine this volume increase with improperly installed equipment and severe problems are sure to follow.

2. The use of random packing and pan distributors should be limited in fouling services, especially quench towers.

3. A small injection of py gas to a quench water tower greatly improves operation.

Ethylene Saturator and Dilution Steam Generator Towers

The ethylene industry recognizes that saturator and dilution steam generator tower designs are a challenge. Many technology suppliers will provide multiple or small dual units to be utilized when the large tower is being refurbished.

Typically challenges include;

1. Corrosion due to pH issues

2. Erosion due to coke particles

3. Fouling due to polystyrene and pyrolysis tars,

4. Emulsion issues due to carry over of hydrocarbons from quench water towers and hydrocarbon removal systems, which can be a stripper tower or Dispersed Oil Extraction - DOX Unit.

Environmentally this is an important tower, which has two environmental functions. The first is water conservation. The Saturator and Dilution Steam Generators recycle the water needed in the cracking furnace, thus reducing the total amount of wastewater generated. The second is hydrocarbon losses to the wastewater treatment system, partially phenols. This tower strips the hydrocarbons from the tower bottoms that are sent to the wastewater treatment facility.

Typical Design of a DSG System

The effluent from a pyrolysis furnace can be introduced to a quench oil tower and then to a quench water tower for cooling. The effluent is a full range mixture of hydrocarbons and water.

Water for the dilution steam system is withdrawn from the circulating quench water loop of the quench water tower and can be pretreated in filters and coalescers to remove most of the entrained hydrocarbons and coke fines.

The filters and coalescers can be an elaborate system utilizing anthracite and multiple coalescers sometimes called a Dispersed Oil Extraction (DOX). Some systems utilize a low pressure water stripper using steam to strip-off the volatile hydrocarbons back to the quench water tower.

The partially stripped water is introduced into the dilution steam generator tower. The water feed stream is typically introduced on the top tray and counter flow with the rising vapor, which allows the steam to leave at saturation.

Steam is added to raise dilution steam temperature to slightly above superheat, preventing condensation along the dilution steam piping from the tower to the pyrolysis furnace.

Steam is also used in the dilution steam generator reboiler as the heating medium. Blow downs from furnace steam drums and boilers can also be introduced to the DSG tower. Dual flow trays or other non-fouling distillation equipment should be utilized to providing resistance to fouling, which is typically in this service. Styrene polymers have been know to foul the top of DSG Towers. Heavy compounds leave the dilution steam system in the tower bottom blow down to the wastewater treatment.

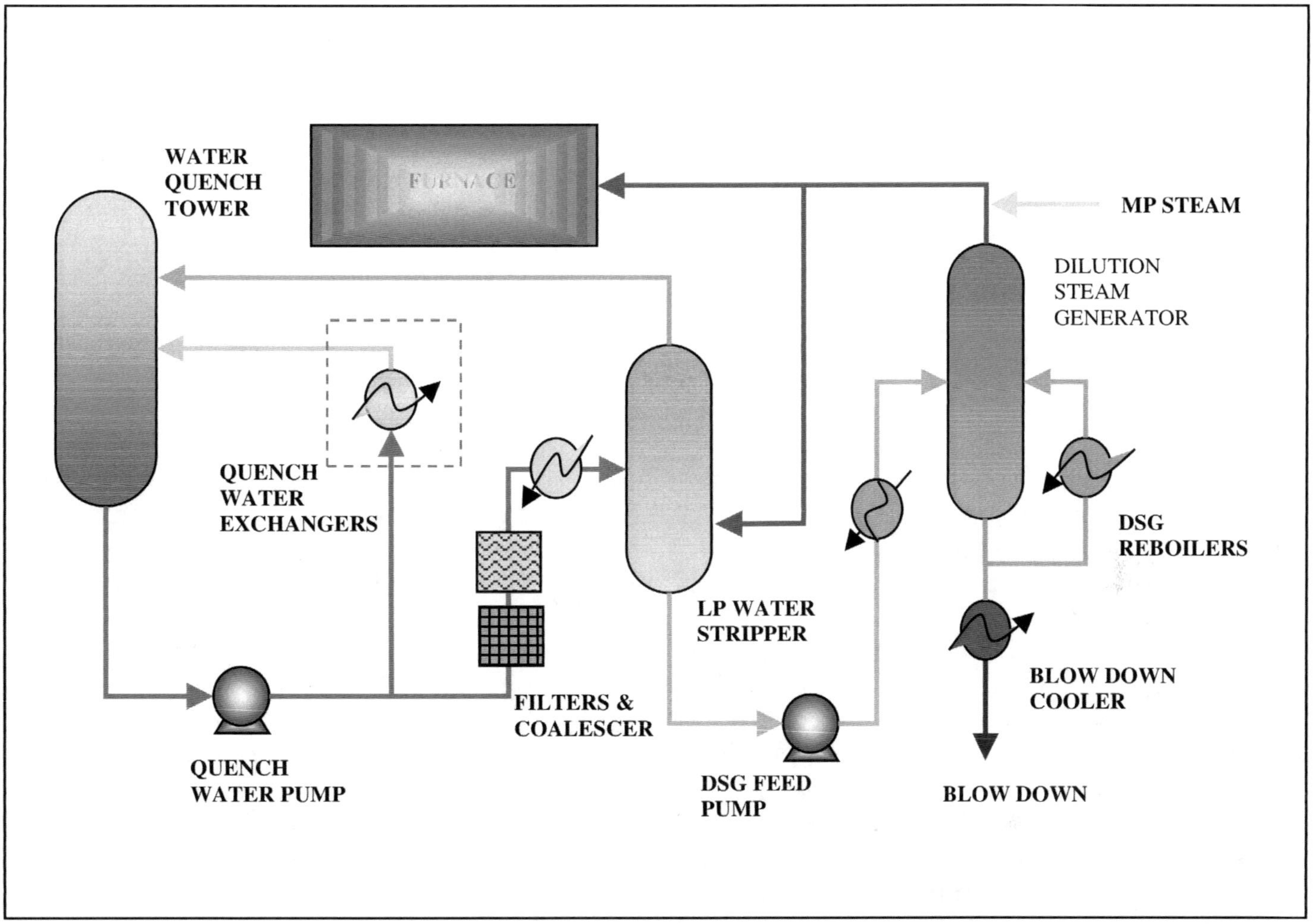

WATER
QUENCH
TOWER
FURNACE
MP STEAM
DILUTION
STEAM
GENERATOR
QUENCH
WATER
EXCHANGERS
DSG
REBOILERS
LP WATER
STRIPPER
BLOW DOWN
COOLER
FILTERS &
COALESCER
QUENCH
WATER PUMP
DSG FEED
PUMP
BLOW DOWN

Typical Design of a Saturator System

The effluent from a pyrolysis furnace can be introduced to a Quench Oil Tower and then to a Quench Water tower for cooling. The effluent is a full range mixture of hydrocarbons and water.

Water for the dilution steam system is withdrawn from the circulating quench water loop of the Quench Water Tower and can be pretreated in filters and coalescers to remove most of the entrained hydrocarbons and coke fines.

The filters and coalescers can be an elaborate system utilizing anthracite and multiple coalescers sometimes called a Dispersed Oil Extraction unit (DOX). Some systems utilize a Low Pressure Water Stripper using steam to strip-off the volatile hydrocarbons back to the Quench Water Tower.

The then partially stripped or extracted water is introduced into the Saturator Tower. The water feed steam is typically introduced on the top of the tower and the hydrocarbon vapor feed steam is typically introduced at the bottom of the tower. The counter flow of the rising vapor generated from the reboilers and the hydrocarbon allows the stream to leave at saturation.

Steam can be added to raise dilution steam temperature to slightly above superheat, preventing condensation along the dilution steam piping from the tower to the pyrolysis furnace. Knock out pots can also be utilized to remove any water that condenses. Additionally a second pass through the furnace can superheat the saturated stream to super heated conditions.

Steam is also used in the reboiler as a heating medium. Blow downs from furnace steam drums and boilers can also be introduced to the Saturator tower. Dual flow trays or other non-fouling distillation equipment should be utilized to providing resistance to fouling, which is typically in this service. Heavy compounds leave the dilution steam system in the tower bottom stream. This stream is typically blown down to the wastewater treatment system.

Design Of a Typical Saturator Tower

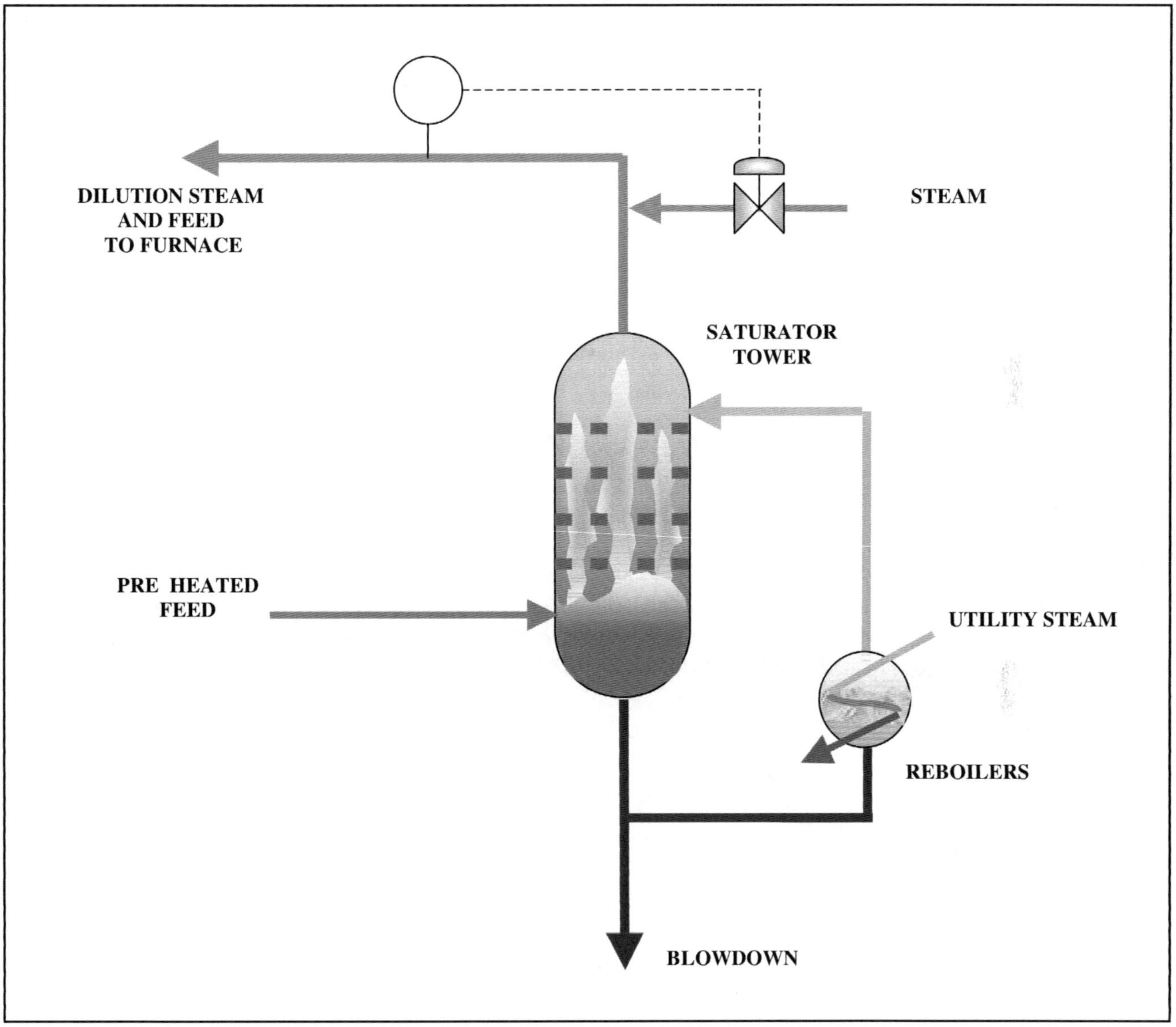

Saturator Tower Design Options

Baffle Trays

Baffle trays, particularly shed decks have been used successfully in saturator towers. The positives of shed deck are that the fouling potential of the decks is almost zero because there are no stagnate zones and low residence time. The negatives of baffle trays and shed decks are the low efficiency of the tray.

Random Packing

A small negative of random packing is the horizontal component, which is an area for the fouling phenomena to begin to propagate.

Grid Equivalents

Grid fits the design guide lines for packing in fouling service, which include low-pressure drop, smooth surface, low residence time.

Dilution Steam Generator Tower Design Options

Baffle Trays

Baffle trays could be utilized in DSG Tower design and might be the preferred option. The positives of baffle trays are that the fouling potential of the decks is almost zero because there are no stagnate zones and low residence time. The negatives of baffle trays are the low efficiency that can be seen if not designed properly

Ripple Trays

An Asian producer has seen good service from Ripple Trays except for the emulsion issues, which are related to pH control.

No Internal Distillation Devices

Several Producers have chosen to remove the distillation devices and have only a spray tower. This option is the most operationally reliable, but may not be the best environmentally.

Examples

1. US Gas Cracker. A saturator tower was commissioned with a combination of spray nozzles and random packing with pan distributors. The random packing was replaced in this tower in 1999 and 2000 because of high-pressure drop and loss of capacity. The tower was experiencing high-pressure drop and reduction in capacity only a few months after the packing was replaced. Attached is a table of pressure drop during this period of operation.

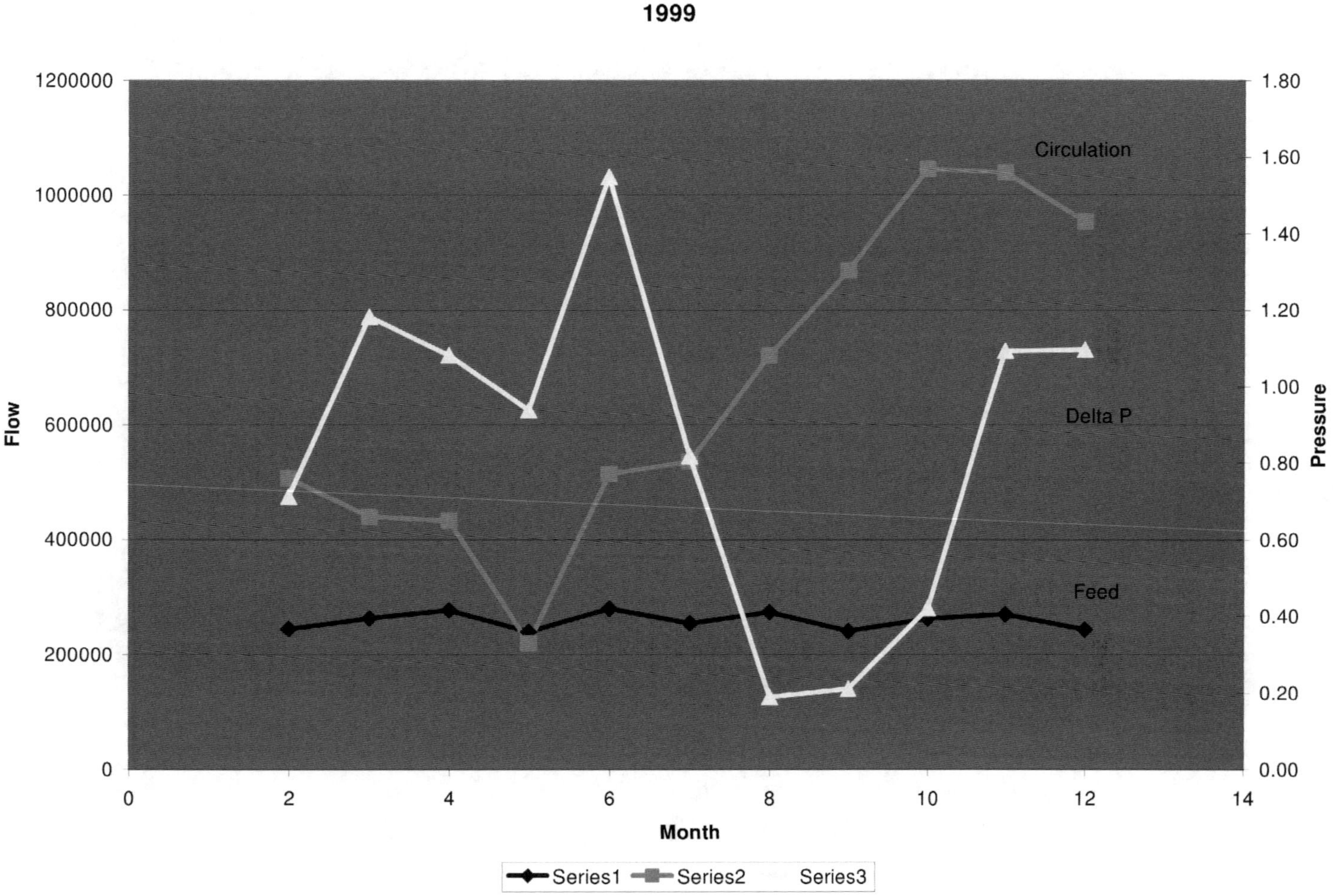

The pan distributor was replaced with a v-notch distributor without parting boxes to redistribute the flow. Because this tower is mainly for heat exchange and not mass transfer the parting boxes were not installed to reduce fouling potential.

Installed V-Notch Pan Distributor in Saturator Tower

2. Asian Ethylene Producer

During the commissioning of a furnace sulfur is added to reduce the catalytic coking of the metal based tubes. This sulfur can reduce the pH of the quench water tower and form an emulsion. Because water is polar and styrene is polar the water / styrene emulation can be difficult to break. This emulsion can carry through to the Dilution Steam Generator and form styrene polymer when introduced to the higher temp of the DSG Tower.

Ethylene Unit Caustic Towers

Ethylene caustic tower have many of the fouling phenomena: polymerization, sedimentation, precipitation, crystallization, chemical reactions, and corrosion. Caustic towers qualify as a complex fouling system with more than two fouling phenomena occurring at the same time while being mutually reinforcing.

A certain amount of polymer formation in a caustic tower is expected in most systems. Caustic tower polymer fouling is typically thought of as a function of acetaldehyde concentration and aldol condensation reactions. The acetaldehydes can be polymerized by the caustic to form first a yellow oil polymer, and then a red oil polymer by aldol condensation reactions, which can then lead to emulsions.

Caustic tower sedimentation results in the salt by-product being deposited on the underside of the tray. Over time this by-product can product a high-pressure drop resulting in reduced rates and increase energy consumption in the Cracked Gas Compressor. Some producers utilize a small continuous py gas wash, followed by a larger py gas wash when the pressure drop increases.

Caustic Tower Design Options

Standard Sieve Deck Trays

Standard sieve deck trays can be utilized in caustic tower design. The positives of sieve decks are that the chemical reaction ability of the process can be maintained over time. The negatives of trays are the higher-pressure drop as compared to packing. Caustic Towers have seen excellent service from standard sieve deck trays. Run lengths have exceeded five years.

Random Packing

Caustic Towers have seen good service from random packing. A small negative of random packing and grid equivalents is the horizontal component, which is an area for the fouling phenomena to begin to propagate. Run lengths have exceeded five years.

Lessons Learned

1. The use of random packing and pan distributors should be limited in fouling services.

2. A small injection of py gas to a caustic tower greatly improves operation.

3. Internal overflows need to be carefully considered during the design phase of a revamp. High-pressure drops can result if internal overflows are undersized.

Ethylene Unit Butadiene Fouling

An ethylene unit can experience Butadiene Polymer fouling in the deethanizer, depropanizer, and debutanizer columns. Butadiene polymerizes to poly butadiene, sometimes called popcorn polymer because it is pyrophoric and makes a popping sound when exposed to air. The polymer formation is temperature dependent and recommendations are to keep temperatures below 200 degrees F.

Example of Butadiene Polymer Fouling in a depropanizer.

Example of depropanizer Tower Bottoms

Butadiene Polymer Tower Design Options

Standard Sieve Deck Trays

Standard sieve deck trays can be utilized in butadiene polymer tower design. Many Producers have seen excellent service from standard sieve deck trays. Run lengths have exceeded five years. Electro-polished trays have been used in this service to prevent the popcorn from sticking to the tray.

High Capacity Trays

With the short residence times of high capacity trays, particularly the elimination of dead zones on the tray by flow directors, polymer production could be reduce in the tower bottoms where higher temperatures are necessary.

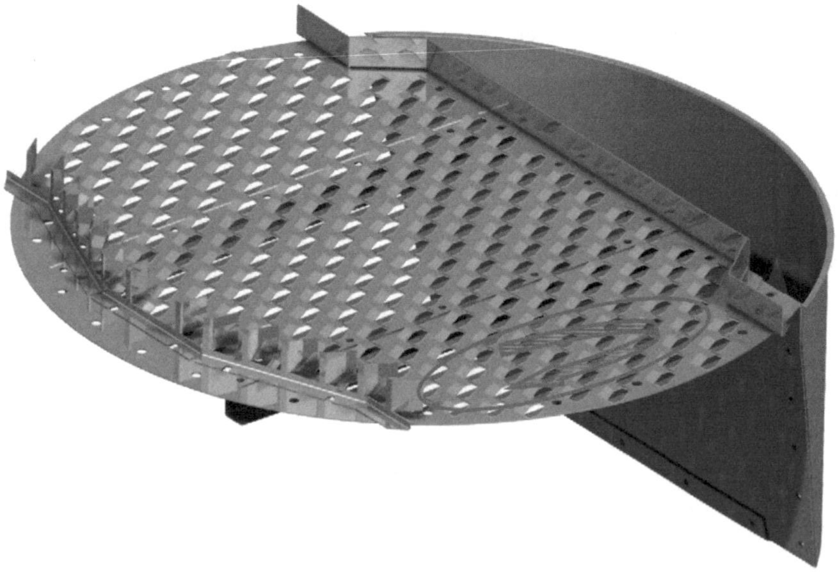

Conclusions

There are many different opinions on designing tower internals for use in fouling services, and because distillation is part art and part science differing opinions can be proven to be serviceable in the field. Trays and packings can both be utilized for fouling services. The key to a successful design is the proper application of internals, trays and or packings. This paper is a starting point for producers to develop, based on their operating experience, practical, serviceable columns for fouling service.

References

1. Andrew W. Sloley, Gary R. Martin, "Subdue Solids in Towers", Chemical Engineering Progress, January 1995

2. G A Mansoori, "Arterial Blockage Fouling" September 2001

3. Muller Steinhangen, Heat Exchanger Fouling – Mitigation and Cleaning Technologies, IchemE UK 2000

4. Bott, TR, Fouling of Heat Exchangers, Elsevier, 1995

5. Timothy M. Zygula, Karl Kolmetz, Randy Sommerfeldt, "Revising An Ethylene Feed Saturator for Increased Reliability", AIChE 2003 Spring National Meeting, April 1, 2003, New Orleans, Louisiana

6. Kolmetz K, Zygula T, "Resolving Process Distillation Equipment Problems", Prepared for The 5th Annual Regional Olefins Conference, October 31st - November 3rd, 2000, Johor Bahru, Malaysia

7. Kister, H. Z. "Distillation Design", McGraw-Hill Book Company Inc., New York, 1992.

8. Kister, H. Z., "Troubleshoot Distillation Simulations", Chemical Engineering Progress, 63-75, June 1995.

9. Liem V. Pham, Mike Binkley, P.E., J. Y. Jang ,Timothy M. Zygula, Robert M. Garner, HIGH-PERFORMANCE ANTI-FOULING TRAY TECHNOLOGY, Paper presented at the AIChE Spring National Meeting, Houston, TX March 12, 1997

Developments in Splitter Revamps

by

Waldo de Villiers & Jose Bravo, Shell Global Solutions (US), Inc.
Peter Wilkinson, Shell Global Solutions International, Amsterdam, NL
Daniel Summers, Sulzer Chemtech, Tulsa, OK

Abstract: This presentation will focus on developments in ethylene plant fractionator revamps. It will discuss the use of high capacity distillation trays to increase capacity in fixed tower sizes. It will present an overview of Shell high capacity mass transfer technology including the HiFi† Plus and ConSep† trays. Additionally it will present revamp case studies and performance data illustrating the potential of these devices.

Ethylene plant distillation trains

An ethylene plant makes extensive use of distillation, especially in the cold section illustrated in figure 1. Any revamp project will also need to consider the required column internals for the columns shown below to allow the desired throughput.

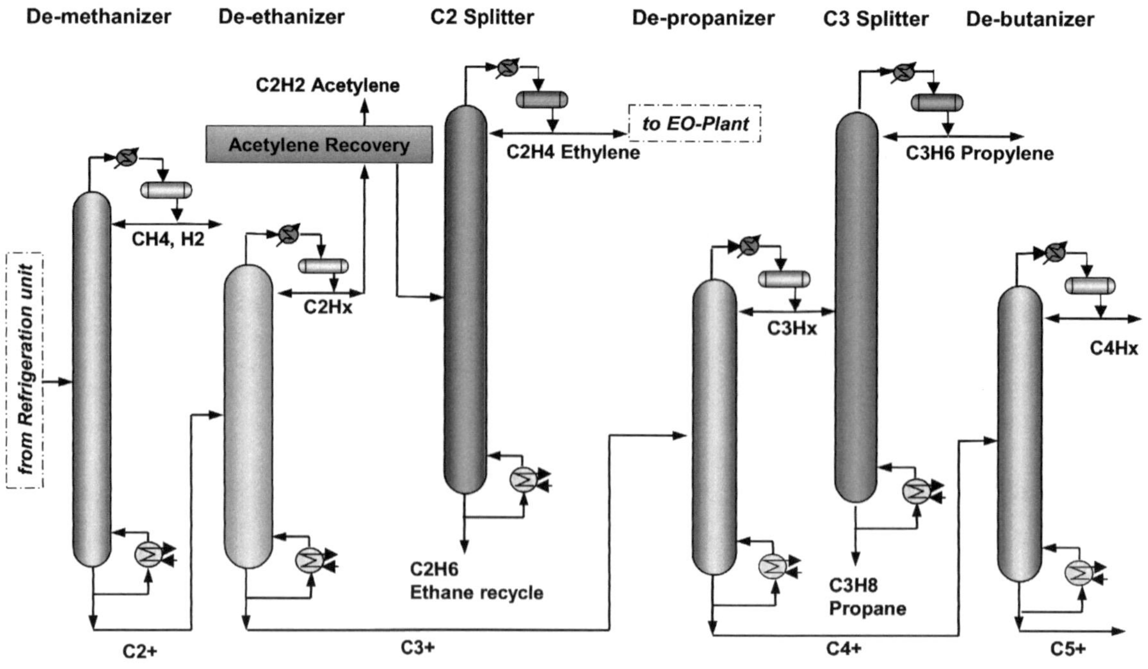

Figure 1: Ethylene Plant Cold Section

† Shell Group trademarks

Principles of High Capacity trays

Revamps of high pressure distillation columns such as those found in the cold section of ethylene plants can effectively make use of high capacity tray technology. Features of high capacity trays are:

- Truncated downcomers that maximize the bubbling area
- Low entrainment bubblers such as Sulzer MVG[††] valves
- Long weir lengths providing high liquid handling capacity
- Good tray efficiency ensured by providing adequate flow path length and avoiding stagnant areas through good liquid distribution
- Ability to function at low tray spacing
- Devices at the inlet of downcomers that produce a lower and more uniform froth height

The Shell family of high capacity devices such as the Calming Section Plus (CS Plus) and HiFi Plus trays make use of these features. Figure 2 illustrates the expected capacity and range of application of these devices relative to conventional trays and packing. The design principle is that for every flow parameter ϕ there is an optimal tray layout – one that maximizes vapor capacity whilst maintaining an acceptable efficiency. At high system pressures downcomer choking may occur resulting in increased froth height and an earlier limitation on tray capacity. Hence the maximum useful capacity is really that determined by the minimum of jet flood, downcomer choking and downcomer inlet velocity restrictions. The concept of "maximum useful capacity" (MUC) rather than a hydraulic flood is used to indicate the maximum tray capacity. The column may still be operable beyond 100% of MUC but the efficiency will be severely degraded.

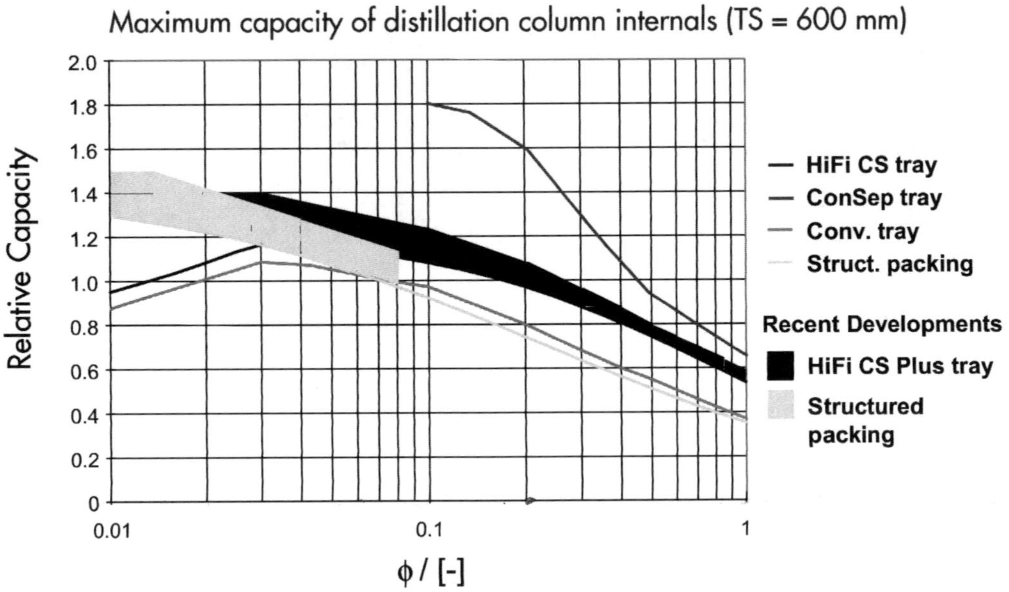

Figure 2: Comparison of column internal capacities

[††] Sulzer Chemtech trademarks

The flow parameter is defined as: $\phi = \frac{L}{V}\sqrt{\frac{\rho_g}{\rho_l}}$, where L/V represents the liquid to vapor mass flow ratio. The density correction factor is the ratio of vapor to liquid density.

Sulzer MVG valves (see figure 3) are utilized in CS Plus and HiFi Plus trays to offer increased vapor capacity and reduced entrainment. The lateral discharge of the vapor jets helps to collapse the froth and reduce entrainment. The directional nature of the valves help to sweep liquid across the tray deck. Turndown capability is between that of sieve and moving valves. MVG tray panels are mechanically more robust than sieve or moving valve tray panels.

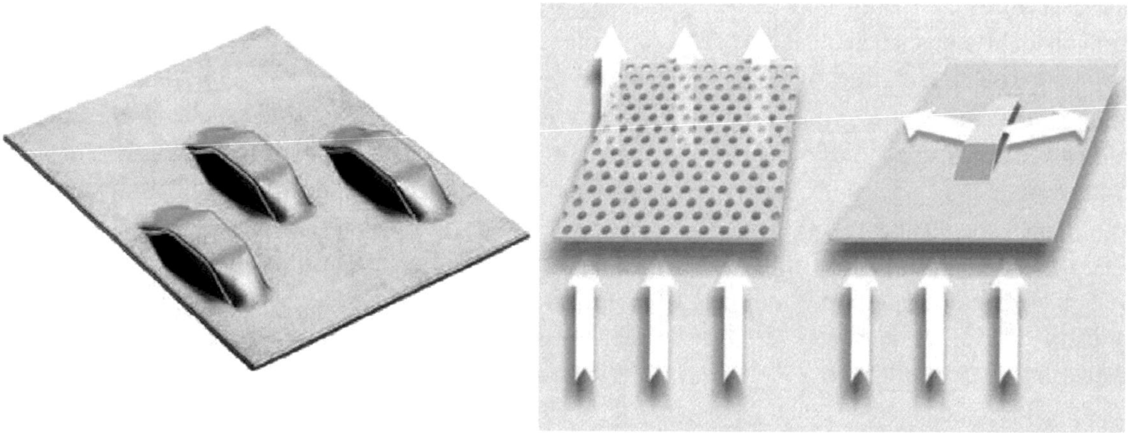

Figure 3: MVG valves

In the case of the HiFi Plus tray, use can also be made of a special downcomer inlet device, called a Crown Inlet Device (CID[†]), to provide additional downcomer liquid handling capacity for high pressure columns.

Typical tray layouts are illustrated in figures 4 and 5.

Figure 4: An 8 downcomer Calming Section Plus tray utilizing MVG™ fixed valves

Calming Section† (CS) trays are typically used for flow parameters below 0.2 such as Main Fractionators, De-isobutanizers, C6/C7/C8 fractionations, atmospheric and mild vacuum chemical columns and in general any place where entrainment is a concern. A single tray design is required for odd and even trays.

Figure 5: An 8 downcomer HiFi† tray utilizing sieve perforations

HiFi trays are typically used for flow parameters above 0.1 such as Deethanizers, Depropanizers, Debutanizers, Rectified Absorbers, C2 and C3 splitters, pump-around sections in fractionators and in general any place where liquid capacity is a concern. Again a single tray design is required for odd and even trays.

Since the 1980s, there have been more than 2000 applications of Shell Calming Section and HiFi trays world-wide.

Minimize installation time and effort

When revamping C2 and other splitters the downtime required to install new internals can often become the critical path. The length of the shutdown can have a significant impact on the revamp economics. Techniques that reduce installation effort are therefore of particular interest. Shell high capacity trays use envelope downcomers that are fully supported by the tray deck. The downcomers are 'dropped' into place with no need for modifications to downcomer bolting bars. Missing support ring segments can be filled in using expansion rings thus avoiding welding to the vessel wall.

Shell Global Solutions has formed an alliance with Sulzer Chemtech to market and sell the Shell mass transfer technology world-wide. Sulzer Chemtech has developed a number of time and cost saving techniques that can be utilized with Shell high capacity trays [1]. These are summarized below.

The Lip-slot[††] panel connection (fig. 6) eliminates the need for bolting at panel overlaps. This reduces the time required to fasten panels to each other dramatically. Only the last panel is required to be bolted or connected with wedge type connections. Split-wedge[††] tray support ring and downcomer connections further reduce the need for bolting. After installation the wedges are secured by bending them outwards.

Figure 6: Lip-slot and Split-wedge type connections

Existing support rings can be used to support new expansion rings with vertical struts and lattice beams to avoid having to weld to the column wall. This practice (fig. 7 left hand) eliminates the need for post welded heat treatment or pressure testing after the work has been completed. In addition up to two HiFi trays can be supported by a single lattice beam at low tray spacings (fig. 7 right hand).

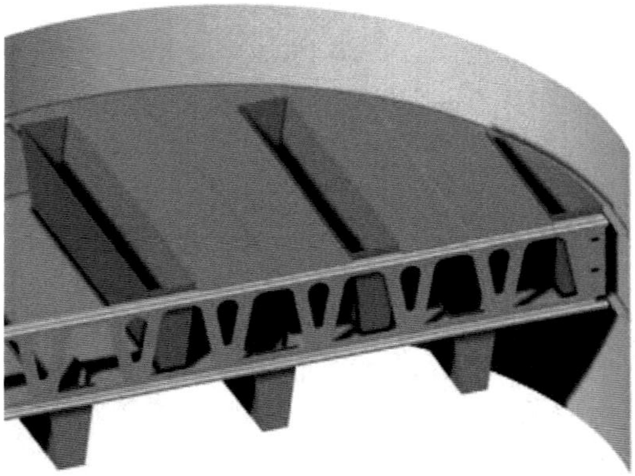

Figure 7: Avoid welding to vessel wall (left). Supporting two trays from a single lattice beam (right).

The use of expansion rings, vertical bars and lattice beams combined with Lip-Slot panel connections may save up to 35% of the installation time, resulting in a significant money savings over conventional methods.

Debottlenecking opportunities

Table 1 illustrates the Shell internals that can be considered for revamping ethylene plant columns (mostly cold-side).

Table 1: Internals options for revamps

Column	Pressure [bar abs]	High Capacity Internals	Super-high Capacity Internals
Caustic Scrubber	> 30	CS	
Demethanizer	> 40	HiFi	
Deethanizer	> 30	CS/HiFi	ConSep
Depropanizer	> 20	CS/HiFi	ConSep
Debutanizer	> 10	CS/HiFi	ConSep
C2 Splitter	10-40	HiFi	ConSep*
C3 Splitter	10-40	HiFi	

(* Non heat-pumped)

The ConSep tray is a relatively new device offering extremely high capacity. It will be described in more detail later on.

Splitter Revamps

Bravo, Sikkenk and Roza [1] reported on the relationships between operating pressure, relative volatility, number of stages and required reflux ratio for a given set of feed and product specifications for low (heat-pumped) and high pressure C2 Splitters. For low pressure C2 Splitters the ability to increase reflux is often limited by heat pump or refrigeration limits. They indicated that lower pressure in C2 Splitters yields lower reflux requirements however that is offset by the increase in vapor velocities due to the lower vapor density. The revamp then becomes a trade-off between the effects of pressure reduction on relative volatility (beneficial) and tray vapor capacity (negative). In such cases the use of high capacity trays is of great benefit. For high pressure C2 Splitters the detrimental effect of higher pressure on volatility dominates and there is a clear preference for lower pressure.

Two case studies covering a C2 and C3 Splitter revamp for a Shell lower-olefins facility in Europe are described below. The C2 splitter revamp involved a 1 for 1 retray of 4-pass conventional trays by 130 HiFi sieve trays (tray spacing 18 inch). The column was a high pressure C2 Splitter operating at 20 barg with a diameter of 4.4m. The throughput of the Splitter before and after the revamp is shown in figure 8. A capacity increase of approximately 45% was witnessed. Figure 9 illustrates the required reflux to feed ratio before and after the revamp confirming that the tray efficiency remained the same. A reduction in pressure drop of approximately 30% was also observed.

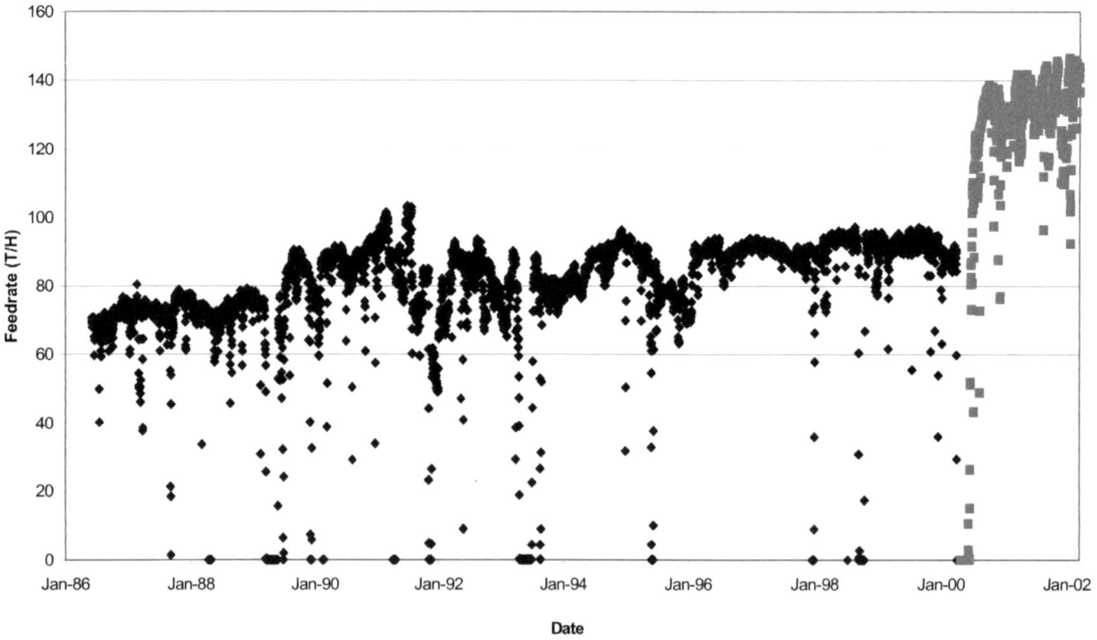

Figure 8: C2 Splitter capacity before and after revamp

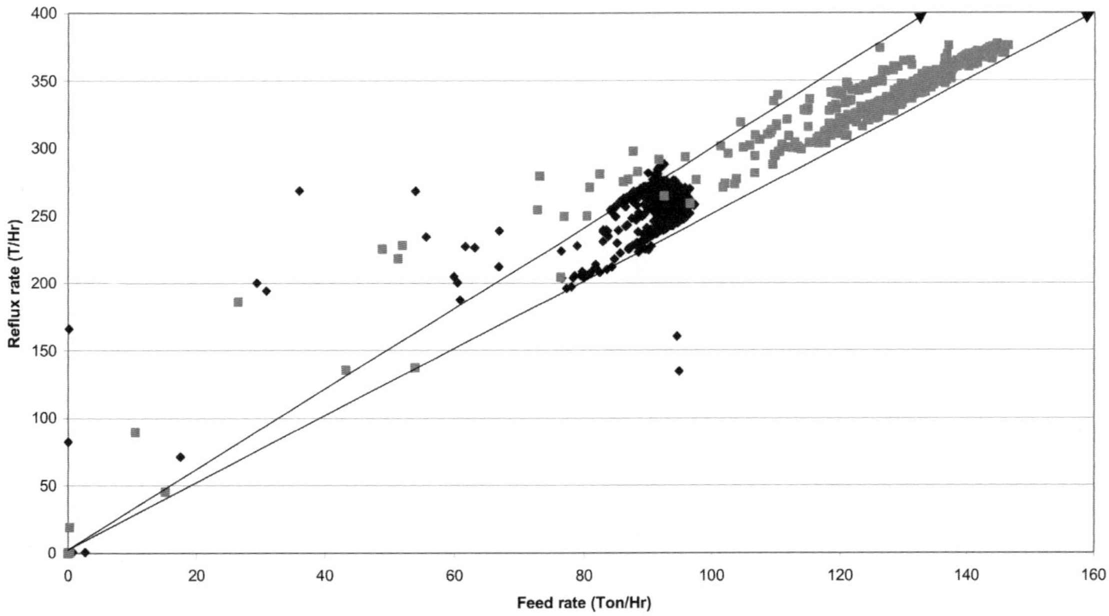

Figure 9: C2 Splitter reflux before and after revamp

The C3 Splitter revamp involved the replacement of 150 conventional 4-pass trays at 18 inch tray spacing with 211 HiFi sieve trays at 12 inch tray spacing. The column is a high pressure C3 Splitter operating between 11-18 bara (due to environmental fluctuations) with a diameter of 6.0m. The throughput of the Splitter before and after the revamp is shown in figure 10. A capacity increase of approximately 30% was witnessed.

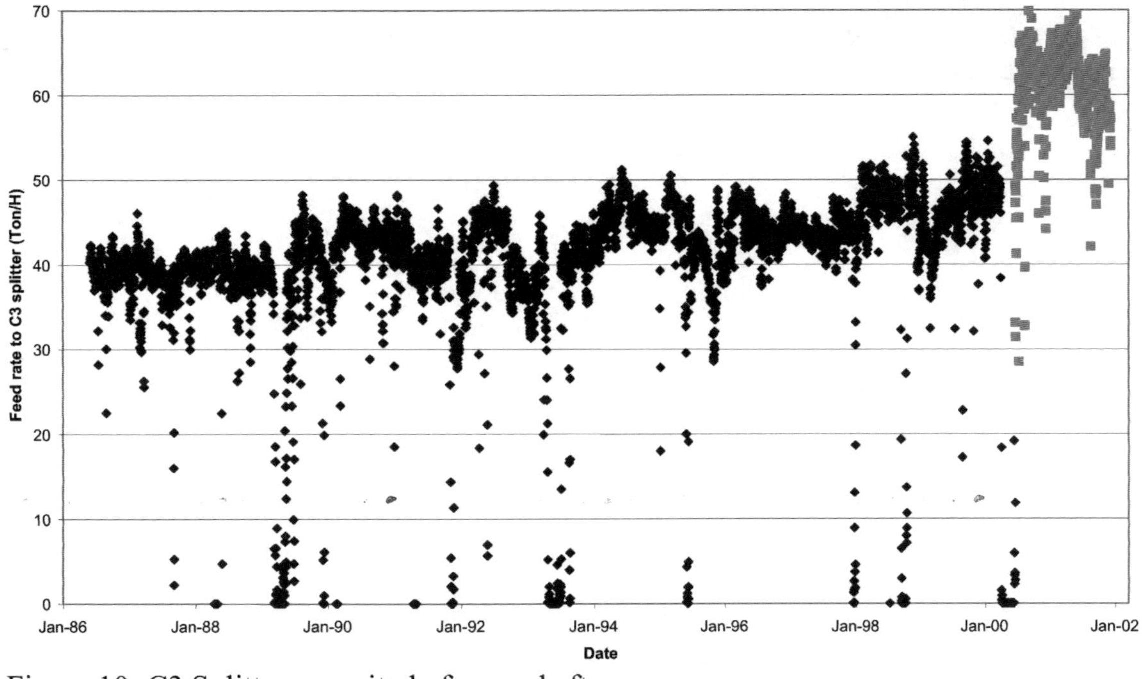

Figure 10: C3 Splitter capacity before and after revamp

Moving beyond High Capacity Trays

FRI defines the 'Ultimate Capacity' of a column to be the vapor load at which liquid would be entrained upwards independent of the column hardware. The Shell ConSep tray is designed to operate at or beyond the FRI ultimate capacity. It combines a conventional contacting deck that operates under conditions of severe entrainment with a separation deck utilizing centrifugal force to separate the entrained liquid from the vapor. The entrained liquid is returned to the tray below. See Figure 11.

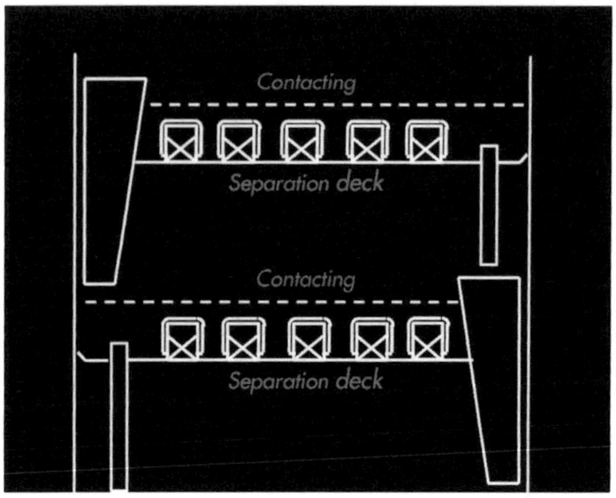

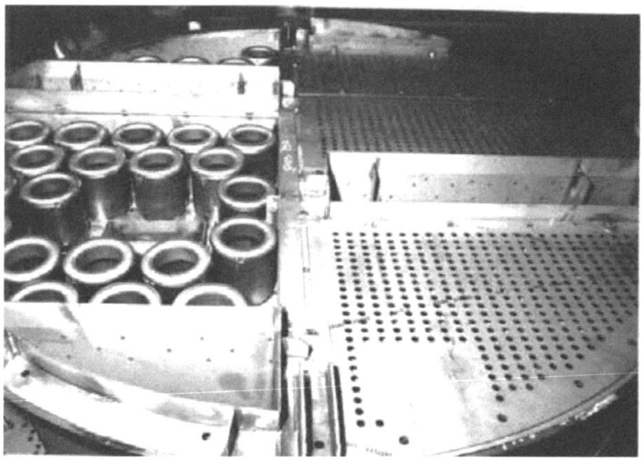

Figure 11: Schematic and photo of the Shell ConSep™ tray

The ConSep tray represents an enhancement of the HiFi tray in that it allows operation above the jet flood point, but permits normal downcomer flow as well. It is the highest capacity device in the Shell family of fractionation devices. It offers increased capacity over conventional trays of approximately 50-80% and 30-50% over CS or HiFi trays. It maximizes jet flood and liquid handling capacity. It can be retrofitted to existing towers utilizing tray spacings of 18-32 inches and provides comparable efficiency to conventional trays.

Table 2 provides a summary of ConSep applications in various services. Note that the Shell Calming Section/HiFi trays that were replaced already achieved a capacity increase of 10-40% over conventional multipass trays [2].

Table 2: Applications of Shell ConSep trays to replace existing HiFi/CS trays

Country	**Plant**	**Column diameter, m**	**Year**	**Max capacity increase achieved**[1]
UK	NGL debutanizer	1.9	1995	22%[2]
Australia	FCCU debutanizer	1.9	1996	30%[3]
Germany	HCU main fractionator	2.2	1999	50%
Australia	NGL debutanizer	1.7	1999	15%[3]
Singapore	FCCU debutanizer	2.5	2000	-[4]

1) Post startup performance testrun - capacity achieved on top of the existing HiFi/CS trays.
2) Limited by reboiler duty. 3) Limited by feed to the column. 4) No test run data available.

A Main Fractionator at a European refinery was retrofitted with ConSep trays to realize a capacity gain opportunity offered by an optimized process [2]. Previously the top section of the main fractionator had been the bottleneck of the hydrocracker. Startup was trouble-free and the new design capacity was met and exceeded in quick time. Figure 12 illustrates the increase in throughput that was achieved as a result of the revamp.

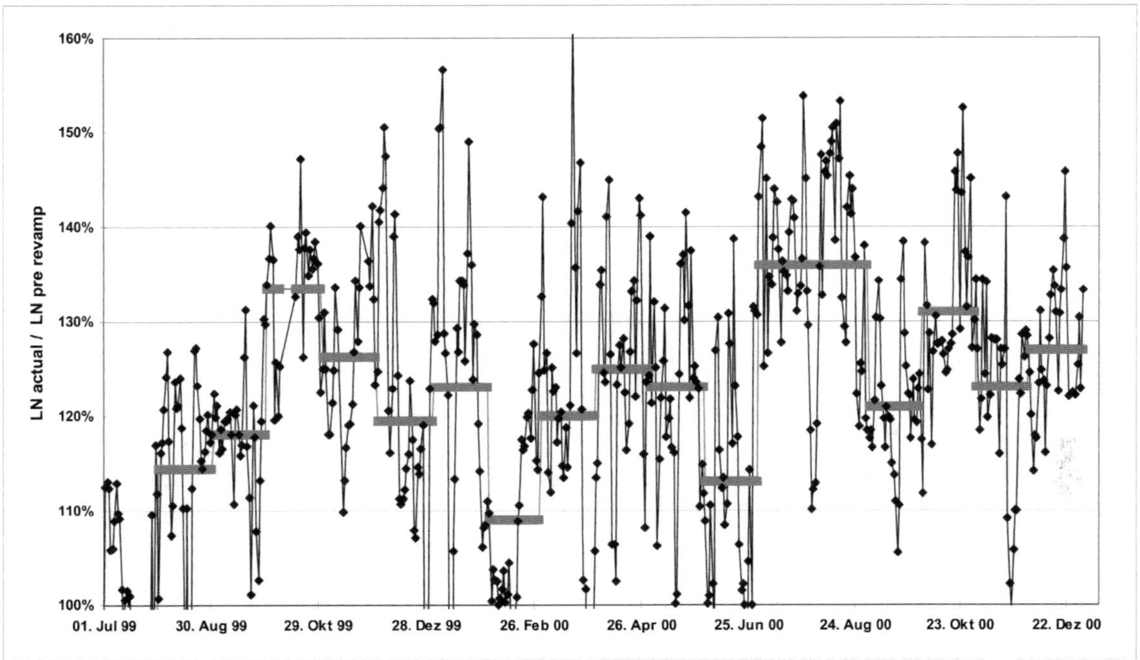

Figure 12: Production increase after revamping with ConSep™ trays

A second ConSep case study involves a NGL debutanizer revamp. In this Shell plant there were two identical debutanizers both receiving the same feed from a depropanizer. In 1995 one of these debutanizers was revamped with ConSep trays (1 for 1 tray replacement of all trays). This has increased the capacity by more than 20% (relative to high capacity Shell Calming Section valve trays). At the highest reboiler duty shown in figure 13 the calculated maximum column load factor is 0.12 m/s. Operating at even higher loads is not possible due to reboiler constraints (the reboiler has not been revamped).

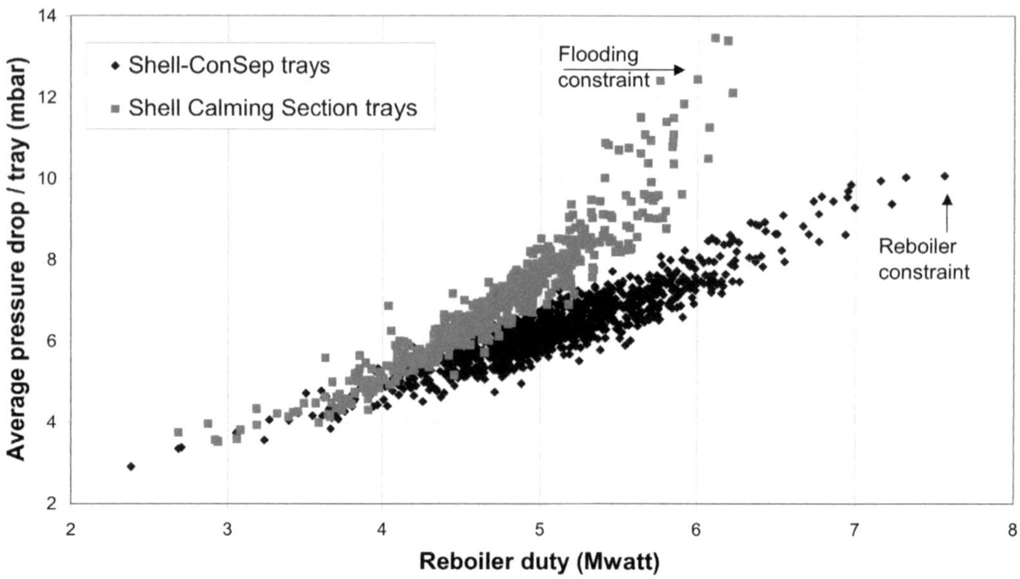

Figure 13: Comparison between debutanizer with and without ConSep trays.

Conclusions

1. The use of high capacity trays with sensible mechanical design and installation features offer an attractive means of revamping ethylene plant splitters and other fractionation columns.
2. Shell Calming Section and HiFi trays offer an attractive combination of performance enhancing features to maximize jet flood and liquid handling capacities with over 2000 documented installations.
3. Shell ConSep trays represent a breakthrough in distillation tray technology offering exceptional debottlenecking potential.

References

[1] "Design and revamp of modern C2 Splitters using high capacity internals and fast installation techniques" presented by J.L. Bravo, J.H. Sikkenk and M Roza at the ERTC Petrochemical Conference, February 2002, Amsterdam, Netherlands

[2] "The Shell ConSep tray technology provides unparalleled distillation capacity" presented by C. Groenendaal, B. Trautrims, K. Kusters and J.L. Bravo at the EFChE-Conference April 2001 in Bamberg, Germany

---

The technical information contained in this presentation is controlled for export by the US Government under EAR99. Please do not forward the contents without obtaining authorization from the author/presenter.

HYDRAULIC ANALYSIS IS KEY TO EFFECTIVE LOW-COST DEMETHANIZER DEBOTTLENECK

By
Steven P. Bellner, Bill Ege, Eastman Chemical Company, Longview, TX,
and Henry Z. Kister, Fluor, Aliso Viejo, CA

Summary

Eastman's demethanizer was bottlenecked by flooding, that was apparent just above the tower swage. A hydraulic analysis based on simulation of plant operating data was used as a basis for a detailed hydraulic analysis. The hydraulic analysis indicated absence of a bottleneck above the swage, but diagnosed the bottleneck to be caused by downcomer backup flooding just below the swage. It also revealed that all the trays below the swage operated close to the downcomer backup flooding limit.

The analysis identified the root cause of the downcomer backup flooding to be low open hole areas on the trays below the swage. Increasing hole area by replacing all tray panels below the swage with new panels containing more holes completely eliminated the flooding and enhanced column capacity. Demonstrated capacity went up by more than 20%, and no limit has been reached. After the hydraulic limitation was eliminated, the temperature in the coldest tower feed drum could be significantly lowered, which reduced ethylene losses in the tower overhead.

A good understanding of the equipment limitations, and a correct hydraulic analysis combined with field measurements and simulation, held the key for successfully identifying and eliminating the tower bottleneck.

Introduction

In the preface to a recent book (1), Norman Lieberman states: "The general knowledge as to how process equipment really functions is disappearing from the process industries. This is not only my opinion, but the general view of senior technical managers, in many large corporations."

A consequence is that hydraulic analysis in search of equipment bottlenecks is becoming a lost art. When a problem arises, the approach taken is often replacing some or all the trays by specialty trays or packing. This sledge-hammer solution often, but not always, cracks the nut. When successful, the bottleneck is swept away at the price of an unnecessarily expensive sledge-hammer solution. However, should the sledge hammer miss the nut, the bottleneck will remain. A few years ago we reported a case (2) where a flashing feed entering a demethanizer downcomer bottlenecked an entire olefins plant for 17 years. This bottleneck survived three failed fix attempts before a good hydraulic evaluation finally led to correct bottleneck diagnosis. Once correctly diagnosed, the bottleneck was easily and inexpensively eliminated.

This paper continues our swim against this tide. The case history presented here is of debottlenecking three similar demethanizers. The successful debottleneck did not involve specialty trays nor packing. Instead, it was achieved by applying good engineering principles to identify and diagnose the equipment bottleneck. The result was an inexpensive, successful debottleneck readily implemented in a turnaround.

It is this good understanding of process equipment, and a correct hydraulic analysis combined with field measurements and simulation, that hold the key for identifying and eliminating tower bottlenecks.

Process and Bottleneck Description

In the demethanizer chiller system, cracked gas is chilled and condensed in four stages by progressively colder levels of refrigerant. Each chilling stage produces a liquid stream which is flashed to the demethanizer pressure and fed to the tower. The feeds enter on trays 7, 14, 26, and 29, with those condensed at colder stages entering progressively higher up the column. The tower had a small diameter upper section, containing 14 sieve trays, and a larger diameter bottom section containing 59 sieve trays.

The demethanizer (Figure 1) separates hydrogen and methane from ethylene and heavier components. The primary objective for the demethanizer is to minimize the loss of ethylene in the overhead while preventing methane from entering the bottom product. At normal operating conditions, overhead ethylene losses were at about 1.5 mole percent, and methane in the column bottom was less than 1 mole ppm.

The tower operated well up to its bottleneck point. The normal tower differential pressure was 5-6 psi. Once the tower feed rates were raised above the bottleneck point, flooding was observed in the top section of the tower. Tower pressure drop rose to 8 psi. If no corrective action was taken, the pressure drop kept climbing, reaching as high as 10-12 psi. The corrective action alleviating the flood was to cut back reflux. This brought the pressure drop down, but at the expense of greater ethylene losses in the overhead. An alternative corrective action was to warm the drum that supplied the feed to tray 7. This lowered the feed rate to the upper section of the tower and lowered the pressure drop, but again at the expense of greater ethylene losses.

The differential pressure system in Figure 1 indicated that the pressure drop rise occurred above the swage. The appearance was of a bottleneck in trays 7-13. With tray 7 feed predominantly liquid, the section between it and the swage usually bottlenecks the capacity in the upper section of a demethanizer. Some vendors proposed replacing trays 7-13 by specialty high-capacity trays. With these trays being at small spacing (18 inches), there were also concerns that the tray spacing could be too small, and a proposal to replace the trays by random packing.

It was, however, appreciated that for correct bottleneck identification, it was necessary to perform a hydraulic analysis. This approach was chosen. The very first step was to prepare a simulation that reflects plant data. This simulation became the basis for our hydraulic analysis.

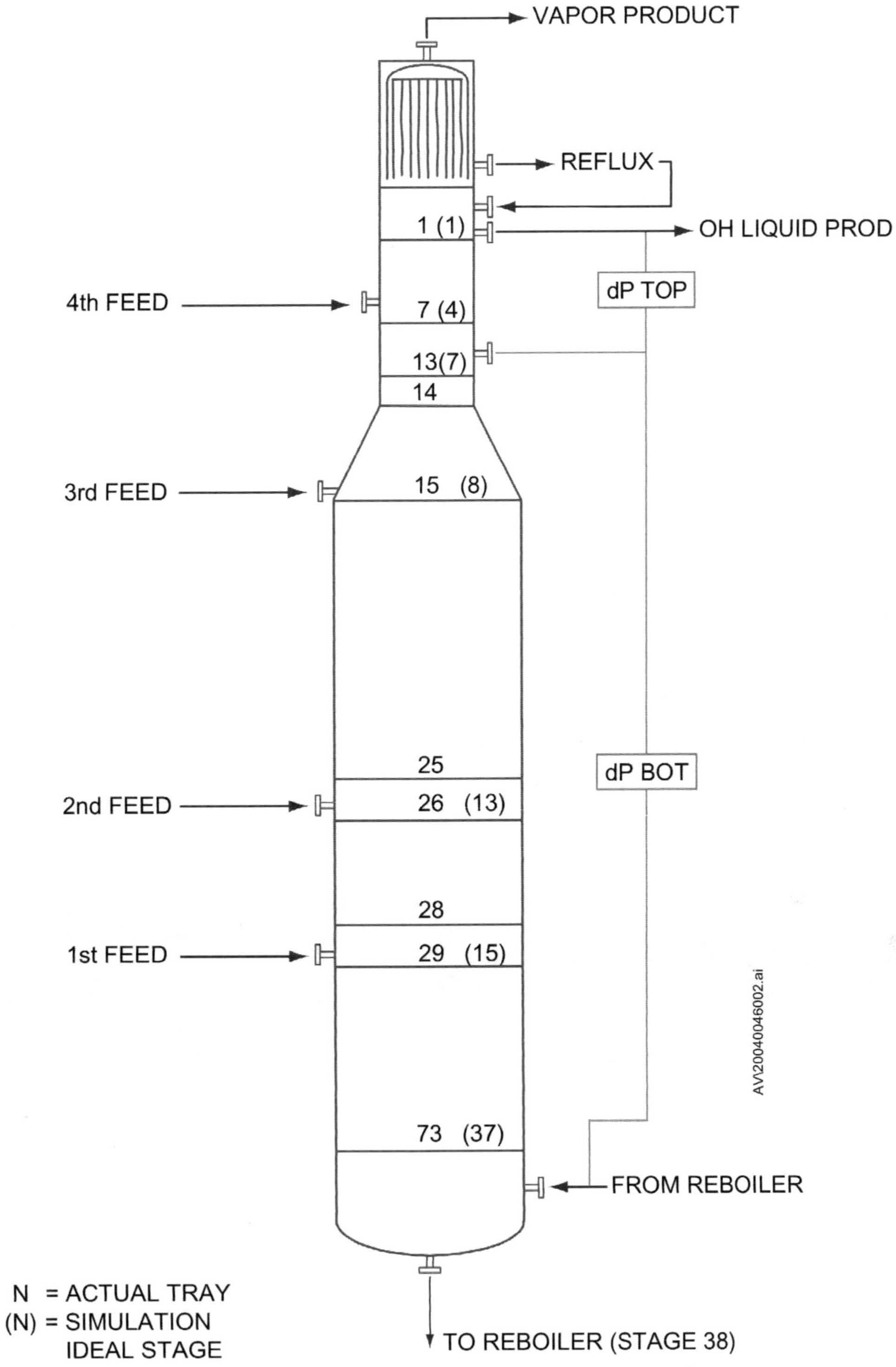

Figure 1
EASTMAN'S DEMETHANIZER:
TOWER TRAYS AND STAGES

Simulation

Feed Train In order to correctly determine the feed flow rates and compositions, it was necessary to rigorously simulate the feed train. Operating data were therefore collected around the feed train as well as the demethanizer and formed the basis to the simulation.

During the simulation preparation, it became apparent that some of the feed train operating data were inconsistent. Initial data collected included pressure, temperature, and liquid flowrates in and out of the feed drums. When temperature and pressure were matched, large discrepancies were apparent between calculated and measured resultant flow rates. Additional operating data were then collected to help resolve these inconsistencies. Vapor samples from the feed drums were analyzed along with more accurate temperature and pressure measurements. Liquid feed rates were then calculated to close material balances around the individual drums. This approach resulted in a better agreement between simulated data and plant data.

Despite attempts to manipulate pressure and temperature, closure of the material balance around the flash drum feeding tray 15 was unsuccessful. The measurements were checked and found to be repeatable. In order to close this balance, it was necessary to simulate 8.4 percent of liquid entrainment in the vapor leaving the drum. The material balance error around this drum indicates that either the vapor analysis from this drum was in error or that actual entrainment was occurring.

Tower Simulation The tower simulation was based on plant data. The Peng-Robinson Equation of State on a commercial simulator was used for vapor-liquid equilibrium (VLE) prediction, as is common practice in demethanizers. It was recognized that the VLE accuracy could have been improved by tuning the equation to closely match VLE data, as done in a previous analysis (3). However, in the previous analysis the objective was to correctly predict separation and efficiency, which requires accurate VLE predictions. Here the objective was to develop vapor and liquid loads for a hydraulic analysis, and our experience is that for this purpose the version available on the commercial simulation is adequate. Because of uncertainties in the measurement of reflux flow, it was difficult to reliably calculate an overall column efficiency. Based on experience with columns in similar service, we selected a 50 percent overall column efficiency and this gave us an excellent match to product compositions, and was therefore used in the column simulation

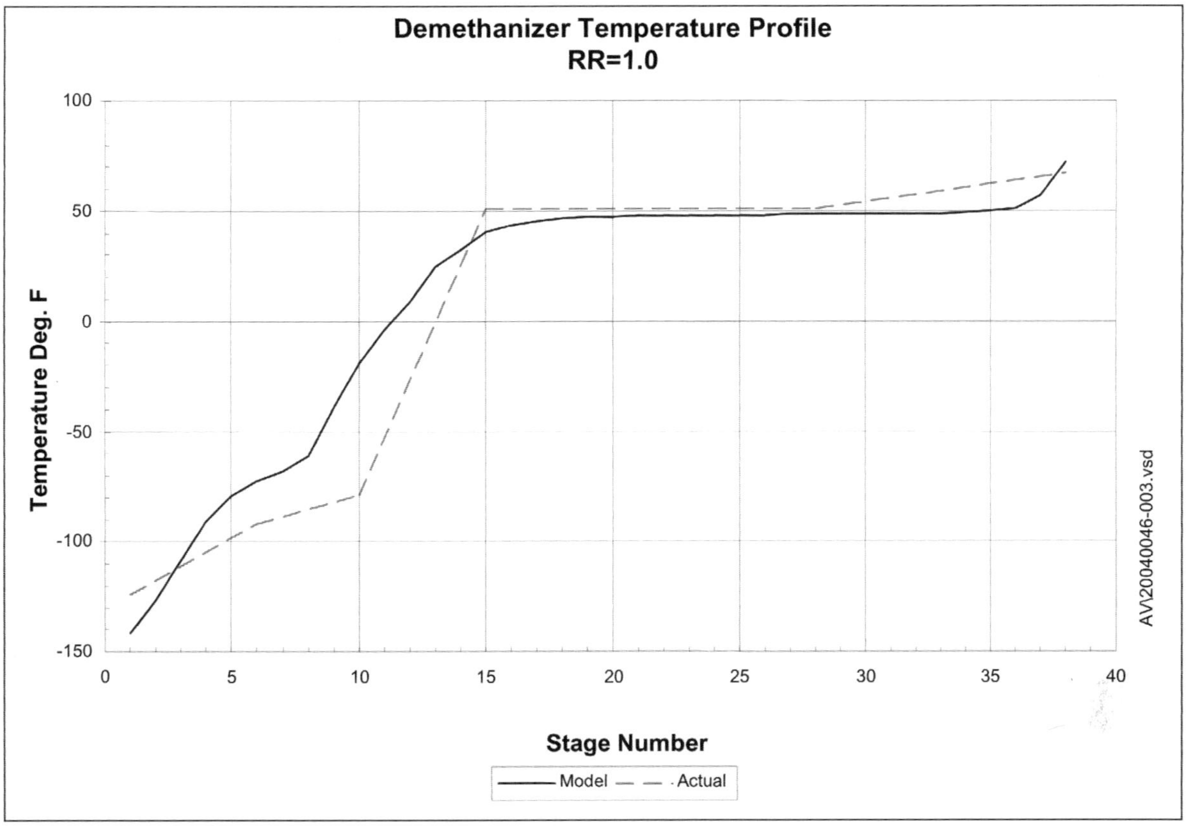

Figure 2 Demethanizer Temperature Profile

Figure 2 compares the simulated temperature profile to actual plant data. There was a good match from near equilibrium stage 15 to the base of the column. Above stage 15, there was a significant discrepancy between simulated and measured temperatures. This discrepancy reflects a combination of sampling and measurement errors and modeling deficiencies. Attempts to alter conditions in the simulation to obtain a temperature match in the top 15 stages wreaked havoc with the rest of the profile and were therefore abandoned. It became apparent that mitigating the discrepancy required a time-consuming and costly detailed analysis, both of the simulation input data and the field measurements. The sensitivity of the hydraulic loads to variations in the temperature profile were examined by varying conditions in the simulation, and found to be minimal. It was therefore concluded that the benefit of perfecting the simulation model for the mere purpose of the hydraulic study did not justify the cost.

One lead, however, was followed. In Figure 2, measured temperatures were *lower* than simulated at the cold end of the tower. This was unexpected. In cryogenic services, freeze-ups at thermowells often produce false high measurements, so measured temperatures tend to be higher than simulated. This argued against a measurement problem, and suggested the possibility that the upper feeds in the simulation are richer in high-boilers than in the plant. To reduce the feed boiling points, we lowered the simulated temperatures of the two coldest feed drums. When the temperatures in both drums were lowered by 20 °F, the simulated temperature profile perfectly matched the plant data. Subsequent hydraulic studies were performed for both alternative versions of the simulation and the differences in the hydraulic results were negligible. So the hydraulic analysis and our conclusions were insensitive to even the large difference between the two models.

Graphical Analysis Figure 3 is a Hengstebeck diagram, i.e., a McCabe-Thiele analysis modified for multicomponent systems (4). It had been advocated (5) as an invaluable tool for troubleshooting tower simulations and gaining insight into tower operation.

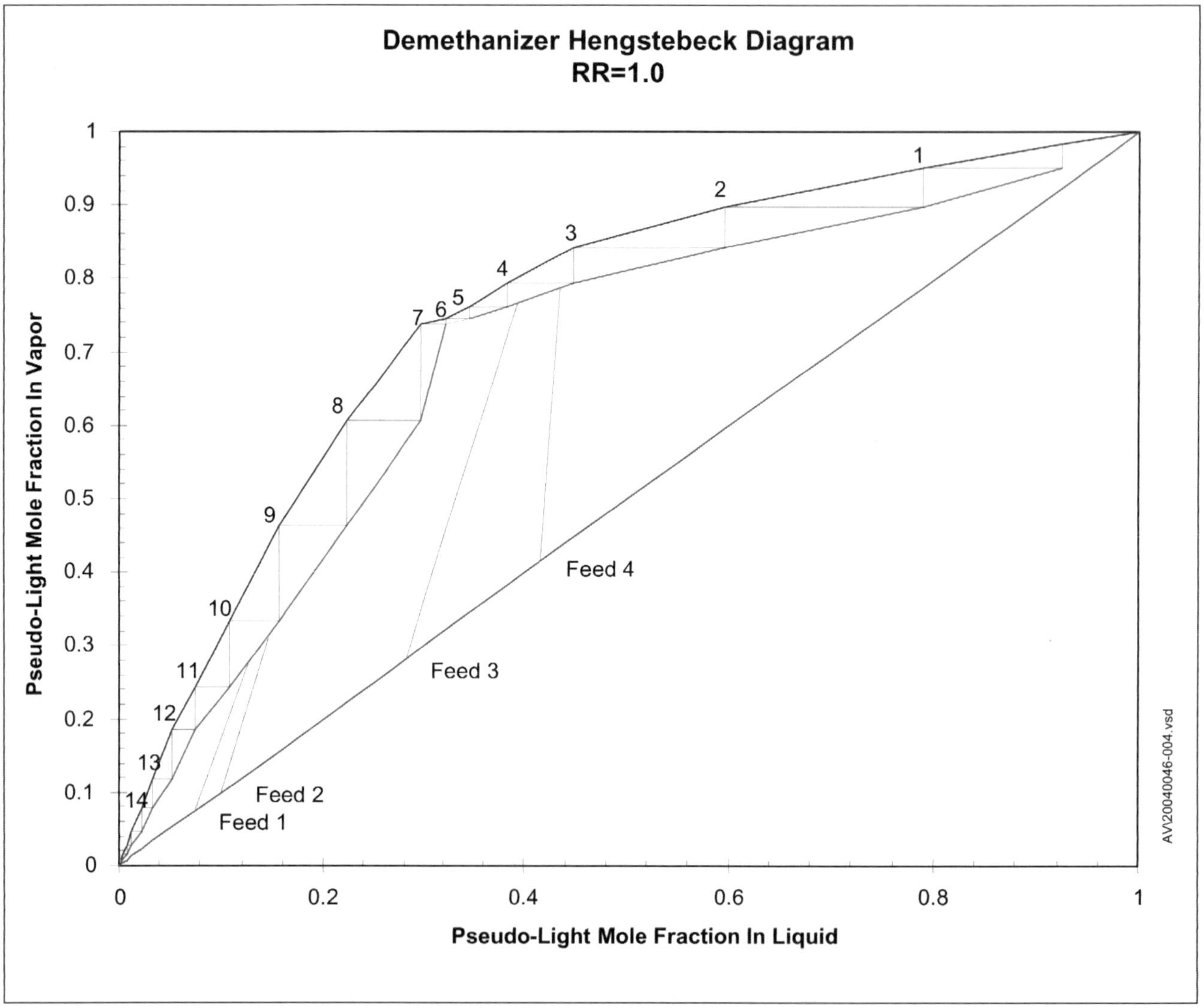

Figure 3 Demethanizer Hengstebeck Diagram

Figure 3 shows that the column is very close to a pinch near stage 6. The operating reflux ratio of 1.0 was therefore close to minimum reflux. The proximity to the pinch and minimum reflux fully explained why the ethylene loss in the tower overhead vapor so rapidly escalated upon reflux rate reduction (while attempting to get out of flood). Similarly, it explained why warming up the coldest feed drum (also in an attempt to get out of flood) also rapidly raised ethylene losses. Finally, it explained why the initiation of flood was not accompanied by a rapid rise in ethylene losses, at least initially. Reflux, not trays, was in short supply, so losing mass transfer on some trays was far less detrimental to ethylene loss than reflux rate cutback.

Figure 3 also shows how the tower operation can be optimized. The key is the liquid to vapor (L/V) ratio just above the pinch. Liquid to this region is supplied both from the reflux and from the upper feed. The two could be balanced to optimize either ethylene recovery or ethylene refrigerant consumption. For minimizing ethylene losses, reflux needs to be maximized and temperature in the upper flash drum needs to be minimized. To minimize the load on the refrigeration compressor, the upper feed drum still needs to be operated as cold as possible, but the reflux (which is condensed by colder refrigerant) can be reduced to lower the load on the ethylene refrigeration system. This reduction is constrained by rapid escalation in ethylene losses in the demethanizer overhead.

Hydraulic Analysis

The simulation supplied vapor and liquid loads and physical properties on which our hydraulic analysis was based. Our hydraulic analysis used FRI's correlations and data. Table 1 summarizes key tray dimensions and results of our hydraulic analysis. Table 1 shows that the tray dimensions vary from one tower section to another, probably tailored to the tower initial design loads.

Table 1 shows that none of the trays operated anywhere near jet-flood. This is not surprising, since the vapor densities in high-pressure demethanizers are usually high, making the vapor velocities low, which keeps most demethanizers well away from jet flood. Table 1 also shows that none of the trays operated close to the downcomer choke inlet velocity, which is a common limit encountered in high-pressure demethanizers. The reason for the comfortable margin from this limit is the large downcomers in the bottom section, which appear nicely optimized to circumvent this limit.

Table 1 shows extremely high froths in the downcomers of the trays below the swage. The results shown in Table 1 for downcomer backup were directly calculated from FRI's correlation. Figure 4 is a graphic representation of the calculated downcomer froth heights. The backup calculations indicate that trays 15 and 25 were operating above the downcomer backup flood point. Below tray 25, the column was very near the downcomer backup flood point. Above the swage, trays 1 to 14 operated a large margin away from the downcomer flood limit.

The investigation focused on the root cause of the high downcomer froth backup. On tray 20, the downcomer length, from the floor to the top of the outlet weir above, was 19.5 inches. The calculated froth height was 20.6 inches. The calculated aeration factor was 0.574, which is well in line (if anything, a bit high) with aeration factors typical for this kind of system (6). This renders the clear liquid backup in the downcomer 11.8 inches, which is extremely high for trays at 18 inch spacing. This backup was the sum of three terms: the inlet clear liquid height on the tray (this was calculated at a normal 3.1 inches), the head loss of liquid exiting the downcomer (this was calculated at quite a normal 1.2 inches), and the tray pressure drop, which was calculated to be an extremely high 7.5 inches.

The root cause of the high pressure drop was an extremely high "dry" pressure drop, i.e., friction through the sieve holes. FRI's correlation gave an extremely high dry tray pressure drop of 6.4 inches of liquid, suggesting shortage of hole area on the tray. Table 1 shows that on trays 15 through 25, the tray hole area was only 3.2 percent of the active area. This hole area is very small compared to normal practices, that make hole area typically 8-10 percent of the active area. Further, the trays have very large downcomers, making the active areas themselves quite small, about 40% of the tower cross section area. This small hole area, coupled with large downcomers, created a large dry-tray pressure drop, raised downcomer backup, in turn initiating tower flooding.

Table 1

DEMETHANIZER HYDRAULIC ANALYSIS

	Top Section		Bottom Section		
	Trays 1-6	Trays 7-14	Trays 15-25	Trays 26-28	Trays 29-731
Tower Diameter, ft	3	3	6	6	6
Tray spacing, inches	18	18	18	21	23
Hole area, % of active area	6.5	6.5	3.2	5.6	7
DC inlet area, % of tower area	14	14	30	30	30
DC clearance, inches	1.3	1.3	1.5	2.8	2.8
Outlet weir height, inches	1.5	1.5	1.5	2	2
Vapor C-Factor, based on net area, ft/s	0.14	0.13	0.08	0.11	0.14
Liquid load, GPM/inch of outlet weir	2.1	2.8	6.3	11.0	13.5
Downcomer inlet velocity, ft/s	0.13	0.18	0.11	0.20	0.24
% Jet Flood	47	46	62	65	70
% Froth in downcomer	58	53	106	94	87
% Downcomer choke velocity	51	53	38	65	78
Total DC backup, inch liquid	6.0	6.0	11.8	12.3	12.3
Tray inlet clear liquid height, inch	2.4	2.5	3.1	4.4	4.4
Head loss under DC, inch liquid	0.2	0.3	1.2	1.1	1.7
Tray pressure drop, inch liquid	3.4	3.2	7.5	6.8	6.2
Tray dry pressure drop, in liquid	2.3	1.9	6.4	4.7	3.7

Dry-tray pressure drop below tray 25 was still high, but to a lesser extent (Table 1 and Figure 4). On these trays, the larger hole areas somewhat lowered the dry pressure drop, but this was counter-balanced by the high liquid loads and taller weirs that raised the tray inlet clear liquid heights and liquid head pressure drop. On these trays, it was the taller tray spacing that kept the downcomer froth backup some margin away from flood.

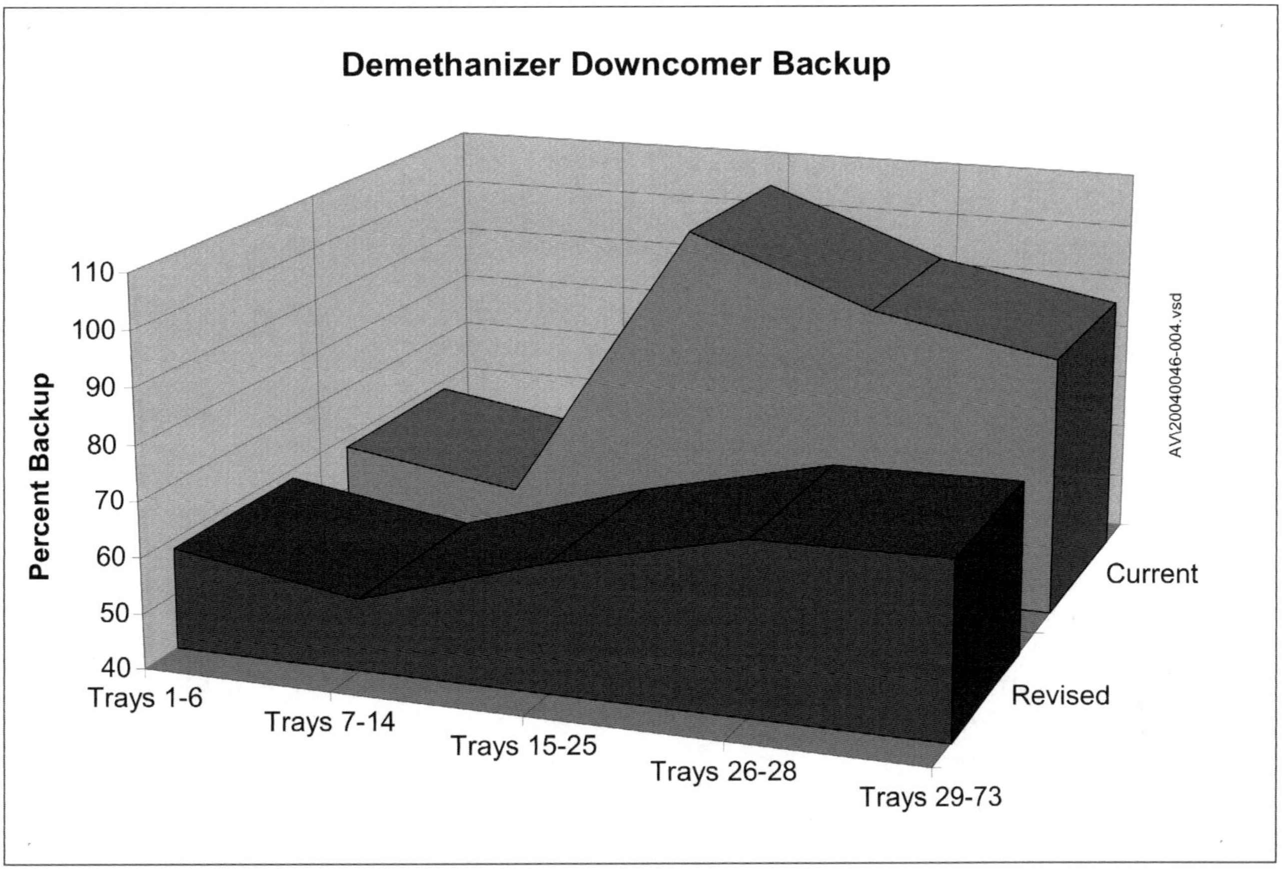

Figure 4 Demethanizer Downcomer Backup

The hydraulic analysis showed that the tower bottleneck was not on trays 7-14 as was initially thought. These trays operated a large margin away from any type of flood (Table 1). Instead, the flood initiated in the section just below. Since the bottleneck occurred near the top of the section below, the appearance was identical with a flood initiating in the section above.

A search of historical data did not reveal the reason for the small hole areas below the swage. Most likely, in the initial years the plant (and tower) operated at much lower loads than currently. A common practice for dealing with low tower loads is to reduce sieve hole area. This prevents tray weeping and permits sieve trays to achieve good efficiencies at high turndown. Since then, tower plant throughput has largely raised, with it so did the hydraulic loads, turndown was not a major issue, but the small hole area became a bottleneck.

Feed Distribution Piping Tower hydraulic analysis remains incomplete until all the relevant points of transition (i.e., any point where a stream enters or leaves the tower) is complete. Our survey of tower failures (7) indicated that points of transition are some of the most severe tower bottlenecks. In another demethanizer (2), a flashing feed entering a downcomer bottlenecked a demethanizer. It was therefore imperative to analyze all the points of transition in the region of flood initiation. Analysis of the feed piping indicated that internal piping was well-designed for potential flashing feed service and was not expected to contribute to the column bottleneck.

Solution

The decks of trays 15-73 were replaced by new sieve decks that had higher hole areas. The new hole areas were increased to the point of incipient weep, and ranged from 8.5% on trays 15-25 to 12% of the active areas on trays 29-73. The downcomers were not modified. The calculated percent froth in the downcomers was lowered from 87%-106% (Table 1) to 63-71%.

Following the retray, the tower flooding and bottleneck completely disappeared. Demonstrated tower capacity was over 20% higher than previous. Pressure drop across the tower declined to 3 psi. No capacity limit has been encountered in the tower since the debottleneck. The cost of repacing the sieve decks was small.

Eastman operates two other plants with similar demethanizers. The analysis described here extended to these plants as well. The demethanizers in these two plants were debottlenecked by similar means, with the same results.

Even though the bottleneck appeared to occur on trays 7-14, it initiated in the section below. Therefore, replacing trays 7-14 by specialty trays or packings, as proposed by some vendors, would have done nothing to alleviate the tower bottleneck. It took developing a good understanding of the equipment limitation, and a correct hydraulic analysis, combined with field measurements and simulation, to correctly identify and diagnose the bottleneck. And this effort was rewarded by a low-cost, effective solution.

References

1. Lieberman, N. P., and E. T. Lieberman, "A Working Guide to Process Equipment", McGraw-Hill, 1997.

2. Kister, H. Z., T. C. Hower, P. R. de Melo Freitas, and J.Nery Souza Neto, "Problems and Solutions in Demethanizers and C_2 Splitters with Interreboilers", 8th Annual Ethylene Producers Conference, New Orleans, LA, February 25-29, 1996.

3. Kister, H. Z., S. G. Chellappan, and C. E. Spivey, “Debottleneck and Performance of a Packed Demethanizer”, Proc. of the 4th Annual Ethylene Producers Conference, New Orleans, Louisiana, March 28 – April 1, 1992.

4. Hengstebeck, R.J. "An Improved Shortcut for Calculating Difficult Multicomponent Distillations", Chem. Eng., p.115, Jan 13, 1969.

5. Kister, H. Z. “Troubleshooting Distillation Simulations”, Chem. Eng. Progr., p.63, June 1995.

6. Kister, H. Z., "Distillation Design", McGraw-Hill, NY, 1992.

7. Kister, H.Z., "What Caused Tower Malfunctions in the Past 50 Years?", IChemE Transactions, p. 5, January 2003.

SESSION 66D – DIVIDED WALL TECHNOLOGY FOR BUTADIENE EXTRACTION
PRESENTED AT: ETHYLENE PLANT TECHNOLOGY
ADVANCES IN DISTILLATION TECHNOLOGY FOR ETHYLENE PLANTS

PRESENTED BY:
Robert J. Brummer (speaker), ABB Lummus Global
Bernd Heida, BASF

Divided wall technology can be used to reduce the investment cost for butadiene extraction plants. BASF, developers of the butadiene extraction process using NMP as solvent, has developed a process scheme to reduce the required piece count by using divided wall technology. BASF has a great deal of experience using divided wall columns in other technologies. This proven divided wall concept can be applied to the BASF butadiene extraction process. The main benefit is the reduction in the pieces of equipment, thereby reducing the investment cost and maintenance cost for the plant. Details of the process were provided in the paper.

AUTHOR INDEX

AUTHOR INDEX

AUTHOR INDEX

Food, Pharmaceutical and Bioengineering Division 2016

Core Programming Area at the 2016 AIChE Annual Meeting

San Francisco, California, USA
13 - 18 November 2016

Volume 2 of 2

ISBN: 978-1-5108-3436-1

Printed from e-media with permission by:

Curran Associates, Inc.
57 Morehouse Lane
Red Hook, NY 12571

CURRAN ASSOCIATES INC. proceedings.com

Some format issues inherent in the e-media version may also appear in this print version.

Copyright© (2016) by AIChE
All rights reserved.

Printed by Curran Associates, Inc. (2017)

For permission requests, please contact AIChE
at the address below.

AIChE
120 Wall Street, FL 23
New York, NY 10005-4020

Phone: (800) 242-4363
Fax: (203) 775-5177

www.aiche.org

Additional copies of this publication are available from:

Curran Associates, Inc.
57 Morehouse Lane
Red Hook, NY 12571 USA
Phone: 845-758-0400
Fax: 845-758-2633
Email: curran@proceedings.com
Web: www.proceedings.com

TABLE OF CONTENTS

VOLUME 1

VOLUME 2

457504 Structural Insights into Protein Splicing Inhibition By Platinum Therapeutics As Potential Anti-Microbials

Abstract:

Abstract

Inteins, self-splicing protein elements, interrupt genes and proteins in many microbes, including the human pathogen *Mycobacterium tuberculosis*. The intein in the mycobacterial recombinase, RecA, is specifically inhibited by cisplatin, *cis*-$[Pt(NH_3)_2Cl_2]$, widely used to treat a variety of cancers. Cisplatin also inhibits mycobacterial growth by targeting inteins. Structural and kinetic analyses of cisplatin binding to the RecA intein were performed to investigate the inhibition of protein splicing by this platinum(II) compound. Mass spectrometric and crystallographic studies of cisplatin binding revealed a complex in which two platinum atoms bind to the intein at N- and C-terminal catalytic cysteine residues in the presence of the reducing agent tris(2-carboxyethyl)phosphine (TCEP). Kinetic analyses of NMR spectroscopic data support a two-step mechanism for cisplatin binding, in which a Pt(II) unit first rapidly interacts at the N-terminus, followed by a slower, first-order binding event involving both the N-terminus (Pt) and C-terminus (Pt+TCEP). Notably, the ammine ligands of cisplatin, which remain bound to platinum in DNA adducts formed as an early step in the mechanism responsible for its anticancer activity, are no longer bound to the metal atom in the intein adduct, as revealed by crystallography. We also explored two additional platinum-based compounds with and without ammine ligands for their ability to inhibit intein splicing. A platinum 4,4,4-trifluorobenzoylacetonate (tfbz) complex, $[Pt(tfbz)(NH_3)_2](NO_3)$, and $K[Pt(Cl_{3\hat{A}}$
to cisplatin, in the low micromolar range. These findings should inform the design of additional platinum-based compounds for targeting inteins in microbial pathogens.

Copyright © American Institute of Chemical Engineers. All rights reserved.

458068 Novel Peptoid Ligands for IgG Purification

Abstract:

A great number of peptides with high affinity for protein targets is currently known and utilized as drugs, affinity ligands, and diagnostic agents. Recently, a broad spectrum of peptide mimetics has emerged, which hold considerable promise for replacing peptides in a variety of biotech and medical applications. Among them, peptoids in particular show excellent biorecognition properties, as well as higher chemical and biochemical stability as compared to peptides. This indicates the need to develop a strategy for translating known protein-binding peptides into peptoid analogues with equal or higher affinity. To begin exploring this field, a small library of peptoid analogues of the IgG-binding peptide HWRGWV was constructed using both amino acid-like and non-natural functional groups, different arrangement of the residues, inclusion of spacing amino acids, and different length. Ligands with affinity for IgG were selected by screening the peptoid variants immobilized on solid supports via IgG binding tests in non competitive and, subsequently, competitive (cell culture supernatant added with bovine serum) conditions. The sequences offering the highest IgG recovery and purity were also tested to select those with the highest chemical and biochemical stability. To this end, the peptoids were challenged with a variety of chemical agents and proteolytic enzymes. Three peptoid-based adsorbents were finally characterized by binding isotherm to determine equilibrium capacity (Q_{max}) and strength of binding on solid phase ($K_{D,S}$). Selected peptoids were also modeled *in silico* to estimate the binding site on IgG and the binding strength in solution. To validate such predictions, one sequence was tested by HDX-MS and isothermal calorimetry, which confirmed both predicted binding site and true affinity ($K_{D,L}$) in solution. This study offers the first insight into the design of bioactive peptoids with equal or better affinity than their parental peptides, and an approach for overcoming the challenges that, despite the structural similarity, the peptide-to-peptoid translation presents.

Copyright © American Institute of Chemical Engineers. All rights reserved.

471698 Anti-*Candida* Peptide Production Using a Recombinant Expression Platform

Abstract:

A smart design for recombinant peptide expression of various lengths is possible with the knowledge of target product, cell line, fermentation strategy, and downstream recovery options. Using a novel anti-Candida peptide as an illustrative example, we show that the combination of a cell line engineered with a reduced host cell proteome (Lotus®), and a mutant version of Green Fluorescent Protein that is chemically resistant confers several advantages for bioprocessing. Specifically, the use of this new*Escherichia coli* strain permitted a 37% increase in column capacity for the primary capture step (ion exchange) and purification from a CNBr-digested mixture was greatly simplified as the number of peptides were reduced. In all, the blending of judicious mutations provide an argument for the recombinant approach to peptide manufacturing, especially as the length of the product increases.

Copyright © American Institute of Chemical Engineers. All rights reserved.

450192 Surface Deae Groups Facilitate Bound Protein Transport on Polymer Chains in Deae-and Deae-Dextran-Modified Sepharose

Abstract:

To explore the contribution of surface ligands (ion exchange groups) to the mass transport of bound proteins on polymer chains in ion exchangers with both surface and polymeric ligands, Sepharose FF was modified with diethylaminoethyl-dextran (DEAE-dextran, DexD) and/or DEAE (D) to fabricate three types of ion exchangers, i.e., FF-DexD (grafting-ligand resin), FF-D (surface-ligand resin), and FF-D-DexD (mixed-ligand resin), for protein adsorption equilibria and kinetics studies. It was found that both adsorption capacity and uptake rate (effective diffusivity, D_e) were significantly enhanced by grafting DEAE-dextran. Notably, the D_e/D_0 (D_0, free diffusivity) values in FF-DexD and FF-D-DexD were larger than 1.4, six times greater than those in FF-D ($D_e/D_0<0.3$). More importantly, the increase of surface-ligand density greatly enhanced uptake kinetics in FF-D-DexD. The results indicate that the surface ligands assisted the transport of bound proteins on polymer chains in the mixed-ligand resins. That is, the surface ligands worked as "transfer stations" between two neighboring chains, resulting in enhanced transport of bound proteins on chains. The research thus disclosed the unique role of surface ligands in facilitating protein uptake kinetics in polymer-grafted porous ion exchangers.

Copyright © American Institute of Chemical Engineers. All rights reserved.

472212 Effects of Histidine and Sucrose on the Physical Properties and Ultrafiltration Performance of a Monoclonal Antibody

Abstract:

Effects of histidine and sucrose on the physical properties and ultrafiltration performance of a monoclonal antibody

2016 AIChE Annual Meeting

Food, Pharmaceutical & Bioengineering Division

Youngbin Baek[1], Nripen Singh[2], Abhiram Arunkumar[2], and Andrew Zydney[1]

[1]The Pennsylvania State University, University Park, PA 16802

[2]Bristol-Myers Squibb, Global Manufacturing and Supply, Devens, MA 01434

Abstract

Monoclonal antibodies (mAb) are currently the dominant class of bio-therapeutics, with important applications in the treatment of cancer and autoimmune disorders. Monoclonal antibodies are concentrated / formulated by ultrafiltration (UF) and diafiltration (DF) to produce the formulated drug substance. There can be significant challenges in the production of the very highly concentrated solutions (often >200 g/L) needed to deliver the appropriate therapeutic dosage by subcutaneous injection [1]. The objective of this study was to examine the effects of buffer conditions on the physical characteristics and UF/DF performance of a mAb provided by Bristol-Myers Squibb. The mAb was

characterized by its hydrodynamic radius (determined by dynamic light scattering), zeta potential, osmotic pressure, and viscosity, using buffers with different pH, histidine concentration, and sucrose (used as an excipient). UF performance was evaluated using both Pellicon 3 (C- and D-screen) and Pall Omega T-series tangential flow filtration cassettes. The hydrodynamic radius was a strong function of both the histidine and sucrose concentrations, likely reflecting the complex interactions between these species and the mAb / water [2,3]. For example, the hydrodynamic radius decreased from 6.4 to 5.6 nm as the histidine concentration increased from 10 to 250 mM, while the addition of sucrose caused a slight increase in the mAb radius. These changes in radius were correlated with changes in both the osmotic pressure and viscosity of the mAb solution, particularly in the presence of high concentrations of sucrose. The change in buffer conditions also affected the UF behavior, with the increase in viscosity causing a significant decrease in filtrate flux (at high mAb concentrations) and maximum achievable mAb concentration. These results provide important insights into the factors controlling the UF behavior of mAb products, including the impact of buffer conditions on the flux and maximum achievable protein concentration.

References

[1] Binabaji E, Ma J, Zydney AL. Intermolecular interactions and the viscosity of highly concentrated monoclonal antibody solutions. *Pharm Res*. 2015:32:3102-3109.

[2] Liad SM, Du QS, Meng JZ, Pang ZW, Huang RB. The multiple roles of histidine in protein interactions. *Chem Cent J.* 2013:7:44-44

[3] Nain AK, Pal R, Sharma RK. Physicochemical study of solute–solute and solute–solvent interactions of l-histidine in water+sucrose solutions at different temperatures. *J Mol Liq.* 2012:165:154-160

Copyright © American Institute of Chemical Engineers. All rights reserved.

472083 Host Cell Protein Removal Via Flow-through Chromatography: A Proteomics-Based Approach

Abstract:

Manufacturing costs represent a major barrier limiting the access to biotherapeutics by the global patient population. As purification processes particularly contribute to biomanufacturing costs, the development of affordable and more efficient purification technologies is a topic of growing interest. Platform processes in current downstream processing rely heavily on "bind-and-elute" chromatography, an inherently batch process that requires long processing times, significant buffer storage capacity, and often expensive chromatography resins. As an alternative to this methodology, we aim to develop a new purification paradigm for biotherapeutics hinged on a "truly continuous" process comprising only flow-through or continuous mode steps. In applying this strategy, removal of host cell protein is a primary concern due to the wide variety in sequence and structure. In this work, we explored methods for removing host cell protein (HCP) by chromatography in flow-through mode. Commercially available chromatography resins commonly used in flow-through mode for polishing applications were evaluated over a range of different loads of HCP obtained from CHO-S culture harvest to (1) determine order of magnitude of protein removal and (2) identify and characterize what protein species are removed with each technology. A high-throughput protein detection process by mass spectrometry enabled us to identify and quantify the HCPs in both cell supernatant and process fluids. Our results indicate that commercially available technologies are not sufficient for removing extracellular HCP when operating strictly in flow-through mode, as the maximum log reduction observed under the conditions tested was ~1 LRV. HCP clearance was then evaluated over an immobilized combinatorial peptide library to determine the expected binding capacity range using affinity interactions of the library with the HCP mixture, a promising strategy for capture of host cell proteins.

Copyright © American Institute of Chemical Engineers. All rights reserved.

471294 An Affinity Chromatography Method for Purification of Antibodies Utilizing Nucleotide Binding Site Targeting with a Small Molecule

Abstract:

We describe a novel affinity membrane chromatography technique (m-NBS^{Tryptamine}) for the purification of monoclonal and polyclonal antibodies from complex protein mixtures such as cell culture media as well as ascites fluids. m-NBS^{Tryptamine} method utilizes the nucleotide-binding site (NBS) that is located on the Fab variable domain of all antibodies to facilitate capturing antibodies on a membrane affinity column via a small molecule, tryptamine, which has a moderate binding affinity to the NBS. Regenerated cellulose membrane was selected as a matrix due to the major advantages over traditionally used resin-based affinity systems. In this study, we used the pharmaceutical antibody Rituximab in our proof of concept experiments. Antibody capture was accomplished by injection of samples while running equilibration buffer (50 mM sodium phosphate pH 7.0) and elution of the antibody was achieved by running a gradient of mild elution buffer (3M NaCl in 50 mM phosphate pH 7.0). Our results demonstrate that the m-NBS^{Tryptamine} column efficiency for Rituximab was >98%, with a purity level of >98%. This small molecule affinity purification method is further tested for a number of necessary parameters such as: injection concentrations, volumes, wash/bind time, elution gradient, antibody/protein-contaminant combinations, effects of injection buffer, post-purification antigen binding activity of antibodies, and column reusability and stability. The m-NBS^{Tryptamine} method provides a superior alternative to the protein-A affinity purification method that is widely used for purification of humanized and chimeric antibodies. We hope that in the long term, this new method will help to increase the availability of antibody based treatment and diagnostic systems to patients, as well as reducing the cost of antibody molecules used in scientific research.

Copyright © American Institute of Chemical Engineers. All rights reserved.

475035 Intein-Based Bioseparations: Development of a Disruptive Technology for a Conservative Industry

Abstract:

Over the past 20 years, my laboratory has worked on the development self-cleaving affinity tags for the capture and initial purification of recombinant proteins. The goal of this work is a truly general platform for the traceless and tagless affinity purification of virtually any target protein expressed in any recombinant host. Although tag-based affinity methods are common at laboratory scale, the expense and complexity of tag removal have prevented these methods from being adopted for the manufacture of therapeutic recombinant proteins. Thus, the development of a simple and inexpensive self-cleaving system constitutes a disruptive advance, with the potential to resolve tag removal issues and provide a new platform unit operation for the preparation of therapeutic products. Concurrent with our research, the biopharmaceutical industry has undergone major transformations. These include explosive increases in recombinant protein titer, the anticipation of biosimilars as patents expire, the increasing adoption of disposable single-use technology, and the growing interest in continuous processes in both upstream and downstream processing. In this presentation, I will discuss our development of intein-based self-cleaving tag methods in the context of the changing biopharmaceutical industry. This will include the basic approaches used to engineer new self-cleaving tags, and how those approaches relate to industry pressures in the highly regulated and conservative biopharmaceutical industry. In our most recent work, we have demonstrated the ability to purify fully processed glycoproteins from mammalian cell culture hosts, and we have incorporated this method into a portable biopharmaceutical factory platform. In both of these cases, we have completely eliminated premature cleavage of the intein tag through the adoption of a split-intein architecture, and we have introduced several additional mutations to impart highly pH-sensitive cleavage for tag removal. This platform is now capable of providing highly purified targets, in a traceless and tagless form using a single column step, from virtually any expression system capable of producing the folded protein. The simplicity of this approach suggest a strong future in the rapid development of new biosimilars, as well as additional applications in pure research and the production of novel non-mAb targets.

Copyright © American Institute of Chemical Engineers. All rights reserved.

475293 Multi-Organ on Chip Models to Advance Drug Development and Model Disease

Abstract:

There have been tremendous advancements in multi-organ on a chip models, motivated largely by shortcomings of existing simplified in vitro culture models and lack of translation of observations in animal models to humans. Multi-organ on a chip models could be valuable for simulating the human body response to drugs and for modeling disease. The components required for these systems to be useful include: 1. organ models capturing key cell types, architectures, and functions of the organ they are intended to model, here termed microphysiological systems (MPSs), 2. culture platforms enabling controlled maintenance, stimulation, and analysis of MPSs in fluidic communication for multiple weeks, and 3. computational frameworks for relating observations and measurements in these culture systems to function in vivo. This talk will describe advances in these areas, and highlight insight provided by even relatively simple single-(gut) and two-(gut-liver) MPS cultures when integrated with computational systems analysis of soluble factors. The gut, the largest immune organ in the body, is of particular interest given observed relationships between gut homeostasis and varied, numerous conditions including obesity, diabetes, neurodegenerative disorders, and depression. We have developed an immune-competent human intestinal model incorporating representative cellular components of the intestinal mucosal environment (enterocytes, goblet cells and immune cells) to begin to develop a quantitative understanding of how coordinated communication of multiple cell types regulates intestinal homeostasis and its impact on other organ systems, including the liver.

Copyright © American Institute of Chemical Engineers. All rights reserved.

472323 A Lab-on-a-Chip Platform of Mono-and Poly-Microbial Biofilms for High-Throughput Downstream Applications

Abstract:

BACKGROUND: Most hospital-acquired infections are caused by microbial colonization that is polymicrobial in nature. Some adapt to biofilm mode of growth which complicates treatment and contributes to unacceptably high mortality rates. These manifestations are common in patients with compromised immunity. Hence there is a dire need for novel and effective antimicrobial agents in eradicating pathogenic communities and limit the emergence of drug-resistant strains.

METHODS: We have developed a novel lab-on-a-chip platform of mono-and poly-microbial biofilms of *Staphylococcus aureus*, *Pseudomonas aeruginosa* and *Candida albicans* for high-throughput antimicrobial screening. The nano-biofilm chip (*nBio*Chip) is robotically printed, robustly handled and scanned using a standard microarray reader. The platform is â€œtrulyâ€ high-throughput and serves as a rising alternative to conventional modes of drug screening.

RESULTS: The biofilm chip consisted of identical nano-biofilms embedded in a hydrogel and exhibited resistance to washing steps involved in the drug screening assay. The biofilm, although nano-scale in nature, demonstrated i) the morphology and three-dimensional architecture of a macroscopic biofilm; ii) elevated drug resistance to antibiotic treatment; and iii) synthesized idiosyncratic exo-polymeric matrix material of a true biofilm. Based on our results, the *nBio*Chip can generate reliable high-throughput antimicrobial susceptibility testing (HT-AST) in 12-18 h. The chip serves as a proof-of-concept platform for high-throughput drug screening against any microorganism and furthers the understanding of microbial interactions in mixed-species communities.

Copyright © American Institute of Chemical Engineers. All rights reserved.

467576 Microfluidic Cell Isolation Technology for Drug Testing of Single Tumor Cells and Their Clusters

Abstract:

There is a growing interest in conducting drug assays with patient-derived cells such as circulating tumor cells (CTCs) to predict better cancer patient outcomes. Recent advances in separating CTCs and their clusters from blood have opened up new opportunities for downstream assays of rare CTCs for drug discovery and probing drug resistance. However, conducting drug assays with CTCs can be challenging due to potential loss of the rare cells (1- 100 CTCs per mL of blood) while handling. Although *ex vivo* culture methods are beginning to emerge to culture CTCs for drug assays, the molecular heterogeneity of individual CTCs and clusters is often lost during the bulk expansion process making it difficult to identify drug resistant cells.

In this study, we present a pipette-based *microfluidic cell isolation* (*MCI*) technology that is capable of conducting single cell resolution drug assays with a small number of tumor cells or their clusters present in small sample volumes (e.g. 10 â€“ 100 cells in 10 mL). A micropipette is used to deliver both the carrier fluid as well as the assay fluid through a microchannel network to nanoliter-sized fluidic traps. The method can digitize the assay fluid producing an array of stationary droplets laden with single cells or their clusters. To establish proof-of-principle of our pipette-based *MCI* method, we use breast cancer cells (MCF-7) and a chemotherapy drug, doxorubicin. We demonstrate that (i) our method can efficiently isolate single MCF-7 cells without appreciable loss during the pipetting and digitization steps (ii) our method can isolate clustered tumor cells with a distribution of 2-22 cells/cluster (iii) the droplet surrounded by an immiscible carried fluid does not adversely affect the viability of anchorage-dependent breast cancer cells (iv) although individual tumor cells display significant heterogeneity in the kinetics of uptake of doxorubicin, they all undergo apoptosis when a critical amount of drug is accumulated (v) the overall viability of clusters is improved simply due to a greater number of cells per cluster suggesting that a CTC cluster has a better chance of surviving chemotherapy drug treatment than a single CTC. Thus, our microfluidic cell isolation technology has significant

potential as a simple yet efficient tool for time-resolved analysis of rare tumor cells and their clusters.

Copyright © American Institute of Chemical Engineers. All rights reserved.

469692 Versatile Microfluidic Platform for Single-Cell Culture and Analysis

Abstract:

Long-term tracking of single cells is essential to interrogate cell-to-cell heterogeneity in clonal cell populations, cell fate regulation, and cellular responses to environmental signals. Compared to classical experimental approaches using microscope coverslips, slides, multi-well plates, or agarose plates, microfluidic technology has revolutionized the monitoring and analysis of individual cells due to its benefits of comparable dimension to a cell, precisely controlled microenvironment, and capability of simultaneously visualizing multiple single cells over prolonged periods. Despite the power of previously developed microfluidic cell culture platforms, they are specially designed for either microchemostat or compartmentalized cell growth and for limited cell types.

Here we present a highly versatile microfluidic platform providing various functionalities for single-cell growth and comparative analysis of diverse types of unicellular organisms ranging from yeasts (*Yarrowia lipolytica*) and bacteria (*Shewanella oneidensis* MR-1) to microalgae (*Chlamydomonas reinhardtii*). This is the first demonstration of an all-in-one microfluidic platform for the trapping and growth of cells with two cultivation formats using the same chip architecture. The device features an array of 30 picoliter-scale circular cell growth chambers, each of which is interconnected with two main channels through two narrow connecting channels arranged in parallel. The serpentine main channels transect the parallel connecting channels for the delivery of medium, oil, dye solutions, or other reagents with the guidance of multiple pneumatic valves. A pair of offset castellated microvalves under the main channels is actuated to enable cells to flow through all of the growth chambers, thereby enhancing trapping efficiency. Microchemostat is created by introducing medium into the main channels and rapidly transporting nutrients into the growth chambers via molecular diffusion, while compartmentalized growth is achieved by infusing oil into the main channels and closing the isolation valve under the connecting channels to form stationary droplet arrays in situ. Coupled with live cell imaging, we have probed single-cell division and growth, long-term cellular dynamics under nutrient switching, and intracellular organelle tracking by dye staining. We envision that our broadly applicable device is extensible to other biological systems (i.e., adherent mammalian cells) and functions (i.e., chemotaxis and analysis of extracellular events such as secretion), and could stimulate additional cell biological explorations.

Copyright © American Institute of Chemical Engineers. All rights reserved.

473652 Point-of-Care Monitoring of Red Blood Cell Adhesion As a Marker of Disease Severity and Treatment Response in Sickle Cell Disease

Abstract:

Abnormal adhesion of blood cells to endothelium are pathophysiologically central to Sickle Cell Disease (SCD). Despite recent advances in identifying and targeting cellular adhesion in SCD, knowledge of abnormal cellular adhesion has not been integrated into routine clinical care or trial design, due to a requirement for complicated custom-designed systems, highly trained personnel, and extensive sample manipulation. The SCD Biochip is a novel point-of-care (POC) microfluidic assay that allows rapid, preprocessing-free, and standardized interrogation of red blood cell (RBC) adhesion to endothelium components (fibronectin, FN, or laminin, LN) in whole blood.

After a complete blood count as a standard of care, 15 μL of surplus whole blood was injected, without processing, on to FN- or LN-functionalized Biochips at or above the physiological shear stresses of post-capillary venules (1-5 dyne/cm^2). Serial quantitative and qualitative evaluations of RBC adhesion and deformability, using standardized protocols, were performed on >100 SCD adult subjects with correlative clinical data.

Greater RBC adherence per unit area to FN or LN was seen in HbSS compared with compound heterozygous SCD. HbSS samples with a low fetal hemoglobin (HbF) displayed greater adhesion to FN or LN than did samples with a high HbF (8% HbF as the cutoff). RBC adhesion was also significantly greater in samples from HbSS patients with higher LDH levels compared to those with lower LDH levels (500 IU/L as the cutoff). Qualitative analyses revealed deformable and non-deformable HbSS-containing RBCs. Deformable RBCs adhere less to FN when exposed to higher shear stresses, and the most tightly adherent cells were non-deformable RBCs. At highest shear stress (50 dyne/cm^2), 75±17% of adhered HbSS RBCs were non-deformable (n=9). Overall, there was a significant association between adhered non-deformable RBCs (%) and serum LDH levels (n=21, Pearson correlation coefficient of 0.79, p<0.005).

Longitudinal analyses performed on individual subjects â‰¥1 month apart showed stable RBC adhesion to FN or LN being treated for SCD with transfusions or hydroxyurea (stable levels of HbA or HbF). In two patients monitored over 4 months, adhesion to FN or LN dropped >60% from baseline after episodes of transfusion (elevated levels of HbA) or initiation of hydroxyurea (elevated levels of HbF).

The SCD Biochip evaluates, simply and with small sample volumes, complex adhesion properties, which reflect clinical phenotypes, including hemoglobin composition, hemolysis, and treatment status. Applied serially and under varied clinical scenarios, this adaptable POC technology will yield a more precise characterization of abnormal adhesive events in a given individual and a more accurate assessment of response to therapy overall.

Copyright © American Institute of Chemical Engineers. All rights reserved.

469308 Enhancing Ex Vivo Platelet Production through Shear Forces within Defined Bioreactors

Abstract:

Currently, platelet transfusions possessing profound clinical importance are entirely derived from human volunteer donors. However, this approach is limited by a 5-day shelf life and differences in donor/recipient histocompatibility. *In vivo*, platelets are formed when bone marrow megakaryocytes (Mks) extend long, cytoplasmic projections, called proplatelets (PPL), into the sinusoid where shear forces accelerate PPL elongation and release platelets into circulation. Methods of *in vitro* platelet production have yielded less than 10 platelets/megakaryocyte (plts/Mk), compared to >1000 plts/Mk *in vivo*. There is a need for a clinically relevant process for platelet production, but much is unknown about what initiates and regulates PPL formation and how to maximize platelet release. We are interested in the generation of ex-vivo platelets through the optimization of Mk maturation, following in the production of functional platelets through a microfluidic bioreactor that utilizes shear forces on Mks to generate platelets. Microfluidic devices have emerged as a valuable tool for cell culture studies. There are numerous advantages to utilizing a microfluidic platform, including a low input cell requirement, the ability to screen multiple conditions in parallel, compatibility with time-lapse imaging, and tight control of microenvironment conditions. In addition, device fabrication is straightforward and inexpensive using soft photolithography.

Through the use of computational fluid dynamic simulations and microfluidic device fabrication, a design â€“ test â€“ build methodology is used to present a dual-flow microfluidic bioreactor system in which aspects of the bone marrow niche are recapitulated to both study the platelet formation process and enhance *in vitro* yields. Experimental studies are conducted to validate the simulations in terms of streamline profiles and flow patterns with and without cell capture. Microenvironment characteristics include shear profiles and extracellular matrix (ECM) protein coatings. Furthermore, the design of the bioreactor generates uniform shear profiles on Mks and PPLs and allows for exploration of a wide range of physiological shear rates. Our results indicate that our bioreactor design produces 21 ± 3 plts/Mk,

and we are currently working to further increase production. Bioreactor-derived platelets are shown to be functional and retain characteristics similar to those of fresh blood platelets. Characterization of the collected platelets from the bioreactor includes surface markers expression, activation in the presence of the thrombin agonist, and morphological/cytoskeletal changes before and after activation. With this microfluidic reactor and further experimental plans, we aim to understand the factors required for initiation of PPL formation to improve *in vitro* platelet yields.

Copyright © American Institute of Chemical Engineers. All rights reserved.

474897 Real-Time Monitoring of Metabolic Function in Liver-on-Chip Microdevices Tracks the Dynamics of Mitochondrial Dysfunction

Abstract:

Microfluidic organ-on-a-chip technology aims to replace animal toxicity testing, but thus far has demonstrated few advantages over traditional methods. Mitochondrial dysfunction plays a critical role in the development of chemical and pharmaceutical toxicity, as well as pluripotency and disease processes. However, current methods to evaluate mitochondrial activity still rely on end-point assays, resulting in limited kinetic and prognostic information. Here, we present a liver- on-chip device capable of maintaining human tissue for over a month in vitro under physiological conditions. Mitochondrial respiration was monitored in real time using two-frequency phase modulation of tissue-embedded phosphorescent microprobes. A computer-controlled microfluidic switchboard allowed contiguous electrochemical measurements of glucose and lactate, providing real-time analysis of minute shifts from oxidative phosphorylation to anaerobic glycolysis, an early indication of mitochondrial stress. We quantify the dynamics of cellular adaptation to mitochondrial damage and the resulting redistribution of ATP production during rotenone-induced mitochon- drial dysfunction and troglitazone (Rezulin)-induced mitochondrial stress. We show troglitazone shifts metabolic fluxes at concentrations previously regarded as safe, suggesting a mechanism for its observed idiosyncratic effect. Our microfluidic platform reveals the dynamics and strategies of cellular adaptation to mitochondrial damage, a unique advantage of organ-on-chip technology.

Copyright © American Institute of Chemical Engineers. All rights reserved.

472562 Identifying Circadian Drug Targets for Maintained Oscillatory Precision

Abstract:

The mammalian circadian clock is a complex dynamic system comprised of interlocked transcription-translation feedback loops. Approximately 10% of all mammalian transcripts are under circadian regulation, highlighting the importance of this system. Disruptions such as shift work, high-fat diets, or aging have been shown to reduce the amplitude of circadian oscillation, and these diminished amplitudes have been associated with a higher incidence of metabolic disorders such as diabetes and obesity. There has therefore been significant recent interest in identifying pharmaceutical modulators of the clock to support robust circadian function. The cellular circadian oscillator is subject to intrinsic molecular fluctuations caused by the low copy number of biomolecules, resulting in cycle-to-cycle variability in period length for cells in the absence of synchronizing cues. Recent work has shown that potential therapeutic treatments affect the precision of the clock, in addition to the period, phase, or amplitude. Furthermore, the vast majority of clock hits within a recent genome-wide siRNA screen resulted in either decreased oscillatory amplitudes or decreased clock precision, indicating that identifying suitable drug targets in the circadian network is a significant challenge.

Rational design of circadian modulators is desirable; however, computational screening for changes in oscillator precision generally requires discrete stochastic simulation of hundreds or thousands of trajectories and then fitting a phase diffusion coefficient for each possible drug target. This approach is prohibitively computationally expensive. Here, we present an efficient analytical method for determining the precision of the circadian oscillator subject to intrinsic molecular noise. Briefly, we construct a model-reduced oscillator with relaxation dynamics governed by the Floquet multipliers of the system. By calculating the phase sensitivity to perturbation and approximating the noise distribution, we are able to analytically calculate a phase diffusion coefficient. We demonstrate agreement between this method and full stochastic simulation. Finally, we apply this method to a computational model of the circadian clock, and identify potential therapeutic targets which allow manipulation of clock period without diminishing precision.

Copyright © American Institute of Chemical Engineers. All rights reserved.

469181 A Systems-Biology Approach to Investigate the Antimicrobial Activity of Oleuropein

Abstract:

In recent years, there is a considerable interest regarding the use of natural antimicrobial agents as food additives to inhibit the growth of food pathogens. Oleuropein and its hydrolysis products, including 3,4-dihydroxy-phenylethanol (hydroxytyrosol), elenolic acid and aglycon [1], are olive phenolic compounds that have antimicrobial effects to a variety of bacteria and fungi such as *Staphylococcus aureus, Pseudomonas aeruginosa, Escherichia coli* and *Candida albicans* [2-3]. The proposed antimicrobial mechanism of oleuropein is due to its surface activity at the bacterial cell membrane: molecular interactions between oleuropein and the phosphate group might modify the phospholipid/water interface properties [4]. In particular, oleuropein was proven to cause leakage of intracellular components such as potassium, phosphate and glutamate from bacteria along with the degradation of the cell membrane[5]. Additionally, oleuropein showed the ability to enhance nitric oxide production in macrophage cells, thus to fight against endotoxins and facilitate the elimination of pathogens[6]. From this, we hypothesize that intracellular metabolic reactions associated with nitric oxide should have larger fluxes for the antimicrobial function.

In this study, a genome-scale modeling approach which can help predict systems-level changes of intracellular metabolisms was adopted to investigate the effect of oleuropein on *S. aureus*, a common food-borne pathogen. Genome-scale modeling approaches, which incorporate genomic genes and metabolites in a metabolic reaction network, have been proved of excellent performance in predicting the systems-level changes of intracellular metabolisms in organisms under specific nutrient environment or treated by specific stimuli. Among all the pathogens that are inhibited by oleuropein, *S. aureus* has been widely studied experimentally [1], [7] as well as by modeling [8]–[10]. In this work, the model developed by Becker et al.[8] was first extended to incorporate the nitric oxide reaction into the intracellular metabolism based on experimental research. The stress that oleuropein imposes on *S. aureus* was mimicked by regulating the export of potassium, phosphate and glutamate out of the cell. With decreased phosphate uptake rate and increased glutamate secretion rate, *S. aureus* growth fits experimental results [1] in the presence of oleuropein at various concentrations. In addition, the reactions and enzymes that are highly related to oleuropein stress are identified by flux balance analysis. This study provides the first genome-scale model for oleuropein metabolism study, and the results indicates that genome-scale modeling is a promising method for revealing the intracellular metabolisms of oleuropein biological properties and thus help to find the food and pharmaceutical applications of oleuropein.

References:

[1] C. C. Tassou and G. J. E. Nychas, "Inhibition of Staphylococcus aureus by Olive Phenolics in Broth and in a Model Food System," *J. Food Prot.*, vol. 57, no. JANUARY, p. 5, 1994.

[2] H. P. Fleming, W. M. Walter Jr., and J. L. Etchells, "Antimicrobial properties of oleuropein and products of its hydrolysis from green olives," *Appl Microbiol*, vol. 26, no. 5, pp. 777–782, 1973.

[3] A. P. Pereira, I. C. F. R. Ferreira, F. Marcelino, P. Valentão, P. B. Andrade, R. Seabra, L. Estevinho, A. Bento, and J. A. Pereira, "Phenolic compounds and antimicrobial activity of olive (Olea europaea L. Cv. Cobrançosa) leaves.," *Molecules*, vol. 12, no. 5, pp. 1153–1162, 2007.

[4] N. Caturla, L. Pérez-Fons, A. Estepa, and V. Micol, "Differential effects of oleuropein, a biophenol from Olea europaea, on anionic and zwiterionic phospholipid model membranes," *Chem. Phys. Lipids*, vol. 137, no. 1–2, pp. 2–17, 2005.

[5] B. Juven, Y. Henis, and B. Jacoby, "Studies on the Mechanism of the Antimicrobial Action of Oleuropein*," *J. Appl. Bacteriol.*, vol. 35, no. 4, pp. 559–567, 1972.

[6] F. Durlu-özkaya and M. T. Özkaya, "Oleuropein using as an Additive for Feed and Products used for Humans," *Food Process. Technol.*, vol. 2, no. 3, pp. 1–7, 2011.

[7] J. Casas-Sanchez, A. M. Alsina, M. K. Herrlein, and C. Mestres, "Interaction between the antibacterial compound, oleuropein, and model membranes," *Colloid Polym. Sci.*, vol. 285, no. 12, pp. 1351–1360, 2007.

[8] S. A. Becker and B. Ø. Palsson, "Genome-scale reconstruction of the metabolic network in Staphylococcus aureus N315: an initial draft to the two-dimensional annotation," *BMC Microbiol.*, vol. 5, no. 1, p. 8, 2005.

[9] M. Heinemann, A. Kümmel, R. Ruinatscha, and S. Panke, "In silico genome-scale reconstruction and validation of the Staphylococcus aureus metabolic network," *Biotechnol. Bioeng.*, vol. 92, no. 7, pp. 850–864, 2005.

[10] D. S. Lee, H. Burd, J. Liu, E. Almaas, O. Wiest, A. L. Barabasi, Z. N. Oltvai, and V. Kapatral, "Comparative genome-scale metabolic reconstruction and flux balance analysis of multiple Staphylococcus aureus genomes identify novel antimicrobial drug targets," *J. Bacteriol.*, vol. 191, no. 12, pp. 4015–4024, 2009.

Copyright © American Institute of Chemical Engineers. All rights reserved.

472684 VEGFR1 Mediates Cell Migration through Activation of PI3K and PLC

Abstract:

VEGFR1 mediates cell migration through activation of PI3K and PLC

Introduction:

VEGF-VEGFR regulate neovascularization; despite this, efforts to control this signaling axis towards directing neovascularization have yet to be achieved [1]. This challenge underlies a continuing need to increase our understanding of signal-to-response. Indeed, there is significantly less knowledge of the VEGFR1 role, which is often described as a decoy receptor due to its high VEGF affinity but low kinase activity [2], relative to the strongly pro-angiogenic VEGFR2. However, our recent computational models have identified that high-VEGFR1 levels may negate the effect of a VEGF inhibitor [3]. Thus, to understand the VEGFR1 signaling role, we develop computational models of the initial intracellular signaling mechanisms connected to VEGFR1: tyrosine site-specific VEGFR1-adapter interactions. Furthermore, we examine how cooperativity in adapter binding to specific VEGFR1 tyrosine sites, and adapter-adapter interactions, direct VEGFR1 signaling. We introduce a novel weighing scheme that correlates adapter-VEGFR1 binding to a cell response (e.g., proliferation, migration, or ubiquitination). We validate our computational predictions using wound healing assays, in vitro.

Materials and Methods:

Computational methods: We develop and compare four endothelial VEGFR1-adapter interaction models in MATLAB

Simbiology: each increases in complexity to gauge how each mechanistic component affects the physiologic response (Fig. 1A). The four models are as follows: Model 1, simplest: adapters bind to one C-terminal Tyr site; Model 2: Model 1 + adapter-adapter binding: adapter-adapter interactions can occur, with binding to one Tyr site; Model 3: Model 1 + Tyr site specificity: adapters bind to physiologically known Tyr sites; Model 4, most physiologically accurate: adapter-adapter interactions occur with binding to known Tyr sites. Therefore, the following comparisons provide insight into VEGFR1 signaling: Model 1 vs. Model 3 = site-specific binding, Model 1 vs. Model 2 = adapter cooperativity, and Model 1 vs. 4 = site-specific binding + cooperativity. We take a data-driven approach towards model development: our experimental qFlow (quantitative) cytometry provides membrane-receptor concentrations while literature mining provides adapter concentrations via western blots, kinetics via surface plasmon resonance, and protein-protein interactions via co-immunoprecipitation. We mine literature to determine how the cell response (migration, proliferation, or ubiquitination) changes when each adapter is inhibited. We quantify these changes in migration, proliferation, and ubiquitination, and assign a corresponding weight to each adapter and linearly correlate adapter contributions to cell response.

Experimental methods: Wound healing assays are performed using mouse RAW 264.7 macrophages: their high VEGFR1 levels (4,820 ± 112 VEGFR1/cell) relative to VEGFR2 (1,770 ± 132 VEGFR2/cell) allows us to focus on VEGF-VEGFR1 response. RAW macrophages are seeded into a 12-well plate and cultured as monolayer to ~90% confluence. The cells are serum starved overnight with DMEM supplemented with 0.5% FBS and 1% PS. The monolayer is scratched with a 100 μL pipette tip and washed once with PBS to remove floating cells. Wells are treated with 750 μL of the serum starved growth factor media, VEGF-A_{164} (50 ng/mL), 10 μM Wortmannin (Anti-PI3K, IC50 = 3 nM), 10 μM U73122 (Anti-PLC , IC50 = 1 μM), 10 μM Imatinib Mesylate (Anti-Abl, IC50 = 600 nM), or a combination of VEGF-A_{164} and an inhibitor. Images of the wounded cell monolayer are taken using a brightfield microsocope at 0 h and 24 h after scratching. All experiments are independently carried out in triplicate. Cell migration is quantified as the number of cells contained within the total gap area relative to the number of cells immediately after the scratch, using Image J.

Results and Discussion:

Adapter negative cooperativity decreases VEGFR1-mediated migration. We find that VEGFR1 regulates HUVEC migration: in Model 1, the integrated migratory response is 2.4-fold greater than the proliferation, and 4.6-fold greater than the ubiquitination, integrated responses. Examining how adapter cooperativity (Model 2) affects VEGFR1-mediated cell responses reveals two key findings. First, VEGFR1 still primarily promotes migration when accounting for adapter cooperativity, evidenced by the migration pathway having the greatest integrated response, 2.1-fold and 3.1-fold greater than proliferation and ubiquitination, respectively. Second, accounting for adapter cooperativity significantly alters the phosphorylation of multiple adapters. Specifically, the integrated response and phosphorylation amplitude of the adapter Nck decrease 88% and 86%, respectively, when adapter-adapter interactions occur (Model 2) compared to no adapter cooperativity (Model 1). Thus, cooperativity in adapter binding mechanistically preventing VEGFR1-mediated Nck activation, implying that Nck negatively cooperates with the other adapters.

VEGFR1 Tyr sites preferentially activate the PI3K and PLC adapters. Accounting for VEGFR1 tyrosine site specificity directs the preferentiality of adapter-VEGFR1 binding and subsequent adapter phosphorylation. Specifically, we find that accounting for VEGFR1 tyrosine sites (Model 3) increases PI3K and PLC signaling; we observe a 1.7-fold and 1.5-fold increase in PI3K and PLC signaling, respectfully, when accounting for VEGFR1 tyrosine sites (Model 3) compared to no tyrosine site specificity (Model 1). Thus, when accounting for specific VEGFR1 tyrosine sites (Model 3), we identify that VEGFR1 is structured to preferentially activate PI3K and PLC .

Tyr site specificity with adapter cooperativity amplifies VEGFR1 through increased PLC activation. By comparing VEGFR1 signaling in the simplest (Model 1) and most physiologically relevant (Model 4) cases, we find that VEGFR1. First, the cell response is unchanged: VEGFR1 signaling primarily promotes cell migration, followed by proliferation, with ubiquitination having the lowest activation. Second, both adapter cooperativity and binding specificity are necessary to enable signal amplification. We arrive at this conclusion because all integrated responses increase in Model 4 compared to Model 1: 6.5-fold for proliferation and 3.5-fold for both migration and degradation.

These increased cell responses do not occur with only Tyr site specificity (Model 3) or adapter-adapter interactions (Model 2). Furthermore, we find that signal amplification primarily results from the preferential activation of PLC ; the PLC integrated response is 2.5-fold greater than the next most activated adapter. We believe this high PLC activation is due to coordination between VEGFR1 Tyr sites and adapter cooperativity. Indeed, PLC has the highest number of available Tyr binding sites on VEGFR1, in addition to undergoing cooperative binding with

Wound healing assays validate VEGFR1-induced migration predictions. Our results indicate that PI3K and PLC are the two adapters primarily regulating cell migration within the VEGF-VEGFR1 signaling axis. To test the robustness of this prediction, we extend our model to macrophages, changing model parameters to match literature-minded adapter and receptor concentrations (e.g., [VEGFR1] >> [VEGFR2]). Again, we predict that PI3K and PLC regulate macrophage migration. Moreover, we rank-predict that PLC inhibition (max=58%/min=24%) > PI3K inhibition (max=51%/min=15%) >> Abl inhibition (< 1% decrease) (Fig. 1B). We validate these predictions to wound healing assays (Fig. 1C).

Conclusions:

Our findings further define VEGFR1 as an active controller in angiogenesis-rather than a mere decoy receptor. Moreover, we define the key proteins mediating this migration role, as PI3K and PLC . Since these pathways are linked to Ca^{2+}signaling [4], these models offer a path towards further understanding VEGFR1-migration via calcium signaling. Future studies can also apply this modeling methodology to understand and control VEGFR1-signaling in diseases where VEGFR1 concentrations are high, such as in breast cancer [3] and ischemia [5].

References:

[1] Eichmann A. *Curr Opin Cell Biol* (2012) 24:188. [2] Zhang Z. *Cell Death Differ* (2010) 17:499-512. [3] Weddell JC. *PLOS ONE* (2014) 9:e97271. [4] Rameh LE. *J Biol Chem* (1998) 273:23750. [5] Imoukhuede PI. *Am J Physiol Heart Circ Physiol* (2013) 304:H1085.

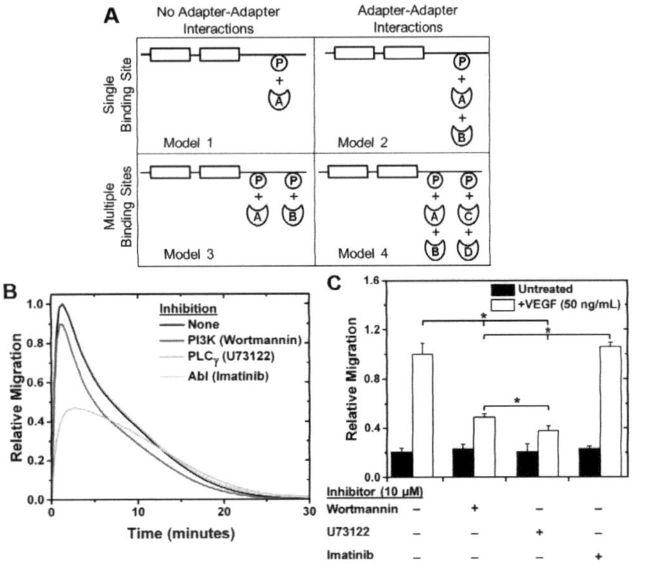

Figure 1. (A) The four models of adapter interactions with VEGFR1. (B) Simulations in RAW macrophages predicting mean migration with adapter inhibition. (C) RAW migration was experimentally quantified with wound healing assays.

Copyright © American Institute of Chemical Engineers. All rights reserved.

469683 Developing a Stochastic Model of LPS-Induced TNF-α Production in Macrophages

Abstract:

Developing a stochastic model of LPS-induced TNF-α production in macrophages

Macrophages are immune cells that are present throughout the body and are involved in a variety of physiological responses. One of its most important roles is its regulation of innate and adaptive immune responses [1]. Toll-like receptors (TLR) embedded in the cell membrane mediate these responses by recognizing microbial components in the body. One such microbial component is LPS (lipopolysaccharide), which is a major constituent of bacterial membranes [2]. Binding of LPS to TLRs leads to the initiation of intracellular signaling pathways and downstream gene expression to rapidly eliminate the bacterial components. One of the major signaling pathways activated by TLR signaling is through the transcription factor NF-κB, which leads to the production of the pro-inflammatory cytokine TNF-α (tumor necrosis factor-α). TNF-α, in turn, initiates signaling through the TNF signaling pathway and leads to propagation of the inflammatory response to microbial components. Hence, LPS-induced signaling pathway is an important part of the early immune response.

Previous studies [3, 4, 5] have formulated population-level models to study the dynamics of signaling through the NF-κB pathway. In these models, the numerical model predicts population-level activation of NF-κB and downstream TNF-α gene expression. However, predictions from the population-level model could be misleading since the model masks the behaviors of individual cells. It has been shown that the effect of LPS stimulation can be highly stochastic, especially to low concentrations of LPS where only a fraction of cells respond to stimulation [6]. Therefore, we developed a stochastic model of LPS-induced NF-κB signaling and TNF-α production in macrophages. Specifically, in addition to gene activation and cellular membrane receptor activation, enzymes and proteins involved in the intermediate stages of LPS-induced NF-κB signaling were modeled as states. Then, the corresponding kinetic parameters for intracellular signaling pathway were identified via the sequential approach of model reduction and inverse modeling. Model simulations were validated and refined using experimental data on LPS stimulated macrophages. Specifically, the activation of NF-κB, TNF-α and IκB in RAW macrophages upon stimulation with different LPS doses was studied using flow cytometry and intracellular staining at the single-cell level. These models are expected to predict the diversity of cellular responses that cannot be captured by deterministic and population-based models. Furthermore, it was possible to simulate intercellular interactions with this stochastic model to analyze cell-to-cell communications - a step toward constructing the multicellular or tissue model. More importantly, the developed stochastic model was used to design a feedback control system that helps producing TNF-α at the desired rate through the autocrine regulation of the LPS-induced intracellular signaling pathway in macrophages.

References:

[1] Parameswaran N, Patial S. 2010. Tumor Necrosis Factor-α Signaling in Macrophages. Crit Rev Eukaryot Gene Expr. 20: 87-103.

[2] Takeuchi O, Hoshino K, Kawai T, Sanjo H, Takada H, Ogawa T, Takeda K, Akira S. 1999. Differential Roles of TLR2 and TLR4 in Recognition of Gram-Negative and Gram-Positive Bacterial Cell Wall Components. Immunity 11: 443-451.

[3] Choudhary S, Kalita M, Fang L, Patel K, Tian B, Zhao Y, Edeh C.B, Brasier A.R. 2013. Inducible TNF receptor associated factor-1 expression couples the canonical to non-canonical NF-κB pathway in TNF stimulation. Journal of Biological Chemistry 288:14612-14623.

[4] Lipniacki T, Paszek P, Brasier A.R, Luxon B, Kimmel M. 2004. Mathematical model of NF-κB regulatory module. Journal of Theoretical Biology 228: 195-215.

[5] Shih V, Kearns D, Basak S, Savinova O, Ghosh G, Hoffmann A. 2009. Kinetic control of negative feedback regulators of NF-κB/RelA determines their pathogen- and cytokine-receptor signaling specificity. Proceedings of the National Academy of Sciences 106: 9619-9624.

[6] James C, Moorman M, Carson B, Branda C, Lantz J, Manginell R, Martino A, Singh A. 2009. Nuclear translocation kinetics of NF-κB in macrophages challenged with pathogens in a microfluidic platform. Biomed Microdevices 11: 693-700.

Copyright © American Institute of Chemical Engineers. All rights reserved.

468915 Parameter Estimation for Sparse Biological Data: Glucose-Dependence of Renin-Angiotensin System in Podocytes Cells during Diabetic Kidney Disease

Abstract:

Diabetic kidney disease (DKD) is the primary cause of end-stage renal failure. Hyperglycemia is known to initiate and exacerbate the pathophysiology of DKD. Podocytes are terminally differentiated glomerular epithelial cells that play a key role in maintaining the structure and function of glomerular filtration barrier in kidneys. Podocytes express a local renin-angiotensin system (RAS) that is altered in hyperglycemia. Studies have shown that angiotensin II (ANG II), which is a RAS peptide, is modulated in hyperglycemic conditions and triggers podocyte injury and apoptosis. The progression of DKD could be slowed by controlling the ANG II levels to prevent irreversible podocyte loss. However, experimental evidence for glucose-dose-dependency of ANG II is scarce in the literature. Hence, for better understanding of the underlying mechanism, we use mathematical modeling to describe the glucose-stimulated RAS signaling in podocytes that produces ANG II.

We have formulated a mathematical model to describe the glucose-sensitive reaction network that triggers the synthesis of ANG II. The local podocyte RAS signaling pathway is represented by a system of ordinary differential equations to track RAS peptides, enzymes, and receptors without explicit glucose-dependence. The system is assumed to be at steady state, thus the model reduces to a system of nonlinear equations. Previous experimental studies were used to estimate the unknown parameters and kinetic constants for the model. Due to sparse experimental data on RAS at sufficiently many glucose states, two different approaches were used to parameterize the model to add glucose-dependency through the parameters: (1) parameterizing the RAS model at normal and high glucose states where all the RAS components are known or estimated and approximating the parameters for intermediate glucose states where no RAS component information is available by considering glucose-dependent linear mathematical functions for each of the sensitive parameters, and (2) fitting the parameters for the RAS model with different proposed glucose-dependent mathematical functions to experimental data from the literature for ANG II vs. glucose [1]. In both the approaches, only the sensitive parameters were varied whereas the non-sensitive parameters were assumed to be independent of glucose.

The model was used to study the change in ANG II concentrations in hyperglycemic conditions with varying glucose levels. The results showed a rise in ANG II levels with increasing glucose concentrations consistent with experimental observations. Sensitivity analysis was conducted. The two parameter estimation approaches resulted in multiple candidate models that are reasonable for glucose-dependence of podocyte intracellular RAS.

Using this model, we were also able to discriminate between possible models for the glucose dose-dependency of ANG II production in podocytes. The key RAS-modulating biomarkers can be identified by studying their effect on ANG II levels. The model can also be used to study the effect of different combinations of various ANG II modulating therapies, which could be useful for drug development.

References: [1] Durvasula & Shankland. 2008. Am J Physiol-Renal Physiol, 294, F830-F839.

Copyright © American Institute of Chemical Engineers. All rights reserved.

461205 A Method for Learning a Sparse Classification Model in the Presence of Missing Data

Abstract:

This work addresses two issues in building classification models: finding a subset of a large number of possible predictors and learning in the presence of missing data. Computational and interpretability concerns promote interest in learning sparse models. Modern dataset often have many measurements but few replicates which can lead to numerical issues. Furthermore, models with many parameters may not provide insight to the analyst. This setting is particularly common in biological or medical studies. Missing data can occur for a wide variety of reasons and is not atypical in the analysis of real datasets. This work focuses on supervised generative binary classification models, specifically linear discriminant analysis (LDA). The parameters are found using an expectation maximization [1] (EM) algorithm. EM allows both missing observations to be handled more naturally as well as the introduction of priors to promote sparsity. An algorithm that finds a sparse LDA model for datasets with and without missing data is presented.

To test the algorithm, a case study for the classification of two types of acute leukemia is presented. The dataset is gene expression data from a microarray. It is a public benchmark problem and has been widely studied [2]. Missing data is artificially added to be representative of missing data in microarrays [3]. The proposed approach is compared to the nearest shrunken centroids algorithm [4] and sparse linear discriminant analysis [5]. Missing data is handled with complete case analysis, mean imputation and k-nearest neighbor imputation, all common approaches in the field. The proposed approach outcompetes alternative methods.

[1] A. P. Dempster, N. M. Laird and D. B. Rubin. Maximum likelihood from incomplete data via the EM algorithm. Journal of the Royal Statistical Society. Series B Methodological, 39: 1-39, 1977.

[2] T. R. Golub, D. K. Slonin, P. Tamayo, C. Huard, M. Gaasenbeek, J. P. Mesirov, H. Coller, M. L. Loh, J. R.

Downing, M. A. Caligiuri, C. D. Bloomfield, and E. S. Lander. Molecular classification of cancer: Class discovery and class prediction by gene expression monitoring. Science, 286: 531-537, 1999.

[3] O. Troyanskaya, M. Cantor, G. Sherlock, P. Brown, T. Hastie, R. Tibshirani, D. Botstein, and R. B. Altman. Missing value estimation methods for DNA microarrays. Bioinformatics, 17: 520-525, 2001.

[4] R. Tibshirani, T. Hastie, B. Narasimhan, G. Chu. Diagnosis of multiple cancer types by shrunken centroids of gene expression. PNAS, 99: 6567-6572, 2003.

[5] K. Sjöstrand, L. H. Clemmensen, R. Larsen, B. Ersbøll. SpaSM: A Matlab toolbox for sparse statistical modeling. Journal of Statistical Software, 2012.

Copyright © American Institute of Chemical Engineers. All rights reserved.

456695 Optimal Experiment Design for Uncertain Biological Systems

Abstract:

Computational modeling has become an indispensable tool to aid discovery in biological systems, which are commonly characterized by highly nonlinear and uncertain behavior. The predictive power of the constructed model largely depends, among others, on the accuracy of the estimates of model parameters. Model-based Optimal Experiment Design (OED) techniques are being increasingly adopted for the purpose of parameter estimation in systems biology [6], as they allow for designing experiments that yield informative data for parameter estimation while accounting for the practical system considerations. However, classical OED approaches cannot account for the intrinsic heterogeneity and uncertainty of biological systems [4]. In addition, uncertainty in the estimates of model parameters cannot be accounted for. In OED for biological systems, the disregard of uncertainty originating from the intrinsic dynamics of the biological system under study as well as inaccuracy in the estimates of the model parameters can lead to severely suboptimal experimental designs [1].

This work considers OED for uncertain biological systems. Two different approaches to OED are comparatively investigated - OED based on deterministic descriptions of uncertainties and OED based on probabilistic descriptions of uncertainties. Deterministic experiment design with respect to all uncertainty realizations, including worst-case uncertainties, can potentially lead to overly conservative designs that underperform when maximizing information content of experiments. On the other hand, the probabilistic OED approach allows for taking into consideration the probability associated with each uncertainty realization. In this case, hence, the uncertainty realizations that have a very small probability of occurrence can be accounted for in the design of experiments, thus resulting in possibly less conservative designs. In the Probabilistic OED approach, the generalized polynomial chaos framework [11] is employed along with Galerkin projection [3] to propagate the uncertainty of the parameter estimates through the system dynamics, resulting in the full probability distribution of system states and outputs at each time point [9]. The deterministic OED approach is formulated through a max-min optimization problem [5]. The D-optimality criterion

[10], which corresponds to the determinant of the Fisher Information (FI) Matrix, is used as a quantitative measure for information content of experiments in both approaches.

The efficacy of both approaches is compared for the canonical Wnt (Wnt/Î²-catenin) signaling pathway, which is ubiquitous in mammalian cell systems and orchestrates critical processes such as cell growth and differentiation [2,7]. A previously reported compartmentalized stochastic model of the Wnt/Î²-catenin pathway is adopted in this work [8]. The OED approaches are intended to design an optimal Wnt stimulation profile (input), which aims at maximizing the information content of the response of the signaling pathway.

The use of a stochastic model that quantitatively accounts for all sources of uncertainty enables obtaining a more representative description of the naturally occurring behavior in a given cell population, thus resulting in a superior performance of experimental design. Stimulation of the signaling pathway with the optimal Wnt profiles show that the probabilistic OED approach yields more accurate parameter estimates, as compared to the deterministic OED. However, the higher accuracy obtained with the probabilistic approach comes at the expense of allowing for a pre-specified probability of constraint violation.

References:

[1] Asprey, S.P. and Macchietto, S. (2002). Designing robust optimal dynamic experiments. *Journal of Process Control*, 12, 545-556.

[2] Clevers, H. and Nusse, R. (2012) Wnt/Î²-Catenin signaling and disease. *Cell*, 149, 1192-1205.

[3] Ghanem, R.G. and Spanos, P.D. (1991) Stochastic Fine Elements: A Spectral Approach. Springer-Verlag, New York.

[4] Kaern, M., Elston, T.C., Blake, W.J., Collins, J.J (2005). Stochasticity in gene expression: From theories to phenotypes. *Nature Reviews Genetics*, 6 , 451- 463.

[5] Korkel, S., Kostina, E., Bock, H.G., Schloder, J.P. (2004). Numerical methods for optimal control problems in design of robust optimal experiments for nonlinear dynamic processes. *Optimization Methods and Software Journal*, 19, 327-338.

[6] Kreutz, C. and Timmer, J. (2009). Systems biology: Experimental design. *FEBS Journal*, 276, 923-942

[7] MacDonald, B., Tamai, K., and He, X. (2009). Wnt/beta-catenin signaling: Components, mechanisms, and diseases. *Developmental Cell*, 17, 9-26.

[8] Martin-Casas, M. and Mesbah, A. (2016). A Stochastic model for the canonical Wnt Signaling pathway. *Conference in Foundations of Systems Biology and Engineering*, Submitted.

[9] Mesbah, A. and Streif, S. (2015). A probabilistic approach to robust optimal experiment design with chance constraints. In *Proceedings of the 9th International Symposium on Advanced Control of Chemical Processes*, 100 -105.

[10] Pronzato, L. (2008). Optimal experiment design and related control problems. *Automatica*, 44, 303-325.

[11] Xiu, D. and Karniadakis, G. (2002). The Wiener-Askey polynomial chaos for stochastic differential equations. *SIAM Journal on Scientic Computing*, 24, 619-644.

Copyright © American Institute of Chemical Engineers. All rights reserved.

464761 Data-Driven Parameter Reduction in Sloppy Models

Abstract:

The past decades have seen significant research progress towards developing reduced, low-dimensional descriptions of complex dynamic models, leading to simpler systems that are easier to interpret. However, the parallel field of reducing the number of parameters required to capture relevant system behavior is relatively nascent: the Manifold Boundary Approximation Method and active subspaces are two promising approaches to this problem of model â €œsloppinessâ€. In this talk we present advances in our own, data-based technique that is capable of discovering sloppy directions in a high-dimensional parameter space without recourse to the analytical equations that constitute the model. We illustrate its effectiveness in the context of biological networks, and show how these methods can be used to perform selective optimization over important parameter combinations, ignoring sloppy directions.

Copyright © American Institute of Chemical Engineers. All rights reserved.

472415 Streamlined Design for Continuous Production of a Pharmaceutical Drug Product

Abstract:

The pharmaceutical industry is beginning to overcome historical reliance on batch processing and embrace the use of continuous processing - a mainstay of other process industries [1]. Continuous unit operations offer many production, efficiency and quality control advantages, and if designed to address the specific risks of a particular process, can also be achieved at lower cost, smaller footprint and more rapid development. One key to this outcome is the participation of a cross-functional team of development and process engineers; operations and supply chain staff; analytical, PAT, and statistical/modeling specialists; equipment vendors; along with inclusion of representatives from Quality and Regulatory functions early in the design process.

We illustrate how functional perspectives have been used to reach balance in the design of a new facility for continuous manufacture of a direct compression and film coated drug product. Design decisions explored include the definitions of a batch in this context, the use of loss in weight feeders, process-model based controls, continuous blending, semi-batch coating pans and automated on-line tablet testing for real time release. The needs of compliant and robust commercial manufacture as well as new product development flexibility are explored. A risk based analysis of drug product characteristics forms the basis for selection among feasible design, control, and operational choices. We present a proof-of-operations case for an initial implementation of continuous manufacturing for the efficient manufacture of drug product batches, and present considerations related to constraints around global regulatory agency acceptability, future flexibility, the competing needs of commercial manufacturing and process development, complexity of process controls, and adaptability to changing market demand.

The presentation is intended to illustrate our belief that for a well-characterized drug product, a streamlined, robust design can be achieved in the absence of multiple complex layers of redundancy (in instrumentation, controls, qualification expense, etc) while still achieving the goals of robust assurance of product quality and batch size flexibility.

[1] Sau L. Lee, Thomas F. Oâ€™Connor, Xiaochuan Yang, Celia N. Cruz, Sharmista Chatterjee, Rapti D. Madurawe, Christine M. V. Moore, Lawrence X. Yu , Janet Woodcock â€œModernizing Pharmaceutical Manufacturing: from Batch to Continuous Productionâ€, *J Pharm Innov, 10*(3), 191-199.

Copyright © American Institute of Chemical Engineers. All rights reserved.

468879 Flowsheet Modeling of a Continuous Direct Compression Process at Production Scale

Abstract:

A flowsheet model for a continuous direct compression and tablet coating and drying process is developed and applied to process data generated at production scale. The flowsheet model contains feeding, blending, tablet compaction and tablet coating and drying units. The model is constructed using gSOLIDS process modelling software (PSE Ltd). The feeding units are assumed to have a constant flowrate, the blending units are based on a tanks-in-series (TIS) methodology and the tablet press model produces tablet hardness as a critical quality attribute. A thermodynamic model for the coating and drying process considers mass and energy conversation to describe five segregated sub-processes (loading, preheating, coating, drying and discharging). Process data is collected by Merck & Co using a GEA Group (AG) continuous direct compression and coating and drying process designed for a six ingredient formulation. Parameter estimation is used determine the mean residence time and lag time for the tanks in series representing the blending process. A number of tablet porosity models are added to the tablet press model increasing model accuracy and flexibility. The coating and drying model describes the process exhaust temperature and relative humidity within a relative error of 2.5%. The flowsheet model performs well using the existing data set and will be used for further process optimization as the process approaches commercial production. In addition, the model is used to investigate process variability in the feeding flowrate and compaction force of the tablet press. A major application of the model is to be used in conjunction with process analytical technology (PAT) for process control.

Copyright © American Institute of Chemical Engineers. All rights reserved.

474389 Characterizing a Drug Product Continuous Manufacturing Platform's Residence Time Distribution Using Major and Minor Unit Formula Excipient Step Changes and Soft Sensors

Abstract:

A continuous manufacturing (CM) line's residence time distribution (RTD) must be appropriately characterized to provide product traceability in the event of a process upset or when switching raw material batches. Lilly's current process to determine the CM line's RTD is to characterize the final blend concentration versus time as a result of intentional step changes to ingredient concentration in the formula. Currently, the ingredient concentration is determined using a Near-infrared (NIR) application located in the tablet press feed frame. This presentation explores 1) using step changes in either the major and/or minor excipient ingredients in the formula to produce similar concentration response curves in the feed frame using the NIR application and 2) the option to use soft sensors in lieu of the NIR application to characterize the CM line's response to the ingredient step changes and ultimately determine its RTD.

Copyright © American Institute of Chemical Engineers. All rights reserved.

466624 Establishing a Procedure for Transferring HME Processes Using Physical Similarities

Abstract:

Introduction

Hot Melt Extrusion (HME) attracted increasing interest in the pharmaceutical industry during recent years. The majority of newly developed APIs (active pharmaceutical ingredients) is hardly soluble in aqueous systems, thus, requires adequate formulations and manufacturing technologies to achieve sufficient levels of bioavailability. HME is a potent pathway to overcome this challenge by establishing solid solutions. Since the process parameters (e.g., screw design, temperature, screw speed) are affecting the morphology of the API a transfer of a HME process from one line to another needs proper considerations. In polymer processing such scale up procedures are common practice especially for single-screw extruders by using physical similarities [1]. Although several guidelines for scaling up of twin-screw extruders exist, it is necessary to know the limitation factors (volume scale-up, energy scale-up, temperature scale-up) [2]–[5].

The objective of this study is therefore to establish a suitable transfer of a HME process from a model extruder (*Pharma Extruder ZSK 18* from *Coperion GmbH*) to a target extruder (*PHARMALAB 16* twin-screw extruder from *Thermo Fisher Scientific Inc.*), where the properties of the intermediate products of both processes should be as identical as possible.

Theoretical Basis

As appropriate transfer methods, the existing method from Menges and Feistkorn [3] as well as the one from Rauwendaal [2] were chosen. The first one is based on the model theory and allows a fully worked out transfer of twin-screw extruders (TSEs). The second method does not allow a complete transfer of the extrusion process since it states to hold critical process parameters, (e.g., the specific mechanical energy consumption (SMEC)) constant. By using the suggested equations, the throughput, the screw speed and the SMEC of the target extruder can be calculated. The two methods are compared in this study.

In pharmaceutical HME, product temperature and shear rates can be critical process parameter since pharma polymers and API can be strongly sensitive to both. Due to the high shear rates in the extrusion die and the impact on the processed material, it is essential to consider the extrusion die properly in the transfer procedure, to avoid different pressure or temperature development. This can be achieved by calculating the die conductance of the die of each extruder [6].

Experimental Methods

Materials

The intermediate products consist of a binary material system. As API Fenofibrate *F6020* from *Sigma-Aldrich* and as polymer matrix *Soluplus®*, a graft copolymer from *BASF*, was chosen. A premix (1:9) was blended before the extrusion process.

Experiments

The set-up of the trials on the model extruder is shown in Table 1. The screw and die configuration were not changed during the investigations.

Run	Temperature profile [°C]	Screw speed [1/min]	Feed rate [kg/h]
1	60/80/100/100/100/110/110/110/110/120	120	2
2	60/80/100/100/100/110/110/110/110/120	200	2
3	60/80/115/115/115/125/125/125/125/135	120	2
4	60/80/115/115/115/125/125/125/125/135	200	2

The set-up of the trials on the target extruder is shown in Table 2. The process input parameters have been calculated according to the two existing transfer methods. As scale-factor for the assembly of the screw configuration of the target extruder, the unrolled screw length Z was used.

Run	Temperature profile [°C]	Screw speed [1/min]	Feed rate [kg/h]
1_M	60/80/100/100/110/110/110/120	145	1.6
1_R	60/80/100/100/110/110/110/120	140	1.7
2_M	60/80/100/100/110/110/110/120	240	1.6
2_R	60/80/100/100/110/110/110/120	240	1.8
3_M	60/80/115/115/125/125/125/135	145	1.6
3_R	60/80/115/115/125/125/125/135	140	2
4_M	60/80/115/115/125/125/125/135	240	1.6
4_R	60/80/115/115/125/125/125/135	240	2

The residence time distribution (model and target extruder) was investigated by means of a coloured tracer and video analysis.

Characterization methods

The intermediate products, obtained from the HME process on the model and target extruder, were characterized by differential scanning calorimetry (DSC), dissolution test according to Ph.Eur. – 2.9.3 with apparatus 2 and content uniformity test according to Ph.Eur. – 2.9.40.

Table 3 shows the different glass transition temperatures (T_g) for the runs 1-4 and 1_M. As can be seen especially the shear rate (represented through the screw speed) causes a shift in glass transition to higher temperature, which indicates that the formulation is sensitive to shear. The T_g of Run 1_M is very close to Run 1, which indicates similar thermal properties.

Table 3: Glass transition temperature of different runs

Run	Glass transition temperature [°C]
1	41.2
2	49.3
3	41.6
4	48.8
1_M	40.5

Figure 1 shows the DSC curves of run 1 in comparison to run 1_M, which have been extruded according to the transfer method of Menges and Feistkorn. As can be seen the shape of the two curves are similar with similar T_g (see Table 3) indicating that the transfer process was successful regarding thermal properties.

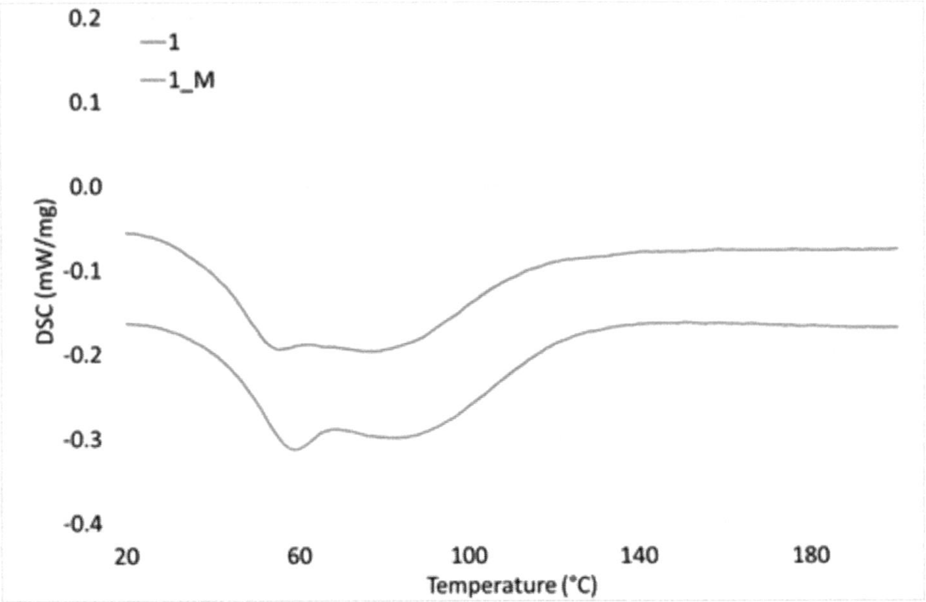

Figure 1: DSC curves (first heating) of Run 1 and Run 1_M

Conclusion and Outlook

By comparing the process output parameters and product characteristics from the model and target extruder, it will be shown if the selected transfer methods are suitable. First results showed that the transfer regarding thermal properties was successful for run 1 using the method of Menges and Feistkorn. Planned analysis will reveal if the transfer was successful regarding dissolution behaviour and content uniformity for all runs. The gained results and the established procedure will help to build up profound knowledge and understanding about scale up of hot melt extrusion processes in pharmaceutical applications.

References

[1] M. Zlokarnik, "Dimensionsanalyse," in *Scale-up: Modelluebertragung in der Verfahrenstechnik*, Second Edition., Weinheim: Wiley-VCH Verlag GmbH & Co. KGaA, 2005, pp. 3–15.

[2] C. Rauwendaal, "10 - Twin Screw Extruders," in *Polymer Extrusion (Fifth Edition)*, Fifth Edition., C. Rauwendaal, Ed. Hanser, 2014, pp. 697 – 761.

[3] G. Menges and W. Feistkorn, "Scale-Up of twin screw extruders application and verification with the example of PVC," *Adv. Polym. Technol.*, vol. 4, no. 2, pp. 123–129, 1984.

[4] A. Dreiblatt, "13 - Technological Considerations Related to Scale-Up of Hot-Melt Extrusion Processes," in *Hot-melt Extrusion: Pharmaceutical Applications*, Chichester: John Wiley & Sons Ltd., 2012, pp. 285–300.

[5] K. Kohlgrueber, *Co-rotating Twin-screw Extruders: Fundamentals, Technology, and Applications*. Munich: Carl Hanser Publishers, 2008.

[6] W. Michaeli, *Extrusionswerkzeuge für Kunststoffe und Kautschuk: Bauarten, Gestaltung und Berechnungsmoeglichkeiten*, 2. Auflage. Munich, Vienna: Carl Hanser, 1991.

468639 Investigation of the Influence of Continuous Blending Process Parameters on the Lubrication of Pharmaceutical Powder Blends

Abstract:

The lubrication of blends of pharmaceutical powders containing 1% magnesium stearate was evaluated in a continuous blender across a broad range of mass flow rates and blender speeds. Significant changes in both the bulk density and tensile strength of powder compacts were observed as a function of blending conditions. The results demonstrate that for a given blender configuration, various final degrees of blend lubrication can be achieved. An optimal combination of mass flow rate and blender speed for maximum lubrication is found to reside at intermediate process values, consistent with results from other studies. The relationship between blending conditions and blend lubrication state are subsequently described by a numerical model, which describes changes in final blend attributes as a function of key blending process parameters. This model, however, collapses at higher blender speeds, where powder conveyance dominates and mixing dynamics are significantly diminished.

Copyright © American Institute of Chemical Engineers. All rights reserved.

473846 "Correlating Raw Material Properties to Critical Processing Parameters to Critical Quality Attributes in a Direct Compaction Continuous Line."

Abstract:

In the last few years, Pharmaceutical manufacturing is experiencing unprecedented scientific and technological innovation as it has been involved in a fundamental transformation from batch to continuous. This revolution has the potential of reducing both the manufacturing time and cost, increase process control and improve overall process quality.

This study demonstrates a systematic methodology to find the critical processing parameters affecting the critical quality attributes of tablets produced in the continuous direct compaction manufacturing line. The methodology is demonstrated in a case study where the API is fixed, the formulation is varied along the DOE but keeping the API at low concentrations (2 to 10%).

The approach follows a screening bi-block two level fractional factorial DOE to elucidate the critical process parameters and corresponding failure modes of the continuous line which depends on the material properties of the formulation. The screening study involves eight processing and two formulation parameters but it only consists on twenty four experiments. A three level factorial DOE on the remaining parameters is then run for characterization and optimization purpose with the ultimate goal to relate material properties to the critical processing parameters and finally to the critical quality attributes of the tablets produced.

The study is run in Rutgerâ€™s direct compaction continuous line and consists of two feeders (API and excipient) connected to a mill and to a blender with a third feeder (lubricant) connected also to the blender, bypassing the mill to

prevent blendâ€™s over lubrication. After the blender, the mix flows through a chute to the feed frame of the rotary tablet press where tablets are compacted. The quality attributes measured on tablets are content uniformity, weight variability, dissolution and hardness, but only the first is expected to have a relevance in this case study at low API concentration (2 to 10%).

Copyright © American Institute of Chemical Engineers. All rights reserved.

461124 A Novel Analysis of Residence Time Distributions for Continuous Powder Mixing

Abstract:

Continuous pharmaceutical manufacturing has attracted the attention of industry and academia for a wide variety of reasons. Because this type of manufacturing can utilize smaller manufacturing facilities, it can alleviate bottlenecks related to scale-up and improve control of product quality. With regards to materials, this type of process can eliminate the isolation and storage of intermediate products as well as decrease raw-material and final-product waste. By reducing the number of manufacturing steps, continuous pharmaceutical manufacturing can shorten manufacturing cycle times, provide greater response capacity, increase production yields, and lead to an overall product manufacturing efficiency (Lee et al., 2015).

The residence time distribution (RTD) is a probability distribution function that describes the length of time a material spends in the system. The RTD can be used to address the traceability of raw materials in continuous pharmaceutical manufacturing processes (Engisch & Muzzio, 2016), providing useful data for cases relating to out of specification product investigations, consumer complaints, product recalls or any other situations that may have public health impact. Knowledge of the RTD enables the prediction of when an affected material will reach a point downstream in the process, as well as how much of the adjacent material it may have affected through dispersion. The RTD of convective powder mixers has previously been experimentally studied and correlated to lumped parameters such as the Peclet number and mean residence time in the Taylor dispersion model (Gao, Vanarase, Muzzio, & Ireapetritou, 2011) . However, the RTD measured/fitted for one set of operating conditions may not predict the system RTD when the conditions are altered.

In this work, we conducted an analysis using previous experimental RTD data characterizing continuous powder mixing (Marikh, Berthiaux, Mizonov, Barantseva, & Ponomarev, 2006; Vanarase & Muzzio, 2011). Vanarase and Muzzio (2011) conducted experimental RTDs on a Gericke GCM 250 continuous mixer for eight different conditions (four different impeller rotation rates: 40, 100, 160, and 250 rpm at two throughput levels: 30 and 45 kg/h). Similarly, Marikh et al. (2006) measured eight experimental RTDs on a Gericke GCM 500 continuous mixer (four different throughputs: 40, 60, 80, 100 kg/h at two different impeller rotation rates: 15 and 60 rpm). The experimental RTDs were converted to RTDs in dimensionless time by scaling to each distribution with its respective mean residence time. It is observed that RTD in dimensionless time is invariant with throughput at different impeller rotation rates under certain conditions. In other words, the shape of RTD is the same but the RTD itself is just shrunk or stretched based on the mean residence time. Next, we introduced a novel concept of a time-dependent RTD to contrast with RTDs measured under steady-state conditions. This approach predicts transient changes in the RTD as throughput varies arbitrarily with time. It is anticipated that this effort will enhance process understanding to facilitate extending empirically measured RTDs for solid oral continuous drug product manufacturing processes in a straightforward manner.

References

Engisch, W., & Muzzio, F. (2016). Using Residence Time Distributions (RTDs) to Address the Traceability of Raw Materials in Continuous Pharmaceutical Manufacturing. *J Pharm Innov, 11*(1), 64-81.

Gao, Y., Vanarase, A. U., Muzzio, F., & Ireapetritou, M. (2011). Characterizing continuous powder mixing using residence time distribution. *Chemical Engineering Science, 66*, 417-425.

Lee, S. L., Oâ€™Connor, T. F., Yang, X., Cruz, C., Chatterjee, S., Madurawe, R., . . . Woodcock, J. (2015). Modernizing Pharmaceutical Manufacturing: from Batch to Continuous Production. *J Pharm Innov, 10*(3), 191-199.

Marikh, K., Berthiaux, H., Mizonov, V., Barantseva, E., & Ponomarev, D. (2006). Flow Analysis and Markov Chain Modelling to Quantify the Agitation Effect in a Continuous Powder Mixer. *Chemical Engineering Research and Design, 84*(11), 1059-1074.

Vanarase, A. U., & Muzzio, F. J. (2011). Effect of operating conditions and design parameters in a continuous powder mixer. *Powder Technology, 208*(1), 26-36.

Copyright © American Institute of Chemical Engineers. All rights reserved.

471433 Model Based Loss-in-Weight Feeder Design and Optimization

Abstract:

Over the past year and half the Pharma Industry has seen the first two approvals for the continuous manufacturing of Oral Solid Dosages. This being further evidence that it is no longer a question of "if" continuous manufacturing will come to the pharma, but rather "when/how" the transformation will take place. From a technical point of view it has been acknowledged by most that Loss-in-weight (LIW) feeding is the most critical operation to continuous Oral Solid Dosage products. In recent years a much deeper fundamental understanding of the LIW feeding process has been developed. This understanding includes a material property driven modeling frame work for predicting performance. The goal of this talk is to present a framework for model based design and optimization of feeders. The key deliverable of this approach would be the ability to tailor each feeder for the specific material properties of the raw material it is feeding.

Copyright © American Institute of Chemical Engineers. All rights reserved.

460739 Screening for Promising Microorganism-Produced Bio-Chemicals

Abstract:

Screening for Promising Microorganism-Produced Bio-Chemicals

Wenzhao Wu, Matthew Long, Jennifer Reed, Christos T. Maravelias*

Dept. of Chemical and Biological Engineering, University of Wisconsin-Madison, Madison, WI 53706

The last decade has seen tremendous progress in metabolic engineering and synthetic biology[1,2]. These advancements enable the use of engineered microorganisms such as *Escherichia coli*, yeast and algae for the production of chemicals that are currently derived mainly from fossil fuel feedstocks[3,4]. However, which chemicals have the highest economic potential remains unclear. Toward this end, we develop a framework for the identification of promising chemicals for bio-based production.

We first examine the US high-production-volume (HPV) chemicals[5], which are manufactured in or imported into the United States in amounts equal to or greater than 454 metric tonnes (MT) per year. HPV chemicals include all commodity chemicals and a portion of fine chemicals. We establish an HPV chemical database (3574 chemicals) by compiling several HPV lists published by the EPA over the past two decades. Then we intersect the HPV chemical database with the KEGG database, which includes most of chemicals produced by characterized reactions in biological systems, and thus 613 overlapping chemicals are found. These chemicals are then imported into a genome-scale metabolic model, and 168 chemicals are identified to be producible by microorganisms, often with the addition of heterologous reactions. These 168 chemicals are the complete pool of candidate targets for bio-based production. In addition, market volume and price data is collected for each of them.

Next, we develop three screening criteria to quantify economic potential.

<u>Criterion 1: "separation cost margin"</u>. The largest the difference between a chemical's selling price and its production cost is, the largest its economic potential is. However, the downstream separation cost is highly product-dependent and difficult to estimate. Therefore, we quantify the economic potential using the "separation cost margin", which is the difference between the price and the upstream cost (including raw material supply cost and bio-conversion cost). We calculate the upstream cost using base cost data from the literature, and theoretical productivity and titer calculated using our metabolic model.

<u>Criterion 2: market volume</u>. The market volume should be greater than an expected production capacity. The specific capacity is estimated based on the type of the bio-conversion system. For example, the capacity for an open pond system can be estimated based on the area of a typical open pond facility.

<u>Criterion 3: market value</u>. The market value should be large enough to attract investment and recover capital cost within an expected time horizon.

Note that the specific values adopted in these criteria depend on the type of supply sources (e.g. flue gas, CO_2, or sugar as the carbon source), bio-conversion type (photosynthesis or fermentation-based), specific reactor types (continuous or batch), and classification of chemicals (commodity or fine chemical). We investigate several benchmark scenarios. For example, we apply the three criteria on all the commodity chemicals in the candidate pool, assuming photosynthetic bio-conversion in a batch reactor.

Finally, we present detailed results for a specific system: cultivation of photosynthetic bacteria in a continuously operating open pond system, supplied with captured CO_2 from flue gas, and nutrients and water from a waste water treatment facility. We set a uniform separation cost margin for all chemicals to be greater than 1.8 $/kg, which is an estimated margin for a benchmark product - polyhydroxyalkanoate (PHA). The volume was set to be greater than 31751 MT/year, calculated from an expected open pond area of at least 1000 ha. Finally, the market size was assumed to be greater than 200 million $/year to be attractive enough for investment, and to recover capital cost in ~5 years supposing a 50% market share. Applying all the three criteria on the 168 chemical candidates, as a preliminary result, we identify four chemicals as the most promising ones: glutaric acid, acrylamide, propanal, and octanoic acid.

References

[1] Gavrilescu, M. & Chisti, Y., 2005. Biotechnology- a sustainable alternative for chemical industry. *Biotechnology advances,* Volume 23, pp. 471-499.

[2] Bornscheuer, U. T. & Nielsen, A. T., 2015. Editorial overview: chemical biotechnology: interdisciplinary concepts for modern biotechnological production of biochemicals and biofuels. *Current Opinion in Biotechnology,* Volume 35, pp. 133-134.

[3] Wilson, S. A. & Roberts, S. C., 2014. Metabolic engineering approaches for production of biochemicals in food and medicinal plants. *Current Opinion in Biotechnology,* Volume 26, pp. 174-182.

[4] Zhang, X., Tervo, C. J. & Reed, J. L., 2016. Metabolic Assessment of E. coli as a Biofactory for Commercial Products. *Metabolic Engineering,* Volume 35, pp. 64-74.

[5] US EPA, 2004. High-production-volume (HPV) chemicals status and future directions of the HPV Challenge Program. *Office of Pollution Prevention and Toxics*, Washington DC.

Copyright © American Institute of Chemical Engineers. All rights reserved.

472332 Acceleration of Microbial Metabolic Engineering with Crispri

Abstract:

The field of metabolic engineering is entering a new era in which readily addressable synthetic transcription factors like RNA-guided dCas9 are opening new possibilities for dynamic regulation and for rapid assessment of multiplexed genetic interventions to alter the metabolic landscape in both workhorse and previously intractable microbes. Toward this end, our group has focused on several applications of CRISPRi-enabled metabolic engineering. Previously, we developed a modular assembly method for traditional restriction-ligation cloning of type II-A CRISPR array libraries for multiplex, combinatorial dCas9-mediated transcriptional repression. Importantly, we demonstrated for the first time that dCas9 can be utilized for metabolic engineering in *E. coli*, leading to enhancement of flavonoid titers through simultaneous repression of an endogenous transcription factor and several central carbon enzymes, including partial downregulation of a synthetic lethal pair. Furthermore, we showed that this system could be utilized to generate complex phenotypes, and we also demonstrated that a single plasmid can be used to attenuate virulence genes in disparate *E. coli* lineages by repressing capsular polysaccharide secretion in both probiotic strain Nissle 1917 and pathogenic strain K5. Expanding upon this work, we have recently developed a novel, streamlined cloning strategy to assemble natural type II-A CRISPR arrays, enabling simultaneous repression of combinations of genetic targets in only a few days. We showcase this method by improving production of a panel of natural products through repression of several novel knockdown targets. Comparison of CRISPRi strains with analogous deletion strains shows that CRISPRi can lead to production improvements comparable to gene deletions, potentially obviating the role of cumbersome combinatorial gene deletions during hypothesis testing stages early in the strain development pipeline.

Copyright © American Institute of Chemical Engineers. All rights reserved.

473033 Enhancing Pathway Synthesis Using Genetic Information

Abstract:

The engineering of non-native synthesis pathways in microbial hosts has shown promise in the production of commercially useful biomolecules. In some cases, a choice for the synthesis pathway may be obvious. For example, there is only one known pathway for biosynthesis of 1,3-propanediol from glycerol. This pathway consists of two reactions, each catalyzed by a singular enzyme. More generally, the number of alternative pathways for a given product may be too large for experimental exploration. To facilitate the identification of possible synthesis pathways for a desired target biomolecule, current computational synthesis methods identify one or more possible synthesis pathway that can be added to the host. Each path consists of a series of non-native reactions that can be catalyzed by a number of possible genes. While there currently is a lack of data and consensus on the best pathway scoring methods, proposed metrics have been based on metabolic objectives such as yield, metabolic burden, and thermodynamic feasibility.

Despite progress, it is still difficult to identify the best synthesis pathway. Current synthesis methods assume uniform activity levels for all identified enzymatic reactions despite differences in kinetic parameters and solubility. To date, more than 1,000 prokaryotic genomes have been fully sequenced and annotated. Partial or draft genomes are available for more than 6,000 species. The total number of reactions listed in the Kyoto Encyclopedia of Genes and Genomes (KEGG) currently exceeds 8000. To exploit the diversity of metabolic enzymes across many different organisms, it becomes necessary for synthesis methods to efficiently compare the various genes available to catalyze a reaction along the identified synthesis pathway.

This work addresses the problem of identifying viable synthesis pathways and their gene implementation. In addition to identifying reaction steps required to construct each synthesis pathway, we identify the best organisms from which to cull the enzymes required to catalyze each reaction step. The major steps in our approach are as follows. First,

ProbPath (Yousofshahi, M., K. Lee, and S. Hassoun, Metabolic Engineering, 2011) is utilized to construct non-native synthesis pathways for a desired product metabolite. ProbPath is graph-based pathway construction algorithm that utilizes probabilistic selection of reactions to construct pathways between a desired metabolite and a metabolite within the host. Second, synthesis pathways are ranked based on their yield using Flux Balance Analysis. Third, the solubility of each possible enzyme from each available organism is evaluated. To accomplish this, we create a database with reviewed protein sequences. The database contains information about specific activity for all proteins from specific organisms present in the BRENDA database. Our contribution is in creating a method for efficiently evaluating solubility and specificity for a large number of genes. We validate the workflow using several test cases.

The work described here advances the state of pathway synthesis by exploring the underlying gene design space, thus leading to more profitable design options when compared to using pathway synthesis tools that only explore the metabolic network design space. Importantly, advancing the design and engineering of synthetic biological systems will lead to reduced development cost, time, and effort, which in turn will enable new discoveries that have a positive impact on human health and the environment.

Copyright © American Institute of Chemical Engineers. All rights reserved.

461672 Engineering of Hydroxytyrosol Production in *Escherichia coliÂ* for Industrial Application

Abstract:

Hydroxytyrosol (HTy) is one of the most powerful antioxidants and has a wide range of potential applications in industry. Currently, HTy is produced from enriched olive extracts after chemical or enzymatic hydrolysis, but microbial fermentation can be a promising solution of an increasing demand for stable and sustainable production.

We have previously reported a synthetic pathway in *E. coli* capable of producing HTy from tyrosine. Our synthetic HTy pathway consists of five heterologous genes for tyrosine hydroxylation and downstream conversion of L-DOPA to HTy. Although we have shown that this pathway is functional, the low titer from tyrosine or from glucose remained as an issue to be addressed. Here we report three metabolic engineering efforts to improve the efficiency of the synthetic HTy pathway. First, we have engineered the co-factor biosynthetic pathway for tyrosine hydroxylation. Next, we have identified inhibition of the tyrosine hydroxylation by downstream chemicals and applied co-culture strategy to overcome it. Lastly, the gene expression of pathway enzymes was optimized using metabolomics and proteomics data. With these engineering, we were able to improve the product yield from tyrosine about 3-fold. In addition, we have engineered the host strain to overproduce tyrosine. Using this engineered strain, we confirmed the production of L-DOPA from glucose without any external supplementation of tyrosine, and the resulting L-DOPA was further converted into HTy by co-culture strategy with an improved yield from the previous result.

Copyright © American Institute of Chemical Engineers. All rights reserved.

468637 Metabolic Engineering for the Production of Medium Chain Fatty Acids and Alcohols in *Escherichia coli*

Abstract:

Medium chain length alcohols are useful in several commercial products, such as cosmetics, plastics, and surfactants. While short chain alcohols, like ethanol and butanol, are available as fermentation products, and longer chain alcohols are readily available as products from the Ziegler process, medium chain alcohols are produced in relatively small amounts. The biochemistry of oleochemical production is analogous to polymerization reactions, including initiation, elongation, and termination steps. We have identified different enzymatic modules that comprise these three steps and assembled metabolic pathways in the bacterium *Escherichia coli* for the production of medium-chain fatty alcohols and fatty acids. We further to seek to improve yield, titer, and selectivity of our fatty acid and alcohol producing strains by optimizing strain design and fermentation conditions.

Copyright © American Institute of Chemical Engineers. All rights reserved.

470198 Engineering Production of Monomers from the Lysine Pathway

Abstract:

There is great interest in developing bio-based aromatic chemicals to replace phthalates, one of the highest volume monomers used in industrial and consumer products. We envision dipicolinic acid (DPA) could be a useful phthalate analog. In addition, the nitrogen atom in the aromatic ring confers new possibilities, as a metal chelator leading to its recent use in specialty polymers, with applications in heavy metal detection[1], catalysis[2], and even as the primary functional component of a self-repairing polymers[3]. DPA is naturally produced in *Bacillus subtilis* cells during sporulation. The biosynthetic pathway, however, has not yet been well-characterized. We have successfully identified the relevant genes for production of DPA in *E. coli*, including two uncharacterized genes *spoVFA* and *spoVFB* which are instrumental for production[4]. Initial fermentations yielded up to 1.4 g/L production of DPA in 48 hours. The productivity was also examined for *E. coli* where threonine and lysine production was reduced. Finally, engineered wild-type *E. coli* was able to produce DPA at maximum of 4.3 g/L with minimal media optimization. DPA production for all fermentations was monitored by HPLC and verified using LC-MS fragmentation. LC-MS of unpurified fermentation media also show production of several by-products, including homoserine and HTPA.

1. Darbha, G. K., Ray, A. & Ray, P. C. *ACS Nano* **1,** 208–214 (2007).
2. Li, C.H. *et al. Nat. Chem.* 1–7 (2016).
3. Kirillova, M. V., Kirillov, A. M., Guedes Da Silva, M. F. C. & Pombeiro, A. J. L. *Eur. J. Inorg. Chem.* 3423–3427 (2008).
4. Daniel, R. A. & Errington, J. *J. Mol. Biol.* **232,** 468–483 (1993).

Copyright © American Institute of Chemical Engineers. All rights reserved.

474962 Invited Presentation: Seawater Based Microbial Synthetic Biotechnology

Abstract:

Industrial biotechnology based on microbial fermentations aims at producing chemicals, materials and biofuels to partially replace petrochemical routes. However, industrial biotechnology cannot compete with the petrochemical industries as it suffers from high consumptions on energy and freshwater, discontinuous processing, low product concentration and thus high product recovery cost as well as low substrate to product conversion efficiency. Therefore, to make industrial biotechnology competitive, we need develop fermentation platform technology which is energy and fresh water saving and run in a continuous way instead of a batch way.

To overcome these disadvantages of the traditional fermentation industries, seawater based fermentation technology has been developed based on *Halomonas* spp. Taking production of microbial bioplastics polyhydroxyalkanoates (PHA) as an example to show the advantages of this new technology. To lower the production cost of PHA, we have developed a seawater technology based on *Halomonas* spp. grown in seawater in an unsterile and continuous process. This new blue water technology achieves fresh water saving, energy saving and reduces process complexity combined with mixed substrates like kitchen wastes, it significantly reduces the production cost of PHA and its biofuels.

In this lecture, the most updated technology to reprogram the *Halomonas* spp. will be introduced. We have been able to assemble several multiple genes containing pathways in *Halomonas* spp. for production of various PHA molecules. The growth patterns of the *Halomonas* spps. have been changed to influence the shapes of *Halomonas* spps. from short rods to long fibers, even further to large spheres, allowing more PHA accumulation and easy product recovery. The technology has been pilot tested in industrial setting, it has been proven to be able to reduce PHA production cost at least 100%. The seawater technology should be able to improve the economic competitiveness of industrial biotechnology. We expect the technology can play an increasingly important role in the fermentation industries.

References

1) Chen, G. Q., et al. Chemical Reviews, 2012. 112, 2082-2099

2) Fu, X. Z., et al. Metabolic Engineering, 2014, 23, 78-91

3) Tan, D., et al. Metabolic Engineering, 2014, 26, 34â€“47

4) Wang, Y., et al. Metabolic Engineering, 2014, 25, 183-193

5) Wang, Y., et al. Current Opinion in Biotechnology, 2014, 30, 59-65

6) Wu, H., et al. Applied Microbiology and Biotechnology, 2014, 98, 10013-10021

7) Yin, J., et al. Applied Microbiology and Biotechnology, 2014, 98, 8987â€“8997

8)Â Chen, G. Q., et al. Trends in Biotechnology, 2015, 33, 559-564

9)Â Jiang, X. R., et al. Metabolic Engineering, 2015, 29, 227-237

Copyright © American Institute of Chemical Engineers. All rights reserved.

458840 Use of Real-Time Analytics and Apoptosis Assays for Enhanced Understanding of Cell Growth and Viability in a CHO-Based Process for Monoclonal Antibody Production

Abstract:

In a cell culture process for biopharmaceutical manufacturing, monitoring cell growth and viability is crucial for understanding cell behavior, as well as for determining cell culture duration and timing of harvest. Standard industry practice typically includes daily sampling of bioreactors for cell count and metabolite profiles. However, daily measurements are often too infrequent to detect rapid changes in metabolite concentrations and viability, and may not capture early indications of decline in cell health. We have evaluated the use of apoptosis assays as well as in-line monitoring techniques to gain a more complete understanding of cell growth, health and viability throughout the cell culture process.

In this study, we have applied markers of early, middle and late apoptosis to evaluate changes in cell growth, health and viability during a fed-batch process in 5L bioreactors. The markers that were monitored included 7-AAD, Caspases, and TUNEL assay for DNA nicking. Multiple early and mid-stage apoptotic markers were compared for their effectiveness in evaluating cell health and predicting cell growth trends in a bioreactor. Additionally, to provide more frequent cell growth measurements and metabolite readings to align with apoptotic assays, in-line probes were incorporated for real-time data analysis. Through calibration and method development, a capacitance-based Incyte probe was used for real-time monitoring of viable cell density (VCD), and an in-line Raman spectroscopy probe was used for real-time monitoring of glucose and lactate.

The results demonstrated that apoptotic markers can provide additional insight into cell health beyond what is learned from daily cell counts, thereby greatly deepening our understanding of cell culture conditions and process parameters during development. Markers of early apoptosis were observed several days prior to a drop in viability as measured by a commercial cell counter, while late-stage apoptotic markers were observed hours in advance or concurrently with the VCD and viability drop. Further insight into the health and productivity of the culture may be obtained by correlation of in-line metabolite and cell growth profiles with cell apoptotic progress.

In summary, optimization and application of real-time analytics and apoptosis assays in a CHO cell process can provide a rich and complete data set for better visualization of cell growth and health throughout the culture process, thereby enabling process improvement to maximize product titer and quality.

Copyright © American Institute of Chemical Engineers. All rights reserved.

468928 Providing a Molecular Basis to Modeling Protein Adsorption to Multimodal Carbohydrate-Based Ligands

Abstract:

Recognition and binding of proteins to carbohydrate-based multimodal ligands plays a critical, but poorly understood, role in both fundamental cellular (e.g., cell glycocalyx and viral protein binding) and industrial processes (e.g., antibody drug bioseparations). Manufacturing costs of biopharmaceutical drugs can be drastically reduced through development of inexpensive multimodal (or mixed-mode) chromatographic bioprocesses for protein isolation and purification. However, unlike classical affinity or ion exchange based bioseparations that have been well studied, multimodal chromatographic (MMC) systems are still in their developmental stages [1]. Inexpensive and biocompatible MMC matrices/ligands derived from carbohydrates have tremendous potential for industrial bioseparations. However, we currently have a poor understanding of the molecular mechanisms driving protein binding to traditional carbohydrate (e.g., sepharose) and/or glycoprotein (e.g., concanavalin A) derived chromatographic matrices. This has been further stymied by the lack of suitable experimental tools and predictive adsorption models that explain the heterogeneous nature of protein-carbohydrate interactions. One of the challenges with traditional methods (e.g., solution depletion) used for monitoring protein adsorption is that bulk ensemble measurements provide limited insight into the underlying molecular mechanisms driving protein-ligand binding. Studying binding interactions of proteins and carbohydrate-based ligands at the single-molecule level is critical to explaining observed macroscale measurements, which will ultimately lead to the development of mechanistically-relevant protein macroscale adsorption models.

Here we study the adsorption of carbohydrate-binding proteins to two distinct model polysaccharide substrates with identical molecular composition but varying surface hydrophilicity using complementary bulk and single-molecule based adsorption techniques. We have modified a recently developed optical tweezers force spectroscopy technique [2] to characterize the single molecule protein binding-unbinding force dynamics to polysaccharide surfaces and probed the protein-ligand bond lifetimes at the single molecule level. Analysis of our records was used to identify trends in the force-binding lifetime datasets to find that the unbinding force and total binding lifetimes reflects the underlying heterogeneity of the available binding sites for carbohydrate-derived ligands. This distribution reflects the spectrum of varying affinity sites present on the polysaccharide surface and ultimately provides a molecular basis for using various multi-site Langmuir-type adsorption models to characterize protein adsorption to polysaccharide surfaces. Interestingly, simple regression analysis of bulk adsorption data using multi-site Langmuir-type models reveals that traditional protein bulk adsorption measurements are unable to provide data with sufficiently high resolution to discriminate between the low and high affinity protein binding sites for polysaccharides if the difference in the protein-carbohydrate ligand unbinding force is relatively small. In summary, we now have the necessary tools to systematically characterize protein adsorption to multimodal carbohydrate-derived ligands for optimizing industrial bioseparations using a first-principles approach.

References

[1] Pinto, I. F., Aires-Barros, M. R. & Azevedo, A. M. Multimodal chromatography: debottlenecking the downstream processing of monoclonal antibodies. Pharm. Bioprocess. 3, 263â€"279 (2015).

[2] Brady, S. K., Sreelatha, S., Feng, Y., Chundawat, S. P. S. & Lang, M. J. Cellobiohydrolase 1 from Trichoderma reesei degrades cellulose in single cellobiose steps. Nature Communications. 6, 10149 (2015).

Copyright © American Institute of Chemical Engineers. All rights reserved.

472914 Operational Control Strategy Development Via First Principles Mapping

Abstract:

Product and Operational control strategies are essential elements in efficient and effective manufacturing of quality medicines. Many forms of guidance exist (e.g., PQLI) to help define and structure documented control strategies. Most control strategies are good at identifying the controlled variables and critical parameters that are essential to ensure product quality. What is often missed, however, is identifying the underlying first principle transformations that are the core of every deliberate step that is taken in making medicine. Each process step has an objective to be met. In fact, when we look at each process step, there are multiple objectives to be met. We must ensure the product quality is delivered, but we must also ensure that we do it cost effectively, protect the operator, protect the community, protect the environment, meet demand, etc. To manufacture the Product, we must design a Process Map as a collection and sequence of Process Stages and Process Operations that combine to deliver certain features and requirements of the product. The process operation is simply stated as an action (or transformation) on an object with a qualifier, e.g. *Disperse powder to less than 1% variation.* The mechanisms for transformation can always be traced to first principles of science. Thus, the Control Strategy is based upon key first principles and associated attributes and variables, at the most fundamental level. To satisfy the criteria of the transformation we must choose and implement effective Control Methods. For the example of "disperse powder", we might choose a vertical agitated tank that will deliver an adequate environment for the first principles attributes of shear and fluid velocity. For each process operation, we must consider objectives and control methods for all Features, such as quality, safety, cost, etc. In the case of Safety, for example, we can provide control methods for preventing dust explosions based upon the first principles of ignition source control.

This presentation will define and demonstrate the means to construct and implement a structured product control strategy via process mapping using first principles-based transformations and chosen control methods.

Copyright © American Institute of Chemical Engineers. All rights reserved.

460708 Nanoextrusion: A Platform Enabling Comparative Assessment of Nanocomposites Vs. Amorphous Solid Dispersions for Drug Dissolution Enhancement

Abstract:

A relatively novel nanoextrusion process was used to produce extrudates of griseofulvin (GF), a poorly water-soluble drug, in a polymeric matrix with the goal of enhancing the dissolution rate and thus the bioavailability. Unlike a traditional hot melt extrusion (HME) process, nanoextrusion process (Baumgartner et al., 2014; Khinast et al., 2013) uses a wet-milled suspension of the drug as a feed along with an extrusion polymer and disperses the drug nanoparticles in the polymeric matrix while evaporating water, thus yielding extrudates in the form of nanocomposites. Simply, nanoextrusion process acts as a continuous drier unlike the traditional HME process. Being a continuous process and having the capability to handle viscous drug suspensions, it could be advantageous over various pharmaceutical drying operations in the manufacturing of solid nanoparticle-based formulations. In this study, we propose that as another potential advantage of the nanoextrusion, extrudates with different forms of the drug can be prepared by the same process via use of different extrusion polymers: nanocrystalline drug dispersed in the polymeric matrix (nanocomposites) or amorphous drug molecularly dispersed with the polymer (amorphous solid dispersion, ASD). It is hypothesized that the drug nanocrystals in the nanocomposites dissolve faster than the amorphous drug in the ASD when the drug dose is so low not to cause supersaturation in the dissolution medium (low drug dose). Specifically, hydroxypropyl cellulose (HPC) and Soluplus were used to stabilize wet-milled suspensions and to form the matrix of the extrudates. Feed suspensions of GF were prepared by wet stirred media milling using HPC and Soluplus in the presence of sodium dodecyl sulfate (SDS), an anionic surfactant. The milled suspensions along with additional polymer (HPC/Soluplus) were fed to a co-rotating twin-screw extruder, which dried the suspensions and formed various extrudates. The extrudates were milled into powders via mortar–pestle for further analysis. The effects of drug particle size in the feed suspensions, stability of drug suspensions, and polymeric matrix type–size on the drug dissolution from the milled extrudates were examined. The suspensions were characterized via laser diffraction, while the extrudates and their milled powders were characterized by laser diffraction, SEM, and XRD. Digital microscopy was used to visualize the changes of different polymeric matrices when exposed to water. Statistical analysis with Korsmeyer-Peppas model helped us to identify the drug release mechanisms. By using polymers with different interactions/miscibility with the drug, two different forms of the drug were formed: extrudates with nano/micro-crystalline drug particles dispersed in the HPC matrix as a secondary phase (nanocomposites or microcomposites) and extrudates with amorphous drug molecularly dispersed within the Soluplus matrix (amorphous solid dispersion). In the composites, drug particle size was the dominant factor that affects the dissolution performance, whereas the matrix size appears to play a significant role in the drug dissolution performance in the amorphous solid dispersion. We have demonstrated that for low-dose drugs, e.g. highly potent poorly soluble drugs, drug nanoparticles in the nanocomposites may actually dissolve faster than amorphous form of the drug in an amorphous solid dispersion, depending on the size of the polymeric matrix.

References

Baumgartner, R., Eitzlmayr, A., Matsko, N., Tetyczka, C., Khinast, J., Roblegg, E., 2014. Nano-extrusion: A promising tool for continuous manufacturing of solid nano-formulations. Int. J. Pharm. 477, 1 11.

Khinast, J., Baumgartner, R., Roblegg, E., 2013. Nano-extrusion: a one-step process for manufacturing of solid nanoparticle formulations directly from the liquid phase. AAPS PharmSciTech 14, 601 604.

Copyright © American Institute of Chemical Engineers. All rights reserved.

468257 Evaporation of Binary Mixtures and Shell Formation in Spray Dried Droplets

Abstract:

Evaporation of binary mixtures and shell formation in spray dried droplets

P.C. Valente[1], Í. Duarte[1], T. Porfirio[1], X. Liu[2], F. Zhang[2] & M.N. Temtem[1]

[1] R&D Drug Product Development, Hovione Farmacência S.A, Sete Casas, 2674-506 Loures, Portugal;

[2] College of Pharmacy, University of Texas at Austin, U.S.A;

Amorphous solid dispersions (ASDs) of poorly water soluble active pharmaceutical ingredients (APIs) in hydrophilic polymeric matrices are a commonly used strategy to increase the solubility and dissolution rate of oral-dosage forms [1]. However, since the polymer is hydrophilic and the API is poorly water soluble, it is often difficult to find a common solvent for both and a mixture of co-solvents is used instead. Solvent mixtures also offer the possibility of influencing particle morphology and drying rate [2] and have been reported to offer advantages in the solid state drug-polymer miscibility and on the stability of the ASDs [3,4].

To fully explore the potential of spray dried ASDs from co-solvent drug-polymer solutions, a fundamental understanding on the solvent chemistry, droplet composition history throughout drying and its impact on the drug-polymer interaction, stability and phase behavior are required. These mechanisms are, in turn, strongly influenced by the process conditions, such as thermodynamic layout, atomizing characteristics and spray drier geometry.

The focus of the present work lies on the evaporative dynamics of binary solvent droplets and how they are influenced by process conditions. Unlike the single component counterpart, modelling the evaporation of binary mixtures depends not only on the process parameters, but also on the mass (and momentum) transfer within the droplet [5]. For example, if one considers solely the molecular diffusion for solvent mixtures with starkly different vaporization enthalpies, the evaporation rate of each solvent will be proportional to their initial mass fraction and, counter-intuitively, will be independent of the vaporization enthalpies [6]. However, whenever there is a non-negligible relative velocity between the droplet and the surrounding drying gas, the internal circulation within the droplet can lead to a steep increase in the effective mass diffusion rate. The evaporation rate of each solvent will also strongly depend on their vaporization enthalpy (see Figure 1). This leads to the more intuitive case where the more volatile solvent tends to evaporate first.

We show that these two limits lead to dramatically different drying kinetics and can strongly affect the onset of drug/polymer precipitation and shell formation. After shell formation the drying kinetics tend to be strongly modified, highlighting the importance in predicting the onset of shell formation to bound the applicability of the drying models. These two limiting cases of infinitely slow or infinitely fast droplet component diffusion may be used to formulate scalar (0D) evaporation models [5,6]. In practical spray drying situations, both limits can be attained depending on the binary mass diffusion coefficients, atomization characteristics, temperature profile or drying gas flow velocity, to name a few parameters. These evaporation models are particularly useful when coupled to a numerical solver to predict stability and phase separation of amorphous solid dispersions [7,8] or a numerical solver of the Navier-Stokes equations, commonly denoted as computer fluid dynamics (CFD) [9,10].

We extend the previously developed platform for phase separation prediction of drug-polymer-solvent systems to systems with binary solvent blends in an effort to generalize a screening methodology for ASDs and compare the data against experimental results.

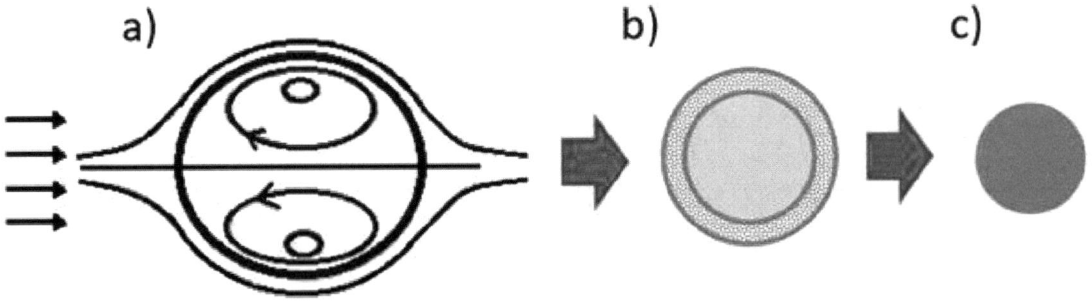

Figure 1: Sketch of various stages throughout the droplet drying history: a) droplet with internal circulation caused by the relative velocity between the droplet and the surrounding gas, b) shell formation and c) the final particle.

References

[1] Perrie Y., Rades T. "Pharmaceutics - Drug Delivery and Targeting", Pharmaceutical Press, 2010.

[2] Wulsten, E., Kiekens, F, van Dycke, F., Voorspoels, J. & Lee, G., "Levitated single-droplet drying: Case study with itraconazole dried in binary organic solvent mixtures", Int. J. Pharmaceutics, 2009, 378(1-2) pp. 116-121

[3] Janssens, S, Anné, M., Rombaut, P. & Van den Mooter, G., "Spray drying from complex solvent systems broadens the applicability of Kollicoat IR as a carrier in the formulation of solid dispersions", Eur. J. Pharm. Sci. 2009, 237, pp. 241-248.

[4] Paudel, A. & Van den Mooter, G., "Influence of Solvent Composition on the Miscibility and Physical

Stability of Naproxen/PVP K 25 Solid Dispersions Prepared by Cosolvent Spray-Drying", Pharm. Res. 2012, 29, pp. 251-270.

[5] Sirignano, W. A. "Fluid dynamics and transport of droplets and sprays", Cambridge University Press (2nd Ed.), 2010.

[6] Law, C. K. "Recent advances in droplet vaporization and combustion", Prog. Energy Combust. Sci, 1982, 8 pp. 171 – 201, 1982.

[7] Duarte, Í., Santos, J. L, Pinto, J. G & Temtem, M., "Screening methodologies for the development of spray-dried amorphous solid dispersions", Pharm. Res., 2015, 32 pp. 222 – 237, 2015.

[8] Prudic, A., Ji, Y. & Sadowski, G., "Thermodynamic Phase Behavior of API/Polymer Solid Dispersions", Mol. Pharmaceutics, 2014, 11(7), pp. 2294-2304

[9] Sazhin, S.S et al., "A simplified model for bi-component droplet heating and evaporation", Int. J. Heat Mass Transfer, 2010, 21 pp. 4495-4505

[10] Maqua, C., Castanet, G. & Lemoine, F., "Bicomponent droplets evaporation: Temperature measurements and modelling", Fuel 2008, 87(13-14), pp. 2932-2942.

Copyright © American Institute of Chemical Engineers. All rights reserved.

455382 Impact of Tooling Design during Development and Risk Evaluation of Tri-Layer Tablet Formulations

Abstract:

Multi-layer tablet is one of the formulation strategies utilized to mitigate chemical and/or physical incompatibility between different drug substances when delivering fixed dose combination drugs. Compressing multi-layer tablets has significant challenges because the interface(s) between neighboring layers are subject to defects such as delamination. It is therefore important to understand the factors that contribute to the strength of the interfaces. During early stages of drug development, feasibility studies are often conducted by making multi-layer compacts with round flat-faced tooling as the exact size and shape for the commercial tablets may not be finalized at that point. However, the link between different tooling designs and multi-layer tablet performance is not well established. This study examines the major engineering considerations when using mini-piloting tools to gauge the risk of interfacial defects in a tri-layer tablet. The first layer (layer A) of the tri-layer tablet was a wet granulation formulation of 56% drug load. The middle layer (layer M) was the inert layer consisting of a placebo blend. The third layer (layer B) was a dry granulation formulation of 33% drug load. The impact of tablet weight and dimensions (diameter and thickness) were evaluated to gain understanding during scale up/down as needed in early feasibility assessment and in troubleshooting. The factors in tooling selection, including tablet shape (round vs. oval), cup depth (0 vs. 0.054â€), and embossing (small vs. large), were evaluated to gain insight on the impact of tooling design on the interface strength of the tri-layer tablet. Tablets were compressed under different main compression forces while keeping the first and second layer tamping forces constant. Axial test was performed to quantify the interface strength under tensile stress. Â The thickness, diameter, and aspect ratio of the tablet were found to have a profound impact on the interface strength at a given compression pressure and on the compression pressure under which the interface start to fail, causing tablet delamination. These were the results of combined factors of the compressibility of the individual layers, the force transmission through the layers during compression (effect of tablet thickness), and the ejection force (effect of aspect ratio). The presence of a 0.054â€ cup depth was found to lower the interface strength to less than 50% of that of a flat tablet. This cup depth effect is consistent with the observations made in the monolithic tablet regime and can be explained by the non-uniform density distribution caused by the curvature of the tablet interface. The oval shaped tablets were found to reduce the interface strength by 30 N or more compared to the round shaped tablets, suggesting that compacts made with round flat faced toolings may provide false positive results when used to assess the interface risks of a multi-layer tablet. Large embossing on the tooling appeared to help enhancing the interface strength, probably by providing more surface areas for the neighboring layers to interlock. This study elucidates the effects of scale-related factors like tablet weight and dimensions on the interface strength of a tri-layer tablet. These effects, not previously well studied or even realized during development, can significantly impact the feasibility evaluation of a multi-layer tablet. It also highlights the importance of tooling design selection on the performance of a tri-layer tablet that must be accounted for when using minipiloting tools for accelerated development of fixed dose combination products.

Copyright © American Institute of Chemical Engineers. All rights reserved.

471306 End-Point Determination and Scale-up/Scale-Down of High Shear Wet Granulation Process Using an in-Line Di-Electric Probe

Abstract:

The main goal of this study is to demonstrate anovel in-line process analytical technology (PAT) for detecting end-point of granulation in a high shear wet granulation (HSWG) process. This technique enabled us to monitor liquid distribution and granule densification across different scales. While many reports in the literature have focused on indirect techniques such as power consumption or torque (which are scale and geometry dependent), in this method a physical property (dielectric permittivity of medium) dependent on the liquid amount and granule porosity (or density) is monitored directly during the course of granulation.

A di-electric probe was modified for recording complex dielectric permittivity of wet mass. The output of the probe was monitored and recorded during liquid addition and wet massing, and samples were taken at corresponding time intervals to correlate the final property of granules to the instrument output. Dependence of the real part of the effective dielectric permittivity on solid fraction of moist powder was suggested by Louge et. al. (1997) for non-spherical particles (snow). We adopted this model for high shear wet granulation process and monitored (in-line)Â density of wet mass during granulation.

Use of this in-line characterization tool enabled us to successfully transfer a 50% drug load API to a 600 L top-driven granulator without the need for any development batch by monitoring granulation end-point based on physical properties of wet mass. Specifically, we could understand the impact of changing process parameters (impeller speed, spray time and wet massing) and API properties (particle size /density) on final granule attributes and tablet CQAâ€™s which was consistent during scale-up/down to geometrically similar or non-similar granulators.

M.Y. Louge, R.Â Steiner, S.C. Keast, R. Decker, J. Dent, and M. Schneebeli, "Application of capacitance

instrumentation to the measurement of density and velocity of flowing snow", *Cold Regions Sci. & Tech.*, 25, 47-63 (1997).

Copyright © American Institute of Chemical Engineers. All rights reserved.

464179 Programmable Immunotherapeutic Biomaterials to Potentiate Chemotherapy

Abstract:

Introduction: Acute Myeloid Leukemia (AML) has a very poor prognosis, with a 5-year survival of about 25%. While cytotoxic chemotherapy is successful at inducing remission of acute myeloid leukemia (AML), the disease has a high probability of relapse with chemo-resistant cells. One strategy to prevent relapse is to vaccinate mice against AML. Key to this strategy is to sustain a cytotoxic $CD8^{+}$T-cell response to eliminate residual cells after debulking by chemotherapy. In this study, we explored the development of a modular biomaterial-based vaccine with AML-specific antigens to prevent AML relapse.

Materials and Methods: Injectable macroporous alginate hydrogels were synthesized by low-temperature redox-induced free-radical polymerization. Granulocyte macrophage colony stimulating factor (GM-CSF), CpG ODN 1826, (5'-TCC ATG ACG TTC CTG ACG TT-3') and ovalbumin (OVA), cell lysate or an immunogenic peptide of Wilms' tumor protein-1 (WT-1) were added to the mixture prior to cryopolymerization. A GFP-luciferase expressing MLL-AF9 AML knock-in cell line from a C57BL/6J mouse was used to induce AML without irradiation. One version of the cell line was transduced to express ovalbumin. Cytosine arabinoside (Ara-C 100mg/kg) and Doxorubicin (Dox 4mg/kg) were used for induction chemotherapy. Hydrogels were injected subcutaneously and the recruitment of immune cells to the scaffold was monitored by periodically harvesting the scaffolds. Antigen specific $CD8^{+}$T-cells were monitored in blood. Secondary transplants were performed to test treatment efficacy.

Results and Discussion: In the MA-Alg/MA-PEG hydrogel, we observed an infiltration of neutrophils, dendritic cells and macrophages as early as 2 days after subcutaneous injection. There was an upregulation in activated dendritic cells at the site of the hydrogel, as well as in the draining lymph nodes. When used as an antigen prophylactically, the lysate of the bone marrow cells of leukemic mice prevented the engraftment of AML cells in mice. A western blot analysis of vaccinated mouse serum revealed the production of antibodies against a broad range of targets. When OVA was used as the antigen, paired with the OVA expressing AML cell line, the onset of leukemia was delayed significantly in prophylactically vaccinated mice, but not prevented altogether. SIINFEKL tetramer positive $CD8^{+}$ cells were detected in peripheral blood and were maintained over the duration of the study, until the mice were terminal. An analysis of the hematopoietic compartments revealed the gradual loss of OVA expression on AML cells, suggesting immune escape through loss of antigen. A bolus subcutaneous injection transiently increased SIINFEKL tetramer positive $CD8^{+}$, and delayed the onset of AML when challenged within that window – however it did not translate into increased survival. A known MHC class 1 restricted peptide fragment of a clinically defined AML antigen, WT-1, was incorporated in the vaccine as an antigen. When used prophylactically, WT-1 extended survival significantly, more than OVA, but not prevent AML associated mortality entirely. To mimic the clinical therapeutic management of AML, an induction chemotherapeutic regimen was used before the administration of the WT-1 vaccine. The chemotherapy significantly debulked the tumor, whereas the WT-1 vaccine prevented the relapse of AML in mice. WT-1 tetramer positive $CD8^{+}$cells were detected in peripheral blood over the duration of the study. Secondary transplants were performed after 150 days and confirmed the eradication of AML.

Conclusions: The results indicate the promise of a programmable biomaterial vaccine with clinically defined leukemia antigen targets. When used prophylactically, the vaccine alone can extend survival by sustaining the cytotoxic $CD8^{+}$T-cell response. When used in conjunction with cytotoxic chemotherapy, the biomaterial vaccine can potentially prevent the relapse of AML in a clinical setting.

Acknowledgements: The work was supported by the National Institutes of Health through grants U19HL129903 and R01EB014703 and R01EB015498.

Copyright © American Institute of Chemical Engineers. All rights reserved.

464448 Surface Presentation of Adjuvants on Protein Nanoparticle Vaccines

Abstract:

Nanoparticle vaccines leverage the inherent immunogenicity of particulate matter to generate specific, adaptive immune responses to poorly immunogenic antigens. To examine the fundamental properties of effective nanoparticle vaccines, we have made nanoparticles directly from a crosslinked model antigen, ovalbumin (OVA), and examined the effects of nanoparticle size and coating on antigen uptake and processing, as well as dendritic cell (DC) inflammation and maturation. We observed that OVA nanoparticles enhance antigen uptake, experience attenuated endosomal acidification, and upregulate the DC maturation marker CD86 as compared to soluble protein. Dendritic cell inflammation was nanoparticle size- and coating-dependent. Small particles (270 nm) enhanced TNF-Î± production as compared to larger (350, 560 nm) nanoparticles, and 270 and 350 nm nanoparticles coated with an extra layer of OVA enhanced IL-1Î² production more than large coated nanoparticles, uncoated nanoparticles, or soluble OVA did. Based on these results, we have chosen coated, 270 nm nanoparticles as a platform for exploring other enhancements to nanoparticle vaccine adjuvancy.

In addition to coating with antigen, coating nanoparticles with molecular adjuvants can further enhance immunostimulation through several pathways. Molecular adjuvants can trigger immunostimulatory intracellular pathways, activate the extracellular complement system or target vaccines to specific cell types. Immunoglobulins can enhance immunity through all three of these mechanisms, and immunoglobulin G (IgG) and immunoglobulin M (IgM) have been identified as promising classes of adjuvants. Unlike traditional toll-like receptor ligands, such as flagellin, immunoglobulins are host-derived proteins, which greatly reduce the risk of an off-target immune response.

Building off of our prior work, we have coated OVA, flagellin, IgM, and IgG onto OVA nanoparticles. The activity of the coatings was confirmed by TLR-5/NFÎºB activation and complement activation studies *in vitro* for flagellin and immunoglobulin, respectively. Using an *in vivo* mouse immunization model, we examine the serum antibody responses and splenic T cell cytokine secretions in response to pathogen- and host-derived protein adjuvants. Our current work addresses three fundamental questions surrounding protein adjuvant use: (1) The benefit of antigen-adjuvant co-delivery, (2) The effectiveness of host-derived immunoglobulin in enhancing nanoparticle-based immune responses, and (3) Whether different immunoglobulin isotypes can trigger different types of adaptive immune responses. Our results contribute to current design strategies for molecular adjuvant incorporation onto nanoparticle vaccines.

Copyright © American Institute of Chemical Engineers. All rights reserved.

467357 Acetalated Dextran Microparticulate Subunit Anthrax Vaccine Formulated Using Coaxial Electrospray Preserves Toxin Neutralization and Enhances Protection

Abstract:

Protein-based subunit vaccines are considered the safest type of vaccine, and their typically poor immunogenicity can be enhanced with well-characterized, potent Th1-skewing adjuvants. The use of polymer particles is an established approach to efficiently deliver both antigens and adjuvants to target antigen-presenting cells. However, particles composed of polyanhydrides and polyesters such as poly(lactic-co-glycolic acid) have detrimental acidic hydrolytic byproducts and relatively slow release kinetics. These substantial drawbacks emphasize the need for new polymer particles that exhibit pH-neutral hydrolytic byproducts and efficient intraphagosomal antigen/adjuvant delivery. Acetalated dextran (Ace-DEX) particles are a novel, acid-degradable formulation that fulfills these preferred design parameters. We have previously demonstrated Ace-DEX particlesâ€™ efficacy in several subunit vaccine formulations. These particles were formulated using the common batch solvent evaporation technique, and thus, some of the resultant in vivo immune responses were likely hampered due to the particle fabrication methodâ€™s denaturing characteristics.

An alternative polymer particle fabrication method such as coaxial electrohydrodynamic spraying (electrospray) is amenable to overcoming these formulation barriers. Coaxial electrospray processes do not create harsh denaturing conditions and have demonstrated gentle encapsulation of proteins in model studies. Despite these promising results, coaxial electrospray methods have yet to be applied to notable therapeutic or vaccine applications utilizing polymer particles. In our current work, we use coaxial electrospray to formulate an anthrax vaccine. We efficiently and gently encapsulate the labile recombinant protein antigen (antigen from *Bacillus anthracis*, causative agent of anthrax) and Th1-skewing resiquimod adjuvant (toll-like receptor 7/8 agonist) either in separate particle sets or co-encapsulated in

the same particle. We compare our electrospray formulations in parallel with particles fabricated by solvent evaporation and the FDA-approved anthrax vaccine (BioThrax). Our electrospray formulations administered parenterally lead to increased Th1-skewing antibody production, preserved toxin neutralization, and enhanced protection against an inhaled challenge of the clinically virulent *B. anthracis* Ames strain. These promising results support electrosprayâ€™s potential to be a powerful platform technique for fabricating subunit vaccines and other protein-based drug delivery formulations.

Copyright © American Institute of Chemical Engineers. All rights reserved.

469783 Engineering Lymph-Node Targeting Vaccines for Type 1 Diabetes

Abstract:

Vaccine approaches to restore antigen-specific immune tolerance to pancreatic beta-cell antigens without global immune suppression are potential therapeutic interventions for Type 1 diabetes (T1D). However, a method for realizing both efficacy and safety is yet to be developed. A major challenge in the development of effective vaccines is efficient delivery of vaccine components to antigen presenting cells (APCs) in lymphoid organs, where the orchestrations of immune cells are initiated.

We recently demonstrated that both the efficacy and safety of subunit vaccines can be dramatically enhanced by rational molecular design which target lymphoid organs via â€˜albumin-hitchhikingâ€™, a mechanism by which structurally optimized molecular vaccines are engineered with a lipophilic albumin-binding tail and follow subcutaneous injection, exhibit dramatic increases in LN accumulation via in situ complexation and transport with endogenous albumin, leading to 30-fold augmentation in CD8+ T-cell priming while greatly reducing systemic toxicity. Here we translate these findings to deliver immune suppressive vaccines to LNs, where they can be efficiently filtered by resident phagocytes and accumulate, promoting antigen-specific tolerance. Treatment of NOD mice with albumin-hitchhiking peptide vaccines significantly delayed and reduced the incidence of diabetes. This is associated with significant suppression of antoreactive CD4 T cells and increased frequency of regulatory T cells, as well as increased suppressive cytokine productions. Our findings suggest antigen delivery to lymph nodes via albumin-hitchhiking might be broadly applicable to treat autoimmune diseases.

Copyright © American Institute of Chemical Engineers. All rights reserved.

466796 Silk Microneedle Skin Patches for Tunable HIV Subunit Vaccine Delivery

Abstract:

Development of a prophylactic HIV vaccine requires novel immunogen design and vaccine delivery strategies. Recent studies in our laboratories have shown that the kinetics of vaccine release modulate humoral immune responses. To control kinetics of prophylactic HIV vaccine, we developed a submicron-scale dissolving microneedle patch (MN) with release kinetics dependent on MN composition. We utilized silk protein to encapsulate an HIV envelope trimer (BG505 SOSIP) immunogen and adjuvants with poly(acrylic acid) (PAA) base. Upon dermal application, rapid dissolution of PAA is followed by silk beta sheet-dependent sustained release of encapsulated vaccine. In mice, whole animal imaging showed higher model antigen retention over one week in the skin with minimal local inflammation for silk MNs compared to equivalent intradermal injections. Antigenicity of SOSIP was maintained in silk as determined by binding to broadly neutralizing antibodies (bNAbs: PGT151, PGT145, PGT121, VRCO1) without binding to non-bNAb 14e. In mice, vaccine delivery in silk MNs promoted and sustained germinal center (GC) responses in lymph nodes with significantly higher GC B cells, plasmablasts and T follicular helper (Tfh) cells for upto 14 days following immunization compared to intradermal injections. Significant increases in MN induced anti-SOSIP IgG titers post boost correlated with Tfh induced increase in plasma CXCL13, a biomarker for strong antibody responses. Taken together, these results indicate the potential of silk microneedles for tunable HIV subunit vaccine delivery.

Copyright © American Institute of Chemical Engineers. All rights reserved.

467119 Enhanced Cancer Immunotherapy By Microneedle Patch-Assisteddelivery of Anti-PD1 Antibody

Abstract:

Enhanced Cancer Immunotherapy by Microneedle Patch-Assisted Delivery of Anti-PD1 Antibody

Yanqi Ye[a, b], Chao Wang[a, b], Gabrielle M. Hochu[a] , Hasan Sadeghifar[c], Zhen Gu[a, b, d *]

[a] Joint Department of Biomedical Engineering, University of North Carolina at Chapel Hill and North Carolina State University, Raleigh, NC 27695, USA;

[b] Division of Molecular Pharmaceutics and Center for Nanotechnology in Drug Delivery, Eshelman School of Pharmacy, University of North Carolina at Chapel Hill, Chapel Hill, NC 27599, USA;

[c] Department of Forest Biomaterials, North Carolina State University, Raleigh, NC 27695, USA;

[d] Department of Medicine, University of North Carolina School of Medicine, Chapel Hill, NC, 27599, USA

Introduction: For skin cancer treatment, immunotherapies have been intensively studied over the past several years).[1,2,3] The anti-PD-1 antibodies that target the inhibitory receptors have shown striking antitumor activity in phase II and III clinical trials of advanced melanoma). Despite the exciting clinical results of anti-PD-1 antibodies for the treatment of melanoma, the e cacy of the approach remains to be improved.

Materials and Methods: Here we present the physiologically self-degradable MN patch-assisted cancer immunotherapy for controlled delivery of aPD1 toward melanoma (Figure 1). Each MN is composed of biocompatible hyaluronic acid (HA) integrated with NPs that encapsulate aPD1 and glucose oxidase (GOx).[4, 5] With the GOx/CAT enzymatic system immobilized inside the NPs, the enzyme-mediated generation of gluconic acid promotes the gradual self-dissociation of NPs and results in the sustained release of aPD1.

Results and Discussion: We show that a single administration of the MN patch induces robust immune responses in B16F10 mouse melanoma model exceeding MN in the absence of the trigger element (GOx) or intratumor injection of free aPD1. Moreover, we demonstrated that the MN coloaded with aCTLA-4 and aPD1 resulted in synergistic treatment of melanoma. These results demonstrate that the MN patch-assisted system provides an innovative delivery strategy of aPD1 via a simple and safe technique that improves cancer immunogenicity and facilitates the clinical treatment of melanoma.

Conclusions: We have developed the MN patch-assisted immunotherapy that delivers aPD1 for the enhanced treatment of the skin cancer. Additionally, MN severs as a platform for combined therapy with other immunomodulators to enhance immunotherapy e ciency

Acknowledgements: This work was supported by NC TraCS (grant 550KR51307), NIH's Clinical and Translational Science Awards (CTSA, 1UL1TR001111) at UNC-CH, the NC State Faculty Research and Professional Development Award, and start-up package from the Joint BME Department of UNC and NCSU to Z.G.

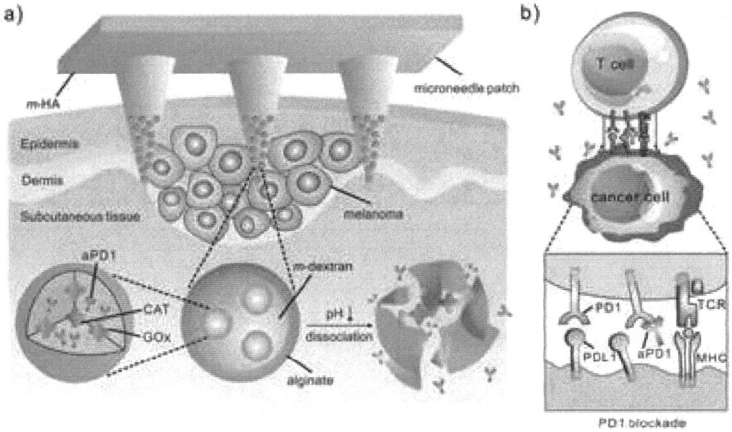

Figure 1. Schematic of the MN patch-assisted delivery of aPD1 for the skin cancer treatment. (a) MN patch loaded with self-dissociated NPs. (b) The blockade of PD-1.

References:

[1] Wang, C.; Ye, Y.; Gu, Z. *et al. Nano Letters* **2016**.

[2] Sullivan, S. P.; Prausnitz, M. R. *et al. Adv. Mater.* **2008**, 933−938.

[3] Gu, L.; Mooney, D. J. *Nat. Rev. Cancer* **2015**, 56−66.

[4] Yu, J.; Zhang, Y.; Ye, Y., Gu, Z. *et al. Proc. Natl. Acad. Sci.* **2015**, 8260−8265.

[5] Gu, Z.; Langer, R. S.; Anderson, D. G. *et al. ACS Nano* **2013**, 4194−4201.

Copyright © American Institute of Chemical Engineers. All rights reserved.

464254 Targeting and Altering *In Vivo* macrophage Responses with Modified Polymer Properties

Abstract:

Targeting and altering *in vivo* macrophage responses with modified polymer properties

Kaitlin Bratlie

Departments of Materials Science & Engineering and Chemical & Biological Engineering, Iowa State University

Introduction: Two pathways for activating macrophages (MΦs) exist. One of these routes is termed the classically activated M1 pathway and is achieved through exposure to lipopolysaccharide (LPS). M1 MΦs are known as pro-inflammatory cells. The other pathway is reached through interleukin-4 (IL-4) and is known as the alternatively activated M2 pathway. M2 MΦ produce pro-angiogenic factors. Here, we successfully use *in vivo* imaging and histological analysis to identify the MΦ response and activation. We demonstrate the ability to induce various MΦ phenotypes with a change in material functionality as well as identify certain materials parameters that seem to correlate with each phenotype. This suggests the potential to develop materials for specific applications and predict the outcome of MΦ activation in response to new surface chemistries.

Ketone

Sulfone

Control

Acetal

Ketal

Oxime

Figure 1. *In vivo* fluorescence images of cathepsin activity in response to implanted polymer nanoparticles. (Left) Fluorescence image of cathepsin activity 7 days after implantation of 600nm pNIPAm particles. (Right) Implantation scheme.

Figure 2. *Ex vivo* arginase:iNOS levels for modified pNIPAm particles.

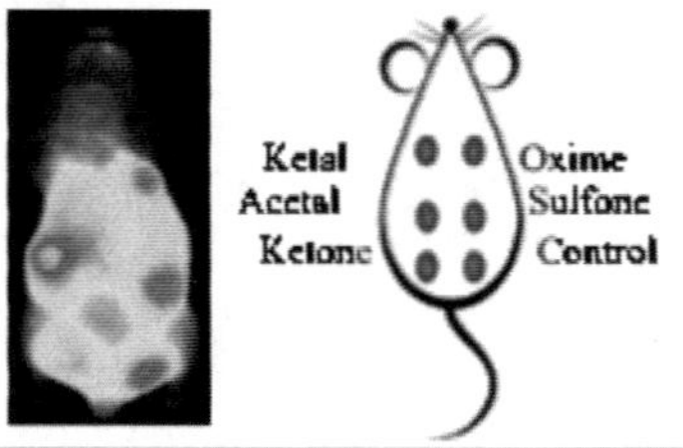

Figure 1. *In vivo* fluorescence images of cathepsin activity in response to implanted polymer nanoparticles. (Left) Fluorescence image of cathepsin activity 7 days after implantation of 600nm pNIPAm particles. (Right) Implantation scheme.

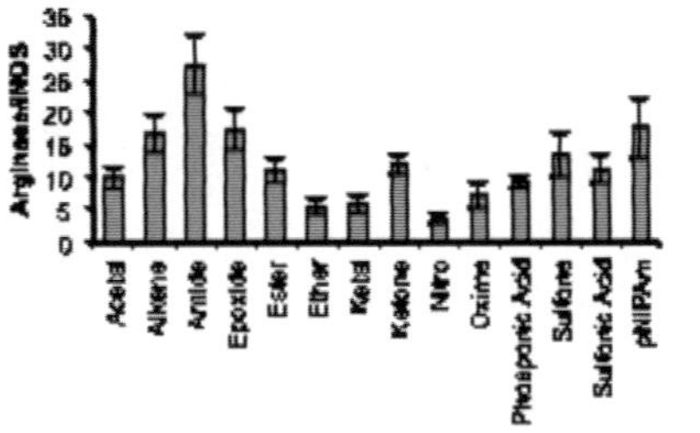

Figure 2. *Ex vivo* arginase:iNOS levels for modified pNIPAm particles.

Materials and Methods: The poly(N-isopropylacrylamide) (pNIPAm) (~600nm) particles were synthesized using established methods and modified through conjugation with surface modifiers through carbodiimide chemistry. MΦs were polarized to the M1 and M2 phenotypes and internalization of functionalized pNIPAm particles were measured. These materials were also subcutaneously injected in SKH1-E mice. Arginase:iNOS profiles have been consistently used as a measure of MΦ phenotype. Arginine activity was measured indirectly through a urea assay in which the lysate was activated, exposed to arginine, and conversion to urea was quantified. Nitrites were measured through a Griess reagent assay to indirectly determine the levels of inducible nitric oxide synthase (iNOS). Material properties of the modified particles including water contact angle, melting temperature, and zeta-potential were also measured.

Results and Discussion: SKH1-E mice were injected with the polymer particles and imaged for cathepsin activity (**Figure 1**). A range in cathepsin activity was observed for the different surface modifiers. The tissue sections from the mice in **Figure 1** were excised and homogenized. The arginase:iNOS displayed a spectrum between the M1 and M2 phenotypes for different surface modifiers as shown in **Figure 2.**

Distinct material parameters influence MΦ phenotypes. For example, increasing the hydrogen bonding of polymeric particles can polarize MΦs towards the M1 phenotype. The induction of multiple MΦ phenotypes confirms the ability to control MΦ activation *in vivo* and suggests the importance of chemical properties in the biocompatibility of materials. The results obtained suggest that changing properties can result in opposite MΦ phenotypes.

Conclusions: With this work, useful insight was gained about *in vivo* MΦ response to various material functionalities. Overall, this *in vivo* study was successful at drawing conclusions about the biomaterial parameters that will allow for tuning MΦ polarization responses to biomaterials and targeting MΦs.

463503 Chemokine Releasing Polymer Implants to Direct Immune Cell Migration in the Setting of Colon Cancer

Abstract:

Colorectal cancer (CRC) is the third leading cause of cancer related death in both men and women. The major cause of death is metastasis and frequently, the target organ is the liver. Successful metastasis depends on a microenvironment conducive to seeding and colonization by metastatic cells. Inflammatory immune cells are implicated in the formation of this microenvironment and interfering with current treatment options including chemotherapy. The objective of the current work is to control immune cell migration and accumulation *in vivo*, directing these cells away from target organs for metastatic disease. In an orthotopic model of colon cancer in which CT26 cells are injected into the cecum, liver metastasis occurs in 90% of mice. Histological analysis indicated that immune cells accumulate within the liver prior to the arrival of metastatic cells. Microarray analysis of pre-metastatic livers from mice injected with CT26 cells indicated that inflammatory chemokines CXCL1 and CCL6 were increased 5.6 and 6.3 fold, compared to mice receiving saline injection. This presentation will focus on the development of biocompatible polymer implants designed for sustained release of these chemokines from the subscapular space, a site where immune cells do not generally accumulate and colon cancer metastasis does not occur. We will discuss the ability of these implants to attract immune cells expressing CXCR2 and CCR1, prototypical receptors for CXCL1 and CCL6, respectively. We will identify the immune cell populations that reside within the implant site and report their numbers in the pre-metastatic liver and primary tumor in mice with and without polymer implant. Finally, the efficacy of these immunomodulatory implants to decrease liver metastatic load will be presented. Collectively, we expect the outcome of this work to impact our understanding of the underlying immunological mechanisms that promote metastatic cancer and aide in the efforts to treat it.

Copyright © American Institute of Chemical Engineers. All rights reserved.

465003 Selection Support Methodology Between Batch and Continuous Technologies for Pharmaceutical Tablet Manufacturing Processes

Abstract:

In solid drug product manufacturing, continuous technology is being developed as an alternative to the conventional batch technology. There are many differences between these two technologies, which may affect critical objectives such as cost-effectiveness, quality, process safety or occupational health. However, general tools are still missing, which can support the choice between batch and continuous technologies, and furthermore, process design of solid drug product manufacturing.

This work presents the selection support methodology between batch and continuous technologies for pharmaceutical tablet manufacturing processes. The methodology consists of four steps and includes iteration: (I) process modeling, (II) evaluation, (III) sensitivity analysis, and (IV) interpretation. The initial input of step I is a design situation, such as a launch of a new product to be manufactured in either technology that needs to be decided. In step I, the processes are modeled with covering the main processes such as weighing, granulation, blending, compression, and coating, and supporting processes such as testing, disposal, cleaning, maintenance of process analytical technology, and heat ventilation and air conditioning. Models are created so that the characteristics of two technologies in the actual manufacturing situations are reflected, such as numbers of operators, product yield or required area for manufacturing. Using these models, evaluation is conducted in step II, which produces results of the objective functions such as net present cost (NPC) after the launch. In step III, sensitivity analysis is conducted in order to quantify the effects of varying input parameters in the evaluation results. Finally in step IV, the results of evaluation and sensitivity analysis are interpreted, and the necessity of iterating the previous steps is investigated. The final outcome of the methodology is the appropriate technology to be chosen, i.e., batch or continuous, for manufacturing, for the tablet manufacturing.

A case study was performed for demonstrating the methodology. The design situation was the launch of a new product for which the estimation is available regarding the API price as well as the maximum demand quantity over the production period of 30 years. It was also assumed that the tablet product can be produced by either technology which needs to be specified based on the economic evaluation. After performing the steps I and II, the quantitative relation between demand quantity/API price and NPC over 30 years was clarified. The NPC did not include the investment cost, since the plant was assumed to have both technologies already. The result showed that continuous manufacturing was more lucrative in the areas of high demand quantity with low API price or low demand quantity with high API price. On contrary, batch manufacturing showed lower NPC in the areas where both API price and demand quantities are both high. Step III identified the number of campaign production in batch manufacturing as one of the most sensitive parameters in the model, for which the step II was iterated. In step IV, an economically appropriate technology was suggested based on the interpretation of the evaluation and sensitivity analysis results.

Copyright © American Institute of Chemical Engineers. All rights reserved.

467693 Development of a Continuous Production Line for Solid Dosage Forms Via Dry Granulation

Abstract:

During the last years, academia, industry and regulators have increasingly demonstrated and addressed the advantages of continuous manufacturing. More and more, this is leading to a shift in paradigm for pharmaceutical manufacturing, towards advanced continuous processing based on real-time quality assurance and away from batch processing. As a result, turn-key solutions of continuous manufacturing systems are provided by various equipment suppliers. The available equipment is ranging from raw material feeding, to blending, granulation as well as tableting and coating. Even though several systems have been studied and discussed in the literature there are still some remaining questions open. Now with the first continuous plants being implemented for continuous production of solid oral dosage forms, interest is shifting towards system robustness, integration of unit operations, in-line measurement and process control. Such â€advancedâ€ turn-key solutions are required to fulfill a wide variety of demands, when processing different raw materials with varying properties.

In this talk we will highlight a control strategy that was validated by system tests of the integrated plant. The objective of the control strategy was to consider all possible process settings and raw material properties, thereby guaranteeing constant product quality. The described manufacturing line is dedicated to the continuous and automated production of solid-oral-dosage forms via a dry granulation route. In order to enable continuous process control and real-time-release, several PAT instruments were implemented. Proper control actions are based on actual process information from the in-line monitoring systems. Process parameters were screened for utilization as soft sensors, to enable even better process understanding and control. Out-of-spec material is detected and removed from the process at different stages. The control strategy enables continuous production even if out-of-spec material has to be removed. By setting the proper process parameters the mass flow can be modified, while still producing in-spec products. A concept will be presented, which allows variation of the mass flow of single unit operations to keep the entire production line within specified limits.

Copyright © American Institute of Chemical Engineers. All rights reserved.

464713 Considerations in Developing a Control Strategy for Continuous Direct Compression

Abstract:

In this work a continuous direct compression (CDC) process was investigated for a low dose drug product for an API with cohesive material properties. The CDC process was developed on the GEA CDC-50 system. A risk based approach was followed during the entire process development of CDC. The process parameters for feeding and blending were selected such that the feeding variability is sufficiently filtered by the available residence time distribution (RTD) of the blending system. The holistic control strategy for CDC involved specifications for input materials, in-process controls for various drug product attributes including blend concentration, tablet weight and hardness, and a product release strategy. The blend concentration was monitored using an in-line NIR method and a RTD model of the CDC line as two redundant methods. In the event when the blend concentration falls outside the given specification limits, non-conforming tablets were diverted. For using the RTD model as a process monitoring method, sensitivity of RTD model to material variability and method of feed rate measurement were identified as the main risks. In this presentation, the development studies conducted to address these risks will be discussed. In addition the pros and cons of using each process monitoring method as well as considerations in designing a diversion system will be discussed.

Copyright © American Institute of Chemical Engineers. All rights reserved.

463948 Applying Supervisory Process Control Practices to Continuous Pharmaceutical Manufacturing

Abstract:

Successful supervisory process control implementations in the pharmaceutical industry have been hampered due to the batchwise nature of most process operations. In fact, batch processes present some specific challenges because the main steps operate discontinuously. Consequently, temperatures, concentrations and other properties vary in time which requires complex non-linear and/or time-varying models and control strategies. In contrast, continuous manufacturing is one step closer to the "holy grail" for process control engineers, i.e. a linear time invariant (LTI) system. The availability of common practices and numerous examples for such a LTI system representation significantly reduce the complexity of process identification and control implementation. Hence, dynamic process operation with feedback and automatic manipulation of quality attributes is proposed as the way forward for continuous manufacturing [1]. A technical framework for this approach is already available in continuous oil refining, and in the chemical and food industries.

In order for the pharmaceutical industry to embrace fully continuous manufacturing with supervisory process control, numerous challenges still exist. This said, the current work tries to face them and apply existing classical and modern continuous control laws to manipulate quality attributes of a commercial continuous wet granulation unit within a continuous from-powder-to-tablet installation, being the ConsiGma™-25. The idea is to have stable automatic control of species concentration in the granules formed during twin-screw wet granulation. In this case the water content is considered, as it is directly related to both chemical stability and physical characteristics of the granules.

A first encounter in this effort was the inline estimation of moisture concentration from the particulate stream using the possibilities offered by high-speed NIR measurements. This includes process-instrument interfacing, industrial data communication, time-alignment of instrumentation and process data, latent variable modeling and signal filtering. These steps are typically highly customised to the situation, but some general comments will be presented. In addition, control of a process requires some form of model, and here a black-box mathematical model, of the type auto-regressive with moving average and exogenous inputs (ARMAX), was identified to predict the fast dynamic behaviour from system actuator to the actual measurement. This allowed us to tune, simulate and validate empirical (proportional-integral) as well as model-based (extended prediction self-adaptive) controllers for the granule moisture concentration loop under study. Another key point that will be addressed is that real-time calculations, here for analysis and control, require computationally inexpensive methods to keep up with the dynamics of the system under study.

[1] Myerson, A. S., Krumme, M., Nasr, M., Thomas, H. and Braatz, R. D. (2015), Control Systems Engineering in Continuous Pharmaceutical Manufacturing. May 20–21, 2014 Continuous Manufacturing Symposium. J. Pharm. Sci., 104: 832–839. doi: 10.1002/jps.24311

Copyright © American Institute of Chemical Engineers. All rights reserved.

456403 Control of a System of Loss-in-Weight Feeders in a Continuous Manufacturing Drug Product Application

Abstract:

Dispensing materials using loss-in-weight feeders is a critical unit operation in most continuous drug product manufacturing processes. As feeding is typically one of the furthest upstream operations, its output has a large influence on the rest of the process. Through experience continuously processing several pharmaceutical formulations, several common feeding problems have been identified including high variability feeding, drifting from set point, and transient disturbances. This work discusses the control of a system of feeders in addition to individual feeder control, showing examples of how a system level ratio control loop can be used to mitigate these common feeding issues and provide consistent output to the downstream process. In addition, identifying special cause transient disturbances from the feeding operation using different types of feeder data will be discussed.

Copyright © American Institute of Chemical Engineers. All rights reserved.

461837 Dynamic Optimization of Continuous Manufacturing of Pharmaceuticals

Abstract:

Continuous manufacturing (CM) of pharmaceuticals is being explored as an alternative to traditional batch-wise production. This shift holds great promise to increase production efficiency, enable smaller production facilities, minimize waste, energy consumption, and raw material use and to enable drug quality monitoring on a continuous basis. However, implementing CM in the pharmaceuticals industry implies short operational campaigns, with significant transient phases (start-up and shutdown), constituting up to 30% of the entire time horizon. This significantly hampers the plant's economic feasibility. Efforts to optimize the overall production process involve mathematical modeling and solution of hybrid (discrete-continuous) systems embedding differential-algebraic equations (DAEs), along with the associated parametric sensitivity trajectories to be used in rigorous gradient-based optimization algorithms.

In this contribution, these efforts will be described, addressing the numerical issues related to the hybrid and non-smooth nature of the mathematical model. A case-study pilot plant model incorporating several synthesis, separation and purification steps to produce final tablets [1] is used for demonstration. The dynamic optimization approach that is employed maximizes the accumulated on-spec production directly over the entire time horizon (so-called Economic-Optimization), while guaranteeing differentiability with respect to the controls [2]. Optimization results will be presented followed by a discussion on the process design and formulation and the intelligent choice of the open-loop control decision variables.

[1] S. Mascia et al., "End-to-End Continuous Manufacturing of Pharmaceuticals: Integrated Synthesis, Purification, and Final Dosage Formation", *Angew* Chem. Int. Ed. Engl., **2013**, 52(47), 12359-63

[2] A. Sahlodin and P.I. Barton, "Optimal Campaign Continuous Manufacturing", *Ind. Eng. Chem. Res.*, **2015**, 54(45), 11344-59.

Copyright © American Institute of Chemical Engineers. All rights reserved.

468095 Extended Iterative Optimzation Technology for Mixture Composition Estimation in Moving Powders

Abstract:

The calibration and maintenance effort for a near infra-red sensor and calibration based solution can be a time, labor and material-intensive exercise. A calibration-free method (iterative optimization technique or IOT) was recently published and applied onto mixture component prediction.[1]. IOT estimates fractions of individual components via optimization by rigorously imposing Beer-Lambert's law and mixture constraints to the measured spectra, given the known spectrum for the pure materials. Advances in IOT theory propose the use of massive blind-search evolutionary algorithms to select spectral regions that can improve the performance of IOT in a real-time scenario.

The application of IOT onto powder flows is severely affected by scattering and uneven sample presentation (among others) which results in large deviations from Beer-Lambert in the mixture spectra. Other application-specific artifacts (like peak overlap) can also influence the performance of IOT, resulting in predictions with trends and biases that can range from mild, to completely unreasonable and inaccurate. This work presents an extension to IOT that preserves the calibration-free nature of the method while improving its real-time performance in the estimation of fractions of individual components in the powder stream of a continuous drug product manufacturing line.

Although still calibration free, EIOT requires at the very least measured spectra at a minimum of two levels of concentration for the active ingredient – this data set is readily available if a step change study is done in the line to determine mixing behavior of the system. The power of the method resides in the spectral differentiation across materials, as such, a data set consisting of spectral measurements at several levels of the individual components is much preferred [2].

EIOT fundamentally challenges the application of Beer-Lambert in relating the fractions of each material through the mixture spectra in the process and pure spectra collected in a static sample. As such, EIOT starts with the estimation of the dynamic pure component spectra that is mathematically necessary such that, the concentrations and the dynamic mixture spectra in the "training set" is explained by Beer's law. Multiple methods are presented to produce this estimate. It has been observed that the dynamic pure spectra estimated differ only in small features from the measured pure spectra. Once these pure spectra are estimated, the procedure continues as IOT would. These differences, although small play an important role in the performance of the method. Several examples are presented, contrasting the performance of a PLS model, IOT and EIOT in estimating the fractions of ingredients from spectra measured with a NIR instrument installed in the feed frame of a tablet press, at the end of a continuous drug product manufacturing line. The trends produced by EIOT lack undesired trends that are attributed to process effects onto the spectra, and represent a useful metric to monitor the real-time fraction of the materials in the feed-frame. Such an estimate can be used in combination with a deterministic model of the system to confirm mixture composition and confirm the state of control of the process.

[1] Muteki, K., Blackwood, D.O., Maranzano, B., Zhou, Y., Liu, Y.A., Leeman, K.R.and Reid, G.L. Industrial & Engineering Chemistry Research, 52, 12258 (2013).

[2] Eriksson, L.; Johansson, E.; Wikstrom, C. Mixture design--design generation, PLS analysis, and model usage. Chemometrics and Intelligent Laboratory Systems, 43, 1-2, 1-24. (**1998**)

Copyright © American Institute of Chemical Engineers. All rights reserved.

475462 *Activatable Nanoparticles and Hydrogels: Integration of Synthetic Polymers, Polypeptides, and Oligonucleotides for the Detection and Treatment of Cancer*

Abstract:

Polymeric nanoparticles and hydrogels that integrate a mechanism of user- or disease-controlled activation can be utilized in the development of contrast or therapeutic agents that provide increased selectivity for the targeted diseased tissues. Our group developed conductive polymer nanoparticles that absorb light in the near infrared range and which have capabilities for photothermal conversion. These particles are being investigated as agents for localized, laser-induced photothermal ablation of cancer cells as well as for combined phototherapy and chemotherapy. Similarly, we have developed of enzymatically activatable near infrared fluorescent polymeric nanoparticles as highly specific contrast agents for optical imaging and chemotherapeutic treatment of cancer. These particles provide an off-to-on fluorescence switch that is triggered by the cleavage of polypeptides that are anchored on the surface of the nanoparticles by proteolytic enzymes that are overexpressed in tumor tissue. More recently our group has been working on the development of DNA-enabled hydrogels that can have a wide range of functionality due to the ability of DNA to act as a biodegradable polymer, while also providing specificity for complementary strands and molecular targets. We have demonstrated DNA-crosslinked poly(ethylene glycol) hydrogels that can be degraded upon interaction with disease-related nucleases and their potential as controlled release systems. We are currently working to develop similar hydrogels that can degrade upon interaction to molecular targets through the inclusion of hybridized aptamer/DNA complexes as crosslinkers. This talk will describe the synthesis, characterization, and *in vitro* evaluation of the nanoparticle and hydrogel systems described above, and provide a glance into the enormous potential of responsive polymeric biomaterials in biomedicine.

Copyright © American Institute of Chemical Engineers. All rights reserved.

475463 *Direct Molecular Design for Chemical & Materials Engineering*

Abstract:

Many design problems in engineering and materials science can be cast in terms of finding molecular structures and mixtures that satisfy thermodynamic criteria. Such common criteria, e.g. binding energy, solubility and partitioning, frequently demand selectivity among components in a mixture. Engineers using molecular modeling to solve these design problems typically proceed in the same, forward fashion as Edisonian experimentalists. Structures are guessed, properties are predicted and the results are compared to the desired outcomes. Learning from the outcomes, this loop is iterated until the results match the desired outcome. In contrast, a direct design method allows the engineer to derive the required molecular structure(s) starting with only the desired properties. Direct computational design may be the best approach when Edisonian trial and error testing is not feasible. Furthermore direct design methods may be faster and provide new molecular structures, previously unknown to the engineer.
COSMOdesign is a new direct design methodology to derive optimal compounds or molecular structures that satisfy specified bounds on thermodynamic properties. The method is based on performing genetic Monte Carlo transformations on the solvation surfaces of molecules under design. The solvation surface provides a simple, two-dimensional surface from which thermodynamic properties can be calculated and appropriate chemical structures can be derived. The evolution of mutations enables the engineer to both assess the opportunity for improving further the design as well as identify the improved molecular structure(s). This methodology has been applied to the design of solutions to many industrial problems involving gas-liquid, liquid-liquid and solid-liquid inverse problems to accelerate the development of green products and processes. The COSMOdesign methodology is illustrated in an example optimization of ionic liquids for more efficient refrigeration based on absorption cooling. Ionic liquids form a broad class of solvents which can often be applied to develop green processes. However, the diversity of options for charge centers, chemical group functionality, valency and molecular configuration presents the most significant obstacle to their selection.

Copyright © American Institute of Chemical Engineers. All rights reserved.

475464 40 Years of Research, Student Education, and Fun

Abstract:

We will discuss the beginnings of the research program in polymers, biomaterials, and drug delivery. We will point out some lesser known studies from the lab that led to the development of new theories and important guidelines for the development of new medical products.

Copyright © American Institute of Chemical Engineers. All rights reserved.

458141 Enhancing Process Understanding and Control with Real-Time Analytics for a Challenging Solid Form System of a Pharmaceutical Compound

Abstract:

In the recent years, process analytical technology (PAT) has been widely used on the pharmaceutical development and manufacturing for better process understanding and control. For example, the information of the in-situ particle size distribution by Particle Track (Lasentec ® FBRM) is invaluable on the crystallization development and scale-up. This paper discussed a case of solid state development of a drug candidate that has high tendency to form solvates in most process solvent systems. The only feasible pathway to produce the desired non-solvated form was through a slurry conversion after adjusting the solvent composition through solvent swap distillation. Multiple PAT tools have been evaluated for process understanding and in-line monitoring of a form conversion process, for example, 1) the use of PVM to monitor the conversion of morphology, 2) the use of the FTIR to monitor the progress of de-solvation 3) the use of the combination of Particle Track and Raman for solid form control.

Copyright © American Institute of Chemical Engineers. All rights reserved.

471843 Simultaneous Fast Measurements of Tablet Porosity/Moisture: Real-Time Release Testing for QC Dissolution Prediction

Abstract:

The motivation for this work comes from the dissolution/disintegration sensitivity of a solid dispersion formulation with an erosion based mechanism of dissolution to compression force (which affects tablet solid fraction and hardness), moisture, and the interaction between the two. Solid dispersions were processed by spray drying and subsequent roller compaction which can generate granules with a wide range of quality attributes that impact compactability and compressibility of granules during compression and can be a challenge for obtaining similarities in dissolution rate for proven acceptable range of tablet hardness. We found that for this case dissolution strongly correlates to disintegration time and can be controlled by tablet moisture and density (solid fraction or porosity), but these properties as well as the disintegration test traditionally require separate off-line long measurements.
A novel at-line process analytical technology (PAT), di-electric probe, was developed for simultaneous measurement of solid fraction and moisture content for the tablets. The basic measurement principle of the probe is to estimate the physical properties of bulk solids (such as solid fraction and moisture) by standard correlation with the measured electrical properties (i.e. dielectric permittivity). It was demonstrated that while both real and imaginary component of di-electric permittivity depend on moisture and solid fraction, the phase lag (ratio of imaginary to real part of di-electric permittivity) and di-electric constant of the active formulation depend only on moisture content or relative humidity (RH%). A statistical model was then developed to show dependence of dissolution rate and disintegration time to solid fraction/moisture measured by di-electric probe.

Copyright © American Institute of Chemical Engineers. All rights reserved.

467706 Real Time Prediction of Tablet Tensile Strength By Ultrasound in Continuous Manufacturing

Abstract:

Tablet tensile strength is a critical property that needs to be well controlled in order to produce an acceptable pharmaceutical product. Methods that allow real time and non-destructive prediction of tablet properties can greatly help the development efforts in a continuous manufacturing line. Real-time measurement of tablet strength by ultrasound increases assurance of product quality and is economically beneficial in continuous manufacturing. A model for predicting tablet strength by ultrasound will enable rapid techniques in continuous assessment of tablets ensuring the desired quality attributes.

The objective of this research project was to develop a statistical model to predict tablet tensile strength in a direct compaction continuous manufacturing line. Tablets consisted of model API, Acetaminophen (APAP), lactose monohydrate as the filler, and magnesium stearate as lubricant. Multivariate effects were analyzed using Quality-by-Design methods. The experiment was based on a fractional factorial design with 30 experiments, including 3 repeated center points, to study the influence of four process variables â€“ drug concentration (%), compaction force (KN), blending speed (rpm), and feed frame speed (rpm) â€“ on the final product. Ultrasound transmission speed was determined by measuring the sound wave time-of-flight (TOF) using transducers with a frequency of 2.25 MHz. The speed of sound (SOS) was calculated by dividing the measured tablet thickness by the TOF. The same tablets were characterized for diametrical crushing strength using a mechanical hardness tester. The ANOVA results indicated that compaction force was the main parameter affecting tensile strength and SOS of the tablets in this process. However, ultrasound was sensitive to drug concentration and feed frame speed. Therefore, a predictive model was developed to correlate tensile strength with SOS and relative density. The model was verified using an independently manufactured validation set. The predicted values obtained from the model were in agreement with the experimental results with a correlation coefficient of 0.92. It was further concluded that by using the ultrasound measurement technique it is possible to assess the quality characteristics of tablets non-destructively and in real time.

Copyright © American Institute of Chemical Engineers. All rights reserved.

472375 Toward Predicting Tensile Strength and Characterizing the Acoustic Properties of Pharmaceutical Tablets on Continuous Manufacturing Platforms

Abstract:

Purpose. An ultrasound measurement system was employed as a non-destructive method to evaluate its reliability in predicting the tensile strength of tablets and investigate the benefits of incorporating it in a continuous line, manufacturing solid dosage forms.

Methods. Cylindrical tablets were prepared either continuously or in batch. The formulation was kept constant, while the compaction force and level of shear strain varied. Ultrasound testing was used to evaluate the strength of tablets by measuring the time of flight. The tensile strength of the same tablets was then determined using a mechanical hardness tester.

Results. The effect of two processing parameters, compaction force and level of shear strain were examined. Ultrasound speed of sound was found to be sensitive to the relative density and the level of shear strain. Ultrasound testing could detect even small differences between tablets that a hardness tester failed to do so.

A strategy for hardness prediction is proposed that uses the existing models for Youngâ€™s modulus and tensile strength of porous materials. A clear correlation between Youngâ€™s modulus and tensile strength at zero porosity is presented.

Conclusions. Ultrasound testing is a good candidate to be placed on/at-line to measure the mechanical integrity of tablets non-destructively. These results provide information about the behavior of processing parameters on the

performance of tablets and the ability to engineer product properties.

Copyright © American Institute of Chemical Engineers. All rights reserved.

466106 Overcoming the Challenges of Noisy Data during the Calibration of NIR As a PAT Technique to Measure Roller Compacted Ribbon Density

Abstract:

INTRODUCTION

A number of research groups have investigated the use of NIR spectroscopy for monitoring ribbon density [1-4]. These studies have taken advantage of the fact that NIR spectra are affected by the density of the material being analysed and offer the potential of measuring/monitoring roller compacted ribbon density in real time [1-3, 5]. In order to develop a calibration equation relating density to the NIR spectra Gupta [2] fitted a regression line through all the scanned wavelengths and used the fitted slope of the line. However, this approach does not take into account issues which arise when there are multiple scans on multiple samples of the same type of material, and each scan records simultaneous NIR spectra for several probes. The spectral dataset can contain erroneous outliers. Data processing in real time is challenging as sampling error outliers must be either filtered or cleaned from the data set before further analysis. [6]. Visual inspection of the data and removal of sampling error outliers though justifiable is not practical for real time data processing. Existing data cleaning and filtering models are complex and are essentially off-line operations [6]. A simple solution is required.

This study aims to highlight how variability in roller compacted ribbon quality can impact on NIR spectral measurement and proposes a simple method of data analysis to deal with this.

<>Materials and Methods

Roller compaction: A Fitzpatrick CCS220 roller compactor was used for this study. The roll speed was kept constant at 2 rpm, roll gap at 0.5 mm and the roll pressure was adjusted for each run to 2, 4, 6, 8, 10, 12, 14 and 16 kN. Blends of microcrystalline cellulose, lactose monohydrate and magnesium stearate were compacted. Each of the three blends differed in MCC % moisture content. MCC was stored at 43% for blend 1, blend 2 at 11% and blend 3 at 75% relative humidity. The variation in moisture was introduced to assess its impact on the ribbon quality and thus challenge the NIR technique. Ribbon envelope density was determined by the GeoPyc™ 1360.

NIR spectral collection: Four MultiEye™ NIR probes scanned samples at-line in the manner in which they were produced by the roller compactor i.e. samples were not cut to a specific size. This set up ensured outliers were detected in a similar manner to which they would be detected in-line when monitoring the process. The Perkin Elmer Spotlight 400 FT-IR was also used and the off-line results were compared to the at-line MultiEye™ data. Samples were cut to a specific size to facilitate off-line Perkin Elmer Spotlight analysis.

Statistical data analysis: Three different calibration methods were developed between spectral slopes and envelope density and compared; (1) using the entire data set unfiltered, (2) a visual discard method and finally (3) a 33% Trimmed Mean method.

RESULTS AND Discussion

Calibration of the NIR method to ribbon density was challenging due to the presence of spectral data from broken, split and curved ribbons in the data set particularly for blend 3 which had the highest moisture content (6.96 % w/w). All initial processing of the full data set indicated that some form of data cleaning was required in order to process the MultiEye™ at-line data set. The visual discard method was applied to the full data set and was found to be particularly successful for blends 1 and 2 (**Figure 1**). After using the visual discard data cleaning method MulitEye™ data the calibration correlation was found to be as good as the off-line higher scan resolution Perkin Elmer spectrometer. Blend 1 off line r=0.92, visual discard r=0.95.

However it is necessary to replace the visual method of spectra cleaning with a non-subjective method which was capable of screening for these erroneous probe readings. The trimmed mean eliminates a specified percentage of the data from both the high and the low ends of the dataset before evaluating the standard mean of the remaining data. For this data set the trimmed mean method sets a limit on how data is cleaned from the data set allowing for the removal of a faulty probe reading (25% of data) or a poor sample (33% of data). The 33% Trim Mean in place of the Full Mean reduced the impact of spectral variation or misreads between samples or probes. The Trim Mean method was found to be at least as successful as the Visual Discard method at cleaning the data set prior to development of the calibration equation (**Table 1 and Figure 2**). The Trim Mean method was preferred as the decision on data to include in the calibration model as it is not subjective and simple to apply. Major variation is apparent in the spectral data for Blend 3 (75%) which renders it unsuitable for prediction and calibration. These processing variations were attributes to the high moisture of MCC in this compacted blend.

CONCLUSIONS

The 33% Trim Mean method offers a simple and practical solution to dealing with the NIR spectral challenges when applying this PAT technique to roller compacted ribbon in-line/at-line. Further development of this method when used for calibration could investigate the optimal number of ribbon samples to % data trimmed to achieve a statically robust balance of data cleaning. Increasing the number of samples used to during calibration may allow for a greater retention of data in the TrimMean e.g. 5 samples and keeping 3 giving a 20% Trim Mean. The Trim Mean method is therefore no less successful then other simple calibration methods but offers the advantage of a simple, easy to follow and non-biased means of removing NIR data collected due to probe sampling error at-line/in-line.

ACKNOWLEDGMENTS

Research is funded by the Synthesis and Solid State Pharmaceutical Centre (SSPC) under grant number 12/RC/2275, the Mathematics Applications Consortium for Science and Industry funded by Science Foundation Ireland Investigator Award 12/IA/1683 and the Pharmaceutical Manufacturing Technology Centre, Ireland. Microcrystalline cellulose PH102 donated by FMC Corporation. Innopharma Laboratories for use of the MultiEye^{TM} NIR Spectrometer.

REFERENCES

1. Acevedo, D., et al., *Evaluation of Three Approaches for Real-Time Monitoring of Roller Compaction with Near-Infrared Spectroscopy.* AAPS PharmSciTech, 2012. **13**(3): p. 1005-1012.

2. Gupta, A., et al., *Nondestructive measurements of the compact strength and the particle-size distribution after milling of roller compacted powders by near-infrared spectroscopy.* J. Pharm. Sci., 2004. **93**(4): p. 1047-1053.

3. Gupta, A., et al., *Influence of ambient moisture on the compaction behavior of microcrystalline cellulose powder undergoing uni-axial compression and roller-compaction: A comparative study using near-infrared spectroscopy.* J. Pharm. Sci., 2005. **94**(10): p. 2301-2313.

4. Lim, H., et al., *Assessment of the critical factors affecting the porosity of roller compacted ribbons and the feasibility of using NIR chemical imaging to evaluate the porosity distribution.* Int. J. Pharm., 2011. **410**(1-2): p. 1-8.

5. McAuliffe, M., et al., *The Use of PAT and Off-line Methods for Monitoring of Roller Compacted Ribbon and Granule Properties with a View to Continuous Processing.* Organic Process Research & Development, 2014. **19**(1): p. 158-166.

6. Liu, H., S. Shah, and W. Jiang, *Online outlier detection and data cleaning.* Comput. Chem. Eng., 2004. **28**(9): p. 1635-1647.

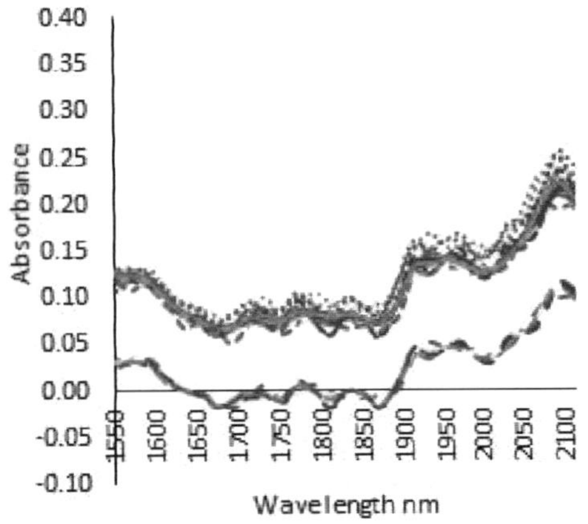

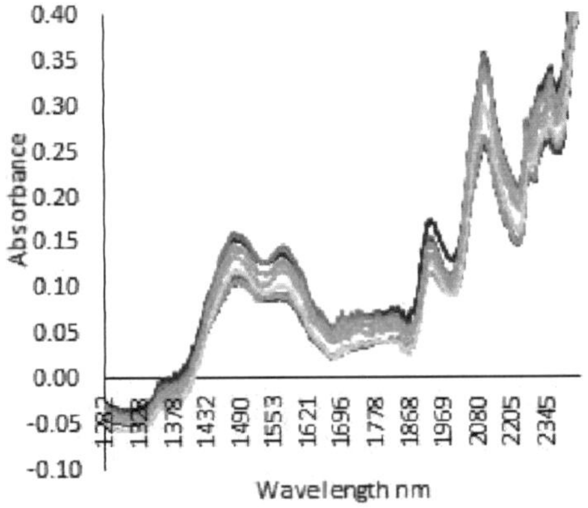

Figure 1 Left panel - Representative spectral data for one piece of ribbon as scanned by the MultiEye™ NIR probes at-line. Probe 1 (continuous line), Probe 2 (dashed line), Probe 3 (dotted line) and Probe 4 (dashed line in lower absorbance region), each of the 5 scans for each probe is seen as a different color i.e. red, yellow, blue, green and purple. When applying the visual discard method to this spectral data set, probe 4 was removed from the data set. Right panel –Representative spectral data for ribbon samples from the same batch as measured by Perkin Elmer Spotlight off-line.

Table 1. Calibration models and the respective correlation coefficients (r) for each of the three blends as calculated by the three different data discard methods. ED indicates envelope density.

Calibration Method	Blend	Calibration model	r
Trim Mean	Blend 1	ED=0.416 + 3584 x Average Spectral Slope	0.96
Visual Discard	Blend 1	ED=0.409 + 3628 x Average Spectral Slope	0.95
Full Mean	Blend 1	ED=0.420 + 3788 x Average Spectral Slope	0.93
Trim Mean	Blend 2	ED=0.319 + 3788 x Average Spectral Slope	0.97
Visual Discard	Blend 2	ED=0.170 + 4257 x Average Spectral Slope	0.88
Full Mean	Blend 2	ED=-0.041 + 5587 x Average Spectral Slope	0.89
Trim Mean	Blend 3	ED=0.465 + 3021 x Average Spectral Slope	0.74
Visual Discard	Blend 3	ED=0.734 + 1515 x Average Spectral Slope	0.92
Full Mean	Blend 3	ED=0.474 + 3115 x Average Spectral Slope	0.74

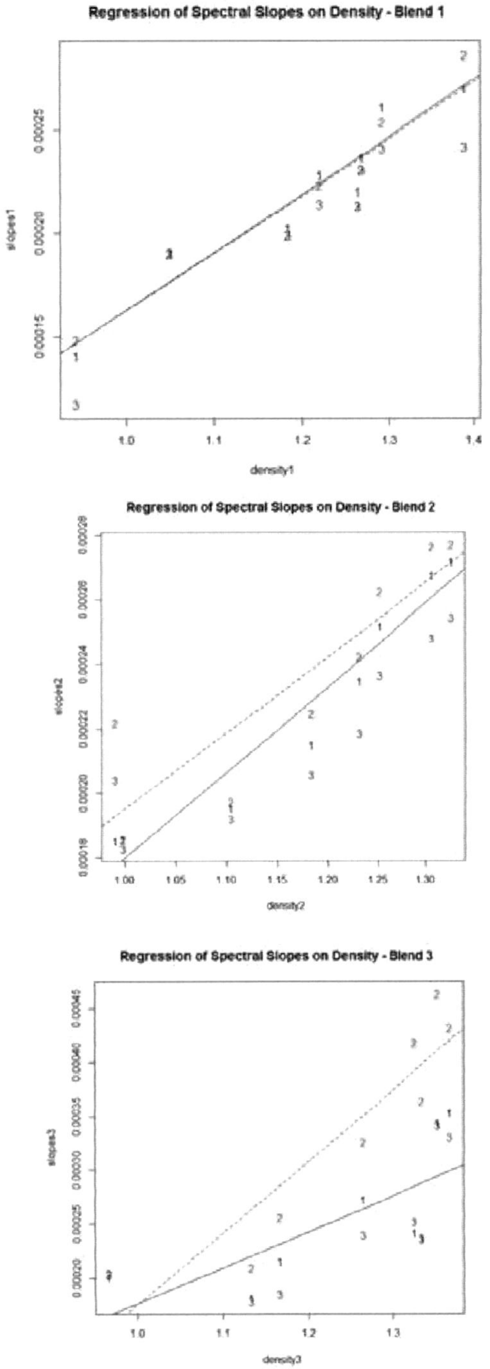

Figure 2. Regression of NIR spectral slopes on Envelope Density. Note: symbols 1=Trim Mean samples, 2= Visual Discard samples and 3 = Full Mean samples. Solid line is fitted line for Trim Mean samples, dotted line is for Visual Discard samples.

Copyright © American Institute of Chemical Engineers. All rights reserved.

457254 PAT Based Batch Statistical Process Control for Pharmaceutical SEMI-SOLID and Liquid Manufacturing Processes

Abstract:

PAT BASED BATCH STATISTICAL PROCESS CONTROL FOR PHARMACEUTICAL SEMI-SOLID AND LIQUID MANUFACTURING PROCESSES

N. Bostijn[1,*], C. Vervaet[2], J.P. Remon[2], T. De Beer[1]

[1]Laboratory of Pharmaceutical Process Analytical Technology, Ghent University, Ottergemsesteenweg 460, 9000 Ghent, Belgium

[2]Laboratory of Pharmaceutical Technology, Ghent University, Ottergemsesteenweg 460, 9000 Ghent, Belgium

Introduction

Pharmaceutical semi-solids and liquids are produced in a batch-wise manner with off-line time-consuming and less efficient quality control (1). The aim of this study is to simulate the batch process on a laboratory scale of five commercial available semi-solid and liquid formulations and to control the critical quality attributes of the process by monitoring the chord length distribution (CLD) of the particles, viscosity, temperature and active pharmaceutical ingredient (API) concentration/homogeneity in-line. For each formulation, reference batches were used to develop a reference batch partial least squares (PLS) regression model. From this model, score control charts were created describing the average batch trajectory. These control charts were used to monitor test batches and to detect any deviations from the expected batch trajectory. This approach allows early fault detection and real-time process adjustments, hence avoiding batch loss.

MATERIALS AND METHODS

Materials

For this study, five commercial available pharmaceutical formulations were selected. Two liquids (shampoo and suspension) and three semi-solids (ointment, cream and gel).

Methods

All experiments were performed using a customized IKA LR2000 modular mini plant reactor system (IKA, Staufen, Germany). The mixing vessel was equipped with a double wall for controlling the temperature of the process. A peristaltic pump (Watson-Marlow, Falmouth, Cornwall, UK) was used for pumping water from a water bath (Type 1032, GFL, Burgwedel, Germany) to the mixing vessel. In-line Raman spectra were recorded using a Raman Rxn1 spectrometer (Kaiser Optical Systems, Ann Arbor, MI, USA), equipped with an air-cooled CCD detector and a fiber-optic immersion probe. The laser wavelength was the 785 nm line from an Invictus NIR diode laser. Every 30 s a spectrum was acquired in the 0–1900 cm^{-1} region with an exposure time of 15 s and a resolution of 4 cm^{-1}.

There were also Raman spectra recorded using a Raman Rxn2 spectrometer (Kaiser Optical Systems, Ann Arbor, MI, USA), equipped with a fiber-optic PhAT probe. The laser wavelength was 785 nm. Spectra were collected every 30 s with an exposure time of 15 s and a resolution of 5 cm^{-1}. The spectral range of the system was from 150–1890 cm^{-1}. Viscosity was measured using an in-line viscometer (Sofraser, Villemandeur, France), equipped with an internal temperature probe. The viscometer had a range from 0.1–1000 mPa.s. Each second the viscosity and temperature were recorded. A focused beam reflectance measurement (FBRM) probe (model C35, Mettler-Toledo, Columbus (Ohio), USA) was used for monitoring the chord length distribution of particles in the system. The probe window was cleaned every 10 s by a pressurized air activated scraper and each 2 s the averaged chord length distribution was saved. The batch model was developed and evaluated using the SIMCA software (Version 14, Umetrics, Umeå, Sweden).

RESULTS AND DISCUSSION

To investigate whether the reference batch models were able to distinguish good from bad batches, process and formulation deviations were induced in the test batches. Only the results of the ointment will be discussed in this abstract.

For test batch 1 the API was added in three different steps. For all the reference batches, API was added in one step after ± 1200 s. In this test batch, 50 % (w/w) of the API was added after 1200 s, 45 % (w/w) after 2520 s and 5 % (w/w) after 3780 s. This induced error resulted in a decrease of the PLS component 1 (PLSC 1) scores in the middle part of the batch evolution scores plot. To monitor and evaluate the overall performance of new batches, a Hotelling's T2 plot can be calculated. When detecting deviations in the Hotelling's T2 plot, it is possible to identify the variables responsible for the deviations via contribution plots. The contribution plots at 1800 s (50 % API) and 3030 s (95 % API), showed that the deviation was determined by a lower viscosity and API concentration. After addition of all the API, the viscosity and API concentration recovered to the average values of the reference model.

In test batch 2 the set point of the water bath was set 10 °C higher compared to the reference batches and corrected after 2580 s. The PLSC 2 scores fell out of the control limits but evolved towards the average batch trajectory at the end of the process when the water bath temperature reached it correct value. At 1800 s the temperature was higher and the viscosity lower compared to the average observations, as seen from the contribution plot.

CONCLUSION

The batch model, developed from four reference batches, was able to detect both the process and formulation induced deviations in the test batches by monitoring the temperature, viscosity, CLD and API concentration in-line.

CHALLENGES AND FUTURE WORK

A next step is to investigate the possibility to produce the same formulations with a continuous production system and to monitor the process using the same and/or new PAT tools.

REFERENCES

1. De Beer T et al. Near infrared and Raman spectroscopy for the in-process monitoring of pharmaceutical production processes. Int. J. of Pharm. 2011; 417: 32–47.

Copyright © American Institute of Chemical Engineers. All rights reserved.

***** Formatting Issue - Best Available Paper/Graphic *****

472690 Real Time PAT Monitoring of Bioreactor Mab IgG3 Cell Culture Process Dynamics Via Fourier Transform Infrared Spectroscopy: Implications and Challenges

Abstract:

Compared to small molecule Process Analytical Technology (PAT) applications, biotechnology product PAT applications have certain unique challenges and opportunities. Understanding the process dynamics of bioreactor cell culture processes is essential to establish an appropriate control strategy for biotechnology product PAT applications. Inline spectroscopic techniques for real time monitoring of bioreactor cell culture processes are highly amendable to PAT approaches to monitor the manufacturing of biotechnology drug products. However, the use of inline Fourier Transform Infrared (FTIR) spectroscopic techniques for bioreactor cell culture process monitoring has not been reported. In this work, real time inline FTIR Spectroscopy is applied to a lab scale bioreactor mAb IgG3 cell culture fluid biomolecular dynamic model. The technical feasibility of using FTIR Spectroscopy for real time tracking and monitoring four key cell culture metabolites (including glucose, glutamine, lactate, and ammonia) and protein yield at increasing levels of complexity (simple binary system, fully formulated media, actual bioreactor cell culture process) is evaluated via a stepwise approach. The FTIR fingerprints of the key metabolites are identified. The multivariate Partial Least Squares (PLS) calibration models have been established to correlate the process FTIR spectra with the concentrations of key metabolites and protein yield of in-process samples, either individually for each metabolite and protein or globally for all four metabolites simultaneously. Applying the 2nd derivative pre-processing algorithm to the FTIR spectra helps to significantly reduce the number of PLS latent variables needed and thus simplify the interpretation of the PLS models. The validated PLS models show promise in predicting the concentration profiles of glucose, glutamine, lactate, and ammonia as well as protein yield over the course of the bioreactor cell culture process. Therefore, the technical feasibility of real time monitoring of the bioreactor cell culture process via FTIR spectroscopy has been demonstrated in this w

Copyright © American Institute of Chemical Engineers. All rights reserved.

475688 Food, Pharmaceutical & Bioengineering Division Award: Control, Sensing, and Chemical Analysis using Microfluidic Devices

Abstract:

The potential uses of microfluidic devices in chemical processing and sensing is almost unlimited. Construction of these devices is currently relatively easy, and there are a large number of published "lab on a chip" systems constructed from a variety of substrates using different actuation, sensing, and control components. In our work, we have focused on components and integrated systems that can be used to measure physical, chemical, and biochemical properties. The applications for these devices have ranged from disease diagnosis to pneumatic computing, and the construction complexity for each device can be very different. For instance, one of our simplest devices is a Newtonian and non-Newtonian viscometer that uses a single microfabricated channel made from a single substrate material. In contrast, our most complex device requires multiple substrates, metal depositions, and ion implants to form a complete system that detects and subtypes the influenza virus. Also, for some of our devices we have needed to redefine the typical low volume and velocities normally seen in microfluidic systems: we have made chips to monitor gallons of fluid per minute or which have fluid velocities approaching a Mach number of 1. In my talk, I will describe some of these devices and discuss the successes and challenges we have encountered. In addition, I will also describe some new areas we are exploring such as reducing water use in food production and continuous flow monitoring of process fluids.

Copyright © American Institute of Chemical Engineers. All rights reserved.

475689 Area 15A, Food - Plenary: Bioelectronic Nose: Integration of Biotechnology and Nanotechnology

Abstract:

Human have five senses. Among the five senses, three of them are physical senses, which are vision, hearing, and touch. The other two senses are chemical senses, which are smell and taste. The science and technology for the physical senses have been advanced enormously, and electronic devices have been developed for these senses, e.g. camera, audio recorder, and tablet PC. However, we do not have any device which can capture smell or taste. In order to mimic the function of human nose and tongue a novel concept for sensor devices functionalized with odor and tastant-recognizing biomolecules was suggested. They are the bioelectronic nose and tongue. A bioelectronic nose consists of primary and secondary transducers. The primary transducer is a biological recognition element such as olfactory receptors. The secondary transducer is a highly sensitive nanodevices that convert biological events into measurable signals. In this presentation, the basic concept and principles of bioelectronic nose will be presented. In addition, applications of bioelectronic nose and the current issues will be discussed.

Copyright © American Institute of Chemical Engineers. All rights reserved.

475690 Area 15B/PD2M, Pharmaceuticals - Plenary: Human Multi-Organ-on-a-Chip Platform for Pharmaceutics

Abstract:

Drug discovery is a lengthy, risky, and expensive process hampered by systemic side effects. Thus, identification of potential liabilities of new molecular entities has become essential in safety assessment throughout drug discovery and development. However, there is no effective in vitro assay currently available for accurately detecting cytotoxicity at early stages of drug development, whereas animal models used at later stages are often ineffective at predicting human responses to candidate drugs. Therefore, there is an increasing demand for improved in vitro models that can better predict the physiological responses of the human body to novel pharmaceutical compounds, particularly with respect to tissue/organ toxicity. With the development of the microfluidic organs-on-a-chip systems, multiple physiological and biochemical parameters of the organ models representing their in vivo counterparts can now be simulated. However, while many drugs can create chronic cellular reactions or delayed cell responses, there is a serious unmet demand of analytical methods for long-term, continual, sensitive analysis of cell-secreted biomarkers from an organ model. In tackling these major challenges facing the pharmaceutical industry, we have innovated a multi-organ-on-a-chip platform integrated with a novel sensing system capable of continual monitoring of physicochemical parameters and soluble biomarkers to enable long-term analysis of drug toxicity in microfluidic organs-on-a-chip. This talk will focus on the development of such an integrated platform, from engineering biomimetic organoids using bioprinting/microtechnologies to sensor integration and automated system operations over extended periods of time.

Copyright © American Institute of Chemical Engineers. All rights reserved.

475691 Area 15C, Bioengineering - Plenary: Status of Biocatalysis: Protein and Process Engineering

Abstract:

Biocatalysts are often highly active and extremely selective but are only slowly gaining general acceptance for catalyzing most applicable transformations. However, there has been significant progress: (*a*) novel protein engineering tools have significantly enhanced biocatalyst development; (*b*) The deactivation of biocatalysts under various stresses can be i) described quantitatively via rational models and ii) remedied via accumulating all stabilizing mutations found in earlier rounds; (c) A host of novel biocatalysts in the past few years has considerably broadened the toolbox for synthetic chemistry.

The presentation discusses both status of and perspectives for biocatalysis. Examples are drawn both from othersâ€™ and the presentersâ€™ own work, for applications ranging from transformations of renewable resources to highly selective synthesis in life science applications.

Copyright © American Institute of Chemical Engineers. All rights reserved.

475692 Area 15D/E, Engineering Fundamentals in Life Science - Plenary: Cellular Bioengineering: Past, Present, and Future

Abstract:

Availability of cultured mammalian cells in the late 70â€™s provided *in vitro* models that enabled pursuit of novel scientific directions, promoted new trends in biomedical research, and initiated the field of â€œcellular bioengineeringâ€.

Recognition that the cellular/molecular events at the microenvironment of the tissue/biomaterial interface dictate the fate of implants *in vivo* resulted in use of cellular *in vitro* models in search of elucidating the underlying mechanisms associated with successful implant-related outcomes. Pertinent findings were subsequently incorporated in the design and formulation of the next generations of biomaterials. Moreover, cellular, *in vitro* models proved valuable alternatives to animal models in examining, for example, the cytocompatibility of biomaterials and in determining the conditions/stimuli (*i. e*., chemical, mechanical and electrical) needed to promote cell functions pertinent to new tissue formation and/or tissue regeneration.

Advances in the medical sciences as well as the need for new therapeutic methodologies in health-care practice expanded the scope of cellular bioengineering applications which now include (but are not limited to) the regenerative capabilities of functional tissues, use of pluripotent stem cells, and changing concepts of biocompatibility. Developments in biology, molecular biochemistry, genetics, and wound healing which are pertinent to new tissue formation/regeneration provide challenges and, at the same time, great opportunities for biomedical engineers. These areas are a substantive part of the present cutting-edge cellular bioengineering milieu and, undoubtedly, will be a major component of future endeavors in biomedical engineering. The scope and creativity of these endeavors are impressive. Researchers working at these frontiers are blazing new trails and are exploring approaches which were undreamed of in the past.

This presentation will briefly review past accomplishments and current advances as well as highlight trends in cellular bioengineering which had an impact but also have the potential for further contributions to the clinical milieu as well as to the biotechnology, tissue engineering, and tissue regeneration fields.

Copyright © American Institute of Chemical Engineers. All rights reserved.

469392 Engineering *Escherichia coli* strains for Optimal Performance in Co-Culture with *Trichoderma Reesei* for Consolidated Bioprocessing

Abstract:

Consolidated bioprocessing, in which enzyme production, hydrolysis, and fermentation occur in a single step, is predicted to offer the lowest cost configuration for cellulosic biofuel production given a sufficiently efficient process. Use of synthetic microbial consortia can help to address some of the limitations of monoculture bioprocessing, including high metabolic burden on the cells and the need to optimize multiple pathways or functions in the same species. Design of synthetic microbial consortia, however, brings new challenges including identifying or engineering strains that can perform optimally within a mixed population.

We are working on a synthetic co-culture of cellulase hyper-producer *Trichoderma reesei* strain RUT-C30 and *Escherichia coli* carrying a pathway for the drop-in biofuel isobutanol from glucose. This co-culture can produce isobutanol directly from cellulosic substrate in one consolidated bioprocess [1]. The *E. coli* strain NV3 [2] initially employed in this system performs suboptimally, achieving lower yield and titer, higher byproduct formation, and notably higher plasmid loss compared to monoculture conditions. Here we present our work to generate *E. coli* strains with superior co-culture performance. We have taken a directed evolution approach, selecting for *E. coli* strains with improved growth on valine analog norvaline in co-culture with *T. reesei*. Additionally we have integrated and optimized the copy number of the isobutanol production pathway genes in the *E. coli* genome to achieve a more stable process.

[1] Minty JJ, Singer ME, Scholz SA, Bae C-H, Ahn J-H, Foster CE, Liao JC, & Lin XN (2013) Design and characterization of synthetic fungal-bacterial consortia for direct production of isobutanol from cellulosic biomass. *Proc Natl Acad Sci USA* **110**(36): 14592-14597.

[2] Smith KM & Liao JC (2011). An evolutionary strategy for isobutanol production strain development in *Eschericia coli*. *Metab Eng* **13**(6): 674-81.

Copyright © American Institute of Chemical Engineers. All rights reserved.

472567 PHA Production Using Yarrowia Lipolytica and Alternative Feedstocks

Abstract:

Polyhydroxyalkanoates (PHAs) are naturally-occurring polyesters produced by a number of different bacterial species as a means of carbon and energy storage. PHAs have been suggested as possible replacements and supplements to petroleum-derived plastics due to PHAsâ€™ tunable mechanical properties and high potential for feedstock flexibility. Feedstock flexibility allows production of a desired product even under conditions of environmental or economically-induced scarcity. Natural producers of PHAs are often highly sensitive to the substrates and media on which they are grown. This variability precludes feedstock flexibility. In order to address this issue, we have engineered a far more metabolically-robust organism, the oleaginous yeast *Yarrowia lipolytica*, to produce PHAs from a number of different feedstocks.

To accomplish this, the PHA-producing enzymes, called PHA Synthases, from the natural PHA producers *Pseudomonas aeruginosa* and *Ralstonia eutropha* were incorporated into the yeast. Flux to this enzyme was increased by overexpression of the native acetyl-CoA carboxylase (ACC1) and modification of Multi-Function Enzyme (MFE1). ACC1 facilitates the synthesis of longer carbon chains, which is desirable when building a specific larger product from smaller substrates. The modification to MFE1 prevents this double-acting enzyme from degrading a highly desired PHA precursor. *Yarrowia lipolyticaâ€™s* native catabolic pathways allow it to be an effective metabolizer of fats, sugars, and proteins. Our engineered organisms demonstrated PHA production using pure and waste fats, crude glycerol, and hexose and pentose sugar feedstocks. The final PHA products were also characterized for their structure and composition: necessary information for the tuning of the mechanical properties.

Copyright © American Institute of Chemical Engineers. All rights reserved.

473357 Rewiring Native Metabolism of Yarrowia Lipolytica for Enhanced Assimilation of Complex Biomass-Derived Sugars and High-Yield Production of Organic Acids

Abstract:

While the model oleaginous yeast *Yarrowia lipolytica* is well known for decades to grow mainly on C6 sugars, its native metabolic capability to assimilate xylose and cellobiose, dominant sugars derived from lignocellulosic biomass, is poorly understood. In this study, we elucidated and activated native xylose and cellobiose pathways of *Y. lipolytica* through comprehensive metabolic and transcriptomic analyses. We identified 7 glucose-specific transporters, 16 xylose-specific transporters, and 4 cellobiose-specific transporters that were transcriptionally upregulated for growth on respective single sugars. While *Y. lipolytica* is capable of using xylose as a carbon source, xylitol dehydrogenase (XYL2) is the key bottleneck of xylose assimilation. *Y. lipolytica* has a set of 5 extracellular and 6 intracellular beta-glucosidases (BGLs) to assimilate cellobiose via both extra- and intra-cellular mechanisms, but the latter being dominant for growth on cellobiose as a sole carbon source. By tuning the expression of native metabolic enzymes and transporter genes, we engineered a *Y. lipolytica* mutant with enhanced assimilation of xylose and cellobiose as well as their mixtures with glucose. Further, we demonstrated the performance of *Y. lipolytica* mutant for high-yield production of high-value organic acids (e.g., alpha-ketoglutaric acid) using ionic liquid-pretreated lignocellulosic biomass in a simultaneous saccharification and fermentation process.

Copyright © American Institute of Chemical Engineers. All rights reserved.

467715 Using Protein Design to Evaluate the Relationship Between Protein Surface Potential and Protein-Lignin Binding for the Eventual of Low Lignin Binding Cellulases

Abstract:

Increasing total turnover number or recycling cellulases used in deconstruction is a key step in making microbial biofuel production from lignocellulosic feedstocks economically viable. Nonspecific binding interactions between cellulases and lignin denature cellulases and hinder cellulase recycling. Previous literature supports that this binding event is influenced by the surface potential characteristics of the cellulases themselves, namely surface hydrophobicity and electrostatic surface potential. To further investigate the relationship between surface potential and nonspecific protein-lignin binding, we used a hybrid computational/experimental approach to (i.) computationally redesign surface potentials spanning a large range of net charge, total charge density, and surface hydrophobicity on a model protein; and (ii.) experimentally characterize the resultant designs for function, stability, and protein-lignin binding. In all, we tested sixteen designs spanning the physiological ranges of the above potentials. We found that designs were, on average, better expressed and more stable that the parental protein, showing that the computational design process is robust. We characterized variant binding to AFEX pretreated, dioxane extracted cornstover lignin with two different experimental methods â€“ quartz crystal microbalance with dissipation (QCM-D) and fluorescence based subtractive mass balance assay. Comparing the binding capacity of variants between characterization methods highlights that both methods isolate the same low, medium, or high lignin binders. Analysis of determined binding capacity with known surface potential properties suggests a weak relationship between protein net charge and protein:lignin binding capacity, while other parameters were not statistically significant. We then applied this method to redesign surfaces of different catalytic domains on endocellulases. Implications for improving the deconstruction step in the next generation of biorefineries will be discussed.

Copyright © American Institute of Chemical Engineers. All rights reserved.

454097 Towards Development of *Rhodococcus Opacus* As a Microbial Cell Factory: Conversion of Lignin-Derived Aromatic Compounds into Lipids

Abstract:

Lignocellulosic biomass is a renewable feedstock that can be converted into biofuels and biochemicals using microorganisms. However, pretreatment of lignocellulosic biomass releases toxic compounds that inhibit microbial growth. Additionally, lignin has been an untapped carbon source due to its recalcitrance and toxic monomeric units (e.g., aromatic compounds). To achieve sustainable and economic utilization of lignocellulosic biomass, we must develop technologies that utilize the complete biomass, including lignin, which is currently considered as waste. To this end, our work focuses on developing *Rhodococcus opacus* as a new chassis for conversion of aromatic compounds into triacylglycerols (TAGs), biodiesel precursors. *R. opacus* is a promising host due to its high tolerance to and utilization of aromatics as a sole carbon source. Importantly, *R. opacus* can accumulate TAGs up to ~80% of cell dry weight under nitrogen-limiting conditions. To enhance its innate aromatic-degrading capacity, we applied adaptive evolution, a growth-based strain selection method, by sequentially sub-culturing cells in diverse combinations of lignin-derived aromatic compounds as sole carbon sources. Our adapted strains demonstrated higher growth rates and higher lipid accumulation compared to the wild type strain. Whole genome sequencing, RNA-seq, and ^{13}C-fingerprinting analysis have identified possible aromatic tolerance and utilization mechanisms such as upregulation of degradation pathways and putative transporters for aromatic compounds. In addition, synthetic biology tools have been developed to enable engineering of the identified tolerance genes, regulators, and pathways. We will present progress towards development of *R. opacus* as a microbial cell factory.

Copyright © American Institute of Chemical Engineers. All rights reserved.

471737 Defining the Minimal Set of Microbial Genes Required for Valorization of Lignin Biomass

Abstract:

Lignin is the second-most abundant biopolymer in the world, but substantial previous efforts to break it down into useful carbon-based platform chemicals in a reliable and scalable manner have been largely unsuccessful, rendering lignin substantially underutilized. By contrast, numerous fungi and bacteria readily degrade this exceptionally recalcitrant biopolymer using a milieu of specialized enzymes, and are thus able to metabolize lignocellulose as an energy source. Reconstituting the lignin-degrading machinery of these specialized microbes in a genetically-tractable model organism may therefore enable a route to efficient valorization of lignin biomass. To date, however, no engineered heterologous host has been shown to have degradation activity toward whole lignin biomass. Identification of the minimum necessary set of enzymes required for the conversion of lignin biomass into valuable feedstock chemicals has likewise not yet been realized. This study investigates the co-expression of putative lignin-degrading enzymes in the model host organism *Saccharomyces cerevisiae.* Specifically, recombinant strains were engineered to drive efficient export of peroxide-generating enzymes as well as ligninase candidates identified through transcriptomic and secretomic data from lignin-degrading fungi. A combinatorial approach was employed to develop an enhanced understanding of the interconnected roles of putative ligninases in degrading lignin biomass.

Copyright © American Institute of Chemical Engineers. All rights reserved.

474992 Invited Presentation: Lignin Conversion By Biological Funneling and Chemical Catalysis

Abstract:

Lignin is an underutilized feedstock for producing renewable fuels and chemicals, where it is instead burned for process heat because its inherent heterogeneity and recalcitrance make it difficult to selectively valorize. Despite many decades of lignin depolymerization research, most catalytic strategies to break down lignin yield a highly heterogeneous slate of aromatic compounds. In Nature, some organisms have evolved metabolic pathways that enable the utilization of lignin-derived aromatic molecules as carbon sources. Aromatic catabolism typically occurs via upper pathways that act as a â€œbiological funnelâ€ to convert heterogeneous substrates to central intermediates, such as protocatechuate or catechol, which undergo ring cleavage and are further converted via the Î²-ketoadipate pathway to central carbon metabolism. We are using a natural aromatic-catabolizing organism, *Pseudomonas putida* KT2440, to demonstrate that these metabolic pathways can be harnessed and engineered to convert both aromatic model compounds and heterogeneous, lignin-enriched streams into value-added compounds. To make this concept a reality will require tailoring the lignin chemistry of the feedstock to the catalytic deconstruction strategy and the biological pathways for upgrading. Overall, this work demonstrates that the use of microbial aromatic catabolism may one day enable an approach to valorize lignin by overcoming its inherent heterogeneity to produce fuels, chemicals, and materials.

Copyright © American Institute of Chemical Engineers. All rights reserved.

475260 Next-Generation Testing Platform for Strain Optimization, Screening, and Fermentation Testing

Abstract:

Amyris is a renewable products company that engineers yeast to convert sugar into valuable chemicals. We have successfully produced four isoprenoid molecules at commercial scale, replacing petroleum, animal or plant-sourced chemicals with reliable, sustainable alternatives. Underlying this commercial success is an automated pipeline for the construction, testing, and deployment of our yeast strains that has made each strain development timeline shorter with each successive molecule. We are now in the midst of expanding our footprint on the metabolic map into chemical targets outside isoprenoids, and consequently, expanding the capabilities of our strain construction pipeline. This talk will focus on our strain testing technologies, and describe the next-generation testing tools we are developing to enable testing of strains targeting diverse chemical space and multiple global markets.

Copyright © American Institute of Chemical Engineers. All rights reserved.

475263 Harnessing Biological Complexity and Diversity through Metabolic Engineering: The Manus Bio Approach

Abstract:

Specialized metabolites represent a class of compounds with unique and complex chemistries and which possess a myriad of commercial applications, from pharmaceuticals and food products to industrial chemicals. Despite their intrinsic value, however, sourcing remains a bottleneck to more widespread use due to their low abundance in nature and the inherent difficulties in applying metabolic engineering and industrial biotechnology approaches to produce these complex natural chemistries in microbial hosts. At Manus Bio, we focus on addressing the challenges of microbial fermentation of plant secondary metabolites through the combined application of our proprietary Multivariate Modular Metabolic Engineering (MMME) and Pathway Integrated Protein Engineering (PIPE) technologies for optimizing heterologous enzymes and pathways. With this approach, we have successfully demonstrated the potential of using a single microbe to catalyze several complex chemical reactions, including polymerization, cyclization, multiple oxidations, and up to six glycosylations, all while producing the target metabolite at greater than 95% purity. In this presentation, we will highlight some of the important insights and guiding principles used to engineer complex biochemical pathways in simple bacterial systems, such as *Escherichia coli*.

Copyright © American Institute of Chemical Engineers. All rights reserved.

475265 The Emerging Organism Engineering Industry

Abstract:

There is an emerging demand for sourcing plant-derived extracts (nutraceuticals, flavors, fragrances, sweeteners, etc.) from engineered microbes. While recent advances in synthetic biology and metabolic engineering provide feasible approaches to engineering such organisms, commercial success for developing these â€œculturedâ€ ingredients presents specific challenges. Unlike biofuels, where efforts can be focused on one particular molecule given the enormous market size, cultured ingredients require developing different organism lines in a rapid and low cost fashion. This requires a scalable solution for bio-manufacturing of organisms, which is provided by our state of the art foundry that continues to grow. I will describe how organism development at Ginkgo leverages our foundry to accelerate the design/build/test using specific examples. In particular, I will highlight the value of combining computer-aided engineering software tools, cheap gene synthesis and high resolution-accurate mass LCMS to develop engineered microbes. Finally, Iâ€™ll touch on how our improvements in manufacturing organisms lend to opportunities outside of cultured ingredients.

Copyright © American Institute of Chemical Engineers. All rights reserved.

475269 General Purposing Synthetic Biology: a Biofoundry in a Box

Abstract:

What could you build with unlimited DNA writing? Productivity of DNA synthesis continues to increase at exponential rates. The oligonucleotide synthesis market alone is forecast to reach nearly $2B USD by 2020 with an annual growth rate in excess of 10%. This increase in oligo synthesis capacity and the associated reduction in cost is transforming synthetic biology; cheap oligo synthesis will remake the impossible to the new routine. Muse's foundational technology is a proprietary synthetic biology platform that scales directly with multiplex oligo synthesis, enabling designer DNA construction at 1000X lower cost than the current state of the art. To date, Muse has demonstrated the ability to generate and track over 50,000 designer mutants simultaneously. Applications of this technology include sequence to activity mapping for site saturation mutagenesis, protein and pathway engineering, and strain reconstruction of complex evolved strains to name a few. Gene editing capabilities have been demonstrated in various industrially relevant organisms, including applications across biotechnology. Further, this technology is being automated in the form of a bench-top device that allows a single user to perform a broad range of molecular biology endeavors at rates 1000X beyond the current state of the art.

Copyright © American Institute of Chemical Engineers. All rights reserved.

475259 Making Silk without Silkworms: Using Industrial Biotechnology to Make Performance Protein-Based Fibers

Abstract:

Bolt Threads is a Bay Area startup that believes that the combination of molecular biology and materials science can revolutionize how people think about and use natural materials. Bolt's materials can dramatically improve the performance and environmental impact of a broad range of consumer and industrial materials. Specifically, Bolt is developing the next generation of sustainable performance apparel by leveraging the latest advances in industrial biotechnology, automation, and materials science. Derived from our knowledge of spider silk protein, the company has developed a high-throughput platform to generate thousands of different silk proteins, all from a scalable microorganism host. Using industrial biotechnology processes and fiber spinning, Bolt is able to make commercial volumes of protein and fiber. Precision control from primary protein sequence through fiber extrusion allows the company to produce materials that combine the benefits of both natural and synthetic fibers, and to go beyond even what is currently available in nature and the marketplace. Leveraging the unique properties of its fibers, Bolt Threads will be delivering its own line of consumer apparel made with its Engineered Silk(TM) as well as working with partners across industry verticals.

Copyright © American Institute of Chemical Engineers. All rights reserved.

475548 The Development of Platform-Based Technologies for the Optimization of Sustainably Produced Chemicals

Abstract:

Genomatica has established an integrated computational/experimental bioengineering platform to design, create, and optimize novel high-producing organisms and bioprocesses. Here we present the use of our platform to develop *E. coli* strains for the production of the industrial chemical 1,4-butanediol (BDO) from carbohydrate feedstocks. These efforts have culminated in commercial scale production of over 5M lbs of BDO from dextrose. BDO is a chemical intermediate produced commercially through energy-intensive petrochemical processes from hydrocarbon feedstocks and is used in a variety of products including automotive, electronics and apparel. Therefore, this product represents an opportunity to make a significant impact on the replacement of traditional petrochemical processes with benign bioprocesses using renewable feedstocks.

We are continuing to enhance our technology platform, enabling faster development of new processes with increased chance of success. Here we describe the application of cell-free transcription-translation (TX-TL), towards improving the design-build- test-learn cycle for the construction of a high-performing microorganism capable of producing BDO and other industrial chemicals from carbohydrates. We used systems biology approaches including 13 C-flux analysis, proteomics, and metabolomics to characterize TX-TL applied to different engineered metabolic pathways, including BDO, and compared the results to the analogous in vivo processes. These studies provide critical insight into the active metabolism of TX-TL and mechanisms of redox and energy regeneration. Learnings from these studies were applied towards improving the TX-TL platform for the production of intermediate and specialty chemicals. The presentation will highlight the successful integration of the TX-TL technology into our rational and high-throughput strain engineering platform, and will show the potential advantages for using cell-free systems towards the optimization of sustainably produced chemicals.

Copyright © American Institute of Chemical Engineers. All rights reserved.

478302 Imagine, Design, Create Cloud Based Software for Biological Design

Abstract:

Over the past 10 years the ability to sequence, edit, and build life has grown exponentially. Software in this space has not kept pace with the rapid pace of biological innovation resulting in a tremendous gap between the modeling and manufacturing of biological systems. The Autodesk BioNano team works to bring design solutions to biology, making the interaction of scientists with software more seamless and efficient. Our goal is to allow scientists to focus on what matters, the science. Autodesk has experience in designing and displaying complex, man-made mechanical structures in the cloud. It is time to apply this knowledge to biology to show how proteins and molecules interact to create life. Autodeskâ€™s BioNano group hopes to empower scientists to Imagine, Design, and Create a better world.

Copyright © American Institute of Chemical Engineers. All rights reserved.

463022 Applications of Omics Technologies for Enhanced Process Robustness in a Chemically Defined Fed-Batch Industrially Relevant CHO Cell Culture Process

Abstract:

We have previously reported on an undesirable high lactate phenotype, caused by overfeeding a mAb-producing CHO cell line, in a chemically-defined process. To better understand this phenomenon, we utilized metabolic flux analysis (MFA) using isotopic labeling with multiple tracers (glucose and glutamate) and untargeted metabolomics. The results from these analyses provided new insights into our cell culture process. We first identified two supplements capable of delaying the onset of the high lactate phenotype by ~3 days. The addition of the first supplement increased the peak viable cell density by 15% and harvest titer by 25%. The second supplement increased both peak viable cell density and harvest titer by 25%. Following this observation, untargeted metabolomics was performed to investigate the mechanism of action for these supplements by comprehensively characterizing the impact of the supplements on the levels of intracellular and extracellular metabolites. Focusing on central carbon metabolism, we conducted isotopic labeling experiments with glucose and glutamate tracers, and analyzed the results using MFA to characterize the fluxes in and around the TCA cycle. We identified overflow of malate and citrate from the mitochondria as a potential mechanism underlying growth inefficiency, which was attenuated by the two medium supplements. In addition, we assessed the redox potential in our cell culture, identified supplements capable of preventing a redox imbalance, and confirmed their ability to delay the high lactate phenotype in our production culture. Utilizing the insight gained through our systems biology approach, we evaluated the proposed mechanism of action by reformulating our nutrient feed to implement the inhibition of mitochondrial metabolite overflow independent of the two supplements.

Copyright © American Institute of Chemical Engineers. All rights reserved.

471844 Scale Related Variables and Their Potential Impact on Process and Product Quality in the Production of Biologics

Abstract:

Cell culture technologies for the production of recombinant protein therapeutics have evolved over the years and span from batch to fed batch, perfusion and other production methods. Process intensification has resulted in new challenges during process scale-up. We assessed dissolved oxygen and pH control strategies at 2-L, 100-L and 2000-L scale to evaluate process and product consistency across the three scales. Differences in mass transfer characteristics and offline sample handling across the scales caused up to 20% differences in integral viable cell density. The changes in the cell growth profiles resulted in shifts in product glycan profiles and associated effector function attributes. Understanding the cell culture operating space based on the selected production method and associated control variables is key to successful process scale-up.

Copyright © American Institute of Chemical Engineers. All rights reserved.

458796 Design and Simulation of a Continuous Rotating Annular Bioreactor with an Internal Spiroid

Abstract:

This research introduces a continuous bioreactor designed to increase cell oxygenation, growth rate and overall high productivity through its novel geometry. The bioreactor consists of an outer shell cylinder and inner core cylinder, each with adjustable rotational velocities to control the mixing in the reactor and provide a variety of shear environments. The reactor may be operated with or without headspace. Inlet flows are controlled to provide medium and oxygen to the reactor volume. A spiroid tube is attached to the inside of the bioreactor shell wall to increase the gas-liquid contact area and thus improve oxygen transfer. The complex flow patterns were successfully analyzed by using computational fluid dynamics (CFD) simulations. Simulations were conducted for multiphase fluids with various operating conditions to predict the actual reaction environment in the bioreactor prior to experimental studies. Various experiments along with mathematical models were used to calculate the volumetric mass transfer coefficients at different operating conditions for bioreactor models with or without the spiroid tube. The comparison of the computational with experimental results was used to support the improved oxygen transfer performance and to predict the higher cell cultivation performance of this bioreactor due to the addition of the spiroid tube.

Copyright © American Institute of Chemical Engineers. All rights reserved.

468411 Cell Culture of the Termite Gut Microbiome Using a 3D-Printed Synthetic Microhabitat

Abstract:

A community of microbes living in association with a host organism is sometimes called a microbiome. The interactions among members of a microbiome can give rise to important communal functionality that cannot be achieved by its constituent organisms alone. Examples of microbiome function include normal human digestion, productive and resilient plant root systems, and the degradation of lignocellulose in lower termites such as *Reticulitermes flavipes*. A microbiome lives in a complex and often dynamically changing microhabitat. No single set of cell culture conditions is suitable for the entire microbiome. As a result, microbiomes are often studied in ways that destroy the community such as the use of metagenomic analysis. However, emerging technology in microfabrication is leading to novel cell culture systems that better emulate complex features of microhabitats. Here, the steep oxygen gradient that naturally exists in the lower termite hindgut was emulated in a microfluidic device using a rapid prototyping and iterative design approach. 3D-printed devices with a 200-µm resolution were used as casting molds for oxygen-permeable, polydimethylsiloxane (PDMS) cell culture chambers. The microbiota from *Reticulitermes flavipes* hindguts were inserted into synthetic microhabitats under anaerobic conditions before allowing diffusive flux from an 8 ppm dissolved oxygen reservoir to reach one side of the culture chamber, thereby closely emulating oxygen gradients in the real hindgut. Initial results indicate spatial reorganizations of protists within this synthetic microbiome as micro-oxygen gradients evolve. Several enabling technologies are under development that will further enhance complex microhabitat cell culture systems. One such technology is semipermeable membranes incorporated directly into disposable and easy to use 3D printed cell culture chambers for the delivery of oxygen and other nutrients. Our work contributes to a better understanding of the *Reticulitermes flavipes* hindgut microbiome by enabling hypothesis-driven research of how culture conditions and perturbations influence community structure and microbiome function. Eventually, gradient bioengineering technology may enable efficient, adaptive conversion of waste biomass into valuable chemicals and fuels.

Copyright © American Institute of Chemical Engineers. All rights reserved.

460147 Single High-EDR Exposure of Cultured Human Cells Reveal Effects of Prolonged Doublings and Microcarrier Attachment on Cell Shear Susceptibility

Abstract:

Background: Human Mesenchymal Stem Cells (hMSCs) are very important for cell therapy use, and are able to avoid immune responses when used in allogeneic therapies [1-3]. hMSCs are cultured at centralized manufacturing centers before being shipped to where they are needed and then differentiated. Currently there are no standardized protocols for the isolation, expansion and harvesting of the hMSCs[4] Additionally there is concern about the effects of shear stress on hMSCs, as damage during scale up and manufacturing have been reported. Based on these reports, we expected that hMSCs would be more sensitive to shear stress than cells typically used in large scale cell culture, specifically compared to Vero cells. Additionally, scale-up issues are present with the large scale production that would be needed for therapeutic use, where up to a trillion cells could be necessary. Microcarriers are one method that could be used to produce cells, but different microcarriers can have different effects on growth [5], and it is important to determine if the microcarrier type has an impact on the shear sensitivity of cells grown on that carrier.

Materials and Methods: Beginning with cells grown in T-flaks, a contractional flow device, called a "torture chamber" was used, flowing hMSC and Vero cells through with flow rates, subjecting them to previously defined hydrodynamic forces[6]. The Vero and hMSC cells were suspended in their respective media and then were flowed at various flow rates through the torture chamber. Following this the cells were collected and centrifuged. A cytotoxicity assay (Promega) was used to quantify the cell damage due to the shear stress. T-flasks were reseeded using cells that were not passed through the torture chamber. An approximate count of the cells that attached one day after seeding was gathered to determine the generation count. The number of cell divisions was estimated for each hMSC batch run through the torture chamber. These were placed into bins of generation counts for comparison to Vero cells using ANOVA, each bin representing what is traditionally considered a "passage". Following this microcarrier effects on shear sensitivity were studied. Human Embryonic Kidney cells (HEK-293) were grown on two different types of microcarriers, collogen coated (Solohill) and dextran matrix (GE cytodex). The HEK-293 cells were grown on the carriers in spinner flasks under protocols typically used for growth on the microcarrier, and then the carriers were flowed through the torture chamber at varying concentrations. These different concentrations were formed by collecting different amounts of media with carriers from the spinner flaks and diluting them to a constant volume, with the highest concentration run being 4 times the lowest concentration, and intermediate concentrations run at 2 and 3 times the lowest concentration. The cells and carriers were then collected, centrifuged, and tested using the same cytotoxicity assay to assess the cell damage due to the shear stress. After taking a sample of the supernatant for the cytotoxicity assay, the cells were then lysed, putting the LDH level on a per cell basis.

Results and Discussion: For suspended cells, the amount of LDH released after shear stress exposure for the hMSC cells after 20.5-24.5 and 24.5-28.5 were not significantly different than the vero cells. However, the hMSCs that had generation counts over 28.5 released significantly ($p<0.05$) more LDH than vero cells at energy dissipation rates above $6.0*10^6$ W/m^3. Specifically, the vero cells and the hMSC cells at generation counts below 28.5 show less than 10 percent LDH, compared to around 20 for the higher generation hMSC cells. At EDR rates of around $4.0*10^7$ W/m^3, the hMSC cells with a generation count of 32.5 to 36.5 exhibit about 60% LDH, indicating a high amount of cell damage and death. The generational effect could be due to changes in the cytoskeleton that have been previously found, possibly contributing to tears and LDH release[7]. The work with the micorcarriers is currently ongoing, but preliminary results show that the percent LDH increases with increasing micorcarrier concentration. The percent LDH increases sharply at higher concentrations, indicating a concentration where the shear effects on the cells become a significant factor. In our preliminary results, the LDH levels increase dramatically between the 3X concentration trial 4x concentration trial.

Conclusions: Our results indicate that the hMSCs might be just as tolerant to sheer stress as other cells currently used for manufacturing, as long as the cells are used when they are still young and aren't entering the process of senescence. The results also indicate that the torture chamber can be used to screen conditions that would allow for optimal growth of cells. Computational modeling has been previously done on the chamber to determine EDR rates for the different flow rates used, and data is available for EDR in spinner flaks and bioreactors based on agitation speed, so the torture chamber could be used to determine the agitation speed that allows for minimal sheer stress effects. The torture chamber can also be used to determine the concentration of carriers that can be used with minimal cell death. More carriers provide more surface area for the cells to grow on, but too many carriers introduces cell damage due to collisions. The use of the chamber allows for the determination of the concentration that would have the most surface area without introducing excessive cell damage. The torture chamber can be used to screen conditions for production of various products without needing to waste production runs to try out conditions to determine viability.

1. Wei, X., Yang, X., Han, Z., Qu, F., Shao, L., Shi. Y., 2013, "Mesenchymal stem cells: a new trend for cell therapy", Acta Pharmacologica Sinica, Vol 34, pp. 747-754.
2. Kim, N., Cho, S., 2013, "Clinical applications of mesenchymal stem cells", Korean Journal of Internal Medicine, Vol. 28, pp. 387-402.
3. Ankrum, J., Ong, J., Karp, J., 2014, "Mesenchymal stem cells: immune evasive, not immune privileged", Nature Biotechnology, Vol. 32, pp. 252-260.
4. Ikebe, C., Suzuki, K., 2014, "Mesenchymal Stem Cells for Regenerative Therapy: Optimization of Cell Preparation Protocols", BioMed Research International, http://dx.doi.org/10.1155/2014/951512
5. Rowley J., Abraham, E. Campbell, A., Brandwein, H., Oh S., 2012, "Meeting Lot-Sized Challenges of Manufacturing Adherent Cells for Therapy", Bioprocess International, Volume 10, pp. 16-22.
6. Shenkman, R., Godoy-Silva, R., Papas, K., Chalmers, J., 2009, "Effects of Energy Dissipation Rate on Islets of Langerhans: Implications for Isolation and Transplantation", Biotechnology and Bioengineering, Vol. 103, pp. 413-423.
7. Maloney, J., Nikova, N., Lautenschlager, F., Clarke, E., Langer, R., Guck, J., Van Vliet, K., 2010, "Mesenchymal stem cell mechanics from the attached to the suspended state", Biophysical Journal, Vol 99, pp. 2479-2487.

Copyright © American Institute of Chemical Engineers. All rights reserved.

463688 Metabolite Pathway Analysis for Analyzing Consumption of Amino Acids in Hybridoma Cell in Different Types of Growth Media and Cell Culture Process Conditions

Abstract:

In hybridoma cell cultures, amino acid and glucose consumption, cell growth and antibody production are key considerations during upstream process optimization, particularly media optimization. Gaining knowledge on their interrelations and the relevant cellular physiology could provide good insight for higher and more consistent antibody production. The amino acid consumption pattern was analyzed in hybridoma cell cultures producing monoclonal antibody (mAb). The amino acid concentrations, product concentrations, and cell growth profiles were measured in three different cell cultures subject to an amino acid supplementation strategy. In this research, we examine the amino acid metabolism and its impact on cell growth and mAb production during the different type of cell culture processes. We found that a group of key amino acids and metabolic pathways, the highest mass fraction of the product, and these could become limiting for product expression. In this study, metabolic pathway analysis was applied to show that certain metabolic pathways are significantly different between the culture process and media composition. Metabolic pathway analysis was helped to better evaluate the impact of amino acid supplementation on the most relevant pathways in the culture process.

Copyright © American Institute of Chemical Engineers. All rights reserved.

474761 Confessions of a Cell Culture Engineer: Process Impact on Protein Products in a QbD World

Abstract:

In producing therapeutic proteins from cell culture processes, it is necessary to develop processes capable of delivering consistent protein characteristics, particularly those that could affect the structure or function of the biomolecule and potentially impact the product safety or efficacy. Over the past thirty years, cell culture scientists and engineers have navigated the waters of process understanding in tackling protein product quality challenges. In many cases, the root cause of protein product attribute variation is related to cell culture process conditions.

The impacts of process conditions on protein attributes may be due to biosynthetic (biological origin) or degradative (chemical or physical origin) effects, and these effects may be driven by intracellular or extracellular factors. This presentation will review industrially relevant experiences, and discuss approaches toward better understanding and control of recombinant protein product characteristics through cell culture manipulations. Specific protein attribute variation addressed may include amino acid sequence, glycosylation (micro and macroheterogeneity), and charge variant distributions, and their potential for impact on various aspects of protein quality. The linkage between this process knowledge and todayâ€™s Quality by Design (QbD) approaches to process and product development will be discussed.

Copyright © American Institute of Chemical Engineers. All rights reserved.

470265 Constructing Predictive Kinetic Models of Metabolism for Guiding Strain Design

Abstract:

Existing computational strain design tools operating at a genome-scale primarily rely on a stoichiometric description of metabolism. Despite many successes, designed mutants often fail as stoichiometric models do not directly account for enzyme level, metabolite concentration and substrate-level regulatory barriers. Developing kinetic models of metabolism at a genome-scale that faithfully recapitulate the effect of multiple genetic interventions would be transformative in our ability to reliably design novel overproducing microbial strains. Here, we introduce k-ecoli457, a genome-scale kinetic model of *Escherichia coli* metabolism that satisfies fluxomic data for a wild-type and 25 mutant strains for different substrates (glucose, acetate, and pyruvate) and growth (aerobic vs. anaerobic) conditions [1]. Model parameterization is carried out using a machine-learning genetic algorithm by simultaneously imposing all available fluxomic data. Overall, the predicted product yields by k-ecoli457 achieve significantly higher value of correlation with experimental data (i.e., Pearsonâ€™s correlation coefficient of 0.84) than FBA, MOMA or maximization of product yield (i.e., 0.18, 0.37 and 0.11, respectively).

In addition, we reconstructed a second-generation genome scale metabolic model (*i*Cth446) for *C. thermocellum* by correcting cofactor dependencies, restoring elemental and charge balances and updating GAM and NGAM values to improve phenotype predictions. The *i*Cth446 model was next used as a scaffold to develop k-ctherm118, a core kinetic model of *C. thermocellum* metabolism [2]. k-ctherm118 parameterization was carried out by simultaneously imposing fermentation data in lactate, malate, acetate and hydrogen production pathways for 19 measured metabolites spanning a library of 22 distinct single and multiple gene knock-out mutants. k-ctherm118 pointed at a systemic effect of limiting nitrogen resulting in increased yields for lactate, pyruvate and amino acids and an increase in ammonia and sugar phosphates concentrations (~1.5 fold) due to down-regulation of fermentation pathways under ethanol stress. The changes in the concentrations for fourteen out of eighteen metabolites in a *â^†ldh* mutant compared to wild-type were

correctly predicted by k-ctherm118. These results quantitatively demonstrate that the developed kinetic models can reliably be used to predict genetically perturbed phenotypes under different growth conditions with a higher accuracy than any other earlier modeling effort.

[1] Khodayari, A., Maranas, C. **k-ecoli457: A genome-scale Escherichia coli kinetic metabolic model satisfying flux data for multiple mutant strains**. submitted.

[2] Dash, S., Khodayari, A., Lynd, L., Maranas, C. **Characterization of Clostridium thermocellum strains with disrupted fermentation end-product pathways**. in preparation.

Copyright © American Institute of Chemical Engineers. All rights reserved.

461252 Optimizing Ensemble Modeling Framework to Generate Kinetic Models of Metabolism

Abstract:

Enabling kinetic and regulatory modeling of cellular metabolism is a major challenge in metabolic engineering and systems biology. Constraint-based stoichiometric modeling greatly aids in characterizing and improving strain designs but without kinetic information it is difficult to identify rate limiting steps and interrogate regulatory behavior. Unfortunately, *in vitro* derived kinetic parameters for enzymes do not necessarily reflect true *in vivo* behavior and are often determined under varying experimental conditions. Consequently, a single kinetic model combining these *in vitro* derived parameters is often unable to resolve experimentally observed *in vivo* data. The ensemble modeling (EM) framework was previously developed to address these hurdles by sampling kinetic parameters for the entire metabolic network simultaneously and screening them against a single experimental dataset.[1] Furthermore, the EM method constrains the large kinetic parameter sample space using readily available thermodynamic, stoichiometric, and steady state flux data. However, despite its numerous advantages, EM becomes computationally limiting with increasing network size and complexity. As we increase the availability of larger stoichiometric metabolic models for various organisms, it is imperative to improve our ability to generate larger kinetic models of these systems as well.

In this work we build on the previous developments in EM by optimizing parameter screening techniques and introducing methods to reduce structural model complexity. We have investigated the trade-offs between model fitness and computation time associated with additional parameter sample space constraints, alternative model screening methods, and reduced kinetic rate law forms. We have also implemented a conservation analysis step to eliminate linear dependency in our models and reduce the stiffness and screening time of our kinetic parameter sets. Most importantly, our work toward optimizing the existing EM method can directly plug into and benefit concurrent EM efforts to develop genome-scale kinetic models across our field.[2] Specifically, through reducing computation time and optimizing parameter sampling and model screening, we are better equipped to more extensively explore the larger parameter sample spaces associated with larger metabolic networks. Lastly, we hope our combined findings will serve as a useful starting point for future EM users to select the best sampling and screening method options to use in their own, unique applications.

[1] Tran, L. M., Rizk, M. L., & Liao, J. C. (2008). Ensemble modeling of metabolic networks. *Biophysical journal*, *95*(12), 5606-5617.

[2] Khodayari, A., Zomorrodi, A. R., Liao, J. C., & Maranas, C. D. (2014). A kinetic model of Escherichia coli core metabolism satisfying multiple sets of mutant flux data. *Metabolic engineering*, *25*, 50-62.

Copyright © American Institute of Chemical Engineers. All rights reserved.

468512 Integrated Analysis of the Trade-Off Between Thermodynamics and Yield in Non-Natural Chemical Biosynthesis Pathways

Abstract:

Metabolic pathway design is essential to enabling the production of valuable non-natural chemicals by microorganisms. There are usually multiple pathways capable of producing a chemical of interest. However, there is likely one or a small set of pathways that is optimal from the perspective of yield, titer and productivity.

Computational pathway prediction methods have become increasingly common in metabolic engineering. Given some reactants and a desired final product, these methods enumerate all possible biosynthetic pathways capable of taking the former to the latter. They are combinatorial in nature and result in a multitude of pathways to the product. For example, the Biopathway Predictor developed by Genomatica identified 10,000 routes from a central metabolite to 1,4-butanediol in four to six steps. The brute force approach to pathway design is experimental characterization. This can be done when the number of pathways is small, but it becomes unmanageable when the number of pathways is greater than a few hundred. In addition, there is a significant time and monetary investment associated with several parallel experimental pathway implementations. Hence, there is a requirement for in silico pathway prioritization methods to identify from the multitude of pathways, the optimal one that balances the objectives of yield, titer and productivity.

The most common metric for the in silico comparison of metabolic pathways is maximum theoretical yield. However, cells have evolved to maximize their growth rather than the production of non-natural chemicals, so the yield of implemented pathways often falls short of the maximum. This restricts the use of maximum theoretical yield to the identification of pathways able to reach yield values above a threshold. However, in addition to the maximum theoretical yield, thermodynamic driving force and kinetic bottlenecks can also limit the flux through these non-natural pathways and hence, additional metrics that consider thermodynamics and kinetics are required to further reduce the number of candidate metabolic pathways.

We apply an integrated metric that leverages pathway thermodynamic and kinetic data for the analysis of non-natural pathways, using 1,4-butanediol pathways as a case study. We develop methods for assessing the variation in the enzyme kinetics of these non-natural chemical biosynthesis pathways and quantify the sensitivity of the metric to errors in this estimation. Furthermore, we use the results from the application of this metric to identify infeasible or rate limiting factors which can be used to direct engineering strategies for pathway optimization. We believe that this metric will enable the efficient prioritization of non-natural chemical biosynthesis pathways and accelerate the development of commercially viable microbial strains.

Copyright © American Institute of Chemical Engineers. All rights reserved.

470815 Thermodynamic Analysis of the Rigidity of Metabolic Nodes Via a Dynamic Flux Balance Approach

Abstract:

The knowledge of operation and regulation of the fluxes in metabolic networks enables rational design and manipulation of cell metabolism for medical or industrial purposes. To understand the functions and flexibility of metabolic networks, quantification of time-dependent changes of intracellular metabolite concentrations and metabolic turnovers under various internal and external perturbations are crucial. However, experimental determination of intracellular metabolite concentrations is tedious and challenging.

Our approach enables the elucidation of the dynamic relationship among the reaction driving forces (Gibbs free energies), metabolite pool sizes and flux rates. This allows us to quantify the relative rigidity of different metabolite nodes as a function of the net thermodynamic driving forces. We demonstrate our method using a published genome scale *E. coli*model (iAF1260: http://systemsbiology.ucsd.edu/InSilicoOrganisms/Ecoli/EcoliSBML) by following steps. First, we perform a dynamic flux balance analysis. This step employs a pseudo steady state technique using simple kinetic growth experimental data as constraints, which predicts the time-rate change of intracellular concentrations. Second, the metabolite concentration bounds can be narrowed by using Gibbs free energy constrained flux balance analysis for each pseudo steady state. Third, the dynamics of intracellular metabolite concentrations and the time-rate change of Gibbs free energy are resolved using a set of self-consistent calculations based on the thermodynamic equilibrium relationship.

Our computational framework employs Gibbs free energies for revealing metabolite pool sizes and node rigidity under different genetic and environmental perturbations. The simulations are largely consistent with the reported experimental measurements of intracellular metabolites. Our methodology used in conjunction with other constraint based techniques provides guidelines for metabolism analysis and rational metabolic engineering.

Copyright © American Institute of Chemical Engineers. All rights reserved.

467256 Analyzing Protein Glycosylation in Mammalian Perfusion Cell Culture Using Glycosylation Flux Analysis

Abstract:

In recent times, perfusion cultures of mammalian cells, particularly Chinese hamster ovarian (CHO) cells, have been drawing a growing interest as an avenue for continuous biopharmaceutical production. Perfusion cultures provide several benefits over traditional fed-batch processing such as higher yield and improved product quality. In the production of monoclonal antibodies (mAb), N-linked glycosylation is among the most important product quality attributes. Such post-translational modification has strong effects on the biological activity and therapeutic efficacy of these drugs. While the macroscopic process variables such as the viable cell density and the concentrations of main metabolites and products in the media are measured throughout the cell culture operation, there is little or no direct measurements on intracellular variables. For this reason, model-based data analysis such as metabolic flux analysis is often required to understand intracellular changes during cell culture production.

In this work, we extended the traditional metabolic flux analysis (MFA) to study N-linked glycosylation process during the operation of a perfusion CHO cell culture reactor to produce immunoglobulin gamma (IgG) antibodies. Like MFA, this novel analysis, that we called glycosylation flux analysis (GFA), is a flux-based technique that relies on a stoichiometric model of the reaction network in the protein glycosylation to predict reaction rates or fluxes, under the assumption of pseudo steady-state condition. In dynamic GFA (DGFA), the steady-state assumption is relaxed by allowing fluxes to vary linearly between time points, as done in dynamic MFA [1]. We employed GFA as well as DGFA to link the extracellular measurements of cell viability, IgG titer and fractions of glycoforms to the intracellular changes in the glycosylation reaction network during long term continuous cultivation.

The overall goal of our study was to understand the regulation of the glycosylation process in CHO cells during perfusion culture, particularly during the initial phase and when going through transitions over different operating conditions. The results of (D)GFA showed that while macroscopic variables such as the viable cell density could be controlled at a desired set-point, intracellular processes such as glycosylation varied over time. This time dependent change in fluxes through the glycosylation network suggested that the viable cell density might not be the most appropriate controlled variable for optimizing the perfusion culture performance while maintaining a constant product quality. By computing flux conversions, defined as the ratio between an efflux and the total influx around a glycoprotein in the network, we linked changes in the activities of the pertinent enzymes to external macroscopic variables [2]. The acquired knowledge about the dynamics of glycosylation process and the enzymes involved will be applied for further process optimization and the control of mAb product quality.

[1] Leighty, R.W., Antoniewicz, M.R., 2011. *Dynamic metabolic flux analysis (DMFA): a framework for determining fluxes at metabolic non-steady state.* Metab. Eng.

[2] Hang, I., Lin, C., Grant, O.C., Fleurkens, S., Villiger, T.K., Soos, M., Morbidelli, M., Woods, R.J., Gauss, R., Aebi, M. (2015.) *Analysis of site-specific N-glycan remodelling in the ER and the Golgi.* Glycobiology.

Copyright © American Institute of Chemical Engineers. All rights reserved.

474039 Modeling Cell Signaling Control of Epithelial-Mesenchymal Transition

Abstract:

Epithelial-mesenchymal transition (EMT) is a normal cellular developmental program that occurs aberrantly in many cancers and has been associated with metastasis and chemoresistance. This understanding has motivated efforts to identify the central molecular drivers of EMT. This is a challenging undertaking, however, because EMT is clearly regulated by complex multivariate processes, including signaling perturbations due to oncogenic mutations, complex transcriptional regulatory programs, and mechanical stimuli. Growth factor-mediated signaling is also centrally involved, with the combined activation of growth factor receptors in different families typically leading to the most complete EMT. This complexity suggests that it is unlikely that any individual signaling pathway exerts exclusive control over EMT. We undertook an unbiased and systematic analysis of the cell signaling processes leading to EMT. Specifically, we developed a partial least squares regression (PLSR) model to predict the quantitative relationships between multivariate signaling events and EMT-associated phenotypes using a cell culture model of EMT in response to transforming growth factor beta, epidermal growth factor, hepatocyte growth factor, or combinations of these ligands. Using multiplexed Luminex measurements, the phosphorylation states or expression levels of 32 distinct signaling pathway proteins (initially identified by antibody microarray) were measured at six time points over a 48 hr time course in a human pancreatic cancer cell line treated with five different ligand treatment conditions. In parallel, phenotypic measurements were made including the expression or localization of vimentin and E-cadherin and measurements of cell scatter from epithelial clusters. Using these data, PLSR was implemented, with the initial model being refined through several rounds of trimming redundant or minimally informative information. The final PLSR model identified ten distinct signaling measurements predicted to be most important for EMT. The importance of these signaling proteins in the human cell line used to build the model was validated in two-dimensional cell culture and three-dimensional cell spheroid studies using targeted inhibitors and siRNA reagents. The results were further validated in primary mouse pancreatic carcinoma cell lines not used in model development. In both human and mouse cell lines, targeting many of the individual signaling proteins identified by the model partially inhibited EMT, but combinations of inhibitors were substantially more effective. Indeed, specific combinations of inhibitors demonstrated an extraordinary ability to antagonize spheroid invasion in collagen in response to ligands. EMT inhibition using the same combinations of inhibitors also increased the sensitivity of murine cell lines to the EGFR inhibitor gefitinib by an order of magnitude. Our results identify a network of critical cell signaling processes required for efficient EMT. The new understanding gained provides a basis for multiple lines of future investigation, including the analysis of cell-to-cell variability in EMT and the rational design of durable combination therapies for cancer.

Copyright © American Institute of Chemical Engineers. All rights reserved.

453122 Developing Therapeutic Medicines through Chemical Engineering Innovations

Abstract:

Over the past 40 years, Genentech scientists and engineers have pioneered and exploited the use of genetic engineering to develop and manufacture recombinant proteins and monoclonal antibody therapeutics that have revolutionized medicine. Building on this legacy of innovation, chemical engineers in the laboratory, in pilot plants, and in manufacturing continue to play a key role in transforming the latest scientific breakthroughs into innovative products for patients. This presentation highlights various applications of chemical engineering in the development of technologies to enable novel and breakthrough therapeutic medicines and products.

Copyright © American Institute of Chemical Engineers. All rights reserved.

The Automated Design of Multi-Enzyme Pathways for Efficient and Scalable Optimization in High-Dimensional Expression Spaces

Sean Halper

Pennsylvania State University

AICHE 2016

PennState

A Grand Challenge in Metabolic Engineering

How do we best design heterologous pathways for maximum downstream productivity?

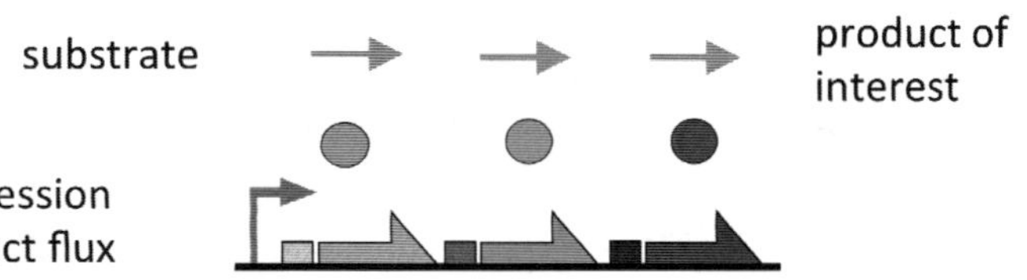

Tune enzyme expression to maximize product flux

State of the Art:
Iterative Design
Build Test Cycle

ANALYZE

DESIGN libraries of genetic parts to vary enzyme expression

BUILD pool of pathway variants with varying enzyme expression

TEST pathway variants for optimal productivity

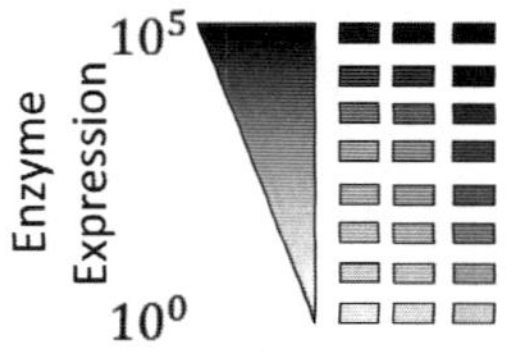

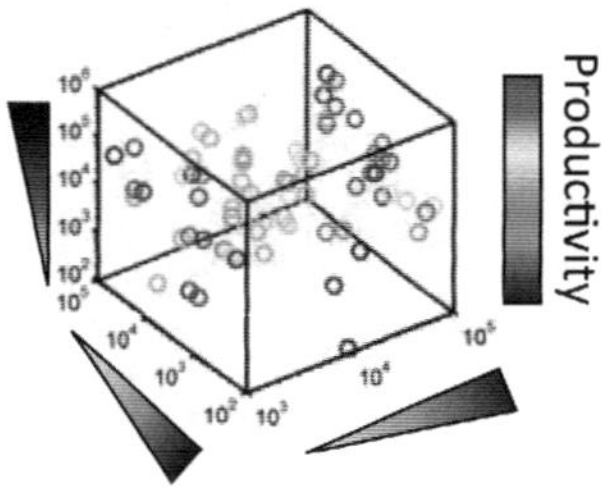

A Grand Challenge in Metabolic Engineering

How do we best design **large** heterologous pathways for maximum downstream productivity?

3 enzymes

9 enzymes

As number of enzymes increases...

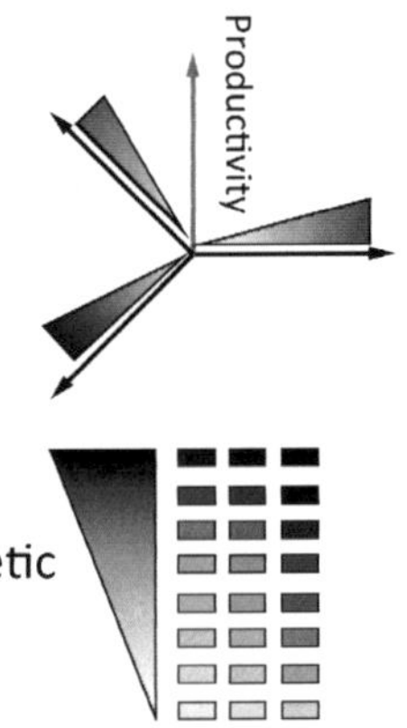

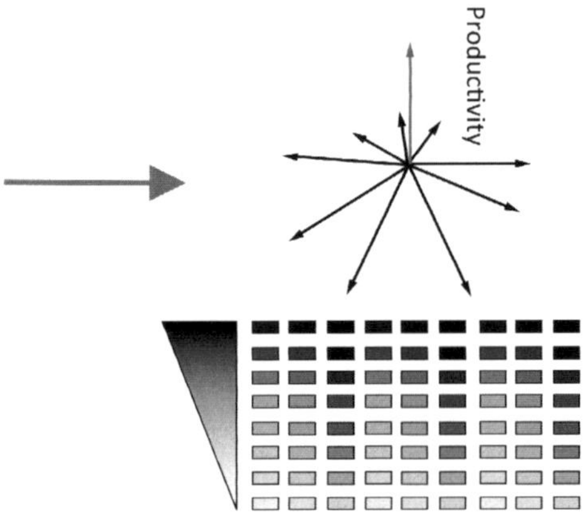

Number of genetic parts required increases

512 possible pathway variants

134 billion possible pathway variants

High dimensionality of larger pathways provides experimentally challenging problem for traditional DBT

3

A Grand Challenge in Metabolic Engineering

How do we address this combinatorial problem?

Biophysical models linking promoter/ RBS sequences to enzyme expression

Kinetic modeling approaches linking enzyme expression to productivity/titer

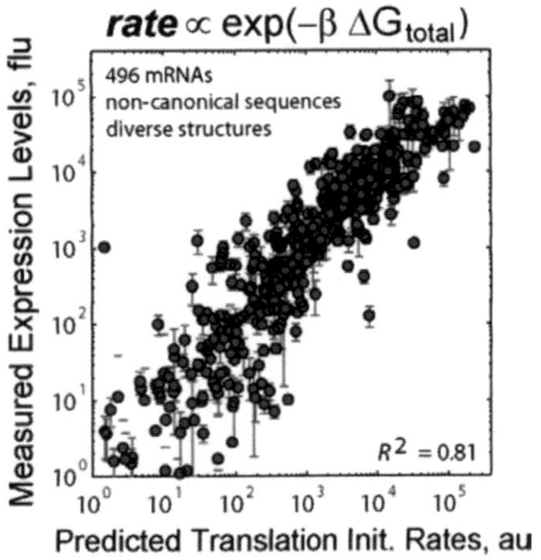

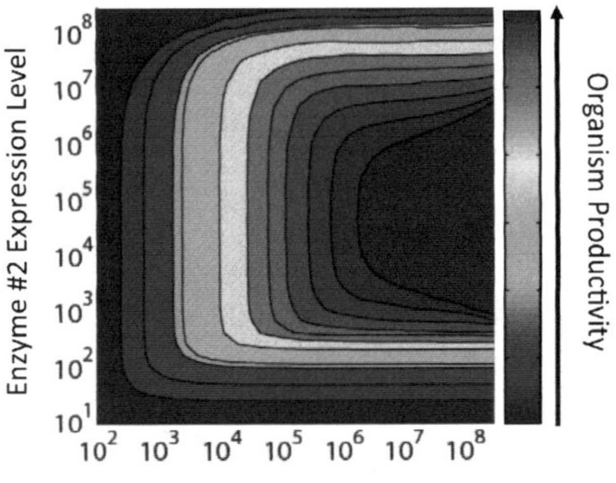

Espah Borujeni & Salis, *JACS*, 2016

Farasat et al, *Molecular Systems Biology*, 2014

4

An Automated Computational Pipeline for Pathway Optimization

By incorporating rational design approaches into a unified workflow, computational pathway design can yield sequence-optimized pathway variants in 2 DBT iterations

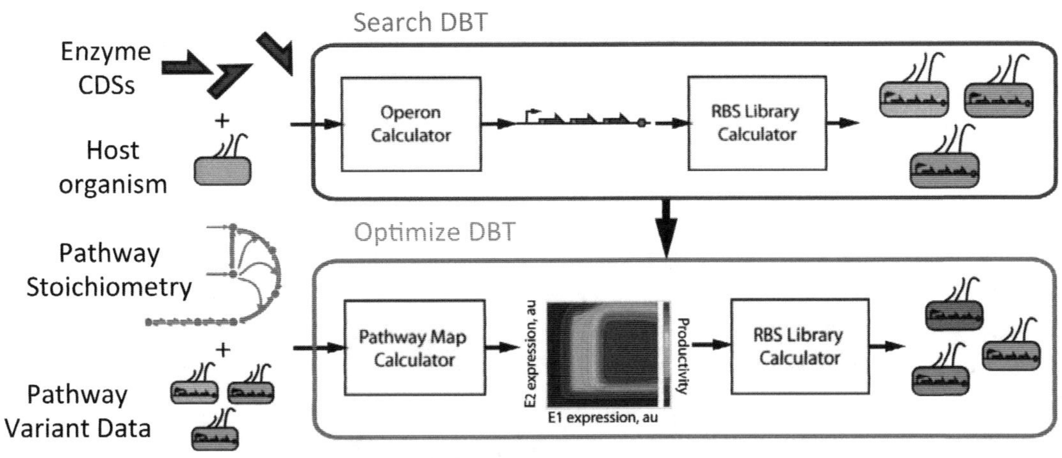

5

An Automated Computational Pipeline for Pathway Optimization

We have applied the pipeline to two metabolic pathways, engineering the sequence of each for maximum productivity using <100 pathway variants

Example 1: crtEBI Pathway
Terpenoid production

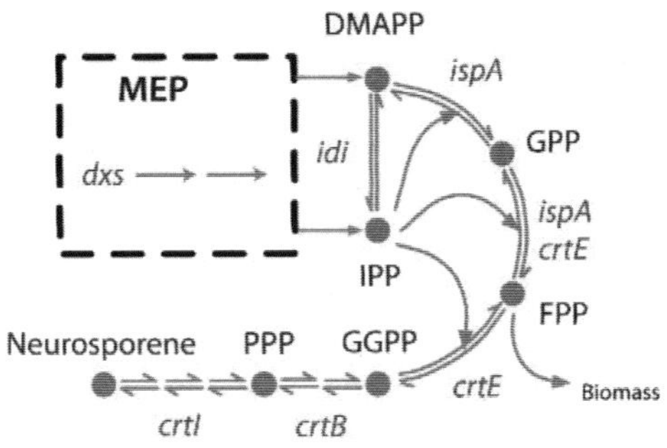

Farasat et al, *Molecular Systems Biology*, 2014

Example 2: Entner Doudoroff Pathway
NADPH Regeneration

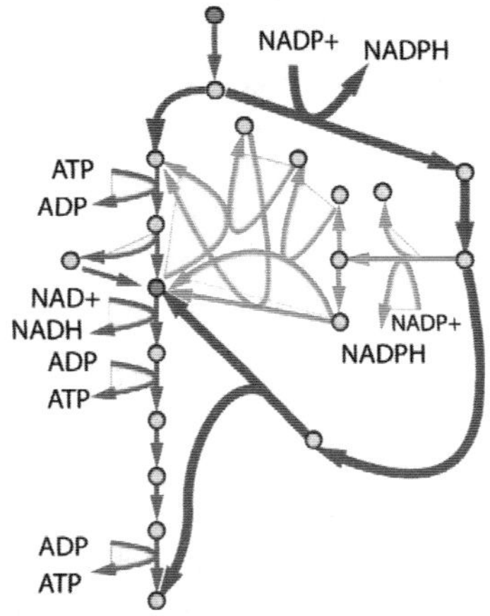

Ng et al, Metabolic Engineering, 2015

6

The Operon Calculator

Objective: Design an operon sequence for your genetic system of interest with reliable expression of a desired compound

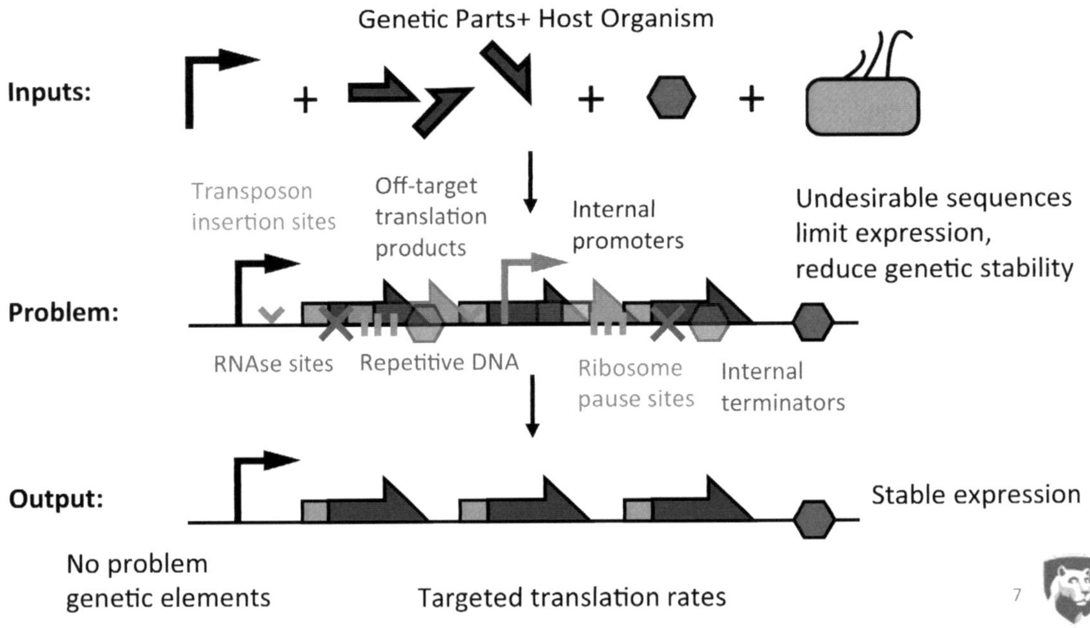

7

The Operon Calculator

Using a genetic algorithm and rational design rules, the algorithm optimizes the sequence of the operon for maximum productivity and longevity

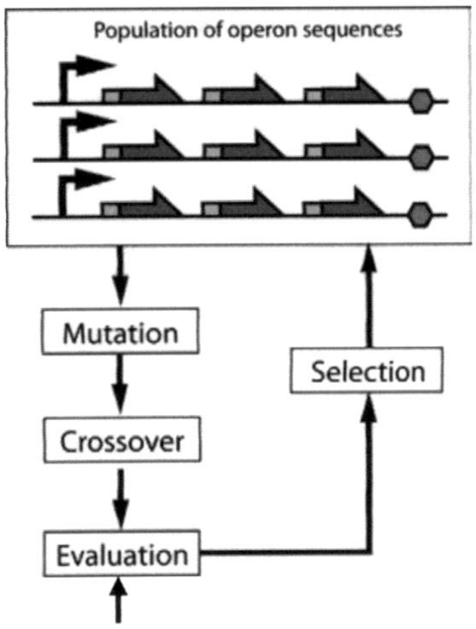

15 Design rules:
- Evolutionary Robustness
- mRNA Stability
- High translation rates

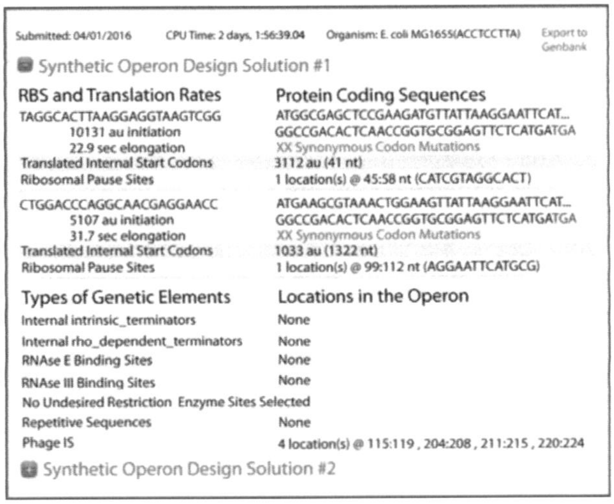

Multiple optimized operon solutions

8

The Operon Calculator

Assembled ED pathway of Z mobilis for stable, inducible expression on two operons

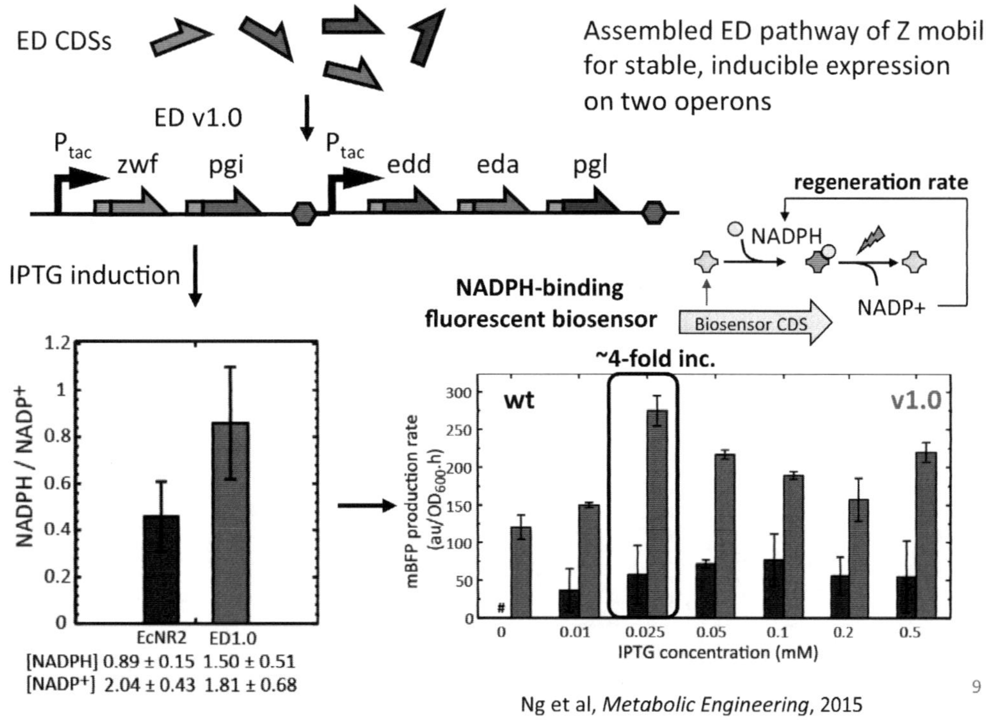

Ng et al, *Metabolic Engineering*, 2015

9

The RBS Library Calculator

Objective: Vary the expression of each enzyme in an operon over a wide dynamic range using as few cloning steps as possible

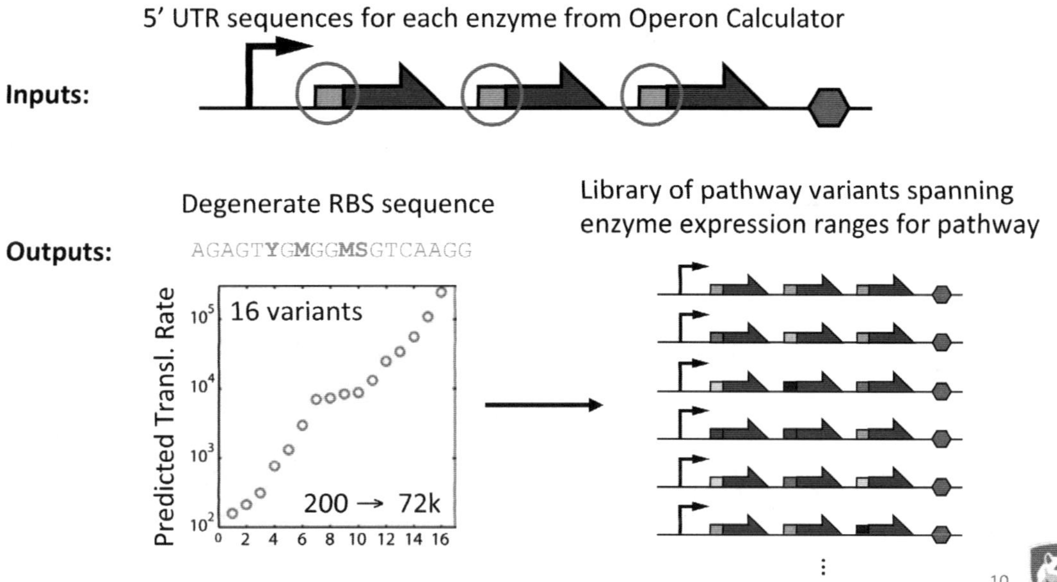

10

The RBS Library Calculator

Using a biophysical model of ribosome binding and a genetic algorithm, the RBS Library Calculator generates minimally degenerate RBS libraries spanning a wide range of translation initiation

Ribosome-mRNA free energy model

$$\Delta G_{total} = \Delta G_{mRNA:rRNA} + \Delta G_{spacing} + \Delta G_{start} + \Delta G_{standby} - \Delta G_{mRNA}$$

$$r_{translation} \propto \exp(-\beta \Delta G_{tot})$$

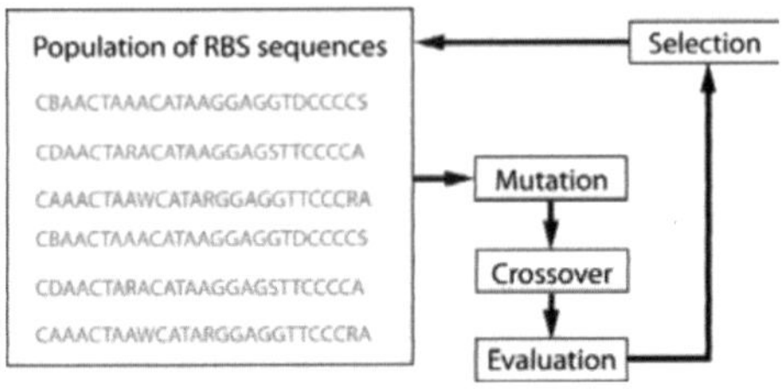

Reliable in diverse organisms

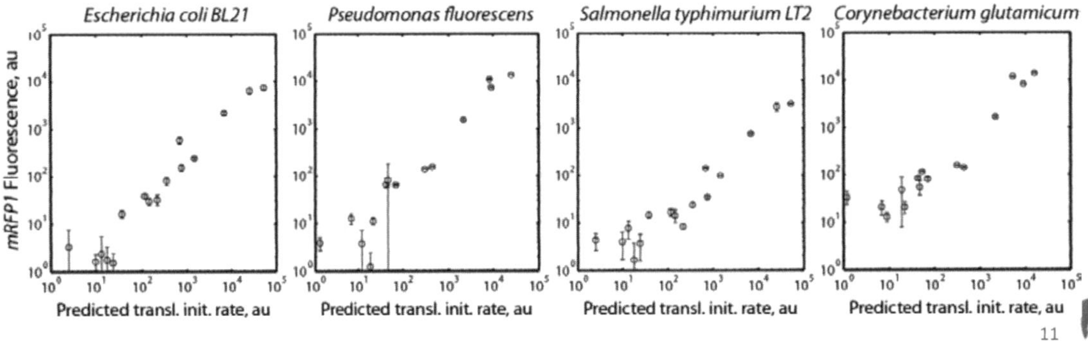

11

The RBS Library Calculator

Using the RBS Library Calculator, we produced RBS libraries spanning wide translation initiation rates, resulting in diverse pathway activities

crtEBI Pathway

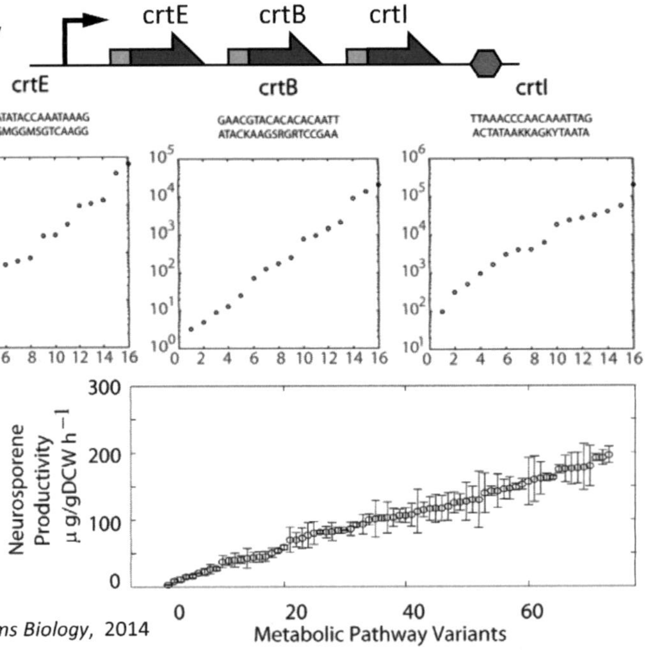

16 member degenerate libraries

73 member pathway variant library

Farasat et al, *Molecular Systems Biology,* 2014

12

The RBS Library Calculator

Using the RBS Library Calculator, we produced RBS libraries spanning wide translation initiation rates, resulting in diverse pathway activities

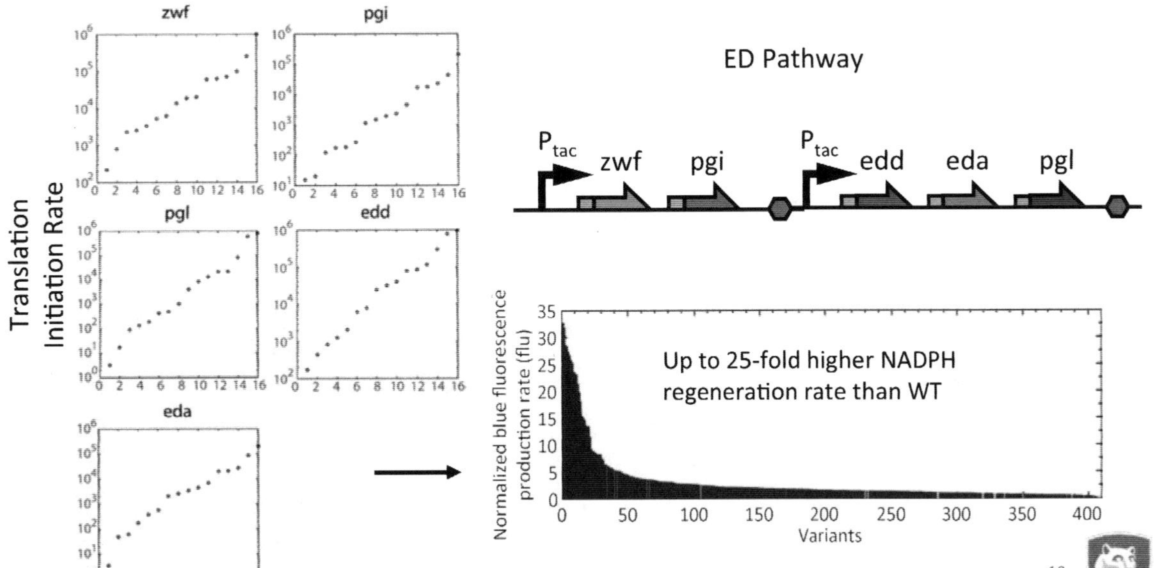

Ng et al, *Metabolic Engineering*, 2015

13

The Pathway Map Calculator

Objective: Map the complex relationship between expression of each enzyme in a pathway and the pathway's final productivity

Inputs:

Expression levels:
RBS translation initiation rates from RBS Library Calculator

Productivity Measurements:
GC-MS
LC-MS
HPLC
Colorimetric/fluorescent measurements

Reaction Network:
Stoichiometric and mechanistic data for the pathway

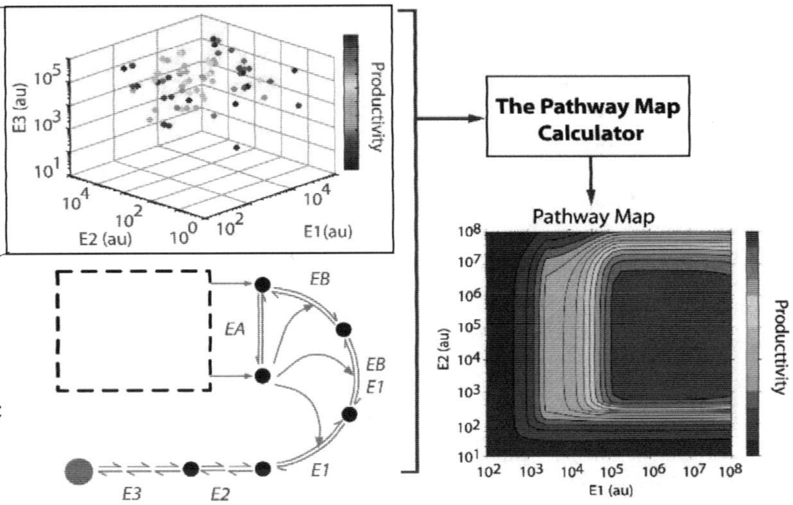

Halper, Farasat & Salis, *in prep*

Output: Pathway Map linking enzyme expression and productivity

14

The Pathway Map Calculator

Using a genetic algorithm and dedimensionalization, the Pathway Map Calculator predicts the kinetic parameters to best fit the kinetic model to the pathway variant data

$$V_{f,1} = k_{f,1}[X][E_1] = k_{f,1}[X^{ref}][E_{1,tot}^{ref}]\frac{[X]}{[X^{ref}]}\frac{[E_1]}{[E_{1,tot}^{ref}]} = \hat{k}_{f,1}[\hat{X}][\widehat{E_1}]$$

$$V_{r,1} = k_{r,1}[E_1X] = k_{r,1}[E_{1,tot}^{ref}]\frac{[E_1X]}{[E_{1,tot}^{ref}]} = \hat{k}_{r,1}[\widehat{E_1X}]$$

$$R_1 = \frac{V_{f,1} - V_{r,1}}{V_{f,1} + V_{r,1}} = \frac{V_{net,SS,1}}{V_{f,1} + V_{r,1}}$$

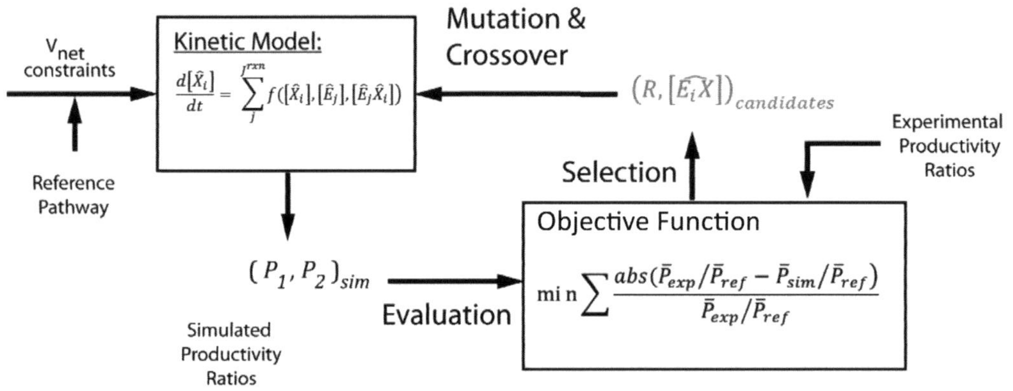

15

The Pathway Map Calculator

The Pathway Map Calculator was used to optimize the productivity of the crtEBI pathway for neurosporene production

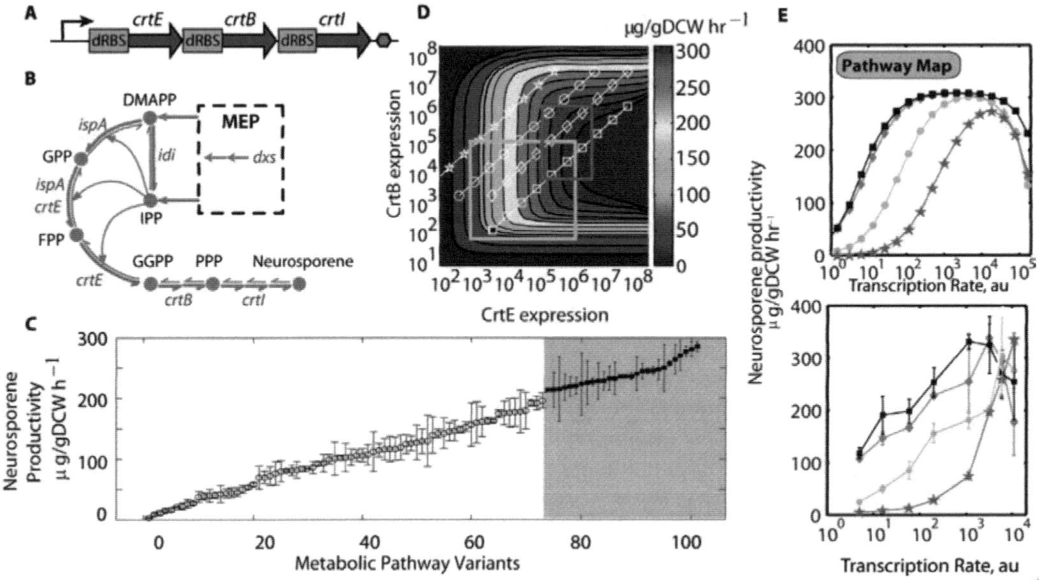

Final productivity of 517 ug/gDCW/hr

Farasat et al, *Molecular Systems Biology*, 2014

16

The Pathway Map Calculator

The Pathway Map Calculator was used to map the Entner Douroff pathway to maximize the NADPH regeneration rate of *E coli*

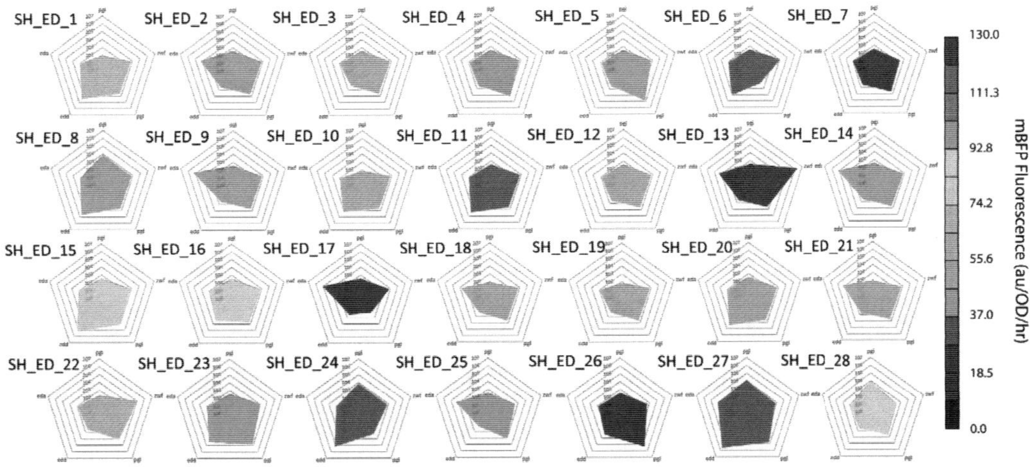

17

The Pathway Map Calculator

The Pathway Map Calculator was used to map the Entner Douroff pathway to maximize the NADPH regeneration rate of *E coli*

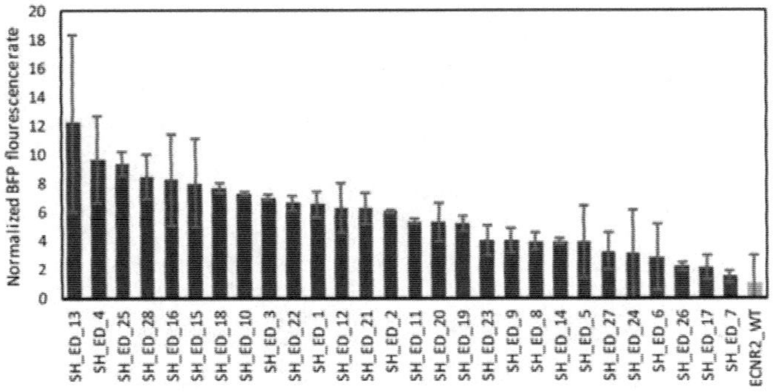

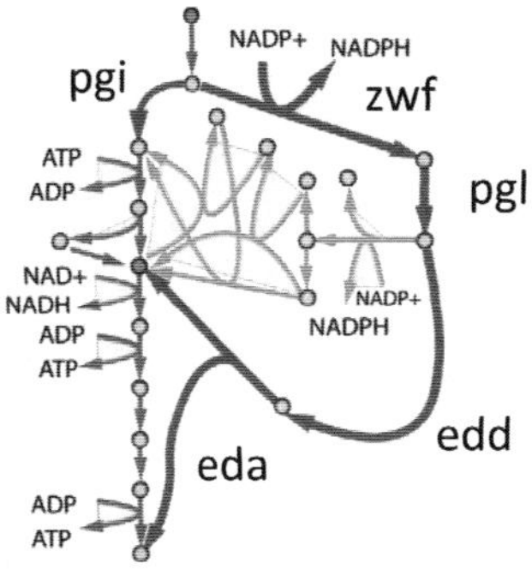

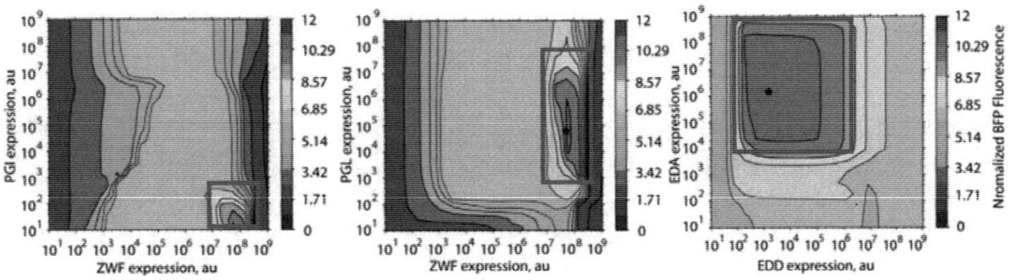

18

Accessing the Pipeline

Interested in our pipeline? Check us out online

Academic: salislab.net/software

Commercial: denovodna.com/software

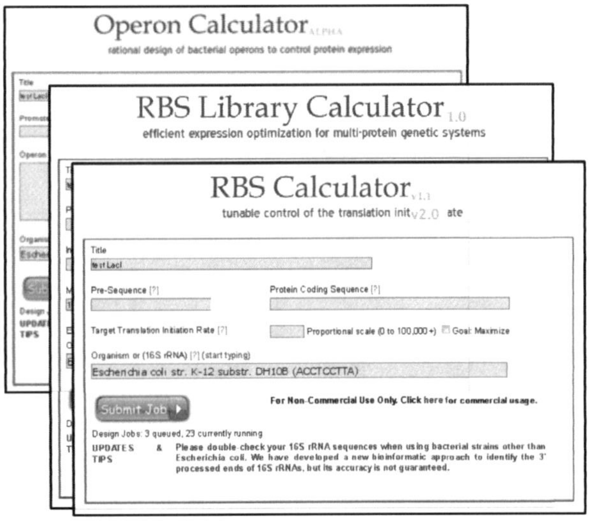

Pathway Optimization & Circuit Engineering

Xylanase and glucanase expression (Liu et. al., 2012)
Nitrile hydratase expression (Kang et. al., 2014)
Secreted protein expression (Heggeset et. al. 2013)

Third–party validation (Biggs et. al., 2014)

FR900098 biosynthesis (Freestone et. al., 2015)
6-ACA biosynthesis for nylon (Zhou et. al., 2015)
Xylitol biosynthesis (Su et. al., 2015)
Riboflavin biosynthesis (Lin et. al., 2014)
Terpenoid biosynthesis (Zhou et. al., 2014)
Amorphadiene biosynthesis (Nowroozi et. al., 2014)
2,3-butanediol biosynthesis (Oliver et. al., 2014)
Engineering memory genetic devices (Yang et. al., 2014)
Butanol biosynthesis (Lim et. al., 2013)
Fatty acid biosynthesis (Lennen et. al., 2013)
Spectinabilin biosynthesis (Shao et. al., 2013)
Nitrogen fixation (Temme at. al., 2012)
Isoprene biosynthesis (Zurbriggen et. al. 2012)

19

Acknowledgements

Graduate students
Sean Halper
Grace Vezeau
Alexander Reis
Daniel Cetnar
Melissa Vidal
David Samson

Undergraduate students
Phillip Clauer, Danielle Hand,
Kavya Vaseekaran, Lipika Gadila

Alumni
Dr. Tian Tian
Dr. Manish Kushwaha
Dr. Amin Espah Borujeni
*Dr. Iman Farasat
Long Chen
*Chiam Yu Ng

Current Funding

ONR Revolutionary Research... Relevant Results

AFOSR AIR FORCE OFFICE OF SCIENTIFIC RESEARCH

NSF

psiee penn state institutes of energy and the environment

464856 Engineering Transcriptional Interference for Designing Synthetic Gene Circuits

Abstract:

An important goal of synthetic biology involves the extension and standardization of novel biological elements for applications in medicine and biotechnology. Transcriptional interference, occurring in sets of convergent promoters, offers a promising mechanism for building elements for the design of tunable gene regulation. Here, we investigate the transcriptional interference mechanisms of antisense roadblock and RNA polymerase traffic in a set of convergent promoters as novel modules for synthetic biology. We show examples of elements, including antisense roadblock, relative promoter strengths, inter-promoter distance, and sequence content that can be tuned to give rise to repressive as well as cooperative behaviors, therefore resulting in distinct gene expression patterns. Our approach will be useful towards engineering new biological devices and will bring new insights to naturally occurring *cis*-antisense systems. Therefore, we are reporting a new biological tool that can be used for synthetic biology.

Copyright © American Institute of Chemical Engineers. All rights reserved.

458544 Repurposing Endogenous Crispr-Cas Bacterial Immune Systems for Programmable Gene Repression

Abstract:

CRISPR-Cas systems offer powerful and versatile biomolecular tools with many applications including genome editing, gene regulation, sequence-specific antimicrobials, and imaging. Because these RNA-directed immune systems are found in most prokaryotes, an opportunity exists to harness endogenous CRISPR-Cas systems as convenient tools in these organisms. We report that the Type I-E CRISPR-Cas system in the bacterium *Escherichia coli* can be co-opted for programmable transcriptional repression. Deletion of the signature *cas3* gene converted this immune system into a programmable gene regulator capable of reversible and multiplexable silencing of heterologous and endogenous genes.

Utilizing this platform, we explored how the length of the CRISPR RNA impacts the form and function of the Type I-E effector protein complex called Cascade. We discovered that the length of the Type I-E CRISPR RNA molecule is not fixed and can be extended, altering the stoichiometry of the Cascade effector complex while preserving functionality. Interestingly, altering RNA length can elicit significant improvement in transcriptional silencing efficiency, but only for particular target locations. These findings heighten our understanding of CRISPR-Cas systems as they continue to increase in popularity and may offer insight into designing tunable transcriptional regulators.

Copyright © American Institute of Chemical Engineers. All rights reserved.

451929 Reprogramming MHC Specificity of Immune Cells By Crispr-Cas9 Assisted Cassette Exchange

Abstract:

Allogeneic cellular transplantation is widely employed to treat various genetic diseases and hematological malignancies. Locating suitable donors for these procedures is often challenging, as MHC/HLA gene alleles need to be matched in order to prevent transplant rejection. We report here a proof of concept study for an ex-vivo CRISPR-Cas9-based approach for the precise exchange of MHC alleles (~5kb) at the native genomic locus. For initial evaluation, the method was performed on immortalized mouse antigen-presenting cells (RAW264.7 macrophages) expressing high levels of surface MHC (H2-Kd). Genomic exchange at the H2-K locus was achieved via co-transfection of cells with a plasmid carrying Cas9 and guide RNA targeting the H2-Kd locus and a homology donor repair template encoding an alternate MHC allele (H2-Kb). Following homology directed repair, modified cells were isolated by flow cytometry based on expression of new MHC. Engineered cells expressing alternate H2-Kb allele were able to present a model antigen peptide (OVA-SINFEKLL) and activate a cognate T cell hybridoma line, demonstrating they were active for downstream immune functions. We envision this approach could be used in the future to improve MHC/HLA matching in cellular transplantation.

Copyright © American Institute of Chemical Engineers. All rights reserved.

466806 Pmpes: A Modular Peptide Expression Vector for the Production of Antimicrobial Peptides from Probiotic *E. coli*

Abstract:

Antimicrobial peptides (AMPs) are a diverse class of small proteins with antibacterial activity against a wide array of pathogens. AMPs have long been used in the food industry as preservatives and are currently being explored for use in the medical realm. Unfortunately, AMP delivery has been a challenge; AMPs may be toxic when administered intravenously, or they quickly degrade in the stomach and the gastrointestinal (GI) tract upon oral administration.

To overcome this challenge, we are engineering probiotic bacteria to produce and secrete AMPs inside host GI tracts. AMPs are chosen or engineered to target medically-relevant pathogens, which reside inside GI tracts, such as Salmonella and Vancomycin-resistant enterococcus (VRE).[1–3] These probiotics are safe to administer orally, are bile-resistant and will reach the site of infection, where they will eliminate the targeted pathogen.

A goal of ours is to develop a library of antibiotic probiotics that can target a wide variety of GI tract pathogens. The development of a bacterial vector that can powerfully express and secrete a wide array of AMPs would be an important breakthrough in order to efficiently engineer this library of probiotics for AMP delivery. However, the issue of secretion poses a major challenge in implementing and testing new peptides from a single bacterial production system. In most naturally-occurring bacterial AMP production systems, each AMP has its own dedicated secretion machinery. For our application, it becomes challenging to construct unique secretion machineries for each new peptide. These systems often comprise several large, poorly characterized proteins. This translates to additional, extensive molecular biology research before they can be employed. Additionally, simultaneous expression of multiple large secretion systems could quickly become a burden on the producer organism.

We have created a powerful AMP expression vector, pMPES, which employs a single previously-characterized *E. coli* AMP secretion system to produce and secrete a wide-array of AMPs from probiotic *E. coli* Nissle 1917. To achieve this, we have remodeled a Microcin V production plasmid, pHK22, so as to create a vector that contains the entire MicV secretion machinery. This plasmid also contains ProTeOn+, a powerful synthetic promoter system we have built and an AMP molecular cloning site.[4] We refer to this vector as pMPES (Modular Peptide Expression System).

In this presentation, we will describe the development of this system and will demonstrate that using this vector, we are able to produce and secrete several AMPs of diverse structure, function, and origin from Nissle 1917. Thus far, we have tested secretion of *E. coli*-derived Microcin V, Microcin N, and Microcin L and of *Enterococcus*-derived Enterocin A, Enterocin P, Hiracin JM79, and Enterocin B. With the exception of Enterocin B, sufficient quantities of all peptides were secreted so as to allow detectable inhibition of the target pathogen. This new expression system enables the rapid development and testing of single or multiple peptide constructs against any given pathogen of interest.

1. Geldart, K., Borrero, J. & Kaznessis, Y. N. A Chloride-Inducible Expression Vector for Delivery of Antimicrobial Peptides Against Antibiotic-Resistant Enterococcus faecium. *Appl. Environ. Microbiol.* **81,** 3889–3897 (2015).

2. Borrero, J., Chen, Y., Dunny, G. M. & Kaznessis, Y. N. Modified Lactic Acid Bacteria Detect and Inhibit Multiresistant Enterococci. *ACS Synth. Biol.* (2014). doi:10.1021/sb500090b

3. Volzing, K., Borrero, J., Sadowsky, M. J. & Kaznessis, Y. N. Antimicrobial peptides targeting Gram-negative pathogens, produced and delivered by lactic acid bacteria. *ACS Synth. Biol.* **2,** 643–50 (2013).

4. Volzing, K., Biliouris, K. & Kaznessis, Y. N. proTeOn and proTeOff, New Protein Devices That Inducibly Activate Bacterial Gene Expression. *ACS Chem. Biol.* **6,** 1107–1116 (2011).

Copyright © American Institute of Chemical Engineers. All rights reserved.

467499 Development of a Quantitative Framework for Tuning the Performance of Cell-Based Device

Abstract:

Mammalian cell-based devices represent a promising emerging technology that leverages the natural capabilities of cells – including sensing, secretion, and biosynthesis – for the treatment of disease. There now exists a substantial set extracellular and intracellular sensor "parts" suitable for engineering mammalian cells to produce a defined output in response to a physiological input. We recently developed one such tool, termed MESA, which is a self-contained receptor-signal transduction platform that enables detection of soluble extracellular cues (via scFv antibody fragment-based ligand binding domains) and in response, modulates the activity of endogenous genes (via a Cas9-based transcription factor). As a proof-of-concept experiment, we utilized these receptors to functionally "rewire" T cells to secrete an immune-potentiating factor when these cells are exposed to an immunosuppressive cue – a functionality that is not observed in nature. Thus, this work demonstrated the possibility of engineering or custom-rewiring cellular input/output for a range of applications. However, how one would best implement such a technology to enable robust performance across various contexts, which of particular important for translational applications, remains an open and important question that applies to MESA as well as many other cellular engineering technologies.

Here, we report the development of a systematic framework for evaluating MESA implementation to facilitate tuning biosensor performance in new contexts and for new applications. We first evaluate the effect of genetic context by utilizing Cas9 to stably integrate MESA constructs at specific genomic locations known to be safe harbor loci, comparing biosensor performance between integration sites. Beyond enabling an examination of the effects of integration location on biosensor function, this approach also enables us to systematically examine biosensor performance characteristics within a genetically identical population of cells. We next investigate how relative expression of MESA chains influences biosensor performance. Utilizing synthetic upstream open reading frames to systematically vary translation rates, we evaluate how variations in protein expression influence receptor function. Collection of this quantitative data could in turn

enable dynamic modeling of the MESA system, facilitating future analysis and design of MESA systems. Overall, this study addresses the general goal of predictably linking implementation methodologies to the performance of engineered cell-based devices, ultimately enabling the development of safer and more robust therapies.

Copyright © American Institute of Chemical Engineers. All rights reserved.

475218 Harnessing Crispr/dCas9 and Polyculture Strategies for Enhancing Natural Product Production in Recombinant Hosts (Featured Presentation)

Abstract:

The field of metabolic engineering is entering a new era in which synthetic transcription factors such as RNAguided-dCas9 are providing the means for dynamic regulation and for rapid assessment of multiplexed genetic interventions to alter the cellular metabolic landscape.

In this talk we will first present a modular assembly method for traditional restriction-ligation cloning of type IIA CRISPR array libraries for multiplex, combinatorial dCas9-mediated transcriptional repression and its application in metabolic engineering of *E. coli* for enhancing flavonoid titers through simultaneous repression of an endogenous transcription factor and several central carbon enzymes, including partial downregulation of a synthetic lethal pair. We will also demonstrate how this method can be expanded by adapting Golden Gate Shuffling for assembly of natural type IIA CRISPR arrays, enabling randomized one-pot assembly of array libraries simultaneously repressing all combinations of a user-defined set of target genes. This approach was tested for improving production of glycosylated natural productsâ€"including the pharmaceutically and nutraceutically valuable polysaccharides heparin and chondroitin and the colorful class of plant natural pigments known as anthocyaninsâ€"through repression of novel knockdown targets and partial downregulation of an essential glycolytic enzyme. Direct comparison of CRISPRi strains with analogous deletion strains (obtained through Î»-red recombineering) shows that CRISPRi can lead to better production improvements compared to gene deletions. Finally, we will demonstrate the construction of orthogonal variants of the classic T7lac promoter using site-directed mutagenesis, generating a panel of inducible hybrid promoters regulated by both LacI and dCas9 and covering a wide expression range. Remarkably, dCas9 orthogonality in our system is mediated by only 23 nucleotide mismatches in a narrow window of the RNA:DNA hybrid, neighboring the protospacer adjacent motif (PAM). We demonstrate that, contrary to many reports, one PAMproximal mismatch is insufficient to abolish dCas9-mediated repression. A subset of these refactored promoters were incorporated into the highly branched violacein biosynthetic pathway, where they act as orthogonal, dCas9-dependent valves capable of throttling and selectively redirecting carbon flux in *E. coli*.

Finally, the presentation will also cover our work on the development and optimization of polycultures (three or more strains in co-culture) for the extension of recombinant pathways, such as the flavonoid branch pathway, *in vivo*. This technology has enabled, for the first time, the *de novo* production of flavan-3-ols and anthocyanins in *E. coli*. Utilizing a computationally guided optimization approach, we were able to demonstrate up to a 970-fold improvement over previously published monoculture titers.

Copyright © American Institute of Chemical Engineers. All rights reserved.

475268 Technology Transfer and Scale-Down Models for Mammalian Cell Culture Manufacturing the Process Knowledge Questions, and the Practical Ones

Abstract:

Fed-batch processing using mammalian cell culture is the most widely used technology for large-scale manufacturing of monoclonal antibodies. Process optimization for the production of monoclonal antibodies is a continued process, potentially involving multiple process generations, and manufacturing sites.

Manufacturing Science groups are a key part of process optimization, as part of their responsibilities is to perform technology transfer from development groups to commercial manufacturing sites, or from a commercial manufacturing site to another. Technology transfer may involve the utilization of lab-scale systems to build process knowledge and to understand performance before manufacturing initiates at large scale. Once commercial scale production is in place, scale-down models are used to monitor and provide real time troubleshooting.

This presentation will focus on the establishment of a scale-down model to support the technology transfer of a new generation of a commercial product to Genentech Vacaville, a multiproduct commercial manufacturing site with the capability of producing monoclonal antibodies in large quantities. Work was performed to optimize the scale-down model, while at the same time provide answers to questions related to manufacturability (e.g., differences in process and impact to run rate, and limitations due to equipment capability). Scale-down model optimization tested the impact of process parameters on performance and product quality, while scale-down model utilization provided answers that enable process fit at large scale.

Copyright © American Institute of Chemical Engineers. All rights reserved.

475264 Metabolic Engineering of *Saccharomyces Cerevisiae* for Increased Production of Opioids

Abstract:

Opioids are the primary drugs used in Western medicine for pain management and palliative care. Farming of opium poppies remains the sole source of these essential medicines despite diverse market demands and uncertainty in crop yields due to weather, climate change, and pests. Researchers at Stanford University recently demonstrated production of the opioid compounds thebaine and hydrocodone from sugar in engineered yeast strains. This milestone achievement combined enzyme discovery, enzyme engineering, and strain optimization to assemble a functional pathway containing over 20 different enzyme activities from plants, mammals, bacteria, and the native yeast host. At Antheia, Inc. we are further developing yeast strains and fermentation processes to translate this proof-of-concept work into a commercially viable venture. The result will be a sustainable, lower cost, and more secure and stable supply of opioid pharmaceutical ingredients.

Copyright © American Institute of Chemical Engineers. All rights reserved.

475253 Mining for Therapeutics from the Microbiome

Abstract:

Bacteria that live in and on us (i.e., the microbiome) make thousands of molecules that have drug-like properties. There is a new revolution afoot to mine for metagenomic functions of interest, clone and express such genes, and test the function of the bioactive (protein/peptide, etc.). in appropriate in vitro and in vivo models. Since many of these bioactives never previously been purified, characterized, and tested in therapeutic doses, this has potential to open up entirely new biology which can be exploited for finding new therapies for many diseases. Building a scalable microbiome drug discovery platform based on this premise requires overcoming key big data and engineering challenges that will be the main focus of this presentation.

Copyright © American Institute of Chemical Engineers. All rights reserved.

475254 Antibiotic Probiotics Reduce Salmonella Enterica in Pre-Harvest Poultry

Abstract:

Chicken has become the most popular type of meat in the world. In the US, per capita consumption has doubled to nearly 90 lbs/year since 1980. The value of chicken broilers produced during 2014 was $32.7 billion, up 6 percent from 2013.

Chicken is also the leading cause of salmonella food poisoning. Despite the vast arsenal of technologies employed to control salmonella in poultry products, these products are the most significant sources of Salmonella enterica that sickens people. In the US, Salmonella enterica is the leading cause of foodborne disease outbreaks, with approximately 1 million infections annually, and calculated health care and lost productivity costs exceeding $3 billion every year. These disconcerting numbers may increase after the FDA-directed phasing out of the use of medically important antimicrobials in food animals for production purposes.

Herein, we report on a new antibiotic technology that can reduce Salmonella enterica in poultry. We have engineered the probiotic strain E. coli Nissle 1917 to produce and secrete the antimicrobial peptide microcin J25. Employing synthetic biology techniques we engineered a DNA promoter for strong, constitutive expression of the peptide. In vitro tests demonstrate the efficacy of the system against Salmonella enterica. The reported animal experiments provide strong proof-of-concept results. Salmonella enterica clear the ceca of poultry faster than in negative control experiments. Approximately 97% lower salmonella carriage is measured in the treated group, by end of the experiment, 14 days post-challenge, compared to the untreated group.

Probiotic bacteria are generally regarded as safe to consume, are bile-resistant and can plausibly be modified to produce many out of the thousands of antimicrobial peptides now known. The reported systems may provide a foundation for platforms to launch antimicrobials against any one of many gastrointestinal tract pathogens, including ones that are multi-drug resistant.

Copyright © American Institute of Chemical Engineers. All rights reserved.

475392 Use of Synthetic sgRNA to Improve CRISPR Editing Efficiency

Abstract:

CRISPR has made genome editing accessible for a wide range of cell-types. However, obtaining consistent editing efficiencies remains a challenge. Synthego has developed novel RNA synthesis technology to produce 100-mer sgRNA for CRISPR at a practical scale and price. We demonstrate that Synthego sgRNA produces consistent and improved genome editing.

Copyright © American Institute of Chemical Engineers. All rights reserved.

481290 From Mind to Molecule: Building the Information Technology Backbone for Synthetic Biology

Abstract:

Biology is now an information science. Our ability to read, write and edit biology has undergone a radical transformation in just the last few years. At TeselaGen we are building software systems that help scientists and managers keep up with our rapidly expanding ability to sequence, synthesize and edit biology. In collaboration with Dow, TeselaGen is building an enterprise platform for synthetic biology that is secure, scalable and expandable and powers rapid, precise and accurate DNA modification. With automatic protocol generation using patented *j5* technology from Berkeley Lab and the best available CRISPR-Cas based techniques, TeselaGen is advancing in our ability to design and build in the biological realm in much the same way that electronic design automation enabled design and build in the electronic realm. TeselaGen is an early stage company that makes its home in the software hub of San Francisco, CA. Our customers include medical, industrial and agricultural companies from around the world that leverage biotechnology to make medicines, enzymes, bio-based chemicals, and help farmers with plants that can better withstand pests and environmental stress.

Copyright © American Institute of Chemical Engineers. All rights reserved.

470239 Discovery and Manipulation of a Native Polyketides Influencing ABE Production in *Clostridium Acetobutylicum.*

Abstract:

The production of ABE (acetone, butanol, ethanol) by the anaerobic bacteria *Clostridium* is a bioprocess capable of converting various renewable substrates into high value solvents and fuels. *Clostridium acetobutylicum* ATCC 824 has long served as a model organism for clostridial ABE production, with an extensive body of literature ranging from fundamental cellular biology to industrial-scale fermentation. Despite the abundance of knowledge concerning this organism, the series of cellular events responsible for the dramatic and somewhat mysterious transition from acid- to solvent- production (deemed the â€œmetabolic switchâ€) remain unclear. Towards addressing this phenomena, we investigated a polyketide synthase (PKS) gene from *C. acetobutylicum* ATCC 824, which has previously been shown to be upregulated directly preceding the metabolic switch in ABE fermentation of this organism. Through targeted gene knockouts, HRMS-guided comparative metabolomics, and extensive structural characterization, we identified a previously undiscovered family of three polyketides synthesized by the PKS gene product. Batch fermentations of PKS-modified strains revealed that while the polyketides are not essential for the metabolic switch (and hence, ABE production), the presence or absence of these compounds could influence both the solvent titers and productivities of *C. acetobutylicum*. Through this work, we hope to highlight the potentially significant role of secondary metabolites in clostridial fermentations, and provide a relatively unexplored strategy for manipulating these industrially-relevant organisms.

Copyright © American Institute of Chemical Engineers. All rights reserved.

463439 Machine Learning-Driven Pathway Optimization: Application of Automation in Industrial Biotechnology

Abstract:

Microbial biosynthesis of fuels and chemicals often requires heterologous expression of metabolic pathways. To achieve the highest efficiency, the pathway flux must be carefully tuned by changing the expression level of each enzyme in the pathway coupled with a high-throughput screening. However, the high-throughput screening is unavailable for most targets. Here, we present a versatile platform for pathway optimization where a machine learning algorithm directs the experimentation to find the highest producing combination with a minimum number of experiments. Compared with other methods like Exterior Derivative Estimation (EDE) used by Lee et al., [1], our machine learning algorithm was able to find the maximum of the test data with less than half of evaluations required by EDE model. As proof of concept, we sought to optimize the lycopene biosynthesis pathway with 24 different expression levels for each gene resulting in more than 13,000 different combinations. The expression level gradient was achieved by mutating both the T7 promoter and ribosome binding sites (RBS). Although we have developed the iBioFAB platform [2], assembly and characterization of all these combinations is very difficult and expensive, if not impossible. By adding the machine learning component and developing an automation friendly lycopene extraction assay, we can achieve an unprecedented level of automation in pathway optimization. This machine learning driven pathway optimization strategy can be applied to optimize any pathway of interest, especially when a high throughput screening method is not available.

1. Lee, M.E., et al., *Expression-level optimization of a multi-enzyme pathway in the absence of a high-throughput assay.* Nucleic Acids Res, 2013. **41**(22): p. 10668-78.

2. Chao, R., Y. Yuan, and H. Zhao, *Building biological foundries for next-generation synthetic biology.* Sci China Life Sci, 2015. **58**(7): p. 658-65.

Copyright © American Institute of Chemical Engineers. All rights reserved.

461757 Engineering Methanogenesis Pathway of Methanogenic Archaea for Biofuel Production

Abstract:

Engineering methanogenesis pathway of methanogenic archaea for biofuel production

Maryam Raeeszadeh-Sarmazdeh, Jacqueline Gonzalez, Wilfred Chen

Department of Chemical and Biomolecular Engineering, University of Delaware, Newark, DE 19716

Methane gas is the second most prevalent greenhouse gas and an abundant source of energy in the United States. Methane is both produced and consumed in marine sediments through anaerobic methanogenesis. Methanogenic archaea primarily use one-carbon substrates such as CO_2, acetate, or methanol for growth and methanogenesis through CO_2 reduction, aceticlastic, or methylotrophic pathways, respectively. The key step in the methanogenesis pathway, reducing methyl-coenzyme M (CH_3-S-CoM) and coenzyme B (HS-CoB) to methane, is catalyzed by Methyl-coenzyme M reductase (MCR). This reaction, which leads to the formation of a heterodisulfide substrate (CoM-S-S-CoB), is exergonic. Therefore, for reverse methanogenesis to occur, the methane oxidation reaction needs to be coupled with an endergonic reaction that provides a thermodynamically favorable situation.

The genetic toolbox in methanogenic archaea is limited. Therefore, we have investigated Nanoluciferase expression, as a sensitive reporter, through episomal expression and integration into the genome in *Methanosarcina acetivorans* . In order to engineer anaerobic Methanogenesis pathway to produce biofuel, we have expressed functional Methyl-coenzyme M reductase (MCR) from *Methanothermobacter marburgensis* in *Methanosarcinals*. The recombinant strains showed improvement in methane formation yield compared to the wild-type strain. Furthermore, the effect of soluble electron acceptors such as Fe(III), and the regeneration of a heterodisulfide substrate on methane oxidation were investigated.

Copyright © American Institute of Chemical Engineers. All rights reserved.

459970 Production of Wax Esters As High-Value Compounds in Yeast *Yarrowia Lipolytica*

Abstract:

Wax esters have a lot of applications in lubricant, cosmetics, inking, and coating industries. Currently the main biological source for wax esters is the seeds of jojoba. Mainly due to the limited supply of jojoba, wax esters are currently produced at a high cost. To alleviate this issue, we introduced the wax ester biosynthetic pathway into the oleaginous yeast, *Yarrowia lipolytica*, which is well known for its superior ability to accumulate triacylglycerol (TAG) under appropriate culturing condition and genetics combinations. To divert the metabolic flux from TAG biosynthesis to wax ester biosynthesis, the genes involved in the wax synthetic pathway from difference sources with unique substrate specificities were tested, and several metabolic engineering strategies have been applied to increase precursor availability and remove substrate competition. The resulting strain produced wax esters at a titer 5-fold higher than the previous record in literature. With a comprehensive optimization on medium recipe, we anticipate a further significant improvement in wax ester titer.

Copyright © American Institute of Chemical Engineers. All rights reserved.

471627 A Microbial Factory for Sustainable Production of Polymer Precursors: Chain Length-Specified ω Hydroxy Fatty Acids and α,ω-Dicarboxylic Acids

Abstract:

In the hope of advancing sustainable plastic production, great progress has been made toward efficient microbial conversion of renewable feedstock to short-chain (C2-C8), bifunctional polymer precursors (e.g. succinic acid, cadaverine, 1,4-butanediol). Relatively little attention has been given to medium-chain (C12-C14) monomers such as ω-hydroxy fatty acids (ω-OHFAs) and α,ω-dicarboxylic acids (α,ω-DCAs), which nonetheless are precursors to high performance polyesters and polyamides. In our work, we are engineering complete microbial conversions of glucose to various medium chain ω-OHFAs and α,ω-DCAs, with precise control of product chain length. We combine bioinformatics and *in vivo* production to screen a wide range of enzymes across phyla and identify combinations that yield complete conversion of intermediates to product α,ω- DCAs. This approach, coupled with optimization of culture conditions has allowed us to achieve near g/L titers of specific medium chain α,ω- DCAs. In particular, our results indicate potential for this microbial factory to enable commercially relevant, renewable production of C12 α,ω- DCA – a valuable precursor to the high-performance plastic nylon-6,12.

Copyright © American Institute of Chemical Engineers. All rights reserved.

472557 Model-Guided Metabolic Engineering of Increased 2-Phenylethanol Production in Plants

Abstract:

2-Phenylethanol (2-PE) is a naturally occurring aromatic with properties that make it a candidate oxygenate for petroleum-derived gasoline. In plants, biosynthesis of 2-PE competes with the phenylpropanoid pathway for the common precursor phenylalanine. The phenylpropanoid pathway in plants directs approximately 30% of carbon flux towards the biosynthesis of lignin, a major constituent of plant cell walls that impedes the process of biofuel production. Therefore, we propose a genetic engineering strategy at the phenylalanine branch point, whereby a portion of the carbon flux towards lignin biosynthesis is diverted towards the production of an economically valuable product, 2-PE. Transgenic *Arabidopsis thaliana* were generated that overexpress aromatic aldehyde synthase (AAS) in tandem with phenylacetaldehyde reductase (PAR) from tomato introducing a pathway for the production of 2-PE. The primary stem was selected as the experimental system for analyzing competition of Phe between phenylpropanoid metabolism and the engineered 2-PE pathway. Excised 5-week-old stems were exogenously fed different concentrations of $^{13}C_6$-ring labeled Phe. Both the amount and isotopic enrichment of downstream intermediates was quantified using LC-MS/MS at multiple time points after feeding. A kinetic model of the phenylpropanoid network was constructed, and the parameters were identified through non-linear optimization with training datasets, and validated with data from an independent experiment. *In silico* analysis of the results from our model predicted that the endogenous cytosolic Phe pools limit the 2-PE production in these transgenic plants. This prediction was tested by combining overexpression of PAR and PAAS with overexpression of a feedback-insensitive 3-deoxy-D-arabino-heptulosonate 7-phosphate (DAHP) synthase, the latter of which has been previously shown to have hyper-induced phenylalanine biosynthesis in Arabidopsis. This transformation led to a significantly increased accumulation of 2-PE in transgenic *Arabidopsis*. The use of kinetic modeling combined with time-course *in vivo* metabolite profiling is shown to be a promising approach to rationally engineer plants that accumulate high-value commodity chemicals.

Copyright © American Institute of Chemical Engineers. All rights reserved.

474734 The Emerging Organism Engineering Industry (Featured Presentation)

Abstract:

There is an emerging demand for sourcing plant-derived extracts (nutraceuticals, flavors, fragrances, sweeteners, etc.) from engineered microbes. While recent advances in synthetic biology and metabolic engineering provide feasible approaches to engineering such organisms, commercial success for developing these "cultured" ingredients presents specific challenges. Unlike biofuels, where efforts can be focused on one particular molecule given the enormous market size, cultured ingredients require developing different organism lines in a rapid and low cost fashion. This requires a scalable solution for bio-manufacturing of organisms, which is provided by our state of the art foundry that continues to grow. I will describe how organism development at Ginkgo leverages our foundry to accelerate the design/build/test using specific examples. In particular, I will highlight the value of combining computer-aided engineering software tools, cheap gene synthesis and high resolution-accurate mass LCMS to develop engineered microbes. Finally, I'll touch on how our improvements in manufacturing organisms lend to opportunities outside of cultured ingredients.

Copyright © American Institute of Chemical Engineers. All rights reserved.

465663 Proteomics Analysis of Antibody Producing CHO DG44 Cell Lines in Different Fed-Batch Processes

Abstract:

Fed-batch cell culture process is the most dominate operation mode for the production of recombinant therapeutic proteins. However, the limited systems biology understanding of the effect of different fed-batch processes on cellular mechanisms has hampered the effective process engineering for high protein productivity and quality. In this study, four basal media, four concentrated feeding supplements, and different feeding schedules were studied. The CHO DG44/IgG1 cell line produced high volumetric titer (1.3 g/L) and cell growth (19.6×10^6 cells/mL). The comparative proteomics was performed using gel LC-MS/MS to analyze the host cell protein expression in different fed-batch processes. The CHO DG44 host cell and CHO DG44/IgG1 cultivated in both batch and simple fed-batch processes were used as controls. The expression levels of the proteins associated with transcription, translation, post-translational modification, carbon metabolism and cell growth were highlighted and investigated. In addition, the effect of glucose, glutamine, lactate, and amino acids were also evaluated. This proteomic study enabled the identification of the potential regulators of cell culture process engineering and cell engineering.

Copyright © American Institute of Chemical Engineers. All rights reserved.

464449 Dendrinow: Automated Dendritic Cell Production

Abstract:

In the last five years, autologous dendritic cell therapies have become one of the most promising candidates for anti-cancer treatment. Dendritic cells represent the bodyâ€™s immunological memory and the means by which the immune system targets various diseases. Autologous cell therapies utilize patient-specific cells, which is advantageous to minimize the chance of host-rejection. Prior studies have proven the effectiveness of these treatments. Phase II clinical studies in India were able to prolong refractory cancer patientsâ€™ lives from an expectancy of less than three months to a median overall survival of 397 days. With nearly 15 million people in the United States affected by cancer, the need for additional cancer therapies has increased the expected market value to $2.2 billion by 2017. Currently, dendritic cells are generated in a small lab-scale setting which presents major regulatory, documentation, and quality concerns to be addressed during scale up. In addition, dendritic cells do not reproduce, therefore many manufacturing lines or unit operations are necessary to produce the desired amount of product. The primary goal of the DendriNow system is to culture dendritic cells from $CD14^+$ peripheral blood mononuclear cells (PBMCs). This system includes two components: a skeleton incubator system and a disposable cell culture kit. Key process changes were also incorporated into the DendriNow system to maximize the yield of dendritic cells. DendriNow will provide research institutions and companies the means to easily produce dendritic cells in a more cost-effective and timely manner, with the ultimate goal of accelerating the further investigation of dendritic cell therapies.

Copyright © American Institute of Chemical Engineers. All rights reserved.

453681 Elucidating and Engineering the Role of Arabinogalactan Proteins in Somatic Embryogenesis

Abstract:

There is a critical need for efficient reproduction of high-value crops including timber crops, biofuel feedstocks and medicinal plants. Our work focuses on the loblolly pine (*Pinus taeda*), a critical crop to the timber industry, currently occupying 30 million acres of U.S. timberlands and making up one half of the U.S. standing pine volume. Somatic embryogenesis (SE) provides a means to create a clonal crop and enables selection for plants with ideal characteristics (*e.g.* fast growth rate, disease resistance, superior wood quality), and our goal is to identify and manipulate predictors of embryo yield in the earliest stage of development to streamline SE and enable process commercialization. SE is an *in vitro* cellular process that leads to the formation of a fully developed and viable embryo from a single cell. The three main stages that lead to the formation of seedlings from a plant cell culture are: 1) embryonic suspensor masses (ESMs) are propagated and maintained in liquid suspension culture; 2) ESMs are plated onto solid development media to induce SE; and 3) embryos are isolated and germinated to generate seedlings. The primary limitation to large-scale implementation of SE is that the process is not fundamentally understood, leading to large variability in embryo yieldâ€"a number that cannot be determined until 3-4 months from SE initiation. Arabinogalactan proteins (AGPs) are a promising biomarker that have been implicated in other SE systems, providing many avenues for manipulation to improve yields. In the work presented here, targeted treatments were applied to maintenance cultures to evaluate influence on yield. For example, we demonstrated that AGPs are influential on loblolly pine somatic embryo yield by inhibiting AGP function *in vitro* using the AGP-binding Î²-glucosyl Yariv reagent. We also introduced generic, exogenous AGPs (*e.g.*, gum arabic, from the Acacia tree) to cultures, supplementing the endogenous AGPs to quantify the molecular response and effect on SE. We further extend this approach using novel characterization techniques to elucidate the role of AGPs in culture, including new colorimetric assays, ELISAs using antibodies with AGP subclass specificity and UPLC characterization and fractionation. Understanding the complex properties of AGPs and their influence on SE will enable a better fundamental understanding of SE and identification of further strategies to promote high embryo yield.

Copyright © American Institute of Chemical Engineers. All rights reserved.

455380 Elucidating Cellular Metabolism through Untargeted Omics Methods Towards Improving Process Robustness

Abstract:

In this work, an untargeted metabolomics approach was applied to an industrially relevant system wherein high peak densities of CHO cells were attained, using a fed-batch process with chemically defined media. This study was used to extend the scope of the analysis beyond well-known metabolic byproducts such as lactate and ammonia, which inhibit growth upon accumulation. Cell supernatant samples were collected from the exponential growth phase and at peak density from multiple cell lines, spanning two parental origins, varying growth rates and monoclonal antibodies produced. An untargeted metabolomics workflow using a combination of liquid chromatography-mass spectrometry (LC-MS) methods was used in conjunction with a novel, biological context-driven data processing and annotation pipeline to characterize extracellular metabolites resulting cellular metabolic activity. About 60 metabolites were identified, with approximately 40% being confirmed against the METLIN database. The metabolite profiles across all samples were then systematically compared to identify those that correlate with cell growth arrest. Comparison of trends from the exponential and peak samples as well as low and high growing cells, within and across both parental cell lines was performed. About 8 metabolites accumulating at higher peak densities compared to the growth phase and in low growing cell lines were of potential interest. These represent metabolic byproducts that could be toxic or growth inhibitory and warrant further investigation. Toxicity dosing studies were subsequently performed to validate the detrimental effects of these compounds on cell growth. A survey of the metabolic pathways involved resulted in design of strategies to mitigate their generation. Overall, using an untargeted â€˜omics approach provided important insights into cellular metabolism and allowed for further process optimization and robustness.

Copyright © American Institute of Chemical Engineers. All rights reserved.

458435 Optimization of a Mammalian Cell Perfusion Culture in Productivity and Product Quality

Abstract:

Optimization of a mammalian cell perfusion culture in productivity and product quality

Contact person: Moritz Wolf, moritz.wolf@chem.ethz.ch

The continuous production of therapeutic proteins using mammalian cell perfusion cultures combines several advantages compared to traditional batch-wise operations applied in industry. So far, this processing mode has been used in particular for the production of labile proteins. However, the achievement of long-term stable cultures, higher productivities, and enhanced product quality are pathing the way to a more general application of continuous cell perfusion cultures [1]. Continuous supply of fresh nutrients as well as removal of waste metabolites combined with the retention of cells allow steady state cultures at very high cell densities [2]. Stable and robust operations further minimize product heterogeneities and allow the simple modulation of critical quality attributes.

This study aimed at the optimization of productivity and product quality in a stirred tank perfusion bioreactor (3) by tuning key cell culture variables for a CHO cell line producing a monoclonal antibody (mAb). The applied sequential screening approach of most influential process parameters including, media composition, cell specific perfusion rate (CSPR), and temperature allowed a rapid evaluation of multiple operating points. While varying one parameter at a time, measurements of extra- and intracellular metabolites, product concentration as well as quality attributes were used to characterize their effect on cellular growth, productivity and product quality at steady state.

The tuning of key cell culture parameters led to a superior performance of the perfusion culture compared to the established fed-batch platform. Especially, the decrease of the CSPR resulted in higher product amounts. Whereas the

optimization of media composition lead to an increase of cell density while maintain perfusion rate low. At steady state operation of the perfusion bioreactor naturally occurring microheterogeneities of the produced mAb (e.g. N-linked glycosylation pattern and charge isoforms) were reduced compared to fed-batch runs. Low byproduct levels and the decoupling of tested variables allowed the deliberate shaping of these attributes. Overall, this study highlights the potential of mammalian cell perfusion cultures for the production of therapeutic proteins.

Literature

[1] V. Warikoo, R. Godawat, K. Brower, S. Jain, D. Cummings, E. Simons, T. Johnson, J. Walther, M. Yu, B. Wright, J. McLarty, K.P. Karey, C. Hwang, W. Zhou, F. Riske, K. Konstantinov, Integrated continuous production of recombinant therapeutic proteins, Biotechnol. Bioeng. 109 (2012) 3018â€“3029.

[2] M.-F. Clincke, C. MÃ¶lleryd, Y. Zhang, E. Lindskog, K. Walsh, V. Chotteau, Very high density of CHO cells in perfusion by ATF or TFF in WAVE bioreactorTM. Part I. Effect of the cell density on the process., Biotechnol. Prog. 29 (2013) 754â€“767.

[3] D. Karst, E. Serra, T. Villiger, M. Soos, M. Morbidelli, Characterization and comparison of ATF and TFF in stirred bioreactors for continuous mammalian cell culture, Biochem. Eng. J. 110 (2016) 17-26.

Copyright © American Institute of Chemical Engineers. All rights reserved.

471216 A Metabolic Flux Based Dynamic Model for Antibody Producing Mammalian Cells

Abstract:

The demand for high-value biopharmaceuticals is on a rapid rise due to their therapeutic and diagnostic applications. Monoclonal antibodies (MABs) are important reagents used in biomedical research, diagnosis and treatment of diseases, affinity production systems, and in vivo imaging. Monoclonal antibodies account for nearly 40% of the sale of biologics in the U.S. and are produced in animal cell bioreactors at a variety of scales. Mammalian cell cultures have been the favored production hosts for MABs, since microbial systems are not able to carry out the complex post-translational and functional modifications of these proteins, such as glycosylation. Chinese Hamster Ovary (CHO) cells and hybridoma cells, which share many similar metabolic characteristics, have been the popular cell types for production of MABs. The large scale production of MABs occurs via in vitro cultivation of mammalian cell lines in bioreactors using techniques similar to those used for microbial cultivation. The simplest approach for producing MABs in vitro is to grow the mammalian cells in batch and fed-batch cultures, since these are easy to operate and scale up, and recover and purify the MABs from the culture medium. However, under batch conditions, mammalian cells exhibit an inefficient glucose and glutamine metabolism, causing increased production of lactate and ammonia, major waste products of metabolism of glucose and glutamine, respectively. Both lactate and ammonia inhibit growth of mammalian cells and deteriorate MAB production. To reduce formation of these toxic by-products, fed-batch operations with intermittent supplementation of principal carbon and nitrogen sources have been performed with these cell lines, resulting in a higher volumetric productivity and extended cell viability. A cost-effective production of MABs requires an understanding of the effects of bioreactor process variables on the physiology of mammalian cells. The development of methods for optimization of cell growth and cell productivity has become a crucial issue to enhance MAB yield in vitro production. Although fed-batch operations have been successfully employed for microbial systems, kinetics of mammalian cell cultures is still under investigation to determine quantitatively as well as qualitatively cost-effective production strategies. Creating these strategies requires understanding of the effects of cell metabolism on the system dynamics. Substantial cell death and sensitivity of MAB productivity to culture conditions are important issues.

Mathematical models have been significant tools for combining cell physiology and engineering to predict the behavior of cell metabolism and optimize culture conditions by identifying process parameters that significantly impact cell growth and target metabolite productivity. The cell metabolism can be described and characterized using a metabolic network that accounts for biotic phase reactions. Metabolic flux analysis is a widely used approach to characterize the state of cellular metabolism and activities of various metabolic pathways. The analysis enables computation of intracellular fluxes from information on experimentally determined metabolite uptake and excretion rates by using the stoichiometry of an identified metabolic network. The analysis enables identification of all significant fluxes and significant metabolites influencing cell metabolism so that it is possible to obtain a set of macro-reactions linking the substrates to the end-products, the so-called elementary flux modes. In this work, a novel dynamic model is developed on the basis of these macro-reactions involving extracellular substrates and products for a mammalian cell culture with a MAB as the target product. The processes accounted for in the model are cell growth and death, uptake of key nutrients, glucose and glutamine, accumulation of lactate and ammonia, and MAB synthesis. In the end, the cell-specific kinetics of growth and death of mammalian cells, utilization of glucose and glutamine, and generation of MAB, ammonia, and lactate are expressed in terms of concentrations of glucose, glutamine, lactate, and ammonia. The model incorporates a structured kinetic representation of MAB synthesis. The model accounts for the impact of glutamine availability and ammonia accumulation on translation rates and stability of mRNAs involved in assembly of MAB. Glutamine provides remarkable metabolic energy for cell growth and protein synthesis and is an important precursor of proteins and peptides, as well as amino sugars, purines and pyrimidines. It has been observed that glutamine depletion in the culture greatly influences the overall antibody synthesis rate and increases the apoptotic death rates of mammalian cells. This is accounted for in the dynamic model. Performance of batch and fed-batch cultures with intermittent or sustained addition of glucose and/or glutamine is simulated. The model predictions are in good agreement with experimental data reported in the literature for Immunoglobulin G (IgG) antibodies.

Copyright © American Institute of Chemical Engineers. All rights reserved.

456010 Response Surface Approach to Optimize CO_2 Fixation Using *Chlorella vulgaris* Culture

Abstract:

Microalgae based biofuels have been reported as an attractive alternative for fossil fuel, since they constitute a renewable energy source that reduces greenhouse gas emissions to the atmosphere. the adeptness to optimizely model microalgae productivity and aswell CO_2 biofixation especially under the varying conditions is crucial for evaluating the profitability and sustainability of their cultivation at large scale for biofuel industries. In this study, In this study, the response surface methodology in conjunction with the central composite design (RSM-CCD) was used to model specific growth rate (SGR), CO_2 uptake rate and biomass productivity of microalgae *chlorella vulgaris*. The non-linear models developed were used to optimize CO_2 concentration in air and cultivation time for SGR, CO_2 uptake rate and biomass productivity in tubular batch photobioreactor. The optimal set (3% CO_2 and 7 days culture) was found to occur at SGR of 2.23 per day, CO_2 uptake rate of 0.065 g/L/day and productivity of 0.092 g/L/day. The optimum values agree well with the literature values.

Acknowledgement

The author would like to gratefully acknowledge the support provided by King Abdulaziz City for Science and Technology (KACST) through the Science & Technology Unit at King Fahd University of Petroleum & Minerals (KFUPM) for funding this work through project No. KACST ARP-A-T-32-62 as part of the National Science, Technology and Innovation Plan.

Copyright © American Institute of Chemical Engineers. All rights reserved.

469260 A Regulatory Perspective for Manufacture of Complex Drug Substances

Abstract:

The quality of chemically synthesized, small-molecule drug substances usually can be characterized by routine analytical testing. In contrast, for drug substances that are polymeric molecules (e.g. peptides), heterogeneous mixtures (e.g. heparins, glatiramer acetate, etc.), or molecular complexes (e.g. iron sucrose), the structure and impurity profile are more complex than small-molecule drugs, and generally cannot be characterized easily by analytical means. As a result of this complexity, quality control of these complex drug substances cannot solely rely on analytical testing. The identity, impurity profile, and content of complex drug substances are highly dependent on their manufacturing process as well as raw material and in-process controls. This presentation will discuss a regulatory perspective for manufacture of complex drug substances in support of both new drug and generic drug applications.

Copyright © American Institute of Chemical Engineers. All rights reserved.

473006 CQA Control Strategy for a Synthetic Peptide Therapeutic Candidate Etelcalcetide

Abstract:

Application of innovative continuous improvement strategies to ensure consistency and robustness of pharmaceutical drug substance (DS) manufacturing processes can add significant value over a productâ€™s lifecycle. A case study from etelcalcetide, a synthetic peptide therapeutic candidate, will be presented highlighting Amgenâ€™s modality-independent approach to process design. Application of QbD and QRM principles were fundamental in establishing a solid understanding of the impact of variability on DS critical quality attributes (CQAs) and overall process performance. Holistic frameworks to proactively identify, assess, and mitigate CMC risks was leveraged during the process design phase. Examples of specific DS CQAs will be used to show how establishment of appropriate controls required a combination of material input controls, procedural controls, in-process testing, and release testing.

Copyright © American Institute of Chemical Engineers. All rights reserved.

457444 Development and Implementation of a Mathematical Model As a Key Component of a Palladium Control Strategy for a Semi-Continuous Process to Produce a Small Molecule Drug Substance

Abstract:

The manufacture of a small molecule drug substance employs a coupling reaction catalyzed with soluble palladium in the final synthetic step of the sequence. The control strategy around palladium removal involves selective adsorption of the palladium onto a functionalized silica resin in a continuous fashion followed by a continuous deprotection reaction and the selective crystallization of the drug substance. During the registration stability campaign, at-line and off-line samples were collected. To minimize cost associated with the selective adsorbent, and to ensure a robust control strategy for palladium, a predictive model was developed to aid in the appropriate design, scaling and setup of the manufacturing equipment. Use of the model and the selected control strategy for palladium, alleviated the need for continuous sampling for palladium during processing and enabled a robust operation during validation with significant production cost reduction.

Copyright © American Institute of Chemical Engineers. All rights reserved.

467794 Monitoring the Trace Levels of Free Sulfate Release to Terminate a Depolymerization Reaction Using ATR-FTIR

Abstract:

The synthesis of a pharmaceutical product involves a critical depolymerisation reaction which determines the final product quality attributes. Hence, termination of this reaction at the desired end point is utmost important to achieve a consistent product with required specifications. However, monitoring this reaction is challenging due to low chemical specificity of the reaction species involved. In this work, we developed a novel PAT method based on ATR-FTIR combined with chemometrics to monitor this reaction. To achieve this, we first identified the reaction mechanism and the rate-controlling step and monitored the reaction progress by tracking the free sulfate release which forms as a by-product during the reaction. The PAT method has been validated at lab and kilo scales and subsequently transferred to the plant and successfully implemented at manufacturing scale.

Copyright © American Institute of Chemical Engineers. All rights reserved.

474335 FDA Review Chemist's Participation in Pre-Approval Inspections: Case Study from Drug Substance Facility

Abstract:

In 2014, FDA initiated a new approach to incorporate review chemist's expertise in the inspection of drug substance/product manufacturing facilities. A pilot was launched with the objective to evaluate the feasibility, effectiveness, and efficiency of Office of Pharmaceutical Quality's vision for the integrated team-based quality assessment of A/NDAs. A risk analysis was conducted jointly by the compliance officer, drug substance, drug product, and facility reviewers to determine the need for a pre-approval inspection (PAI) and prioritize the issues to be covered on the inspection. The integrated team-based approach resulted in improved knowledge sharing among the reviewers and compliance officers, and in better identification and evaluation of high-risk areas during the application review. These high-risk areas identified by the review team were covered during the PAI. Collaboration between the review chemist and investigators helped uncover significant issues and resulted in the issuance of deficiencies to the firm.

Based on outcomes of the Agency's pilot integrated PAI program, other inspections were conducted using this paradigm including the recent PAI of a complex drug substance manufacturing facility. Participation on this inspection provided the reviewer with an opportunity to examine the equipment and process closely, to gain a better understanding of the development and optimization of manufacturing process and controls, and in turn allowed the inspection team to identify the CMC-related observations. Furthermore, the firm provided positive feedback regarding the presence of a review chemist on inspection deemed particularly helpful to better understand and improve regulatory submission of future applications. In summary, implementation of this new integrated approach for conducting a PAI has enhanced the efficiency and quality of review process, and improved communication between FDA and industry.

Copyright © American Institute of Chemical Engineers. All rights reserved.

473022 Panel Discussion: Current Issues and Challenges for CQA Monitoring and Control for Drug Substance Manufacturing

Abstract:

In this panel discussion, all of the speakers in this session and invited FDA leaders will join for an interactive Q& A session. It will highlight current issues and challenges in the CQA monitoring and control for drug substance manufacturing, from development, manufacturing, to regulatory review and inspection, and life cycle management.

Copyright © American Institute of Chemical Engineers. All rights reserved.

464765 Modeling the Sexual Dimorphism in Feedback Regulation of the HPA Axis

Abstract:

Significant sex differences are observed in disease prevalence. In general, it is observed that there is a greater prevalence of autoimmune diseases in women, while men are more prone to developing infectious diseases and coronary heart disease (1). Furthermore, a number of studies show evidence for sex differences in the endocrine response to acute stress (2). The primary component of this stress response is the hypothalamic-pituitary adrenal (HPA) axis, which is composed of the important neuroendocrine mediators corticotrophin releasing hormone (CRH), adrenocorticotrophic hormone (ACTH), and cortisol (corticosterone in rodents). In response to stress (both physical and psychological) there is increased hypothalamic CRH release, which induces the secretion of ACTH from the pituitary, inducing the release of the adrenal cortisol, the primary effector molecule of the HPA axis, which promotes adaptation to stress through both receptor-mediated genomic as well as non-genomic mechanisms. Finally, cortisol binds to the glucocorticoid receptor to regulate its own synthesis by inhibiting the expression of CRH and ACTH, respectively, thus, completing the feedback loop.

Rodent studies consistently report sex differences in the HPA axis in response to variety of stressors such as immune challenges, foot shocks and forced running. In general, studies have demonstrated that administration of stressors results in a greater secretion of both ACTH and corticosterone (CST) in females than in males. Interestingly, basal levels of CST also exhibit sexual dimorphism (3). These differences are most apparent in the circadian rhythms of the CST. In general, female rats exhibit higher plasma CST than male rats throughout the circadian period, with differences being most prominent at the circadian peak. Moreover, the amplitude of the circadian rhythm in females is twice as high compared to that in males (4). Furthermore, studies have also shown that both the circadian characteristics and its response to stress vary with estrous cycle in rats (5). Given these observations, it is hypothesized that sexual dimorphism in the HPA axis might be due to differential feedback regulation in the HPA axis under the possible influence of gonadal steroids, primarily estrogen in females and testosterone in males.

In such circumstances mathematical modeling can be particularly insightful as it can be used to formulate and test hypotheses, when it is not as feasible to do so through experimental means. In the present work, we develop a mathematical model that can study the possible differential feedback regulation of the circadian dynamics of the HPA axis in female and males rodents both in a baseline homeostatic state. We focus our current modeling efforts on sexual dimorphism in rodents since; the literature on human studies is much less consistent in comparison due to differences in demographic composition of study subjects and methodological considerations. Furthermore, modeling with a focus on rodent data supports broader experimental efforts in studying sexual dimorphism in the pharmacological response to glucocorticoids.

Our model builds on earlier work by Sriram et al. and our group, and is comprised of three components (6-8); a central oscillatory compartment that simulates the circadian secretion of CST in the HPA axis, and a peripheral compartment that captures the downstream effects of secreted CST on the expression of proinflammatory cytokines, which we intend to use as a physiological mediator for an immune stressor. The oscillations in the central compartment are generated due to the negative feedback between CST and corticotropin-releasing hormone (CRH)/adrenocorticotropic hormone (ACTH) using a modified Goodwin oscillator model that incorporates Michaelis-Menten kinetics to express the degradation terms in the hypothalamus, pituitary, and adrenals. The HPA axis is positively modulated by the expression of cytokines in the peripheral compartment, which in turn is balanced by the negative feedback of CST. We estimated kinetic parameters to calibrate our model to the relatively high amplitude oscillations observed in female CST circadian data as obtained by Atkinson et al (5). As a first approximation, we consider the influence of the gonadal steroids on the HPA axis to be implicit in the kinetic parameters of the differential equations describing CRH, ACTH and CST dynamics. As proof of concept in evaluating the prevailing hypothesis regarding the differential regulation of the HPA axis in females and males, we have identified an initial set of physiologically important parameters that the amplitude of circadian oscillation is dependent on, including the strength of negative feedback of CST on CRH and ACTH, and additionally, the rate of production of CST. We hypothesize that these parameters are important in accounting for the observed sex differences in basal HPA axis activity and its response to stress. The parameter space is explored to determine possible transitions in model output and response to immune stressor.

This work provides a quantitative framework for improving our mechanistic understanding of the critical processes responsible for the observed differences in stress response in females and males and will have broad application to efforts in quantitative systems pharmacology, which seek to address issues relating to sexual dimorphism.

References:

1. Oertelt-Prigione S. The influence of sex and gender on the immune response. Autoimmun Rev. 2012;11(6-7):A479-85. doi: 10.1016/j.autrev.2011.11.022. PubMed PMID: 22155201.

2. Kudielka BM, Kirschbaum C. Sex differences in HPA axis responses to stress: a review. Biol Psychol. 2005;69(1):113-32. doi: 10.1016/j.biopsycho.2004.11.009. PubMed PMID: 15740829.

3. Goel N, Workman JL, Lee TT, Innala L, Viau V. Sex differences in the HPA axis. Compr Physiol. 2014;4(3):1121-55. doi: 10.1002/cphy.c130054. PubMed PMID: 24944032.

4. Seale JV, Wood SA, Atkinson HC, Bate E, Lightman SL, Ingram CD, et al. Gonadectomy reverses the sexually diergic patterns of circadian and stress-induced hypothalamic-pituitary-adrenal axis activity in male and female rats. J Neuroendocrinol. 2004;16(6):516-24. doi: 10.1111/j.1365-2826.2004.01195.x. PubMed PMID: 15189326.

5. Atkinson HC, Waddell BJ. Circadian variation in basal plasma corticosterone and adrenocorticotropin in the rat: sexual dimorphism and changes across the estrous cycle. Endocrinology. 1997;138(9):3842-8. doi: 10.1210/endo.138.9.5395. PubMed PMID: 9275073.

6. Mavroudis PD, Corbett SA, Calvano SE, Androulakis IP. Mathematical modeling of light-mediated HPA axis activity and downstream implications on the entrainment of peripheral clock genes. Physiological Genomics. 2014;46(20):766-78. doi: 10.1152/physiolgenomics.00026.2014.

7. Sriram K, Rodriguez-Fernandez M, Doyle FJ. Modeling Cortisol Dynamics in the Neuro-endocrine Axis Distinguishes Normal, Depression, and Post-traumatic Stress Disorder (PTSD) in Humans. PLoS Computational Biology. 2012;8(2). doi: 10.1371/journal.pcbi.1002379.

8. Ramakrishnan R, DuBois DC, Almon RR, Pyszczynski NA, Jusko WJ. Fifth-generation model for corticosteroid pharmacodynamics: application to steady-state receptor down-regulation and enzyme induction patterns during seven-day continuous infusion of methylprednisolone in rats. J Pharmacokinet Pharmacodyn. 2002;29(1):1-24.

Copyright © American Institute of Chemical Engineers. All rights reserved.

460104 Chemical Kinetic Modeling of Intracellular Viral Processes in VSV-Infected Single Cells

Abstract:

Viral infectious diseases, such as Influenza, pose an important health concern. Even a single virus particle can infect a cell, hijack its resources, and manufacture thousands of new virus particles. The production of viruses from infected cells is a complicated process involving a large number of intracellular reactions. Study of these intracellular viral processes can provide us with vital clues that can help us develop new vaccines, antiviral medications, and even help us engineer viruses to treat cancer. Intracellular reactions that depend on small numbers of molecules and the interaction between viral and cellular components result in highly heterogeneous infection outcomes. Single-cell studies have shown that virus yields from individual cells can vary widely, as much as 300-fold [1]. It is possible that a few infected cells produce exceptionally high numbers of virus progeny relative to the average or majority of the infected cells. Behavior of these rare cells may have important long-term consequences towards the progression of the infection [1]. The throughput of single-cell studies can be increased through the use of fluorescent reporter strains of viruses. We have shown that these strains can be used to analyze viral protein production within cells over time, and that viral reporter expression is correlated with virus yields [2]. In this paper, we use quantitative viral reporter measurements from infected single cells and stochastic and deterministic reaction models to study intracellular viral processes.

Baby Hamster Kidney (BHK) cells were infected with a strain of vesicular stomatitis virus (VSV), encoding the gene for DSRed-Express protein, at various multiplicities of infection (MOIs) and then seeded into microwells. Wells containing one cell were identified and the fluorescent expression from these wells analyzed. Overall, we obtain RFP measurements at 30 minute intervals from 500-1000 individual VSV-infected BHK cells for MOI values of 1, 10, 40, and 80. This quantitative data allows us to build chemical kinetic models and estimate the kinetic rate parameters involved.

Such a wide variation in biological systems is often considered intrinsic (or aleatory) [3] and described via stochastic chemical kinetic models [4]. Stochastic chemical kinetic models describe the behavior of reacting systems when the number of molecules are small (1-1000) and the continuum assumption is not suitable. Stochastic reaction modeling involves probabilistically simulating the integer-valued molecule counts instead of deterministically simulating the concentration of reactive species, which is the case in a classical, deterministic reaction modeling approach. In this paper, we developed new chemical kinetic models -- both deterministic and stochastic -- to describe the relevant intracellular viral reproductive processes such as activation, transcription, replication, translation, and degradation. For both kinds of models, we estimate kinetic rate parameters for the available data using statistical techniques such as maximum likelihood estimation, Bayesian inference, and Monte Carlo methods [4]. Finally, we compared the fit of a deterministic reaction model against the fit of a stochastic reaction model to the same data. This analysis is done in order to discover whether the uncertainty described by the data is more aleatory or epistemic [3].

References:

[1] Timm, Andrea and Yin, John. Kinetics of virus production from single cells, Virology, Volume 424, Issue 1, 1 March 2012, Pages 11-17, ISSN 0042-6822, http://dx.doi.org/10.1016/j.virol.2011.12.005.

[2] Jay Warrick, Andrea Timm, Adam Swick, and John Yin, Tools for single-cell kinetic analysis of virus-host interactions, PLoS One, January 2016.

[3] Simeone Marino, Ian B. Hogue, Christian J. Ray, Denise E. Kirschner, A methodology for performing global uncertainty and sensitivity analysis in systems biology, Journal of Theoretical Biology, Volume 254, Issue 1, 7 September 2008, Pages 178-196, ISSN 0022-5193, http://dx.doi.org/10.1016/j.jtbi.2008.04.011.

[4] Gupta, Ankur and Rawlings, James B. Comparison of parameter estimation methods in stochastic chemical kinetic models: Examples in systems biology. AIChE Journal, Volume 60, Issue 4, Pages 1253--1268, ISSN 1547-5905, 2014. http://onlinelibrary.wiley.com/doi/10.1002/aic.14409/abstract

Copyright © American Institute of Chemical Engineers. All rights reserved.

461333 Seq2Flux: A Web-Based Platform to Bridge High-Throughput Sequencing and Metabolic Flux Analysis

Abstract:

High-throughput sequencing techniques, such as whole-genome sequencing and RNA-Seq, have been widely applied on over 3,000 species for uncovering the mysteries of biological systems. With the rapid advancement of the sequencing techniques, the cost of sequencing service has been dramatically decreased to $50-60/genome. Currently, with the access to over 100 TB sequencing data, the interpretation of the sequencing data, however, becomes the bottleneck for developing advanced sequencing service. This is mainly because of the fact that few detailed phenotypic readouts, e.g., metabolic flux distribution, could be directly linked with sequencing data for intuitive interpretation of the sequencing results. To this end, we developed Seq2Flux as a web-based platform to bridge the high-throughput sequencing data and metabolic flux analysis. We implemented Seq2Flux for a proof-of-concept study to predict metabolism of *E. coli* from customized genome sequencing. In general, we used the whole-genome sequencing data of *E. coli* strain provided by users as the input, and run the logic operation of the gene-protein-reaction association to map the gene annotation into a genome-scale metabolic model of *E. coli* (iJO1366). We next developed a constraint-based flux analysis algorithm to integrate large-scale fluxomics data with the sequence-curated metabolic models to achieve accurate prediction of metabolic fluxes as well as other phenotypes (e.g., growth rate). Taken together, Seq2Flux could reveal valuable metabolic information directly from genome sequencing data, which paves the ways for advanced applications of high-throughput sequencing data such as personalized medicine.

Copyright © American Institute of Chemical Engineers. All rights reserved.

464265 Mathematical Modeling of White Blood Cell Population Dynamics for Disease Prognosis

Abstract:

The complete blood count (CBC) is one of the most frequently ordered clinical tests for diagnosis and disease monitoring. CBCs are measured using automated hematology analyzers, usually based on principles of flow cytometry. Conventionally, the total number of each blood component (e.g., red blood cells, white blood cells) from the CBC informs diagnosis and monitoring. This approach, while broadly useful, discards a variety of additional data such as the distribution of different components that can be obtained from the analyzers and used for more accurate and earlier detection of diseases. The approach proposed in this project focuses on a case study consisting of patients with acute myocardial infarction (heart attack or AMI), the number one cause of deaths in the world. Early detection of AMI can result in appropriate therapeutic intervention, which can prevent fatal outcomes.

There is sufficient evidence in the literature highlighting the correlation of white blood cells (WBCs) and WBC subpopulations with AMI. This evidence motivates us to develop additional prognostic tools for examining the WBC dynamics in patients with AMI. In this framework, we aim to extract information from the raw flow cytometry measurements and develop a dynamic mathematical description of WBC subpopulation behavior. For this purpose we are using two cohorts of clinical datasets containing troponin, the gold standard measure of heart muscle cell death, and CBC measurements of patients suspected with AMI. More broadly, our objective is to use patient data to derive insight into the dynamics of human pathophysiology and to develop related clinical tools. This methodology is associated with no additional laboratory testing costs and has immense potential to advance clinical diagnosis, monitoring, and optimization of treatment.

Copyright © American Institute of Chemical Engineers. All rights reserved.

470524 A Pareto Optimality Explanation of the Glycolytic Pathways in Nature

Abstract:

The Entner-Doudoroff (ED) and Embden-Meyerhof-Parnas (EMP) glycolytic pathways are generally well-conserved across saccharolytic species in all three domains of life (i.e. Eukarya, Bacteria and Archaea). Although both pathways convert glucose to pyruvate, reactions constituting these two pathways span different intermediates resulting in varying ATP yields for different organisms. One could construct a multiple of possible ways of converting glucose to pyruvate with sometimes higher ATP yield. This raises the question as to why nature generally employs either EMP or ED glycolysis. Is this a coincidence, convergent evolution or there exists a driving force towards either of the two pathway designs? We addressed this question by first employing the loopless-optStoic algorithm to prospect for multiple possible pathways between glucose and pyruvate at different pre-determined stoichiometric yields of ATP (i.e., from 0.5 to 5 mol ATP/mol glucose). Subsequently, we analyzed the thermodynamic feasibility of all the pathways at physiological metabolite concentrations and quantified the protein cost of the feasible solutions. We identified the pareto optima for the trade-off between energy efficiency and protein cost for each of the feasible routes. Interestingly, we found that both the naturally evolved ED and EMP pathways are among the most protein cost-efficient pathways when compared to other pathways generating one and two mol ATP/mol glucose, respectively. Although at standard conditions, conversion of glucose to pyruvate could generate up to 5 mol ATP/mol glucose with favorable thermodynamic driving force (Î”$_r$GÂ° < 0), we found that most pathways designed for higher ATP production (> 2 ATP) are thermodynamically infeasible at physiological metabolite concentration ranges. In addition, while there exist several thermodynamically feasible pathways with up to 3 ATP production, they required approximately 4-fold higher investment on catalytic machinery to drive the pathway flux, which provides one possible explanation why these high ATP-yielding pathways are not found in nature.

Copyright © American Institute of Chemical Engineers. All rights reserved.

472184 Genome-Scale Metabolic Model of a Medicinal Plant Culture System

Abstract:

The taxane diterpenoids are a class of compounds first isolated from species of the genus *Taxus*, all possessing the taxadiene core. Three taxanes, the natural product paclitaxel and its two synthetic analogs docetaxel and cabazitaxel, are currently FDA approved for the treatment of a wide variety of cancers. A number of production platforms have been used to supply market quantities of paclitaxel and semi-synthetic analog precursors dating back to the initial paclitaxel clinical trials; however, industrial scale plant cell culture has proven to be the most economic and environmentally friendly. Plant cell cultures are not optimized for taxane synthesis, plagued by slow growth rates and inconsistent product formation, necessitating novel ways to analyze and understand cellular metabolism.

Genome-scale models (GSMs) are *in silico* stoichiometric representations of all or part of the metabolism of a cell, tissue or organism of interest. GSMs have become indispensable tools for the study of bacterial metabolism and physiology, assessing gene essentiality and designing rational metabolic engineering strategies. However, the tool is under-utilized in the plant community, with GSMs having been constructed only for model plant systems such as *Arabidopsis,* maize and sorghum. Plant metabolic reconstruction is made complicated by compartmentalization, pathway duplication, and the fact that the functions of many plant genes have yet to be elucidated.

A draft *Taxus* GSM has been constructed and will be discussed here. Although several automated platforms have been developed for genome annotation and model construction, it was found that these platforms are not appropriate for under-characterized systems, and hence, *Taxus* metabolism was reconstructed using custom programs to identify and extract a set of reactions from the KEGG database that correspond to genes in an annotated *Taxus* transcriptome, a collection of more than 48,000 contigs present during various stages of cellular growth and metabolism. Manual curation is ongoing and comprises the addition of subcellular compartments, intracellular transporters and exchange reactions, all of which can be obtained from various databases and primary literature. Once fully curated, quantitative analysis of the constructed model can be performed with the incorporation of condition-specific transcriptome data (i.e., elicited and nonelicited, various time points, small and large sized aggregate cultures). Simulating biomass growth under various conditions allows for the study of cellular metabolism and physiology during different stages of paclitaxel synthesis and can provide key information to devise strategies to improve process performance. The final GSM will represent the first such construction of both a medicinal plant and non-model plant system.

Copyright © American Institute of Chemical Engineers. All rights reserved.

472778 Modeling the Spatio-Temporal Dynamics of Calcium Signal Propagation in Liver Lobules

Abstract:

The liver performs a wide variety of physiological functions such as regulation of intermediary metabolism, lipid synthesis, bile production and xenobiotic detoxification. Normal liver function requires tight regulation of intracellular processes as well as intercellular coordination. Free Ca^{2+} in the intracellular domain participates in the regulation of hepatic functions such as energy metabolism, bile secretion, proliferation and apoptosis. Regulation of cytosolic Ca^{2+} is particularly important in hepatocytes, the cells responsible for the bulk of metabolic and detoxification activities in the liver. Consequently, disruption of Ca^{2+} dynamics can potentially lead to pathological conditions. Extracellular stimulants such as hormones elicit a signaling cascade involving phospholipase C (PLC) cleavage of PI-4,5-P_2, IP_3 generation, and subsequent Ca^{2+} release from intracellular stores. Gap-junction mediated transfer of IP_3 molecules between adjacent hepatocytes induces, in neighboring hepatocytes, Ca^{2+} oscillations that spread through the tissue. This cell-cell interaction leads to a coordinated response at the tissue level. Measuring lobular Ca^{2+} responses upon treatment of perfused rodent livers with PLC-activating agonists has revealed a remarkable spatial organization. For low stimulus levels, Ca^{2+} signals in liver lobules start near a specific lobular locus, in the pericentral region, and propagate towards the periportal region in a wave-like fashion.

Although the behavior of Ca^{2+} signal propagation at the lobular scale is understood fairly well, the contribution of hepatocyte-specific Ca^{2+} signaling components to the lobular scale Ca^{2+} wave propagation remains largely unknown. In the present study, we performed a causality analysis of time series data on Ca^{2+} signals in single hepatocytes to identify a cell-cell influence network underlying the Ca^{2+} wave propagation in liver lobules. We also developed a

computational model of spatio-temporal Ca^{2+}signaling and cell-cell interactions to predict the coordinated changes in the intracellular Ca^{2+} signaling parameters sufficient to yield the lobular scale Ca^{2+} propagation dynamics identified in the time series data analysis.

We acquired time-resolved images of Ca^{2+} spikes caused by different vasopressindoses in the isolated perfusedrat liver at the lobular level. We segmented the acquired images to obtain hepatocyte-specific cytosolic Ca^{2+} profiles. For a given time lag, we estimated pairwise transfer entropies (TEs) for Ca^{2+} signal information received *by* a hepatocyte *from* all the other hepatocytes within the two-dimensional lobular visual field to obtain hepatocyte- and lag-specific empirical TE distributions. For each hepatocyte, we identified the set of statistically significant causal influences by comparing the TE values from its immediate neighbors with the hepatocyte-specific empirical TE distribution. Our analysis revealed that at the lobular scale, for the vasopressin levels employed in the experiment, Ca^{2+} waves propagate from the pericentral to the periportal region along the hepatocyte cords. However, the transfer entropy analysis revealed that the direction of causal influence between successive neighboring hepatocytes may not be consistently in the pericentral-to-periportal direction, i.e., in some instances, Ca^{2+} signal information may â€œflowâ€ in the direction opposite to lobular scale Ca^{2+} wave propagation between neighboring hepatocytes lying within a hepatocyte cord.

We utilized a computational modeling approach to explore how such local switches in the direction of Ca^{2+} signal influence could arise, and evaluated their influence on the overall lobular scale wave propagation. We developed an ODE based computational model of cytosolic Ca^{2+} spiking in hepatocytes to identify the spatial patterns of cell-intrinsic Ca^{2+} signaling parameters that can reproduce the experimentally observed lobular Ca^{2+} wave propagation. Our model consists of individual hepatocytes in which the intracellular Ca^{2+} signals are elicited via a G-protein coupled receptor mediated pathway. The hepatocytes are oriented along a cord and exchange IP_3 with the adjacent hepatocytes via gap junctions. Model simulations revealed that 1) gap junction coupling results in coordination of Ca^{2+} spikes in hepatocytes along a cord, despite variability in intracellular Ca^{2+} signaling parameters across cells, and 2) correlated spatial gradients of intracellular Ca^{2+} signaling parameters are required for yielding unidirectional flow of Ca^{2+} waves at the lobular scale. In the absence of gap junction coupling, Ca^{2+} spikes did not occur sequentially along a hepatocyte cord in response to stimuli that lasted over extended periods of time. Disruption of spatial gradients of intracellular signaling parameters resulted in local deviations from the overall direction of Ca^{2+} wave propagation along hepatocyte cords.

Next, we tested the effect of local disruptions in cell-specific Ca^{2+} signaling ability of hepatocytes on the lobular scale Ca^{2+} wave propagation. Inability of hepatocytes to respond to Ca^{2+} agonists may contribute to aberrant bile trafficking and promote cholestasis. We initialized our model of a hepatocyte cord in which a randomly distributed subset of hepatocytes was insensitive to Ca^{2+} agonists. Simulations revealed that Ca^{2+} waves do not propagate throughout the lobule when Ca^{2+} signals are not elicited in a subset of hepatocytes. Furthermore, increase in the strength of gap junction coupling was sufficient to restore the lobular scale Ca^{2+} signaling pattern even in the presence of local disruptions in Ca^{2+} response, suggesting that efficient intercellular communication can yield robust lobular scale Ca^{2+} response despite local insufficiencies in Ca^{2+} signaling.

In summary, we have developed and illustrated a computational model based approach to analyze and interpret time series data on Ca^{2+} signals in single cells in intact liver tissue. Our results led to new predictions on the role of gap junctions in overriding the cell-intrinsic variability towards a robust and spatially-organized lobular scale Ca^{2+} response to external stimuli.

Copyright © American Institute of Chemical Engineers. All rights reserved.

473169 Bioprocess Optimization Under Uncertainty Using Ensemble Modeling

Abstract:

Mathematical modeling has become an indispensible tool in bioprocess design and optimization [1], especially after the implementation of Quality by Design for biopharmaceuticals by the US Food and Drug Administration's (FDA) office [2]. Since the majority of bioprocesses involve batch or fed-batch culture and are therefore dynamic in nature, kinetic modeling using ordinary differential equation (ODE) is the predominant framework to model these processes. However, the development of first principle dynamic bioprocess models often faces significant challenges due to high degree of process nonlinearity and limited datasets for parameter estimation (model training), the combination of which leads to model identifiability problem. This problem causes uncertainty in the modeling, and as a consequence, there often exist a family of equivalent kinetic models and model parameters, which are consistent with the available data and information on the bioprocess. Such uncertainty, if not addressed properly, could lead to suboptimal bioprocess design and operation.

In this work, we investigated several strategies for bioprocess optimization under uncertainty using ensemble modeling. The bioprocess optimization under uncertainty involved the maximization of mean-standard deviation objective function:

$$U(c) = E_\theta(f(c, \theta)) + a\sqrt{V_\theta}(f(c, \theta))$$

where c denotes the vector of operating conditions, θ denotes the vector of model parameters, f(c,θ) describes the process objective function (e.g. yield or product titer), a is a weighting factor, and E_θ (f(c,θ)) and V_θ (f(c,θ)) are the expected and variance of $f(c,\theta)$ over the posterior probability density function of the parameters $P(\theta)$. Similar objective functions have been proposed previously for bioprocess optimization under uncertainty [3, 4]. Our work differed from the previous studies in the generation of the parameter ensemble, allowing efficient evaluation of the objective function, especially for high dimensional parameters. We previously created an ensemble modeling toolbox for MATLAB, called REDEMPTION (Reduced Dimension Ensemble Modeling and Parameter Estimation) [5], which offered a user-friendly interface for kinetic ODE modeling of bioprocesses using time series process data, including the identification of an ensemble (family) of statistically equivalent model parameters. In particular, we used a combination of an out-of-equilibrium Adaptive Monte Carlo and multiple ellipsoids-based uniform sampling, called HYPERSPACE [6], to obtain the ensemble of parameters whose negative log-likelihood function values were lower than a given threshold. In the bioprocess optimization above, we approximated the expectation function over the posterior distribution of parameters using the following:

$$E_{\theta}(g(\theta)) = \Sigma_{\theta \quad \Omega} g(\theta) P(\theta)$$

where Ω denotes the parameter ensemble.

We evaluated the performance of the proposed strategy using a case study of monoclonal antibodies (mAb) production. We employed the kinetic bioprocess model proposed by Kontoravdi et al. [1, 7] to simulate *in silico* time-series concentration data, contaminated with Gaussian noise at 50% coefficient of variation. We subsequently used this dataset to obtain the maximum likelihood (ML) parameter estimate. We further generated an ensemble of ~400,000 parameter combinations using a threshold of 1.2 times the negative log-likelihood value of the ML estimate. The bioprocess optimization involved finding the optimal starting glutamine concentration that maximizes the mAb concentration at the end of batch culture, for different initial glucose concentrations. We implemented two strategies: $a = 0$ (i.e. maximization of expected value of $f(c,\theta)$) and $a = -1.282$ (corresponding to 90% lower confidence bound of $f(c,\theta)$). We compared our strategy to maximizing $f(c,\theta)$ directly using the ML parameter estimate. The final mAb concentration for each optimal glutamine concentration was computed using the model used to generate the data above. The comparison in Table 1 showed that the optimal glutamine concentrations using the proposed bioprocess optimization under uncertainty led to higher final mAb concentration than those using the ML parameter estimate, especially at high initial glucose concentration. In particular, the strategy using $a = 0$ consistently outperformed the optimization using the ML model. The result thus demonstrated the benefit and importance of considering parameter uncertainty in bioprocess optimization.

Table 1. Final mAb concentrations from different optimization strategies.

	mAb concentration (10^3 mg/L)		
Initial glucose concentration (mM)	Proposed method (a=0)	Proposed method (a=-1.282)	Optimization using ML model
25	1.3127	1.2742	1.3020
50	1.3238	1.3105	1.1856
75	1.3116	1.3158	1.1390
100	1.3042	1.3183	1.1347

1. Kiparissides, A., et al., *'Closing the loop' in biological systems modeling - From the in silico to the in vitro.* Automatica, 2011. **47**(6): p. 1147-1155.

2. Rathore, A.S. and H. Winkle, *Quality by design for biopharmaceuticals.* Nature Biotechnology, 2009. **27**(1): p. 26-34.

3. Mandur, J. and H. Budman, *Robust optimization of chemical processes using Bayesian description of parametric uncertainty.* Journal of Process Control, 2014. **24**(2): p. 422-430.

4. Kim, K.K.K. and R.D. Braatz, *Probabilistic Analysis and Control of Uncertain Dynamic Systems: Generalized Polynomial Chaos Expansion Approaches.*2012 American Control Conference (Acc), 2012: p. 44-49.

5. Liu, Y., E. Manesso, and R. Gunawan, *REDEMPTION: reduced dimension ensemble modeling and parameter estimation.* Bioinformatics, 2015. **31**(20): p. 3387-3389.

6. Zamora-Sillero, E., et al., *Efficient characterization of high-dimensional parameter spaces for systems biology.* BMC Systems Biology, 2011. **5**: p. 142.

7. Tatiraju, S., M. Soroush, and R. Mutharasan, *Multi-rate nonlinear state and parameter estimation in a bioreactor.* Biotechnology and Bioengineering, 1999. **63**(1): p. 22-32.

Copyright © American Institute of Chemical Engineers. All rights reserved.

461930 A Rapid, Materials-Sparing Approach to the Exploration of a Roller Compaction Process Design Space

Abstract:

Roller compaction (RC) is a time and material intensive process commonly used to improve the overall processing behavior of formulations. The typical approach to identifying target processing conditions on a pilot- or production scale system can easily consume kilograms of formulation, and the material consumption associated with system startup and shutdown further preclude effective in-process measurements. These issues can be magnified during efforts to establish a design space according to the principles of QbD, where, for example, a broad range of material property inputs may be evaluated to assess their impact on both downstream process behavior and product performance.

Here we demonstrate the application of a well-established scale down approach to roller compaction simulation in order to evaluate the influence material attributes of a spray dried dispersion (SDD) incur on RC processing and tablet compression behavior. The approach couples a compaction simulation of the ribbon formation process with a ¼-scale replica of an in-line rotary mill to yield granules with densities and size distributions analogous to what is produced on a Gerteis Mini-Pactor roller compactor. Intragranular blends of a constant composition were prepared using SDD lots of varying bulk densities and particle size distributions, and each blend was compressed to form simulated RC ribbons of varying target porosities. All ribbons were subsequently milled under common granulation conditions, and tablets were prepared from these granulations to evaluate both the compression behavior and *in vitro* dissolution performance of each final system. This materials-sparing methodology enables the exploration of the relationship between SDD properties and simulated roller compaction conditions across a broad design space, resulting in the establishment of statistical models which describe the influence of these factors on meaningful attributes of the final drug product.

Copyright © American Institute of Chemical Engineers. All rights reserved.

452650 Advances in Discrete Element Modeling of High-Shear Wet Granulation Process Using Rocky-DEM

Abstract:

High-shear wet granulation process is widely used in the pharmaceutical industry to improve the bulk density and flow characteristics of a formulation. There are different particle-level phenomenon and mechanisms at play during a granulation process, such as nucleation, agglomeration and breakage. The challenge is to ensure that these mechanisms occur similarly at different scales of the granulator to keep consistency in the quality of the granules. The ability to accurately predict the stress and collisional behavior in a granulator, would not only help our fundamental understanding of the process but also help alleviate experimental costs during scale-up studies.

Discrete element method (DEM) is a valuable tool to model such particle-level interactions, but in the past, has suffered from practically restrictive run times. In order to circumvent this, assumptions on particle size and shape (modeled as spheres or glued-spheres) were routinely made. However, more recently, advances in DEM technology, such as that provided by software Rocky-DEM, have addressed some of these limitations. Rocky-DEM provides a step-change in the ability to perform large scale DEM simulations significantly faster by utilizing the large number of cores available in NVIDIA® graphics card (GPU processing). The simulation run times using the GPU were compared with traditional and more expensive CPU solvers. In addition, several capabilities unique to Rocky-DEM in comparison with other commercially available DEM codes were utilized in this study; a few of them include the representation of Avicel using non-spherical particle shapes (polyhedral shape representation) and the ability to include cohesive forces and breakage without loss of mass or volume.

Rocky-DEM was used to predict the movement of Avicel particles in a Diosna® granulator at 50% volume fill (Figure 1). The simulations were validated by comparing results to experimentally-obtained surface velocity profile measurements using video-imaging. The validated DEM model was then used to establish a relationship between various parameters on simulation measurements of bulk properties (such as average velocity and stress distributions in the powder). Some of the variables included the particle stiffness (low and high), particle shape (spherical vs Avicel-shaped), particle size (0.8, 1.8 mm), impeller speed (Froude number vs tip speed), granulation type (wet vs dry particles), and granulator scale (1L, 10L and 150L). Thus, by numerically predicting granular pressures, velocities, and stresses in different regions of the granulator, this study provides guidelines on achieving reproducible particle environments with changes in scale and impeller speeds, and establishes the effect of particle shape and size. The Rocky DEM platform enabled the utilization of actual particle shapes and sizes in these studies and therefore are a closer reflection of a real-case scenario.

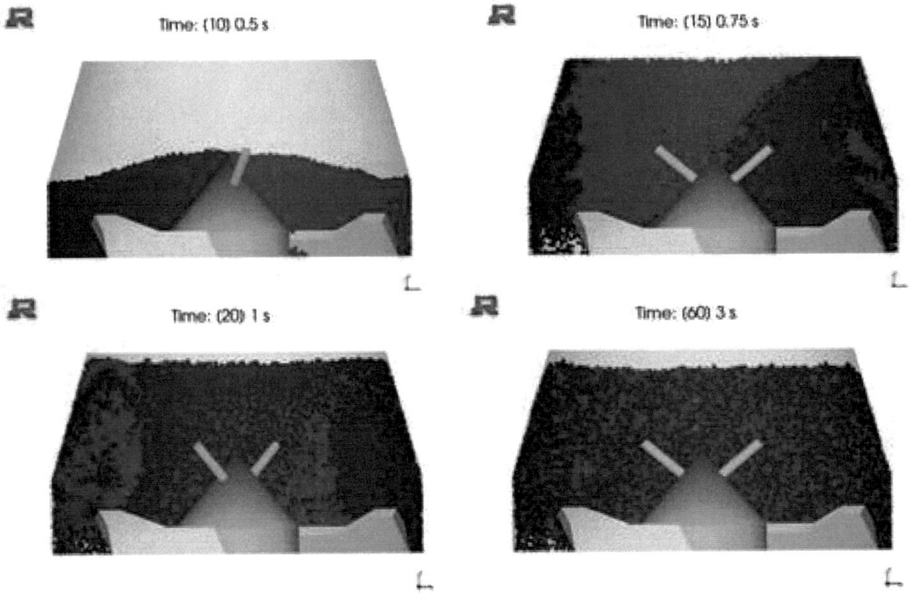

Figure 1. A cross-section of the granulator colored by the particle initial fill position to show mixing profiles

Copyright © American Institute of Chemical Engineers. All rights reserved.

458336 Scale-up of HME Process from a Trilobed 16mm to a Bilobed 27mm Extruder; SPH Simulations and 1D Mechanistic Modelling

Abstract:

Introduction

Hot melt extrusion (HME) is a continuous manufacturing process primarily using co-rotating intermeshing twin-screw extruders (TSE). The continuous manufacturing nature of the process allows for a steady product quality and a decrease in production costs. Furthermore, the process has the potential of increasing the bioavailability of poorly soluble drugs by enabling manufacturing of amorphous solid dispersions, making it interesting to the pharmaceutical industry. Although the process has been used in the polymer and rubber industry for years it is still not well enough understood and fairly new in the pharmaceutical industry. Especially the technical transfer between different extruder scales (scale-up) is still an open question.

Methodology

Simulation of extruders via traditional CFD is prohibitively complex, due to the small gaps, the complex rheology and, especially, due to the free surfaces that occur in the not-fully-filled screw elements. In order to better understand the process conditions and product quality relationships, 1D mechanistic and 3D smoothed particle hydrodynamics (SPH)[1] simulation approaches and the established scale-up theory were used. Standard 3D mesh-based computational fluid dynamics (CFD) models struggle with the complex rotating intermeshing TSE geometry, and lack the ability to investigate partially filed TSE elements[2]. The 1D mechanistic model is able to simulate the HME process as a whole, giving information about the filling degree, pressure and temperature distribution, as well as the specific mechanical energy input (SMEC) across the screw configuration[3]. In order to setup the 1D model, the individual screw elements have to be characterized by determining the inherent conveying capacity and the pressure build up capacity[4]. The

characterization is done using the mesh-free SPH simulation approach, giving additional insight into the mixing and sheer rate distribution for every screw element, allowing for a comparison between triple- and double-threaded extruder screw elements.

Results

An HME process on a trilobed 16mm and a bilobed 27mm extruder, with five different materials, was investigated, providing recommendations for the process setup. Additionally, a scale-up scenario was investigated, using the proposed scale-up laws with an in-silico process DoE, with an experimental validation.

Literature

[1] J. J. Monaghan, "Smoothed particle hydrodynamics," *Reports Prog. Phys.*, vol. 68, no. 8, pp. 1703–1759, Aug. 2005.

[2] M. Bierdel, "Computational Fluid Dynamics," in *Co-Rotating Twin-Screw Extruders*, K. Kohlgrüber, Ed. Munich: Carl Hanser Publishers, 2008.

[3] A. Eitzlmayr, G. Koscher, G. Reynolds, Z. Huang, J. Booth, P. Shering, and J. Khinast, "Mechanistic Modeling of Modular Co-Rotating Twin-Screw Extruders," *Int. J. Pharm.*, vol. 474, no. 1–2, pp. 157–176, 2014.

[4] K. Kohlgrüber, *Co-Rotating Twin-Screw Extruders*. Munich: Carl Hanser Publishers, 2008.

Copyright © American Institute of Chemical Engineers. All rights reserved.

455147 Scale-up and Blender Change Model for the Pharmaceutical Lubricated Mixing Process

Abstract:

In the pharmaceutical lubricant mixing process, lubricity of granules after mixing was evaluated by measuring the contact angle with water and tablet hardness. The granules were prepared using three different types of popular tumble mixers; V-blender (effective volumetric capacity: 70 litters), Container blender (36 litters) and Bin blender (20 litters) with different powder loading rates (10-88%), blender rotation speeds (18-37 rpm) and mixing times (5-120 min) for model placebo formulation. Contact angle measurement and tablet hardness are useful as alternative characteristics for evaluation of granule lubricity. Based on the experimental data, Mixing Performance Index (MPI), which is an empirical equation including blender type, powder loading ratio, mixing rotation speed and batch size as functional parameters, were developed. Mixing Ability (MA) was defined by the combination of MPI and mixing time. Calculated MA exhibited good correlation with tablet hardness (correlation coefficient is 0.89) in all datasets. MA was verified by nine different drug product data with different manufacturing scales, to enable the MA model to support the formulation researcher to set mixing process parameters when the batch size or blender type changes.

Copyright © American Institute of Chemical Engineers. All rights reserved.

462399 Scaling Intra-Tablet Coating Variability

Abstract:

Pharmaceutical tablets usually exhibit variability in coating thickness, both across individual tablets as well as over a single tabletâ€™s surface. In this work, the coefficient of variation in coating thickness, referred to as intra-tablet coating variability, is examined as a function of processing scale. The discrete element method (DEM) is used to simulate the dynamics of two different tablet shapes in coating pans of four different sizes.Â The asymptotic intra-tablet coating variability is observed to increase with increasing pan size. Studies designed to explain the observed trend are also presented.

Copyright © American Institute of Chemical Engineers. All rights reserved.

454707 Improving Workflow for Modeling of High-Shear Rotor-Stator Wet Milling for Pharmaceutical Applications

Abstract:

Wet milling is an important unit operation for the size reduction of pharmaceutical powders. The benefits of wet milling to the pharmaceutical industry relies on its ability to control particle size distributions (PSDs) while managing containment concerns and avoiding dry powder handling of potent active pharmaceutical ingredients (APIs). Due to the high-value nature of APIs, a trial-and-error experimentally-driven scale-dependent approach to obtain optimal milling conditions would not be cost-effective.

The use of modeling-aided scale-up of high-shear rotor-stator wet milling processes using a standardized population balance equation (PBE) approach has been reported by this group in the past.[1-2] In this work, recent model improvements as well as the creation of templates for easy data collection and the actions taken to improve the workflow will be discussed. A few case studies from early to late phase small molecules and multiple scales of mills from lab to manufacturing will be used to exemplify the approach.

[1] Luciani et al. PBE-Aided Design of Slurry Milling Process to Reduce Particle Size of Intermediate and Active Pharmaceutical Ingredients, AIChE Annual Meeting 2012 (366b)

[2] Luciani et al. PBE-Aided Design of Slurry Milling Process to Reduce Particle Size of Intermediate and Active Pharmaceutical Ingredients. *Org. Process Res. Dev.*, 2015, *19* (5), 582â€"589

Copyright © American Institute of Chemical Engineers. All rights reserved.

468475 Characterizing Mixing Time Scales in an Exothermic Mixing-Limited Reaction

Abstract:

Exothermic mixing-limited reactions are commonly encountered in the pharmaceutical and fine chemicals industry, and are usually executed in semi-batch reactors. Their scale-up, however, requires a keen understanding of the heat and mass transfer aspects of the reaction and can be influenced by process parameters such as reagent dosing rate and agitation. In this contribution, we will present a case study where scale-up of a highly exothermic reaction resulted in variable yields in different reactor geometries due to heat and mass-transfer limitations. A systematic analysis of local meso-mixing rates with correlations by Bourne et al.[a] and computational fluid dynamics in laboratory and pilot plant reactors revealed presence of local temperature hot-spots, which is attributed as the main cause of variable reaction yields. As a result, a scale-independent meso-mixing rate based parameter was applied to obtain a high reaction yield at different scales.

[a]Bourne, John R. "Mixing and the selectivity of chemical reactions." *Organic Process Research & Development* 7.4 (2003): 471-508.

The work presented here was sponsored by AbbVie. AbbVie contributed to the design, research, and interpretation of data, writing, reviewing, and approving the publication. All authors are employees of AbbVie.

Copyright © American Institute of Chemical Engineers. All rights reserved.

471225 Synergistic Effect of Low Power Ultrasonication and Antimicrobial Peptide Action in Deactivation of *Listeria Monocytogenes*

Abstract:

Ultrasound is a relatively new method of food sterilization. Low intensity ultrasound can form transient pores in cell membranes due to shock waves generated by collapse of bubbles that are formed by cavitation. In conventional sterilization using ultrasound, the threshold of power needed for deactivation of microorganisms by thinning and rupture of cell membranes is very high. As a result, conventional ultrasonic sterilization may not be cost effective and may also result in loss of texture of food product. Since antimicrobial peptides kill bacteria by pore formation in cell membranes, transient pores formed by low power ultrasonication should result in enhancement of antimicrobial activity. Because of its relatively gentle action (low intensity), low frequency ultrasound is expected to have no adverse effect on food texture. Experiments were conducted for deactivation of *L. monocytogenes* using a naturally occurring antimicrobial peptide (AMP) melittin in the absence as well as in the presence of ultrasonication. In the absence of AMP melittin, ultrasonication has very small effect on cell density upto a power level of 40. However, at a higher power level of 60, we observed a dramatic decrease in cell density which implies cell lysis. At low AMP concentration, low power ultrasonication does not improve the antimicrobial activity. At high AMP concentrations, however, AMP was found to completely inactivate *L. monocytogenes.* The synergistic effect of AMP with ultrasonication was found to be the maximum at AMP concentration of 0.78 Î¼g/ml of melittin. A dramatic decrease in 2 orders of magnitude in cell density was observed for ultrasonication in the presence of 0.78 Î¼g/ml of melittin compared to either ultrasonication alone or AMP action alone. This clearly demonstrates synergistic effect of ultrasonication and AMP for pore formation in cell membrane. A mathematical model was developed to describe the formation of transient pores in cell membranes as a result of cavitation and its effect on pore formation and rupture of cell membrane due to antimicrobial peptide insertion to explain this synergy.

Copyright © American Institute of Chemical Engineers. All rights reserved.

467749 Membrane Based Process for Polyphenols Recovery from Winery Effluents

Abstract:

Introduction:

Winery effluents are rich in added-value compounds, among which the polyphenols stand out by their antioxidant properties. In wine-producing regions, these effluents are abundant and a cheap and attractive source for the recovery of these natural antioxidants [1]. The recovery of these compounds reduces the environmental impact of wineries, representing a significant enhancement in maintaining the environmental balance, with gains in economic and environmental issues. In this regard, the development of efficient and environmentally friendly processes for the recovery and concentration of polyphenols from agro-industrial effluents is required.

In this sense, due to their inherent characteristics (low energy requirement, no additives, mild operating conditions, separation efficiency and easy scaling up), membrane processes like ultrafiltration and nanofiltration have been employed to recover and concentrate polyphenols from agro-food extracts and effluents.

In a previous work [2], a sustainable process of aqueous extraction associated to microfiltration (MF) for the recovery of natural antioxidants from wine lees, generated in the second racking of red wine making, was developed. More than 20% of the total content of polyphenols present in the raw effluent were recovered in the MF permeates. These are complex mixtures mainly composed of polyphenols, anthocyanins, organic acids, polysaccharides and minerals.

Literature data [3] report that membrane processes applied to concentrate complex mixtures containing polyphenols lead to drastic permeation fluxes decline and severe problems of membrane fouling. The present work addresses the ultrafiltration(UF) and the nanofiltration (NF) of the MF permeates obtained in a previous work [2] and evaluates the effect of operating conditions of transmembrane pressure and feed flow rate on the productivity (permeation fluxes) and on the selectivity (solute rejections). The quantification of this effect and the assessment of concentration polarization is carried out through the film theory model.

Materials and Methods

Permeation experiments were realized in laboratory flat-cell units with a membrane surface area of 14.5 x 10^{-4} m^2, using the NF270 and ETNA01PP membranes, with molecular weight cut-off (MWCO) of 300 and 1000 Da, respectively. Experiments were conducted in total recirculation mode, where the permeate and the retentate streams were recirculated to the feed tank to study the variation in the permeate fluxes and the solute rejection coefficients. The stabilization time for each experimental run was 30 min, after which permeate samples were taken for chemical analysis. These permeation runs were performed at feed flow rates of 100, 150 and 200 L h^{-1}, transmembrane pressures of 3, 5, 7 and 15 bar and using as feed solution the MF permeates with the composition characteristics displayed in Table 1.

Table 1. Physico-chemical characteristics of the feed solution.

Parameter	**MF Permeate**
Total Organic Carbon (mg L^{-1} C)	716 ± 10.8
Turbidity (NTU)	< 1.0
Conductivity (μS cm^{-1})	241 ± 2.0
Total Polysaccharides (mg L^{-1} Glucose)	10.1 ± 0.4
Total Polyphenols (mg L^{-1} GAE)	26.1 ± 0.1
Monomeric Anthocyanins (mg L^{-1} Mv3g)	4.20 ± 0.1
Total Antioxidant Activity (mM TEAC)	10.1 ± 0.1

Results and Discussion

Nanofiltration permeation fluxes with the NF270 membrane increase linearly with transmembrane pressure and are independent of the feed circulation flow rate.

The ETNA01PP membrane displayed typical patterns of UF permeation fluxes, where at low transmembrane pressures (up to ~5 bar) the permeation fluxes increase linearly and at higher transmembrane pressures the limiting flux is reached. In fact, working at feed flow rates of 100 and 150 L h^{-1}, the ETNA01PP membrane do not yielded substantial difference in permeation fluxes, while at 200 L h^{-1}higher permeation fluxes were achieved. Besides, for the investigated membranes, the solution permeability decreases with increasing the membrane MWCO. This drop in permeability is boosted by increasing pressure and lowering feed circulation flow rate, that is, such conditions favor the occurrence of concentration polarization phenomenon, a precursor of the membrane fouling.

The tighter membrane (NF270) displayed higher rejection coefficients, i.e., 90% for polyphenols and close to 100% for anthocyanins, providing a permeate with very low antioxidant activity. On the other hand, the ETNA01PP showed moderate polyphenols and anthocyanins rejections (40-50%), showing that the polyphenols preferably permeate this UF membrane. Besides, for both the membranes, the rejection coefficients for polyphenols increase with increasing transmembrane pressure and are virtually independent of feed circulation flow rate. This behavior is attributed to the fact that at higher transmembrane pressures concentration polarization phenomenon is more severe, leading to the formation of a more selective layer on the membrane surface, and therefore increasing the rejection coefficients. The occurrence of concentration polarization was quantified by the film theory model, showing that the concentration of solutes on the boundary layer adjacent to the membrane (C_{am}) increased with adverse hydrodynamics (low values of feed circulation rates and high values of transmembrane pressure).

Thus, according to the aforementioned data, the integration of different membrane separation processes is an alternative for the recovery and concentration of antioxidants (phenolic compounds and anthocyanins) from the second racking effluents. In such a process, a permeate with antioxidant compounds can be obtained by means of microfiltration associated with water dilutions (aqueous extraction), and can subsequently be concentrated by nanofiltration, resulting in a product with high antioxidant properties.

Acknowledgements

The authors are grateful to the Brazilian funding agencies (FAPERGS, SEBRAE/RS, CAPES and CNPq) and to the Portuguese funding agency (FCT) for their financial support, as well as to the Vinícola Almaúnica (Brazil) by the financial support and for providing the effluents.

References

[1] Balboa, E. M., Soto, M. L., Nogueira, D. R., González-López, N., Conde, E., Moure, A., Vinardell, M. P., Mitjans, M., Domínguez, H., Potential of antioxidant extracts produced by aqueous processing of renewable resources for the formulation of cosmetics, Industrial Crops and Products, Vol. 58, No. 0, pp 104-110, 2014.

[2] Giacobbo, A., do Prado, J. M., Meneguzzi, A., Bernardes, A. M., de Pinho, M. N., Microfiltration for the recovery of polyphenols from winery effluents, Separation and Purification Technology, Vol. 143, No. 0, pp 12-18, 2015.

[3] Minhalma, M., Dias, C., De Pinho, M., Membrane fouling in ultrafiltration of cork processing wastewaters, Advances in Environmental Research, Vol. 3, No. 4, pp 539-549, 2000.

Copyright © American Institute of Chemical Engineers. All rights reserved.

462144 Impact of Post-Harvest Drying and Extraction Method upon Total Phenol, Antioxidant Potential, and Inhibitory Activity of Black Walnut Husks

Abstract:

The fruit of the black walnut consists of the kernel, shell and husk. The kernel is often used for consumption and the shells as an abrasive and filler, however the husks are often discarded as compost or landfill. Walnut husks contain tannins and polyphenolic compounds which exhibit antimicrobial properties and can treat inflammation. The naphthoquinones in walnut husks can be used as broad spectrum biocides, bio-herbicides, and have been demonstrated to have cytotoxic effects against melanoma cells. Supercritical fluids, which have high diffusivity are excellent solvents for extracting chemical species from plant matter. Supercritical carbon dioxide, by itself has a relatively low dielectric constant and ethanol can be added as a modifier to help improve the extraction of polyphenolic compounds. For this study, black walnuts were extracted either using ultrasonic assisted extraction using ethanol or supercritical carbon dioxide extraction with an ethanol modifier. The effects of temperature and ethanol content with constant density supercritical carbon dioxide with ethanol upon the extraction of ground walnut husks was evaluated using a factorial design of experiments. Temperature was varied between 50 and 70C and ethanol mole fraction between 10 and 20%. Both dried and damp walnut husks were evaluated with damp walnut husks exhibiting higher antioxidant potential. The optimal extraction conditions were found to be 68 °C and 20 wt-% ethanol in supercritical carbon dioxide. At these conditions, the antioxidant potential as measured by the ferric reducing ability of plasma (FRAP) assay was 0.027 mmol trolox equivalent/g (mmol TE/g) for dried walnut husk and 0.054 mmol TE/g for walnut husks that were not dried. The antioxidant potential was also evaluated using the total phenolic content (TPC) assay and DPPH assay. The antimicrobial inhibition, quorum sensing inhibition, as well as chemical composition were also evaluated and will also be presented.

Copyright © American Institute of Chemical Engineers. All rights reserved.

Superheated Steam in Spray Drying of Milk

Amanda Lum, Shahnaz Mansouri, Karen Hapgood, Meng Wai Woo
Department of Chemical Engineering, Monash University, Clayton, Victoria, Australia
Corresponding email: meng.woo@monash.edu

Introduction

Superheated steam, also known as the condition of steam above its vaporising point has a wide range of application. Its application extends from cleaning, sterilization, heat treatments, production of energy, drying and many more. Along with its increasing attention, the exploitation of superheated steam and its advantages in the drying technologies have been actively studied in the recent years.

In the last decade, the incorporation of superheated steam in various drying processes have been well established with several existing commercial superheated steam dryers in the industry. The integration of superheated steam in place of hot air such in various conventional dryers have been successfully exploited to a vast variety of applications. In principle, any conventional hot air dryer can be transformed into a superheated steam dryer through various modifications although most reformations differs from another and are in general based the different drying imperative required[6].

To this date, the manipulation of superheated steam in the drying technologies is substantially ascertained. However it is noted that its applications are targeted towards the dring of solids or sludge materials. Its application on liquid state materials have yet to be realised and limited research as well as field experiences are found in the literature. Therefore, the use of superheated steam and its effect on solid formation during the liquid state drying such as those containing dissolved solids or droplet suspensions in spray drying have yet to be uncovered. This gap in research understanding have been recognised by a few authors which have affirmed the need to understand the potential of superheated steam in liquid drying processes.

In this report, superheated steam was used in place of hot air in the drying of milk droplets to understand how superheated steam affects element migration and segregation in a multicomponent system. One of the main challenges of powdered milk produced by current production processes are the poor powder reconstitution properties such as powder wettability and flowability due to the accumulation of surface fats during the powder production process. Among the different stages of reconstitution process, it was noted that the wettability of powders is usually the rate-limiting step [4] thus placing effective wetting properties to have more importance as compared to the rest of the reconstitution stages. Wettability of powders which are also commonly represented by Washburn's equation is reported to be largely affected by the powder surface composition[2].

It was therefore proposed that the use of superheated steam as the drying medium would lead to a change in elemental migration pattern during the particle formation process. With a hydrophilic nature, the incorporation of steam during the particle formation process was hypothesized to modify and control the movements of the individual elements within the multicomponent system. This would potentially allow for the engineering of dairy powders according to their specific needs.

Materials and Methods

This study was conducted via a single droplet analysis with set up as shown in Figure 1 to closely observe the behaviour of milk droplets throughout its drying period. A 2.5 ± 0.05 μL of fresh full cream milk (Pura Milk, Australia) droplet was suspended on a glass filament designed for such study. The temperature of the initial fresh milk droplet was maintained at 23°C throughout the study. Saturated steam generated from a steam generator was allow to be further heated into its desired superheated condition of 110°C by a prototype single pass tube and shell heat exchanger with hot air. Similarly, studies on hot air drying was performed as a basis of comparison with superheated steam drying.

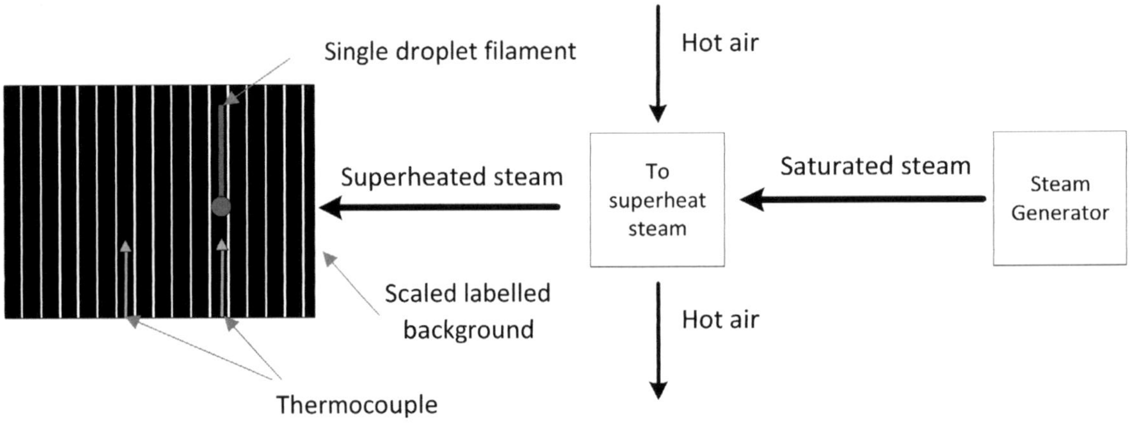

Figure 1: Superheated steam single droplet drying setup

A prototype steam trap was also designed to capture condensed water droplet in the free flowing stream to ensure that the steam is adequately superheated throughout the experiments. A digital HD video camera recorder (Sony HDR- PJ430) with 3 full sets of close up-lens (Hoya 46mm Multicoated Close-up Lens Set including +1, +2, +4 Diopter Strengths) were used to record the entire drying process.

The study was carried out in an open system setting instead of the required close system analysis from the literature due to the indubitably condensing steam which would hinder the observation and recording of the drying process. The temperature of the desired drying condition is closely monitored by thermocouples placed in the vicinity of the suspended droplet. Droplets were left to be fully dried prior to the post experimental analysis.

In situ wetting study

One of the common direct elemental surface analysis on particle surface is via its surface wettability. In situ wetting test was conducted to investigate the wetting ability which is also the hydrophlicity of the particle surface after drying. The analysis was carried out by resuspending a 2 ± 0.05 μL milli Q water (Millipore, Australia) droplet onto the tip of the dried milk droplet. The subsequent changes after was being further recorded which was then used to measure for its initial apparent projected contact angle (APCA). Due to no observable difference in solubility between the two drying mediums, the dissolution test was not further pursued.

Control study

A set of control study was implemented to define the extreme cases of wettability. Lactose which are water soluble are dried under identical hot air drying conditions and were used to represent surfaces which have a high degree of wettability. On the other side, Vaseline, a petroleum based product was coated onto the single droplet glass filament to denote a poor wettability surface. Both control runs provide a bench mark to the extreme cases in the wettability scale which results obtained from this study was compared to.

Results and Discussions

Full cream milk experiments

The recorded drying history of both superheated steam as well as hot air drying shows similar gradual shrinkage as expected in the literature. With close observation, the corresponding slight enlargement of droplet size as reported in the literature was also noted [3]. This phenomena is also known as the initial condensation of superheated steam which occurs as they come into contact with a cooler sample.

The sequence of the respective drying progression was analysed based on a 10 seconds interval as shown in figure 2. It was observed that hot air drying leads to an earlier final shape as compared to those dried under superheated steam. A constant shape was observed after nearly 40s into the drying process under hot air condition whereas steam-dried milk droplet was delayed until after 80s of drying. Given both having a similar drying period, the superheated steam drying scheme was reflected to cause the droplets to have a slower gradual shrinkage and thus leading to a longer shrinkage period. Such trends was postulated to be due to the occurrence of case hardening also known as crust formation at the outer layer of the droplet which is one of the common traits in hot air drying [7]. The formation of crust over the surface of the droplets would likely cause the droplet to hold a fixed figure in a much earlier stage which thus leads to the inhibition of internal moisture transfer. On the other hand, it was reported that superheated steam is less susceptible and often does not cause case hardening thus possibly leading to the delay in forming a fixed final shaped [1].

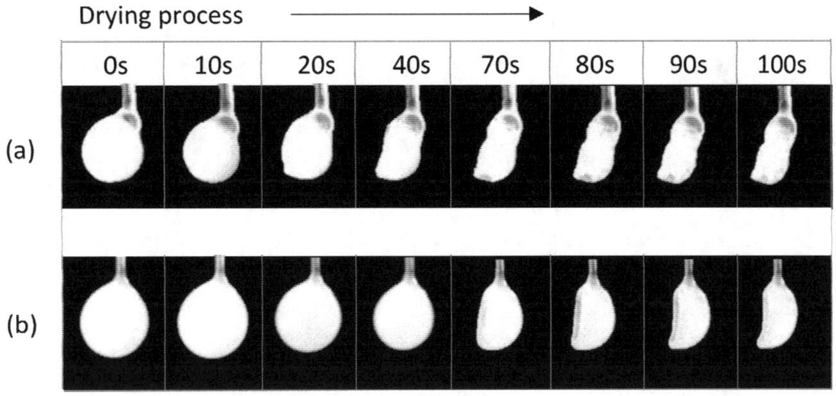

Figure 2: Drying of milk droplets at 110°C with 10s time lapse. (a) Hot air drying (b) Superheated steam drying

Based on the analysis of the droplet size for the superheated steam drying runs, an initial condensation with an increase of 4.8% in droplet size was observed in the first 10s into the drying process. The gain in moisture was quickly evaporated as the droplet size shrank to its initial size approximately 3s after. Besides that, a relatively constant size reduction in steam dried droplet was also detected in the first 50s of evaporation before they could not be presumed as spherical while hot air dried droplets were seen to deform after 20s. This observation was supposed to be due to the effect of crust formation during the hot air drying process at which surface moisture evaporates at a much faster rate when compared to the outwards diffusion from its core such that the surface experiences a lack of moisture thus solidifying. The layer of solid shell being form inhibits the diffusion of moisture to the drying front.

Examining the drying behaviour of superheated steam and hot air, it was realised that the drying temperature experience by the droplets in the constant rate period was different between the two despite an identical drying temperature was being delivered by the mediums. In the case of hot air, the droplet experiences a constant wet bulb temperature which are interdependent between the dry bulb temperature and relative humidity of the ambient condition. Being independent of the atmospheric conditions, a constant boiling temperature at the specific operating pressure is being encountered by the droplets. This further denotes that milk droplets being dried under superheated steam conditions experiences a much greater temperature at its drying front when compared to hot air drying.

Furthermore, an immediate elevation of the product temperature to the boiling point upon contact with superheated steam was also reported [8]. Comparing to the gradual increase in product temperature in the hot air drying process, steam dried milk droplets was perceived to experience evaporation at a higher temperature for a longer period in a given time frame. This may have contributed to the difference in observation in the drying history of hot air and superheated steam. It was proposed that the lack of mass transfer resistance at the droplet drying front, as moisture evaporates into superheated steam have prevented the formation of crust and thus allowing for a more uniform evaporation as observed in Figure 2(b).

Wetting Test

An immediate examination on the surface hydrophilicity on the dried milk droplets was done via an in situ wettability test. The apparent projected contact angle (APCA) was used as a rough measurement to particle surface wettability and thus indicating surface hydrophilicity. A large APCA signifies higher degree of coverage of the apparent resuspended water over the dried droplet, thus representing a more wettable surface, directing towards greater surface hydrophilicity, and vice versa. APCAs are evaluated based on the contact angle between the tangent to solid and liquid interface at the vapour –liquid-solid intersection or as indicated as alpha (α) in figure 3. A smaller α contact angle imposes a more hydrophilic surface which the inverse for hydrophobic surfaces. APCA provides an approximated quantification to the surface wettability of the dried particle.

The set of control wettability measurements were generated as a baseline for comparison; an absolute hydrophobic surface and complete hydrophilic surface which are used as a standard to assess the surface wettability of the dried milk surfaces from this study. Resuspended water droplet was observed to have a much lesser coverage over the hydrophobic glass filament coated with Vaseline compared to the hydrophilic surface. Such observation was as predicted, with the hydrophobic surface resulting in the highest α APCA of 90.7^{o} and hydrophilic surface obtaining the lowest of 18.5^{o} in the spectrum.

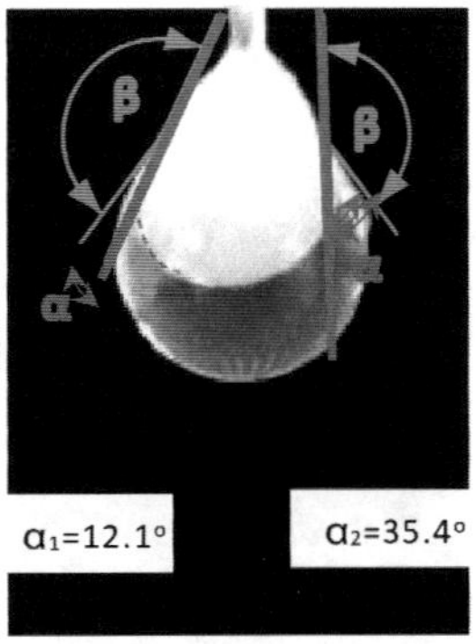

Figure 3: APCA measurements for wetting analysis.

Based on the evaluated results between hot air and superheated steam drying, hot air dried milk droplets was reported to retain an overall greater APCA as compared to superheated steam dried milk droplets. Evaluating from 30 individual trials, an overall average α APCA of approximately 70.4° was obtained from hot air drying while 28.8° from superheated steam drying.

It is apparent that both hot air dried milk droplet and superheated steam dried milk falls within the range of the extreme cases of the control spectrum. By evaluating the α APCA generated from the study, milk dried under hot air was observed to position towards the hydrophobic end of the scale while the steam dried milk surfaces are towards the hydrophilic end. Such observation have primarily suggested the possibility of a shift in component migration when milk droplets are dried under superheated steam as compared to hot air. A lower α APCA suggested that the surface of the steam dried milk droplet are more wettable and thus are more hydrophilic relative to those dried under hot air.

Lactose and Whey Protein Experiments

In view of the interesting observations with the full cream milk experiments, additional experiments were undertaken with one of the main components of milk: lactose. Lactose solution with 10%wt solid was dried under different superheated steam temperature, 110°C, 130°C, and 145°C. It was observed that complete drying of lactose at 110°C was not possible. Lactose droplets were seen to shrink until it reaches a saturated state at which the droplet is sticky and rubbery as shown in Figure 4 (a). Further evaporation was not achievable despite a longer drying period. It was postulated that such phenomena was due to the insufficient drying force between the medium and the droplet as the drying rate was mainly governed by the temperature difference between the drying medium and surface temperature (T_m - T_s). Given the drying medium operates at 110°C, a 10°C difference may not have provided a sufficient driving force with the temperature at the droplet surface is at its boiling temperature, 100°C.

Drying was proceeded with an increase in the superheated steam drying temperature to 130°C and 145°C. Similar observations were observed at which the size of the droplets were seen to reduce until its saturated state. A disturbance was attempted with the intention of examining the state of the drying droplet. Interestingly, the saturated lactose droplet was seen to bubble as soon as the droplet was interrupted. After the vigorous bubbling, the droplet was observed to turn opaque and thus solidify. Such observation was suspected to be caused by the inhibition of further moisture extraction from the droplet due to a saturated layer being form around the surface of the droplet as soon as the condition of the surface is in equilibrium with the medium. The saturated layer does not solidify such like air but sufficient to prevent and hold moisture from further evaporation.

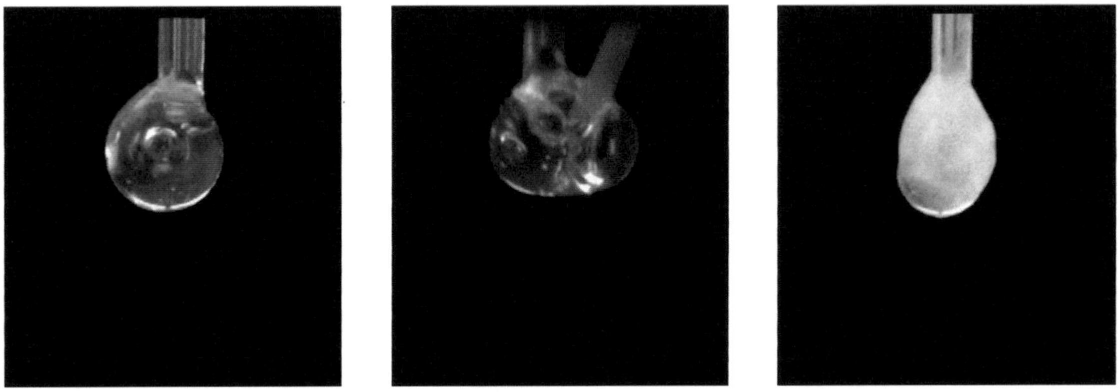

(a) (b) (c)

Figure 4: Lactose drying under superheated steam condition. (a) Lactose at saturated state, (b) Disturbance to the drying droplet and (c) Final state of lactose after disturbance.

Superheated steam dried lactose were observed under SEM and have shown possible formation of crystals structures as presented in Figure 5. Although further investigation is required to confirm the degree of crystallinity of these lactose crystals, the preliminary study have identified the possibilities of lactose crystal formations and thus indicating the possible in situ crystallisation of lactose under a superheated steam spray drying condition.

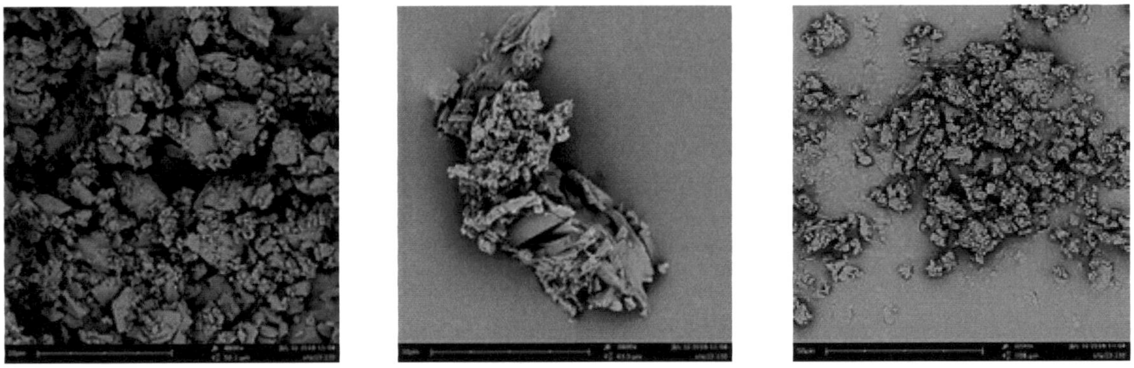

(a) (b) (c)

Figure 5: SEM image of crystallized lactose dried under superheated steam condition.

Similar drying procedure was conducted in the preliminary study of WPI at which solutions are dried under both hot air and superheated steam at 110°C. It was observed that the droplet immediately became opaque upon contact with steam followed by progressive shrinkage of the droplet (results not shown here for brevity). Shown in Figure 6 are the SEM images of WPI dried under hot air as well as superheated steam conditions. Hot air dried protein molecules are seen to cluster and forms a continuous network like structure. While define spherical protein molecules were observed when dried under superheated steam. The cause of such observations are still under investigation but is preliminary suspected to be due to the precipitation of protein particles when undergoing high heat treatment from superheated steam.

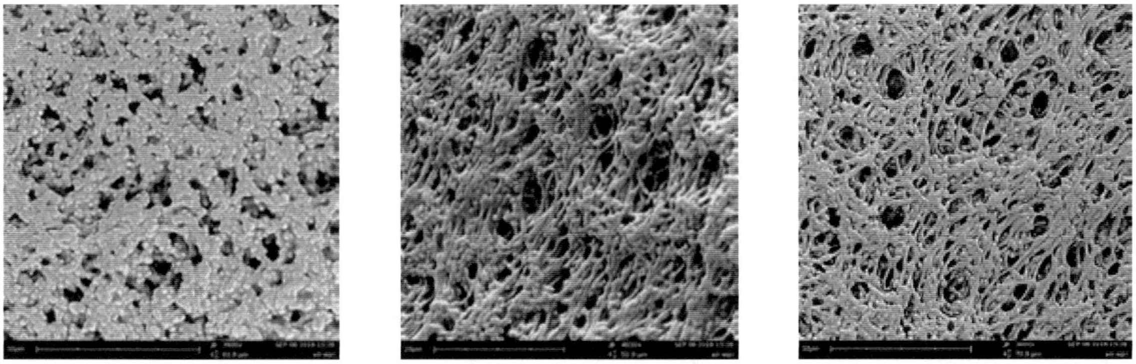

(a)

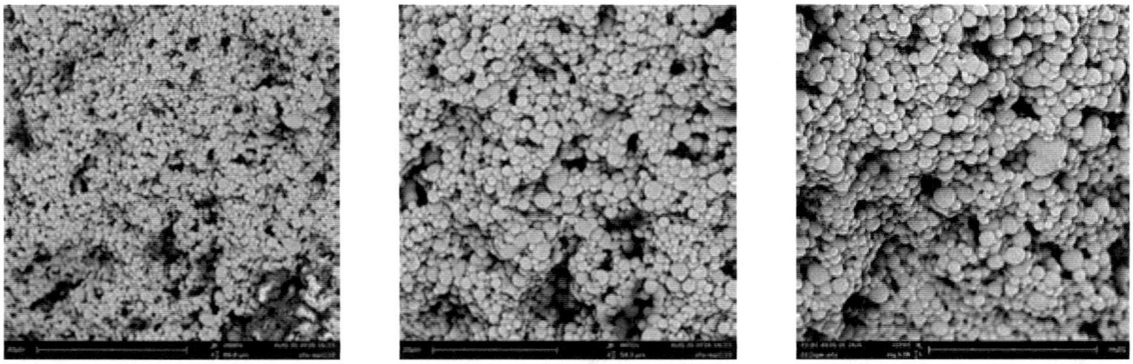

(b)

Figure 6: WPI dried under (a) Hot air and (b) superheated steam conditions

Conclusion

The potential effect of superheated steam on the relocation of milk components during the spray drying of milk was observed in this study. A preliminary investigation on surface wettability between air dried milk droplet and steam dried milk droplet were conducted. It was found that milk dried under superheated steam conditions are seemingly more wettable due to the observed lower APCA obtained. The outcomes suggested the possibility of shift in the migration of individual components in milk during the particle formation process. Initial study on the distinct component found in milk suggested possible crystallisation behaviour as well as interesting changes in protein molecules morphology when undergoing drying under superheated steam conditions. Future work will be carried out in a prototype superheated steam spray dryer for further studies and evaluation to understand the impact of the medium on milk powder formation.

References

[1] BHATTACHARYA, S. 2014. *Conventional and Advanced Food Processing Technologies,* National Technical University of Athens, Zografou, Greece, Wiley.

[2] FÄLDT, P. & BERGENSTÅHL, B. 1996. Spray-dried whey protein/lactose/soybean oil emulsions. 2. Redispersability, wettability and particle structure. *Food Hydrocolloids,* 10, 431-439.

[3] IYOTA, H., NISHIMURA, N., YOSHIDA, M. & NOMURA, T. 2001. SIMULATION OF SUPERHEATED STEAM DRYING CONSIDERING INITIAL STEAM CONDENSATION. *Drying Technology,* 19, 1425-1440.

[4] J. KIM, E. H., CHEN, X. D. & PEARCE, D. 2005. Melting characteristics of fat present on the surface of industrial spray-dried dairy powders. *Colloids and Surfaces B: Biointerfaces,* 42, 1-8.

[5] LANGRISH, T. A. G. 2008. Assessing the rate of solid-phase crystallization for lactose: The effect of the difference between material and glass-transition temperatures. *Food Research International,* 41, 630-636.

[6] MUJUMDAR, A. S. 2014. *Handbook of Industrial Drying, Fourth Edition*, Taylor & Francis.

[7] RAHMAN, M. S. 2007. *Handbook of Food Preservation, Second Edition*, CRC Press.

[8] VAN DEVENTER, H. C. & HEIJMANS, R. M. H. 2001. DRYING WITH SUPERHEATED STEAM. *Drying Technology,* 19, 2033-2045.

466920 Mechanical Steam Treatment of Plant Material

Abstract:

Since approx. 15 years it is known by the research work of the Lemgo University of Applied Sciences that bacterial cells can be separated efficiently from plant surfaces by flash evaporation. A short steam treatment is followed by an extremely fast evacuation which results in an intensive evaporation that reduces the cell count of microbes.

First investigations were undertaken in a special lab-scale VSV-apparatus (Vacuum-Steam-Vacuum) with a treatment chamber of 0.7 L. The plant material as a fixed bed and the model systems (inoculated wooden lamellae or small glas plates) were treated with saturated steam. Even heat resistant bacterial spores are reduced up to 4 decades in only 10 s steam treatment of approx. 120 °C /1/. The formation of a thin layer onto the surfaces of the model material is important to reduce the bonding forces between the cells and the surface. The obtained type of kinetic is unusual and can be divided in two steps: a first phase with a steep linear spores reduction; with an activation energy E_A of about 10 kJ/mol and a second step with a flatter linear diminution with an E_A of ca. 200 kJ/mol (typical for heat-influenced inactivation). The 2nd step is not necessary for an industrial application, resulting in a short batch time of an industrial process. There must be a minimum temperature of the saturated steam depending on the specific bonding force in the range of 110 to 125 Â°C /2/.

A lot of research was done in the last years to decontaminate dry plant materials (spices, herbs and drugs) which confirms the mild process parameters giving a good quality concerning active substance concentration. Additional experiences are existing of an industrial scale decontamination with a solid mixer as treatment chamber.

In this contribution a summary of the research work of this â€œMechanical Steam Decontaminationâ€ is given.

/1/ MÃ¼ller U, Wilhelm P.: Lebensmitteltechnik 2002; 9: 58-59

/2/ Lilie M, Dissertation, University of Bielefeld/D. 2009

Copyright © American Institute of Chemical Engineers. All rights reserved.

Influence of Seed Pre-Soaking and Oil Body Washing Regime on the Physical Quality of the Final Suspension

Zarani Mat Taher[1, 2], Tim Foster[1], David Gray[1*]

1. Division of Food Science, School of Bioscience, University of Nottingham, Sutton Bonington Campus, LE12 5RD United Kingdom
2. Institute of Bioproduct Development, Universiti Teknologi Malaysia, Skudai 81310 Johor Malaysia

Introduction

Oil bodies (oleosomes) are tiny subcellular (diameter around 0.5μm to 3μm) spherical organelles that provide energy during germination or post-germination. Oil bodies or also known as oleosomes have been extensively studied during the last decades (Garcia *et al*.,1988, Murphy and Cummins, 1989, Tzen and Huang 1992; Huang, 1996), in different plant species such as maize germ (Nikiforidis *et al*.,2013), sunflower seeds (Beisson *et al*.,1996, White *et al*.,2006, Fisk *et al*.,2008) , rice bran (Nantiyakul *et al*.,2012), oat (White *et al*.,2006), soybean (Iwanaga *et al*.,2008, Waschatko *et al*.,2012), rape seed (Tzen *et al*,1993, Katavic *et al*.,2006, Jolivet *et al*.,2011, Jolivet *et al*.,2009,) and *Echium plantagenum* (Gray *et al*.,2010).

From the structure point of view, oil bodies are mainly composed of triacylglycerol (TAG) in the core, surrounded by monolayer of phospholipids (PL) and the surface of the oil body is covered with a protein known as oleosin along with minor proteins such as caleosin and steroleosin (Tzen *et al*., 1993). Oleosins are alkaline proteins with a molecular weight in the range of 16-25kDa; they confer a net negative charge on the surface of oil bodies at neutral pH by evaluating the surface charge of your OBs in comparison with literature behaviour. (Tzen and Huang, 1992, Tzen *et al*., 1990). Oleosin is synthesised during seed development and maturation; and is believed to play an important role on the physically stabilising these organelles *in vivo* (Fisk *et al*., 1998). Due to its physical and chemical stability, oil bodies have received attention as natural oil droplets that can be recovered from seeds through wet-milling with potential use in a range of food and non-food applications such as pharmaceutical and cosmetic (Iwanaga *et al*., 2007). Furthermore, these natural emulsions offer versatility as a food ingredient.

Despite have been characterised, the effect of recovery regimes and post-recovery (downstream) processing on the quality of oil bodies extracted from sunflower seeds has not been extensively reported. The aim of this work is to determine the effect of seed pre-soaking on the quality of recovered oil bodies. Physical measurements such as droplet size and charge were made, and washing recovered oil bodies was used as a diagnostic of oil body integrity.

Material and methods

2.1.1. Materials

Sunflower seeds were purchased from Cargill Inc. The seeds, cultivated in Europe, were harvested in October 2013, transported to Nottingham and de-hulled using hammer mill (Glen Creston, UK). All chemicals used were of analytical grade (Sigma Aldrich, UK).

2.1.2. Recovery of oil bodies

Oil bodies were extracted from the seeds that have been soaked in deionised water for 2, 6 and 12 hours; pre-soaked seeds were drained prior to wet-milling. 100 gram of partially de-hulled sunflower seeds ('dry' or pre-soaked) was wet-milled in deionised water (1:4 w/w solids to milling water) using a kitchen blender (Krupss,UK) on full power for 2 minutes. The homogenate was then filtered through three layers of cheese cloth under vacuum. The slurry was centrifuged at 10,000 rcf, for 20 minutes at 4°C. The cream fat pad was collected to obtain the crude oil body cream and labelled as crude oil bodies (COB).

2.1.3. Washing Crude Oil Bodies

10g of crude cream was re-suspended (40ml) either in deionised water, or 0.1M sodium bicarbonate (pH 8.3). The process was repeated three more times using the same solution. For urea washed, oil body was re-suspended once in 40ml of 9M urea solution and rinse three times with deionised water. All the slurry was homogenised using a Potter Elvehiem (Wheaton, USA) at 700 rpm for ten passes for uniform dispersion followed by centrifugation (Thermo Jouan CR3i Multifunction, UK) at 2400 rcf for 20 minutes at 4°C. The final washed material called washed oil bodies (WOB) was kept in 4°C for a maximum of 48 hours prior to use for analysis.

2.2. Physical measurement

2.2.1. Zeta potential

The zeta potential (ζ-potential) represents the indicator of colloidal dispersion stability. This property was determined using Delsa Nano Particle Analyser (Beckman Coulter Inc. USA). This method calculates ζ-potential by measuring the direction and velocity of the droplets in an applied electric field. Emulsions were then injected into the measurement chamber of the particle electrophoresis instrument (Delsa Nano Particle Analyzer, Beckman Coulter,Inc., USA).

2.2.2 Particles size

The droplet diameters of oil body suspension were determined by using a laser diffraction particle size analyser equipped with universal liquid module (LS 13 320 Beckman Coulter, High Wycombe, UK) using distilled water as fluid. The particle size value reported for each sample type as the surface mean diameter D (3, 2). The particle size reported for each sample type was determined from the average of three samples.

2.2.3. Microscopy

The microstructure of emulsions and sunflower seed sections were determined using optical microscopy (Nikon microscope Eclipse E400, Nikon Corporation, Japan). A drop of the emulsion was placed on a glass slide and covered with a cover slip. The prepared glass slide was observed under the microscope at a magnification of 40 x. Transmission Electron Micrograph (TEM) was prepared as described previously by White *et al.*, 2006.

2.2.4 Statistical analysis

Statistical analysis was performed in Minitab version 16.2.2 software. Significant differences were identified by Post-Hoc Tukey's test ($P < 0.05$).

3.0 . Results and discussion

3.1 Imaging

Electron micrograph showing oil bodies was displayed in Figure 1 and Figure 2. Figure 1 shows the features of oil bodies when viewed using a Transmission Electron Microscope

(TEM). It can be seen that oil bodies were tightly packed within the cell however, they still remained as individual organelles. However, image of oil bodies was flattened under TEM resulting from dehydrated during sample preparations (Cabra and Samso, 2015). Image from microscopy indicates that the size of oil body from sunflower seeds approximately 0.5-3.5μm. It can be seen that oil body was appeared in shining spherical structure and occupied within the cytoplasm region and agreement with other author (Nikiforidis *et al.*, 2013, Neuberger *et al.*, 2008).

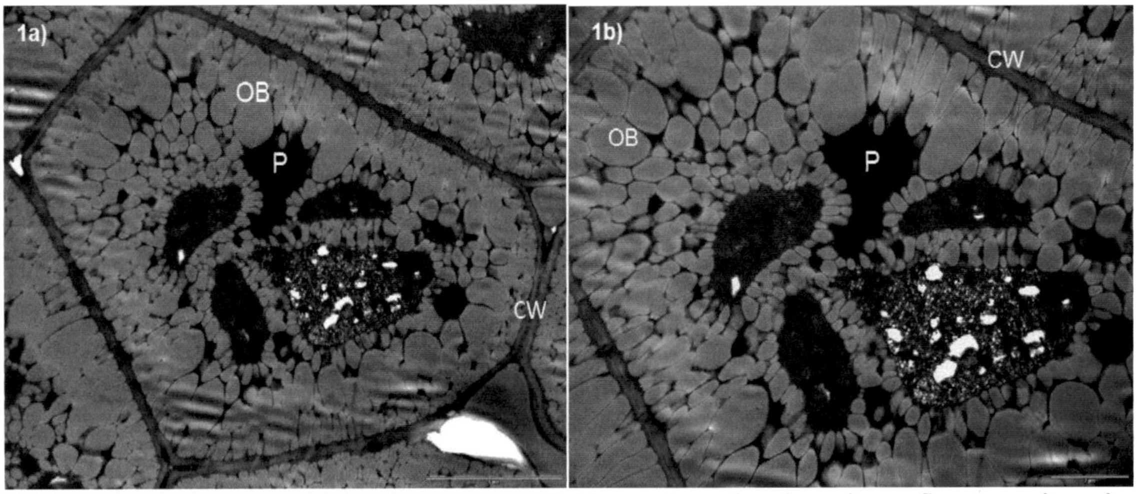

Figure 1: The structure of oil bodies appear as light grey irregular shape in sunflower seeds under TEM. **OB** = oil bodies, **CW** = cell wall **P**= plastid. Scale bars represent 1a) 5 μm 1b) 2 μm

3.2 Effect of seed pre-soaking on the size of recovered oil bodies

It was hypothesized that seed pre-soaking prior to wet milling would soften the tissues and minimise the breakage of oil bodies experienced during grinding process. The influences of high mechanical forces experienced during grinding, resulting changes in sunflower seeds chemical surface chemistry or led to damage of oil bodies housed within the cytoplasm as damaged seeds may potentially reduce the oil yield and quality (Gunasekaran *et al.*, 1988, Parde *et al.*, 2002).

The influence of seed pre-soaking times (0h, 2h, 6h, and 12h) on the physical stability of oil bodies was studied. The physical state of seeds during wet milling had a significant impact on the quality of the oil body suspension. The volume mean diameter of oil bodies measured after the pre-soaking is presented in Table 1.

Table 1: Effect of seed soaking time on the volume mean diameter of oil bodies.

Seed soaking time (hrs)	Crude oil bodies (μm)	Washed oil bodies (μm)
0	11.06 ± 0.5^{a}	5.08 ± 0.3^{c}
2	11.20 ± 0.2^{a}	5.12 ± 0.4^{c}
6	9.08 ± 0.6^{b}	3.13 ± 0.0^{d}
12	9.01 ± 0.6^{b}	3.15 ± 0.0^{d}

Note: $^{a\text{-}d}$ Means values that do not share the letter are significantly different. Tukey 95% Simultaneous Confidence Intervals

Results shown in Table 1 suggest that the average volume mean diameter ($D_{4,3}$) of crude oil bodies has decreased with the increasing of pre-soaking time. The size of crude oil bodies in samples pre-soaked for 6 hours was slightly bigger (9.08 ± 0.6 μm). The size of the crude sample was slightly bigger due to carry-over of cell contaminants or any interference of non-solubilise pigments or aggregation. Since 12 hours of soaking showed similar trend to 6 hours thus, it is suggested that 6 hours for seed pre-soaking is an optimum time. The results indicates that plasticisation of the seed through pre-soaking in water reduces the damage of oil bodies during wet-milling. The results indicates that larger oil bodies droplets (above normal size of oil bodies in plant, 3.5 μm) may be due to coalescence or hybrid formed during grinding process, which potentially weaken the oil bodies.

3.2.1 Impact of washing COBs from ALL seed preparations on protein/lipid composition

Crude oil bodies, prepared from seed that had been soaked for 6 hours, were washed repeatedly with a solution to remove extraneous protein and seed contaminants; a range of washing solutions was tested. The proximate composition of these preparations is shown in Table 2. The mean moisture content of oil bodies in this study was 45.7 ± 1.3 %. In general, an increase in washing power resulted in removal of proteins from the crude preparation and a consequential relative increase in lipid content of the extract. Therefore, urea-washed was the promising washing regime to remove protein leaving behind integral protein content.

Table 2: Lipid and protein content (weight dry basis) of oil bodies prepared using different washing regimes

Sample	Lipid content (%)	Protein content (%)
Crude	87.0 ± 0.0^{b}	8.06 ± 0.1^{a}
Water-washed	84.9 ± 0.8^{c}	5.16 ± 0.0^{b}
Urea-washed	96.3 ± 1.2^{a}	1.42 ± 0.0^{d}
Alkali-washed	94.4 ± 0.7^{a}	1.96 ± 0.1^{c}

Note: [a-d] Means values that do not share a letter are significantly different. Tukey 95% Simultaneous Confidence Intervals.

Figure 4 shows the reduction of protein content when oil body samples were repeatedly washed either in water, urea or sodium bicarbonate. It is clearly seen that the protein content progressively decreased after the first and second washes but any subsequent washing did not show any protein reduction in any of the samples. Lower protein contents were observed after application of a urea-wash or alkali-wash than protein content for water-washed.

Impact of washing of COBs on surface charge

Zeta potential (ζ-potential) is a key parameter to assess emulsion or suspension stability. The application is widely used for the determination of magnitude of electrical charges when exposed to the oscillating voltage across a measurement cell (McClement, 1999). The average of surface charge obtained in the crude oil body sample was -30.3 mV, water-washed was approximately -29.4 mV and alkali-washed was - 28.4 ± 1.2 mV in final oil body suspensions. However, a statistically significant reduction in surface charge was discovered (approximately -24mV) after washing with urea. Zeta potential values in urea-washed were significantly ($P<0.05$) reduced from -30.3 ± 0.6 mV to - 22.4 ± 1.9 mV. This suggests that, the absence of some proteins, which cover the entire surface of oil bodies, would contribute to the loss of surface charge after washing.

Impact of washing COBs on size of OBs

The size of oil bodies under different washing regimes was observed using light microscopy and typical micrographs are shown in Figure 6. The mean particle size of oil bodies washed using different solutions is shown in Table 3. In practice, the average size of droplets: remained the same on washing with water (3.13 ± 0.01 μm for COB compared with 3.17 μm for WWOB); decreased after washing with 0.1M sodium bicarbonate to 2.20 ± 0.08 μm; but increased on exposure to 9M urea to 7.20 ± 0.29 μm. Removal or extraneous protein by the harshest agent (9M urea) triggered widespread droplet coalescence. When the protein is removed using 0.1M sodium bicarbonate, the reduction in the apparent particle size could be explained by a loss of this aggregation effect rather than by the loss of a set of larger droplets, although the latter cannot be discounted. The results indicate that purification with 9M urea triggered oil body droplets to form aggregations and to coalesce.

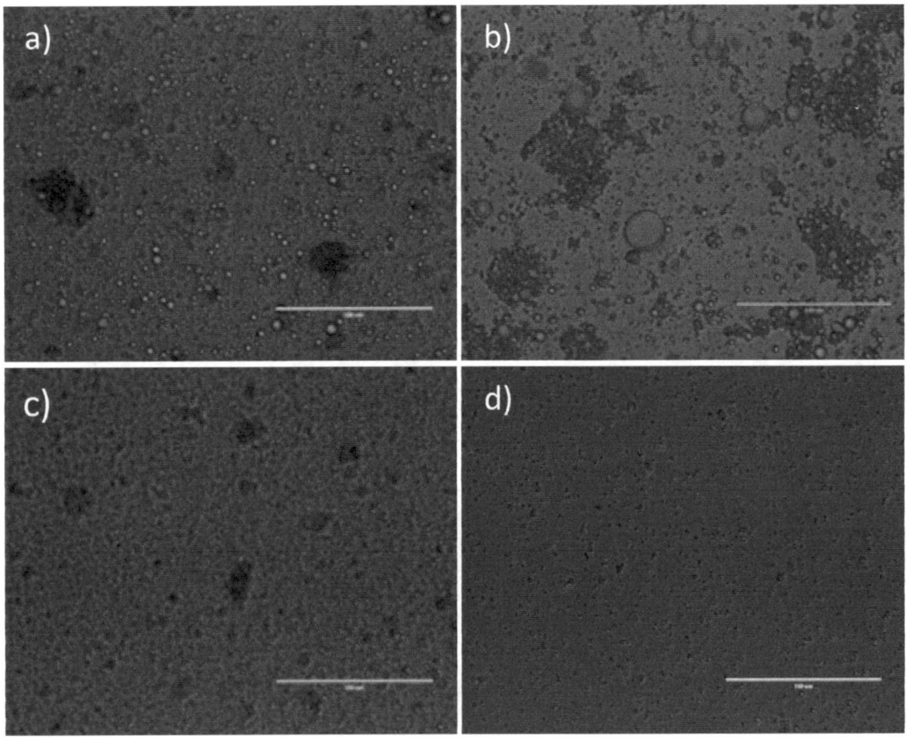

Figure 2: Microstructure of crude and washed oil bodies isolated from partially de-hulled sunflower seeds. Images of **a)** Crude oil bodies, **b)** 9M urea washed oil bodies, showing coalescence and leading to larger droplets **c)** oil bodies washed with deionised water **d)** alkali-washed (0.1M sodium bicarbonate) showed intact oil bodies with small droplets. Scale bars represent 100 μm.

Conclusion

Pre-soaking softens sunflower seed and a 6 hour pre-soak of our batch of seed provided good quality oil bodies. The particle size of oil bodies is important to detect any sign of aggregation or coalescence in the suspension system. Therefore, different type of washing treatment was carried out to test the stability of oil bodies under each treatment. Depending on the nature of plant itself, surface membrane of sunflower seeds oil body was damaged during urea-washed causing coalescence of oil bodies.

Acknowledgement

The authors would like to thank Islamic Development Bank for the financial support, Division of Food Sciences, University of Nottingham for laboratory facilities.

References

1. Beisson, F., Ferte, N. & Noat, G. (1996). Oil-bodies from sunflower (*Helianthus annuus L.*) seeds. Biochemical Journal 317: 955-956.
2. Cabra, V., Samsó, M. 2015).Do and Don'ts of Cryo-electron Microscopy: A Primer on Sample Preparation and High Quality Data Collection for Macromolecular 3D Reconstruction. *J. Vis. Exp.* (95), e52311 doi: 10.3791/52311

3. Fisk, I. D., White, D. A., Lad, M., and Gray, D. A. (2008). Oxidative stability of sunflower oil bodies. *European Journal of Lipid Science and Technology,* 110 (10), 962-968.
4. Garcia M.J, Quintero C., Mancha.M (1988). Oil bodies and lipid synthesis in developing soybean seeds. Phytochemistry, Vol.27, No.10 pp 3083-3087.
5. Huang, A. H. C. (1996). Oleosins and oil bodies in seeds and other organs. Plant Physiology, 110, 1055-1061.
6. Iwanaga, D., Gray, D, Decker, E. A., Weiss, J., and McClements, D. J. (2008) Stabilization of soybean oil bodies using protective pectin coatings formed by electrostatic deposition. 26; 56(6):2240-5.
7. Iwanaga, D., Gray, D. A., Fisk, I. D., Decker, E. A., Weiss, J., and McClements, D. J. (2007). Extraction and characterization of oil bodies from soy beans: a natural source of pre-emulsified soybean oil. Journal of Agricultural and Food chemistry, 55 (21), 8711-8716.
8. Jolivet. P, Boularda. C , Bellamy.A , Valotc.B , d'Andréaa.S , Zivyc .M , Nesib.N, Chardota.T.(2011). Oil body proteins sequentially accumulate throughout seed development in Brassica napus.Journal of Plant Physiology 168: 2015-2020.
9. Katavic, V., Agrawal, G. K., Hajduch, M., Harris, S. L. & Thelen, J. J. (2006). Protein and lipid composition analysis of oil bodies from two Brassica napus cultivars. Proteomics 6(16): 4586-4598
10. Lacey, D. J., Wellner, N., Beaudoin, F., Napier, J. A., and Shewry, P. R. (1998). Secondary structure of oleosins in oil bodies isolated from seeds of safflower (Carthamus tinctorius L.) and sunflower (Helianthus annuus L.). *Biochemical Journal,* 334, 469-477.

463259 Gentle Microfluidic Cell Capture and Release: PLL Stabilization of the Secondary Anchor Targeted Cell Release System

Abstract:

Gentle Microfluidic Cell capture and Release: PLL Stabilization of the Secondary Anchor Targeted Cell Release System

Ali Ansari, P. Imoukhuede, Ph.D.

University of Illinois at Urbana-Champaign

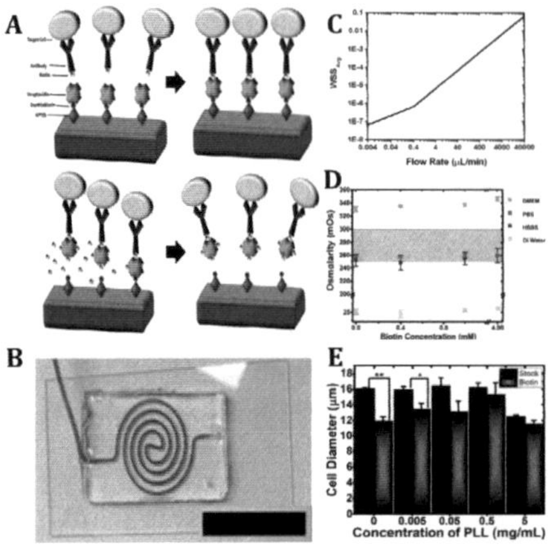

Figure 1. SATCR System Diagram (A) Spiral Mixer Design. Scale bar is 33 µm (B) Wall Shear Stress through spiral across input flow rates. (C) Biotin titration & Osmolarity (D) Cell Diameter & PLL concentration (E).

Introduction: Patients exhibit differential responses to drugs and therapeutics due to their unique genetic, proteomic, and cellular characteristics[1–3]. Advancing personalized medicine thereby requires methods of isolating and profiling these differences to individualize therapeutic regimens. However, some current cell isolation techniques are slow, disrupting physiological receptor quantities[4] while several of the more rapid "lab-on-a-chip" methods apply force or chemical digestion to release selected cells[5–9], which causes irreversible changes to biomarkers [6,7,10–13]. Thus, there is distinct lack of gentle and rapid cell isolation technologies for clinical translation [13]. Towards these aims of translatable, fast and gentle cell isolation, we describe adaptation of our novel, Secondary Anchor Targeted Cell Release System[14] (SATCR) to microfluidics and optimization of the microfluidic spiral design and buffer solutions to ensure gentle, but effective cell capture.

Materials and Methods: We have previously described the SATCR surface functionalization for selective isolation of cells in suspension[14] (Fig. 1A). Briefly, 24-well glass bottom plates are oxygen plasma cleaned, 2% APTES is applied, d-Desthiobiotin (DSB) is solubilized, activated, and dissolved in pH 6.0 MES buffer and mercaptoethanol quenches. Following overnight incubation, excess DSB is washed with PBS, and PBS-reconstituted streptavidin is applied overnight at 4°C. The surface is rinsed with PBS, rewetted, and stored at 4°C until use. Here we integrate the SATCR within a spiral microfluidic device consisting of a polydimethylsiloxane (PDMS) spiral mold (Fig. 1B), cast with a 3D printed master, which is bonded to glass through oxygen plasma treatment (Fig.1A). MCF7GFP cells, a luminal breast cancer cell line, were obtained from Cell Biolabs (San Diego, CA). MCF7GFP cells were grown in high glucose Dulbecco's modified Eagle medium (DMEM) supplemented nonessential amino acids (University of Illinois Cell Media Facility, School of Chemical Sciences, Urbana, IL), 10% fetal bovine serum (Invitrogen, Carlsbad, CA), and 1% Penicillin–Streptomycin (Invitrogen). These cells were then fixed with 4% formaldehyde and then labeled with previously biotinylated h-HLA-ABC antibody. FEAP software was used to simulate spiral wall shear stress. Towards optimizing gentleness, buffer osmolarity readings were taken on the Wescor Vapor Pressure Osmometer Vapro 5520 and Human Umbilical Vein Endothelial Cell (HUVEC) sizes and concentrations were measured on the Countess II Automatic Cell Counter. Flow rates for cell capture were 300 µL/min and cells were released via 4 mM biotin solution.

Results: Our computational simulation of the spiral revealed that the experimental flow-rate of 400 µL/min gave a 0.1 mPa wall shear stress (Fig. 1C), a gentle shear stress as it is 4 orders of magnitude below arterial wall shear stress[15]. The rationale for a spiral design was to increase mixing, and we experimentally observe 12% more cell release for recollection in a SATCR spiral micromixer versus a SATCR straight micro-channel design, indicating the efficacy of the SATCR spiral. We observed a significant, 26% (ANOVA Fisher test, $p<0.001$), decrease in HUVECs size following capture and release from the SATCR spiral (Fig. 1E). This decrease was not due to the biotin, as increasing biotin concentrations did not affect buffer osmolarity (Fig. 1D). Instead, it was likely due to sequestration of anions, due to an acidic isoelectric point of the SATCR spiral: we observed an entering buffer pH =7.2 and an exit pH =6.4. We determined that buffer supplementation with cationic PLL, stabilized cell diameter: optimal [PLL] = 0.05-0.5 mg/mL (Fig. 1E).

Conclusions: Reducing cell isolation induced stresses will advance personalized therapeutic regimens by retaining physiological biomarkers. Here we have identified optimizations for the adaption of the Secondary Anchor Targeting Cell Release System to capture cells for downstream analysis and recollection by reducing both wall shear and osmolarity induced stresses. These optimizations will allow us to further reduce the stresses experienced by the cells and help maintain physiological receptor and biomarker expression of cells released from the device. Future work will apply these optimizations to spiked blood samples to ensure translatability towards circulating cell capture.

References: **1.**Bergers, G. & Hanahan, D. *Nat Rev Cancer* **8,** 592–603 (2008).**2.**Ramos, P. & Bentires-Alj, *Oncogene* (2014). **3.**Batchelor, T. T. *et al. Cancer Cell* (2007).**4.**Imoukhuede, P. I., et al. *Am J Physiol Hear. Circ Physiol* **4,** H1085–93 (2013).**5.**Li, P. *et al.* (2015). doi:10.1073/pnas.1504484112 **6.**Mizuarai, S., et al. *Histol. Histopathol.* (2005). **7.**De Spiegelaere, W. *et al. Histol. Histopathol.* (2011).**8.**Imoukhuede, P. I. & Popel, A. S. *Cancer Med.*(2014).**9.**Imoukhuede, P. I., et al. S. *Am. J. Physiol. Heart Circ. Physiol.* (2013). **10.**Wynick, D. & Bloom, S. *Neuroendocrinology* (1990). **11.**Blann, A. D. *et al.* (2005). doi:10.1160/TH04 **12.**van Beijnum, J. R., et al. *Nat. Protoc.* (2008). **13.**Alunni-Fabbroni, M. & Sandri, M. T. *Methods* (2010). **14.**Ansari, A. et al. *Biotechnol. Bioeng.* n/a–n/a (2015). **15.** Morita, A et al. *Stroke* (2014)

Copyright © American Institute of Chemical Engineers. All rights reserved.

461540 Population-Based Detection of Cell Penetrating Peptide Uptake in a Microfluidic Droplet Trapping Array

Abstract:

Droplet-based microfluidic devices have emerged as a highly efficient technique for performing high-throughput, single cell analysis of heterogeneous samples such as tumor biopsies. The underlying principle involves compartmentalizing single cells in picoliter-sized monodisperse aqueous droplets in an inert continuous oil phase. Traditionally, this technology has required the use of off-chip incubation of samples coupled with high speed cameras for static measurements of cells or extracellular products. Incorporation of a trapping array downstream of the flow focusing junctions allows for the dynamic measurement of cellular behavior or intracellular enzyme activity. In this study, a microfluidic droplet trapping array was fabricated using polydimethylsiloxzne (PDMS) by established soft lithography and PDMS replication techniques, in order to encapsulate and trap single cells in aqueous droplets and assess the single cell uptake of cell penetrating peptides (CPPs) across an entire population. This technology is capable of delivering a more quantitative and complete understanding of the dynamic permeability efficiency of select CPPs. The continuous phase was fluorinated oil (Novec 7500, 3M) with a 0.2% w/w 008-fluorosurfactant (RAN Biotechnologies). Oil and water flow rates were optimized to maximize single cell encapsulation and droplet trapping efficiency in the array. Cell viability was determined in entrapped droplets to verify that encapsulation and extended trapping did not affect cellular viability. CPP permeability efficiency was determined across 767 droplets using both commercially available and novel CPPs. The microfluidic droplet trapping array consisted of a flow-focusing junction upstream of a droplet trapping array to allow for on-chip incubation and visualization of the encapsulated cells. The geometry was designed to produce droplets with diameters ranging from 60-140 microns by adjusting the flowrates of the inlet oil and water streams. The droplet trapping array consisted of 70 μm circles imprinted 40 μm into the PDMS directly above the downstream trapping channel. It was observed that cell viability was not affected in the trapped droplets for the duration of the experiment (~3-6 hours). Dynamic analysis of CPP uptake was easily visualized across the entire population of three different types of cancer cells (HeLa, OPM2, and HCT116). This study demonstrated a novel application of microfluidic droplet trapping array specifically relevant to cancer detection and diagnostics. Single cancer cells were encapsulated in aqueous droplets that were subsequently trapped and analyzed in a time-dependent manner. While the CPP studies are proof of concept, they demonstrate the capabilities of the technology to quantify the intracellular fluorescence in single cells across a heterogeneous sample.

Copyright © American Institute of Chemical Engineers. All rights reserved.

473659 Point-of-Care Diagnosis of Hemoglobin Disorders with a Mobile Electrophoresis Chip

Abstract:

In developing countries, diagnostic tests for homozygous (HbSS) or compound heterozygous (HbSC or HbS-Beta thalassemia) sickle cell disease (SCD) are not readily available at the point-of-care (POC). Very few infants are screened in Africa for SCD because of the high cost, time for sample transfer to a central laboratory (2-6 weeks), and level of skill needed to run traditional tests. The World Health Organization recognizes a crucial need for early detection of SCD in newborns, since it is estimated that 70% SCD-related deaths in Africa are preventable with early cost-effective interventions. The diagnostic barrier can be broken with affordable, POC tools that facilitate early detection immediately after birth or at the time of immunization. To address this unmet clinical need, we have developed a mobile electrophoresis platform (HemeChip) for reliable, affordable, and rapid diagnosis of SCD.

The HemeChip uses a microfabricated platform, with a material cost less than $0.87 per device, housing cellulose acetate electrophoresis to rapidly separate hemoglobin (Hb) types. Less than 5 microliters of blood, which can be obtained through a finger stick or heel stick, is processed on a piece of cellulose acetate paper via an applied electric field in alkaline buffer within 10 minutes. We clinically tested and benchmarked HemeChip against standard clinical methods using 51 blood samples from 14 patients.

The HemeChip reliably identifies and discriminates amongst Hb C/A2, S, F and A0. The HemeChip hemoglobin concentration results were correlated (Pearson Correlation Coefficient (PCC) of â‰¥0.96 for all Hb types tested) with standard clinical hemoglobin screening methods, including high performance liquid chromatography (HPLC). The agreement between the HemeChip and HPLC results were assessed using the Blandâ€“Altman plot, which showed a strong agreement between estimated (HemeChip) and actual (HPLC) hemoglobin percentages. The majority (95.5%) of the differences between actual and estimated hemoglobin percentages were within the limits of agreement.

Furthermore, the receiver Operating-Characteristic (ROC) curves showed more than 0.89 sensitivity and 0.86 specificity for identification of hemoglobin types using the HemeChip, based on the band travelling distance from the sample application point.

We developed a web-based image processing application for automated and objective quantification of HemeChip results at the POC using cloud computing resources. This intensity-based mobile phone image quantitation method showed high correlation with HPLC results for tested patient blood samples (PCC=0.95). Moreover, the Bland-Altman analysis showed strong agreement between the HemeChip results analyzed with the mobile user interface and HPLC. The majority (91%) of the differences between actual (HPLC) and estimated (mobile user interface) were within limits of agreement.

HemeChip technology offers a low-cost, easy to use, rapid approach and an innovative solution to POC diagnosis of SCD and other hemoglobin disorders. HemeChip can distinguish between different patient phenotypes, including HbSS (HbS only), transfused HbSS (HbS and HbA), and Hemoglobin SC disease (HbS and HbC). In summary, the HemeChip identification and quantification of hemoglobin phenotypes, as a POC technique, were comparable to standard clinical methods. This platform has clinical potential in under-served populations worldwide, in which SCD is endemic.

Copyright © American Institute of Chemical Engineers. All rights reserved.

473064 Nanopore-Based Sequence-Specific Nucleic Acid Detection at 1aM

Abstract:

A PCR-free, optics-free platform has been developed for the detection of *E. coli*16S rRNA at 1 aM. This sequence-specific nucleic acid (NA) detector is based on electromechanical signal transduction using a nanopore-based scheme. Rapid, sensitive, and cost-effective detection of NAs of specific sequence is in high demand for diagnosing pathogenic diseases, detection of food contaminants, patient screening during epidemics, or oncological status assessment during surgery. In order to detect target NAs, our device utilizes peptide nucleic acid (PNA) capture probes conjugated to polystyrene beads. Since PNA is charge neutral, the bead-PNA conjugates are designed to be charge neutral until they hybridize to target NAs at which point the complex becomes negatively charged and mobile in an electric field. If the electric field is oriented through a pore that is too small for the bead-PNA conjugate with hybridized target to pass through, it will at least partially block the pore thereby causing an easily measured step reduction in ionic current. In this way, the selective, nucleic acid hybridization event is electromechanically transduced.

Previously, we introduced a first-generation device based on a drawn pipette tip as the â€œporeâ€ with a 10 fM limit of detection (LOD). In this system, the bead-PNA conjugates and the nucleic acid-containing sample were placed inside the buffer-filled drawn glass pipette tip and an electric field was applied over the length of the pipette. This sensor could distinguish between complementary and non-complementary NA sequences and no false positives were observed with 1613-base DNA oligomers as targets. [1, 2]

In current work, we replaced the drawn glass pipette tip with a glass nanopore in a thin (1 Î¼m or thinner) glass membrane. The planar device with a nanopore is compatible with microfluidic integration and also makes high throughput manufacturing possible. Smaller bead-PNA conjugates become mobile enough to block the nanopore at lower concentrations of target NAs. This device has proven capable of detecting *E. coli* 16S rRNA (target) at 1 aM against a 1 pM background of RNA from *Pseudomonas putida*. In the absence of *E. coli* rRNA, no false positive

signals were obtained in the presence of 1 pM *P. putida* 16S rRNA. Also, a universal PNA probe complementary to both a portion of the 16S rRNA target from *E. coli* and rRNA from *P. putida*was used as a positive control, and as expected, a response to rRNA from both bacteria was detected using the positive control PNA probe. This detection scheme shows promise for integration into portable, low-cost systems for rapid detection of pathogenic bacteria in food, water and body fluids.

[1] Esfandiari L, Monbouquette HG, Schmidt JJ. Sequence-specific Nucleic Acid Detection from Binary Pore Conductance measurement. J. Am. Chem. Soc. 2012; 134: 15880-15886.

[2] Esfandiari L, Lorenzini M, Kocharyan G, Monbouquette HG, Schmidt JJ. Sequence-Specific DNA Detection at 10fM by Electromechanical Signal Transduction. Anal. Chem. 2014; 86: 9638-9643.

Copyright © American Institute of Chemical Engineers. All rights reserved.

462494 An Integrated Microfluidic System Enabling Real-Time Autonomous Field Surveillance for Vector-Borne Pathogens

Abstract:

Mosquito-borne viruses such as West Nile, dengue, chikungunya, and Zika viruses are important causes of emerging and re-emerging diseases worldwide. We report on a fully integrated, portable, and inexpensive microfluidic chip and chip-handling system that can be deployed in a network to enable real-time autonomous field surveillance for vector-borne pathogens over an arbitrarily sized geographic region. Our system attracts mosquitos to feed on sugar baits and then analyzes mosquito saliva using reverse transcription loop mediated isothermal amplification (RTLAMP) coupled to our recently published QUASR detection method to test for a panel of mosquito-borne viruses, including West Nile virus, western equine encephalitis virus, and St. Louis encephalitis virus. Prototype devices selectively and specifically detect RNA viruses in the laboratory and will be field-tested this summer. The chip handling system, or "smart trap," enables daily, autonomous sample processing and wireless data transmission to a cloud-based biosurveillance system.

The microfluidic chip's design minimizes the need for external manipulation and reduces the risk of system contamination. Chips are patterned from acrylic and bonded a thin backing onto the chip. Passive, normally closed microfluidic check valves with adjustable opening pressures are installed using a press fit. An onboard storage reservoir holds rehydration buffer. The chip's channel geometries allow for dry-storage of all RT-LAMP assay reagents into 8 parallel reaction channels. After desiccation, channel fill and vent holes are sealed off with press fit plug strips that create a gas-tight seal. The chip's feeding well presents a solid sugar bait to mosquitos, which are attracted to feed by a floral scent. Upon feeding on the sugar bait, mosquitoes deposit saliva which can be tested for the presence of viral RNA. The onboard rehydration buffer dissolves the sugar bait prior to testing by RT-LAMP. The headspace above the sample is then pressurized, pushing it through the second check valve and rehydrating the dried assay reagents. Cylindrical air pockets positioned at the ends of each reaction channel limit the extent of filling.

The smart trap loads one chip daily from a stack of chips that enable autonomous operation for 1 month at a time. A belt drive system moves the chip through different stations in the trap. Linear actuators are used to load chips, dispense onboard fluids, and create a face seal for pneumatic chip filling. A small pump provides the pressure needed to push the sample into the chip. A heater maintains the chip at 65°C for the 30 minute RT-LAMP reaction. Once the chip cools, an inexpensive fluorometer determines whether RT-LAMP reactions are positive or negative. The traps relay data to a master receiver trap, which uploads daily test results to the cloud via text message. A web-based application loads real-time biosurveillance data and overlays test results with vector abundance maps.

Sandia National Laboratories is a multi-program laboratory managed and operated by Sandia Corporation, a wholly owned subsidiary of Lockheed Martin Corporation, for the U.S. Department of Energy's National Nuclear Security Administration under contract DE-AC04-94AL85000.

Copyright © American Institute of Chemical Engineers. All rights reserved.

471810 Capture and Concentration of Pathogens in Chaotic Flows

Abstract:

Infectious diseases of both viral and bacterial origin continue to be a health threat to millions of people in developed and underdeveloped countries. One successful treatment strategy for cases of sepsis or viral infections, such as Ebola disease and HIV, is the effective capture and removal of the pathogenic agent from the bloodstream (i.e., pathogen blood cleansing). We are presently developing filter-less technologies for the direct capture of bacteria (i.e., *Escherichia coli*) and eventually viruses (i.e., Ebola virus-like-particles) from the bloodstream. We use a portable/disposable system for the continuous capture of pathogens circulating through a microfluidic chamber. The system is based on the specific recognition of proteins on *E. coli* membranes; it integrates the use of (a) anti-*E. coli* polyclonal antibodies, (b) magnetic nanoparticles (MNP), (c) a microfluidic chaotic flow system, and (d) a neodymium magnet. Anti-*E. coli* antibodies are covalently immobilized within commercial magnetic nanoparticles to fabricate nanoparticles that will bind *E. coli* bacteria. Our experiments compare the performance of different immobilization strategies (amino-carboxylic covalent binding and streptavidin-biotin binding) and different magnetic nanoparticle sizes (range 30â€“800 nm). The heart and distinctive feature of our system is a microfluidic chamber in which the *E. coli* binding particles and the bacteria are mixed by the action of a laminar chaotic flow produced by the alternating rotation of two cylinders. The intimate contact induced by this chaotic laminar flow promotes the capture of the bacteria by individual nanoparticles or nanoparticle clusters. The trapped *E. coli* are then concentrated by a simple magnet located downstream from the microchamber.

Â This platform has key advantages over currently available methods, which are mostly based on the use of microfluidic channels or filtering membranes: (a) It is faster (overall capture times in the order of 1â€“5 minutes); (b) It offers superior capture due to the intimate mixing induced by the chaotic flow; (c) It is easy to use, which reduces labor efforts and eliminates the need for dedicated and costly infrastructure.

Copyright © American Institute of Chemical Engineers. All rights reserved.

474749 Invited Presentation: Sensor Technologies for High Throughput Bioprocess Monitoring

Abstract:

Dr. Raoâ€™s research has focused on applications of fluorescence spectroscopy to bioprocess engineering. His lab has developed next-generation sensors for low-cost non-invasive monitoring of oxygen, pH and pCO2 in bioreactors. In addition, novel sensors for glucose and glutamine have been developed. These sensors have led to a paradigm shift in bioprocess technology, as they collectively enable High Throughput Bioprocessing.Â In addition, the Rao lab has focused on developing novel applications of Green Fluorescent Protein in Bioprocessing. Its unique fluorescent properties and ease of use in a variety of culture systems have provided new insights into heterologous protein production. Recently, the Rao lab has started developing the next generation of sensors based on surface plasmon coupled fluorescence. These promise to lead to ultra-sensitive and ultra-fast sensors but at low-cost.

Copyright © American Institute of Chemical Engineers. All rights reserved.

462192 Pharmaceutical Crystallization: A Manufacturing Process Perspective

Abstract:

Typically a number of critical quality attributes (CQAs) such as crystal structure and crystal component identification, purity, crystallinity, crystal size distribution (CSD), polymorphism and impurity, and stability are evaluated to assess the robustness, controllability, repeatability, and consistency of a pharmaceutical crystallization process. There are different ways to achieve in-depth process understanding and to realize rationale process control. This presentation provides several FDA research case studies to demonstrate the technical feasibility of real time process monitoring approaches for better understanding, design and control of pharmaceutical crystallization processes and product characterization. The case studies are: 1) Combined real-time PAT monitoring and process chemometrics for mapping the state of a pharmaceutical crystallization process; 2) Combined real-time PAT process monitoring and first-principle modeling for elucidating the nucleation mechanisms of a dynamic pharmaceutical crystallization process; 3) Combined real-time PAT process monitoring, Design of Experiments (DOE), and General Linear Modeling (GLM) to establish a hybrid approach for process characterization and process design space development; and 4) Integrated PAT approach for nucleation induction time measurement. This presentation will briefly address some of the current opportunities in pharmaceutical crystallization, including switching from batch to continuous crystallization and several promising, novel technologies in the area of crystallinity monitoring, from a process engineering perspective.

Copyright © American Institute of Chemical Engineers. All rights reserved.

456484 Cooling Crystallization within Micromixers: Mathematical Modeling, Theoretical Analysis, and Experimental Validation

Abstract:

A key objective in crystallization is to control crystal size distribution (CSD), to improve product quality (e.g., pharmaceutical and biopharmaceutical) and process efficiency. Micromixers have been an effective technology to enhance the control of nucleation, which is the key step towards final product CSD in continuous [1] and semi-continuous crystallization processes [2]. Common micromixers combine solution and anti-solvent streams such as in a dual-impinging-jet (DIJ) micromixer (e.g., [3]–[5]).

Rather than anti-solvent crystallization as commonly used, a DIJ mixer has been demonstrated to combine hot and cold saturated solutions of the same solute/solvent to generate small uniform seed crystals [2], which appears to be the first reported cooling DIJ nucleation. The product crystals were less than 10 microns in length, placing them in the correct size range for direct application in inhalers [6]. Theoretically, if the mixing was perfect (that is, concentration and temperature were completely mixed at the molecular scale in the mixer), the average supersaturation level would be too low to nucleate crystals in a cooling DIJ mixer. The surprising result of nucleating crystals by combining hot and cold saturated solutions in a DIJ mixer motivated the theoretical analysis of the system, which showed that nucleation was enabled by the much faster energy transfer than mass transfer rates near the interface. This difference in rates enabled the temperature of the hot solution to drop to approximately the average temperature of the two solutions before its solution concentration had significantly changed, resulting in a supersaturation sufficiently high to nucleate crystals.

The theoretical analysis is supported with process simulations. The two-dimensional energy and solute mass balances are solved using COMSOL input with an analytical solution for the velocity field, to generate spatial distributions of temperature, concentration, and supersaturation near the hot-cold interface within a cooling DIJ mixer. In the most important spatial region of importance for characterization of nucleation, the two-dimensional fields are shown to be very close to analytical solutions derived from a one-dimensional approximation of the energy and molar balances. This simplification applied the derivation of design criteria to facilitate quick assessment of whether any particular pharmaceutical-solute combination will nucleate crystals in a cooling DIJ mixer, based on the physicochemical properties. These criteria could save time and material by avoiding or reducing trial-and-error experiments, which is helpful at the early stage of pharmaceutical process development. The process simulations are then expanded to include the full solution of the spatially-varying population balance model [5] to calculate the crystal size distributions throughout the spatial domain. The mathematical models and methodologies are validated for the crystallization of L-asparagine monohydrate (LAM) in aqueous solution.

References

[1] M. Jiang, Z. Zhu, E. Jimenez, C. D. Papageorgiou, J. Waetzig, A. Hardy, M. Langston, and R. D. Braatz, "Continuous-flow tubular crystallization in slugs spontaneously induced by hydrodynamics," *Cryst. Growth Des.*, vol. 14, no. 2, pp. 851–860, 2014.

[2] M. Jiang, M. H. Wong, Z. Zhu, J. Zhang, L. Zhou, K. Wang, A. N. Ford Versypt, T. Si, L. M. Hasenberg, Y. E. Li, and R. D. Braatz, "Towards achieving a flattop crystal size distribution by continuous seeding and controlled growth," *Chem. Eng. Sci.*, vol. 77, pp. 2–9, 2012.

[3] B. K. Johnson and R. K. Prud'homme, "Chemical Processing and Micromixing in Confined Impinging Jets," *AIChE J.*, vol. 49, no. 9, pp. 2264–2282, 2003.

[4] M. Midler Jr., E. L. Paul, E. F. Whittington, M. Futran, P. D. Liu, J. Hsu, and S. H. Pan, "Crystallization method to improve crystal structure and size," US5314506 A, 24-May-1994.

[5] X. Y. Woo, R. B. H. Tan, and R. D. Braatz, "Precise tailoring of the crystal size distribution by controlled growth and continuous seeding from impinging jet crystallizers," *CrystEngComm*, vol. 13, no. 6, pp. 2006–2014, 2011.

[6] T. G. D. Capstick and I. J. Clifton, "Inhaler technique and training in people with chronic obstructive pulmonary disease and asthma," *Expert Rev. Respir. Med.*, vol. 6, no. 1, pp. 91–103, Feb. 2012.

Copyright © American Institute of Chemical Engineers. All rights reserved.

452797 Real Time PAT Monitoring of the Solvent-Mediated Transformation of Ciclesonide Acetone Solvate into Methanol or Cyclohexane Solvate

Abstract:

Due to structural differences, pseudopolymorphs of a given drug substance exhibit different physicochemical properties, which impacts drug dissolution, bioavailability, and efficacy. Therefore polymorphic transformation has been a vital aspect of pharmaceutical crystallization process engineering. In this work, three solvates of ciclesonide were characterized by Differential Scanning Calorimetry ï¼ˆDSCï¼‰, Thermogravimetry (TG), and X-ray Powder Diffraction (XRPD) analysis. The driving force of solvates transformation process was investigated via measuring solubility of ciclesonide solvates in different mixed solvent systems. To follow the crystallization process kinetics and understand the phase transformation mechanisms, , several Process Analytical Technology (PAT) techniques such as Fourier Transform Infrared Spectroscopy (FTIR), Raman Spectrum (RAMAN), Particle Video Microscope (PVM) and Focused Beam Reflectance Measurement (FBRM) were employed to monitor the phase transformation from acetone solvate into methanol or cyclohexane solvate in real time. In addition, the effects of temperature, solvent composition, and stirring rate on phase transformation process were discussed. Furthermore, it was demonstrated that real time PAT process monitoring can help to realize better process control and achieve desired crystal product quality attributes.

Copyright © American Institute of Chemical Engineers. All rights reserved.

472470 Crystal Morphology and Size Control in Pharmaceutical Industry

Abstract:

Crystal Morphology and Size Control in Pharmaceutical Industry

Junbo Gong[1,2*], Yan Wang[1,2], Jingkang Wang[1,2]

[1] National Engineering Research Center of Industry Crystallization Technology, School of Chemical Engineering and Technology, Tianjin University, Tianjin, China;

[2] The Co-Innovation Center of Chemistry and Chemical Engineering of Tianjin, Tianjin University, Tianjin, China;

E-mail address: junbo_gong@tju.edu.cn * wangyan57@tju.edu.cn

Abstract

Crystal shape and size engineering is an integral part of pharmaceutical crystallization. It aims to design and control crystal habit and particle size distribution (PSD) of drug substance. This program is of great significance due to the strong impact of product quality, including bioavailability, dissolving rate, flowability, etc. Downstream particle handling is also affected, such as filtration, drying and compaction. The factors influencing crystal morphology and size come from two groups: the one is the operating condition, which includes supersaturation, temperature, the hydrodynamics of the crystallizer (e.g., stirring rate, mixing conditions) and external field (e.g., magnetic field, ultrasound field), and the other is chemical composition of the solution which consists of solvents and additives. In order to control crystal shape and PSD, experiments should be designed and carried out to monitor crystal growth process. The influence on crystallization thermodynamics and kinetics should also be investigated. During the research,

microscopy techniques, like polarizing optical microscope, scanning electron microscopy (SEM) and atomic force microscope (AFM) will be applied to analyze crystal shape. Particle sizing techniques are also used to count for PSD, like the laser diffraction (LD) and focused beam reflectance measurement (FBRM). On the other hand, the use of additives has attracted increasing attention in the drug manufacturing process. The presence of additives could substantially affect the solid forms and morphology. High throughput screening (HTS) techniques to additives should be developed to accelerate the screening efficiency. The effects of additives on crystal shape may be resulted from the specificity of interaction between the additive and the solute molecule, charge distribution, hydrophilic-hydrophobic property, stereoselectivity as well as diffusion rate and competitive adsorption, etc. With fast advances made in computational techniques, modeling and prediction of crystal morphology can be realized in the absence or presence of additives. They not only provide powerful tools to design and screen effective additives, but lead to further understanding of crystal growth mechanisms.

Fig. 1. SEM of the drug substance without (a) with (b) the modification of the crystallization environment

Copyright © American Institute of Chemical Engineers. All rights reserved.

472381 Liquid Liquid Phase Separation during the Cooling Crystallization of Pyraclostrobin

Abstract:

Liquid−Liquid Phase Separation during the Cooling Crystallization of Pyraclostrobin

Crystallization, as a method of isolation and purification, is widely used in fine chemicals, food and pharmaceutical industries. It is a powerful mean to improve the properties of particle, such as changing crystal form, narrowing crystal size distribution (CSD), modifying crystal habit and so on. However, sometimes droplets like oil drops are generated instead of crystals are appeared from the supersaturated solution during the cooling and/or solventing-out crystallization process. This phenomenon is always termed as liquid¨Cliquid demixing, liquid¨Cliquid-phase separation (LLPS) or oiling out.

The main purpose of this investigation was to investigate the cooling crystallization process of pyraclostrobin in isopropanol/cyclohexane solvent mixtures. It was found that liquid−liquid phase separation will occurred under some certain conditions. In order to understand the mechanism of the liquid−liquid separation and the influence of cooling rate on oiling out, the phase diagram and cloud point curves at different cooling rate were experimentally measured with the aid of in situ tools, such as PVM, FBRM, ATR-FTIR and Raman. The phase diagram includes the solubility curve, the liquid−liquid separation curve, supersolubility curves, and clear point. It was found from the phase diagram that a normal single-phase crystallization would change to a oiling-out crystallization once the concentration of pyraclostrobin exceeded a certain value during crystallization. Contrasting with cloud point curves of different cooling rate, it is obvious that the cooling rate has a significant effect on the metastable state critical oiling out point. The concentration corresponds to the metastable state for critical oiling out point decreases with the increase of cooling rate. In the cooling and heating cyclic process, the solution has experienced five stages. During cooling down the solution from high temperature, oiling out including stable LLPS and metastable LLPS occurred, then nucleation and growth of crystals happened; upon heating up, the crystals dissolved gradually, then the solution with a small amount of particles transformed into transient oil phase and finally it formed a stable and clear solution.

Considering the crystal properties of final products, some experiments designed to investigate the influence of cooling rate and initial concentration. And the result showed that the higher cooling rates, the more serious agglomeration and poorer crystal habit of final products would get. Another result is that initial concentration has little influence on the crystal properties. High quality of products can be obtained by designing route of cooling crystallization according to the phase diagram.

Figures

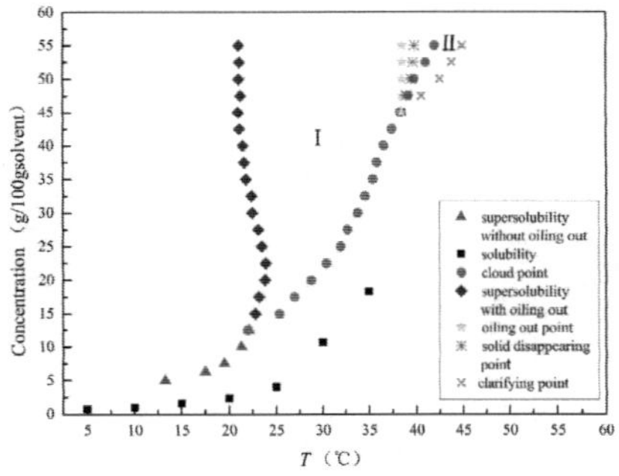

Figure 1 Phase diagram of pyraclostrobin in 10 wt % isopropanol/cyclohexane solvent mixtures.

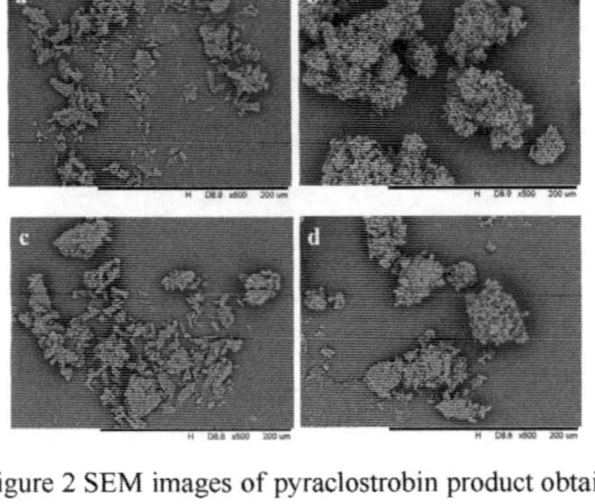

Figure 2 SEM images of pyraclostrobin product obtained from different experiments

a refer to pyraclostrobin products at the cooling rate of 0.1 K/min under concentration of 10g/100g solvent; b refer to pyraclostrobin products at the cooling rate of 0.5 K/min under concentration of 10g/100g solvent; c refer to pyraclostrobin products at the cooling rate of 0.1 K/min under concentration of 15g/100g solvent; d refer to pyraclostrobin products at the cooling rate of 0.5 K/min under concentration of 15g/100g solvent.

Copyright © American Institute of Chemical Engineers. All rights reserved.

472527 Discriminate the Differences Between Surface Water and Channel Water in the Formation of Solvates

Abstract:

Discriminate the differences between Surface water and Channel water in the formation of solvates

Yumin Liu[1,2], Dandan Han[1,2], Jingkang Wang[1,2], Junbo Gong[1,2*]

[1] National Engineering Research Center of Industry Crystallization Technology, School of Chemical Engineering and Technology, Tianjin University, Tianjin, China;

[2] The Co-Innovation Center of Chemistry and Chemical Engineering of Tianjin, Tianjin University, Tianjin, China;

E-mail address: liuyumin@tju.edu.cn junbo_gong@tju.edu.cn *

Abstract

Solvates are defined as Active Pharmaceutical Ingredients (APIs) solids which incorporate one or more solvent molecules in the crystal lattice or channel to form new crystalline substances. Converting the drugs into hydrophilic solvates can help to improve the solubility in water and increase drug bioavailability. Meanwhile, it also can improve the stability of drugs, thus facilitating the formulation and storage of the drugs. There are four binding modes between compounds and solvents, just presented in the following Figure 1. Liquid inclusions and incorporation into the lattice are composed of solvates. With the increased incidence of solvate formation, it is essential to get a comprehensive understanding of their structure, physical stability, and the implication of de-solvation on the properties of the APIs.

Hydrate are the most abundant solvates, owing to the high propensity of water to be entrapped in the lattice of APIs, with its small size and tendency to form multi-directional hydrogen bonds. Pharmaceutical solids may come in contact with water during processing steps, such as crystallization, lyophilization, spray-drying, wet granulation, or aqueous film-coating, and they may be exposed to humid air during storage. Absorbed water molecules may reside on crystal surfaces, in crystal channel, or in crystal lattice structures which can change the packing in the unit cell. Surface water and channel water is quite different from each other. Without a doubt, we can utilize SXRPD to discriminate that differences between them. But in our laboratory, several other offline analytical techniques, such as XRPD, TGA and DVS, were also adopted to distinguish it and achieved remarkable results.

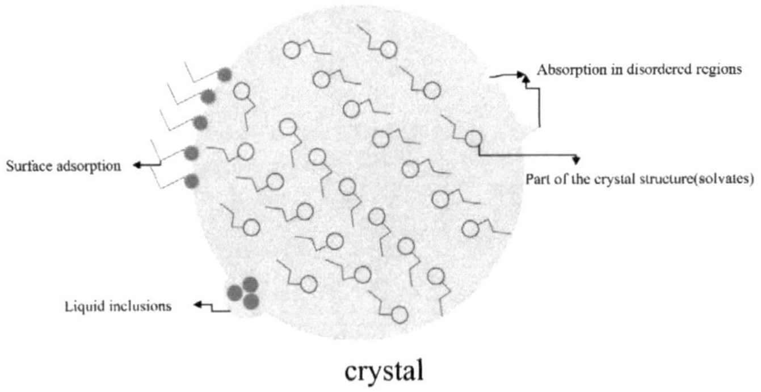

Fig.1 Different principles of solvent associations with crystalline solids

Fig.1 Different principles of solvent associations with crystalline solids

Copyright © American Institute of Chemical Engineers. All rights reserved.

472078 Influence of Ultrasonic Wave on the Crystallization of Lysozyme

Abstract:

Due to the highly sensitivity of proteins to the crystallization conditions, a slight change in operating parameters might affect the outcome of the crystallization. In this study, the effects of ultrasonic wave on the crystallization of lysozyme were investigated by applying different levels of ultrasound during the purification process of lysozyme from the aqueous solutions of hen egg white through the solvent freeze out (SFO) technology. Tetragonal lysozyme crystals were obtained during the process, which were then verified by X-Ray Powder Diffraction (XRPD). The quality of lysozyme crystal (composition and purity, size and shape, enzyme activity) was measured using SDS-PAGE, polarized light microscopy and spectrophotometer, respectively. It was found that the purity and size of lysozyme crystal were affected significantly by ultrasonic wave, while the enzyme activity of lysozyme crystal werenâ€™t. In addition, the influence mechanism of ultrasonic wave on lysozyme crystallization was discussed. The results obtained by this work will provide essential information for the purification of other proteins.

Copyright © American Institute of Chemical Engineers. All rights reserved.

471180 Site-Specific Conjugation of a Drug to an Antibody Via Nucleotide Binding Site

Abstract:

Antibody-drug conjugates (ADCs) are recent pinnacle of immunotherapeutic agents that enhance the antitumor effects of antibodies and reduce adverse systemic effects of potent drugs. To obtain an efficient ADC, the drug should be conjugated to the antibody with a linker while preserving antibody activity and high potency of the drug with an optimized drug-antibody ratio (DAR). Conventional ADCs are typically produced by conjugating a drug to an antibody through the side chains of either surface-exposed lysines or free cysteines generated through reduction of inter-chain disulfide bond. Many obstacles have been associated with the conventional drug conjugation methods to antibodies, such as aggregation of antibodies, high number of the DAR yielding rapid *in vivo* clearance, heterogenous modifications causing weak performance of the ADC, and requirement of harsh chemical modification resulting with damaged antibodies. In this study, we describe a site-specific method for conjugation of drugs to monoclonal therapeutic antibodies utilizing nucleotide-binding site (NBS). We used an NBS ligand â€"ring structured small moleculeâ€" that has moderate binding affinity to the NBS located between heavy and light chains on the variable region of Fab fragment to conjugate a drug with a PEG linker to the antibody. We have used previously described UV-NBS method that utilizes UV-energy to initiate a photo-cross-linking reaction between the NBS and the NBS-ligand, to obtain controlled site-specific conjugation of the drug to the antibody. We have assessed the effectiveness of the ADC generated with the UV-NBS method utilizing various drugs such as paclitaxel and bortezomib in combination with various antibodies including Rituximab, anti-CD133 and anti-CD38. We further characterized *in vitro* efficacy, toxicity, and the DAR of the ADCs as well as activity and sensitivity of the antibodies post-conjugation with the drugs.

Copyright © American Institute of Chemical Engineers. All rights reserved.

461822 Acid-Sensitive Silylated or Acetalated Polysaccharides for Vaccine Applications

Abstract:

Acid-degradable polymers are well-suited for use as drug delivery vehicles because numerous physiological sites (e.g., intracellular endocytic pathway) are acidic. In particular, our work has concluded that acid-sensitive polymers are advantageous for vaccine and other immune modulatory therapies because once phagocytosed by immune cells, the low pH environment of the phagosome leads to triggered release of the encapsulated cargo. Here we report the synthesis of acid-sensitive silylated or acetalated polysaccharides derived from either dextran or inulin. For the silyl chemistry, various alkyl substitutions on the silicon center were performed, including trimethylsilyl dextran (TMS-DEX), ethyldimethylsilyl dextran (EDMS-DEX), and triethylsilyl dextran (TES-DEX) on dextran. Because dextran and acetalated dextran are not inherently immunogenic, we chemically modified a naturally-occurring, immunostimulatory polysaccharide, inulin, to produce tunable, acid-sensitive biopolymers acetalated inulin (Ace-IN) and trimethylsilyl inulin (TMS-IN). These polymers were then fabricated into microparticles (MPs) via emulsification followed by solvent evaporation. Our data indicates the MPs are stable at extracellular pH 7.4 and display a wide range of degradation half-lives (fifteen minutes to greater than nine days) when exposed to acidic pHs of 2.0 and 5.0. The half-lives are dependent on the extent of silylation/acetalation and steric crowding on the silicon center (trimethyl to ethyldimethyl to triethyl). Silyl-DEX Ace-IN, and Silyl-IN MPs exhibit cytocompatibility when cultured with macrophages. TES-DEX, Ace-IN, and TMS-IN MPs, composed of highly hydrophobic moieties or the parent immunostimulatory inulin, elicited substantial in vitro production of tumor necrosis factor alpha (TNF-Î±), a cytokine associated with an innate immune response. In vivo immunization with a model ovalbumin antigen encapsulated in silylated or acetalated polysaccharide MPs, without a separate adjuvant, results in a dual humoral and cellular response that is superior to an alum-adjuvanted formulation. Overall, we present Silyl-DEX, Silyl-IN, and Ace-IN as members of the acid-degradable polymer family for potential use in subunit vaccines and other drug delivery applications.

Copyright © American Institute of Chemical Engineers. All rights reserved.

467628 The Effect of Particle Physico-Chemical Properties on Phagocytosis

Abstract:

The effect of particle physico-chemical properties on phagocytosis
Anusha Garapaty[1], Julie A.Champion[1], (1) School of Chemical and Biomolecular Engineering, Georgia Institute of Technology, Atlanta, GA

The tunability of polymeric particle properties makes them attractive for a wide range of applications in drug delivery, imaging and diagnostics. Their physico-chemical properties such as size, shape, surface chemistry and stiffness can be altered to influence their interactions with cells. It has become imperative to establish a correlation between these physico-chemical properties and biological behavior to design targeted delivery systems. During these delivery applications, particles introduced into the body interact with macrophages of the reticuloendothelial system. These macrophages rapidly clear them through a process known as phagocytosis thus limiting their potential in biomedical applications involving either phagocytic inhibition or macrophage targeting.

The influence of material parameters size, shape and mechanical stiffness on phagocytosis has been established previously [1-3]. However, the combinatorial effects of these material biophysical parameters on phagocytosis is still unknown. Our objective in this work is to understand how the combined effects of particle physico-chemical properties size, shape and stiffness affect the process of phagocytosis using particles which can be independently tuned in each of these properties while keeping the chemistry constant.

Tunable particles have been fabricated by the development of a template based layer-by-layer (LbL) assembly process. Polystyrene particles of varied sizes were chosen as the template. Shape is imparted to the polystyrene particles by heat liquefaction while stretching. The polystyrene template particles are adsorbed with polyelectrolytes poly (vinyl pyrrolidone) (PVP), poly (ethyleneimine) (PEI), and poly (acrylic acid) (PAA) by LbL self assembly. Mechanical stiffness can be varied either by maintaining the solid template core or by making hollow capsule particles through tetrahydrofuran treatment and thiol crosslinking of the polyelectrolyte layers. To study specific Fc interaction with macrophages, the particles were coated with IgG antibody. Our results show that for a spherical shape, when the size is altered cores are internalized slightly more than corresponding capsules during phagocytosis. However, when the shape is altered to a rod for a particular size, capsules exhibit enhanced internalization compared to corresponding cores. These particles thus demonstrate that shape and stiffness play a crucial role in tuning the specificity of IgG functionalized particle interactions with macrophages. This novel polymer particle platform can also be further extended to functionalization of active ligands for phagocytic inhibiton or modulation of macrophage activity and phenotype in immunological therapeutic applications.

References

1. Beningo, K.A. and Y.-l. Wang, Fc-receptor-mediated phagocytosis is regulated by mechanical properties of the target. Journal of cell science, 2002. 115(4): p. 849-856.

2. Champion, J.A. and S. Mitragotri, Role of target geometry in phagocytosis. Proceedings of the National Academy of Sciences of the United States of America, 2006. 103(13): p. 4930-4934.

3. Champion, J.A., A. Walker, and S. Mitragotri, Role of particle size in phagocytosis of polymeric microspheres. Pharmaceutical research, 2008. 25(8): p. 1815-1821.

Copyright © American Institute of Chemical Engineers. All rights reserved.

473372 Process Development and Scale-up of an ADC Conjugation Process

Abstract:

Antibody-drug conjugates (ADCs) are a class of pharmaceuticals that use an antibody to selectively deliver a cytotoxic drug to cancer cells. Critical quality attributes (CQAs) are controlled by means of the input quality of antibody and small molecule drug/linker components as well as execution of the conjugation within defined process parameter ranges. Fairly minor changes in ADC process parameters may have the potential to result in changes to the drug substance CQAs. Furthermore, the limitations of in-process sampling may not offer immediate detection nor the potential for corrective action. Thus, in-depth understanding of the conjugation process, determination of scale-dependent parameters, and operating knowledge of manufacturing equipment are essential elements necessary to develop a control strategy. This presentation will provide an overview of the experiences learned from a successful scale-up campaign.

Copyright © American Institute of Chemical Engineers. All rights reserved.

469000 *In Situ* crosslinked Endosomolytic Polymeric Vesicles for Immunotherapeutic Targeting of the Sting Pathway

Abstract:

Cyclic dinucleotides (CDNs) are an emerging class of immunomodulators that have shown particular promise in cancer immunotherapy for their ability to elicit the production of type I interferons and trigger the generation antitumor leukocyte phenotypes in immunosuppressive tumor microenvironments (TMEs). The endogenous ligand of the stimulator interferon genes (STING), 2â€™3â€™ cyclic guanosine monophosphate-adenosine monophosphate (cGAMP) has recently shown promise as an immunotherapeutic capable of eliciting tumor regression and immune memory. However, a number of drug delivery barriers limit the therapeutic efficacy and translation of CDNs, including nuclease degradation, poor intracellular uptake, and inefficient cytosolic delivery. To address this challenge, we have developed polymeric vesicles that actively enhance endosomal escape of cyclic dinucleotides for efficient delivery to the cytosolic STING receptor.

Reversible addition-fragmentation chain transfer (RAFT) polymerization was used to synthesize poly[(ethylene glycol)-b-(butyl methacrylate-co-diethylamino ethyl methacrylate-co-pyridyl disulfide ethyl methacrylate)] (PEG-b-DBP) amphiphilic diblock polymers that self-assemble into endosomolytic pH-responsive vesicles with $d < 100$ nm in aqueous conditions with neutral surface charge. Vesicles encapsulated cGAMP with efficiencies of 38% and significantly enhanced potency *in vitro*, manifesting in multiple order of magnitude decreases in cGAMP EC50 when encapsulated. Nanoparticle cGAMP formulations demonstrated significant inhibition of tumor growth and increased mouse survival times in a murine melanoma model. Collectively, these data demonstrate that endosomolytic polymer nanoparticles significantly improve CDN potency by enhancing delivery to the cytosolic STING sensor and suggest that encapsulated cGAMP formulations may prove to be a powerful cancer immunotherapeutic.

Copyright © American Institute of Chemical Engineers. All rights reserved.

472945 Impact of Dose, Route, and Composition on the Immunogenicity of Polyelectrolyte Multilayers Assembled on Gold Templates

Abstract:

Novel biomaterial vaccines are promising in treating challenging diseases such as cancer and infectious diseases. We recently developed a vaccine platform based on self-assembly of immune signals into immune polyelectrolyte multilayer (iPEMs). These iPEMs vaccine can be delivered on gold nanoparticle template to stimulate effector cytokines *in vitro* and expand antigen-specific T cells *in vivo*. Here we investigate the impact of vaccine dose, route, and adjuvant choice on the generation of T cell immunity and anti-tumor response. Three injection routes (i.e. intradermal, subcutaneous and intramuscular injections) and three iPEM dosing levels were employed. Intradermal injection induced the most potent antigen-specific T cell responses, and in all cases, the level of response was dose dependent. Importantly, we discovered that iPEMs generate durable memory T cells responses, indicated by potent, antigen-specific $CD8^+$ T cells responses in mice receiving a vaccine challenge 70 days after a priming immunization. In a common exogenous antigen melanoma model, iPEM vaccines inhibited tumor growth more effectively than equivalent admixed formulations consisting of antigen and one of several toll like receptor agonists used as adjuvants. This study reveals the ability of iPEMs to enhance response to several different classes of vaccine cargos using common injection routes, creating a new vaccine and adjuvant delivery platform based entirely on immune signals.

Copyright © American Institute of Chemical Engineers. All rights reserved.

473160 Diminution of Macrophage Attachment on CD47-Functionalized and Polydopamine-Coated Surface

Abstract:

CD47 is a ubiquitously expressed protein known as "marker of self". Recently, CD47 functionalized surface has shown the ability to reduce inflammatory response and platelet adhesion. Polydopamine is an ultrathin but robust film formed by pH-induced polymerization. Polydopamine has been used as a universal surface modification agent capable of further surface functionalization. The purpose of this study is to use polydopamine as a platform to immobilize CD47 onto substratum in order to prevent macrophage attachment. The surface of 6-well cell culture plates was modified with biotin molecules through polydopamine coating. CD47-streptavidin fusion protein was then immobilized onto the biotin-functionalized surface through the high affinity between biotin and streptavidin. The effect of CD47-SIRPα interaction on cell attachment was studied by inoculating J774A.1 macrophages and human mesenchymal stem cells (hMSCs) on CD47 immobilized cell culture plates. Macrophage cells have high amount of SIRPα expressed on the cell surface, while hMSCs do not have any SIRPα. CD47 modified surface deferred macrophage attachment for up to 8 hours. However, hMSCs attached to CD47 immobilized surface within 2 hours. Our results demonstrate that CD47-functionalized surface can prevent macrophage attachment due to CD47-SIRPα interaction.

Copyright © American Institute of Chemical Engineers. All rights reserved.

467092 Predictive Multi-Scale Computational Modeling Environment for Simulating Multicellular Dynamics

Abstract:

Multicellular development depends in large part on the growth, patterning and morphogenesis of epithelial sheets. How individual epithelial cells coordinate multicellular processes is still poorly understood due to the inherent complexity of emergent systems-level behavior. Generating and testing hypothetical novel biophysical mechanisms combining multiple spatial and time scales requires the creation of multi-scale computational models that can span subcellular to tissue levels. However, the task of including detailed descriptions of interactions between the cytoplasm, cortically enriched cytoskeleton and intercellular adhesion has been challenging due to the prohibitively high computational costs. Here, we introduce a multi-scale modeling environment called Epi-Scale for simulating epithelial tissue dynamics using a modified subcellular element method. Epi-Scale explicitly and in detail simulates the separate mechanical interactions between multiple subcellular components. Computational implementation of the model is based on an efficient parallelization algorithm that utilizes clusters of Graphical Processing Units (GPUs) for simulating large numbers of cells within a reasonable computational time. Epi-Scale naturally recapitulates cellular and tissue-scale properties observed in epithelial systems such as cell neighbor relationships and cell geometries. As a demonstration of the predictive power of the model, detailed simulations cell-cell rearrangement outcomes as a result of changes in tissue growth rates and dynamic mitotic rounding are described. A particular advantage for the Epi-Scale environment is its extensibility toward investigating complex biological processes of multiple cell types and interactions between cells and extracellular matrix, including morphogenesis, wound healing and metastasis. Identifying the control systems for tissue homeostasis is important for understanding the underlying causes of birth defects as well as diseases that occur due to misregulated growth such as cancer.

Copyright © American Institute of Chemical Engineers. All rights reserved.

471522 Physical Modeling of the Spreading of Epigenetic Modifications through DNA Looping

Abstract:

Epigenetics refers to variations in gene expression that are caused not by changes in the sequence of DNA, but by chemical modifications in nucleosomes. These epigenetic marks often spread along the chromatin fiber through enzymatic processes that require genomically distant DNA segments to come into close spatial proximity. Such DNA looping is a common phenomenon in many critical biological processes within the nucleus, such as recombination events and gene regulation. Gaining knowledge of the dynamics of loop formation and the kinetics of mark transfer is therefore an essential step towards a better understanding of epigenetic processes.

Loops in DNA occur at a range of length scales, from short DNA lengths on the order of 100 basepairs for the binding of regulatory proteins, to intermediate lengths of kilobases for chromosome folding, to long lengths of hundreds of kilobases for genetic recombination. It is important to recognize that different physics govern looping at different length scales. At short lengths, the rigidity of the DNA must be handled explicitly, whereas at long lengths, the overall chain dynamics dominate. Because of the range of length and time scales involved in the motion of DNA, gaining insight into looping dynamics has proved a particularly complicated pursuit. Previous theoretical studies have addressed the problem through physical modeling and simulation of DNA as a Rouse or semiflexible polymer and have obtained qualitative agreement with experimental results.

To build upon this research, we are working to develop a comprehensive kinetic model for the spreading of epigenetic modifications through DNA looping. We use both analytical theory and computer simulations to calculate the probabilities that various lengths of DNA will loop, and in both cases, we account for the dynamics of DNA looping over the entire range of relevant length scales. We consider the spreading of methyl marks along chromosomal DNA, as dictated by these looping probabilities and rates of methylation and demethylation. Given an initial methylation profile, we determine the steady-state probabilities that other sites will become marked. We examine the effects of the type of transfer (i.e. local vs. at-a-distance), the initial profile (i.e. presence or absence of a nucleation site), the overall topology of the DNA (i.e. linear vs. ring), and the DNA configuration (i.e. static vs. dynamic).

Copyright © American Institute of Chemical Engineers. All rights reserved.

456775 Clostridium Difficile Biofilms: In silico Metabolic Analysis and Antibiotic Treatment

Abstract:

Clostridium difficile is an anaerobic bacterium responsible for recurring in fections in the gastrointestinal tracts of patients previously treated with oral antibiotics that disrupt the healthy gut microbiome. Recent in vitro experiments have demonstrated the ability of C. difficile to form biofilms on surfaces. We developed a metabolic model of C. difficile biofilms to investigate the effects of biofilm formation on antibiotic treatment in vivo. The model was formulated by combining a genome-scale reconstruction of C. difficile primary metabolism with reaction-diffusion type equations for key nutrients (glucose and six essential amino acids) and the common oral antibiotic vancomycin. A very simple model of vancomycin pharmacokinetics was used to predict the efficacy of a typical treatment schedule under the assumption of a fixed thickness of the mature biofilm. Our model predicted that vancomycin will effectively eradicate biofilms of sufficiently small thickness. Once the thickness passed a critical threshold, the model predicted vancomycin treatment would fail catastrophically due to insufficient antibiotic penetration into the biofilm caused by the combination of limited diffusion and vancomycin binding to cell wall precursors. This critical biofilm thickness was shown to be very sensitive to model parameters associated with the stool levels and killing efficiency of vancomycin.

Copyright © American Institute of Chemical Engineers. All rights reserved.

462830 Modeling of a Recombinant Protein Production System: Cell Culture Dynamics and Dynamic Metabolic Flux Analysis

Abstract:

Process development for new antibody products typically starts with selecting candidate cell lines that fit an existing platform process and then tailoring this process to improve antibody productivity, cell vitality and product quality with respect to critical quality attributes (CQAs). Process optimization to achieve the desired outputs relies on in house expertise and historical datasets, and usually consists of a trial-and-error approach. Besides requiring a large investment in time and resources, this approach can lead to only sub-optimal solutions. On the contrary, systems biotechnology can provide a more systematic and rational approach to optimization and aspire to predict how one system changes over time and under varying conditions.

In theory, first-principles models can be used for prediction. However, even for first-principles models the extent of prediction capability greatly depends on the information content and range of the experimental information the model is based on. This is especially true for unstructured models.

In this work, we aim to capture the physiological response of Chinese hamster ovary (CHO) cells to extracellular changes during antibody-producing fed-batch culture as well as to elucidate resource utilization under three different feeding strategies and during the different cell culture phases. To reach these goals we have developed a new model structure comprising two levels. At the extracellular level, cell culture dynamics (CCDyn) are described using a Monod-type kinetic model. Segregation among proliferating, resting and dead cell subpopulations is incorporated into this kinetic model, unlike the classic approach which is usually unsegregated. At the intracellular level, cell metabolism is described using a stoichiometric approach. This coupling is based on the assumption that intracellular reaction kinetics are much faster than the changes in the extracellular process (at least between two subsequent feeding times) so that pseudo-steady state can be assumed to apply for all intracellular reactions. The intracellular compartment is modeled through a fully determined dynamic metabolic flux analysis (dMFA) which describes, in a parameter-free manner, the most significant metabolite fluxes in both the cytoplasm and the mitochondrion. The flux model includes glycolysis, TCA cycle, amino acid metabolism, NS metabolism, as well as mAb glycosylation. One novelty compared to the existing models in CHO systems, is that the MFA includes the Aspartate-Malate shuttle, which directly links amino acid metabolism with TCA. The inputs to this model are the nutrients uptake and bio-products secretion rates calculated by the CCDyn model, while the outputs are the intracellular metabolite fluxes varying with respect to time as a function of changes in the extracellular environment.

Generally speaking, a stoichiometric modelling approach cannot be quantitatively predictive, even in the dynamic version we adopted. However, it is particularly useful to elucidate a metabolic network structure in the different cell culture phases and inform the structure of the CCDyn model. In particular, by including the Asp-Mal shuttle, the dMFA provides additional information on how amino acids (namely, asparagine and glutamate) are consumed towards energy metabolism. Moreover, our dMFA strategy directly links asparagine consumption with energy metabolism and ammonia secretion. Such information is particularly useful in the proposed approach because the CCDyn model is developed as dependent on the extracellular concentrations only. As a consequence, all the experimental data required to develop and validate the model come from the analysis of the supernatant only.

We will present our evaluation of the prediction capability of this new model structure against an independent dataset from fed-batch CHO cell cultures. We will show that the proposed CCDyn model can accurately predict key performance indicators without the need for parameter re-estimation for the new experimental conditions. We can conclude that the proposed approach, that is coupling a segregated Monod-kinetic model with a parameter-free dMFA, is able to give enough information at a reduced price in terms of computational and experimental effort. For this reason, this study represents a significant step towards predictive modeling of animal cell cultures.

Copyright © American Institute of Chemical Engineers. All rights reserved.

471236 A Novel Four-Dimensional Omics Approach for Mapping Large Targetomes of Global Regulatory Networks

Abstract:

Manipulation of global regulators is a new frontier for metabolic engineering. Before global regulators can be explored fully for this application, however, a deep and more cohesive understanding of both the extent to which they exert their cellular effects and their mechanisms of action is required. High-throughput omics experiments have been used to begin to gain this understanding, revealing vast yet diverse and variable targetomes for global regulators. Sorting through the high volumes of target information associated with omics experiments is challenging and false positives are common. For these reasons, we hypothesize that applying only a single omics experiment is insufficient to capture a high-confidence quantification of a global regulatorâ€™s targetome.

In this work, we have proposed and demonstrated an integrative four-dimensional omics (InFO) approach that evaluates changes in gene expression after an (i) **imposed stress** as a function of (ii) **time** in (iii) **multiple genetic backgrounds** by (iv) **multiple omics methods** (transcriptomics, proteomics, and high throughput sequencing crosslinking immunoprecipitation). We apply this approach to the Carbon storage regulator (Csr) system in *Escherichia coli* as proof of concept. Collecting and integrating 126 sets of omics data revealed a well defined set of genes that likely have direct regulatory interactions with CsrA, the main Csr protein regulator. These systems-scale predictions were confirmed via the use of two *in vivo* follow-up experiments that tested both regulation effect of the Csr regulator on its presumed identified targets as well as the ability for physical interactions. A key element of this analysis is the classification of omics-detected genes by potential regulatory mechanisms (up- or down-regulation at the transcript or protein level). Additionally, this novel approach has served to demonstrate the variety of ways that a global protein regulator can regulate its mRNA targets at the systems-level as well as the selectivity that it can exhibit in regulation to yield directional control of biological pathways. Overall, this highly integrative approach has led to a core set of genes that can be confidently called true Csr targets, illustrating the advantages of combining multi-dimensional omics experiments in elucidating targetomes of global regulators.

Copyright © American Institute of Chemical Engineers. All rights reserved.

469419 Combined Omics Approach Reveals the Effects of a Phthalate Pollutant on Adipocyte Metabolism and Inflammation

Abstract:

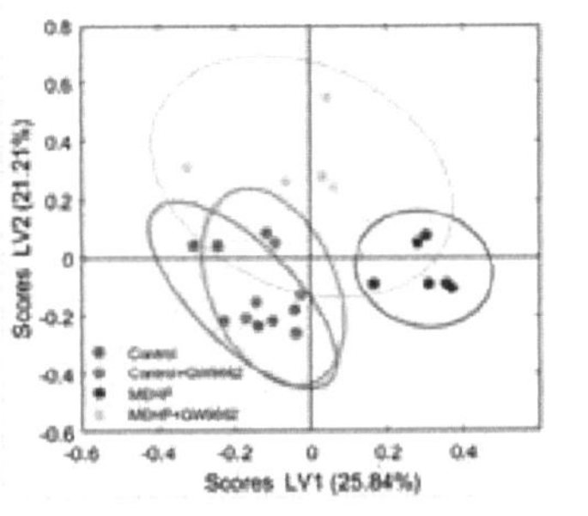

Fig. 1: Latent variable projections from PLS-DA of untargeted proteomic data.

Introduction: Contamination of the environment with organic pollutants has emerged as a significant public health concern due to the pervasive nature of these contaminants. Epidemiological studies have linked chronic endocrine disrupting chemical (EDC) exposure to adverse effects on reproduction, development, and more recently, metabolic diseases [1]. A growing number of studies have reported that perinatal exposure to certain EDCs, termed obesogens, could contribute to weight gain through an adipogenic effect that leads to increased fat mass. To date, studies have mainly focused on the impact of suspected obesogens on stem cell fate and tissue development, sometimes yielding conflicting results. Less attention has been paid to clarifying whether these chemicals can directly disrupt metabolic regulation in differentiated cells of adult tissue. For example, EDCs could interfere with endogenous regulatory pathways to impair the ability of adipocytes to sequester fatty acids or activate pro-inflammatory pathways, both hallmarks of obesity-related metabolic diseases [2]. Studies have shown that EDCs can bind to nuclear receptors (NRs) in a variety of tissues to disrupt the normal function of the endocrine system [3]. One such NR, peroxisome proliferator activated receptor gamma (PPARγ), is a particularly promising link between EDCs and their metabolic effects. This NR recognizes structurally diverse ligands, and regulates the transcription of genes involved in lipid storage homeostasis. Importantly, this NR interacts with other nutrient sensors through feedback mechanisms [2], thus coupling control over metabolism and signaling in adipocytes.

Objective: Our aim in the current work was to study the effects of low, physiologically relevant doses of EDCs on differentiated murine adipocytes, working under the hypothesis that EDCs act via both metabolic and inflammatory signaling pathways to elicit their effects.

Methods: Due to their exogenous origin, EDCs cannot be readily placed into the context of a canonical biochemical or signaling pathway. In this light, a data-driven (e.g., multi-omic) approach could provide valuable clues in determining the pathways impacted by the chemical, which in turn could lead to mechanistic insights. In this study, we combined mass spectrometry metabolomic and proteomic methods with gene expression analysis to study the biochemical changes elicited by a pervasive EDC, monoethylhexyl phthalate (MEHP), in differentiated 3T3-L1 adipocytes. To broadly assess the metabolic effects, we profiled the levels of ~50 intracellular metabolites, and performed a partial least squares discriminant analysis (PLS-DA) on the metabolite data. We also used untargeted proteomics methods to profile the levels of ~1000 cellular proteins and performed PLS-DA, pathway/network analysis, and targeted quantification on the data.

Results: To establish that EDCs elicit an inflammatory response in mature adipocytes, we screened for the effects of 3 representative EDCs - tributyltin (TBT), bisphenol A (BPA), and MEHP - by first inducing adipogenic differentiation of 3T3-L1 cells, and then treating the cells with the chemicals at doses reported in human exposure studies. At nanomolar doses, the chemicals did not affect cell morphology or fat accumulation, but significantly increased expression of inflammatory gene markers such as monocyte chemo-attractant protein 1 (MCP-1). Latent variable score projections from PLS-DA of the metabolite data showed clear differences in the metabolite profiles across the treatment groups. The differences relative to control cells treated with only the vehicle (DMSO) were most pronounced for MEHP treated cells. The corresponding loadings showed that multiple metabolites contributed to the differences, with the majority comprising free fatty acid (FFA) species. Focusing on MEHP as a potent EDC, we performed a dose response experiment to obtain a more detailed understanding of the metabolic and inflammatory responses. MEHP treatment increased the expression of an array of chemokines and cytokines, similar to the effects of the prototypical pro-inflammatory cytokine TNFα. Exposure to MEHP also altered the intracellular FFA profile, raising the levels of both saturated and unsaturated FFAs. We next performed untargeted proteomics experiments with targeted quantification to profile the impact of MEHP exposure on the levels of cellular proteins. The proteomic data indicated that the chemical exposure broadly and significantly altered the levels of enzymes and other proteins involved in regulating the lipid balance and inflammatory pathways in adipocytes. These data were consistent with our hypothesis that MEHP impacts transcriptional/translational regulation of metabolic pathways. Several of the up-regulated proteins were downstream targets of PPARγ (e.g. GPDH, FABP4), which supports the involvement of this key transcriptional regulator. Pathway enrichment analysis using the String database and Ingenuity Pathway Analysis found that lipid metabolism and PPAR signaling, respectively, were the most significantly overrepresented metabolic and signaling pathways in the discriminatory protein dataset. To further investigate the involvement of PPARγ in the observed metabolic and inflammatory effects, we tested whether inhibiting PPARγ activity would abrogate the effects of MEHP exposure. A comparison of protein expression profiles showed that co-treating the cells with the selective synthetic antagonist GW9662 during MEHP exposure attenuated MEHP-induced differences, taking the adipocytes to a phenotypic state more similar to vehicle control (Fig 1). Additionally, co-treatment with MEHP and GW9662 prevented the significant increase in chemokine and cytokine expression observed when the cells were treated with MEHP alone.

Conclusions and ongoing work: Taken together, our results support the hypothesis that the metabolic and inflammatory responses are coupled, and that PPARγ acts, at least in part, as a key mediator. Our results also support a mechanistic link between EDC exposure and adipocyte inflammation, which underpins many of the metabolic dysfunctions of obesity. The next step of this research is to apply another dimension of omics analysis, fluxomics, to better understand the metabolic alterations that lead to increased levels of FFAs, the hypothesized link between metabolic and inflammatory signaling. To this end we are repeating MEHP exposure experiments in adipocytes fed an isotopically labeled substrate, and using isotopic labeling based metabolic flux analysis to gain insight into which central carbon metabolism and lipid synthesis/breakdown pathways are perturbed. We anticipate that the results of this analysis will help guide focused loss-/gain-of-function experiments to further clarify the role PPARγ and other regulator molecules play in the observed effects.

References:

[1] Heindel (2003) Toxicol Sci 76: 247-9.

[2] Manteiga et al. (2013) Wiley Interdiscip Rev Syst Biol Med 5: 425-47.

[3] Grun et al. (2007) Rev Endocr Metab Disord 8: 161-71.

Copyright © American Institute of Chemical Engineers. All rights reserved.

474646 Towards Whole-Cell Models of Higher Organisms

Abstract:

Our lab recently reported completion of the first "whole-cell" model, explicitly accounting for the known functions of every gene product and molecule in the simplest culturable bacterium, *Mycoplasma genitalium*. In this talk I will discuss our efforts to expand this new technology to model more complicated organisms, as well as the applications that we see for such models.

Copyright © American Institute of Chemical Engineers. All rights reserved.

475309 The Power of Prediction: Rapid Optimization of Metabolic Pathways without High-Throughput Screening

Abstract:

The Salis Lab and De Novo DNA have developed an integrated computational-experimental pipeline that rapidly designs and optimizes many-protein genetic systems, such as metabolic pathways & networks, inside engineered micro-organisms, while requiring only a small number of characterization experiments. Our pipeline can be applied to any large genetic system to optimize its performance, using only two design-build-test-learn cycles. To do this, we've combined several predictive models, design rules, and optimization algorithms to map the genetic system's sequence-expression-activity relationship, predict its optimal protein expression levels, and prioritize protein engineering efforts. Our platform is web-accessible at http://www.denovodna.com/software, and has been used by over 6000 registered researchers to design over 100,000 synthetic DNA sequences. We present case studies to demonstrate the pipeline's process.

Copyright © American Institute of Chemical Engineers. All rights reserved.

475294 Reimagine Sequence Space

Abstract:

The applications enabled by perfect DNA sequences from a highly scalable synthesis platform are limitless. With Twist Bioscience's innovative silicon-based platform, reaction volumes are miniaturized by a factor of 1000 enabling high quality DNA production at reduced price and turn-around time. Delivering DNA as genes, ready-to-use oligonucleotide pools or variant libraries simplifies experiments by circumventing the need to clone, allowing use of sequence space to be reimagined on a grander scale, ultimately accelerating discoveries.

Copyright © American Institute of Chemical Engineers. All rights reserved.

475262 Rapid Optimization of Industrial Microbes Via High-Throughput Strain Engineering

Abstract:

Zymergen has built a technology platform for high-throughput engineering of a wide variety of microbes. We use this technology platform to build and test many design ideas and rapidly improve strain performance. In this presentation, an overview of our technology platform for strain design, strain construction and strain testing will be given. In addition, selected case studies will be discussed.

Copyright © American Institute of Chemical Engineers. All rights reserved.

475251 Optimizing Protein Production Strains for Downstream Purification

Abstract:

Boston Mountain Biotech is working on simplifying the removal of host cell proteins (HCPs) that unfavorably bind during protein purification. Despite major advancements in biopharmaceutical manufacturing, protein purification remains one of the most costly and restrictive aspects of production. BMB seeks to directly address this challenge by developing the Lotus® manufacturing platform that identifies the HCPs that exert the greatest burden on downstream processing and then silences these genes in the production strain. This platform could be readily implemented across much of the biomanufacturing industry, with the potential to reduce costs to consumers in the ever-burgeoning healthcare market and ease the time and investment required to bring new therapeutics to market.

Copyright © American Institute of Chemical Engineers. All rights reserved.

475258 Chemically Synthesized Modified Guide RNAs Enhance Crispr/Cas Genome Editing

Abstract:

Using an advancement in RNA synthesis technology (Dellinger et al. J. Am. Chem. Soc. 2011, 133, 11540â€“11556), we find it straightforward to chemically synthesize RNAs of more than 100 nt in length and to incorporate modified nucleotides at any position. Using this technology, we have demonstrated that chemical alterations to synthesized single guide RNAs (sgRNAs) enhance genome editing efficiency in human primary T cells and CD34+ hematopoietic stem and progenitor cells (Hendel et al. Nat. Biotech. 2015 doi:10.1038/nbt.3290). Here, we extend this work characterizing the performance of chemically modified sgRNAs for genome editing in additional cell types including human primary cells (e.g. Fibroblasts, hepatocytes) and induced pluripotent stem cells. The ability to chemically synthesize high quality sgRNAs in a scalable manner enables their widespread application in all phases from basic research to therapeutic use.

Copyright © American Institute of Chemical Engineers. All rights reserved.

470203 Self-Assembly of Antibody-Polymer Conjugates into Novel Sensing Materials

Abstract:

Affinity biosensors based on antibody binding represent one of the most successful and widely investigated categories of protein materials, having yielded technologies as ubiquitous in the US today as home pregnancy tests and often forming the basis for efforts at the development of low cost diagnostics. The sensitivity of these materials is fundamentally limited by the density of binding sites on the material surface, with the possibility of an order of magnitude boost in sensitivity of a transformative increase in surface binding sites could be achieved.

Here, we show that such an increase in binding sites can be achieved through antibody self-assembly into three-dimensional nanostructured materials. Using a site-selective chemistry, bioconjguates are prepared of IgG antibody with the synthetic polymer poly(N-isopropylacrylamide) (PNIPAM). Lyotropic self-assembly of the materials leads to the formation of large lamellar nanostructures that can be observed by both X-ray scattering and transmission electron microscopy. Studies of the self-assembly as a function of bioconjguate concentration and polymer length provide insight into the phase behavior of the system. This is the first such demonstration of antibody self-assembly into large structured arrays.

Using a simple flow coating and crosslinking process, the antibody bioconjguates can be immobilized in nanostructured thin films on different substrates. These films show a 3-4 times higher capacity for protein G binding than adsorbed antibody monolayers. As the size of the binding protein is increased, however, the difference is reduced, presumably because the large binder has trouble penetrating the polymer channels within the protein nanostructure, indicating a need for additional engineering of nanostructure size and transport rates. In addition to this, ongoing work is focusing on developing protein materials from smaller engineered protein binders such as nanobodies.

Copyright © American Institute of Chemical Engineers. All rights reserved.

458100 Structural Studies to Determine the Mechanisms of Biobased Nanoparticle Synthesis

Abstract:

Protein and peptide directed synthesis of inorganic nanomaterials offers several advantages over traditional synthesis methods; however, despite being used to great effect, the fundamental mechanisms of how biomolecules control and drive nanomaterial synthesis is not well understood. We are using high-resolution structural biology techniques on a model gold binding and mineralizing peptide system to investigate how biomolecules are able to influence inorganic nanomaterial nucleation and growth. In this presentation we will present our current results, as well as discuss how this model gold mineralizing system is being used to develop an improved colorimetric ionizing radiation sensor for therapeutic use in cancer treatment.

Copyright © American Institute of Chemical Engineers. All rights reserved.

460477 Biopolymers from a Thermophile: Production, Characterization, and Application

Abstract:

Exopolysaccharides (EPSs) are high molecular weight carbohydrate biopolymers that are composed of sugar residues linked together through glycosidic bonds, and secreted by microorganisms into the surrounding environment with certain properties and functions. The EPSs are one of the adaptations developed by various kinds of extremophiles in order to survive in the environments with extreme conditions, such as extreme temperature, salinity, or pH. Extremophiles are considered as potential producers of EPSs with novel and unusual characteristics which may hardly be found in more traditional polymers. Recently some extremophilic EPSs have already been characterized, and the properties and functions of EPSs from related extremophiles were also partly elucidated. However the accumulated knowledge concerning the structural and functional properties of EPSs secreted by extremophiles is still limited. In this research, a thermophilic strain, *Geobacillus* sp. (WSUCF-1) which produced unique type of EPSs has been studied. WSUCF-1 could utilize glucose as optimum carbon and energy source to produce significant amount of EPSs (382 mg/L). Two purified EPSs were obtained through ion-exchange and gel filtration columns, and both of them demonstrated higher degradation temperature (319.20 and 314.85 °C) than the thermophilic EPSs recently reported. Monosaccharide composition analysis showed that these two EPSs were composed by mannose and glucose with various relative molar ratios (1/0.2141 and 1/0.0270). Spectrum analysis also showed several unique features of the EPSs. EPSs from the relatively unexplored WSUCF-1 strain will likely provide a valuable resource for exploitation in novel biotechnological processes and opportunities in various industrial fields including bio-based polymeric materials.

Copyright © American Institute of Chemical Engineers. All rights reserved.

471023 Genetically Encodable Acoustomagnetic Reporters for Background-Free Molecular and Cellular MRI

Abstract:

Technologies for non-invasive whole-organ imaging of molecular targets and cellular processes such as gene expression are essential to support biological studies, medical diagnostics, and the development of cellular therapeutics. While optical methods provide high-resolution images, they are fundamentally limited by the penetration depth of light in opaque tissues. In comparison, magnetic resonance imaging (MRI) and ultrasound are powerful and complementary non-invasive imaging modalities and have been widely used in clinics. Yet they have never been paired for molecular imaging. Such a pairing, especially in a way that one modality can control the image content from the other modality, could take advantage of the unique features of each modality to provide information not available via either technology alone. **Here, we describe the first contrast agents for acoustomagnetic imaging. These agents are based on gas vesicles (GVs), a unique class of genetically encoded gas-filled protein-only nanostructures formed by photosynthetic microbes as a means to regulate buoyancy.** These agents are genetically encodable, robustly detectable using susceptibility-dependent MRI sequences and "erasable" at specific ultrasound pressures. Thus they provide, for the first time, a direct method to eliminate the endogenous background, a major drawback for susceptibility imaging including functional MRI and superparamagnetic iron oxide (SPIO) nanoparticle imaging. In addition, clustering of these agents based on the presence of a specific analyte produces changes in MRI contrast that could enable the use of these agents as functional sensors.

We reasoned that the gaseous interior of GVs would have a different magnetic susceptibility from the surrounding aqueous media, and would thus induce a local magnetic field gradient that would alter the precession of nearby aqueous ^{1}H nuclear spins. Indeed, we show that GVs can be imaged at nanomolar concentration using gradient echo MRI and quantitative susceptibility mapping. Furthermore, GVs are "erasable" MRI contrast agents, since ultrasound pressure applied in a spatially selective manner can collapse these nanostructures, eliminating their gas compartment and thus the MRI contrast. This is very useful because, many endogenous tissue properties, such as the presence of blood vessels and the interface between different types of tissue, can produce hypointense contrast in gradient echo images, and the complication of endogenous contrast has been a major drawback to the design of susceptibility-dependent T2* MRI contrast agents. Due to their ability to be imaged before and after acoustic collapse, GVs enable, for the first time, the acquisition of background-free T2* contrast. We demonstrate the concept of acoustically erasable background-free imaging using GVs injected into mouse brains and collapsed using ultrasound.

Furthermore, acoustic collapse occurs at different specific pressures for GVs from different organisms, enabling multiplexed "acoustomagnetic" imaging using a pressure-scan paradigm. The paradigm could enable the tracking of multiple GV-labeled cell types using MRI. Finally, GVs can potentially be engineered as sensors of specific biological signals because the clustering of GVs produces a marked change in gradient echo images, as demonstrated ***in vitro*** by streptavidin-biotin cross-linking.

Overall, these results support the introduction of GVs as the first acoustomagentic reporters, which use the combination of ultrasound and MRI to enable background-free and multiplexed imaging. The ability of GVs to be genetically encoded and engineered opens the possibility of using this new form of contrast in a wide range of biological applications, especially in synthetic biology and cellular therapeutics.

Copyright © American Institute of Chemical Engineers. All rights reserved.

474994 How Bolt Threads Is Using Chemical and Process Engineering to Grow Its Business of Engineering Silk (Featured Presentation)

Abstract:

Bolt Threads is a startup that believes that the combination of molecular biology, chemical engineering, and materials science can revolutionize how people think about and use natural materials. Bolt is developing the next generation of sustainable apparel by leveraging the latest advances in industrial biotechnology, automation, and materials science. The company is producing spider silk, a notoriously strong and lightweight material that scientists have tried to commercialize for decades. Boltâ€™s secret weapon is chemical and process engineering, which allows it to do what its predecessors couldnâ€™t: scale up its technology. From bench scale to industrial scale, process engineering is the foundation of building this high growth startup business. Leveraging the unique properties of its fibers, Bolt Threads will be delivering new products to consumers and brands worldwide.

Copyright © American Institute of Chemical Engineers. All rights reserved.

470235 Efficacy of 5-Aminolevulinic Acid (5-ALA)-Mediated Photodynamic Therapy (PDT) Using Cold Atmospheric Plasma (CAP) As a Light Source for Anti-Tumor Applications

Abstract:

Introduction: Photodynamic therapy (PDT) is a treatment modality that takes advantage of the cytotoxic effects induced by a light source and a photosensitizer to generate singlet oxygen species [1]. It is well known that 5-ALA has been widely used in photodynamic therapy, which is a natural biochemical precursor for heme synthesis in living mammalian cells. In this study, cold atmospheric plasma (CAP) was used for the first time as a light source in photodynamic therapy. CAP is an ionized gas where the ion temperature is close to room temperature. It contains electrons, reactive oxygen species (ROS), reactive nitrogen species (RNS), radicals, and UV light [2]. The wavelengths for CAP are widely covered from 250 nm to 800 nm, which makes CAP a possible light source for photodynamic therapy.

Materials and Methods: 5-ALA and gold nanoparticles 2-4 nm in diameter were encapsulated inside polymersomes by self-assembly. After being encapsulated, TEM, DLS and SEM were used to measure particle size, zeta potential and particle morphologies. Both fibroblasts (ATCC, PCS-201-011) and melanoma A375 (ATCC, CRL-1619) cells were seeded at $2*10^4$ cells/ml on a 96-well plate maintained in DMEM (ATCC) supplemented with 10% FBS (ATCC) and 1% penicillin/streptomycin (Sigma). After 24 hours of incubation, polymersomes at different concentrations (from 100 to 500 Î¼g/ml) were added to the cells and were further co-cultured overnight. Then, both cells with nanoparticles were exposed under a CAP light source for 90s. Cell viability of the two cell lines was quantified by an MTS assay, and cell morphology was observed under confocal microscopy. All cells were incubated in a standard incubator at 5% CO_2at 37 Â°C. The cellular experiments were run in triplicate and repeated three times for each group.

Results and Discussion: Cell viability results with the 5-ALA mediated photodynamic therapy indicated that using CAP as light source can significantly decrease melanoma cell number when compared with non-CAP treatment. Specifically, at concentrations of 200 Î¼g/ml, the melanoma cell number reduced by more than 50% when compared to controls. More importantly, this treatment was non-cytotoxic to fibroblasts. In addition, confocal images indicated that the 5-ALA/PDT treated melanoma cells lost their cellular protrusions and showed a polygonal, nearly epithelial cell-like appearance in contrast to the elongated cells with multiple protrusions in untreated controls. Fibrillar actin formation in growth cones was reduced after ALA/PDT treatment, which showed a rearrangement of the actin cytoskeleton, resulting in a loss of cellular polarization. Also, focal adhesions decreased significantly observed from confocal images, which is related with cell integrin adhesion, signaling and actin cytoskeleton formation after ALA/PDT treatment.

Conclusions: 5-ALA/PDT with CAP as a light source treatment induced melanoma cancer cell apoptosis as well as suppression of cell growth and cell morphological changes. All of these results showed that 5-ALA/ PDT with CAP as a light source is a novel effective PDT method for anti-tumor applications deserving further study.

Acknowledgement: This study was supported by the Northeastern University.

References: [1] L. M. Rossi, P. R. Silva, L. R. Vono, et al. Langmuir, 2008, 24, 12534-12538.

[2] M. Wang, X. Cheng, W. Zhu, et al. Tissue engineering: part A, 2014, 5(20), 1060-1071.

Copyright © American Institute of Chemical Engineers. All rights reserved.

455670 Repeatable and Adjustable on-Demand Local Anesthesia By Phototriggerable Liposomes

Abstract:

Current treatment of pain relies heavily on opioids and conventional local anesthetics. The former is associated with systemic side effects and addictionÂ (1, 2), while the latter provides short durations of anesthesia (normally 6 â€" 12 h when typical post-operative pain lasts for a timescale of days) (3). Although sustained release systems can provide longer nerve block durations in animal models (up to days) (4), they have a major limitation: once administered, their drug release profile cannot be changed, regardless of the patientâ€™s changing needs. This is particularly troublesome when physical therapy or movement are desired, as motor block often accompanies the desired sensory block (4). An on-demand local anesthesia system where patients and physicians could noninvasively control the timing, intensity and duration of local anesthesia therefore highly desirable.

Here we demonstrate a phototriggerable local anesthesia system that can realize this goal. A near infrared (NIR) light-absorbing photosensitizer was encapsulated within liposomes composed of lipids labile to reactive oxygen species (ROS) (6). Upon irradiation with NIR light, the photosensitizer released singlet oxygen, inducing lipid peroxidation and enhancing the permeability of liposomal lipid bilayers, resulting in the release of encapsulated drug. A potent local anesthetic, tetrodotoxin (TTX), was loaded into these phototriggerable liposomes (Lipo-PS-TTX). *In vitro* experiments showed repeatable phototriggered TTX release with 730-nm laser irradiation (7). *In vivo* injections of Lipo-PS-TTX percutaneously onto the rat sciatic nerve induced an initial sciatic nerve block, and subsequent irradiations at 730 nm induced repeated sciatic nerve block. The timing, intensity and duration of nerve block were adjustable by the parameters of irradiation. Nerve block duration showed a clear relationship with irradiation energy density, demonstrating the controllability of the system. On-demand phototriggerable nerve block systems may allow pain treatment to be personalized and substantially improve pain management.

1. J. L. Apfelbaum, C. Chen, S. S. Mehta, T. J. Gan, *Anesth Analg* **97**, 534 (2003).

2. M. E. Lynch, C. P. N. Watson, *Pain Res Manag* **11**, 11 (Spring, 2006).

3. M. J. M. D. F. Fredrickson, A. F. Abeysekera, R. M. A. White, *Regional Anesthesia & Pain Medicine September/October* **37**, 495 (2012).

4. H. Epstein-Barash *et al.*, *Proc Natl Acad Sci USA* **106**, 7125 (2009).

5. A. V. Soldatova *et al.*, *Inorg Chem* **50**, 1135 (2011/02/07, 2010).

6. Y. Yorihiro, N. Etsuo, K. Yoshio, S. Hiroyuki, *Biochim Biophys Acta* **795**, 332 (1984).

7. A. Y. Rwei *et al.*, *Proceedings of the National Academy of Sciences* **112**, 15719 (December 22, 2015, 2015).

Copyright © American Institute of Chemical Engineers. All rights reserved.

472751 Engineering Polymeric Nanostructures in Silicone Hydrogel Contact Lens Biomaterials for Controlled Release to Treat Glaucoma

Abstract:

Topical eye drops are the current standard of care for ocular therapeutics. Despite dominating the market, eye drops are only marginally effective with typically less than 1-7% of the applied drug being productively absorbed. Although the eye is easily accessible for treatment, there exist highly effective physiological barriers that make topical treatment extremely challenging. Low bioavailability and short residence time are combated with a rigorous eye drop regimen, making treatment highly susceptible to patient compliance issues. A number of ocular conditions rely heavily on prescribed topical treatment, sometimes leading to devastating consequences if treatment is inadequate. This is the case for glaucoma, the world's leading cause of irreversible blindness which is estimated to affect over 60 million worldwide and currently has no cure. If untreated or poorly treated, individuals experience progressive, permanent vision loss due to an increase in intraocular pressure that damages the optic nerve. Vision loss and ocular disease costs the United States approximately $139 billion per year, with an estimated 4.3 Americans estimated to develop glaucoma by 2030.

Our solution to replace eye drops for ocular therapeutics is the therapeutic contact lens. Macromolecular memory is one of the most promising concepts for therapeutic lens design, and our lab has shown (both *in vivo* and *in vitro*) this technology is capable of delivering drug at a controlled rate for a wide range of drug classes and molecules, lens types, lens modalities, and lens wear schedules. Engineering macromolecular memory involves molecular imprinting of flexible polymer chains. Functional monomers are selected to have high affinity to the drug via non-covalent interactions. Molecular complexes comprised of the functional monomers and drug molecules are crosslinked into the hydrogel matrix during polymerization. These nanostructures remain after removal of the drug and provide interaction points throughout the polymer that delay release. Lenses are specifically designed for the selected drug because the chemistry and structure of the drug molecule dictate which types of non-covalent interactions are utilized. This study highlights a selected combination of functional monomers interacting with two forms of latanoprost, a prominent glaucoma medication prescribed today.

Silicone hydrogel and hydroxyethyl methacrylate (HEMA) contact lens formulations were developed to release the prostaglandin analogue, latanoprost, in both its active form (LPA) and prodrug form (LP). The major components of our silicone hydrogel lenses were methacryloxypropyl-tris-(trimethylsiloxy) silane (TRIS), dimethyl acrylamide (DMA), and methacryloxypropyl terminated polydimethylsiloxane (DMS-R11). HEMA was the main constituent of the HEMA lenses. Poly(ethylene glycol) 200 (PEG200) and ethylene glycol dimethacrylate (EGDMA) served as crosslinkers. Functional monomers were selected to specifically cater to the chemistry of both forms of latanoprost. For LP, hydrogen bonding was the focus with acrylic acid (AA) and methacrylic acid (MAA) to form variations of a poly(AA-*co*-MAA-*co*-TRIS-*co*-DMS-R11-*co*-DMA-*co*-EGDMA-*co*-PEG200DMA) network and poly(AA-*co*-MAA-*co*-HEMA-*co*- PEG200DMA) network. For LPA, ionic bonding potential was taken advantage of with 2-(diethylamino)ethyl methacrylate (DEAEM) and diallyldimethylammonium chloride (DADMAC) for the development of poly(DEAEM-*co*-DADMAC-*co*-TRIS-*co*-DMS-R11-*co*-DMA-*co*-EGDMA-*co*-PEG200DMA) network and poly(DEAEM-*co*-DADMAC-*co*-HEMA-*co*-PEG200DMA) network. UV free-radical photopolymerization was used to form the contact lenses from solution. Drug binding studies were performed to show successful incorporation of macromolecular memory. Dynamic release studies were done in two ways: a large volume release system (200 mL at 34°C) and microfluidic devices designed to mimic physiological volumes and flow rates (3μL/min).

One of the most significant variables to consider in these formulations is the molar ratio of the functional monomers (M) to the template drug molecule (T)—the M/T ratio. In this study, various combinations of the M/T ratio were tested for both drug types (20/1, 50/1, 100/1) in the silicone hydrogel and HEMA lenses. Ratios between the monomers varied between 1 and 3 to show how the monomer content and proportions directly impact the nature of the memory sites formed and the resultant release properties from the lenses. Previous studies have shown that the release rate and overall release profile can be controlled through the adjustment of the M/T ratio and the diversity of functional monomers used. For both HEMA and silicone hydrogel lenses produced, higher M/T ratios resulted in a slower release of drug over time. Both LP and LPA were successfully released for over a week under *in vitro*physiological conditions in both lens types. Optical clarity and mechanical properties were also tested to show that the polymers had appropriate physical properties for use as contact lenses.

There exists a great need for an improved, more efficient and effective method of topical drug delivery to the eye. Macromolecular memory has the potential to revolutionize ocular therapeutics for a wide range of ocular conditions and diseases through the incorporation of nanostructured memory sites specifically catering to the chosen drug molecule. This technology has been proven to provide tailorable release rates with versatility in application, while also being capable of fitting seamlessly into current manufacturing schemes. Our study shows how this technology is applied for the first time to the global issue of glaucoma as a possible remedy for current drug delivery methods. Improved treatment via macromolecular memory could ultimately reduce incidence of glaucomatous blindness and greatly improve patients' quality of life.

Copyright © American Institute of Chemical Engineers. All rights reserved.

465938 Core Shell Chitosan Microcapsules for Programmed Sequentialdrug Release

Abstract:

A novel type of core-shell chitosan microcapsules with programmed functions of sequential drug release are designed and developed by microfluidic technology for acute gastrosis therapy. The microcapsule is composed of a cross-linked chitosan hydrogel shell and an oil core containing both free drug molecules and drug-loaded poly(lactic-co-glycolic acid) (PLGA) nanoparticles. The resultant microcapsules can keep structural integrity in neutral medium without leakage of encapsulated substances. Based on the acid-triggered decomposition of chitosan shell and the sustained release of drug from PLGA nanoparticles, our proposed microcapsules successfully achieve the programmed sequential drug release performance, i.e., burst release first and then sustained release. In acid medium, the encapsulated free drug and drug-loaded PLGA nanoparticles can be rapidly released within 60 s to achieve the first burst-release. Later, the drugs loaded in PLGA nanoparticles are slowly released for a few days to achieve the second sustained-release due to the synergistic effect of drug diffusion and PLGA degradation. Such core-shell chitosan microcapsules with programmed functions of sequential drug release are promising to achieve a more rational drug delivery and controlled-release for the treatment of acute stomach illness. In addition, the designed strategy of programmed sequential release in this work is of versatility for various biomedical applications.

Copyright © American Institute of Chemical Engineers. All rights reserved.

471991 Engineering of Virus-like Particles Functionalized with CD47 Extracellular Domain to Improve Targeted Delivery By Avoiding Macrophages

Abstract:

Targeted drug delivery is important to both improve the drug efficacy and reduce side effects. Virus-like particles (VLPs) have been engineered to be used as targeted drug delivery vehicles. VLPs are self-assembled, homogeneous nanoparticles derived from the viral capsid and contain no genetic material. The Swartz Lab has engineered Hepatitis B core VLPs to be stable during manufacture and storage, but to lose stability under cytoplasmic conditions, allowing us to release the drug cargo once the VLP is inside the cells. Additionally, the Swartz Lab has developed a process to functionalize the VLP surface by incorporating non-natural amino acids (nnAAs) using cell-free protein synthesis technology. By incorporating a nnAA with an azide group on the VLP and a nnAA with an alkyne group on the ligands, we can decorate the VLP surface with multiple ligands using the copper(I)-catalyzed azide-alkyne cycloaddition ("click") reaction. With this approach, each VLP can accommodate up to 120 ligands including targeting molecules such as antibody fragments on the surface.

One major challenge in using VLPs as a drug delivery vehicle is avoiding phagocytic clearance before reaching the targeted cells. For this purpose, the CD47 extracellular domain will be displayed on the VLP surface. CD47 is known as the "marker of self", because its extracellular domain interacts with SIRPα on phagocytes to inhibit phagocytosis. Several CD47 extracellular domain variants have been produced using cell-free protein synthesis and displayed on the VLP surface using the "click" reaction. These VLPs will be further functionalized with targeting molecules for specific delivery to prostate cancer cells. The CD47-functionalization will enhance the delivery efficiency of the VLPs by avoiding macrophage engulfment.

Copyright © American Institute of Chemical Engineers. All rights reserved.

460473 Zwitterionic Gel Encapsulation Promotes Protein Stability, Enhances Pharmacokinetics Andreduces Immunogenicity

Abstract:

As an emerging class of therapeutics, protein drugs offer the advantages of higher specificity and potency than small molecules. However development of protein therapeutics is hindered by their poor stability, inadequate pharmacokinetic (PK) profiles, and immunogenicity. Polyethylene glycol conjugation (PEGylation) is the most successful strategy to date to overcome these shortcomings, and more than ten PEGylated proteins have been brought to market. However, anti-PEG antibodies induced by treatment raise serious concerns about the future of PEGylated therapeutics. Based on extensive studies of PEGylation, we believe that two issues are responsible for its immune response: the immunogenicity/antigenicity of PEG itself, and limited PEG coverage on the protein surfaces. By solving these two issues, we demonstrate a zwitterionic polymer network encapsulation (ZPNE) technology that effectively enhances protein stability and pharmacokinetics while reducing the immune response. Highly immunogenic fungal-derived uricase was used as a model therapy. In this study, uricase modified with a comprehensive zwitterionic polycarboxybetaine (PCB) gel network exhibited exceptional stability and a greatly prolonged circulation half-life. More importantly, the PK behavior was unchanged and neither anti-uricase nor anti-PCB antibodies were detected after three weekly injections in a rat model. This technology is applicable to a variety of proteins and unlocks the possibility to adopt highly immunogenic proteins for therapeutic or protective applications.

Copyright © American Institute of Chemical Engineers. All rights reserved.

472682 Drug-Bearing Supramolecular Filament Hydrogels for Local Treatment of Diseases

Abstract:

We report here the direct use of therapeutic agents to create self-assembling supramolecular filaments that can form hydrogels in physiologically relevant conditions. Our strategy is to chemically conjugate a short peptide onto a therapeutic agent to form a drug amphiphile for self-assembly into one-dimensional (1D) nanostructures. Through the choices of the peptide sequences and the manipulation of the assembly kinetics, we are able to tune the gelation conditions, the drug release rates, and other properties important for local delivery. In one specific example, we found that rationally designed camptothecin (CPT) amphiphiles could spontaneously form a viscous solution in water containing self-assembled CPT filaments. This solution gels immediately upon contact with body fluids or cell medium. We carefully studied their drug release kinetics and in vitro efficacy against brain primary cells, and evaluated its in vivo efficacy using a resection brain tumor mouse model. Our results suggest that these drug-bearing supramolecular hydrogels have enormous potential to improve the current standards for local treatment of many diseases including brain tumor.

Copyright © American Institute of Chemical Engineers. All rights reserved.

455388 Drug Release Study of a Tripartite Gold Nanoconjugate for Spinal Cord Injury Treatment

Abstract:

Gold nanoparticles (AuNPs) show great potentials for biomedical applications with regard to both diagnostic and therapeutic utilities. As a drug delivery carrier, AuNPs enable versatile conjugation chemistry with drugs and proteins with the ease of tuning particle size, shape, and surface chemistry. Our work focuses on the engineering of AuNPs/drug conjugates for treating respiratory dysfunction resulted from high-level of spinal cord injury (SCI). Respiratory problems are common and often lead to fatal consequences caused by SCI. In SCI patients, systemic administrations of drugs that improve respiratory function also cause harmful side effects. By chemically conjugating the AuNP drug carrier to a transporter protein (wheat germ agglutinin coupled to horseradish peroxidase or WGA-HRP) and to theophylline (a SCI drug or THP), a single administration of the nanoconjugate improved diaphragmatic activity at a much lower dosage than the free drug dosage. This talk will describe the synthesis and characterization of the nanoconjugate, particularly, the use of thermogravimetric analysis (TGA) and laser desorption ionization mass spectrometry (LDI-MS) to determine the numbers of drug molecules conjugated to AuNP. This talk will further describe the drug release profiles of the gold nanoconjugate in buffer and *in vivo*. For the *in vivo* study, recovery was found 72 hours after injection, which was 24 hours after the detection of the nanoconjugate in the targeting site. The *in vivo* results are correlated with the results from the drug release study in buffer and artificial spinal cord fluid using mass spectrometry. This study demonstrates a novel nanoconjugate capable of targeted drug delivery for the treatment of respiratory dysfunction associated with SCI.

Copyright © American Institute of Chemical Engineers. All rights reserved.

471398 Stimuli-Responsive Nanoscale Hydrogels for the Co-Delivery of Chemotherapeutic Agents

Abstract:

The lack of specificity in traditional chemotherapeutic administration typically leads to significant dose-limiting toxicities and requires patients to wait for long periods between treatments. During this time, cancer cells have an opportunity to recover from the treatment and develop multi-drug resistance[1]. Our work seeks to improve treatment specificity through the use of intelligent nanoscale hydrogels (nanogels) to localize the chemotherapeutic agents at targeted disease sites via the enhanced permeation and retention effect, ultimately limiting the toxicity to healthy tissues.Â Here, we show that the nanogel molecular architecture can be tailored to carry a variety of cargos with widely varying physicochemical properties, and release the cargo only in response to an acidic intracellular environment. We also optimized the nanogel surface properties to avoid the bodyâ€™s natural clearance mechanisms and increase circulation time, while still maintaining the necessary characteristics to promote cellular uptake. Nanoparticle-mediated combination therapy offers many advantages including the ability to signal different pathways in the cancer cells, maximize the therapeutic efficacy against specific targets, target different phases of the cell cycle, and overcome efflux-driven mechanisms of resistance[2]. Further, it allows PK/PD to be dictated by the in vivo distribution and cellular uptake of the nanogels rather than the physicochemical properties of the free chemotherapeutic agents, ensuring optimal synergistic ratios are delivered to the cytosol.

Prepared nanogels are comprised of: (i) a hydrophilic, cationic monomer 2-(diethylamino)ethyl methacrylate (DEAEMA) that imparts the pH-response by ionization of amine pendant groups; (ii) a tetraethylene glycol dimethacrylate (TEGDMA) crosslinker to improve chemotherapeutic agent retention; (iii) an alkyl methacrylate monomer to improve chemotherapeutic agent-polymer interactions; and (iv) surface-grafted poly(ethylene glycol) methacrylate (PEGMA) to impart serum stability. Nanogels were synthesized using a UV-initiated oil-in-water emulsion polymerization with a 2.5 mol% crosslinking density[3]. Hydrogel properties were investigated through systematic variation of monomer functionality and chain length. The physical properties of the resulting nanogels were compared using dynamic light scattering, zeta potential, titration, pyrene fluorescence, and red blood cell hemolysis as a function of pH to elicit the influence of polymer composition on swelling ratio, surface charge, pKa, relative hydrophobicity and hydrophile-hydrophobe phase transition, and erythrocyte membrane disruption capability. The therapeutic delivery potential was analyzed using hydrophobic and hydrophilic chemotherapeutic agents.

The nanogels resulted in well-defined and controllable particle size, morphology, and composition. We demonstrated the tunability of our multicomponent nanogel systems to entrap varied molecular cargos, and showed that the molecular architecture can be rationally designed to respond intelligently to different environments. Inclusion of a hydrophobic monomer significantly altered the resulting nanogel physical properties. Varying both the chain length and steric bulk allowed for precise control over the thermodynamic response (relative swelling ratio), dynamic behavior (nanogel pKa and membrane disruption potential), and drug-polymer interaction (therapeutic delivery potential). We also demonstrated that the zeta potential changes drastically as a function of grafted PEGMA coverage on the surface of the nanogel, while the isoelectric point of the network core remains constant. The volume swelling ratio (VSR) of the nanogels was significantly influenced by the amount of PEGMA, with an increased percentage relating to a decreased VSR. Ultimately, we found that optimizing both the grafted monomer and hydrophobic monomer composition leads to varying nanogel surface and core characteristics, respectively, that may be exploited to enable long-circulation of the nanogels and effective transport of the drugs to the tumor site.

Acknowledgements: This work was supported by a grant from the National Institutes of Health (R01-EB-000246-22). AMW is supported by an NSF Graduate Research Fellowship. AS, NAS, and BC equally contributed to the work.

Author Correspondence: angelamwagner@utexas.edu, peppas@che.utexas.edu

References: **[1]** Kolishetti, N., et al. (2010). "Engineering of self-assembled nanoparticle platform for precisely controlled combination drug therapy." PNAS 107(42): 17939-17944. **[2]** Jabr-Milane, L. S., et al. (2008). "Multi-Functional Nanocarriers to Overcome Tumor Drug Resistance." Cancer Treat Rev 34(7): 592-602. **[3]** Fisher, O. Z., et al. (2009). "Enhanced core hydrophobicity, functionalization and cell penetration of polybasic nanomatrices." Pharm Res 26(1): 51-60.

Copyright © American Institute of Chemical Engineers. All rights reserved.

472439 Exploiting Serum Interactions with Cationic Biomaterials Enables Label-Free Circulating Tumor Cell Isolation

Abstract:

Exploiting Serum Interactions with Cationic Biomaterials Enables Label-Free Circulating Tumor Cell Isolation

Michael J. Mitchell[1], Carlos A. Castellanos[2], Michael R. King[2]

[1]Department of Chemical Engineering, Koch Institute for Integrative Cancer Research, MIT, Cambridge, MA 02139

[2]Department of Biomedical Engineering, Cornell University, Ithaca, NY 14853

Introduction: The ability to isolate circulating tumor cells (CTCs) from patient blood at high purity will facilitate the development of effective personalized medicine regimens for patients with metastatic cancer. Our lab has developed microscale flow devices with immobilized cell adhesion molecules to both isolate CTCs from patient blood and deliver therapeutics to cancer cells (1-3). Improving upon current isolation methods is challenged by (1) the lack of adhesion receptors common to a broad range of CTCs, and (2) low CTC capture due to leukocyte bio-fouling onto capture surfaces. Herein, we evaluated the effect of charged polymer coated surfaces on differential capture of CTCs under flow.

Materials and Methods: Microscale flow devices were functionalized with cationic polymer by incubation with 0.025 - 0.1% (w/v) aqueous poly-L-lysine (PLL). Functionalized surfaces were characterized using dynamic light scattering, AFM, TEM, circular dichroism spectroscopy, and contact angle goniometry. Cells were suspended in PBS buffer containing up to 1% (w/v) human albumin and perfused through microscale devices at a flow rate of 0.008 mL/min for 10 minutes using a motorized syringe pump and monitored via an inverted microscope. Micrographs and video of cell adhesion were used to measure cancer cell capture and purity.

Results and Discussion: PLL-functionalized surfaces were able to differentially capture cancer cells and leukocytes suspended in albumin-containing buffers (Fig. 1A - B). Leukocyte capture significantly decreased on surfaces when cells were suspended in 1% (w/v) albumin solution, while the number of cancer cells captured remained unchanged. Cell suspensions from metastatic prostate cancer patients were processed on PLL-coated surfaces and putative CTCs identified on basis of DAPI, cytokeratin, and CD45 staining (Fig 1 C).

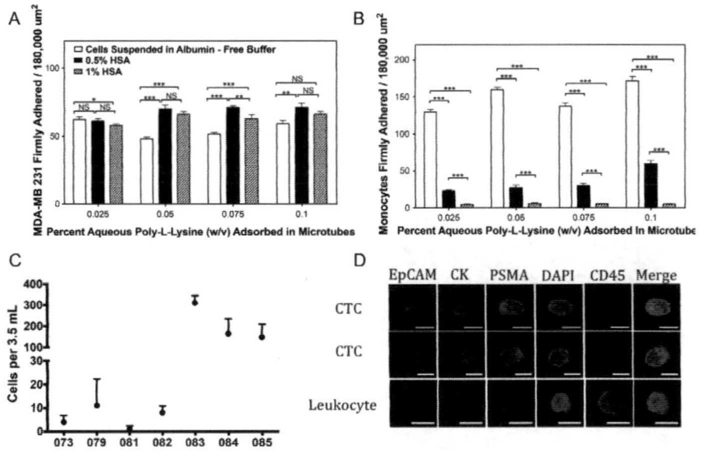

Figure 1: PLL functionalized surfaces differentially capture cancer cells and leukocytes under flow. Number of firmly adhered cancer cells (A) and leukocytes (B) per 180,000^2 um. ***P<0.0001. *P <0.01. NS: not significant. (C) Number of prostate cancer cells captured from blood of metastatic cancer patients and representative images of CTCs. CTCs were confirmed on basis of DAPI, cytokeratin (CK), and CD45 staining. Scale bar: 5 um.

Conclusions: We have shown that PLL functionalized surfaces contribute to differential capture of cancer cells under flow. While both cancer cells and leukocytes adhesively interact with polylysine, the addition of albumin to buffer solution acts to repel leukocytes while maintaining cancer cell capture, which has the potential to improve applications for the isolation of rare CTC populations of interest, for the development of effective personalized medicine regimens.

References:

1. Mitchell MJ, Castellanos CA, King MR. Surfactant functionalization induces robust, differential adhesion of tumor cells and blood cells to charged nanotube-coated biomaterials under flow. *Biomaterials*. 2015; 56:179-186.

2. Mitchell MJ, Wayne E, Rana K, Schaffer CB, King MR. TRAIL-coated leukocytes that kill cancer cells in the circulation. *PNAS*. 2014; 111: 930-935.

3. Mitchell MJ, Castellanos CA, King MR. Immobilized surfactant nanotube complexes support selectin mediated capture of viable circulating tumor cells in the absence of capture antibodies. Journal of Biomedical Materials Research Part A. 2015; 103:3407-3418.

Copyright © American Institute of Chemical Engineers. All rights reserved.

463508 Lipid Exchange Envelope Penetration (LEEP) of Nanoparticles for Plantengineering: A Universal Localization Mechanism

Abstract:

Many important plant functions, such as carbon dioxide reduction or energy generation is carried out within the chloroplast – a plant organelle that appears greatly under explored as an engineering material. Here, we examine the subcellular uptake and kinetic trapping of a wide range of nanoparticles for the first time, using the plant chloroplast as a model system, but validated in vivo in living plants. Confocal visible and near-infrared fluorescent microscopy and single particle tracking of gold-cysteine- AF405 (GNP-Cys- AF405), streptavidin-quantum dot (SA-QD), dextran and poly(acrylic acid) nanoceria, and various polymer-wrapped single-walled carbon nanotubes (SWCNTs), including lipid-PEG- SWCNT, chitosan-SWCNT and 30-base (dAdT) sequence of ssDNA $(AT)_{15}$ wrapped SWCNTs (hereafter referred to as $ss(AT)_{15}$-SWCNT), are used to demonstrate that particle size and the magnitude, but not the sign, of the zeta potential are key in determining whether a particle is spontaneously and kinetically trapped within the organelle, despite the negative zeta potential of the envelope. We develop a mathematical model of this lipid exchange envelope and penetration (LEEP) mechanism, which agrees well with observations of this size and zeta potential dependence. The theory predicts a critical particle size below which the mechanism fails at all zeta potentials, explaining why nanoparticles are critical for this process. LEEP constitutes a powerful particulate transport and localization mechanism for nanoparticles within the plant system.

Reference: Wong MH, Misra R, Giraldo JP, Kwak SY, Son YW, Landry MP, Swan JW, Blankschtein D, Strano MS. 2016. Lipid Exchange Envelope Penetration (LEEP) of Nanoparticles for Plant Engineering: a Universal Localization Mechanism. Nano lett, 16 (2), pp 1161–1172

Copyright © American Institute of Chemical Engineers. All rights reserved.

470276 Nanoparticle Uptake in Flow-Conditioned Endothelial Cells Depends Â on Glycocalyx Structure and Nanoparticle Shape and Charge

Abstract:

The development of atherosclerosis in arterial blood vessels preferentially occurs at branched, constricted and curved arteries and spans decades. The condition is difficult to treat due to its location and slow progression. The glycocalyx is a sugar-rich extracellular matrix, composed of glycosaminoglycans such as heparan sulfate and sialic acids, with pore sizes of 7 nm or less. These components line the endothelial cells of blood vessels and are critical in maintaining vascular health. The glycocalyx regulates mechanotransduction, chemokine and signaling molecule interactions, and cell-cell communication (Ebong, 2011). In atherosclerotic patients this glycocalyx layer is found to be shed, which suggests that glycocalyx may be linked to cardiovascular disease. Our interest is to leverage unhealthy glycocalyx and its characteristics, combined with nanoscale particle drug delivery vehicles, to develop a preventative approach to atherosclerosis development. To achieve our goal we looked into how the surface and interface properties of the glycocalyx differs under various conditions, and how each component of the matrix affects nanoparticle uptake by blood vessel endothelial cells. A monolayer of human endothelial cells were exposed to shear stress at 15 dynes/cm^2 for six hours. For a healthy glycocalyx, the cells were cultured with serum-rich media, while a collapsed glycocalyx was modeled through serum-deficient media. Glycocalyx cleaving enzymes were perfused across the endothelial cells for degraded models (Fels, 2014). Heparan sulfate was degraded using heparinase III enzyme at 45 mU/mL (U is sigma units) and sialic acid residues were removed through neuraminidase at 1215 mU/mL. After degradation, the shed components were introduced back into the culture at concentrations found in a healthy glycocalyx to simulate repairing of the matrix. After exposure to flow and glycocalyx treatments, the endothelial cells were incubated for 16 hours with various gold nanoparticles coated in polyethylene glycol and AlexaFluor 647 fluorophore, then imaged through confocal microscopy to determine glycocalyx structure and nanoparticle uptake. The particles used were gold nanospheres with diameters of 10 nm and nanorods with diameters of around 20 nm, with charges ranging from -10 mV to -40 mV. These characteristics allow for the particles to localize at dysfunctional glycocalyx - a healthy glycocalyx would discourage uptake of particles that are negatively charged and/or larger than 7 nm because of the pore size cutoff and innate negative charge of the matrix. The glycocalyx of the human endothelial cells were found to be more prominent and continuous with flow conditioning, as imaged by bovine serum albumin, heparan sulfate, and sialic acid immunostaining. This matches results by Dewey *et al.* with regards to heparan sulfate regrowth under laminar shear flow. The thickness and coverage of all stains increased after pre-treatment with flow stimulus and coincided with minimal nanoparticle uptake as expected. However, after serum deprivation or enzyme treatments the uptake of gold nanoparticles increased significantly because of glycocalyx disruption. Addition of the degraded components back into the culture during the nanoparticle treatment decreased the uptake. Further, localization of the nanoparticles were observed to be different within the monolayers depending on how the glycocalyx was degraded. These results suggest an importance in consideration of the glycocalyx layer in observing nanoparticle interaction and uptake for vascular drug delivery. The charge and size of the particles have been previously identified as important factors in uptake kinetics. However, here we show the proper experimental modeling of the cellular conditions and glycocalyx in the form of flow stimulus and degradation is also necessary to accurately predict how the cells interact with nanoscale particles. Also, the ability to target compromised glycocalyx that is prone to atherogenesis is an important first step in reducing plaque development and cardiovascular disease prevention.

Copyright © American Institute of Chemical Engineers. All rights reserved.

470342 Determining the Size of Protein Conjugated Nanoparticles By Quantifying Brownian Motion

Abstract:

Determining the Size of Protein Conjugated Nanoparticles by Quantifying Brownian Motion

Katherine N. Clayton[1], Janelle W. Salameh[2], Julia G. Fraseur[2*], Nelda Portalatin-Vasquez[2*], Alyssa Panitch[2], Steven T. Wereley[1], Tamara L. Kinzer-Ursem[2]

[1]School of Mechanical Engineering, Purdue University

[2]Weldon School of Biomedical Engineering, Purdue University

As the field of colloidal science continues to expand, tools for rapid and accurate physiochemical characterization of colloidal particles will become increasingly important. Here, we present Particle Scattering Diffusometry (PSD), a nanoparticle measurement technique that uses low sample volumes that can be integrated into micro- or nanofluidic systems. Based on the fundamental principles of diffusion, gold nanoparticles (100 nm) in a 4 mL sample volume (or less) undergoing Brownian motion are imaged under dark field microscopy and recorded with a CCD camera for ~8 seconds then processed using a MATLAB algorithm. The diffusion coefficient of the particles is calculated by correlating successive particle images (at time *Dt*) to one another (cross-correlation, s_c) and the particle image with itself (autocorrelation, s_a) at a magnification (*M*).

$$D = \frac{s_c^2 - s_a^2}{16M^2\Delta t}$$

The size of the gold nanoparticles is then calculated from the Stokes-Einstein equation.

We initially characterized the size of gold nanoparticles with calmodulin (CaM), bovine serum albumin (BSA), and lysozyme (Lyso) conjugated to the particle surface. PSD was used to calculate the diffusion coefficient, polydispersity, and biomolecule layer thickness on the gold nanoparticles. Post-characterization, the presence of protein monolayers on gold nanoparticle surfaces was measured in order to investigate the sensitivity in diffusion coefficient measurements of PSD. Differentiating the change in the gold nanoparticle diffusion coefficient down to protein enables PSD to be used in lieu of common laboratory workflows. Therefore, we are currently applying PSD for the detection of inflammatory cytokines and characterization of proteins involved in learning and memory.

In summary, we have established a rapid (seconds) and sensitive (protein monolayer) technology for characterizing biomolecular surface modifications in very low sample volumes (microliters) using instrumentation and tools that are common to most laboratory settings. In this presentation we will to discuss the development of the PSD technique, our current results and challenges associated with PSD measurements, as well as future work to size particle samples and as a tool for molecular diagnostics and characterizing protein dynamics.

Copyright © American Institute of Chemical Engineers. All rights reserved.

470665 Evaluation of Vascular Targeted Carriers Designed with Dual Ligand Strategies to Target an Inflamed Endothelium

Abstract:

During inflammation, activated endothelial cells (ECs) upregulate surface leukocyte adhesion molecules (LAMs) such as selectins, intercellular adhesion molecule-1 (ICAM-1), and vascular cell adhesion molecule-1 (VCAM-1). This enables firm capture of circulating leukocytes by activated ECs and assists extravasation into the tissue to perform various functions. These same LAMs can be harnessed to localize drug delivery carriers to the diseased endothelium. Vascular-targeted carriers (VTCs) are designed as leukocyte mimics and decorated with ligands that target leukocyte adhesion molecules (LAMs) to facilitate adhesion. Due to target vessel sizes and the presence of blood flow, VTCs require different design considerations than many other targeted particle therapies; adhesion of VTCs in regions with dynamic blood flow requires multiple ligand-receptor (LR) pairs that provide particle adhesion and disease specificity. Despite the ultimate goal of leukocyte mimicry, the specificity of multiple LAM-targeted VTCs remains poorly understood, especially in physiological environments. Here, we develop a series of VTCs with well-controlled ligand properties to investigate particle binding to inflamed endothelium. Particle adhesion was evaluated *in vitro* to an inflamed human umbilical vein endothelial cell (HUVEC) monolayer using a parallel plate flow chamber (PPFC) and *in vivo* in a model of mesentery inflammation via intravital microscopy. Combining sites of sialyl Lewis A (sLe^A), a ligand for selectin, and anti-ICAM onto a single particle resulted in ~3-7-fold increase of adherent particles at the endothelium over single-ligand particles. At a constant total ligand density, a particle with a ratio of 75% sLe^A: 25% anti-ICAM resulted in more than 3-fold increase over all over other ligand ratios tested in our *in vivo* model. Combined with *in vitro* and *in silico* data, we find the best dual-ligand design of a particle is heavily dependent on the surface expression of the endothelial cells. Interestingly, the greatest adhesion was observed for VTCs with more ligand for the lesser-expressed EC receptor. These results establish the importance of characterizing endothelial expression associated with particular diseases and of considering LR-kinetics in intelligent VTC ligand design for future

therapeutics.

Copyright © American Institute of Chemical Engineers. All rights reserved.

465562 Toward Biological Membrane-Inspired Nanotechnology

Abstract:

Since the invention of the optical microscope in the 17th century it has been observed that biological membranes compartmentalize the cellular machinery of life. Then, nearly 100 years ago, it was concluded, through surface science, that the cell membrane is a lipid bilayer. Since that time, many observations and theories have been put forward to try to explain how complex behavior emerges in living systems from the ubiquitous lipid bilayer structure and its integrated proteins and carbohydrates. We are now in a period of time when some of the most important paradigms posited in the last several decades for emergence of complex behavior in living cell membranes are being tested and questioned. These include the membrane (or lipid) raft hypothesis and mechanisms for generation of curvature and lipid asymmetry. An important outcome of this work may be the engineering of new nanoscale self-assembled systems and composites. This talk will focus upon our work in biomembrane mechanical properties, nanoscale phase separation, crowding-induced mixing, nanoscale curvature generation, and bionanocomposites that may contribute toward the design of biological membrane-inspired nanotechnology.

Copyright © American Institute of Chemical Engineers. All rights reserved.

457963 De Novo Design of Skin-Penetrating Peptides for Enhanced Transdermal Delivery of Peptide Drugs

Abstract:

Skin-penetrating peptides (SPPs) are attracting increasing attention as a noninvasive strategy for transdermal delivery of therapeutics. Recent work has demonstrated their potential for dermal delivery, and experimental as well as theoretical studies have elucidated their mechanism of action. However, discovery of such peptides is largely driven by experimental screening of large libraries rather than theoretical or computational designs. Recent mechanistic studies have shown that SPPs must possess affinity for both keratin in skin and the drug of interest to be effective enhancers. Based on this finding, we computationally generated and screened virtual libraries of disulfide-bonded, cyclic peptides against keratin and Cyclosporine A (CsA) to identify SPPs capable of enhancing transdermal delivery of CsA. The sequential screening procedure was performed using the docking software HADDOCK and returned a list of sequences that were tested *in vitro* to validate their ability to bind CsA and enhance its transdermal permeation. The selected sequences were experimentally tested and were found to bind keratin and CsA, as determined by mass spectrometry and affinity chromatography, as well as to enhance CsA transport across the skin. The leading heptameric sequences were tested and yielded CsA permeation on par with previously discovered SPP (SPACE). An octameric peptide yielded significantly higher delivery of CsA compared to heptameric SPPs. Further, the safety profile of the selected sequences was validated by incubation with skin keratinocytes. This method represents an effective procedure for the de novo design of skin-penetrating peptides for the delivery of desired therapeutic or cosmetic agents.

Copyright © American Institute of Chemical Engineers. All rights reserved.

472586 Ligand-Targeted Conjugate Systems for Delivery of siRNA to Tumors

Abstract:

Small interfering ribonucleic acids (siRNA) are a class of nucleic acid-based drugs that have the potential to knock down any gene with a high degree of specificity. They harness a natural gene regulation and antiviral mechanism for efficient destruction of any mRNA. This is useful for both the understanding of and treatment of diseases ranging from genetic disorders and viral infections to cancer. However, to be effective, the siRNA must travel from the injection/infusion site to the tissue of interest and then reach the cytoplasm of the desired cell type. Nucleic acids are highly charged and cannot cross the cell membrane unaided. Exposure to serum endonucleases also accelerates the destruction of the siRNA payload. Methods to cross these barriers include targeted delivery using vehicles ranging from lipid nanoparticles to polyplexes. Unfortunately, the delivery materials tend to be toxic, limiting the efficacy in patients, and also tend to localize to immune cells, the endothelium, or the liver. Currently, there are very few biomaterials that can successfully deliver siRNA to tissues outside of this group, and none are in Phase II clinical trials.

Recently, advances in siRNA stabilization technology have made unencapsulated siRNA conjugates viable, as chemical modifications to the RNA backbone have been used to reduce susceptibility of the siRNA to endonuclease degradation. These conjugates also have desirable treatment properties, including an in-human safety profile of up to 300 mg/kg and a clear regulatory path due to a well-defined chemical structure. Furthermore, due to their small size and thus ability to diffuse relatively quickly through extracellular matrix, the conjugates can be injected subcutaneously, a home procedure familiar to many diabetics on insulin, as opposed to nanoparticles, which must be infused intravenously at a clinic or hospital.

We have developed an azide-alkyne cycloaddition based system that allows for the delivery of ligand-targeted siRNA conjugates to models of glioblastoma (a form of brain tumor) *in vitro* and *in vivo*. The ligand used is based on chlorotoxin. *In vitro* studies using U87 (glioblastoma) cells show a similar degree of knockdown to commercially available transfection materials relative to a scrambled siRNA control. Confocal microscopy studies using derivatives of these ligands with fluorescently-labeled siRNA demonstrate a high degree of uptake relative to unconjugated siRNA in both cases. These conjugates are also able to traverse the bloodstream, deliver a nucleic acid payload into the tumor cells, and efficiently knock down expression of a housekeeping gene. Using fluorescent derivatives of our siRNA, we are able to observe localization of the siRNA to subcutaneous or intracranial tumor xenografts. Intravital imaging experiments demonstrate the ability of these conjugates to pass the blood-brain barrier and permeate the tumor. *In vivo* knockdown studies show a moderate degree of tumor-specific knockdown of a housekeeping gene in an intracranial model of glioblastoma using U87 cells.

These results demonstrate that siRNA knockdown of genes in tumors using these modular conjugate systems are feasible and present a way forward for using siRNA in the treatment of cancer. Future work involves using biologically-relevant siRNA to demonstrate anti-tumor effects in this and syngeneic tumor models.

Copyright © American Institute of Chemical Engineers. All rights reserved.

472815 Design,Development and Optimization of pH-Responsive Hydrogels for the Oral Delivery of Human Growth Hormone

Abstract:

Protein therapeutics have the potential to treat a variety of debilitating disorders ranging from diabetes, growth hormone deficiencies, and even certain forms of cancer. However, due to their large and low stability, their administration has been traditionally limited painful and embarassing injections. An orally delivered carrier capable of protecting the delicate cargo during its transit through the gastrointestinal tract and undergoing targeting delivery to the absorptive region of the upper small intestine has the potential to overcome limits of protein stability, improve oral bioavailability and provide a more patient tolerated means of protein administration. To this end, a pH-responsive hydrogel carrier composed of methacrylic acid (MAA), N-vinyl pyrrolidone (NVP), and poly(ethylene glycol), P((MAA-co-NVP)-g-EG), was developed to deliver higher molecular weight protein therapeutics via the oral route. By varying crosslinking density and length, the swelling behavior of these gels could be modulated and adjusted to accomodate the larger protein, with loading efficiencies up to 60% for human growth hormone. Additionally, due to their pH-responsive behavior, the protein was not released under simulated gastric conditions but was preferentially released in simulated intestinal conditions. The hydrogel carriers also proved to be non-toxic in both model intestinal cell lines - Caco-2 and HT29-MTX - and in vivo upon administration to C57Bl/6 mice. Finally, when administered to the intestine, human growth hormone loaded P((MAA-co-NVP)-g-EG) hydrogels successfully delivered their protein payload with oral bioavailabilities of 0.5 - 5%, which is similar to other reported oral delivery stratgies for protein therapeutics.

Copyright © American Institute of Chemical Engineers. All rights reserved.

460422 Design of Biodegradable Layer-By-Layer Films Capable of Localized, Efficient, and Sequential Gene Delivery

Abstract:

Layer-by-layer (LbL) films containing cationic polyelectrolyte and anionic bioactive molecules such as DNA are promising biomaterials for controlled and localized gene delivery for a number of biomedical applications such as cancer DNA vaccine delivery and tissue engineering. The LbL technique allows a wide variety of biological molecules including DNA, RNA, protein, and peptides to be incorporated into surface coating with exact amount and spatial distribution. Additionally, the release time and dosage are controllable by the LbL film-based gene delivery vehicle. Although nonviral gene delivery vectors generally show lower cytotoxicity than their viral counterpart, their low transfection efficiency prevents their clinical translation. In order to improve the transfection efficiency, we synthesized a novel bioreducible poly(amido amine) (PAA). We incorporated a highly transfecting monomer, 5-amino-1-pentanol (APOL) into the PAA molecular structure. We also modified this new type PAA with different amount of disulfide bond. The disulfide bond here can be reduced in target cell thus enabling DNA release from the polymer/DNA complex. The buffering capacity of the new APOL-containing PAA in the relevant intracellular pH range was higher than poly(ethyleneimine) (PEI) as well as the PAA without APOL. The high transfection efficiency of the APOL-PAA was confirmed by in vitro experiments using HEK 293, NIH 3T3, and MC 3T3 cells. We chose the highest transfecting APOL-PAA and applied it into our LbL film design. We observed increased transfection in the APOL-PAA containing LbL films. We also modified the LbL interior structure by incorporating non-biodegradable PEI as a barrier layer to achieve sequential disassembly and sustained DNA release. Moreover, by using two different DNA plasmids embedded in different layers of the LbL film, the two types of DNA were shown to be released at different times. Our study contributes new molecular and LbL film designs for improved DNA delivery based on the LbL film technique.

Copyright © American Institute of Chemical Engineers. All rights reserved.

466875 Aminoglycoside-Derived Liposomes for Drug Delivery

Abstract:

Breast cancer is one of the leading cancers in women, and is responsible for an estimated 50,000 deaths every year in the US. Triple-negative breast cancer (TNBC) is diagnosed in 15-30% of all breast cancer cases, and dvelopment targeted therapeutics are challenging due to lack of estrogen, progesterone, and HER2 receptors in this disease. Aminoglycosides contain multiple sugar groups with with amines and hydroxyls, and have been investigated as antibiotics in the clinic. Aminoglycoside-derived lipid-containing polymers were formulated into nanoparticles (liposomes) for encapsulation and delivery of mitoxantrone, an FDA-approved DNA damaging drug, to different cancer cells. Five different liposomes were synthesized, formulated and characterized for their size, surface charge, and stability over time. The efficacy of liposomal mitoxantrone for ablation of TNBC, bladder, and prostate cancer cells was evaluated both as a single-agent treatment and in combination with TRAIL (Tumor Necrosis Factor-apha Related Apoptosis-Inducing Ligand), which is a protein that has been shown to selectively kill cancer cells compared to normal cells. In addition the combination of liposomal mitoxantrone with PARP (poly ADP ribose polymerase enzyme) inhibitors was also evaluated in cancer cells. Liposomal mitoxantrone demonstrated 6-10 fold higher efficacy for ablation of cancer cells compared to unencapsulated drug in vitro. In vivo biodistribution and efficacy of liposomal mitoxontrone is being studied using TNBC tumor-bearing mice.

Copyright © American Institute of Chemical Engineers. All rights reserved.

462047 Water Soluble Enteric Polymer

Abstract:

Water Soluble Enteric Polymer

Jin Zhao[1], Oliver Petermann[2], Matthias Knarr[2], Josh Imperial[1], Debbie Holbrook[1], Harold Bernthal[1]

(1) Pharmaceutical Excipients R&D, Dow Pharma & Food Solutions, Midland, MI 48674

(2) Pharmaceutical Excipients R&D, Dow Pharma & Food Solutions, Bomlitz, ND 29699, Germany

Enteric dosage forms have been developed to protect either acid sensitive active pharmaceutical ingredients (API) from gastric conditions or the gastric tissue from an irritating APIs. Current technology typically involves coating multi-particulates, capsules or tablets with a known enteric polymer. Such coating processes usually involve organic solvents that are flammable, toxic, and explosive, or involve aqueous dispersions that are hard to handle, tend to block coating nozzles and have poor stability ADDIN EN.CITE ADDIN EN.CITE.DATA (*1-3*). To overcome these drawbacks, a new class of enteric polymer has been developed that can be directly dissolved in water and used to manufacture enteric dosage forms such as intrinsic soft and hard shell capsules, as well as enteric films and coatings.

Films and capsule shells were prepared using water soluble (WS) hypromellose acetate succinate (HPMCAS) aqueous solutions. The rheology of aqueous solutions of WS HPMCAS showed a reversible thermal gelation behavior which can be adjusted to fit various film forming manufacturing requirements by varying the polymer type, molecular weight, polymer concentration and the usage of plasticizers. The resulting films demonstrated comparable mechanical strength to films made from organic solutions. The API diffusion across the film also demonstrated acid resistance and release performance that satisfied USP enteric requirement.

Water soluble HPMCAS capsules were produced by a dipping process where the cold aqueous solution can be dipped with relatively low pin temperatures and less aggressive drying process (Figure 1).

The findings from this research can help pharmaceutical scientists manufacture intrinsic enteric dosage forms such as soft and hard capsules, chewable films, as well as apply enteric coatings with cost effective process. The enteric capsules can be manufactured without additional coating steps. This will not only enable controlled oral delivery of acid sensitive biomolecules, vaccines, and live organisms, but also enable rapid prototyping of New Chemical Entities (NCEs) and New Biological Entities (NBEs) that require enteric protection. Such enteric dosage form can provide not only enteric release but also delayed or sustained release in both pharmaceutical and nutraceutical applications.

Drying

Figure 1. Enteric capsules made from WS HPMCAS aqueous solutions.

Acknowledgements

The authors wish to thank Robert Schmitt and Shrikant Khot for their insights and suggestions throughout this research.

References

ADDIN EN.REFLIST 1. M. A. Osman, R. B. Patel, D. S. Irwin, P. G. Welling, Absorption of theophylline from enteric coated and sustained release formulations in fasted and non-fasted subjects. *Biopharmaceutics & drug disposition* **4**, 63-72 (1983).

2. M. A. Osman, R. B. Patel, D. S. Irwin, P. G. Welling, Absorption of theophylline from enteric coated and sustained release formulations in fasted and non-fasted subjects. *Biopharmaceutics & drug disposition* **4**, 63-72 (1983)

3. P. Mura, F. Maestrelli, M. Cirri, M. L. Gonzalez Rodriguez, A. M. Rabasco Alvarez, Development of enteric-coated pectin-based matrix tablets for colonic delivery of theophylline. *Journal of drug targeting* **11**, 365-371 (2003).

4. D. Cade, H. straub. (Capsugel France SAS, WO2013/164121).

468054 Preparation of a Drug Delivery System (DDS) of Hydropylic and Hydrophobic Drugs: In Vitro Validation in Human Lung Cancer Cells

Abstract:

PREPARATION OF A DRUG DELIVERY SYSTEM (DDS) OF HYDROPYLIC AND HYDROPHOBIC DRUGS: IN VITRO VALIDATION IN HUMAN LUNG CANCER CELLS

Fernando Mesías-Recamán, Jesús A. Rodríguez-Rodríguez, Miguel A. Galán, Eva Martin Del Valle

It is well known that current therapies present many side effects due to their low tumor cells-recognition capacity (low specificity), low solubility, low specificity, low bioavailability, prolonged circulation time in the body (1, 2) and high toxicity (3). In this regard, developing new medical strategies which could solve these disadvantages is necessary. Therefore, in this work, two different drugs with different hydrophobicity were combined to obtain an important synergic effect on the tumor cell line H460 (human lung cancer).

Our strategy is based on the combination of the carboxymethyl-β-cyclodextrin (CMCD) covalently attached to cisplatin as paclitaxel (PTX) carrier. In this manner, cisplatin and PTX could be delivered at the same time obtaining a synergic effect. CMCD was selected due to it presents low toxicity, small size, non-immunogenicity, excellent biocompatibility and high conjugation capacity with a variety of lipophilic compounds (4, 5, 6).

Furthermore, this molecule presents a hydrophilic ring and a hydrophobic cavity (7) that allows the link of cisplatin in its surface and forms the inclusion complexes with PTX. Cisplatin was covalently attached to the outside oligosaccharide ring through activation of the carboxylate functional groups of the CMCD, while PTX was included inside the oligosaccharide ring forming an inclusion complex. In addition, the solubility of both drugs was increased in aqueous conditions when drugs were bind to/formed the inclusion complex with

CMCD.

This system was characterized by NRM and FTIR, and thermodynamic studies. The PTX inclusion was determined by HPLC.

Later, their stability over time and the paclitaxel and cisplatin release kinetics were also studied. Finally, the efficacy of the both drugs delivery system has been validated *in vitro* with the H460 cell line by MTT assays and promising results have been obtained. Thus, this system could be a novel strategy to avoid the side effects that current antitumor therapies generate.

References

[1] Ganta S., Devalapally H., Shahiwala A., Amiji M. A review of stimuli-responsive nanocarriers for drug and gene delivery. J. Control Release. 2008, 126(3)187-204.

[2] Barreto J. A., Malley W. O., Kubeil M., Graham B., Stephan H., Spiccia L. Nanomaterials: Applications in Cancer Imaging and Therapy. Adv. Mater. 2011, 23(12):H18-40.

[3] Kesharwani P., Jain K., Jain N. K. Dendrimer as nanocarrier for drug delivery. Prog. Polym. Sci. 2014, 39(2):268–307.

[4] Zhang J. and Ma P. X. Host-Guest Interaction Mediated Polymeric Core-Shell Assemblies: Versatile Nanocarriers for Drug Delivery. Angew. Chem. Int. Ed. Engl. 2009, 48(5):964-968.

[5] Davis M. E., Brewster M. E. Cyclodextrin-based pharmaceutics: past, present and future. Nat. Rev. Drug. Discov. 2004, 3(12):1023-1035.

[6] Zhang J. X., Ma P. X. Host-guest interactions mediated nano-assemblies using cyclodextrin-containing hydrophilic polymers and their biomedical applications. Nanotoday. 2010, 5(4):337-350.

[7] Álvarez-Parrilla E., De la Rosa L. A., Torres-Rivas F., Rodrigo-García J., González-Aguilar G. A. Complexation of apple antioxidants: chlorogenic acid, quercetin and rutin by β-cyclodextrin (β-CD). J. Inclusion Phenom. Macrocycl. Chem. 2005, 53:121-129.

Copyright © American Institute of Chemical Engineers. All rights reserved.

471693 Community Data-Driven Metabolic Network Reconstruction for Predicting Interspecies Interactions in a Photoautotroph-Heterotroph Consortium

Abstract:

Microbial communities exert a significant impact on our environment, human health and industry. A mechanistic understanding of complex interplay between member species is essential for engineering microbial communities. Metabolic network models that provide comprehensive community-wide predictions on cross-species metabolite exchanges can serve as a useful tool for that purpose. Compared to single species analyses, community metabolic network reconstruction is more complex because it has to account for interspecies interactions. Current approaches focus on reconstruction of high-quality individual networks so that combined networks can subsequently lead to the prediction of interspecies interactions and community behaviors. This method becomes ineffective, however, for environmental communities whose members cannot be sufficiently characterized due to the difficulty in isolation and axenic cultivation. In order to address this limitation, we here tested a new approach that uses community-level data as a critical input for network reconstruction. Incorporation of community data is advantageous because it provides information on interspecies metabolic interactions, which is not necessarily obtainable from axenic cultures. To evaluate the validity of the proposed method, we reconstructed community networks based on alternative strategies of gapfilling (i.e., individual vs. community-level gapfilling). We implemented all processes for network reconstruction and refinement using the DOE Systems Biology Knowledgebase (KBase) platform (www.kbase.us). In the case study of a photoautotroph-heterotroph binary consortium, metabolic networks that were gapfilled and refined using community data provided experimentally validated predictions of how a photoautotrophic cyanobacterium provides organic carbon and nitrogen sources to support the growth of an obligate heterotrophic species. Further developments are in progress to extend the proposed method to more complex communities beyond binary consortia.

Copyright © American Institute of Chemical Engineers. All rights reserved.

470013 Model Communities to Study Mechanisms of Plant-Microbiome Function

Abstract:

The microbiome of plants is comprised of thousands of taxonomically and functionally diverse microbiota, the majority of which are uncultivable. Current methods for studying how the genetics of community members contribute to microbiome function include shotgun metagenome sequencing and microbiome replacement studies. In this work we present the design and analysis of a model community of cultivable, genome-sequenced representatives of taxa in the microbiome of *Populus* plants, a second generation biofuel feedstock. Results from one-on-one studies show that individually, strains induce diverse changes in root architecture and have unique metabolic capabilities; and as a community lead to increased root growth in *Populus* hosts. Across 20 plants from two host genotypes the community is dominated by two taxa, with *Burkholderia* and *Pantoea* isolates representing 99% of detected organisms. Within the remaining 1%, the *Streptomyces* and *Rhizobium* representatives are the most abundant, followed by *Sphingobium* and *Duganella* isolates. These abundance results suggest structuring of the community, rather than dominance by a single taxon. We also observe positive and negative correlations in abundance between community members, emergent properties that are not predicted from community design. This work demonstrates the feasibility and analysis of model communities to study microbiome function in plant systems.

Copyright © American Institute of Chemical Engineers. All rights reserved.

456663 Metabolic Modeling of Pseudomonas Aeruginosa and Staphylococcus Aureus Interactions in Multispecies Biofilms

Abstract:

Chronic wounds are often colonized by consortia comprised of different bacterial species growing as biofilms on a complex mixture of wound exudate. The spatial organization of biofilm consortia cause bacteria to exhibit phenotypes distinct from planktonic growth and often render the application of antibacterial compounds ineffective. We developed spatiotemporal models to investigate the multispecies metabolism of a biofilm consortium comprised of two common chronic wound isolates: the aerobe Pseudomonas aeruginosa and the facultative anaerobe Staphylococcus aureus. By combining genome-scale metabolic reconstructions with partial differential equations for metabolite diffusion, the models were able to provide both temporal and spatial predictions with genome-scale resolution. The models were used to analyze the metabolic differences between single species and two species biofilms and to investigate the tendency of the two bacteria to spatially partition in the multispecies biofilm as observed experimentally. The two species system was predicted to support a maximum biofilm thickness much greater than P. aeruginosa alone but slightly less than S.aureus alone, suggesting an antagonistic metabolic effect of P. aeruginosa on S. aureus. Nutrient gradients imposed by supplying glucose at the bottom and oxygen at the top of the biofilm induced spatial partitioning of the two species, with S. aureus most concentrated in the anaerobic region and P. aeruginosa present only in the aerobic region. When each species was allowed to enhance its growth through consumption of secreted metabolic byproducts assuming identical uptake kinetics, the competitiveness of P. aeruginosa was further reduced due to the more efficient glucose oxidative metabolism of S. aureus. Lysis of S. aureus by a small molecule inhibitor secreted from P. aeruginosa and/or P. aeruginosa aerotaxis were predicted to substantially increase P. aeruginosa competitiveness in the aerobic region, consistent with in vitro experimental studies.

Copyright © American Institute of Chemical Engineers. All rights reserved.

472731 Environmental Perturbations Lead to Correlated Changes in Cecum Microbiota Composition and Metabolite Profile

Abstract:

The adult mammalian gastrointestinal (GI) tract is colonized by ~10^{14} bacteria belonging to at least several hundred species, which collectively outnumber the genes of their host by ~100-fold. This gut microbiota impacts an array of physiological functions in the GI tract. For example, it participates in the digestion of complex carbohydrates. Experiments comparing germ-free (GF) and conventionally raised (CONV-R) mice have shown that the gut microbiota dramatically impacts the level and diversity of bile acids [1]. A significant alteration in the microbial populations, or dysbiosis, correlates with not only GI diseases, but also other chronic diseases such as cancer and asthma. Dysbiosis also correlates with alterations in the profile of metabolites in the intestine and systemic circulation. These metabolites can act as signaling molecules to engage various host receptors and regulatory molecules [2, 3]; however, specific functions have been identified for only a few of these metabolites [4]. There is a pressing need to characterize the metabolic functions of the microbiota and better understand how these functions depend on microbiota composition.

A major challenge in resolving the enzymatic pathways and organisms responsible for particular metabolic functions lies in the complexity of interactions between the host and microbiota as well as interactions within the microbiota community. Thus, experimentally tractable model systems are needed to characterize these chemicals and how these chemicals depend on the microbiota composition. Gnotobiotic mice are powerful animal models that capture physiological interactions between the host and microbiota species, but the *in vivo* setting limits the ability to manipulate microbiota functions without also perturbing the balance of resident commensal microbes. Cultures of individual species are useful for determining whether a species can carry out a particular metabolic function, but do not capture community-level interactions. More advanced *in vitro* systems such as mixed cultures of microbiota isolates can preserve these interactions, while also offering control over the environment and stimuli. However, it remains to be seen whether these systems can faithfully capture the microbiota composition and chemical profile.

In this study, we describe a simple and robust model system in the form of an anaerobic batch culturing system capable of representatively culturing the microbiome isolated from the cecum. We use a multi-omic approach to validate this system by profiling the community composition and metabolites present in the cecum culture. The metabolic products of cecum culture are also compared against the differences in metabolite profiles between the cecum contents isolated from CONV-R and GF mice. We then utilize the system to investigate the impact of an environmental perturbation from xenobiotic chemical exposure.

Results from 16S rRNA sequencing analysis show that the cecum culture community composition mimics the microbiome structure *in vivo*. At the phylum level, the Operational Taxonomic Units (OTUs) present in the cultured microbiota exactly match the cecal microbiota freshly isolated from 6 week old CONV-R C57BL/6 mice. At the species levels, the similarity is approximately 75%. The major phyla detected in the culture samples were Bacteroidetes, Firmicutes and Proteobacteria. While there were differences in the abundance of individual groups like the Clostridia, this did not affect the representation of the communities in the culture, and all the major genus reported previously from the fecal samples in other studies [5] were represented. The freshly isolated and cultured microbiota also show similar diversity based on inverse Simson diversity index analysis (0.9 for CONV-R cecal contents and 0.8 for culture), Shannon diversity index (2.8 and 2.0) and Schao1 scores (156 and 113).

Using an untargeted metabolomics, we characterized the time evolution of metabolite profiles in the cecal culture. Samples were collected on day 1 and 7 of the culture and extracted using a solvent based method. Metabolites were analyzed on a high-resolution time-of-flight (TOF) mass analyzer operated in an information dependent acquisition (IDA) mode. Clustering the detected features (precursor ion m/z) show distinct profiles for metabolites that are produced and consumed by the microbiota. Based on their time profiles, the features clustered into four groups: rapidly consumed (depleted in 1 day), slowly consumed (depleted over 7 days), rapidly produced (significantly elevated in 1 day), slowly produced (significantly elevated after 7 days). The levels of these features in the culture medium remain unchanged when the culture is not inoculated with the cecal isolates, confirming that the changes are due to microbial metabolism. Features that were significantly elevated or depleted compared to medium control (without inoculation) were identified using one-way ANOVA ($p < 0.05$) and annotated using MS/MS libraries or an *in silico* fragmentation tool. Approximately one-third of the metabolites significantly elevated in the cecum culture were significantly depleted in GF mouse compared to CONV-R mouse, suggesting that these microbiota metabolites are also produced *in vivo*.

We next perturbed the cecum culture using a pair of representative environmental chemicals â€“ bisphenol A (BPA, 1 and 10 uM) and diethylhexyl phthalate (DEHP, 10 and 100 uM) â€“ to investigate the dependence of the metabolite profile on culture community composition. Metabolomics analysis showed that both BPA and DEHP brings about significant changes in the metabolite profile in a dose-dependent manner. Using hierarchical clustering, we identified groups of features that were significantly altered in the DEHP or BPA condition compared to vehicle control. Out of 1,044 metabolic products detected in our metabolomics analysis, 122 and 176 were significantly altered with DEHP and BPA treatment, respectively. Fifty-seven of the significantly altered metabolites were common to both BPA and DEHP treatment. These metabolite profile alterations correlated with significant shifts in the culture community composition. Interestingly, the compositional changes were not evident at the phylum level. Rather, significant changes were detected at family, genus and species levels.

In conclusion, our results to date show that the anaerobic batch culture system mimics the microbial community composition expected in the cecum *in vivo*, and can capture significant correlations in the structural and functional changes resulting from an external perturbation. Ongoing work investigates the culture systemâ€™s response to an endogenous hormone, and compares this response to *in vivo* trends to further establish the culture system as a robust model of the murine cecal microbiota.

References:

1. Sayin SI, Wahlstrom A, Felin J, Jantti S, Marschall HU, Bamberg K, et al. Gut microbiota regulates bile acid metabolism by reducing the levels of tauro-beta-muricholic acid, a naturally occurring FXR antagonist. Cell Metab.

2013;17(2):225-35. doi: 10.1016/j.cmet.2013.01.003. PubMed PMID: 23395169.

2. Laparra, J.M, Sanz Y. Interactions of gut microbiota with functional food components and nutraceuticals. Pharmacol Res. 2010;61:219-25.

3. Sanz Y, De Palma G. Gut microbiota and probiotics in modulation of epithelium and gut-associated lymphoid tissue function. Int Rev Immunol. 2009;28:397-413.

4. Vogt SL, Pena-Diaz J, Finlay BB. Chemical communication in the gut: Effects of microbiota-generated metabolites on gastrointestinal bacterial pathogens. Anaerobe. 2015;34:106-15. doi: 10.1016/j.anaerobe.2015.05.002. PubMed PMID: 25958185.

5. Goodman AL, Kallstrom G, Faith JJ, Reyes A, Moore A, Dantas G, et al. Extensive personal human gut microbiota culture collections characterized and manipulated in gnotobiotic mice. Proceedings of the National Academy of Sciences of the United States of America. 2011;108(15):6252-7. Epub 2011/03/26. doi: 10.1073/pnas.1102938108. PubMed PMID: 21436049; PubMed Central PMCID: PMC3076821.

Copyright © American Institute of Chemical Engineers. All rights reserved.

462900 Engineered Microconsortium of Lactic Acid Bacteria to Combat Wound Infections

Abstract:

Up to 60% of all wounds and 80% of chronic wounds are infected with biofilms â€" surface-associated microbial communities residing within an extracellular matrix â€" that prevent healing and are responsible for other major health complications. For dermal wounds, noninvasive topical antimicrobial treatments, such as silver, iodine, medicinal honey, and antibiotics, are common yet can be ineffective at overcoming the protection conferred by the biofilm matrix and the associated growth paradigm. Disrupting the structural elements of the biofilm is an interesting strategy for re-sensitization of biofilm associated microbes to antimicrobials, an essential step in noninvasive infection clearance. Recent work on alternative wound treatments has shown that topical application of probiotic lactic acid bacteria (LAB) inhibits pathogenic bacterial growth on wounds and promotes wound healing. However, LAB activity against mature biofilms has not been characterized. In this talk, we will describe the design of a synthetic microconsortium of engineered LAB (*Lactobacillus plantarum* and *Lactobacillus reuteri*) that serve as a topical wound treatment and can provide therapeutic benefit through synergistic combination of various therapeutic mechanisms. This LAB microconsortium is designed to detect the presence of pathogens and execute appropriate responses to disassemble mature biofilm matrices and inhibit nascent biofilm formation through enzymatic disruption, while also inhibiting microbial growth through production of antimicrobial compounds. We will discuss our recent efforts in development of this system against *Pseudomonas aeruginosa* â€" a major opportunistic biofilm-forming pathogen found in most chronic wound infection sites.

Copyright © American Institute of Chemical Engineers. All rights reserved.

470107 Invasion and Co-Existence of Phototrophic Microbes in a Synthetic Model System Mediated By Nitrogen Availability

Abstract:

Microalgae are widely regarded as a promising renewable and sustainable feedstock for the production of various valuable molecules, including biofuels. However, biological contamination has detrimentally affected biomass production and is therefore considered as a major obstacle for open-system microalgae cultivation[1]. New methods are needed to greatly enhance microalgae resistance to biological invasion. One underexplored strategy is to leverage working principles established in ecology research, such as those governing invasion and species co-existence in complex natural biological systems[2].

In this work, we investigate how competition and resource availability affect the outcomes of mutual invasion in a synthetic phototrophic microbial system. The system consists of *Chlorella vulgaris*, a microalga studied extensively in biofuel research, and *Synechocystis* sp. PCC6803, a model cyanobacterium. In each experiment, a resident species is established in a semi-continuous culture and the other species is introduced at low cell density as a potential invader. It has been found that the availability of nitrate, the sole nitrogen source in this system, determines the composition of resulted new steady state. Specifically, with abundant nitrate, *Synechocystis* successfully establishes itself in *C. vulgaris* culture, becoming the numerically dominant species while driving the microalga into low cell density. When nitrate is provided at low concentrations, *Synechocystis* fails to invade *C. vulgaris*. Similarly, *C. vulgaris* can invade established *Synechocystis* culture when nitrate is limited. Interestingly, mutual invasion can occur and lead to stable co-existence of the two species under certain conditions. Together, these results provide insights on the driving forces and dynamics of phototrophic ecosystems, which could have important implications for the development of new biological control methods.

Keywords: Microalgae, cyanobacteria, nitrogen limitation, invasion, co-existence

References

1 Wang H, Zhang W, Chen L, Wang J, Liu T. The contamination and control of biological pollutants in mass cultivation of microalgae. Bioresour Technol, 2013, 128, 745-750.

2 Chesson P. Mechanisms of maintenance of species diversity. Annu. Rev. Ecol. Syst., 2000, 31, 343-366.

Acknowledgements

Chun Wan is grateful to Chinese Government Scholarship Council (CSC) for financial support with his study in the US.

Copyright © American Institute of Chemical Engineers. All rights reserved.

471894 Invited Presentation: Engineering the Microbiome for Form and Function

Abstract:

Microbial communities hold great potential for a range of biotechnological applications where dividing labor between multiple organisms can enable increased productivity, novel functions or combinations of functions, or simplified process optimization and flexibility by treating the different organisms as modules that can easily be swapped and recombined. An ongoing challenge remains the development of strategies and tools for controlling community composition, feedback between the organisms, and how to optimally divide labor between multiple species. We have developed new set regulatory components and systems that can be used to turn gene expression on or off in response to a cell-cell communication signal, and enable interspecies communication between representative Gram-negative (*Escherichia coli*) and Gram-positive (*Bacillus megaterium*) organisms. We are also employing new strategies to control the growth and coexistence of species in co-cultures by manipulating ecological interactions; a key challenge for applications with microbial consortia. We will describe the use of synthetic ecology approaches, where both the environment and cells are manipulated, to control community composition in batch, fed-batch and continuous culture. Finally, we will describe our recent efforts to distribute functions between these microbes in a model bioprocess and our broader efforts to understand the links between form and function in natural and engineered microbial communities.

Copyright © American Institute of Chemical Engineers. All rights reserved.

458195 Surface Display-Enabled Directed Evolution of Stabilized Alpha Helix Peptides

Abstract:

Surface display-enabled directed evolution of stabilized alpha helix peptides

Keywords: stapled peptides, surface display, directed evolution, PPI inhibitors

The majority of interfaces between interacting proteins include an alpha helix, the most common protein secondary structure motif. Such interfaces have posed attractive drug targets due to their importance in a wide range of diseases, in which inhibition, by blocking interactions, or activation, by triggering function, can result in therapeutic applications. However, due to the large size of protein-protein interaction surface areas, small molecules have seen little success in drug development. Here, we describe an application of bacteria surface display using directed evolution to screen for potent stabilized peptide binders of intracellular targets.

As done in reported development of stabilized peptide inhibitors, the peptideÕs sequence is generally determined from the bound residues observed in a co-crystal structure of the target and one of its known binders. However, when designing a stabilized peptide, introduction of the staple may abrogate binding, as the modified residues and bulk of the staple may interfere with the peptide-target interaction. This has been illustrated by several recent examples, and optimization generally requires painstaking staple scanning. The directed evolution method presented here allows for rapid screening of $>10^8$ unique stabilized peptide sequences.

By incorporating azidohomoalanine (AHA) residues spaced 7 residues apart in solid phase peptide synthesis, we and others have demonstrated stabilization of peptides through double-click chemistry using a bifunctional alkyne-containing linker. Such design allows for the inclusion of fluorophores for imaging applications or pharmacologic enhancers like PEG chains. Our lab and others have demonstrated that the resulting stabilized peptides have increased protease resistance, subcutaneous bioavailability, and binding affinities. Taking advantage of the ability of methionine auxotrophic *E. coli* to efficiently incorporate AHA as a methionine surrogate, we constructed libraries of randomized peptides containing two AHA residues for double-click surface stabilization. We demonstrate efficient display of AHA incorporated peptides using the eCPX scaffold, an artificial protein based on endogenous OmpX, and robust reaction yields and specificity. Following surface display and chemical double-click stabilization, the resulting peptides show increased protease resistance. Here, we demonstrate the development of the technique and its application in screening for novel stabilized alpha helix binders.

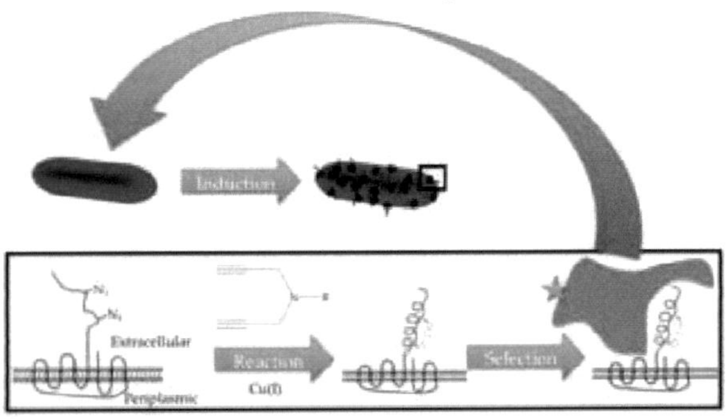

Figure 1- the directed evolution process for stabilized peptides involves inducing bacteria with AHA non-natural amino acids 7 residues apart, followed by CuAAC for stabilization and selection. This process can be repeated to enrich forstrongly binding clones and affinity maturation can also be done to further improve affinity.

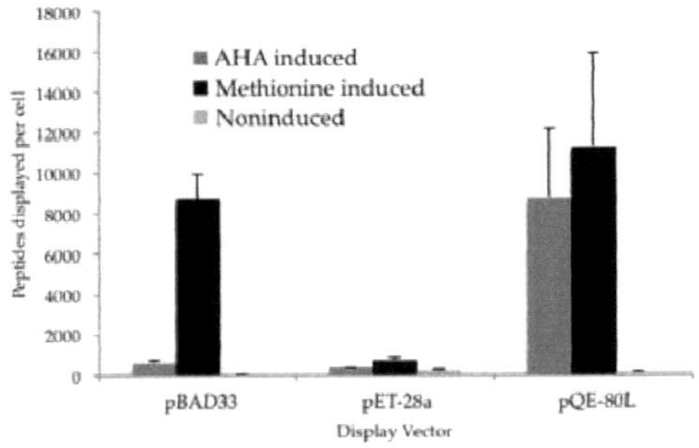

Figure 2- surface display levels of exendin-4 , a model peptide, using eCPX with different vectors. pQE-80L shows optimal display of peptide under both methionine and AHA incorporation conditions.

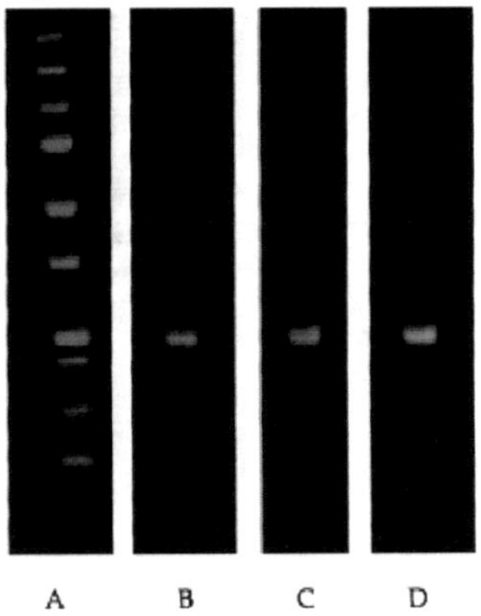

Figure 3- AHA incorporation is exendin-4 specific. Cells were reacted with SulfoCy5-alkyne after induction and run on a reducing SDS-PAGE gel. A - molecular weight ladder . B - 700 nm scan (SulfoCy5 channel) showing one bright band near the expected molecular weight (20 kDa). C – 800 nm scan (anti-exenatide-4 channel) showing one band for the exenatide-4-eCPX fusion. D – overlay showing co-localization of the two channels indicating specific reaction.

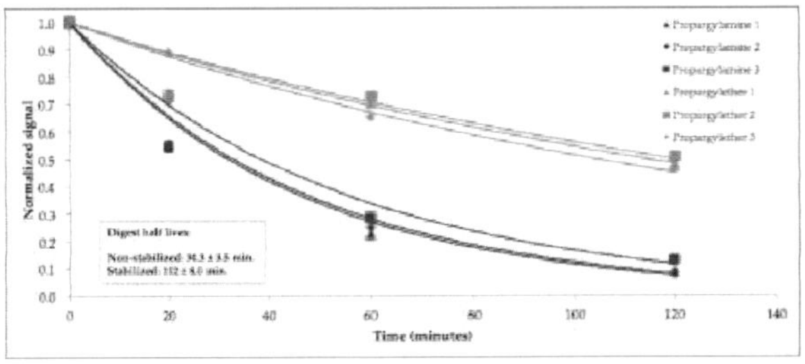

Figure 4 - digest trajectories of surface displayed exendin-4 reacted with either propargylamine (single alkyne, non-stabilizing) or propargyl ether (bisalkyne, stabilizing) showing significant increase in protease resistant upon stabilization on the surface of bacteria.

Copyright © American Institute of Chemical Engineers. All rights reserved.

470904 Fitness Landscapes for an Enzymes on Multiple Substrates

Abstract:

The emerging fields of biomanufacturing and biopharmaceuticals require enzymes with improved or novel activities. Yet, our lack of total understanding of the intricacies of how enzymes function has constrained our ability to robustly engineer substrate specificity[1–3]. Furthermore, the mechanisms of natural evolution leading to improved or novel substrate specificities are not wholly defined[4]. In this talk we will present the use of a synthetic biology approach and deep mutational scanning to generate three comprehensive fitness landscapes for hydrolysis activity of an amidase with different aliphatic amide substrates: acetamide, propionamide, and isobutyramide. We assessed the function of >6500 point mutations and found that 20.6% of all variants from the isobutyramide selection had above wild-type fitness. Consistent with existing theories of adaptive molecular evolution, we also found that beneficial mutations were exponentially distributed[5]. Remarkably, there was essentially no correlation between the fitness of beneficial mutations identified in the isobutyramide selection to the other two selections. In this talk we will also discuss the evaluation of fitness landscapes with respect to enzyme structure. Our results show that, at least for this enzyme, fitness effects for a mutation on one substrate cannot be predicted based on known effects for a different substrate.

1. Arnold, F. H. Combinatorial and computational challenges for biocatalyst design. ***Nature* 409,** 253–257 (2001).

2. Bornscheuer, U. T. ***et al.*** Engineering the third wave of biocatalysis. ***Nature* 485,** 185–194 (2012).

3. Liberles, D. A. ***et al.*** The interface of protein structure, protein biophysics, and molecular evolution. ***Protein Sci.* 21,** 769–785 (2012).

4. Khersonsky, O. & Tawfik, D. S. Enzyme promiscuity: a mechanistic and evolutionary perspective. ***Annu. Rev. Biochem.* 79,** 471–505 (2010).

5. Orr, H. A. The Population Genetics of Adaptation: The Distribution of Factors Fixed During Adaptive Evolution. ***Evolution (N. Y).* 52,** 935–949 (1998).

Copyright © American Institute of Chemical Engineers. All rights reserved.

471035 High-Throughput Screening of Biological H_2 Production

Abstract:

Hydrogenases, ferredoxins (Fd), and ferredoxin-$NADP^+$ reductases (FNR) are redox proteins that often mediate electron metabolism *in vivo*. These proteins are also potential components for biological H_2 production technologies, but have not necessarily evolved for this application. Here, we report a High-throughput H_2 Production Assay Device (H_2PAD) that enables simultaneous evaluation of 96 individual H_2 production reactions. Using a CCD camera, H_2PAD senses the chemo-optical response of Pd/WO_3 thin films to H_2 produced in the individual wells of a 96-well plate. Image analysis software has been developed to interpret these time-lapse photographs. The resulting rates of color change indicate H_2 production rates by the biological assays in the wells. We used H_2PAD to discover hydrogenase and FNR mutants that can enhance biological H_2 production. From a library of 10,080 randomly mutated *Clostridium pasteurianum* [FeFe] hydrogenase (CpI), we found a mutant with nearly 3-fold higher H_2 production specific activity across a range of pH values. From a library of 400 semi-randomly mutated *Oryza sativa* (rice root) FNR (RrFNR), the top hit exhibited a 2-fold higher catalytic activity within an NADPH-driven H_2 production pathway. In addition to accelerating the pace of directed evolution in biological H_2-producing systems, H_2PAD has the potential to elucidate fundamental biochemical mechanisms of these systems. As such, it represents a tool to allow research to more rapidly meet the growing global demand for sustainable H_2.

Copyright © American Institute of Chemical Engineers. All rights reserved.

472199 Combinatorial Pairwise Assembly Efficiently Generates High Affinity Binders and Enables a "Mix-and-Read" Detection Scheme

Abstract:

We show that a combinatorial library constructed by random pairwise assembly of low affinity binders can efficiently generate binders with increased affinity. Such a library based on the Sso7d scaffold, from a pool of low affinity binders subjected to random mutagenesis, contained putative high affinity clones for a model target (lysozyme) at higher frequency than a library of monovalent mutants generated by random mutagenesis alone. Increased binding affinity was due to intramolecular avidity generated by linking binders targeting non-overlapping epitopes; individual binders of $K_D \sim 1.3$ μM and 250 nM produced a bivalent binder with apparent $K_D \sim 2$ nM. Furthermore, the bivalent protein retained thermal stability (T_M= 84.5 °C) and high recombinant expression yields in *E. coli*. Finally, it is important to note that efficient identification of binders with non-overlapping epitopes on the target is implicit to our strategy; this in turn can be exploited in biosensing applications. In particular, we show that the two subunit proteins comprising the high affinity bivalent binder can be used to design a simple "mix-and-read" format assay for target quantification. In our scheme, binders comprising the bivalent protein are fused to two out of the three fragments of tripartite split GFP. In the presence of the target, assembly of GFP and fluorescence reconstitution occurs in concentration-dependent manner. Mix-and-read assays, wherein a signal can be obtained by simply mixing reagents with the target analyte, have significant advantages over conventional assay formats such as an enzyme-linked immunosorbent assay (ELISA), wherein time-consuming wash steps and protein immobilization are required.

Copyright © American Institute of Chemical Engineers. All rights reserved.

467804 Affinity-Enhancing Mutations in Different Antibody Complementarity-Determining Regions Display Unique Affinity/Stability Trade-Offs

Abstract:

The natural process by which germline antibodies acquire somatic mutations to mediate specific recognition of target antigens must be balanced by mutations that compensate for the destabilizing effects of remodeling the complementarity-determining regions (CDRs) for high-affinity binding. However, little is known about affinity/stability trade-offs for mutations that accumulate within the CDRs and framework regions of antibodies during affinity maturation. We have evaluated the contribution of mutations that were acquired during directed evolution of an antibody variable (V_H) domain specific for the Alzheimerâ€™s AÎ²42 peptide that was initially grafted with AÎ²Â residues 33-42 in CDR3. We find that the most important binding mutation is located in the grafted AÎ² peptide in CDR3. The other key binding mutations were also located in the CDRs. Importantly, the binding mutations in CDR3 did not significantly impact stability, while those binding mutations in other CDRs reduced stability. The destabilizing effects of the binding mutations were compensated for by multiple stabilizing mutations. Our findings demonstrate that affinity-enhancing mutations in CDR3 of an autonomous human V_H domain typically lack affinity/stability trade-offs, while mutations in the other CDRs and framework regions typically display such trade-offs. We expect that these findings will lead to improved methods for designing antibodies and libraries thereof with reduced affinity/stability trade-offs, which is important for the reliable generation of antibody fragments and full-length antibodies with high stability in addition to high affinity.

Copyright © American Institute of Chemical Engineers. All rights reserved.

456142 Evolution of Antibody Agonists That Functionally Recapitulate Natural Cytokines

Abstract:

Ligand-mediated receptor dimerization is the most prevalent signaling mechanism used by secreted growth factors to activate their cognate cell surface receptors. The canonical model for ligand-mediated dimerization involves activation of receptor subunits when oriented into specific receptor dimer geometries. Helical cytokines constitute a particular class of ligands that signal through dimerization of cytokine receptors to elicit a wide range of biological activities, especially those relevant to immune regulation. Cytokines bind to receptor extracellular domains (ECDs) and enforce dimerization of monomeric receptor subunits, activating constitutively associated Janus kinases (JAKs), which in turn recruit and activate signal transducer and activator of transcription (STAT) proteins to orchestrate cell behavior. Due to their critical role in determining immune cell fate, there has been a great deal of interest in harnessing the agonistic potential of cytokines for immunotherapy, but progress has been hampered by several practical limitations. First, poor pharmacokinetic properties mandate frequent injection or continuous infusion and the high serum concentrations of cytokines required for therapeutic response induce adverse off-target effects. Second, cytokines are inherently pleiotropic, often activating a wide range of cell types expressing shared receptors, which hinders efficacy and can lead to systemic toxicity. Third, cytokines are difficult to re-engineer or modify without concerns about immunogenicity. Consequently, there exists a need for novel dimerizing agents that utilize protein scaffolds with improved drug-like properties.

The monoclonal antibody platform represents a robust, engineerable scaffold that benefits from extended in vivo half-life and we thus aimed to engineer antibody-based constructs that dimerize receptors in their native conformations induced by endogenous cytokines to functionally recapitulate cytokine signaling. We devised a novel yeast display-based evolutionary strategy to isolate antibodies that use a single binding site to bridge the interface between two different receptor subunits within a dimeric signaling complex. Our approach involved screening antibody libraries against receptor-ligand complexes that served as 'molecular casts' of the active cytokine-receptor complex conformation and performing iterative rounds of selections to evolve 'staplers' that recognize a composite epitope between heterodimeric receptor subunits. Using this new engineering strategy, we identified staplers for two different cytokine systems that bound fully assembled cytokine-receptor complexes but not any of the individual components thereof. Total internal reflection fluorescence (TIRF) microscopy co-localization studies demonstrated that these molecules efficiently dimerized receptor subunits in living cell membranes in the absence of cytokine. Determination of the crystal structure of the stapler bound to the cytokine-receptor complex ECD provided biophysical evidence that staplers recognize a composite epitope formed by the conjunction of two receptor subunits and that they bind heterodimeric receptors in their active, cytokine-bound signaling conformation. Further, we demonstrated that staplers activated JAK/STAT pathway signaling in human cell lines with equivalent intensity to native cytokines and showed that staplers recapitulated functional responses such as immune cell proliferation and immunoglobulin class switching. Our evolutionary strategy can be applied to isolate stable and readily engineerable agonistic antibodies that serve as staplers for other immune receptor complexes and, more generally, for any dimeric receptor system to address a broad range of disease applications.

Copyright © American Institute of Chemical Engineers. All rights reserved.

474615 New Approaches for High-Throughput Analysis and Engineering of Enzymes and Protein Therapeutics (Featured Presentation)

Abstract:

Our group engineers proteins as therapeutics and diagnostic agents for applications including cancer and regenerative medicine. We are also creating new technologies for protein engineering. Our team has developed a high-throughput screening platform that enables massively parallel, quantitative biochemical measurements to be performed on millions of protein variants expressed from yeast or bacteria. We used this technology to rapidly discover an antibody against a clinical cancer target, evolve an orange fluorescent protein biosensor, and engineer an enzyme with reduced sensitivity to an inhibitor. These three distinct protein engineering applications, each with unique assay requirements and different host organisms, highlight the flexibility and broad technical capabilities of this new technology platform.

Copyright © American Institute of Chemical Engineers. All rights reserved.

472207 A QbD Approach for Identifying Critical Process Paramerters in Drug Substance Manufacturing

Abstract:

The Vertex approach to critical process parameter identification via an effect threshold concept will be discussed and accompanied by examples applying the method to drug substance development and manufacture.

Copyright © American Institute of Chemical Engineers. All rights reserved.

481150 QbD and Continuous: A Step Change in Performance

Abstract:

Traditionally the majority of active pharmaceutical compounds have been manufactured at large plant scales using batch methodologies. Continuous manufacturing is becoming an increasing attractive alternative however due to apparent benefits such as smaller manufacturing footprints, increased process safety, lower manufacturing costs and increased supply-chain flexibility. Realising all of the above benefits using conventional continuous equipment is however quite challenging and sometimes not achievable. In this project, a new approach to the design of continuous manufacturing platforms is being presented which is not restricted to using such conventional equipment but which is based around the bespoke design and manufacture of all unit operations, optimised to fit the specific process being examined. Processes are quantified in detail using methods such as computational fluid dynamics (CFD) leading to tailored equipment designs which can be manufactured with high precision using metallic 3D printing.

In order to demonstrate the suitability of applying this design approach to the entire manufacturing process, a case study is also presented in which two reactions are coupled together, followed by an extraction and wash step and a liquid-liquid separation. All unit operations within this platform are designed using fundamental engineering principles combined with numerical analyses (based upon the requirements of the specific process), prior to being 3D printed and combined to create a bespoke, sealed, hard-piped manufacturing system. The optimised design of the units enables extremely high throughput rates to be achieved corresponding to 580,000 kg/L/day (compared to an approximate batch productivity of 525kg/L/day).

Copyright © American Institute of Chemical Engineers. All rights reserved.

472550 A QbD Approach to Experimental Designs with Kinetic Parameters

Abstract:

Kinetic parameters are commonly ignored when applying traditional DoE approaches to chemiscal reactions, resulting in a failure to accurately capture and describe the time-dependent nature of the system. An alternative method utilizes such parameters as responses in the traditional DoE, allowing the impact of process parameters on the chemical kinetics to be evaluated and communicated within a standardized QbD framework.

Copyright © American Institute of Chemical Engineers. All rights reserved.

455311 New Trends in Materials-Sparing and Modelling Approaches Towards Pharmaceutical Development

Abstract:

Rising development costs and longer development times that cut into exclusivity periods have pushed the pharmaceutical industry into re-examining current development processes. Historically, the development of pharmaceutical products relied heavily on trial-and-error approaches. Over the last couple of decades, the concept of QbD (Quality by Design) has been incorporated to build robustness and consistent quality into the final product. More recently, there has been a significant push to develop materials-sparing techniques and predictive modelling approaches to develop a sound framework of product and process knowledge at small scales in order to detect risks early in development and facilitate efficient diagnosis and mitigation strategies for challenges at the commercial scale. Such knowledge promotes both the development of robust products and also enables an acceleration path for development activities when costly API (active pharmaceutical ingredient) is limited.

In this talk, examples of how materials-sparing approaches have been used for the development of both an oral solids drug product (OSD) and a biologic drug product will be presented. For the OSD case, plastic/elastic material properties were leveraged to develop an empirical heuristic used to predict the feasibility of two materials to successfully yield a bilayer tablet. The primary tool used to develop this assessment was a compaction simulator fitted with an instrumented die. The materials-sparing work-flow efficiently guided the formulation development of bilayer tablets (using <50g of material); the approach was validated at pilot and commercial-scales and during accelerated stability testing. For the biologic drug product, an FEM-based model was used to optimize commercial lyophilization cycles by leveraging knowledge of heat and mass transfer principles. Lyophilization, which involves freezing a product and then reducing the surrounding pressure to remove the frozen solvent via sublimation, is used to increase the shelf-life of biologic products by enabling products to be stored as dry powders instead of in solution. Thus, heat and mass transfer rates during both freezing and sublimation processes impact product quality and stability. However,

technology transfer and scaling approaches typically utilize large DOEs (design of experiment) to determine cycle parameter ranges through a trial-and-error approach, which is costly both in terms of material needs (API) and resources (time, utility costs). By developing an understanding of how heat and mass transfer rates change with scale, the transfer of a product from lab to commercial scale was successfully completed using only two engineering runs. The presentation will focus primarily on key decisions made during the development process that were enabled by the use of materials-sparing laboratory techniques or modelling predictions and discuss how the propagation and acceptance of these capabilities can streamline the development of future products.

Copyright © American Institute of Chemical Engineers. All rights reserved.

461149 The Journey of Mechanistic Models from Design to Optimization/ Launch and Beyond

Abstract:

This work describes a diversity of technical needs that led to the development of several mechanistic models to guide the design of specific drug substance processes at early phases of development. Some of these former models were also used for process optimization to define design spaces, predict process acceptable ranges and normal operating ranges and define conditions for process validation. Considerations on the requirements for models classified as medium impact and the approach to include mathematical expressions in the common technical document will be discussed. Finally, some details on the planned integration of mechanistic models to the manufacturing environment post-validation will be provided.

Copyright © American Institute of Chemical Engineers. All rights reserved.

471359 Scale-up of Reaction and Crystallization Steps of a Commercial Manufacturing Process in a Multi-Purpose Plant Using Mechanistic Modeling

Abstract:

This work presents an approach on modeling a manufacturing process in a multi-purpose plant, that comprises multiple reaction and crystallization steps, from the laboratory to production scale. In a multi-purpose plant, more often than not, the equipment used might not be the most suited for a particular process (e.g., mass transfer limitations due to poor mixing capabilities of an impeller) and thus it becomes crucial to use modeling tools and first principle methodologies to increase process understanding and further optimize the process to highest performance possible.
The case-study to be shown comprises a manufacturing multi-step, multi-phase process already in commercial phase, in which the reaction and crystallization unit operations are both modeled in laboratory and large scales. It is assessed if the conditions used in today's manufacturing process are optimized and if there is room for improvement either in yield of final product (reduce impurity formation), or particle size distribution (narrower interval and more suited range for further particle engineering). The data source for the modeling is both offline (HPLC analysis) and online (FBRM) which can be used as-is in modeling tools such as Dynochem (no data processing needed) to adjust the kinetics and crystallization constants. Both models are derived from laboratory scale experiments and subsequently are used to predict the behavior of the large scale process. After having the models validated in large scale batches, they are used to propose new conditions for the large scale process, using narrower ranges inside the Design Space of the current manufacturing process, for either yield of final product or particle size distribution.
In parallel to the kinetics and crystallization modeling, potential mass transfer limitations are assessed with Dynochem and CFD (Computational Fluid Dynamics) tools, depending on the complexity of reactor and impeller designs (lab and large scale designs). The mixing efficiency and mass transfer coefficients are derived from this study and are added to the kinetics model to increase its robustness and improve the overall understanding of the process.

Copyright © American Institute of Chemical Engineers. All rights reserved.

469184 Case Study: QbD Approach to Design Space Verification

Abstract:

Design Space Verification was completed for a recent regulatory filing using the Quality by Design paradigm for a drug substance process. A risk analysis approach was utilized to identify the key issues to address in order to verify that the design spaces derived from laboratory scale experiments were appropriate for commercial scale production.

Copyright © American Institute of Chemical Engineers. All rights reserved.

472822 Novel Method for Protein Stability and Delivery through the Formation of Complex Coacervates

Abstract:

Polyelectrolyte complexes and complex coacervation offer a novel means for the delivery of proteins and other bioactive materials. Polypeptides are biocompatible materials whose tunable properties lend themselves as viable vehicles for the delivery of drugs and, in particular, proteins. We also investigated the encapsulation of green fluorescent protein (GFP), bovine serum albumin (BSA), and lysozyme as model protein systems into polypeptide-based complex coacervates. These polypeptides are capable of mimicking the milieu found in cells, particularly cellsâ €™ natural characteristics of compartmentalization, crowding and soft interactions. The phase separation that occurs with coacervate formation gives way to compartmentalization and the high concentration of molecules simulates the biomolecules content, which also favors soft interactions between involved species. Additionally, these coacervates are aqueous, just like cells, and have tunable chemistry for furthering their cell-like characteristics. As such, the ability for peptide based coacervate systems to partially mimic this biological environment presents a viable means for potentially allowing proteins to exist in a wide array of environmental conditions without concern for their degradation. To further investigate protein encapsulation, we propose the creation of polypeptide-based block copolymers built from poly-lysine and polystyrene. We will focus on the characterization of polyelectrolyte complex micelle formation between various combinations of polypeptide and polystyrene. Previous work has shown that different ratio of hydrophobic to hydrophilic parts has a large impact on the geometry of micelle formation. By controlling the molecular weight of the block copolymer, the behavior of micelle stability will be investigated, which is very useful in protein encapsulation. It has been shown that the ionic strength and pH of the environment heavily influences coacervate formation and affects the encapsulation efficiency of proteins, thus, various mixtures of polycation/polyanion/protein mixtures were investigated. Trends from these environmental factors will then used to design delivery systems in the future. Key features of these mixtures that were varied were total monomer concentration, the geometry of micelle formation when using complex coacervate micelles, ratio of polycation to polyanion and initial concentration of protein. The stability

and geometry of micelle will be determined by light scattering and transmission electron microscopy. Encapsulation efficiencies were determined using a Bradford assay and UV/Vis spectroscopy and stability data were acquired using circular dichroism. Polypeptide-based coacervates are a powerful platform technology to understand the role of the chemical environment on the stability of a protein formulation. Ultimately, we look to harvest these materials to enable the development of highly stable therapeutics and biologics such as refrigeration-free vaccines. This novel system is a major stepping-stone for the effective treatment of therapeutic regiments, providing not only a milieu that mimics the biological environment, but a means for stabilization of bioactive materials.

Copyright © American Institute of Chemical Engineers. All rights reserved.

472681 Silica Based pH-Responsive Nanocomposite Nanoparticles As Controlled Drug Release Carriers

Abstract:

Biocompatible drug carriers with the ability of controlled therapeutic agent delivery is a promising way to overcome the limitations of conventional therapies. Among all types of carriers, silica-based nanocomposites emerged as an encouraging candidate because of the biocompatibility, high thermal stability of silica materials. Moreover, silica based pH-responsive nanocomposite materials can be produced by encapsulation of natural or synthetic pH-responsive polymers, such as alginate, chitosan, poly (methacrylic acid) and poly (acrylic acid). In this study, the pH-responsive silica based nanocomposite nanoparticles were made by silica sol-gel method in water in oil microemulsion system with different polymer (alginate, chitosan, etc.) solutions as water droplets. Nano sized particles between 20-80 nm were detected by transmission electron microscopy. Different surface zeta potentials were showed by dynamic light scattering with different silica polymer nanocomposites. Silica-alginate nanocomposites showed negative zeta potential in acidic solution, however, silica-chitosan nanocomposites had positive zeta potential in the same solution. The polymer weight ratio was detected by thermogravimetric analysis. The results showed increased polymer weight ratio with higher concentration polymer solution used in the water in oil microemulsion. Different drug molecules, such as timolol maleate and nadolol, were used to test the release profile of the silica-polymer nanocomposite nanoparticles at different pH buffer solutions. For silica-alginate nanocomposites, more drugs were released at physiological pH solution compared to acidic solution. However, the release profile of silica-chitosan nanocomposites was opposite compared to the silica-alginate one. Consequently, the aim of this study was to investigate the different silica-polymer nanocomposite nanoparticles and tested their pH-responsive controlled release property.

Copyright © American Institute of Chemical Engineers. All rights reserved.

454276 Porous Organic Cages for Drug Delivery

Abstract:

Porous materials have shown significant promise in drug delivery applications due to their high surface areas for adsorption, tunable pore size and chemistry, and regular, well-defined porosity. While several inorganic or hybrid organic-inorganic porous materials have been studied as drug carriers, concerns exist regarding the potential toxic effects of metals present in these systems. Porous organic cages are individual organic molecules with intrinsic porosity that, under specific conditions, self-assemble window-to-window with close packing to yield all-organic crystals with continuous, interconnected channels. In this work, we present the first-ever evaluation of these porous organic cages, specifically CC3 and CC5, for their drug delivery properties and biocompatibility. Crystalline particles of CC3 and CC5 have been loaded with cancer therapeutics (dichloroacetate and Î±-cyano-4-hydroxycinnamic acid), caffeine (useful for medicinal and cosmetic applications), and calcein (a model fluorescent molecule), with loadings of up to 25% by mass achieved, as determined by thermogravimetric analysis. Drug release from loaded CC3 and CC5 over several days was tested under simulated biological conditions in phosphate buffered saline (PBS), and drug release curves were obtained via UV-vis spectrometry to determine the kinetics and efficiency of drug delivery. The materials have also been shown to be stable in both water and PBS for at least one week, with no loss in crystallinity observed in powder x-ray diffraction patterns. Finally, two cytotoxicity assays have been conducted using HeLa cells to determine the effect of these materials both on cell membrane permeability and on metabolic activity. Neither CC3 nor CC5 compromised cell membrane integrity up to a concentration of 3 mg/mL, and while no reduction in metabolic activity was observed after incubation with CC3, the effects of particle size and potential breakdown products of CC5 on cytotoxicity are being investigated.

Copyright © American Institute of Chemical Engineers. All rights reserved.

461136 Contact Lenses Prepared By Using the Layer-By-Layer Method for Sustained Drug Delivery

Abstract:

It is known that a lot of the people are suffering from eye diseases. Approximately 90% of the drug formulation is implemented as eye drops in topical implementation[1]. However, bioavailability of the eye drop is too low and only 1-5% of the drug can reach the target tissue. The remaining drug is thrown away from the nasal cavity and discarded drugs can cause serious side effects[2]. Researchers has developed various approaches to eliminate these disadvantages of the use of eye drops in recent years. Increasing the viscosity and permeability of eye drops or loading drugs into micro and nano particles or into contact lenses are some of the approaches that have been studied. Among the others, contact lense based drug delivery systems have two basic advantages such as minimizing the side effects of drugs used excessively and elimination of drug loss/waste[3].

A Synthetic polymer p-HEMA is used frequently for controlled release applications as well as being used as a soft material. There are few studies that focus on usage of p-HEMA contact lenses to achieve prolonged drug release. Even though, they are more successful than eye drops and ointment, sustained drug release cannot be provided as desired. Drug loaded p-HEMA contact lenses release the drug in few hours. There are some methods used to slowdown the drug release[4]. One of these methods is utilizing the layer by layer method in order to modify the surface of the p-HEMA hydrogels. This method is based on the affinity between complementary functional groups[5].

In this study it is aimed to use the drug loaded p-HEMA contact lenses for the treatment of the conjunctivitis, which is one of the common eye diseases. Conjunctivitis is inflammation of conjunctiva and the membrane layer of eyelid on inner surface. Dexamethasone is a hydrophobic drug used for conjunctivitis treatment which was added into hydrogels. The surface of hydrogels was modified via "Layer by Layer" (LbL) technique. By using this technique, it is possible to obtain up to ten layers. Moreover, a thick layer of drug free biodegradable materials were added on the top of the contact lenses, which already had five layers formed via LbL technique, as mentioned before. The formation of each layer was monitored by measuring absorbance, light transmittance and water contact angle of the lenses. Hydrogels were characterized by using FT-IR (Fourier Transform Infrared Spectroscopy), DSC (Differential Scanning Calorimetry) and SEM (Scanning Electron Microscope). Drug release concentration was measured by using UV-vis spectroscopy. It was determined that, p-HEMA hydrogel released around 60 % of the drug in 8 h. Hydrogels prepared by LbL methods (5 layers) have sustained drug release in which only 30 % of drug was released in 50 hours. It was demonstrated that the formed biodegradable material layers act as a barrier for drug release. Furthermore, addition of the drug-free layers on top of this surface modified hydrogels at different thickness gave rise to further reduction at the drug release. As a unique result, for the first time in literature, it is seen that addition of biodegradable material layer on the top of the surface modified hydrogels had allowed us to observe a "delay time". The observed delay time varied with the changing thickness of the layer applied. The delay time increased from 15 minutes to 45 minutes with increasing thickness. Thus, controlled drug release was accomplished. Moreover, addition of biodegradable materials layer on top of the contact lenses allowed us to see a delay time before the release of drug as well as it prevented the burst release. Also, side effects of the drug were eliminated which was released at the beginning.

This study shows clearly that combining the distinct properties of different materials by using LbL method has been an effective approach for the controlled release of ophthalmic drugs. The dose of drug, which each patient must receive according to his/her drug regimen, can be adjusted by varying the thickness or the changing the number of the layers applied.

REFERENCES

[1] C. Le Bourlais, L. Acar, H. Zia, P.A. Sado, T. Needham, R. Leverge, 1998, 'Ophthalmic Drug Delivery Systems - Recent Advances', Progress in Retinal Eye Research, 17(1),33-58.

[2] J.C. Lang, 1995, 'Ocular Drug-Delivery Conventional Ocular Formulations', Adv Drug Deliver Rev, 16(1), 39-43.

[3] Y. Kapoor, J.C. Thomas, G. Tan, V.T. John, A. Chauhan, 2009, 'Surfactant-laden soft contact lenses for extended delivery of ophthalmic drugs', Biomaterials, 30, 867–878.

[4] D. Gulsen, A. Chauhan, 2004,'Ophthalmic Drug Delivery through Contact Lenses' Investigative Ophthalmology & Visual Science, 45(7), 2342-2347.

[5] X. H. Hu, Y. S. Zhang, 2014, Synthesis and surface modification of chitosan containing hydrogel for ophthalmic drug delivery, 'Materials Technology: Advanced Performance Materials', 29(3).

Copyright © American Institute of Chemical Engineers. All rights reserved.

468570 Chitosan-Coated Selenium Nanoparticles for Anti-Acne Applications

Abstract:

Introduction: The skin is the bodyâ€™s strongest form of protection against infectious and pathogenic microbes, and when this barrier is perforated by wounds, lesions or other chronic skin conditions, the body is vulnerable to more serious diseases. Among these skin diseases, acne vulgaris is one of the most prevalent, affecting up to 95% of all adolescents and persisting in up to 10% of adults. *Proprionibacterium acnes* has been suggested as the main contributor to acne vulgaris. Our lab has synthesized selenium nanoparticles to combat bacterial infection, and have found them to be effective against several bacterial strains, and relatively non-toxic to healthy fibroblasts and other mammalian cells. Here, to make such particles suitable for skin applications, we employed a new synthesis of chitosan-coated selenium nanoparticles as an effective topical treatment for *P. acnes*, as well as common skin denizens, such as *S. aureus*, while preserving the healthy skin tissue surrounding the acne lesions.

Materials and Methods: Chitosan-coated selenium nanoparticles (Ch-SeNP) were synthesized in a precipitation reaction, in which chitosan (dissolved in 1% acetic acid) was added dropwise to sodium selenite solution, followed by an addition of ascorbic acid. The resulting Ch-SeNP are separated via centrifugation and resuspended in DI water. Characterization was performed via dynamic light scattering and transmission electron microscopy. *P. acnes* (ATCC 11827) were cultured in modified reinforced clostridial broth medium or on tripticase soy agar plates with defibrinated sheep blood, while *S. aureus* was cultured in 3% tryptic soy broth or on agar plates. For *S. aureus*, kinetic growth curves were tracked by measuring the optical densities of treated cultures via spectrophotometry over 24 hrs. *P. acnes*, an anaerobic bacterium, required anaerobic conditions to grow, and were thus tested via a colony-forming unit assay in anaerobic gas chambers over 4 days. All experiments were repeated three times, and statistical significance were analyzed via a two-talked Studentâ€™s T-test.

Results and Discussion: Ch-SeNP were found to be as effective, if not more, as antibacterial agents as the previously studied BSA-coated SeNP on *S. aureus*, with lag times lengthening as Ch-SeNP concentrations increased. At a mid-range concentration of Ch-SeNP (0.023 mg/ml), *P. acnes* growth was completely absent compared to untreated colonies, suggesting that Ch-SeNP were strongly effective at preventing *P. acnes* proliferation.

Conclusions: Chitosan-coated selenium nanoparticles were found here to be exceedingly more stable in solution than their BSA-coated predecessors without sacrificing their antibacterial efficacy. Additionally, we found the Ch-SeNP to be effective in preventing the proliferation of the anaerobe *P. acnes*, suggesting a novel SeNP application in topical skin infections.

Acknowledgements: The authors thank Northeastern University and the College of Engineering for funding.

Copyright © American Institute of Chemical Engineers. All rights reserved.

471959 Regulating the Timing and Sequence of Pro- and Anti-Inflammatory Cytokine Deliveries from Magnetically Responsive Biomaterials for Use in Wound Healing Applications

Abstract:

In the United States, approximately 6.5 million patients are diagnosed with chronic wounds. The treatment of chronic wounds costs more than $25 billion annually and this is expected to grow due to the increasing cost of healthcare, an aging population, and a rise in the occurrence of diabetes and obesity worldwide [1]. The process of wound healing includes four major steps: hemostasis (blood clothing), inflammation, proliferation, and tissue maturation [2]. Even a slight perturbation in this process can disrupt proper healing, leading to chronic wounds. Chronic wounds are often a result of arrest in the inflammation phase of healing [2]. Although inflammation critically initiates repair and helps clear infections, a prolonged inflammatory reaction can cause considerable harm to the injury site. After an appropriate duration of inflammation, this inflammatory response can be shifted to a more pro-healing response through the delivery of cytokines like interleukin 4 (IL-4) and interleukin 10 (IL-10). These anti-inflammatory cytokines alter the phenotype of macrophages from pro-inflammatory (M1) to anti-inflammatory (M2), suggesting a potentially powerful drug delivery strategy if these cytokines can be delivered in a delayed manner. We hypothesize that the transition of macrophage phenotype from pro-inflammatory (M1) to anti-inflammatory (M2) can be controlled through sequenced delivery of interferon gamma (IFN-γ), followed by IL-4 and/or IL-10. The goal of this research was to develop a wound-healing hydrogel system that initially delivers IFN-γ, followed by magnetically triggered delivery of IL-4 and/or IL-10.

Our biomaterial system was composed of two-compartments: (1) a porous gelatin outer compartment designed to recruit macrophages and establish an initial pro-inflammatory (M1) phenotype, and (2) a magnetically responsive alginate inner compartment which was designed to deliver IL-4 and/or IL-10 when magnetically triggered to shift the response to anti-inflammatory by promoting (M2) phenotype. This magnetically responsive inner compartment was made by casting 7 wt% Fe_3O_4in 1 wt% alginate with 2.5 mM adipic acid dihydrazide (AAD) crosslinking in the presence of a magnet to generate biphasic gels. These biphasic gels were then swollen in water, frozen at -20°C, and lyophilized to generate porous, magnetically compressible biphasic gel structures. Gelatin outer compartments were loaded with 1000 ng of IFN-γ and designed to diffusively release IFN-γ early on. Magnetically deformable inner compartment gels were loaded within 100-1000 ng of IL-4 or IL-10 and were designed to deliver these anti-inflammatory proteins to macrophages that were recruited to the outer compartment at a time dictated by the application of the magnetic stimuli. A pulsatile 5 kGauss graded magnetic field was used in these experiments and applied at 1.5 Hz. Validation of these protein delivery profiles over time were quantified for IL-4, IL-10, and IFN-γ using ELISA.

We have demonstrated this system's ability to initially release a pro-inflammatory cytokine (IFN-γ). We have found that more than 95% of the loaded IFN-γ was released from the outer compartment in the first 6 hours. We have also shown the ability to retain anti-inflammatory cytokines (IL-4 and IL-10) for several days and magnetically trigger their release from the inner compartment. Furthermore, we were able to magnetically control the timing and rate of these deliveries. For example, we showed that when we load 1000 ng of IL-4 in the inner compartment, we were able to retain IL-4 in the gel for 4 days with low release rates prior to magnetic stimulation (29.6 ± 3.3 ng/day for 4 days) and magnetically trigger enhanced release rates on day 4 (345.8 ± 103.29 ng/day over 4 hours). Using gels loaded with 500 ng of IL-10, we again demonstrated excellent retention for 3 days prior to magnetic stimulation (0.17 ± 0.04 ng/day for 3 days) and magnetically enhanced release rates on day 3 (6.99 ± 4.24 ng/day over 3 hours). The timing of IL-10 delivery could be stimulated on day 5 instead of day 3. In this case, IL-10 was retained at a low release rate prior to magnetic stimulation (0.11 ± 0.04 ng/day for 5 days) and magnetically released at enhanced rates on day 5 (3.67 ± 0.776 ng/day for 4 hours). Finally, the rate of IL-4 release could be controlled by using different magnetic stimulation regiments. For example, the release rate of IL-4 when magnetically stimulated at 1.5 Hz for 1 hour every 12 hours for over days was only 20.23 ±7.33 ng/day whereas continuous magnetic stimulation at 1.5 Hz for 4 hours resulted in a much higher release rate (345.85 ±103.29 ng/day). We believe that this biomaterial system will be highly useful for understanding how the timing and sequence of cytokine deliveries impact wound healing outcome.

Keywords: Wound healing, Macrophages, Cytokine delivery, Biomaterial, Magnetic stimulation.

REFERENCES

1. Sen, Chandan K. et al. "Human Skin Wounds: A Major and Snowballing Threat to Public Health and the Economy." *Wound Repair and Regeneration* 17.6 (2009): 763–771.

1. Frykberg, Robert G., and Jaminelli Banks. "Challenges in the Treatment of Chronic Wounds." *Advances in Wound Care* 4.9 (2015): 560–582.

Copyright © American Institute of Chemical Engineers. All rights reserved.

471748 Evaluation of Poly(curcumin) Microparticle Degradation and Activity in the Presence of Free Radical Generating Systems

Abstract:

Under a variety of pathophysiological conditions (e.g., radiation injury, inflammation, acute lung injury), cells are no longer able to maintain a balance between antioxidants and exogenous or endogenous radical generation, resulting in free radicals overproduction and oxidative stress.

Curcumin, a potent antioxidant, has three potential scavenging sites, but due to its hydrophobicity, poor solubility and bioavailability, it has possessed a low therapeutic effect *in vivo*. Free antioxidant drug delivery systems can also pose strain to the environment that can lead to pro-oxidant effects rather than the latter. To overcome these limitations, our group has previously synthesized a covalently crosslinked poly(curcumin beta amino), poly(curcumin), film, where the hydrolytically degradable backbone of the polymer network increases curcuminâ€™s bioavailability, utilizing a controlled release mechanism, whichÂ gives great potential to show positive efficacy on oxidative stress treatment and cellular protection. Poly(curcumin) films can be cryomilled into microparticles to employ versatility in size and for localized delivery of interest. An advantage of the poly(curcumin) network is its comonomer system composed of a hydrophilic (polyethylene glycol (400) diacrylate) and hydrophobic (curcumin multiacrylate (CMA)) component. The ratios of these components can be varied to modify the rate of drug release. The design of microparticle blends can provide unique regulation to future dosage and treatments to specific oxidative environments (e.g. different stages of oral mucositis) to maintain a constant therapeutic release of curcumin. The use of kinetic and predictive modeling can be used to model poly(curcumin) microparticles as a two compartment system to verify the contributing factors to microparticle degradation as well as the consumption or scavenging rates of the released curcumin. We hypothesize that curcumin release from microparticles of different compositions that are introduced to reactive oxygen species to the microparticle system will respond and consume differently based on different degradation rates. and accelerate the hydrolytic cleavage of the weak ester bonds, further accelerating the release and consumption of curcuminâ€™s

antioxidant activity due to its scavenging capabilities. Predictive models can allow for us to evaluate and understand our complex polymer network mechanistically and better understand the contribution of volume, hydrolysis, and drug affinity in our system.

Copyright © American Institute of Chemical Engineers. All rights reserved.

470891 Simulating Intestinal Molecular Transport Using Tissue Engineered in Vitro Models

Abstract:

Introduction: The intestinal mucosal barrier lining the gastrointestinal tract â€" composed of an epithelial cell monolayer and the mucins they secrete â€" plays a critical role in maintaining metabolic and immunological homeostasis. Recently developed microfluidic gut-on-a-chip devices have significantly advanced the capability of in vitro models in recapitulating in vivo intestinal tissue physiology (1). However, remaining key challenges for such devices include (i) supporting both anaerobic and aerobic intestinal milieu, (ii) simultaneous human cell co-culture, and (iii) real-time in situ observation of their dynamic crosstalk.Â Here, we introduce a new bioengineered platform technology with high analytical power that can potentially allow for anaerobic and aerobic culture by creating a transverse compartmentalized device that retains previously identified anatomical and mechanical tissue complexities. Peristaltic motion that is critical in promoting gut epithelial cell growth and function was further simulated via pneumatic actuation.

Materials and Methods: Type I collagen infused polyacrylamide hydrogel precursor solution was casted in a 500μm thick PDMS sheet having pre-cut villous mimicking micropatterns on a silanized glass slide. (2) An actuating layer made with Ecoflex that has been used in soft-robotics (3) was attached outside of a hydrogel layer and sealed with another glass slide. We then increased the cellular and extracellular complexity by sequentially introducing (i) HT-29 intestinal epithelial cells, (ii) mucin extract from pig intestines, (iii) bacteria, and (iv) human peripheral blood-derived mononuclear cells (PBMC). By using a solenoid valve in tandem with a microcontroller (μC), the chamber was expanded and contracted at controlled pressures and frequencies using in house vacuum and air.

Results and Discussion: Covalent attachment of polyacrylamide hydrogels to silanized glass slides was achieved in a sealed, compartmentalized microfluidic device. Molecular diffusion tests using rhodamine dye showed that the hydrogels are permeable to gas and allow for small molecular diffusion. Different spatially patterned hydrogels can be microfabricated with villous-like structures that recapitulate the physiology of the gastrointestinal tract. The luminal compartment of the microfluidic was seeded with HT-29 colorectal cancer cells that preferentially adhered to the hydrogel, forming an epithelial barrier while the tissue compartment was filled with PBMCs derived from blood donors. Additionally, mucin and bacteria were introduced in the luminal compartment and continuously perfused in order to reduce bacterial density while cell-to-cell interactions were imaged in real-time. Actuation was controlled by an Arduino-based μC that can have very diverse control of operations.

Conclusions: We have developed a new type of microfluidic device that integrates soft biomaterials. We envision that such a modular in vitro microphysiological intestinal tissue model can serve as a translational platform to discover the biophysical etiology of disruption of the mucosal barrier.

Copyright © American Institute of Chemical Engineers. All rights reserved.

467964 Development of Injectable Microgels for Galns Enzyme Replacement Therapy

Abstract:

Introduction: In this study, we developed injectable microgels comprised of multiarm polyethylene glycol acrylate (PEG-Ac) with tunable degradation and controlled release features. We employed the microgels for the encapsulation and delivery of N-acetylgalactosamine 6 sulfate sulfatase (GALNS) enzyme with the broader goal of developing an effective enzyme replacement therapy. Deficiency of GALNS causes accumulation of glycosoaminoglycans and ultimately Morquio A disease. Although effective in alleviating some disease symptoms, current enzyme replacement therapy (ERT) has many limitations, such as a need for recurrent infusions and occurrence of immune responses.[1] Delivery of GALNS encapsulated in PEG microgels could offer an improvement over existing therapies, as it preserves enzyme activity, leads to longer circulation times and attenuates immune responses, while requiring a single injection.

The PEG microgels in this study were prepared by electrospraying by Michael's addition reaction. We were able to obtain microgels of size range 70-300 μm by varying the electrospraying set-up parameters. The microgels could be injected through a needle, while still preserving their integrity upon injection. Further, the degradation of the microgels could be tuned to control release of encapsulated proteins. Release kinetics for several model proteins varying in size and hydrodynamic radii was found to be dependent on the protein size, as well as mesh size and degradation rate of the hydrogel. Furthermore, we observed preservation of GALNS enzyme activity in the upon encapsulation in the microgels. The nanoporous microgel mesh size precluded protein release until sufficient microgel degradation occurred.

Conclusions: The developed drug delivery approach has potential as a modified-ERT system to improve treatment outcomes and quality of life of Morquio A patients.

References: 1. Montaño, A. M.; Tomatsu, S.; Gottesman, G. S.; Smith, M.; Orii, T. International Morquio A

Registry: clinical manifestation and natural course of Morquio A disease. *Journal of inherited metabolic disease* **2007,** 30, (2), 165-74.

Copyright © American Institute of Chemical Engineers. All rights reserved.

471705 Computationally-Driven Design of a Heterobivalent and Gate to Improve Targeting Agent Selectivity

Abstract:

Targeted molecular imaging agents have greatly advanced the sensitivity and specificity of detection methods and increased the ability of clinicians to detect and characterize cancers. However, these imaging agents often target biomarkers that are not exclusively expressed on cancerous cells, resulting in false positives and high signal background. Many cancer types do not have a single biomarker that can consistently distinguish them from the normal tissue. Yet the overexpression of two biomarkers is a more unique phenomenon that could enable selective targeting for a diverse array of patients. The goal of this study is to develop a modular molecular targeting system, labeled a biological AND gate, that can discriminate between cells that have a moderate overexpression of two biomarkers over cells that have moderate or even high expression of only one.

The AND gate molecule is a heterobivalent ligand comprising two binding domains linked via a flexible peptide. The two binding domains target two disease-associated biomarkers with a weak affinity that precludes monovalent binding to singly overexpressing cells while enabling robust bivalent binding to dually overexpressing cells. Optimal AND gate design – including individual affinities of the binding domains, linker length, and linker structure – depends on several properties of the both the tumor and the biomarker, such as average cell surface expression, internalization rate, homogeneity across cell membrane, and potential agonistic/antagonistic effects. To guide quantitative design, a geometric, cell-surface binding model was developed and integrated with a literature biodistribution model (Thurber and Wittrup *J. Theor. Biol.* 2012). The hybrid model indicates that, given a surface expression of 100,000 biomarkers per cell (each), a 100-fold increase in specificity can be achieved over cells expressing 100,000 of only one biomarker with individual ligand affinities of 125-575 nM. Stronger affinities hinder selectivity while weaker affinities preclude even bivalent binding. The optimal flexible linker length, to maximize local ligand concentration after binding of the first domain, is 22 amino acids for a receptor radius of 20 Å. Differential results are observed for flexible and rigid linkers.

These optimal ligand properties were then engineered into an experimental heterobivalent ligand. In our model system, the binding domains are two non-antibody protein scaffolds (10th type III domain of fibronectin (Fn3) domains) that have been previously engineered to bind either epidermal growth factor receptor (EGFR, clone A') or carcinoembryonic antigen (CEA, clone C7) with initial affinities of 0.42 ± 0.06nM and 15 ± 15 nM, respectively. To obtain ligands with affinities within the 125-575 nM range, yeast surface display libraries, created by mutagenesis of each parental gene, were sorted by flow cytometry for decreased binding to the proper target. Mutants from the sorted sub-populations were produced and titrated on cancer cells, which revealed almost no correlation between yeast-displayed affinity to recombinant target and soluble affinity to the natively-expressed target. Nevertheless, several mutants for each target were identified; C7 mutants ranged from 2.5 to 5,600 nM and A' mutants from 15 to 3,000 nM. To determine scaffold modularity, *i.e.*, whether their affinities were affected by the fusion of a second Fn3 domain, the parental ligands A' and C7 were fused on either the N- or C- termini with a non-binding Fn3 clone. The fusion of a non-binding Fn3 to the C-termini (distal to the binding site) decreased the affinity of A' 50-fold to 21 ± 12 nM and C7 affinity held at 16 ± 10 nM. When the non-binding Fn3 was fused to the N-termini (near the binding site), A' affinity decrease was similar (18 ± 6 nM), while C7 showed a decrease in affinity to 89 ± 97 nM. Two mutants of A' were also tested in the fusion context; A' mutants A'.1 and A'.2, which have affinities of 28 ± 3 nM and 17 ± 4 nM, respectively, as single domains, displayed affinities of 28 ± 12 nM and 125 ± 93 nM, respectively, when fused to the non-binding Fn3 via their C-termini. An additional high-affinity Fn3 binder to EGFR (clone D) demonstrated no effect when fused via its C-terminus (0.4 ± 0.1 nM) compared to the monovalent affinity of 0.25 nM. These variations in fusion modularity demonstrate the importance of quantifying the *in vitro* affinity of fusion components – in the fusion context – and highlight dependence on unknown binding paratope, receptor binding epitope, binding orientation, and structural folding variations. Ongoing efforts include identifying the structural effect of the fusion via circular dichroism and SEC analysis.

For initial AND gate testing, three heterobivalent ligands were synthesized: A'-C7, A'.1-C7, and A'.2-C7, where the N-terminus of C7 was fused to the C-terminus of the three EGFR binding Fn3 via a glycine-rich linker to test their selectivity on native and modified cell lines of varying EGFR and CEA expression. Preliminary data of the A'-C7, A'.1-C7, and A'.2-C7 fusions on A431 cells (<2,000 CEA, 4,000,000 EGFR per cell) show that the fusion of C7 via the C-terminus further decreases the EGFR affinity of the fusion compared to the non-binding Fn3 fusion to 102 ± 31 nM, 443 ± 238 nM, and 1200 ± 1000 nM, respectively. However, titration of A'.1-C7 on lentiviral-transfected HEK-293T cells (250,000 CEA, <2,000 EGFR receptors per cell) results in an affinity of 69 ± 10 nM, showing that the C7 affinity is indistinguishable when fused to either A'.1 or the non-binding Fn3. This data shows that the effect of the fusion can be dependent on the clone you are fusing, which is crucial when accurately measuring the properties of the system, and the importance of cell lines with accurately characterized expression to quantify individual domain affinities within the heterobivalent ligand. Ongoing studies, evaluating binding selectivity and sensitivity on cell lines with varying levels of CEA and EGFR, will be discussed. Additional results to be shared include evaluation of ligands of varying affinities, additional scaffolds, and alternative fusion methods to create a more modular system.

Copyright © American Institute of Chemical Engineers. All rights reserved.

461410 Supramolecular Pegylation of Biopharmaceuticals

Abstract:

Introduction: The practice of medicine has been transformed by the use of protein drugs, but these also introduce an array of attendant complications arising from poor chemical and/or structural stability, thus requiring extensive effort in the development of stabilizing excipients and formulation methods. Direct chemical modification with poly(ethylene glycol) (PEG) has demonstrated improvements in protein stability and circulation half-life, a practice that has been used clinically for a number of protein therapies. Herein, we demonstrate for the first time a platform approach to PEGylation of biopharmaceuticals using only supramolecular interactions by leveraging high-affinity host-guest interactions in order to endow PEG-like functionality on proteins non-covalently.

Materials and Methods: A supramolecular macrocycle with high binding affinity was synthesized and functionalized with PEG chains of controlled molecular weight *via* "click" chemistry. Leveraging host-guest binding of this macrocycle with native residues on proteins, this PEGylated compound was added in formulation with insulin, glucagon, and a therapeutic antibody. Functional performance of all three biopharmaceuticals was assayed both *in vitro* and *in vivo.*

Results and Discussion: The supramolecular PEGylation (SM-PEG) of insulin extends *in vitro* stability and preserves potency when aged under agitating conditions, extending protein life from approximately 14 hours alone to over 100 days when in formulation with SM-PEG. Moreover, in a diabetic mouse model the duration of insulin activity can be tuned by selection of PEG molecular weight. Extending this approach to glucagon and a therapeutic anti-CD20 antibody, the SM-PEG lends significant stability and preservation of activity to these proteins as well. This approach to supramolecular modification of therapeutic proteins provides an excipient-only strategy to modulate protein activity and stability without the need for covalent modification of the protein drug molecule, and has broad possible use as a platform for biopharmaceutical formulation.

Conclusions: In biopharmaceutical practice, assuring stability and modulating pharmacokinetics are two primary areas of interest, and covalent PEGylation has been among the more successful strategies evaluated in this regard. Using high binding affinities afforded by specific supramolecular host-guest interactions, proteins can be modified non-covalently with PEG chains of various molecular weights. This supramolecular PEGylation strategy dramatically extended the stability of three therapeutic proteins, and also prolonged the activity of insulin in a modular fashion dependent on molecular weight. This approach may be broadly applicable as a platform of "smart" excipients for formulation and shelf-life extension of biopharmaceuticals.

Copyright © American Institute of Chemical Engineers. All rights reserved.

462185 *Non-Invasive Screening of Diseases Using an Orally Available Near-Infrared Imaging Agent*

Abstract:

Early detection of diseases is crucial to improving the chances of successful treatment, such as curative surgery of cancer or early intervention with disease-modifying treatments to prevent irreversible damage in rheumatoid arthritis. Current methods of detection involve either minimally invasive blood tests or medical imaging modalities. Blood tests can provide large amounts of molecular information but lack spatial information about the site of the disease, and imaging modalities can provide spatial information but only offer anatomical data. Molecular imaging modalities can bridge the gap between the aforementioned techniques, but the drawbacks include safety concerns with ionizing radiation and the associated high cost.

We aimed at developing a near-infrared orally available imaging agent to detect diseases non-invasively. This technique employs near-infrared fluorescence as a method of detection that makes the process safe (non-ionizing radiation) but limits the detection of diseases within 1 to 2 cm from the surface of the skin. This method of detection is easily portable, cheap and the use of an orally available imaging agent makes it completely non-invasive. Importantly, oral delivery avoids associated risks with parenteral delivery and could improve compliance in a screening setting.

As proof of concept, we chose a single-digit nanomolar integrin binding peptidomimetic as a model system. The small molecule ligand has the highest affinity for Î±$_v$Î²$_3$ integrin, which is highly expressed on certain forms on cancer and on macrophages present at sites of inflammation such as arthritic joints. The binder was conjugated to several near-infrared dyes in order to make a series of agents with varying physicochemical and optical properties to ascertain what properties are important for orally available imaging agents. *In vitro* and preliminary *in vivo* experiments were conducted to find the agent that had the best mix of oral bioavailability, *in vivo* optical properties and specificity. Based on these results the IRDye 800CW conjugate was used in an orthotopic breast cancer model to show that this method of detection is effective following oral delivery. We were able to produce target-to-background ratios (TBR) of 4.7 and tumor-to-muscle ratios over 20 using our method of delivery and detection. A mechanistic model based on partial differential equations was also made to correlate the properties of the agents to TBR, in order to highlight the most important properties from a quantitative pharmacokinetic and transport perspective.

Copyright © American Institute of Chemical Engineers. All rights reserved.

467547 Investigation of the Cellular Uptake and Cytotoxicity of Cell-Penetrating Peptides in *Candida* Fungal Pathogens

Abstract:

Cell-penetrating peptides (CPPs) have been widely studied as tools for delivering a variety of molecular cargo into cells, including DNA, RNA, proteins, and nanoparticles. Previous work with CPPs has mainly focused on their uptake in mammalian cells, but CPPs also have potential as drug delivery and research tools in other organisms, including fungi. We have investigated 12 different peptides for their ability to function as CPPs for the important human fungal pathogens *Candida albicans* and *C. glabrata*. The peptides were commercially synthesized with an N-terminal 5-carboxyfluorescein label serving as small-molecule cargo and as a fluorescent probe to report cellular uptake. We have used flow cytometry to screen the CPPs for their potential to efficiently enter the *Candida* cells. Among the 12 CPPs, we have identified several promising CPPs, including pVEC and SynB, that can enter both *C. albicans* and *C. glabrata* cells, and the overall uptake process depends on both the salt concentration in the buffer and on the concentration of the peptides. We have also identified CPPs with some specificity; for example, Hst-5 can only translocate into *C. albicans,* whereas TP-10 has higher uptake in *C. glabrata*. We found that generally the CPPs with higher uptake also have higher toxicity toward both *Candida* species, but SynB showed almost no antimicrobial activity at a concentration where nearly 100% of cells have high levels of peptide uptake. Furthermore, we explored the subcellular localization of CPPs at different concentrations with fluorescent microscopy and flow cytometry. Most of the promising CPPs showed a 3-stage behavior, where CPPs evenly distributed in the cytosol at high concentrations, localized inside vacuoles at moderate concentrations, and gathered around the membranes at lower concentration. Cecropin B, in contrast, shows â€œall-or-nothingâ€ behavior, with the peptide either present in the cytosol or exhibiting no apparent association or uptake. In addition, we have also investigated the uptake at different temperatures to aid understanding of the mechanism of uptake for the peptides. Some peptides (e.g., MPG and $(KFF)_3K$) are significantly affected by temperature, indicating a potential ATP-dependent uptake mechanism, whereas temperature has minimal impact on

other peptides (e.g., pVEC and Cecropin B). Based on these results, we are now exploring the structure-function relationship for translocation of CPPs into fungal cells.

Copyright © American Institute of Chemical Engineers. All rights reserved.

470913 Catalase-Laden Microdevices for Cell-Mediated Enzyme Delivery

Abstract:

Catalase-Laden Microdevices for Cell-Mediated Enzyme Delivery

Junfei Xia[1], Zhibin Wang[1], Yuanwei Yan[1], Zhijian Cheng[2], Li Sun[2], Yan Li[1], Yi Ren[2], Jingjiao Guan[1,*]

[1] Department of Chemical and Biomedical Engineering, FAMU-FSU College of Engineering, Florida State University

[2] Department of Biomedical Sciences, College of Medicine, Florida State University

Abstract

Enzymes have been used as drugs to treat a wide variety of human diseases and traumas. However, therapeutic utility of free enzymes is generally impeded by short circulation time, lack of targeting ability, immunogenicity, and inability to cross certain biological barriers. Various strategies have been developed to overcome these limitations. Among them is cell-mediated drug delivery, which is featured by integrating drugs with live cells and taking advantage of the unique capabilities of the cells to achieve controlled drug delivery. We developed simple and inexpensive methods based on soft lithography and layer-by-layer (LbL) assembly for fabricating disk-shaped microparticles termed microdevices and generating the cell-microdevice complexes. [1, 2] We herein applied these methods to the fabrication of enzyme-laden microdevices and corresponding cell-microdevices complexes for the first time. We used catalase as a model enzyme in this study. It is a water soluble protein with an isoelectric point of 5.8. It is thus negatively charged at pH 7. Poly(diallyldimethyl ammonium chloride) (PDAC) was chosen here as a polycation to form a bilayer with catalase. Catalase in the microdevices was catalytically active and active catalase was slowly released from the microdevices. The catalase-laden microdevices were attached to the external surface of live suspension and adhesion cells respectively to form cell-microdevice complexes. This technique is applicable to other therapeutic enzymes and therapeutic cells, and thus promises to find clinical applications for treating various human diseases and traumas.

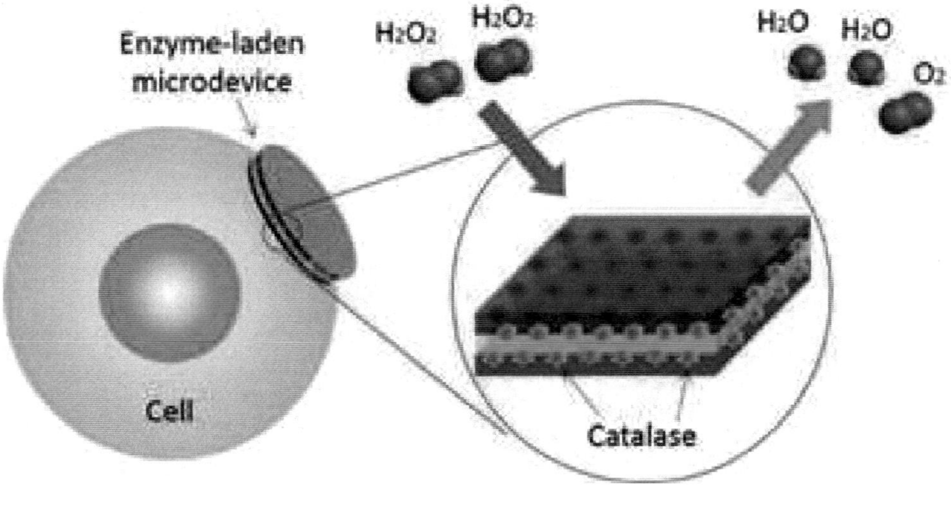

References

1. Junfei Xia, Zhibin Wang, Danting Huang, Yuanwei Yan, Yan Li, Jingjiao Guan*,*Asymmetric Biodegradable Microdevices for Cell-Borne Drug Delivery*, **ACS Applied Materials & Interfaces** (2015) 7: 6293-6299.

2. Zhibin Wang, Junfei Xia, Yuanwei Yan, Ang-Chen Tsai, Yan Li, Teng Ma, Jingjiao Guan*,*Facile functionalization and assembly of live cells with microcontact-printed polymeric biomaterials*, **Acta Biomaterialia** (2015) 11: 80-87.

Copyright © American Institute of Chemical Engineers. All rights reserved.

472660 Ragweed Pollen-Based Oral Vaccine Delivery

Abstract:

Oral vaccine delivery although highly desirable, continues to be a difficult target. This is because the harsh acidic and enzymatic environment together with physical and biological barriers in the gastrointestinal (GI) tract pose unique challenges to deliver vaccines orally. We have developed pollen grains (PGs) as a novel method to deliver antigens across the GI mucosa. In this study ragweed pollens were used to deliver ovalbumin (OVA) orally.

Ragweed pollens were chemically processed to remove native biomolecules and to obtain clean pollen particles. Mice were then immunized with a formulation of OVA and ragweed pollens. Immune responses were characterized by determining the presence of serum anti-OVA IgG, IgG1, IgG2a, IgE, fecal IgA, salivary IgA, and vaginal IgA antibodies using enzyme-linked immunosorbent assay (ELISA). Our study shows that vaccine formulations with ragweed pollens stimulated significantly higher antibodies compared to control mice that received only OVA without ragweed pollens. Using ragweed pollens, high OVA-specific systemic immune response was induced compared to control mice. Furthermore, anti-OVA IgA antibodies were stimulated not only in the intestinal mucosa, but at the distal vaginal mucosal site. Overall, our study demonstrates the potential of ragweed pollen for a simple method of oral vaccination that can induce systemic and mucosal immunity. Further study is required to optimize the antigen dose and frequency of the delivery system.

Copyright © American Institute of Chemical Engineers. All rights reserved.

472952 the Roles of Competing Dissolution, Diffusion and Transient Marangoni Convection Fluxes in Surfactant-Enhanced Spreading of Aerosols for Pulmonary Delivery

Abstract:

We are formulating aerosols for pulmonary delivery with surfactants to overcome aerodynamic limitations when treating obstructive lung diseases such as cystic fibrosis. The added surfactant is designed to establish transient Maragoni flows which can convect drug laterally after aerosol droplets are deposited onto airway surface liquid (ASL) and disease-associated mucus plaques. We have used a combination of aerosol droplet deposition experiments with mock systems (entangled (bio)polymer solution subphases mimicking the ASL, fluorophores mimicking drugs, multiple surfactant species) and numerical transport simulations (via COMSOL) to examine the competing roles of surfactant dissolution, diffusion and transient Marangoni convection fluxes in determining the extent of drug spreading. Large dissolution and diffusion fluxes lead to significant net drug transport, but poor spreading while small dissolution and diffusion fluxes lead to poor drug transport and poor spreading; extensive drug spreading occurs when dissolution and diffusion fluxes are similar in magnitude to the transient Marangoni convection flux. Several combinations of mock subphases, surfactants, and mock drugs will be discussed along with implications for delivery via both liquid and solid aerosols.

Copyright © American Institute of Chemical Engineers. All rights reserved.

472776 Improving Single Use Bioreactor Design and Process Development, New Research Towards Intensifying Seed-Train and Scale-up Methods Using 5:1 Turn-Down

Abstract:

Operating bioreactor vessels at low working volumes (low turn down ratio) is often desirable but brings about challenges in regard to mixing, mass transfer, and process control. Research done towards optimizing cell culture has provided methods to improve performance and control when operating under these conditions. Implementing bottom heat exchange, making changes to impeller angle and height, taking advantage of the unique Thermo Fisher drilled hole sparge design, and implementing a new cross flow sparge into the headspace have yielded reliable mass transfer and cell culture results with the additional benefit of lower turn down mixing when draining the reactor.

Copyright © American Institute of Chemical Engineers. All rights reserved.

462181 Characterization and Design of a 3L Single-Use Bioreactor to Support Process Development and Tech Transfer

Abstract:

The design and characterization of a 3L rigid single-use bioreactor (SUB), to replace existing 3L glass bioreactors, was pursued to further enhance the efficiency of development operations in Merckâ€™s Kenilworth biologics facility.

A myriad of bench-scale, rigid, single use bioreactors are commercially available at this time. We required one to replace our current 3L glass bioreactor vessels and meet the diverse requirements of our biologics facility, which supports a continually increasing number of cell culture projects. The new SUB needed to perform like our current bioreactors, so that process development activities would not be impacted. Overall, the SUB was expected to be versatile, robust, and simple to use.

A multi-faceted, iterative approach was used in the development of a new custom 3L SUB, with the following requisites:

- Similar configuration to larger vessels (e.g. drilled tube sparger and rigid baffles)
- Magnetic bottom drive agitation system
- Perfusion ready
- Expanded number of sensor ports and addition lines for flexibility
- Compatible with Automated sampling systems

To characterize the designs, we performed gas transfer (k_La) studies, computational fluid dynamics (CFD) modeling, and cell growth studies with both fed-batch and perfusion processes. The data generated from these studies was used to refine the design and confirm comparability to the glass vessel controls.

Implementation of the customized 3L SUBs has enabled enhanced process development capability by introducing technology to improve operational efficiency.

Copyright © American Institute of Chemical Engineers. All rights reserved.

461864 Foam Control in the Mammalian Cell Culture Process through Extrinsic and Intrinsic Methods

Abstract:

During biopharmaceutical manufacturing, the formation of foam within bioreactors has been an industry-wide issue for decades. As the product throughput has increased over the years, higher viable cell densities have increased oxygen demand. To support this higher uptake, agitation and aeration rates have also increased, leading to more foam generation. This results in a myriad of problems, including cell entrapment within foam, cell damage caused by bubbles bursting at the surface, interruption of gas transfer across the air/liquid interface, and equipment failure due to clogged vent filters. To optimize bioreactor conditions, Biogen has been working on different methods of monitoring, removing, and controlling foam in bioreactors to dramatically reduce antifoam addition and prevent the regeneration of foam. While focusing on extrinsic methods of monitoring and removing foam through automation, the intrinsic characteristics of foam generation and stability are also being studied to find alternative methods of eliminating the foam in the cell culture process. New methods for monitoring and controlling the foam in bioreactors include using a digital camera to image the culture surface within the vessel. Antifoam is then programmed to be fed automatically based on the percentage of foam coverage. Studies of the delivery of antifoam have also shown that spraying antifoam as a mist reduces the required dose to remove the foam layer, while not significantly interrupting the mass transfer. It not only reduces the foam regeneration rate, but also protects cells from shear damage by allowing for a lower gas entrance velocity from the sparger. The foam generated during cell culture manufacturing processes has been characterized as stable Pickering foam, with slower foam decay. The foam stability was correlated to the accumulation of biomass within the culture. This stability reduces the efficacy of the traditional antifoam remediation strategy over time. Therefore, alternative methods of foam removal are now being investigated to improve the cell culture processes. Overall, the systematic ways of controlling foam have notably reduced the likelihood of equipment failure and improved bioreactor success rates.

Copyright © American Institute of Chemical Engineers. All rights reserved.

460729 Morphological and Life Cycle Changes Exhibited at the up-Scaling of a *Saccharomyces Cerevisiae* High-Cell Density Fed-Batch Biopharmaceutical Process

Abstract:

In a previous study (Fu et al, 2014), we observed a clear difference in yeast cell volumes (determined by wet cell weight (WCW)) grown at 10L laboratory scale versus 10,000L manufacturing scale. That study also included metabolite analysis which suggested hypoxia during scale up. We hypothesized that hypoxia weakens the yeast cell wall during the scale up, leading to changes in cell permeability, and/or cell mechanical resistance.Â Cells treated with the zymolyase showed a more sensitive cell wall later in the production process at manufacturing scale, although cell viability from samples obtained at both scales were similar. In the current study we further examine the relationship between relative volumes of yeast cells to cell viability, size, morphology, and life cycle at laboratory scale (10L) and manufacturing scale (10,000L). Vi-cell and flow-cam methodologies demonstrate a larger sub-population of cells with aspect ratio 0.5 and â‰¥6 micron-diameter at 10,000L scale, corresponding to more one- and two-budded mothers in the larger scale process. In addition, results from a cell cycle study using flow cytometry and DAPI fluorescence microscopy demonstrated multi-budded cells having increased chromosomal content with a long residence in the M phase without completing cytokinesis. In summary, these results collectively suggest a possible relationship between the level of cell wall compromise with cell volume and the presence of a major sub-population of larger budding cells reflecting a longer residence in M phase.

Copyright © American Institute of Chemical Engineers. All rights reserved.

456526 Accelerating Manufacturing in a Clinical Drug Substance Plant through Application of Plant Simulation

Abstract:

Given the major investment required to build or expand biopharmaceutical manufacturing facilities, intensification of existing assets is a highly valuable proposition. However, drug substance processes have many complex operational constraints. Increasing the run rate of a facility should only be undertaken when a thorough evaluation of all operational constraints has been performed.

In this case study, plant simulation was applied to an existing clinical drug substance manufacturing facility executing a wide variety of processes to evaluate opportunities to increase facility run rate. The modeling effort identified an opportunity to increase capacity by 40% with minimal capital investment. At current time, with partial implementation of the identified changes, a 20% increase in run rate has been realized.

Copyright © American Institute of Chemical Engineers. All rights reserved.

461029 Model Predictive Control: A Process Development Tool for Mammalian Cell Culture Processes

Abstract:

In recent years, there has been a growing necessity to increase the speed, efficacy and information content associated to the development and scale-up of biopharmaceutical processes, particularly given the regulatory authorities current shift in philosophy towards quality-by-design (QbD). This QbD requirement for increased process understanding intensifies the work involved in the earlier phases of the drug-product lifecycle. The benefit, however, is that the increased process knowledge can speed the technical transfer from development into manufacturing, deliver a more optimised, robust process with higher titres and greater reproducibility and aid in troubleshooting and root-cause analysis of deviations during production.

A model predictive control (MPC) technology platform incorporating Raman spectroscopy was developed as a tool to streamline upstream bioprocess development and applied to a fed-batch CHO culture. In this talk, the core elements of the MPC will be discussed, namely the development of the Raman PAT method for the online quantification of cell density, substrate and metabolite concentrations, the process model and the MPC control algorithm.

The application of the MPC to process development will be demonstrated through a case study of its application to a CHO fed-batch culture. The MPC enabled the transition from traditional bolus to a PAT-enabled continuous feeding strategy in order to prevent nutrient depletion and deliver a stable macro-environment for the cells. Overall, application of the glucose set-point control strategy resulted in a 1.5 fold increase in peak viable cell density (VCD) and the integral of the viable cell density (IVC) which is directly related to increased titre.

Copyright © American Institute of Chemical Engineers. All rights reserved.

466474 Accurate Determination of B_{22} Data â€“ Multivariate Analysis to Improve B_{22} As a Predictive Tool

Abstract:

ABSTRACT

Protein-protein molecular interactions are known to be involved in protein solution aggregation behaviour and are a common issue for the manufacturing of therapeutic proteins such as mAbs. Much effort has been employed to gain a better understanding of aggregation, however the mechanisms leading to protein aggregation are still not fully understood. The osmotic second virial coefficient (B_{22}) is a fundamental physiochemical property that describes protein-protein interactions in solution, which can be a useful tool for not only predicting the aggregation propensity of proteins but also protein solubility, crystallisation and phase behaviour.

A rapid and high-throughput way of determining B_{22} is self-interaction chromatography (SIC). However, the method is still critiqued to be less accurate than its predecessors such as static light scattering. This paper is evaluating the impact of the different experimental strategies and theoretical estimations on the parameters involved in the second virial coefficient determined by SIC. This has been done to help focus on the small variations and combinations that have the largest impact on B_{22} and can directly determine a critically different value.

A number of journal papers recently published describe important experimental considerations for more accurate B_{22} estimations. Many of these small differences in the experimental procedure could result in wrongly estimated B_{22}

values and the interpretation that the protein-protein interactions are repulsive when they are actually attractive and an instable condition for the protein. To easily identify what initial parameters that have the greatest impact an extensive multivariate and sensitivity analysis has been performed to report the impact of a single or multiple parameters on B_{22}. The values that were found to be the most accurate representations of B_{22} were compared to aggregation data to identity that B_{22} was a useful and powerful predictor of aggregation propensity.

Apart from proving to be an accurate predictor for stability in pre-formulation and formulation studies, this tool has also shown to be a good indicator for suitable conditions that can be used for purification and manufacturing of mAbs to enable better product stability while not compromising recovery and quality.

In summary, this work is providing a good insight into important considerations when determining protein-protein interactions and virial coefficients to obtain accurate values that can be used as a powerful tool for biopharmaceutical development.

REFERENCES

Hedberg, S. H. M., Heng, J. Y. Y., Williams, D. R. & Liddell, J. M. (2015). Self-Interaction Chromatography of MAbs: Accurate Measurement of Dead Volumes. *Pharmaceutical Research,* 32, 3975-3985.

Quigley, A., Heng, J., Liddell, J. & Williams, D. (2013). The Accurate Measurement of Second Virial Coefficients Using Self Interaction Chromatography: Experimental Considerations. *European Journal of Pharmaceutics and Biopharmaceutics,* 85, 1103-1111.

Copyright © American Institute of Chemical Engineers. All rights reserved.

458703 Multivariate Data Analysis (MVDA) for Cell Culture Process Development

Abstract:

Multivariate Data Analysis (MVDA) for Cell Culture Process Development

John Bowers, Maria Khouzam, Louis Obando, John Higgings, and Balrina Gupta.

Analyzing mammalian cell culture results has traditionally been done after batch completion by creating and reviewing univariate plots of each variable separately. This approach results in a large amount of data that is not easily interpreted. MVDA Multivariate Data Analysis (MVDA) resolves these because it analyzes parameters and their covariance of large data sets to summarize the performance of the process. Real-time MVDA allows for online process monitoring, and provides immediate understanding.

To apply MVDA during cell culture development, several hurdles were overcome. The number and variety of bioreactors generated our first challenge. Experiments were performed in systems ranging from an automated set of twenty four 250 mL mini-bioreactors to a manufacturing skid with two 500 L stainless steel bioreactors. Since batch context was not available in a consistent format across the bioreactor controllers, special software was developed to enter it into PI. Furthermore each reactor had its own configuration, so a universal configuration structure was develop and the actual configuration of each bioreactor entered into it.

In addition, continuous process data were collected into the data archiving system while off-line results were collected into a separate electronic notebook (ELN) system, making it harder to use combined results to holistically analyze new batches. To combine offline measurements with continuous bioreactor data, SIMCA-Online was configured to extract the daily offline measurements from E-Notebook BioAssayÂ® (PerkinElmer) database. Batch information and the continuous data were taken from a PI data historian. A â€œgrab and holdâ€ of the last entered value was sufficiently representative of daily offline measurements to account for the sampling rate difference between the daily and continuous measurements in the model. Models based on these combined modes significantly better described culture behavior, and were much better predictors of batch quality measures, such as titer or charge variants.

Furthermore, modeling approaches were developed to account for scale changes and intentional parameter adjustments. Â Most parameters used were intrinsic, such cell and metabolite concentrations, pH, DO, temperature etc. To account for scale effects in useful extrinsic parameters, such as gas flows or feed amounts, a volume based scaling was introduced. With these, we were able to use results from automated 250 mL mini-bioreactor system Ambr® 250 (Sartorius Stedim Biotech) to build models for 3-L and larger scale bioreactors.

In order to account for intentional changes in parameters, such as pH or temperature settings, variables were tracked as deviations from setpoints. The deviation from setpoint allowed for diagnostics of equipment performance, without the multivariate statistics being dominated by an intentional setpoint change. We applied this approach to process ranging studies to determine the impact of changing a parameter on overall process performance live, giving immediate feedback on what a parameter change impacted. When needed, the setpoints were then included in the models to quantify the magnitude of the change to performance or quality attributes.

We successfully applied MVDA using SIMCA-13 and SIMCA-Online® (Umetrics) to several development activities, such as assessing the impact of the parameter changes on the process performance, confirming similarity between master and working cell banks, and demonstrating process comparability across scales. Applying the techniques described here, as much information as possible was obtained from development activities, even in the context of continuing process changes. Six to eight initial similar batches can be used to create models for future work. These models, in turn, can be used to improve process understanding to guide future work. The new results can then be included in models to further improve understanding and model performance.

Copyright © American Institute of Chemical Engineers. All rights reserved.

467886 Flux Balance Analysis and Crispr/Cas9-Facilitated Genetic Modification for the Increased Production of Beta-Carotene in Recombinant *Saccharomyces Cerevisiae*

Abstract:

Carotenoids are an important group of molecules found in many biological systems. They are a precursor to vitamin A and are used in vitamins, supplements, and food dyes. Carotenoids are produced naturally in plants in low quantities, and chemical synthesis is not a viable option, so overproducing this molecule in organisms such as yeast is the most economical choice in an increasing global market. Previous work has been done to insert genes from yeast species *Xanthophyllomyces dendrorhous* into *Saccharomyces cerevisiae* and to determine that *S. cerevisiae* is an appropriate host to produce a particular carotenoid, beta-carotene[1]. This *S. cerevisiae* strain has been further developed using directed evolution and an increase in beta-carotene production has been seen[2]. This work will focus on two metabolic engineering strategies to continue to increase the production of beta-carotene and improve understanding of the yeast strain and the production process.

First, the substrate utilization and beta-carotene production through the use of a metabolic flux model of *S. cerevisiae* is examined[3]. This model is modified to include the additional beta-carotene-producing reactions as well as experimentally determined substrate utilization. Flux variance analysis is used to further refine the model to accurately depict the glucose and ethanol metabolism in the yeast strain. The results from this model, along with previously obtained experimental bioreactor data, will be used to determine the system state that exhibits the highest production of beta-carotene. In addition, this refined intracellular model is combined with a previously developed extracellular kinetic model leading to dynamic predictions of all intracellular metabolite fluxes.

A second strategy looks to improve beta-carotene productivity through the targeted deletion of cellular pathways.

Utilizing the OptKnock functionality of the COBRA Matlab toolbox in conjunction with the modified metabolic flux model, it is possible to identify individual genes and groups of genes whose absence could lead to higher beta-carotene productivity[4]. Potential knockout genes targeted by OptKnock include several genes involved in the conversion of pyruvate and acetate, key intermediates in both the growth and beta-carotene production pathways. In order to confirm these results experimentally, several of these genes are disrupted using a CRISPR/Cas-9 method[5]. The effects of these genetic disruptions, particularly regarding cellular growth and beta-carotene productivity will be emphasized.

[1]Verwaal et al., *App. and Env. Microbiology*, **73**, 4342-4350 (2007)

[2]Reyes et al., *Metabolic Engineering*, **21**, 26-33 (2014)

[3]Herrgård et al, *Nat Biotechnol.*, **26**, 1155-1160 (2008)

[4]Burgard et al., *Biotech. and Bioeng.*, **84**, 647-657 (2003)

[5]Bao et al., *ACS Synthetic Biology*, **4**, 585-594 (2015)

Copyright © American Institute of Chemical Engineers. All rights reserved.

465087 Biobutanol Production By *C. Tyrobutyricum* through Proteomics Guided Carbon and Redox Rebalance

Abstract:

With similar energy intensity, biobutanol is an environmentally friendly substitute for fossil fuel. *Clostridium tyrobutyricum* is an excellent host cell that can be engineered to produce butanol because of its high butanol tolerance and conversion yield from various rate to butyryl-CoA. Our comparative proteomics study revealed the limiting steps in the central carbohydrate pathway of *C. tyrobutyricum*. In this study, the rational host cell engineering was performed to redistribute the metabolic flux of carbon and redox to produce butanol. The heterologous NAD^+-dependent formate dehydrogenase (*fdh*) was synthesized to intracelluarly generate extra reducing power. The homologous thiolase (*thl*) was up-regulated to direct more flux from C2 to C4. A new mutant ACKKO-*thl-fdh-adhE2* was constructed and characterized using free-cell fermentation in a 2-L bioreactor. It was found that the butanol titer was improved from 15 g/L to 19.5 g/L using the new mutant. The proteomics was used to further investigate how the cell engineering changed the expression of enzymes involved in the butanol formation. This study showed that the proteomics-guided rebalance of carbon and redox is an effective strategy to improve butanol production.

Copyright © American Institute of Chemical Engineers. All rights reserved.

Optimization of Production Medium for High Cell Mass Production and the Functional Characteristics Study of *Bifidobacterium Longum* WICC B34

Roslinda Abd Malek[1], Muhammad Khairuddin Malek[1], Nor Zalina Othman[1], Solleh Ramli[1], Ramlan Aziz[1] and Hesham A. El Enshasy[1,2,3]

[1]Institute of Bioproduct Development, Universiti Teknologi Malaysia, Skudai, Johor, Malaysia, [2]Department of Chemistry of Natural and Microbial Products, National Research Cener, Dokki, Giza, Egypt. [3] Bioprocess Development Department, City of Scientific Research and Technology Application, New Burg Al Arab, Alexandria, Egypt.

Introduction

Microorganism of the genus *Bifidobacterium* constituting one of the predominant microorganisms in the colon during the early stages of life (Harmsen *et al.,* 2000; Lay *et al.,* 2005), at a time when the immune system is not fully developed and the observed antagonistic has stimulated a great deal of interest in the role of *Bifidobacteria* in human health. Increasing health concerns and knowledge of intestinal microflora has led to the interest in using probiotic bacteria especially from genus *Bifidobacterium* as supplements in various types of foods products. *Bifidobacteria* considered as a probiotic as they contribute in desirable changes and produces protective effect in intestinal tract (Lay *et al.,* 2005). *Bifidobacteria* plays a role in inhibition growth of pathogenic bacteria by producing organic acids as it will reduce the pH of the colon (Ballongue, 1998). The ability of adherence to colon mucosa also helps to prevent the pathogen adherence as well as colon cancer induction (Ballongue, 1998). Apart from that, *Bifidobacteria* also important in the prevention and treatment of diarrhoea by participating in competitive exclusion against diarrheagenic bacteria (Bernet *et al.,* 1993). A lot of media have been proposed for the detection of these bacteria however none can give complete selective effect on the media proposed (Rada and Petr, 2002). Media for *B. longum* which are simple and cost-effective have been established, but the influence of each medium composition and the interaction between each of media component still unknown. Growth of *B. longum* influenced much by the composition of the medium used such as carbon sources, nitrogen sources, micronutrients, and macronutrients. Each medium composition plays a major part in terms of cost especially in large scale production process. Screening method was carried out before getting an optimized medium composition. During screening, each composition with best concentration will be select for further optimization process. Thus, in this study, the effect of each media component on the growth of *B. longum* was evaluated by using one factor at a time (OFAT) approach. This approach is important to accurately predict the influence of each component to get optimum cell mass production at shake flask level (Hun *et al.*,2013, Malek *et al.*, 2012, Elmarzugi *et al.,* 2010, Othman *et al.*, 2009, Azaola *et al.,* 1999,). Despite of medium optimization study, there are some others important factors such as acidity, bile, hydrophobicity, antimicrobial sensitivity, and antibiotics susceptibility which can affect the probiotic cells to remain and exert their potential functionalities in a host. The tolerance of probiotic bacteria toward acid and bile salt was determined by the study growth of the cells in extreme environment. Thus, a study on the growth of *B. longum* towards acidity and bile environment was carried out, followed by antimicrobial test with various pathogenic bacteria, antibiotics susceptibility, and hydrophobicity test with different types of solvent.

Materials and methods

Culture Medium and Inoculum Preparation

The probiotic strain that was used in this study is *B. longum* WICC-B34 obtained from the Wellness Industries Culture Collection (WICC), Institute of Bioproduct Development (IBD), Universiti Teknologi Malaysia. *B. longum* was stored as frozen working culture at -80 °C in deManRagosa Sharpe (MRS) supplemented with 1g L^{-1} L-cysteine HCL-H_2O, and 13% v/v glycerol. One ml stock cultures were inoculating with deManRagosa Sharpe (MRS)+ L-cysteine and incubated in 20 ml cap test tube for 48h, at 37 °C. One ml cell suspensions were inoculated with 50 ml fresh deManRagosa Sharpe (MRS) and incubated in screw capped shake flasks for 48 h and 37 °C.

Screening of Different Types of Cultivation Media in Shake Flask Level.

Nine different types of media were used in this study for primary evaluation for selection of highly productive medium. Fifty millilitre (ml) of each medium types was dispensed into 250 ml screwed cap Erlenmeyer flask and autoclaved at 121 °C for 15 min. The carbon source of each medium type was sterilized separately and added to the medium before inoculation. After cooling, it was inoculated with 48 h culture growing in MRS medium at 10% inoculums. The pH value was adjusted to 6.0 before being autoclaved. The culture was then incubated anaerobically at 37 °C for 48 h with static conditions. All media that were used were reported before for their ability to support cell growth of *Bifidobacteria* sp. The composition of the media used for screening *B. longum* was shown in Table 1.

Table 1: Composition of nine different types of media used for growth screening of *B. longum*.

Medium No.	Composition g L^{-1}/ %	References
1	MRS medium: peptone (10), yeast extract (5), beef extract (10), glucose (20), polysorbate-80 (1), ammonium citrate (2), sodium acetate (5), magnesium sulfate (0.1) manganese sulfate (0.05), dipotassium hydrogen phosphate (2), L-cysteine (0.5), pH 6.0	Her *et al.* (2004)
2	Reconstituted RSM (skimmed milk (12 % w/w)) supplemented with % ; glucose (2), yeast extract (1), and 0.05 L-cysteine-HCL contain 10 % glycerol, pH 6.5-7.0	Kamaly, (1997)
3	Peptone Yeast (PY medium): yeast extract (10), peptone (20), glucose (10) + lactose (5 %), pH 5.5-6.0	Andaloussi *et al.* (1995)
4	MRS broth supplanted with L-cystein-HCL (1), glycerol (13 % v/v), yeast extract (35), and glucose (20), pH 6.2	Kiviharju, (2005)
5	MRS broth supplanted with L-cystein-HCL (1), and glycerol (13 % v/v), yeast extract (35), and raffinose (10).	Garro *et al,* (2006)
6	Lactose (50), beef extract (10), peptone (10), yeast extract (5), sodium acetate (7), ammonium citrate (2), disodium phosphate (2), polysorbate-80 (1), manganese sulfate pentahydrate (0.2), L-cysteine (0.5)	Song *et al.* (2003)
7	TPY medium + glucose (0.5 %): trypticase (BBL) (10), phytone (BBL) (5.0), yeast extract (2.5), polysorbate-80 (1), L-cysteine (0.5) dipotassium hydrogen phosphate (2.0), Magnesium Chloride hexahydrate (0.5), zinc sulfate heptahydrate (0.25), calcium chloride (0.15), Ferum (III) chloride (0.005), pH 6.3	Yamazaki and Dilawari, (1990)
8	Skim milk (28), and yeast extract (22), pH 6.5	Stephanie *et al.* (2007)
9	skim milk (12 %) containing (5 %) lactose or molasses, pH 6.3	Reyed, (2006)

Optimization of medium composition using one-factor-at-a-time (OFAT) method.

Medium that contained suitable composition for the cell growth was further optimized by using one-factor-at-a-time approach. By using this approach, the selected medium was optimized by changing different concentration of media components by varied one component but others were remain constant. Different glucose concentrations (0-100 g L^{-1}) were added individually in the medium. The concentration of peptone added was varied from 1-14 g L^{-1}whereas the concentration of yeast extract was added in ranges of 0-12 g L^{-1}. Furthermore, the concentration of meat extract, sodium acetate and K_2HPO_4 were added varied from 0-14 g L^{-1}, 0-9 g L^{-1}, and 0-5 g L^{-1} respectively.

Bioreactor cultivations

B. longum cultures was cultivated in pilot scale bioreactors (BioEngineering, Wald, Switzerland) of 16-L with working volume of 6 litres. Batch cultivations involved two different strategies which were controlled and uncontrolled pH. Cultivation was conducted at different parameter, agitation speed at 200 rpm, temperature at 37 °C and initial pH at 6.0 by automatic addition of 3M NaOH and 3M HCL based on cultivation strategy. Anaerobic conditions were maintained by sparging the culture with filter sterilized nitrogen gas at 0.3 v/v/m (air volume, per liquid media volume, per minute).

Analytical Methods

Cell Dry Weight and Optical Density Determination

Samples were withdrawn every 4 hours and further analyzed. Cell dry weight was measured by gravimetrical method and optical density by spectrophotometer. Supernatant was stored at -20 °C for subsequent chromatographic analyses (glucose and lactic acid determination). The cells then were dried in the oven at the temperature of 65 °C until constant weight. For Optical Density (OD) method, cell dry weight was determined by using UV-VIS single-beam spectrophotometer (HACH DR/2500, Germany) at wavelength of 600 nm. The OD culture then converted to dry cell mass through pre-determined standard curve between the OD value and cell dry weight.

Lactic Acids Analysis by HPLC

Concentration of organic acids (lactate and acetate) in fermented medium in batch cultures was determined by HPLC (Waters, Milford, MA). One ml of supernatant sample was diluted 1:5 in deionized water and filtered through a 0.45 μm nylon membrane filter. Separation of compounds was carried out using a 300 mm x 7.8 mm ORA (8% cross linked resin hydrogen ionic form) column (Waters, Milford, MA) and a UV detector Waters 966 photodiode array detector at 210 nm were used. The mobile phase was 0.005N H_2SO_4 as eluent at flow rate of 0.5 ml min^{-1} at 55 °C and using a pump waters 600 controller. Compounds were detected using a refractive index detector. Reported data are means of duplicate analysis.

Glucose Determination

Glucose determination was conducted in order to determine the glucose uptake pattern by *B. longum* during the fermentation. The glucose concentration in broth should decrease gradually over the cultivation time. The maximum glucose consumption rate was calculated using the measured concentrations and sampling times. Glucose was analysed from cell free samples by using colorimetric method by DNS reagent and instrument YSI 2700 Select (Yellow Springs Instrument Inc., U.S.A.). The colorimetric method was performed based on the colour developed after the reaction between reducing sugar and DNS (3, 2-Dinitrosalisylic acid).

Survival of B .longum Cells in Bile Salt Tolerant

Preparation of Bile Salt Solutions

The survival of *B. longum* was evaluated at different concentration of bile salts oxgal (Difco, Lab, Detroit, MI, USA).Concentration of 0.5, 1.0, 2.0, 3.0, and 4.0 % of bile salts were prepared in 250 mL Erlenmeyer flasks containing MRS medium (Aslim and Alp, 2009).

Survival of B. longum in Simulated Gastric Juice (SIG)

Preparation of Simulated Gastric Juice at Different pH

The simulated gastric juice was prepared according to the method of Kim *et al.* (2007). Artificial gastric juice was prepared by suspending 3.5 g D-glucose, 2.05 g NaCl, 0.6 g $KHPO_4$, 0.11 g $CaCl_2$, 0.37 g KCl, 0.05 g ox gall bile (Merck) and 13.3 g of pepsin (was added after sterilization) in 1 L distilled water. The pH value of artificial gastric juice prepared was adjusted to different pH value at 1.0, 2.0, 3.0, and 4.0 by using 1 M HCl. The prepared solution was sterilized by autoclaving at 121°C for 15 minutes.

The Viability of B. longum Cells Towards Acidic pH (1.0-4.0) Within Certain Period of Time

Artificial gastric juice solutions with different pH concentration (1.0-4.0) that was previously prepared were transferred into sterile universal bottle containing 0.1 mL of concentrated cells. The survival rate of the cells after 2 hours was determined by incubating the mixture at time intervals of 0, 30, 60, 90, and 120 minutes. The mixture was pipetted (0.1 mL) into the MRS agar at every time intervals and spread plate technique was carried to grow the cell. The agar plates were incubated for 48 hours in anaerobic jar at 37 °C. Following incubations, colony forming unit (CFU) were counted and recorded. Three replicates were conducted with each concentration of bile salts.

Hydrophobicity Assays of B. longum in Different Solvents

In hydrophobicity assays, *B. longum* cells after 3 days activation in MRS broth at 37 °C was centrifuged at 8000 x g for 10 minutes and were twice washed in phosphate buffer (KH_2PO_4-Na_2HPO_4 50 mM, pH 6.0), suspended in KNO_3(0.1 M, pH 6.2). The optical density (ODA) was determined at 600nm. In this experiment, 3 different types of solvent were used to test the adhesion of the cells which were xylol (apolar solvent), chloroform (acidic solvent), and ethyl acetate (basic solvent). Four ml of the cells suspension were added to 1 ml of each solvents, and after 5 minutes the phase were vortexes for 2 minutes and the optical density (ODB) of the aqueous phase were determined at 600nm after one hour. The hydrophobicity of the cells was measured as follow H % = [(OD of A – OD of B)/OD of A] x 100 (Perez *et al.,* 1998).

Determination of Antibiotic Susceptibility

In the study of antibiotic susceptibility, five different types of antibiotic (erythromycin, rifampicin, tetracycline, streptomycin and gentamycin) in ranges between 5 $\mu g\ ml^{-1}$ to 20 $\mu g\ ml^{-1}$ were used on *B. logum* cells. Each of the antibiotic were carefully weight, dissolved, and diluted in appropriate diluents factor. Evaluation was followed according to the method by Kirk-Bauer scale (Bauer *et al.,* 1966) which showed that most of the cells more susceptible to the antibiotic testing.

Antimicrobial Activity of B. longum Towards Other Pathogenic Species.

Antimicrobial resistance testing was done by using agar disc diffusion method. In this method, cells that grown in MRS broth supplemented with 0.05% (w/v) L-cystein hydrochloride at 37° C for 48 h was used. Pathogenic bacteria culture (0.1 ml) was added to MRS agar and pour into plate to promote the growth of pathogenic bacteria (Domig *et al.,* 2007).

Results and discussion

Optimization of high cell mass of Bifidobacterium longum in the shake flask cultivation

The potential effect of each medium composition towards the production of high cell mass of *B. longum* has been tested in this study. Among different media that was screened (Table 1), the highest mediums that contribute to the highest cell mass were selected and further optimized. The medium selected composed of (g L^{-1}): peptone, 10; yeast extract, 5; beef extract, 10; glucose, 20; polysorbate-80, 1; ammonium citrate, 2; sodium acetate, 5; magnesium sulphate, 0.1; manganese sulphate, 0.05; dipotassium phosphate, 2 and L-cystein, 0.5 (Her *et al.,* 2004). The cultivation *B. longum* in this media yielded the cell mass of 2.235 g L^{-1} in shake flasks study. The growth of *B. longum* was further studied by cultivation in the different types of carbon sources to determine which one give the best cell mass production. Utilization of different carbon sources by *B. longum* was determined by cultivating the probiotic cells in media with 5 different carbon sources as follows; glucose, mannitol, glycerol, lactose and sucrose in shake flasks culture. Utilization of glucose in shake flask gives the maximum cell mass of 2.7 g L^{-1} with the production of lactate and acetate was 2.4 g L^{-1} and 1.28 g L^{-1} respectively. Thus, in this study glucose which gave higher cell mass was selected to be used for the medium optimization. In this study, the effect of different concentrations of glucose ranging from 0 to 100 g L^{-1} on the growth of *B. longum* was further investigated. The initial cell mass without addition of glucose was at 0.64 g L^{-1} and increasing until it reached maximal cell mass of 4.66 g L^{-1} at glucose concentration of 70 g L^{-1}. The best concentration of glucose for the highest cell mass was at 70 g L^{-1} with reasonable acetate and lactate production about 1.6 g L^{-1} and 7.55 g L^{-1} respectively. Growth kinetics comparison study between un-optimized and optimized classical was done in shake flasks level. *B. longum* was cultivated in a media that contains original media composition and the media composition obtained from classical optimization method. Cultivation period in this experiment was 96 hours and samples were taken every 4 hours for analysis. The medium composition of these two media is shown in Table 2. Finally, the medium that has been optimized classically (Table 2) gave the cell mass of 6.03 g L^{-1} when compared growth in un-optimized media gave only 2.58 g L^{-1} after 72 hours cultivation in the shake flask level. The cell mass of *B. longum* in optimized classical media (Table 2) showed an increase approximately 22.4 % over the cell mass obtained in un-optimized media.

Table 2: Summarizes the growth kinetics of *B. longum* cultivated in shake flasks culture by using un-optimized, and optimized OFAT media.

Parameters	**Shake flasks**	
	Un-optimized (g L^{-1})	**Classical Optimized (g L^{-1})**
Maximum Biomass, X_{max}(g L^{-1})	2.58	6.03
Specific growth rate,μ(h^{-1})	0.0231	0.0338
Biomass productivity, P_X ($gL^{-1}h^{-1}$)	0.036	0.083
Doubling time,d_t(h)	30.00	20.51
Medium components:	g/L	g/L
Glucose	20	70
Yeast Extract	5	5
Sodium acetate	5	7
Peptone	10	12
Meat extract	10	14
K_2HPO_4	2	3
Ammonium citrate	2	2
Polysorbate-80	1	1
Magnesium sulfate	0.1	0.1
Manganese sulfate	0.05	0.05
L-cystein	0.5	0.5

Batch cultivation to increase high cell mass of B. lungum in 16-L stirred tank Bioreactor

The cell growth, production of lactate and pH changes during cultivations of *B. longum* using optimized media was studied in 16-L stirred tank bioreactor under controlled (a) and uncontrolled (b) batch fermentation, and was illustrated in Figure 1. Bioreactor batch fermentation with 6-L working volume was carried out with the cultivation of 5 % (v/v) inoculum size. Experiments were performed in anaerobic conditions by sparging the cultures with nitrogen gas at the rate of 0.3 volumes per volume per minutes (v/v/m). As shown in the Figure 1(b) (uncontrolled pH), the cells grew exponentially for 72 hours until the cells grow constantly and reach the maximum cell mass at 7.5 g L^{-1} at the growth rate of 0.127 h^{-1}. Comparing with the cell mass obtained in the un-optimized shake flasks media, the value of cell mass yield shows an increment of approximately 189 % higher during cultivations in uncontrolled batch bioreactor. The pH of the production medium decreased from 6.0 until 4.43 during the growth of the cells due to the increase production of lactate from 0.4 g L^{-1} and achieved the maximum at 4.38 g L^{-1}. High production of lactate concomitantly with cell growth in the culture will lyse the cells as it acts as microbial inhibitor which later decrease the cell mass.

Contrary with the cultivation in pH controlled bioreactor, the cell mass produced was higher and higher growth rate when compare to the uncontrolled pH bioreactor. Figure 1(a) shows that at the controlled pH at 6.0, the maximum cell mass yield was 11.97 g L^{-1} at 52 hour with the growth rate of 0.245 h^{-1}. The cells then constantly grow until 60 hour before it starts to drop to 10.91 g L^{-1} at 64 hours. The value was 363 % higher than the cell mass obtained in un-optimized shake flasks media. The results showed that the growth of *B. longum* was greatly influenced by the pH controlled. Previous bioreactor cultivation resulted low cell mass with higher production of lactate when the pH was left uncontrolled. The growth performance of *B. longum* was dramatically improved when the culture pH was constantly controlled at 6.0. The concentration lactate was lower and biomass increasing when the cultivation medium was controlled to pH 6.0. The production of lactate in controlled pH bioreactor gave the value approximately between 0.69 g L^{-1} and 2.61 g L^{-1} as the environment culture pH was controlled by the addition of 2 M sodium hydroxide (NaOH).

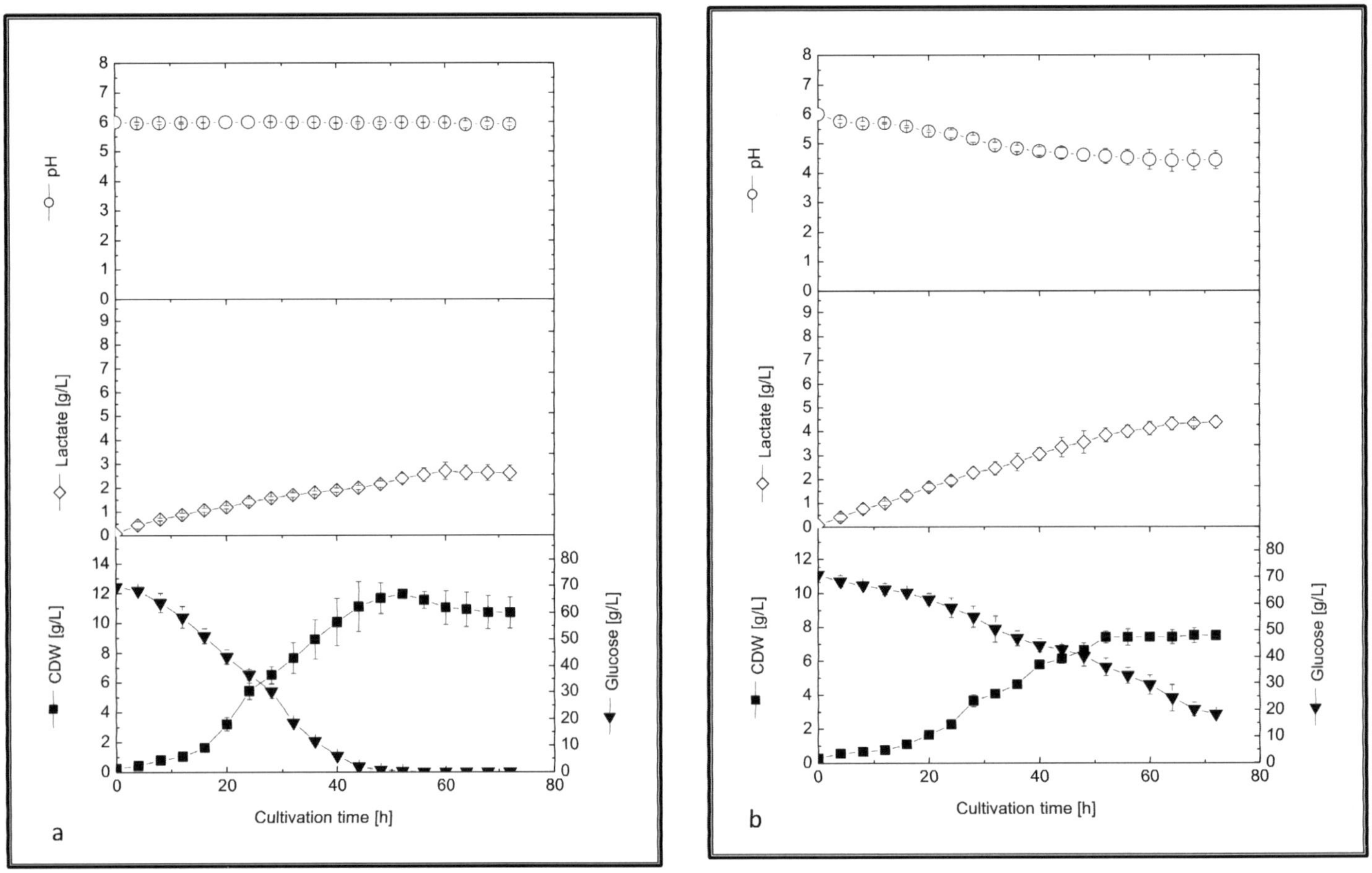

Figure 1: Kinetics of cell growth, glucose consumption, lactate and pH changes in batch cultivation of *B. longum* in 16-L stirred tank bioreactor at 37°C for 72 hours with pH control (a) and Un-controlled pH (b).

Functional Characteristics Study of Bifidobacterium Longum WICC B34

The cells harvested were then proceeding for to functional characteristics test. In this study, *B. longum* cells were tested for its survival under different ranges of low pH value and different concentration of bile salt in order to be good probiotics. The high percentage of survival was observed in bile salt concentration ranges of 0.05% to 0.3% which give almost 80 % survivals of the cells (Table 3). It can be observed that, *B. longum* exhibit higher tolerance at pH 3.0 and 4.0 and can withstand lower pH value in the certain limit The percentage of survival of the cells at pH 1, 2, 3, and 4 were recorded at 20.6 % (in 60 min), 18.8 % (in 120 min), 39.7 % (in 120 min), and 41.7 % (in 120 min) respectively (Table 4). The ability of the cells to adhere was also tested by carried out hydrophobicity test. Lowest percentage of cell adhesion can be seen when exposed to Ethyl acetate (44.12 %), and cells exposed to chloroform gave the highest result of 86.1 %. The cells were then tested for antibiotics and antimicrobial resistance. *B. longum* cells were resistant to streptomycin and susceptible to other antibiotics. In antimicrobial study, *B. longum* cells showed resistance effect towards the entire pathogenic microorganism in this study, *Bacillus sp, Staphylococcus Aureus, Salmonella enterica, Candida albicans, E. coli, Aspergillus brasiliensis, Pseudomonas aeroginosa* and *Staphylococcus epidermidis.*

Table 3: Survival of *B. longum* in MRS at varied of bile salt concentration at 37° C, as determined by viable counts

MRS + varied of bile salt concentration	**Survival of cells after expose in different concentration of bile salt solution at different times (log cfu ml^{-1})**					**Percentage of inhibition (%)**	**Percentage of survival (%)**
	0 min	30 min	60 min	90 min	120 min		
MRS + 0.0 5%	9.826	9.612	8.88	8.66	8.59	14.39 %	85.61 %
MRS + 0.1 %	9.973	9.71	8.79	8.64	8.51	17.19 %	82.81 %
MRS + 0.2 %	9.591	9.279	8.5	8.43	8.04	19.29 %	80.71 %
MRS + 0.3 %	8.94	8.73	7.87	7.75	7.40	20.81 %	79.82 %

% of inhibition = cfu ml^{-1} initial - cfu ml^{-1} final / cfu ml^{-1} final x 100

Table 4: Viability of *B. longum* in synthetic gastric juice in pH ranges between pH 1 to pH4 after 0 min, 30 min, 60 min, and 90 min, in 2 hours' time

MRS + varied pH value incubation at 37 °C	**Survival of cells after expose in different ranges of pH at different times (log cfu ml^{-1})**				
	0 min	30 min	60 min	90 min	120 min
MRS + pH 1	6.30 (100 %)	1.7 (26.9 %)	1.3 (20.6 %)	No growth	No growth
MRS + pH 2	6.38 (100 %)	2.6 (40.7%)	2.4 (37.6 %)	1.9 (29.8 %)	1.2(18.8 %)
MRS + pH 3	6.81 (100 %)	5.13 (75.3 %)	4.34 (63.7 %)	3.41 (50.1 %)	2.71(39.7 %)
MRS + pH 4	6.70 (100 %)	5.21 (77.7 %)	4.42 (65.9 %)	3.51(52.3 %)	2.8(41.7 %)

Acknowledgement

The authors would like to thanks Universiti Teknologi Malaysia (UTM) for Contract research grant no R.J130000.7344.4B200 and R.J130000.7646.4C101.

References

1. Andaloussi, S.A., Talbaoui, H., Marczak, R., and Bonaly, R. (1995). Isolation and Characterization of Exocellular Polysaccharides Produced By *Bifidobacterium longum. Journal of Applied Microbiology and Biotechnology*. 43, 995-1000.
2. Aslim, B. and Alp, G. (2009). Relationship between the Resistance to Bile Salts and Low Ph with Exopolysaccharide (EPS) Production of *Bifidobacterium* sp. Isolated from Infants Faeces and Breast Milk. *Journal of Food Microbiology*. 16, 101-105.
3. Azaola, A., Bustamante, P., Huerta, S., Saucedo, G., Gonz'alez, R., Ramos, C. and Savali, S. (1999). Use of Surface Response Methodology to Describe Biomass Production of *Bifidobacterium Infantis* in Complex Media. *Journal of Biotechnology Techniques.*13,93-95.
4. Ballongue, J. (1998). *Bifidobacteria* and Probiotic Action. In Lactic Acid Bacteria (Salminen, S. & vonWright, A., eds.). Marcel Dekker. New York, U.S.A. 519-587.
5. Bauer, A.W., Kirby, W.M., Sherris, J.C., and Turck, M. (1966). Antibiotic Susceptibility Testing By a Standardized Single Disk Method. *American Journal of Clinical Pathology.* 45 (4), 493-496.
6. Bernet, M.F., Brassart, D., Neeser, J.R. and Servin, A.L. (1993). Adhesion of Human *Bifidobacterial* Strains to Cultured Human Intestinal Epithelial Cells and Inhibition of Enteropathogen-Cell Interactions. *Journal of Applied and Environmental Microbiology.* 59(12), 4121-4128.
7. Canzi, E., Guglielmetti, S., Mora, D., Tamagnini, I., and Parini, C. (2005). Conditions Affecting Cell Surface Properties of Human Intestinal *Bifidobacteria. Antonie van Leeuwenhoek.* 88, 207–219.
8. Corre, C., Madec, M.N. and Boyaval, P. (1992). Production of Concentrated *Bifidobacterium bifidum*. *Journal of Chemical Technology Biotechnology*.53, 189-194.
9. Domig, K. J., Mayrhofer, S., Zitz, U., Mair, C., Petersson, A., Amtmann, E., Mayer, H.K. and Kneifel, W. (2007). Antibiotic Susceptibility Testing of *Bifidobacterium thermophilum* and *Bifidobacterium pseudolongum* Strains: Broth Microdilution vs. Agar Disc Diffusion Assay. *International Journal of Food Microbiology*. 120, 191-195.
10 El Marzugi N., El Enshasy H.A., Abd.Malek R., Othman Z., Sarmidi M.R., Abdul Aziz R. 2010. Optimizati.on of cell mass production of the probiotic strain *Lactobacillus lactis* in batch and fed-batch culture in pilot scale levels. In: Médez-Vilas A, ed. Current Research, Technology and Education Topics in Applied Microbiology and Microbial Biotechnology, Vol. 2, Spain: Formatex Research Centre, Badajoz 873-79
11. Garro, M.S., Aguirre, L. and de, Giori, G.S. (2006). Biological Activity of *Bifidobacterium longum* in Response to Environment pH. *Journal of Applied Microbiology Biotechnology*.70,612-617.
12. Harmsen, H.J.M., Wildeboer-Veloo, A.C.M., Raangs, G.C., Wagendorp, A.A., Klijn, N., and Bindels, J.G. (2000). Analysis of Intestinal Flora Development in Breast-Fed and Formula-Fed Infants by Using Molecular Identification and Detection Methods. *Journal of Pediatric Gastroenterology Nutrition*. 30, 61–7.

13. Hennessy, A.A., Ross, R.P., Devery, R., and Stanton, C. (2009). Optimization of a Reconstituted Skim Milk Based Medium for Enhanced CLA Production by *Bifidobacteria. Journal of Applied Microbiology*. 106 (4), 1315-1327.
14. Her, S.L., Duan, K.J., Sheu, D.C. and Lin, C.T. (2004). A Repeated Batch Process for Cultivation of *Bifidobacterium longum*. *Journal of Industrial Microbiology Biotechnology*. 31, 427-432.
15 Hun, C.H., Mohd Sueb, M.S., Malek, A., Othman, Z., Elsayed, E.A., Ramili, S, Elmarzugi, N.A., Sarmidi, M.R.., Aziz R., El Enshasy, H.A (2013). Bioprocess development for high cell mass production of the probiotic yeast *Kluyveromyceslactis*. IOSR J Pharceut Biol Sci, ; 8: 49-59.
16. Hopkins, M.J., Cummings, J.H. and Macfarlane, G.T. (1998). Inter-Species Differences in Maximum Specific Growth Rates and Cell Yields of *Bifidobacteria* Cultured on Oligosaccharides and Other Simple Carbohydrate Sources. *Journal of Applied Microbiology*.85, 381-386.
17. Kamaly, K.M. (1997). *Bifidobacteria* Fermentation of Soybean Milk. *International Journal of Food Research*. 30 (9), 675-682.
18. Kim, P., Jung, M. Y., Chang, Y. H., Kim, S., Kim, S. J. and Park, Y. H. (2007). Probiotic Properties of *Lactobacillus* and *Bifidobacterium* Strains Isolated From Porcine Gastrointestinal Tract. *Journal of Applied Microbial and Cell Physiology*. 7, 1103-1111.
19. Kiviharju, K. (2005). Optimization of *Bifidobacterium longum* Production Process. *Journal of Biotechnology*. 117, 299-308.
20. Lay, C., Rigottier-Gois, L., Holmstrom, K., Rajilic, M., Vaughan, E.E., de Vos, W.M., Collins, M.D., Thiel, R., Namsolleck, P., Blaut, M. and Dore, J.(2005). Colonic MicrobiotaSignatures across Five Northern European Countries. *Journal of Applied and Environmental Microbiology*. 71, 4153–4155.
21. Makras, L., and Vuyst, L.D. (2006). The in vitro Inhibition of Gram-Negative Pathogenic Bacteria by *Bifidobacteria* is caused by the Production of Organic Acids. *International Dairy Journal*. 16 (9), 1049-1057.
22 Malek, R.A., Hamdan, S., El Enshasy, H., Othman, N.Z., Zainol, N.A., Sarmidi, M.R., Aziz, R.. (2010). Production of *Lactobacillus salivarius*, a new probiotic strain isolated from human breast milk, in semi-industrial scale and studies on its functional characterization. In: Médez-Vilas A, ed. Current Research, Technology and Education Topics in Applied Microbiology and Microbial Biotechnology, Vol. 2, Spain: Formatex Research Centre, Badajoz1196-1204
23 Othman, N.Z, El Enshasy, H.A., Abdul Malek, R., Sarmidi, M.R., Aziz, R.A (2009). Kinetics of cell growth and functional characterization of probiotic strains *Lactobacillus delbrueckii and Lactobacillus paracasei* isolated from breast milk. Deutche Lebensmittel Rundschau, 105: 444-450
24. Perez, P.F., Minnaard, Y., Disalvo, E.A., and Antoni, G.L. (1998). Surface Properties of *Bifidobacteria* Strains of Human Origin. *Journal of Applied and Environmental Microbiology*. 64 (1), 21-26.
25. Perrin, S., Warchol, M., Grill J.P., and Schneider, F. (2001). Fermentations of Fructo-Oligosaccharides and their Components by *Bifidobacterium infantis* ATCC 15697 on Batch Culture in Semi-Synthetic Medium. *Journal of Applied Microbiology*. 90, 859-865.
26. Rada, V. and Petr, J. (2002). Enumeration of *Bifidobacteria* in Animal Intestinal Samples. *Journal of Vet. Med.-Czech,* 47.(1), 1-4.
27. Reyed, R., and Diwany, A. (2006). Molasses as Bifidus Promoter on *Bifidobacteria* and Lactic Acid Bacteria Growing in Skim Milk. *The Journal of Microbiology*. 5 (1).

28. Rodríguez, E., Arqués, J.L., Rodríguez, R., Peirotén, A., Landete, J.M., and Medina, M. (2012). Antimicrobial Properties of Probiotic Strains Isolated from Breast-Fed Infants. *Journal of Functional Food*. 4 (2), 542-551.
29. Scardovi, V. (1986). *Bifidobacterium* in Bergey's Manual of Systematic Bacteriology. 9th Ed. Volume 2, P.H. Sneath, N.S. Mair, M.E. Sharpe and J.G. Holt (Eds.) Williams and Wilkins Publishers, Baltimore, MD. 1418.
30. Shene, C., Mardones, M., Zamora, P. and Bravo, S. (2005). Kinetics of *Bifidobacterium longum* ATCC 15707 Fermentations:Effect of the Dilution Rate and Carbon Source. *Journal of Applied Microbiology Biotechnology*. 67, 623-630.
31. Song, S.H., Kim, T.B., Oh, H., and Oh, D.K. (2003). Optimization of *Bifidobacterium longum* Growth by Use of Calcium Carbonate-Alginate Beads. *Journal of Microbiology and Biotechnology*. 19, 727–731.
32. Stephanie, B.S., Mercenier, A., and Rochat, F. (2007). Use of *Bifidobacterium longum* for the Prevention and Treatment of Inflammation. *U.S. Patent No. 279/472*. Washington D.C.
33. Yamazaki, H., and Dilawri, N. (1990). Measurement of Growth of *Bifidobacteria* on Inulofructosaccharides. *Journal of Applied Microbiology*. 10 (6), 229-232.

455871 Combined Effects of Dissolved Oxygen and Alcohol Dehydrogenase Inhibitor for Enhancing Lactate Production in *Rhizopus Oryzae*

Abstract:

Lactate fermentation by *Rhizopus oryzae* usually suffers from ethanol production as a result of limited local oxygen concentration during cultivation of filamentous fungi. High oxygen supply was introduced during fermentation to maintain sufficiently high dissolved oxygen (DO) level in broth; however, ethanol was still observed. Our previous work showed metabolic fluxes shifted toward lactate production with lower ethanol formation in the fermentation containing enzyme inhibitors including 1,2-diazole, 2,2,2-trifluoroethanol, 4-methylpyrazole, and 3-hydroxypyruvate.

In this study, we further investigated the dynamic responses of the living *R. oryzae* cultivated in the static bed bioreactor to the effects of enzyme inhibitors and DO levels together. Either increasing DO level or adding inhibitor alone limited ethanol and promoted lactate production while cell biomass remained unchanged. Lactate yield was increased by 24% with the increasing DO from 40% to 80% during the fermentation. Among 4 inhibitors studied, 2,2,2-trifluoroethanol gave the maximum improvement in fermentation efficiency. Approximately, 32% increasing lactate yield was obtained in the fermentation at 40% DO with 0.1 mM 2,2,2-trifluoroethanol. Together, inhibitor and high DO level better controlled metabolic flux toward lactate production. Dramatic increasing lactate yield of 54% was obtained in the fermentation containing 2,2,2-trifluoroethanol at 80% DO control with 100% improved productivity. Findings from this work exhibited the simple but effective way to improve lactate fermentation by *R. oryzae*.

Copyright © American Institute of Chemical Engineers. All rights reserved.

467452 Important Role of Class I Heat Shock Genes on the Butyrate Production and the Response to Low pH Condition By Clostridium Tyrobutyricum ATCC 25755

Abstract:

Class I heat shock genes (HSGs) code for molecular chaperones which play a major role in the bacterial response to sudden increases of environmental stress by assisting protein folding. The objective of this study was to identify the expression profile of HSGs of *Clostridium tyrobutyricum* under butyric acid stress, and explore the effect of overexpressing these genes on butyric acid production. The qRT-PCR analysis shows that five Class I HSPs genes of the *C. tyrobutyricum* significantly up-regulated in response to butyric acid or butyrate. At a butyric acid concentration of 6g/L, the expressions of *grpE, DnaK, DnaJ, groESL, htpG* increased 2.3, 1.8, 0.7, 2.7, 2.6, 1.0- fold. Meanwhile, the expressions of those genes increased 6.2, 8.6, 4.7, 8.9, 9.2, 2.6-fold at a butyrate concentration of 10g/L, respectively. Overexpression of *DnaK*, *groESL*, *htpG* significantly improved the tolerance to butyric acid. At a butyric acid concentration of 10g/L, the *groESL*-overexpressing strains grew faster, and can achieve higher cell density (OD_{600} 2.0 vs OD_{600} 1.4 for the wild type). Under butyrate stress, the *groESL*- and *htpG*-overexpressing strains had higher inhibition rate constant (*Kp*=3.52g/L, *Kp*=2.56 vs *Kp*=1.75g/L for the wild type), suggesting that both of them have higher tolerance to butyrate. The free-cell fermentation results show that *groESL*-overexpressing strains stopped producing butyric acid at a butyrate concentration of 44.25±2.60 g/L, which was higher compared to the wild type (34.52±1.92 g/L). The results showed that metabolic engineering to improve the butyric acid tolerance and production has a potential application prospect.

Copyright © American Institute of Chemical Engineers. All rights reserved.

469930 Production of Propionic Acid from Soy Molasses By High-Cell Density Fermentation with *Propionibacterium Acidipropionici*

Abstract:

The three-carbon fatty acid propionic acid has wide industrial applications such as food and feed preservatives, perfumes and artificial flavorings. Its current industrial production is mainly through petrochemical processes, which are not sustainable and result in environmental concerns. In contrast, propionic acid production through fermentation is sustainable and can fulfill consumer demands for natural product. Although the best propionic acid producer *Propionibacterium acidipropionici* can produce propionic acid with high yield, purity and productivity, further production cost reduction is required to make fermentation process economically competitive against petrochemical process. In this study, low-cost soy molasses, a byproduct from soybean refinery, was used as the sole carbon source to produce propionic acid by *P. acidipropionici* in batch and fed-batch fermentations. After simple pretreatments to remove undesirable inhibitors, soy molasses, which contained sucrose, stachyose, raffinose, glucose, xylose, galactose, fructose and lactic acid, could be readily fermented by *P. acidipropionici*to produce propionic acid at a yield of >0.4 g/g sugar consumed, confirming that soy molasses is a promising feedstock for economical production of propionic acid.

Copyright © American Institute of Chemical Engineers. All rights reserved.

457386 Integrated Biocatalytic Process for Trehalose Production and Separation from Maltose

Abstract:

To develop cost-effective, bio-based, industrial trehalose production from maltose, an integrated bioprocess for the trehalose production and separation from maltose with recombinant trehalose synthase in permeabilized cells is proposed in this study. We have successfully established an efficient production system for recombinant trehalose synthase (TreS) in a 50 L fermentor and suitable processes for separation and purification to achieve comprehensive utilization of the raw material and products. Maltose syrup (30 wt %) was considered as the feedstock and was hydrolyzed into trehalose in one step using an efficient TreS (9,234 U/mL) by whole-cell catalysis, and a trehalose conversion rate of 75% at 30 °C can be reached in the bioreactor system. Industrial process optimization analysis was also employed to improve the technology. Trehalose separation and purification processes based on simulated moving bed chromatography (SMB) were performed, and the product was crystallized with 97.6 ± 0.02% purity and 95.9% recovery yield.

Copyright © American Institute of Chemical Engineers. All rights reserved.

Screening for Ligninolytic Enzyme Production by White Rot Fungi from Malaysia Using Different Medium

Hanapi, S.Z[1]., Hatti-Kaul, R[3]., Aziz, R[1].,Othman, N.Z[1]., El-Enshasy, H.A[1,2]

1. Institute of Bioproduct Development (IBD, Universiti Teknologi Malaysia, (UTM), 81310 UTM Skudai, Johor, Malaysia;
2. Bioprocess Development Departments, City for Scientific Research (CSAT), New Burg Al Arab, Alexandria, Egypt.
3. Department of Biotechnology, Lund University, Chemical Centre, P. O. Box 124, SE-221 00 Lund, Sweden.

Introduction

In soil biosphere, white rot fungi was devoted as the best groups of organisms in degrading complex lignin polymers with their key components of extracellular ligninolytic enzymes (Jebapriya and Gnanadoss, 2013). The evidence for white rot fungus is borne out by researches that show its ability in many bioremediation processes (Andrade et al., 2013). The growth patterns of fungi are closely related to environmental condition with the association with the changes in mycelium morphology in response to oxygen (Rahardjo et al., 2005), carbon dioxide (Bowen et al., 2007; McIntyre and McNeil, 1997) and nutrient concentrations in the medium (Harris, 2008). This study therefore was set out to identify white rot fungi associated in local virgin soil and assess its best extracellular ligninolytic enzymes production using different medium.

Material and Methods

Chemicals and Culture

All the chemicals used in the study were in high purity and analytical grade. Kraft-lignin, Guaicol, ABTS (2,3'-azino-bis(3-ethylbenzothiazoline-sulfonic acid), Poly R-478 and Veratryl alcohol were purchased from Sigma. Reactive Phanerochaete chrysosporium (DSMZ 6909) was obtained from Leibnitz Institute DSMZ-German Collection of Microorganisms and Cell Cultures, Braunschweig, Germany and used as a control.

Maintenance of Fungus and Preparation of Spore Inoculum

Fungi are grown on supplemented agar as described by Tien and Kirk (1988) and incubated for 7 days at 30 °C. Five discs of mycelium were transferred to a liquid growth medium (50 ml, pH 4.5) and shaken at 150 rpm at 30°C for 7 days. Mycelium obtained from each flask was blended in three cycles of 5s (Warring Blender) and used as initial inoculum for each experiment carried out (Pointing et al., 2000).

Assessment of Ligninolytic Enzymes production in Different Medium

Isolation and Identification

Forest cover and protected area system in Johore, south peninsular Malaysia were chosen as the sampling locations for the isolation of white rot fungi from soil. Isolation method was carried accordingly (Ishaq and Khan, 2011) and identified by macro-morphology and micro-morphology. Characteristics of ligninolytic producer were identified using PMM medium with guaiacol, ABTS and Poly R-478 (Kiiskinen *et al.*, 2004) and lignin medium as a substrate (Renugadevi *et al.*, 2011). Molecular identification of fungal isolates was conducted by First BASE Laboratories Sdn Bhd using primer pJET1.2/blunt (Accession No. EF694056.1). Fungi were identified by searching sequences obtained using Blast2 in GenBank through the National Center for Biotechnology Information (NCBI) as described by Brown *et al.* (2015). The Boot-strap Test of Phylogeny was performed by using software MEGA version 6 (Tamura *et al.,* 2013). The accession numbers for all sequences derived from the GenBank database are included in the phylogenetic trees.

Enzymes Assay

Lignin peroxidase (LiP) activity was assayed according to Kheiralla *et al.* (2013). Mn-dependant peroxidase (MnP) was measured by oxidation of 2,6-dimethoxyphenol (Terrazas-Siles *et al.*, 2005) and Laccase (Lac) activity as described by Kumar *et al.* (2011). All calculations of enzymes were according to Baltierra-Trejo et al. (2015).

Medium Screening

The triplicate of 250 ml conical flasks each containing 50 ml of growth media; Medium 1, MEA (Kheiralla *et al.,* 2013), Medium 2 (Tien and Kirk, 1988), Medium 3 (Marco-Urrea *et al.*, 2009), Medium 4 (Yang and Yu, 1996), Medium 5 (Dhouib et al., 2005), Medium 6 (Zheng *et al.*, 1999) and medium 7 (Sharari *et al.*, 2011) were selected. Each medium is inoculated with 5% of each fungal inoculum and incubated at 30 °C in a shaking incubator of 150 rpm for 7 days.

Result and Discussions

A total of 77 soil samples were collected from 11 substrata comprising of 118 white rot fungi isolates and only two were selected further. Analysis from rDNA sequences illustrated that isolate WICC F38 is almost identical to that of *Phanerochaete chrysosporium* (Figure 1). Analysis of 18s region (accession no: GU190187.1 and GU190189.1) strengthen this conclusions, thus, it was termed *Phanerochaete* sp. WICC F38.The second most potent strains, WICC F39 then was found identical with *Cerrena unicolor* (accession no: JN546140.1) then was estimated as *Cerrena* sp. WICC F39.

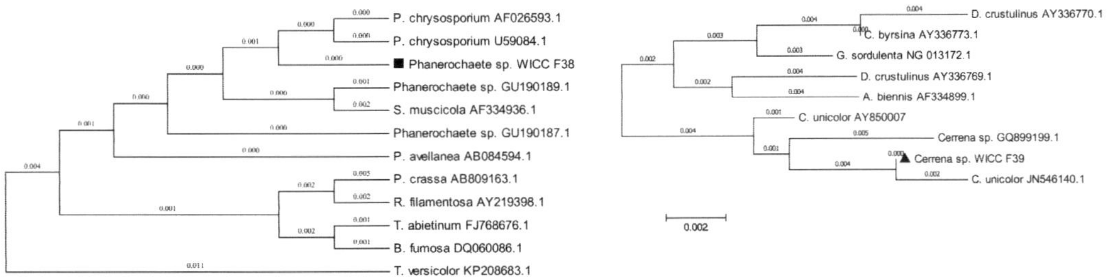

Figure 1: Molecular Phylogenetic analysis by Maximum Likelihood method of white rot fungi on virgin soil inferred from the pJET1.2/blunt (Accession No. EF694056.1), in which the *Trametes versicolor* (KP208683.1) (a) was used as outgroup.

In general, productions of multiple ligninases were observed in high titres for *Cerrena* sp. WICC F39 in most of the medium tested especially laccase. This strain touched highest yields in medium 4 with activity of laccase at 56.34±6.4 U L^{-1} (Figure 2) at pH 5.4, respectively. The reason for this is not clear but it may be explained by the fact that laccases are usually the first ligninolytic enzymes secreted to the surrounding media by the fungus (Maciel and Ribeiro, 2010). Laccase differs from LiP and MnP in that it does not require H_2O_2 during oxidation of substrates (Yaropolov et al., 1994). In other words, the laccases from fungi can, in particular, to degrade lignin in the absence of peroxidases (Eggert et al., 1997). The secretion of lignin peroxidase vary in pace for both P*hanerochaete* sp. for each medium. Dekker et al. (2001) stated that the secretion of both enzymes strongly depends on the addition of veratryl alcohol as substrates. This is in agreement for data depicted from *P. chrysosporium* DSMZ 6909. Based on the results, lignin peroxidase and yield derived from medium 5 were 56.34±6.4 U L^{-1} and 19.42 g, which the only medium using glycerol as a sole carbon source. However, no effect has been observed for *Phanerochaete* sp. WICC F39. According to Cancel *et al.* (1993), the presence of veratryl alcohol in a medium more likely acts as a stabilizer for lignin peroxidase but not the enhancer for their synthesis. It was proven here in medium 1 which lignin peroxidase was not influenced directly with the addition of veratryl alcohol. In case for manganese peroxidase, Maciel and Ribeiro (2010) stated that manganese peroxidase is usually expressed in almost all white-rot fungi and more widely spread than the lignin peroxidase. However, this finding cannot be extrapolated to both *Phanerochaete* sp. in this study. While this is the case, results for *Cerrena* sp. WICC F39 are in line with those reported. This study also supports the claim that *Phanerochaete* sp. is lack in laccase. Preliminary studies by Kirk and Farrell (1987) reported that laccase was not involved in degradation of lignin due to lack activities of laccase observed in *P. chrysosporium*. This study was also confirmed by Hatakka (1994b) and Novotný et al. (1997). Hatakka (1994a) also suggested that white rot fungi that appear not to produce laccase may not be essential for lignin decay. However, a report by Srinivasan *et al.* (1995) contracted the findings. According to Larrondo *et al.* (2003), there are no conventional laccase were found in genome search for *P. chrysosporium*. Therefore, suggested that *P. chrysosporium* does not have the capacity to produce laccases.

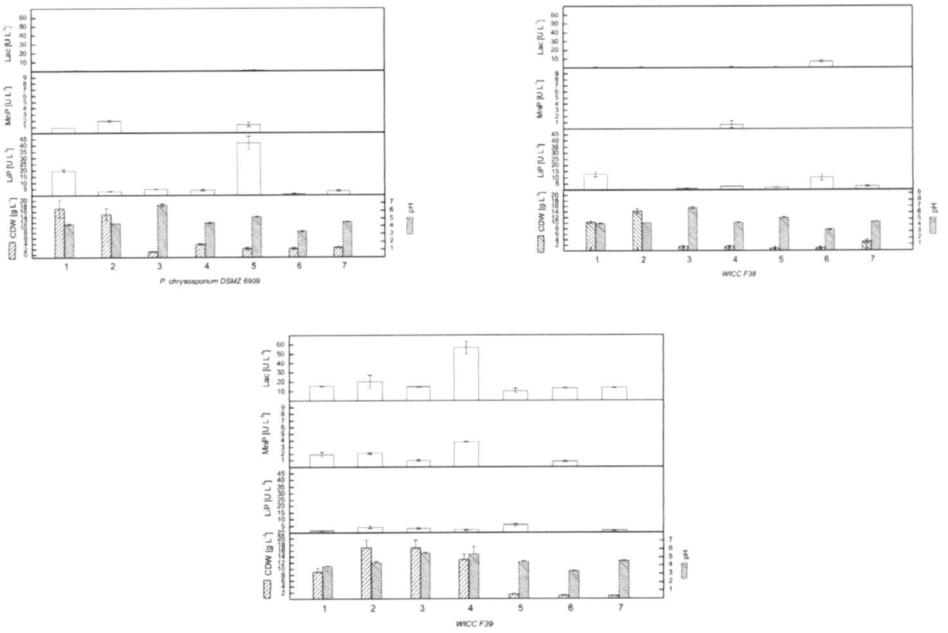

Figure 2: Production of ligninolytic enzymes from *P. chrysosporium* DSMZ 6909, WICC F38 and WICC F39 isolates subjected to different medium compositions

Conclusion

High activities of ligninases especially laccase was obtained for *Cerrena* sp. WICC F39 compared with other isolate, *Phanerochaete* sp. WICC F38. However, the use of different medium for each strain clearly effects the production of each ligninolytic enzymes. In this study, while the ability to secrete three lignases was shown for both isolates, though the activities of some of the enzymes were low; however, this does not fully show their capability. Results from this comparison growth medium apparently showed the sensitivity of ligninase enzymes to culture compositions. Furthermore, despite to these promising results, questions remain. Therefore, to develop full picture of ligninolytic enzymes production by these strains, additional studies in characterization and optimization of medium are needed.

Acknowledgement

The authors would like to thanks the MOHE and Universiti Teknologi Malaysia (UTM) for contract research Grant no R.J130000.7646.4C101.

References

1. Andrade, M. V. F., Silva, K. M. L. D., Siqueira, J. P. d. S., Wanderley, C. R. P., Araújo, R. d. S., Marinho, G., and Rodrigues, K. (2013). Azo dye degradation by Phanerochaete chrysosporium in the medium enriched with nitrogen in the presence of primary co-substrate. Brazilian Archives of Biology and Technology 56: 867-874.
2. Baltierra-Trejo, E., Márquez-Benavides, L., and Sánchez-Yáñez, J. M. (2015). Inconsistencies and ambiguities in calculating enzyme activity: The case of laccase. Journal of Microbiological Methods 119: 126-131. doi: http://dx.doi.org/10.1016/j.mimet.2015.10.007.
3. Bowen, A. D., Davidson, F. A., Keatch, R., and Gadd, G. M. (2007). Effect of nutrient availability on hyphal maturation and topographical sensing in Aspergillus niger. Mycoscience 48: 145-151. doi: 10.1007/s10267-007-0352-x
4. Brown, G. R., Hem, V., Katz, K. S., Ovetsky, M., Wallin, C., Ermolaeva, O., Tolstoy, I., Tatusova, T., Pruitt, K. D., and Maglott, D. R. (2015). Gene: a gene-centered information resource at NCBI. Nucleic acids research 43: D36-D42.
5. Cancel, A. M., Orth, A. B., and Tien, M. (1993). Lignin and veratryl alcohol are not inducers of the ligninolytic system of Phanerochaete chrysosporium. Appl Environ Microbiol 59: 2909–2913.
6. Dekker, R. H., Vasconcelos, A.-F., Barbosa, A., Giese, E., and Paccola-Meirelles, L. (2001). A new role for veratryl alcohol: regulation of synthesis of lignocellulose-degrading enzymes in the ligninolytic ascomyceteous fungus, Botryosphaeria sp.; influence of carbon source. Biotechnology Letters 23: 1987-1993. doi: 10.1023/A:1013742527550.
7. Dhouib, A., Hamza, M., Zouari, H., Mechichi, T., Hmidi, R., Labat, M., Martinez, M., and Sayadi, S. (2005). Screening for Ligninolytic Enzyme Production by Diverse Fungi from Tunisia. World Journal of Microbiology and Biotechnology 21: 1415-1423. doi: 10.1007/s11274-005-5774-z.
8. Eggert, C., Temp, U., and Eriksson, K.-E. L. (1997). Laccase is essential for lignin degradation by the white-rot fungus Pycnoporus cinnabarinus. FEBS Letters 407: 89-92.
9. Harris, S. D. (2008). Branching of fungal hyphae: regulation, mechanisms and comparison with other branching systems. Mycologia 100: 823-832.
10. Hatakka, A. (1994a). Lignin-modifying enzymes fungi: production and role. FEMS Microbiology Reviews 13: 125-135.
11. Hatakka, A. (1994b). Lignin-modifying enzymes from selected white-rot fungi: production and role from in lignin degradation. FEMS Microbiology Reviews 13: 125-135. doi: http://dx.doi.org/
12. Ishaq, F., and Khan, A. (2011). Isolation, Identification and Comparative Study of Fungal and Bacterial Strains Found in Organic and Inorganic Soils of Different Agricultural Fields. Recent Research in Science and Technology 3: 30-36.
13. Jebapriya, G. R., and Gnanadoss, J. J. (2013). Bioremediation of textile dye using white rot fungi: A review. International Journal of Current Research and Review 5: 1-13.
14. Kheiralla, Z. H., El-Din, S. M. B., Malek, S. M. A., and Aziz, D. H. A. (2013). Optimization of cultural conditions for lignin peroxidase production by Phanerochaete chrysosporium and Pleurotus ostreatus. . Academia Journal of Biotechnology 1: 87-95.
15. Kiiskinen, L. L., Ratto, M., and Kruus, K. (2004). Screening for novel laccase-producing microbes. J Appl Microbiol 97: 640-646. doi: 10.1111/j.1365-2672.2004.02348.x
16. Kirk, T. K., and Farrell, R. L. (1987). Enzymatic" combustion": the microbial degradation of lignin. Annual Reviews in Microbiology 41: 465-501.

17. Kumar, V. V., Kirupha, S. D., Periyaraman, P., and Sivanesan, S. (2011). Screening and induction in laccase activity in fungal species and its applications in dye decolorization. Afr. J. Microbial. Res 5: 1261-1267.
18. Larrondo, L. F., Salas, L., Melo, F., Vicuña, R., and Cullen, D. (2003). A novel extracellular multicopper oxidase from Phanerochaete chrysosporium with ferroxidase activity. Applied and Environmental Microbiology 69: 6257-6263.
19. Maciel, M. J. M., and Ribeiro, H. C. T. (2010). Industrial and biotechnological applications of ligninolytic enzymes of the basidiomycota: A review. Electronic Journal of Biotechnology 13: 14-15.
20. Marco-Urrea, E., Pérez-Trujillo, M., Vicent, T., and Caminal, G. (2009). Ability of white-rot fungi to remove selected pharmaceuticals and identification of degradation products of ibuprofen by Trametes versicolor. Chemosphere 74: 765-772. doi: http://dx.doi.org/10.1016/j.chemosphere.2008.10.040
21. McIntyre, M., and McNeil, B. (1997). Dissolved carbon dioxide effects on morphology, growth, and citrate production in Aspergillus niger A60. Enzyme and Microbial Technology 20: 135-142. doi: http://dx.doi.org/10.1016/S0141-0229(96)00108-1
22. NCBI http://www.ncbi.nlm.nih.gov/. In http://www.ncbi.nlm.nih.gov/.
23. Novotný, Č., Vyas, B., Erbanova, P., Kubatova, A., and Šašek, V. (1997). Removal of PCBs by various white rot fungi in liquid cultures. Folia Microbiologica 42: 136-140.
24. Pointing, S., Jones, E., and Vrijmoed, L. (2000). Optimization of laccase production by Pycnoporus sanguineus in submerged liquid culture. Mycologia: 139-144.
25. Rahardjo, Y. S. P., Sie, S., Weber, F. J., Tramper, J., and Rinzema, A. (2005). Effect of low oxygen concentrations on growth and α-amylase production of Aspergillus oryzae in model solid-state fermentation systems. Biomolecular Engineering 21: 163-172. doi: http://dx.doi.org/10.1016/j.bioeng.2005.01.001
26. Renugadevi, R., Ayyappadas, M. P., Preethy , P. H., and Savetha, S. (2011). Isolation, screening and induction of mutation in strain for extra cellular lignin peroxidase producing bacteria from soil and its partial purification. Journal of Research in Biology 4: 312-318.
27. Sharari, M., Jahan Latibari, A., Guillet, A., Aurousseau, M., Mouhamadou, B., Rafeiee, G., Mirshokraei, A., and Parsapaghouh, D. (2011). Application of the white rot fungus Phanerochaete chrysosporium in biotreatment of bagasse effluent. Biodegradation 22: 421-430. doi: 10.1007/s10532-010-9415-3
28. Srinivasan, C., Dsouza, T., Boominathan, K., and Reddy, C. (1995). Demonstration of Laccase in the White rot Basidiomycete Phanerochaete chrysosporium BKM-F1767. Applied and Environmental Microbiology 61: 4274-4277.
29. Tamura, K., Stecher, G., Peterson, D., Filipski, A., and Kumar, S. (2013). MEGA6: molecular evolutionary genetics analysis version 6.0. Molecular biology and evolution 30: 2725-2729.
30. Terrazas-Siles, E., Alvarez, T., Guieysse, B., and Mattiasson, B. (2005). Isolation and characterization of a white rot fungus Bjerkandera sp. strain capable of oxidizing phenanthrene. Biotechnol Lett 27: 845-851. doi: 10.1007/s10529-005-6242-4
31. Tien, M., and Kirk, T. K. (1988). Lignin peroxidase of Phanerochaete chrysosporium. In Methods in enzymology, ed. Scott T. Kellogg Willis A. Wood, 238-249: Academic Press.
32. Yang, F.-C., and Yu, J.-T. (1996). Development of a bioreactor system using an immobilized white rot fungus for decolorization. Bioprocess Engineering 15: 307-310. doi: 10.1007/BF02426439
33. Yaropolov, A., Skorobogat'Ko, O., Vartanov, S., and Varfolomeyev, S. (1994). Laccase. Applied Biochemistry and Biotechnology 49: 257-280.

34. Zheng, Z., Levin, R. E., Pinkham, J. L., and Shetty, K. (1999). Decolorization of polymeric dyes by a novel Penicillium isolate. Process Biochemistry 34: 31-37. doi: http://dx.doi.org/10.1016/S0032-9592(98)00061-2.

451591 Engineering the Shell Proteins of a Bacterial Microcompartment to Control Small Molecule Transport

Abstract:

Many non-native reactions in bacterial hosts have unacceptably low yields due to pathway toxicity or competition with other cellular processes. Sequestration of pathways within a diffusion barrier is a potential solution to these problems. Compartmentalization can prevent escape of volatile or toxic intermediates, prevent off-pathway reactions, and create private cofactor pools. Bacterial microcompartments are a class of proteinaceous organelles comprising a characteristic protein shell enclosing a set of enzymes. To enhance the function of heterologous pathways by encapsulation within synthetic microcompartments, we must understand how to control diffusion in and out of the microcompartment organelle. To this end, we explored how small differences in shell protein structure result in changes in the diffusion of metabolites through the shell by comparing the ethanolamine utilization (Eut) and the1,2-propanediol utilization (Pdu) microcompartments. We found that the Eut protein EutM properly incorporates into the Pdu microcompartment to form chimeric MCPs. This substitution alters native metabolite accumulation and the resulting growth on 1,2-propanediol. Further, we identified a single pore-lining residue mutation that confers the same phenotype as substitution of the full EutM protein, indicating that small molecule diffusion through the shell is the cause of growth enhancement. Finally, we show that the hydropathy index and charge of pore amino acids are important indicators to predict how pore mutations will affect growth on 1,2-propanediol, likely by controlling diffusion of one or more metabolites. Toggling pore properties will allow metabolic engineers to selectively control small molecule diffusion for the optimization of heterologous pathways.

Copyright © American Institute of Chemical Engineers. All rights reserved.

465723 Modifying Histatin 5 for Enhanced Resistance to *Candida albicans* Secreted Aspartic Proteases

Abstract:

Histatin 5 is a highly cationic antimicrobial peptide found in human saliva and it has potential as an antifungal therapeutic to treat *Candida albicans* infections. *C. albicans* is a commensal organism that is part of the normal flora of oral cavity, but it is also an opportunistic fungal pathogen that can cause infections, especially in immunocompromised patients. Although histatin 5 has strong and specific antifungal activity against *C. albicans*, the pathogen produces a family of ten secreted and cell-wall bound enzymes called secreted aspartic proteases (Saps), some of which can degrade histatin 5. Several of the Saps have been previously shown to inactivate histatin 5 by cutting the peptide at either side of a positively charged lysine residue. To improve the therapeutic potential of histatin 5 and better understand its interaction with various Saps and *C. albicans* cells, we have designed and tested variants of histatin 5 that have each of the lysine residues substituted with either an arginine or a leucine. All variants were found to have antifungal activity similar to the parent histatin 5, which shows that the peptide is tolerant of arginine and leucine modifications without affecting its fungicidal capabilities. On the other hand, several of the peptide variants showed a significant difference in susceptibility to proteolysis by purified Sap2 and Sap9. Replacing Lys17 with either arginine or leucine led to almost complete resistance to proteolysis by both Saps, while only 45% and 10% of the unmodified histatin 5 remained intact after exposure to Sap2 and Sap9, respectively. Interestingly, substitution at Lys13 to either leucine or arginine made the peptide more prone to proteolysis by Sap9, with Lys13Leu resulting in complete degradation by Sap9. Similar results were obtained after incubation of histatin 5 and the modified peptides with *C. albicans* cells. Both of the Lys17 substitutions showed less degradation while Lys13Leu mutation showed more degradation compared to histatin 5. These results show that peptide modifications could provide insight into the interaction between histatin 5 and the Saps and aid in designing more robust histatin 5 analogs with enhanced therapeutic potential.

Copyright © American Institute of Chemical Engineers. All rights reserved.

470417 Engineering New Allosterically - Regulated Protein Interactions for Applications in Biotechnology

Abstract:

In nature, there are many proteins and peptides that exhibit allosteric regulation. These natively disordered proteins gain structure when they interact with other proteins or small molecules, such as ions. Using protein-engineering tools, we have investigated one of such peptide: the β-roll. This peptide is isolated from the Block V repeats-in-toxin (RTX) domain of Adenylate Cyclase from Bordetella Pertussis. It is disordered in the absence of calcium and it folds into a β-roll secondary structure composed of two parallel β-sheet faces upon binding to Ca^{++} ions. This way, the peptide can transition between its unfolded state and the β-roll structure in a reversible way. We have utilized the allosteric regulation of this domain as a tool to engineer new protein-protein interactions. In its folded state, the peptide has two faces, serving as binding surfaces available for interaction with other proteins. Each beta strand in the β-roll face has two residues with solvent exposed side chains. Our work involves alteration of the residues, which form these faces upon calcium binding, via combinatorial protein design techniques.

We evaluated the potential of this peptide as a cross-linker for hydrogel formation. We showed that calcium dependent hydrogel formation can be achieved by designing the two faces of the folded beta helix to contain leucine residues. When there is no calcium, β-roll domains remain unstructured, delocalizing the leucine rich patches. After calcium binding, the β-roll folds and the leucine rich faces are exposed creating a hydrophobic driving force for self-assembly[1]. We created various constructs where we used the leucine rich β-roll mutant as a hydrogel building block. In addition, we showed that this peptide can serve as a new scaffold for biomolecular recognition. We randomized the amino acid residues on one of the beta sheet forming faces and selected mutants, which demonstrated an affinity for various targets, including lysozyme. Isothermal titration calorimetry (ITC) was utilized to quantitatively determine the thermodynamic parameters of interactions in solution between the mutants and their target. In the case of the lysozyme library, we achieved sub-micromolar / mid-nanomolar (0.52μM - 63nM) dissociation constants. We also

performed affinity chromatography experiments and showed that the best mutant is capable of capturing its target, in the presence of calcium[2]. The reversibility of the calcium binding allows for the engineered molecular recognition to be controllable. The captured target is easily eluted upon removal of the calcium ions. This will be of great interest in the bio-separation area as well as other areas of biotechnology, where binding events can be easily and reversibly controlled.

References:

1. K. Dooley, B. Bulutoglu and S. Banta, *Doubling the cross-linking interface of a rationally-designed beta roll peptide for calcium-dependent proteinaceous hydrogel formation.* Biomacromolecules, 2014, 15(10): 3617-3624.
2. B. Bulutoglu, K. Dooley, M. Blenner and S. Banta, *Catch & Release: Fishing for target proteins with an evolved β-roll peptide exhibiting allosteric regulation*, submitted.

Copyright © American Institute of Chemical Engineers. All rights reserved.

472551 Enhancing Protein Solubility By Rationally Programming Co-Translational Folding Using Synonymous Codons

Abstract:

Biological production of active proteins requires soluble expression. However, few techniques exist to improve protein solubility, largely because solubility optimization is believed to be protein-dependent, difficult to predict computationally, and more challenging for more complex proteins. As it relates to complex multidomain proteins, it has been hypothesized that strategic translational pauses between domains and between distinct individual structural motifs can prevent interactions between nascent chain fragments that generate kinetically trapped misfolded peptides and thereby enhance protein yields. In this work, we have explored the introduction of synthetic transient pauses between structural domains in a heterologous model protein based on computationally designed patterns of synonymous codons that affect the affinity between the mRNA and the anti-Shineâ€"Dalgarno (aSD) sequence on the ribosome. Although synonymous codon sequences do not affect the amino acid identity of a protein, they have demonstrated profound effects on protein function. Specifically, we have demonstrated that optimizing translation attenuation at domain boundaries can predictably affect solubility patterns in bacteria; in these studies attenuation is based on computationally designed increased affinity (i.e. binding energies) between the ribosomal anti-Shine-Dalgarno sequence and the messenger RNA. Exploration of the affinity space showed that modifying less than 1% of the nucleotides in short interdomain amino acid linkers can vary soluble protein yields up to â^¼7-fold without altering the primary sequence of the protein. Given our findings that interdomain ribosomal affinity can affect protein solubility in a predictable manner, we have also explored computational design of protein solubility by exploiting ribosomal affinities in combination with other parameters that include mRNA structure, and tRNA abundance, also known to influence nascent protein fate. Based on a proteinâ€™s secondary structure, we hypothesized key regions where transient pausing can provide the greatest enhancement in solubility. Highly structured, high affinity, and/or low tRNA abundance synonymous mutations were implemented in these regions. Conversely, relatively unstructured, low affinity, and high abundance codons were used in regions where enhanced translation would not lead to detrimental interactions between nascent chain fragments. In this talk our results on the rational design of key factors that modulate translation timing as a novel protein engineering strategy will be discussed.

Copyright © American Institute of Chemical Engineers. All rights reserved.

473766 Development of Thermostable Affinity Reagents for Low-Cost, Point-of-Care Diagnostics

Abstract:

The timely diagnosis and treatment of disease in resource-constrained settings requires the development of robust point-of-care (POC) diagnostics, which provide accurate results and can be employed by users with minimal medical training and limited access to basic infrastructure. One of the most common POC diagnostic formats is the immunochromatographic rapid diagnostic test (RDT), which traditionally uses nitrocellulose-immobilized IgG antibodies as affinity reagents for the capture of disease-relevant biomarkers from patient samples. However, these antibodies are prone to thermal denaturation and non-specific binding events that can render test results unreliable, and in infrastructure-poor contexts where continuous cold chain storage may be infeasible, this silent test invalidation can occur with regularity. This serves to diminish device performance, resulting in adverse patient outcomes and eroding patient trust in the diagnostic format.

We here report the development of thermostable affinity reagents based on a DNA-binding protein isolated from the hyperthermophilic bacteriaÂ *S. solfataricus*. This scaffold was used to develop a model binding protein, which has been assessed for its surface-bound activity, its utility within the dilute antigen regime, and its activity retention under thermal challenge. The demonstrated kinetic stability of this species is nearly three orders of magnitude greater than that of standard polyclonal antibodies, and the species can be facilely produced in high molar yields via bacterial expression and purified in a single chromatographic step. Using yeast surface display and flow cytometric analysis, rationally-designed mutagenesis libraries have also been screened for potential binders to four TB biomarkers identified in the urine of infected patientsÂ via mass spectrometry. These studies have confirmed the scaffoldâ€™s potential applicability to a broad suite of disease biomarkers, and have validated the suitability of this scaffold as a replacement for antibodies in *in vitro* point-of-care diagnostic devices.

Copyright © American Institute of Chemical Engineers. All rights reserved.

472606 Redesign of *E. coli*Â Water Channel Protein, Ompf, Using Iterative Protein Redesign and Optimization (IPRO) Suite

Abstract:

The water footprint for food production, processing and energy production is increasing everyday with the world population and freshwater is becoming an increasingly limited resource. Ultra-permeable membranes are rapidly emerging as a widely implemented technology to utilize and efficiently recycle marginal water sources. Existing membranes are mainly dense solution or diffusion-based or channel- based. Aquaporins (AQPs) are the most popular biological channel-based membranes due to their short pore lengths and high water permeabilities of billion molecules per second. Another particular member of the â€œporinâ€ class of beta-barrel proteins is Outer Membrane Protein F (OmpF). Unlike AQPs, OmpF is exceptionally stable with the ability to survive contact with solvents such as 100% ethanol. While the stability and mutation tolerance of OmpF makes it a suitable candidate for computational design and subsequent experimental validation for performing separations at the angstrom scale, it can also be easily assembled into stable block-copolymer membrane sheets. To this end, we employed the Iterative Protein Redesign & Optimization (IPRO) suite (1) of programs to perturb the OmpF protein backbone, mutate and recursively repack the sidechains in order to obtain the selective internal structure of AQP inside the stable beta-scaffold of OmpF.

We started with the wild-type *E.coli* OmpF (elliptical pore constriction with major and minor axes of ~1.1 nm and 0.7 nm respectively). Pore sizes between 0.3 nm and 0.5 nm are key to several industrial and environmental separations such as methane/CO_2 and salt/water, which is accessible to the less stable AQPs. We first performed ensemble analysis on molecular dynamics (MD) simulations of water permeation through *E. coli* AQP1 using *k*-means clustering technique. Next, we fixed the relative coordinates of water molecules in a water-wire from each ensemble and put them inside the OmpF pore. Subsequently, we identified the OmpF pore-constricting residues and allowed them to be mutated to only hydrophobic residuesto eliminate any interactions between the water-wire and the pore wall. In

addition, we extended the IPRO suite with the ability to select designs based on a geometric criterion in which only designs that have both the pore-constriction axes smaller than 0.35 nm, were accepted. Interestingly, we identified three main types of mutants from our results. The first type comprises mutants where the design positions were only allowed to mutate to tryptophan (longest hydrophobic side chain). This was performed to test out the physical limit up to which the OmpF pore area can be reduced. We have identified a mutant with 25 tryptophan mutations having pore axes of 0.224 nm and 0.18 nm, respectively. The other two types of mutants comprise one with typical hour-glass shaped constriction (pore axes lengths: 0.319 nm and 0.2552 nm respectively) and another which appears as two stacks of ellipses, one arranged on top of the other with the major axes of the lower stack nearly at a right angle to that of the ellipses in the stack above, thereby making the pore very small (axes lengths: 0.236 nm and 0.194 nm respectively). We will perform subsequent experimental validation in establishing a platform for designing precisely tuned membrane transporters. These engineered membrane proteins will serve as important industrial workhorses in energy-efficient water treatment.

1. Pantazes RJ, Grisewood MJ, Li T, Gifford NP, Maranas CD. The Iterative Protein Redesign and Optimization (IPRO) suite of programs. J Comput Chem. 2015;36(4):251-63.

Copyright © American Institute of Chemical Engineers. All rights reserved.

475003 Invited Presentation: Adventures in Rational Protein Design: Goldilocks and the Bears

Abstract:

Nature's principal machinery (proteins) offers great promise toward the future development of useful biological electronic devices, novel therapeutics and conditional enzymatic function. We seek to engineer novel, non-natural proteins of tailored function for high impact applications — *while advancing our fundamental knowledge of protein structure function relationships*. Protein engineering applies the fundamental principles of biophysics and biochemistry toward design, and is best achieved through an integrated (iterative) experimental and computational framework. Our current protein engineering efforts toward enzymes that function under non-physiological conditions includes, temperature-adpated enzymes, protein scaffolds capable of biological charge transfer, proteins that regulate gene expression networks, and proteins that are structurally and functionally resistant to oxidative stress.

Copyright © American Institute of Chemical Engineers. All rights reserved.

466051 Uncertainty Analysis As Essential Step in the Establishment of the Dynamic Design Space of Primary Drying during Freeze-Drying

Abstract:

Since large molecules are considered the key driver for growth of the pharmaceutical industry, the focus is shifting from small molecules to biopharmaceuticals. The list of approved biopharmaceutical drug products from the Food and Drug Administration (FDA) and European Medicines Agency (EMA) consists of approximately 50% freeze-dried products (Constantino and Pikal, 2004). Therefore, freeze-drying is the preferred way to stabilise biopharmaceutical drug products which are unstable in aqueous solution. However, freeze-drying requires a lot of energy and has a long processing time, making it a costly process (Burns, 2009; Gieseler, 2012). The largest fraction of freeze-dried biopharmaceutical products are therapeutic protein formulations and vaccines (Pikal, 2002).

Conventional pharmaceutical freeze-drying (lyophilisation) is a batch process that consists of three consecutive steps: freezing, primary drying (ice sublimation) and secondary drying (desorption of unfrozen water). During the primary drying step the temperature at the sublimation front is critical, and should not exceed a critical value (dependent on the freeze-dried product) to prevent cake collapse. In this project, the focus is on the modelling of the primary drying step. The purpose of the model is to use it to support future optimisation of the process parameters during operation of the freeze-dryer in a dynamic way. During primary drying only two process parameters are adaptable, i.e. the shelf temperature and the chamber pressure. In conventional pharmaceutical manufacturing, both process parameters are fixed, resulting in a conservative cycle. However, by dynamically updating both variables (i.e. resulting in a time variant Design Space), a substantial decrease in processing time can be obtained.

The modelling approach can benefit from incorporating model parameter uncertainty to prevent that the critical temperature at the sublimation front is exceeded. Once the time variant Design Space is created for a certain set-up (i.e. a combination of a specific product and a piece of freeze-drying equipment), the values for both process variables can be fed to the equipment and a significant decrease in processing time can be obtained. As a result, the cost of processing will also be decreasing. Moreover, end product quality is guaranteed.

Burns, L. (2009). The biopharmaceutical sectorâ€™s impact on the US economy: analysis at the national, state and local levels.

Constantino HR, Pikal MJ. Lyophilization of Biopharmaceuticals. Arlington (VA): AAPS Press; 2004.

Gieseler, H. (2012). Insights in lyophilization 2012. Current best practices & research trends. Antwerp, 2012.

Pikal, M. J. (2002). Freeze Drying. Encyclopedia of pharmaceutical technology, 1299-1326.

Copyright © American Institute of Chemical Engineers. All rights reserved.

472886 Tunable Atomization of Viscous Liquids in the Production of Amorphous Solid Dispersions

Abstract:

Atomization of bulk liquids into droplets underlies several critical properties of amorphous dispersions of pharmaceuticals such as dry particle size, morphology, bulk density, physical stability, and dissolution behavior. Because these properties must often be maintained throughout the life cycle of a compound, a comprehensive understanding and control of atomization is required over a range of process scales. Droplet sizes of low viscosity Newtonian fluids are relatively well predicted by several empirical and semi-empirical correlations in the literature. However, viscous and / or non-Newtonian fluids, such as the polymer-rich solutions used to create amorphous solid dispersions, deviate significantly from these historical correlations. This work will present an atomization model for high viscosity liquids that derives its functional form from first principles of fluid dynamics for a pressure-swirl atomizer. The model is subsequently tuned to match experimental data using a small number of empirically determined free parameters. This methodology allows droplet size to be reliably predicted from controllable process parameters such as atomization nozzle geometry, solution flow rate, and liquid properties.

Copyright © American Institute of Chemical Engineers. All rights reserved.

468023 Secondary Drying Scale-up Methodology: Eliminating a Bottleneck with a Lean Development Approach

Abstract:

Secondary drying scale-up methodology: eliminating a bottleneck with a lean development approach

T. Porfirio[1,2*], P. Valente[1], I. Matos[1], J. Moreira[1], J. Vicente[1], M. Temtem[1], V. Semiao[2]

[1] Hovione Farmaciencia SA, Sete Casas, 2674-506 Loures, Portugal; *tporfirio@hovione.com

[2]LAETA, IDMEC, Mechanical Engineering Department, Instituto Superior Tecnico, Universidade de Lisboa, Av. Rovisco Pais 1, 1049-001 Lisbon, Portugal

Owing to the short residence time of the droplets/particles within the spray drying chamber and process constraints to maintain product quality, most spray drying processes include a secondary drying (post-drying stage) to reduce residual solvent content to values below ICH guidelines (International Conference on Harmonization of technical requirements for registration of pharmaceuticals for human use). In spite of the widespread use of secondary drying process the physical mechanisms governing it are not yet fully understood [1, 2]. When using vacuum dryers, this complementary step is often the bottleneck of spray drying process and thus there is an increasing need to optimize process conditions to maximize throughput and ensure product quality. The process optimization for spray dried dispersions (SDD) is normally challenging due to the presence of an amorphous polymeric matrix with a glass transition temperature that limits the operating conditions in order to avoid the crystallization of the product.

A systematic methodology is proposed to allow an early recognition of scale-up and drying issues before process implementation on industrial scale to support the SDD development. The methodology includes a model that describes the drying mechanism of free-flowing powders under-vacuum considering the penetration theory [3,4]. This theory assumes a continuous mixing process with an alternate fictitious sequences of static periods and perfect mixing periods [5]. The understanding of the drying mechanism is crucial to optimize the process where the fundamental drying mechanism and mass transfer principles play an important role. The proposed methodology was created under the Quality by Design and Development by Design paradigms where the scale-independent relations (such as drying rate curve) and product/process understanding are found on lab-scale trials in order to achieve a lean development (Figure 1).

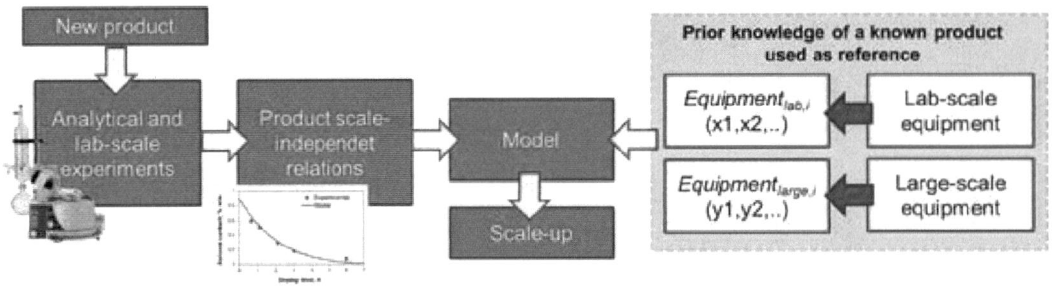

Figure 1: Proposed methodology framework

The method was constructed including the prior knowledge and the data of a known product used as reference. Several testes at lab- and large-scale units were performed for this reference case in order to establish the relevant equipment's parameters.

With that, when a new product is introduced, the proposed methodology provides an analytical assessment to evaluate a glass transition temperature, density, particle size and other important quality attributes that are relevant for the drying performance and understanding, e.g. the plasticization curve of the SDD to assure the amorphous state during the process since the risk of crystallization grows when the drying temperature is too close to the glass transition temperature. Succeeding, the product and process behavior and understanding are attained on lab-scale experiments to predict scale-independent relation. The lab-scale experiments are also important to understand the product behavior in terms of agglomeration and accumulation.

With the obtained scale-independent relation, it is possible to fit the model parameters and extrapolate for the large-scale unit where the model parameters of the equipment were previous stablished.

The intended benefits of this method are:

- Reduce secondary drying process time through process optimization;
- Support secondary drying development and scale-up;
- Tech transfer between secondary drying units.

A case study comprising the secondary drying process development of a SDD will be presented by following the proposed methodology. In this case study, operating conditions (namely the temperature and rotation speed) were optimized by the model without compromising the amorphous state of the temperature in order to reduce the drying time.

[1] Gianfrancesco, 2012. New method to assess water diffusion in amorphous matrices during storage and drying. Food Chemistry, 132, 1664-1670

[2] Kougoulos, 2011. Impact of agitated drying on the powder properties of an active pharmaceutical ingredient. Powder Technology, 210, pp. 308-314

[3] Schlunder and Mollekopf, 1984. Vacuum contact drying of free flowing mechanically agitated particulate material. Chemical Engineering and Processing, 18, 93-111

[4] Tsotsas and Schlunder, 1986. Contact drying of mechanically agitated particulate material in the presence of inert gas. Chemical Engineering and Processing, 20, 277-285

[5] Shani and Chaudhuri, 2012. Contact drying: a review of experimental and mechanistic modeling approaches. International Journal of Pharmaceutics, 434, 33-348
Copyright © American Institute of Chemical Engineers. All rights reserved.

471328 Scientific and Regulatory Considerations on Particle Engineering for Dry Powder Inhaler (DPI)

Abstract:

Dry Powder Inhaler (DPI) is a leading pulmonary drug delivery technology for the treatment of asthma, chronic obstructive pulmonary disease (COPD), and respiratory infections. For DPI drugs, due to significant impact of the dry powder properties on the inter-particle interactions as well as aerosol performance, particle engineering provides an effective approach for development and optimization of particulate formulation to ensure efficient drug delivery as well as optimal therapeutic outcomes. An advanced particle engineering technique can further improve the pharmaceutical manufacturing process to minimize all possible sources of the product variability and to ensure manufacturing consistency as well as drug product quality.

Recently, a variety of particle engineering techniques have been adopted for formulation development and commercial manufacturing of DPI drugs. For each particle engineering technique, in-depth understanding and control of the particulate and solid-state properties for dry powders is critical. This presentation will discuss various particle engineering techniques available for manufacture of DPI and will focus on scientific and regulatory considerations for characterization and control of dry powder properties for DPI drugs. This presentation will also include a discussion of regulatory submission expectations with regard to characterization and control of the particulate and solid-state properties for dry powders based on commonly seen deficiencies in new drug applications (NDAs) and abbreviated new drug applications (ANDAs). The objective of this presentation is to illustrate how to use a risk based approach as advocated by the Quality by Design (QbD) paradigm for developing particle characterization methods as well as establishing control limits for key dry powder properties of DPI drugs to ensure reproducible commercial manufacturing for consistent product quality.

Copyright © American Institute of Chemical Engineers. All rights reserved.

469151 Production of Pharmaceutical Tablets By FDM 3D Printing

Abstract:

Production of pharmaceutical tablets by FDM 3D printing

Production of tablets using 3D printers is a novel method of pharmaceutical formulation. Although still in the research&development phase, since recently it already has its first commercial application. The main advantage of this approach is that every single tablet can be printed to meet the needs of a specific patient, the dosage and dissolution rate can be controlled by the composition and porosity of the tablet. Furthermore, two or more drugs can be present in one tablet by using multiple printer nozzles with varying feed materials. The topic of this work is fabrication of pharmaceutical tablets, using Fused Deposition Modeling (FDM), a commercially available 3D printing technology, which utilizes thermoplastic polymer filaments as feed material. Biodegradable filaments were produced from mixtures of pharmaceutical polymers with additions of plasticizers and defined contents of a model drug, processed by hot-melt extrusion. For this purpose, the DIY hot-melt extruder Filastruder v1.5 was employed, which surpassed standard pharmaceutical extruders in terms of price, maintenance time and replaceability of its components, making it ideal for the purposes of an extensive parametric study of filament composition. This study resulted in production of drug-laden filaments with mechanical properties suitable for 3D printing, which were then successfully used as feed material for FDM 3D printer to produce tablets of varying shape and porosity. Furthermore, the effect of various printing parameters on the quality of the printed tablets was documented. Composition analysis by HPLC has confirmed that the drug contents in the tablets correspond with the initial drug loadings of the filaments. XRPD and DSC analyzes have shown, that the drug in the tablets exists in the form of an amorphous solid solution. Dissolution analysis has illustrated the effect of tablet shape, porosity and printing settings on the dissolution profile.

Although some cases of 3D printed tablets through Fused Deposition Modeling were already reported in the past, to the best knowledge of the author, this work presents the only case, where a conventional drug (with a relatively low melting point), which did not contribute to the printability of the material, was processed. Also, the extrusion and printing temperature is the lowest achieved so far in the field of FDM tablet printing. This makes the documented method more universal (applicable for a wider range of drugs) and presents potential for future research, which will involve incorporation of multiple drugs into one tablet, increasing resolution of the printing and expanding control over dissolution behavior.

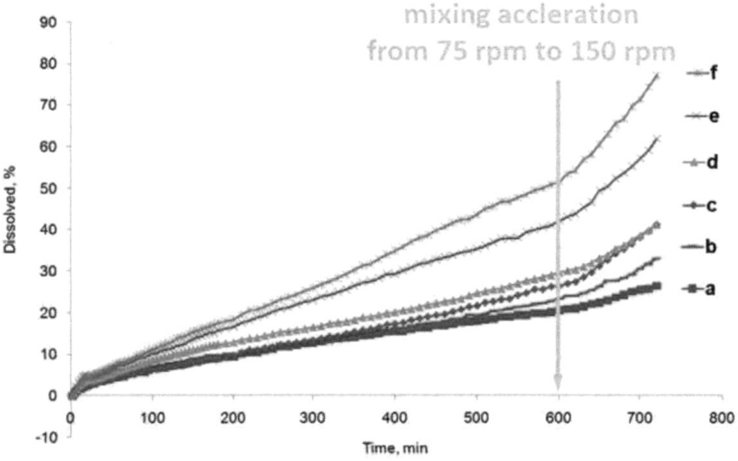

a) A disc-shaped tablet with deformations, resulting from using an unsuitable model, **b)** A disc-shaped tablet from a more suitable model, printed with a slight drop-shaped deformation on one side, **c)** A disc-shaped tablet from the same model as b), but with no deformation (due to better printing settings), **d)** A caplet-shaped tablet, **e)** A caplet-shaped tablet with micropores (pore height of 0.6 mm), which were most probably at least partially clogged during dissolution due to the slight swelling of the tablet, **f)** A caplet-shaped tablet with larger pores (pore height of 1.2 mm), which most probably stayed at least partially open during dissolution, which explains the highest dissolution rate of this tablet

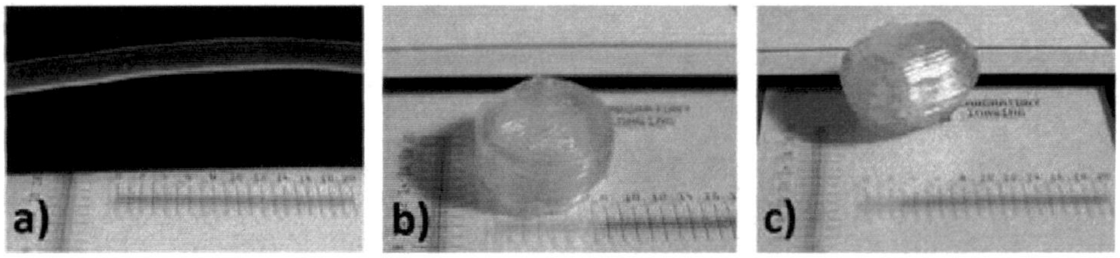

a) A biodegradable drug-laden filament, suitable for 3D printing purposes, **b)** A cylinder-shaped tablet,

c) A caplet-shaped tablet with predefined partially visible micropores

Copyright © American Institute of Chemical Engineers. All rights reserved.

453052 Experimental Analysis of Pharmaceutical Twin-Screw Wet Granulation Using Hydrophilic and Hydrophobic Formulations

Abstract:

Introduction

Traditional pharmaceutical solid-dosage manufacturing exists of batch-wise processes which are rather inefficient and expensive. Today, efforts are made to shift from batch to continuous processing. Twin-screw granulation is a promising continuous wet granulation technique for solid dosage manufacturing. Nevertheless, little is known about how process and equipment settings affect granule formation mechanisms and kinetics.

Objective

The aim of this study was to evaluate the twin-screw granulation process for both hydrophilic and hydrophobic formulations to understand the role of the individual screw modules and process settings, as well as their interaction upon granule formation. This involved investigating the influence of process settings on granule properties at different locations in the screw barrel.

Results from this experimental study will be used for the development, calibration and validation of a two-dimensional Population Balance Model (PBM) involving particle properties (granule size and liquid content or density/porosity) relevant for twin screw granulation. The 2D PBM will incorporate the rate processes which are considered dominant in the wetting and mixing zones of the granulator, such as aggregation and breakage, based on an inflow and outflow granule size distribution in each zone. The experimental data from different locations inside the granulator will be used to obtain the unknown model parameters for individual screw modules.

Material and methods

Materials

Two hydrophobic model drug formulations were used. Therefore, two poorly water soluble API's (60% w/w) were separately blended with 16% (w/w) lactose monohydrate (Lactochem® Regular, DFE Pharma, Goch, Germany), 16% (w/w) microcrystalline cellulose (Avicel® PH 101, FMC, Philadelphia, USA), 3% (w/w) hydroxypropylcellulose (Klucel® EXF, Ashland, Covington, Kentucky, USA) and 5% (w/w) crosscarmellose sodium (Ac-Di-Sol®, FMC, Philadelphia, USA).

Because a poor flowability was observed for one model formulation, 0.5 % (w/w) colloidal silicon dioxide (CAB-O-SIL® M-5, Cabot Corporation, Alpharetta Georgia, USA) was added to that formulation in order to ensure adequate dosing of the dry ingredients into the granulation module. Additionally, a hydrophilic formulation existing of 97.5% (w/w) lactose monohydrate (Lactochem® Regular, DFE Pharma, Goch, Germany) and 2.5% (w/w) Polyvinylpyrrolidone (Kollidon®30, BASF, Ludwigshafen, Germany) was studied. All formulations were processed using demineralized water as granulation liquid.

Methods

Granules were produced using the high shear wet granulation module of the ConsiGma^{TM}-25 system (GEA Pharma systems, Collette, Wommelgem, Belgium).

To develop and calibrate the 2D PBM model, a 3-level full factorial Design of Experiments (DoE) (MODDE 11, MKS Umetrics, Umeå, Sweden) was performed for three formulations by varying the following parameters: screw speed (450-900 rpm), material throughput (5-25 kg/h) and liquid-to-solid ratio (formulation dependent). The temperature of the cooling water around the granulator barrel was kept constant at 25°C. This DoE was executed for one screw arrangement existing of 2 kneading zones of each six kneading elements ordered at a stagger angle of 60°. Several responses were assessed and are described below.

In order to gain insight in the granulation progress along the length of the granulation barrel, samples were collected in well-defined zones of the granulator unit:

- Location 1: Before the first kneading zone, where the granulation liquid is added (wetting)
- Location 3: After the first kneading zone
- Location 5: After the second kneading zone
- Location 6: After the chopper section being a kneading block at the end of the screw (i.e. granulator outlet).

Granules collected at these locations were oven dried (24h, 40°C, 25% RH) and granule size distribution (QICPIC®, Sympatec GmbH, Clausthal-Zellerfeld, Germany) was measured. Furthermore, all samples were sieved in different fractions (i.e. >2000, 1000-2000, 850-500, 150-500, 75-150 and <150 µm) using a Retsch VE 1000 sieve shaker (Haan, Germany). True density (Helium-pycnometry, Accupyc 1330 Pycnometer, Micrometrics, Norcross, USA) and specific surface area (Brunauer–Emmett–Teller (BET)-theory) was measured for every fraction.

Moreover, distribution of moisture among the different sieve fractions was assessed using a colored tracer (0.5% w/w cochenillered) dissolved in the granulation liquid. UV spectroscopy was used to measure tracer content in every sieve fraction, allowing to visualize the moisture content per sieve fraction at the different locations of the granulator. This experimental data about granule properties collected at intermediate locations inside the screw barrel will help to describe the dynamics during twin-screw granulation to be included in the aimed 2D PBM.

The granules collected at the end of the barrel (location 6) were oven dried (24h, 40°C, 25% RH) and analyzed for several responses: torque, granule size distribution (QICPIC®, Sympatec GmbH, Clausthal-Zellerfeld, Germany), friability (Friabilator, PTFE Pharma Test, Hainburg, Germany), flowability (Ring Shear

Tester RST-XS, Dietmar Schulze, Wolfenbüttel, Germany), mean residence time and Peclet number (determined with residence time distribution – SentroPAT FO system, Sentronic, Dresden, Germany).

Conclusion:

This experimental study proved the ability to capture the stepwise formation and transition of granules at the different locations inside the granulator in terms of granule properties (e.g. granule size, moisture, density/porosity).

It was affirmed that process parameters like liquid-to-solid ratio can dictate the formation of granules and their corresponding quality attributes. Furthermore it should be emphasized that differences are observed between hydrophilic and hydrophobic formulations. This could be derived from the measured moisture content in the different sieve fractions along the length of the granulator barrel. Results showed a more heterogeneous moisture distribution among the different size fractions in the hydrophobic formulations where moisture is less likely to be found in the smaller size regions. This could be explained by a significant fraction of un-wetted particles.

Copyright © American Institute of Chemical Engineers. All rights reserved.

470232 Model-Based Identification and Evaluation of Abdominal Aortic Aneurysm Surveillance and Surgery Strategies on the Population Level

Abstract:

Abdominal aortic aneurysm (AAA) is when the diameter of the abdominal aorta is larger than 30 mm. The primary risk associated with AAA is an increased risk of aortic rupture, which is fatal in 68-90% of cases [1]. Once a patient is diagnosed with AAA, the AAA is monitored via abdominal ultrasound. The rationale for the regular surveillance is that the risk of rupture is low for AAA, less than 55 mm in size, but increases dramatically in diameter larger than 55 mm [2]. Early surgery on patients with smaller AAA diameters (lower risk of rupture) has a higher mortality rate than taking no action. Despite numerous researches done about prediction of AAA size, there is a lack of a design that quantifies the risk of surgery and rates of rupture and mortality at surveillances and integrates it with the process of decision making. This research addresses the necessity of integrating the rupture rate in different time periods.

A Monte-Carlo simulation technique was applied to a growth model based on Bayesian Analysis to simulate 10,000 hypothetical patients [3]. To ensure that the generated data correlated to the original data, the Cholesky decomposition was determined from the patient cohort data and applied to generation of characteristics of the hypothetical patients. The probability of each possible growth trajectory and cumulative risk of rupture is computed by Bayesian Analysis for each patient. Mortality and rupture rates are calculated individually, applying the Monte-Carlo simulation on meta-analysis paper [2] and National Vital Statistics System data for 2014 [4]. The risk of rupture increases in patients with increase in the size and the mortality rate increases with the time.

Different protocols regarding the surgical intervention threshold, risk of surgery, and observation time limits were designed, and the effects of life expectancy simulated. Simulating all 10,000 hypothetical patients, and comparing the results for different designed protocols and current available protocols in different countries, gave us a unique opportunity to analyze the effect of the surveillance and surgery decisions on patients' mortality.

Applying 120 designed protocols on 10,000 different individuals provided us with an overview on the differences within the designed protocols and also with current available protocols. There are surveillance protocols available in different countries for abdominal aortaâ€™s screening [2]. These population based protocols use the current size observed at screening, to recommend the time for the next surveillance. After determining the non-dominated protocols using the individual based simulations, we analyzed the results of all patients to build a general surveillance protocol solely based on the AAA size in the time of observation. Simulating the future outcome of 10,000 patients by using the population based protocols distinguished the efficiency of the protocols in terms of ruptures, deaths, and surgeries.

References:

[1] Sherer, E. A., R. R. Bies, et al. Growth of Screen-Detected Abdominal Aortic Aneurysms in Men: A Bayesian Analysis. *CPT: pharmacometrics and systems pharmacology*. **1**: e12. (2012).

[2] Bown MJ, Sweeting MJ, Brown LC, Powell JT and Thompson SG. Surveillance Intervals for Small Abdominal Aortic Aneurysms; A Meta-analysis. *The Journal of American Medical Association*, **309**: 806-813 (2013).

[3] Shah BH, Borwanker JD and Ramkrishna D. Monte Carlo Simulation of Microbial Population Growth. *The Journal of Mathematical Biosciences*, **31**:1-23 (1976).

[4] Arias E and Division of Vital Statistics. United States Life Tables. *National Vital Statistics Report,* **62:** 1-63 (2014).

Copyright © American Institute of Chemical Engineers. All rights reserved.

463048 Model-Based Novel Strategy for Individualized Treatment of Sickle Cell Disease with Hydroxyurea

Abstract:

Sickle Cell Disease is a hereditary disease caused by a single nucleotide mutation in the Î² subunit of hemoglobin producing abnormal, rigid and sickle shaped red blood cells. Current treatment method involves lifelong treatment with daily oral dosage of hydroxyurea. Hydroxyurea is known to induce fetal hemoglobin synthesis that inhibits hemoglobin polymerization and thus relieves the symptoms associated with the disease. The major side effect of the drug is it inhibits enzyme ribonucleotide reductase, causing bone marrow toxicity.

The main challenges associated with hydroxyurea are substantial interpatient variability in pharmacokinetic and pharmacodynamic profiles, cytotoxicity, non-compliance, lack of an effective biomarker to predict treatment efficacy and consequently lack of optimal dosing regimen for individual patients. Clinical studies have shown that mean cell volume (MCV) of red blood cell (RBC) increases with hydroxyurea treatment and is strongly correlated to treatment outcome. Therefore the MCV of RBC is used as a surrogate marker to monitor treatment progression. However, the effect of the drug on RBC is a manifestation of its effect on the early precursor cells containing nucleus. Red blood cells circulate in the blood for around 120 days and therefore the effect of hydroxyurea can only be seen after such time period. The slower RBC dynamics delays the clinical decision making process. Reticulocyte, an immediate precursor to red blood cell, circulates in blood for a relatively shorter time period before developing into mature red blood cell. Thus, reticulocytes offer faster dynamics and MCV of reticulocytes is proposed in this work as an effective biomarker.

A pharmacodynamic (PD) model is developed to describe reticulocytes dynamics as a function of hydroxyurea dosage. The model uses the population balance approach which uniquely describes the evolutionary distribution of various cell types in the bone marrow and their emergence in the periphery. In order to reduce the myelosuppression and maximize drug efficacy, the drug dosage is optimized based on the response obtained from individual patient pharmacodynamic model. This study shows the feasibility of using MCV of reticulocytes as an effective surrogate biomarker. The model can aid clinicians in timely prediction of individual patient dosage and improve the quality of life among patients. For model validation, the data is collected from Riley Hospital for Children in Indianapolis.

Copyright © American Institute of Chemical Engineers. All rights reserved.

468376 Broadly Neutralizing Antibodies Enhance Clearance of HIV-Infected Cells ̂ In Vivo: Pharmacodynamic Analysis of Viral Load Kinetics with Validation in Humanized Mice

Abstract:

The primary role of antibodies typically involves neutralizing free virus particles in circulation by binding to specific regions of their cognate antigen. In the case of HIV, the causative agent of AIDS, the antigen is the envelope (Env) glycoprotein found on the surface of the virus particles. Broadly neutralizing antibodies (bNAbs), such as 3BNC117, are capable of neutralizing a large diversity of mutational variants of Env, leading to a rapid but transient reduction in viral load when passively administered. In vitro experiments have previously demonstrated the ability of antibodies to not only block new infection and reduce circulating virus concentrations, but also to engage particular classes of host immune effector cells by recognizing Env found on the surface of infected cells. This interaction is mediated by receptors on the host cell surface that bind to constant (Fc) regions on the antibodies. The host cells then mediate the death of these infected cells through mechanisms such as antibody-dependent cellular cytotoxicity (ADCC) or antibody-dependent cellular phagocytosis (ADCP). However, this interaction has not previously been demonstrated in vivo.

This work represents the first evidence of enhanced infected-cell clearance in vivo. Based on results from the first-in-man dose escalation phase 1 clinical trial of the bNAb 3BNC117, we found that the antibody kinetics upon passive administration are well-described by a simple two-compartment model where antibody can transport between blood plasma and lymph tissue (LT) compartments. We then coupled this transport model, with parameters inferred from healthy individuals, to a popular model of viral infection dynamics, including birth of virus from infected cells (which reside primarily in LT), and transport of virus between plasma and LT. This coupling took the form of an additional kinetic term corresponding to antibody binding free virus particles, leading to their subsequent elimination from the system. This simple model was unable to recapitulate the observed kinetics of viral load decay; in particular, the time scale and extent of decay could not both be fit simultaneously. However, by adjusting our model to include a mechanism whereby antibody causes death of infected cells, we substantially improved the fit to the viral load data because it causes a higher-order decay in the viral load over a longer time scale. Thus, the clinical patient data can not be explained when antibody exclusively neutralizes free virus particles. Motivated by these findings, our experimental collaborators demonstrated in a humanized mouse model that the clearance rate of infected cells is indeed increased, via a mechanism that involves engagement of Fc receptors (FcRs). These results have implications for passive immunotherapy (treatment or prevention) targeted against HIV-infected cells.

Copyright © American Institute of Chemical Engineers. All rights reserved.

467238 Computational Methods at the Intersection of Tuberculosis Immunology and Therapy

Abstract:

Tuberculosis (TB) poses a global public health threat, causing 9.6 million new infections and 1.5 million deaths in 2014 (WHO Global tuberculosis report 2015). While most drug-susceptible TB is treatable with a combination of four antibiotics, the treatment lasts 6-9 months and has failed to prevent drug-resistant TB worldwide. Three of the challenges facing new TB treatment strategies are: 1) complex disease pathology (i.e. spherical lesions called granulomas form in the lung), 2) variable bacterial susceptibility to antibiotics, and 3) combination therapy (necessary to avoid drug resistance). It is difficult to address these challenges in a single experimental system; the costly non-human primate model of TB is the only animal model that recapitulates human-like pathology, and *in vitro* systems do not capture the spatial or long-term temporal aspects of TB pathology.

We use a computational approach to integrate spatial and temporal data from multiple *in vitro* and *in vivo* experimental systems. The experimental data encompass a large variety of temporal scales, spatial scales, biological compartments and experimental methods. The biology, experimental data type and model structure determine the model calibration workflow. Quantitative approaches in complex biological systems require well-defined strategies to incorporate a variety of experimental data. We illustrate one such strategy for TB immunology and therapy. Our computational model captures host immunity, bacterial dynamics (including differential antibiotic susceptibility), and antibiotic dynamics in plasma, lung tissue and TB granulomas. The model comprises a combination of agent-based, ODE and PDE models. We calibrate each of the model components to non-human primate microbiological and immunological data, rabbit and human pharmacokinetic (PK) data, and *in vitro* and animal pharmacodynamic (PD) data. In some instances multiple experimental data types are available for a single model component. E.g. antibiotic distribution in granulomas is measured using LCMS as well as MALDI-MS. In such situations the model is calibrated to both data sets in parallel, weighting the contribution of each data set based on known experimental measurement constraints.

Our model bridges and connects multiple experimental data sets, and predicts efficacy of multiple antibiotics in the context of host immune dynamics. This model is applied to efficiently and systematically evaluate potential TB treatment regimens containing five anti-TB antibiotics (isoniazid, rifampin, moxifloxacin, gatifloxacin and levofloxacin), to complement and inform future animal or human trials.

Copyright © American Institute of Chemical Engineers. All rights reserved.

452313 Cell Population-Based Mathematical Model of Liver Homeostatic Renewal

Abstract:

Many chronic diseases impair the self-renewal capability of hepatocytes in the liver. This impaired capacity for renewal may even be one of the major factors contributing to the progression of chronic liver disease through steatohepatitis and fibrosis to cirrhosis. As evidence of this, in the intermediate and late stages of chronic liver disease hepatocyte death leads to scar tissue formation rather than hepatocyte renewal. The harmful effects of chronic diseases on liver regeneration have been widely studied using experimental and computational methods, however, the effects of chronic diseases on homeostatic liver renewal are less well understood.

Even though it is widely recognized that much of liver cellular turnover is accomplished through a continuous process of liver homeostatic renewal, partial hepatectomy followed by liver regeneration is often still the methodology of choice used to investigate dynamic liver function and tissue renewal capability. Regeneration, however, provides a picture of liver function under abnormal operating conditions rather than during normal function; it is therefore of limited utility for investigating everyday tissue function. We showed earlier that an integrated network of multiple cell types is required for liver regeneration following partial hepatectomy and that chronic ethanol use causes an imbalance between populations of pro-regenerative and anti-regenerative hepatic stellate cells prior to hepatectomy. We further showed that this homeostatic imbalance in hepatic stellate cell populations, coupled with the well-documented increase in cytokine production in alcoholic liver disease (ALD), could explain impaired regeneration caused by chronic ethanol abuse. Because diseases such as ALD cause homeostatic imbalances to cell populations, it is possible that such imbalances affect homeostatic renewal in addition to impairing regeneration. Therefore, using a cell population-based mathematical modeling framework, this current study extends that previous work by investigating how homeostatic renewal may be regulated in the healthy liver and how chronic insults and disease adaptation could affect liver homeostatic renewal.

We develop a mathematical model of liver homeostatic renewal that takes into account two recently described populations of hepatocytes: Axin2+ and Axin2-. Hepatocyte renewal occurs predominantly by replication of Wnt-responsive Axin2+ â€œstem-cell-likeâ€ hepatocytes, which have been shown to cluster pericentrally and which respond to factors produced by pericentral endothelial cells (i.e., WNT-2a). As Axin2+ hepatocytes renew, some of the daughter cells migrate from the pericentral region, become Axin2- cells, and populate the remainder of the liver. We investigate the dynamic behavior of this two-state system by simulating tissue renewal, first assuming no feedback between populations, and subsequently investigating the effect of combinations of feedbacks, including product inhibition, cellular competition, and others. We use a systematic design of experiment-based approach to investigate the system behavior under all possible combinations of feedback configurations, specifically characterizing system recovery by calculating total deviation from nominal cell levels and overall recovery time in response to transient perturbations in apoptosis rates and proliferation rates. This exercise allows us to identify which of the investigated models of liver homeostatic renewal responds most robustly to multiple types of disturbances. We assume that nature selects for robust behavior, so that the most robust model indicates the most biologically plausible configuration.

In the presentation, we will discuss our modeling and analysis approach in detail, especially our investigation of the dynamic behavior of our homeostatic renewal model using frequency-based analyses, sensitivity analyses, and stability analyses. We will also discuss our model simulations of chronic insults to healthy and diseased livers. Our simulations and model analyses results lead us to predict that liver homeostatic renewal is governed by a combination of intrinsic (passive) system stability and active control by non-parenchymal cell networks. We further predict that chronic diseases likely impair both the hepatocyte homeostatic renewal system and control by non-parenchymal cell networks.

Research Support: R01 AA018873, R21 AA022417, T32 AA007463, F31 AA023445

Copyright © American Institute of Chemical Engineers. All rights reserved.

469109 Quantitative Dissection of Bacterial Nitric Oxide Stress Networks

Abstract:

Nitric oxide (NO) is an important antimicrobial produced by the immune system to neutralize pathogens. Many pathogens, such as *Mycobacterium tuberculosis*, *Neisseria meningitides*, *Pseudomonas aeruginosa*, and pathogenic *Escherichia coli*, harbor NO detoxification and repair systems that have been linked to their virulence, and thus represent attractive targets toward which novel anti-infective therapies could be identified. Unfortunately, the few compounds that have been found to inhibit bacterial NO detoxification systems suffer from poor transport or toxicity, which warrants searches for other compounds or strategies to impair those processes. To complicate matters, the broad reactivity of NO and its reaction products give rise to an expansive and interconnected biochemical reaction network whose experimental measurables are difficult to interpret. To facilitate quantitative analyses of bacterial NO stress networks, we constructed and experimentally validated a kinetic model of the *E. coli* NO biochemical network, including relevant processes such as enzymatic NO detoxification, iron-sulfur cluster damage and repair, DNA deamination and repair, transcriptional regulation, and NO autoxidation. Here, I will discuss how we have used this computational tool in an iterative workflow with experimentation to mechanistically investigate how a variety of chemical, genetic, and environmental perturbations alter the *E. coli* NO stress response, as well as how this tool could facilitate the development of next-generation antibiotics that inhibit bacterial NO defenses.

Copyright © American Institute of Chemical Engineers. All rights reserved.

462356 Informing the Structural Model of the Mammalian Voltage-Gated Sodium Ion Channel: Probing Protein Conformational Dynamics Using Small-Molecule Neurotoxins

Abstract:

Voltage-gated sodium ion channels (Na_Vs) are critical to life, being responsible for generating action potentials in skeletal muscle, nerve, and cardiac cells.[1] The abnormal expression and/or malfunction of Na_Vs is tied to epilepsy, arrhythmia, intractable acute and chronic pain, metastatic cancers, erythromelalgia, and congenital insensitivity to pain.[1,2] Despite their importance to human function and disease, a complete understanding of the conformational dynamics of Na_Vs, their responsiveness to changes in membrane potential, and the influence of small molecules on ion gating remains elusive. Critical to the understanding of Na_V function is the construction of a detailed model of the tertiary structure of the pore forming and voltage-sensing domains. As such, ligand docking studies using the small molecule neurotoxins veratridine [partial agonist, 3] and batrachotoxin [agonist, 4] with a homology model of Na_V, based on the crystal structure of a prokaryotic Na_V [5], have been performed. These computational modeling studies have allowed for the assignment an overlapping putative binding site for both toxins within the ion pore. Chemical synthesis has enabled the construction of de novo synthetic and semisynthetic neurotoxin analogs. In combination with protein mutagenesis, the systematic exploration of the inner pore of Na_V has confirmed the binding site proposed by the model. Although these results are in agreement with other reported experimental findings, they directly refute a previously proposed pharmacophore model for veratridine and batrachotoxin. Additionally, these studies have revealed a previously unreported state-dependent functional duality of veratridine as either a partial agonist or antagonist. The continued exploration of Na_V-ligand interaction will inform Na_V homology modeling and improve the understanding of Na_V dynamics.

[1] Nardi, A.; Damann, N.; Hertrampf, T.; Kless, A. *ChemMedChem* **2012**, *7*, 1712.

[2] Martin, F.; Ufodiama, C.; Watt, I.; Bland, M.; Brackenbury, W. J. *Front. Pharamcol.* **2015**, *6*, 273.

[3] (a) Rando, T. A. *J. Gen. Physiol.* **1989**, *93*, 43. (b) Codding, P. W. *J. Am. Chem. Soc.* **1983**, *105*, 3172.

[4] (a) Khodorov, B. I. *Prog. Biophys. Molec. Biol.* **1985**, *45*, 57. (b) Daly, J. W.; Witkop, B.; Bommer, P.; Biemann, K. *J. Am. Chem. Soc.* **1965**, *87*, 124.

[5] McCusker, E. C.; Bagneris, C.; Naylor, C. E.; Cole, A. R.; Dâ€™Avanzo, N.; Nichols, C. G.; Wallace, B. A. *Nat. Commun.* **2012**, 3, 1102.

Copyright © American Institute of Chemical Engineers. All rights reserved.

465683 Computational Modeling of Amyloid Aggregation Kinetics

Abstract:

The aggregation of amyloidogenic proteins through cross -sheet interactions depends upon two main factors: the primary sequence and environmental conditions such as presence of ions in solution[1,2]. The ordered fibrous aggregates formed as a result of this process are involved in a number of neurodegenerative diseases. The morphology of these aggregates is influenced by the aggregation kinetics and governs disease patterns in mammals and phenotypic differences in yeast.

The process of amyloid formation involves a two-step pattern of initial nucleation corresponding to a lag phase followed by a fiber elongation phase. This behavior has been modeled empirically using mathematical functions such as the logistic function, and other kinetic models, such as the Finke-Watzky (F-W) 2-step mechanism, which capture the near sigmoidal shape of the aggregation curve. However, experimental aggregation data does not always mirror the simple S-shape of a sigmoidal function. In this work we have developed a model for amyloid aggregation kinetics following a nucleation, elongation and fragmentation scheme. The effect of factors influencing aggregation kinetics is studied by changing parameters, such as fragmentation rate, initial concentrations of seeds, or critical nucleus size. Our results show that non-sigmoidal aggregation kinetics can result from varying some of the above mentioned parameters.

[1] Rubin, J., Khosravi, H., Bruce, K. L., Lydon, M. E., Behrens, S. H., Chernoff, Y. O. and Bommarius, A. S., 2013. Ion-specific effects on prion nucleation and strain formation. Journal of Biological Chemistry, 288(42), pp.30300-30308. doi: 10.1074/jbc.M113.467829

[2] Yeh, V., Broering, J. M., Romanyuk, A., Chen, B., Chernoff, Y. O. and Bommarius, A. S. (2010), The Hofmeister effect on amyloid formation using yeast prion protein. Protein Science, 19: 47–56. doi: 10.1002/pro.281

Copyright © American Institute of Chemical Engineers. All rights reserved.

460826 The Swim Pressure: Forces and Stresses on Living Materials

Abstract:

To understand complex biological systems like the interior of a cell, it is essential to characterize the motion of molecular motors and the interaction of active constituents with other cellular components.Â Active elements like motor proteins generate internal forces and stresses that enable them to control their own behavior and that of their surroundings, driving the cell far from equilibrium.Â In an effort to model the cell, we developed a method to encapsulate swimming Janus particles, microorganisms, and other nano/micron-sized objects inside of a phospholipid vesicle. We use a new â€œswim pressureâ€ perspective to analyze the mechanical deformation and lysis of cells triggered by the motion of active elements and pathogenic bacteria. We further fabricate soft, flexible materials that can expand/shrink, elongate, move, and steer on command by loading the material with bacteria that can move in a directed, coherent manner. These soft materials may be used as micro- or nanomechanical devices and motors that could have multiple applications in medicine (e.g., focused drug delivery), biophysics, and other fields. In addition to being a useful rheological tool to quantify the mechanical forces, stresses, and (thermo)dynamics of active elements inside a cell, the swim pressure concept may engender new discoveries in cellular morphology, function, and the complex intracellular transport within the cytoplasm.

Copyright © American Institute of Chemical Engineers. All rights reserved.

472487 Mechanical Amplification of Tumor Death Using Polymeric Nanoparticles

Abstract:

Mechanical Amplification Of Tumor Death Using Polymeric Nanoparticles

Michael J. Mitchell[1], Robert Langer[1]

[1]Chemical Engineering and David H. Koch Institute for Integrative Cancer Research, MIT, Cambridge MA 02139

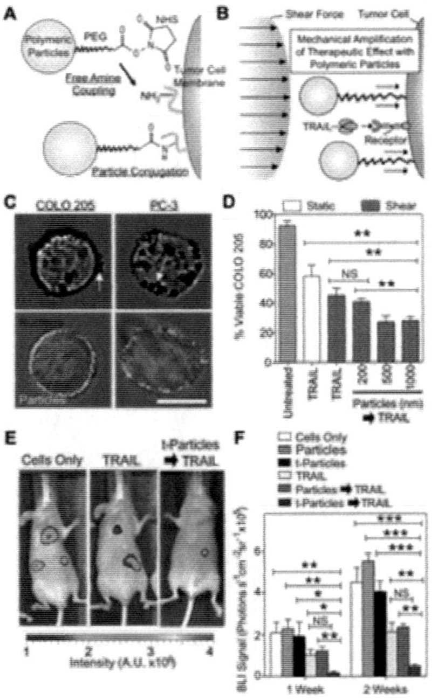

Introduction: Regulation of receptor-mediated phenomena at the nanoscale in a controlled manner remains a challenge for a broad range of fields, including cell biology, medicine, and bioengineering. Fluid forces play a crucial role in receptor-mediated signaling necessary for physiological function, while materials such as nanoparticles (NPs) have demonstrated potential for spatiotemporal control of cellular signaling. Here, we have developed polymeric mechanical amplifiers that tether to the cell surface and increase receptor-mediated apoptosis in the presence of physiological fluid flow.

Materials and Methods: Polymeric NPs (size: 100-1000 nm) were conjugated to free amines on tumor cells via NHS crosslinker chemistry (**Fig. 1A**). Nondegradable polystyrene and degradable (PLGA, PCL) NPs bound to tumor cells were assessed using flow cytometry, brightfield, and confocal fluorescence microscopy. A cone-and-plate viscometer was used to apply a fluid shear force (2.0 dyn/cm^2) to tumor cell suspensions and to amplify the force exerted by polymeric particles on tumor cells (**Fig.1B**). Tumor cells (COLO 205, PC-3, MCF-7) were treated with 0.1 µg/mL of a TNF-related apoptosis-inducing ligand (TRAIL) to assess receptor-mediated apoptosis. *In vivo* mechanical amplification of TRAIL apoptosis was assessed in nu/nu mouse models of lung metastasis.

Figure 1: (**A**) Conjugation of polymer amplifiers to tumor cell surface via NHS crosslinker chemistry. (**B**) Polymeric particles amplify force exerted on the tumor cell surface under fluid forces to increase mechanotransduction and therapeutic efficacy. (**C**) Confocal and brightfield images of 1 µm polymeric particles conjugated to the surface of tumor cells. Scale bars: 10 µm. (**D**) Particle-functionalized tumor cell viability after TRAIL treatment in the presence of shear force. **P<0.01. NS: not significant. Bioluminescence imaging (**E**) and quantification (**F**) of COLO 205 tumor cell burden in mice post-injection of targeted (t)-particles followed by TRAIL. ***P<0.001.

Results and Discussion: NHS crosslinker chemistry stably bound polymeric NPs to the tumor cell surface (**Fig.1C**). In the presence of fluid forces, polymeric NPs increased TRAIL apoptosis (**Fig.1D**). Increasing the number of NPs conjugated to the tumor cell surface enhanced TRAIL apoptosis under fluid shear exposure, while the therapeutic effect under static conditions was not altered. Inhibition studies indicated the response is caspase signaling-dependent, and increased tumor death receptor expression in the presence of shear forces was also observed. Targeted anti-EPCAM polymeric NPs delivered to tumor cells *in vivo* mechanically amplified a subsequent treatment of soluble TRAIL, and reduced both circulating tumor cells in blood and overall tumor cell burden *in vivo* by over 90% (**Fig. 1E,F**).

Conclusions: We conclude that surface-bound polymeric nanoparticles enhance receptor-mediated apoptosis in the presence of physiological shear forces, and represents a potentially new application for a broad range of nanotechnologies to maximize the capacity of receptor-mediated signaling and function in the presence of a ligand.

Copyright © American Institute of Chemical Engineers. All rights reserved.

469931 A Multimodally Regulated Microenvironment for Mechanotransduction Studies of Organ Development

Abstract:

The emergence of diseases is known to depend on both genetic and biomechanical factors. However, much work remains to be done to define the complex regulatory interactions responsive to environmental stimuli. More specifically, the impact of exogenous forces, such as mechanical compression, on the regulation of these signaling cascades, which effect basic cellular processes such as proliferation and apoptosis, remains unresolved. Although the mechanisms of regulation are unknown, mounting experimental evidence has increasingly implicated mechanical forces in the regulation of cell cycle, cell survival and cellâ€"cell communication. However, probing these relationships experimentally remains problematic, owing to the unique challenges involved in mechanically manipulating tissues both *in* and *ex vivo*. Here we describe the fabrication and implementation of a scalable microfluidic culture chip (the **R**egulated **E**pithelial **M**icroenvironment **Chip**) for studying the impact of exogenous forces on development and gene expression in organ cultures. The chip consists of individually addressable culture chambers. Each chamber allows control over the chemical perfusion of culture media, while the precise application of compressive forces exerted on the disc is achieved via a pneumatically operated membrane on the chamberâ€™s ceiling. Enabled by the REM-Chip, we demonstrate that Ca^{2+}Â signaling is inhibited during mechanical compression, but once compression is removed a Ca^{2+}Â wave cascades throughout the disc. This response is dependent on the presence of specific serum conditions indicating synergy between chemical and mechanical factors. Further, specific pharmacological or genetic manipulation of specific morphogen pathways modulates the frequency and amplitudes of long-distance Ca^{2+} waves. The REM-Chip enables quantitative studies into cross-talk between genetic and environmental factors at the tissue and organ scale.

Copyright © American Institute of Chemical Engineers. All rights reserved.

462514 Mechanosensing Dynamics of Human Red Blood Cells

Abstract:

Mechanical stress-induced deformation of human red blood cells (RBCs) plays important physiopathological roles in oxygen delivery, blood rheology, transfusion, and malaria. Recent studies demonstrate that, in response to mechanical deformation, RBCs release adenosine-5'-triphosphate (ATP), suggesting the existence of mechanotransductive pathways in RBCs. Most importantly, the released ATP from RBCs regulates vascular tone and impaired release of ATP from RBCs has been linked to diseases such as type II diabetes and cystic fibrosis. To date, however, the mechanisms of mechanotransductive release of ATP from RBCs remain unclear. Given that RBCs experience shear stresses continuously during the circulation cycle and the released ATP plays a central role in vascular physiopathology, understanding the mechanotransductive release of ATP from RBCs will provide not only fundamental insights to the role of RBCs in vascular homeostasis but also novel therapeutic strategies for red cell dysfunction and vascular disease. This talk describes the main research in my group on integrating microfluidic-based approaches to study the mechanosensing dynamics of RBCs. Specifically, I will introduce a microï¬‚uidic approach that can probe the dynamics of shear-induced ATP release from RBCs with millisecond resolution and provide quantitative understandings of the mechanosensitive ATP release processes in RBCs. Furthermore, I will also describe our recent findings about the roles of the Piezo1 channel, a newly discovered mechanosensitive cation channel in the mechanotransductive ATP release in RBCs. Last, possible functions of RBCs in the regulation of cerebral blood flow will be discussed.

Copyright © American Institute of Chemical Engineers. All rights reserved.

460424 Combinatorial ECM Arrays Reveal the Role of Biomechanics in Liver Progenitor Differentiation

Abstract:

Combinatorial ECM arrays reveal the role of biomechanics in liver progenitor differentiation

Andreas P. Kourouklis, Kerim B. Kaylan, Gregory H. Underhill

Department of Bioengineering, University of Illinois at Urbana-Champaign, Urbana, IL

Recent findings suggest that biomechanical signals within the liver microenvironment can regulate the differentiation of mature hepatocytes **[1]**. However, the role of ECM biomechanics in liver progenitor differentiation remains primarily unexplored, despites its potential importance in the processes of liver morphogenesis and regeneration. To examine these mechanisms, we created high-throughput cellular arrays with the capacity to reiterate combinatorial ECM cues and characterize the corresponding phenotypic expression. Moreover, we combined these arrays with substrates of modular stiffness and integrated them with traction force microscopy (TFM) to assess the associated traction stress. This strategy provides a novel avenue to examine cell differentiation and elucidate the role of combinatorial ECM cues in cellular fate.

With this objective, we fabricated high throughput BMEL cell arrays on soft (4kPa) and stiff (30kPa) substrates presenting all single and pairwise combinations of 5 ECM molecules (Fig. 1A). We found that the ECM composition along with the substrate stiffness coordinately control cholangiocytic differentiation (e.g. OPN+ cell fraction). Most notably, collagen IV (C4) and fibronectin (FN) exhibited distinct effects (Fig.1B). BMEL cells on FN expressed stiffness-mediated cholangiocytic differentiation, indicated by the higher OPN+ fraction on stiff substrates (Fig.1B), while the cells on C4 exhibited cholangiocytic differentiation indifferent of substrate stiffness. Moreover, BMEL cells on FN developed higher traction stress on stiff substrates (Fig.1B), while cells on C4 exhibited significant traction on both substrates. These data are suggestive of a mechanism by which liver progenitor differentiation is regulated by traction stress at the cell-ECM interface.

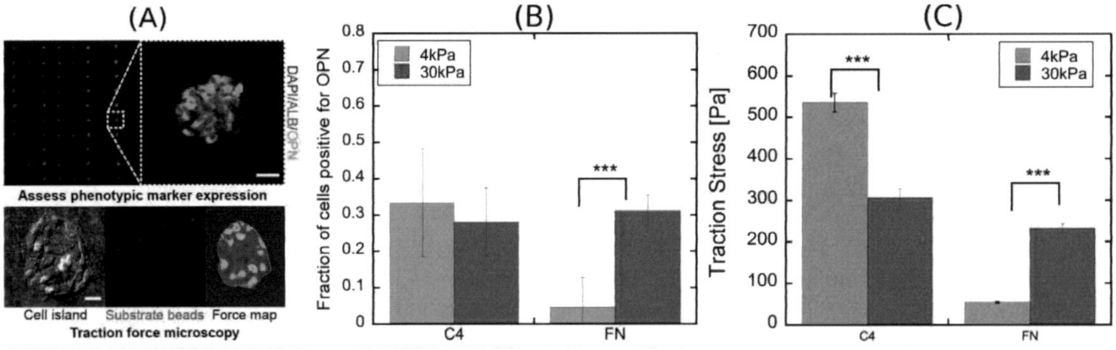

Figure 1: (A) Successful fabrication of cell arrays with BMEL domains of a diameter of ~150um. The BMEL arrays are immunostained for hepatocytic (albumin: ALB) and cholangiocytic (osteopontin: OPN) markers for high-throughput analysis. Lower raw demonstrates bead embedded gels for TFM with the corresponding phase contrast and heat maps. OPN expression (B) and traction stress (C) in response to stiffness and ECM composition.

In conclusion, we integrated microarray fabrication with TFM in order to reveal the role of cell biomechanics in the differentiation of liver progenitor cells. We showed that cells integrate combinatorial ECM cues using traction stress before they commit to their differentiated phenotype. Future experiments will assess the intracellular pathways that direct this differentiation response. Overall, this strategy provides new insight into liver progenitor fate decisions that could aid the investigation of liver disease mechanisms and the development of cell-based therapies.

References:

[1] Wells, R.G., Hepatology, 2008. **47**(4): p. 1394-400.

Copyright © American Institute of Chemical Engineers. All rights reserved.

471179 Loss of Giant Obscurins Enhances Migration and Cell Dynamics in Pancreatic Ductal Epithelial Cells

Abstract:

Loss of Giant Obscurins Enhances Migration and Cell Dynamics in Pancreatic Ductal Epithelial Cells

Daniel J Shea[1], Konstantinos Konstantopoulos[1], and Aikaterini Kontrogianni-Konstantopoulos[2]

[1]Department of Chemical and Biomolecular Engineering, Johns Hopkins University, [2]Department of Biochemistry and Molecular Biology, University of Maryland School of Medicine.

Introduction: Obscurins are giant cytoskeletal proteins with structural and regulatory roles that are encoded by the gene *OBSCN*. Depletion of giant obscurins from breast epithelial cells has been correlated with increased apoptotic resistance[1], epithelial to mesenchymal transition, tumorigenicity, and metastasis[2]. Additionally, mutations in the *OBSCN* gene have been associated with pancreatic ductal adenocarcinoma[3, 4]. Since pancreatic cancer has the lowest (5-year) survival rate of any major malignancy, and metastasis is the single most important contributor to this prognosis, we looked to assess the role of obscurins in pancreatic cancer metastasis. We investigated how depletion of obscurins from human pancreatic ductal epithelial (HPDE) cells affects cell migration and cytoskeletal dynamics using a series of microchannel migration assays and real-time measurements of actin, microtubule, and focal adhesion (FA) turn-over rates.

Materials and Methods: Immunoblotting was utilized to assess the obscurin levels in six pancreatic cell lines. A pancreatic tissue microarray was stained for obscurin alongside with obscurin-expressing HPDE cells, and imaged via confocal microscopy to determine obscurins localization and expression. HPDE cells were transfected with shRNA constructs targeting the *OBSCN* gene or control (scramble) shRNA, and stable cell lines were produced. Cross-shaped PDMS microchannels with a base channel (10μm) bifurcating to three-branch channels with varying widths of 3, 6 and 10μm (the 10μm device shown in Fig 1A) were fabricated using photolithography. Cells were tracked as they entered the channel bases until they made a decision at the bifurcation point and continued to the branch regions. Additionally, actin dynamics at cell-cell junctions were measured using fluorescence recovery after photobleaching (FRAP) for LifeAct-RPF transfected HPDE cells. Microtubule growth was examined using time-lapse confocal microscopy to image EB1-GPF transfected HPDE cells. FA density was measured via TIRF microscopy. RhoGTPase pull-down was performed to assess the levels of active rhoA, given that the RhoGEF motif of obscurins specifically binds and activates RhoA. All experiments were performed using obscurin-KD and control HPDE cells.

Results and Discussion: Pancreatic tissue microarray analysis shows high obscurin expression in normal pancreatic tissue and reduced expression in pancreatic adenocarcinoma. Using immunoblotting, giant obscurins were identified in epithelial but not mesenchymal pancreatic cell lines, and HPDE cells were found to express the highest levels of obscurins. Confocal images of HPDE cells show that obscurins are localized in the cytosol, and are absent from cell-cell junctions when grown in the absence of FBS. In the cross-shaped microchannels, obscurin-KD HPDE cells migrated with a higher velocity in the channel base, made a decision in the bifurcation region more quickly (Fig 1A), were more protrusive (lower solidity), and less circular than control cells, indicating that they were more dynamic. Coinciding with a pronounced decrease in active rhoA levels (Fig 1C), obscurin-KD HPDE cells were found to have faster (more dynamic) microtubule growth (Fig 1B) and decreased FA density. Furthermore, actin dynamics at cell-cell junctions were only increased for obscurin-KD cells in the presence of 10% FBS, where obscurin localization shifted from cytosol to the cell-cell junctions.

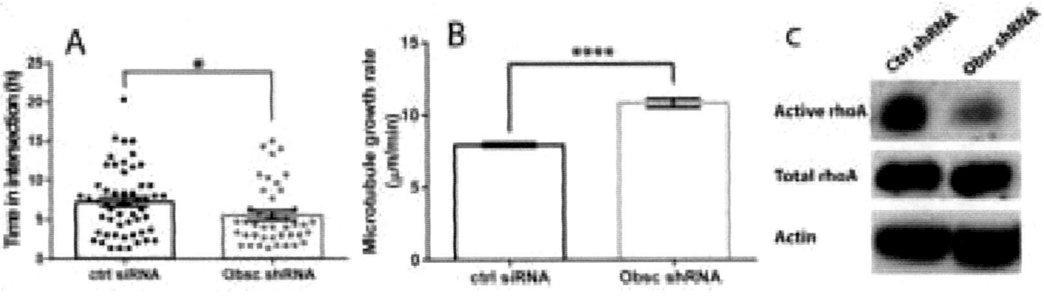

Figure 1. (A) Obscurin-KD HPDE cells exhibit faster decision-making time in cross-shaped device, **(B)** increased microtubule dynamics and **(C)** have less active rhoA when compared to control cells. (*P<.05, ****P<.0001)

Conclusions: Given that one of the major hallmarks of pancreatic cancer is its extensive local tumor invasion, and its dissemination to distant organs, identifying biomarkers that play a role in pancreatic cancer metastasis is vital to the treatment and prevention of the disease. We found that the loss of giant obscurins from epithelial pancreatic cells results in a more migratory and dynamic cell phenotype. Furthermore, in addition to comprehensively exploring cell migration and cytoskeletal properties, we have also begun exploring the effects of obscurinsÕ depletion *in vivo* using orthotopic pancreatic implantation of cells in mice.

References: (1) Perry, N.A. *et al., FASEB J,* 2012, 26,7. (2) Shiver M. *et al., Oncogene*, 2014, *34*(32), 4248-4259 (3) Hocker, J. R., et al., *Cancer Lett,* 2015, *359*, 2. (4) Murphy, S. J., *et al., Gastroenterology* 2013, *145*, 5.

Copyright © American Institute of Chemical Engineers. All rights reserved.

474396 Cell Motility and Traction Force Generation in Three-Dimensional Fibrin Hydrogels

Abstract:

The ability of cells to exert force on their surroundings underlies cell migration and extracellular matrix (ECM) remodeling, and is thus an essential aspect of complex, metazoan life. Due to technical challenges, few studies have examined how living cells transduce cytoskeletal force to the ECM in soft, three-dimensional (3D) environments. We used timelapse 3D imaging, quantitative image analysis, and theory to determine how fibroblasts embedded in a 3D fibrin matrix exert force on the surrounding ECM. Under these circumstances, cytoskeletal force generation and transmission are dominated by contractile actin bundles (stress fibers) attached at their ends to large, stable integrin-based adhesions (focal adhesions). Stress fiber contraction is counterbalanced by continuous elongation at focal adhesions. The resulting dynamic equilibrium can explain how cells maintain stable, contractile connections to the ECM during cell migration, and provides a probable means by which fibroblasts contract provisional matrices during wound healing.

Copyright © American Institute of Chemical Engineers. All rights reserved.

472245 Alginate/Chitosan Microparticles for Gastric Passage and Intestinal Release of Therapeutic Protein Nanoparticles

Abstract:

Proteins have significant potential to treat diseases due to their specific mechanisms of action compared to small molecule drugs. Protein nanoparticles offer a means to turn soluble protein, which has very low cellular uptake, into a colloidal protein carrier capable of enhanced uptake and subsequent intracellular activity. Previous work produced a stable therapeutic nanoparticle formulation encapsulating a bacterial effector protein, AvrA, which is capable of modulating inflammatory signals in an in vivo inflammatory bowel disease (IBD) mouse model. A major drawback is that these nanoparticles are instilled transrectally, limiting delivery to lower regions of the colon. To implement our protein nanoparticles as a clinically viable treatment for IBD with access to the entire intestinal tract and improved patient compliance, an oral formulation must be designed.

Localized intestinal delivery of protein therapeutics via the oral route is a highly attractive approach, however, the harsh acidic and enzymatic conditions in the gastrointestinal tract challenge retention of protein function. The approach is to deliver protein nanoparticles within gastro-protective microparticles made from alginate and chitosan that would release the nanoparticles in the intestine. A flow focusing microfluidic device was used to control alginate droplet diameters by tuning the continuous and dispersed phase flow rates. Alginate droplets encapsulating protein nanoparticles were then simultaneously crosslinked with calcium and layered with chitosan to form uniform, coated microparticles with high loading efficiency. Protein nanoparticles encapsulated within crosslinked alginate/chitosan microparticles retained approximately 50% of their activity after 1 hr incubation in simulated gastric fluid (SGF). Subsequent incubation in simulated intestinal fluid (SIF) released all remaining active protein within 3 hrs. Alginate vs. chitosan droplets, chitosan concentration, alginate concentration, number of layers, and crosslinker type were tuned to optimize protein protection in SGF and controlled release in SIF. Altogether, these hydrogel microparticles could serve as a potential oral delivery vehicle for protein therapeutics.

Copyright © American Institute of Chemical Engineers. All rights reserved.

458095 Photodynamic Enhancement of Drug Delivery to Tumors

Abstract:

Pancreatic cancer is associated with one of the poorest prognosis of all malignancies, with an unchanged five- year survival rate of 5% or less. The complex desmoplastic microenvironment impeding effective drug delivery to pancreatic tumors and the compensatory ability of cancer cells to adapt alternative survival pathways suggest that treatments will need to attack several targets. Photodynamic therapy (PDT) may have a specific role in enhancing drug delivery to pancreatic tumors and ameliorating drug resistance mechanisms by the targeted destruction of specific pathways. We have developed a bi-directionally cooperative combination regimen of PDT and irinotecan for pancreatic cancer using two clinically approved, nanotechnology-enabled therapies with a clear molecular basis. In an orthotopic pancreatic cancer mouse model, a combination of photodynamic therapy (PDT) and FDA-approved liposomal irinotecan (nal-IRI, MM-398) synergistically reduced the primary and metastatic tumor burden, and prolonged animal survival. This presentation will discuss how PDT increases tumor permeability and irinotecan transport by several mechanisms including disruption of efflux pumps, vascular and stromal integrity. Our findings provide a framework for the translation of mechanistic insights into rational new therapies for several cancers. This study is supported by the following NIH grants: 1K99CA194269 (HH); 5K99CA175292/ 4R00CA175292 (IR); 4R01CA15841505, 5R01CA16099805 5P01CA08420312 (TH). JL was supported by the Herchel Smith fellowship and Harvard College Research Program.

Copyright © American Institute of Chemical Engineers. All rights reserved.

471199 Highly Efficient Encapsulation of Small-Molecule N-Acetylcysteine within PLGA Nanoparticles to Restore Redox Balance in Oxidant-Stressed Environments

Abstract:

Introduction: Overproduction of reactive oxygen species (ROS) is a major factor in the pathogenesis of multiple sclerosis[1]. Localized antioxidant delivery is therapeutically relevant to counteract and neutralize ROS in such oxidant-stressed systems. Poly(lactic-co-glycolic acid) (PLGA) nanoparticles can provide a localized, controlled release of encapsulant, though reported encapsulation efficiencies of small-molecule therapeutics have been low: 12%[2] and 16%[3] for dopamine and N-acetylcysteine (NAC), respectively. Here we investigate how modifying the traditional double-emulsion, solvent evaporation method by doping external water phases with the small-molecule drug affects encapsulation efficiency, temporal release profiles, released drug activity, and released drug protective effect on oligodendrocyte progenitor cell (OPC) viability after H_2O_2 stress. We aim to deliver NAC, a potent, cheap, FDA-approved antioxidant, with high efficiency and temporal control.

Materials and Methods: NAC-loaded PLGA nanoparticles with varying lactide:glycolide (L:G) ratios and inherent viscosities were prepared by the double-emulsion, solvent evaporation method. The solvent for all water phases was deionized water. Briefly, the internal water phase (10% w/v NAC) was dispersed via sonication within the oil phase (15% PLGA in dichloromethane). The primary emulsion was dispersed in the outer water phase (1% PVA with or without 2.5% NAC) via sonication. The secondary emulsion was stirred in 0.3% PVA with or without 5% NAC for 4 hours. Particles were washed 3x in either DI water or 2.5% NAC and lyophilized. Release was determined by incubating particles at 37°C in phosphate-buffered saline and measuring supernatant NAC via Ellman's Reagent. LAP photoinitiator radical concentrations were measured via luminol to determine the dose response of NAC radical scavenging. OPC viability in response to oxidative conditions with and without the NAC antioxidant was measured via the CellGlo reagent.

Results and Discussion: Encapsulation efficiencies (relative to theoretical load without doped external phases) for NAC-loaded PLGA nanoparticles were greatest with external phase doping – reading greater than 300% of the case when undoped. With the doped particles, NAC was released for 14 days. NAC ranging from 0.5-1000 μM (the particle supernatant concentration range for each sample time) scavenged photoinitiated LAP radicals in a dose-dependent manner, with 0.5 μM and 1000 μM NAC scavenging 6.82% and 95.5% of radicals, respectively. Ongoing experiments show NAC-loaded nanoparticle antioxidant activity and protective effect on OPC viability.

Conclusions: Drug-doping of external water phases during the double emulsion, solvent evaporation method gave NAC-loaded PLGA nanoparticles with significantly higher encapsulation efficiencies than without doping, and up to 20-fold higher than previously reported values. NAC scavenged LAP radicals in a dose-dependent manner and was released from PLGA nanoparticles for at least two weeks. This indicates that these particles will provide a larger and more continuous small molecule dose longer than previously reported methods. Efficiently loaded antioxidant nanoparticles capable of controlled and localized delivery could have therapeutic potential in oxidant-stressed systems and bias progenitor cells such as OPCs toward a self-renewing fate.

References: [1]Lehnardt S. ***J. Neurosci,*** 2002**.** 22:2478-86. [2]Pahuja R. ***ACS Nano,*** 2015**.** 9:4850-71. [3]Lancheros R. ***Univ. Sci.,*** 2014. 19:161-168

Copyright © American Institute of Chemical Engineers. All rights reserved.

472474 Fabrication and Characterization of PLGA Nanoparticle-Bacteria Conjugate Nanobeads for Increased Tumor Penetration

Abstract:

Cancer is the second cause of overall mortality in the world and systemic chemotherapy is a major therapeutic approach for nearly all types and stages of cancer. Grand challenges of limited efficacy and systemic toxicity of existing chemotherapeutic treatments can be addressed through better drug carrier targeting and enhanced penetration of drug carriers. Attenuated strains *Salmonella* Typhimurium have previously been shown to have the ability to preferentially colonize tumor tissues. However, clinical success in immunocompetent hosts has been rare due to insufficient tumor colonization by bacteria[1–3]. We hypothesize that a combinatorial therapy approach based on integrating bacteria with chemotherapeutics-loaded nanoparticles will amplify both forms of therapy. Thus, we have developed a Nanoscale Bacteria-Enabled Drug Delivery System (NanoBEADS)[4] in which the functional capabilities of bacteria are interfaced with and augmented by chemotherapeutic-loaded nanoparticles. Each NanoBEADS agent is constructed by interfacing a *Salmonella* Typhimurium VNP20009 (i.e. a live engineered tumor-targeting bacterium) with an ensemble of poly(lactic-co-glycolic acid) nanoparticles loaded with chemotherapeutics for the purpose of using the bacterium to transport the particle deep into the tumor. To create the NanoBEADS, *S.* Typhimurium VNP20009 bacteria were cultured in Luria-Bertani broth (1% tryptone, 0.5% yeast extract and 0.5% sodium chloride) and coated with biotin-conjugated rabbit polyclonal anti-*S.* Typhimurium antibody. The nanoparticles were fabricated using a modified nanoprecipitation method[5] and fluorescently tagged with 6,13-Bis(triisopropylsylylethynyl)pentacene (TIPS pentacene). With a size of approximately 110 nm in diameter and a zeta potential of about -30 mV, the carboxyl end groups allow for EDC coupling to streptavidin. The biotin-coated bacteria were then mixed with streptavidin coated nanoparticles for 30 minutes at room temperature to promote self-assembly of the particles to bacteria. Scanning electron microscopy (SEM) images of NanoBEADS show the average number of nanoparticles

attached to each bacterium to be 35. In-vitro assays were carried out to compare the penetration of passively diffusing PLGA nanoparticles and NanoNBEADS in multicellular tumor spheroids.

1. Heimann, D. M. & Rosenberg, S. A. Continuous intravenous administration of live genetically modified salmonella typhimurium in patients with metastatic melanoma. *J. Immunother.* **26,** 179–80 (2003).

2. Toso Gill, V., Hwu, P., Marincola, F., Restifo, N, Schwartzentruber, D, Sherry, R, Topalian, S, Yang, J., Stock, F., Freezer, L., Morton, K., Seipp, C., Haworth, L., Mavroukakis, S., White, D., MacDonald, S., Mao, J., Sznol, M., and Rosenberg, S., J. & Toso, J. Gill, V., Hwu, P., Marincola, F., Restifo, N, Schwartzentruber, D, Sherry, R, Topalian, S, Yang, J., Stock, F., Freezer, L., Morton, K., Seipp, C., Haworth, L., Mavroukakis, S., White, D., MacDonald, S., Mao, J., Sznol, M., and Rosenberg, S. Phase I study of the intravenous administration of attenuated Salmonella typhimurium to patients with metastatic melanoma. *J Clin Oncol.* **20,** 142–152 (2002).

3. Pawelek, J. M., Low, K. B., Bermudes, D. & Pawelek, J. M., Low, K. B., Bermudes, D. Bacteria as tumour-targeting vectors. *Lancet Oncol.* **4,** 548–556 (2003).

4. Traore, M. A., Damico, C. M. & Behkam, B. Biomanufacturing and self-propulsion dynamics of nanoscale bacteria-enabled autonomous delivery systems. *Appl. Phys. Lett.* **105,** 173702 (2014).

5. Niu, X., Zou, W., Liu, C., Zhang, N. & Fu, C. Modified nanoprecipitation method to fabricate DNA-loaded PLGA nanoparticles. *Drug Dev. Ind. Pharm.* **35,** 1375–83 (2009).

Copyright © American Institute of Chemical Engineers. All rights reserved.

460403 Interstitial Release of Cisplatin from Triggerable Liposomes Enhances Efficacy Against Triple Negative Breast Cancer Solid Tumor Analogues

Abstract:

In solid tumors, including Triple Negative Breast Cancer (TNBC), limited penetration of therapeutics leads to heterogeneous intratumoral distributions reducing efficacy. Liposomal cisplatin decreases toxicities associated with free cisplatin. However, the heterogeneous intratumoral distributions of liposomes combined with limited release of cisplatin from liposomes may also reduce efficacy. This study uses pH-responsive liposomes designed to release cisplatin within the acidic tumor interstitium (7.0 < pH â‰¤ 6.0) with the aim (1) to improve the penetration of the free drug within tumors on the assumption of greater diffusivities based on the free drug's much smaller size and (2) to increase the availability of the free agent near cancer cells.

Introduction: In advanced solid tumors, including Triple Negative Breast Cancer (TNBC) solid tumors, the limited penetration of therapeutics results in limited therapeutic effects. TNBC is a subgroup of breast cancer associated with a high chance of cancer reoccurrence outside the breast and has been found to exhibit sensitivity to DNA-damaging platinum-based compounds, which is the basis for increased clinical use of platinum-derived agents. To enable selective and effective treatment of TNBC solid tumors, we study a drug delivery carrier of CDDP that is expected to result in potentially deep penetration and homogeneous distribution of the drug within the tumor interstitium. The carrier is a tunable (pH-releasing) liposome encapsulating CDDP. Tumor selectivity by these liposomes is mediated by the EPR effect. Efficacy is mediated by the pH-triggered release of contents within the tumor interstitium (7.0 < pH < 6.5) by extravasated liposomes. The small size of CDDP relative to its nanometer-sized carrier should exhibit greater diffusivities and deeper tumor penetration.

Methods: To study the efficacy of these liposomes on TNBC cells two different TNBC cell lines were studied: MDA-MB-231 and MDA-MB-468. The efficacy of pH-releasing liposomes (LD_{50}) loaded with CDDP was compared to the efficacy of other constructs including conventional (non-p-releasing) liposomes, empty liposomes, and free cisplatin. Following incubation of 6 and 24 hours, in addition to cell viability, cisplatin per cell, cisplatin intercalated to DNA and (if applicable) retention of cisplatin within the liposomes, were measured. In multicellular spheroids exhibiting intracellular pH gradients measured in situ by confocal microscopy using a membrane impermeant pH-indicator, we evaluated the spatiotemporal distributions of liposomes and contents. In dose-response studies, upon completion of incubation with all constructs, spheroids were washed and their volume change (volume at time t/volume at initiation of treatment x100%) was monitored in time followed by an outgrowth assay.

Results: On MDA-MB-231 and MDA-MB-468 cell monolayers, extracellular acidification resulted in lower LD_{50} values only for pH-releasing liposomal cisplatin (101.9 ± 24.1 and 14.5 ± 3.6 μg/mL, respectively) which were significantly less than for non-pH-releasing liposomal cisplatin (374.2 ± 20.0 and 33.7 ± 5.0 μg/mL, respectively). In contrast, at neutral pH, the LD_{50}values of both liposome types were comparable. Killing efficacies correlated strongly with the nuclear DNA-associated cisplatin independent of the delivery modality or the extracellular pH. pH-releasing liposomal cisplatin decreased the volumes of multicellular spheroids by 46.4-70.0 % more than non-pH-releasing liposomal cisplatin. pH-releasing liposomes resulted in improved penetration and more homogeneous profiles within spheroids of a cisplatin-surrogate cell-permeant fluorophore compared to non-pH-releasing liposomes.

Conclusions: These findings demonstrate the potential of intratumorally CDDP-releasing nanometer-sized carriers to selectively and effectively control TNBC solid tumors.

Funding Sources:

American Cancer Society (ACS): RSG-12-044-01

National Science Foundation (NSF): DMR-1207022

AAUP-AFT Professional Development Fund

Copyright © American Institute of Chemical Engineers. All rights reserved.

457602 Bioresponsive Polymer Coating on Targeted Drug Nanorods

Abstract:

Non-specific delivery and ineffective drug release at the target site are among the top challenges to effective treatment. The first challenge reflects the fact that interactions with carriers (*i.e.*, nanoparticles) can denature the active ingredients of drugs by shear stress. The second challenge is low affinity for specific receptor proteins overexpressed on the target cell surface. The third challenge of ineffective drug release can be overcome by exploiting the disease microenvironment. Here, we report a one-step process to prepare bioresponsive polymer coated drug nanorods (NRs) from liquid precursors using the solvent diffusion method. A thin layer (10.3 ± 1.4 nm) of poly(e-caprolactone) (PCL) polymer coating was deposited on the surface of camptothecin (CPT) anti-cancer drug NRs. The mean size of PCL-coated CPT NRs was 500.9 ± 91.3 nm length x 122.7 ± 10.1 nm width. The PCL polymer coating was biodegradable at acidic pH 6 as determined by Fourier Transform Infrared (FT-IR) spectroscopy. CPT drugs were released up to 51.5% when PCL coating dissolved into non-toxic carboxyl and hydroxyl groups. Trastuzumab (TTZ), a humanized IgG monoclonal antibody, was conjugated to the NR surface for breast cancer cell targeting. Combination treatments using CPT and TTZ decreased the HER-2 positive BT-474 breast cancer cells by 66.9 ± 5.3 % *in vitro*. These results suggest highly effective combination treatments of breast cancer cells using bioresponsive polymer coated specific drug delivery.

Copyright © American Institute of Chemical Engineers. All rights reserved.

466655 Internalization of Nanoparticles Functionalized with Low Molecular Weight Protamine into Erythrocytes

Abstract:

Erythrocytes have been exploited extensively for their potential applications as delivery vehicles due to their remarkable biocompatibility, biodegradability and prolonged life-span in circulation. The mature human erythrocyte is a terminally differentiated cell that does not possess an endocytic uptake mechanism. As a result, the efficient migration of bioactive molecular cargos such as large macromolecular drugs, genetic material, or proteins through the plasma membrane of erythrocytes remains as a major challenge for intracellular delivery applications. A variety of techniques have been successfully developed to entrap proteins, drugs, and enzymes within erythrocytes. The majority of these techniques, however, require disruption of the cell membrane to create large pores or perturbations through which biomolecules can diffuse. These techniques include chemical methods such as drug-induced endocytosis, mechanical methods such as hypotonic osmosis/dialysis, and electrical methods such as electroporation. The main disadvantages associated with these methods are the loss of structural integrity and cellular components of the cells due to weakened cell membranes. As a consequence, these modified erythrocytes are easily recognized by the host immune system. To achieve the benefits associated with erythrocytes as a drug delivery vehicle it is vital to maintain complete structural and functional integrity of erythrocytes.

Toward this goal, we introduced a novel method for transporting nanoparticles (NPs) into intact erythrocytes by combining the advantages of both cell-penetrating and targeting peptides. The method employs an ERY1 (WMVLPWLPGTLDGGSGCR) peptide that has a high affinity for the erythrocyte membrane protein glycophorin A and a low molecular weight protamine (LMWP) peptide that has been shown to transport proteins and small molecules into erythrocytes. These peptides are conjugated to a NP formed between fluorescently-labeled bovine serum albumin and a grafted copolymer of poly(ethylene) glycol and poly-L-lysine.

The formation of nanoparticles was observed using SEM, TEM, and dynamic light scattering. The nanoparticles had a core/shell structure and their size ranged from approximately 20-100 nm. Confocal microscopy and flow cytometry data confirmed that uptake of LMWP functionalized NPs was significantly higher (approximately > 3-5 fold) than that of other NPs at 6 h. Under these conditions, synthesized NPs were not toxic to the erythrocytes and no significant hemolysis was observed. Catalytic activity of the encapsulated protein was not negatively affected upon encapsulation or crosslinking. In conclusion, NPs were efficiently internalized into RBCs without adversely affecting cellular function and this method has the advantage over existing methods in that it does not appear to alter the structural integrity of the cell. The above approach of utilizing a surface modified NP architecture to improve internalization into erythrocytes may serve as a universal platform to deliver therapeutic proteins/drugs into RBCs in future. Furthermore, this nanoparticle system capable of entering red blood cells and extending circulation time has enormously broad potential, including improved treatment of blood-specific diseases such as malaria as well as treatment of acute lymphoblastic leukemia.

Copyright © American Institute of Chemical Engineers. All rights reserved.

466720 Enzyme Replacement Therapy Extended to the Brain through Nano-Polymersomes

Abstract:

Approximately 1 in 5,000 to 8,000 children are born annually with a lysosomal storage disease (LSD). Current treatment of LSDs, called enzyme replacement therapy (ERT), involves weekly to monthly intravenous(IV) infusions of missing endogenous enzymes. While ERT has proven effective in LSDs without central nervous system (CNS) involvement, the brain has remained untreatable due to the presence of the blood-brain barrier (BBB). The BBB is made up of endothelial cells with tight junctions, possessing a width between connections of 0.2 nm in size, which prevents passage of 98% of small molecule drugs, including enzymes, from the blood into the brain. Patients with neuropathic LSDs, which account for 50-70% of diseases in this category, present with severe CNS degeneration, ataxia, and premature death, with no treatment available on the market. The disease of focus, GM1 gangliosidosis, causes premature death between the ages of two and four with no hope available for parents of patients. Encapsulating enzyme into polymeric vesicles, called polymersomes, facilitates the transport of this missing enzyme into the brain through IV injections, effectively extending ERT therapy to the brain for the first time.

Polyethylene glycol-b-poly(lactic acid) (PEGPLA) has been proven to self-assemble into vesicles via both passive mixing and the solvent injection, using dimethyl sulfoxide (DMSO) and water. The mixing enzyme in GM1 gangliosidosis, β-galactosidase (βgal), is fluorescently tagged with reactive Alexa Fluor 488 at an efficiency of 13 dye molecules per mole of protein, with a finally protein concentration of 0.35 mg/mL attainable. Fluorescently tagged βgal is dissolved in stirring water prior to injection of DMSO and PEGPLA. Apolipoprotein E (ApoE) and CF350 attachment is facilitated through the introduction of homobifunctional PEG during injection. Polymersome diameter is characterized by dynamic light scattering (DLS). Self-assembly was stopped by freezing prior to lyophilization, with mannitol as a lyoprotectant. βgal loaded polymersomes were released *in vitro* in buffers of pH 7.4 and 4.8. Attachment was confirmed with DLS and Lowry assays. An immortalized feline fibroblast line, GM1/SV3, was used in *in vitro* studies to confirm particle uptake and treatment efficacy.

Previous studies indicated that incorporating 2 wt%/v mannitol during formation led to PSDs 54 ± 0.8% smaller after lyophilization. Therefore, 2 wt%/v mannitol was used to aid in the formation of brain-deliverable polymersomes encapsulated with βgal. PEGPLA polymersomes loaded βgal at 0.08 ± 0.03 mg βgal/mg polymersome with activity maintained at a value of 452 ± 82 a.u. DLS data indicates that polymersomes with βgal had diameters statistically similar to control. Release of βgal was doubled over 6 hours when loaded polymersomes are placed into citrate buffer (pH=4.1) when compared to tris buffer (pH=7.4), indicating increased enzyme release expected in the lysosome. First order kinetics of release were followed for three hours when βgal was released from polymersomes in acidic buffer prior to a plateau in mass released, indicating the potential for sustained long term release from polymeresomes. Activity of βgal was maintained after release. CF350 Amine was attached to polymersome surface, with an increase in diameter from 147.2 ± 24 nm to 210.5 ± 8.6 nm (Fig 1A). Fluorescence microscopy confirmed attachment (Fig 1C), with 79 ± 21% of polymersomes labeled with blue fluorescence. Furthermore, our carriers were capable of both encapsulating Alexa Fluor 488 and attaching CF350 at a high efficiency, with 87 ±12% of polymersomes presenting with both colors in flow cytometry. Attachment of ApoE was confirmed with an increase in polymersome diameter to 194.4 ± 3.5 nm and an increase in protein content from 0.06 ± 0.11 mg/mL to 1.03 ± 0.71 mg/mL. GM1SV3 cells, an immortalized cell line from felines with GM1 gangliosidosis, present low density lipoprotein receptors (LDLR) on the surface, as shown in confocal microscopy, indicating that they are a good model for ApoE mediated delivery through the BBB. ApoE increases the uptake of polymersomes into GM1/SV3 fibroblasts when compared to unlabeled control polymersomes.

This work highlights our ability to create polymersomes capable of protecting βgal in the blood stream, shown by limited release in a pH 7.4 environment, targeting the nanocarrier to the BBB, shown by the facilitated attachment of ApoE, and after transcytosis, releasing βgal in the lysosome of neural cells, shown by extended release in a pH 4.1 enviornment and cell studies in GM1SV3 cells. This combination of ERTand nanotechnology, demonstrating the capability of our carrier to transport enzymes to the brain while maintaining their activity, will create a paradigm shift in the treatment of CNS disease, providing treatment for currently untreatable and fatal diseases like GM1 gangliosidosis.

Copyright © American Institute of Chemical Engineers. All rights reserved.

473905 Regulatory Considerations for Introducing New Technologies during Drug Development

Abstract:

This presentation will cover regulatory aspects to consider when introducing new technologies during the development phase, with emphasis on changes made during the latter stage of product development. Some case studies of new technologies will be also be discussed.

Copyright © American Institute of Chemical Engineers. All rights reserved.

466568 Viscoelasticity andÂ Molecular Mobility of Celecoxib/Pvp/Tpgs Melt: A New Approach for Amorphous Drug Dispersion Assessment during Hot Melt Extrusion Product Design

Abstract:

Innovations at the Product Design and Discovery Interface

Abstract # 466568

Title: Viscoelasticity and molecular mobility of celecoxib/PVP/TPGS melt: A new approach for amorphous drug formulation assessment during hot melt extrusion (HME) product design.

Introduction/Motivation/Objectives:

Formulating drugs (active pharmaceutical ingredients) as amorphous solid dispersions is an alternative for improving the solubility of poorly water soluble drugs. Hot melt extrusion (HME) has demonstrated to be the adequate manufacturing unit process for compounding such amorphous solid dispersions. However, the risk of a potential event of inadequate amorphous content in the extrudate is always present, which is of critical significance when assessing the risk to the quality of such formulations during developmental product design phases.

Celecoxib is a poorly water soluble (BCS class II) non-steroidal anti-inflammatory drug. When compared to other BCS class II compounds, it has relatively low molecular mobility (high viscosity) and crystal growth rate upon cooling from the undercooled melt state [1,2]. The presence of surfactants or polymers has demonstrated to significantly increase or inhibit the radial crystal growth rate of Celecoxib, respectively [3].

We have previously studied the impact on the Celecoxib crystallization by PVP (polymer) and TPGS (surfactant) in terms of molecular mobility. Over all it was demonstrated that the polymer decreases the mobility of the drug, whereas the surfactant acts as a plasticizer. The dynamic viscosity of the Celecoxib with polymer and/or surfactant follows Arrhenius temperature dependence near the melting point (10°C of undercooling), however the VTF model results in higher agreement as the undercooling degree increases.

Even though, viscosity has been assessed for HME processability of melt formulations, it has not been associated to the amorphous content in the extrudate when processed using HME. Herein we present the link between viscoelasticity and amorphous state and content in the extrudate of mixtures of celecoxib, PVP and/or TPGS. We expect these results to complement the knowledge available for when assessing a drug as a candidate for a solid dispersion formulation during developmental stages of a drug product.

Sample Preparation:

Physical samples were prepared by mixing the drug with the corresponding amount of polymer and/or surfactant. Samples were mixed manually following a mixing protocol for homogeneous material distribution, and hence reproducibility.

Binary and ternary mixtures consisted of 50 wt.% TPGS or 50 wt.% PVP dispersed in a 50 wt.% Celecoxib, and ternary mixtures of 25 wt.% PVP, 25 wt.% TPGS and 50 wt.% Celecoxib. Samples where extruded using a Thermo Scientific MiniLab II HAAKE Rheomex CTW5 II HME. Extruded amorphous content and state was characterized using X-Ray Powder Diffraction (XRD), and Raman spectroscopy combined with a microscope for area of sample selection to obtain the spectra. The latter allowed for testing of drug homogeneous content throughout the extruded material.

Methodology/Data Analysis:

Extruded drug samples where analyzed in terms of amorphous content and amorphous state. The results where compared with our previous data on viscoelasticity and molecular mobility of the melt. Together we linked the molecular mobility (viscoelasticity) of the melt with the amorphous content in the extrudate of celecoxib mixed with PVP and/or TPGS.

Results and Analysis:

The samples were extruded using a bench-top HME (Thermo Haake MiniLab II) at a temperature of 140°C and a speed of 25 1/min (rpm). Raman spectroscopy of the extrudate samples resulted in a significantly less amorphous content when celecoxib was combined with PVP only, than when combined with both PVP and TPGS. This is evidenced with a noticeable larger Raman intensity of the celecoxib/PVP/TPGS mixture of each of the most important functional groups of celecoxib (CF_3, SO_2 and NH_2) compared to celecoxib mixed with PVP.

Powder diffraction characterization (XRD) showed a significant reduction in the crystalline state of the drug when mixed with PVP compared to PVP/TPGS, evidenced by the disappearance of the characteristic peaks of Celecoxib between 13 and 25 degrees (2Θ). On the other hand, TPGS increased the intensity of these peaks, and hence the crystalline state.

SEQ Table * ARABIC 1. Comparison between the off-line rheological and Raman data. A decrease in molecular mobility of the melt results in a higher amount of amorphous content in extrudate.

Samples	Off-line rheology			**HME**	Raman
	Molecular mobility (viscosity)*	Viscoelastic behavior*	Viscoelastic behavior* (140°C)		Amorphous content in extrudate*
Celecoxib/PVP	Decreased (viscosity increase)	Liquid to Elastic	Liquid		Increased
Celecoxib/PVP/TPGS	Decreased *(Decreased much less than with PVP alone)*	Liquid to Elastic	Liquid	**140°C, 25 min^{-1}.**	Increased *(Increased much less than with PVP alone)*
Celecoxib/TPGS	Increased (viscosity decrease)	Liquid to Elastic	Liquid		Decreased

*Compared each to pure Celecoxib.

The table summarizes the molecular mobility behavior and the amorphous content of the extrudate. When analyzed together, scientifically justifies the inclusion of viscoelasticity when designing a drug product formulation for HME. A decrease in viscosity results in an increase of the amorphous content of the drug, whereas an increase in molecular mobility results in a decrease in amorphous content in the extrudate. However, it must be highlighted that this correlation is conditional to the materials present in the melt, polymers (PVP) or surfactants (TPGS) as when these are combined less amount of amorphous content may result.

Conclusions:

We have elucidated the relationship between the molecular mobility of celecoxib when in presence of PVP (polymer) and/or TPGS (surfactant), the elastic/liquid (viscoelasticity) behavior, and the resulting amorphous content of the celecoxib in the extrudate. This work provides the proof-of-concept that viscoelasticity characterization combined with analytical techniques in HME could be used to tailor an amorphous drug with reproducible and adequate amorphous content manufactured using HME, being this correlation currently unknown and highly beneficial for when designing an amorphous drug dispersion.

References

[1] Baird, J. A.; Santiago-Quinonez, D.; Rinaldi, C.; Taylor, L. S. Role of Viscosity in Influencing the Glass-Forming Ability of Organic Molecules From the Undercooled Melt State. *Pharm Res* **2011**, *29*, 271–284.

[2] Baird, J.A.; Van Errdenbrugh, B.; Taylor, L.S. A Classification Tendency of Organic Molecules from Undercooled Melts. *J Pharm Sci* **2010**, 99, 3787-3806.

[3] Mosquera-Giraldo, L. I.; Trasi, N. S.; Taylor, L. S. Impact of Surfactants on the Crystal Growth of Amorphous Celecoxib. *International Journal of Pharmaceutics* **2014**, *461*, 251–257.

[4] Trachenko, K. The Vogel-Fulcher-Tammann Law in the Elastic Theory of Glass Transition. *Journal of Non-Crystalline Solids* **2008**, 354, 3903-3906.

[5] Gupta, P.; Chawla, G.; Bansal, A.K. Physical Stability and Solubility Advantage from Amorphous Celecoxib: The Role of Thermodynamic Quantities and Molecular Mobility. *Molecular Pharmaceutics* **2004**, 1, 406-413.

Copyright © American Institute of Chemical Engineers. All rights reserved.

472436 Miniaturized Purification Platform for Automated Screening of Flow Synthesis Using Extraction and Crystallization

Abstract:

R&D for continuous flow chemistry (e.g. end-to-end continuous manufacturing and automated synthesis lab) necessitates a modular and reconfigurable chemical synthesis platform capable of screening of a variety of chemical pathways. Flow microreactors are now commercially available, and more recently, drop-on-demand formulation device, such as ink- and droplet-based pharmaceutical printers, have made it to market. While these devices can be naturally combined to construct a modular end-to-end continuous flow platform, the platform cannot be fully realized without its purification counterpart. Nevertheless, microscale purification platform is not yet well established; Although a number of innovative separation techniques have been reported in the microfluidics literature, including evaporation, extraction, crystallization, filtration and chromatography, most of these processes have been designed as a standalone for lab-on-a-chip applications. An automated and reconfigurable purification device which can seamlessly integrate with other parts of the flow synthesis platform in a flexible and scalable manner is therefore desirable. In this work, a microscale purification platform which can virtually be constructed using off-the-shelf parts is described. Specifically, our device consists of a combination of commercially available parts, such as capillaries, chips, mixers, fittings, etc. as well as custom parts, including tubing heaters and zero-dead volume Y-junctions. In addition, the device support multiple reservoirs for use with aqueous and organic solvents, a Peltier-based heating and cooling system, a couple of process analytical tools, including UV and non-contact Raman spectroscopy and an auto-sampler for sample collection of offline analysis using techniques such as MS and UPLC. The device operation is controlled using a LabView software which integrates, controls and monitors electronic actuators, including syringe pumps, valves, pressure sensors, etc., and executes a sequence of instructions to perform priming, loading and cleaning. The capability of the device is demonstrated via automated screening of continuous and segmented (air and liquid) flow extraction and crystallization under various favorable and unfavorable conditions for the purification of model APIs including diphenhydramine and atropine.

Copyright © American Institute of Chemical Engineers. All rights reserved.

464011 Novel Validation Method Development for a Drug Delivery Device Used during Breastfeeding

Abstract:

4.5 million infants die every year. Many of these deaths can easily be prevented, with existing medications, if hygienic and effective paediatric formulations were available. In low-resource settings, the need for medications, which do not require refrigeration, can easily be accurately dosed, and can be hygienically administered is especially high, due to electricity, potable water, and literacy limitations. Breastfeeding infants are especially in need of appropriate drug and delivery device combinations, considering few drug delivery systems are commercially available specifically for this population, and infant compliance can be low.

We are responding by developing a novel device for administering medications or nutrients to infants, and *in vitro* validation methods for characterising itâ€™s efficacy. The device, when worn by a mother during breastfeeding, releases active pharmaceutical ingredients contained in the device into the milk directly consumed by the infant.Â It therefore has many potential advantages. For example, by suspending the medication or nutrient into the breast milk directly, and since it could be disposable, it overcomes the need for potable water as a suspension or cleaning fluid. It contains a solid dosage form, and therefore does not require refrigeration. Since the device is pre-loaded with medication a mother does not need to be literate to measure out a dose. Furthermore, since it is used during breastfeeding, infants may find it to be a more natural means of receiving medicine than conventional spoons or syringes.

Several validation methods have been used and developed in optimising the device design and potential corresponding dosage forms. These include a breastfeeding simulation apparatus, a novel infant tongue peristalsis simulation apparatus, and a novel tablet disintegration characterisation method using texture analysis.

The breastfeeding simulation apparatus was used to assess the release from the device of several dosage forms containing zinc, used in the treatment for diarrhoeal disease. The apparatus mimics maternal lactation, by flowing human milk through a silicone breast mimic. The milk then passes through the device loaded with the therapeutic insert, and then is sucked by a modified breast pump, at pressures physiologically representative of breastfeeding, into an exit line from which samples are collected with a fraction collector. The generated breast milk samples supplemented with zinc, collected over a simulated breastfeed are then analysed via inductively coupled optical emission spectrometry to test drug release rates from the drug delivery system. Zinc release over the course of 20 minutes, more than the time typically used during breastfeeding, in the simulations for four rapidly disintegrating tablet formulations ranged from 32-51%, indicating further formulation optimisation is needed.

An infant tongue mimic system was developed to simulate *in vitro* the peristaltic motions and associated pressures exerted from an infant tongue during breastfeeding. The system was used to characterise the potential effects of tongue movement on the disintegration of model tablets from the drug delivery system. The tongue mimic system includes a modified peristaltic pump equipped with a novel tongue mimic, a piezoresistive sensor for pressure measurements, a tube for holding the tablet under model breastfeeding flow conditions, and an adjustable palate mimic, allowing for pressure to the tablet to be modified based on clinical data. The release profiles of the model compound, sulforhodamine B, measured at two rotational values possible during breastfeeding for two potential compression values, representative of two potential tongue strengths of infants were statistically different ($P=0.0013$ and $P<0.001$, using 2-way ANOVA with matching for the lower and higher compression values respectively). For breastfeeding rotational rates representing non-nutritive breastfeeding, compression amount was found to be a significant factor ($P=0.0162$). In addition to being useful for characterising dosage forms for the device under the influence of tongue motion during suckling, this system could be useful in characterising other oral dosage forms or products.

Disintegration characterisation in human milk of potential tablet formulations is critical for screening dosage form appropriateness for the device. Therefore, a texture analysis disintegration method was also developed to analytically characterise tablet disintegration in opaque media. This method was developed to overcome visibility challenges which would be associated with use of the USP disintegration test with human milk. In this method, a texture analyser applies a load to a rapidly disintegrating tablet immersed in opaque media, and load data is collected over time as the tablet disintegrates, with fluctuations in the load indicating tablet fracturing. An in-house tablet formulation of a rapidly disintegrating tablet, and a commercially available tablet formulation were characterised in water, bovine, and human milk. These were found by one-way ANOVA to have statistically different disintegration times in each media ($P<0.0001$). This system is useful in characterising other tablet formulations for the device, and may be useful in the characterisation of other materials in opaque media.

Further work aims to continue testing various designs of the drug delivery system, as well as other formulations of corresponding dosage forms, to optimise the release of active pharmaceutical agent from the device.

Copyright © American Institute of Chemical Engineers. All rights reserved.

470420 Photothermal Therapy of Bladder Cancer and Melanoma Via Targeted Single-Walled Carbon Nanotubes

Abstract:

Single-walled carbon nanotubes (SWNTs) are unique in that they strongly absorb near-infrared (NIR) light, while biological systems have very low levels of absorption of NIR light. This project focuses on targeting the SWNTs using conjugation with the human annexin A5 protein. Annexin A5 (molecular weight of 36,000) is a monomeric protein that binds with high affinity to phophatidylserine (PS) in phospholipid bilayers. PS is the most abundant anionic phospholipid of the plasma membrane and is tightly segregated to the internal side of the plasma membrane in most cell types. Anionic phospholipids are largely absent from the external surfaces of resting mammalian cells under normal conditions. PS is expressed on the external surface of tumor cells and also of endothelial cells that line the blood vessels in tumors, but it is not expressed on the outside surface of the vascular endothelium in normal organs. The SWNTs serve as an NIR target resulting in tumor heating and ablation.

This project focuses on the photothermal therapy treatment of two types of cancer that initially occur on surfacesâ €"bladder cancer and melanoma. Approximately 80% of bladder cancer patients have recurrent tumors starting with stage I superficial and quickly progressing towards more muscle invasive and metastatic malignancies with each recurrence. The high incidence of reappearance is believed to be due to the residual tumor left behind in a significant number of patients. The current form of treatment is a transurethral resection for lower grade tumors and cystectomy for higher grade masses. Novel treatments are needed for eliminating all bladder tumor cells with a minimum number of treatments, ideally one. One type of skin cancer, malignant metastatic melanoma (CMM), has a high mortality rate. When CCM becomes metastatic, patient survival rates fall as low as 5%. Over the last half-decade rapid advances in biologic therapeutics and nanotechnology have spawned a new generation of treatment options proving a rich opportunity to investigate combinatorial therapies.

In the study of this therapy for treating bladder cancer, *in vitro* binding tests confirmed a strong binding affinity of SWNT-annexin A5 to MB49 mouse and J82 human bladder cancer cells. Combining SWNT-annexin A5 heating via NIR light confirmed significant cell death as compared to untreated controls for both cell lines. The *in vitro* tests provided statistically significant validation for the potential of this targeted ablation therapy. *In vivo* testing on immune competent mice was conducted to confirm the efficacy of this treatment even further. A biodistribution study followed by FT-Raman analysis confirmed no non-specific accumulation of SWNT-annexin A5 in any organs. Using a 360° diffuser to deliver NIR light to the inside of mouse bladders, NIR power tolerance tests confirmed that no healthy tissue damage occurred at 50 J/cm^2. *In vivo* therapy testing of MB49 cells implanted in the bladder of immune competent mice confirmed a promising treatment modality which is currently being further investigated with more testing using bioluminescence imaging of the cells in combination with fluorescence microscopy.

The work on treating melanoma used B16F10-luciferase mouse melanoma cells injected into immune competent mice. Plasma levels of SWNT-annexin A5 were monitored for up 24 hours after i.v. injection in B6 immune competent albino mice; plasma levels of the conjugate were negligible 1 hour after injection. The minimum injected dose of SWNT-annexin A5 used in combination with photothermal therapy using NIR irradiation was found for obtaining complete eradication of primary tumors (implanted subcutaneously into the flank). Having established the necessary photothermal therapy treatment conditions, we are currently undertaking an evaluation of the synergism between this photothermal therapy and the immunostimulants anti-PD1, anti-CTLA-4, anti-CD73, and cyclophosphamide to ascertain possible mechanisms of harnessing photothermal antigen production to eradicate distant metastases following primary tumor ablation.

Copyright © American Institute of Chemical Engineers. All rights reserved.

471289 Bioproduct Design through Market Assessment: NSF Innovation Corps (I-Corps^{TM}) Participation Guides the Development of New *Escherichia coli* Strains

Abstract:

Host cell proteins (HCPs) complicate the recovery of biologically-derived products due to their presence as the target product is isolated by various unit operations that include chromatography, precipitation, and other common bioseparation techniques. While the technology exists to rewrite the genome of the host cell used for expression, few have attempted to use these technologies to reduce HCP burden on the bioseparation steps. Such a space to investigate is quite large because of the number of different techniques and conditions used for bioseparation. Nevertheless, our team set out to elucidate a product design that could gain traction with academia and industry. Participation in the I-Corps^{TM} program defined the characteristics of a minimum viable product through a combination of hypothesis-driven discovery and potential end-user comments.

To wit, a target bioseparation technique (diethylaminoethyl or DEAE ion exchange chromatography) and improvement threshold was identified based on extensive discussions with potential end-users, and a minimum viable product was constructed. An engineered *Escherichia coli* cell line (Lotus®) was built via homologous recombination that displays excellent fermentation and expression characteristics. Moreover, DEAE column capacity was improved by 37% based on chromatography breakthrough data. This presentation will describe the fundamental research efforts, I-Corps^{TM} participation, minimum viable product development, and finally the assessment of the effort.

Copyright © American Institute of Chemical Engineers. All rights reserved.

451290 Recombinant Snake Antivenom Peptide

Abstract:

Tens of thousands of people die by snakebite every year, primarily in developing countries. Many more survivors continue to suffer permanent physical and mental disabilities from a bite. Cytotoxic snake venoms are present in many countries including from American rattlesnakes. Current Fab-based treatment for US snakes costs over $1K per vial but is covered by some insurance policies for humans but not normally for animals. A small ribosomal peptide was reported to neutralize snake venom that was identified as the N-terminus of a serum protein in the North American Opossum (*Didelphis virginiana*). We have found that the chemically synthesized peptide is able to neutralize the hemorrhagic toxins in different species of snake venoms when tested in mice. In addition to activity studies, we have explored the possibility of making the peptide recombinantly by a process that would be amenable to a large scale production. The goal of our work is to propose a large scale, low cost process for the peptide that can be implemented in developing countries where snakes are prevalent and the population cannot afford expensive horse serum-based antivenom treatments. Our process obviates the use of snake venoms that are currently required to inject horses for the production of antibodies. We have successfully produced the peptide in *E. coli* and purified sufficient quantity to demonstrate activity in mice. Process optimization is currently underway, in addition to expanded activity studies. The presentation will cover details of our process studies and the activity of the *E. coli*-produced peptide.

Copyright © American Institute of Chemical Engineers. All rights reserved.

453622 Ultrasensitive Microfluidic Assay for Genome-Wide DNA Methylation Analysis and Precision Medicine

Abstract:

Epigenetics refer to any process that alters gene activities without changing the DNA sequence. Previous studies have demonstrated that epigenetic regulation, such as DNA methylation, plays a critical role in many biological procedures, such as embryo development and learning process. Genome-wide study of dynamics of DNA methylation at single-base resolution provides a comprehensive understanding of methylome configuration. Bisulfite sequencing is regarded as the gold-standard technology for analyzing DNA methylation. This technology is based on the different consequences after cytosine and 5-methycytosine (5-mC) treated with sodium bisulfite. However, conventional protocols required more than 300 ng DNA due to the significant sample loss during bisulfite treatment (up to 98% [1, 2]). This limits its application for studying precious samples and understanding the underlying heterogeneity of clinic samples. To broaden the applications of bisulfite sequencing, it is necessary to develop ultrasensitive tools for epigenomic study, particularly DNA methylation analysis. **We reported the first microfluidic DNA methylation assay at genome-wide scale and single-base resolution with >3 orders of magnitude higher sensitivity than conventional protocols**.

To improve the sensitivity of the assay for studying subsets of neurons, we developed a diffusion-based microfluidic assay that preserved significantly more DNA than conventional assays. There were two main factors contributing to the DNA preservation: 1) Taking advantage of manipulating solution with extremely small volume (<1 μl), bisulfite treatment (e.g. dosing, temperature and time) were well controlled. 2) Column-based DNA desalting in conventional assay was replaced by diffusion purification [3]. The diffusion scheme also increased the concentration of bisulfite mix which shortened the reaction time from 4~16 hours to 1-2 hours.

We used our microfluidic protocol to perform reduced representative bisulfite sequencing (RRBS) with various amounts of genomic DNA extracted from GM12878 cells. For all 4 sample sizes, replicate experiments were highly correlated (r = 0.995, 0.992, 0.979 and 0.912 for 300, 10, 1 and 0.3 ng, respectively). With similar sequencing depth, we identified averagely 1.82 million CpGs with 5x coverage using 0.3ng DNA, compared to 0.94 million in ENCODE data (deposited standard data using 1-2 μg DNA). When 300 ng samples were used as a reference, 92.1, 83.8, and 57.9% of methylation was observed in 10, 1, and 0.3 ng samples, respectively. This indicates that >50% of the genome was covered when the amount of starting DNA was decreased by 3 orders of magnitude.

References

1. M. Ehrich, S. Zoll, S. Sur and D. van den Boom, *Nucleic Acids Res.*, 2007, **35**, e29-e29.

2. C. Grunau, S. J. Clark and A. Rosenthal, *Nucleic Acids Res.*, 2001, **29**, E65-65.

3. S. Ma, D. N. Loufakis, Z. Cao, Y. Chang, L. E. K. Achenie and C. Lu, *Lab Chip*, 2014, **14**, 2905-2909.

Copyright © American Institute of Chemical Engineers. All rights reserved.

460310 Recurrent Patterns of DNA Copy Number Alterations in Tumors Reflect Metabolic Selection Pressures

Abstract:

Human tumors exhibit recurrent patterns of DNA amplifications and deletions across diverse cancer types. These patterns are suggestive of conserved selection pressures during tumor evolution but cannot be fully explained by known oncogenes and tumor suppressor genes. Applying principal component analysis (PCA) to genome-wide DNA copy number alteration (CNA) from 17 tumor types, we identified previously unreported CNA genomic sub-signatures. Combining these human signatures in a cross-species comparison with signatures from mouse models of cancer allowed for synteny-based elimination of passenger genes. Examination of the cross-species refined CNA signatures revealed coordinate enrichment of copy number alterations in core glycolysis enzymes. Notably, these PCA-defined CNA signatures are predictive of glycolytic phenotypes, including FDG-avidity of patient tumors. Using an experimental immortalization system, we show that exogenous expression of metabolic enzymes alters the copy number status of the corresponding endogenous loci, supporting the hypothesis that these metabolic genes act as drivers within the conserved CNA amplification regions. Finally, using mass spectrometry-based metabolomics, we demonstrate that cells with glycolytic CNA signatures channel an increased proportion of glucose-derived carbon into pentose phosphate-associated biosynthetic pathways including nucleotide synthesis. Taken together, our experimental and computational data support a model in which glycolysis-linked selective pressures encountered during tumorigenesis (eg, redox stress and senescence) shape the highly recurrent DNA copy number alterations found in aneuploid human tumors.

Copyright © American Institute of Chemical Engineers. All rights reserved.

451812 Comparative Mapping of Dengue Virus-Host Interactions Using Global Proteomics

Abstract:

Introduction: Dengue virus (DENV) is transmitted by mosquitoes and infects nearly 400 million people annually. Like other viruses, DENV interacts with the host to orchestrate myriad biochemical processes required for replication. A unique characteristic of vector-borne viruses such as DENV is that they use cellular machinery in two very different hosts to successfully replicate. This raises the question, how does DENV hijack cellular processes for replication in two hosts that have diverged over hundreds of millions of years.

Materials and Methods: We investigated this question using complementary systems biology techniques. We first sought to systematically identify DENV-human and DENV-mosquito protein interactions for all 10 DENV proteins using affinity purification/mass spectrometry (AP/MS). To identify DENV-human and DENV-mosquito protein interactions, individual Strep-tagged DENV proteins were over-expressed in human HEK293T or *A. aegypti* Aag2 cells, and subjected to AP/MS. High confidence interactions were determined using a statistical algorithm called MS Interaction STatistics (MiST) to filter the resulting MS data based on specificity, reproducibility and abundance of the identified peptides. We then sought to systematically link physical interactions to functional outcomes during DENV replication using RNAi knockdown and a reporter virus system in human Huh7 cells.

Results and Discussion: From our proteomic work, we recovered many previously reported physical interactions, included interactions between NS1 and Complement C4, NS5 and FAM192A, and NS5 and Importin-a/b. We also found a significant enrichment for orthologous interactions that are shared between the human and mosquito hosts, potentially providing conserved physical interfaces by which functional modules can be manipulated during replication. The functional genetic characterization of 349 DENV-interacting human factors identified 95 genes that significantly impacted viral replication, including previously reported replication factors SEC61G and UBE3A. To

further understand how shared interactions facilitate replication in mosquitoes, we are testing the impact DENV-interacting insect factors on DENV replication in the mosquito *A. aegypti* Aag2 cells. This comparative approach will elucidate how conserved functional modules are manipulated by a virus to facilitate replication two evolutionarily distant hosts. Upon integration of the proteomic and genetic data, we identified several nuclear human factors that impact viral replication and physically interact with DENV. We are currently investigating how these DENV-host interactions impact global gene expression using ChIP-seq and RNA-seq. Our preliminary data suggest that DENV inhibits the expression of antiviral interferon stimulated genes through a protein interaction between the viral polymerase and transcription elongation machinery.

Conclusions: Using proteomics and genetics, we identified DENV-host protein interactions in both human and mosquito systems. We further determined the impact of these host factors on viral replication, and identified interactions between DENV proteins and host nuclear factors involved in the regulation of gene expression. Our data support a model in which a physical interaction between the DENV polymerase and transcription elongation machinery inhibits the antiviral innate immune response.

Acknowledgements: We thank Dr. Michel Tassetto and Dr. Jason Wojcechowski for feedback. This work was supported by DARPA 10-93-Prophecy-PA-008 to RA and NJK, and NIH F32AI112262-01A1 to PSS.

Copyright © American Institute of Chemical Engineers. All rights reserved.

462527 A Capture-Based Technique for Enhancing RNA-Seq Analysis of Bacterial Transcriptomes during in Vitro and In Vivo Infections

Abstract:

To fully understand the dynamic spatial and temporal interactions of a pathogen with its host, it is necessary to analyze the transcriptomes of both the host and the pathogen throughout the course of an infection. Even using massively parallel techniques such as microarrays or RNA-Seq, analysis of the pathogen transcriptome is complicated by the overwhelming amount of host RNA (>99.9%) associated with an infected sample.

We have developed a solution-phase hybridization-based capture approach to bind specifically to pathogen transcripts, followed by a spin column-based chromatography step to recover the bound transcripts. This protocol dramatically increases the number of reads mapping to pathogen genes by >100-fold. We demonstrate application of the technique to in vitro tissue culture infections with a multi-drug resistant isolate of Klebsiella pneumomiae, and demonstrate high-level constitutive expression of numerous drug resistance genes even in the absence of antimicrobial drugs. We further detect expression of putative virulence-associated genes throughout the course of the infection. We also present preliminary investigations into in vivo infections with Klebsiella pneumoniae or Pseudomonas aeruginosa in the murine lung, demonstrating the capability of the technique for simultaneous host and pathogen expression profiling. Further, we are developing consumables and automation to increase the throughput of the capture technique, in an attempt to reduce the hands-on time and improve the reproducibility of the technique. We discuss possible applications of the technique to targeted drug discovery, as well as potential applications to other (non-infectious disease) systems where RNA of a species of interest is greatly outnumbered in a background of non-informative RNA.

Copyright © American Institute of Chemical Engineers. All rights reserved.

467629 Biologically Consistent Annotation for Metabolite Identification from Metabolomics Data

Abstract:

Research on biological systems increasingly relies on analytical techniques that generate large datasets which can be difficult to process. Metabolomics, which deals with the global analysis of low molecular weight metabolites in a biological system, has emerged as a key "omics" field that can provide detailed information about cellular reactions. Metabolite analysis has become increasingly relevant for diverse applications, including process intensification in bioprocess development, identifying diagnostic biomarkers, and understanding cell-cell interactions in complex microbial consortia, to name a few.

Advances in detection technology have greatly expanded the number and type of metabolites that can be detected. Coupled with appropriate chromatographic separation methods, high-resolution mass spectrometry (MS) can be used to perform sophisticated untargeted experiments to achieve broad coverage of the chemical species in a sample, while also obtaining rich information on the species' molecular structure through their fragmentation patterns and corresponding MS/MS spectra. On the other hand, datasets resulting from untargeted experiments are very complex, and can contain $10^3 \sim 10^4$ spectral signatures or "features." Analyzing such datasets to obtain meaningful biological information depends on reliably and efficiently resolving the chemical identities of the detected features.

Despite progress, the identification of metabolites from spectral data remains a major bottleneck. In untargeted metabolomics, the objective is to characterize as many compounds as possible. The gold standard is to annotate the data by referencing a high-purity chemical standards library generated using the same analytical method as the samples. However, this is often intractable, since the sample could contain thousands of unknown metabolites. A common alternative is to search publicly available spectral libraries for entries that match the observed MS/MS spectra. However, the number of pure chemical standards covered by spectral MS/MS libraries is very small compared to the number of metabolites occurring in nature. Two commonly used libraries, METLIN and HMDB, contain MS/MS data for only ~16% of metabolites catalogued in KEGG.

In recent years, a number of computational tools have been developed that utilize *in silico* fragmentation, molecular fingerprinting, and machine learning. While these tools can greatly increase the number of features that can be assigned a putative identity, difficulties remain in functionally annotating the data. Given that many compounds share the same chemical formula (and hence mass) and fragment similarly, associating a spectral signature with a chemical identity can be ambiguous. This is also reflected in the outputs of *in silico* fragmentation tools, which typically suggest many possible matches for a given experimentally observed MS/MS spectrum.

This paper describes a novel computational workflow that utilizes the biological context of a sample to annotate and interpret the metabolomics data. The basis for this context-driven approach is that the metabolites present in a sample reflect enzyme-catalyzed biochemical reactions active in the corresponding biological system. We develop this workflow to specifically address the comparison or differential analysis of samples representing two or more different experimental conditions, a typical problem in metabolomics.

The major steps in the workflow are as follows. First, a metabolic model is constructed from genome annotation data in KEGG to define a set of metabolic reactions that could be expressed in the biological system of interest. Second, the detected features are mapped to candidate compounds in the model, creating a graph that connects the candidate compounds via edges representing reactions. The nodes in this graph are weighted by scores reflecting the confidence in the mapping of detected features to candidate compounds. These scores are calculated based on matches of features to MS/MS spectra in databases and predicted spectra generated using *in silico* fragmentation tools. The edges in the graph are weighted by the confidence that the corresponding reactions are expressed in the biological system. Third, a local neighborhood analysis is performed on the weighed graph to further refine the confidence calculations. If a large fraction of connected metabolites in a subset of the graph, or neighborhood, consists of high confidence metabolites, this increases confidence in the assigned identities of all metabolites in the neighborhood. Features associated with multiple candidate metabolites are preferentially mapped to such high confidence neighborhoods, since there is a greater likelihood that the reactions are actively engaged. The results of this analysis also inform pathway engagement, as pathways covering many high confidence neighborhoods are also likely engaged.

We validate the workflow using experimental data from two case studies. The first case deals with metabolomic data collected on a single, well-characterized cell type, whereas the second case deals with data on a complex microbial community harboring only partially characterized member species. Untargeted metabolomics experiments were run using multiple chromatography methods in combination with both positive and negative ionization modes, which broadened coverage of unique masses by >50%. Use of an accurate metabolic model is critical; up to 90% of the candidate compounds that match to KEGG compounds based on mass (precursor m/z) could be eliminated as likely false matches by considering only compounds present in the model.

The annotation generated using our workflow was compared against annotation results collected from available databases (METLIN and HMDB) and recently published *in silico* prediction tools (CFM-ID and MetFrag). For selected metabolites with available chemical standards, the predicted annotation was also confirmed experimentally. Our results to date indicate that all three steps in the workflow greatly contribute to improving the annotation yield and accuracy.

Copyright © American Institute of Chemical Engineers. All rights reserved.

467602 Deductive Determination of Dynamic Cellular Objectives from Biological Data

Abstract:

Cybernetic models describe metabolic changes by means of simplifying intricate cellular regulation mechanisms with the concept of "dynamic metabolic goals." In prior work, these goals have taken the form of inductive assumptions about the nature of metabolism and are explicitly stated in forms like "maximization of growth rate" or "maximization of the rate of carbon uptake." Also, often cellular phenotypes resulting from specific stimulus can serve as context-specific objective functions. For instance, in the case of macrophage metabolism in response to stimulation by LPS, the inflammation phenotype represented by cytokine readout profiles can serve as an objective function. These inductively determined metabolic goals have yielded numerous predictions of metabolic phenomena, but a procedure to determine the most descriptive cybernetic objective function for a given system has yet to be proposed. To address this, we develop a method to deductively resolve metabolic objectives from a combination of metabolomic and transcriptomic data.

This analysis simultaneously relates dynamic metabolic changes with the regulation of enzymes that influence those metabolic changes. It provides objective functions that take the form of optimal weightings which represent the return on investment for various metabolic pathways that are competing for same pool of resources. To give biological context to these returns on investment, a method to mine gene expression data and explain the objective function using pathway enrichment analysis is proposed. We use this approach to analyze three systems: two diauxic growth scenarios and prostaglandin metabolism in a mammalian cell line. Pathway enrichment analysis of the genes mined using this procedure provides objective functions that show agreement with already established experimental knowledge related to the behavior of these systems in their respective conditions. The ability to determine objective functions deductively from the data allows for the formulation of robust cybernetic descriptions of systems where objective functions are difficult to determine inductively. This approach has applications in specifying objective functions for more complicated metabolic systems in multi-cellular organisms.

Copyright © American Institute of Chemical Engineers. All rights reserved.

467852 'Omics Profiling of a Continuous Manufacturing CHO Production Platform for Monoclonal Antibodies

Abstract:

The manufacturing of biotherapeutics such as recombinant proteins is usually performed in fed-batch mode using Chinese hamster ovary (CHO) cell lines as main expression system. Currently, continuous manufacturing is only preferred if the protein is unstable and prone to chemical or enzymatic degradation. Nevertheless, using perfusion reactors is beneficial for the reduction of quality inconsistencies due to the short residence and steady operation, which is equally favourable for more stable products as monoclonal antibodies (mAbs). Realizing process and product steady-state by systematic process development instead of using empirically driven approaches is one of the key points within the Quality by design (QbD) concept. Thereby, the increase in knowledge about cell physiology is a cornerstone within this development strategy, where 'omics technologies can be helpful to address the regulatory behaviour of the used cell lines.

In this work, transcriptome and proteome profiling of a CHO cell line producing a mAb was performed to investigate the intracellular variability during steady-state of a perfusion reactor. Daily samples were analysed by high-throughput methods depicting the time evolution of several thousand transcripts and proteins. Data mining by different clustering methods allowed the analysis of each individual 'omics study. Furthermore up- and down regulated intracellular processes and pathways responsible for the quality of the secreted mAb were investigated. By the use of bioinformatics databases the biological function was assigned and moreover key functions were characterized by functional annotation. Additionally, to identify differences between the mRNA and protein expression in time, an integrated data analysis of both datasets was completed, as the correlation of the proteomic and transcriptomic expression response is not necessarily given for every regulatory cell process. In conclusion, useful insights into cellular behaviour can be gained by the joint application of 'omics technologies to address the intracellular process consistency of perfusion cell culture.

Copyright © American Institute of Chemical Engineers. All rights reserved.

457484 CFD MODEL Based Near-Infrared Spectroscopy Implementation for in-LINE ASSAY Monitoring of a Pharmaceutical Suspension

Abstract:

INTRODUCTION AND PURPOSE

During bottle filling of a pharmaceutical suspension formulation, a decreasing trend in assay is observed at the end of the batch. This leads to approximately 17% of the batch content which cannot be commercialized. The root cause can most likely be found in foam present in the top layer of the suspension in the holding tank which contains less API. The aim in this study is to implement a PAT solution for establishing a batch-dependent endpoint (I.e. assay â‰¤ 95%) determination while reducing in-process sampling drastically.

MATERIALS AND METHODS

Materials

Tests were performed with a high-dose (i.e. 100 mg/mL) pharmaceutical suspension formulation. Lab scale samples were prepared in a 10L mixing vessel. Full scale batches are manufactured in a 150L compounding vessel following a one-pan process. After the transfer to a holding tank, the full scale batches are gradually discharged to a specific bottle filling system with the help of a peristaltic pump. During this process step, the suspension is stirred continuously.

Methods

NIR spectra were recorded using a Fourier-transform Near-Infrared (NIR) spectrometer (Bruker Matrix-F Duplex) equipped with a TE-InGaAs detector, a quartz halogen lamp and a fiber-optic non-contact probe. Every 30

seconds, a spectrum was acquired in the 12000-4000 cm^{-1} region with a resolution of 8 cm^{-1}, averaged over 64 scans. Ultra Performance Liquid Chromatography (UPLC) was selected as reference technique for API content determination. All UPLC measurements were performed at Janssen Pharmaceutica, Beerse, Belgium, using their developed and validated method (confidential).

RESULTS AND DISCUSSION

Samples in a range of 85-115% target concentration were analyzed with both NIR spectroscopy and UPLC. A lab-scale calibration model was developed which shows acceptable feasibility for NIR spectroscopy as potential PAT tool for the API assay determination in suspension.

For approaching the production measurement setup as good as possible, the calibration model was optimized by using a circular system with a similar flow as in production (i.e. 800 mL/min) with the help of a customized T-piece. Samples with concentrations varying between 90-103% were measured and included in the prediction model.

At production scale however, the developed model showed a weak performance with an unexpected variation in NIR assay results. It was assumed that the signal variation was created by disturbed flow patterns/turbulence in front of the lens due to possible back pulse formation at the system output. A Computational Fluid Dynamics (CFD) simulation of the flow pattern of the suspension inside the measuring interface confirmed this assumption.

CFD simulations demonstrated that an extension of the original T-piece could optimize the measurements by transferring the back pulse formation further away from the lens. This mechanical solution allowed a constant sample volume per time unit in front of the NIR laser. This was confirmed with additional lab-scale tests and the follow-up of several production scale batches.

CONCLUSIONS

Despite the dominantly present water absorption bands, NIR spectroscopy is feasible to establish good API assay predictions of this high solid-load suspension formulation.

Experiments indicated that a fine-tuning of the interfacing device is crucial to achieve adequate in-line NIR measurements. Therefore, sufficient process knowledge, including fluidum mechanics, is desired to rationalize the choice for an optimal measurement setup.

CHALLENGES AND FUTURE WORK

In the future it is aimed to investigate if NIR spectroscopy in combination with this specific interfacing device is applicable for other liquid pharmaceutical formulations.

Copyright © American Institute of Chemical Engineers. All rights reserved.

466854 Using Optical Coherence Tomography (OCT) to Analyze in Real-Time the Tablet Coating Variability of a Continuous Drum Coating Process

Abstract:

This study demonstrates an in-line measurement system used in a continuous drum coater (Driaconti-T lab, Driam Anlagenbau GmbH). Process monitoring is performed by an optical coherence tomography (OCT) sensor (OCT Pharma 1D, Phyllon GmbH) providing cross-sectional images of film-coatings in a non-destructive and contactless manner.

The continuous coater has three chambers, each with a product volume of around 2 liters of tablets. In continuous operation mode, the tablets are inserted in chamber 1 and transported to chamber 3, while being coated in each of them. Thus, the product is handled in small batches, without exchange of product between the chambers, resulting in a defined residence time for each tablet.

OCT measurement and subsequent automatic data analysis is a method to determine the coating thickness from OCT images without the need of chemometric calibration models, based on a known refractive index of the coating materials. OCT was used to obtain the temporal evolution of the quality parameters, i.e., mean coating thickness, inter-tablet coating variability and intra-tablet coating variability during the coating process. Thereby, the influence of spray rate, air temperature and spray pattern pressure on the quality parameters was studied based on a full-factorial DoE. Additionally, the variability of single drum segments of an extended period of time was studied.

Continuous coating, here in the form of small batches, is an important enabler for continuous manufacturing of solid dosage forms. In combination with an in-line OCT system the coating process can be monitored and actively

controlled.

Copyright © American Institute of Chemical Engineers. All rights reserved.

467352 Advanced Flexible Control System Implementation into Direct Compaction Continuous Pharmaceutical Manufacturing Pilot-Plant

Abstract:

Currently, pharmaceutical companies are going under paradigm shift from batch to continuous manufacturing. Real time monitoring and supervisory control system are necessary part of efficient continuous manufacturing. However, the minimum level of control necessary to produce the product with desired critical quality attributes depends on several factors including drug content (low dose, high dose), formulation properties (e.g. flowability, segregation), and product properties. Nevertheless, different control strategies such as feedforward (FF), feedback (FB), or combined FF/FB[1]as well as different control algorithms such as PID, MPC, or hybrid MPC-PID can be potentially used. Selection of a right control strategy and algorithm is mandatory task both for efficient manufacturing as well as regulatory point of view. However, because of different levels of complexities and unavailability of systematic method and tools, the identification of required level of control, type of control strategy and control algorithm are still a challenging task.

In this work, a flexible control system has been implemented into a direct compaction continuous tablet manufacturing pilot-plant. The flexible control system consists of three control strategies (FF, FB, combined FF/FB[1]) and three control algorithms (PID, MPC, hybrid MPC-PID) giving options for multiple control schemes. The flexible control system is designed in such a way that any control scheme can be easily selected. The control system is also modular in nature that can be easily transferred to any other control platform and manufacturing plant. The continuous tablet manufacturing pilot-plant runs under each control scheme and their performance are evaluated using standardized performance indicators. Based on the results obtained, a systematic methodology is proposed to assist the selection of control strategy and algorithm for continuous pharmaceutical manufacturing process.

Direct compaction continuous tablet manufacturing process is considered to demonstrate the implementation of flexible control strategy. The flexible control strategy consist of the five supervisory feedback control loops and three supervisory forward control loops among which the required number of control loops can be selected. Feedback loops are for drug concentration, powder level in chute, main compression force, tablet weight and hardenss control while the feedforward loops are for proactively mitigate the effects of process and raw materials variabilities. For example, measured powder bulk density is considered in real time in tablet press to proactively mitigate the effect of powder bulk density on tablet weight and hardness via changing the fill depth. Similarly, the feed forward loop to mitigate the effects of lubricant concentration variability on tablet hardness and thereby on tablet dissolution is added. As well as a feed forward control loop to mitigate the effects variation on blend composition on tablet potency. All feedforward control loops can not be used simultaneously because of limitations of number of actuators and therefore the methodology plays an important role to identify the appropriate control loops for a particular pharmaceutical product. The flexible control system are implemented using industry standard control platform (e.g. DeltaV), PAT data management tool (synTQ), OPC communication protocol, fieldbus devises, chemometric tools and real time monitoring tools.

Reference

1. Singh, R., Muzzio, F., Ierapetritou, M., Ramachandran, R. **(2015)**. A combined feed-forward/feed-back control system for a QbD based continuous tablet manufacturing process. PROCESSES Journal, 3, 339-356.

Copyright © American Institute of Chemical Engineers. All rights reserved.

468862 Raman Spectroscopy- Towards the Prediction of Quality Attributes and Application in Cell Culture Process Development

Abstract:

In recent years, the United States Food and Drug Administration (FDA) and the European Medicines Agency (EMA) have stressed the need of improved process understanding and control in the area of pharmaceutical development, manufacturing and quality assurance, summarized in the Process Analytical Technology (PAT) initiative. Lately, Raman spectroscopy has emerged as a promising online monitoring and control tool in bioprocesses. The advantages of this technology are the invasive and nondestructive operations, the simultaneously detection of multiple process variables, as well as the online operation mode and real-time information release.

Raman spectroscopy is mainly used for monitoring purposes at production stage, where variations are small and the process is already well defined. However, the exploitation of Raman technology for process development, where different process conditions, media, products or cell lines are tested resulting in a large variability and few experiments at similar conditions, could bring very important benefits, in terms of process understanding, speed and cost reduction. This work presents several ideas, experimental results and recommendations to overcome those challenges in process development.

Different fed-batch cultures, varying in cell lines, products and platform media, were monitored online. In addition, diverse spiking strategies were performed to artificially increase the calibration range as well as the quality of the prediction models. The evaluation of advanced modeling techniques to derive optimal results from the obtained spectral data plays a central role in this work. Those techniques included an in-depth analysis of the pretreatment procedure, wavelength selection and outlier elimination with regards to robust calibration and prediction of the components of interest.

Besides classical parameters such as glucose, lactate, titer and viable cell density, all 20 amino acids could be very well predicted. More than that, even the prediction of quality attributes like aggregates and glycans showed promising results. In addition, the potential of generic models, which can be applied independently of the cell line, process platform and medium, could be proven. This enables the Raman spectroscopy as an online monitoring tool of multiple process variables and its implementation into a supervisory control of a continuous integrated bioprocess to ensure consistent process efficiency and shaping the product qualities.

Copyright © American Institute of Chemical Engineers. All rights reserved.

470108 Hydrodynamics and Mass Transfer Effects in Mini Vessel Dissolution Testing Systems

Abstract:

In the pharmaceutical industry, dissolution testing is routinely conducted to evaluate the *in vitro*performance of solid dosage forms in pharmaceutical development, determine the dissolution behavior of formulations and provide an optimization of drug release dosage forms, and as a quality control tool to insure that solid dosage forms have consistent dissolution properties. The USP Dissolution Testing Apparatus 2 is the device most commonly used for this purpose. In a typical test, a tablet is added to the vessel and samples are manually withdrawn from the vessels over time and analyzed for their drug content.

The USP 2 drug dissolution testing method is not always appropriate, especially for dissolution testing involving very small tablets or when not enough drug substance may be available for proper dissolution testing in the USP 2 Apparatus, especially during the initial stage of drug development. Small-volume dissolution testing apparatuses (mini vessel systems) can be advantageously used instead of the standard USP 2 system to overcome these limitations. The most common of these mini vessel systems is similar in shape and operation to the USP 2 apparatus, but it only requires a small volume vessel (200 mL) and a mini paddle impeller.

Before mini vessel systems can gain wide acceptance in industry, their hydrodynamics needs to be studied, and their operating conditions must be defined in order to compare the velocity distribution and mass transfer coefficient in a mini vessel system with those in the standard USP2 system. Very few studies on mini vessel systems are available. Therefore, the objective of this work was to quantify the hydrodynamics and drug mass transfer rates in mini vessel systems, and determine the operating conditions under which mini vessel systems can be reliably used instead of USP 2 system for dissolution testing purposes.

The velocity distribution in a minivessel was obtained here using Computational Fluid Dynamics (CFD) at different agitation speeds. The predictions were compared to the velocity vector maps obtained using Particle Image Velocimetry (PIV). Four horizontal iso-surfaces were selected to be studied in detail. The CFC predicted and PIV measured flow patterns were found to be in good agreement. As expected, the 3D flow velocity magnitude increased as the agitation speed increased. However, the velocity in the inner core region below the impeller always remained very low.

The tablet-medium mass transfer process was additionally modeled, and the mass transfer coefficients at four different agitation speeds were obtained and compared to the standard system. The mini vessel system has a flow pattern similar to that of the standard USP 2 system. It was found that similar velocities at the tablet surface (on the iso-surface at y=7mm, i.e., at the vessel bottom) in the mini vessel system and in the USP 2 system were obtained when the mini vessel was operated a 100 mL, 125 rpm and the USP 2 system was operated at 900rpm and 100rpm, indicating that these two systems should result in similar dissolution profiles when operated under these conditions.

Copyright © American Institute of Chemical Engineers. All rights reserved.

471592 3D Raman Spectroscopy – a Method for Mapping Ingredient Distribution throughout a Tablet Volume

Abstract:

Blending of powders can often result in agglomeration when one or more raw materials in a blend are cohesive, hydrophilic or have the tendency of acquiring electrostatic charges. Agglomeration often leads to failure modes in pharmaceutical tablets, particularly "low dose" formulation tablets where the active pharmaceutical ingredient (API) is present in small amounts. When powders agglomerate, the uniformity of the final tablet is adversely affected, leading to variation in the tablet potency and dissolution profiles, which are usually undesirable. New technology enables the quantitative examination of the microstructure within tablets to determine the extent of agglomeration and to assess its effects on product performance. This can be useful as a means for understanding how the process causes agglomeration. If process parameters can be correlated to the tablet microstructure, this knowledge can assist in designing tablets of desirable quality thereby reducing product failure drastically.

Raman imaging is a powerful tool for studying tablet structure. It has higher chemical specificity than NIR and is also less affected by physical characteristics of the sample. The main disadvantage of Raman, its low quantum yield, requiring long integration times, is vanishing quickly, due to vast advances made recently in laser and detector technology. This is causing a significant renewed interest in using chemical mapping to examine the microstructure of pharmaceutical tablets.

This talk will focus on the use of a new Raman imaging platform, the mPAT system from H_2Optx, which is equipped with a 10x10 micron 2D spatial resolution along with the ability to shave layers off a tablet for depth analysis. This level of spatial resolution enables the user to extract detailed structural data that reveals the relationships between process, product structure, and product performance. This technique has the ability to do a quantitative content determination within a sample and, combining this for each pixel, generate 3D maps w.r.t. the individual constituents.

Tablets manufactured from traditional batch processing methods and continuous manufacturing methods were examined, including tablets made via Direct Compaction (DC), Wet Granulation (WG) and Roller Compaction (RC). Tablets using several formulations for each of the three methods were made with various levels of induced agglomeration and were analyzed using the mPAT system. The effects of the different processing conditions on agglomeration were characterized and their effect on tablet microstructure and dissolution were quantified. These observations will be discussed in this presentation.

Copyright © American Institute of Chemical Engineers. All rights reserved.

463044 Real-Time Monitoring and Modeling of Milling in a Continuous Tableting Line Via Dry Granulation

Abstract:

In the Pharmaceutical Industry, batch operation is the traditional mode of manufacture. During batch production, the quality of the batch usually is subject to batch-to-batch variability and thus quality is controlled by statistical sampling and if necessary rejection of the entire batch. Conversion to the continuous processing mode offers the potential for quality improvement as well as potential cost savings. However, continuous processing requires on-line measurements and active process control and thus the development of Process Analytical Technology (PAT) tools is crucial [1].

In the continuous dry granulation tableting line there are a number of variables that relate to the final product quality. For instance, composition and tablet hardness are among the CQAs in this process. In the dry granulation line, the roller compactor serves as a granulation method. In this unit, the powders are fed into a hopper and then conveyed by the feed screw to two counter rotating rolls. The powder is gripped by the rolls and pulled into the nip area of the compacting rolls before being compacted into a ribbon. Then the ribbon is milled into granules by a hammer mill [2]; generating a granular feed for the tablet press, which has favorable flow and compaction properties.

The milling step is an important unit operation because the granules particle size distribution (PSD) will affect compaction properties and subsequently the tablet hardness. Therefore, the granules PSD is also a CQA. In the roller compaction process, an increase in roll pressure will translate to an increase in ribbon density. The ribbon density has a relation to granule PSD [3]. As a consequence, the ribbon density can be considered an intermediate CQA. The relationship between ribbon density and granule PSD is difficult to establish from first principles, but a correlation can be established by conducting a series of experiments. In this design of experiments, the roll pressure is varied, the ribbon is milled at different speeds and the PSD measured through either off-line means such as sieve analysis or else on-line using laser based methods.

In continuous operations, it is desirable to measure all of the CQAs online, so as to permit the implementation of appropriate process control logic. In the tableting line installed at Purdue University, the roller compacted (RC) ribbon properties such as density and composition can be measured online using either a microwave (MW) and/or a Near-Infrared (NIR) sensor [4]. The PAT tool for online PSD measurement that we have adopted in our work is a high speed imaging camera (Innopharma Eyecon Camera), with which our team has had some previously experience in wet granulation applications [5]. This equipment counts and images particles with a sampling time of 2s and has a scanning rate of $10 m.s^{-1}$.

The milling unit operation is commonly modeled using a population balance model [6]. The key model component of the population balance model (PBM) is the breakage kernel. This component represents the process of creation of smaller particles, as larger particles are impacted by the mill hammers or pins which causes the fracture. A wide range of breakage kernels have been proposed, most of purely empirical or semi-empirical nature. Each of these models requires the estimation of model specific parameters which will depend on feed material properties as well as mill operating variables [7].

This project has three objectives. The first one is to calibrate the Innopharma Eyecon camera. The second objective is to establish the parameters of the breakage kernel and thus the PBM model and the third objective is to develop the empirical relation between the ribbon density and granules PSD both measured using the on-line PAT tools. The first objective is achieved by generating samples of granules prepared under different conditions then relating the Innopharma Eyecon Camera output to the sieve analysis results. Since the sieve analysis measures fractions mass while the camera measures particle size, there is a conversion required which depends on particle density. The data for meeting the needs of the second and third objectives is determined through a series of experiments in which the mill speed, the composition of the powder and the density of the ribbon in the roller compactor and PSD are measured on-line. This data is gathered in runs when the process reaches steady state.

The results of this three-part study enable the capability for on-line monitoring of the combined compactor-mill subsystem. The PBM model and ribbon density-PSD relationship are important components of a data reconciliation model for the entire line. In this paper we report the results of these studies and demonstrate their utility in data reconciliation.

References

[1] S. L. Lee, T. F. O. Connor, X. Yang, C. N. Cruz, L. X. Yu, and J. Woodcock, "Modernizing Pharmaceutical Manufacturing : from Batch to Continuous Production," no. 1, pp. 1–25, 2015.

[2] R. W. Miller and P. J. Sheskey, "Roller Compaction Technology for Pharmaceutical Industry." Marcel Dekker Inc., 2003.

[3] S. Yu, "Roll Compaction Of Pharmaceutical Excipients," University of Birmingham, 2012.

[4] A. Gupta, J. Austin, S. Davis, M. Harris, and G. Reklaitis, "A novel microwave sensor for real-time online monitoring of roll compacts of pharmaceutical powders online - A comparative case study with NIR," *J. Pharm. Sci.*, vol. 104, no. 5, pp. 1787–1794, 2015.

[5] A. S. El Hagrasy, P. Cruise, I. Jones, and J. D. Litster, "In-line size monitoring of a twin screw granulation process using high-speed imaging," *J. Pharm. Innov.*, vol. 8, no. 2, pp. 90–98, 2013.

[6] E. B. Litster, J. D., *The Science and Engineering of Granulation Processes*, vol. 53, no. 9. 2013.

[7] J. K. Pettersen and K. L. Sandvik, "Estimating the breakage and selection functions for a continuous mill," *Int. J. Miner. Process.*, vol. 35, no. 3–4, pp. 149–158, Aug. 1992.

Copyright © American Institute of Chemical Engineers. All rights reserved.

460448 Targeted Anticoagulation Therapy for the Prevention of Venous Thrombosis

Abstract:

Background: Deep vein thrombosis (DVT) and its complications are a leading cause of cardiovascular morbidity and mortality. Despite recent advances in anticoagulation therapy, the rate of bleeding complication remains high, especially for patients with two or more risk factors. In this study, we constructed a fusion protein SCE5-TAP that combines potent FXa inhibition with platelet targeting. Using this unique approach, we illustrate enrichment of targeted TAP at sites of venous thrombosis without altering systemic coagulation profiles.

Method and Results: We used genetic engineering to fuse an activation-specific single chain antibody (scFv) with a potent, direct FXa inhibitor, tick ant-coagulant peptide (TAP). Specific antibody binding of the fusion molecules SCE5-TAP to activated platelets was proven in flow cytometry and anti-FXa activity was demonstrated in chromogenic assays. Intravital microscopy of localized laser-induced injury was used to demonstrate the capacity of SCE5-TAP to target and prevent thrombus formation to murine cremaster venules. Furthermore, anticoagulation efficacy was evaluated in a murine model of inferior vena cava (IVC) thrombosis induced by electrolytic injury, as representative of non-occlusive thrombus. SCE5-TAP was effective in limiting acute venous thrombosis event, by decreasing 32% of thrombus weight, 75% of neutrophil, and 55% of monocyte infiltration. In contrast to Enoxaparin (LMWH) and Rivaroxaban, SCE5-TAP revealed anti-thrombotic effects at low doses at which bleeding time was not prolonged and anti-FXa activity in the plasma was below 0.1 IU/mL.

Conclusions: We present *in vivo* data suggesting unique properties of SCE5-TAP in limiting thrombosis burdens associated with DVT. In summary, SCE5-TAP represents a promising therapeutic candidate that reduce the risk of DVT without an attendant risk of bleeding.

Copyright © American Institute of Chemical Engineers. All rights reserved.

472383 Diffusion Altering Reporter Genes for Magnetic Resonance Imaging

Abstract:

Background: The development of robust reporter molecules for imaging gene expression in the context of living, optically opaque animals is critical for facilitating research in the emerging fields of genetic and cell based therapeutics. Despite their wide prevalence in cell biology, fluorescent and luminescent reporter genes have limited utility in this respect due to the poor penetration of light into deep tissues. In contrast to optical techniques, magnetic resonance imaging (MRI) offers excellent depth penetration and high spatial and temporal resolution. However, existing reporter genes for MRI are primarily based on metal-binding proteins, which are limited by their reliance on metal ion bioavailability, or derived from acidic and basic proteins with exchangeable protons, which tend to have relatively low sensitivity. In this work, we introduce an entirely new class of MRI reporter genes that works by altering water diffusivity in cells. Specifically, we show that the human water channel aquaporin 1 (AQP1) is a sensitive, nontoxic, and metal-free reporter that produces robust MRI contrast by increasing diffusion of water molecules across the cell membrane.

Methods: We used lentiviral transfection to construct stable CHO, U87, and N2A cell lines expressing AQP1 and control cells expressing GFP. AQP1 and GFP-expressing cells were imaged using diffusion-weighted MRI to map effective water diffusivity across a broad range of gene expression levels and in mixed cell populations. Finally, AQP1 was used to image gene expression in a murine intracranial tumor model. All animal protocols were approved by the Institutional Animal Care and Use Committee of the California Institute of Technology.

Results: AQP1-expression was found to enhance water diffusivity in several mammalian cell lines by 82 187%, resulting in enhanced contrast compared to GFP-expressing control cells imaged using diffusion-weighted MRI. Next, using doxycycline-dependent AQP1 expression, we demonstrated that AQP1-based increase in molecular diffusion of water can be used to reliably image gene expression across a broad, dynamic range spanning 0.01 to 10 µg/mL inducer concentration. Notably, AQP1 expression at submicromolar concentrations (~ 457 µM estimated) was found to be sufficient for producing significant MRI contrast, thereby placing AQP1 among the most sensitive MRI reporters. Furthermore, we developed Monte Carlo models for water diffusion in mixed cell populations, which predicted that AQP1-based increase in water diffusion should be evident even in heterogeneous populations comprising only a fraction of AQP1-labeled cells. Consistent with these predictions, our experiments revealed that as few as 10% AQP1 expressing cells is sufficient to produce MRI contrast in diffusion weighted images. Finally, we demonstrated that AQP1 can be used for imaging induced gene expression in mouse brain tumor xenografts, with the average intensity in AQP1 tumors decreasing by 41% following doxycycline administration.

Conclusions: Our work lays the foundation for AQP1 as the first in a class of MRI reporter genes that works by altering water diffusion in cells without impacting cell size or viability. As a sensitive, metal-free, and nontoxic reporter, AQP1 is well suited for dynamic imaging of gene expression *in vivo*. We anticipate that the wide prevalence of diffusion weighted MRI, together with the high performance, biocompatibility, and engineering capacity of aquaporins will enable the pervasive application of diffusion altering reporter genes in basic biological research as well as support the clinical development of genetic and cell-based therapies.

Copyright © American Institute of Chemical Engineers. All rights reserved.

470954 Pepdarts: Novel Peptide Vectors for Drug Delivery to the Brain

Abstract:

Delivery of biological cargo across the blood brain barrier (BBB) for therapeutic interventions is critical in treating neurological and neurodegenerative diseases but remains a formidable challenge. The BBB poses a strong physical and metabolic barrier for exchange of xenobiotics between systemic circulation and the central nervous system. Strategies employing the transcellular transport pathway of circulatory proteins such as transferrin have shown some success in this respect. We aim to employ the complementary paracellular pathway between the endothelial cells of the brain capillary microvasculature for transporting therapeutics to the central nervous system. Conventionally regarded as being reserved exclusively for the transport of small molecules and ions, there is some evidence that pathogens and immunomodulatory agents such as cytokines and immune cells can indeed access this pathway by disrupting the tight junctions (TJs) between the brain endothelial cells. TJs are non-covalent protein assemblies formed by extracellular loops (ECLs) of transmembrane proteins on endothelial cell surfaces. Claudin 5 (cldn5) has been identified as the key component of BBB TJs. Animal models representing cldn5-knockouts exhibit an easily accessible central nervous compartment to molecules in systemic circulation but fail to survive beyond the neonatal stage.

Here, we report on the design and development of PEPDARTs, a suite of novel peptide vectors, designed specifically to target and transiently disrupt the cldn5 TJ assemblies of the brain capillary endothelial cells. PEPDARTs (**P**ermeability **E**nhancing **P**eptides for **D**isruption, **A**ttenuation & **R**ecovery of **T**ight junctions) disrupt cldn5 TJ assemblies by binding to cldn5 ECLs and interfering with the homophilic cis- and trans-interactions of ECLs on the neighboring endothelial cells to enhance permeability across the BBB. Furthermore, since PEPDARTs are relatively small (10-15 amino acids) their circulation half-life is short (~30-45 minutes) causing quick clearance from the TJ interface and systemic circulation. Any disruption of the BBB is therefore anticipated to be transient under *in vivo* conditions.

We will present our preliminary studies from peptide microarrays and *in vitro* cell culture models that have led to fundamental understanding of the interactions between cldn5 ECLs (responsible for the formation of TJs). Additionally, these studies have led to the design of several PEPDART candidates that increase the BBB permeability at levels relevant for therapeutic applications. We will conclude with *in vivo* results in animal models demonstrating the transient and reversible disruption of the BBB in the presence of PEPDARTs. We are currently investigating the potency and safety of PEPDARTs in their efficacy and utility as drug delivery vectors for CNS-relevant therapeutics for delivery to the brain. This work is expected to not only have a significant impact on the development of new therapeutic modalities for treatment of NDs, but also elucidate the fundamental mechanisms of molecular transport across the BBB.

Copyright © American Institute of Chemical Engineers. All rights reserved.

470906 Nanoallergens: A Liposome Based Platform for Modeling Allergenic Proteins

Abstract:

Allergies, also known as type I hypersensitivity, is an immunological disorder of certain immune cells where allergen proteins bind multiple cell surface bound immunoglobulin Eâ€™s (IgEs), triggering the release of inflammatory agents and enzymes.Â The release of inflammatory compounds, known as the degranulation reaction, occurs on mast cells and basophils; and can result in severe symptoms including a potentially fatal anaphylactic shock. Due to the growing prevalence of allergies in the general population, there is an increased need to model the allergen proteins for accurate identification of the most important epitopes that carry critical roles during activation of the degranulation reactions to elucidate the details of this complex cellular response. Recently, our laboratory has developed a new platform for modeling allergen proteins we call nanoallergens. By displaying peptide mimotopes of potential IgE binding epitopes on the surfaces of liposomal nanoparticles, we can stimulate degranulation responses in a similar fashion to natural allergen proteins. Liposomes offer precise control over epitope loading ratios and heterogeneity, and allow a systematic study of an individual patient reactivity to each epitope. Furthermore, nanoallergen platform allows a much needed prediction of a potential clinical response to allergens in a personalized therapy setting that is currently unavailable using current allergy testing methods. Nanoallergens are also extremely versatile; so far we have evaluated the potency of IgE epitopes in several major peanut proteins using clinical samples and likewise used nanoallergens as an allergy testing model for penicillin allergies. Given the ease of synthesis and wide range of applications, we demonstrate that nanoallergens have potential as valuable tools for studying degranulation responses, determining and ranking allergy reactive epitopes as well as clinical allergy testing.

Copyright © American Institute of Chemical Engineers. All rights reserved.

469395 Lasso Peptide Unthreading: A Model System for Knotted Proteins

Abstract:

Lasso peptides are a class of protein-derived natural products with a curious threaded structure which endows the peptides with protease resistance and often thermostability. Lasso peptides have an array of therapeutically relevant activities ranging from antimicrobial function to receptor binding. Lasso peptides are also of interest because of their relatively simple entangled structure which can serve as a model system for larger knotted protein systems. In this talk I will present data on the kinetics and energetics of lasso peptide unthreading including experiments to precisely probe the unthreading mechanism.

Copyright © American Institute of Chemical Engineers. All rights reserved.

462327 T Cells Expressing CD19/CD20 Bispecific Chimeric Antigen Receptors Eradicate CD19⁻ Lymphoma and Prevent Antigen Escape *In Vivo*

Abstract:

The adoptive transfer of T cells expressing chimeric antigen receptors (CARs) has demonstrated clinical efficacy in the treatment of advanced cancers, with anti-CD19 CAR-T cells achieving up to 90% complete remission among patients with relapsed B-cell malignancies. However, a substantial fraction (>10%) of these patients eventually relapse due to antigen escape, a process by which tumors evade T-cell detection by eliminating expression of the targeted antigen. Here, we report the design and optimization of bispecific, OR-gate CARs that can trigger robust T-cell activation in response to target cells that present either CD19 or CD20, thus preventing malignant B cells from escaping T-cell therapy by simply losing CD19 expression. We demonstrate that optimized OR-gate CAR-T cells—but not conventional, single-input CD19 CAR-T cells—can eliminate pre-existing $CD19^-$ mutant tumors as well as prevent the spontaneous emergence of antigen-negative tumor growth *in vivo*. This technology presents an effective measure against antigen escape in CD19 CAR-T cell therapy that is fully compatible with current clinical practice and manufacturing processes.

Copyright © American Institute of Chemical Engineers. All rights reserved.

475004 Rapid Antibody Discovery Platform for Antibody Drug Conjugates and Bispecific Antibodies

Abstract:

The development of novel antibody therapeutics, especially antibody-drug conjugates (ADCs) and bispecific antibodies, is often a challenging and time-consuming process. Utilizing Sutroâ€™s cell-free protein synthesis (CFPS) system, we describe an antibody discovery platform for the selection, screening, and characterization of antibodies for novel drug targets in approximately four weeks. Using ribosome display technology, we can select binders from billion-member libraries and directly screen the clonal output in CFPS. Because CFPS allows the rapid expression of variants at high titers, we have the ability to screen thousands of variants for desirable antibody characteristics such as cell binding, cell internalization, binding affinity, thermostability, and epitope. Top leads can be further optimized for non-natural amino acid incorporation and drug conjugation for the generation of homogeneous ADCs. Alternatively, variants can be combined and evaluated as bispecific antibodies. Sutroâ€™s antibody discovery platform enables rapid candidate generation and characterization for making the most efficacious and safe ADCs and bispecific antibodies.

Copyright © American Institute of Chemical Engineers. All rights reserved.

464240 Deciphering Lignin Deconstruction Mechanism of Lower Termite

Abstract:

Although fascinating, the exact process of plant cell wall degradation developed by nature and implemented in lower termites remains unclear. We have been conducting systematic studies on the lignin deconstruction mechanism over the last seven years as an effort to develop nature-inspired technologies for lignocellulose processing. A series of experimental and analytical techniques from various disciplines were used in our studies. Scientific questions explored included mircrochemical environment of the termite guts, structural and composition changes of lignocellulose along the termite digestion track, critical enzyme systems involved in lignin deconstruction, chemical reactions that modifies lignin structure, and disassociation energy levels of different chemical bonds that form lignin structure. Integrating these results makes it possible to propose a chemical reaction mechanism and associated lignin degradation pathways. The mechanism and the pathway will be presented and discussed in the proposed paper. The results of the study provide new insights for developing more sustainable processes for plant cell wall deconstruction required in biorefineries.

Copyright © American Institute of Chemical Engineers. All rights reserved.

455682 Production of Food, Energy, and Fuel from Microalgae in Holograhic Diffractive Optic-Solar Glass Reactors

Abstract:

This project is focused on developing technology platforms for the solar-assisted production of food, fuel, and clean water from municipal wastewater sources. This novel food, energy water (FEW) platform employs new light management strategies to simultaneously: (1) generate electrical energy; and (2) optimize algae growth for food and fuel. This is accomplished by use of a holographic diffractive optic-solar glass (HDO-SG) reactor. The glass transmits the visible light spectrum that is used for photosynthesis and internal reflects the IR and UV light that is directed toward photovoltaic (PV) cells that are integrated into the reactor. The PV- generated energy can be used to operate the photobioreactor. Preliminary results show that *Chlorella* sp and *Scendesmus* sp. algae grow well in reactors containing HDO-SG. Water loss is minimized compared to open systems which is key when growing in arid climates. Growth of *Chlorella* sp on secondary and tertiary wastewater has been demonstrated, as well as growth using of flue gas as a carbon source. No observed changes in lipid productivity were found compared to control cultures grown in laboratory media.

Copyright © American Institute of Chemical Engineers. All rights reserved.

474778 BUTANOL PRODUCTION BY *CLOSTRIDIUM ACETOBUTYLICUM* IN A SERIES OF PACKED BED BIOFILM REACTORS

Francesca Raganatia, Dipartimento di Ingegneria Chimica, dei Materiali e della Produzione Industriali - Università degli Studi di Napoli Federico II, Napoli, Italy
Alessandra Procentese, Dipartimento di Ingegneria Chimica, dei Materiali e della Produzione Industriali - Università degli Studi di Napoli Federico II, Napoli, Italy
Giuseppe Olivieri, Dipartimento di Ingegneria Chimica, dei Materiali e della Produzione Industriali - Università degli Studi di Napoli Federico II, Napoli, Italy
Maria Elena Russo, Istituto di Ricerche sulla Combustione, Consiglio Nazionale delle Ricerche, Napoli, Italy
Peter Gotz, Department of Life Sciences and Technology / Bioprocess Engineering, Beuth University of Applied Sciences Berlin, Berlin, Germany
Piero Salatino, Dipartimento di Ingegneria Chimica, dei Materiali e della Produzione Industriali - Università degli Studi di Napoli Federico II, Napoli, Italy
Antonio Marzocchella, Dipartimento di Ingegneria Chimica, dei Materiali e della Produzione Industriali - Università degli Studi di Napoli Federico II, Napoli, Italy

The reactor design plays a key role in the fermentative production of biobutanol. The high cell concentration that may be reached in confined – biofilm, membrane, recycling - cell reactors offers high conversion rates. The concentration of solvents in the broth from biofilm reactors reported in literature is not particularly high and it negatively affects the successive stages for butanol recovery (Qureshi et al., 2005). The low concentration of solvents in the produced stream is typically due to the inhibitory effect of solvents on the fermentation. Therefore, the butanol bioreactor productivity is as low as the bioreactor behaviour approaches the CSTR limit.

Another major element of the research on ABE fermentation is the mathematical modeling of this process. The experimental investigations in any typical fermentation process are not only time consuming due to slow kinetics of the process, but also involve significant cost. The performance of ABE fermentation process, in terms of selectivity and yield of solvents, is dependent on many physical and design parameters, in addition to the *Clostridium* strain involved and basic mode of fermentation operation. Thus, optimization of the process involves intense experimentation with different combinations of physical and design parameters, which is rather impractical. Simulation of the fermentation process with mathematical models gives an insight into the characteristics of the process, and also helps to identify the influence of each variable on the overall process.

This contribution reports recent advance on ABE fermentation at Napoli. The study regarded an innovative immobilized cell reactor system: an experimental campaign and the development of a model.

Clostridium acetobutylicum was adopted for the fermentation process. The conversion was carried out in 4 packed bed reactors connected in series: the first reactor of the series was kept under acidogenesis and the successive reactors were kept under solventogenesis. Tests were carried out feeding the reactor system with solutions bearing both glucose as carbon source.

A mathematical model of the PBBRs system was also formulated using kinetic data assessed experimentally. The proposed model was an unstructured-unsegregated model and summarizes biochemical as well as physiological aspects of growth and metabolite synthesis by the production strain.

Introduction

The development of biotechnological processes to produce butanol from renewable resources as eco-sustainable alternatives to the petrochemical production routes (Kumar et al., 2012) is still an open challenge. Acetone-Butanol-Ethanol (ABE) fermentation by clostridia is drawing new interest as a way to turn renewable resources into valuable base chemicals and liquid fuels. Nevertheless the high industrial potential interest for the butanol production by the biotechnological route, some features of the ABE fermentation process hinder its success on an industrial scale. Indeed, the ABE fermentation is characterized by low yield, the acid-solvent two phase feature, and low concentration of butanol due to its inhibiting effect on fermentation. Moreover, the cost of traditional feedstocks - starch and molasses – is a larger fraction of the butanol sale price. Therefore, the use of a low-cost renewable resource is a pre-requisite for the industrial success of biotechnological process to produce butanol. Key issues for the success of the ABE fermentation process include the detailed metabolic analysis of the clostridia coupled with the development of improved engineered strains too (Papoutsakis 2008).

Reactor design and operating conditions play a key role in fermentative productions. The main factors that hinder the commercial development of the traditional ABE batch fermentation processes include low cell density, cell – and relative feedstock - lost at the end of the fermentation, low reactor productivity, high down-times, nutritional limitations and severe product inhibition (Chen and Blaschek, 1999). Process intensification may be obtained by increasing cell concentration in the reactor: cell immobilized reactors and retention membrane reactors are two potential solutions (Qureshi et al., 2005). These reactors configuration takes also the advantage to be operated continuously (Qureshi et al., 2000; Procentese et al., 2015). Continuous bioconversions are characterized by several advantages with respect to batch cultures in biofilm reactors (Qureshi et al., 2000). The main advantages are: the high cell concentration, the reactor operation at high dilution rates – high throughout - without cell washout, high yield, high butanol concentration that enhance butanol recovery performances. Moreover, the biofilm support can be reused.

The reactor design and its optimization for ABE fermentation may take advantages form reactor modelling The performance of ABE fermentation process, in terms of selectivity and yield of solvents, depends on several physical and design parameters, on the Clostridium strain, and on the mode of fermentation operation. Simulation of the fermentation process with mathematical models gives an insight into the characteristics of the process, and also helps to identify the influence of each variable on the overall process.

This contribution reports recent advance on ABE fermentation at Napoli. The study regarded an innovative immobilized cell reactor system: an experimental campaign and the development of a model.

Apparatus and model

The anaerobic solventogenic commercial *Clostridium acetobutylicum* DSM 792 was used for the fermentation process. The conversion was carried out in 4 packed bed biofilm reactors (PBBRs) connected in series: the first reactor (fed with stream bearing the the carbon source) was operated under acidogenesis conditions, and the three successive reactors were operated under solventogenesis conditions. The two phases of the ABE fermentation were operated in separate vessels: acids were produced in first section of the system, the produced acids and the residual sugar were converted in solvents in the second section. The PBBR system performance was characterized in terms of final butanol concentration and productivity as a function of the dilution rate.

The apparatus was made of reactor system, liquid pumps, heating apparatus, device for pH control, and on-line diagnostics (sketch in Fig. 1). The reactor system was made of four fixed beds. Each bed was at the bottom of a 100 mL glass lined pipe (4 cm ID, 8 cm high) jacketed for the heat exchange. Water from an external circulating water bath (Julabo heating circulator MA4) was fed into the jacket of each reactor to keep the operating temperature at 37 °C. The liquid head was controlled by the overflow duct in each reactor: the working volume of each reactor was set at 40 mL. Nitrogen was sparged at the bottom of each reactor to ensure anaerobic conditions. The pH control device – one for each reactor - included a pH-meter, a peristaltic pump, a vessel with NaOH 0.3 M solution and a controller.

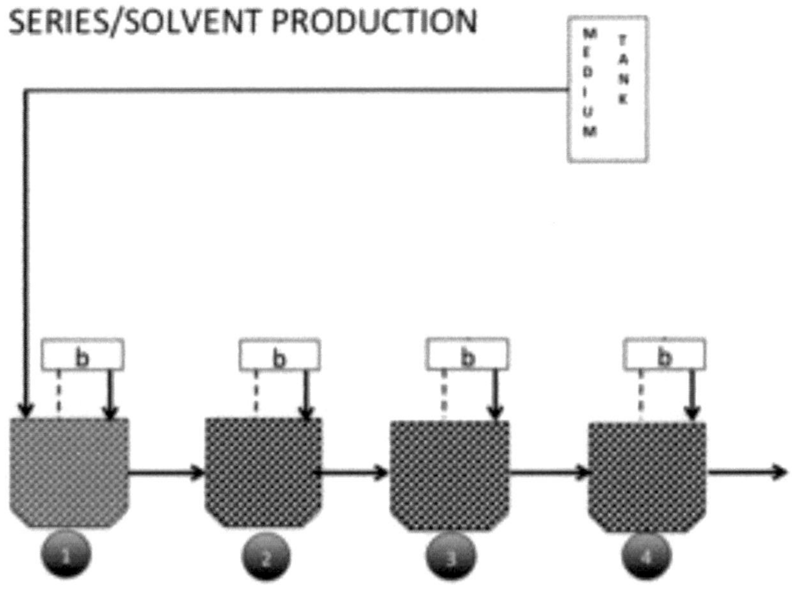

Fig.1: Outline of the apparatus used for the continuous process: A) PBBRs connected in parallel during the start-up phase; B) PBBRs connected in series during the butanol production phase. b: pH measure/control device.

Tygon rings (3/1 mm OD/ID) were chosen as biofilm carriers.
Clostridium acetobutylicum DSM 792 was used. Details regarding the reactivation and pre-inoculum procedures are reported in Raganati et al. 2013.

The composition of the medium fed to the PBBR system is reported in (Procentese et al. 2015a). Glucose (100 g/L concentration) was used as carbon source.

A mathematical model of the PBBRs system was formulated using glucose kinetic data assessed experimentally according to the procedure proposed by Procentese et al. (2015a; 2015b). The proposed model was an unstructured-unsegregated model and summarizes biochemical as well as physiological issues of growth and metabolite synthesis by the production strain. The key fermentation rates were expressed and evaluated with regard to substrate consumption and butanol end-product inhibitory effects.

Kinetic data regarding the growth rate (acidogenesis) and the butanol production rate (solventogenesis) were from specific experimental campaigns (details in Procentese et al. 2015a and 2015b). The main assumptions of the proposed model are reported hereinafter (Raganati et al., 2016).

- the PBBR system was assumed as a series of CSTR;
- the biomass present in each PBBR as free cells and immobilized cells was a heterogeneous cell population consisting of: acidogenic cells, solventogenic cells and spores ;
- the kinetics of cell growth and butanol production of biofilm-cells were assumed equal to those of the free cells;
- cells attachment and detachment processes were considered.

Experimental results

Two PBBR configurations were used (Raganati et al., 2016).

Parallel configuration) The four fixed beds were operated in parallel. This configuration was used during start-up to promote biofilm formation.

Series configuration) The four fixed beds were connected in series (Fig. 1). This configuration was used during butanol production after a biofilm layer had formed in each unit in parallel mode.

Butanol production tests were carried out with the 4 PBBRs connected in series (Fig.1) and operated at pre-set conditions. The pH of the reactor 1 (Fig.1) was set at 5.5 to promote acidogenesis conditions. The pH of the reactors 2, 3, and 4 (Fig.1) was set at 4.7 to promote solventogenesis conditions.

The overall dilution rate (D) – ratio between the feeding flow rate and the total volume of the 4 fixed beds – ranged between 0.05 and 1.4 h^{-1}. After setting the dilution rate, the reactor system was operated until steady state conditions were reached: metabolite and glucose concentration in each reactor staying constant for at least 5 times the space-time of the reactor.

Fig. 2 reports the concentration of butanol and glucose in the four PBBRs measured under steady state conditions for D=0.15 h^{-1}. It is interesting to analyse the substrate and metabolite concentration in the four reactors from the PFR behaviour point of view. The gradual decrease in glucose concentration (Fig. 2B), and the gradual increase in solvent concentration (Fig. 2A) along the PBBR series reproduce the typical behaviour of batch fermentation. As expected, the formation of a stable biofilm in the four units and the choice of the proper operating conditions lead to a successful reactor system that behaves like a PFR with continuous inoculum and constant biofilm concentration.

The PBBRs system was successfully operated for more than three months to produce butanol. The reactor performances – butanol productivity and butanol concentration – were assessed as a function of the reactor system dilution rate. By tuning the D it was possible: i) to totally convert the carbon source (D <0.15 h^{-1}); ii) to maximize the concentration of butanol in the produced stream (about 15 g/L at D = 0.65 h^{-1}); iii) to maximize the butanol productivity (about 12.6 g/Lh at D = 0.9 h^{-1}).

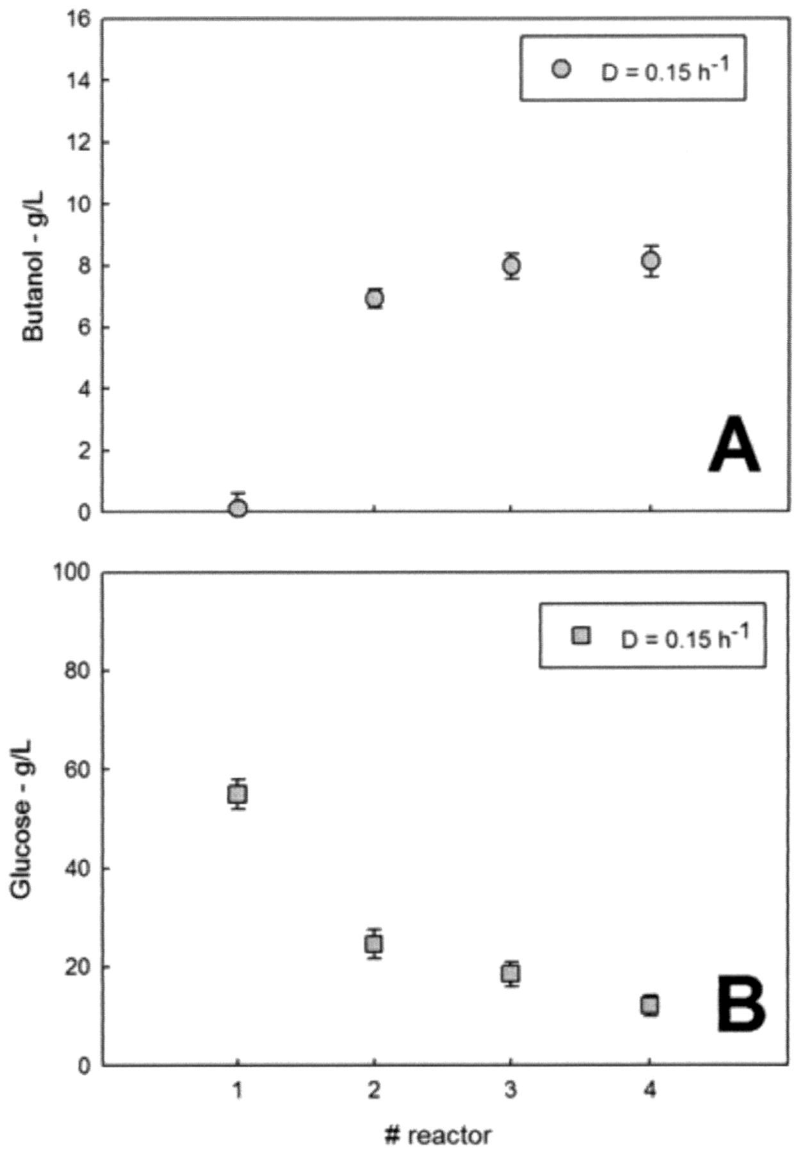

Fig.2: Butanol and glucose concentration measured along the PBBR series at D=0.15 h^{-1}.

Model results

Fig. 3 reports the estimated concentration of butanol and glucose in the four PBBRs for D=0.15 h^{-1}. It is evident that the developed model succeeded in reproducing the behavior of the PBBRs system.

Acknowledgements

The authors thank the European Union's Horizon 2020 research and innovation program for their financial support to the project Waste2Fuels (grant agreement No 654623).

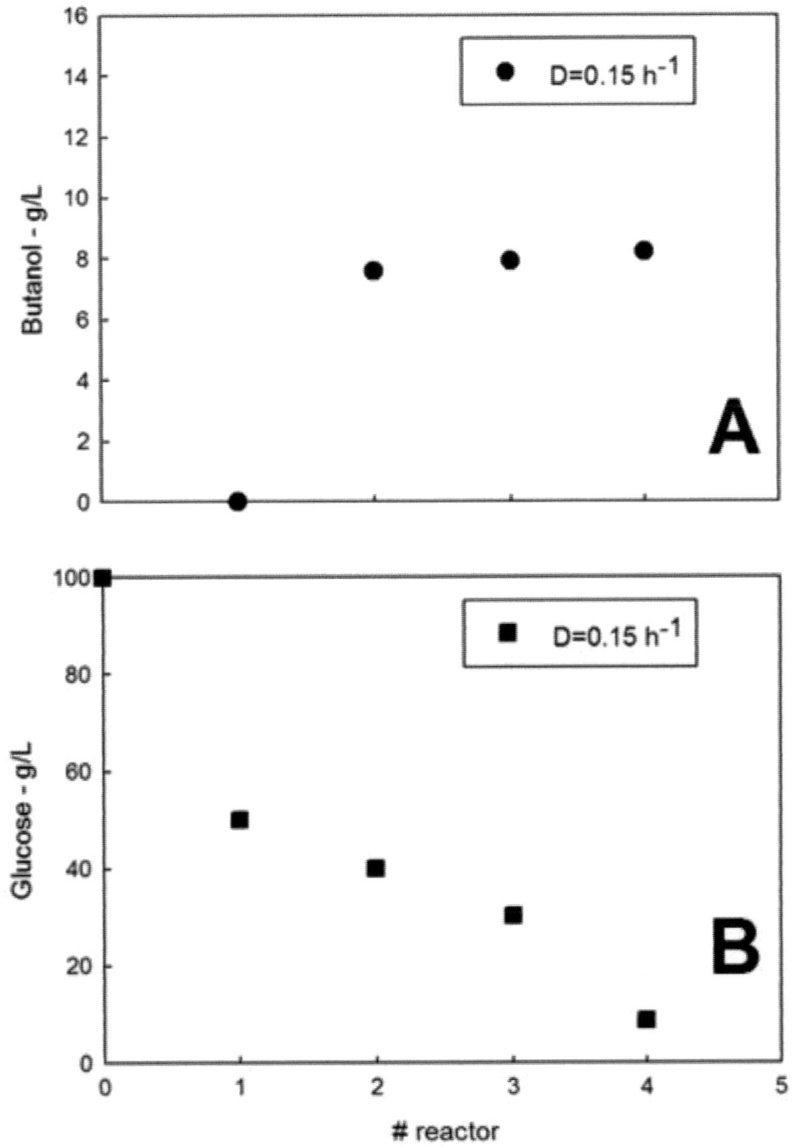

Fig.3: Estimated butanol and glucose concentration along the PBBR series at $D=0.15\ h^{-1}$.

References

1. Chen CK, Blaschek HP. Acetate enhances solvent production and prevents degeneration in *Clostridium beijerinckii* BA101. Appl Microbiol Biotechnol 1999; 52:170–173
2. Kumar M, Goyal Y, Sarkar A, Gayen K. Comparative economic assessment of ABE fermentation based on cellulosic and non-cellulosic feedstocks. Appl Energ 2012; 93:193–204.
3. Papoutsakis, E. T. Engineering solventogenic clostridia. Current Opinion in Biotechnology 2008 19(5), 420–9.

4. Procentese, A., Raganati, F., Olivieri, G., Russo, M. E., Salatino, P., Marzocchella, A. Continuous xylose fermentation by *Clostridium acetobutylicum* – Kinetics and energetics issues under acidogenesis conditions. Bioresource Technol 2015a 164, 155–161.
5. Procentese A, Raganati F, Olivieri G, Russo ME, Salatino P, Marzocchella A. Continuous lactose fermentation by *Clostridium acetobutylicum* – Assessment of solventogenic kinetics. Bioresour Technol 2015b; 180:330–337.
6. Qureshi N, Schripsema J, Lienhardt J, Blaschek HP. Continuous solvent production by *Clostridium beijerinckii* BA101 immobilized by adsorption onto brick. World J Microb Biot 2000; 16:337-382.
7. Qureshi, N., Annous, B. A., Ezeji, T. C., Karcher, P., & Maddox, I. S. (2005). Biofilm reactors for industrial bioconversion processes : employing potential of enhanced reaction rates. Microbial Cell Factories, 21, 1–21.
8. Raganati, F., Olivieri, G., Procentese, A., Russo, M. E., Salatino, P., Marzocchella, A. Butanol production by bioconversion of cheese whey in a continuous packed bed reactor. Bioresour Technol 2013; 138:259–265.
9. Raganati, F., Procentese, A., Olivieri, G., Russo, M. E., Götz, P., Salatino, P., Marzocchella, A. Butanol production by *Clostridium acetobutylicum* in a series of packed bed biofilm reactors. Chemical Engineering Science 2016; 152; 678–688.

466294 Effect of Environmental Factors Towards Patchoulol Production By *Chlamydomonas Reinhardtii*

Abstract:

Patchoulol ($C_{15}H_{26}O$) is a sesquiterpene alcohol and a valuable compound in the fragrance and pharmaceutical industry. Traditionally, patchouli oil is extracted from the leaves of a plant, *Pogostemon cablin*. This approach faces the threat of limited and intermittent supply as these plants only grow in select regions of the world. But now with the advance of genetic engineering, it is possible to genetically modify microorganisms such as yeasts and bacteria to produce patchoulol. Photosynthetic microorganisms, such as microalgae in particular, would be of high interest as they have the advantage of directly converting carbon dioxide to high-value products through carbon fixation and photosynthesis. In this work, we have cultivated a patchoulol-producing strain of *Chlamydomonas reinhardtii* and elucidated the effect of light and temperature towards cell growth and patchoulol production. We also considered the effect of light-and-dark cycles towards patchoulol productivity. The findings from these environmental factors not only increase our understanding of patchoulol expression in microalgae, but also represent a step towards advancing this production platform to other target molecules. It is envisioned that this platform would be applicable to other terpenoids such as bisabolene and limonene, both hydrocarbon precursors to a range of industrially relevant chemicals.

Copyright © American Institute of Chemical Engineers. All rights reserved.

470048 Nutrient Capture By a Sustainable Symbiotic Biofilm: Simultaneous Phosphorous and Nitrogen Recovery By Attached Fungi and Microalgae Biofilm

Abstract:

Nutrient pollution, a condition of excess phosphate (P) and nitrate (N) from point and non-point sources, is a major water quality concern worldwide due to the tremendous volume of urban, agricultural and industrial wastewater production [1]. The elimination of phosphorus from effluents before discharging is made mandatory as even traces of P (above 0.02 mg/L) can cause eutrophication, also P level is regulated to be lower than 0.05 mg/L, if streams discharge into lakes or reservoirs. On the other hand, the P reserves are being depleted swiftly which apparently is linked to global food security, with more than 90% of phosphorus used for agricultural purpose [2]. Currently a curative solution to combat water pollution by recuperating N and P, recycling the P for food production by closing the nutrient loop to effectively manage dwindling rock P reserve are the major challenges. The most common phosphorus removal techniques for municipal and industrial wastewater are biological treatment and chemical precipitation. Bio-valorization offers an excellent alternative to chemical treatment technologies as the biomass produced can be a source of protein and other valuable bio-chemicals also it is an eco-friendly and low cost technology.

In this present work, a novel composite biofilm called mycoalgae biofilm [3] was developed using the isolated strains of polyphosphate accumulating fungi and microalgae and we want to apply this platform for the efficient recovery and recycling of nutrients in waste stream. The effect of various process conditions (fungal strain, carbon levels, temperature, P/N ratio) on the nutrient recovery efficiency and the changes in the biofilm biomass composition was studied. Since the biofilm will be used for nutrient recovery from an array of samples a synthetic medium for the cell growth with the initial P (40 mg/L), N (200 mg/L) and other nutrients were used as the medium for the biofilm development. Depending on the nature of the waste water and its composition, and the fungi-microalgae combination intended to use for bio-remediation, the developed mycoalgae biofilm can be a sustainable microbial process for food, feed or energy source.

Methods: Water and soil samples were collected from Sarita wetland near UMN to isolate the native fungi for the biofilm formation. Fungal isolates were identified by genetic identification of fungal internal transcribed spacer (ITS) regions. Maximum Phosphorous and Nitrogen removal was observed in strain Sa7 (*Mucor sp.)* and *Chlorella sp,* combination which was considered for the biofilm formation and Nutrient recuperation. The morphology of the biofilm was also studied using CLSM and SEM at the optimized conditions to study the interaction between the cells. The nutrient removal efficiencies were calculated using the equation: $RE_i = (S_{i0}$ â€“ $S_{it})/S_{i0}$; where RE_i is the removal efficiency of the nutrients *i* (TN, TP, PO_4-P) or the pollutants (*i*) using the mycoalgae lichen type biofilm; S_{i0} and S_{it} are the initial and final concentration of component *i* after *t* days. The rate of nutrient and pollutant removal by mycoalgae biofilm is an important factor which will be calculated by; $dS/dt = r_i = (S_{i0}$ â€“ $S_{it})/ (t_i$ â€“ $t_0)$; Where: r_i (g/L/d) is the removal rate of substrate *i* (TN, TP, PO_4-P).

Conclusion: The composite biofilm shows promising results in waste water treatment especially in removing the N, P, suspended solids and other organics from nutrient polluted water in a single-step reactor with easy biomass recovery and high efficiency, when compared to the individual pure cell cultures. The biofilm attaches to the polymer matrix completely leaving behind the clean water for recirculation, and the composite biomass for nutrient recycle. The biofilm composition can also be tailored based on the influent stream components, feasibility of the strains to grow together and its nutrition value for further use. The commercial expansion of this biofilm system could be a step ahead in having an integrated and sustainable water treatment systems. **References**

1. E.K. Wilson, Danger from microcystins in Toledo water unclear, Chem. Eng. News, 92 (32) (2014), p. 9
2. Ye, Y., Gan, J., & Hu, B. (2015). Screening of Phosphorus-Accumulating Fungi and Their Potential for Phosphorus Removal from Waste Streams. Applied biochemistry and biotechnology, 177(5), 1127-1136.
3. Bo Hu, Aravindan Rajendran, A Novel Microalgae Cultivation Technology Using Fungi and Lichen Biofilm, UMN invention disclosure case # 20140274

Copyright © American Institute of Chemical Engineers. All rights reserved.

461292 Incorporating Hydrodynamics into Spatiotemporal Metabolic Models of Bubble Column Gas Fermentation

Abstract:

A promising route to renewable liquid fuels and chemicals is the fermentation of carbon monoxide containing gas streams to synthesize desired products such as ethanol and 2,3-butanediol. We recently developed a spatiotemporal metabolic model for bubble column reactors with the CO fermenting bacterium *Clostridium ljungdahlii* as the microbial catalyst [1]. This first generation model was based on the simplifying assumptions of ideal plug flow for the vapor phase and plug flow plus axial dispersion for the liquid phase. However, industrial bubble column reactors are complex multiphase processes in which spatial variations in the gas superficial velocity, volumetric holdup and interfacial area can have profound e ects on fermentation performance.
We discuss the development of a second-generation bubble column model that accounts for the e ects of gas phase hydrodynamics on cellular growth, nutrient update and byproduct synthesis rates in a temporarily and spatially resolved manner. The modeling approach involves the coupling of a genome-scale metabolic reconstruction with mass and momentum balances that allow the calculation of local gas metabolite concentrations, superficial velocity, volumetric holdup and interfacial area as well as local concentrations of biomass and liquid phase metabolites. The resulting set of partial di erential equations with embedded linear programs is spatially discretized and numerically integrated using the MATLAB based code DFBAlab. Simulations results are shown for two CO fermentating organisms: the research model *C. ljungdahlii* and the industrially relevant *Clostridium autoethanogenum*. Model predictions are compared to those generated without gas phase hydrodynamics and to experimental data collected from a laboratory scale bubble column reactor with respect to CO conversion, biomass production, ethanol titer and ethanol-to-acetate ratio.

1. Chen, J., J. A. Gomez, K. Ho ner, P. I. Barton and M. A. Henson, "Metabolic Modeling of Synthesis Gas Fermentation in Bubble Column Reactors," Biotechnology for Biofuels, 8, 89, doi:10.1186/s13068-015-0272-5 (2015).

Copyright © American Institute of Chemical Engineers. All rights reserved.

464191 A Commercial Demonstration of Biorefinery of Lipids--Coproduction of Biodiesel and 1,3-Propanediol

Abstract:

Lipase-catalyzed transesterification (also esterification) from oils or fats for biodiesel production has some advantages over chemical-catalyzed transesterification, such as environmental friendship, lower energy consumptionï¼Œwidely suitable for low quality vegetable oils. However, the low stability of the lipase due to alcohol has been being regarded as the main hurdle to the industrialization of lipase-catalyzed biodiesel production. Tsinghua University has developed a novel process to eliminate the inhibition of methanol on lipase, thus the operational life of the lipase could be improved into more than 300 cycles from several (1-5) cycles while traditional process applied. The patented novel process was successfully demonstrated in a pilot plant (200kg/day) and commercial plant (50,000ton/year) in China. It has also been being demonstrated in a pilot-plan in Brazil.

As a by-product, glycerol will be produced at about 10% of biodiesel during the process of biodiesel production. More and more concerns were pay on how to utilize crude glycerol. It could be a promising way to produce 1,3-propanediol (PDO) by fermentation with crude glycerol as feedstock. PDO is a valuable monomer to synthesize polytrimethylene terephthalate (PTT). PTT has excellent properties for textile fibre or plastic, and it was found to have some better properties than other polyesters such as PET, PBT. Tsinghua University has developed a novel flexible process for PDO production by fermentation from glycerol or glucose, and it was successfully commercialized with capacity of 20,000 tons/year PDO .

Copyright © American Institute of Chemical Engineers. All rights reserved.

459025 Cell-Free Mixing of *Escherichia coli* Crude Extracts to Prototype Isoprenoid Biosynthesis

Abstract:

Cell-free metabolic engineering (CFME) is advancing a powerful paradigm for prototyping and designing biochemical synthesis pathways. To date, most CFME efforts use either purified enzymes or a crude extract from a source strain containing the entire enzymatic pathway. In this work, we used mixing of crude lysates, or extracts, from *Escherichia coli* to reconstitute a portion of the isoprenoid pathway; isoprenoids are a diverse class of compounds with applications as pharmaceuticals, fragrances, pesticides, biofuels, and chemical feedstocks. Specifically, we individually overexpressed three enzymes, mixed the resulting lysates, and utilized the native glycolytic enzyme cascade to convert glucose to mevalonate, an intermediate of isoprenoid synthesis. Avoiding the synthesis of cytotoxic HMG-CoA in the source strain, this plug-and-play approach was used to screen enzyme variants and optimize enzyme ratios which facilitated the construction of an all-in-one strain. The open nature of CFME allowed exploration of cofactor landscapes at varying ATP, NAD^+, and CoA concentrations. Genomic modification to the source strain confirmed that metabolism of acetate salts via the *ackA-pta* pathway is critical for cofactor recycle and high titers (>10 $g \cdot L^{-1}$). Building on these results, we have extended the biosynthetic pathway by adding six additional enzyme-enriched extracts to generate the monoterpene limonene. Our results highlight the high cofactor turnover of cell-free lysates and the ability to rapidly prototype and debug enzymatic pathways *in vitro*.

Copyright © American Institute of Chemical Engineers. All rights reserved.

467112 Biosensor-Guided Improvement of Microbial Alkaloid Production

Abstract:

Monoterpene indole alkaloids (MIAs) have important therapeutic value, including as anticancer and antimalarial agents. Because of their chemical complexity, therapeutic MIAs, or advanced intermediates thereof, are often isolated from the native plants. Here, I present our recent work in the microbial production of the modified MIA 10-hydroxystrictosidine from glucose and the monoterpene secologanin via a pterin-dependent mono-oxidation strategy. This work opens the door to the scalable production of alkaloids as well as the production of modified MIAs to serve as late intermediates in the semisynthesis of known and novel therapeutics. In order to reach scalable levels, we are currently focusing on creating G-protein coupled receptor (GPCR)-based biosensors to increase microbial production of biogenic amines. This will allow for high-throughput development of higher yielding alkaloid production strains and serve as an example of the utility of GPCR-based biosensors in the development of more productive chemical-producing microbial platforms.

Copyright © American Institute of Chemical Engineers. All rights reserved.

471718 Towards the Elucidation of the Limonin Biosynthetic Pathway, a Triterpene Derived Natural Product in Citrus sinensis

Abstract:

Triterpene derived plant natural products have essential properties in plant immunity and wide uses in the food, health and biotechnology sectors. The bioactivities of these molecules are attributed to the vast range of enzymatic tailoring on the triterpene scaffold including oxygenations, glycosylations, and acylations. These tailoring reactions modify triterpenes into complex molecules that are outside our chemical synthesis capabilities. Thus, the discovery and characterization of enzymes involved in triterpenoid biosynthesis is necessary to uncover novel chemistries and for metabolic engineering efforts towards the production of intricate triterpene derivatives.

One important group of triterpene derivatives are the limonoids. Limonoids display potent insect-deterrent properties including anti-feedant, growth disruptive, and reproduction inhibitive activities. Limonoids exhibit high degrees of oxygenation and complex intermolecular rearrangements resulting from an estimate of a dozen unknown genes. Herein, we aim to elucidate the biosynthetic pathway of limonin, a prototypical limonoid found in *Citrus sinensis* (sweet or Valencia orange). We selected gene candidates via a homology and transcriptome analysis, and we developed a platform for rapid pathway discovery combining transient expression in the host *Nicotiana benthamiana* with liquid chromatography mass spectrometry (LC-MS) metabolomics. The utilization of this platform allows for fast screening of gene candidates potentially involved in the limonin biosynthetic pathway.

Copyright © American Institute of Chemical Engineers. All rights reserved.

470509 Organelles By Design: Custom Nanobioreactors in Bacteria

Abstract:

Despite the vast array of enzymatic activities available to metabolic engineers, many biosyntheses are impractical due to the production of toxic intermediates and off-target products, poor kinetic performance, and cofactor imbalances. We aim to address these roadblocks by organizing biosynthetic pathways inside custom organelles within cells. Bacterial microcompartments, porous proteinaceous organelles found in *Salmonella* and other bacteria, provide an ideal starting point for the engineering of subcellular bacterial organelles. To this end, we develop tools for the control of microcompartment formation, enzyme encapsulation, and small molecule transport across the compartment shell, as well as computational tools to study native compartment function and identify promising target pathways for encapsulation.

In order to tune the catalytic activity of the microcompartments, we orchestrate microcompartment formation and cargo expression to control when the organelles are formed and how much heterologous cargo is encapsulated. We also create new localization signals that allow fine control over the loading of multiple enzymes to the compartments simultaneously. We are investigating the role of pore residues in small molecule transport across the shells of virus-like particles and the microcompartments themselves, as well as using protein engineering approaches to examine virus-like particle assembly as a model system. Lastly, we use computational tools to model microcompartment function in order to facilitate the selection of appropriate pathways for encapsulation. Together, these tools allow the creation of custom biosynthetic organelles with user-defined enzyme content and transport properties, enhancing flux and decreasing the toxicity of heterologous biosynthetic pathways.

Copyright © American Institute of Chemical Engineers. All rights reserved.

In Situ Labeling of Natural Products with Alkyne Functionality and Its Applications

Xuejun Zhu, Wenjun Zhang
University of California, Berkeley

The terminal alkyne is a functionality widely used in organic synthesis, pharmaceuticals, and material science. It is a particularly useful moiety in azide-alkyne click chemistry reactions, which have recently emerged as one of the most powerful tools in drug discovery and chemical biology. Despite the importance of the terminal alkyne and its prevalence in synthetic compounds, the biosynthetic routes to terminal alkynes are not well understood. We recently identified and characterized two biosynthetic pathways in microbes that lead to the generation of terminal alkynes through unprecedented carrier protein-dependent pathways. These pathways were further exploited for tagging various natural products with terminal alkyne functionalities through starter unit or extender unit engineering. For example, pyrones (polyketides), antimycins and eponemycins (polyketide-non ribosomal peptide hybrids) have been successfully labeled with alkyne functionality. The natural product tagging strategy also enabled the development of a fluorogenic bioorthogonal platform to quantify the newly generated alkyne tags. Using this quantification platform as a screening method, we applied directed evolution strategy to improve the activity of a membrane-bound bifunctional desaturase/acetylenase. One round of mutagenesis along with the newly developed screening method yielded a desaturase/acetylenase variant with increased activity in *E. coli*. This study not only provides alkyne biosynthetic tools that are useful for chemical biology, but it also has applications in which enzymatic generation of terminal alkynes and *in situ* bioorthorgonal chemical transformation are required or preferred.

474995 Invited Presentation: Tandem Catalysis: From Biosynthesis to Biocatalysis

Abstract:

Tandem or sequential catalysis is a hallmark of natural product biosynthesis in which a series of enzymes act together to convert a simple precursor metabolite to a structurally complex product. In contrast, biocatalysis typically involves single enzymes and single-step transformations. In recent years, there is a growing trend to explore the principle of tandem catalysis for biocatalysis. In this presentation, I will first discuss our recent work on the discovery and engineering of natural product biosynthetic pathways for drug discovery and development (1) and then discuss our attempts to develop tandem catalytic processes that combine biocatalysts (enzymes or engineered whole cells) with chemical catalysts with an ultimate goal of exploring the synergy among different types of catalysis (2,3).

1. Y. Luo, H. Huang, J. Liang, M. Wang, L. Lu, Z. Shao, R. Cobb, and H. Zhao. "Activation and Characterization of a Cryptic Polycyclic Tetramate Macrolactam Biosynthetic Gene Cluster." *Nature Communications*, 4: 2894 (2013).
2. C. A. Denard, J. Hartwig, and H. Zhao. "Multistep one-pot reactions combining biocatalysts and chemical catalysts for asymmetric synthesis." *ACS Catalysis*, 3, 2856–2864 (2013).
3. C. Denard, H. Huang, M. J. Bartlett, L. Lu, Y. Tan, H. Zhao, and J.F. Hartwig. "Cooperative Tandem Catalysis by an Organometallic Complex and a Metalloenzyme." *Angewandte Chemie,* 53, 465-469 (2014).

Copyright © American Institute of Chemical Engineers. All rights reserved.

466273 Current Status and Prospects of Industrial Xylitol Production in China

Abstract:

Xylitol is an alternative sweetener that has many beneficial health properties including regulating insulin-independent metabolism in humans, preventing dental caries and a number of pediatric diseases. Based on these promising advantages, xylitol has been commercially developed as a valuable ingredient for use in food and pharmaceutical industries. In China, xylitol is mainly produced industrially by the reduction of xylose derived from hemicellulose hydrolysates of biomass materials in the presence of Raney nickel catalysts. Currently, xylitol market lies between 20,000 and 40,000 tons per year with an economic value ranging between 90 and 340 million dollars and the xylitol market is expect to increase by 15% per year. The mainly Chinese companies including FUTASTE, LONGLIVE and HUAKANG PHARMA can produce about 40,000 tons per year. While chemical synthesis of xylitol has some disadvantages including a high energy requirement, extensive purification steps, and a high cost of product. Accordingly, the biological production of xylitol has recently attracted much attention. Many efforts have been made for the biological conversion of xylitol using microorganisms from hemicellulosic hydrolyzate by Chinese scientists. It is noteworthy that the xylitol productivity of recombinant *Escherichia coli* (1.84 $g\ l^{-1}\ h^{-1}$ and 143.8 $g\ l^{-1}$ yield) by our previous work and the xylitol productivity of recombinant *Candida tropicalis* (2.01 $g\ l^{-1}\ h^{-1}$ and 75.1 $g\ l^{-1}$yield) by Dr. Yuan's group are the highest xylitol productivity rates among recombinant bacteria and fungi from hemicellulosic hydrolyzate, respectively. Further, the industrial bio-production of xylitol and arabinose using *C.tropicalis* by Thomson Biotech from xylose mother liquor has been reported. All of above processes have been studied in an effort to provide one that is high yielding and competitive with chemical processes.

Copyright © American Institute of Chemical Engineers. All rights reserved.

457047 Rational Substitution of Firefly Luciferase from *Photinus Pyralis* for Improvement of Thermostability

Abstract:

Firefly luciferase is widely used in many areas of biotechnology and molecular biology, but the application is restricted due to its poor thermal stability. Random mutagenesis and directed evolution paly an important role in the thermal stability of protein; nevertheless, large number of experimental attempts appear to be time- and labor-consuming. In this paper, to enhance the thermal stability of firefly luciferse, we got two mutants (E325K, E354K) by reversing the charges of individual residues at active sites with opposite interaction energies on the basis of molecular dynamic (MD) simulation. Root-mean-square-deviation(RMSD) and radius of gyration were used to evaluate the thermostability of the mutants. The residual activities of wide-type luciferase, E325K, E354K were 4.52%,7.22% and 30%, respectively, after being heated at 37 â„ƒ for 5 h, which showed the same trends with the computational predictions (E354K > E325K > WT). It was worth nothing that the T_{50} and T_m of E354K were 4.45 and 3.2 â„ƒ higher, respectively, than that of the native luciferase. Rational substitution of residue in regulation of thermostability using MD simulation would be a faster, less resource-intensive and more effective than random mutagenesis and directed evolution for instructional design to enhance thethermostability.

Copyright © American Institute of Chemical Engineers. All rights reserved.

466981 Improving Cellular Robustness and Butanol Titers of *Clostridium Acetobutylicum* ATCC824 By Introducing the Heat Shock Proteins from Extremophilic Bacteria

Abstract:

In recent years, increasing concerns over environment, energy and climate have renewed interest in biotechnological production of butanol. However, growth inhibition by fermentation products and inhibitory components of raw biomass hindered the development of ABE fermentation. Improving cellular robustness of Clostridium acetobutylicum is of great importance for efï¬cient ABE production. In this study, we attempt to improve the robustness and butanol titers of C. acetobutylicum ATCC824 by overexpressing the groESL and dnaK from the extremely radioresistant bacteria Deinococcus wulumuqiensis R12 and C. acetobutylicum ATCC824. Three recombinants were obtained and designated as ATCC824(groESLR12), ATCC824(dnaK R12) and ATCC824(groESL824). These three recombinants were found to signiï¬cantly improve diverse types of stress-tolerances including furfural, butanol, oxidative, acid and osmotic-tolerance. Meanwhile, the butanol titers increased to 11.22g/L, 13.04g/L and 10.68g/L, which were 29.7%, 51% and 23.5% higher compared with the wild type, respectively. Especially for ATCC824(dnaK R12), the production of acetic and butyric acid decreased about 50% and 100% compared with the wild strain. These results indicated that overexpression of groESL and dnaK enhanced the metabolic rate of EMP pathway and improved acids re-uptake efficiency. In addition, ATCC824(groESLR12) was superior to ATCC824(groESL824) in diverse types of stress-tolerances and butanol titers because GroESL from an extreme bacterium could more efficiently perform their function in the heterologous host. Our study provided a new clue and a theoretical basis for improve the robustness and butanol titers of C. acetobutylicum ATCC824.

Copyright © American Institute of Chemical Engineers. All rights reserved.

455883 Polymalic Acid Fermentation: A Platform for Production of Biopolymers and Chemicals

Abstract:

Polymalic acid (PMA) is a novel polyester polymer that has been broadly used in the medical and food industries. Its monomer, L-malic acid, is also a potential C4 platform chemical and extensively used in the food industry while it is currently supplied with the racemic mixture (DL-Malic acid) through chemical synthesis. Recently, great researches have been attempted to develop biotechnological process for L-malic acid production. In our study, a novel route for production of L-malic acid and its derivatives based on PMA platform was developed. Here, a specific fluorescence probe for detecting acidity strength was designed and used for high-throughput screening. Based on the genome-scale metabolic model *i*ZX637, Engineered the strain with overexpressed the gene of pyruvic carboxylase resulted in a PMA titer increase by 30% in shake flask. During the fermentation process, nitrogen limitation was found to be beneficial for PMA biosynthesis, which could upregulate the expression of key genes involved in the PMA biosynthesis pathway. Disruption or overexpression of the nitrogen transcriptional regulator Gat1 controlled by the TOR signaling pathway could obviously regulate PMA biosynthesis. Finally, the downstream process and progress in industry in China are presented.

Copyright © American Institute of Chemical Engineers. All rights reserved.

476737 Engineering Yeast with Minicellulosome and Cellodextrin Pathway for Co-Utilization of Cellulose and Mixed Sugars

Abstract:

Consolidated bioprocessing (CBP), which integrates cellulase production, cellulose saccharification, and fermentation into one step, has been widely considered as the ultimate low-cost configuration for producing second-generation bioethanol from lignocellulose. However, the requirement of a microbial strain able to hydrolyze cellulosic biomass and convert the resulting sugars into high-titer ethanol limits CBP application. In this work, cellulolytic yeasts were developed by engineering ***Saccharomyces cerevisiae*** with a heterologous cellodextrin utilization pathway and bifunctional minicellulosomes. The cell-displayed minicellulosome was two-scaffoldin derived, and contained an endoglucanase and an exoglucanase, while the intracellular cellodextrin pathway consisted of a cellodextrin transporter and a β-glucosidase, which mimicked the unique cellulose-utilization system in ***Clostridium thermocellum*** and allowed ***S. cerevisiae*** to degrade and use cellulose without glucose inhibition/repression on cellulases and mixed-sugar uptake. Consequently, only a small inoculation of the non-induced yeast cells was required to efficiently co-convert both cellulose and galactose to ethanol in a single-step co-fermentation process, achieving a high specific productivity of ~62.61 mg cellulosic ethanol/g cell·h from carboxymethyl cellulose (CMC) and ~56.37 mg cellulosic ethanol/g cell·h from phosphoric acid-swollen cellulose (PASC). Our work provides a versatile engineering strategy for co-conversion of cellulose-mixed sugars to ethanol by ***S. cerevisiae***, and the achievements in this work may further promote the application of CBP in cellulosic biofuel production.

Copyright © American Institute of Chemical Engineers. All rights reserved.

476734 Integrated Chemical and Bio-Catalysis for Production of High-Value Liquid Transportation Fuels from Agricultural Residues

Abstract:

Catalytic conversion of lignocellulosic biomass to high-valued transportation petrol, jet and diesel fuels is of great importance to develop versatile renewable energy and boost rural economy, thus receiving worldwide attention. However, the state-of-the-art processes still suffer from low efficiency in yield and productivity, which hinders the progress of its commercialization. Thus, it is essential to explore the consolidated catalytic reaction and refinery process. Here we report an advanced approach/process that could continuously produce long-chain (C5-C15) ketones from lignocellulosic biomass in a cascade mode by integrating several bio- and chemical catalysis reactions, which exhibits high efficiency and stable conversion rate (>70 %) in alkylation reaction. This technology consolidated the acetone-butanol-ethanol (ABE) fermentation, palladium catalyzed alkylation and in situproduct recovery, allowing the catalytic reaction to proceed consistently and smoothly. The remarkable conversion property is mainly attributed to excellent compatible linkage of chemical catalysis with biosynthesis and process design to stabilize the activity of Pd catalyst. This validated strategy for rational process design and integration of chemical- and bio-catalysis offers great demonstration in producing high-valued transportation biofuels derived from agricultural residues.

Copyright © American Institute of Chemical Engineers. All rights reserved.

457314 Cholesteryl Hyaluronic Acid-Coated, Reduced Graphene Oxidenanosheets for Anti-Cancer Drug Delivery

Abstract:

Here, we report hyaluronyl reduced graphene oxide (rGO) nanosheets as a tumor-targeting delivery system for anticancer agents. Hyaluronyl-modified rGO nanosheets were prepared by synthesizing cholesteryl hyaluronic acid (CHA) and using it to coat rGO nanosheets, yielding CHA-rGO. Compared with rGO, CHA-rGO nanosheets showed increased colloidal stability under physiological conditions and improved in vivo safety, with a survival rate of 100% after intravenous administration of 40 mg/kg in mice. The doxorubicin (Dox) loading capacity of CHA-rGO was 4-fold greater than that of rGO. Uptake of Dox by CD44-overexpressing KB cells was higher for CHA-rGO than for rGO, and was decreased in the presence of hyaluronic acid through competition for CD44 receptor binding. After intravenous administration in tumor-bearing mice, CHA-rGO/Dox showed higher tumor accumulation than rGO/Dox. The in vivo antitumor efficacy of Dox delivered by CHA-rGO was significantly increased compared with free Dox or rGO/Dox. In CHA-rGO/Dox-treated mice, tumor weights were reduced to 14.1% 0.1% of those in untreated mice. Our findings indicate that CHA-rGO nanosheets possess greater stability, safety, drug loading capacity, and CD44-mediated delivery of Dox than rGO nanosheets. These beneficial properties of CHA-rGO improved the distribution of Dox to tumors and facilitated the cellular uptake of Dox by CD44-overexpressing tumor cells, resulting in enhanced anticancer effects.

Copyright © American Institute of Chemical Engineers. All rights reserved.

471641 Pocket Centrifuges: Building a 20-Cent Centrifuge for Global Health Applications

Abstract:

Centrifugation is critical to most diagnostic assays, but the infrastructure requirements (cost, electricity) of centrifugation present a significant barrier to low-cost, point-of-care diagnostics. I will describe the design and implementation of an ultra-fast centrifuge that is low-cost (20-cents), light-weight, and operates on human-power, to deliver up to 30,000xg.

The instrument can be easily disposed (use-and-throw centrifugation) and can be carried to remote locations in the field (fits in a pocket).

Using centrifugal microfluidics, we are able to isolate plasma from whole blood in less than 2 mins. I will also highlight easy-incorporation of quantitative buffy coat (QBC) technique for Malaria diagnostics.

Copyright © American Institute of Chemical Engineers. All rights reserved.

463450 Patterned Biofilm Formation to Investigate Biofilm-Associated Antibiotic Resistance

Abstract:

Biofilm-related high-level tolerance to antimicrobials is a leading cause of chronic infections in humans and persistent biofouling in industry. However, the mechanism of such tolerance is still not fully understood due to the intrinsic heterogeneity in biofilm structure and associated spatial variation in cell physiology. Previously, we demonstrated that patterned biofilms with well-defined size and shape of micron-scale cell clusters and distance between adjacent cell clusters can be obtained using gold surfaces with patterned functional alkanthiols to control structural heterogeneity. Using this system, we found that multicellular connections can be formed between cell clusters depending on the size of interacting clusters and the distance between them.

These results led to our hypothesis that antibiotic tolerance of biofilm cells may change during biofilm growth and the cells involved in inter-cluster interactions are more sensitive to antibiotics. To test this hypothesis, antibiotic treatment of *Escherichia coli* biofilm cells during biofilm formation was monitored using live/dead staining and colony forming units (CFU) viability assays. The results revealed that the antibiotic susceptibility of biofilm cells increased in the first two hours after incubation and then gradually decreased to the level of overnight planktonic cultures. In addition, *E. coli* RP437 cells attached between micron-scale square patterns were found more sensitive to ampicillin compared to the cells within the square-shaped cell clusters. These findings are consistent with our earlier report that active interaction between all clusters is involved during biofilm formation, and provide new information for maximizing the outcome of antibiotic treatment.

Copyright © American Institute of Chemical Engineers. All rights reserved.

471540 Elucidating Bacterial Interactions in the Vaginal Microbiome Using Microfluidic Droplet Technology

Abstract:

Elucidating Bacterial Interactions in the Vaginal Microbiome Using Microfluidic Droplet Technology

Corine Jackman[1] (corinemj@umich.edu), **Xiaoxia (Nina) Lin**[1,2] (ninalin@umich.edu)

[1] *Department of Chemical Engineering, University of Michigan, Ann Arbor, MI 48109*

[2] *Department of Biomedical Engineering, University of Michigan, Ann Arbor, MI 48109*

There is increasing evidence suggesting that the bacteria in the human vagina, referred to collectively as the vaginal microbiome (VMB), plays fundamental roles in womenÕs health and susceptibility to diseases. For instance, bacterial vaginosis (BV), the most common vaginal infection in women of reproductive age in the United States, is characterized by disturbances in the VMB where lactic acid bacteria level is reduced dramatically by an overgrowth of various other bacteria. BV is associated with an increased risk of preterm birth, which has accounted for approximately 75 percent of perinatal deaths in the United States[1]. Women with BV also have an increased susceptibility for contracting sexually transmitted infections, such as HIV. The question of how the diverse microorganisms in VMB interact with one another and with their host functionally remains largely unanswered. This is due in part to the immense ecological complexity and the challenge in isolating microbes using conventional cultivation methods.

Microfluidic droplet technology has emerged as a new approach for compartmentalizing cells from microbial communities to perform assays, measure growth, and detect interactions in individual nano- or pico-liter droplets.

Advantages of this approach include high-throughput experimentation, and reduced space, time, chemical reagent and labor requirement. In this work, we use microfluidic droplet co-cultivation to investigate pairwise interaction between two vaginal bacteria, specifically *Gardnerella vaginalis* and a *Lactobacillus* species. *G. vaginalis* is associated with BV and *Lactobacillus* species are believed to promote health by producing lactic acid and reducing pH in the vagina. A previous study demonstrated a severe killing effect on *G. vaginalis* after incubation in supernatant of *Lactobacilli.* We investigate this interaction by co-cultivating these species in micro-droplets and comparing growth in the bi-culture verses monocultures. Droplets after incubation under appropriate condition are analyzed through Fluorescent *In Situ* Hybridization (FISH) using specific probes that hybridize to the 16s rRNA of targeted species. More cell growth in droplets with monocultures in comparison to droplets with bi-cultures indicate a negative interaction. This work demonstrates the effectiveness of using droplet co-cultivation for investigating microbial interactions in the VMB.

Reference:

1. McCormick, Marie C. "The contribution of low birth weight to infant mortality and childhood morbidity." *New England journal of medicine* 312.2 (1985): 82-90.

Copyright © American Institute of Chemical Engineers. All rights reserved.

453912 Microfluidic Medip-Seq for Low-Input Epigenomic Analysis and Personalized Medicine

Abstract:

Epigenetic modifications, such as DNA methylation and histone modifications, play pivotal roles in gene expression and regulation, and are highly involved in cellular processes such as stem cell pluripotency/differentiation and tumorigenesis. In mammals, DNA methylation is the best studied epigenetic modification. It refers to the addition of a methyl group at the carbon-5 position of cytosine residues within CpG dinucleotides, and it is heavily associated with a large number of human diseases. For example, transcriptional silencing of tumor-suppressor genes by CpG-promoter hypermethylation plays an important role in cancer development. Therefore, understanding of epigenetic regulations will help to improve various aspects of biomedicine. For instance, personalized medicine can be tailored based on epigenetic profile of certain patient to specifically control gene expression in the disease treatment. However, the technology for profiling epigenetic modifications, i.e. methylated DNA immunoprecipitation (MeDIP), suffers from serious limitations. The key limitation is the sensitivity of the assay. Conventional assay requires a large number of cells ($>10^6$ cells per MeDIP). This is feasible when using cell lines. However, such requirement has become a major challenge when primary cells are used because very limited amounts of samples can be generated from lab animals or patients. Population heterogeneity information may also be lost when a large cell number is used.

In this project, we developed an automated ultrasensitive microfluidic methylated DNA immunoprecipitation followed by next-generation sequencing (MeDIP-Seq) technology for profiling DNA methylomes. We extensively optimized design parameters for each and every step of MeDIP (e.g. sonication/crosslinking time, antibody concentration, washing conditions) in order to reach highest sensitivity of 0.1 ng DNA (or ~50-100 cells) as starting material for IP, which is roughly 4-5 orders of magnitude higher than the prevailing protocol and 2-3 orders of magnitude higher than the-state-of-the-art(~50 ng). With such sensitivity, we were able to study temporal dynamics in the DNA methylomes during the various stages of mammary cancer development from a transgenic mouse model.

Copyright © American Institute of Chemical Engineers. All rights reserved.

465273 High-Throughput Measurements of Cancer Cell Deformability

Abstract:

We designed a high-throughput deformability cytometer for assessing the mechanical properties of cancer cells. Changes in cancer cell mechanics impact the likelihood of cancer to metastasize within the body, where highly invasive cells are typically more deformable than non-malignant cells or primary tumor cells. Cellular mechanics are often assessed using low throughput techniques, such as atomic force microscopy or micropipette aspiration. The rapid assessment of cellular properties would be particularly useful for characterizing cancer cells due to the high heterogeneity of tumor cells. Here we present a microfluidic device for the high-throughput assessment of cancer cell deformability. Our device consists of an obstacle course of restriction channels for cells to squeeze through. Measurements of entrance time, transit time, and elongation index within the channels are quantified to evaluate the mechanical behavior of cells. We use ovarian cancer cells of variable metastatic potential from a murine model to assess the changes in deformability that relate to the site of cell origin. We consider primary ovarian cancer cells, cells derived from a tumor on the hepatic lobe, and cells from a metastatic tumor found in the lung [Zhang, L. et al., Oncogene, 2016]. We consider the contributions of intracellular structure to the measured changes in cell mechanics using quantitative fluorescence microscopy. This work advances the understanding of how deformability and intracellular structure contribute to the metastasis of ovarian cancer.

Copyright © American Institute of Chemical Engineers. All rights reserved.

468929 Deterministic Encapsulation of Single Cells in Thin Tunable Microgels for Niche Modeling and Therapeutic Delivery

Abstract:

Deterministic Encapsulation of Single Cells in Thin Tunable Microgels for Niche Modeling and Therapeutic Delivery

Angelo S. Mao[1], Jae-Won Shin[1,2], David A. Weitz[1], David J. Mooney[1]

[1]School of Engineering and Applied Sciences, Harvard University, Cambridge, MA

[2]Department of Pharmacology, University of Illinois College of Medicine, Chicago, IL

Introduction: High-throughput encapsulation of single cells into thin hydrogel layers has the potential to abet regenerative medicine therapies and basic science applications, for example enabling the construction of microscale assemblages *in vitro* and improving existing cell therapy techniques. Current methods for encapsulating cells into microscale hydrogels generally follow a Poisson distribution and possess limited control over hydrogel properties. We report a microfluidic, water-in-oil based method for deterministic encapsulation of single cells in an ~6 micron thin layer of alginate hydrogel that was used for microscale tissue construction *in vitro* and improved cell delivery *in vivo*.

Materials and Methods: A cross-junction polydimethylsiloxane (PDMS) microfluidic device was used to fabricate cell-containing alginate microgels using calcium carbonate as the polymer cross-linker. Different polymer molecular weights were tested, as were combinations of alginate with extracellular matrix proteins. Encapsulated cells were cultured in PDMS microwells with exposure to osteogenic induction media or delivered intravenously into mice.

Levels of donor-derived soluble factors were measured from blood plasma of injected mice.

Results and Discussion: By adsorbing calcium carbonate nanoparticles to cells prior to encapsulation, a high percentage of alginate microgels were found to contain cells without decreasing the fraction of cells that were encapsulated. Cells exhibited viability near 90% over a three-day period. Microgels were found to be mechanically tractable and capable of forming interpenetrating hybrid microgels. Moreover, microscale cell assemblages of encapsulated mesenchymal stem cells (MSCs) were constructed with a PDMS-based microwell system. Singly-encapsulated cells were observed to differentiate more robustly than multiply-encapsulated cells in larger microgels. Finally, singly-encapsulated human MSCs were found to respond more strongly to stimulatory factors, experience more delayed clearance kinetics, and sustain greater donor-derived soluble factors in blood plasma following intravenous delivery into mice, compared to multiply-encapsulated MSCs in larger gels.

Conclusions: The improved differentiation capacity of singly-encapsulated MSCs and delayed clearance kinetics of intravenously delivered singly-encapsulated MSCs indicate the potential of this approach in a variety of regenerative medicine and tissue engineering applications.

Copyright © American Institute of Chemical Engineers. All rights reserved.

474644 Advances in Microfabricated Sensors for Health and Environmental Monitoring

Abstract:

Accurate detection of hazardous chemicals is critical to the health of the public and the environment, and to the safe and efficient operation of many industrial processes. This presentation will discuss recent advances in low-power chemical sensing for health, environmental and process monitoring. It will also present our approach, which leverages the microfabrication technology to realize highly efficient microheater-based platforms with fast response and recovery times. The integration of a number of nanotechnology-based sensing materials with the microheater platform enables fast and sensitive detection of several health and environmental pollutants. The talk will end with current limitations and plans for future directions.

Copyright © American Institute of Chemical Engineers. All rights reserved.

461681 Application of PAT-Based Feedback Control for the Crystallization of Pharmaceuticals in Porous Media

Abstract:

The development of new drug delivery methods to tailor the desired drug release have been of major interest such as substrates, polymer films and microspheres. The drug release can be affected by various attributes of the polymer properties such as molecular weight, size and size distribution, porosity, morphology and make-up. However, the purpose of this work is to evaluate the benefits of controlling the bulk crystallization in the presence of alginate porous particles. The main objective is to increase the drug loading (DL) in the alginate beads and minimizing the heterogeneous nucleation occurring in the bulk system. Two types of controlled cooling profiles and the model-free Direct Nucleation Control (DNC) strategy were evaluated. A mean DL from 49 to 69 % was achieved through the various scenarios. The application of DNC increased or maintained the mean DL achieved compared to the cooling only scenarios, while a decrease of about 75 % was observed in the batch-to-batch variability. SEM images showed that crystals of different crystal size distribution (CSD) were obtained for the various cooling scenarios, which can be related to the different operating trajectories in the phase diagram achieved during the feedback control approaches. The possibility of manipulating the CSD of particles within the polymer beads can be used to tailor dissolution profile and bioavailability. The work demonstrates the benefits of using feedback control approaches to increase DL, decrease batch-to-batch variability and control dissolution behavior of drug-loaded polymer microspheres.

Copyright © American Institute of Chemical Engineers. All rights reserved.

470315 Optimization of the Bottom-up Precipitation of Nano-Particles of an Ultra-Fast Crystallizer through the Qualitative Characterization of API-Polymer Interactions

Abstract:

Formulations of poorly soluble drugs, seeking to improve bio-availability by API size reduction into the nano-range, typically rely on top-down comminution techniques, e.g. media-milling. While effective, media-milling is costly, time and energy-intensive as well as presenting some isolation challenges. Alternative methods, aiming to access the nano-range through control of the precipitation process, usually rely on very high-intensity mixing in specialty devices like impinging jets and multi-inlet vortex mixers.

For a special class of compounds, crystal nucleation and growth are so fast that even very rapid solvent-antisolvent mixing is insufficient to limit the final crystal size to the nano range.

This work explores the use of polymeric additives to quench the crystal growth of a very fast crystallizer in an attempt at bottom-up nano-crystal precipitation. Lack of first principles understanding of the exact process, through which polymers suppress crystal growth typically leads to high-throughput screening methods for the identification of the optimal additive. However, the large number of potential polymeric candidates, coupled with the parameters controlling the complex high-shear precipitation process lead to a design space too large to explore through brute-force techniques.

Instead, chemometric analysis of the API-polymer interactions was correlated to the performance during precipitation, decoupling polymer optimization from the precipitation process. Identification of specific functional group interactions between the polymer and API also sheds some light on the nature of the crystal growth suppression effect. The

combination of high-shear solvent-antisolvent mixing in a rotor-stator wet-mill and polymer additive effect enabled the scalable bottom-up production of nano-particles.

Copyright © American Institute of Chemical Engineers. All rights reserved.

469039 Core-Shell Nanocomposites for Fast Release of Hydrophobic Active Pharmaceutical Ingredients: Impact of Carrier Size

Abstract:

The bioavailability of a large percentage (estimated to be between 40% and 70%) of newly discovered active pharmaceutical ingredients (APIs) is limited by their poor water solubility–hydrophobicity and slow dissolution rate. Over the last decade, major research efforts have been focused on the development of API nanocrystals in suspension form and subsequently converting it into dry core-shell nanocomposite microparticles (NCMPs) to overcome the aforementioned challenges. Preparing NCMPs via fluidized bed (FB) coating of the API nanosuspensions onto inert excipient particles has been used in the manufacturing of nanoformulation based solid dosages. Recent work shows that if the suspensions are properly stabilized, core-shell NCMPs having a water-soluble core can enhance the dissolution rate of poorly water soluble APIs signficantly[1,2]. However, the impact of the carrier particle size has not been investigated. In this study, formation of core-shell nanocomposites of Fenofibrate and Itraconazole, model poorly water soluble APIs, via FB coating of their well-stabilized high drug loaded nanosuspensions is investigated. Specifically, the extent of dissolution enhancement, when fine carrier particles (sub-50 mm) as opposed to the traditional large carrier particles (>300 µm) are used, is examined. This allows testing the hypothesis that greatly increased carrier surface area and corresponding thinner shell at the same API loading for finer carriers can significantly increase the dissolution rate. Fine sub-50 µm lactose (GranuLac® 200) carrier particles were made fluidizable via dry coating with nano-silica, enabling decreased cohesion, fluidization and nanosuspension coating. For both APIs, 30% loaded suspensions were wet-stirred media milled along with a polymer and a surfactant as stabilizers. The stabilizer amounts were screened to achieve stable suspensions with small particle size. The suspensions were FB coated onto hydrophilic nano-silica (M5P) dry coated sub-50 micron lactose (GranuLac® 200) carrier particles or larger carrier particles of median size >300 µm (PrismaLac®40). The resulting finer composite powders (sub-100 mm) based on GranuLac®200 were freely flowing, had high bulk density, and exhibited much faster, immediate release of the poorly water-soluble drugs, in particular for Itraconazole. This is attributed to a much higher specific surface area of the carrier leading to corresponding thinner coating layer for fine carriers as opposed to those for large carrier particles.

Keywords: Active Pharmaceutical Ingredients/drugs, Poorly Water Soluble, Wet-Stirred Media Milling, Fluidized Bed Coating, Fine and large Carrier Particles, Fast Release/Dissolving Nanocomposite Powders

References:

1. C. Knieke, **M. Azad**, D. To, E. Bilgili, R. Dave, Sub-100 micron Fast Dissolving Nanocomposite Drug Powders, **Powder Technol.** 271 (2015) 49–60.
2. A. Bhakay, **M. Azad**, E. Bilgili, R. Dave, Redispersible Fast Dissolving Nanocomposite Microparticles of Poorly Water-Soluble Drugs, **Int. J. Pharm. Sci.** 461(2014) 367–379.

Copyright © American Institute of Chemical Engineers. All rights reserved.

469735 *'designer'* Hydrogel Particles for Controlled Crystallization and Release of Hydrophobic Active Pharmaceutical Ingredients

Abstract:

Nearly 40% or more of drugs currently in the pipeline of the pharmaceutical companies are poorly water soluble, which introduces a significant hurdle for formulating such major class of drugs with an effective dosage. Utilization of innovative biocompatible materials in formulations of hydrophobic drugs to enhance their solubility-limited oral bioavailability and absorption is a rapidly growing area in pharmaceutical materials research. In this work, we develop a novel bottom-up approach for producing and formulating *nanocrystals* of poorly water-soluble active pharmaceutical ingredients (APIs) using core-shell composite hydrogel beads. Organic phase nanoemulsion droplets stabilized by polyvinyl alcohol (PVA) and containing a model hydrophobic API (fenofibrate) are embedded in the alginate hydrogel matrix and subsequently act as crystallization reactors. Controlled evaporation of this composite material produces core-shell structured alginate-PVA hydrogels with drug nanocrystals (500 - 650 nm) embedded within the porous confinement environment of the core. Adjustable loading of API nanocrystals up to 83% by weight is achieved with improved dissolution behavior. We also develop a quantitative model and experimentally validate that the drug release patterns of the fenofibrate nanocrystals can be modulated by controlling the thickness of the PVA shell and drug loading. In summary, our composite hydrogel enabled encapsulation system allows for controlled crystallization, adjustable drug loading, and customizable dissolution profiles ranging from fast to delayed release by utilizing the unique microstructures of coreâ€"shell hydrogels, thus provides a versatile oral delivery system.

Copyright © American Institute of Chemical Engineers. All rights reserved.

Dissolution of Carbamazepine Crystallised directly onto Excipients

Vivek Verma, Clare M Crowley, Peter Davern, Sarah Hudson & B. Kieran Hodnett
Synthesis and Solid State Pharmaceutical Centre, Department of Chemical and Environmental Sciences, Materials and Surface Science Institute, University of Limerick, Ireland
Email ID: vivek.verma@ul.ie

For BCS Class II drugs, such as carbamazepine (CBMZ), *in vivo* bioavailability is significantly affected by dissolution rate [1]. Conventional approaches to improving the latter have relied on mechanical crystal size reduction techniques, such as attrition milling, but this can lead to degradation of active pharmaceutical ingredients (APIs), non-uniform crystal size distributions and the incorporation of impurities [2]. To circumvent these issues various 'bottom-up' approaches (*e.g.*, spray drying, antisolvent precipitation, sonoprecipitation, *etc.*) to crystal size reduction have been investigated [3].

A potential alternative approach to reducing crystal size is the heterogeneous nucleation of APIs in the presence of dispersed excipient particles. Chadwick *et al* [4] showed an increase in the nucleation rate of acetaminophen (AAP) in the presence of α-Lactose Monohydrate (α-LMH) and β-D-Mannitol (β-D-Man). Increasing nucleation rates should concomitantly reduce the crystal size distribution of the APIs and may, therefore, obviate the need to mill API batches. In this study CBMZ was selected as the model API and the influence of the process conditions (*i.e.,*time and supersaturation) and dispersed, insoluble excipients on its crystallisation from metastable CBMZ-methanol solutions and its subsequent dissolution was examined. Unlike the AAP system studied by Chadwick et al, the presence of excipients in a range of supersaturation and crystallisation time had no impact on the final crystal size distribution (CSD) of CBMZ.

PXRD and Raman spectroscopy of the isolated solids fractions confirmed the formation of CBMZ FIII from metastable CBMZ-methanol solutions (supersaturations (S) of 1.22, 1.34 and 1.55) in the absence and presence of excipients (α/β-Lactose (α/β-Lac), β-D-Mannitol (β-D-Man), microcrystalline cellulose (MCC) and carboxymethyl cellulose (CMC)). The crystal size distribution (CSD) of the FIII crystals, assessed by image analysis of the SEM micrographs, showed similar CSD in the presence and absence of dispersed excipient (Table 1 and Figure 1). However, the dissolution rate of CBMZ FIII was enhanced 2-fold upon crystallisation in the presence of dispersed excipients as compared to CBMZ FIII recrystallised in the absence of excipients under the same conditions (Figure 2). The dissolution rate of CBMZ FIII crystallised in the presence of dispersed excipients is higher than the physical mix of CBMZ FIII and excipients, mostly evident with α/β-Lac and β-D-Man (Table 2) reflecting the superiority of the process. SEM-Raman indicates the presence of low concentration of CBMZ on the surface of excipient facilitating faster dissolution rate at initial times. Figure 2 also presents the comparitive dissolution of CBMZ FIII crystallised in the presence of dispersed β-D-Man and the physicsal mix of CBMZ FIII and β-D-Man indicating the improved dissolution rate throughout the dissolution profile.

Table 1: Mean crystal size distribution (CSD) (μm) of CBMZ FIII crystals formed in the presence and absence of excipients

S	Excipient				No Excipient
	α/β-Lac	β-D-Man	MCC	CMC	
1.22	16±11	13±14	17±11	13±7	22±10
1.34	23±14	13±7	14±7	32±22	23±10
1.55	17±9	18±10	19±10	20±10	27±12

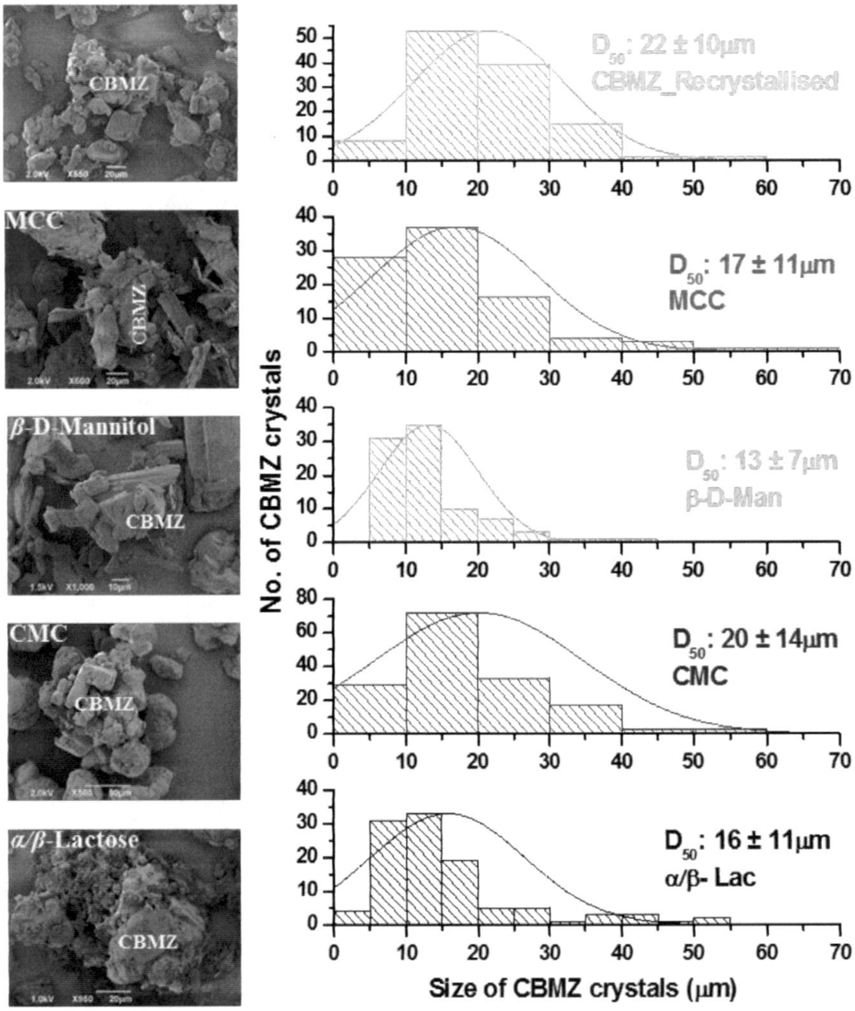

Figure 1: SEM micrographs and CSDs of CBMZ FIII crystallised in the absence and presence of excipients from metastable CBMZ-MeOH solution (t=3 hr aged samples; S = 1.22)

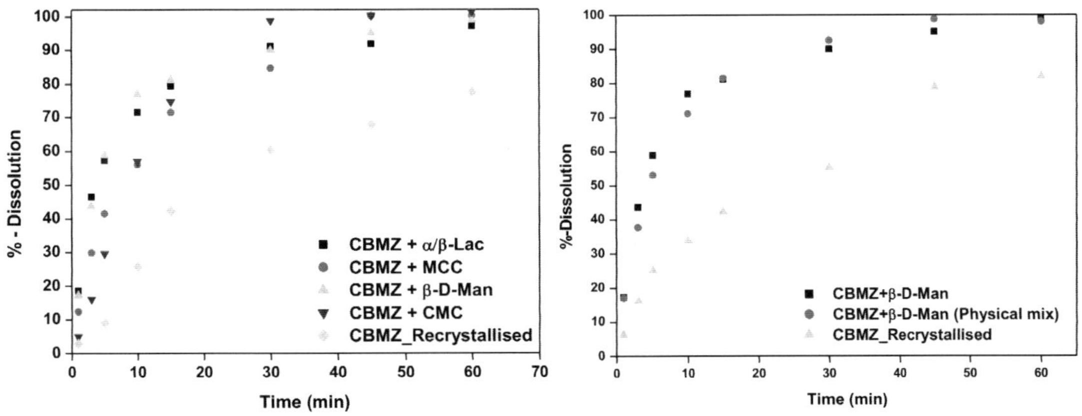

Figure 2: (Left) Comparison of the dissolution rates of CBMZ FIII crystallised in the absence and presence of excipients from metastable CBMZ-MeOH. (Right) Comparision of the dissolution rates of CBMZ crystallised in presence of β-D-Man and the physical mix of CBMZ FIII and β-D-Man; (S=1.22, t=3hr)) (dissolution media: PBS, pH = 7.4; sink conditions ($[CBMZ]_{max}$= 40 mg/L))

Table 2: Dissolution rate of CBMZ FIII over 5 min in the presence and absence of excipients

Excipients	**S**	**Dissolution rate over 5 min**
None	Recry	3.47
None	Recry	3.47
None	Recry	3.47
None	Recry	3.47
CMC	1.22	5.75
CMC	Phymix	6.01
CMC	1.55	6.12
CMC	1.34	6.45
MCC	1.34	8.36
MCC	1.22	8.85
α/β-Lac	Phymix	9.84
MCC	Phymix	10.79
β-D-Man	Phymix	11.31
β-D-Man	1.34	11.97
MCC	1.55	12.2
α/β-Lac	1.34	12.3
β-D-Man	1.22	12.65
α/β-Lac	1.22	12.72
α/β-Lac	1.55	13.62
β-D-Man	1.55	15.74

Acknowledgements
This work was funded by Science Foundation Ireland under Grant 12/RC/2275. This work was also conducted under the framework of the Irish Government's Programme for Research in Third Level Institutions Cycle 5, with the assistance of the European Regional Development fund.

References

1. Lindenberg M., S. Kopp and J.B. Dressman, European Journal of Pharmaceutics and Biopharmaceutics, 2004, **58**(2): p. 265-278.
2. Horn D. and J. Rieger, Angewandte Chemie International Edition, 2001, **40**(23): p. 4330-4361.
3. Poornachary S. L. et al., Crystal Growth and Design, 2016, **16**(2):p. 749-758.
4. Chadwick, K.; Chen, J.; Myerson, A. S.; Trout, B. L., Toward the Rational Design of Crystalline Surfaces for Heteroepitaxy: Role of Molecular Functionality. *Crystal Growth & Design* 2012, **12** (3), 1159-1166.

*** Formatting Issue - Best Available Paper/Graphic ***

467271 Modeling Spray Dried Dispersion: From Droplet to Particle

Abstract:

Modeling Spray Dried Dispersion: From Droplet to Particle

Jaime L. Curtis-Fisk[1], Priti Jain[2], Jodi Mecca[1], Bart Rijksen,[3] Shrikant Khot[1], William W. Porter III[1]
[1]The Dow Chemical Company, Midland, MI USA

[2]Dow Chemical International Private Limited, Mumbai, India

[3] Dow Benelux BV, Terneuzen, Netherlands

***Purpose*:**

Poorly soluble APIs are often formulated as a solid amorphous dispersion in combination with an excipient polymer to enhance their solubility. A common route to producing the dispersion is through spray drying. This process involves spraying a solution of polymer, active pharmaceutical ingredient (API) and solvent into a heated drying gas where the liquid spray droplets dry into solid particles of the amorphous dispersion. Hydroxypropyl methylcellulose acetate succinate (HPMCAS), such as AFFINISOL™ HPMCAS, is one polymer shown to significantly enhance API solubility when in an amorphous dispersions. The purpose of this work is to develop a droplet drying model connecting the fundamental properties of HPMCAS polymer solutions to the spray drying particle formation kinetics and the resulting particles' morphology and compositional uniformity. With this modeling capability, it may be possible to predict and gain insight into the observed properties of amorphous spray dried dispersions.

Approach and Methods:

A particle formation model was developed considering the balance between solvent evaporation and solute diffusion to predict particle morphology and composition. The model, as illustrated in Figure 1, considers the diffusion of API and excipient towards the core of the droplet and surface recession of the solvent as the two main counteracting mechanisms in the drying of droplets. Methods for predicting particle formation are known, but do not fully address the complexities of producing spray dried dispersions from a solution containing API and polymer.[i] This model framework builds upon these references and addresses the non-ideality of the liquid phase due to the use of polymeric excipients, including the interaction between solutes and the change in evaporation rate due to non-ideal vapour liquid equilibrium. This added complexity will provide a more accurate route to predicting the impact of formulation and spray drying conditions on particle morphology.

To provide inputs to the model, fundamental properties including viscosity and diffusivity of HPMCAS-API solutions were characterized in a wide range of solvents and solvent blends. Additionally methods to characterize HPMCAS-API interactions were used to understand potential effects on mass transport within the solution droplet.

*** Formatting Issue - Best Available Paper/Graphic ***

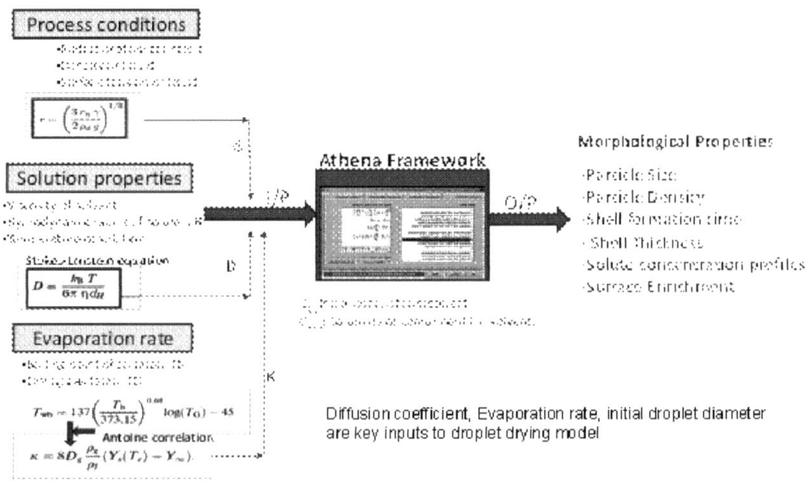

Figure 1. **The model framework combines process conditions, solution properties, and evaporation rate in an Athena framework to generate predictions of particle morphological properties**

Results:

This particle formation model was shown to compare well against experimental results in the literature. Figure 2 illustrates a comparison of the model predictions of spray dried trehalose against results from Vehring et al.[ii] The samples evaluated cover a low Peclet number range of 0.44 to 1, with an initial droplet diameter of 20 µm resulting in a dry particle diameter of 1.7 to 1.8 µm. The model accurately predicts the measured dry particle density, which is near the density of amorphous trehalose indicating solid particles.

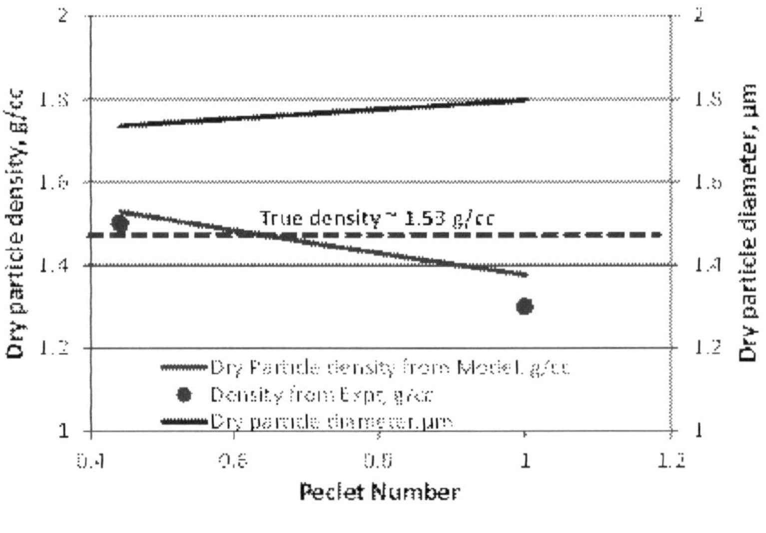

Figure 2. Comparison of experimental results from Vehring et al and model predictions of trehalose particles formed through spray drying.[ii]

It was also found to predict well the experimental results from Vehring et al.[ii] for spray drying of glycoprotein solution as shown in Figure 3. As the Peclet number increases, the particle diameter increases while the density decreases. Both the experimental and modeled particle densities (0.1 to 0.5 g/cc) are significantly lower than the density of solid glycoprotein. This indicates hollow particles with void space, a result confirmed through microscopy of the generated particles.

***** Formatting Issue - Best Available Paper/Graphic *****

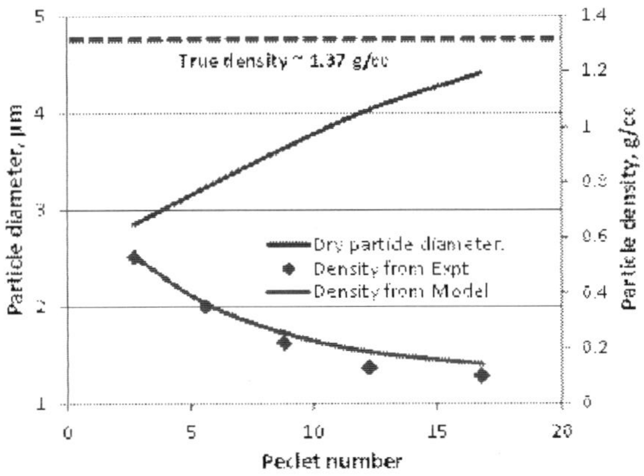

Figure 3. Comparison of experimental results from Vehring et al and model predictions of glycoprotein particles formed through spray drying.[ii]

To apply the droplet drying model to spray drying API-HPMCAS dispersions, fundamental solution properties were characterized. The viscosities of such solutions were found to vary strongly based on the specific HPMCAS polymer and spray drying solvent. As shown in Figure 4, at similar concentrations a wide range of viscosities were observed indicating very different polymer configurations in solution. Similar measurements were conducted to quantify the diffusivity of polymer and API solutes in solution and the surface tension of these solutions. Having quantified these properties, predictions around the droplet drying process and resulting particle properties will be shared. It will be shown that depending on the solvents selected for spray drying, diffusion controlled or surface recession controlled drying kinetics can occur resulting in very different particle morphologies and compositional uniformity.

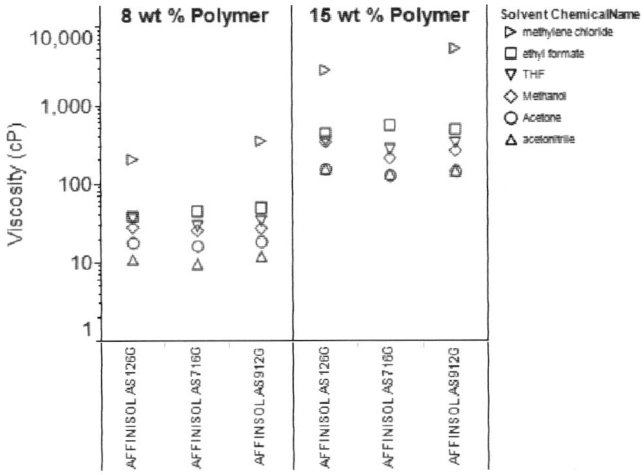

Figure 4: Viscosity of Polymer/Solvent Solutions that serve as inputs for particle formation modeling

Conclusions:

Experimental analysis of formulation solution properties and polymer chemistry provide a solid framework for building modeling capabilities to predict the formation of solid particles from liquid droplets in the spray dried dispersion process. The two examples presented highlight the dramatically different potential outcomes from a spray

*** Formatting Issue - Best Available Paper/Graphic ***

drying process, from solid particles of high density to low density hollow particles. The ability to predict the nature of the particles formed based on formulation and process properties provides the formulator with a valuable tool to optimize this process to hone in on a targeted outcome prior to conducting time consuming experiments. Particle morphology can play a critical role in downstream processing and ultimately the application performance. Combining polymer chemistry and formulation expertise with modeling capabilities provides formulators with a streamlined route to product development.

---

[i] Chen, Xiao Dong; Sidhu, Harvinder; Nelson, Mark; "Theoretical probing of the phenomenon of the formation of the outermost surface layer of a multi-component particle, and the surface chemical composition after the rapid removal of water in spray drying," Chem. Eng. Sci. 66 (2011) 6375-6384.

[ii] Vehring, Reinhard; Foss, Wiillard R.; Lechuga-Ballesteros, David; "Particle formation in spray drying," Aerosol Science 38 (2007) 728-746.

Copyright © American Institute of Chemical Engineers. All rights reserved.

458912 Modeling Dissolution of Polymer Strip Films Loaded with Poorly Water-Soluble Drug Nanoparticles

Abstract:

Thanks to a wide variety of excipients, APIs, and their interactions, reliable prediction of the performance of any solid dosage form is a significant challenge. This is especially true for swellable polymeric strip films containing poorly water-soluble drug nanoparticles, as several different mechanisms may potentially control the rate of drug release. These limiting mechanisms include diffusion of the dissolved drug out of the swollen polymer gel layer, erosion of the polymer matrix, and dissolution of the poorly water-soluble drug particles. The objective of this work is to develop a predictive model for drug release from nanoparticle-laden polymer strip films based on various formulation parameters and inherent drug/excipient properties. This model utilizes first principles to estimate evolution of film thickness, dissolved drug concentration, drug particle size, and water concentration throughout the film over time during the dissolution process. Diffusion of dissolved drug and solvent, drug particle dissolution, polymer erosion, and film swelling/shrinkage are all taken into consideration. Results from this model will be used to better understand the fundamental driving forces behind drug release from poorly water-soluble drug particle-laden polymer strip films, allowing for more intelligent formulation design of targeted applications.

Copyright © American Institute of Chemical Engineers. All rights reserved.

468415 A Novel Method for Grindability Assessment of Pharmaceutical Powders

Abstract:

In the early stages of development of a new Active Pharmaceutical Ingredient (API), insufficient material quantity is usually available for addressing process engineering issues, such as milling. Mechanical properties that are accountable for size reduction are hardness, toughness and stiffness (for semi-brittle failure), and these can be characterised using a small sample quantity by nanoindentation. However, the use of nano-indentation may not be appropriate for strain-rate sensitive solids, and hence impact testing, backed up by breakage analysis, can give better representative information on grindability. Unfortunately, there is no commercially-available single particle impact testing for fine particulate solids. In contrast, dry powder dispersers, such as the Scirocco powder disperser of Malvern Mastersizer 2000, are widely available. In this device, particles are dispersed by an air jet in a Venturi eductor configuration, impacted at a bend and presented to the laser light beam for size analysis by laser light diffraction. So this device has the potential to be used for testing the breakage propensity (grindability) of particles. However, particles of different sizes accelerate to different velocities, due to short acceleration length, and break to different extents; hence the results will be 'fuzzy', unless the relation between particle size and impact velocity and nozzle pressure is established.

Three dimensional multiphase CFD simulations of the Scirocco disperser have been carried out to analyse the air flow field, following which particle trajectories and impact velocities have been calculated. These calculations are then used in the estimation of the breakage propensity parameter. The extent of breakage and change in the surface area are then related to the breakage propensity parameter, for which a good correlation is shown for a range of particle sizes of several organic particles. The obtained information shows the ease of grindability of different materials relative to each other.

Copyright © American Institute of Chemical Engineers. All rights reserved.

471687 Feedforward Control of Continuous Pharmaceutical Manufacturing Process

Abstract:

Continuous manufacturing has been in use in many industries for a long time, however the interest in pharmaceutical continuous manufacturing is relatively new and most pharmaceutical companies are now moving from batch to continuous. With continuous manufacturing we have shorter processing times, increased efficiency and increased product quality assurance in real time. The product quality is of utmost importance in pharmaceutical manufacturing and in order to control that, we need to make sure that effect of input disturbances are proactively mitigated before they affect end product quality. Therefore it is desired that a combined feedforward/feedback control system be implemented over a traditional feedback control system to dissipate and control the disturbances from entering the process. The feedforward controller measures and takes corrective actions for the disturbances before they affect the process while the feedback controller ensures the consistency of the output [1].

In this work, feedforward control loops have been designed for continuous pharmaceutical manufacturing process via wet granulation (WG), roller compaction (RC) and direct compaction (DC) and their performance have been evaluated through a flexible integrated process flowsheet model. For WG line, a combined feedforward/feedback control system has been developed to control the drug concentration at twin screw granulator (TSG) outlet. This process involves a batch pre-blending step where API and intragranular components are mixed, and subsequently fed into a continuous wet granulation (TSG) processing step, together with second stream containing excipients. The commingling of these two streams in the twin screw granulator leads to an overall low API concentration. This concentration of the API is very small and hence, difficult to be measured by a Near Infrared (NIR) sensor. Thus, the role of the residence time distribution (RTD) model in predicting the concentration at the outlet of the TSG as well as to determine the disturbance dissipation along the process becomes significant. Along with this, RTD can also be used for the raw material traceability and scale-up requirements. A major source of drug concentration variability is the potential lack of homogeneity of the stream entering the granulator. Pre-blends discharged by a tumbling blender are known to fluctuate fairly significantly in composition and due to limited back mixing in the twin screw granulator, these fluctuations are propagated to the granulator exit, affecting the content uniformity of API in the granules. In turn, such variability can affect product content uniformity (CU). Since the composition of the stream discharged by the feeder can be monitored instantaneously using PAT method, it is therefore possible to adjust the ratio of the API-bearing stream and the excipient-only stream, and their flow rates, to compensate for composition fluctuations entering, and therefore exiting, the granulator. This has been achieved through a feedforward control mechanism combined with a feedback control loop. For RC line, a feed forward control loop has been developed to mitigate the effect of lubricant concentration. This loop has been integrated with feedback control loop for ribbon density which is cascade with hydraulic pressure feedback (slave) control loop, and a feedback control loop for throughput. For DC line, three feedforward control loops have been developed to proactively mitigate the effects of variations of powder bulk density, lubricant concentration and drug concentration on tablet weight & hardness, tablet hardness, and tablet potency respectively. However, maximum two feed forward control loops can be activated at a time. These feedforward loops have been integrated with feedback control loops [2].

The aim of this presentation is two-fold. First to highlight the feedforward control loops involved in continuous pharmaceutical manufacturing via WG, RC and DC and second to demonstrate the performance of combined feedforward/feedback control system over only feedback control system.

References:

[1]. Singh, R., Muzzio, F., Ierapetritou, M., Ramachandran, R. **(2015)**. A combined feed-forward/feed-back control system for a QbD based continuous tablet manufacturing process. PROCESSES Journal, 3, 339-356.

[2]. Singh, R., Sahay, A., Karry, K. M., Muzzio, F., Ierapetritou, M., Ramachandran, R. (2014). Implementation of a hybrid MPC-PID control strategy using PAT tools into a direct compaction continuous pharmaceutical tablet manufacturing pilot-plant. International Journal of Pharmaceutics, 473, 38â€"54.

Copyright © American Institute of Chemical Engineers. All rights reserved.

471742 Simultaneous Measurement of Granule Disintegration and API Dissolution Profiles As a Tool for Maintaining Product Quality during Process Scale-up

Abstract:

The release rate of an active pharmaceutial ingridient (API) is a crucial atribute of a solid dosage form. In vitro dissolution tests with clearly defined conditions are used to assess quality of the solids. A dissolution profile provides information about the amount of dissolved API from the solid dosage form in time. Comparison of the different profiles could serve as a tool to assess a success of a transfer or scale-up of a given process. Granulation proces is a key step during the manufacture of solid dosage forms, therefore the granules are tested by in vitro dissolution experiments. However, single dissolution profile of API from the granules does not provide deep insight into dissolution process. While classical in vitro dissolution tests are not always discriminatory, their combination with the simultaneous monitoring of other variables, such as the mechanical propertier of disintegrating granules before and during dissolution, may be more sensitive to in consisitncies in the product quality â€" either between batches or during scale-up or transfer.

The aim of this work is to investigate the dissolution mechanism of the granules consisted of 6-component formulation with a high load of API (58 % wt.) and produced at qualitatively different conditions of the granulation process (various main impeller speed, expressed by Froude number ranging from 0.26 to 2.04) and on different scales (granulator volume: 4, 25 and 400 L). A novel metodology which combines granule disintegration online monitoring with a measurement of API dissolution profile was developed. An evolution of a granule size distribution is analyzed online using a laser particle size analyzer (Horiba Partica LA950) by static light scattering. A 300ml circuit with flow cell and input cone for load and to take the samples for offline measurement on UV spectrophotometer in order to obtain dissolution profile coupled with information about granule disintegration. The inner structure of the granules

was investigated using x-ray micro-CT tomography and their mechanical properties were analyzed by CT3 Texture Analyzer. Three possible mechanisms of API release from granules were proposed: leaching, surface erosion and granule breakage. These mechanisms were identified in different stages of the dissolution process and different contribution of each mechanism was observed for granules produced at different Froude number granulation set-up. In a more detailed part of study, where granules sieved into coarse (over 400 μm) and fine (under 400 μm) fractions were investigated, breakage of the coarse granules into considerably stable sub-granules was observed, where the prevailing mechanism of release was leaching of API. This demonstrates that a sub-population of granules exists in each batch regardelss of the process conditions, which shows very robust and repeatable properties. The challenge for future QbD research will be to identify mechanisms that lead to the formation of this sub-population.

Copyright © American Institute of Chemical Engineers. All rights reserved.

471422 Particle Characterization Technique to Enable Rapid Pharmaceutical Formulation and Process Optimization

Abstract:

Pharmaceutical processing consists of several unit operations that transform active ingredients and excipients to improve the processing and handling characteristics for downstream processing into the desired solid dosage form. These same processes can impart physical changes to an active an ingredient that can potentially affect the in vivo performance of the formulated product.

The current in vitro analytical methodology used by industry to predict in vivo absorption is often times poorly correlated to true product performance. The standard USP/Ph Eur in vitro dissolution methods do not allow for a complete understanding of dissolution kinetics and underlying particle-engineering principles that may be at the root cause of a potential processing issue.

Understanding the performance of the drug product is of utmost importance to establish the fundamental mechanisms of drug release. The drug release is governed by the formulation composition and to a larger extent the processing conditions for a given formulation. If a drug substance has low solubility, for example, then the particle size will be a governing factor of drug absorption.

The focused beam reflectance measurement (FBRM) probe is an integral tool for formulation and process development at Nucleo Life Sciences. The FBRM is implemented in formulation optimization, qualification of API particle size, evaluation of formulation performance, and scale-up qualification. The measurements are simple and quick to analyze and provide a wealth of information on the particulate system.

Evaluating formulations using the FBRM particle size characterization provides valuable insight to formulation development. The ability to monitor tablet and granule disintegration behavior allows for an understanding of the underlying mechanisms of product performance. All of these dynamic events can be monitored in situ. Our goal is to assist companies with implementing the data generated by FBRM into regulatory submissions to support the selection of the in vitro dissolution method development or even use the FBRM methodology as a routine in process release test.

Copyright © American Institute of Chemical Engineers. All rights reserved.

*** Formatting Issue - Best Available Paper/Graphic ***

468287 HPMC Cloud Point: Exploring Hydroxypropylmethyl Cellulose Behavior in Pharmaceutical Formulations

Abstract:

HPMC cloud point: Exploring hydroxypropylmethyl cellulose behavior in pharmaceutical formulations

Maryam Mohammadpour[1,2,*], Brian Kerins[1], Mary Ellen Crowely[1], Gavin W. Walker[2], Abina M. Crean[1,2]

School of Pharmacy, University College Cork, Ireland[1]

Synthesis & Solid State Pharmaceuticals Centre (SSPC), Bernal Institute, University of Limerick, Ireland[2]

*Tel: + 353 (21) 4901683, email: marym.mohammadpour@ul.ie

INTRODUCTION

Hydroxypropylmethyl cellulose (HPMC) is widely used in pharmaceutical applications as a sustained release agent (Viriden, Wittgren et al. 2011), tablet coating, nanoparticulate stabilisers and nucleation inhibitor (Ghosh, Bose et al. 2011). In these applications its gelation behaviour is key to its performance and therefore needs to be understood within each system. Aqueous solutions of HPMC display two characteristic temperatures: a thermal gelation temperature (TGT) and a lower critical solution temperature (LCST). The gelation behaviour of HPMC is influenced by the HPMC degree of substitution, formulation additives (such as salts, surfactants, drug molecules) and temperature (Zhang, Wang et al. 2015). [i]

HPMC begins to precipitate out of solution at the LCST due to the loss of water of hydration at increased temperatures (Viridén, Wittgren et al. 2011). This phase separation into a polymer rich and polymer depleted phase, caused by association of hydrophobic groups leading to aggregation, allows for the cloud point to be determined as the polymer rich phase is capable of scattering light. Though there seems to be no direct relationship between TGT and cloud point, it seems that both properties are affected by electrolytes in the same manner i.e. if an electrolyte increases the TGT, it will also increase the cloud point (Mitchell, Ford et al. 1990).

The objective of this paper is to demonstrate the ability of cloud point measurements to identify the impact of formulation additives on the behaviour of HPMC within formulations during dissolution. The additives investigated include; a surfactant, docusate sodium (DOSS) and a model drug carbamazepine. The goal of this work is to use cloud point measurements to understand formulation interactions with HPMC which will increase understanding of variability in formulation behaviour during dissolution.

Materials and Methods

The HPMC grade investigated was HPMC substitution type 2910, label viscosity 3 cp. In comparison a 2nd HPMC derivative was also investigated, HPMCAS Grade AS-MG viscosity type 3 cp. HPMCAS is a cellulose polymer which does not exhibit a cloud point. The goal of the study was to investigate both polymers in aqueous systems with and without docusate sodium and carbamazepine to assess the additive effect on cloud point behaviour.

Aqueous polymeric solutions were prepared by adding HPMC to an aqueous solution of additive while stirring with a magnetic stirrer at 500 rpm. The solutions were made up to the required volume with ice cold deionised water and left mixing for 20 minutes. The solution was degassed by a sonicator for one hour and then moved to a refrigerator to hydrate overnight.

A Chirascan circular dichroism spectrometer performed cloud point assays. Percentage transmission was examined at 800nm over a range of 25 – 91 °C at a rate of 1.0°C/min. Sample was present in a 0.5mm quartz cuvette. An air background calibration was carried out prior to testing each day.

Results

To ensure the accuracy of the produced results a centre point (6% HPMC, 0.5% DOSS) was ran in triplicate over 3 separate days. The average CP was calculated to be 73.25 ± 0.725 °C. The low standard deviation indicates the Chirascan instrument is capable of generating reproducible CP measurements.

There is a clear trend evident that as the concentration of DOSS increases the the CP increases. This we attribute to the increased energy needed to overcome the solubilisation of the hydrophobic groups of the HPMC by the surfactant (Figure 1). When ionic surfactants and non-ionic polymers are mixed, two competing phenomena take place: 1) Self aggregation of surfactant molecules to form micelles or 2) interaction of individual surfactant molecules with monomer units of the polymer chain (Acevedo, Takhistov et al. 2014). The free surfactant molecules bind to the hydrophobic parts of the polymer through adsorption until a state of saturation is reached. The priority binding of the surfactant produces polar outshells, increasing the energy barrier for the crosslinking of the HPMC chains to occur, by hindering the free access to HPMC chains. The thermal energy requirements for dismantling this shell and for the aggregation of the polymer is determined by the chemical structure and electrostatic interaction between the surfactant and HPMC (Joshi 2011).

*** Formatting Issue - Best Available Paper/Graphic ***

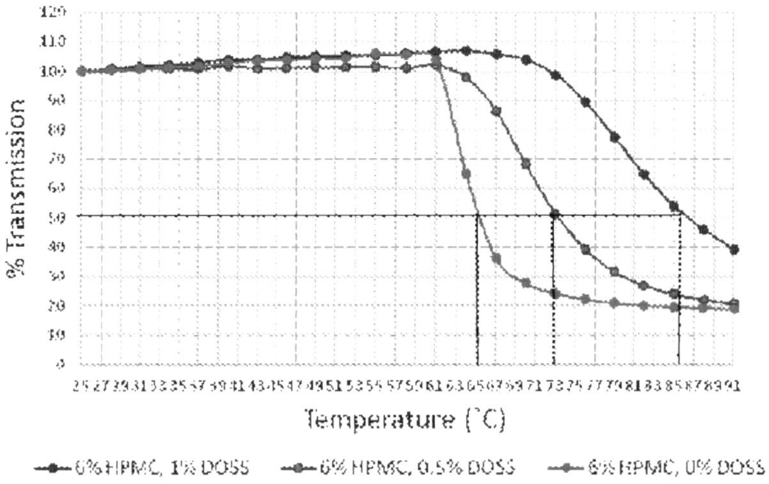

Figure 1: The effect of varying concentration of DOSS on HPMC cloud point (Cloud point was reached when he overall % transmittance of light reaches 50% of that compared to the same sample at 25°C (Silva, Pinto et al. 2008).

Additional results will be presented looking at the effect of carbamazepine level on HPMC cloud point. Also the corresponding behaviour of HPMCAS polymer will be presented. Studies to date have not detected a cloud point for aqueous solutions of HPMCAS alone. To show the relevance of these cloud point studies, dissolution studies was be present from a range of these formulation compositions.

Conclusions

Formulation additives such as docusate sodium were shown to positively influence the cloud point of HPMC. Future work will explore how this alteration of the polymer self-assembly behaviour and cloud point can inform the HPMC-formulation behaviour during dissolution studies.

References

Acevedo, A., et al. (2014). "Thermal gelation of aqueous hydroxypropylmethylcellulose solutions with SDS and hydrophobic drug particles." Carbohydrate polymers **102**: 74-79.

Joshi, S. C. (2011). "Sol-Gel behavior of hydroxypropyl methylcellulose (hpmc) in ionic media including drug release." Materials **4**(10): 1861-1905.

Mitchell, K., et al. (1990). "The influence of additives on the cloud point, disintegration and dissolution of hydroxypropylmethylcellulose gels and matrix tablets." International Journal of Pharmaceutics **66**(1): 233-242.

Silva, S. M., et al. (2008). "Aggregation and gelation in hydroxypropylmethyl cellulose aqueous solutions." Journal of colloid and interface science **327**(2): 333-340.

Viriden, A., et al. (2011). "The consequence of the chemical composition of HPMC in matrix tablets on the release behaviour of model drug substances having different solubility." European Journal of Pharmaceutics and Biopharmaceutics **77**(1): 99-110.

Zhang, L., et al. (2015). "Rheological and gel properties of hydroxypropyl methylcellulose/hydroxypropyl starch blends." Colloid and Polymer Science **293**(1): 229-237.

Copyright © American Institute of Chemical Engineers. All rights reserved.

452573 QbD for Generic Modified Release Drug Products

Abstract:

As stated in the White paper recently published on FDA website - FDA Pharmaceutical Quality Oversight: One Quality Voice, a number of challenges to pharmaceutical quality continue to manifest. Among all the problems, product recall and defect reporting data is listed as the number one problem. The high occurrences of problems can be attributed to inherent defects in product and process design; these data further indicate failures in the implementation of manufacturing process scale-up as well as routine production.

FDA's approach to quality oversight has evolved in recent years, from an emphasis placed on quality control using empirically derived specifications, moving towards an increasing emphasis on in-depth product and process understanding. However, for a while, a disproportionate amount of regulatory attention is devoted to low-risk products and issues, diverting resources needed for the assessment of high-risk products. Therefore, risk based Quality by Design (QbD) approach is necessary. For simply designed immediate release oral products, traditional quality control approach is adequate. However, for more complex modified (oral) release products, risk based QbD approach should be adopted.

More and more drug products have been developed as modified release dosage forms in order to maintain therapeutic activity for an extended time, reduce toxic effects, protect the active substance against degradation due to low pH, etc. As these pharmaceutical products have increased their design complexity, approaches adopted to ensure the equivalence and quality of generic drug products, must likewise evolve. This presentation will focus on the quality assessment of modified release drug products based on their risk level, product design, and process understanding.

Copyright © American Institute of Chemical Engineers. All rights reserved.

461268 Utilization of Quality By Design Principles to Define Formulation Best Practices for a Direct Compression Excipient

Abstract:

Utilization of Quality by Design Principles to Define Formulation Best Practices for a Direct Compression Excipient

Kathryn O. Hewlett (KOHewlett@Dow.com), True L. Rogers (TLRogers@Dow.com), Karen M. Balwinski (KMBalwinski@Dow.com), Jaime L. Curtis-Fisk (JLCurtisFisk@Dow.com), Robert L. Schmitt (SchmitRL@Dow.com), Shrikant N. Khot (SNKhot@Dow.com)

Introduction: Quality by Design (QbD) initiatives enable pharmaceutical manufacturers to proactively design quality and performance into drug products, with the ultimate goal to maximize safety and efficacy for the patient. Hydrophilic matrix tablets represent one of the most prevalently utilized modified-release oral drug delivery systems, which are relatively straightforward to develop and cost-efficient to manufacture. Hydroxypropyl methylcellulose (HPMC) is a common rate-modifying polymer incorporated into hydrophilic matrices. The water-soluble HPMC particles are uniformly distributed throughout the matrix tablet. Upon contact with aqueous gastrointestinal media, the particles hydrate and swell. The particles coalesce to form a swollen polymer network layer. The release of active pharmaceutical ingredient (API) from the tablet is modulated by both diffusion through the swollen network and erosion of the outermost surface.

Matrix tablet performance can be impacted by a number of variables, such as API physicochemical properties (solubility, crystalline morphology), physicochemical properties of the rate-modifying polymer (particle shape and size, viscosity grade), formulation composition (filler, glidant, lubricant selection) and manufacturing methodology (blending, direct compression vs. granulation, tableting conditions). The Dow Chemical Company has recently begun commercial-scale production of a new direct compression grade HPMC, METHOCEL™ DC2, to complement the finer particle size, but less flowable controlled release grade, METHOCEL™ CR. We used QbD principles to characterize properties and performance of matrix tablets containing METHOCEL™ DC2 in order to determine and balance optimal performance regimes for tablet physical properties and modified-release (MR). The results of these investigations will be presented along with formulation guidances and best practices. Particular emphasis will be placed on filler selection, HPMC molecular weight and morphology grade, and polymer concentrations utilized to ensure robust MR performance.

Methods: Gliclazide is a poorly soluble sulfonylurea used to treat non-insulin dependent *diabetes mellitus* and was used as model API. HPMC concentrations of 10, 20, 30, 40, and 50% (w/w) were investigated to cover the performance design space. The new K100LV, K4M, and K100M DC2 grades of HPMC were utilized as rate-modifying polymers, and the corresponding CR grades were used as comparative controls. Fillers investigated were Flow-Lac lactose (Meggle), Avicel PH102 microcrystalline cellulose (FMC), Starch 1500 pregelatinized starch (Colorcon), DiTab dicalcium phosphate (Rhodia), and Manogenn powdered mannitol (SPI Polyols). Tablets were produced via direct compression on a pilot-scale Manesty Beta Press at 50-rpm turret speed. Tablet physical properties, such as weight and tensile strength, were characterized using a Sotax HT 100 - 500NV instrument. Modified-release performance was characterized using the USP II paddle method in either a Varian VK 7010 dissolution system equipped with a Varian Cary 50 UV visible diode array spectrophotometer (Agilent Technologies, Santa Clara, CA USA) or a Distek 2100 single-bath dissolution system (North Brunswick, NJ USA) equipped with an Agilent 8453 diode array spectrophotometer (Agilent). Each tablet tested for MR performance was positioned in a hanging basket to minimize dissolution measurement error.

Results: Figure 1 shows the modified release performance of METHOCEL™ K100LV CR and DC2 as a function of time for formulations with differing HPMC content. Burst release occurred at low HPMC concentrations (10 and 20%) with both morphology grades. Increasing the level of HPMC in the formulation resulted in a shift into a robust MR performance regime. Above this robustness transition, modified release performance from both CR and DC2 matrix tablets was consistent. Further increase of HPMC level minimally impacted API release rate once the robust regime was reached, so there is a diminishing impact with further increase in HPMC concentration once inside the robust regime.

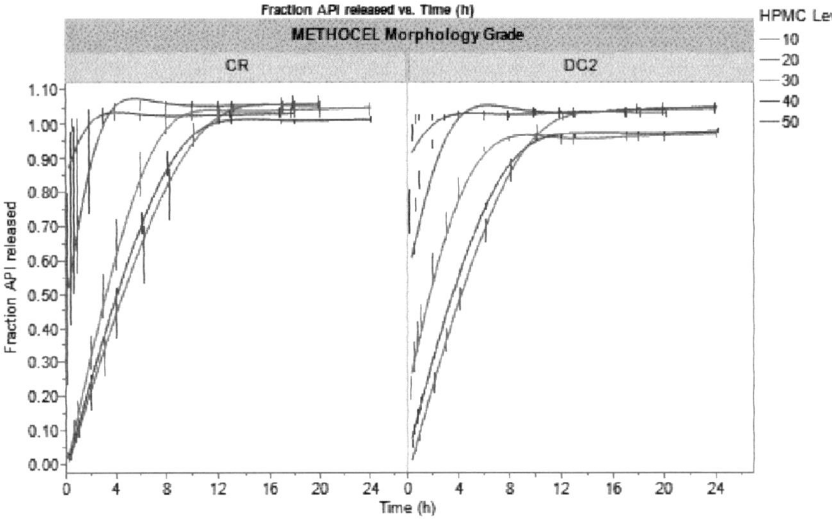

Figure 1. Release of gliclazide over time, as a function of METHOCEL grade and formulation level.

Optical imaging reflected the trends observed from dissolution testing. Figure 2 shows images taken after 4 hours of swelling for tablets comprising 10% (a) and 40% (b) METHOCEL™ K100LV CR. At lower HPMC concentration, the polymer content was insufficient to achieve a continuous swollen network. As a result, the swollen network was unable to maintain structural integrity, causing burst API release as the tablet rapidly eroded. Higher HPMC concentration delivered a continuous swollen network. Dark regions in the figure represent water and areas devoid of swollen network, and the dry or partially hydrated matrix appears as grey or white. The white line in approximately the center of each image represents the location of the initial dry tablet surface. After 4 hours, the swollen network of the 10% HPMC formulation was noncontiguous (Fig. 2a). A large region near the top of the image was devoid of swollen network. The presence of voids in the swollen network is consistent with the observed burst release at lower HPMC concentrations. By contrast, the 40% formulation had a contiguous, swollen network layer surrounding the tablet and delivered robust MR performance (Fig. 2b).

(b)

(a)

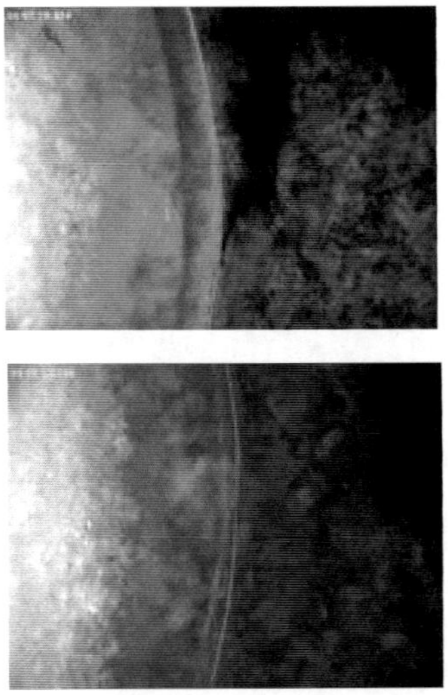

Figure 2. Optical images of (a) 10% and (b) 40% CR tablets after 4 hours of swelling.

Tablet filler selection also impacts MR performance. Figure 3 shows MR profiles of the gliclazide formulation containing 30% METHOCEL™ K4M DC2 with a range of soluble and insoluble fillers. Depending upon filler solubility, the API release rate can be increased or decreased. Once inside the robust regime, filler solubility can be used to adjust MR performance to the desired target.

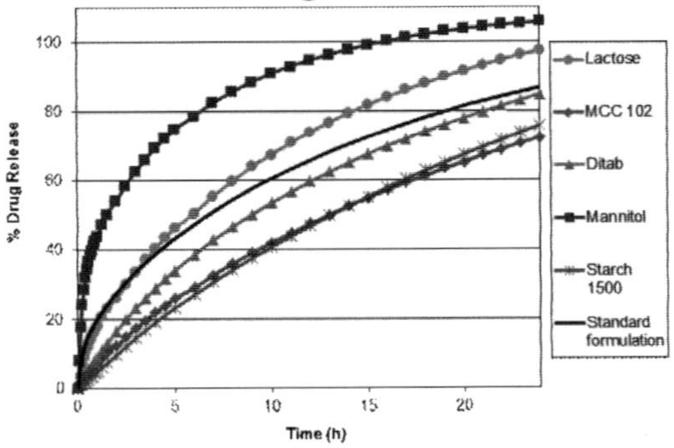

Figure 3.The effect of filler solubility on gliclazide release rate. Tablets were formulated with 30% METHOCEL DC2.

Conclusions: When working with matrix tablets, many factors must be considered to obtain robust MR performance. This work focused on examining the role of filler selection, the molecular weight and morphology grade of HPMC, and polymer concentration necessary to ensure robust MR performance. A new, direct-compression grade HPMC, METHOCEL™ DC2, was compared to METHOCEL™ CR, and both morphology grades enabled robust MR performance at comparable levels. As the HPMC concentration was increased, the formulation transitioned into a robust regime of consistent MR performance. Further increase in HPMC concentration had diminishing impact once inside the robust MR regime. Filler solubility also impacted MR performance and could be used to fine-tune the desired release rate. The systematic framework utilized while conducting this study enabled us to define formulation guidances and best practices for the new direct compression grade of HPMC. There is now a clear understanding of the influences of HPMC grade, polymer level, and filler selection on MR performance.

Copyright © American Institute of Chemical Engineers. All rights reserved.

455234 Microcrystalline Cellulose   Does Wood Pulp Source Impact the Direct Compression Performace of This Excipient?

Abstract:

INTRODUCTION

In a quality by design (QbD) environment knowledge and understanding of raw material critical quality attributes (CQAs) are essential [1]. Excipient variability is an important consideration particularly when blended with challenging APIs where manufacturing may occur close to the edge of failure, i.e. close to the acceptance limits for critical quality attributes. In such cases small variations in an excipients attribute could possibly impact the end product quality, even if the variation is within the excipient's certificate of analysis release specifications. Continuous process improvement, continuous manufacturing and regulatory requirements for improved process understanding as covered by ICH Q10 [2] drive a need to understand the impact of excipient variability on finished product performance.

Microcrystalline cellulose (MCC) was chosen as the excipient of study due to its widespread use as a diluent/filler or binder in solid oral dosage form manufacturing. MCC has been available as a pharmaceutical excipient since 1964 [3] and consequently there has been much previous research into its material attributes. Moisture [4], crystallinity [5-7], degree of polymerisation (DP) [3], surface area and porosity [4] have been suggested in the literature as material attributes for consideration when processing MCC.

In order to understand the possible variability in MCC it is important to understand its manufacturing process. Pharmaceutical cellulose is derived from wood pulp. Trees are broken down into wood chips which undergo a chemical process known as the 'Kraft process'[8] which removes the lignin that links cellulose chains together as wood. Wood pulp is the starting material bought by MCC manufacturers. The type of wood source (soft/hard woods) will affect the chemical content (the hemicellulose content and crystallinity) of the wood pulp.

MCC manufacturers treat the wood pulp with acid to breakdown the long cellulose chains into smaller chain lengths (reduce the degree of polymerization). Acid hydrolysis cleaves the long cellulose chains at amorphous sites where hemicellulose sugar chain branches are located. Therefore how the cellulose is hydrolysed is dependent on its chemical content and the number of amorphous regions. The length of the hydrolysis time and final DP of a batch of MCC is influenced by the wood pulp type as different wood pulps have characteristic leveling off degrees of polymerization e.g. 180-210 for hard woods and 210-250 for soft woods [3]. By blending different wood pulps together before hydrolysis the DP can be predicted by manufactures. This is an important control as it can influence the MCC final bulk density[3]. The resulting shorter cellulose chains are thus more crystalline as the amorphous regions have been cleaved, hence the name microcrystalline cellulose. Post hydrolysis steps (spray drying/air stream drying and sieving) offer manufactures the opportunity to manipulate the final lot bulk density, PSD and mean particle size to produce product within specification.

The study presented aims to understand if samples of MCC grade PH102 produced from different wood pulp mixes show differences in physicochemical properties and hence compaction performance during a direct compression process.

<>Materials and Methods

Three batches of MCC PH102 produced from a range of pulp sources were supplied by FMC. The pulp types have not been disclosed due to commercial sensitivity. All samples were produced using the commercial scale process and meet with supplier and pharmacopeia specifications.

- Sample 1 - 75% Pulp X: 25% Pulp Y
- Sample 2 - 50% Pulp X: 50% Pulp Z
- Sample 3 - 100% Pulp X

MCC PH102 lots were characterized and compared for differences in particle size distribution (sieve analysis), moisture content (thermogravimetric analysis), crystallinity (FTIR) and surface area (BET N_2 adsorption), degree of polymerization and scanning electron microscopy. Two compaction studies were carried out on an instrumented Piccola™ rotary tablet press. The first compaction study was on samples 'as received'. The second compaction study was performed on a sieved size fraction (106-250 μm) of each lot in order to remove the confounding factor of particle size distribution differences. Compacted material hardness, weight and thickness were measured on a Pharmatron™ Smart Test 50 and compaction profile generated.

RESULTS AND DISCUSSION

The particle size distribution (PSD) profiles of each sample as received did indicate a significant difference in PSD particularly the % fines. The sieved size fraction (106-250 µm) of each sample was then compared in order to eliminate the influence of PSD and % fines as a possible reason for differences in the compressibility of the lots.

Characterization of the 106-250 µm size fractions for each sample identified differences in the surface area, crystallinity and bulk density (Table 1) and the PSD (Figure 1). All lots had a moisture content determined by loss on drying in the range 3.74 - 4.86% w/w. It was assumed that within this range moisture was not an influencing factor [4].

When comparing compression profiles of samples (Figure 2), Samples 1 and 2 which were manufactured from pulp mixes were similar. However Sample 3, manufactured form 100% of pulp X, showed a significantly higher compressibility.

The study results show that particle size is a major factor that influences MCC compressibility. Following correction for particle size, Sample 3 produced from a single pulp source showed superior compression properties. The particle properties resulting in this increase in compaction are currently inconclusive and being explored. Further studies will focus on degree of polymerization and particle mechanical parameters during compression.

CONCLUSIONS

Previous studies [9, 10] have discussed wood pulp as a source of variability between batches of MCC. Mixing wood pulps minimizes this variability. In this study a MCC sample produced from 100% of one particular wood pulp during manufacture of MCC produced material which performed differently during direct compression compared to samples from pulp mixes when corrected for difference in particle size between samples.

The impact of this variability in MCC will be formulation specific. It is envisaged for blends with a high % API loading which is difficult to compact, this variability may be significant to drug product manufacturers. This also highlights the importance of using a number of different lots of excipients during drug development to capture excipient variation where it is potentially a risk factor.

ACKNOWLEDGMENTS

This research is funded by the Synthesis and Solid State Pharmaceutical Centre (SSPC) and Science Foundation Ireland (SFI) under grant number 12/RC/2275. PH102 was donated by FMC Corporation.

REFERENCES

1. Chen, Z., D. Lovett, and J. Morris, *Process analytical technologies and real time process control a review of some spectroscopic issues and challenges.* Journal of Process Control, 2011. **21**(10): p. 1467-1482.

2. ICH, *Pharmaceutical Quality System Q10*, in *ICH Q10*, I.C.o.H.o.T.R.f.R.o.P.f.H. Use, Editor 2008, International Conference on Harmonisation of Technical Requirements for Registration of Pharmaceuticals for Human Use.

3. Thoorens, G., et al., *Microcrystalline cellulose, a direct compression binder in a quality by design*

environment-A review. Int J Pharm, 2014. **473**(1-2): p. 64-72.

4. Sun, C.C., *Mechanism of moisture induced variations in true density and compaction properties of microcrystalline cellulose.* Int. J. Pharm., 2008. **346**(1-2): p. 93-101.

5. Park, S., et al., *Cellulose crystallinity index: measurement techniques and their impact on interpreting cellulase performance.* Biotechnol Biofuels, 2010. **3**: p. 10.

6. Mihranyan, A., et al., *Moisture sorption by cellulose powders of varying crystallinity.* Int. J. Pharm., 2004. **269**(2): p. 433-442.

7. Rowe, R.C., A.G. McKillop, and D. Bray, *The effect of batch and source variation on the crystallinity of microcrystalline cellulose.* Int. J. Pharm., 1994. **101**(1-2): p. 169-72.

8. Chakar, F.S. and A.J. Ragauskas, *Review of current and future softwood kraft lignin process chemistry.* Industrial Crops and Products, 2004. **20**(2): p. 131-141.

9. Landin, M., et al., *Effect of country of origin on the properties of microcrystalline cellulose.* Int. J. Pharm., 1993. **91**(2-3): p. 123-31.

10. Landin, M., et al., *Effect of batch variation and source of pulp on the properties of microcrystalline cellulose.* Int. J. Pharm., 1993. **91**(2-3): p. 133-41.

Table 1. Characterization results of the 106-250 μm size fraction for each of the three samples before compression.

	SAMPLE 1	**SAMPLE 2**	**SAMPLE 3**
Moisture %w/w	4.41 ± 0.68	4.86 ± 0.64	3.74 ± 0.36
Bulk density g/cm^3	0.32	0.32	0.34
IR Crystallinity (***Nelson et al 1964,)***	0.86	0.90	0.94
Surface Area m^2/g	1.232	1.148	1.289
D50 μm	174 ± 0.00	212 ± 3.46	186 ± 1.53

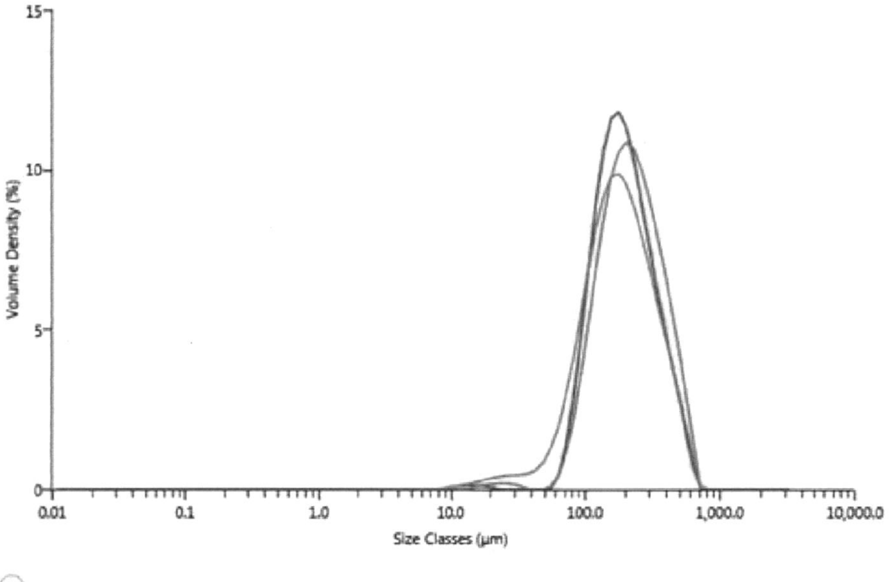

Figure 1. Malvern PSD of 106-250 μm size fraction comparison for the three samples. Sample 1 (green), sample 2 (blue) and sample 3 (red).

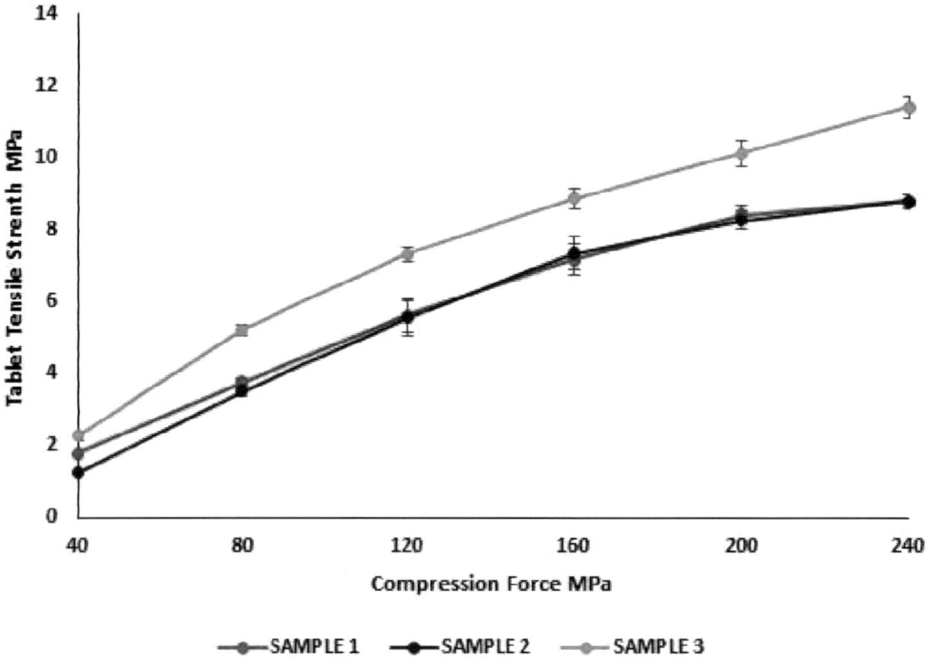

Figure 2. Compression profiles for the three samples. Tensile strength for N=20 tablets

Copyright © American Institute of Chemical Engineers. All rights reserved.

472553 Human Embryonic Stem Cell Models of Early Human Placental Development

Abstract:

The placenta is a complex organ at the maternal-fetal interface, and is critical for fetal development during pregnancy. The trophectoderm layer of the blastocyst stage embryo is the precursor for all trophoblast (TB) cell types in the placenta. Abnormalities in TB development lead to placental pathology, and are associated with adverse outcomes during pregnancy such as preeclampsia and recurrent loss of pregnancy. Moreover, it is now generally accepted that effects on fetal development due to improper placentation may result in adverse health outcomes in later life. Yet, despite its importance to maternal and fetal health, early human TB development remains poorly understood. Mechanistic studies on TB development are difficult due to challenges in research with human embryos and limited availability of placental samples from early gestation. Furthermore, human placental samples collected during the first trimester â€“ typically the earliest samples are ~ 6 weeks (~ 28 days post-fertilization) â€“ are associated with heterogeneity, containing multiple TB and non-TB cell types, and have likely undergone adaptation to the maternal environment due to the influence of cytokines and chemokines present in the maternal decidua.

The use of human embryonic stem cells (hESCs) to develop TB models has gained significant interest since it was first reported that hESCs treated with Bone Morphogenetic Protein (BMP) differentiate into TB-like cells. The use of hESC-derived TB as a model system is attractive because it overcomes several challenges associated with studies on TB development discussed above. However, lack of understanding of the mechanistic basis of TB differentiation of hESCs is a major impediment for widespread acceptance of hESC-derived TB as a model system for early placental development. Why hESCs, believed to be epiblast-like, can access TB fate remains an enigma. In contrast to mouse embryonic stem cells (mESCs), hESCs have been reported to readily differentiate to TB-like cells upon treatment with Bone Morphogenetic Protein (BMP) and/or inhibitors of the Activin/Nodal. However, whether hESCs can give rise to bona fide TB has been a subject of intense debate. A role for BMP in trophectodermal differentiation has not been identified in the mouse. It has been postulated that mESCs are epigenetically restricted from TB differentiation through hypermethylation of the *Elf5* promoter region. Analogously, hESCs are hypermethylated at the *ELF5-2b* promoter locus, whereas early placental TB is hypomethylated, suggesting a similar epigenetic restriction. Furthermore, unlike placental TB, hESC-derived TB has been reported to not downregulate HLA class I antigens.

We have shown that *bona fide* TB can indeed be obtained from hESCs, and they possess key markers of placental TB such as hypomethylation of the ELF5 promoter locus and downregulation of HLA class I antigens. Whether the hESC-derived TB can function as *bona fide* trophoblasts during placental development cannot be experimentally tested. Nevertheless, a surrogate criterion is whether the response of hESC-derived TB towards their signaling environment is similar to those of primary TB. Transforming Growth factor-beta (TGFÎ²) suppresses epithelial to mesenchymal transformation (EMT) and invasive TB formation from placental TBsÂ and epidermal growth factor (EGF) triggers syncytialization of placental TBs. Our studies show that hESC-derived TB behave similarly, i.e. EGF treatment leads to formation of multinucleate syncytiotrophoblast (STB) and inhibition of Activin/Nodal signaling leads to EMT and formation of invasive TBs. Importantly, we have also identified conditions for selective differentiation of hESC-derived TB to invasive TB or STB lineages in monolayer cultures. Here we specifically discuss the role of BMP signaling in TB fate acquisition in hESCs, and subsequent differentiation to specialized TB subtypes.

Copyright © American Institute of Chemical Engineers. All rights reserved.

470260 Engineering Xeno-Free Microcarriers with Recombinant Vitronectin, Albumin and UV Irradiation for Human Pluripotent Stem Cell Bioprocessing

Abstract:

The development of technologies for the expansion and directed differentiation of human pluripotent stem cells (hPSCs) [1] in large quantities under xeno-free conditions is a critical step toward enabling envisioned stem cell-based therapies. Stirred suspension microcarrier bioreactors afford great surface-to-volume ratio and scalability but their use in the expansion and differentiation of hPSCs require the engineering of xeno-free microcarriers and media. In this study, a simple and low-cost approach was established for generating microcarriers without animal-derived components.

Microcarriers were coated with recombinant vitronectin (VN) and/or human serum albumin (HSA) with or without UV illumination. The expression of genes and proteins was determined by quantitative PCR, flow cytometry, western blotting and immunostaining as described [2]. Fourier transform infrared (FTIR) spectroscopy, absorbance and ELISA methods were employed to characterize the surface of the engineered beads [3]. Directed hPSC differentiation regimens to definitive endoderm (DE), cardiac mesoderm (MS) and neuroectoderm (NE) were also reported [2, 4].

Polystyrene beads coated with VN, which is routinely used as a substrate for 2D culture of hPSCs resulted in 24.3±3.1% seeding efficiency of H9 human embryonic stem cells (hESCs) but the cells detached from the carriers and their concentration declined over 5 days of stirred suspension cultivation. The inclusion of HSA did not improve significantly the efficiency (27.8±1.3%, $p=0.15$) and growth of cells but increased their retention on the beads under agitation. However, UV irradiation resulted in enhanced seeding efficiency (30.5±1.6%, $p=0.013$) and retention while hPSCs grew 20.5±2-fold per passage over multiple successive passages and cell viability remained above 80%. More

than 90% of the cells were $SSEA4^+$ and $OCT4^+$. After multi-passage culture, hPSCs formed embryoid bodies displaying tri-lineage specification markers. Similarly, hPSCs propagated on engineered microcarriers were coaxed to DE, MS or NE and expressed pertinent genes and proteins. The observed performance of this xeno-free system is comparable to that of cultures involving Matrigel-coated beads.

In conclusion, the microcarriers engineered in this study supported the growth and potential for multi-lineage specification of self-renewing hPSCs. Such culture modalities will be critical for the realization of the potential of stem cells in areas ranging from regenerative medicine to drug discovery.

Acknowledgements: Funding to EST is acknowledged from the National Institutes of Health (NHLBI R01-103709) and the National Science Foundation (CBET-1547785).

References:

1. Fan, Y., Wu, J., Ashok, P., Hsiung, M., Tzanakakis, E.S., *Stem Cell Rev.*, 2015, 11:96-109.

2. Fan, Y., Hsiung, M., Cheng, C., Tzanakakis, E.S., *Tissue Eng. Part A*, 2014, 20:588-99.

3. Fan, Y., Zhang, F., Tzanakakis, E.S., *ACS Biomaterials Science and Engineering*, 2016.

4. Wu, J., Tzanakakis, E.S., *PLoS One*, 7:e50715, 2012.

Copyright © American Institute of Chemical Engineers. All rights reserved.

474200 Biomaterial Based Engineering of Vascularized Pancreatic Islet Organoids Derived from Human Pluripotent Stem Cells (hPSCs)

Abstract:

Organoids are 3D engineered constructs comprising multiple organ specific cell populations capable of recapitulating an in vivo organ systemâ€™s specific structure and function in an in vitro setting. Development of such systems requires an organ specific parenchyma cell source and a platform for self-organization and lineage specific induction of the chosen cell types. Human and murine hPSCs, unique in their abilities for self-renewal and capacity to differentiate into organ specific cell types, have been used to produce organoids representing all three germ layers. In addition to the necessary organ specific cell populations, attaining functional vasculature is a critical component, vital for maintaining nutrient supply and appropriate organ function. This is particularly critical for pancreatic islet organoid development, since generating a specialized, fenestrated vasculature is crucial for glucose delivery and insulin response. Recent success in generating pancreatic b-cells from hPSCs has set the ultimate stage for engineering vascularized islet organoids from a regenerative cell source.

We have thus developed a novel hydrogel promoting spontaneous aggregation of pre- differentiated hESC derived pancreatic progenitor cells (hESC-PPs) into a 3D organoid when cultured directly on the hydrogel surface without embedment. The resulting spheroids are amenable for size and cellular composition tuning. Following 3D organoid formation, the hESC- PP cells exhibited enhanced PP gene and protein expression throughout the cell population. Furthermore, using isolated microvessel elements from adipose, a new microvasculature was integrated within the hESC-PPs organoids during initial seeding and spheroid formation. Additionally, inclusion of a supporting mesenchyme within the organoid promoted microvascular expansion and vascular specification within the organoid in 4 days. Upon continued culture, the gene expression of key pancreatic maturation markers (NKX6.1, PDX1, and INS) was enhanced in the vascularized organoids even in the absence of specific chemical induction. Moreover, the intra-organoid vasculature demonstrated an increase in islet endothelial specific API gene expression and PLVAP, a marker noted for increased endothelial diaphragm and fenestration. In conclusion, we believe the results represent a major step in the in vitro production of hPSC islet organoids with further microvasculature, a development indirectly extendable to other organ-specific applications.

Copyright © American Institute of Chemical Engineers. All rights reserved.

471525 The Role of FGF2 in Reprogramming of Epidermal Keratinocytes Toward Neural Crest Fat

Abstract:

Neural crest cells (NC) play a central role in forming the peripheral nervous system, the craniofacial skeleton and pigmentation of the skin during development. Due to their differentiation diversity these cells have been a focal point of research for disease therapeutics via cell transplantation. However, an easily accessible autologous cells source for therapeutic cell transplantation remains one of the main challenges facing the field. Recently our lab discovered that human epidermal inter-follicular keratinocytes (KC) can turn into neural crest (NC) stem cells without ectopic expression of transcription factors or reprogramming to the pluripotent state. Here we studied the role of FGF2 in this novel phenomenon.

Keratinocytes were isolated from glabrous foreskin of 1-3 day old neonates and exposed to KC-NC induction medium containing FGF2 growth factor. FGF receptor knockdown through lentiviral transfection was used, as well as chemical inhibitors for various signaling pathways. Clonal density experiments were performed using single keratinocyte cells to create a pure KC population that can give rise to NC. NC were further differentiated into Schwann cell, peripheral neuron, melanocyte and smooth muscle cell lineages. Gene expression was assessed with RNA-Seq and validated through quantitative real time RT-PCR. On protein level immunocytochemistry was used in order to detect markers in NC and differentiated lineages.

Our results show that FGF2 is necessary for KC-NC reprogramming (Fig:1A), as evidenced by use of chemical inhibitors as well as gene knockdown using shRNA. We identified the FGF receptor that mediates the signaling as well as subsequent pathways including Akt. Ultimately, FGF signaling leads to expression of the NC-specific transcription factor, *SOX10*, which is necessary for NC reprogramming. The newly produced NC express NC genes including *SOX10, Nestin, Pax3, FoxD3* (Fig:1B) and could be coaxed to differentiate into peripheral neurons, Schwann cells, melanocytes and smooth muscle cells as shown by molecular as well as functional assays.

In conclusion, we provide mechanistic insight into the process of keratinocyte reprogramming to neural crest stem cells. This work represents a paradigm shift in stem cell biology as it demonstrates the unusual plasticity of human adult KC that can turn into many different cells types without genetic modification or reprograming to the pluripotent state. The dearth of cell sources for treatment of neurogenic disorders, combined with the accessibility and growth potential of human epidermal cells suggest that the proposed work could have tremendous implications for the use of cell therapy for treatment of many debilitating diseases.

Copyright © American Institute of Chemical Engineers. All rights reserved.

463233 Dexamethasone-Releasing Polyurethane Microfiber Meshes to Facilitate Human Cardiac Stem Cell Proliferation and Differentiation

Abstract:

c-kit positive human cardiac stem cells (hCSCs) have demonstrated great promise for treating ischemic heart failure. However, therapeutic efficiencies have been low, likely due to limited cell survival, retention, engraftment, and uncontrolled differentiation following transplantation. Cardiac tissue engineering has offered new strategies to overcome these problems. Cell or biomaterial cardiac patches, 3D constructs, and bioreactor-conditioned scaffolds have been previously utilized to repair cardiac damage. Factors necessary to enhance cell retention, survival rate, and ultimate cardiac differentiation include mimicking the properties of the cardiac extracellular matrix (ECM), providing a high surface area for display of adhesive ligands, and presenting a nano-scale architecture with topographical features to guide cell morphology and orientation. We propose that elastomeric microfiber meshes that release anti-inflammatory corticosteroids could facilitate cardiomyocyte differentiation of hCSCs, and enhance cell retention and engraftment *in vivo*.

As a first step, in this study we examined aligned electrospun polyurethane (PU) microfiber meshes for engineering of cardiac patches using human cardiac stem cells (hCSCs). To this end, thin (~5 μm) elastomeric meshes consisting of aligned PU (1.7 μm diameter) electrospun micro-fibers were electrospun and suspended above glass slides using PDMS strips for supports (Figure 1). hCSCs were then seeded on these meshes and cultured for up to 31 days within differentiation media and compared to hCSC seeded in tissue culture polystyrene plates (TCPS). The viability, morphology, proliferation rate, gene, and protein expression of the hCSCs were then analyzed. Briefly, we found that cells on aligned microfiber meshes displayed an elongated morphology aligned in the direction of fiber orientation, lower proliferation rates, but increased expressions of genes and proteins majorly associated with cardiomyocyte phenotype. The early (*NK2 homeobox 5*, *Nkx2.5*) and late (*cardiac troponin I, cTnI*) cardiomyocyte genes were significantly increased on mesh (*Nkx2.5* = 56.2 ± 13.0, *cTnI* = 2.9 ± 0.56,) over TCPS (*Nkx2.5* = 4.2 ± 0.9, *cTnI* = 1.6 ± 0.5, n = 9, $p<0.05$ for both groups) after 31 days of differentiation (Figure 2). In contrast, expressions of smooth muscle markers *Gata6* and smooth muscle myosin heavy chain (*SM-MHC*) were significantly decreased on microfibers. Concurrently, immunocytochemical analysis with cardiac specific antibody exhibited a similar pattern of cardiac lineage differentiation.

As the next step, we are currently incorporating the corticoid dexamethasone within the core phase of core/sheath coaxially electrospun fiber meshes and characterizing its release as a function of the thicknesses of the core and sheath phases. Our hypothesis is that the sustained release of this anti-inflammatory agent will increase hCSC proliferation and cardiomyocyte differentiation and mitigate hCSC apoptosis. Therefore, our plan is to show increased viability of doxorubicin-treated hCSCs, as well as to characterize mRNA markers of hCSC differentiation into cardiomyocytes (*Nkx2.5, cTnI*), smooth muscle cells (*Gata6, SM-MHC*), and endothelial cells (*CD-31*).

Copyright © American Institute of Chemical Engineers. All rights reserved.

460307 Oligodendrocyte Survival, Proliferation, and Intracellular Redox State Is Dependent on 3D Hydrogel Mechanics

Abstract:

Oligodendrocyte precursor cells (OPCs) differentiate into oligodendrocytes, which in the central nervous system are responsible for generating the myelin sheath, an electrically insulating layer around neuronal axons. This myelin sheath enables neurons to send quick and efficient electrical signals to surrounding neurons. Diseases or injuries to the myelin sheath, such as multiple sclerosis, are highly debilitating diseases that have a large impact on the patientâ€™s quality of life. Multiple sclerosis is an autoimmune disorder where inflammation occurs and oligodendrocyte myelin segments are disrupted and impaired. Following relapses, stem cells are capable of migrating to the damaged area, however OPCs are often unable to survive or differentiate into functional, myelinating oligodendrocytes. Biomaterials may be a potential method of repair following these insults or injuries, however little is known about what biomaterial factors stimulate OPC proliferation, differentiation, and functional myelination. Research indicates that there is a correlation between a cellâ€™s ability to proliferate or mature and its intracellular redox state, which can be modulated by both soluble antioxidants and microenvironmental stiffness.

Here, we investigate how simply tuning the mechanical stiffness of a polyethylene glycol (PEG) based hydrogel affects the proliferation and intracellular redox state of two different oligodendrocyte precursor cell lines. PEG-dimethacrylate hydrogels with storage moduli from 230 to 1000 Pa were formed by tuning the concentration and molecular weight from 6% to 10% (wt/v) and 4600 to 8000 Da, respectively. When cells were encapsulated in the hydrogels, they proliferated in a stiffness dependent trend, where the largest increases in ATP and DNA concentrations were found in the most compliant hydrogel formulation. In gels with storage moduli of 230 Pa, the concentration of ATP was found to increase 12 fold over seven days while DNA increased 44% over seven days. To test the influence of antioxidants, we incorporated lactic acid into the hydrogel both as a soluble factor or into the polymer backbone where it can be released through hydrolytic degradation. As a measure of intracellular redox state, glutathione content within cells was measured in its reduced GSH form and in its oxidized GSSG disulfide form. Results from determining the ratio of reduced GSH to total glutathione (GSH and GSSG) suggest that incorporating lactic acid further reduces the intracellular redox state and increases cellular proliferation. These findings suggest the potential use of tunable PEG hydrogel systems to promote OPC growth, increase oligodendrocyte maturation, and repair the myelin sheath.

Copyright © American Institute of Chemical Engineers. All rights reserved.

475271 Engineering the Morphogenesis of Pluripotent Stem Cells

Abstract:

Pluripotent stem cells (PSCs) exhibit a unique ability to recapitulate dynamic morphogenic processes that mimic analogous developmental biology events to yield complex tissue structure and function that results from coordinated multicellular interactions. The recent success in generating a diverse set of â€œorganoidsâ€ from PSCs demonstrates their tremendous inherent potential for complex tissue formation. However, the inability to robustly and predictably generate organoids currently limits their use for drug discovery and development, modeling of human development and disease, and regenerative medicine therapies. Thus, we have sought to develop methods to reproducibly and robustly direct the morphogenesis of pluripotent stem cell aggregates by controlling physical and biochemical elements of the 3D microenvironment. We have systematically examined the effects of numerous factors, such as aggregate formation methods, hydrodynamic forces and presence of various types of biomaterials, on the differentiation and morphogenesis of PSCs. We are also now combining these approaches with novel cell engineering methods in PSCs to spatially and temporally control the initiation of morphogenic processes and complement our experimental studies with computational modeling approaches. We anticipate that these studies will lead to a greater mechanistic understanding of stem cell biology as well as scalable and translatable technologies for manufacturing of PSC-derived products for regenerative medicine.

Copyright © American Institute of Chemical Engineers. All rights reserved.

467293 Engineering an Environmentally-Isolated Bacterium for Continuous Biofuel Production and Recovery Under Supercritical CO_2

Abstract:

Model organisms such as *E. coli* and *S. cerevisiae* are the workhorses of biotechnological product generation due to their well understood and easily manipulated genetics, fast doubling times, and low cost fermentation media. However, bioprocesses utilizing these hosts are prone to contamination and are negatively affected by product toxicity. Environmental strain isolation provides an opportunity to discover alternative organisms with unique growth characteristics and physiological traits to overcome these long-standing bioprocess challenges. Using a targeted bioprospecting approach by sampling fluid from a deep subsurface supercritical carbon dioxide ($scCO_2$) well, a bacterium was isolated that is rarely able to demonstrate consistent, robust growth in the presence of $scCO_2$. Due to the broad microbial lethality and solvent chemistry of $scCO_2$, we hypothesize that a dual-phase reactor of growth media and $scCO_2$ will simultaneously provide a sterile growth environment and the capacity to continuously strip off strain-produced biofuels (short chain alcohols) to alleviate product toxicity. Additionally, $scCO_2$ is a sustainable, labile solvent that can be separated from desired products through depressurization, leaving these products at high concentrations and providing advantages over other co-solvent extraction systems. We have begun characterizing this unique host through genome sequencing and by determining optimal growth conditions in various environments. Transformation is possible using a protoplast-based method, which has permitted the identification of promoters capable of inducible heterologous protein expression in both aerobic and anaerobic conditions. We engineered the $scCO_2$ tolerant strain to produce isobutanol by introducing a two-enzyme (2-ketoisovalerate decarboxylase (KivD) and alcohol dehydrogenase (Adh)) pathway. The two-step conversion occurs at approximately 70-80% from 2-ketoisovalerate when grown aerobically; however, the intermediate aldehyde was found to accumulate at short culture times. Due to the high partition coefficient for the aldehyde to the $scCO_2$ phase, five alternative alcohol dehydrogenases were tested. A variant was identified that lowered the build-up of isobutyraldehyde and resulted in conversion of 2-ketoisovalerate to isobutanol above 85% in aerobic cultures. Efforts to scale up this process and to develop tools to integrate the biofuel pathway into the genome are currently underway.

Copyright © American Institute of Chemical Engineers. All rights reserved.

464608 A Whole-Plant Genome-Scale Model of Maize

Abstract:

A whole-plant metabolic model of maize was used to analyze the transport of metabolites between organs, as well as evaluate the unique metabolic properties of the major organs. The model links the root, stalk, leaf, kernel, and tassel organs using the phloem as the main tissue for metabolite transport among the organs. The network of metabolic reactions in each organ was determined using whole-plant transcriptomic data to determine organ or tissue specific gene protein reaction (O-GPR or T-GPR) relationships. 3017, 2714, 3372, 3242, and 3206 genes corresponding to metabolic reactions were expressed in the root, shoot, leaf, tassel, and kernel organs, respectively. Once reactions with known activity were allocated into each organ, the flow through the biomass reaction of each organ was restored by preferentially adding reactions known to occur in maize with no organ or tissue localization information over reactions corresponding to low expression in the focus organ. Organ-specific biomass equations were defined and ATP requirements were determined for photosynthetic vs. non-photosynthetic organs based on literature. Three growth stages of maize development representing the vegetative leaf growth, tassel development, and kernel filling stages were modeled based on the change in dry weight of each organ over time. Additionally, the transport between organs was normalized based on the proportion of dry weight at each growth stage modeled. Using parsimonious analysis, the total number of reactions required during leaf growth, tassel development, and kernel filling stages were identified with approximately 80% of the reactions in the vegetative leaf growth stage common among all growth stages. As expected, sucrose is transported to the kernel during the kernel filling stage from all other organs. Using a whole-plant model, we aim to further enhance the understanding of the metabolites transported between organs, as well as identify the reactions underpinning nitrogen utilization efficiency.

Copyright © American Institute of Chemical Engineers. All rights reserved.

471323 Plants 2.0: Towards Metabolic Engineering of Natural Product Biosynthesis into Crops

Abstract:

By 2050, 9.7 billion people will inhabit the earth. To feed this growing population, we need to combat plant pathogens, which cause 65% of crop damage in the United States alone. One solution is to engineer production of natural defense metabolites from wild, or even domesticated, plants into crops. When attacked by pathogens, a few cruciferous plants, such as Chinese cabbage (***Brassica rapa)*** and watercress (***Nasturtium officinale***), produce a group of secondary metabolites called phytoalexins. Indole phytoalexins (brassinins) and benzyl phytoalexins (nasturlexins) are especially affective against bacterial and fungal pathogens[1]. They can improve human health as well; in vitro studies show brassinin is highly affective against cancer cell lines[2].

By engineering brassinin synthesis into a crop, we introduce not just a single anti-fungal, anti-cancer compound, but potentially over a dozen brassinin derivatives, each with unique defense and therapeutic properties. Towards this end, we recently discovered a 10 gene pathway in ***B. rapa*** for the synthesis of brassinin and two more enzymes responsible for its derivatization. Furthermore, preliminary results show this pathway is active in planta when transiently and constitutively expressed in tobacco (***Nicotiana benthamiana***). However, transient engineering is unacceptable for a crop trait that must be passed through generations and constitutive production of brassinin enzymes and intermediates appears to have detrimental effects on plant health. Therefore, our ultimate goal is creation of a stable engineered crop line which produces brassinin only under pathogen stress.

Here we present our progress towards introduction of phytoalexin biosynthesis in planta. We first focus on initial transient expression of these pathways in tobacco and the model plant ***Arabidopsis thaliana.*** Then we discuss our progress toward engineering of ***A. thaliana*** stable lines which both synthesize brassinin and, with careful promotor design, regulate brassinin biosynthesis. We chose to create an ***A. thaliana***

prototype both as a proof of concept and to study the effect of this pathway on plants and their commensal microbial communities. Finally, we present our outlook for the characterization of brassinins on the plant-microbe system and the introduction of this pathway into a crop, such as soybean (***Glycine max).*** Understanding these phenotypes and initial engineering of these pathways will be extremely important for successful integration of aromatic defense molecules into crop plants and improved food yield.

[1] Pedras, M. Soledade C., Estifanos E. Yaya, and Erich Glawischnig. "The phytoalexins from cultivated and wild crucifers: chemistry and biology."***Natural product reports*** 28.8 (2011): 1381-1405.

[2] Kim, Sung Moo, et al. "Brassinin Induces Apoptosis in PC 3 Human Prostate Cancer Cells through the Suppression of PI3K/Akt/mTOR/S6K1 Signaling Cascades." ***Phytotherapy Research*** 28.3 (2014): 423-431.

Copyright © American Institute of Chemical Engineers. All rights reserved.

467483 Metabolic Model of *Trichodesmium Erythraeum,* a Marine Diazotroph

Abstract:

It is estimated that 42% of the annual global nitrogen fixation is performed by the marine cyanobacterium, *Trichodesmium.* It is also known to leak 30 â€“ 50 % of the nitrogen it fixes, thus acting as an important source of biologically available nitrogen in the open ocean. *Trichodesmium* is a unique diazotrophic cyanobacterium because it can simultaneously fix carbon and nitrogen by differentiating the metabolic pathways in two subsets of cells while maintaining similar cellular structure. Despite the prominence of this organism in the nitrogen cycle, little work has been done to investigate how carbon and nitrogen is directed through metabolism within and between cells. We have reconstructed the metabolic network of *T. erythraeum*, which is the one species of *Trichodesmium* with a sequenced genome. We have used this model to predict fluxes between cells based on constraints which we have measured in the lab or *in situ* data found in literature. Our model successfully predicts equilibrium cell compositions (carbon fixing versus nitrogen fixing) which have been measured in the open ocean. This model will be further refined and improved to enable more advanced modeling of cultures with distinct metabolisms.

Copyright © American Institute of Chemical Engineers. All rights reserved.

471639 Developing a Tunable and Tightly Controlled Gene Expression System in *Synechocystis* sp. PCC6803

Abstract:

Cyanobacteria are photosynthetic microbes that can derive energy from sunlight and carbon from atmospheric carbon dioxide, as well as other sources. In the past two decades, there were numerous efforts geared towards developing cyanobacteria as bioproduction platforms. Several of these organisms are amenable to genetic manipulation, thus facilitating synthetic biology/metabolic engineering applications. The unique demands of living on light as a primary nutrient make the use of many established gene expression control systems difficult or impossible to apply. Among other approaches, researchers have used endogenous promoters such as P*psbA2* and P*cpcB* (promoters that drive expression of major photosynthetic components), and synthetic riboswitches with limited/mixed success. Hence, developing an effective genetic toolkit to regulate gene expression in cyanobacteria is of utmost importance. In this work, as a first case study, we combined a previously developed theophylline-responsive riboswitch with the T7 control system to drive expression of a yellow fluorescence protein (YFP) reporter in *Synechocystis* sp. PCC 6803. We then demonstrated, as a second case study, the application of this system by using P*nblA* to drive expression of this same reporter. Note that, under environmental stresses such as nitrogen or sulfur starvation, cyanobacteria utilize phycobilisomes (that account for 30-40% of total cellular protein) as a nutrient source through a degradation process that leads to yellowing or bleaching of the culture. The *nbl* (nonbleaching) genes are involved in this process, with *nblA* interacting directly with the phycobilisome, possibly causing rod instability and the beginning of the disassembly process. Overall, the utility of the designed gene expression control system was clearly shown via reduced baseline expression and higher ratio of induction. Thus, our work shows the utility of a tightly controlled and tunable gene expression control system that can exploit the differences in light/dark life cycle and expedite biotechnological applications in cyanobacteria.

Copyright © American Institute of Chemical Engineers. All rights reserved.

467630 A Strain Engineering Strategy for the Containment of Genetically Modified Cyanobacteria

Abstract:

Cyanobacteria are photosynthetic microorganisms whose metabolism can be manipulated through genetic engineering for the production of many molecules of interest to industry directly from CO_2 and sunlight[1,2]. However, the costs of constructing and maintaining photobioreactors for scaled up cultivation of these genetically modified cyanobacteria are cost prohibitive[3]. A major concern for the cultivation of cyanobacteria in more affordable open pond reactors is the escape and subsequent environmental contamination by these genetically modified cyanobacteria. Modelling and preliminary experimental results suggest that a potentially useful strategy for containment of genetically modified cyanobacteria could involve abolishment of function the native CO_2-concentrating mechanism (CCM) through gene knockouts, thereby eliminating the ability to fix CO_2 in ambient air but causing no reduction in CO_2-fixation rate under an artificially elevated CO_2 concentration[4]. In the first part of this work, the efficacy of this method was studied for a model cyanobacterium, *Synechococcus sp.* PCC7002. The genes eliminated in this method have homologs across a wide diversity of cyanobacteria. Thus, a probable mechanism for subversion of containment would involve horizontal gene transfer to restore CCM function. In the second part of this work, elimination of genes associated with the natural competency of *Synechococcus sp.* PCC7002 was examined as a method to decease the likelihood of this subversion. The final result of this work is a strain of cyanobacteria engineered to have high fitness in an industrial environment containing high-CO_2, an inability to propagate in the surrounding environment, and a low likelihood of restoring viability in ambient air.

1. Angermayr, S. A., Gorchs Rovira, A. & Hellingwerf, K. J. Metabolic engineering of cyanobacteria for the synthesis of commodity products. *Trends Biotechnol.* **33,**352–361 (2015).

2. Sarkar, D. & Shimizu, K. An overview on biofuel and biochemical production by photosynthetic microorganisms with understanding of the metabolism and by metabolic engineering together with efficient cultivation and downstream processing. *Bioresour. Bioprocess.* **2,**17–36 (2015).

3. Jones, M. E. & Banholzer, W. F. Solar flux, water, and land impose limits on biology. *Biotechnol. Bioeng.* **111,**1059–61 (2014).

4. Clark, R., Cameron, J., Root, T. & Pfleger, B. Insights into the industrial growth of cyanobacteria from a model of the carbon-concentrating mechanism. *AIChE J.* **60,** 1269–1277 (2014).

Copyright © American Institute of Chemical Engineers. All rights reserved.

457458 Understanding the Role of Hpmcas in Phenytoin Solid Dispersion Formulation Using Coarse Grained Molecular Dynamics Simulations

Abstract:

Many orally administrated drug candidates currently being developed are highly lipophilic, which necessitates the use of an excipient to stabilize and increase their solubility in the gastrointestinal tract. Hydroxypropyl methylcellulose acetate succinate (HPMCAS) has been identified to be a very effective excipient; however, to date very little understanding has been established on how HPMCAS interacts with drug molecules at a molecular level, and how different structural variables such as molecular weight and substitution pattern affect the performance of HPMCAS as an excipient. In this study, we employed coarse-grained (CG) molecular dynamics (MD) simulations to model HPMCAS and a drug candidate, phenytoin, in an implicit water environment. We parameterized our CG model using the radial distribution functions (RDFs) from atomistic simulations of short HPMCAS oligomers and applied the model to simulate long chains. We studied the effect of individual substitution type (e.g. methyl, hydroxypropyl, acetate etc.) by simulating homogenous polymer chains with drug molecules. We found the conformation and the structural details of the polymer-drug complex are affected by monomer substitution type, the amount of polymer loading, and polymer length. Specifically, bulky hydroxypropyl group results in a more loosely packed polymer matrix, thus allowing drug molecules to diffuse through the matrix more easily. The use of high polymer concentration (>80 wt%) can significantly reduce the size of the drug aggregates in the complex. We determined the drug release rate for different polymer-drug complex by running simulations for over 50Î¼s. We also modeled the heterogeneous copolymer and investigated the effect of deprotonated succinate substitution on the release behavior. We found that the polymer-drug complex swells when succinate groups switch from protonated to deprotonated state, thus facilitates the release of the drug molecules. Our simulation box size is comparable to the size of polymer-drug complex identified from the experiment. This study provides insights to help optimizing the composition of HPMCAS (e.g. ratio of acetate to succinate group) for any given drug.

Copyright © American Institute of Chemical Engineers. All rights reserved.

464851 Assessing the Efficacy of Thermo-Responsive Poly(N-isopropylacrylamide) in Phenytoin Solid Dispersion Formulation Using Atomistic Molecular Dynamics Simulations

Abstract:

Abstract for AICHE Annual Meeting 2016

Submission Deadline: May 9th, 2016

Session: 26004 Amorphous Solid Dispersion for Drug Product

Category: Pharmaceutical Discovery, Development, and Manufacturing Forum

Assessing the Efficacy of Thermo-Responsive Poly(N-isopropylacrylamide) in Phenytoin Solid Dispersion Formulation Using Atomistic Molecular Dynamics Simulations

Soroush Moghadam[1] and Ronald G. Larson[2]

[1]Department of Mechanical Engineering, University of Michigan, Ann Arbor, MI, 48109

[2]Department of Chemical Engineering, University of Michigan, Ann Arbor, MI, 48109

For an active pharmaceutical ingredient (API) to be absorbed from the gastro-intestinal tract into the bloodstream, it must be sufficiently soluble in the GI fluid. However, more than 40% of new chemical entities in the pharmaceutical industry are poorly soluble, leading to drug aggregation and crystallization, which necessitates the use of an excipient to stabilize and increase the solubility. Solid dispersions can maintain the API in an amorphous form and result in a supersaturated API solution upon its dissolution in the GI tract. As the hydrophilic component, thermo-sensitive poly(N-isopropylacrylamide), namely pNIPAAm, has been suggested to provide desired controlled release properties due to its lower critical solution temperature (LCST~32°C), biocompatibility and also large design space for its functional groups. Although many simulation studies analyzed the LCST behavior of pNIPAAm chains in water, to date there has not been a computational analysis of how pNIPAAm interacts with drug molecules at molecular levels as a function of temperature and how we can tune these properties to reach desired stability of polymer-drug complex. In this study, we performed all-atom molecular dynamic (AA-MD) simulations to model pNIPAAm and a drug candidate, phenytoin, in an explicit water environment. AMBER forcefield and solvated ensemble averaging of partial charges have been implemented. After validating the forcefield parameters using the well-known lower critical solution behavior of pNIPAAm, we simulated the polymer-drug complex in water and its behavior at temperatures below (295K) and above the LCST (310K). The effect of copolymerization with hydrophilic elements (e.g. Dimethylacylamide (DMA)) on the excipient behavior has been studied. We found that existence of hydrophilic comonomers in a pNIPAAm chain dramatically affects its temperature transition dependence. However, It affects affinity of the excipient functional groups to interact with drug molecules as well. Using radial distribution functions, we found that there is an optimum comonomer molar fraction of around 20-30% DMA at which interaction with phenytoin drug molecules is strongest. Furthermore, maintaining the weight percents of the components, using shorter polymer chains close to pNIPAAm reported persistence lengths can significantly influence the drug-polymer interaction upon heating. Although the length and time scales of our simulations are not high enough to be conclusive, this study provides insights to help optimizing the composition of pNIPAAm for any given drug.

KEYWORDS: Polymer Drug Aggregates, Molecular Dynamics, pNIPAAm

Copyright © American Institute of Chemical Engineers. All rights reserved.

474782 Solid State Form Understanding Applied to the Design of a Spray Dried Dispersion Process: A Quality By Design Approach to Optimization of Feed Solution Loading

Abstract:

Spray drying is a common unit operation across many industries due to the ability to control bulk powder properties and isolate and dry product from a liquid feed in a single operation. For the pharmaceutical industry, spray drying offers the control of physical properties of crystalline or amorphous active pharmaceutical ingredients (APIs), as a neat or formulated drug product that can lead to improved bioavailability of the drug candidate. Drug candidate molecules are often formulated as a solid dispersion, incorporating both API and a polymer excipient for enhanced dissolution performance and improved physical stability. In order to reliably achieve solid form control to the amorphous state during the spray drying operation, the API must be fully solubilized in the spray dryer feed solution, eliminating crystalline API material. The crystalline form of the API may or may not be the most stable form in the selected spray drying solvent. To ensure robust solid state form control of the spray dried dispersion (SDD), not only should the form landscape of the API in the spray drying solvent be well understood the process control capabilities during feed solution preparation must be considered as well. Numerical modeling of the solubility of relevant API forms, in conjunction with feed solution loading and process control capability should be used to determine an optimal feed solution loading. The operating point can be selected through optimization based upon on selected risk-based criteria and control capability. Applying this methodology to Compound A, the throughput of the spray drying process was improved two-fold, while ensuring a robust control strategy to deliver the desired API in an amorphous form as a spray dried dispersion.

Copyright © American Institute of Chemical Engineers. All rights reserved.

468548 Ternary Dispersions of Itraconazole: A Comparison of Hot Melt Extrusion and Spray Drying

Abstract:

Ternary dispersions of itraconazole: a comparison of hot melt extrusion and spray drying

Mark T. Davis[1]*, Catherine B. Kelly[1], & Gavin M. Walker[1,2]

[1]Synthesis and Solid State Pharmaceutical Centre (SSPC), Bernal Institute,

University of Limerick, Limerick, Ireland

[2]Pharmaceutical Manufacturing Technology Centre, PMTC, University of Limerick, Limerick, Ireland

Tel.: +353-(0)-61- 213096, E-mail: mark.davis@ul.ie

ABSTRACT

BCS Class II drug itraconazole (ITR) represents a difficult formulation challenge due to its very low aqueous solubility and poor dissolution kinetics. It has been extensively studied in an attempt to achieve higher bioavailability and therefore better dissolution. The amorphous state represents a disordered structure relative to the crystalline form which has been shown to have an advantage of higher dissolution rates and better oral absorption. Much work has been carried out on the performance of single polymers in the formulation of the amorphous form as an amorphous solid dispersion however evidence exists that ternary amorphous solid dispersions (ASDs) containing two rationally chosen biocompatible polymers may exhibit better amorphous stability and dissolution profiles than their binary

equivalents.

In this study, ITR was formulated into a number of amorphous solid dispersions using hot melt extrusion (HME) and spray drying (SD). The systems comprised ternary formulations of itraconazole with Soluplus® and/or HPMCP. Initial studies within our group have indicated that although HPMCP is an enteric polymer it interacts more favourably with ITR than Soluplus does and so should offer amorphous stabilisation benefits. Soluplus, however, is an excellent aqueous phase solubiliser for ITR so contributes to the dissolution properties of the ASD.

A drug loading of 40% was chosen to produce tablets of 500 mg containing 200 mg of active pharmaceutical ingredient (API). Equivalent physical mixtures were prepared to measure the solubility benefit of the amorphous state. All formulations were shown to be completely free of either liquid-crystalline or crystalline ITR using both wide-angle XRD and hyper differential scanning calorimetry (DSC). Further analysis of physical characteristics was carried out using SEM, laser scattering and density measurements.

Dissolution testing was carried out on both the tablets and the physical mixture using USP dissolution apparatus 2 (paddle at 37 °C) in biorelevant media. 5ml aliquots were removed at regular intervals and, following production and validation of an absorbance - concentration calibration curve for ITR, dissolved drug was quantified using UV-Vis spectrophotometry. Differences in dissolution rate and extent were rationalised in terms of the preparation methods, hydrogen bonding and the material properties.

Copyright © American Institute of Chemical Engineers. All rights reserved.

469561 Tunable Drug Loaded Stable and Fast Release Amorphous Composites Production Using Mesoporous Neusilin ®UFL2

Abstract:

Literature has revealed that there may exist possibility of greatly reduced mobility of drug molecules due to confinement, which could occur through several different manners. Impregnation of poorly water drug solutions in to mesoporous materials is one such attractive approach for nano-confinement that leads not only to creation of amorphous state but also its stability due to what may be the poor molecular mobility. Such as approach is examined here to increase the aqueous solubility and thus the bioavailability of poorly soluble drugs since it allows formation of drug amorphous state which generally exhibit higher apparent solubilities than their crystalline counterparts. Unfortunately, mesoporous materials such as silica are expensive and proprietary, and not easily available in bulk quantities. In this work, Neusilin® UFL2 as commercially available mesoporous amorphous materials was used to prepare amorphous drug composites of the model poorly soluble drugs Fenofibrate and Itraconazole via impregnation. Drug loading could be tuned as low as 1.0% (w/w) to as high as 30%. PXRD, DSC, and Raman spectroscopy were used to assess the amorphousness/crystallinity of the drug in the impregnated composites after formation as well as after storage. It was shown that the impregnated drug composites are stable for three months even stored at 40°C and 75% relative humidity. Re-crystallization of the amorphous drug was completely suppressed in the composites due to the confinement effect provided by the mesopores of Neusilin® UFL2. The amorphous drug composites result in higher apparent solubility and faster release/dissolution rate of the model drugs compared to crystalline counterparts. UV imaging technique, a real-time and sample sparing method, was used to qualitatively differentiate the amorphous and crystalline drug dissolution besides the conventional USP II dissolution. In conclusion, preparation of amorphous drug composites using Neusilin® UFL2 appears to be promising as a stable amorphous drug delivery vehicle providing a rapid drug release and subsequently bioavailability enhancement.

Keywords: Mesoporous Neusilin®, Impregnation, Amorphous Drug Composites, Stability, Dissolution Enhancement

Copyright © American Institute of Chemical Engineers. All rights reserved.

458072 Hot Melt Extrusion of Ternary Solid Dispersions of Itraconazole: Amorphous *Vs* Liquid Crystal Intermediate

Abstract:

Abstract

It has been demonstrated that ternary solid dispersion compositions can be more efficient than binary compositions in terms of their excellent physical stability and ability to inhibit the recrystallization of API's in bio-relevant media. This study is mainly focused on investigating the preparation of ternary solid dispersions of itraconazole, a BSC Class (II) model drug, containing a water-soluble polymer and/or a water-insoluble polymer using hot melt extrusion. The formation, physio-chemical characterization, stability, and phase transformation in the context of drug formulation and development of amorphous/liquid crystals of itraconazole was investigated in detail.

Five combinations of ternary solid dispersions comprising itraconazole, hydroxy propyl methyl cellulose phthalate (HPMCP-HP50) and/or soluplus were prepared by hot melt extrusion (HME) using a 16 mm pharmalab co-rotating twin screw extruder and L/D ratio of 40: 1 (Thermo Fisher Scientific, Karlsruhe, Germany). Information regarding the miscibility of the drug with the excipient, physical state, phase transformation, possible interactions and proper storage conditions were obtained by employing small angle powder X-ray diffractometry (SAXS), differential scanning calorimetry (DSC), Raman spectroscopy and infrared spectroscopic (FT-IR) measurements for specific polymer-polymer and drug-polymer interactions.

The SAXS results showed that itraconazole in the solid dispersion was in the amorphous state and indicated the absence of liquid crystalline itraconazole. The Raman and FT-IR analysis indicated the formation of hydrogen bonding between itraconazole and HPMCP-HP50/soluplus molecules and that itraconazoleâ€'HPMCP-HP50 interactions were stronger than itraconazoleâ€'soluplus interactions. From the DSC scans melt extrudates of itraconazole and HPMCP-HP50/soluplus exhibited one glass transition (Tg) and an absence of an endotherm around the melting point of itraconazole confirming that the API is present only in its amorphous state and that each formulation consists of a single homogenous amorphous phase of Itraconazole and HPMCP-HP50 and/or soluplus.

Copyright © American Institute of Chemical Engineers. All rights reserved.

463448 Adaptable Elastin-like Protein – Hyaluronic Acid (ELP – HA) Hybrid Hydrogels with Tunable Stress Relaxation Rates for Cell-Matrix Interaction Studies

Abstract:

Synthetic hydrogels have been widely used as *in vitro* models to study cell-matrix interactions and elucidate signal transduction mechanisms in biomaterials science. However, these hydrogels are typically elastic with minimal viscous dissipation, whereas native extracellular matrices (ECMs) are viscoelastic with stress relaxation characteristics. Thus, to better mimic the native cellular microenvironment and to understand the role of stress relaxation in guiding cell behaviors, we designed a viscoelastic hybrid hydrogel system consisting of protein-engineered, elastin-like protein (ELP) and chemically modified hyaluronic acid (HA). This family of adaptable ELP-HA hydrogels was formed *via* dynamic covalent hydrazone linkages through the reactions between hydrazine groups on ELP and aldehyde or benzaldehyde groups on HA. By combining the precise sequence control of protein-engineered materials and the kinetic features of dynamic covalent chemistry, we were able to independently tune the stress relaxation rate without causing changes to several other important matrix parameters, including polymer concentration, initial stiffness, and cell-adhesive ligand density. Human intestinal organoids were encapsulated within this family of ELP-HA hydrogels to evaluate their potential as *in vitro* biomimetic scaffolds for cell-matrix interaction studies. While all cultures remained highly viable, matrix stress relaxation rate appeared to influence cell proliferation rate and the ability of the cells to properly polarize and organize into an organoid-like tissue with an epithelial-coated internal lumen.

Copyright © American Institute of Chemical Engineers. All rights reserved.

Targeting Cancer Cells via Tumor-homing Peptide CREKA Functional PEG Nanoparticles

Aysu Ceren Okur[1+], *Pelin Erkoc*[1+], *Seda Kizilel*[1,2]*

[1]*Koç University,* Biomedical Sciences and Engineering, [2]*Chemical and Biological Engineering, Istanbul, 34450, Turkey*

e-mail: skizilel@ku.edu.tr
Fax: 90-212-338-1538

+: equal authorship

*: corresponding author

Keywords: water-in-water emulsion; hydrogel nanoparticles; CREKA; IKVAV

Abstract

Targeting cell microenvironment via nano-particle based therapies holds great promise for treatment of various diseases. One of the main challenges in targeted delivery of nanoparticles for cancer therapy includes reduced localization of delivery vehicles at tumor site. The therapeutic efficacy of drugs can be improved by recruiting delivery vehicles towards specific region of tumorigenesis in the body. Here, we demonstrate an effective approach in creating PEG particles via water-in-water emulsion technique where tumor-homing peptide CREKA was used for functionalization. Simultaneous conjugation of laminin peptide IKVAV into hydrogel network and influence of altered combinations of ligands on intracellular uptake of anticancer drugs by HeLa cells were investigated. CREKA conjugated hydrogel nanoparticles were more effective to improve apoptotic effects of the model drug Doxorubicin (DOX) compared to that of particles conjugated with other peptides. Fluorescence intensity analysis on confocal micrographs suggested significantly higher cellular uptake of CREKA conjugated PEG particles than internalization of nanoparticles in other groups. We observed that fibrin binding ability of PEG particles could be increased up to 94% through CREKA conjugation. Our results suggest the possibility of cancer cell targeting via CREKA-functional PEG nanoparticles.

1. Introduction

Targeting cancer cells via nanoparticle drug delivery vehicles contributed to significant advances in biomedical applications.[1-3] Functionalization of nanoparticles through peptides with specific affinities towards transmembrane proteins and extracellular matrix elements can be an effective strategy for tumor directed drug delivery. To improve internalization of nanoparticles into cancer cells via receptor-mediated endocytosis several peptides have been used that provided advantages such as low immunogenicity, stability in blood fluid and ease of use.[4-9] IKVAV (Isoleucine-Lysine-Valine-Alanine-Valine) is a cell adhesive peptide, derived from α_1 chain of laminin basement membrane protein and was first used to mimic neural microenvironment and trigger neuronal differentiation.[10, 11] However, studies about the role of IKVAV in tumor targeting and imaging are limited. IKVAV is known to interact with different cancer cell types through binding integrins $\alpha_3\beta_1$ and $\alpha_6\beta_1$.[12-15] Recently, $\alpha_6\beta_1$-integrin has been identified as a functionally related receptor in glioma stem-like cells (GSC). GSCs are chemotherapy resistant cell populations in glioblastoma multiforme, and interfere with the binding $\alpha_6\beta_1$ to laminin leading to decreased tumor formation. An IKVAV epitope carrying self-assembling peptide amphiphile was found to increase immobilized β_1-integrin at the GSC membrane, inhibited focal adhesion kinase (FAK) and consequently induced apoptosis in GSCs.[16] In another study, radiolabeled IKVAV was injected intravenously into mice with lung tumors, where peptide contributed to significant localization in lungs compared to other tissues.[10, 12] IKVAV also binds to nucleolin expressed on tumor cell surface, which is a new molecular target for cancer therapy.[17, 18] Besides targeting of cell membrane proteins through the use of integrin binding peptides, a tumor-homing peptide CREKA (Cysteine–Arginine–Glutamic acid–Lysine–Alanine) has been considered for targeting of tumor ECM.[11] CREKA was first selected by *in* vivo phage display peptide libraries that home selectively to tumor blood vessels. Type IV collagen in basement vascular membrane is the receptor for CREKA binding.[6, 19] Due to leaky vasculature of cancer microenvironment, fibrin deposition occurs in tumor ECM, but not in healthy tissues. CREKA has specific affinity towards these deposited fibrin clots.[4-6] Previously, CREKA was conjugated to iron oxide nanoparticles, poly(ethylene glycol)–polylactic acid polimeric nanoparticles, poly(amidoamine) dendrimer nanoparticles and amphiphile micelles.[4, 6, 11, 20, 21] *In vivo* studies suggested significantly improved retention and anti-cancer effects of CREKA conjugated vehicles compared to controls.[6, 11, 21] Even though several studies exist related to tumor-homing potential of CREKA, the knowledge about molecular target of CREKA is limited to *in vivo* phage library screening. Comparison of alternative integrin-binding peptides with CREKA may provide insight into its binding site and may improve its targeting capacity. In addition, tumor-homing function of CREKA peptide in blood may be diminished due to short half-lives as a result of degradation of peptide by endogenous proteases.[22] Hence, finding efficient delivery vehicles for CREKA is important for cancer diagnostics.

Water-in-oil emulsion polymerization has been commonly employed for nano-scale synthesis of hydrogel particles. In these studies, generally hydrogel microparticles were used to encapsulate co-cultures of varying cell types.[23] However, the use of organic solvents, and surfactants used in emulsion processes is not desirable in biological systems.[24] Hydrogel particles can also be prepared in aqueous medium using water-in-water emulsion. For example, PEG hydrogel microparticles can be synthesized in dextran, a natural polysaccharide, utilizing the immiscibility of two aqueous phases of polymers above a certain concentration.[25-27] Application of water-in-water emulsion system as a drug delivery vehicle was limited to only few reports where VEGF and calmodulin proteins were encapsulated in PEG microparticles, and dispersed prepolymer phase were crosslinked in dextran and raffinose continuous phases by UV light photopolymerization. These hydrogel microparticles were then characterized with respect to their chemical compositions, morphologies and size distributions via light microscopy.[24, 25]

Here, our objective was to use water-in-water emulsion for the synthesis of hydrogel particles that have targeting potential towards cancer cells. PEG was chosen as a polymeric network due to its low immunogenicity and low surface charge that extends blood circulation period when used as a delivery vehicle.[28-31] Dextran was selected as the continuous phase in this system, a natural analogue of PEG due to its similar properties such as solubility in water, biocompatibility and anti-fouling. To achieve higher internalization into cancer cells, we reduced the sizes of particles produced by water-in-water emulsion technique down to nanoscale and used visible light photopolymerization. Next, cross-linked PEG particles were recovered from dextran continuous phase after rinsing (**Scheme 1**).

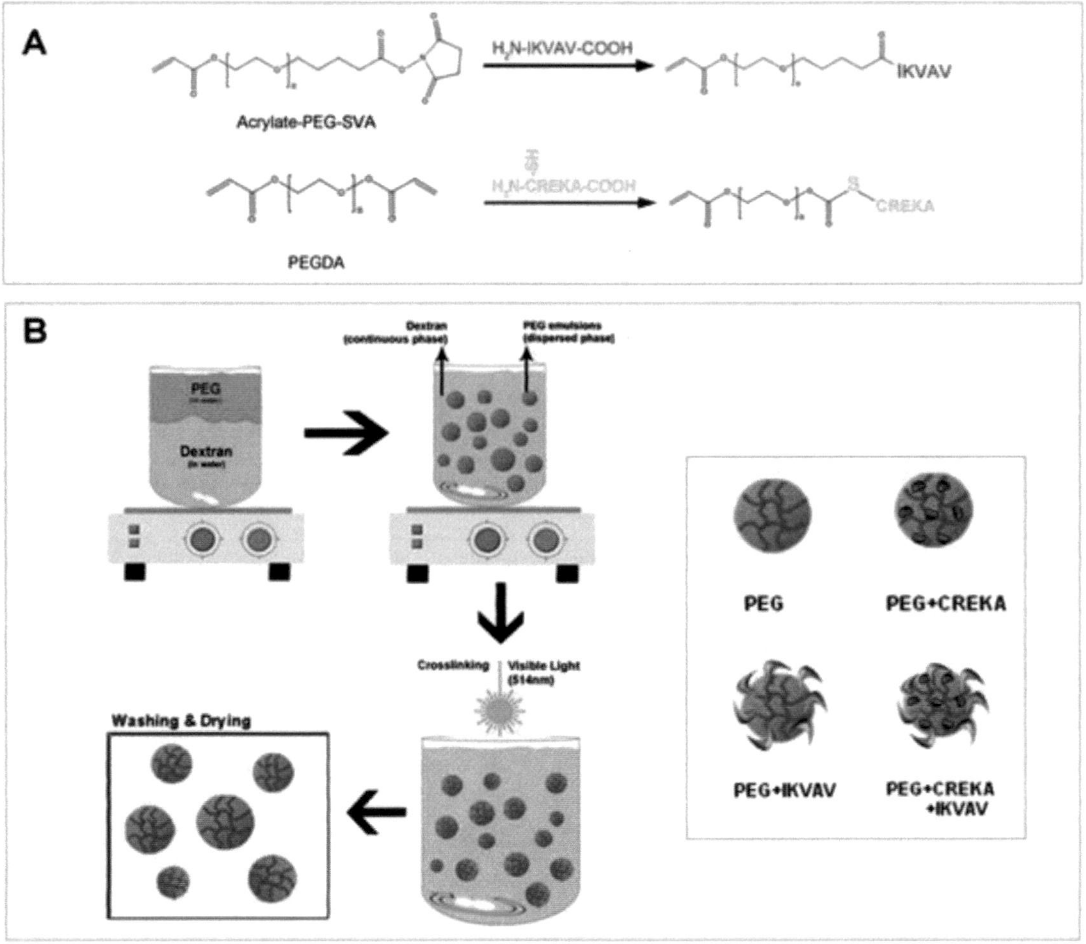

Scheme 1. A) IKVAV peptide conjugated with Acr-PEG-SVA for covalent binding into hydrogel structure, CREKA was bonded to PEGDA polymer chains via thiol groups at the terminal cysteine. **B)** Illustration of water-in-water emulsion approach and functionalization of hydrogel particles with various peptides and peptide groups.

Targeting potentials of PEG particles were investigated through conjugation with IKVAV and CREKA peptides into PEG hydrogel nanocarriers and their targeting efficiencies were further studied in cervical cancer cells (HeLa). Results suggested these delivery vehicles as promising cancer therapeutics to improve cellular uptake and efficacy of drugs. This study is original for simultaneous investigation of combinatory targeting potentials of ECM peptides and utilization of water-in-water emulsion for the synthesis of targeted nanocarriers.

2. Materials and methods

2.1. Materials

Potassium Chloride (KCl), Sodium Bicarbonate (NaHCO3), Fluorescein o-acrylate, Eosin-Y, 1-vinyl- 2-pyrrolidone (VP), Dextran (40 kDa), Poly(ethylene glycol) diacrylate (PEGDA, MW: 575 Da), Triton X-100 and dialysis tubing cellulose membrane were purchased from Sigma–Aldrich (St. Louis, MO). Phosphate Buffer Saline (PBS), Trypsin–EDTA and Penicilin/streptomycin were purchased from Gibco (Life Technologies, Grand Island, NY). Amicon Ultra Centrifugal Filter (NMWL: 60 kDa), cellulose and nylon filters (0.45 μm), in vitro fibrin binding assay, Triethanol-amine (TEA) were obtained from Merck. Doxorubicin Hydrochloride (DOX) was purchased from MP Biomedicals (Solon, OH). Dulbecco's Modified Eagle Medium (DMEM), L-Glutamine and 4',6-Diamidino-2-Phenylindole, dihydrochloride (DAPI) were purchased from Invitrogen (Carlsbad, CA). IKVAV (MW:528 Da), CREKA (MW:605 Da) were purchased from Elim Biopharmaceuticals (Hayward, CA). Acrylate PEG-Succinimidyl Valerate (Acr-PEG-SVA, Mw:3.4 kDa) was purchased from Laysan Bio (Arab, AL). Dialysis cassette (NMWL:3.5 kDa) and Fluoraldehyde™ o-Phthaldialdehyde Reagent Solution (OPA) were obtained from ThermoFisher Scientific (Waltham, Mass). Fetal Bovine Serum (FBS) was obtained by Bioindustries (Cromwell, CT). Cell-Titer-GLO (CTG) viability assay was obtained from Promega (Madison, WI). Cell culture plates were purchased from Greiner Bio-One (Solingen, Germany).

2.2. Preparation of emulsions and synthesis of PEG hydrogel particles

In order to obtain crosslinked and functional PEG hydrogel particles, we used water-in-water emulsion polymerization, where PEG prepolymer was dispersed in dextran phase. First the peptides and peptide-PEG-acrylate conjugates are prepared, then functional hydrogel particles were synthesized with CREKA and IKVAV peptides (**Scheme 1**). For CREKA functionalization, peptide (0.5 mM) was directly added into prepolymer solution, where acrylate groups in PEGDA bound via thiol ether linkage group with sulfur on cysteine. IKVAV (2.5 mM) was conjugated to acr-PEG-SVA from amine terminus in basic pH conditions. The conjugation reaction was carried out for 2 h at room temperature and the solution was dialyzed with dH_2O and lyophilized for storage.

Free amine percentages for both CREKA and IKVAV conjugation reactions were measured to calculate covalent binding efficiency of the peptides using fluorometric fluoraldehyde o-phthalaldehyde (OPA) assay. Briefly, standard solutions of each peptide with altered concentrations were prepared in PBS. Three different concentrations of acr-PEG-IKVAV conjugate was also prepared in PBS. PEGDA prepolymer soultions were prepared with and without CREKA and incubated at room temperature for 1h. All samples and standard solutions were reacted with fluoraldehyde reagent (1:10 v/v) and fluorescence was monitored at 340 nm excitation and 455 nm emission wavelengths using a plate reader (Biotek, Synergy H1).[32, 33] The standard solutions were used to prepare calibration curves.

Prepolymer solution was prepared in dH_2O with 518 mM PEGDA, 225 mM TEA (co-initiator) and 37 mM VP (accelerator) and 0.025 mM Eosin Y (initiator) at pH 8. PEG prepolymer was added to 20% (w/w) dextran and stirred to obtain prepolymer emulsion droplets for 45 minutes. Following exposure to green light (λ: 514 nm, P: 2.5 mW/cm^2, Argon ion laser-Coherent, Innova 70C), gels were rinsed with centrifugal filters and were lyophilized. These gels were then used for DOX loading and subsequent experiments. The conversion of PEG monomer in this system is about 80 %, and crosslinking density is $6x10^{-4}$.[31, 34]

2.3. Structural, chemical and physical characterization of hydrogel particles

Particles were characterized in terms of chemical and physical structure, size and morphology. Field Emission Scanning Electron Microscope (FESEM) (Zeiss Ultra Plus, Bruker, MA) and Atomic Force Microscope (AFM) (The Dimension Icon, Bruker, MA) were used to characterize structure and morphology of the gels. For FESEM characterization, lyophilized particles were placed on conductive carbon layer and coated with 10 nm thick gold. In AFM measurements, Peak Force QNM mode in air, RFESP (MPP 21100-10) cantilever (Veeco probes, USA, k=3 N/m, fo: 86-94 kHz) was used and NanoScope Analysis 1.5 software was used to export images. Dynamic light scattering (DLS) (Zetasizer, Malvern ZS) was used to determine hydrodynamic radii of particles in water. To confirm peptide incorporation, Fourier Transform Infrared Spectrometer (FTIR) (Thermo Scientific iS10 FT-IR), with a single reflection diamond ATR was used. CREKA presence was also confirmed by X-ray Fluorescence Spectrometer (Bruker, Tiger S8).

2.4. Fibrin Binding Assay

Fibrin gels were synthesized in 96-well plates. Each well was treated with fibrinogen and thrombin solutions with 60 % and 40 % (v/v), respectively. Solutions were reacted at 37°C with continuous shaking for 4 hours. Next, 0.5 mM and 1 mM free CREKA peptides, particles without peptides (1 mg/ml), and particles covalently bonded with 0.5 mM and 1 mM CREKA were added on fibrin gels and incubated for 1h. Initial absorbance and absorbance after incubation period was measured at 220 nm with NanoDrop2000 Spectrophotometer (Thermo Scientific, Mass, US). Binding efficiency (%) of molecules to fibrin layer was calculated using Equation (1). First, absorbance of the wells was measured in the beginning, where the wells included CREKA, PEG or CREKA-PEG solutions. This absorbance was defined as total absorbance. Next, these solutions were transferred to the wells that included fibrin gels, and incubated for one hour. Next, fibrin gels were rinsed, and the rinsing buffer was combined with the liquids in the wells before rinsing step. The absorbance of this combined fluid was measured, and this is defined as unbound absorbance.

$$Fibrin\ \ Binding\ \ \ (\%) \ = \ \frac{Total\ \ Absorbance\ - \ Unbound\ \ Absorbance}{Total\ \ Absorbance} \qquad (1)$$

2.5. Drug loading and release from hydrogel particles

Drug-loading was carried out with 10 mg of particles in 0.1 mg/ml DOX solution in an orbital-shaker with vigorous mixing at room temperature. Unloaded DOX was removed with ultracentrifugal filters. Particles were loaded into a cellulose membrane, and incubated in 10 ml PBS supplemented with 10 % FBS at 37°C. Samples were taken from the outside medium of cellulose membrane at predetermined time points for 6 days. DOX absorbance of collected samples was measured with NanoDrop at 480 nm. Loading percentage was calculated from the weight fraction of the drug in doxorubicin-loaded particles using Equation (2), and drug release was calculated using Equation (3) and Equation (4), where C_t is the concentration of DOX in the buffer at time t, V is the total volume of the buffer, V_s is the volume of the sample taken. Using the ratio of M_t values to total weight of DOX released, % mass release was calculated.

$$Loading\ \ efficiency = \frac{(M_0 - M_f)}{M_f} * \ 100 \qquad (2)$$

$$M_t = C_t * V \ + \ \sum C_{t-1} * \ V_s \qquad (3)$$

$$Mass\ \ Release\ \ \ (\%) \ = \ (M_t / M_\infty) * \ 100 \qquad (4)$$

2.6. Cytotoxicity Assay on HeLa and Human Dermal BJ Cells

HeLa cells and human BJ skin fibroblasts were cultured in DMEM supplemented with 10% FBS, 2 mM L-glutamine, and 1% penicillin/streptomycin. Cells were maintained in a humidified incubator with 5% CO_2 at 37°C. To examine the cytotoxicity of gels, cells were seeded to 96-well plates at 5000 cells/well and pre-incubated overnight. Media in wells were removed and particles in different groups were suspended in DMEM at 0.1 mg/ml concentration were added. Effect of hydrogels on cytotoxicity was determined via CTG assay after 24 and 72hours incubation. Luminescence was normalized to control cells which were not treated with gels. Cytotoxicity of DOX-loaded hydrogel particles was also evaluated in HeLa cells, similarly.

2.7. Internalization of DOX- loaded particles

HeLa cells (5000 cells/well) were seeded on previously autoclaved glass slides placed in 24-well plates. Meanwhile, gel groups were incubated with DOX in DMEM with a ratio of 10:1 (w/w) for 24 hours. Next, DOX loading was determined, measuring the absorbance at 480 nm by spectrophotometer and calculating the ratio of absorbance after centrifuge to absorbance before centrifuge. Media of seeded cells were replaced with DOX-loaded particles suspended in DMEM. After 1hour incubation, cells were fixed as described previously.[35] Cell nuclei were stained with DAPI. Cellular uptake of DOX was visualized by laser scanning confocal microscopy (Eclipse 90-i, Nikon). As a sign of cellular internalization efficiency, DOX/DAPI intensity ratio of cell images captured on constant gain and intensities were quantified using Image J. Same method was also applied to human dermal BJ fibroblast cells to quantify DOX internalization in these cells.

2.8. SEM Imaging of Particle Treated HeLa Cells

HeLa cells (5000 cell/well) were seeded on glass slides and incubated with 0.1 mg/ml of hydrogel particles for 1 hour. Cells were fixed with 4% glutaraldehyde and 0.1 % osmium tetroxide at room temperature, following dehydration with increasing concentration gradient of ethanol and slides were dried with a critical point dryer in 40°C and 112 bars for 1 hour. These samples were fixed on carbon tape attached stubs and coated with 10 nm gold layer using a sputter coater (Cressington). Size and morphology of gel treated cells were analyzed via Zeiss Evo LS15 SEM (Bruker) at an acceleration potential of 5 kV.

2.9. Statistical analysis

Data were analyzed using GraphPad Prism5 software package. Oneway ANOVA with Tukey's Multiple Comparison post-test was performed as was indicated in the results. Differences were considered as statistically significant when the p-value was less than 0.05.

3. Results and discussion

3.1. Preparation and characterization of peptide precursors

PEG was covalently bonded to IKVAV peptide using Acr-PEG-SVA through the reaction between SVA and amine groups, where CREKA was bonded to PEGDA polymer chains via Michael addition between thiol groups at the terminal cysteine and acrylate groups in PEGDA (Scheme 1A).[30, 36] Conjugation reaction of IKVAV and functional groups were confirmed by FTIR (Fig. S1a) and amount of free amines present after conjugation was measured by OPA assay. IKVAV conjugation efficiency was calculated as 98% and CREKA binding efficiency to PEGDA was found as 99% (Fig. S1b). The presence of CREKA in the hydrogel network was confirmed by X-ray fluorescence spectroscopy (XRF), where the peak at 2.307 keV demonstrates Kα peak from sulphur atom coming from the cysteine amino acid of CREKA peptide (Fig. S2).

3.2. Synthesis and characterization of particles

Prepolymer solution was emulsified in dextran and was exposed to visible light for crosslinking. This method resulted in the formation of covalently cross-linked hydrogel particles in sub-micron size scale. Briefly, the dispersed droplets of prepolymer solution containing PEGDA, TEA, VP and eosin Y were obtained in dextran continuous phase. Upon visible light exposure dispersed PEG prepolymer droplets were covalently crosslinked to form submicron-scale hydrogel particles. Acr-PEG-IKVAV and CREKA were also added into the prepolymer solution to obtain functional hydrogel particles.[30]

Previous studies about the synthesis of hydrogels via water-in-water emulsion resulted in the formation of micron size particles, where characterization was based on optical microscope imaging.[24, 25, 27] Here, nano-sized hydrogel particles were synthesized through optimization of emulsions, where detailed characterization was carried out using FESEM, AFM and DLS. FESEM and AFM micrographs in **Fig. 1** revealed the size distribution within the range of 140-400 nm. FESEM and AFM images for the IKVAV incorporated particles were also found in submicron size (**Fig. S3-S4**).

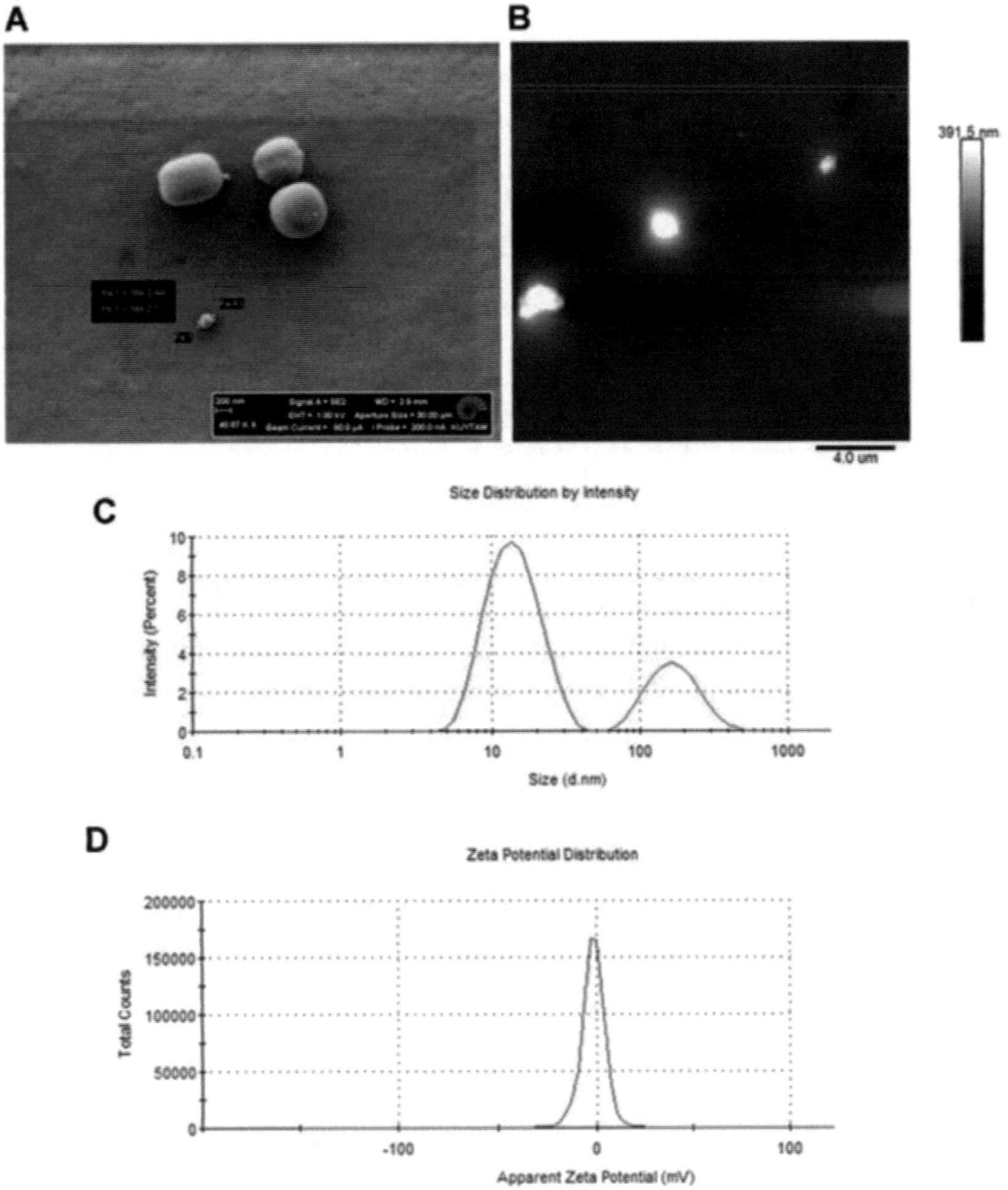

Fig. 1. A) FESEM image of lyophilized particles. The scale bar corresponds to 200 nm. B) AFM image of particles obtained by vacuum drying of aqueous suspension of particles on

glass slides. **C)** DLS characterization of particles depending on intensity percentage. **D**) Zeta potentials of particles.

DLS measurements revealed that hydrodynamic radii of the hydrogel particles were approximately 15 nm for 71 % and 178 nm for 25 % of particles (**Fig. 1C**). Peptide conjugated particles exhibited similar size distribution in DLS measurements (**Fig. S5-S6).** Nanoparticles smaller than 40 nm could not be imaged by FESEM, as this was beyond the limits of imaging capacity of FESEM for polymeric nanoparticles (**Fig. 1A**). Zeta potential of particles in aqueous environment was measured as -3.38 ± 5.79 (**Fig. 1D**). This slightly negative zeta potential value resulted from the lone pair electrons of oxygen atoms in PEG chains, and is consistent with the previous observations in the literature.[35] Previously, it was demonstrated that neutral particles could escape from Kupffer cell clearance and allocate intensely and homogenously in cancer bearing region.[21, 37, 38]

3.3. DOX loading and release from hydrogel particles

DOX molecules were loaded into swollen polymer matrixes, due to the high water uptake capacity of PEG hydrogel particles and electrostatic interaction between slightly negatively charged PEG particles and positively charged DOX, which represents weak base at physiological conditions.[39] The amount of DOX-loaded into particles was calculated as 62 %, based on the measurement of absorbance at 480 nm.

DOX release experiment was carried out to characterize the release profile of the drug from particles. Cumulative release profile of DOX and experimental data was fitted to cumulative Gaussian percentages of nonlinear regression method as was shown in **Fig. 2.**

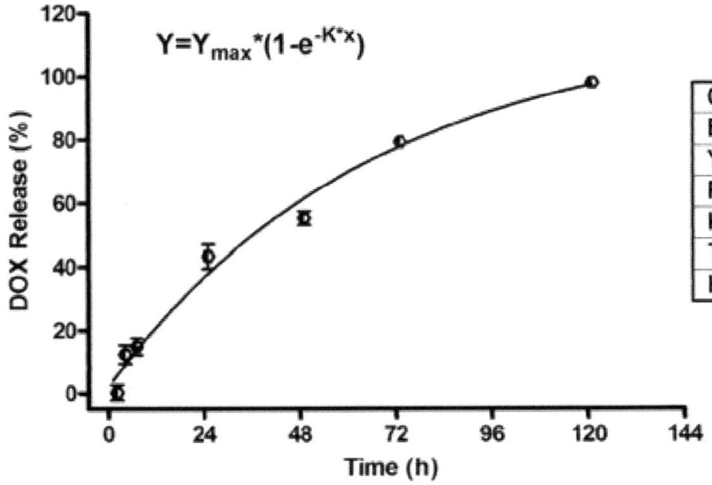

One-phase association	
Best-fit values	
Y0	4.125
Plateau	115.8
K	0.01503
Tau	66.52
Half-time	46.11

Fig. 2. DOX release versus time from DOX-loaded targeted hydrogel particles. Total drug release was calculated from absorbance values of DOX measured at 480 nm at predetermined time points. One-phase exponential association analysis was used for curve fitting to calculate release half-time of the drug. (n=3, R^2=0.9860).

DOX release reached to saturation in 144 hours. Higher release of DOX obtained during initial hours of the experiment may be due to the presence of DOX on hydrogel particle surface, which may result from the static interactions between PEG network and small DOX molecules that might have influenced the release profile.[40] One-phase association equation of nonlinear regression method was applied for the curve fitting of the *in vitro* drug release kinetics as follows:

$$Y = Y_{\max} * \left(1 - e^{-K*X}\right) \tag{5}$$

$$Tau = {}^{1}\!/\!_{K} \tag{6}$$

$$Half - time = \ln(2)/K \tag{7}$$

In equation 5, K represents the rate constant, expressed in inverse minutes; Tau is the time constant. Equation 5 assumes a pseudo-first order release kinetics, where certain fraction of the drug is released during each time interval where fewer number of drug molecules are released as time advances and the curve levels off.[41] Half-time of DOX release from gels was calculated as 46 hours. R^2 value was obtained as 0.9860 suggesting the good level of fitting obtained for experimental data.

3.4. Binding affinity of CREKA and CREKA-hydrogel particles to fibrin clots

In vitro fibrin binding assay was performed to investigate the targeting potential of CREKA functionalized particles (**Fig. 3**).

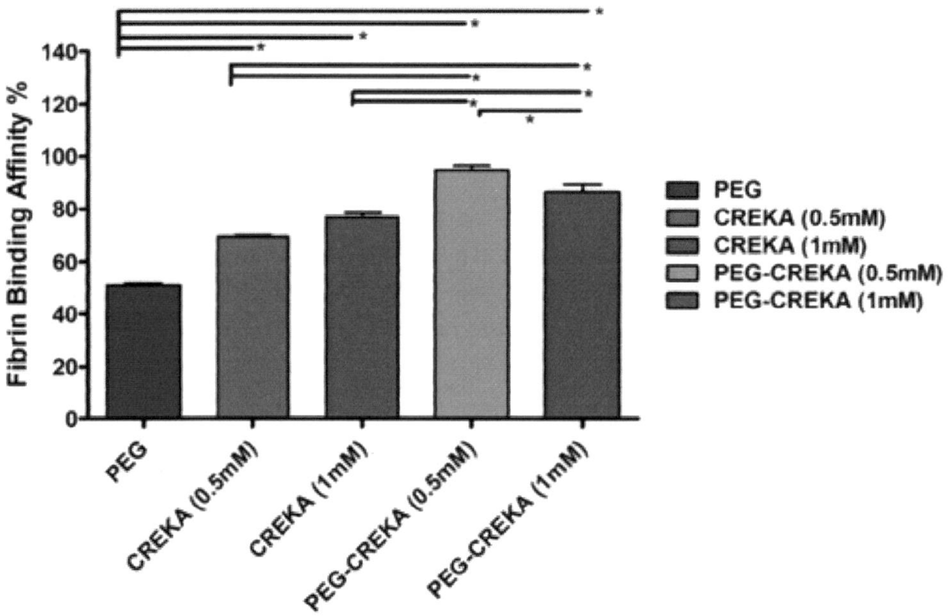

Fig. 3. Fibrin binding affinity of free and PEG bound CREKA peptides. Affinities of particles to fibrin clots were assayed in the presence and absence of two different concentrations (0.5 mM and 1 mM) of CREKA. One-way ANOVA followed by Tukey's multiple comparison test, Results indicated with * show statistical significance between the control group and other groups, $P < 0.05$, n=3.

In a previous study, fibrin-binding ratio related to bound-to-free form of CREKA were measured for CREKA only, CREKA-conjugated iron oxide nanoparticles (CREKA-IONP), and FITC-IONP from FITC absorbance as 0.69, 0.62 and 0.42, respectively.[20] We also obtained similar ratios, where fibrin-binding percentages were found approximately 70 % for free CREKA peptides and 45 % for particles without CREKA. Low level of non-specific affinity measured for PEG particles to fibrin can be explained simply by the transport of particles to underlying fibrin gel due to gravity. CREKA conjugated particles demonstrated high binding affinities towards fibrin for both concentrations of CREKA (0.5 mM and 1 mM) compared to that of CREKA peptide alone. The affinity for binding to fibrin at 0.5 mM was higher than 1 mM CREKA. The slight differences in affinity might suggest that particles reached to saturation at 0.5 mM CREKA during synthesis process. Since particles with 0.5 mM CREKA conjugation demonstrated the highest affinity of 94 % towards fibrin binding, we continued the rest of the experiments with 0.5 mM CREKA.

3.5. Cytocompatibility of particles

HeLa cells and non-cancerous healthy cell line human BJ dermal fibroblasts were incubated with 0.1 mg/ml hydrogel particles with or without peptide groups for characterization of cytocompatibility. Cell-Titer-GLO luminescent viability assay was used to investigate cytotoxicity of particles with altered combinations of IKVAV and CREKA. The luminescence values measured were normalized with respect to untreated control group. The viabilities measured for both HeLa cells and human BJ dermal fibroblasts were above 80 % and that particles were not toxic for both HeLa and BJ cells (**Fig. 4A, Fig. 4B**).

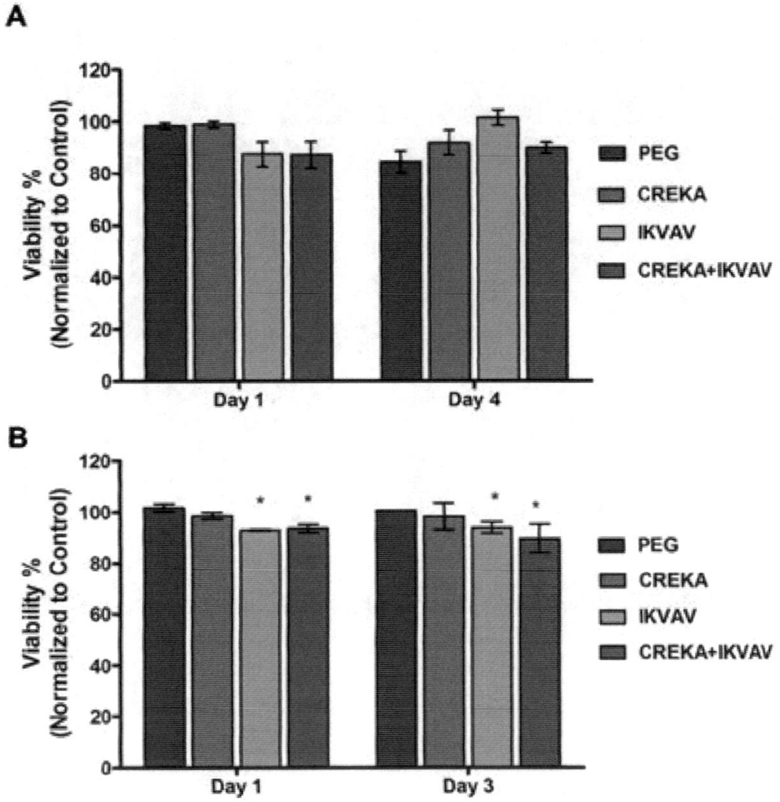

Fig. 4. A) Cell viability of HeLa and **B)** BJ cells. Viabilities were measured on day 1 and day 3 with experimental groups; untreated, PEG particles and peptide functionalized particles treated cells. One-way ANOVA followed by Tukey's multiple comparison test (*indicates statistical significance compared to the control group, $P < 0.05$, n=3).

3.6. Cellular uptake and in vitro cytotoxicity of DOX-loaded particles

Cellular internalization of particles by HeLa cells was also investigated via laser scanning confocal microscopy. Particles were loaded with DOX, a chemotherapeutic drug that has specific excitation at 480 nm and emission maximum at 560-590 nm.[35, 42] DOX-loaded particle treated cells were fixed with formaldehyde and their nuclei were stained with DAPI. Images in DOX and DAPI/DOX merged panels indicated that internalization of CREKA incorporated gels by HeLa cells was significantly higher than internalization of particles in other groups (**Fig. 5A** and **5B**).

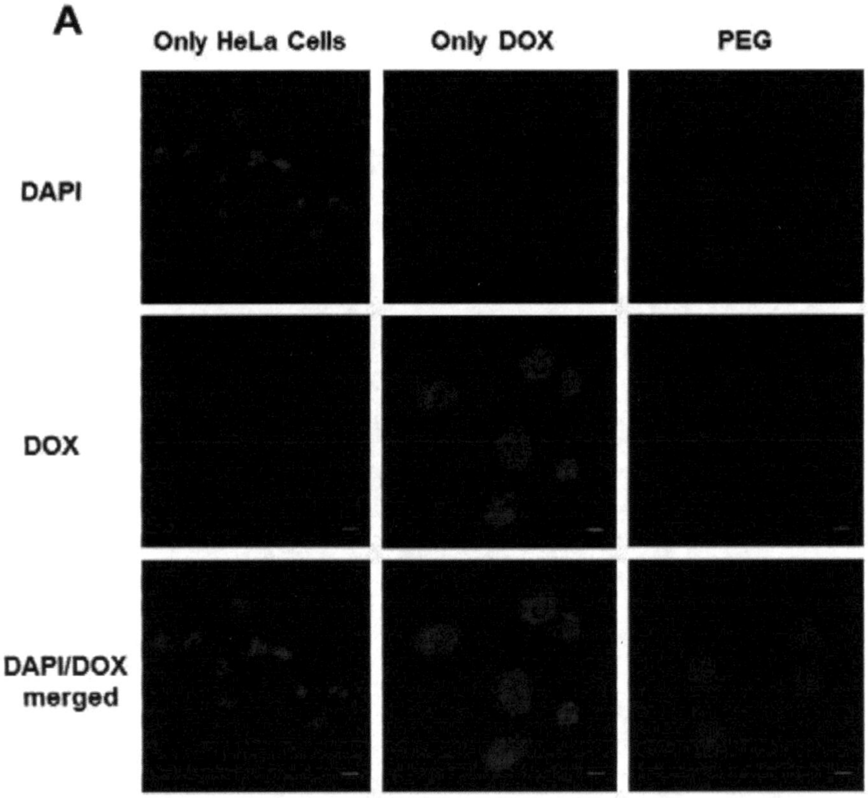

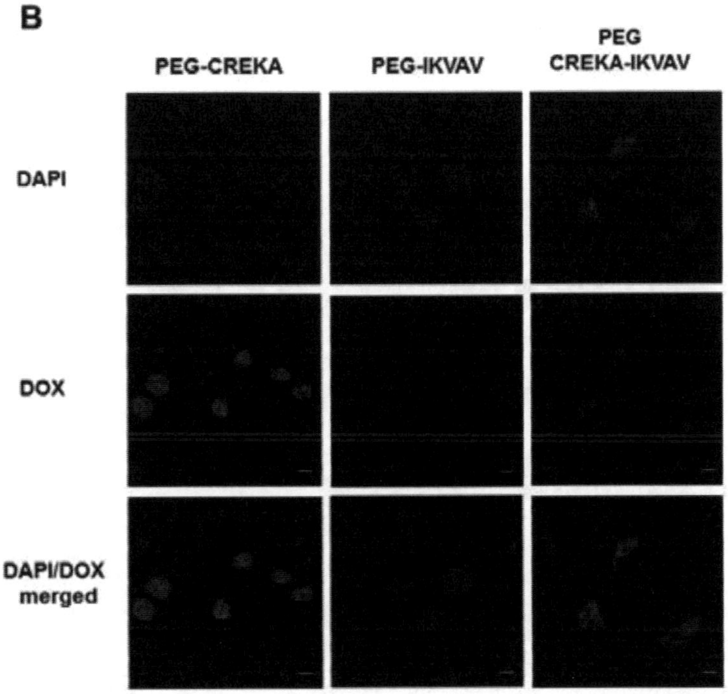

Fig. 5. Cellular internalization of free DOX and DOX-loaded particles visualized by laser scanning confocal microscope. First row illustrates cell nuclei staining with DAPI (blue); middle raw DOX fluorescence and third row, overlay images of first two rows (scale bar =10 μm).

It was also observed that simultaneous conjugation of CREKA with IKVAV did not improve internalization of particles by HeLa cells further (**Fig. 5B**). The reason for higher red fluorescence intensity in free DOX sample than the hydrogel treated samples, except CREKA bound ones, can be explained by the need for a longer uptake period due to larger size of DOX-loaded in hydrogels than its free form. Uptake efficacy was further quantified as the ratio of DOX (red) to DAPI (blue) fluorescence intensity (**Fig. 6**).

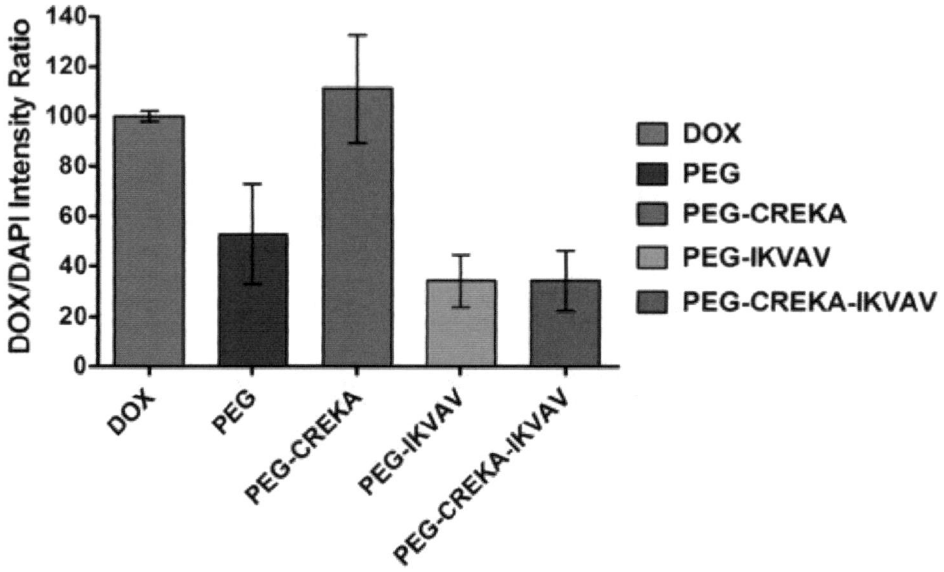

Fig. 6. Quantification of fluorescence intensity based on confocal microscopy images using Image J. One-way ANOVA was followed by Tukey's multiple sample comparison (*indicates statistical significance between PEG-CREKA and other groups, $P < 0.05$, n=3).

DOX-loaded HeLa cells were considered as control and results in the rest were normalized with respect to DOX fluorescence in only DOX incubated HeLa cells. DOX uptake was approximately 50% in the presence of hydrogel particles without any peptide functionality. This high amount of internalization may be related to dextran residues that remained on hydrogel particle surface. The contribution of dextran to internalization of particles was observed previously, in which higher internalization of dextran micelles was explained possibly by the probe role of dextran in fluid phase pinocytosis.[43, 44]

Internalization of particles by HeLa was significantly improved with CREKA conjugation into particles, compared to the conditions of DOX in the cell culture medium alone or IKVAV functional particles or particles with dual peptide functionality. Particle internalization was 111 % for PEG-CREKA, and 34 % for both PEG-IKVAV and PEG-CREKA-IKVAV functionalized particles (Fig. 5 and Fig. 6).

IKVAV target different domains of integrin, where CREKA targets collagen IV in fibrin structure. The differences in targeting domains of these peptides might be the reason for altered efficiencies observed for these peptides. An earlier study showed that HeLa cells expressed collagen IV as an extracellular matrix protein.[45] Hence, increased targeting potential of CREKA can be explained by increased secretion of collagen IV in HeLa cell microenvironment. We observed relatively lower levels of uptake in combinations of CREKA with IKVAV compared to the uptake observed with only CREKA functionalized particles. Reduced internalization of particles despite the presence of CREKA with hydrogel particles of dual peptide functionality could be attributed to the accessibility of CREKA peptide to fibrin. PEGDA chains were conjugated to thiol groups of CREKA peptide via Michael-type addition, where IKVAV was conjugated into the network through the acrylate group of acrylate-PEG-peptide molecules. The presence of long PEG chains in IKVAV-PEG-acrylate and the absence of long PEG chain in CREKA functionalized conjugates might have blocked interactions of CREKA with fibrin, and hence resulted in reduced uptake compared to internalization of CREKA only functionalized particles (**Scheme 1A**).[46] Overexpression of nucleolin in HeLa cells serves as a ligand for IKVAV.[47, 48] Expression of this receptor might have also contributed to the uptake of IKVAV containing hydrogel particles in our study. To evaluate target selectivity of the nanoparticles quantitatively, we carried out control experiments using healthy (non-tumor) cells and the gel nanoparticles with peptides. We investigated cellular uptake of nanoparticles by healthy cells, where we treated healthy human fibroblast cells with PEG nanoparticles that included altered combinations of peptides. Fluorescence images of HeLa and fibroblast cells incubated with DOX and DOX-loaded CREKA-PEG nanoparticles were shown in **Fig. S7**. Pseudo-color maps of the corresponding scattering intensity of cells were generated to highlight intensity differences between samples (**Fig. S7**). When the fluorescence intensities of cell nuclei for DOX-loaded nanoparticle treated HeLa cells and fibroblasts were compared, the cellular uptake was significantly higher in HeLa cells than fibroblasts and that alterations in peptide sequences did not change uptake significantly in fibroblast cells as they did in HeLa cells.

The cellular uptake was also evaluated in HeLa cells via viability assay in the presence of DOX-loaded nanoparticles with altered peptide sequences (**Fig. S8**). The cellular viability was reduced to approximately 20 % in the presence of nanoparticles after 1day treatment, however peptide incorporation to the hydrogels did not significantly affect the viability.

Particle uptake was finally characterized with FESEM for higher spatial resolution (**Fig. 7**). At high magnification it was observed that cells preserved their morphology and integrity after treatment with particles. Focal adhesions were visible around the cell ECM and apoptotic bodies or vesicles known as blebs that started to form near the surface of cells (**Fig. 7B, Fig. S9)**.[49]

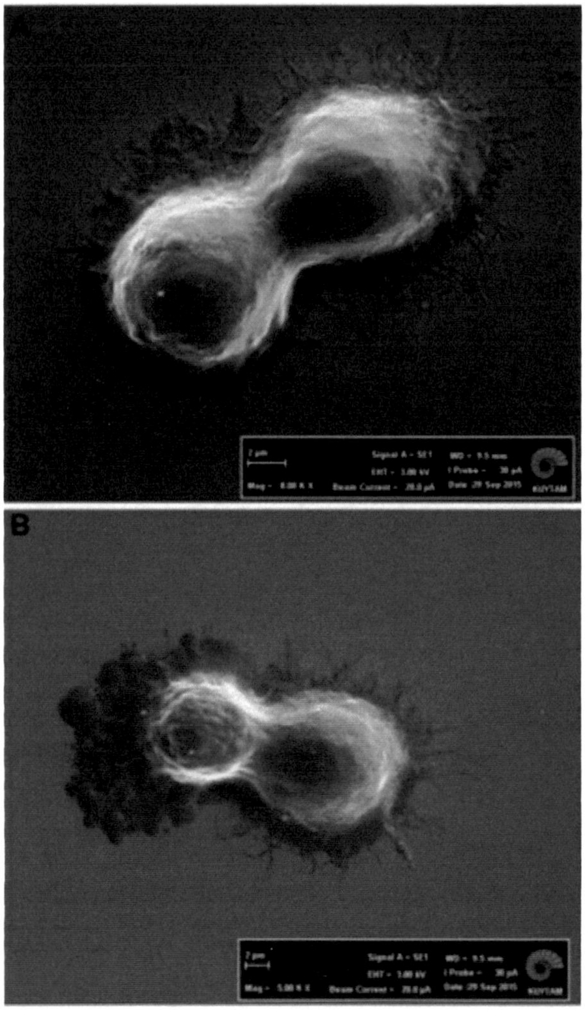

Fig. 7. SEM images of **A**) HeLa cells, **B**) HeLa cells treated with CREKA incorporated particles. Scale bars are equal to 2 μm. Adobe Photoshop CS4 software was used to obtain false colored SEM images of HeLa cells.

The occurrence of blebs might have resulted from the higher cell permeabilization due to the presence of hydrogel particles. However, blebs can also be observed in healthy cells during cell locomotion and division.[50] According to these images, cells preserved their morphology, and the hydrogel particles were observed on the surface of cells.

4. Conclusions

In this study, we developed emulsion-based synthesis of sub-micron biocompatible hydrogel particles without the use of organic solvents and investigated selective targeting potential of ECM derived peptides for cancer therapy. The potential of targeting efficacy of these carriers were studied with cancer cells using an anticancer drug, DOX. The targeting potentials of the two ligands were evaluated, where *in vitro* experiments suggested higher targeting potential of CREKA compared to that of IKVAV. In addition, combination of peptides was not superior with respect to internalization of particles. Incorporation of CREKA into hydrogel network improved internalization of the delivery vehicle, which would be significant to reduce systemic side effects of cancer drugs. This study suggests improved targeting that could be obtained with delivery vehicles towards cancer cells with CREKA functionalization compared to targeting that could be achieved with laminin peptide IKVAV. Targeted PEG nanocarriers developed here is promising to promote the efficacy of drugs, to decrease undesirable side effects of the drug, and to eliminate the resistance of cancer tissue towards drug.

Acknowledgements

The authors would like to thank Cansu Yıldırım and Barıs Yagcı for providing assistance with FESEM, XRF measurements at Koc University Surface Science and Technology Center (KUYTAM), and Tugba Bal for providing help with the preparation of peptides. We thank Dr. Tugba Bagcı Onder at School of Medicine at Koc University for providing human fibroblast cells.

References

[1] C.E. Ashley, E.C. Carnes, G.K. Phillips, D. Padilla, P.N. Durfee, P.A. Brown, T.N. Hanna, J. Liu, B. Phillips, M.B. Carter, N.J. Carroll, X. Jiang, D.R. Dunphy, C.L. Willman, D.N. Petsev, D.G. Evans, A.N. Parikh, B. Chackerian, W. Wharton, D.S. Peabody, C.J. Brinker, The targeted delivery of multicomponent cargos to cancer cells by nanoporous particle-supported lipid bilayers, Nat Mater, 10 (2011) 389-397.
[2] D.J. Irvine, Drug delivery: One nanoparticle, one kill, Nat Mater, 10 (2011) 342-343.
[3] S.E. Jin, H.E. Jin, S.S. Hong, Targeted delivery system of nanobiomaterials in anticancer therapy: from cells to clinics, Biomed Res Int, 2014 (2014) 23.
[4] E.J. Chung, Y. Cheng, R. Morshed, K. Nord, Y. Han, M.L. Wegscheid, B. Auffinger, D.A. Wainwright, M.S. Lesniak, M.V. Tirrell, Fibrin-binding, peptide amphiphile micelles for targeting glioblastoma, Biomaterials, 35 (2014) 1249-1256.
[5] P.P. Karmali, V.R. Kotamraju, M. Kastantin, M. Black, D. Missirlis, M. Tirrell, E. Ruoslahti, Targeting of albumin-embedded paclitaxel nanoparticles to tumors, Nanomedicine : nanotechnology, biology, and medicine, 5 (2009) 73-82.
[6] D. Simberg, T. Duza, J.H. Park, M. Essler, J. Pilch, L. Zhang, A.M. Derfus, M. Yang, R.M. Hoffman, S. Bhatia, M.J. Sailor, E. Ruoslahti, Biomimetic amplification of nanoparticle homing to tumors, Proc Natl Acad Sci U S A, 104 (2007) 932-936.
[7] J.S. Wadia, S.F. Dowdy, Transmembrane delivery of protein and peptide drugs by TAT-mediated transduction in the treatment of cancer, Adv Drug Deliv Rev, 57 (2005) 579-596.
[8] W.C. Chen, A.X. Zhang, S.-D. Li, Limitations and niches of the active targeting approach for nanoparticle drug delivery, European Journal of Nanomedicine, 4 (2012) 89-93.
[9] A. Accardo, D. Tesauro, G. Morelli, Peptide-based targeting strategies for simultaneous imaging and therapy with nanovectors, Polymer Journal, 45 (2013) 481-493.
[10] P.O. Zamora, D. Eshima, D. Graham, L. Shattuck, B.A. Rhodes, Biological distribution of 99mTc-labeled YIGSR and IKVAV laminin peptides in rodents: 99mTc-IKVAV peptide localizes to the lung, Biochim Biophys Acta, 1182 (1993) 197-204.
[11] J. Wu, J. Zhao, B. Zhang, Y. Qian, H. Gao, Y. Yu, Y. Wei, Z. Yang, X. Jiang, Z. Pang, Polyethylene glycol-polylactic acid nanoparticles modified with cysteine-arginine-glutamic acid-lysine-alanine fibrin-homing peptide for glioblastoma therapy by enhanced retention effect, Int J Nanomedicine, 9 (2014) 5261-5271.
[12] Y. Kikkawa, K. Hozumi, F. Katagiri, M. Nomizu, H.K. Kleinman, J.E. Koblinski, Laminin-111-derived peptides and cancer, Cell adhesion & migration, 7 (2013) 150-256.
[13] R.S. Bresalier, B. Schwartz, Y.S. Kim, Q.Y. Duh, H.K. Kleinman, P.M. Sullam, The Laminin Alpha-1 Chain Ile-Lys-Val-Ala-Val (IKVAV)-Containing Peptide Promotes Liver Colonization by human colon-cancer cells, Cancer Res., 55 (1995) 2476-2480.
[14] T.E. King, S.C. Pawar, L. Majuta, I.C. Sroka, D. Wynn, M.C. Demetriou, R.B. Nagle, F. Porreca, A.E. Cress, The role of alpha 6 integrin in prostate cancer migration and bone pain in a novel xenograft model, PLoS One, 3 (2008) e3535.
[15] J.M. Whitelock, M.L. Paine, J.R. Gibbins, R.F. Kefford, R.L. Ogrady, Multiple levels of post-transcriptional regulation of collagenase (matrix metalloproteinase 1) in an epithelial cell line, Immunol Cell Biol, 71 (1993) 39-47.
[16] M. Srikanth, S. Das, E.J. Berns, J. Kim, S.I. Stupp, J.A. Kessler, Nanofiber-mediated inhibition of focal adhesion kinase sensitizes glioma stemlike cells to epidermal growth factor receptor inhibition, Neuro Oncol, 15 (2013) 319-329.
[17] M.C. Kibbey, B. Johnson, R. Petryshyn, M. Jucker, H.K. Kleinman, A 110-kD nuclear shuttling protein, nucleolin, binds to the neurite-promoting IKVAV site of laminin-1, Journal of neuroscience research, 42 (1995) 314-322.

[18] T. Watanabe, K. Hirano, A. Takahashi, K. Yamaguchi, M. Beppu, H. Fujiki, M. Suganuma, Nucleolin on the Cell Surface as a New Molecular Target for Gastric Cancer Treatment, Biol Pharm Bull, 33 (2010) 796-803.
[19] A. Grayson, S.E. Zale, D. DeWitt, Therapeutic particles suitable for parenteral administration and methods of making and using same, Google Patents, 2013.
[20] A.M. Kruse, S.A. Meenach, K.W. Anderson, J.Z. Hilt, Synthesis and characterization of CREKA-conjugated iron oxide nanoparticles for hyperthermia applications, Acta Biomater, 10 (2014) 2622-2629.
[21] J. Zhao, B. Zhang, S. Shen, J. Chen, Q. Zhang, X. Jiang, Z. Pang, CREKA peptide-conjugated dendrimer nanoparticles for glioblastoma multiforme delivery, Journal of colloid and interface science, 450 (2015) 396-403.
[22] D. Zanuy, A. Flores-Ortega, A.I. Jimenez, M.I. Calaza, C. Cativiela, R. Nussinov, E. Ruoslahti, C. Aleman, In silico molecular engineering for a targeted replacement in a tumor-homing peptide, The Journal of Physical Chemistry B, 113 (2009) 7879-7889.
[23] C.L. Franco, J. Price, J.L. West, Development and optimization of a dual-photoinitiator, emulsion-based technique for rapid generation of cell-laden hydrogel microspheres, Acta Biomaterialia, 7 (2011) 3267-3276.
[24] W.J. King, M.W. Toepke, W.L. Murphy, Facile formation of dynamic hydrogel microspheres for triggered growth factor delivery, Acta Biomaterialia, 7 (2011) 975-985.
[25] N.A. Impellitteri, M.W. Toepke, S.K.L. Levengood, W.L. Murphy, Specific VEGF sequestering and release using peptide-functionalized hydrogel microspheres, Biomaterials, 33 (2012) 3475-3484.
[26] O. Franssen, W.E. Hennink, A novel preparation method for polymeric microparticles without the use of organic solvents, International Journal of Pharmaceutics, 168 (1998) 1-7.
[27] M. Parlato, A. Johnson, G.A. Hudalla, W.L. Murphy, Adaptable poly(ethylene glycol) microspheres capable of mixed-mode degradation, Acta Biomaterialia, 9 (2013) 9270-9280.
[28] S. Kizilel, E. Sawardecker, F. Teymour, V.H. Perez-Luna, Sequential formation of covalently bonded hydrogel multilayers through surface initiated photopolymerization, Biomaterials, 27 (2006) 1209-1215.
[29] K. Xiao, Y. Li, J. Luo, J.S. Lee, W. Xiao, A.M. Gonik, R.G. Agarwal, K.S. Lam, The effect of surface charge on in vivo biodistribution of PEG-oligocholic acid based micellar nanoparticles, Biomaterials, 32 (2011) 3435-3446.
[30] T. Bal, B. Kepsutlu, S. Kizilel, Characterization of protein release from poly(ethylene glycol) hydrogels with crosslink density gradients, Journal of biomedical materials research. Part A, 102 (2014) 487-495.
[31] S. Kizilel, Mathematical Model for Microencapsulation of Pancreatic Islets within a Biofunctional PEG Hydrogel, Macromolecular Theory and Simulations, 19 (2010) 514-531.
[32] G.Y. Wu, C.J. Meininger, Analysis of citrulline, arginine, and methylarginines using high-performance liquid chromatography, Method Enzymol, 440 (2008) 177-189.
[33] D.L. Hern, J.A. Hubbell, Incorporation of adhesion peptides into nonadhesive hydrogels useful for tissue resurfacing, Journal of Biomedical Materials Research, 39 (1998) 266-276.
[34] S. Kizilel, V.H. Pérez-Luna, F. Teymour, Modeling of PEG Hydrogel Membranes for Biomedical Applications, Macromolecular Reaction Engineering, 3 (2009) 271-287.
[35] C. Nazli, G.S. Demirer, Y. Yar, H.Y. Acar, S. Kizilel, Targeted delivery of doxorubicin into tumor cells via MMP-sensitive PEG hydrogel-coated magnetic iron oxide nanoparticles (MIONPs), Colloids and Surfaces B: Biointerfaces, 122 (2014) 674-683.
[36] T. Bal, C. Nazli, A. Okcu, G. Duruksu, E. Karaoz, S. Kizilel, Mesenchymal stem cells and ligand incorporation in biomimetic poly(ethylene glycol) hydrogels significantly improve insulin secretion from pancreatic islets, J Tissue Eng Regen Med, (2014).
[37] S. Stolnik, L. Illum, S.S. Davis, Long circulating microparticulate drug carriers, Advanced Drug Delivery Reviews, 16 (1995) 195-214.

[38] T. Neuberger, B. Schöpf, H. Hofmann, M. Hofmann, B. von Rechenberg, Superparamagnetic nanoparticles for biomedical applications: Possibilities and limitations of a new drug delivery system, Journal of Magnetism and Magnetic Materials, 293 (2005) 483-496.
[39] J.O. Kim, A.V. Kabanov, T.K. Bronich, Polymer micelles with cross-linked polyanion core for delivery of a cationic drug doxorubicin, Journal of Controlled Release, 138 (2009) 197-204.
[40] M. Goncalves, P. Figueira, D. Maciel, J. Rodrigues, X. Qu, C. Liu, H. Tomas, Y. Li, pH-sensitive Laponite((R))/doxorubicin/alginate nanohybrids with improved anticancer efficacy, Acta Biomaterialia, 10 (2014) 300-307.
[41] Q. Yang, K.R. Moulder, M.S. Cohen, S. Cai, L.M. Forrest, Cabozantinib Loaded DSPE-PEG Micelles as Delivery System: Formulation, Characterization and Cytotoxicity Evaluation, BAOJ pharmaceutical sciences, 1 (2015).
[42] P.E. Saw, J. Park, E. Lee, S. Ahn, J. Lee, H. Kim, J. Kim, M. Choi, O.C. Farokhzad, S. Jon, Effect of PEG pairing on the efficiency of cancer-targeting liposomes, Theranostics, 5 (2015) 746-754.
[43] R. Raveendran, G.S. Bhuvaneshwar, C.P. Sharma, Hemocompatible curcumin-dextran micelles as pH sensitive pro-drugs for enhanced therapeutic efficacy in cancer cells, Carbohydrate Polymers, 137 (2016) 497–507.
[44] C. Commisso, R.J. Flinn, D. Bar-Sagi, Determining the macropinocytic index of cells through a quantitative image-based assay, Nat. Protocols, 9 (2014) 182-192.
[45] J.J. Furth, T.H. Wroth, S. Ackerman, Genes for collagen types I, IV, and V are transcribed in HeLa cells but a postinitiation block prevents the accumulation of type I mRNA, Exp Cell Res, 192 (1991) 118-121.
[46] J.J. Furth, T.H. Wroth, S. Ackerman, Genes for collagen types I, IV, and V are transcribed in Hela cells but a postinitiation block prevents the accumulation of type I mRNA, Experimental cell research, 192 (1991) 118-121.
[47] C. Nazli, T.I. Ergenc, Y. Yar, H.Y. Acar, S. Kizilel, RGDS-functionalized polyethylene glycol hydrogel-coated magnetic iron oxide nanoparticles enhance specific intracellular uptake by HeLa cells, Int J Nanomedicine, 7 (2012) 1903-1920.
[48] X. Chen, D.M. Kube, M.J. Cooper, P.B. Davis, Cell surface nucleolin serves as receptor for DNA nanoparticles composed of pegylated polylysine and DNA, Molecular therapy : the journal of the American Society of Gene Therapy, 16 (2008) 333-342.
[49] L. Liu, J.K. Zhang, M.L. Li, X.H. Zhang, J.L. Zhang, Z.J. Li, L.K. Wang, J.H. Wu, C. Luo, Inhibition of HepG2 cell proliferation by ursolic acid and polysaccharides via the downregulation of cyclooxygenase-2, Mol Med Rep, 9 (2014) 2505-2511.
[50] L.F. Barros, T. Kanaseki, R. Sabirov, S. Morishima, J. Castro, C.X. Bittner, E. Maeno, Y. Ando-Akatsuka, Y. Okada, Apoptotic and necrotic blebs in epithelial cells display similar neck diameters but different kinase dependency, Cell Death Differ, 10 (2003) 687-697.

469855 Directing Vascular Regeneration *in-Situ*

Abstract:

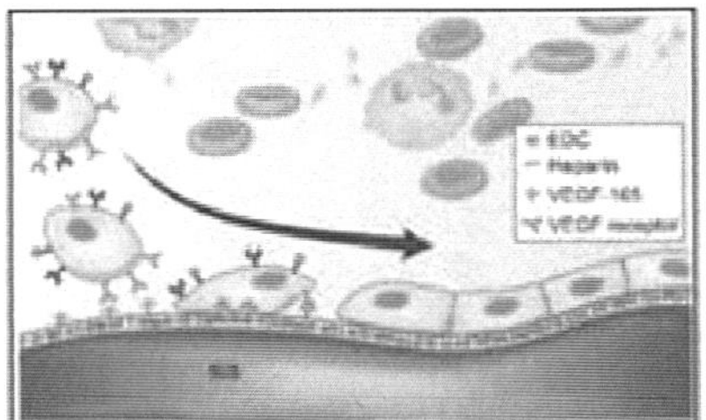

Fig. 1: Schematic indicating the use of VEGF to capture blood borne, VEGFR-expressing cells and repopulate the vascular graft lumen.

Recently our group demonstrated that immobilized VEGF can capture circulating endothelial cells from the blood *in-vitro*. Furthermore, we have demonstrated proof of concept by implanting a-cellular tissue engineered vessels (A-TEVs) comprised of SIS immobilized with heparin and vascular endothelial growth factor (VEGF) into the arterial system of sheep which remained patent (92%, n=12) for 3mo (Fig.1). Upon analysis, the lumen of these grafts was comprised of a fully functional endothelium as early as 1mo post implantation. This study sought to identify the type of cells that are captured by VEGF on the lumen of A-TEVs *in-vivo* and understand how these cells turn into an endothelial (EC) monolayer that is capable of maintaining patency *in-vivo*. A-TEV implantations were performed as previously published. *In-vivo Analysis:* Fixed explants of 1wk, 1mo, 3mo, and 6mo VEGF functionalized A-TEVs are assessed via IHC for MC and EC markers. *In-vitro Capture under Flow:* Capture of VEGFR expressing cells from blood under flow will be assessed in a microfluidic device with a flat surface comprised of chitosan, heparin, and VEGF (CHV) as previously published. *In-vitro differentiation:* Blood borne mononuclear cells that are captured on surface immobilized VEGF are coaxed to differentiate into EC with a combination of soluble and biophysical signals. A-TEVs were implanted as interpositional grafts into the arterial circulation of an ovine animal model. As early as 1mo post-implantation, the graft lumen was fully endothelialized as shown by IHC for EC markers, CD144 and eNOS. At the same time, luminal cells co-expressed leukocyte markers CD14 and CD163 (Fig. 3). To understand these results, we performed cell capture experiments under flow using microfluidic devices. Interestingly, blood mononuclear cells expressing high levels of VEGF receptors were captured on CHV surfaces with high specificity under a range of shear stresses. Initially, these cells expressed high levels of CD14 and CD16. Under the right conditions they were coaxed to differentiate into an EC phenotype as shown by expression of CD144 and eNOS (Fig.2), additional IC analysis, WB, qRT-PCR, and flow cytometry. We will also discuss the role of soluble signals and biophysical forces in transdifferentiation of blood cells into EC that maintain graft patency. We demonstrate the ability of VEGF functionalized surfaces to capture cells directly from the blood *in-vitro* and *in vivo*. In the presence of the right biochemical and biophysical signals these cells differentiate into EC like cells that maintain graft patency and vascular function. Our results shed light into the process of vascular tissue regeneration *in situ* using the body's regenerative capacity.

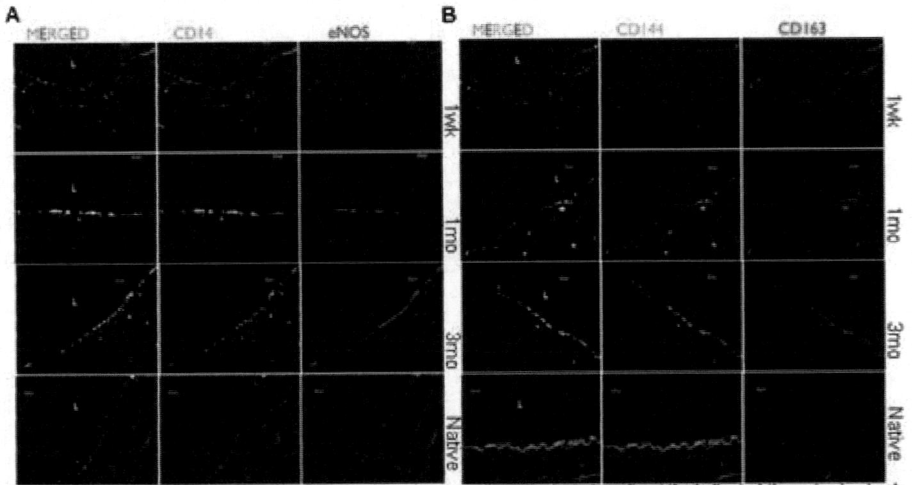

Fig. 2: VEGF functionalized A-TEVs develop a confluent endothelium. Explanted grafts at the indicated times (n=4 animals at each time point) were stained for EC markers eNOS (A) and CD144 (B) and MC/MΦ markers CD14 (A) and CD163 (B). Blue stain=DAPI. Note that the basal lamina of native arteries is highly auto-fluorescent (wavy structure under the EC monolayer).

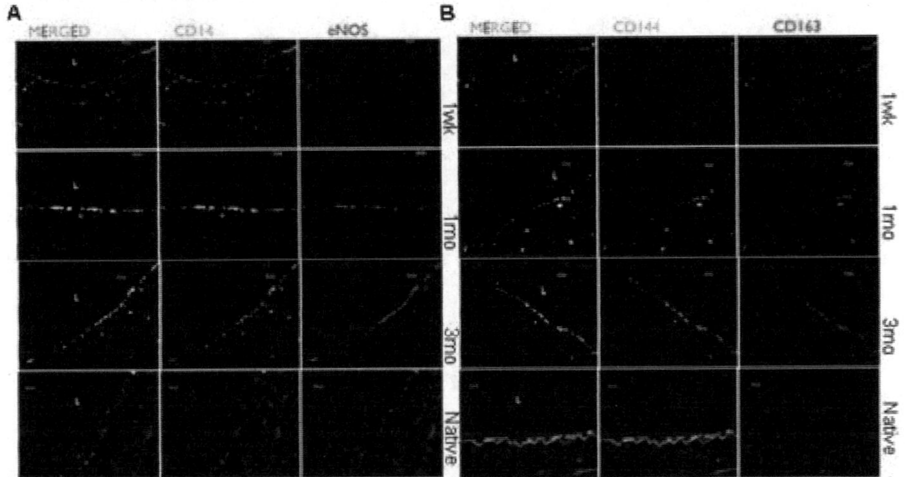

Fig. 2: VEGF functionalized A-TEVs develop a confluent endothelium. Explanted grafts at the indicated times (n=4 animals at each time point) were stained for EC markers eNOS (A) and CD144 (B) and MC/MΦ markers CD14 (A) and CD163 (B). Blue stain=DAPI. Note that the basal lamina of native arteries is highly auto-fluorescent (wavy structure under the EC monolayer).

Copyright © American Institute of Chemical Engineers. All rights reserved.

468119 Heterogeneity in Valvular Interstitial Cell Phenotype Is a Predictor of Cell Activation and Acquisition of Degenerative Properties

Abstract:

Valvular interstitial cells (VICs), the resident cells in heart valves, play major roles in valvular pathophysiology. VICs are quiescent (qVIC) in healthy heart valves and maintain tissue homeostasis. They, however, become contractile and activated (aVIC) under chemically- or mechanically-induced degenerative conditions. It is well-known that qVIC transform to aVIC on stiffer substrates, at higher passage numbers and in the presence of paracrine signals. Regardless of quiescence or activation states, investigations of culture heterogeneity can provide insights into culture likelihood for degenerative transformation. In this study, we quantify six different VIC morphologies – rhomboid, spindled, tailed, round, triangle and multi-extension. We then correlate the distributions of these six morphologies with cell activation state. The main goal of this study is to demonstrate that the distribution of different VIC morphologies changes significantly with culture conditions and those changes, in turn, correlate with the amounts of degenerative cells (aVIC).

Porcine aortic VICs from passage P2 to P4 were cultured on soft polyacrylamide gels and on plastic culture dishes. In a separate experiment, the same cells were grown in the presence or absence of high numbers of progenitor VICs (pVICs). pVICs have the ability to differentiate in culture, and potentially become either qVIC or aVIC. Cell monolayers were imaged under phase and fluorescence microscopy of actin fibers. Three cell morphology parameters were analyzed: maximum cell span, span ratio and cell area. All six distinct morphologies of VICs were represented and their relative proportions in culture were calculated. Immunoblotting of protein lysates and immunofluorescence staining of monolayers were also employed in the determination of qVIC and aVIC content in all experiments.

An increase in α-smooth muscle actin (α-SMA) expression, the main marker for aVICs, was detectable with increasing passage number, increasing substrate stiffness and upon contact with pVICs. Round, rhomboid and triangle VICs were prominent in conditions favoring aVIC transformation (later passages, stiffer substrates and pVICs signals) whereas spindled and tailed morphologies were prominent in conditions favoring VIC quiescence. Span ratio, maximum span and cell area were found to be significantly different ($p<0.5$) between each pair of conditions tested.

The substantial increase of α-SMA expression, or aVIC content, indicates that VICs can sense their environment and become activated in response to increasing matrix stiffness. The substantial increase in α-SMA due to passage number indicates that VICs become activated over time as they multiply *in vitro*. The increase in α-SMA due to pVIC contact indicates that pVIC will differentiate to aVIC phenotype due to paracrine communication. Since earlier passages and soft substrates contained more spindled and tailed morphologies, these can be associated with physiological tissue conditions. Later passages and stiffer substrates expressed more aVICs, characterized by high numbers of round, rhomboid and triangle morphologies.

Our conclusion is that characterization of these six morphologies is arguably more important for engineering of valvular tissues than the traditional qVIC and aVIC classification. Follow-up experiments in progress include 3D characterization studies of heterogeneous populations as well as isolations of these six morphological cases. These studies will lead the way to proper maintenance of spindled and tailed morphologies, guide future differentiation studies of pVICs and potentially provide unique ways to target and inhibit degenerative proteins in aVIC.

Copyright © American Institute of Chemical Engineers. All rights reserved.

470442 *A Novel Integrated Intestine-Liver-Brain Model to Investigate Ethanol Metabolism*

Abstract:

To date, *in vitro* models of the gastrointestinal (GI) tract and liver have been developed separately with very limited efforts to integrate them into a single system. The communications between the GI and liver modulate multiple metabolic functions, inflammatory responses and the biotransformation of chemicals. Ethanol-induced toxicity can have severe consequences on all three of these organs. In the liver, excessive ethanol consumption can lead to fibrosis, while significant cell death has been reported in intestines exposed to ethanol. Excessive alcohol consumption can be toxic to neurons in the brain. There is a need to understand how these organs function cohesively to metabolize ethanol. More importantly, the effects of the metabolites secreted by these organs on the brain are not fully understood.

We describe a novel integrated GI-liver model that is designed by integrating a primary rat ileum explant with a 3D organotypic multi-cellular liver model. Primary ileum tissues were surgically excised from rats. These explants were cultured in oxygenated culture medium that contained antibiotics. The ileum cultures remained stable with virtually no shedding of the intestinal epithelium. Ethanol (EtOH) was administered over 100-1000mM concentration range. These concentrations ranged from sub-lethal to highly toxic. Upon exposing the integrated ileum-liver cultures to 100mM EtOH for 24 hours, significant changes in hepatic function were observed. Albumin levels were approximately 2-fold lower in the integrated ileum-liver models when compared to liver models. A similar trend was also observed in urea production. The integrated ileum-liver models exhibited significantly lower urea levels compared to liver models alone.

EtOH administration is known to result in cell death and in the production of reactive oxygen species (ROS). Hepatocytes in the integrated cultures and liver models were imaged for ROS production upon EtOH administration. Even at a sub-lethal dose of 100mM EtOH, ROS production was significantly higher in the integrated cultures when compared to cells in the liver models alone. Interestingly, ROS production by the integrated ileum-liver culture was higher than liver models that had been treated with a combination of LPS (lipopolysaccharide) and 100mM EtOH.

Astrocyte cultures were exposed to EtOH in culture medium or spent medium obtained from the integrated ileum-liver models. The viabilities of astrocytes exposed to condition medium obtained from EtOH treated gut-liver models were approximately 2-fold lower compared to cultures that were exposed only to EtOH. These trends clearly demonstrate that the integrated gut-liver models exhibit increased sensitivity to EtOH.

Ongoing efforts are focused upon measuring the levels of inflammatory cytokines and production of reactive oxygen species when this multi-organ model (gut-liver-brain) is treated with EtOH. Integrating the GI, liver and brain will serve as a powerful platform to provide holistic information on toxicity, inflammatory response and organ crosstalk.

Copyright © American Institute of Chemical Engineers. All rights reserved.

474170 Alginate Capsule Composition Influences the Pancreatic Differentiation of Human Embryonic Stem Cells

Abstract:

Introduction: Type 1 diabetes (T1D) is an autoimmune disease affecting millions of people worldwide wherein the beta cells of the pancreas are destroyed, resulting in insulin dependence. Shortage of donor islets combined with immune rejection limits islet transplantation from becoming a viable therapy. We have shown previously that calcium alginate encapsulation for differentiation of human embryonic stem cells (hESCs) results in efficient differentiation to insulin-producing cells. The unlimited supply of hESCs and the immunoisolation capability of alginate capsules, suggest this approach as an alternative treatment for T1D. Islet encapsulation studies demonstrate that capsule composition has a significant effect on immunoisolation. However, capsule composition can modify the differentiation of encapsulated cells. Hence in this study we are evaluating the effect of material properties on pancreatic differentiation of encapsulated hESCs.

Materials and Methods: A single cell suspension of ROCK inhibitor treated hESCs were suspended in 1.1% alginate and 0.2% gelatin followed by encapsulation by drop wise addition into a divalent cation bath. hESCs were encapsulated in barium alginate, crosslinked with 10, 15, 20, 50 or 100 mM $BaCl_2$. The hESCs were differentiated entirely under encapsulation and characterized after completion of pancreatic differentiation. Encapsulated hESCs were induced to the pancreatic islet like state following a stagewise differentiation protocol. Definitive endoderm (DE) was induced with ActivinA and Wnt3Afor 4 days followed by pancreatic progenitor induction with KAAD-cyclopamine and KAAD-cyclopamine with retinoic acid for 2 days each, respectively. DE stage differentiation was characterized by gene and protein expression of SOX17, FOXA2, EOMES, and CER. Pancreatic commitment was analyzed by PDX1 gene and protein expression.

Results: Increasing the concentration of $BaCl_2$ from 10 to 100 mM $BaCl_2$ resulted in an increase in substrate stiffness from 4 to 75 kPA, as determined by AFM analysis. Viability and proliferation of encapsulated hESCs were very sensitive to capsules stiffness, with only 4-7 kPa capsules supporting cell growth and colony formation. Diffusional differences were confirmed to be minimal under the tested conditions. DE stage differentiation was found to be enhanced as capsule stiffness was increased, as shown by both gene and protein analysis of SOX17, FOXA2, EOMES, and CER. Pancreatic commitment was analyzed by PDX1 gene and protein expression. *PDX1* gene expression was highest in the 4 kPa stiffness gel (10 mM $BaCl_2$), and was significantly reduced as stiffness increased. The same trend was observed with PDX1 protein expression, which showed the highest number of cells positive for PDX1 in the 10 mM $BaCl_2$ condition.

Discussion: Our study demonstrates that capsule materials properties can significantly modulate the efficiency of chemically induced pancreatic differentiation. Specific encapsulation conditions can significantly enhance the efficiency of chemical induction. In general, higher crosslinking was found to be detrimental to the viability and differentiation of hESCs. Research is currently underway for encapsulating hESCs in a high throughput alginate array platform. This will allow for a more thorough investigation of the effect of substrates stiffness on differentiation, as well as delineate the signaling pathways which govern pancreatic differentiation. Results of these studies will be presented and discussed.

Copyright © American Institute of Chemical Engineers. All rights reserved.

464495 Smart Biomaterials to Study, Mimic, and Exploit Cellular Mechanobiology

Abstract:

Living cells are capable of processing a variety of mechanical signals encoded within their microenvironment, which can in turn act through the cellular structural machinery to regulate many fundamental behaviors. In this sense, cells may be regarded as "smart biomaterials" that dynamically and locally modulate their physical properties in response to environmental stimuli. I will discuss our recent efforts to understand and engineer these living microsystems, and to create new, bio-inspired materials that mimic sequence/structure/function relationships of cytoskeletal networks. Key areas of emphasis will include: (1) Applying material and genetic strategies to probe the timing and molecular mechanisms of stem cell mechanobiology; (2) Using single-cell micropatterning and femtosecond laser ablation to precisely quantify the microscale mechanics of actomyosin stress fibers, and (3) Engineering stimulus-sensitive intrinsically disordered protein brushes based on neuronal cytoskeletal networks.

Copyright © American Institute of Chemical Engineers. All rights reserved.

464453 Podocalyxin Knockdown Impairs Migration of Pancreatic Cancer Cells By Influencing Cytoskeletal Remodeling and Surface Adhesion

Abstract:

Podocalyxin Knockdown Impairs Migration of Pancreatic Cancer Cells by Influencing Cytoskeletal Remodeling and Surface Adhesion

Bin Sheng Wong, Daniel J Shea, Robert A Law, Konstantinos Konstantopoulos

Department of Chemical and Biomolecular Engineering, Johns Hopkins University

Introduction: Podocalyxin (PODXL) is a transmembrane sialomucin glycoprotein normally expressed in kidney podocytes[1]. While typically absent in normal pancreas, PODXL is often overexpressed in invasive pancreatic cancers and is associated with more aggressive tumor phenotypes[2]. Overexpression of PODXL has also been correlated to poor clinical prognosis of other cancers from breast[3], colorectal[4], renal[5] and bladder[6] origins. In this study, we aimed to investigate the effects of PODXL on the morphology, morphodynamics, migration and cytoskeletal dynamics of pancreatic cancer, and to elucidate the mechanism in which PODXL exerts its functions.

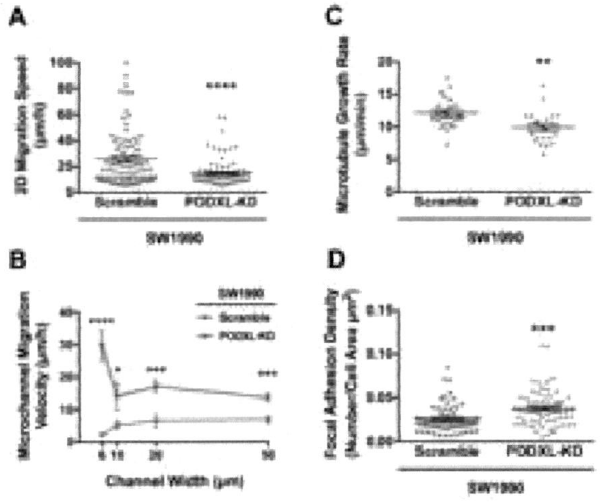

Materials & Methods: Stable clones of metastatic pancreatic adenocarcinoma SW1990 with reduced PODXL expression were generated using small hairpin RNA[2]. Knockdown (KD) efficiency was verified using western blot, flow cytometry and immunofluorescence staining. Random 2D and chemotactic-driven migration of PODXL scramble control (SC) and PODXL-KD SW1990 were assessed on collagen-I coated glass coverslips and microfluidic devices with microchannels of varying width respectively using time-lapse live microscopy. The cellsÕ migration, morphology and morphodynamic parameters were quantified using imageJ and a custom-written MATLAB program. Fluorescence recovery after photobleaching was conducted on cells transfected with LifeAct-GFP to quantify actin dynamics. Microtubule growth rates were examined via time-lapse confocal microscopy in cells transfected with end-binding protein 1-GFP. Cells were immunostained for phosphorylated paxillin and imaged with total internal reflection florescence microscopy to quantify focal adhesion size and density.

Figure 1: PODXL knockdown in SW1990 decreases their migration speed on 2D collagen-I coated glass surfaces (A) and reduces their velocity in microchannels across all channel widths (B). PODXL-KD slows down microtubule growth and dynamics (C). TIRF microscopy revealed diminished focal adhesion density in PODXL KD compared to SC (D). *^{2}0.05, **^{2}0.01, ***^{2}0.001, ****^{2}0.0001 by t-test.

Figure 1: PODXL knockdown in SW1990 decreases their migration speed on 2D collagen-I coated glass surfaces (A) and reduces their velocity in microchannels across all channel widths (B). PODXL-KD slows down microtubule growth and dynamics (C). TIRF microscopy revealed diminished focal adhesion density in PODXL KD compared to SC (D). *≤0.05, **≤0.01, ***≤0.001, ****≤0.0001 by t-test.

Results & Discussions: PODXL was successfully knocked down (>90%) in SW1990 cells, resulting in decreased PODXL surface coverage and membranous localization. Migration of PODXL-KD SW1990 is significantly reduced compared to the SC on both 2D (**Fig. 1A**) and in microchannels of all widths (**Fig. 1B**). PODXL-KD cells are generally smaller, less elongated, more circular, less protrusive and exhibit much slower morphodynamics compared to the SC. While displaying similar actin dynamics, PODXL knockdown significantly slows down the rate of microtubule growth (**Fig. 1C**), thereby contributing to diminished morphodynamics and reduced migration. Furthermore, PODXL-KD cells possess significantly higher focal adhesion density compared to the SC despite having comparable focal adhesion size (**Fig. 1D**). Elevated focal adhesion density may lead to increased traction forces in the PODXL-KD cells which enhances their adhesion to the surfaces, thereby impeding efficient migration.

Conclusion: PODXL knockdown markedly impairs migration of metastatic pancreatic cancer cells SW1990 on collagen-I coated 2D surfaces and in microchannels. These migratory deficiencies could be a result of interplay between reduced cellular morphodynamics, slower microtubule growth and higher focal adhesion density. Co-immunoprecipipation and mass spectroscopy assays failed to identify enrichment of ezrin, a widely reported binding partner of PODXL, in the immunoprecipitates of PODXL in pancreatic cancer cells. Experiments are underway to verify a novel cytoplasmic binding partner of PODXL that we have identified in pancreatic cancer cells and to uncover the exact molecular mechanisms by which PODXL exerts its pro-migratory effects.

References: [1]Kerjaschki, D., et al. (1984) J Cell Biol 98(4): 1591-1596. [2]Dallas, M. R., et al. (2012). Am J Physiol Cell Physiol 303(6): C616-624. [3]Somasiri, A., et al. (2004). Cancer Res 64(15): 5068-5073. [4]Larsson, A., et al. (2011). Br J Cancer 105(5): 666-672. [5]Hsu, Y. H., et al. (2010). Am J Pathol 176(6): 3050-3061. [6]Boman, K., et al. (2013). Br J Cancer 108(11): 2321-2328.

Copyright © American Institute of Chemical Engineers. All rights reserved.

470668 Palmitate Induces the Loss of Desmosomes through Endoplasmic Reticulum (ER)-Stress to Promote Cancer Progression

Abstract:

Long chain free fatty acids (FFAs), especially palmitate (PA), a saturated FFA, is known to induce endoplasmic reticulum (ER) stress in many types of cells and contribute to diseases, including cancers. ER stress is highly induced in various tumors and is closely associated with cancer cell survival and resistance to anti-cancer treatments. PA, abundant in high-fat diets, is believed to contribute to or increase the risk of cancers. Thus it was surprising that there was no direct evidence to support that elevated FFA levels are linked to the progression of cancer, i.e. epithelial to mesenchymal transition (EMT) and tumor metastasis, until now [Nath et al, 2015]. We recently found hepatocellular carcinoma (HCC) progression via induction of EMT is closely associated with elevated levels of FFA [Nath et al, 2015]. In addition to exogenous FAs from the diet, it is well known that FA synthesis is upregulated in cancer [Currie et al, 2013], thereby providing another source of elevating FFA levels. Currently the mechanism by which elevated levels of fatty acids promote the progression of cancers is unclear.

One of the first steps of EMT is the disassembly of epithelial cell-cell contacts (e.g. adherens junction, desmosomes, etc.), and the loss of cell polarity. Previously, through an integrative analysis (computational and experimental), we identified that the desmoplakin (DSP) protein is perturbed by PA[Wang et al, 2011]. The DSP protein is an obligate component of functional desmosomes; intercellular junctions responsible for cell-cell adhesion. DSP together with plakoglobin and plakophilin forms the intracellular desmosomal plaque, which also serves as an anchor for keratin intermediate filament attachment. Reduced DSP protein expression can lead to loss of cell-cell adhesion through desmosomes, disruption of the keratin intermediate filament network, and has been observed in liver carcinoma and in cell lines established from metastatic patient tumor samples. A reduction in desmosomes has been observed in cells undergoing EMT, whereas during the reversal of mesenchymal to epithelial transition desmosomes were restored. While loss of desmosomes and DSP has been reported in HCC [Wang et al, 2011], the role of FFAs in modulating DSP expression and the role of DSP in EMT and metastasis is unknown. Our research focuses on uncovering the mechanism by which elevated saturated fatty acids induce the loss of cell adhesion and enhance migration, mediated through IRE1Î± protein and ER stress. In contrast to prior studies which do not explain how PA induces ER-stress and the unfolded protein response (UPR) signaling pathways, we pursued a novel line of investigation, and assessed whether PA interacts directly with the kinase proteins in the UPR (specifically IRE1Î±) to activate ER stress and its downstream signaling to induce EMT mediated by the loss of DSP.

References:

Currie E, Schulze A, Zechner R, Walther T, Farese R(2013) Cellular Fatty Acid Metabolism and Cancer. Cell Metabolism 18(2): 153-161. doi:10.1016/j.cmet.2013.05.017

Nath A, Li I, Roberts L, Chan C (2015) Elevated Free Fatty Acid Uptake Via CD36 Promotes Epithelial-Mesenchymal Transition In Hepatocellular Carcinoma. Scientific Reports. 5:14752. doi:10.1038/srep14752

Wang X, Nath A, Yang X, Portis A, Walton SP, Chan C (2011) Synergy Analysis Reveals Association between Insulin Signaling and Desmoplakin Expression in Palmitate Treated HepG2 Cells. PLoS ONE 6(11): e28138. doi:10.1371/journal.pone.0028138

Copyright © American Institute of Chemical Engineers. All rights reserved.

468319 Novel Role of Cadherin 11 in Extracellular Matrix Synthesis and Muscular Physiology

Abstract:

Cell-cell adhesion through cadherins has been identified as an important player in stem cell fate decisions. Recently, our lab study has recognized that Cadherin 11 (CDH11) is intimately involved in the process of MSC to smooth muscle cell (SMC) differentiation, and further that engagement of cadherin 11 is a significant event in the regulation of extra cellular matrix (ECM) synthesis. Here we report on the mechanism of CDH11 mediated ECM synthesis and furthermore, we provide in vivo data supporting our in vitro results and link our findings on cellular differentiation to tissue organization, muscle function and neuromuscular physiology. Aorta, bladder and skin were obtained from 8-week old mice (*Cdh11*$^{-/-}$ and WT-wildtype) and tested for mechanical properties. shRNA targeting CDH11 was used in human dermal fibroblasts to study the effect of CDH11 deficiency. An engineered surface immobilized with a fusion protein containing extracellular CDH11 with the Fc domain (CDH11-Fc) was used as a tool to study immediate downstream effects and identify the mechanism of ECM production following CDH11 engagement. To study the effect of CDH11 on muscle function *in-vivo*, 6-month-old male *Cdh11*$^{-/-}$ and WT (n=12) were acclimated to assessment protocols, and then assessed bi-monthly over a 6-month period. Assessments includes an inverted grid hanging test and overnight activity wheel monitoring. Body composition (total body fat and lean mass) was also measured by quantitative magnetic resonance.

Â We demonstrate that CDH11 regulates collagen and elastin synthesis and thereby contributes to tissue integrity both *in vivo* using a *Cdh11*$^{-/-}$ mouse model and *in vitro* using knockdown studies as well as gain of function approaches. We also identified the underlying mechanism and discovered that CDH11 engagement leads to swift and direct activation of both the TGF-Î²1 and ROCK pathways. This is followed by activation of transcription factors, MRTF-A and MYOCD leading to expression of collagen and elastin genes. In summary, our findings indicate that cell-cell contact by cadherin-11 is a novel regulator of ECM synthesis. Notably, these findings have important implications in skeletal muscle function and overall physical strength of the mice. In tests of physical endurance, *Cdh11*$^{-/-}$ mice failed to maintain their body position in an inverted grid hanging test, failing in just 10 seconds as compared to WT (1-3 minutes). Furthermore, *Cdh11*$^{-/-}$ mice have little to no activity wheel usage in contrast to control mice. Finally, our data also indicate that *Cdh11*$^{-/-}$ mice have significantly higher body fat relative to WT as well as lower lean mass.

Â Overall, we discovered a novel role for Cadherin-11 in ECM synthesis and tissue structure and identified the underlying mechanism involving direct TGF-Î² and ROCK activation. Notably, CDH11-mediated molecular events are linked to changes in muscle mass and organization, ultimately affecting neuromuscular function and physiology.

Copyright © American Institute of Chemical Engineers. All rights reserved.

472281 Spatially Separated Chemokine Gradients and Their Effects on Single Cell Migration

Abstract:

A variety of physiological and pathophysiological processes including immune responses, metastasis, and stem cell differentiation are governed by an intricate network of endogenously produced cytokines. *In vivo*, these signals are often present as gradients permeating through the three-dimensional (3D) tissue and their concentrations as well as spatial diffusion are critical in providing adequate guidance for directional cell migration. Several *in vitro* approaches such as Boyden chambers or microfluidic platforms are well established for quantitative as well as qualitative analysis of cell motility; however, require a predetermined optimum chemokine concentration for each cell type. Additionally, these devices are prone to temporally diminishing gradient or provide chemotactic observations in response to a fixed slope of the gradient limiting the ability to recreate *in vivo* like conditions. To date, there is no efficient method to determine the effects of gradient concentrations (e.g. chemotaxis vs fugetaxis) on three dimensional cellular migration. Here, we demonstrate a novel microfluidic strategy that allows creation of gradients with varying slopes in the y-direction as a result of orthogonal gradient in the x-direction. This approach allows real-time observations of transendothelial migration (TEM) and chemotaxis under several different levels of chemotactic stimulation. The multifold advantages of the proposed strategy include the ability to: 1) observe cytokine activated stimulation of endothelial cells in response to different chemokine concentrations and the subsequent effects on homing of immune/cancer cells, 2) identify the optimum chemoattractant concentration governing chemotaxts and TEM, 3) correlate effects of the slope of gradient on cell motility, and 4) observe real-time migration behaviors through a 3D ECM environment at the single cell level. More interestingly, the orthogonal gradient technique allows the possibility of investigation chemotaxis and chemorepulsion (fugetaxis) in a single platform. Using this device, we exposed metastatic breast cancer cells to a known chemoattractant, stromal derived factor 1Î± (SDF-1Î±). We show that the tumor cells only extravasate and migrate towards gradients that are generated with a maximum concentration of 350-300 ng/mL, which closely resembles previously reported concentrations. This device is versatile and may be

implemented in a variety of studies, including leukocyte recruitment, stem cell differentiation and migration, and pharmaceutical screening.

Copyright © American Institute of Chemical Engineers. All rights reserved.

469577 Single Integrin Force Measurements Reveal the Distribution of Forces Exerted By Living Cells

Abstract:

Integrins are heterodimeric transmembrane proteins that mediate cell attachment to the extracellular matrix (ECM) and play key roles in interpreting mechanical cues from a cell's environment. In particular, mechanical forces sensed and transmitted through integrin-ligand interactions are a critical aspect of integrin-based signaling that affects migration, proliferation, and differentiation. However, the forces that individual integrins experience in living cells have yet to be fully characterized. Here, we use Förster resonance energy transfer (FRET)-based molecular tension sensors (MTSs) to measure the distribution of mechanical forces exerted by individual integrin heterodimers in living cells at the single-molecule level. We engineered two MTS variants: MTS_{GPGGA}, to probe a low force range (0-8 pN), and MTS_{HPst}, for examining intermediate forces (8-12 pN). We find that a large fraction of integrins exert modest forces of <12 pN, even within defined integrin adhesion complexes. This finding is intriguing given that the load-bearing capacity of integrins has been reported to be on the order of tens of piconewtons. For integrins specifically within adhesion complexes, we see that the measured force per integrin falls into distinct subpopulations. These subpopulations may correspond to specific adhesion structures or modes of attachment, and may be crucial to understanding how mechanical forces activate downstream signaling. In ongoing work, we take advantage of the unique capabilities of MTSs to determine how subpopulations of load-bearing integrins are altered in response to external mechanical perturbations, and how (and whether) distinct integrin subtypes bear differing levels of mechanical load.

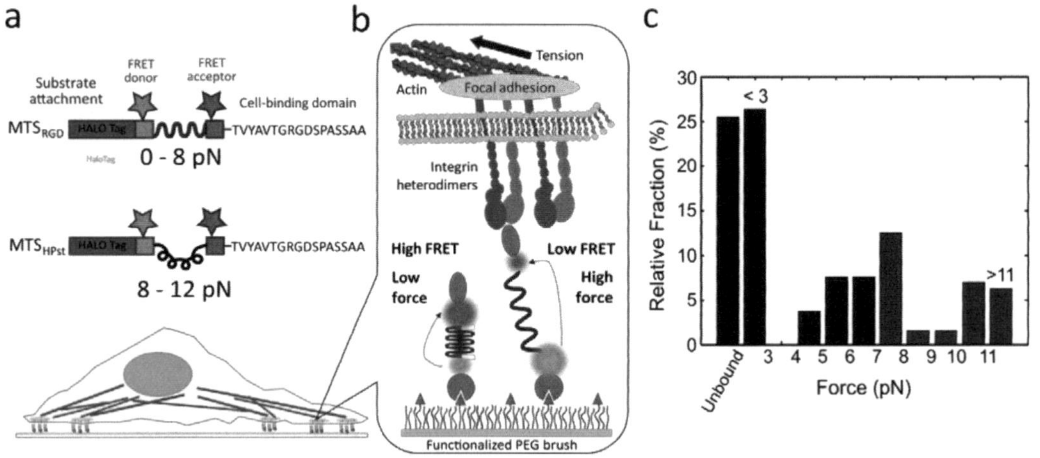

Figure 1. (a) Molecular tension sensors (MTSs) present an RGD-based cell-binding domain and use different force-sensitive linkers flanked by organic fluorophores. MTS_{RGD} measures forces in the 0 to 8 pN range, while MTS_{HPst} measures forces in the 8 to 12 pN range. (b) MTSs are attached to a functionalized PEG coverslip. Integrin heterodimers bind and exert force on the cell-binding domain, stretching the force-sensitive domain and decreasing the FRET efficiency. (c) Relative fraction of molecules measured under adhesions in live cells as measured from MTS_{RGD} (Blue) and MTS_{HPst} (Red).

Copyright © American Institute of Chemical Engineers. All rights reserved.

472486 Cell-Cell Contact Promotes Tgfβ1-Induced Epithelial-Myofibroblast Transition Via Notch1

Abstract:

Myofibroblasts aid in wound healing and upon aberrant and chronic activation they can contribute to the development of pathological conditions including fibrosis and cancer. Myofibroblasts can develop from epithelial cells through an epithelial-mesenchymal transition (EMT), a process in which epithelial cells lose cell-cell adhesion, exhibit morphological changes, and upregulate the expression of cytoskeletal proteins including alpha smooth muscle actin (αSMA). Previous studies have suggested that complete disruption of cell-cell contacts is necessary for induction of EMT by transforming growth factor (TGF)-β1. However, it is not clear how partial cell-cell contacts (such as those experienced by cells along a wound edge) impact EMT progression. Here, we employed a microfabrication approach to decouple the effects of cell-cell contact and cell-matrix adhesion in TGFβ1 induction of EMT. We find that when cells are permitted to spread, the presence of cell-cell contacts enhances myofibroblast development from epithelial cells when compared to single cells lacking neighbors. Furthermore, we find that cell spreading and intercellular contacts together control the activation and subcellular localization of Notch1. Blocking cell spreading abrogates TGFβ1-induced Notch1 activation and expression of EMT markers. Moreover, pharmacological inhibition of Notch activation and knockdown of Notch1 abated cell-cell contact mediated αSMA expression. These studies provide insight into how cell-cell and cell-matrix adhesive cues together contribute to the development of myofibroblasts along wound edges and may suggest approaches to promote wound healing or to prevent fibrosis and cancer.

Copyright © American Institute of Chemical Engineers. All rights reserved.

473148 Learning about Ligand-Receptor Interactions, Pollymers and Biological Recognition through Surface Forces

Abstract:

Much experimental and theoretical effort has been directed at elucidating the forces and dynamics of breaking, rupturing or unbinding ligand–receptor bonds, e.g., measuring the bond strength as a function of pulling rate or time using a variety of techniques. In contrast, much less attention has been given to the formation of ligand–receptor bonds. Using a combination of theory (analytical and simulations) and direct experiments (using the Surface Forces Apparatus technique) of a polymer tethered ligand receptor system, the role of tether dynamics, ligand–receptor binding strength, tethered architecture, and grafting density in determining the spatial range, capture time, and subsequent adhesion of complementary surfaces will be explored. In addition, the impact of the tethering chains' nonspecific repulsive interactions with model membranes, adhesion under flow, and implications for targeting polymer–coated particles towards biomimetic surfaces and cells will be discussed. Methods for tailoring specific adhesion without altering the ligand–receptor bond strength will be highlighted.

Copyright © American Institute of Chemical Engineers. All rights reserved.

470461 Investigation of the Dfferences in Material Properties Between Amorphous Dispersions Created By Conventional Methods and High-Shear Co-Precipitation

Abstract:

High-shear co-precipitation is a technique for the manufacturing of API-polymer amorphous solid dispersions, and is an alternative to the typical spray-drying and hot-melt extrusion processes. The technique relies on the rapid contacting of a solution of polymer and API with a common anti-solvent, resulting in a stable suspension where the precipitated amorphous solid dispersion is below its glass transition temperature.

In this work we present the characterization and analysis of a co-precipitated solid dispersion exhibiting significant differences in properties compared to equivalent dispersions, prepared via spray-drying and hot-melt extrusion. These differences extend across the full spectrum of "fundamental" characteristics, including morphology, true density, glass transition temperature and enthalpy, as well as indirect, "performance-based" properties, such as compactability and dissolution. A large collection of "orthogonal" techniques is brought to bear in the analysis of the material and the elucidation of the physical processes underlying these differences in properties.

Copyright © American Institute of Chemical Engineers. All rights reserved.

469250 Melt Rheology of Cellulose Ethers Designed for Hot Melt Extrusion

Abstract:

Melt Rheology of Cellulose Ethers Designed for Hot Melt Extrusion

Authors: Tirtha Chatterjee, Kevin O'Donnell, Yongfu Li, Uma Shrestha, Robert L. Sammler, Shrikant Khot

Formulation of new drug entities is increasingly challenging due to poor solubility of the drug and limited polymers that can produce amorphous solid dispersions (ASDs). One of the leading technologies to manufacture ASDs is hot melt extrusion (HME) in which the solid or melted drug is mixed with a molten polymeric carrier. In order to be processed, the polymeric carrier must have adequate melt rheology at temperatures acceptable for pharmaceutical applications. The AFFINISOL™ cellulosic polymer family has been engineered to have a melt viscosity that enables their use as excipients in HME.

In order to design a successful HME process the formulator must understand the excipient melt rheology as a function of operating conditions. In the present study, the melt rheology of AFFINISOL™ HPMC HME (hydroxypropyl methylcellulose) and HPMCAS HME (hydroxypropyl methylcellulose acetate succinate) is presented. Traditionally the melt rheology is characterized using a rotational rheometer, however, this was found to be inadequate for this family of polymers. To overcome this challenge an alternative method was adapted allowing rheological analysis directly in line with a twin screw extruder. This enables analysis under conditions identical to those a formulator would experience. Melt viscosity of the cellulose ethers was found to be dependent upon temperature and shear rate with all polymers exhibiting a power law shear thinning behavior. The rheology of select formulations containing model active pharmaceutical ingredients will also be presented.

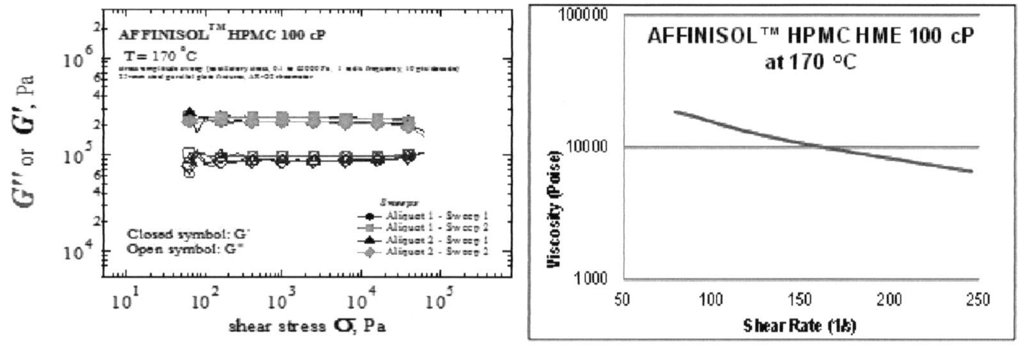

Figure 1: (Left) Rotational rheology of AFFINISOL™ HPMC HME at 170 °C exhibiting solid like behavior despite ability to extrude at temperatures well below this and (Right) alternative in-line method demonstrating melt viscosity as a function of shear rate at 170 °C.

Copyright © American Institute of Chemical Engineers. All rights reserved.

471573 The Water Sorption Behaviour of Amorphous Solids: Experimental and Theoretical Insights Using a Model Protein BSA

Abstract:

This paper considers the water sorption behaviour of complex amorphous solids using advanced experimental investigations as well as a number of thermodynamics models for the formation of solid- solute solutions. The water sorption behaviour of amorphous materials is especially complex kinetically as the loss or gain of water vapour by the solid is controlled by two nominally independent kinetic phenomena. The first is the diffusion kinetics of moisture sorption and the second factor is the relaxation kinetics of the glassy amorphous state.

Experimental evidence for a glassyâˆ'rubbery hydration mechanism of bovine serum albumin (BSA) is reported where BSA is used as an amorphous model protein substrate. Experimental results corroborate that the hydration mechanism in amorphous protein materials are more complex than anticipated. We have experimental evidence indicating that BSA can be experimentally trapped into distinct conformational states during water desorption. In considering the dimension of time during hydration or dehydration of a protein, a significant problem in determining and interpreting a water sorption isotherm is posed by the â€˜sample historyâ€™, a time effect, of a lyophilised amorphous biological products. Its influence, both prior to the sorption experiment as well as produced during the actual measurement, has not been previously clarified.

A water vapour sorption experiment is typically carried out in one of the three modes: (a) interval sorption,where the solvent activity is subsequently raised in a stepwise fashion through one continuous measurement; (b) integral sorption, where the solvent activity is raised from zero up to one specific value and the mass uptake is recorded (typically with a new specimen for each measurement); or (c) differential sorption, where an activity ramp is applied in one continuous measurement. It should be understood that each of these three sorption modes creates a different experiment-induced time dependent behaviour. This paper will provide recommendations in best practise for

determined water sorption properties of amorphous solids.

The final section of this paper will introduce modern equilibrium and non-equilibrium thermodynamics models for analysing water sorption isotherm behaviour for amorphous solids.

Copyright © American Institute of Chemical Engineers. All rights reserved.

469149 A Kinetic Approach to Polymorph Prediction:Â Polymorph Specific Alpha and Beta Glycine Nucleation Rates and Solubility Curves from Molecular Dynamics

Abstract:

Predicting polymorphism, the ability of a molecule to self-assemble into multiple solid-state crystalline forms, continues to elude both experimentalists and molecular modelers alike. Landscape energy minimization techniques, although elegant in their own right, fail to include standard design crystallization variables such as solvent, temperature, or pressure. Furthermore, the effect of crystal size, a critical variable for molecules that display size dependent polymorphism, remains unaddressed in these optimization algorithms. Molecular dynamics (MD), which encapsulates the aforementioned kinetic design variables naturally, has suffered from time scale challenges associated with activated rate processes, such as nucleation. However, with the advent of modern supercomputers, and HPC hardware such as xeon phi coprocessors, these types of simulations are just now becoming within reach of computational scientists. To date, only a handful of attempts to determine the effects of the nucleation pathway on polymorphism of organic molecules have been performed using MD .

In this work, a first principles molecular dynamics approach to the study of homogeneous nucleation rates and solubility of polymorphs is presented for both alpha and beta glycine in water as a model system. To this end, a rare-event molecular dynamics simulation methodology is used in combination with a newly derived stochastic model for one-dimensional nucleation barrier crossings. This works seeks to assess the degree to which polymorph specific nucleation kinetics and solubility, both at the nanoscale and bulk scale (>1000nm), can be used for polymorph prediction. Bulk solubility curves over a 45 K temperature range are predicted directly from molecular dynamics. Both the trend-line estimated and MD predicted alpha glycine solubility values are within 5% to 11% agreement with experimental values. Beta glycine solubility estimation is achieved purely by MD prediction, as the solubility of this polymorph is not measurable in water due to rapid recrystallization kinetics. From the solubility calculation, alpha glycine is shown to be less soluble than beta glycine over the entire 45K temperature window at bulk length sales. This is in agreement with experimental evidence that shows the presence of alpha glycine at equilibrium in aqueous solvent for micron-sized particles. However, it is also shown that beta glycine is less soluble than alpha glycine at the nanoscale. Furthermore, consistent with Ostwaldâ€™s rule of stages, the nucleation kinetics show a strong preference for the formation of beta glycine at ~2nm length scales, due to beta glycine critical clusters being ~40 kcal/mol more stable than alpha glycine. This yields the prediction of beta formation at these length scales as well as the size dependent polymorphism of glycine.

Copyright © American Institute of Chemical Engineers. All rights reserved.

465879 Silica Nanoparticle-Seeded Microfluidic Emulsion Crystalisation - Crystalline Drug Composite Microparticles

Abstract:

We have recently demonstrated the fabrication of co-formulated drug-(polymeric) excipient microparticles with exquisite control over particle size, shape and polymorphism, using microfluidic emulsion crystallization and model hydrophobic drugs such as ROY and Carbamazepine [1]. In some cases, the co-formulation of polymeric excipients and drugs, depending on their mutual interactions, may lead to problems of drug-polymer miscibility and result in amorphous solid dispersions. [2-3] These amorphous solid dispersions may present storage problems, as they can crystallize over time. Here, we demonstrate the use of colloidal silica particles as excipients in microfluidic emulsion crystallization of a model hydrophilic drug. Silica nanoparticles, conventionally incorporated in solid dosage forms at the later stages of formulation, function as glidants or fillers. In contrast, in our work, the silica particles act as heterogeneous seeds for nucleation of the drug, thereby drastically reducing crystallization times, while also acting as an excipient that improve flowability and enable excellent crystallinity. Another notable advantage of using colloidal silica is its general applicability to a wide range of drugs, since the silica particles can easily be functionalized [4]. We envision significant potential for drug-silica composites prepared by microfluidic emulsion crystallisation, which allow rapid crystallization of a variety of drugs and their direct formulation into spherical microparticles without compromising the crystallinity of the drug product.

Word count:213

CONTACT * S. A. Khan, Tel: +65-6516-5133; saifkhan@nus.edu.sg

[1] Leon, R. A. L.; Badruddoza, A. Z. M.; Zheng, L.; Yeap, E. W. Q.; Toldy, A. I.; Wong, K. Y.; Hatton, T. A.; Khan, S. A., Highly Selective, Kinetically Driven Polymorphic Selection in Microfluidic Emulsion-Based Crystallization and Formulation. *Crystal Growth & Design* **2015,** *15*(1), 212-218.

[2]Huang, Y.; Dai, W. G., Fundamental aspects of solid dispersion technology for poorly soluble drugs. *Acta Pharm Sin B* 2014, *4*(1), 18-25.

[3] Marsac, P. J.; Shamblin, S. L.; Taylor, L. S., Theoretical and practical approaches for prediction of drug-polymer miscibility and solubility. *Pharm Res* 2006, *23*(10), 2417-26.

[4] Md. Badruddoza, A. Z.; Toldy, A. I.; Hatton, T. A.; Khan, S. A., Functionalized Silica Nanoparticles as Additives for Polymorphic Control in Emulsion-Based Crystallization of Glycine. *Crystal Growth & Design* **2013,** *13* (6), 2455-2461.

Copyright © American Institute of Chemical Engineers. All rights reserved.

461527 The Effects of Crystal Size and Habit on the Dehydration Induced Polymorphism; The Case of Carbamazepine Dihydrate

Abstract:

The process conditions influence the kinetics and thermodynamics during drying of hydrated materials. The water removal from hydrated crystals induces mechanical stress on the crystal lattice leading to structural deformations. Concurrently, the nucleation of the anhydrous form alters material's properties[1]. Thus, material properties influence dehydration, but also they are influenced by it. This work examines the role of interfacial phenomena of hydrated materials, using carbamazepine dihydrate as model compound.

Three methods were used for the crystallization of carbamazepine dihydrate:

1. 0.075 g*mL^{-1} of anhydrous carbamazepine (as received) were dissolved under heating in a methanol-water (60:40 %v/v) mixture. Then the mixture was left to cool down for 12 hours under stirring (500 rpm). The solution was filtered out and the crystals were collected[2].
2. 0.05 g*mL^{-1} of anhydrous carbamazepine (as received) were dissolved under heating in a methanol-water (50:50 %v/v) mixture. Then the mixture was left to cool down for 12 hours under stirring (500 rpm). The solution was filtered out and the crystals were collected.
3. 0.025 g*mL^{-1} of anhydrous carbamazepine (as received) were dissolved under heating in a methanol-water (40:60 %v/v) mixture. Then the mixture was left to cool down for 12 hours under stirring (500 rpm). The solution was filtered out and the crystals were collected.

X-Ray powder diffraction (XRPD) patterns agreed that all methods result to carbamazepine dihydrate crystals, however optical examination of the crystals, using Scanning Electron Microscope (SEM), revealed differences in their crystal size and habits. Crystals from methods one are smaller in size and they have a smaller aspect ratio, leading to a more compact shape and better flowability and cohesion behaviour, compared with the crystals obtained from the other two methods. It is important, for later on, to mention that the dominant face in the crystals prepared via methods two and three facet (0 2 0) is by far the dominant one.

The three types of carbamazepine dihydrate were exposed in dehydration conditions of 30, 40, 60 and 80°C for 24 hours and the anhydrous material was examined with XRPD and SEM. The material from methods two and three dehydrated to the Triclinic polymorph, the less stable carbamazepine polymorph[3]. The crystals obtained from method one dehydrated concomitantly to the P-Monoclinic and Triclinic polymorphs. Increasing temperature leads to increasing amount of the Triclinic polymorph, as shown from the SEM images. The domination of the Triclinic polymorph, in the high aspect ratio crystals is associated with the formation of well-ordered macroscopic cracks perpendicular to the (0 2 0) facet of the crystal, followed by the growth of Triclinic carbamazepine whiskers. The high aspect ratio of the crystals obtained from methods two and three appears to be quite attractive for crack nucleation and propagation. Triclinic carbamazepine whiskers grow in needle-shaped structures with high surface area to volume ratio to enhance the dissipation of heat from the crystals. In the, smaller and more compact, particles obtained from method one smaller cracks, formed randomly on the crystal facets, seem to be sufficient to dissipate the heat from the system without the need of the whiskers. The smaller size of the particles, probably, contributes to better heat dissipation. From the above, it can be concluded that energy dissipation drives dehydration induced polymorphism in this system.

Material from method one was placed in a glass vial with six milling balls (5 mm diameter)[4-6]left to roll for 12 hours. XRPD examination of the material obtained did not reveal dehydration of the crystals, although the SEM images showed that defects were formed on the crystals. The grinded material was subjected to the aforementioned drying conditions, however the trend in the dehydration-induced polymorphism did not change dramatically, the dried material was again a mixture of the two polymorphs, P-Monoclinic and Triclinic. This highlights the importance of powder's bulk properties (size and habit) over surface properties in the determination of drying induced polymorphism. The effects of stirring were also examined in the context of method one. It was shown that low Reynolds number favors the formation of high aspect ratio dihydrates which give a Triclinic polymorph upon dehydration. Using the same liquid composition as in method one the effect of supersaturation on crystal habit were examined. It was shown that low supersaturation, also, leads to the formation of high aspect ratio dihydrates.

FD-IGC measurements were performed on samples from method 1 dried at the four different temperatures. The shift from lower to higher surface energies is correlated with the increasing amount of Triclinic material observed with SEM. Computational models and FD-IGC measurements on pure P-Monoclinic and Triclinic can be combined to give approximate values of the polymorphic content of each sample.

This work, establishes pathways to obtain pure Triclinic carbamazepine and P-Monoclinic carbamazepine via the dehydration carbamazepine dehydrate. It critically revisits previous studies suggesting that the dehydration follows the Ostwald[2] rule of stages. A mechanism for the dehydration-induced polymorphism, suggesting a concomitant nucleation scheme is proposed. According to this mechanism crystal size and habit determine the dehydration induced polymorphism. The mechanism is driven by the energy dissipation mechanism of the system. This drying mechanism can be mechanistically understood using notions similar to those employed by Machon to describe the influence of defect concentration of nanoparticle's phase transitions using the one dimensional Ginzburg-Landau equation[7,8]. It is, also, demonstrated that FD-IGC measurements can be used to distinguish the polymorphic content of the dried material.

Disclosure:

Imperial College London and AbbVie jointly participated in study design, research, data collection, analysis and interpretation of data, writing, reviewing, and approving the publication. Eftychios Hadjittofis is a graduate student at Imperial College London; Jerry Y. Y. Heng is a professor at Imperial College London. They all have no additional conflicts of interest to report. Geoff G. Z. Zhang is an employee of AbbVie and may own AbbVie stock.

References:

1. Galwey AK. The topochemistry of the reaction interface in decomposition processes in solids. *Reactivity of Solids*. 1990;8(3–4):211-230.

2. Khoo JY, Shah UV, Schaepertoens M, Williams DR, Heng JYY. Process-induced phase transformation of carbamazepine dihydrate to its polymorphic anhydrates. *Powder Technol*. 2013;236:114-121.

3. Grzesiak AL, Lang M, Kibum K, Matzger AJ. Comparison of the four anhydrous polymorphs of carbamazepine and the crystal structure of form I. *Journal of Pharmaceutical Sciences*. 2003;92(11):2260-2271.

4. Wang Z, Shah UV, Olusanmi D, et al. Measuring the sticking of mefenamic acid powders on stainless steel surface. *Int J Pharm*. 2015;496(2):407-413.

5. Shah UV, Olusanmi D, Narang AS, et al. Effect of crystal habits on the surface energy and cohesion of crystalline powders. *Int J Pharm*. 2014;472(1–2):140-147.

6. Shah UV, Olusanmi D, Narang AS, Hussain MA, Hinder SJ, Heng JYY. Decoupling the contribution of surface energy and surface area on the cohesion of pharmaceutical powders. *Pharmaceutical Research*. 2014;32:248-259.

7. Machon D, Melinion P. Size-dependent pressure-induced amorphization: A thermodynamic panorama. *Physical Chemistry Chemical Physics*. 2014;17(2):903.

8. Machon D, Piot L, Hapiuk D, et al. Thermodynamics of nanoparticles: Experimental protocol based on a comprehensive ginzburg-landau interpretation. *Nano Letters*. 2014;14(1):269-276.

Copyright © American Institute of Chemical Engineers. All rights reserved.

462188 Raman Mapping for Analyzing Multi-Component Solid Dosage Formulations

Abstract:

It has always been challenging to use spectroscopic methods to analyze salt disproportionation in a multi-component tablet matrix due to the spectral interference generated by the various excipients. Although combining Raman spectroscopy and chemometrics can be a powerful approach to study the extent of salt disproportionation, it was found in the present study that bulk measurements and chemometric modeling have obvious limitations when the targeted component is present at low levels in the tablet. Hence, a two-step Raman mapping approach was developed herein to investigate salt disproportionation in tablets with a low drug loading (5% w/w). The first step is to locate the area of interest where the drug particles reside throughout the tablet surface by using a statistically optimized sampling method termed deliberate sub-sampling. The second step, referred to herein as close-step mapping, utilize a step by step mapping of the targeted area to find more details of salt disproportionation in the tablet regions where the drug is concentrated. By using this two-step Raman mapping approach, we successfully detected the existence of minor species embedded in multi-component low drug loading tablet matrices, where bulk measurements from routine techniques usually lack of sensitivity. This approach will help formulation scientists detect and understand salt disproportionation and in situ drug-excipients compatibility issues in low dose solid dosage formulations.

Copyright © American Institute of Chemical Engineers. All rights reserved.

472578 Understanding the Roles of Surfactants and Oligomeric Assembly on the Functionality of the Light-Activated Membrane Protein Proteorhodopsin

Abstract:

Membrane proteins fulfill diverse functional roles that include sensing, catalysis and transport that are attractive for technological applications, such as chemical or biological sensing, separations, and energy conversion. The functionalities of membrane proteins depend on interactions with molecular species in the proteinâ€™s local environments, which typically include surfactants (*e.g.*, lipids) and other proteins. An objective for optimally exploiting membrane proteins for technological applications is to understand how protein-surfactant and protein-protein interactions independently influence protein functionalities. However, this goal remains a major experimental challenge because adjustment of protein-surfactant interactions by surfactant quantity or composition also influences protein oligomerization, and thus protein-protein interactions. We hypothesize that membrane protein functionality is mediated primarily by protein-protein interactions because the diversity of amino acid residues at protein-protein interfaces support stronger and more specific interactions than surfactants, which interact with membrane proteins largely by relatively weak hydrophobic interactions. As a model membrane protein for this study we chose proteorhodopsin, which functions as a light-activated H^+-ion pump and can be stabilized in surfactant solutions in monomeric or oligomeric forms, which can be isolated by using size-exclusion chromatography. Solution-state ^1H nuclear magnetic resonance measurements are used to identify and quantify the surfactant species that stabilize proteorhodopsin in solutions, while complementary blue native polyacrylamide gel electrophoresis establishes the distributions of monomeric and oligomeric proteorhodopsin. UV-visible absorption measurements provide information about the light-activated functionalities of proteorhodopsin monomers and oligomers in solutions with different surfactant compositions. The correlation of surfactant composition with functionalities of proteorhodopsin monomers and oligomers yields insight into the functional roles of surfactants and oligomeric assembly. These results provide information about conditions that support the optimal function of membrane proteins, which is broadly important for designing systems that exploit membrane protein functionalities.

Copyright © American Institute of Chemical Engineers. All rights reserved.

464626 Amyloid-β and α-Synuclein: Uncovering Their Commonalities and Differences in Complex with Binding Proteins Using Simulations and Experiments

Abstract:

Alzheimer's disease and Parkinson's disease are the primary neurodegenerative disorders that lead to cognitive and mobility impairment in aging populations (1,2). Amyloid-β peptide (Aβ), and α-synuclein (α-syn) are amyloidogenic proteins that self-assemble and form fibrillar amyloid deposits, pathological features Alzheimer's disease and Parkinson's disease, respectively. In Alzheimer's disease, seline plaques comprise of Aβ proteins, and in Parkinson's disease Lewy bodies, comprise of α-syn proteins. Interestingly, up to 50% of Alzheimer's disease cases exhibit the aggregation of α-synuclein into Lewy bodies, associated with a more aggressive disease course, evidence supporting that Aβ and α-syn interact in vivo to promote the aggregation and accumulation of each other and accelerate cognitive dysfunction (3). Thus, the simultaneous inhibition of aggregation by targeting combinations of the two monomeric proteins involved in the aforementioned diseases may constitute a promising therapeutic strategy (4). We have introduced a combination of computational and experimental studies to investigate the two disease-associated amyloidogenic proteins in complex with binding proteins. Using molecular dynamics simulations, free energy calculations (5-8) in CHARMM (9) and in-house structural analysis programs, we have identified the distinct role of energetic driving forces leading to molecular recognition of the two amyloidogenic proteins, Aβ and α-syn, and the key residue interactions with regard to binding and specificity in the framework of the complex formation. Additionally, our computational studies have identified key commonalities and differences of the two disease-associated amyloidogenic proteins in complex with specific protein counterparts. The results of our study suggest possible key determinants for inhibiting Aβ and α-syn and pave the way for the design and discovery of inhibitors as novel potential therapeutic agents.

1. Alzheimer's Association. Alzheimer's disease facts and figures. Alzheimers Dement. 2011; 7(2) : 208-44.

2. de Lau LM, Breteler MM. Epidemiology of Parkinson's disease. Lancet Neurol. 2006; 5(6) : 525-35.

3. Clinton LK, Blurton-Jones M, Myczek K, Trojanowski JQ, LaFerla FM. Synergistic Interactions between Abeta, tau, and alpha-synuclein: acceleration of neuropathology and cognitive decline. J Neurosci. 2010 May 26; 30(21) : 7281-9.

4. Shaykhalishahi H, Mirecka EA, Gauhar A, Grüning CS, Willbold D, Härd T, Stoldt M, Hoyer W. A β-hairpin-binding protein for three different disease-related amyloidogenic proteins. Chembiochem. 2015; 16(3) : 411-4.

5. Tamamis P, Morikis D, Floudas CA, Archontis G. Species specificity of the complement inhibitor compstatin investigated by all-atom molecular dynamics simulations. Proteins. 2010; 78(12) : 2655-67.

6. Tamamis P, Pierou P, Mytidou C, Floudas CA, Morikis D, Archontis G.Design of a modified mouse protein with ligand binding properties of its human analog by molecular dynamics simulations: the case of C3 inhibition by compstatin. Proteins. 2011; 79(11) : 3166-79.

7. Gorham RD Jr, Forest DL, Tamamis P, López de Victoria A, Kraszni M, Kieslich CA, Banna CD, Bellows-Peterson ML, Larive CK, Floudas CA, Archontis G, Johnson LV, Morikis D. Novel compstatin family peptides inhibit complement activation by drusen-like deposits in human retinal pigmented epithelial cell cultures. Exp Eye Res. 2013; 116 : 96-108.

8. A. Kieslich, Chris; Tamamis, Phanourios; D. Gorham Jr., Ronald; Lopez de Victoria, Aliana; U. Sausman, Noriko; Archontis, Georgios; Morikis, Dimitrios. Exploring Protein-Protein and Protein-Ligand Interactions in the Immune System using Molecular Dynamics and Continuum Electrostatics. Curr Phys Chem. 2012; 2(4) : 324-343(20).

9. Brooks BR, Brooks CL 3rd, Mackerell AD Jr, Nilsson L, Petrella RJ, Roux B, Won Y, Archontis G, Bartels C, Boresch S, Caflisch A, Caves L, Cui Q, Dinner AR, Feig M, Fischer S, Gao J, Hodoscek M, Im W, Kuczera K, Lazaridis T, Ma J, Ovchinnikov V, Paci E, Pastor RW, Post CB, Pu JZ, Schaefer M, Tidor B, Venable RM, Woodcock HL, Wu X, Yang W, York DM, Karplus M. CHARMM: the biomolecular simulation program. J Comput Chem. 2009; 30(10) : 1545-614.

Copyright © American Institute of Chemical Engineers. All rights reserved.

472466 Improving the Activity of a Thermophilic Lipase By Increasing the Flexibility Proximal to the Active Site

Abstract:

Lipases are long chain ester hydrolases used across multiple industries including the detergent, food, flavor, pharmaceutical, fine chemicals, agrochemical, biosensor, bioremediation and cosmetic fields. Cold active lipases are frequently sought after for their selectivity and efficient catalysis as noted by high k_{cat} values. Even though cold-active lipases are widely used in these fields, the vast majority of naturally evolved psychrophilic lipases are too thermolabile for most applications at ambient and moderate temperatures. To take advantage of the intrinsically efficient catalysis of psychrophiles, we must have methods to increase their thermostability; however, prior efforts to improve the thermostability of psychrophilc lipases through rational engineering and directed evolution resulted in weakened cold-activity.

The thermodynamic parameters of reactions catalyzed by psychrophilic enzymes show that they are characterized by either lower transition state free energy values or smaller temperature dependence of activity than their thermophilic counterparts. The latter is achieved by the disruption of fewer enthalpy-driven interactions during the activation process. Therefore, localized increases in flexibility near the active site are responsible for the higher catalytic reaction rates of psychrophiles. Recently, the cold-activity of some thermophilic enzymes has been improved successfully by incorporating the structural elements of their psychrophilic counterparts, which shows that cold activity and thermostability are not mutually exclusive. In this study, we use site directed mutagenesis to increase the flexibility proximal to the active sites of lipase from *Geobacillus thermocatenulatus* by analysis of lipase from psychrophilic *Photobacterium lipolyticum*. Additionally, we compare the thermostability and kinetic parameters between the wild type lipase and the mutated one. In our talk, we will present these results discussing the structural differences and how the activity of thermophilic enzyme is affected.

Copyright © American Institute of Chemical Engineers. All rights reserved.

456927 Understanding Medium Chain Ester Biosynthesis in Yeast through Alcohol-o-Acyltransferase Structure-Function Studies

Abstract:

Fatty acid ethyl esters are secondary metabolites that are produced during microbial fermentation and fruit ripening. Short and medium chain volatile esters have value as natural food additives (e.g. isoamyl acetate) and industrial solvents (e.g. ethyl acetate), and long chain waxy esters have potential for use as diesel fuel substitutes. The biosynthetic pathways in yeast that produce these compounds have a common reaction step, the condensation of an alcohol with an acetyl- or acyl-CoA by alcohol-O-acetyltransferse (AATase, E.C. 2.3.1.84 and 2.3.1.74). Previous activity screening revealed that Eht1 and Eeb1 exhibit high activity towards ethyl ester biosynthesis with acyl-CoAs with C4 to C12 carbon chain length. However, our understanding of medium chain fatty acid ester synthesis in yeast is limited. Here, we present the results of a series of structure-function studies of medium chain fatty acid ester synthase, Eht1 and Eeb1 from *S. cerevisiae*, including 1) analysis of subcellular localization in *S. cerevisiae* and *E. coli* hosts; 2) truncation and alanine scanning studies to identify the protein domains responsible for membrane localization; 3) membrane topology study of Eht1 and Eeb1. Pairwise alignment suggests Eht1 and Eeb1 are paralogs and closely related, with 58% sequence identity and 73% sequence similarity at the amino acid level. Overexpression studies revealed that Eht1 and Eeb1 initially localize to the endoplasmic reticulum. As cell growth progresses to stationary phase, Eht1 traffic to the surface of lipid droplets, while Eeb1 localizes to mitochondria. With overexpression in E. coli, membrane localization of Eeb1 is lost, in comparison to that Eht1 localizes to plasma membrane. The truncation and alanine scanning studies suggested that both N and C termini are important for their localization and trafficking. The membrane topology study of them on endoplasmic reticulum membrane indicated that both termini are exposed to the cytosolic side of ER. Combined, these studies develop our overall understanding of medium chain fatty acid ester biosynthesis in yeast and provide critical information to developing biochemical processes for ester biosynthesis.

Copyright © American Institute of Chemical Engineers. All rights reserved.

460380 Using Poly(EK) Fusion Proteins to Enhance Protein Stability

Abstract:

Protein PEGylation has been used for several decades, and has resulted in the development of numerous PEGylated protein drugs. PEGylation typically improves protein stability at the expense of bioactivity, biodegradability, or immunogenicity by forming a shielding layer around the protein. Recently, ultra low fouling zwitterionic polymers, including poly(carboxybetaine) (pCB), have demonstrated the ability to improve the thermal and chemical stabilities of proteins when chemically conjugated to protein surfaces. Oppositely charged glutamic acid (E) and lysine (K) and residues have been shown to occur on protein surfaces at balanced ratios, and alternating repeating poly(EK) sequences have demonstrated nonfouling zwitterionic characteristics when utilized in functionalizing surfaces and nanoparticles. Thus EK ideally presents itself as a naturally-derived pCB analogue, allowing for the use of recombinant techniques for one-step production of protein-poly(EK) fusion proteins through protein expression techniques.

Genes encoding for poly(EK) tails of 10 kD and 30 kD have been appended to the C-terminus of proteins. The resulting fusion protein products containing the poly(EK) tails were expressed in E. coli strain BL21(de3) and purified. Enzyme kinetics and activity assays and thermal and salt stability assays were performed to demonstrate the effect that poly(EK) had on improving protein stability. The addition of poly(EK) to native Î²-lactamase and a destabilized TEM-19 mutant indicated that the poly(EK) tail was not significantly detrimental to protein activity, with fusion protein products containing the 10 kD poly(EK) maintaining native activity. Additionally the addition of poly(EK) increased the fusion protein productâ€™s affinity to the substrate, with appended 10 kD and 30 kD poly(EK) reducing the Km from 70 to 25 and 45 Î¼M, respectively. Similarly to pCB, the addition of poly(EK) was able to improve the proteinâ€™ thermal stability. The addition of 10 kD and 30 kD poly(EK) increased the midpoint transition temperature by approximately 15 and 20 °C, respectively. Interestingly, the addition 10 kD poly(EK) did not change the profile of the thermal stability curve while increasing stability, whereas the addition of 30 kD poly(EK) generated a modified stability profile. Additionally, proteins containing 30 kD (polyEK) were able to retain more than 30% of initial activity after 1 hour of incubation at 95°C, a nearly 300% improvement over the unmodified wild-type protein. This system can be applied to any protein where increased stability is desired, and is being used to study more application-relevant systems. Additionally further study will allow for probing the effect of poly(EK) on protein structure, allowing for the study of its effects on the protein product's tertiary and quarternary structure and poly(EK)'s secondary structure. Recently it has been demonstrated that PEGylated drugs are beginning to see accelerated clearance, which may be tied to the increased presence of anti-PEG antibodies. This allows for a clear demonstration of the effect of poly(EK) *in vivo* to study protein circulation and immunogenicity.

References:

[1] Liu, E. J.; Sinclair, A.; Keefe, A. J.; Nannenga, B., L.; Coyle, B. L., Baneyx, F, Jiang, S., Biomacromolecules, 2015
[2] Keefe, A. J.; Jiang, S., Nature Chemistry, 2012, 4, 59-63.

Copyright © American Institute of Chemical Engineers. All rights reserved.

465612 Varying Effects of Kosmotropic and Chaotropic Anions on Amyloid beta$_{1-42}$ Aggregation Kinetics and Fibril Formation

Abstract:

Aggregation of amyloid-β peptides to form plaques is considered to trigger the pathogenesis of Alzheimer's disease (AD). A_{1-42}, which is the most toxic amyloid-β peptide, has been widely studied to understand its aggregation properties and role in the pathogenesis of AD. However, the effect of ions on aggregation of A_{1-42} has not been investigated systematically. In this work we have studied the effect of ions of the Hofmeister series on the aggregation of A_{1-42}. Previous work from our group had demonstrated that Sup35NM protein containing the prion domain of yeast prion protein Sup35p exhibited faster aggregation in the presence of strongly hydrated anions (kosmotropes), and slower aggregation in the presence of poorly hydrated anions (chaotropes)[1,2]. The Hofmeister salts also determined the amyloid strain that was preferentially formed, and affected the specificity of species barrier of the Sup35 protein between three closely related *Saccharomyces* species[3]. In this work, we have investigated how Hofmeister ions affect the aggregation kinetics of A_{1-42} peptide. Our experiments show that while the relative effects of kosmotropic and chaotropic ions on aggregation kinetics are similar, the absolute effects of the chaotropic salts on the aggregation of A_{1-42} peptide are different from Sup35NM. Further, the differences in the amyloid strains formed by A_{1-42} in the presence of Hofmeister salts have also been studied by microscopy and electrophoresis techniques to gain better insights into the aggregation mechanism of A_{1-42}.

[1] Rubin, J., Khosravi, H., Bruce, K. L., Lydon, M. E., Behrens, S. H., Chernoff, Y. O. and Bommarius, A. S., 2013. Ion-specific effects on prion nucleation and strain formation. Journal of Biological Chemistry, 288(42), pp.30300-30308. doi: 10.1074/jbc.M113.467829

[2] Yeh, V., Broering, J. M., Romanyuk, A., Chen, B., Chernoff, Y. O. and Bommarius, A. S. (2010), The

Hofmeister effect on amyloid formation using yeast prion protein. Protein Science, 19: 47–56. doi: 10.1002/pro.281

[3] Sharma, A., Bruce, K. L., Chen, B., Gyoneva, S., Behrens, S. H., Bommarius, A. S. and Chernoff, Y. O., 2016. Contributions of the Prion Protein Sequence, Strain, and Environment to the Species Barrier. Journal of Biological Chemistry, 291(3), pp.1277-1288. doi: 10.1074/jbc.M115.684100

Copyright © American Institute of Chemical Engineers. All rights reserved.

475005 The Role of Electrostatics in the Function of a Protein Link Between Innate and Adaptive Immunity

Abstract:

The protein C3d is a proteolytic fragment of C3, the converging protein of three activation pathways of the complement system. Upon cleavage and activation of C3, the fragment C3b covalently attaches to pathogen or damaged host cell surfaces through its thioester domain (TED), acting as (i) an opsonin for recognition of C3b-decorated surfaces by phagocytes, and (ii) as a component of surface-bound convertase complexes that amplify C3 cleavage in a C3b production feedback loop [1]. Further cleavage steps inactivate C3b by converting to C3d, a surface-bound standalone TED. C3d is a biomarker indicating that complement system has been activated on a particular surface. C3d is also a link between innate (complement system based) and adaptive (antigen-antibody based) immunity, through the formation of the C3d-complement receptor 2 (CR2) complex, which is part of the B cell receptor (i.e. antibody) – coreceptor (i.e. CR2) complex. This complex cross-links antigens with B cells and augments antibody production by up to four orders of magnitude.

The association between C3d and CR2 is a two-step process, driven by the long-range electrostatic attraction of protein macrodipoles (recognition step) and the formation of a stable bound complex based on favorable short-range polar and nonpolar pairwise amino acid interactions (binding step) (see [2] and references therein). This model is based on computational studies, such as molecular dynamics simulations, electrostatic calculations, theoretical mutagenesis and alanine scans, and free energy analyses. Also, Brownian dynamics simulations demonstrate that electrostatic steering accelerates the formation of an encounter complex during recognition of C3d and CR2, which is ionic strength dependent [3]. In all cases, the computational data have been compared and are in agreement with existing experimental binding, mutagenesis, and ionic strength studies.

The electrostatic interaction functionality of C3d with CR2, and the appearance of the B cell receptor-coreceptor complex, has been gained throughout evolution, as it is demonstrated by studying the electrostatic character of C3d from a variety of species using the concept of perturbation-resistant electrostatic hotspots [4]. Primitive species have only complement-mediated immunity, whereas more advanced species have both complement-mediated and antigen-antibody-mediated immunity. It is proposed that electrostatic hotspots in C3d have evolved to optimize its dual functionality, i.e. covalent attachment to pathogen surfaces and interaction with CR2.

The CR2-binding site in human C3d is multi-functional, also mediating electrostatic interactions with complement regulators, such as Factor H [5]. Such regulation is important to inhibit complement attack against host tissues. In addition, *S. aureus* bacterial proteins use the CR2-binding site of C3d to disrupt the function of the complement system, through electrostatically-driven binding [6, 7]. Moreover, the C3d (TED) domain of C3b is a site of electrostatic interactions with complement regulators Factor H, MCP, and DAF, as well as with viral proteins that mimic complement regulators to facilitate viral evasion of the complement system [8].

Compromised regulation of the complement system results to autoimmune and inflammatory diseases, such as lupus, rheumatoid arthritis, age-related macular degeneration, and rare kidney diseases. Given the fact that C3d is a biomarker of complement activation, C3d can be used as a diagnostic for early onset of complement-mediated diseases, prior to clinical diagnosis. A virtual screening study of 7 million chemical compounds for binding against C3d has led to the identification of 10 C3d ligands with micromolar affinities and fluorescence properties [9]. These compounds have the potential to become diagnostics for age-related macular degeneration, using *in vivo* fluorescence imaging. They also have the potential for dual therapeutic and diagnostic (theranostic) role, as well as for becoming drug carriers for targeted delivery at sites of local complement-originating inflammation.

[1] Zewde N, Gorham RD Jr, Dorado A, Morikis D (2016) Quantitative modeling of the alternative pathway of the complement system, PLOS ONE 11:e0152337.

[2] Mohan R, Gorham RD Jr, Morikis D (2015) A theoretical view of the C3d:CR2 binding controversy, Molecular Immunology 64:112-122.

[3] Mohan RR, Huber GA, Morikis D (2016) Electrostatic steering accelerates C3d:CR2 association, Journal of Physical Chemistry B, in press – available online.; DOI: 10.1021/acs.jpcb.6b02095.

[4] Kieslich CA, Morikis D (2012) The two sides of complement C3d: evolution of electrostatics in a link between innate and adaptive immunity, PLoS Computational Biology 8:e1002840.

[5] Harrison R, Gorham RD Jr, Morikis D (2015) Energetic evaluation of binding modes in the C3d and factor H (CCP 19-20) complex, Protein Science 24:789-802.

[6] Gorham RD Jr, Rodriguez W, Morikis D (2014) Molecular analysis of the interaction between staphylococcal virulence factor Sbi-IV and complement C3d, Biophysical Journal 106:1164-1173.

[7] Gorham RD Jr, Kieslich CA, Morikis D (2012) Complement inhibition by Staphylococcus aureus: electrostatics of C3d-EfbC and C3d-Ehp association, Cellular and Molecular Bioengineering 5:32-43. [8] Ojha H, Panwar HS, Gorham RD Jr, Morikis D, Sahu A (2014) Viral regulators of complement activation: structure, function and evolution, Molecular Immunology 61:89-99.

[9] Gorham RD Jr, Nuñez V, Lin J-H, Rooijakkers SHM, Vullev VI, Morikis D (2015) Discovery of small molecules for fluorescent detection of complement system activation product C3d, Journal of Medicinal Chemistry 58:9535-9545.

Copyright © American Institute of Chemical Engineers. All rights reserved.

464884 Bottom-up Construction of Synthetic Microbial Pairs Inspired By Nature

Abstract:

Anaerobic fungi in the hindgut of large herbivores are among the most robust organisms at degrading crude lignocellulose. Though gut fungi have been characterized in isolation, such growth does not accurately mimic their native environment. Anaerobic fungi exist in nature as part of much larger consortia, which include archaea, bacteria, and protozoa. In particular fungi form a tight interaction with methane producing archaea (methanogens), which siphon hydrogen and other metabolites from the fungi. This syntrophic partnership allows the fungi to more efficiently produce energy by increasing the flux through their hydrogenosomes. In addition to greatly increasing biomass degradation, fungi-methanogen co-cultures are capable of directly converting crude biomass into methane. Elucidating this specific interaction using â€"OMICS-based approaches will allow for the design of stable consortia capable of efficiently converting biomass to value-added products.

In this work, native co-cultures of anaerobic fungi and methanogenic archaea were isolated from herbivore fecal materials. These co-evolved consortia demonstrated approximately 2-3 fold increased accumulation of fermentation gases and degradation of biomass compared to cultures with the methanogens removed. Natural consortia were remarkably stable in batch culture, surviving together for over 18 months and 150 transfers. Genomic sequencing of the methanogens in this consortia revealed two distinct species of methanogens in the genera *Methanosphaera* and *Methanocorpusculum*. The sequences were assembled and analyzed using the Kbase platform, and then binned into separate genomes. Synthetic co-cultures were created using well-characterized anaerobic fungi and methanogens inspired by the genomic analysis of the native mixture and capable of converting a range of different substrates to methane. The synthetic co-cultures were tested for substrate consumption and stability on a variety of substrates including reed canary grass, avicel, xylan, and pectin. Synthetic co-cultures displayed accelerated growth and degradation of substrates, on par with the native consortia, and enabled fungi to metabolize an expanded array of substrates (e.g. pectin and xylan) that the fungi could not metabolize in isolation. This work demonstrates that mixed-and-matched microbial pairs can perform equally well at converting crude biomass into syngas products as those co-evolved in nature.

Copyright © American Institute of Chemical Engineers. All rights reserved.

466415 Engineered Consortium Overcomes Substrate-Induced Inefficiencies in Hexose Catabolic Pathways

Abstract:

Escherichia coli is an important commercial species used for production of biofuels, organic acids, sugar alcohols, biopolymers, and natural compounds. The growth media and time necessary for production of value added products are important factors for determining the cost effectiveness of a commercial process. Processed biomass and agroindustrial byproducts serve as low cost nutrient sources that contain a variety of carbon sources. However, bioconversion of mixtures of carbon sources by *E. coli* is inefficient due to carbon catabolite repression (CCR), where the transport and catabolic activity of one or more carbon sources is repressed and inhibited by the transport and catabolism of another carbon source. CCR can also result in incomplete bioconversion of carbon sources. Attempts to mitigate the inefficiencies of CCR can be categorized as i) identification or development of a single strain deficient of CCR ii) development of a microconsortia of strains that each catabolize a single sugar, and most work has focused on co-utilization of glucose and pentoses. However, the hexoses mannose and galactose can be found in high concentrations in certain biomass derived nutrient sources as well. We will present a new microconsortium of different *E. coli* strains, each engineered to preferentially catabolize a different hexose - glucose, galactose, or mannose. We modified the specificity and preference of carbon source using a combination of rational strain design and adaptive evolution. The modifications ultimately resulted in strains that preferentially catabolized their specified sugar, overcoming CCR mechanisms (inducer exclusion, transcriptional repression, the phosphorylation state of the adenylate cyclase activator $EIIA^{Glc}$), substrate nonspecific transporters, and complications of resulting from the shared metabolic pathway for hexose catabolism. Finally, we compared the performance of the microconsortium to an existing CCR mutant strain. The microconsortium was capable of catabolizing the total available sugars much faster than a CCR mutant and resulted in much higher final cell density. Further, by including different selectable markers in each strain, we were able to show that by changing the relative sugar concentration in the media and strain composition of the inoculant, we can alter the strain composition of the culture over time. The results indicate that not only have we created a more efficient method for bioconversion of mixtures of glucose, galactose, and mannose, but we have also developed a method to control populations of constituent strains of a microconsortia by simply modifying the sugar composition of the media.

Copyright © American Institute of Chemical Engineers. All rights reserved.

472105 Engineering Microbial Communities for Cellulose Degradation and Modular Production of Target Chemicals

Abstract:

Cellulosic biomass, an abundant biological waste material, can serve as an inexpensive carbon source for producing valuable products using microbial ecosystems. A critical step in developing a system dependent on waste is the degradation of cellulose into glucose monomers, which can be subsequently utilized for growth and product generation. Current approaches rely on the expensive production and purification of enzymes for this purpose, or harsh pretreatment processes. However, the use of a mixed community where the labor is divided between polymer degraders and target product makers has the potential to address this challenge. To this end, we are building a community of *B. megaterium* and *E. coli* for bioproduction from cellulose. We will describe our recent progress engineering the Gram-positive *Bacillus* strain for targeted cellulase secretion. Complete degradation of cellulose requires a combination of three cellulose functions â€" an endoglucanase, a cellobiohydrolase, and a beta-glucosidase. The goal of this work is to generate a system of one or more *B. megaterium* that secrete a combination of requisite cellulases, for synergistic degradation of cellulose in co-culture. We have tagged cellulases with a small library of signal peptides (SPs), found to trigger protein secretion in *B. megaterium*, and demonstrated our novel strategy leads to efficient expression and secretion. Cellulase activity against amorphous cellulose was further confirmed through a series of bioassays, validating that we have engineered strains with significant cellulase secretion and activity. We have also demonstrated that the glucose monomers released through the actions of the cellulases secreted by *B. megaterium* can be used to support both *B. megaterium* and *E. coli* growth in co-culture. We will describe our current efforts to explore the role of induction, expression and inoculation ratios on community composition and the production of target bioproducts by the *E. coli* maker cells.

Copyright © American Institute of Chemical Engineers. All rights reserved.

472543 Extending Natural Product Biosynthesis Pathways through Application of Engineered Modular *E. coli* Polycultures

Abstract:

The past few years have seen an exciting boom in research interest over the use of modular microbial co-cultures for metabolic engineering applications. The ability for these multi-strain systems to spread the metabolic burden of pathway overexpression across the members of the consortia while enabling modular-specific genetic optimization have been highlighted as a key advantages over traditional monoculture approaches. These systems have demonstrated the potential for relatively simple synthetic consortia to out perform their respective monocultures for a wide variety of target products. Until recently, efforts have primarily focused on the use of two microbial strains in co-culture. In this work we will describe our efforts to extend this technology to the use of polycultures, or three or more strains in co-culture, for the extension of natural product production pathways in *E. coli*. Utilizing this technique, we were able to demonstrate the *de novo* production of a variety of flavonoid natural products not previously obtained from monoculture efforts.

To define the production landscape for our system, we identified several key optimization parameters for the individual modules as well as the polyculture as a whole. Through the development of two new high-titer production modules for phenylpropanoic acids from tyrosine and anthocyanins from flavan-3-ols, we were able to realize the *de novo* production of anthocyanidin-3-O-glucosides for the first time from a recombinant cell system. In addition to describing this successful case study, we will discuss our efforts to demonstrate the feasibility of these systems for other target pathways and methods to measure population dynamics in an attempt to address the stability of these systems during extended microbial fermentations. The work presented here highlights the potential of modular microbial co-cultures to revolutionize the future of natural product production from microbial hosts while attempting to address several of the lesser understood aspects of this new and exciting field.

Copyright © American Institute of Chemical Engineers. All rights reserved.

458134 Microbial Synthesis of Plant Monolignols

Abstract:

Monolignols are direct precursors of lignin biosynthesis. Their derivatives also exhibit anti-oxidative, anticancer and canceroprotective activities. To achieve efficient production of the simplest monolignol *p*-coumaryl alcohol, four heterologous enzymes with high catalytic activities were selected for pathway construction. The full pathway was divided into the upstream and downstream partial pathways. Bioconversion study showed that the downstream enzymes were able to efficiently convert the phenylpropionic acids into the corresponding monolignols. *E. coli* strain containing the full pathway produced 501.8 ± 41.4 mg/L of *p*-coumaryl alcohol under optimized conditions. However, when the pathway was extended for the production of caffeyl alcohol, coniferyl alcohol, the promiscuity of the hydroxylase HpaBC led to the production of an instable intermediate *L*-dopa from tyrosine. *In vivo* and *in vitro* studies demonstrated that *E. coli* cell membrane provided a selective barrier toward tyrosine. Based on this, microbial co-cultures were designed. With the optimal inoculum ratio, 401.0 ± 15.3 mg/L of caffeyl alcohol was produced, which is nearly 12 times higher than that of the mono-culture. The same strategy was used for coniferyl alcohol production. Limited by the activity of methyltransferases, the highest titer was 124.9 ± 5.1 mg/L with 232.9 ± 15.1 mg/L of caffeyl alcohol accumulated. In this work, we achieved microbial production of three monolignols. This work also demonstrated the promising potential of microbial co-cultures for prevention of side-reactions.

Copyright © American Institute of Chemical Engineers. All rights reserved.

472380 Mechanisms and Adaptation of the Arabidopsis thaliana Root Microbiota in Iron Mobilization and Uptake from Iron-Limiting Environments

Abstract:

Plants coexist with a diverse community of microbes, some of which assist the plant in overcoming poor nutrient availability in low quality soils. For example, nitrogen fixation is made possible by the nodule-forming rhizobia, and otherwise inaccessible phosphorous is provided to the plant by their root-associated fungi. However, little is known about the extent to which the natural root microbiota facilitates the uptake of the fourth-most limiting nutrient in agriculture: iron. Moreover, a systematic study of how the root-associated microbiota adapts to such low quality conditions, and the impact such adaptation may have on the plant's nutrient status remains poorly understood.

We have identified root-associated bacteria that promote the growth and development of *Arabidopsis thaliana* under conditions simulating low iron availability in vitro. These strains, which represent members within the phyla Proteobacteria and Bacteroidetes, are employed as models to investigate the mechanics facilitating plant iron acquisition by its microbiota. By applying reverse genetics and metabolomics, I investigate the role of the plant's secondary metabolism in mediating plant-microbe signaling in the rhizosphere. In parallel, I investigate adaptation of the A. thaliana root microbiota to calcareous soils, which constitute approximately one-third of the world's arable land. I will apply root-associated bacterial and fungal community profiling methods to determine adaptation in terms of community composition, and cross-reference these results with comparative genomics and plant growth promoting ability of a comprehensive rhizobacterial library derived from this soil type. Finally, I use a plant gnotobiotic system to investigate whether defined bacterial communities obtained from a calcareous soil uniquely display an ability to promote plant growth under iron-limiting conditions.

Copyright © American Institute of Chemical Engineers. All rights reserved.

462699 The Role of the Immune System in Modulating Tumor Cell Migration Following Radiation-Induced Changes in the Tissue Microenvironment

Abstract:

Despite aggressive surgical, radiological, and chemotherapeutic intervention, triple-negative breast cancer patients experience high rates of recurrence. The role of the tumor microenvironment in recurrence, however, is not well understood. We hypothesize that the irradiated tumor microenvironment influences tumor and immune cell behavior. In this study, we characterize the effects of normal tissue irradiation on tumor and immune cell migration to evaluate how tumor-stromal interactions modulate recurrence after therapy. This work represents the first step toward elucidating how stromal radiation response contributes to tumor and immune cell recruitment.

The effect of radiation on tumor cell migration to normal tissues was studied using an orthotopic breast cancer model. Nude mice were inoculated with luciferase-labeled 4T1 murine or MDA-MB-231 human triple-negative breast cancer cells in the mammary fat pad (MFP), and BALB/c mice with depleted CD4+ and/or CD8+ T cells were inoculated with 4T1 MFP tumors. The contralateral normal MFP was irradiated to a dose of 20 Gy with a 250 kVp cabinet x-ray machine when tumors were palpable. Cell migration was monitored with bioluminescence imaging (BLI) 10 d after irradiation. Irradiated and control tissues were evaluated using immunohistochemistry (IHC). Tissue sections were stained with F4/80 to determine the extent of macrophage infiltration. Flow cytometry was also performed on dissociated irradiated and control tissues to characterize immune cell populations. Of particular interest were CD11b+F4/80+ macrophages and CD11b+GR1+ myeloid-derived suppressor cells (MDSCs).

Radiation enhanced tumor cell migration to normal tissues *in vivo*. *Ex vivo* BLI analysis in nude mice revealed that normal tissue irradiation attracted tumor cells to the MFP and surrounding tissues, including the peritoneum and muscle ($p<0.001$). Tumor cells migrated to irradiated normal tissues in BALB/c mice with depleted CD4+ and CD8+ T cells ($p<0.001$). Depletion of CD8+ T cells alone significantly enhanced tumor cell migration in comparison to depleting CD4+ T cells alone ($p<0.01$). This indicates that CD8+ T cells may negatively regulate local recurrence following radiotherapy. An increase in macrophage infiltration was also found in irradiated tissue sections in the absence of T cells. Flow cytometry confirmed increases in macrophage and MDSC infiltration in irradiated areas.

This study establishes that normal tissue radiation response modulates tumor and immune cell migration after radiation. The increase in macrophage and MDSC infiltration after irradiation in the absence of T cells indicates a crucial role of the immune system in tumor cell migration. Our work suggests that the tumor stroma may facilitate tumor cell invasion and tumor regrowth following radiotherapy. Future studies will utilize these results to engineer improved *in vitro* tumor microenvironment models to probe the complex physical, chemical, and biological cues that influence cancer recurrence and metastasis.

Copyright © American Institute of Chemical Engineers. All rights reserved.

470850 Hemica-Hydrogel Encapsylated Micro-Channel Array in Cancer Metastasis

Abstract:

Introduction: Cell migration is a complex process that involves cell interactions with tissues of varying stiffness and porosity. The current state of art for studying cell migration in vitro consists of 2D and 3D matrixes as well as polydimethylsiloxane (PDMS)-based microfluidic devices. These methods cannot be used simultaneously, alter pore-size independent of matrix stiffness or recreate microenvironments of physiologically relevant stiffness. We herein present the first device that bypasses these hurdles by using a polyacrylamide (PA)-based Hydrogel Encapsulated MIcro-Channel Array (HEMICA) and focuses on the study of cancer cell metastasis.

Methods: Silicone wafers with the desired micro-channel design are created via photolithography and used as a mold for PA. Access holes to the PA imprinted design are punched out. Glass cover-slips are activated and PA is cured on them creating a thin (60-100um) layer. Both gel layers are allowed to reach their respective maximum swelling. The upper layer with the design on its lower surface is then sandwiched to the bottom gel/glass via a NHS-ester ligand. The encapsulated micro-channels are coated with collagen type I using Sulfo-SANPAH as a ligand. The device is seeded with cells and subsequently submerged into media for live microscopy. To quantify cell speed, persistence and cell morphology we used FIJI and a custom MATLAB code. Statistical significance was assessed with two-way ANOVA test.

Results and Discussion: MDA-MB-231 breast cancer cells were seeded in HEMICA and studied in a wide range of microenvironments spanning full confinement (*WxH=3x10Î¼m*^{2}) to 2D spaces (*WxH=50x10Î¼m*^{2} *or 200x50Î¼m*^{2}). Cell speed shows a biphasic behavior with maximum speed in 10um width (similar to the dimension of the nucleus) and intermediate stiffness 13kPa (similar to the stiffness of human breast tissue). Cell persistence is lower in the softest gels with higher significance on the larger width channels, suggesting a correlation between stiffness and directed migration. Although large widths allow cells to adhere preferentially to one channel wall (contact guidance), we show that low stiffness (7kPa) reduces contact guidance based on the frequency cells cross the channel center-line and the probability of following a path within the central regions of the channels. In addition, intermediate stiffnesses increase cell polarization, especially on small widths, while low stiffnesses increase cell circularity indicating reduced cell-matrix adhesions. On the other hand, cell projected area increases with increasing channel width independent of stiffness. Moreover, the cell variance of area and circularity is higher in large widths of low stiffness, indicating major morphodynamic changes in the process of migration. In contrast, the variance of solidity (how protrusive cells are) is higher in confinement with intermediate stiffness, which implies that cells in full confinement change the number of their protrusions frequently in order to achieve an optimum mode of migration. Through our ongoing experiments we are identifying the mechanobiology of cancer cell migration by establishing the link between stiffness, myosin II, microtubules and traction forces in confinement.

Conclusion: We have created a novel and versatile device which mimics in vivo microenvironments. HEMICA can be used to study collective migration, cell-cell communication, cell-matrix interactions and cell population dynamics, offers organ-on-a-chip applications and allows post-migration studies. Our results demonstrate that the geometry of the cell microenvironment and tissue compliance promote different modes of migration ultimately affecting all stages of cancer metastasis.

Copyright © American Institute of Chemical Engineers. All rights reserved.

464918 Tissue Origami: Directed Folding of Tissues By Programmed Cell Contractility Networks

Abstract:

A major barrier to the engineering of complex tissues is the difficulty in controlling tissue structure across a wide range of length scales. Further, proper function of synthetic biological tissues is hampered by mass transport limitations, and the lack of extracellular matrix (ECM) physicochemical anisotropies that guide appropriate cell behaviors in space and time. In biology, tissues are typically built over a range of length scales via self-organized morphogenesis processes in which extracellular matrix remodeling occurs at the same time as cell division, migration, and fate specification. We take a biomimetic approach to building tissues by engineering the spatial organization of sheets of ECM gels via contractile networks of cells. Such contractile networks are co-patterned with ÒpassengerÓ cell types that can be guided into prescribed 3D topologies that instruct cell self-organization processes. Thus, the folding trajectory and form is combined with control over the type and placement of different cell types on the sheet, enabling control over 3D tissue architecture in concert with cellular interfaces.

Introduction: Living tissues are characterized by the specific arrangement of biological cells in layers of ECM. Interfaces between cells and their matrix microenvironment are often folded into arrays of functional units such as the epithelial acini of the breast and the crypts and villi of the intestine. In many cases, the curvature and spatial organization of the ECM is relevant to tissue development and function. For example, the specification of endoderm cell fate at crypts in the gut is governed by cell-cell signaling networks that are influenced by the folding state of tissue layers. We seek to build folds of specific geometry in ECM-mimetic gels in order to understand the systems-level interactions between cells in folded tissues. Moreover, we rely on the intrinsic contractility of cells to sculpt the physicochemical anisotropy of the ECM to mimic in vivo tissues.

Materials and Methods: We build arrays of contractile tissues at the upper and lower surface of 250 micron-thick matrigel-collagen ECM-mimetic gels using DNA-patterned assembly of cells (Figure 1a). The type and location of cells on each surface of the gel prescribe the final folded geometry of the ECM. Collagen fibers are pre-labeled with fluorescent dye in order to track curvature of the sheet over time via confocal microscopy.

Results and Discussion: Regular arrays of contractile mammary epithelial cells placed on one side of circular ECM gels yield cap-shaped objects with maximum curvatures that increase over time. The curvature at a given timepoint increases monotonically as the spacing between tissues on the sheet is decreased (and more tissues are added, Figure 1b), scales linearly with cell contractility, and is partially blocked by blebbistatin. We found that anisotropic grids of fibroblast tissues produce consistent anisotropic folds along controlled axes, allowing us to fabricate a range of 3D topographies and predict them with a biophysical FEA model (Figure 1c). For example, we constructed the Miura-ori fold (Figure 1d), which resembles folds of the gut epithelium in development.

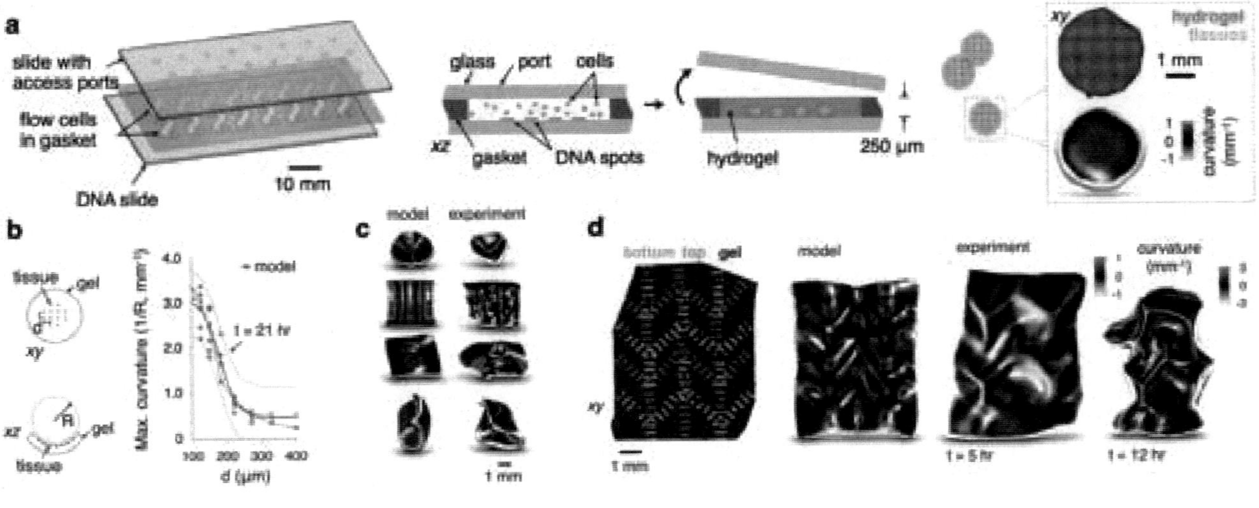

Figure 1. Engineering tissue curvature via cellular contractility. (a) microfluidic apparatus for building thin ECM sheets with contractile tissue actuators at prescribed positions in *xy*. (b) ECM curvature increases with tissue density. (c) Various folds can be prescribed in ECM sheets and predicted via FEA modeling. (d) The Miura-ori fold.

Conclusions: We report engineering control over the 3D shape of biomimetic ECM scaffolds by leveraging tissues as mechanical actuators. Our ability to design specific fold geometries in concert with cell compositions and spatial arrangement in ECM sheets could improve drug toxicity screens and enable studies of the necessary conditions for topological transitions in tissue development.

Copyright © American Institute of Chemical Engineers. All rights reserved.

470860 Engineering the Gut-Liver-Brain Axis to Investigate Chemical Toxicity

Abstract:

The inter-organ crosstalk between the gut, liver and brain has been shown to affect many physiological functions. The accumulation of drugs, chemicals, toxins and their metabolites in the liver can reach the brain and result in impaired cognitive function. Brain diseases such as hepatic encephalopathy result from complex interactions along the gut-liver-brain axis. We have designed and assembled a multi-organ system to obtain fundamental insights into the intricate cross-talk between these three organs.

Primary intestinal tissues were surgically obtained from rat jejunum and cultured in medium containing antibiotics. Gram staining was performed to detect the presence of gut bacteria over a 3-day period. Minimal numbers of Gram-positive and Gram-negative bacteria were released over the 72 h period. Imaging was conducted to determine the viability of intestinal cells that were shed. Enterocyte and goblet cell phenotypes were confirmed via alkaline phosphatase activity and Alcian blue staining, respectively. The intestinal tissues expressed these markers even after 96 h.

Hepatocyte function in the integrated jejunum-liver models was measured through urea secretion and albumin production and was comparable to liver models alone. Acetaminophen, a well-known hepatotoxicant, was administered to integrated gut-liver cultures. Even at a 10mM concentration of acetaminophen, the decrease in urea production and albumin expression increased two-fold in the integrated cultures when compared to liver models only. These trends clearly demonstrate the enhanced sensitivity to acetaminophen exhibited by the integrated cultures, underscoring the need to combine organs.

The gut-liver-brain crosstalk was accomplished by exposing astrocyte cultures to spent medium from the integrated systems and liver models. Specifically, the spent culture medium was obtained from systems exposed to 10mM acetaminophen. The production of reactive oxygen species (ROS) in astrocytes was visualized using the DCFDA assay. ROS production was compared between astrocyte cultures treated with conditioned medium from the integrated gut-liver systems and liver models only. There was approximately 10-fold higher ROS in cultures treated with conditioned medium obtained from only liver models. This trend could be due to increased acetaminophen metabolism by the gut in the integrated cultures through the glucuronidation and sulfation pathways, thus decreasing the production of toxic metabolites from the cytochrome P450 2E1 (CYP2E1) mediated metabolism of this hepatotoxicant.

The design and assembly of multi-organ engineered tissues provides a powerful platform to investigate the interactions and crosstalk that occur during many physiological processes. We are currently focusing on elucidating the signaling pathways between these three organs to better understand how they function.

Copyright © American Institute of Chemical Engineers. All rights reserved.

473906 Modulation of Lipid Composition As a Therapy in Alcohol Injured Liver Cells

Abstract:

Liver cells exposed to ethanol are altered in NADH/NAD$^+$ ratio, generation of reactive oxygen species, excessive fat accumulation and cell apoptosis. These changes in a healthy liver can result in alcoholic liver diseases (ALDs), which are major causes of liver related morbidity in the world. However, despite the well documented metabolism of ethanol in parenchymal liver cells, hepatocytes, current treatment options are limited to cessation of alcohol consumption. Studies in other cell types and organisms have demonstrated that, in addition to changes in metabolism, physical changes to cell membranes occur upon exposure to ethanol and these changes are dependent on cell membrane lipid composition. Here we present an alternative perspective on mitigating ethanol injury on hepatocytes based on modulating the membrane lipid composition. Considering the likely relationship of lipid composition with ethanol tolerance from a yeast system, we use cationic liposomes that contain DOPC (1,2-dioleoyl- sn-glycero-3-phosphocholine) to recover the functionality of ethanol injured primary rat hepatocytes. As a cationic lipid, DOTAP (1,2-dioleoyl-3-trimethylammonium-propane) is used to facilitate liposome cell interactions. Incorporation of the liposomes into the hepatocyte membranes were monitored by measuring DOTAP as a surrogate marker using liquid chromatography / mass spectrometry with a triple quadrupole (QqQ) mass analyzer and an electrospray ionization source (ESI). Cell function was assessed using secreted albumin as a marker. The results of these *in vitro* studies illustrate the potential of using cationic liposomes to mitigate ethanol injury *in vivo* using existing or novel means of delivering lipids to hepatocytes in the liver.

Copyright © American Institute of Chemical Engineers. All rights reserved.

470712 Signal Transduction of the Physical Environment in the Neural Differentiation of Stem Cells

Abstract:

Neural differentiation is largely dependent on extracellular signals within the cell microenvironment. These extracellular signals are mainly in the form of soluble factors that activate intracellular signaling cascades that drive changes in the cell nucleus. However, it is becoming increasingly apparent that the physical microenvironment provides signals that can also influence lineage commitment and very low modulus surfaces has been repeatedly demonstrated to promote neurogenesis. The molecular mechanisms governing mechano-induced neural differentiation are still largely uncharacterized; however, a growing body of evidence indicates that physical stimuli can regulate known signaling cascades and transcription factors involved in neural differentiation. Understanding how the physical environment affects neural differentiation at the molecular level will enable research and design of materials that will eventually enhance neural stem cell differentiation, homogeneity, and specificity. Previous findings have shown that the neural phenotype in stem cells is strongly repressed by the Neuron Restrictive Silencer Factor (NRSF). NRSF coordinately represses up to 300 genes involved in neural differentiation. Recently, we found that the neural induction in mesenchymal stem cells (MSCs) by forskolin and IBMX is dependent on the downregulation of NRSF and derepression of its downstream targets. Therefore we set out to determine whether soft surfaces could induce neural phenotype in MSCs also through downregulation of NRSF. Indeed we found that several neural characteristics induced by soft surfaces including increase in neural marker expression (Tuj), biosynthetic enzymes (TH), and neuronal function are repressed by NRSF.

Copyright © American Institute of Chemical Engineers. All rights reserved.

470794 NANOG Restores the Impaired Contractile Function of Senescent Mesenchymal Stem Cells

Abstract:

Contractility is an essential function of tissue engineered constructs containing smooth muscle cells such as blood vessels and bladders. Although, mesenchymal stem cells (MSC) have been extensively used as a source of smooth muscle cells (SMC), recent studies in our laboratory showed deficits in the myogenic differentiation capacity of senescent MSCs due to decrease in the formation of actin filaments. Unfortunately, current strategies for improving the contractile function have been unsuccessful in restoring this function after senescence, since senescent cells failed to respond to extracellular signaling factors. Interestingly recent work in our laboratory showed that ectopic expression of a single pluripotent transcription factor, NANOG, could reverse aging and restore the impaired contractile properties of aged MSC. In this study, we sought to identify the effects of NANOG on the actin filament organization and force generation capacity of MSC in a 3D microenvironment.

To this end, Human Hair Follicle Mesenchymal Stem Cells (hHF-MSCs) were isolated from a 73 year old donor and senesced by serial passaging in vitro. The cells were transduced with a tetracycline regulatable vector that tightly controls the expression of NANOG by the addition of the tetracycline analogue, doxycycline (Dox). The protein and mRNA levels of Î±-smooth muscle actin (ACTA2) were quantified by western blotting and real time PCR, respectively. Organization of actin filaments were visualized using immunocytochemistry techniques. To measure the biomechanical force and map the biochemical properties of cells in 3D tissues, we utilized micro-fabricated tissue gauges (ÂµTUGs) that contain a pair of flexible micropillars. The statistical significance between two mean values was analyzed using two-tailed unpaired Studentâ€™s t-test values of $p<0.05$ were considered statistically significant.

We found that, cellular senescence impaired the expression of myogenic genes, actin filament organization and the force generation capacity of MSC derived SMC. Notably, expression of NANOG restored all these functions,

ultimately reversing cellular senescence. Interestingly, while senescent cells exhibited impaired TGF-Î²1 pathway, NANOG completely restored this pathway and reversed the loss of contractile force generated by senescent 3D microtissues (Fig. 1). We will show that the mechanism involves activation of SMADs, which were impaired in senescent cells. We also investigated whether NANOG must be continuously present in order to elicit its effects. Despite the short half-life of NANOG protein (2.5Â±0.5 hr), the effects of NANOG on myogenic gene and protein expression and contractile function were sustained up to 10 days after termination of expression. This result may suggest interesting epigenetic regulation that may provide long-term recovery of the senescent phenotype of 3D tissues. We are currently investigating such epigenetic events.

In conclusion, ectopic expression of the embryonic transcription factor, NANOG, restored actin polymerization and the contractile function of senescent MSC-derived SMC, suggesting that NANOG may be used to enhance the contractile properties of tissue engineered constructs containing SMC, such as bladder and arteries. Given the effects of NANOG on SMC differentiation and contractile fucntion, it is possible that NANOG might also affect the contractile function of other types of muscle such as cardiac or skeletal muscle.

Copyright © American Institute of Chemical Engineers. All rights reserved.

472164 Supported Biomembrane Microenvironments of Controlled Composition for Gamma-Secretase Substrate Cleavage Assays

Abstract:

The intramembrane protease gamma-secretase is a current target of therapeutic intervention, with pivotal pathological functions within Alzheimerâ€™s disease and cancer. Our primary objective is to expand methods for *in vitro* studies of the intramembrane proteases (IMPs) with the development of a microsphere-supported (proteolipobead) and planar biomembrane platforms. IMPs are found throughout all branches of life and their functions are extremely broad. Despite extensive studies, understanding of IMP regulation and catalysis has been hampered by membrane-associated enzymology. Rhomboids and Secretases are polytopic membrane proteases that are widely conserved in all organisms. The precise reaction mechanisms of the intermembrane proteases remain to be elucidated and furthermore, rigorous analysis of the kinetics of interfacial catalysis in these systems has not yet been undertaken.

High throughput of screening of drug candidates has been carried out in bulk assay systems using cell membrane fragments, solubilized enzymes and is underway in proteoliposomes. A critical barrier to further progress in the study and HTS of gamma-secretase is that such bulk systems do not allow for the direct in situ quantification of enzyme, substrates, or inhibitors or their relative distributions within the structures under assay. We have expanded in vitro models of gamma-secretase to include supported biomembranes, enabling: 1) characterization and verification of biomembrane loading of substrates, inhibitors and protein effectors, 2) studies of co-localization, phase-partitioning and lateral mobility of membrane-bound assay constituents, and 3) the development of flow cytometry-based assays of cleavage. In this work we have characterized the structures, phase localization and compositions of gamma-secretase proteolipobead and planar systems by employing correlative optical and surface microscopy from the micro- to the nanoscale. We have probed the formation of Lo domains in situ, monitored using 3D FRET phase detection. We have probed the phase partitioning of enzyme, substrates and cleavage products in PLB systems with superresolution microscopy. A combination of biomembrane mobility methods are being used to probe the diffusivities and mobile fractions of lipid, substrates and gamma-secretase. The results of phase partioning and diffusivity measurements are being integrated into bottom-up spatial diffusion-reaction models to correlate microenvironment with enzyme function. As new drug targets in cancer and infectious disease involved in regulated intramembrane proteolysis (RIP) based cell signaling are uncovered, it is expected that other IMP enzymes can be subsequently studied and probed at high throughput with this platform.

Copyright © American Institute of Chemical Engineers. All rights reserved.

470812 Optimizing Process Yields of a Legacy 25,000L Mammalian Cell Culture Process Using Integrated Upstream and Downstream Process Yield Models

Abstract:

Process monitoring is defined as a continuous effort to evaluate manufacturing parameters across multiple lots to ensure commercial manufacturing processes are operating in-control within specifications. As it becomes increasingly important to reduce the cost of goods for cell culture processes, scientists and engineers require digital data-acquisition and aggregation systems to develop a robust process monitoring approach for the entire process. These automated data acquisition tools assist in rapidly contextualizing data over the course of a process lifecycle and can identify correlations across an entire process (cell culture and purification). In particular, advent of digital plant combined with development of informatics tools to generate, acquire, analyze and visualize large data sets for process monitoring has led to increased process understanding, reduced cycle time, and implementation of process improvements based on analysis performed with the tool.

We will present a case study where predictive models of production bioreactor and chromatography steps were used in tandem to identify sources of variability in overall process yield for a 25,000L fed-batch CHO process. From this analysis, bioreactor culture duration was identified to negatively impact glycosylation levels of the recombinant fusion protein produced from the culture. By implementing calibrated, data-driven modifications to the upstream manufacturing schedule, we were able to realize tighter control of upstream protein quality, optimal downstream process performance, and improved overall process yield.

Copyright © American Institute of Chemical Engineers. All rights reserved.

470647 Intelligent Control Strategies Towards the Intensification of Monoclonal Antibody Production Via Continuous Operation

Abstract:

Current industrial trends encourage the development of sustainable, environmental friendly processes with minimal energy and material consumption. Meanwhile, the increasing market demand as well as the tight regulations in product quality, necessitate efficient operation procedures that guarantee products of high purity. In this direction, process intensification via continuous operation paves the way for the development of novel, eco-friendly processes, characterized by higher productivity compared to batch.
This shift could benefit the market of high value biologics, such as monoclonal antibodies (mAbs), as it would lead to shorter production times, decreased costs, as well as significantly less energy consumption compared to batch [1, 2]. In particular, mAb production comprises two main steps: the culturing of the cells (upstream) and the purification of the targeted product (downstream). Both processes are highly complex and their performance depends on various parameters. In particular, the efficiency of the upstream depends highly on cell growth and the longevity of the culture, while the product quality can be jeopardized in case the culture is not terminated timely. Similarly, downstream processing, whose main step is the chromatographic separation, relies highly on the setup configuration, as well as on the composition of the upstream mixture. Therefore, in order to be able achieve a sustainable and efficient continuous operation, it is of vital importance to design intelligent computational tools that form a solid basis for the: (i) execution of cost-free comparisons of various operating strategies, (ii) design of optimal operation profiles and (iii) development of advanced, intelligent control systems that can maintain the process under optimal operation, rejecting disturbances.

This work focuses on the development of advanced control strategies for: (i) a cell culture system in a bioreactor and (ii) a semi-continuous purification process. More specifically, we consider fed-batch culturing of GS-NS0 cells producing a chimeric IgG4 antibody (based on the feeding of five key nutrients) and the semi-continuous Multicolumn Countercurrent Solvent Gradient Purification (MCSGP) [3-5] process for the purification of the upstream mixture. For the controller development we follow the generic framework & software platform PAROC that considers: (i) development of a high-fidelity process model, (ii) approximation of the complex, process model, (iii) design of the multi-parametric controller, (iv) 'closed-loop', in-silico validation of the controller against the process model [6]. PAROC allows 'in-silico' testing of the advanced controllers against the high-fidelity process model and evaluate their performance before operating them online. The results from this study indicate that the in-silico upstream/downstream integration can be achieved and the designed controllers succeed in maintaining both systems under optimal operation. In particular, the upstream system is characterized by a significantly increased mAb titer under the operation of the controller. Moreover, the controller indicates the time point that the culture needs to be terminated in order to minimize the risk of product degradation. Similarly, the downstream control scheme manages to maintain a stable operation throughout the process cycle, tracking efficiently the predefined setpoints. In addition, the controller is immunized against the uncertainty that results from the composition of the upstream mixture by considering the latter as measured disturbance. Finally, the suggested controller input indicates decreased use of utilities, while maintaining process efficiency.

Acknowledgements

The authors would like to thank Mr R. Oberdieck and Miss A. Quiroga-Campano for their contribution in the assessment of the controller validation. The authors would also like to thank Mr. Fabian Steinebach & Prof. M. Morbidelli from ETH Zurich, as well as Dr. Thomas Mueller-Spaeth from ChomaCon AG for their valuable input on the understanding of the MCSGP process. Financial support from the European Commission (OPTICO/G.A. No.280813) & Texas A&M University are also gratefully acknowledged.

References

1. Konstantinov, K.B. and C.L. Cooney, *White Paper on Continuous Bioprocessing May 20-21, 2014 Continuous Manufacturing Symposium.* Journal of Pharmaceutical Sciences, 2015. **104**(3): p. 813-820.

2. Xenopoulos, A., *A new, integrated, continuous purification process template for monoclonal antibodies: Process modeling and cost of goods studies.* J Biotechnol, 2015. **213**(0): p. 42-53.

3. Aumann, L. and M. Morbidelli, *A continuous multicolumn countercurrent solvent gradient purification (MCSGP) process.* Biotechnology and Bioengineering, 2007. **98**(5): p. 1043-1055.

4. Krättli, M., F. Steinebach, and M. Morbidelli, *Online control of the twin-column countercurrent solvent gradient process for biochromatography.* Journal of Chromatography A, 2013. **1293**: p. 51-59.

5. Papathanasiou, M.M., et al., *Advanced control strategies for the multicolumn countercurrent solvent gradient purification process.*AIChE Journal, 2016.

6. Pistikopoulos, E.N., et al., *PAROC—An integrated framework and software platform for the optimisation and advanced model-based control of process systems.* Chemical Engineering Science, 2015. **136**: p. 115-138.

Copyright © American Institute of Chemical Engineers. All rights reserved.

470938 Development of a Scalable Process for Production of a Broadly Protective Influenza Antigen

Abstract:

While most influenza infections cause only mild illness, the World Health Organization (WHO) estimates that, on average, 3 to 5 million people worldwide develop respiratory or heart complications related to seasonal influenza infections each year. The WHO also estimates 250,000-500,000 influenza associated deaths yearly, with a large percentage occurring in developing world children with inadequate access to vaccines. Such mortality would be much worse for a virulent pandemic influenza strain. The current standard of care for influenza immunization is a trivalent whole-inactivated virus vaccine consisting of two influenza A strains (H1N1 and H3N2) and one influenza B strain. Due to antigenic drift, the influenza virus is constantly changing and experts must predict the dominant strains for the upcoming flu season roughly 7-9 months in advance of distribution. As a result, the actual circulating viruses may differ from those included in the vaccine, causing an increase in the number of influenza cases and the number of influenza-related deaths as was the case during the 2014-15 flu season. Once the three strains are selected, the majority of the vaccine is produced in eggs and treated for inactivation. This egg-based method takes roughly 6 months. As a result of the lengthy production times, vaccines cannot be updated to more accurately reflect the circulating strains in the case of a poor prediction.

A broadly protective influenza vaccine that would elicit strong protective responses for most or all possible circulating strains would dramatically reduce influenza-related hospitalizations and deaths by conferring life-long immunity. Our proposed vaccine would address many deficiencies in the current vaccine strategy by targeting a conserved region of the influenza hemagglutinin protein to protect against both seasonal and pandemic influenza strains. We will use a stabilized virus-like particle scaffold to present the antigen in a multivalent manner which more closely mimics the presentation of the antigen on the natural virus. By providing broadly protective immunity worldwide against seasonal and pandemic influenza, we would decrease the cost of medical care for influenza related illnesses, and dramatically reduce the number of influenza related deaths.

We have developed a stabilized hemagglutinin stem antigen which has the potential to provide broad protection against seasonal and pandemic strains of influenza. Here we discuss the development of a scalable refolding and formulation process for the production of our stabilized hemagglutinin stem antigen. By evaluating the kinetic and molecular forces required for the proper folding and trimerization of the hemagglutinin stem antigen, we have transitioned a number of lengthy, industrially unattractive refolding steps to a more reliable, scalable process. We are also working to further stabilize our hemagglutinin stem antigen for improved conformational stability with the ultimate goal of producing a cold-chain independent vaccine which would allow for improved distribution worldwide. We plan to evaluate both lyophilization and spray drying techniques for producing solid drug substance compatible with long term room temperature storage.

Copyright © American Institute of Chemical Engineers. All rights reserved.

467571 Application of Process Monitoring Data Management Systems to Make Multiple Incremental Improvements to Cell Cultures Processes at the Manufacturing Scale

Abstract:

Process monitoring data management systems have been important tools to capture the large number of parameters and conditions that influence both the product quality and productivity of cell culture processes. Recent improvements in process analytical technologies and refinements in process control strategies have improved the biopharmaceutical industryâ€™s ability to gather large, multidimensional data sets that can be used for statistical process monitoring. In this work, we discuss the implementation of a data management platform and its application to build understanding around mammalian cell culture processes at the manufacturing scale. The database comprises of the end to end collection of process information from vial thaw to drug substance release for several commercial biologics produced using mammalian cell culture.

We explore the use of this database to make several incremental improvements to the cell culture process through a series of case studies including responses to health authority requests, technology transfers between multiple facilities, and evaluations of upstream parameters to understand their impact on final drug substance output. These case studies highlight the gains in process knowledge that can be achieved using these data management tools and illustrate some of the challenges in developing these capabilities. While data management tools facilitated the rapid retrieval of available data, there were several obstacles associated with data structure, data acquisition, and the subsequent data analysis. The current work demonstrates the successful application of data management tools to improve the performance and robustness of existing cell culture processes at the manufacturing scale.

Copyright © American Institute of Chemical Engineers. All rights reserved.

471374 Scalable Production of Glioblastoma Cancer Stem Cells in 3 Dimension Thermoreversible Hydrogels for Drug Discovery

Abstract:

There is growing interest in developing drugs that specifically target glioblastoma cancer stem cells. Current cell culture methods, however, cannot cost-effectively produce the large numbers of glioblastoma cancer stem cells required for drug discovery and development. In this paper we report a new method that encapsulates patient-derived primary glioblastoma cancer stem cells and grows them in 3 dimension thermoreversible hydrogels. Our method allows long-term culture (~ 50 days, 10 passages tested, accumulative ~>10^{10}-fold expansion) with both high growth rate (~ 20-fold expansion / 7 days) and high volumetric yield (~ 2.0 x10^7 cells/ml) without the loss of stemness. The scalable method can be used to produce sufficient, affordable glioblastoma cancer stem cells for drug discovery.

Copyright © American Institute of Chemical Engineers. All rights reserved.

470399 Levers to Modulate Product Quality Attributes of Product-a in CHO Cell Culture

Abstract:

Levers to modulate two product quality attributes were investigated as part of process development of Product-A produced in Chinese Hamster Ovary (CHO) cell culture. The addition of a chelating agent to the cell culture fluid was identified as a lever to minimize product fragment formation which indicates that metallo-proteases could potentially be contributing to fragment formation. A study to understand the nature and formation of product size variants identified that aggregate formation was predominantly intracellular and occurred prior to secretion and the fragments were formed in the cell culture fluid after secretion. In addition, media additives and process conditions were evaluated for potential levers to modulate product quality attribute-2. Process parameter-1 was determined to be a robust lever for controlling product quality attribute-2 across various production scales.

Copyright © American Institute of Chemical Engineers. All rights reserved.

468324 Data-Driven Optimization Using an Evolutionary Design of Dynamic Experiments for Biopharmaceutical Processes

Abstract:

Biopharmaceutical processes are substantially complex and are not understood well enough to enable the development of an accurate knowledge-driven model for process optimization purposes. In such a case, a data-driven modelling approach provides a very attractive alternative, which, if designed properly, can quickly optimize the process with a small set of experiments. The recently introduced Design of Dynamic Experiments (DoDE) [1], a generalization of the traditional Design of Experiments (DoE) [2, 3], has demonstrated that it is an effective data-driven modeling and optimization approach [1, 4, 5].

Here we apply the DoDE method in an evolutionary manner to optimize biopharmaceutical processes while satisfying budgetary and developmental time constraints on the number of experiments in two representative biopharmaceutical processes; Penicillin fermentation [6] and Hybridoma cell culture [7]. We maximize the amount of product produced by determining the optimal feeding policy and batch duration.

To minimize the number of experiments we initially consider only a few dynamic sub-factors and the simplest Response Surface Model (RSM). After the initial results are analysed and an approximate RSM is estimated, additional experiments are added sequentially to further improve the data-driven model and the process performance. We demonstrate that with a reasonable amount of evolutionary experiments one can come very close to the model-based optimum. The model-based optimization [8], however, requires a very accurate knowledge-driven model that is rarely at hand.

References

1. Georgakis, C., *Design of Dynamic Experiments: A Data-Driven Methodology for the Optimization of Time-Varying Processes.* Industrial & Engineering Chemistry Research, 2013. **52**(35): p. 12369-12382.

2. Box, G.E.P. and N.R. Draper, *Response Surfaces, Mixtures, and Ridge Analysis*. 2007, Hoboken, NJ: Wiley.

3. Montgomery, D.C., *Design and Analysis of Experiments*. 8 ed. 2013, New York: Wiley. 729.

4. Fiordalis, A. and C. Georgakis, *Data-driven, Using Design of Dynamic Experiments, versus Model-driven Optimization of Batch Crystallization Processes.* Journal Of Process Control, 2013. **23**(2): p. 179-188.

5. Makrydaki, F., C. Georgakis, and K. Saranteas. *Dynamic Optimization of a Batch Pharmaceutical Reaction Using the Design of Dynamic Experiments (DoDE): The Case of an Asymmetric Catalytic Hydrogenation Reaction.* in *Proceedings of the 9th International Symposium on Dynamics and Control of Process Systems*. 2010. Oude Valk College, Belgium.

6. Bajpai, R.K. and M. Reuss, *A Mechanistic Model for Penicillin Production.* Journal Of Chemical Technology And Biotechnology, 1980. **30**(6): p. 332-344.

7. De Tremblay, M., et al., *Optimization of Fed-Batch Culture of Hybridoma Cells Using Dynamic Programming Single and Multifeed Cases.* Bioprocess Engineering, 1992. **7**(5): p. 229-234.

8. Biegler, L.T., *An Overview of Simultaneous Strategies for Dynamic Optimization.* Chemical Engineering And Processing, 2007. **46**(11): p. 1043-1053.

Copyright © American Institute of Chemical Engineers. All rights reserved.

465077 In silico Formulation Development of Oral Solid Dosage Forms Using a Material Property Database

Abstract:

Motivated by the desire to lessen the empirical nature of oral solid dosage formulation and to pursue a quality by design framework, a material property database (MPD) and a material science interface (MSI) were recently developed at Abbvie. The MPD contains critical material properties of common excipients along with the lot-to-lot variability and humidity dependence of these properties. The centralization of these material properties in the MPD, collected in a standardized and reliable manner, also facilitate the use of predictive modeling via the MSI. The MSI is a user-friendly graphical interface which allows access to the MPD and associated modeling tooling used to aid in formulation development. Currently, the MSI contains models to predict blend uniformity, powder flowability, and compactability of pharmaceutical formulations intended for a direct compression process by making use of material properties contained within the MPD. Examples utilizing the MSI models will be given for the formulation development of a model active pharmaceutical ingredient (acetaminophen). The examples given herein show the utility of the MPD and MSI for providing in silico exploration of formulation design space, better guidance and scientific rational for experimentation, and improved understanding in the relationship between material properties, manufacturability, and product performance.

Disclosures:

All authors are employees of AbbVie. The design, study conduct, and financial support for this research was provided by AbbVie. AbbVie participated in the interpretation of data, review, and approval of the publication.

Copyright © American Institute of Chemical Engineers. All rights reserved.

466851 Prediction of Tablet Dissolution from API and Excipient Polymer Properties

Abstract:

Prediction of Tablet Dissolution from API and Excipient Polymer Properties

Jaime L. Curtis-Fisk[1] (JLCurtisfisk@dow.com), True L. Rogers[1] (TLRogers@Dow.com), Karen M. Balwinski[1] (KMBalwinski@Dow.com), Kathryn O. Hewlett[1] (KOHewlett@Dow.com), Robert L. Schmitt[2] (SchmitRL@Dow.com), Shrikant N. Khot[1] (SNKhot@Dow.com)

[1] The Dow Chemical Company, Midland, MI USA

[2] The Dow Chemical Company, Collegeville, PA USA

Purpose: Modified-release (MR) matrix tablets are typically produced using hydroxypropyl methylcellulose (HPMC) as rate-modifying polymer. HPMC effectively sustains release of a broad range of active pharmaceutical ingredients (APIs), but specific dissolution behavior varies depending on the API properties, excipient properties, and formulation. Recently, a new line of HPMC, METHOCEL™ DC2 grades, have been introduced to facilitate direct compression formulations given its enhanced flowability. The ability to predict performance based on properties of components could streamline formulation development and provide a more direct route to troubleshooting challenging APIs. Combined such a capability with a direct compression manufacturing technique would mean faster and less expensive development and manufacturing. The intent of this study was to identify the API and polymer properties that most directly impacted dissolution performance, connect the observed performance to fundamental API and polymer chemistry, and develop predictability that connects this fundamental understanding to efficient formulation design.

Methods: The impact of API and polymer properties on dissolution was investigated by evaluating dissolution performance of APIs covering a broad range of solubility, shown in Figure 1. Tablets contained a standard formulation of 50% API, 30% METHOCEL DC2 HPMC, 19% impalpable lactose, and 1% magnesium stearate. A standard formulation was utilized for all APIs to make a direct comparison of the structure/activity relationship between API properties and dissolution performance. Tablets were prepared by hand-feeding powder formulations into the tablet press to remove variability in die fill due to powder flow differences between API formulation blends.

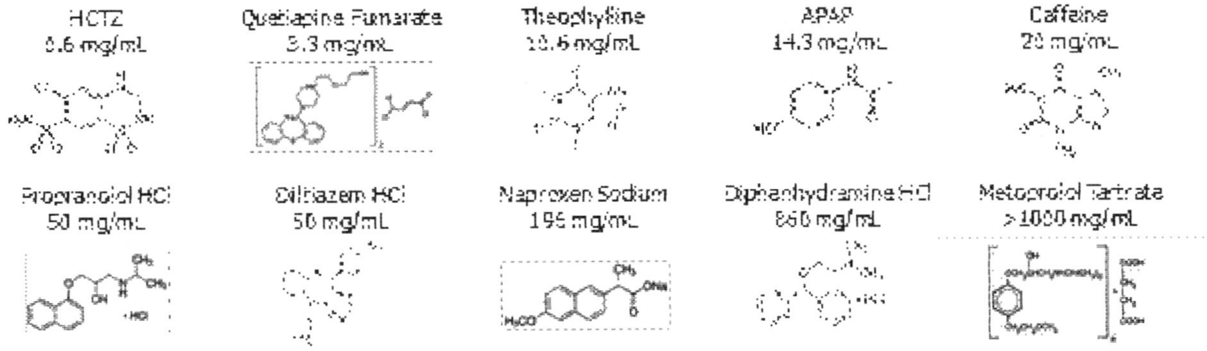

Figure 1. APIs with a broad range of solubility were evaluated in order to connect drug properties to dissolution performance.

Results: API dissolution performance was evaluated for three viscosity grades of METHOCEL DC2 (K100LV, K4M, K100M). Figure 2 displays the results of tablet formulations containing METHOCEL K4M DC2; percent API release at 2, 4 and 12 hours is plotted as a function of API solubility. The amount of API released at a given time point is greater as API solubility increases. The sensitivity to API solubility is most dramatic for intermediate solubility, with very low or high solubility API reaching a plateau where small changes in solubility to do not significantly impact release.

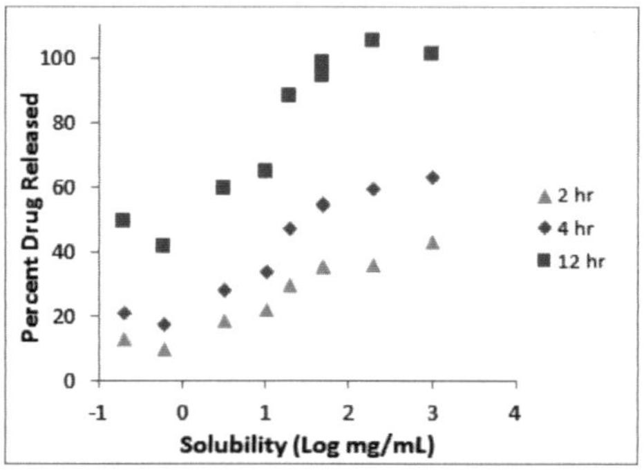

Figure 2. Percent API release as a function of API solubility, demonstrating a similar correlation at 2, 4, and 12 h dissolution time points.

The correlation of API solubility with drug release rate was also evaluated using t_{50} as a measure of release rate. This different perspective on the same datasets provides new insight into dissolution mechanisms of the tablet and the implications to formulation optimization. Figure 3 displays t_{50} values as a function of API solubility for three viscosity grades of METHOCEL DC2. All three viscosity grades demonstrate similar behavior; at low solubility there is a strong correlation between solubility and t_{50}, but at higher solubility the observed t_{50} value is constant. This behavior can be explained considering the competing mechanisms that occur during tablet dissolution. For poorly soluble APIs, the rate limiting step is API exposure to the greater sink of dissolution media as the outermost layer of swollen matrix erodes, followed by API dissolution into the media. When API solubility is greater, the limiting factor becomes diffusion of dissolved API through the swollen polymer layer and subsequent release into the greater sink of dissolution media. While K100M results in greater t_{50} values for all APIs evaluated, followed by K4M and lastly K100LV, the differentiation between grades was greatest for poorly soluble APIs. The difference in performance indicates that for low solubility APIs, polymer viscosity is a critical variable in formulation design.

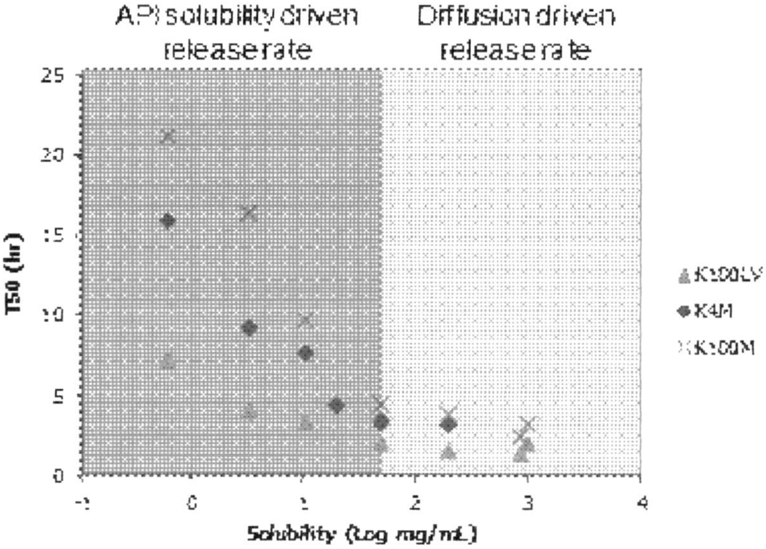

Figure 3. Dissolution rate described by** t_{50} **as a function of API solubility for three viscosity grades of METHOCEL DC2, demonstrating the correlation to API solubility.

Conclusions: A key learning from this study is that API release from matrix tablets prepared with the METHOCEL DC2 product line is strongly impacted by API solubility and polymer viscosity. APIs of low solubility are released through a dissolution mechanism limited by API solubility and erosion of the outermost surface of the swollen polymer layer, but at higher solubility the release is driven by diffusion of dissolved API through the swollen polymer network. While higher viscosity polymers demonstrated greater extended release, the impact of this effect is more pronounced for lower solubility APIs and became negligible for higher solubility APIs. These relationships provide formulation guidance on selecting the appropriate viscosity grade of METHOCEL DC2 based upon API solubility.

Copyright © American Institute of Chemical Engineers. All rights reserved.

463230 Connecting Processed Materials Properties to Process Development Strategy

Abstract:

A quality-by-design (QbD) process development case study would be presented using an example of a high drug load wet granulation formulation. This case study will identify the dependence of drug product performance quality attributes on material properties of intermediate granules. The translation of this mechanistic product and process understanding to process development, process analytical technology (PAT), and process control will be highlighted. The correlation of granule size versus porosity with tablet dissolution will be discussed. The implications of this correlation on the use of focused beam reflectance measurement (FBRM) versus drag flow force (DFF) sensor for process monitoring and control will be discussed. This talk will highlight the importance of material understanding to process development.

Copyright © American Institute of Chemical Engineers. All rights reserved.

458431 Decoupling of a Low-Dose Dosator Capsule Filling Process in Dynamic and Static Mode Tests to Understand the Extend of the Effect of Powder and Process Parameters on Capsule Quality Attributes

Abstract:

Introduction

The precise filling of capsules with small powder quantities is essential for a broad range of industrial operations and a number of low-dose capsule filling machines are currently available. Filling principles can be categorized in either volumetric (e.g., the dosator nozzles, vacuum drum filler, vacuum dosator and tamp filler) or gravimetric (e.g., micro-dosing) methods. Capsule filling using the nozzle dosator technique has been widely investigated and is an important technology applied in the pharmaceutical industry today especially for the filling of capsules for inhalation, as the doses need a controlled degree of compaction to ensure an efficient dose delivery [1] [2] [3]. It is known that a large number of powder and processing parameters affect the quality of filled capsules [4] [5] [6] [7] [8]. However, the precise dosing of small powder quantities still poses significant problems. Thus, the objective of this study is to decouple the dosing event, which consists of multiple steps in order to identify and understand in-depth the complex relationship between material attributes and process parameters and their influence on critical quality attributes (fill weight and fill weight variability). Supported by *in silico* simulation, we will link those material properties to process performance and develop a mechanistic understanding in order to optimize the low-dose dosator capsule filling process.

Materials and Methods

Three different grades of alpha lactose monohydrate (Lactohale100, 200, 220 – DFE Pharma) specially designed for inhalation application were used as received. The following material attributes were characterized in triplicate: particle size (Qicpic OASIS/L dry dispersing system Sympatec, Germany), bulk (BD) and tapped density (TD) (Pharmatest PT-TD200), true density (AccuPyc II 1340, Micromeritics, Norcross, USA). The compressibility (CPL), air permeability (PD), flow function coefficient (FFc), cohesion (C), angle of internal friction (AIF), basic flowability energy (BFE), wall friction angle (WFA) were measured with the FT4 powder rheometer (Freeman Technology, Malvern, United Kingdom). Results are displayed in table 1.

Table 1: Inhalation carrier characterization

	Lactohale 100	**Lactohale 200**	**Lactohale 220**
Particle size and friction properties			
Volumetric mean diameter [µm]	160.02 ±0.62	83.63 ±n.a.	16.38 ±n.a.
Angle of internal friction (AIF) [°]	18.43 ±0.49	21.24 ±0.53	25.73 ±2.66
Wall friction angle (WFA 3 kPa 0.2 Ra) [°]	7.70 ±0.02	11.46 ±0.82	19.97 ±3.82
Bulk powder properties			
Bulk density (BD) [g/ml]	0.6972 ±0.0036	0.6223 ±0.0031	0.3996 ±0.0067
Tapped density (TD) [g/ml]	0.8275 ±0.013	0.9962 ±0.0017	0.7847 ±0.0066
True density [g/ml]	1.5385 ±0.0028	1.5426 ±0.0024	1.5466 ±0.0042
Compressibility (CPL) at 8 kPa (ratio ρ_{comp}/ρ_{BD}) [%]	1.05 ±0.00	12.66 ±0.29	36.95 ±0.16
Air permeability (PD) at 8kPa [mbar]	1.05 ±0.02	7.00 ±0.10	12.04 ±0.55

Flow properties			
Basic flowability energy (BFE) [mJ]	910.67 ±43.00	1722.85 ±27.00	667.32 ±10.80
Flow function coefficient (FFc)	6.58 ±0.01	4.04 ±0.01	1.65 ±0.01
Cohesion (C) [kPa]	0.24 ±0.02	0.39 ±0.01	1.05 ±0.13

Dynamic capsule filling: The powders were filled into hard gelatin capsules of size 3 (supplied by Capsugel) with a lab scale capsule filling machine (Labby, MG2, Bologna) adopted with a special low dose equipment. Two volumes of dosing chamber (2.5, 5 mm), two different dosator sizes (1.9, 3.4 mm), two different powder bed heights (5, 10 mm) and two different filling speeds (500, 3000 cph) were used for the experiments. The study was performed under humidity controlled conditions (45-55 % relative humidity). To create an initial powder bed without densification, the layer was adjusted manually. After setting all machine parameters, capsules were collected at time point 0, after 25, 50, 75, 100 rotations for a total run time of 30 minutes in order to sample from the same position inside the rotating container. Powder feeding was adjusted according to the amount of powder collected during capsule filling. Subsequently, the weight of filled and empty capsules was measured with a Denver (SI-234A) scale.

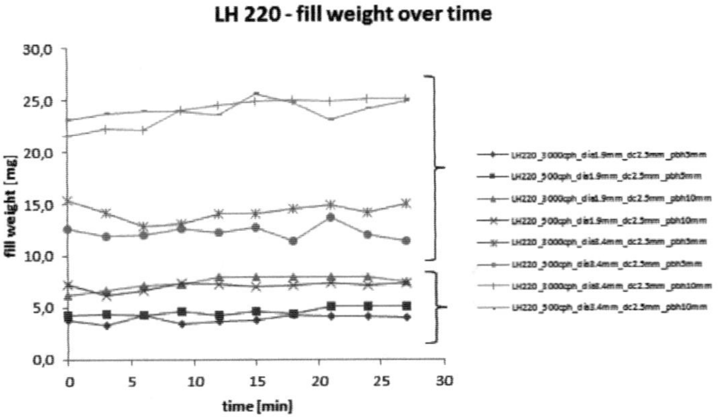

Figure 2: Capsule fill weight of Lactohale220 at different combinations of process settings

Static capsule filling: The static mode capsule filling experiments were conducted under the same combination of process settings and environmental conditions as the dynamic mode tests. The difference here is that the powder is picked up from a static powder bed not a rotating drum in order to study the effect of powder layer homogeneity on capsule quality attributes and to get a mechanistic understanding about the powder compaction behavior inside the dosing chamber.

Simulation and Modeling

In addition to the experiments, DEM simulations of the static capsule filling experiments are performed. Therefore, a model of the dosator is implemented in the DEM software LIGGGHTS® and different particle contact properties are used to describe the filling behavior of different powders. The simulations show the evolution of the powder mass inside the nozzle during filling and therefore help to understand the dosator process in more detail.

Conclusion and Outlook

In this research, various material attributes of different grades of lactose were determined and their performance during capsule filling using a lab scale low dose dosator nozzle system was investigated in dynamic mode. Obtained results will be correlated with experiments conducted with the static stand-alone device and the simulation work to understand the influence of particular process parameters on the filling performance, which will furthermore lead to an improved filling process.

References

[1] B. E. Jones, "The filling of powders into two-piece har+d capsules," *Int. J. Pharm.*, vol. 227, no. 1, pp. 5–26, 2001.

[2] M. Llusa, E. Faulhammer, S. Biserni, V. Calzolari, S. Lawrence, M. Bresciani, and J. Khinast, "The effect of capsule-filling machine vibrations on average fill weight," *Int. J. Pharm.*, vol. 454, no. 1, pp. 381–387, 2013.

[3] M. Llusa, E. Faulhammer, S. Biserni, V. Calzolari, S. Lawrence, M. Bresciani, and J. Khinast, "The effects of powder compressibility, speed of capsule filling and pre-compression on plug densification," *Int. J. Pharm.*, vol. 471, no. 1, pp. 182–188, 2014.

[4] S. B. Tan and J. M. Newton, "Powder flowability as an indication of capsule filling performance," *Int. J. Pharm.*, vol. 61, no. 1, pp. 145–155, 1990.

[5] R. Patel and F. Podczeck, "Investigation of the effect of type and source of microcrystalline cellulose on capsule filling," *Int. J. Pharm.*, vol. 128, no. 1–2, pp. 123–127, 1996.

[6] F. Podczeck and J. M. Newton, "Powder and capsule filling properties of lubricated granulated cellulose powder," *Eur. J. Pharm. Biopharm.*, vol. 50, no. 3, pp. 373–377, 2000.

[7] E. Faulhammer, M. Llusa, C. Radeke, O. Scheibelhofer, S. Lawrence, S. Biserni, V. Calzolari, and J. G. Khinast, "The effects of material attributes on capsule fill weight and weight variability in dosator nozzle machines," *Int. J. Pharm.*, vol. 471, no. 1–2, pp. 332–338, 2014.

[8] E. Faulhammer, M. Fink, M. Llusa, S. M. Lawrence, S. Biserni, V. Calzolari, and J. G. Khinast, "Low-dose capsule filling of inhalation products: Critical material attributes and process parameters," *Int. J. Pharm.*, vol. 473, no. 1–2, pp. 617–626, 2014.

Copyright © American Institute of Chemical Engineers. All rights reserved.

467102 Quantitative Investigation on Segregation Phenomenon Using Discrete Element Method in Large Scale Pharmaceutical Compression Process

Abstract:

Segregation is a common issue in pharmaceutical manufacture of solid dose drug product. Especially during manufacture scale-up, pharmaceutical powder with segregation tendency has potential risk to fail specification of blend or content uniformity in large scale process. It may result in significant waste of drug substance or cause possible quality issues for the drug product. Moreover, experimental investigation on large scale segregation is usually inefficient to determine root cause or provide quantitative improvement plan while the consumption of drug substance was still significant. In this work, quantitative Discrete Element Method (DEM) simulation method was developed and applied on the prediction of segregation tendency of powder blend in large scale pharmaceutical unit operations.

In the large scale drug product manufacture of Project X, segregation was consistently observed in the last quarter of the tablet compression process. For the two major blend components in this project, material properties such as particle size distribution, true density, and flowability were measured and imported into simulation. The previous established correlation between shear flow of experiment and simulation was applied to guarantee that the flow behavior of simulated DEM particles was representative of real pharmaceutical powder [1]. Several slice geometries were developed to represent the dimensions of the current blender- hopper combination and alternative candidates for geometry improvement. The blender discharge and hopper feeding process in the bulk tablet compression of project X was then simulated using the commercial DEM software Star CCM+® v11.02 by CD-Adapco® and with processors of dual CPUs, Intel® Xeon® E52640 @ 2.50 GHz.

Based on the simulation results, difference on true density of the blending components was determined as the root cause of segregation in the compression process of Project X. Geometries show different segregation tendency with the maximum CU range 103% - 120% due to different powder flow and segregation mechanics in these geometries. It was observed that segregation mainly took place on the surface of inclined material bed during the discharging and feeding processes and was significantly influenced by the relative location of outlet. The method was then validated by performing feeding experiment of Project X final blends with 3D printed blender-hopper geometries, which match well with the segregation behavior of corresponding simulations. Suggestions were therefore made on the setting of hopper geometry for further manufacture activity. The case study shows robust applicability of the proposed simulation method on pharmaceutical powder flow and segregation. This method has potential benefit on significant saving of cost and drug substance, as well as risk reduction of batch failure especially in scale-up of solid dose manufacture.

1. Gao Y., Chalasani, S., Mittal, B., Investigation on Key Parameters of Common Pharmaceutical Blending Unit Operations Using Discrete Element Method, AIChE Annual Meeting, Salt Lake City UT, Nov 2015, paper 412e.

Copyright © American Institute of Chemical Engineers. All rights reserved.

454354 Two-Phase Flow (solids: interstitial air) Modeling of a Press Feed System

Abstract:

Production rate limitations can develop when feeding fine powders to a press, granulator or mill. These powder feed rate limitations arise when the powder interacts with entrained air, resulting in adverse â€œtwo-phaseâ€ (powder-air) flow effects. Two-phase flow problems can result in out-of-specification tablets, requiring operating at below target production rates and can occur in a continuous or batch process. A critical step in developing a robust process, including conducting Quality by Design (QbD) studies, is modeling two-phase flow behavior in the press feed system.

Two-phase flow risks can be accounted for during process design by:

- Measuring key flow properties to use as inputs in the two-phase flow model;
- Computational modeling the two-phase flow effects based upon first principle models;
- Evaluating potential corrective actions through two-phase modeling.

The two-phase modeling software uses key powder flow properties, including:

- Permeability
- Compressibility
- Specific gravity of a particle
- Wall friction angles (phi-prime, fâ€™)
- Effective angle of internal friction (delta, d) measured via a cohesive strength test.

The two-phase model as includes the press feed system geometry and process parameters as inputs. The model outputs include parameters that dictate two-phase flow behavior, such as vertical solids stress, interstitial gas pressure, bulk density and normalized gas pressure gradient. These modeling outputs are used to predict whether flow problems will occur and analyze potential corrective actions, including modifying the geometry of the press feed system, increasing the permeability of the powder and adding venting.

Copyright © American Institute of Chemical Engineers. All rights reserved.

462120 Use and Misuse of the Diametrical Compression Test

Abstract:

The diametrical compression (DC) test has been used extensively to derive the tensile strength of pharmaceutical tablets from the measured breaking force. DC-derived tensile strengths provide a good approach to measuring lot-to-lot variability in the mechanical properties of tablets. For these quality control type applications, method precision is required, but accuracy is not. Additionally, DC has been used to calibrate parameters of the Druker Prager Cap (DPC) model, a yield criterion expressing the failure of a powder compact under arbitrary 3D loading conditions. For this application, the DC method must not only provide suitable precision, but also accuracy. In this work we explore the accuracy of the DC method in the measurement of tensile strength and conclude that the true tensile strength of a powder compact is approximately double the DC-derived value. While historical literature assumes that tablets fracture under tension along the centerline of the tablet, analysis of the stress state suggests that tablets are likely to fracture under shear. The impact of this ~50% error should be considered when accuracy of the tensile strength result is required.

Copyright © American Institute of Chemical Engineers. All rights reserved.

470464 Engineering Proteins for Bioorthogonal Surface Conjugation of Gold Nanoparticles

Abstract:

Engineering Proteins for Bioorthogonal Surface Conjugation of Gold Nanoparticles

Julia G. Fraseur[a], Katherine N. Clayton [b], Tamara L. Kinzer-Ursem[a]

[a]Weldon School of Biomedical Engineering, Purdue University[b]School of Mechanical Engineering, Purdue University

As the field of bionanotechnology is rapidly expanding, researchers are taking advantage of nanoparticle properties for the design of vehicles for nanomedicine, therapeutics, and biosensing. Key to the advancement of the field are the development of techniques to bind biomolecules of interest to nanoparticles to produce stable, highly bio-active nanoparticle conjugates. We have developed a novel method for *in situ*conjugation of N-terminal azide-tagged proteins onto gold nanoparticle surfaces with consistent orientation, low aggregation, and high levels of activity.

In this work we selectively label proteins of interest co-translationally in recombinant expression systems using a chemoenzymatic tagging method. This method covalently labels protein with azide functionality that provides a handle for a rapid, bioorthogonal conjugation to nanoparticles via the azide-alkyne cycloaddition reaction, a popular â €œclickâ€ chemistry. Simultaneously, we have developed a method for functionalizing streptavidin-coated gold nanoparticles with a heterobifunctional PEG terminated with biotin and dibenzocyclooctyne (biotin-PEG4-DBCO) in order to conjugate the azide-tagged protein covalently to the nanoparticle surface. This chemistry allows for proteins to be consistently oriented onto the gold nanoparticle regardless of mixture complexity. With the azide tag, our target protein can be conjugated to the particle from a purified state or cell lysate.Â Although aggregation of nanoparticles post-conjugation can be a common occurrence when working with nanoparticles, we have optimized our methods such that protein-conjugated nanoparticles remain monodisperse post-conjugation. Furthermore, we show high levels of protein activity on the nanoparticles, ensuring that neither our tagging method, nor our conjugation methods inhibit protein functionality.

Here we will discuss the underlying methodology of protein tagging and nanoparticle conjugation by click chemistry. This technique streamlines the conjugation process and provides an opportunity for the successful functionalization of target proteins to gold nanoparticles directly from cell lysate. Future applications of these methods include characterization of protein-protein interactions on nanoparticles, biomarker detection, biosensing, and molecular diagnostics.

Copyright © American Institute of Chemical Engineers. All rights reserved.

457546 Detection of Peptide Biomarkers By Engineered Yeast Receptors

Abstract:

Access to low-cost, point-of-care (POC) diagnostics enables screening, diagnosis and treatment monitoring for a number of critical diseases. POC diagnostics, such as lateral flow sandwich immunoassays, nucleic acid amplification tests, and colorimetric chemical reactions, are available for proteins, DNA/RNA, and small molecule disease biomarkers. However, there is a technology gap for assaying peptides and small proteins biomarkers smaller than 15 kDa. These biomarkers are often too small to be simultaneously bound by two antibodies as required for sandwich immunoassays and instead require more sophisticated assays like immunoturbidimetric measurements and mass spectrometry, which can prove impractical and prohibitively expensive at the POC level. Thus, a cost-effective POC peptide diagnostic can improve early detection and treatment management for a range of diseases.

Bakerâ€™s yeast (*Saccharomyces cerevisiae*) is a particularly promising vehicle for a POC peptide diagnostic because it is cheap, robust, and easily engineered with standard biological techniques. Active dry yeast is simple to manufacture and can be freeze-dried and reconstituted to provide easy distribution without requiring a cold-chain. Yeast contains a native G-protein coupled receptor (GPCR) which detects the 13 amino acid a-factor peptide. We hypothesized that Ste2p may be a useful starting point for directed evolution to produce receptors for peptide biomarkers that cannot currently be assayed at the POC.

Directed evolution has been used to evolve GPCRs to detect small molecules, but not peptides. Directed evolution experiments ideally start with proteins that have a low basal activity that can be improved upon. However, most peptide biomarkers have sequences that are very dissimilar from Î±-factor, so Ste2p is not likely to have basal activity for the peptide biomarker. The sequence dissimilarity, representing a large evolutionary distance, is challenging as many simultaneous mutations would likely be required for the receptor to detect the biomarker. To overcome this difficulty, we chose substrate walking as an alternative technique for covering large evolutionary distances. In substrate walking, an enzyme is evolved along a pathway of chimeric substrates which contains features of both the native and desired substrates before being evolved for the desired substrate. Substrate walking with peptides-receptor pairs has not been explored, so we investigated the feasibility and constraints of this strategy.

Here, we chose a peptide biomarker of chronic kidney disease (CKD). CKD, the 12thleading cause of death worldwide, affects 10% of the adult American population and is a significant, emerging burden in low and middle income countries. As CKD worsens, cystatin C accumulates in the serum and urine. Clinically, cystatin C satisfies the criteria for a universal biomarker as it is independent of age, weight, and muscle mass and is more sensitive to early-onset renal failure compared to creatinine, a commonly used biomarker for CKD. Cystatin C is typically measured with a nephelometer, which is costly and requires substantial infrastructure to use and maintain. We used tryptic digest to ensure heterogeneous degradation products are cleaved into standard peptides. Trypsin digestion of cystatin C yields an 11 amino acid peptide that can serve as a CKD biomarker.

In this study, we demonstrate YBBs as a promising, versatile platform for POC diagnostics. We present a generalizable platform for producing peptide biomarker receptors using substrate walking and directed evolution, and validate the approach by developing a YBB for cystatin C, a biomarker for renal failure. Finally, we demonstrate the clinical feasibility of the YBB by showing functionality in pooled human urine, a relevant sample matrix. To our knowledge, this is the first demonstration of engineering a GPCR to detect a fully orthogonal peptide sequence.

Copyright © American Institute of Chemical Engineers. All rights reserved.

472313 Engineering Recombinant Protein Sensors for Probing Epigenetic States in Live Cells

Abstract:

Epigenetics is a term that encompasses heritable changes in gene expression that do not change the underlying DNA sequence of the gene. Epigenetic modifications include chemical modifications such as methylation of DNA and acetylation, methylation or phosphorylation of histone protiens. These modifications play a key role in cellular processes such as differentiation. Aberrant changes in epigenetic modifications have been correlated with various types of cancers and neurological disorders. Current detection techniques of epigenetic changes require tedious sample pre-treatments and can be carried out only using a population of cells. Monitoring epigenetic changes in live cells will enable real-time characterization of cell response to various external factors such as environmental chemicals and drugs which are of primary importance to pharmaceutical industry and regulatory affairs agencies. Combined with state-of-the-art fluorescent techniques, we demonstrated the ability of our recombinant sensors in detecting epigenetic changes using both cellular extracts and *in situ.* The probe is compatible with live cell assays and can sensitively quantify the response of individual cells to drug treatments. The developed platform is ideal for monitoring in live cell changes in the epigenetic state and can be a extremly valuable tool for drug development and toxicology studies.

Copyright © American Institute of Chemical Engineers. All rights reserved.

465795 Purification-Free Antibody Fragment Immobilization Using in *Vivo* Biotinylation

Abstract:

Antibody fragments are attractive tools in research and diagnostics due to their strong antigen-binding ability and their ease of production in *Escherichia coli*. Such applications could benefit from an array of different antibody fragments immobilized on a surface, and constructing antibody arrays would necessitate an immobilization method that is amenable to many antibody fragments. We have established a simple immobilization method for single-chain variable fragment (scFv) antibodies that is based on the biotin-streptavidin interaction and does not require purifying scFvs prior to immobilization. We genetically fused two biotinylation tagsâ€"the biotin carboxyl carrier protein (BCCP) or the AviTag minimal sequenceâ€"to scFvs with different frameworks and antigens. scFvs were site-specifically biotinylated *in vivo* by endogenous *E. coli* biotin ligases during protein expression. All scFv fusions were successfully immobilized directly from cell lysates onto streptavidin-coated plates. We found that scFvs biotinylated via the BCCP tag tended to immobilize better than those biotinylated via the AviTag, even when biotinylation efficiency was improved with the biotin ligase BirA. The immobilized scFvs retained the ability to bind to their antigen, whether the antigen is a purified sample or it is detected in a cell lysate. The immobilization is stable, and the scFvs maintain their functionality to bind to their target for at least 100 days after immobilization and storage at 4 °C or -20 °C. The simplicity and robustness of our method make it a promising approach for future applications that require immobilization of multiple antibody fragments.

Copyright © American Institute of Chemical Engineers. All rights reserved.

466450 A Rapid Diagnostic Assay for Protein Biomarkers Using Whole Cell Biosensors

Abstract:

Synthetic biology is an emerging discipline which seeks to develop new biological systems with user-defined properties by adopting an engineering approach based largely on standardization, characterization, and modularity. Biosensors are one area where synthetic biology has been successfully applied with examples ranging from small molecule metabolites through to toxins and environmental contaminants.

Whole cell biosensors offer a convenient platform for detection as they are self-renewing and easy to transport. However, when trying to sense proteins, designers of whole cell biosensors are faced with an issue of how to transmit information across the cell membrane. To address this challenge, we have developed a class of whole cell biosensors based on the bacterial surface display of an antibody fragments that bind different epitopes of a protein biomarker. The presence of the biomarker induces cell aggregation, which can be transmitted as a downstream signal through optical assays or more sophisticated genetic circuitry, such as activation of bacterial quorum sensing.

Using a dimeric GFP as a model protein biomarker, we developed a semi-quantitative assay procedure that leads to a visual output in as little as 15 minutes. We then showed that the assay can be easily developed for a new biomarker by exchanging the nanobody on the cell surface for one that binds to fibrinogen. Further, we have developed a predictive mathematical model based on the aggregation of colloidal particles to help optimise the procedure for detection within a specific range by modulating nanobody expression level and the number of cells used in the assay. In summary, our 'whole cell ELISA' procedure combines the advantages of traditional antibody detection with the cost effectiveness of whole cell biosensors.

Copyright © American Institute of Chemical Engineers. All rights reserved.

469336 Biosensing Chemicals That Target Nuclear Hormone Receptors: Introducing a Rapid, Versatile Cell-Free Protein Synthesis Approach

Abstract:

The growing concern of the ubiquitous presence of endocrine disrupting compounds in our environment and their effects on human and animals has catalyzed biosensor developement to detect and study the interaction of these chemicals with nuclear hormone receptors (NHRs). Although huge progress has been made, available biosensors remain expensive and time consuming. Here we report a rapid, versatile cell-free protein synthesis platform for detecting chemicals that target NHRs. This new biosensor is based on a multidomain-allosteric fusion protein with a core nuclear hormone ligand binding domain and a reporter domain. This biosensor has ability to produce a near real-time signal. We also report the robustness of cell free system in combination with a â€œjust-add-sampleâ€ technique to detect endocrine disrupting compounds in environmental samples. The cell free system has also been shown to be functional in human samples including blood and urine, which expands the possible portable applications of the biosensor.

Copyright © American Institute of Chemical Engineers. All rights reserved.

472500 Engineering Functionalized Virus-like-Particles with Antibody Fragments to Target Cancer Biomarkers

Abstract:

Functionalized Virus-Like particles (VLPs) are promising nano-particle vehicles for drug delivery or diagnostic applications. Our lab has done extensive work in developing a Hepatitis B core protein based VLP, with an artificially introduced disulfide bond network for added stability. These particles can target cancer cells by functionalizing their surface with antibody fragments, DNA or RNA aptamers using Cu(II) catalyzed click chemistry. The presence of multiple targeting agents on each VLP enhances targeting specificity due to avidity effects. In this paper, we present work on developing antibody fragments with non-natural amino acids (nnAAs), based on the J591 antibody, to target Prostate Specific Membrane Antigen (PSMA), a protein over-expressed on prostate cancer cell-lines. Using our *E. coli* based Cell Free Protein Synthesis (CFPS) platform, we incorporate nnAAs on both the spikes of the VLP (azide side chain) and the antibody fragments (alkyne side chain). Using metabolically engineering cell lines, we are able to express full-length single chain variable fragments (scFvs) and Fragment Antigen-binging (Fab) proteins containing nnAAs. These antibody fragments have been used in developing luminescent nanoparticles to detect and enumerate Circulating Tumor Cells. This work can be extended to target other cancer biomarkers for drug delivery.

Copyright © American Institute of Chemical Engineers. All rights reserved.

AUTHOR INDEX

AUTHOR INDEX

AUTHOR INDEX

AUTHOR INDEX

AUTHOR INDEX

AUTHOR INDEX

AUTHOR INDEX

AUTHOR INDEX

AUTHOR INDEX

AUTHOR INDEX

AUTHOR INDEX

AUTHOR INDEX

AUTHOR INDEX

AUTHOR INDEX

AUTHOR INDEX

AUTHOR INDEX

AUTHOR INDEX

AUTHOR INDEX

AUTHOR INDEX

AUTHOR INDEX

AUTHOR INDEX

AUTHOR INDEX

AUTHOR INDEX

AUTHOR INDEX